l Regions

D0164260

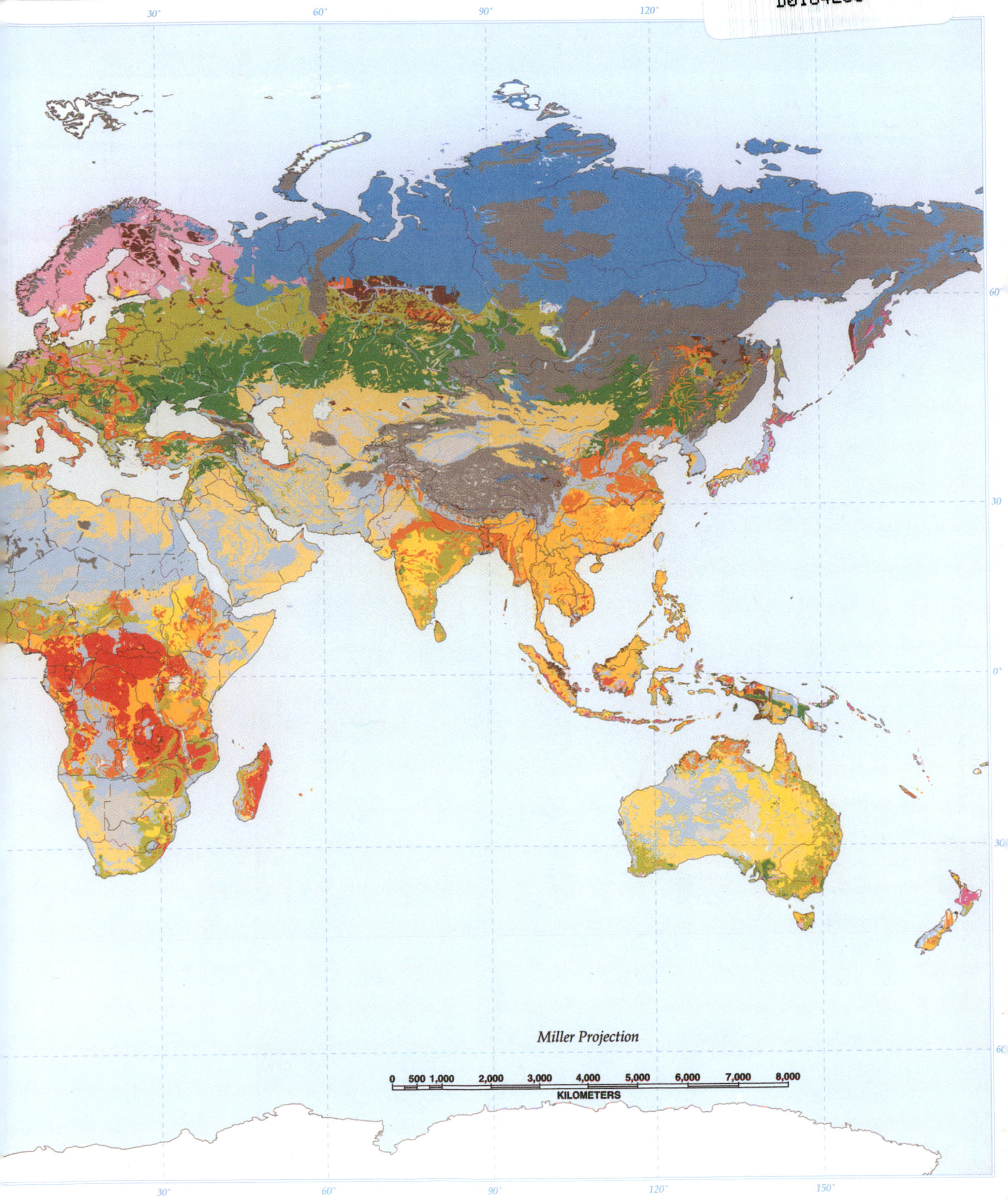

30°
60°
90°
120°
60°
30°
0°
30
60
150°
Miller Projection
0 500 1,000 2,000 3,000 4,000 5,000 6,000 7,000 8,000
KILOMETERS

Elements of the Nature and Properties of Soils

Elements of the Nature and Properties of Soils

Second Edition

Nyle C. Brady
Emeritus Professor of Soil Science
Cornell University

Ray R. Weil
Professor of Soil Science
University of Maryland at College Park

Upper Saddle River, New Jersey 07458

Library of Congress Cataloging-in-Publication Data

Brady, Nyle C.
Elements of the nature and properties of soils / Nyle C. Brady, Ray R. Weil.
p. cm.
Abridged ed. of: The nature and properties of soils. 13th ed. c2002.
Includes bibliographical references and index.
ISBN 0-13-048038-X
1. Soil science. 2. Soils. I. Weil, Ray R. II. Brady, Nyle C. Nature and properties of soils. III. Title.

S591.B792 2004
631.4—dc21 2003043399

Editor-in-Chief: *Stephen Helba*
Executive Editor: *Debbie Yarnell*
Development Editor: *Kate Linsner*
Managing Editor: *Mary Carnis*
Production Management: *Carlisle Publishers Services*
Project Coordinator: *Janice Stangel*
Director of Manufacturing and Production: *Bruce Johnson*
Manufacturing Buyer: *Cathleen Petersen*
Creative Director: *Cheryl Asherman*
Senior Design Coordinator: *Miguel Ortiz*
Formatting: *Carlisle Communications, Ltd.*
Marketing Manager: *Jimmy Stephens*
Marketing Assistant: *Melissa Orsborn*
Printer/Binder: *Courier Westford*
Cover Design: *Carey Davies*
Cover Photography: *Courtesy of Ray Weil*

Pearson Education Ltd.
Pearson Education Australia, Pty, Ltd.
Pearson Education Singapore, Pte. Ltd.
Pearson Education North Asia, Ltd.
Pearson Education Canada, Ltd.
Pearson Educación de Mexico, S.A. de C.V.
Pearson Education—Japan
Pearson Education Malaysia, Pte. Ltd.

10 9 8 7 6 5 4 3
ISBN 0-13-048038-X

To all the students and colleagues in soil science who have shared their inspirations, camaraderie, and deep love of the Earth.

Contents

3 Soil Classification 58

4 Soil Architecture and Physical Properties 94

5 Soil Water: Characteristics and Behavior 134

6 Soil and the Hydrologic Cycle 162

PREFACE

Like mountains or lakes, soils are both natural bodies in the landscape and ecosystems in their own right. Soils are among our most important natural resources. Soil science therefore lies at the heart of terrestrial ecology; understanding the soil system is key to the success and environmental harmony of any human use of the land.

This textbook is designed to help make your study of soils a fascinating and intellectually satisfying undertaking. We are confident that much of what you learn will be of enormous practical value in equipping you to meet the natural resource challenges of the 21st century. Whether your interests lie in forestry, agriculture, ecology, geography, or landscaping, you will soon find that soil science is critical to solving a wide range of problems relevant to these and similar fields. Because soil science integrates concepts from physics, chemistry, geology, and biology, this textbook will provide many opportunities for you to see practical applications for principles from these and other basic sciences.

As is the case for its parent book, *The Nature and Properties of Soils,* 13th edition, this newest edition of *Elements of the Nature and Properties of Soils* explains the fundamental principles of soil science in a manner that you will find relevant to your interests. Throughout, the text emphasizes the soil as a natural resource and highlights the many interactions between the soil and other components of forest, range, agricultural, wetland, and constructed ecosystems. We have sought to craft a book that will serve your needs well, whether your reading this textbook is to be your only formal exposure to soil science or you are embarking on a comprehensive soil science education. This new textbook is meant to provide both an exciting, accessible introduction to the fascinating world of soil science and a reliable reference for your professional bookshelf.

The parent 13th edition of *The Nature and Properties of Soils* maintains the ecological approach to soil science education characteristic of the two previous editions. This approach helps you understand the many critical ecosystem processes that are centered in the soil and that form the basis of interactions between soils and other ecosystem components such as groundwater, air, streams, and vegetation. This *Elements* version preserves the ecological approach of its parent book, while telling the story of soil science in a more condensed form that some students and professors have found more suitable to their needs.

Since 2000, when the first *Elements* edition was published, we have conferred with many students and professors who used the textbook. They told us that they need an updated coverage of soil science principles in approximately 550 pages, rather than the 960 pages of the parent textbook. They also urged that we not diminish either the scientific rigor or the easy readability of the parent book. We have tried to follow their advice.

We have been able to limit *Elements* to about 60% of the number of pages in the parent book by rewriting and condensing many sections, by reorienting much of the artwork, and by reorganizing and reducing application details so as to allow the fundamentals of soil science to be covered in 15 rather than 20 chapters.

New to this edition of *Elements* is expanded coverage of topics that are critically important to the future role of soils in natural resource sciences: wetlands, septic drain fields, salt-affected soils, bioremediation, soil ecology, nutrient and irrigation management, soil hydrology, and new concepts in *Soil Taxonomy.* This edition includes new sections on the pedosphere concept, nonsilicate colloids, inner and outer sphere complexes, effective CEC, proton balance approach to soil acidity, acid and non-acid cation saturation, human-influenced acidity, Ca and Mg in plants and soils, irrigation water quality, biomolecule binding, soil food web ecology, forest nutrient management, a phosphorus pollution site index, indicators of soil quality, and many other topics of current interest in soil science. In response to their popularity in the previous two editions, we have also added new "boxes" that present either engaging examples and applications or technical details and calculations. These boxes both *highlight* material of special interest and allow the logical thread of the regular text to flow smoothly without digression or interruption. In addition to updating many references, we have added a new feature to this edition, a set of World Wide Web links printed in the margins of the relevant chapter sections.

We could not have done all this without the many valuable suggestions, ideas, and corrections sent to us by soil scientists, instructors, and students from around the world. This edition has once again greatly benefited from such contributions. The high level of professional devotion and camaraderie shared by so many students and practitioners of soil science never ceases to inspire us. We appreciate the comments and suggestions concerning the makeup of this abridged edition made by a number of students and professors who responded to inquiries from the field staff of Prentice Hall. We are especially grateful to Lyle Nelson, Professor Emeritus at Mississippi State University for reviewing the entire text and making excellent and constructive comments. Joyce Torio of the American Chemical Society provided very useful background information on the problems and opportunities affecting global soil quality. In addition, Karen Lowell Langholtz provided able research and editorial assistance for this edition.

Special thanks go to the following colleagues who generously reviewed portions of the parent textbook in detail and made valuable suggestions for improvement: Duane Wolf (University of Arkansas); J. Kenneth Torrence (Carleton University); Jessica Davis (Colorado State University); Harold van Es and Martin Alexander (Cornell University); Dan Richter (Duke University); Michael Beug (Evergreen State University); Lee Burras (Iowa State University); Kudjo Dzantor, Delvin Fanning, Robert Hill, Bruce James, Martin Rabenhorst, and Patricia Steinhilber (University of Maryland); Daniel Hillel (University of Massachusetts); Lyle Nelson (Mississippi State University); Jimmie Richardson (North Dakota State University); Darrell Schultze (Purdue University); Murray Milford (Texas A & M University); Rattan Lal (Ohio State University); Mike Swift (UN Tropical Biology Program); Allen Franzluebbers, Jeff Herrick, Scott Lesch, and Jim Rhoades (USDA-Agricultural Research Service); Bob Ahrens, Hari Eswaran, Paul Reich, and Sharon Waltman (USDA/Natural Resources Conservation Service); Fred Magdoff and Wendy Sue Harper (University of Vermont); W. Lee Daniels, S. K. de Datta, and Lucian Zelazny (Virginia Tech.); Clay Robinson (West Texas A & M University); and Tom Siccama (Yale University).

Once again, we express our heartfelt thanks to our wives, Martha and Trish, for their encouragement, understanding, and patience without which we could not possibly have found the time and energy required to make such extensive improvements to this textbook. They bear witness to the fact that our effort to continuously improve *The Nature and Properties of Soils* textbooks is truly a labor of love.

—N. C. B. and R. R. W.

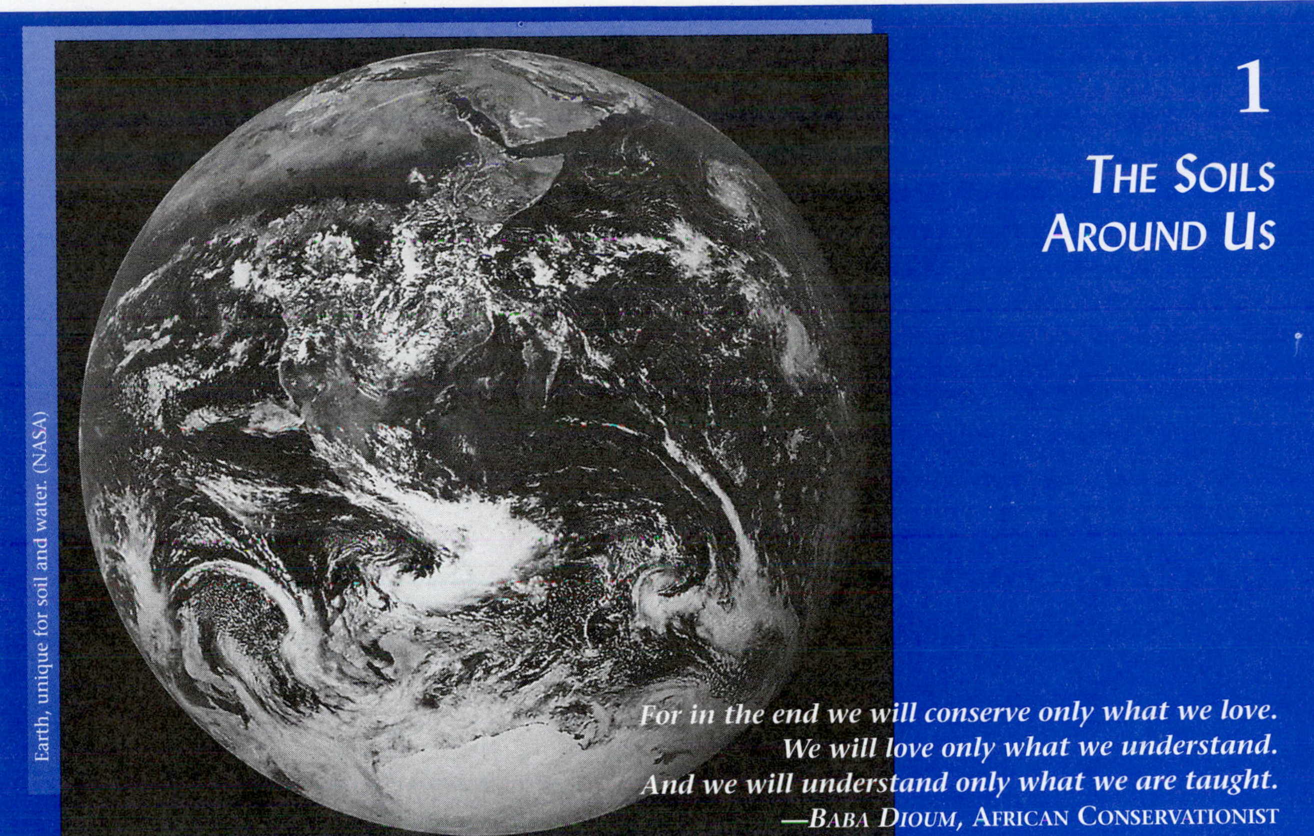
Earth, unique for soil and water. (NASA)

1 The Soils Around Us

For in the end we will conserve only what we love.
We will love only what we understand.
And we will understand only what we are taught.
—Baba Dioum, African Conservationist

The Earth, our unique home in the vastness of the universe, is in crisis. Our planet is covered with life-sustaining air, water, and soil. However, great care will be required in preserving the quality of all three if our species is to continue to thrive.

Depletion of the ozone layer in the upper atmosphere is threatening us with an overload of ultraviolet radiation. Increases of carbon dioxide and methane in the atmosphere are destabilizing our climate. Tropical rain forests, and the incredible array of plant and animal species they contain, are disappearing at an unprecedented rate. Groundwater supplies are being contaminated in many areas and depleted in others. In parts of the world, the capacity of soils to produce food is being degraded, even as the number of people needing food is increasing. It will be a great challenge for the current generation to bring the global environment back into balance.

Soils are crucial to life on earth. From ozone depletion and global warming to rain forest destruction and water pollution, the world's ecosystems are impacted in far-reaching ways by processes carried out in the soil. To a great degree, the quality of the soil determines the nature of plant ecosystems and the capacity of land to support animal life and society. As human societies become increasingly urbanized, fewer people have intimate contact with the soil, and individuals tend to lose sight of the many ways in which they depend upon soils for their prosperity and survival. Indeed, the degree to which we are dependent on soils is likely to increase, not decrease, in the future.

Soils will continue to supply us with nearly all of our food (except for what can be harvested from the oceans). How many of us remember, as we eat a slice of pizza, that the pizza's crust began in a field of wheat and its cheese began with grass, clover, and corn rooted in the soils of a dairy farm? Most of the fiber we use for lumber, paper, and clothing has its roots in the soils of forests and farmland. Although we sometimes use plastics and fiber synthesized from fossil petroleum as substitutes, in the long term we will continue to depend on terrestrial ecosystems for these needs. Besides, on a hot day, would you rather wear a polyester shirt or one made of cotton?

In addition, biomass grown on soils is likely to become an increasingly important feedstock for fuels and manufacturing as the world's finite supplies of petroleum are depleted during the course of this century. The early marketplace signs of this trend can be seen in

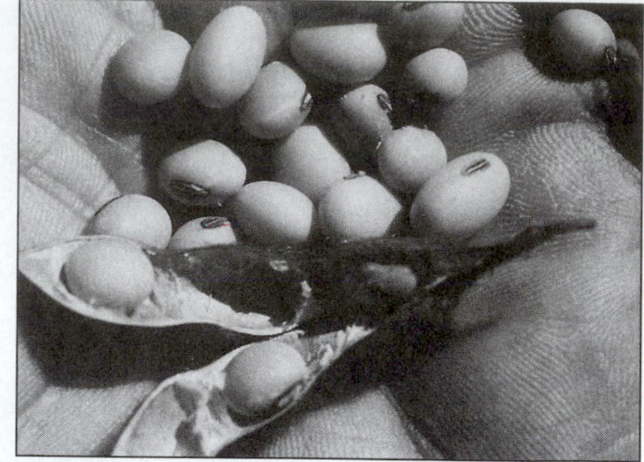

FIGURE 1.1 In the future, we will be increasingly dependent on soils to grow renewable resources that can substitute for dwindling supplies of crude oil. Plastics, fuels, and inks, for example, can be manufactured from soybean oil instead of from petroleum. Soybean oil is edible and far less toxic to the environment. (Photos courtesy of R. Weil)

the form of gasohol and biodiesel fuels made from plant products, printers' inks made from soybean oil, and biodegradable plastics synthesized from cornstarch (Figure 1.1).

One of the stark realities of the 21st century is that the human population that demands all of these products will increase by several billion, while the soil resource base available to provide for them is likely to decline because of soil degradation and urbanization. Thus, to survive as a species, we will have to greatly improve the efficiency and sustainability with which we manage our soil resources.

The art of soil management is as old as civilization. As we meet the challenges of this century, new understandings and new technologies will be needed to protect the environment and, at the same time, produce food and biomass to support society. The study of soil science has never been more important for foresters, farmers, engineers, natural resource managers, and ecologists alike.

1.1 FUNCTIONS OF SOILS IN OUR ECOSYSTEM

In any ecosystem, whether your backyard, a farm, a forest, or a regional watershed, soils have five key roles to play (Figure 1.2). *First,* soil supports the growth of higher plants, mainly by providing a medium for plant roots and supplying nutrient elements that are essential to the entire plant. Properties of the soil often determine the nature of the vegetation present and, indirectly, the number and types of animals (including people) that the vegetation can support. *Second,* soil properties are the principal factor controlling the fate of water in the hydrologic system. Water loss, utilization, contamination, and purification are all affected by the soil. *Third,* the soil functions as nature's recycling system. Within the soil, waste products and dead bodies of plants, animals, and people are assimilated, and their basic elements are made available for reuse by the next generation of life. *Fourth,* soils provide habitats for a myriad of living organisms, from small mammals and reptiles to tiny insects to microscopic cells of unimaginable numbers and diversity. *Finally,* in human-built ecosystems, soil plays an important role as an engineering medium. Soil is not only an important building material in the form of earth fill and bricks (baked soil material), but also provides the foundation for virtually every road, airport, and house we build.

1.2 MEDIUM FOR PLANT GROWTH

Imagine a growing plant. The aboveground portion may be most familiar, but the portion growing below the soil surface, its root system, may be nearly as large as the portion we see above ground. Here is a list of things that plants obtain from the soils in which their roots proliferate:

- Physical support
- Air
- Water
- Temperature moderation
- Protection from toxins
- Nutrient elements

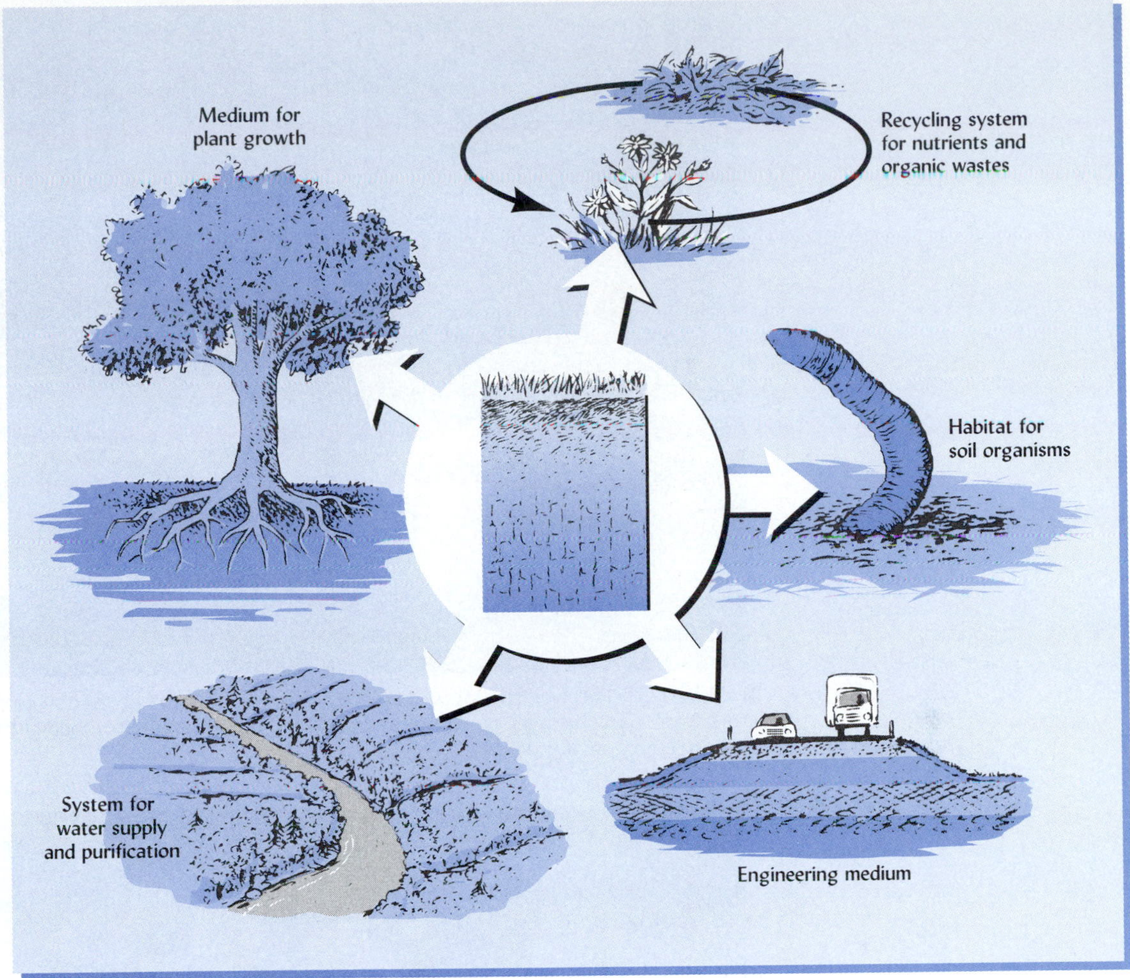

FIGURE 1.2 The many functions of soil can be grouped into five crucial ecological roles.

First, the soil mass provides *physical support,* anchoring the root system so that the plant does not fall over. Occasionally, as in Figure 1.3, strong wind or heavy snow does topple a plant whose root system has been restricted by shallow or inhospitable soil conditions.

Plant roots depend on the process of respiration to obtain energy. Since root respiration, like our own respiration, produces carbon dioxide (CO_2) and uses oxygen (O_2), an important function of the soil is *ventilation*—allowing CO_2 to escape and fresh O_2 to enter the root zone. This ventilation is accomplished via the network of soil pores.

An equally important function of the soil pores is to absorb water and hold it where it can be used by plant roots. As long as plant leaves are exposed to sunlight, the plant requires a continuous stream of water to use in cooling, nutrient transport, turgor maintenance, and photosynthesis. Since plants use water continuously, but in most places it rains only occasionally, the *water-holding capacity* of soils is essential for plant survival. A deep soil may store enough water to allow plants to survive long periods without rain (see Figure 1.4).

As well as moderating moisture changes in the root environment, the soil also *moderates temperature fluctuations.* Perhaps you can recall digging in garden soil on a summer afternoon and feeling how hot the soil was at the surface and how much cooler just a few centimeters below. The insulating properties of soil protect the deeper portion of the root system from the extremes of hot and cold that often occur at the soil surface. For example, it is not unusual for the temperature at the surface of bare soil to exceed 35 or 40°C in midafternoon, a condition that would be lethal to most plant roots. Just a few centimeters deeper, however, the temperature may be 10°C cooler, allowing roots to function normally.

There are many potential sources of **phytotoxic substances** in soils. These toxins may result from human activity, or they may be produced by plant roots, by microorganisms, or by natural chemical reactions. Many soil managers consider a function of a

FIGURE 1.3 This wet, shallow soil failed to allow sufficiently deep roots to develop to prevent this tree from blowing over when snow-laden branches made it top-heavy during a winter storm. (Photo courtesy of R. Weil)

FIGURE 1.4 A family of African elephants finds welcome shade under the leafy canopy of a huge acacia tree in this East African savanna. The photo was taken in the middle of a long dry season and no rain had fallen for almost five months. The tree roots are still using water from the previous rainy season stored several meters deep in the soil. The light-colored grasses are more shallow-rooted and have either set seed and died or gone into a dried-up dormant condition. (Photo courtesy of R. Weil)

good soil to be protection of plants from toxic concentrations of such substances by ventilating gases, by decomposing or adsorbing organic toxins, or by suppressing toxin-producing organisms. At the same time, it is true that some microorganisms in soil produce organic, growth-stimulating compounds. These substances, when taken up by plants in small amounts, may improve plant vigor.

Soils supply plants with inorganic, *mineral nutrients* in the form of dissolved ions. These mineral nutrients include such metallic elements as potassium, calcium, iron, and copper, as well as such nonmetallic elements as nitrogen, sulfur, phosphorus, and boron. By eating plants, humans and other animals usually obtain the minerals they need (including several elements that plants take up but do not appear to use themselves) indirectly from the soil (see periodic table, Appendix B). The plant takes these elements out of the soil solution and incorporates most of them into the thousands of different organic compounds that constitute plant tissue. A fundamental role of soil in supporting plant growth is to provide a continuing supply of dissolved mineral nutrients in amounts and relative proportions appropriate for plant growth.

Interactive periodic table. Look up everything you could want to know about the essential elements: www.webelements.com/

Of the 92 naturally occurring chemical elements, only the 18 listed in Table 1.1 have been shown to be **essential elements** without which plants cannot grow and complete their life cycles. A message that may have been found on a sign hanging on a cafe door may help you remember the 18 elements essential for plant growth. Most of the chemical symbols are obvious, but finding those for copper (Cu) and zinc (Zn) may require some imagination.

C.B. HOPKiNS CaFe, Co.
Closed **M**onday **Mo**rning and **Ni**ght
See You Zoon, the **Mg.**

Essential elements used by plants in relatively large amounts are called **macronutrients;** those used in smaller amounts are known as **micronutrients.** A plant's dry matter consists mainly of carbon, hydrogen, and oxygen, which the plant obtains by photosynthesis from air and water, not from the soil.

Plants can be grown in nutrient solutions without soil (**hydroponics**), but at a high cost of energy, money, and management. Although hydroponic production can be feasible on a small scale for a few high-value plants, the production of the world's food and fiber and the maintenance of natural ecosystems will always depend on the use of millions of square kilometers of productive soils.

TABLE 1.1 Elements Essential for Plant Growth and Their Sources[a]

The chemical forms most commonly taken in by plants are shown in parentheses, with the chemical symbol for the element in bold type.

Macronutrients: Used in relatively large amounts (> 0.1% of dry plant tissue)		*Macronutrients: Used in relatively small amounts (< 0.1% of dry plant tissue)*
Mostly from air and water	*Mostly from soil solids*	*From soil solids*
Carbon ($\mathbf{C}O_2$)	Nitrogen ($\mathbf{N}O_3^-$, $\mathbf{N}H_4^+$)	Iron ($\mathbf{Fe}^{2+}$)
Hydrogen ($\mathbf{H}_2O$)	Phosphorus ($H_2\mathbf{P}O_4^-$, $H\mathbf{P}O_4^{2-}$)	Manganese ($\mathbf{Mn}^{2+}$)
Oxygen ($\mathbf{O}_2$, $H_2\mathbf{O}$)	Potassium ($\mathbf{K}^+$)	Boron ($H\mathbf{B}O_3$)
	Calcium ($\mathbf{Ca}_2^+$)	Zinc ($\mathbf{Zn}^{2+}$)
	Magnesium ($\mathbf{Mg}_2^+$)	Copper ($\mathbf{Cu}^{2+}$)
	Sulfur ($\mathbf{S}O_4^{2-}$)	Chlorine ($\mathbf{Cl}^-$)
		Cobalt ($\mathbf{Co}^{2+}$)
		Molybdenum ($\mathbf{Mo}O_4^{2-}$)
		Nickel ($\mathbf{Ni}^{2+}$)

[a]Many other elements are taken up from soils by plants, but are not essential for plant growth. Some of these (such as sodium, silicon, iodine, fluorine, barium, and strontium) do enhance the growth of certain plants, but do not appear to be as universally required for normal growth as are the 18 listed in this table.

1.3 REGULATOR OF WATER SUPPLIES

There is much concern about the quality and quantity of the water in our rivers, lakes, and underground aquifers. Governments and citizens everywhere are working to stem the pollution that threatens the value of our waters for fishing, swimming, and drinking. For progress to be made in improving water quality, we must recognize that nearly every drop of water in our rivers, lakes, estuaries, and aquifers has either traveled through the soil or flowed over its surface.[1] Imagine, for example, a heavy rain falling on the hills surrounding a river. If the soil allows the rain to soak in, some of the water may be stored in the soil and used by the trees and other plants, while some may seep slowly down through the soil layers to the groundwater, eventually entering the river over a period of months or years as base flow. If the water is contaminated, as it soaks through the upper layers of soil it is purified and cleansed by soil processes that remove many impurities and kill potential disease organisms.

Contrast the preceding scenario with what would occur if the soil were so shallow or impermeable that most of the rain could not penetrate the soil, but ran off the hillsides on the soil surface, scouring surface soil and debris as it picked up speed, and entering the river rapidly and nearly all at once. The result would be a destructive flash flood of muddy water. Clearly, the nature and management of soils in a watershed will have a major influence on the purity and amount of water finding its way to aquatic systems.

1.4 RECYCLER OF RAW MATERIALS

What would a world be like without the recycling functions performed by soils? Without reuse of nutrients, plants and animals would have run out of nourishment long ago. The world would be covered with a layer, possibly hundreds of meters high, of plant and animal wastes and corpses. Obviously, recycling must be a vital process in ecosystems, whether forests, farms, or cities. The soil system plays a pivotal role in the major geochemical cycles. Soils have the capacity to assimilate great quantities of organic waste, turning it into beneficial **humus**, converting the mineral nutrients in the wastes to forms that can be utilized by plants and animals, and returning the carbon to the atmosphere as carbon dioxide, where it again will become a part of living organisms through plant photosynthesis. Some soils can accumulate large amounts of carbon as soil organic matter, thus having a major impact on such global changes as the much-discussed *greenhouse effect* (see Sections 1.12 and 11.2).

1.5 HABITAT FOR SOIL ORGANISMS

When we speak of protecting ecosystems, most people envision a stand of old-growth forest with its abundant wildlife, or perhaps an estuary such as the Chesapeake Bay with its oyster beds and fisheries. (Perhaps, once you have read this book, you will envision a handful of soil when someone speaks of an ecosystem.) Soil is not a mere pile of broken rock and dead debris. A handful of soil may be home to *billions* of organisms, belonging to thousands of species. In even this small quantity of soil, there are likely to exist predators, prey, producers, consumers, and parasites (Figure 1.5).

How is it possible for such a diversity of organisms to live and interact in such a small space? One explanation is the tremendous range of niches and habitats in even a uniform-appearing soil. Some pores of the soil will be filled with water in which swim organisms such as roundworms, diatoms, and rotifers. Tiny insects and mites may be crawling about in other, larger pores filled with moist air. Micro-zones of good aeration may be only millimeters from areas of **anoxic** conditions. Different areas may be enriched with decaying organic materials; some places may be highly acidic, some more basic. Temperature, too, may vary widely.

All about rangeland soil communities:
http://www.blm.gov:80/nstc/soil/communities/index.html

[1] This excludes the relatively minor quantity of precipitation that falls directly into bodies of fresh surface water.

FIGURE 1.5 The soil is home to a wide variety of organisms, both relatively large and very small. Here, a relatively large predator, a centipede (shown at about actual size), hunts for its next meal—which is likely to be one of the many smaller animals that feed on dead plant debris. (Photo courtesy of R. Weil)

Hidden from view in the world's soils are communities of living organisms every bit as complex and intrinsically valuable as their counterparts that roam the savannas, forests, and oceans of the earth. Soils harbor much of the earth's genetic diversity. Soils, like air and water, are important components of the larger ecosystem. Yet only now is soil quality taking its place, with air quality and water quality, in discussions of environmental protection.

1.6 ENGINEERING MEDIUM

"*Terra firma,* solid ground." We usually think of the soil as being firm and solid, a good base on which to build roads and all kinds of structures. Indeed, most structures rest on the soil, and many construction projects require excavation into the soil. Unfortunately, as can be seen in Figure 1.6, some soils are not as stable as others. Reliable construction on soils, and with soil materials, requires knowledge of the diversity of soil properties, as discussed later in this chapter. Designs for roadbeds or building foundations that work well in one location on one type of soil may be inadequate for another location with different soils.

Working with natural soils or excavated soil materials is not like working with concrete or steel. Properties such as bearing strength, compressibility, shear strength, and stability are much more variable and difficult to predict for soils than for manufactured building materials. Chapter 4 provides an introduction to some engineering properties of soils. Many other physical properties discussed will have direct application to engineering uses of soil. For example, Chapter 8 discusses the swelling properties of certain types of clays in soils. The engineer should be aware that when soils with swelling clays are wetted they expand with sufficient force to crack foundations and buckle pavements. Much of the information on soil properties and soil classification discussed in later chapters will be of great value to people planning land uses that involve construction or excavation.

FIGURE 1.6 Better knowledge of the soils on which this road was built may have allowed its engineers to develop a more stable design, thus avoiding this costly and dangerous situation. (Photo courtesy of R. Weil)

1.7 SOIL AS ENVIRONMENTAL INTERFACE

The importance of soil as a natural body derives in large part from its role as an **interface** between the worlds of rock (the **lithosphere**), air (the **atmosphere**), water (the **hydrosphere**), and living things (the **biosphere**). Environments where all four of these worlds interact are often the most complex and productive on Earth. An estuary, where shallow waters meet the land and air, is an example of such an environment. Its productivity and ecological complexity far surpass those of a deep ocean trench, for example (where the hydrosphere is rather isolated), or the upper atmosphere (where rocks and water have little influence). The soil, or **pedosphere**, is another example of such an environment (Figure 1.7).

The concept of the soil as interface means different things at different scales. At the scale of kilometers, soils channel water from rain to rivers and transfer mineral elements from country rocks to the oceans. They also remove and supply vast amounts of atmospheric gases, substantially influencing the global balance of methane and carbon dioxide. At a scale of a few meters (Figure 1.7*b*), soil forms the transition zone between hard rock and air, holding both liquid water and oxygen gas for use by plant roots. It transfers mineral elements from the Earth's rock crust to its vegetation. It processes or stores the organic remains of terrestrial plants and animals. At a scale of a few millimeters (Figure 1.7*c*), soil provides diverse microhabitats for air-breathing and aquatic organisms, channels water and nutrients to plant roots, and provides surfaces and solution vessels for thousands of biochemical reactions. Finally, at the scale of a few micrometers and smaller (less than a millionth of a meter), soil provides ordered and complex surfaces, both mineral and organic, that act as templates for chemical reactions and interact with water and solutes. Its tiniest mineral particles form micro-zones of electromagnetic charge that attract everything from bacterial cell walls to proteins to conglomerates of water molecules. As you read the entirety of this book, the frequent cross-referencing between one chapter and another will remind you of the importance of scale and interfacing to the story of soil.

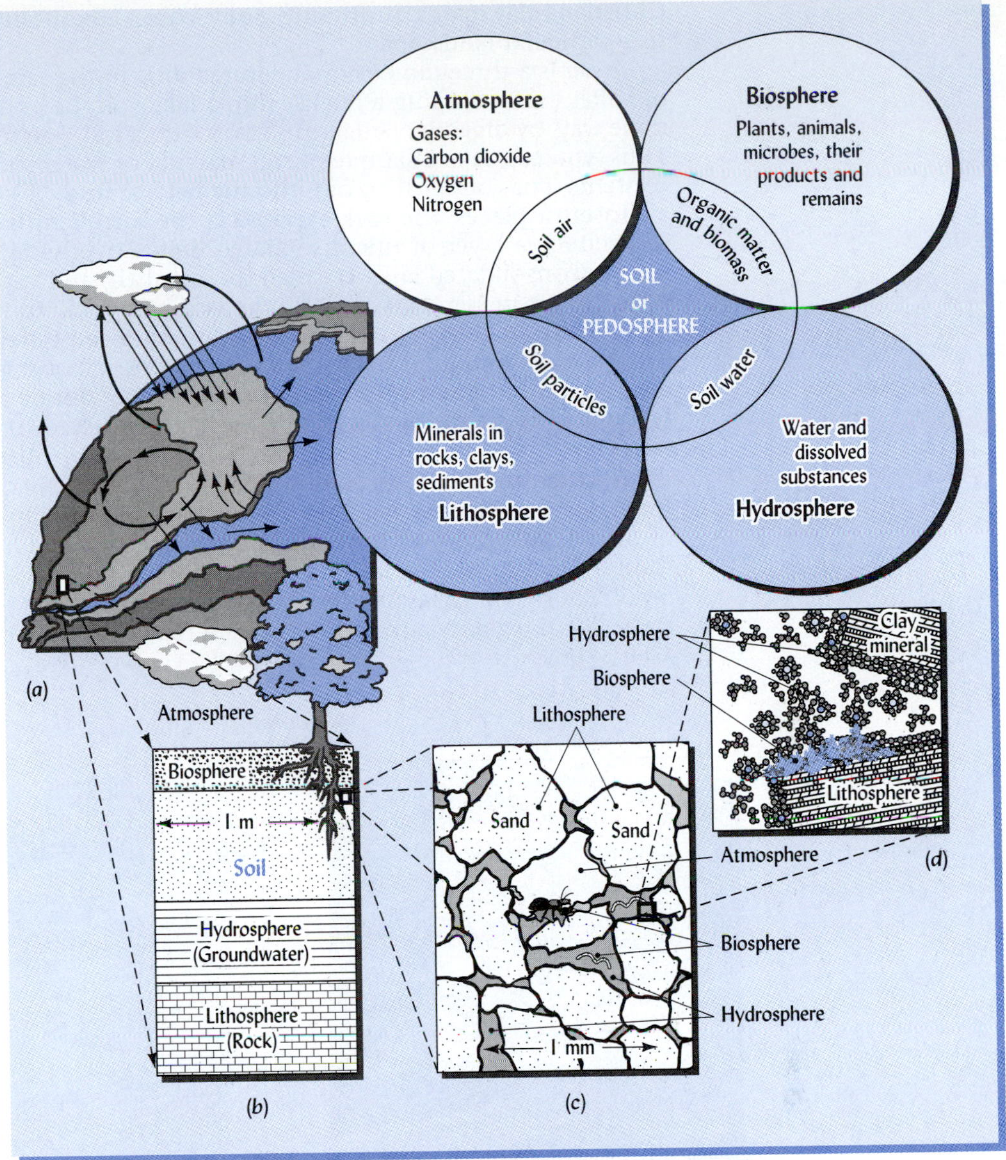

FIGURE 1.7 The pedosphere, where the worlds of rock (the lithosphere), air (the atmosphere), water (the hydrosphere), and life (the biosphere) all meet. The soil as interface can be understood at many different scales. At the kilometer scale (*a*), soil participates in global cycles of rock weathering, atmospheric gas changes, water storage and partitioning, and the life of terrestrial ecosystems. At the meter scale (*b*), soil forms a transition zone between the hard rock below and the atmosphere above—a zone through which surface water and groundwater flow, and in which plants and other living organisms thrive. A thousand times smaller, at the millimeter scale (*c*), mineral particles form the skeleton of the soil that defines pore spaces, some filled with air and some with water, in which tiny creatures lead their lives. Finally, at the micro- and nanometer scales (*d*), soil minerals (lithosphere) provide charges, reactive surfaces that adsorb water and cations dissolved in water (hydrosphere), gases (atmosphere), and bacteria and complex humus macromolecules (biosphere). (Diagram courtesy of R. Weil)

1.8 SOIL AS A NATURAL BODY

You may have noticed that this book sometimes refers to "the soil," sometimes to "a soil," and sometimes to "soils." *The soil* is often said to cover the land as the peel covers an apple. However, while the peel is relatively uniform around the apple, the soil is highly variable from place to place on Earth. In fact, the soil is a collection of individually different soil bodies. One of these individual bodies, *a soil,* is to *the soil* as an individual tree is to the earth's vegetation. Just as one may find sugar maples, oaks, hemlocks, and many other species of trees in a particular forest, so, too, might one find

Christiana clay loams, Sunnyside sandy loams, Elkton silt loams, and other kinds of soils in a particular landscape.

A soil is a three-dimensional natural body in the same sense that a mountain, lake, or valley is. By dipping a bucket into a lake you may sample some of its water. In the same way, by digging or augering a hole into a soil, you may retrieve some soil material. Thus, you can take a sample of soil material or water into a laboratory and analyze its contents, but you must go out into the field to study a soil or a lake.

In most places, the rock exposed at the earth's surface has crumbled and decayed to produce a layer of unconsolidated debris overlying the hard, unweathered rock. This unconsolidated layer is called the **regolith**, and varies in thickness from virtually nonexistent in some places (i.e., exposed bare rock) to tens of meters in other places. The regolith material, in many instances, has been transported many kilometers from the site of its initial formation and then deposited over the bedrock which it now covers. Thus, all or part of the regolith may or may not be related to the rock now found below it. Where the underlying rock has weathered in place to the degree that it is loose enough to be dug with a spade, the term **saprolite** is used (see Plate 11).

Through their biochemical and physical effects, living organisms such as bacteria, fungi, and plant roots have altered the upper part—and, in many cases, the entire depth—of the regolith. Here, at the interface between the worlds of rock, air, water, and living things, soil is born. The transformation of inorganic rock and debris into a living soil is one of nature's most fascinating displays. Although generally hidden from everyday view, the soil and regolith can often be seen in road cuts and other excavations (Figure 1.8).

FIGURE 1.8 Relative positions of the regolith, its soil, and the underlying bedrock. Note that the soil is a part of the regolith, and that the A and B horizons are part of the **solum** (from the Latin word *solum,* which means "soil or land"). The C horizon is the part of the regolith that underlies the solum, but may be slowly changing into soil in its upper parts. Sometimes the regolith is so thin that it has been changed entirely to soil; in such a case, soil rests directly on the bedrock. (Photo courtesy of R. Weil)

A soil is the product of both destructive and creative (synthetic) processes. Weathering of rock and microbial decay of organic residues are examples of destructive processes, whereas the formation of new minerals, such as certain clays, and of new stable organic compounds are examples of synthesis. Perhaps the most striking result of synthetic processes is the formation of contrasting layers called **soil horizons.** The development of these horizons in the upper regolith is a unique characteristic of soil that sets it apart from the deeper regolith materials.

Soil scientists specializing in **pedology** (*pedologists*) study soils as natural bodies, the properties of soil horizons, and the relationships among soils within a landscape. Other soil scientists, called **edaphologists,** focus on the soil as habitat for living things, especially plants. For both types of study it is essential to examine soils at all scales and in all three dimensions (especially the vertical dimension).

1.9 THE SOIL PROFILE AND ITS LAYERS (HORIZONS)

Soil scientists often dig a large hole, called a *soil pit,* usually several meters deep and about a meter wide, to expose soil horizons for study. The vertical section exposing a set of horizons in the wall of such a pit is termed a **soil profile.** Road cuts and other ready-made excavations can expose soil profiles and serve as windows to the soil. In an excavation open for some time, horizons are often obscured by soil material that has been washed by rain from upper horizons to cover the exposed face of lower horizons. For this reason, horizons may be more clearly seen if a fresh face is exposed by scraping off a layer of material several centimeters thick from the pit wall. Observing how soils exposed in road cuts vary from place to place can add a fascinating new dimension to travel. Once you have learned to interpret the different horizons (see Chapter 2), soil profiles can warn you about potential problems in using the land, as well as tell you much about the environment and history of a region. For example, soils developed in a dry region will have very different horizons from those developed in a humid region.

Horizons within a soil may vary in thickness and have somewhat irregular boundaries, but generally they parallel the land surface (Figure 1.9). This alignment is expected, because the differentiation of the regolith into distinct horizons is largely the

FIGURE 1.9 This road cut in central Africa reveals soil layers or horizons which parallel the land surface. Taken together, these horizons comprise the profile of this soil, as shown in the enlarged diagram. The presence and characteristics of particular horizons in this profile distinguish this soil from the thousands of other soils in the world. (Photo courtesy of R. Weil)

result of influences, such as air, water, solar radiation, and plant material, originating at the soil–atmosphere interface. Since the weathering of the regolith occurs first at the surface and works its way down, the uppermost layers have been changed the most, while the deepest layers are most similar to the original regolith, which is referred to as the soil's **parent material.** In places where the regolith was originally rather uniform in composition, the material below the soil may have a similar composition to the parent material from which the soil formed. In other cases, the regolith material has been transported long distances by wind, water, or glaciers and deposited on top of dissimilar material. In such a case, the regolith material found below a soil may be quite different from the upper layer of regolith in which the soil formed.

In undisturbed ecosystems, especially forests, organic materials formed from fallen leaves and other plant and animal remains tend to accumulate on the surface. There they undergo varying degrees of physical and biochemical breakdown and transformation, so that layers of older, partially decomposed materials may underlie the freshly added debris. Together, these organic layers at the soil surface are designated the **O horizons.**

Soil animals and percolating water move some of these organic materials downward to intermingle with the mineral grains of the regolith. These join the decomposing remains of plant roots to form organic materials that darken the upper mineral layers. Also, because weathering tends to be most intense nearest the soil surface, in many soils the upper layers lose some of their clay or other weathering products by leaching to the horizons below. **A horizons** are the layers nearest the surface that are dominated by mineral particles but have been darkened by the accumulation of organic matter. The organically enriched A horizon at the soil surface is sometimes referred to as **topsoil.** Plowing and cultivating a soil homogenizes and modifies the upper 12 to 25 cm (5 to 10 in.) of the soil to form a **plow layer.** In many soils, the majority of fine plant feeder roots can be found in the topsoil or plow layer. Sometimes contractors remove the plow layer from a site and sell or stockpile this topsoil for later use in establishing lawns and shrubs around newly constructed buildings (see Plate 35 after page 498).

In some soils, intensely weathered and leached horizons that have not accumulated organic matter occur in the upper part of the profile, usually just below the A horizons. These horizons are designated **E horizons** (Figure 1.10).

The layers underlying the A and O horizons contain comparatively less organic matter than the horizons nearer the surface and are commonly referred to as **subsoil.** Varying amounts of silicate clays, iron and aluminum oxides, gypsum, or calcium carbonate may accumulate in these underlying horizons. The accumulated materials may have been washed down from the horizons above, or they may have been formed in place through the weathering process. These underlying layers are referred to as **B horizons** (Figure 1.10).

Plant roots and microorganisms often extend below the B horizon, especially in humid regions, causing chemical changes in the soil water, some biochemical weathering of the regolith, and the formation of **C horizons.** The C horizons are the least weathered part of the soil profile. Many of the chemical, biological, and physical processes that characterize topsoils also take place to some degree in the C horizons of soils, which may extend deep into the underlying saprolite or other regolith material. Traditionally, the lower boundary of the soil has been considered to occur at the greatest rooting depth of the natural vegetation, but soil scientists are increasingly studying layers below this in order to understand ecological processes such as groundwater pollution, parent material weathering, and geochemical cycles. The importance of the various soil layers is highlighted in Box 1.1.

In some soil profiles, the component horizons are distinct in color, with sharp boundaries that can be seen easily by even novice observers. In other soils, the color changes between horizons may be gradual, and the boundaries more difficult to locate. However, color is only one of many properties by which one horizon may be distinguished from the horizon above or below it. The study of soils in the field is both a sensual and an intellectual activity. Delineation of the horizons present in a soil profile often requires a careful examination, using all the senses. In addition to seeing the colors in a profile, a soil scientist may feel, smell, and listen to the soil (as in Box 4.1) as well as conduct chemical tests, to distinguish the horizons present.

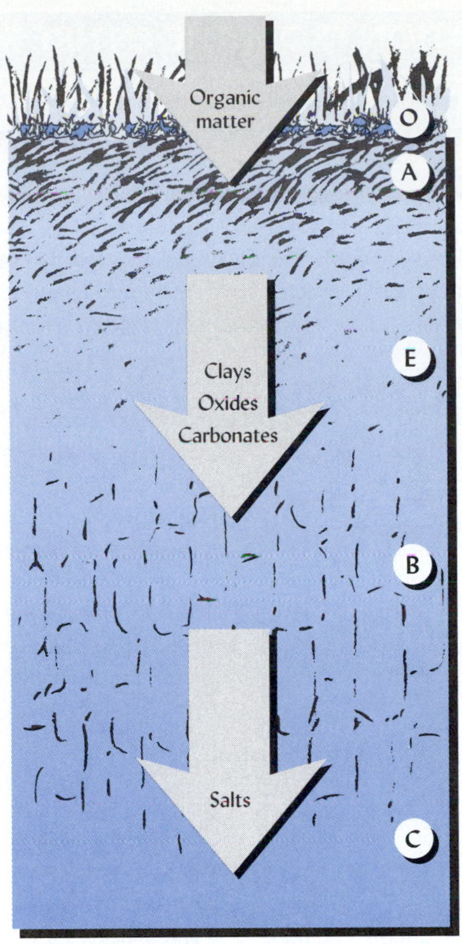

FIGURE 1.10 Horizons begin to differentiate as materials are added to the upper part of the profile and other materials are translocated to deeper zones. Under certain conditions, usually associated with forest vegetation and high rainfall, a leached E horizon forms between organic-matter-rich A and the B horizons. If sufficient rainfall occurs, soluble salts will be carried below the soil profile, perhaps all the way to the groundwater.

1.10 SOIL: THE INTERFACE OF AIR, MINERALS, WATER, AND LIFE

We stated that where the regolith meets the atmosphere, the worlds of air, rock, water, and living things are intermingled. In fact, the four major components of soil are air, water, mineral matter, and organic matter. The relative proportions of these four components greatly influence the behavior and productivity of soils. In a soil, the four components are mixed in complex patterns; however, the proportion of soil volume occupied by each component can be represented in a simple pie chart. Figure 1.12 shows the approximate proportions (by volume) of the components found in a loam surface soil in good condition for plant growth. Although a handful of soil may at first seem to be a solid thing, it should be noted that only about half the soil volume consists of solid material (mineral and organic); the other half consists of pore spaces filled with air or water. Of the solid material, typically most is mineral matter derived from the rocks of the earth's crust. Only about 5% of the *volume* in this ideal soil consists of organic matter. However, the influence of the organic component on soil properties is often far greater than its small proportion would suggest. Because it is far less dense than mineral matter, the organic matter accounts for only about 2% of the *weight* of this soil.

The spaces between the particles of solid material are just as important to the nature of a soil as are the solids themselves. It is in these pore spaces that air and water circulate, roots grow, and microscopic creatures live. Plant roots need both air and water. In an optimum condition for most plants, the pore space will be divided roughly equally among the two, with 25% of the soil volume consisting of water and 25% consisting of air. If there is much more water than this, the soil will be waterlogged. If much less water is present, plants will suffer from drought. The relative proportions of water and air in a soil typically fluctuate greatly as water is added or lost. Soils with much more than 50% of their volume in solids are likely to be too compacted for good plant growth. Compared to surface soil layers, subsoils tend to contain less organic matter, less total pore space, and a larger proportion of small pores (**micropores**) which tend to be filled with water rather than with air.

BOX 1.1 USING INFORMATION FROM THE ENTIRE SOIL PROFILE

Soils are three-dimensional bodies that carry out important ecosystem processes at all depths in their profiles. Depending on the particular application, the information needed to make proper land management decisions may come from soil layers as shallow as the upper 1 or 2 cm, or as deep as the lowest layers of saprolite (Figure 1.11).

For example, the upper few centimeters of soil often hold the keys to plant growth and biological diversity, as well as to certain hydrologic processes. Here, at the interface between the soil and the atmosphere, living things are most numerous and diverse. Forest trees largely depend for nutrient uptake on a dense mat of fine roots growing in this zone. The physical condition of this thin surface layer may also determine whether rain will soak in or run downhill on the land surface. Certain pollutants, such as lead from highway exhaust, are also concentrated in this zone. For many types of soil investigations it will be necessary to sample the upper few centimeters separately so that important conditions are not overlooked.

On the other hand, it is equally important not to confine one's attention to the easily accessible "topsoil," for many soil properties are to be discovered only in the deeper layers. Plant-growth problems are often related to inhospitable conditions in the B or C horizons that restrict the penetration of roots. Similarly, the great volume of these deeper layers may control the amount of plant-available water held by a soil. For the purposes of recognizing or mapping different types of soils, the properties of the B horizons are often paramount. Not only is this the zone of major accumulations of minerals and clays, but the layers nearer the soil surface also are too quickly altered by management and soil erosion to be a reliable source of information for the classification of soils.

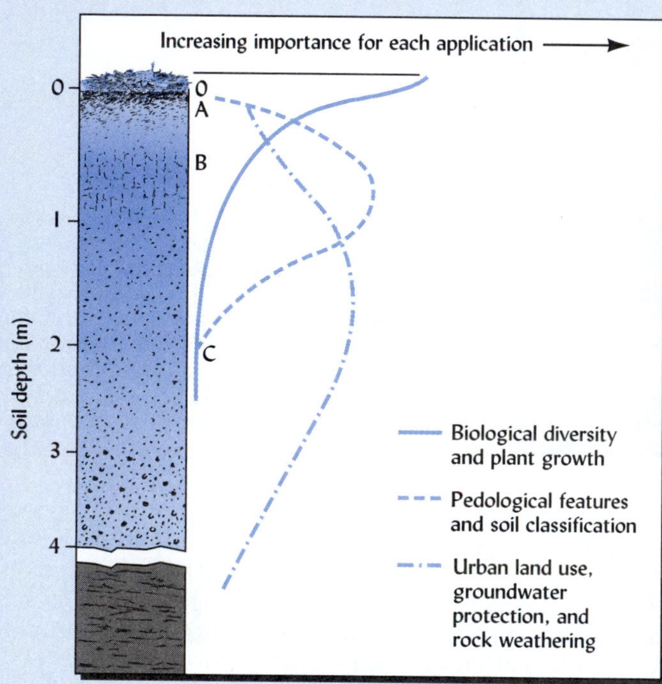

FIGURE 1.11 *Information important to different soil functions and applications is most likely to be obtained by studying different layers of the soil profile. (Diagram courtesy of R. Weil)*

In deeply weathered regoliths, the lower C horizons and saprolite play important roles. These layers, generally at depths below 1 or 2 m, and often as deep as 5 to 10 m, greatly affect the suitability of soils for most urban uses that involve construction or excavation. The proper functioning of onsite sewage disposal systems and the stability of building foundations are often determined by regolith properties at these depths. Likewise, processes that control the movement of pollutants to groundwater or the weathering of geologic materials may occur at depths of many meters. These deep layers also have major ecological influences because, although the intensity of biological activity and plant rooting may be quite low, the total impact can be great as a result of the enormous volume of soil that may be involved. This is especially true of forest systems in warm climates.

1.11 MINERAL (INORGANIC) CONSTITUENTS OF SOILS

Except in the case of organic soils, most of a soil's solid framework consists of **mineral**[2] particles. The larger soil particles, which include stones, gravel, and coarse sands, are generally rock fragments of various kinds; that is, these larger particles are often aggregates of several different minerals. Most smaller particles tend to be made of a single mineral. Thus, any particular soil consists of particles that vary greatly in both size and composition.

---

[2] The word *mineral* is used in soil science in three ways: (1) as a general adjective to describe inorganic materials derived from rocks; (2) as a specific noun to refer to distinct minerals found in nature, such as quartz and feldspars (see Chapter 2 for detailed discussions of soil-forming minerals and the rocks in which they are found); and (3) as an adjective to describe chemical elements, such as nitrogen and phosphorus, in their inorganic state in contrast to their occurrence as part of organic compounds.

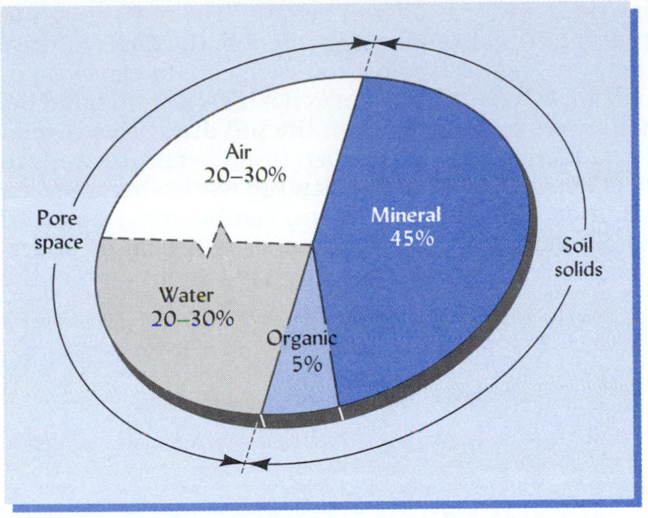

FIGURE 1.12 Volume composition of a loam surface soil when conditions are good for plant growth. The broken line between water and air indicates that the proportions of these two components fluctuate as the soil becomes wetter or drier. Nonetheless, a nearly equal proportion of air and water is generally ideal for plant growth.

The mineral particles present in soils are extremely variable in size. Excluding, for the moment, the larger rock fragments such as stones and gravel, soil particles range in size from **sand** (2.0 to 0.05 mm in diameter) to **silt** (0.05 to 0.002 mm) to **clay** (smaller than 0.002 mm). Each group of particles exhibits unique properties and influences on soil behavior (see Section 4.2).

The smaller particles (< 0.001 mm) of clay (and similar-sized organic particles) have **colloidal**[3] properties and can be seen only with the aid of an electron microscope. Because of their extremely small size, colloidal particles possess a tremendous amount of surface area per unit of mass. Since the surfaces of soil colloids (both mineral and organic) exhibit electromagnetic charges that attract positive and negative ions as well as water, this fraction of the soil is the seat of most of the soil's chemical and physical activity.

The proportion of particles in these different size ranges is called **soil texture.** Terms such as *sandy loam, silty clay,* and *clay loam* are used to identify the soil texture. Texture has a profound influence on many soil properties, and it affects the suitability of a soil for most uses. To understand the degree to which soil properties can be influenced by texture, imagine dressing in a bathing suit and lying first on a sandy beach, and then in a clayey mud puddle. The difference in these two experiences would be due largely to the properties described in lines 5 and 7 of Table 1.2. Other properties related to particle size are also listed therein. Note that clay-sized particles play a dominant role in holding certain inorganic chemicals and supplying nutrients to plants.

TABLE 1.2 General Properties of the Three Major Size Classes of Inorganic Soil Particles

	Property	*Sand*	*Silt*	*Clay*
1.	Range of particle diameters in mm	2.0–0.05	0.05–0.002	Smaller than 0.002
2.	Means of observation	Naked eye	Microscope	Electron microscope
3.	Dominant minerals	Primary	Primary and secondary	Secondary
4.	Attraction of particles for each other	Low	Medium	High
5.	Attraction of particles for water	Low	Medium	High
6.	Ability to hold chemicals and nutrients in plant-available form	Very low	Low	High
7.	Consistency when wet	Loose, gritty	Smooth	Sticky, malleable
8.	Consistency when dry	Very loose, gritty	Powdery, some clods	Hard clods

[3] Colloidal systems are two-phase systems in which very small particles of one substance are dispersed in a medium of a different substance. Clay and organic soil particles smaller than about 0.001 mm (1 micrometer, μm) in diameter are generally considered to be colloidal in size. Milk and blood are other examples of colloidal systems in which very small solid particles are dispersed in a liquid medium.

To anticipate the effect of clay on the way a soil will behave, it is not enough to know only the *amount* of clay in a soil. It is also necessary to know the *kinds* of clays present. As home builders and highway engineers know too well, certain clayey soils, such as those high in smectite clays, make very unstable material on which to build because the clays swell when the soil is wet and shrink when the soil dries. This shrink-and-swell action can easily crack foundations and cause even heavy retaining walls to collapse. These clays also become extremely sticky and difficult to work when they are wet. Other types of clays, formed under different conditions, can be very stable and easy to work. Learning about the different types of clay minerals will help us understand many of the physical and chemical differences among soils in various parts of the world (see Box 1.2).

BOX 1.2 OBSERVING SOILS IN DAILY LIFE

Your study of soils can be enriched if you make an effort to become aware of the many daily encounters with soils and their influences that go unnoticed by most people. When you dig a hole to plant a tree or set a fence post, note the different layers encountered, and note how the soil from each layer looks and feels. If you pass a construction site, take a moment to observe the horizons exposed by the excavations. An airplane trip is a great opportunity to observe how soils vary across landscapes and climatic zones. If you are flying during daylight hours, ask for a window seat. Look for the shapes of individual soils in plowed fields if you are flying in spring or fall (Figure 1.13).

Soils can give you clues to understanding the natural processes going on around you. Down by the stream, use a magnifying glass to examine the sand deposited on the banks or bottom. It may contain minerals not found in local rocks and soils, but originating many kilometers upstream. When you wash your car, see if the mud clinging to the tires and fenders is of a different color or consistency than the soils near your home. Does the "dirt" on your car tell you where you have been driving? Forensic investigators have been known to consult with soil scientists to locate crime victims or establish guilt by matching soil clinging to shoes, tires, or tools with the soils at a crime scene.

Other examples of soil hints can be found even closer to home. The next time you bring home celery or leaf lettuce from the supermarket, look carefully for bits of soil clinging to the bottom of the stalk or leaves (Figure 1.14). Rub the soil between your thumb and fingers. Smooth, very black soil may indicate that the lettuce was grown in mucky soils, such as those in New York State or southern Florida. Light brown, smooth-feeling soil with only a very fine grittiness is more typical of California-grown produce, while light-colored, gritty soil is common on produce from the southern Georgia–northern Florida vegetable-growing region. In a bag of dry pinto beans, you may come across a few lumps of soil that escaped removal in the cleaning process because of being the same size as the beans. Often this soil is dark-colored and very sticky, coming from the "thumb" area of Michigan, where a large portion of the U.S. dry bean crop is grown.

Opportunities to observe soils in daily life range from the remote and large-scale to the close-up and intimate. As you learn more about soils, you will undoubtedly be able to see more examples of their influence in your surroundings.

FIGURE 1.13 *The light- and dark-colored soil bodies, as seen from an airliner flying over central Texas, reflect differences in drainage and topography in the landscape. (Photo courtesy of R. Weil)*

FIGURE 1.14 *Although this celery was purchased in a Virginia grocery store in early fall, the black, mucky soil clinging to the base of the stalk indicates that it was grown on organic soils, probably in New York State. (Photo courtesy of R. Weil)*

Soil Structure

Sand, silt, and clay particles can be thought of as the building blocks from which soil is constructed. The manner in which these building blocks are arranged together is called **soil structure.** The particles may remain relatively independent of each other, but more commonly they are associated together in aggregates of different-size particles. These aggregates may take the form of roundish granules, cubelike blocks, flat plates, or other shapes. Soil structure (the way particles are arranged together) is just as important as soil texture (the relative amounts of different sizes of particles) in governing how water and air move in soils. Both structure and texture fundamentally influence the suitability of soils for the growth of plant roots.

1.12 SOIL ORGANIC MATTER

Soil organic matter consists of a wide range of organic (carbonaceous) substances, including living organisms (the soil **biomass**), carbonaceous remains of organisms that once occupied the soil, and organic compounds produced by current and past metabolism in the soil. The remains of plants, animals, and microorganisms are continuously broken down in the soil and new substances are synthesized by other microorganisms. Over time, organic matter is lost from the soil as carbon dioxide produced by microbial respiration. Because of such loss, repeated additions of new plant and/or animal residues are necessary to maintain soil organic matter.

Under conditions that favor plant production more than microbial decay, large quantities of atmospheric carbon dioxide used by plants in photosynthesis are sequestered in the abundant plant tissues which eventually become part of the soil organic matter. Because carbon dioxide is a major cause of the greenhouse effect which is believed to be warming the earth's climate, the balance between accumulation of soil organic matter and its loss through microbial respiration has global implications. In fact, more carbon is stored in the world's soils than in the world's plant biomass and atmosphere combined.

Even so, organic matter comprises only a small fraction of the mass of a typical soil. By weight, typical well-drained mineral surface soils contain from 1 to 6% organic matter. The organic matter content of subsoils is even smaller. However, the influence of organic matter on soil properties, and consequently on plant growth, is far greater than the low percentage would indicate.

Organic matter binds mineral particles into a granular soil structure that is largely responsible for the loose, easily managed condition of productive soils. Part of the soil organic matter that is especially effective in stabilizing these granules consists of certain gluelike substances produced by various soil organisms, including plant roots (Figure 1.15).

Organic matter also increases the amount of water a soil can hold and the proportion of water available for plant growth (Figure 1.16). In addition, it is a major soil source of the plant nutrients phosphorus and sulfur, and the primary source of nitrogen for

FIGURE 1.15 Abundant organic matter, including plant roots, helps create physical conditions favorable for the growth of higher plants as well as microbes (left). In contrast, soils low in organic matter, especially if they are high in silt and clay, are often cloddy (right) and not suitable for optimum plant growth.

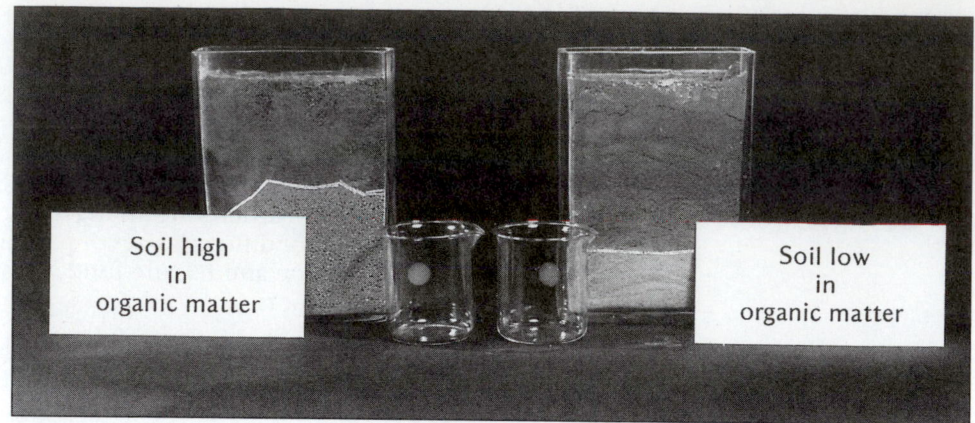

FIGURE 1.16 Soils higher in organic matter are darker in color and have greater water-holding capacities than soils low in organic matter. The soil in each container has the same texture, but the one on the right has been depleted of much of its organic matter. The same amount of water was applied to each container. As the photo shows, the depth of water penetration was less in the high organic matter soil (left) because of its greater water-holding capacity. It required a greater volume of the low organic matter soil to hold the same amount of water.

most plants. As soil organic matter decays, these nutrient elements, which are present in organic combinations, are released as soluble ions that can be taken up by plant roots. Finally, organic matter, including plant and animal residues, is the main food that supplies carbon and energy to soil organisms. Without it, biochemical activity so essential for ecosystem functioning would come to a near standstill.

Humus, usually black or brown in color, is a collection of very complex organic compounds which accumulate in soil because they are relatively resistant to decay. Just as clay is the colloidal fraction of soil mineral matter, so humus is the colloidal fraction of soil organic matter. Because of their charged surfaces, both humus and clay act as contact bridges between larger soil particles; thus, both play an important role in the formation of soil structure. The surface charges of humus, like those of clay, attract and hold both nutrient ions and water molecules. However, gram for gram, the capacity of humus to hold nutrients and water is far greater than that of clay. Unlike clay, humus contains certain components that can have a hormone-like stimulatory effect on plants. All in all, small amounts of humus may remarkably increase the soil's capacity to promote plant growth.

1.13 SOIL WATER: A DYNAMIC SOLUTION

Water is of vital importance in the ecological functioning of soils. The presence of water in soils is essential for the survival and growth of plants and other soil organisms. The soil moisture regime, often reflective of climatic factors, is a major determinant of the productivity of terrestrial ecosystems, including agricultural systems. Movement of water, and substances dissolved in it, through the soil profile is of great consequence to the quality and quantity of local and regional water resources. Water moving through the regolith is also a major driving force in soil formation.

Two main factors help explain why soil water is different from our everyday concept of, say, drinking water in a glass.

1. Water is held within soil pores with varying degrees of tenacity depending on the amount of water present and the size of the pores. The attraction between water and the surfaces of soil particles greatly restricts the ability of water to flow as it would flow in a drinking glass.
2. Because soil water is never pure water, but contains hundreds of dissolved organic and inorganic substances, it may be more accurately called the **soil solution.** An important function of the soil solution is to serve as a constantly replenished, dilute nutrient solution bringing dissolved nutrient elements (e.g., calcium, potassium, nitrogen, and phosphorus) to plant roots.

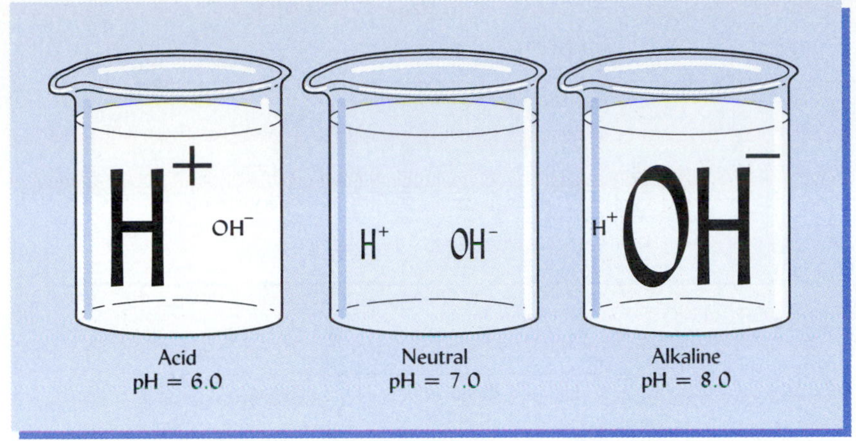

FIGURE 1.17 Diagrammatic representation of acidity, neutrality, and alkalinity. At neutrality the H^+ and OH^- ions of a solution are balanced, their respective numbers being the same (pH 7). At pH 6, the H^+ ions are dominant, being 10 times greater, whereas the OH^- ions have decreased proportionately, being only one-tenth as numerous. The solution therefore is acid at pH 6, there being 100 times more H^+ ions than OH^- ions present. At pH 8, the exact reverse is true; the OH^- ions are 100 times more numerous than the H^+ ions. Hence, the pH 8 solution is alkaline. This mutually inverse relationship must always be kept in mind when pH data are used.

When the soil moisture content is optimal for plant growth (Figure 1.12), the water in the large- and intermediate-sized pores can move about in the soil and can easily be used by plants. As the plant grows, however, its roots remove water from the largest pores first. Soon the larger pores hold only air, and the remaining water is found only in the intermediate- and smallest-sized pores. The water in the intermediate-sized pores can still move toward plant roots and be taken up by them. However, the water in the smallest pores is so close to solid particles that it is strongly attracted to and held on the particle surfaces. This water may be so strongly held that plant roots cannot pull it away. Consequently, not all soil water is *available* to plants. Depending on the soil, one-sixth to one-half of the water may remain in the soil after plants have wilted or died for lack of moisture.

Soil Solution

The soil solution contains small but significant quantities of soluble inorganic compounds, some of which supply elements that are essential for plant growth. Refer to Table 1.1 for a listing of the 18 essential elements and their sources. The soil solids, particularly the fine organic and inorganic colloidal particles, release these elements to the soil solution, from which they are taken up by growing plants. Such exchanges, which are critical for higher plants, are dependent on both soil water and the fine soil solids.

Another critical property of the soil solution is the balance or inverse relationship between H^+ and OH^- ions. When H^+ and OH^- ions are in equal supply, they both have a concentration of 1×10^{-7} moles/L, and the solution is **neutral.** The solution is **acidic** when H^+ ions outnumber OH^- ions (more than 10^{-7} H^+ ions/L and fewer than 10^{-7} OH^- ions/L). It is considered *basic* or **alkaline** when the reverse is true. The acidity or alkalinity of the soil solution is usually expressed as the **pH** or negative logarithm of the H^+ ion concentration. Because of this logarithmic relationship, a difference of one pH unit represents a tenfold difference in the concentrations of H^+ and OH^- ions. For example, the number of H^+ ions present in a liter of solution at pH 5 is 10 times that at pH 6 (see Figure 1.17). For a detailed explanation of pH, refer to Box 9.1.

Often referred to as a *master variable,* pH controls many chemical and biological reactions in soils, including the bioavailability of many nutrients and pollutants. Figure 1.18 shows the descriptive names associated with different pH levels and the pH levels typical for several types of soils. A good grasp of pH relationships is essential for understanding many aspects of soil behavior and management.

1.14 SOIL AIR: A CHANGING MIXTURE OF GASES

Approximately half of the volume of the soil consists of pore spaces of varying sizes (refer to Figure 1.12), which are filled with either water or air. When water enters the soil, it displaces air from some of the pores; the air content of a soil is therefore inversely related to its water content. If we think of the network of soil pores as the ventilation system of the soil connecting airspaces to the atmosphere, we can understand that when

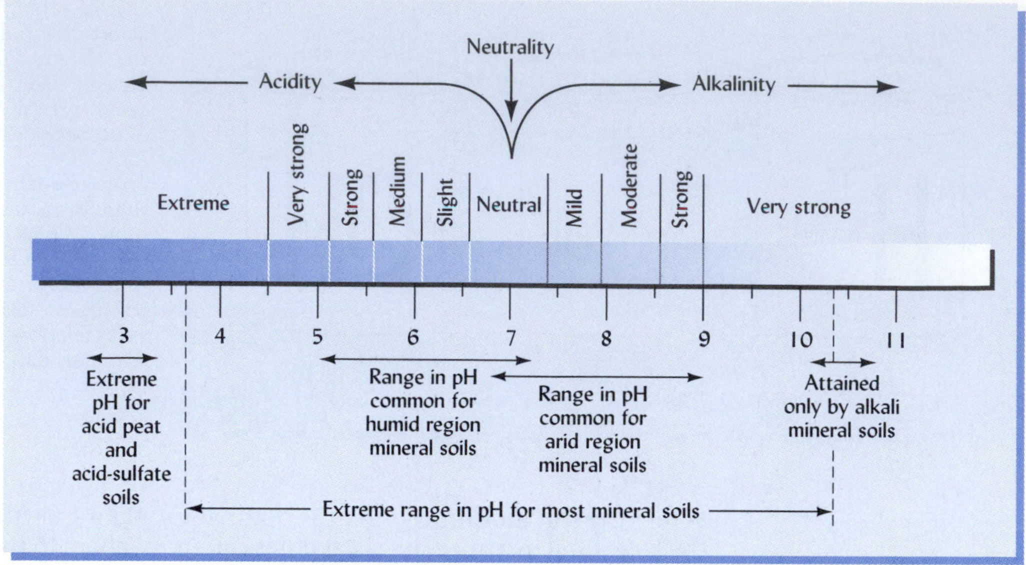

FIGURE 1.18 Value of pH found in various types of soils.

the smaller pores are filled with water the ventilation system becomes clogged. Think how stuffy the air would become if the ventilation ducts of a classroom became clogged. Because oxygen could not enter the room, nor carbon dioxide leave it, the air in the room would soon become depleted of oxygen and enriched in carbon dioxide and water vapor by the respiration (breathing) of the people in it. In an air-filled soil pore surrounded by water-filled smaller pores, the metabolic activities of plant roots and microorganisms have a similar effect.

Therefore, soil air differs from atmospheric air in several respects. First, the composition of soil air varies greatly from place to place in the soil. In local pockets, some gases are consumed by plant roots and by microbial reactions, and others are released, thereby greatly modifying the composition of the soil air. Second, soil air generally has a higher moisture content than the atmosphere; the relative humidity of soil air approaches 100% unless the soil is very dry. Third, the content of carbon dioxide (CO_2) is usually much higher, and that of oxygen (O_2) lower, than contents of these gases found in the atmosphere. Carbon dioxide in soil air is often several hundred times more concentrated than the 0.035% commonly found in the atmosphere. Oxygen decreases accordingly and, in extreme cases, may be 5 to 10%, or even less, compared to about 20% for atmospheric air.

The amount and composition of air in a soil are determined to a large degree by the water content of the soil. The air occupies those soil pores not filled with water. As the soil drains from a heavy rain or irrigation, large pores are the first to be filled with air, followed by medium-sized pores, and finally the small pores, as water is removed by evaporation and plant use. This explains the tendency for soils with a high proportion of tiny pores to be poorly aerated. In such soils, water dominates, and the soil air content is low, as is the rate of diffusion of the air into and out of the soil from the atmosphere. The result is high levels of CO_2 and low levels of O_2, unsatisfactory conditions for the growth of most plants. In extreme cases, lack of oxygen both in the soil air and dissolved in the soil water may fundamentally alter the chemical reactions that take place in the soil solution. This is of particular importance to understanding the functions of wetland soils.

1.15 INTERACTION OF FOUR COMPONENTS TO SUPPLY PLANT NUTRIENTS

As you read our discussion of each of the four major soil components, you may have noticed that the impact of one component on soil properties is seldom expressed independently from that of the others. Rather, the four components interact with each other to determine the nature of a soil. Thus, soil moisture, which directly meets the needs of plants for water, simultaneously controls much of the air and nutrient supply to the plant roots. The mineral particles, especially the finest ones, attract soil water, thus determining its movement and availability to plants. Likewise, organic matter, because of its physical binding power, influences the arrangement of the mineral particles into clusters and, in so doing, increases the number of large soil pores, thereby influencing the water and air relationships.

Essential Element Availability

Perhaps the most important interactive process involving the four soil components is the provision of essential nutrient elements to plants. Plants absorb essential nutrients, along with water, directly from one of these components: the soil solution. However, the amount of essential nutrients in the soil solution at any one time is far less than is needed to produce a mature plant. Consequently, the soil solution nutrient levels must be constantly replenished from the inorganic or organic parts of the soil and from fertilizers or manures added to agricultural soils.

Fortunately, relatively large quantities of these nutrients are associated with both inorganic and organic soil solids. By a series of chemical and biochemical processes, nutrients are released from these solid forms to replenish those in the soil solution. For example, the tiniest colloidal-sized particles—both clay and humus—exhibit negative and positive charges. These charges tend to attract or **adsorb**[4] oppositely charged ions from the soil solution and hold them as **exchangeable ions.** Through ion exchange, elements such as Ca^{2+} and K^{+} are released from this state of electrostatic adsorption on colloidal surfaces and escape into the soil solution. In the following example, a H^{+} ion in the soil solution is shown to exchange places with an adsorbed K^{+} ion on the colloidal surface.

$$\underset{\text{Adsorbed}}{\boxed{\text{colloid}}\,K^{+}} + \underset{\text{Soil solution}}{H^{+}\text{ ion}} \longrightarrow \underset{\text{Adsorbed}}{\boxed{\text{colloid}}\,H^{+}} + \underset{\text{Soil solution}}{K^{+}\text{ ion}}$$

The K^{+} ion thus released can be readily taken up (absorbed) by plants. Some scientists consider that this ion exchange process is among the most important of chemical reactions in nature.

Nutrient ions are also released to the soil solution as soil microorganisms decompose organic tissues. Plant roots can readily absorb all of these nutrients from the soil solution, provided there is enough O_2 in the soil air to support root metabolism.

Most soils contain large amounts of plant nutrients relative to the annual needs of growing vegetation. However, the bulk of most nutrient elements is held in the structural framework of primary and secondary minerals and organic matter. Only a small fraction of the nutrient content of a soil is present in forms readily available to plants. Table 1.3 shows the quantities of various essential elements present in different forms in typical soils of humid and arid regions.

TABLE 1.3 Quantities of Six Essential Elements Found in Upper 15 cm of Representative Soils in Temperate Regions

	Humid region soil			*Arid region soil*		
Essential element	*In solid framework, kg/ha*	*Exchangeable, kg/ha*	*In soil solution, kg/ha*	*In solid framework, kg/ha*	*Exchangeable, kg/ha*	*In soil solution, kg/ha*
Ca	8,000	2,250	60–120	20,000	5,625	140–280
Mg	6,000	450	10–20	14,000	900	25–40
K	38,000	190	10–30	45,000	250	15–40
P	900	—	0.05–0.15	1,600	—	0.1–0.2
S	700	—	2–10	1,800	—	6–30
N	3,500	—	7–25	2,500	—	5–20

[4] *Adsorption* refers to the attraction of ions to the surface of particles, in contrast to *absorption,* the process by which ions are taken *into* plant roots. The adsorbed ions are exchangeable with ions in the soil solution.

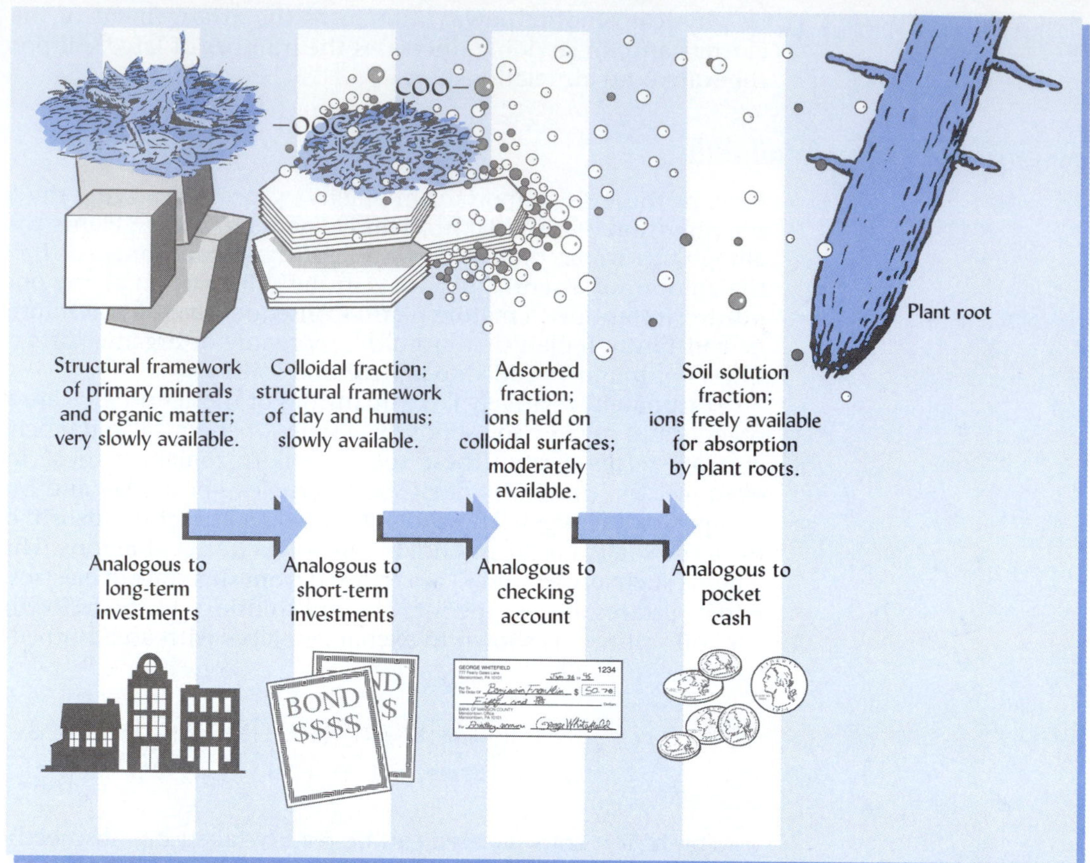

FIGURE 1.19 Nutrient elements exist in soils in various forms characterized by different accessibility to plant roots. The bulk of the nutrients is locked up in the structural framework of primary minerals, organic matter, clay, and humus. A smaller proportion of each nutrient is adsorbed in a swarm of ions near the surfaces of soil colloids (clay and organic matter). From the swarm of adsorbed ions, a still smaller amount is released into the bulk soil solution, where uptake by plant roots can take place. (Diagram courtesy of R. Weil)

Nutrient ion uptake across membranes—an advanced lecture:
http://www.botany.ubc.ca/biol351/351d.htm

Figure 1.19 illustrates how the two solid soil components interact with the liquid component (soil solution) to provide essential elements to plants. Plant roots do not ingest soil particles, no matter how fine, but are able to absorb only nutrients that are dissolved in the soil solution. Because elements in the coarser soil framework of the soil are only slowly released into the soil solution over long periods of time, the bulk of most nutrients in a soil is not readily available for plant use. Nutrient elements in the framework of colloid particles are somewhat more readily available to plants, as these particles break down much faster because of their greater surface area. Thus, the structural framework is the major storehouse and, to some extent, a significant source of essential elements in many soils.

The distribution of nutrients among the various components of a fertile soil, as illustrated in Figure 1.19, may be likened to the distribution of financial assets in the portfolio of a wealthy individual. In such an analogy, nutrients readily available for plant use would be analogous to cash in the individual's pocket. A millionaire would likely keep most of his or her assets in long-term investments such as real estate or bonds (the coarse fraction solid framework), while investing a smaller amount in short-term stocks and bonds (colloidal framework). For more immediate use, an even smaller amount might be kept in a checking account (exchangeable nutrients), while a tiny fraction of the overall wealth might be carried to spend as currency and coins (nutrients in the soil solution). As the cash is used up, the supply is replenished by making a withdrawal from the checking account. The checking account, in turn, is replenished occasionally by the sale of long-term investments. It is possible for a wealthy person to run short of cash even though he or she may own a great deal of valuable real estate. In an analogous way, plants may use up the readily available supply of a nutrient even though the total supply of that nutrient in the soil is very large. Luckily, in a fertile soil, the process described in Figure 1.19 can help replenish the soil solution as quickly as plant roots remove essential elements.

1.16 NUTRIENT UPTAKE BY PLANT ROOTS

To be taken up by a plant, a nutrient element must be in a soluble form and must be located *at the root surface*. Often, parts of a root are in such intimate contact with soil particles that a direct exchange may take place between nutrient ions adsorbed on the surface of soil colloids and H^+ ions from the surface of root cell walls. In any case, the supply of nutrients in contact with the root will soon be depleted. This fact raises the question of how a root can obtain additional supplies once the nutrient ions at the root surface have all been taken up into the root. There are three basic mechanisms by which the concentration of nutrient ions at the root surface is maintained (Figure 1.20).

First, **root interception** comes into play as roots continually grow into new, undepleted soil. For the most part, however, nutrient ions must travel some distance in the soil solution to reach the root surface. This movement can take place by **mass flow**, as when dissolved nutrients are carried along with the flow soil water toward a root that is actively drawing water from the soil. In this type of movement, the nutrient ions are somewhat analogous to leaves floating down a stream. On the other hand, plants can continue to take up nutrients even at night, when water is only slowly absorbed into the roots. Nutrient ions continually move by **diffusion** from areas of greater concentration toward the nutrient-depleted areas of lower concentration around the root surface.

In the diffusion process, the random movements of ions in all directions causes a *net* movement from areas of high concentration to areas of lower concentrations, independent of any mass flow of the water in which the ions are dissolved. Factors such as soil compaction, cold temperatures, and low soil moisture content, which reduce root interception, mass flow, or diffusion, can result in poor nutrient uptake by plants even in soils with adequate supplies of soluble nutrients. Furthermore, the availability of nutrients for uptake can also be negatively or positively influenced by the activities of microorganisms that thrive in the immediate vicinity of roots. Maintaining the supply of available nutrients at the plant root surface is thus a process that involves complex interactions among different soil components.

It should be noted that the plant membrane separating the inside of the root cell from the soil solution is permeable to dissolved ions only under special circumstances. Plants do not merely take up, by mass flow, those nutrients that happen to be in the water that roots are removing from the soil. Nor do dissolved nutrient ions brought to the root's outer surface by mass flow or diffusion cross the root cell membrane and enter the root passively by diffusion. On the contrary, a nutrient is normally taken up into the plant root cell only by reacting with a specific chemical binding site on a large protein carrier molecule that forms a hydrophilic channel across an otherwise hydrophobic lipid (fatty) membrane. Energy from metabolism in the root cell is used to activate this carrier protein so that it will

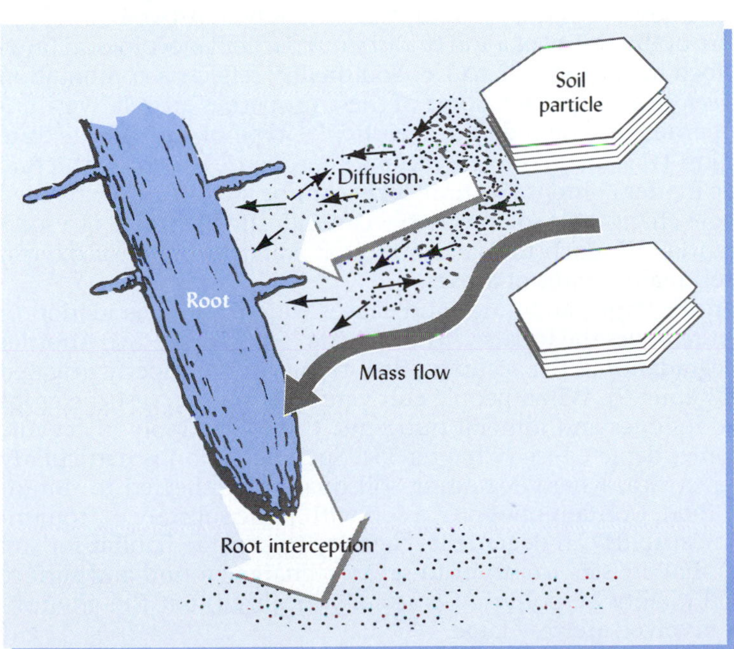

FIGURE 1.20 Three principal mechanisms by which nutrient ions dissolved in the soil solution come into contact with plant roots. All three mechanisms may operate simultaneously, but one mechanism or another may be most important for a particular nutrient. For example, in the case of calcium, which is generally plentiful in the soil solution, mass flow alone can usually bring sufficient amounts to the root surface. However, in the case of phosphorus, diffusion is needed to supplement mass flow because the soil solution is very low in this element in comparison to the amounts needed by plants. (Diagram courtesy of R. Weil)

pass the nutrient ion across the cell membrane and release it into the cell interior. This carrier mechanism allows the plant to accumulate concentrations of a nutrient inside the root cell that far exceed that nutrient's concentration in the soil solution. Because different nutrients are taken up by specific types of carrier molecules, the plant is able to exert some control over how much and in what relative proportions essential elements are taken up.

Since nutrient uptake is an active metabolic process, conditions that inhibit root metabolism may also inhibit nutrient uptake. Examples of such conditions include excessive soil water content or soil compaction resulting in poor soil aeration, excessively hot or cold soil temperatures, and aboveground conditions which result in low translocation of sugars to plant roots. We can see that plant nutrition involves biological, physical, and chemical processes and interactions among many different components of soils and the environment.

1.17 SOIL QUALITY, DEGRADATION, AND RESILIENCE

What is soil quality? Here is an agricultural introduction: http://soils.usda.gov/sqi/sqw.html

Soil is a basic resource underpinning all terrestrial ecosystems. Managed carefully, soils are a *reusable* resource, but in the scale of human lifetimes they cannot be considered a *renewable* resource. As we shall see in the next chapter, most soil profiles are thousands of years in the making. In all regions of the world, human activities are destroying some soils far faster than nature can rebuild them. As mentioned in the opening paragraphs of this chapter, growing numbers of people are demanding more and more from the Earth's fixed amount of land. Nearly all of the soils best suited for growing crops are already being farmed. Therefore, as each year brings millions more people to feed, the amount of cropland per person continuously declines. In addition, many of the world's major cities were originally located where excellent soils supported thriving agricultural communities, so now much of the very best farmland is being lost to suburban development as these cities thoughtlessly expand.

Finding more land on which to grow food is not easy. Most additional land brought under cultivation comes at the cost of clearing natural forests, savannas, and grasslands. Images of the Earth made from orbiting satellites show the resulting decline in land covered by forests and other natural ecosystems. Thus, as the human population struggles to feed itself, wildlife populations are deprived of vital habitat, and overall biodiversity suffers. Efforts to reduce and even reverse human population growth must be accelerated if our grandchildren are to inherit a livable world. In the meantime, if there is to be space for both people and wildlife, the best of our existing farmland soils will require improved and more intensive management. Soils completely washed away by erosion or excavated and paved over by urban sprawl are permanently lost, for all practical purposes. More often, soils are degraded in quality rather than totally destroyed.

Soil quality is a measure of the ability of a soil to carry out particular ecological functions, such as those described in Sections 1.2 to 1.6. Soil quality reflects a combination of *chemical, physical,* and *biological* properties. Some of these properties are relatively unchangeable, inherent properties that help define a particular type of soil. Soil texture and mineral makeup (Section 1.11) are examples. Other soil properties, such as structure (Section 1.11) and organic matter content (Section 1.12), can be significantly changed by management. These more changeable soil properties can indicate the status of a soil's quality relative to its potential, in much the same way that water turbidity or oxygen content indicates the water-quality status of a river.

Mismanagement of forests, farms, and rangeland causes widespread degradation of soil quality by erosion that removes the topsoil, little by little (see Chapter 15). Another widespread cause of soil degradation is the accumulation of salts in improperly irrigated soils in arid regions (see Chapter 9). When people cultivate soils and harvest the crops without returning organic residues and mineral nutrients, the soil's supply of organic matter and nutrients becomes depleted (see Chapter 11). Such depletion is particularly widespread in sub-Saharan Africa, where degrading soil quality is reflected in diminished capacity to produce food. Contamination of a soil with toxic substances from industrial processes or chemical spills can degrade its capacity to provide habitat for soil organisms, to grow plants that are safe to eat, or to safely recharge ground and surface waters. Degradation of soil quality by pollution is usually localized, but the environmental impacts and costs involved are very large.

While protecting soil quality must be the first priority, it is often necessary to attempt to restore the quality of soils that have already been degraded. Some soils have sufficient **resilience** to recover from minor degradation if left to revegetate on their own. In other cases, more effort is required to restore degraded soils (see Chapter 15). Organic and inorganic amendments may have to be applied, vegetation may have to be planted, physical alterations by tillage or grading may have to be made, or contaminants may have to be removed. As societies around the world assess the damage already done to their natural and agricultural ecosystems, the science of **restoration ecology** has rapidly evolved to guide managers in restoring plant and animal communities to their former levels of diversity and productivity. The job of **soil restoration**, an essential part of these efforts, requires in-depth knowledge of all aspects of the soil system.

1.18 CONCLUSION

The earth's soil consists of numerous soil individuals, each of which is a three-dimensional natural body in the landscape. Each individual soil is characterized by a unique set of properties and soil horizons as expressed in its profile. The nature of the soil layers seen in a particular profile is closely related to the nature of the environmental conditions at a site.

Soils perform five broad ecological functions. They act as the principal medium for plant growth, regulate water supplies, recycle raw materials and waste products, and serve as a major engineering medium for human-built structures. They are also home to many kinds of living organisms. Soil is thus a major ecosystem in its own right. The soils of the world are extremely diverse, each type of soil being characterized by a unique set of soil horizons. A typical surface soil in good condition for plant growth consists of about half solid material (mostly mineral, but with a crucial organic component, too) and half pore spaces filled with varying proportions of water and air. These components interact to influence a myriad of complex soil functions, a good understanding of which is essential for wise management of our terrestrial resources.

If we take the time to learn the language of the land, the soil will speak to us.

STUDY QUESTIONS

1. As a society, is our reliance on soils likely to increase or decrease in the decades ahead? Explain.
2. Discuss how *a soil*, a natural body, differs from *soil*, a material that is used in building a roadbed?
3. What are the five main roles of soil in an ecosystem? For each of these ecological roles, suggest one way in which interactions occur with another of the five roles.
4. Think back over your activities during the past week. List as many incidents as you can in which you came into direct or indirect contact with soil.
5. Figure 1.12 shows the volume composition of a loam surface soil in ideal condition for plant growth. To help you understand the relationships among the four components, redraw this pie chart to represent what the situation might be after the soil has been compacted by heavy traffic. Redraw the original pie chart again, but show how the four components would be related on a mass (weight) basis rather than on a volume basis.
6. Explain in your own words how the soil's nutrient supply is held in different forms, much the way that a person's financial assets might be held in different forms.
7. List the essential nutrient elements that plants derive mainly from the soil.
8. Are all elements contained in plants essential nutrients? Explain.
9. Define these terms: *soil texture, soil structure, soil pH, humus, soil profile, B horizon, soil quality, solum,* and *saprolite.*
10. Describe four processes that commonly lead to degradation of soil quality.
11. Compare the pedological and edaphological approaches to the study of soils. Which is more closely aligned with geology and which with ecology?

2 FORMATION OF SOILS FROM PARENT MATERIALS

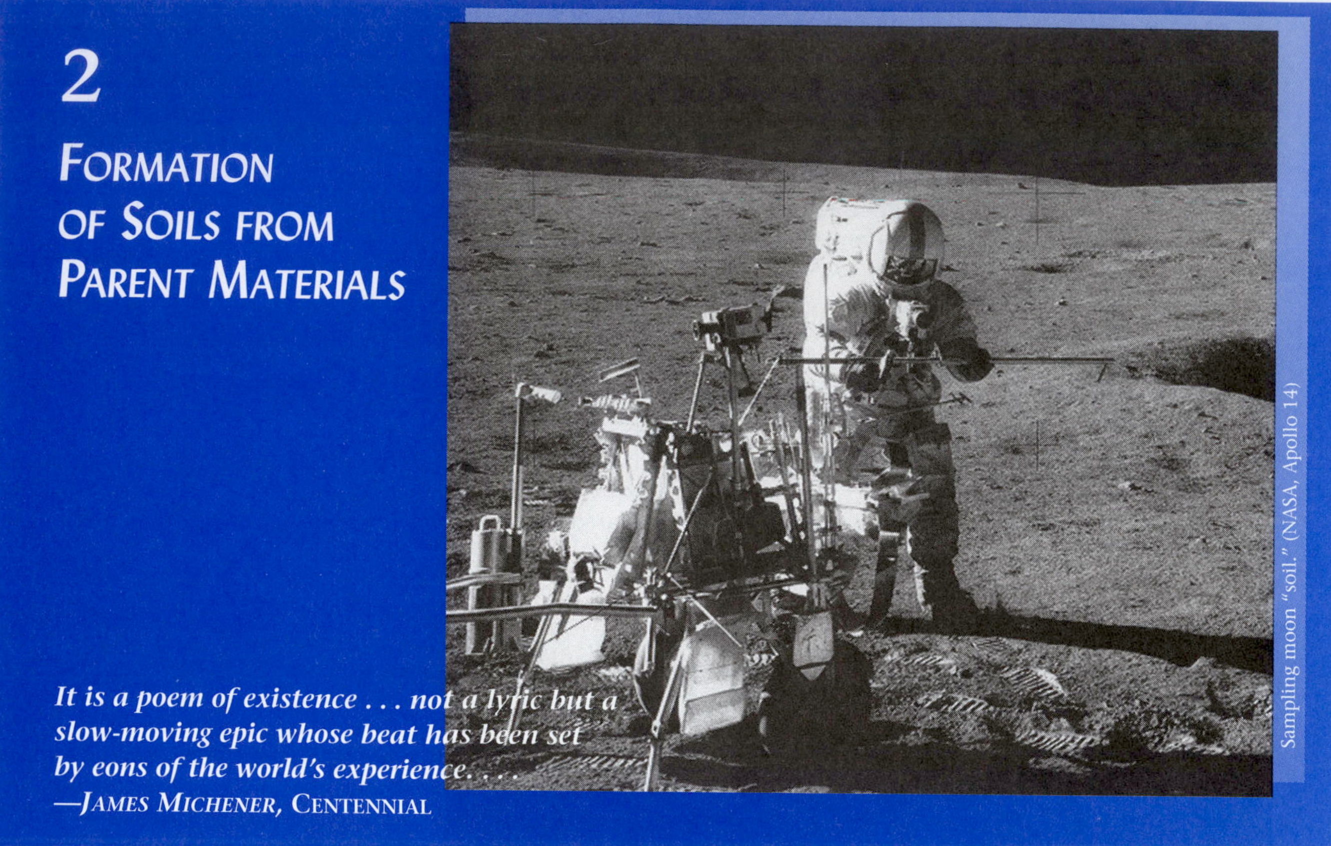
Sampling moon "soil." (NASA, Apollo 14)

It is a poem of existence . . . not a lyric but a slow-moving epic whose beat has been set by eons of the world's experience. . . .
—JAMES MICHENER, CENTENNIAL

The first astronauts to explore the moon labored in their clumsy pressurized suits to collect samples of rocks and dust from the lunar surface. These they carried back to Earth for analysis. It turned out that moon rocks are similar in composition to those found deep in the Earth—so similar that scientists concluded that the moon itself began as a large chunk of molten earth that broke away eons ago, when the young planet nearly melted in a stupendous collision with a Mars-sized object, leaving the Pacific Ocean as a scar. On the moon, this rock remained unchanged or crumbled into dust with the impact of meteors. On Earth, the rock at the surface, eventually coming in contact with water, air, and living things, was transformed into something new, into many different kinds of living soils. This chapter reveals the story of how rock and dust become "the ecstatic skin of the Earth."[1]

We will study the processes of soil formation that transform the lifeless regolith into the variegated layers of the soil profile. We will also learn about the environmental factors that influence these processes to produce soils in Belgium so different from those in Brazil, soils on limestone so different from those on sandstone, and soils in the valley bottoms so different from those on the hills.

Every landscape is a suite of different soils, each influencing ecological processes in its own way. Whether we intend to modify, exploit, preserve, or simply understand the landscape, our success will depend on our knowing how soil properties relate to the environment on each site and to the landscape as a whole.

2.1 WEATHERING OF ROCKS AND MINERALS

The influence of **weathering**, the physical and chemical breakdown of particles, is evident everywhere. Nothing escapes it. Weathering breaks up rocks and minerals, modifies or destroys their physical and chemical characteristics, and carries away the soluble

[1]The apt description of the soil as "ecstatic skin" is from Logan (1995). For some intriguing evidence from Mars Pathfinder that Earth may not be the only planet with a skin of soil, see Rieder et al. (1997).

FIGURE 2.1 Two stone markers, photographed on the same day in the same cemetery, illustrate the effect of rock type on weathering rates. The date and initials carved in the slate marker in 1798 are still sharp and clear, while the date and figure of a lamb carved in the marble marker in 1875 have weathered almost beyond recognition. The slate rock consists largely of resistant silicate clay minerals, while the marble consists mainly of calcite, which is much more easily attacked by acids in rainwater. (Photo courtesy of R. Weil)

products. Likewise, it synthesizes new minerals of great significance in soils. The nature of the rocks and minerals being weathered determines the rates and results of the breakdown and synthesis (Figure 2.1).

Characteristics of Rocks and Minerals

Rocks and the rock cycle illustrated:
http://as.tsud.edu/earth_science/notes/rocks.htm

The rocks in the earth's outer surface are commonly classified as **igneous, sedimentary,** and **metamorphic.** Those of igneous origin are formed from molten magma and include such common rocks as granite and diorite (Figure 2.2).

Igneous rock is composed of primary minerals[2] such as light-colored quartz, muscovite, and feldspars and dark-colored biotite, augite, and hornblende. In general, dark-colored minerals contain iron and magnesium and are more easily weathered. Therefore, dark-colored igneous rocks such as gabbro and basalt are more easily broken down than are granites and other lighter-colored igneous rocks. The mineral grains in igneous rocks are randomly dispersed and interlocked, giving a salt-and-pepper appearance if they are coarse enough to see with the unaided eye.

Sedimentary rock consists of compacted or cemented weathering products from older, preexisting rocks. For example, quartz sand weathered from a granite rock and washed into the ocean may settle on the ocean floor and eventually become cemented into a solid mass called *sandstone.* Later, movements of the Earth's crust may raise this sandstone above sea level, where it could become the material we are so familiar with in certain canyon walls, road cuts, and soil parent materials. Similarly, clays may be compacted into *shale.* Because most of what is presently dry land was at some time in the

Rock texture	Quartz; Light-colored minerals (e.g., feldspars, muscovite)		Dark-colored minerals (e.g., hornblende, augite, biotite)	
Coarse	Granite	Diorite	Gabbro	Peridotite / Hornblendite
Intermediate	Rhyolite	Andesite	Basalt	
Fine	Felsite / Obsidian		Basalt glass	

FIGURE 2.2 Classification of some igneous rocks in relation to mineralogical composition and the size of mineral grains in the rock (rock texture). Worldwide, light-colored minerals and quartz are generally more prominent than are the dark-colored minerals.

[2]Primary minerals have not been altered chemically since they formed as molten lava solidified. *Secondary minerals* are recrystallized products of the chemical breakdown and/or alteration of primary minerals.

past covered by water, sedimentary rocks are the most common type of rock encountered, covering about 75% of the Earth's land surface. The resistance of a given sedimentary rock to weathering is determined by its particular dominant minerals and by the cementing agent.

Metamorphic rocks are those that have formed by the metamorphism or change in form of other rocks. As the Earth's continental plates shift, and sometimes collide, forces are generated that can uplift great mountain ranges. As a result, igneous and sedimentary masses are subjected to tremendous heat and pressure that compress, distort, and/or partially remelt the original rocks. Igneous rocks are commonly modified to form schist or gneiss in which light and dark minerals have been reoriented into bands. Sedimentary rocks, such as limestone and shale, may be metamorphosed to marble and slate, respectively. As is the case for igneous and sedimentary rock, the particular minerals that dominate a given metamorphic rock influence its resistance to chemical weathering (see Table 2.1 and Figure 2.1). A high degree of metamorphism may also physically weaken the rock mass, hastening its breakdown into smaller fragments.

Weathering: A General Case

More on rock weathering processes:
http://uregina.ca/~sauchyn/geog323/weather.html

Weathering is a biochemical process that involves both destruction and synthesis. Moving from left to right in the weathering diagram (Figure 2.3), the original rocks and minerals are destroyed by both *physical disintegration* and *chemical decomposition*. Without appreciably affecting the composition, physical disintegration breaks down rock into smaller rocks and eventually into sand and silt particles that are commonly made of individual minerals. Simultaneously, the minerals decompose chemically, releasing soluble materials and synthesizing new minerals, some of which are resistant end products. New minerals form either by minor chemical alterations or by complete chemical breakdown of the original mineral and resynthesis of new minerals. During the chemical changes, particle size continues to decrease, and constituents continue to dissolve in the aqueous weathering solution. The dissolved substances may recombine into new (secondary) minerals, may leave the profile in drainage water, or may be taken up by plant roots.

Three groups of minerals that remain in well-weathered soils are shown on the right side of Figure 2.3: (1) silicate clays, (2) very resistant end products, including iron and aluminum oxide clays, and (3) very resistant primary minerals, such as quartz. In highly

TABLE 2.1 Selected Primary and Secondary Minerals Found in Soils Listed in Order of Decreasing Resistance to Weathering Under Conditions Common in Humid Temperate Regions

Primary minerals		*Secondary minerals*		
		Goethite	$FeOOH$	Most resistant ↑
		Hematite	Fe_2O_3	
		Gibbsite	$Al_2O_3 \cdot 3H_2O$	
Quartz	SiO_2			
		Clay minerals	Al silicates	
Muscovite	$KAl_3Si_3O_{10}(OH)_2$			
Microcline	$KAlSi_3O_8$			
Orthoclase	$KAlSi_3O_8$			
Biotite	$KAl(Mg,Fe)_3Si_3O_{10}(OH)_2$			
Albite	$NaAlSi_3O_8$			
Hornblende[a]	$Ca_2Al_2Mg_2Fe_3Si_6O_{22}(OH)_2$			
Augite[a]	$Ca_2(Al,Fe)_4(Mg,Fe)_4Si_6O_{24}$			
Anorthite	$CaAl_2Si_2O_8$			
Olivine	$(Mg,Fe)_2SiO_4$			
		Dolomite[b]	$CaCO_3 \cdot MgCO_3$	
		Calcite[b]	$CaCO_3$	↓
		Gypsum	$CaSO_4 \cdot 2H_2O$	Least resistant

[a] The given formula is only approximate since the mineral is so variable in composition.
[b] In semiarid grasslands dolomite and calcite are more resistant to weathering than suggested because of low rates of carbonation weathering (see later in this Section).

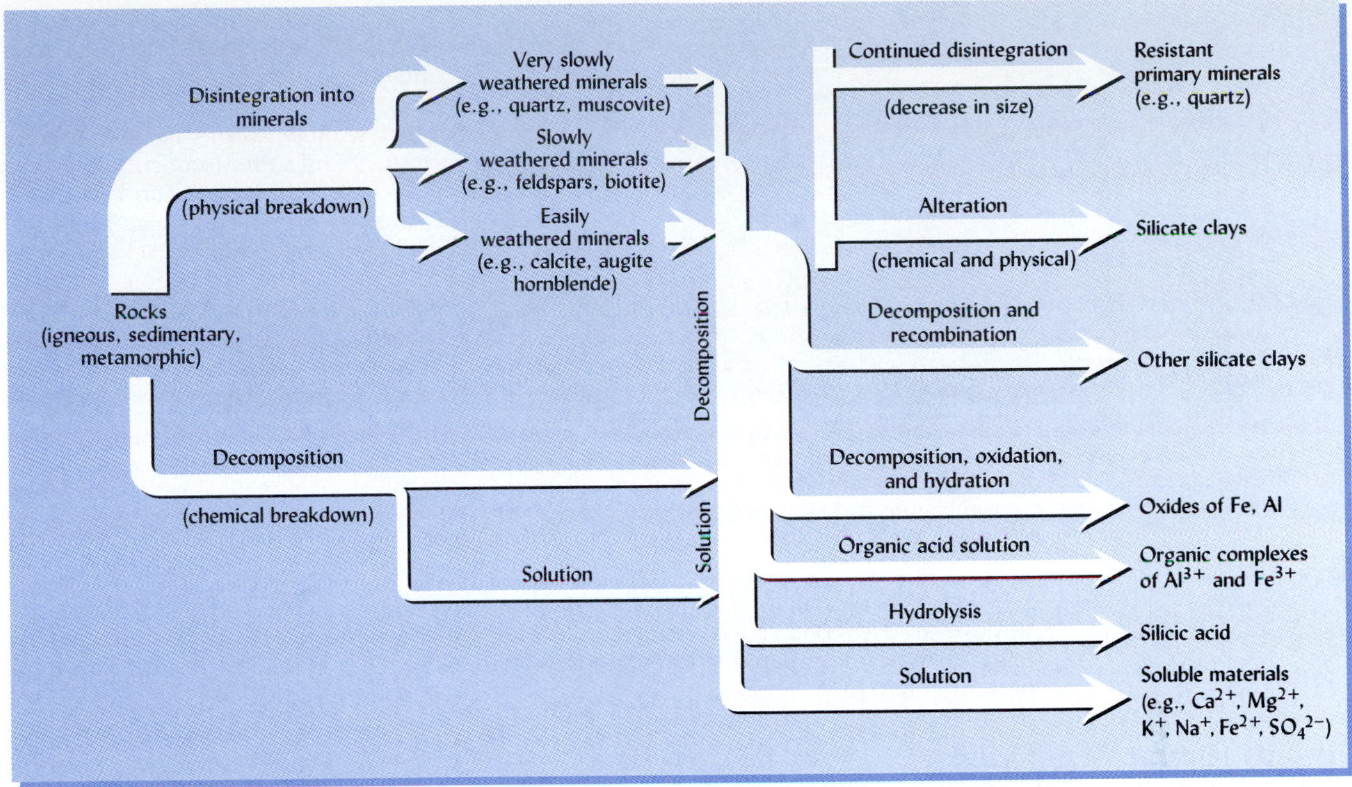

FIGURE 2.3 Pathways of weathering that occur under moderately acid conditions common in humid temperate regions. The disintegration of rocks into small individual mineral grains is a physical process, whereas decomposition, recombination, and solution are chemical processes. Alteration of minerals involves both physical and chemical processes. Note that resistant primary minerals, newly synthesized secondary minerals, and soluble materials are products of weathering. In arid regions the physical processes predominate, but in humid tropical areas decomposition and recombination are most prominent. (Diagram courtesy of N. C. Brady)

weathered soils of humid tropical and subtropical regions, the oxides of iron and aluminum and certain silicate clays with low Si/Al ratios predominate because most other constituents have been broken down and removed.

Physical Weathering (Disintegration)

Temperature. Rocks heat up during the day and cool down at night, causing alternate expansion and contraction of their constituent minerals. As some minerals expand more than others, temperature changes set up differential stresses that eventually cause the rock to crack apart.

Because the outer surface of a rock is often warmer or colder than the more protected inner portions, some rocks may weather by **exfoliation**—the peeling away of outer layers (Figure 2.4). This process may be sharply accelerated if ice forms in the surface cracks. When water freezes, it expands with a force of about 1465 Mg/m^2, disintegrating huge rock masses and dislodging mineral grains from smaller fragments.

Abrasion by Water, Ice, and Wind. When loaded with sediment, water has tremendous cutting power, as is amply demonstrated by the gorges, ravines, and valleys around the world. The rounding of riverbed rocks and beach sand grains is further evidence of the abrasion that accompanies water movement.

Windblown dust and sand also can wear down rocks by abrasion, as can be seen in the many picturesque rounded rock formations in certain arid regions. In glacial areas, huge moving ice masses embedded with soil and rock fragments grind down rocks in their path and carry away large volumes of material (see Section 2.3).

Plants and Animals. Plant roots sometimes enter cracks in rocks and pry them apart, resulting in some disintegration. Burrowing animals may also help disintegrate rock

FIGURE 2.4 Two illustrations of rock weathering. (Left) An illustration of concentric weathering called *exfoliation.* A combination of physical and chemical processes stimulate the mechanical breakdown, which produces layers that appear much like the leaves of a cabbage. (Right) Concentric bands of light and dark colors indicate that chemical weathering (oxidation and hydration) has occurred from the outside inward, producing iron compounds that differ in color. (Right photo courtesy of R. Weil)

somewhat. However, such influences are of little importance in producing parent material when compared to the drastic physical effects of water, ice, wind, and temperature change.

Biogeochemical Weathering

While physical weathering is accentuated in very cold or very dry environments, chemical reactions are most intense where the climate is wet and hot. However, both types of weathering occur together, and each tends to accelerate the other. For example, physical abrasion (rubbing together) decreases the size of particles and therefore increases their surface area, making them more susceptible to rapid chemical reactions.

Chemical weathering is enhanced by such *geological* agents as the presence of water and oxygen, as well as by such *biological* agents as the acids produced by microbial and plant-root metabolism. That is why the term **biogeochemical weathering** is often used to describe the process. The various agents act in concert to convert primary minerals (e.g., feldspars and micas) to secondary minerals (e.g., clays and carbonates) and release plant nutrient elements in soluble forms. Note the importance of water in each of the following six basic types of chemical weathering reactions.

HYDRATION. Intact water molecules may bind to a mineral by the process called **hydration.** Hydrated oxides of iron (e.g., $Fe_{10}O_{15} \cdot 9H_2O$) and aluminum (e.g., $Al_2O_3 \cdot 3H_2O$) exemplify common products of hydration reactions.

HYDROLYSIS. In **hydrolysis** reactions, water molecules split into their hydrogen and hydroxyl components and the hydrogen often replaces a cation from the mineral structure. A simple example is the action of water on microcline, a potassium-containing feldspar.

$$\underset{\text{(solid)}}{KAlSi_3O_8} + \underset{\text{Water}}{H_2O} \xrightarrow{\text{Hydrolysis}} \underset{\text{(solid)}}{HAlSi_3O_8} + \underset{\text{(solution)}}{K^+ + OH^-}$$

The potassium released is soluble and is subject to adsorption by soil colloids, uptake by plants, and removal in the drainage water.

DISSOLUTION. Many minerals under go **dissolution** as water hydrates the cations and anions until they become dissociated from each other and surrounded by water molecules. Salt dissolving in water provides an example.

CARBONATION AND OTHER ACID REACTIONS. Weathering is accelerated by **carbonation** and other acid reactions. For example, when carbon dioxide dissolves in water (a process enhanced by microbial and root respiration), the carbonic acid (H_2CO_3) produced hastens the chemical dissolution of calcite in limestone or marble, as illustrated when the following reaction goes to the right:

$$\underset{\text{Carbonic acid}}{H_2CO_3} + \underset{\substack{\text{Calcite}\\\text{(solid)}}}{CaCO_3} \xrightarrow{\text{Carbonation}} \underset{\text{(solution)}}{Ca^{2+}} + \underset{\text{(solution)}}{2HCO_3^-}$$

Soils also contain other stronger acids, such as nitric acid (HNO_3), sulfuric acid (H_2SO_4), and many organic acids. Hydrogen ions are also associated with soil clays. Each of these sources of acidity is available for reaction with soil minerals.

OXIDATION-REDUCTION. Minerals that contain iron, manganese, or sulfur are especially susceptible to **oxidation-reduction** reactions. Iron is usually laid down in primary minerals in the divalent Fe(II) (ferrous) form. When rocks containing such minerals are exposed to air and water during soil formation, the iron is easily oxidized (loses an electron) and becomes trivalent Fe(III) (ferric). If iron is oxidized from Fe(II) to Fe(III), the change in valence and ionic radius causes destabilizing adjustments in the crystal structure of the mineral.

The oxidation and/or removal of iron during weathering is often made visible by changes in the colors of the resulting altered minerals (see Figure 2.4, right).

COMPLEXATION. Soil biological processes produce organic acids such as oxalic, citric, and tartaric acids, as well as the much larger fulvic and humic acid molecules (see Section 12.4). In addition to providing H^+ ions that help solubilize aluminum and silicon, they also undergo organic **complexation** reactions with the Al^{3+} ions held within the structure of silicate minerals. By so doing they remove the Al^{3+} from the mineral, which then is subject to further disintegration.

Had there been no living organisms on Earth, the chemical weathering processes we have just outlined would probably have proceeded 1000 times more slowly, with the result that little, if any, soil would have developed on our planet.

The various chemical weathering processes occur simultaneously and are interdependent. For example, hydrolysis of a given primary mineral may release ferrous iron [Fe(II)] that is quickly oxidized to the ferric [Fe(III)] form, which, in turn, is hydrated to give a hydrous oxide of iron. Hydrolysis or complexation also may release soluble cations, silicic acid, and aluminum or iron compounds. In humid environments, some of the soluble cations and silicic acid are likely to be lost from the weathering mass in drainage waters. The released substances can also be recombined to form silicate clays and other secondary silicate minerals. In this manner, the biochemical processes of weathering transform primary geologic materials into the compounds of which soils are made.

2.2 FACTORS INFLUENCING SOIL FORMATION[3]

We learned in Chapter 1 that *the soil* is a collection of *individual soils*, each with distinctive profile characteristics. This concept of soils as organized natural bodies derived initially from late-19th-century field studies by a brilliant Russian team of soil scientists led by V. V. Dukochaev. They noted similar profile layering in soils hundreds of kilometers apart, provided that the climate and vegetation were similar at the two locations. Such observations and much careful subsequent field and laboratory research led to the recognition of five major factors that control the formation of soils.

Illustrations of the five soil forming factors: http://www.soils.umm.edu/infoserv/orgs/mapss/soilformingfactors.html#time

1. *Parent materials* (geological or organic precursors to the soil)
2. *Climate* (primarily precipitation and temperature)
3. *Biota* (living organisms, especially native vegetation, microbes, soil animals, and human beings)

[3]Many of our modern concepts concerning the factors of soil formation are derived from the work of Hans Jenny (1941 and 1980) and E. W. Hilgard (1921), American soil scientists whose books are considered classics in the field.

4. *Topography* (slope, aspect, and landscape position)
5. *Time* (the period of time since the parent materials became exposed to soil formation)

Soils are often defined in terms of these factors as *dynamic natural bodies having properties derived from the combined effects of climate and biotic activities, as modified by topography, acting on parent materials over periods of time.*

We will now examine how each of these five factors affects the outcome of soil formation. However, as we do, we must keep in mind that these factors do not exert their influences independently. Indeed, interdependence is the rule. For example, contrasting climatic regimes are likely to be associated with contrasting types of vegetation, and perhaps differing topography and parent material, as well. Nonetheless, in certain situations one of the factors has had the dominant influence in determining differences among a set of soils. Soil scientists refer to such a set of soils as a **lithosequence, climosequence, biosequence, toposequence,** or **chronosequence.**

2.3 PARENT MATERIALS

Illustration of parent material deposition: http://sis.agr.gc.ca/cansis/taxa/genesis/pmdep/atlantic.html

Geological processes have brought to the earth's surface numerous parent materials in which soils form (Figure 2.5). The nature of the parent material profoundly influences soil characteristics. For example, a soil might inherit a sandy texture (see Section 4.2) from a coarse-grained, quartz-rich parent material such as granite or sandstone. Soil texture, in turn, helps control the percolation of water through the soil profile, thereby affecting the translocation of fine soil particles and plant nutrients.

The chemical and mineralogical composition of parent material also influences both chemical weathering and the natural vegetation. For example, the presence of limestone in parent material will slow the development of acidity that typically occurs in humid

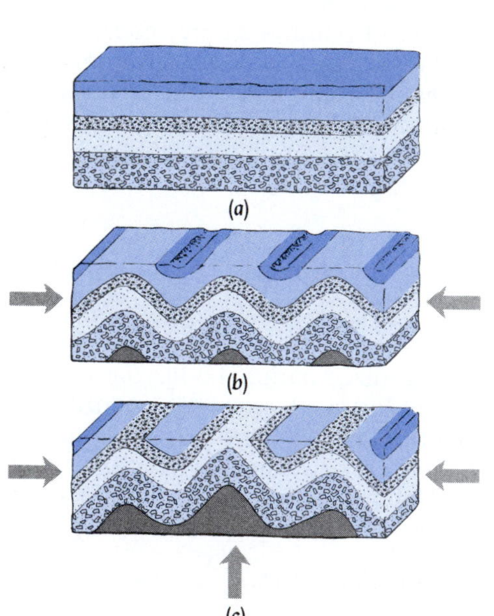

FIGURE 2.5 Diagrams showing how geological processes have brought different rock layers to the surface in a given area. (*a*) Unaltered layers of sedimentary rock with only the uppermost layer exposed. (*b*) Lateral geological pressures deform the rock layers through a process called *crustal warping*. At the same time, erosion removes much of the top layer, exposing part of the first underlying layer. (*c*) Localized upward pressures further reform the layers, thereby exposing two more underlying layers. As these four rock layers are weathered, they give rise to the parent materials on which different kinds of soils can form. (*d*) Crustal warping that lifted up the Appalachian Mountains tilted these sedimentary rock formations that were originally laid down horizontally. This deep roadcut in Virginia illustrates the abrupt change in soil parent material (lithosequence) as one walks along the ground surface at the top of this photograph. (Photo courtesy of R. Weil)

climates. In addition, it greatly influences the kinds of clays that can develop as the soil evolves (see Section 8.5). In turn, the nature of the clay minerals present markedly affects the kind of soil that develops.

Classification of Parent Materials

Inorganic parent materials can either be formed in place as residual material weathered from rock, or they can be transported from one location and deposited at another (Figure 2.6). In wet environments (such as swamps and marshes), incomplete decomposition may allow organic parent materials to accumulate from the residues of many generations of vegetation. Although it is their chemical and physical properties that most influence soil development, parent materials are often classified with regard to the mode of placement in their current location, as seen on the right side of Figure 2.6.

Although these terms properly relate only to the placement of the parent materials, people sometimes refer to the soils that form from these deposits as *organic soils, glacial soils, alluvial soils,* and so forth. These terms are quite nonspecific because parent material properties vary widely within each group, and because the effect of parent material is modified by the influence of climate, organisms, topography, and time.

Residual Parent Material

Residual parent material develops in place from weathering of the underlying rock. In stable landscapes it may have experienced long and possibly intense weathering. Where the climate is warm and humid, residual parent materials are typically thoroughly leached and oxidized, and show the red and yellow colors of various oxidized iron compounds (see Plates 9, 11, and 15 after page 114). In cooler and especially drier climates, the color and chemical composition of residual parent material tend to resemble more closely the rock from which it formed.

Residual materials are widely distributed on all continents. The physiographic map of the United States (Figure 2.7) shows nine great provinces where residual materials are prominent (shades of green on the map).

A great variety of soils occupy the regions covered by residual debris because of the marked differences in the nature of the rocks from which these materials evolved. The varied soils are also a reflection of wide differences in other soil-forming factors, such as climate and vegetation (Sections 2.4 and 2.5).

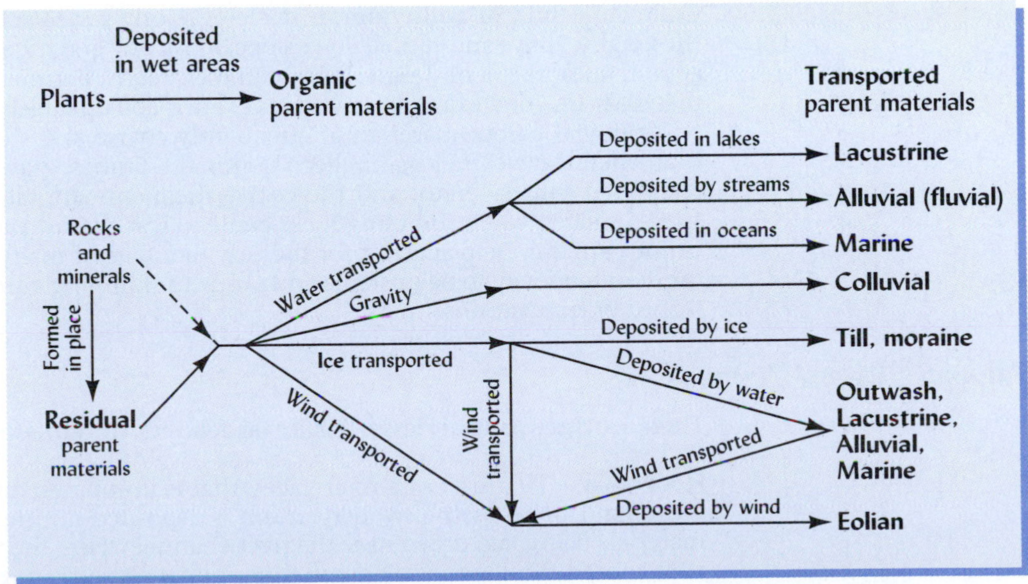

FIGURE 2.6 How various kinds of parent material are formed, transported, and deposited.

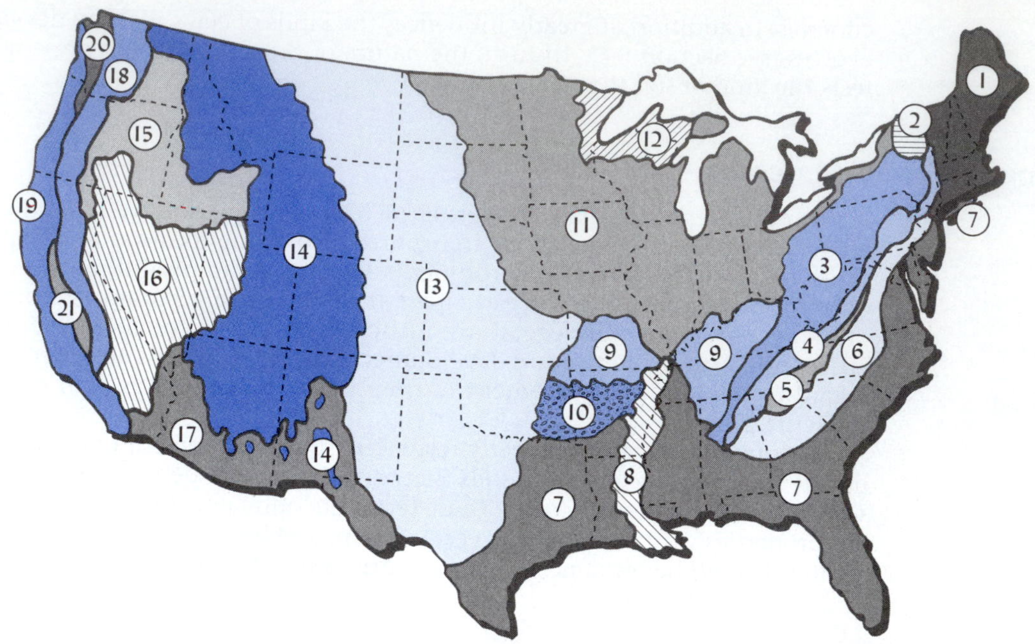

FIGURE 2.7 Generalized physiographic and regolith map of the United States. The regions are as follows (major areas of residual parent material are italicized and shown in shades of orange on the map).

1. New England: mostly glaciated metamorphic rocks.
2. Adirondacks: glaciated metamorphic and sedimentary rocks.
3. *Appalachian Mountains and plateaus:* shales and sandstones.
4. *Limestone valleys and ridges:* mostly limestone.
5. Blue Ridge Mountains: sandstones and shales.
6. *Piedmont Plateau:* metamorphic rocks.
7. Atlantic and Gulf coastal plain: unconsolidated sediments; sands, clays, and silts.
8. Mississippi floodplain and delta: alluvium.
9. *Limestone uplands:* mostly limestone and shale.
10. *Sandstone uplands:* mostly sandstone and shale.
11. Central lowlands: mostly glaciated sedimentary rocks with till and loess.
12. Superior uplands: glaciated metamorphic and sedimentary rocks.
13. *Great Plains region:* sedimentary rocks.
14. *Rocky Mountain region:* sedimentary, metamorphic, and igneous rocks.
15. Northwest intermountain: mostly igneous rocks; loess in river basins.
16. Great Basin: gravels, sands, alluvial fans; igneous and sedimentary rocks.
17. Southwest arid region: gravel, sand, and other debris of desert and mountain.
18. *Sierra Nevada and Cascade mountains:* igneous and volcanic rocks.
19. *Pacific Coast province:* mostly sedimentary rocks.
20. Puget Sound lowlands: glaciated sedimentary.
21. California central valley: alluvium and outwash.

Colluvial Debris

Colluvial debris, or **colluvium**, is made of poorly sorted rock fragments detached from the heights above and carried downslope, mostly by gravity, assisted in some cases by frost action. Rock fragment (talus) slopes, cliff rock debris (detritus), and similar heterogeneous materials are good examples. Avalanches are made up largely of such accumulations.

Colluvial parent materials are frequently coarse and stony because physical rather than chemical weathering has been dominant. Stones, gravel, and fine materials are interspersed (not layered), and the coarse fragments are rather angular. Packing voids, spaces created when tumbling rocks come to rest against each other (sometimes at precarious angles), help account for the easy drainage of many colluvial deposits and also for their tendency to be unstable and prone to slumping and landslides, especially if disturbed by excavations.

Alluvial Stream Deposits

There are three general classes of alluvial deposits: **floodplains, alluvial fans,** and **deltas.**

FLOODPLAINS. That part of a river valley that is inundated during floods is a floodplain. Sediment carried by the swollen stream is deposited during the flood, with the coarser materials being laid down near the river channel where the water is deeper and flowing with more turbulence and energy. Finer materials settle out in the calmer flood waters farther from the channel. Each major flooding episode lays down a distinctive layer of sediment, creating the stratification that characterizes alluvial soils (Figure 2.8).

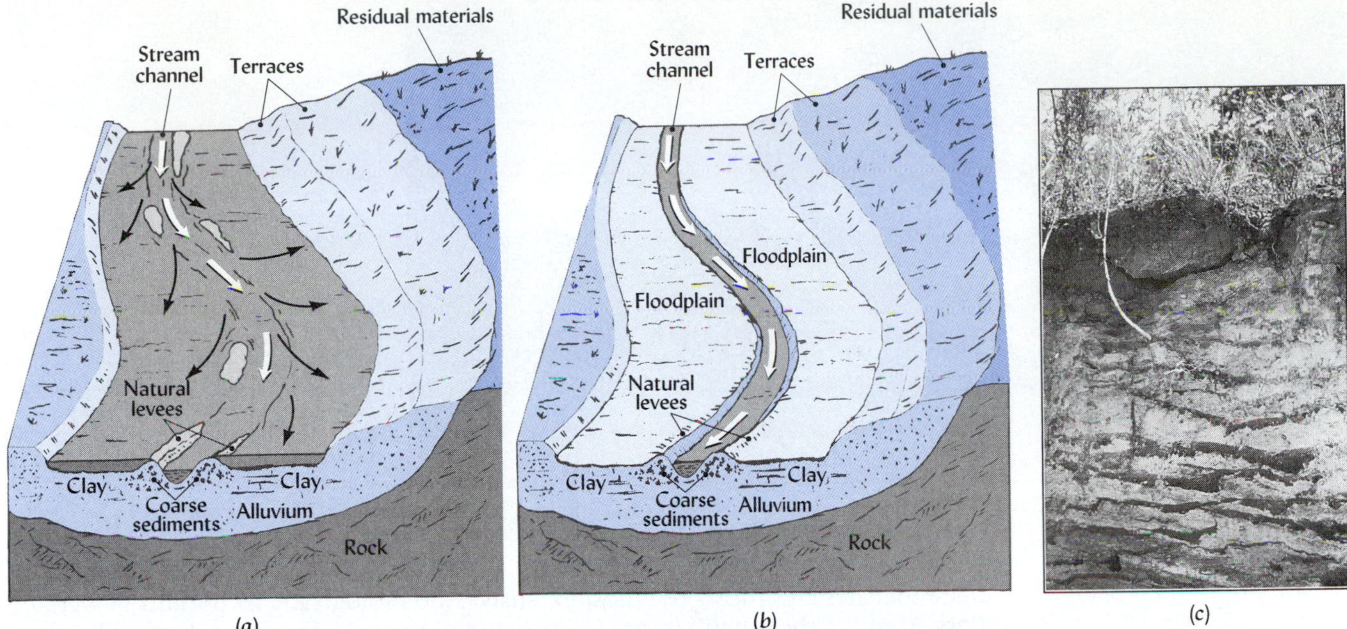

FIGURE 2.8 Illustration of floodplain development. (*a*) A stream is at flood stage, has overflowed its banks, and is depositing sediment in the floodplain. The coarser particles are being deposited near the stream channel where water is moving most rapidly, while the finer clay particles are being deposited where water movement is slower. (*b*) After the flood the sediments are in place and vegetation is growing. (*c*) Contrasting layers of sand, silt, and clay characterize the alluvial floodplain. Each layer resulted from separate flooding episodes. (Redrawn from *Physical Geology,* 2d ed., by F. R. Flint and B. J. Skinner, copyright © 1977 John Wiley & Sons, Inc. Reprinted by permission of John Wiley & Sons, Inc.) (Photo courtesy of R. Weil)

If, over time, there is a change in grade, a stream may cut down through its already well-formed alluvial deposits. This cutting action leaves **terraces** above the floodplain on one or both sides. Some river valleys feature two or more terraces at different elevations, each reflecting a past period of alluvial deposition and stream cutting.

The floodplain along the Mississippi River is the largest in the United States (area 8 in Figure 2.7), varying from 30 to 125 km in width. Floodplains of smaller streams also provide parent materials for locally important soil areas.

Soils derived from alluvial sediments generally have characteristics seen as desirable for human settlement and agriculture. These characteristics include nearly level topography, proximity to water, high fertility, and high productivity. Although many alluvial soils are well drained, others may require artificial drainage if they are to be used for upland crops or for stable building foundations. Alluvial soils can provide natural habitats, such as bottomland forests, which are very productive of timber and support a high diversity of birds and other wildlife.

While alluvial soils are often uniquely suited to forestry and crop production, their use for home sites and urban development should generally be avoided. Building on a floodplain, no matter how great the investment in flood-control measures, all too often leads to tragic loss of life and property during serious flooding.

Alluvial Fans. Streams that leave a narrow valley in an upland area and suddenly descend to a much broader valley below deposit sediment in the shape of a fan, as the water spreads out and slows down (see Figure 2.9). The rushing water tends to sort the sediment particles by size, first dropping the gravel and coarse sand, then depositing the finer materials toward the bottom of the alluvial fan.

Alluvial fan debris is found in widely scattered areas in mountainous and hilly regions. The soils derived from this debris often prove very productive, although they may be quite coarse-textured.

Delta Deposits. In a few river systems, considerable suspended material settles near the mouth of the river, forming a delta. A delta often is a continuation of a floodplain (its front, so to speak). It is clayey in nature and likely to be poorly drained as well.

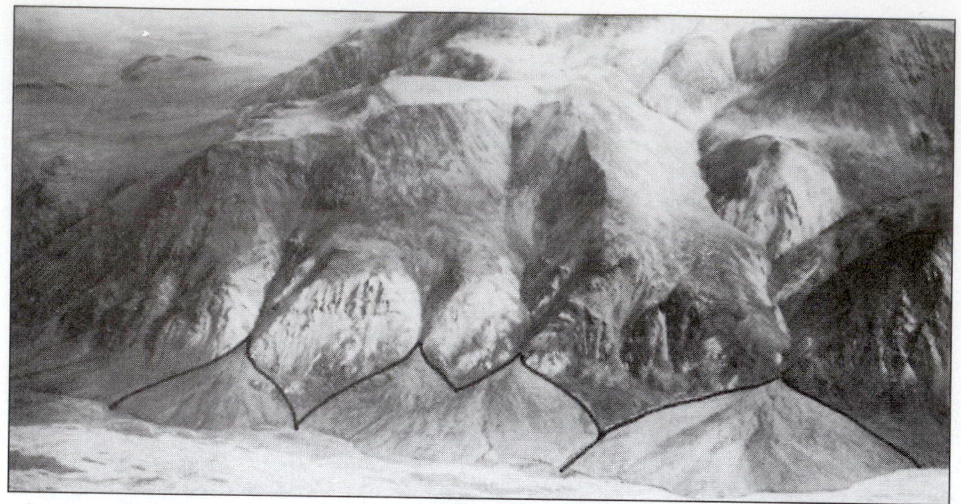

FIGURE 2.9 Characteristically shaped alluvial fans alongside a river valley in Alaska. Although the areas are small and sloping, they can develop into well-drained soils. (Courtesy U.S. Geological Survey)

Delta marshes are among the most extensive and biologically important of wetland habitats. Many of these habitats are today being protected or restored, but civilizations both ancient and modern have also developed important agricultural areas (often specializing in the production of rice) by creating drainage and flood-control systems on the deltas of such rivers as the Amazon, Euphrates, Ganges, Hwang Ho, Mississippi, Nile, Po, and Tigris.

Marine Sediments

Illustrations of parent material deposition in Maritime Provinces:
http://sis.agr.gc.ca/cansis/taxa/genesis/pmdep/atlantic.html

Streams eventually deposit much of their sediment loads in oceans, estuaries, and gulfs. The coarser fragments settle out near the shore and the finer particles at a distance (Figure 2.10). Over long periods of time, these underwater sediments build up, in some cases becoming hundreds of meters thick. Changes in the relative elevations of sea and land may later raise these marine deposits above sea level, creating a coastal plain. The deposits are then subject to a new cycle of weathering and soil formation.

A coastal plain usually has only moderate slopes, being more level in the low-lying parts nearer the coastline and more hilly farther inland, where streams and rivers flowing down the steeper grades have more deeply dissected the landscape. The land surface in the lower coastal portion may be only slightly above the water table during part of the year, so wetland forests and marshes often characterize such parent materials.

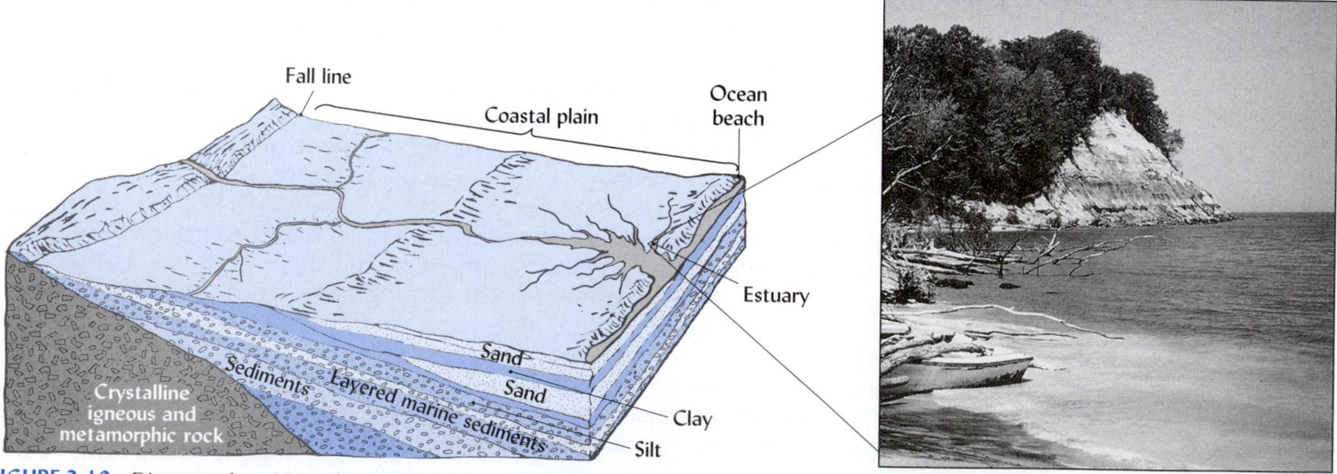

FIGURE 2.10 Diagram showing sediments laid down in marine waters adjacent to coastal residual igneous and metamorphic rocks. Note that the marine sediments are alternate layers of fine clay, silts and coarse-textured sands and gravels. The photo shows such layering on coastal marine sediment. This diagram and photo illustrate the relationship between marine sediments and residual materials in the Coastal Plain of southeastern United States. (Photo and diagram courtesy of R. Weil)

Marine deposits are quite variable in texture. Some are sandy, as is the case in much of the Atlantic seaboard coastal plain. Others are high in clay, as are deposits found in the Atlantic and Gulf coastal flatwoods and in the interior pinelands of Alabama and Mississippi. Where streams have cut down through layers of marine sediments, clays, silts, and sand may be encountered side by side. The properties of the soils that form are heavily influenced by those of the marine parent materials. Because seawater is high in sulfur, many marine sediments are high in sulfur and go through a period of acid-forming sulfur oxidation at some stage of soil formation (see Sections 9.7 and 12.22 and Plate 65).

Parent Materials Transported by Glacial Ice and Meltwaters

During the Pleistocene epoch (about 10^4 to 10^7 years ago), up to 20% of the world's land surface—northern North America, northern and central Europe, and parts of northern Asia—was invaded by a succession of great ice sheets, some more than 1 km thick. Present-day glaciers in polar regions and high mountains cover about a third as much area, but are not nearly so thick as the glaciers of the Great Pleistocene Ice Age. Some scientists predict that if the current global warming trend continues, these present-day glaciers could partially melt, causing an increase in sea level, thus flooding many coastal areas around the world.

In North America, Pleistocene-epoch glaciers covered most of what is now Canada, southern Alaska, and the northern part of the contiguous United States (Figure 2.11).

As the glacial ice pushed forward, the existing regolith with much of its mantle of soil was swept away, hills were rounded, valleys were filled, and, in some cases, the underlying rocks were severely ground and gouged. Thus, the glacier became filled with rock and all kinds of unconsolidated materials, carrying great masses of these materials as it pushed ahead (Figure 2.12). Finally, as the ice melted and the glacier retreated, a mantle of glacial debris or drift remained. This provided a new regolith and fresh parent material for soil formation.

Glacial Till and Associated Deposits. The name **drift** is applied to all material of glacial origin, whether deposited by the ice or by associated waters. The materials deposited

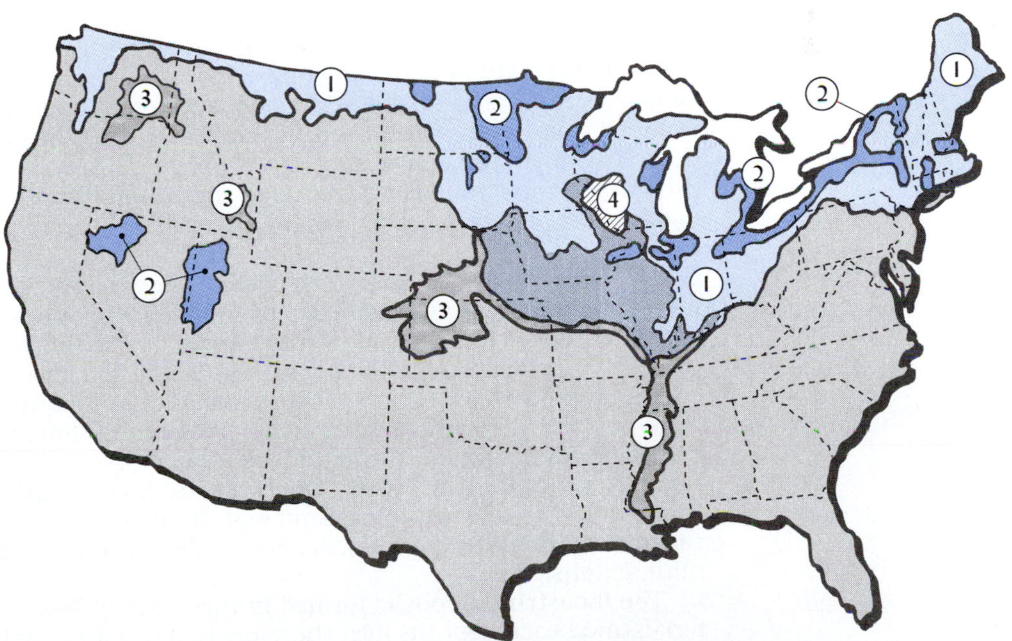

FIGURE 2.11 Areas in the United States covered by the continental ice sheet and the deposits either directly from, or associated with, the glacial ice. (1) Till deposits of various kinds; (2) glacial-lacustrine deposits; (3) the loessial blanket (note that the loess overlies great areas of till in the Midwest); (4) an area, mostly in Wisconsin, that escaped glaciation and is partially loess covered.

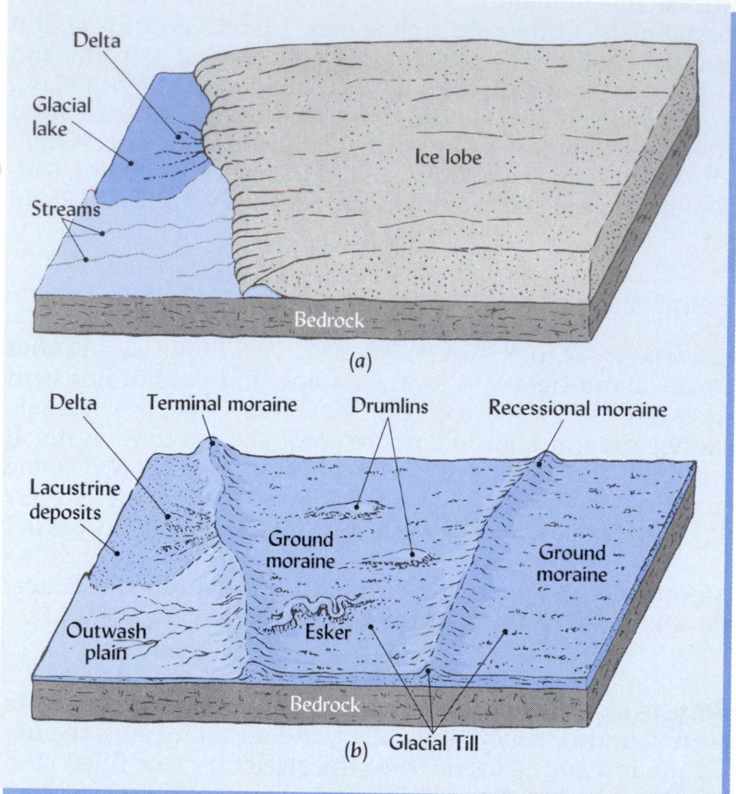

FIGURE 2.12 Illustration of how several glacial materials were deposited. (*a*) A glacier ice lobe moving to the left, feeding water and sediments into a glacial lake and streams, and building up glacial till near its front. (*b*) After the ice retreats, terminal, ground, and recessional moraines are uncovered along with cigar-shaped hills (drumlins), the beds of rivers that flowed under the glacier (eskers), and lacustrine, delta, and outwash deposits. (*c*) The stratified glacial outwash in the lower part of this soil profile in North Dakota is overlain by a layer of glacial till containing a random assortment of particles, ranging in size from small boulders to clays. Note the rounded edges of the rocks, evidence of the churning action within the glacier. Scale is marked every 10 cm. (Photo courtesy of R. Weil)

directly by the ice, called **glacial till**, are heterogeneous (unstratified) mixtures of debris, which vary in size from boulders to clay. Glacial till may therefore be somewhat similar in appearance to colluvial materials, except that the coarse fragments are more rounded from their grinding journey in the ice, and the deposits are often much more densely compacted because of the great weight of the overlying ice sheets. Much glacial till is deposited in irregular ridges called **moraines.** Figure 2.12 shows how glacial sheets deposited several types of soil parent materials.

GLACIAL OUTWASH AND LACUSTRINE SEDIMENTS. The torrents of water gushing forth from melting glaciers carried vast loads of sediment. In valleys and on plains, where the glacial waters were able to flow away freely, the sediment formed an **outwash plain** (Figure 2.12). Such sediments, with sands and gravels sorted by flowing water, are common **valley fills** in parts of North America and Europe.

When the ice front came to a standstill where there was no ready escape for the water, ponding began; ultimately, very large lakes were formed (Figure 2.12). Particularly prominent in North America were those south of the Great Lakes and in the Red River Valley of Minnesota and Manitoba (see Figure 2.11). The latter lake, called Glacial Lake Agassiz, was about 1200 km long and 400 km wide at its maximum extension.

The **lacustrine deposits** formed in these glacial lakes range from coarse delta materials and beach deposits near the shore to larger areas of fine silts and clay deposited from the deeper, more still waters at the center of the lake. Areas of inherently fertile (though not always well drained) soils developed from these materials as the lakes dried.

Parent Materials Transported by Wind

Wind is capable of picking up an enormous quantity of material at one site and depositing it at another. Wind can most effectively pick up material from soil or regolith that is loose, dry, and unprotected by vegetation. Dry, barren landscapes have served, and continue to serve, as sources of parent material for soils forming as far away as the opposite side of the globe. The smaller the particles, the higher and farther the wind will carry them. Wind-transported (**eolian**) materials important as parent material for soil formation include, from largest to smallest particle size: **dune sand, loess** (pronounced "luss"), and **aerosolic dust.** Windblown **volcanic ash** from erupting volcanoes is a special case that is also worthy of mention.

DUNE SAND. Along the beaches of the world's oceans and large lakes, and over vast barren deserts, strong winds pick up medium and fine sand grains and pile them into hills of sand called *dunes*. The dunes, ranging up to 100 m in height, may continue to slowly shift their locations in response to the prevailing winds. Because most other minerals have been broken down and carried away by the waves, beach sand usually consists mainly of quartz, which is devoid of plant nutrients and highly resistant to weathering action. Nonetheless, over time dune grasses and other pioneering vegetation may take root and soil formation may begin.

LOESS. The windblown materials called *loess* are composed primarily of silt with some very fine sand and coarse clay. They cover wide areas in the central United States, eastern Europe, Argentina, and central Asia (Figure 2.13*a*). Loess may be blown for hundreds of kilometers. The deposits farthest from the source are thinnest and consist of the finest particles.

In the United States (Figure 2.13*b*), the main sources of loess were the great barren expanses of till and outwash left in the Missouri and Mississippi River valleys by the retreating glaciers of the last Ice Age. During the winter months, winds picked up fine materials and moved them southward, covering the existing soils and parent materials with a blanket of loess that accumulated to as much as 8 m thick.

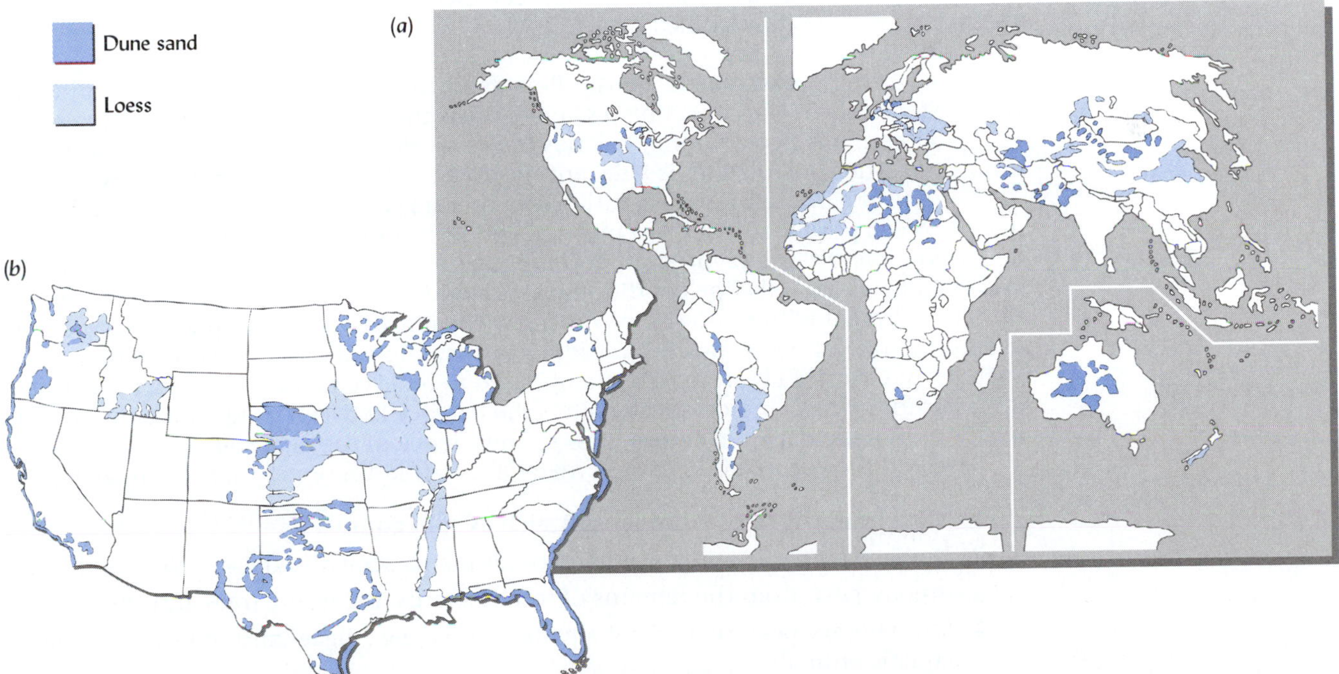

FIGURE 2.13 (*a*) Major eolian deposits of the world include the loess deposits in Argentina, eastern Europe, and northern China, and the large areas of dune sands in north Africa and Australia. (*b*) Approximate distribution of loess and dune sand in the United States. The soils that have developed from loess are generally silt loams, often quite high in fine sands. Note especially the extension of the central loess deposit down the eastern side of the Mississippi River and the smaller areas of loess in Washington, Oregon, and Idaho. The most prominent areas of dune sands are the Sand Hills of Nebraska and the dunes along the eastern shore of Lake Michigan.

In central and western China, loess deposits reaching 30 to 100 m in depth cover some 800,000 km^2. These materials have been windblown from the deserts of central Asia and are generally not associated directly with glaciers. These and other loess deposits tend to form silty soils of rather high fertility and potential productivity.

Aerosolic Dust. Very fine particles (about 1 to 10 μm) carried high into the air may travel for thousands of kilometers before being deposited, usually with rainfall. These fine particles are called *aerosolic dust,* because they can remain suspended in air, due to their very small size. Although this dust has not blanketed the receiving landscapes as thickly as is typical for loess, it does accumulate at rates that make significant contributions to soil formation. Much of the calcium carbonate in soils of the western United States probably originated as windblown dust. Recent studies have shown that dust, originating in the Sahara Desert of northern Africa and transported over the Atlantic Ocean in the upper atmosphere, is the source of much of the calcium and other nutrients found in the highly leached soils of the Amazon basin in South America.

Volcanic Ash. During volcanic eruptions cinders fall in the immediate vicinity of the volcano, while fine, often glassy ash particles may blanket extensive areas downwind. Soils developed from volcanic ash are most prominent within a few hundred kilometers of the volcanoes that ring the Pacific Ocean. Important areas of volcanic ash parent materials occur in Japan, Indonesia, New Zealand, western United States (in Hawaii, Montana, Oregon, Washington, and Idaho), Mexico, Central America, and Chile. The soils formed are uniquely light and porous, and tend to accumulate organic matter more rapidly than other soils in the area. The volcanic ash tends to weather rapidly into allophane, a type of clay with unusual properties (see Section 8.1).

Organic Deposits

Organic material accumulates in wet places where plant growth exceeds the rate of residue decomposition. In such areas residues accumulate over the centuries from wetland plants such as pondweeds, cattails, sedges, reeds, mosses, shrubs, and certain trees. These residues sink into the water, where their decomposition is limited by lack of oxygen. As a result, organic deposits often accumulate up to several meters in depth (Figure 2.14). Collectively, these organic deposits are called **peat.**

Distribution and Accumulation of Peats. Peat deposits are found all over the world, but most extensively in the cool climates and in areas that have been glaciated. About 75% of the 340 million hectares of peat lands in the world are found in Canada and northern Russia. The United States is a distant third, with about 20 million hectares.

The rate of peat accumulation varies from one area to another, depending on the balance between production of plant material and its loss by decomposition. Cool climates and acidic conditions favor slow decomposition, but also slower plant production. Warm climates and alkaline conditions favor rapid losses, but also rapid plant production. Enrichment with nutrients may increase the rate of organic production more than it does the rate of decomposition, leading to very high net accumulation rates. As we shall see in Chapter 11, artificial drainage, used to remove excess water from a peat soil, lets air into the peat and drastically alters the balance between production and decomposition of organic matter, causing loss or subsidence of the peat soil.

Based on the nature of the parent materials, four kinds of peat are recognized:

1. Moss peat, the remains of mosses such as sphagnum
2. Herbaceous peat, residues of herbaceous plants such as sedges, reeds, and cattails
3. Woody peat, from the remains of woody plants, including trees and shrubs
4. Sedimentary peat, remains of aquatic plants (e.g., algae) and of fecal material of aquatic animals

Organic deposits generally contain two or more of these kinds of peats. Alternating layers of different peats are common, as are mixtures of the peats. Because the succession of plants, as the residues accumulate, tends to favor trees (see Figure 2.14), woody peats often dominate the surface layers of organic materials.

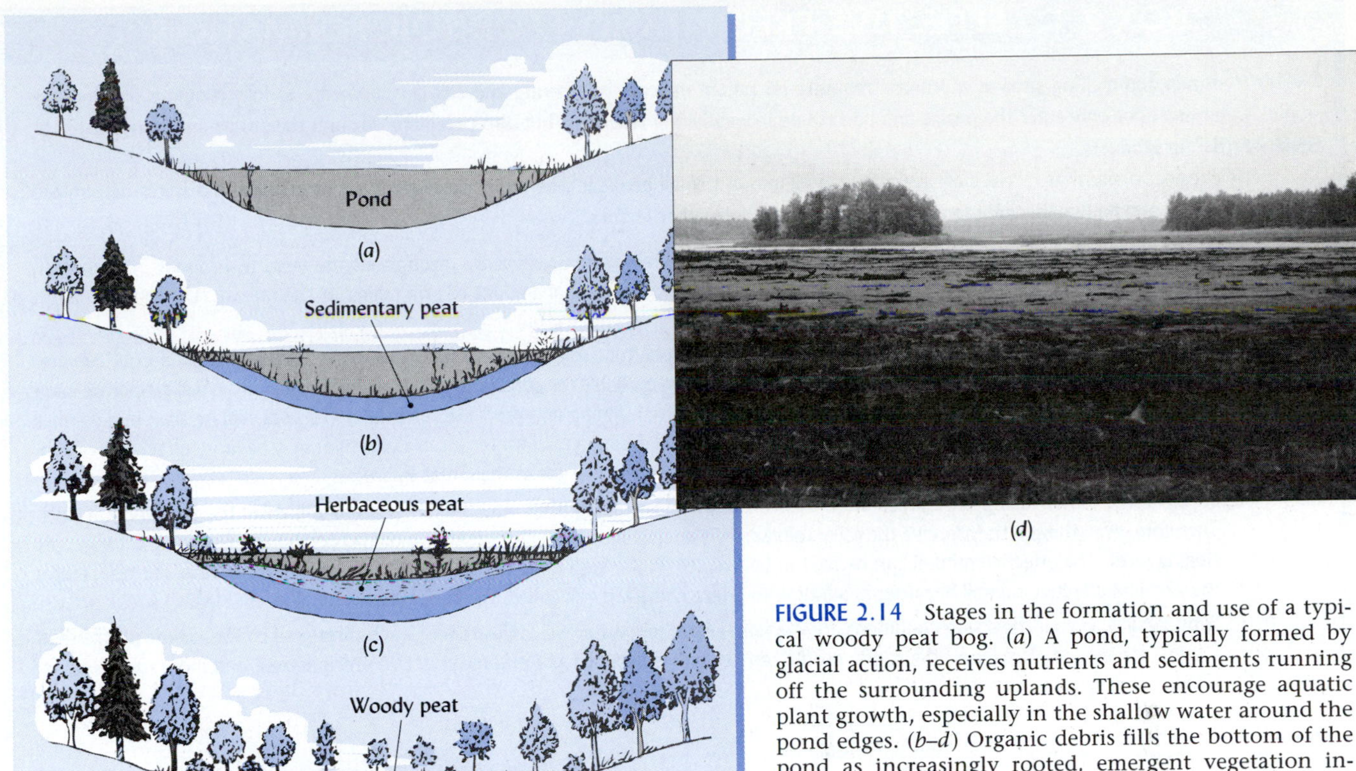

FIGURE 2.14 Stages in the formation and use of a typical woody peat bog. (*a*) A pond, typically formed by glacial action, receives nutrients and sediments running off the surrounding uplands. These encourage aquatic plant growth, especially in the shallow water around the pond edges. (*b–d*) Organic debris fills the bottom of the pond as increasingly rooted, emergent vegetation invades. (*e*) Eventually shrubs and trees take root in the peat and cover the area. Many such bogs have been cleared of trees and drained by ditches to remove some of the water, exposing an organic muck soil that is often highly productive for vegetable crops. The bog in the photo is in Central Michigan. (Photo courtesy of R. Weil)

Histosols (peatlands) of Ireland:
http://www.peatsociety.fi/natcoms/irl/index.htm

The organic material is called **peat**, or **fibric**, if the residues are sufficiently intact to permit the plant fibers to be identified. If most of the material has decomposed sufficiently so that little fiber remains, the term **muck** or **sapric** is used. In mucky peats (**hemic** materials), only some of the plant fibers can be recognized.

Wetland areas are important environmental buffers and natural habitats for wildlife. Drainage of these areas reduces the benefits of wetlands. While organic material is the foundation for some very productive agricultural soils, environmentalists argue that such use is unsustainable because, once drained, the organic deposits will decompose and disappear after a century or so; therefore, these areas might be better left in (or returned to) their natural state (see Section 7.7).

Recognizing that the effects of *parent materials* on soil properties are modified by the combined influences of *climate, biotic activities, topography*, and *time*, we will now turn to these other four factors of soil formation, starting with climate.

2.4 CLIMATE

Climate is perhaps the most influential of the four factors acting on parent material because it determines the nature and intensity of the weathering that occurs over large geographic areas. The principal climatic variables influencing soil formation are **effective precipitation** (see Box 2.1) and *temperature*, both of which affect the rates of chemical, physical, and biological processes.

Effective Precipitation

The greater the depth of water penetration, the greater the depth of soil weathering and development. Surplus water percolating through the soil profile transports soluble and suspended materials from the upper to the lower layers. It may also carry away soluble

BOX 2.1 EFFECTIVE PRECIPITATION FOR SOIL FORMATION

Water from rain and melting snow is a primary requisite for parent material weathering and soil development. To fully promote soil development, water must not only enter the profile and participate in weathering reactions, but also percolate through the profile and translocate soluble weathering products.

Let's consider a site that receives an average of 600 mm of rainfall per year. The amount of water leaching through a soil is determined not only by the total annual precipitation, but also by at least four other factors.

a. **Seasonal distribution of precipitation.** The 600 mm of rainfall distributed evenly throughout the year, with about 50 mm each month, is likely to cause less soil leaching or erosion than the same annual amount of rain falling at the rate of 100 mm per month during a 6-month rainy season.

b. **Temperature and evaporation.** In a hot climate, evaporation from soils and vegetation is much higher than in a cool climate. Therefore, in the hot climate, much less of the 600 mm will be available for percolation and leaching. Most or all will evaporate soon after it falls on the land. Thus, 600 mm of rain may cause more leaching and profile development in a cool climate than in a warmer one. Similar reasoning would suggest that rainfall concentrated during a mild winter (as in California) may be more effective in leaching the soil than the same amount of rain concentrated in a hot summer (as in the Great Plains).

c. **Topography.** Water falling on a steep slope will run downhill so rapidly that only a small portion will enter the soil where it falls. Therefore, even though they receive the same rainfall, level or concave sites will experience more percolation and leaching than steeply sloping sites. The effective rainfall can be said to be greater on the level site than on the sloping one. The concave site will receive the greatest effective rainfall because, in addition to direct rainfall, it will collect the runoff from the adjacent sloping site.

d. **Permeability.** Even if the above conditions are the same, more rain water will infiltrate and leach through a coarse, sandy profile than a tight, clayey one. Therefore, the sandy profile can be said to experience a greater effective precipitation, and more rapid soil development may be expected.

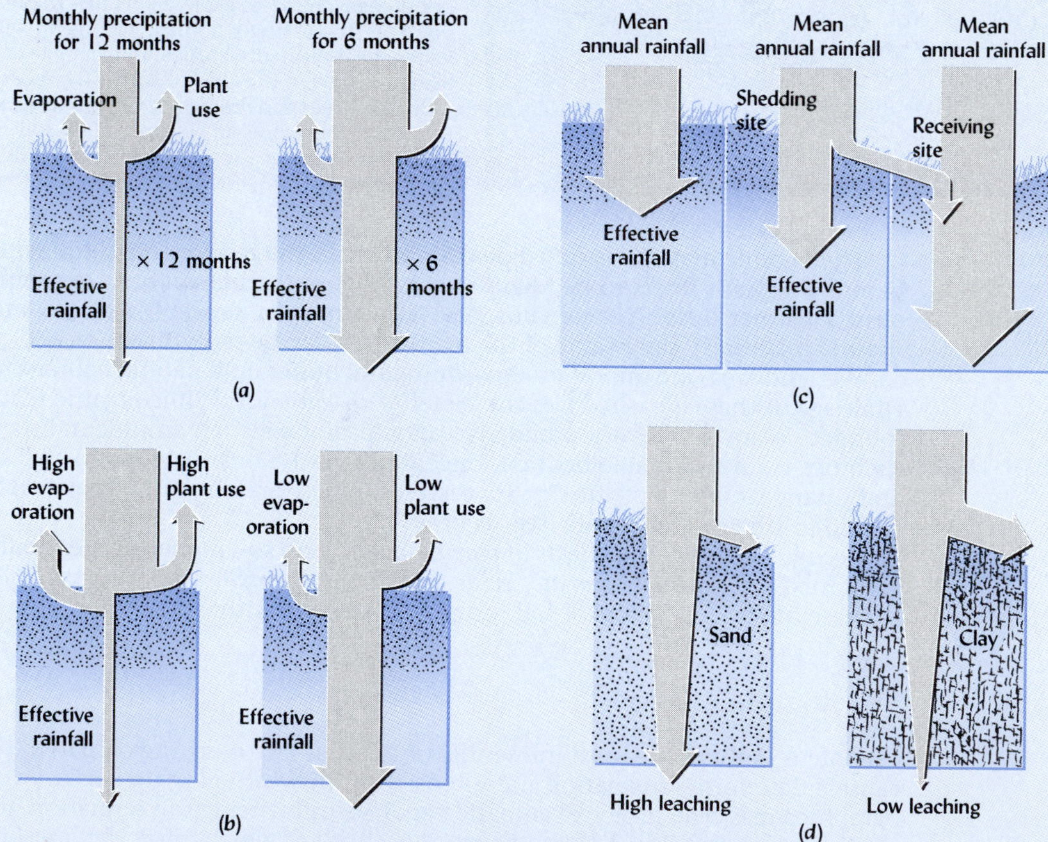

FIGURE 2.15 *The seasonal rainfall distribution (a), evaporative demand (b), site topography (c) and soil permeability (d) interact to determine how effectively precipitation can influence soil formation. (Diagram courtesy of R. Weil)*

materials in the drainage waters. Thus, percolating water stimulates weathering reactions and helps differentiate soil horizons.

Likewise, a deficiency of water is a major factor in determining the characteristics of soils of dry regions. Soluble salts are not leached from these soils, and in some cases they build up to levels that curtail plant growth. Soil profiles in arid and semiarid regions are also apt to accumulate carbonates and certain types of cracking clays.

Temperature

For every 10°C rise in temperature, the rates of biochemical reactions more than double. Temperature and moisture both influence the organic matter content of soil through their effects on the balance between plant growth and microbial decomposition (see Section 11.10). If warm temperatures and abundant water are present in the profile at the same time, the processes of weathering, leaching, and plant growth will be maximized. The very modest profile development characteristic of cold areas contrasts sharply with the deeply weathered profiles of the humid tropics.

Climate also influences the natural vegetation. Humid climates favor the growth of trees. In contrast, grasses are the dominant native vegetation in subhumid and semiarid regions, while shrubs and brush of various kinds dominate in arid areas. Thus, climate exerts its influence partly through a second soil-forming factor, the living organisms.

Considering soils with similar temperature regime, parent material, topography, and age, increasing effective annual precipitation generally leads to increasing clay and organic matter contents, greater acidity, and lower ratio of Si/Al (an indication of more highly weathered minerals. However, many places have experienced climates in past geologic epochs that were not at all similar to the climate evident today. This fact is illustrated in certain old landscapes in arid regions, where highly leached and weathered soils stand as relics of the humid tropical climate that prevailed there many thousands of years ago.

2.5 BIOTA: LIVING ORGANISMS

Organic matter accumulation, biochemical weathering, profile mixing, nutrient cycling, and aggregate stability are all enhanced by the activities of organisms in the soil. Vegetative cover reduces natural soil erosion rates, thereby slowing the rate of mineral surface soil removal. Organic acids produced from certain types of plant leaf litter bring iron and aluminum into solution by complexation and accelerate the downward movement of these metals and their accumulation in the B horizon.

Role of Natural Vegetation

Organic Matter Accumulation. The effect of vegetation on soil formation can be seen by comparing properties of soils formed under grassland and forest vegetation near the boundary between these two ecosystems (Figure 2.16). In the grassland, much of the organic matter added to the soil is from the deep fibrous grass root systems. By contrast, tree leaves falling on the forest floor are the principal source of soil organic matter in the forest. Another difference is the frequent occurrence in the grasslands of fires that destroy large amounts of aboveground plant matter and surface litter. Also, the much greater acidity under many forests inhibits the action of certain soil organisms that otherwise would mix much of the surface litter into the mineral soil. As a result, the soils under grasslands generally develop a thicker A horizon with a deeper distribution of organic matter than in comparable soils under forests, which characteristically store most of their organic matter in the forest floor (O horizons) and a thin A horizon. The microbial community in a typical grassland soil is dominated by bacteria, while that of the forest soil is dominated by fungi (see Chapter 10 for details). Differences in microbial action affect the aggregation of the mineral particles into stable granules and the rate of nutrient cycling. The light-colored, leached E horizon typically found under the O or A horizon of a forested soil results from the action of organic acids generated mainly by fungi in the acidic forest litter. An E horizon is generally not found in a grassland soil.

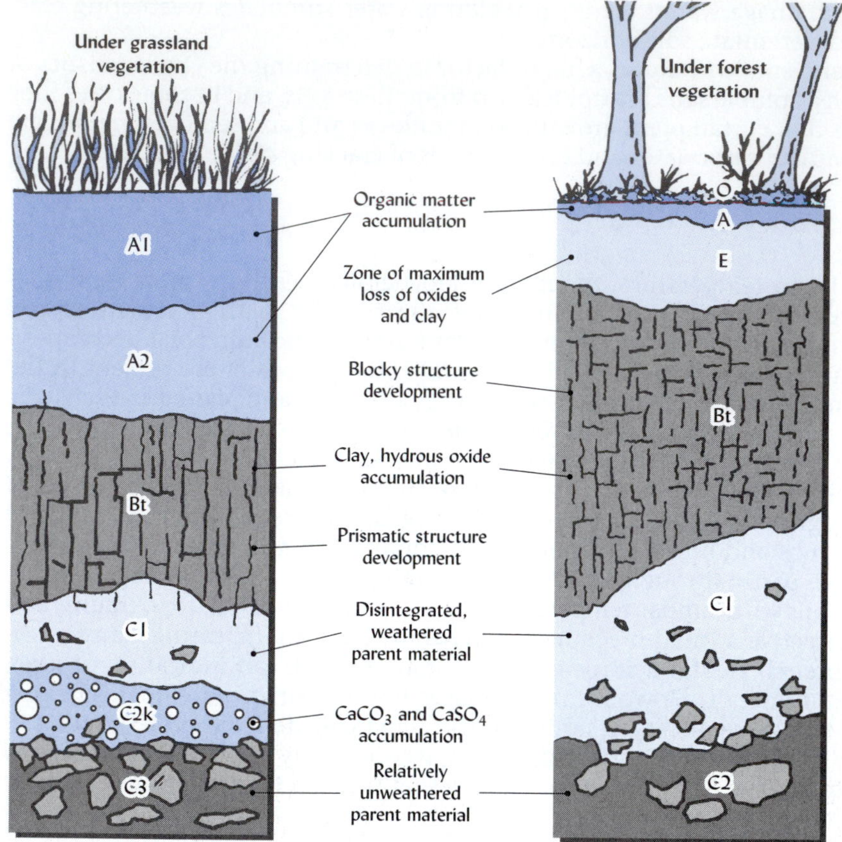

FIGURE 2.16 Natural vegetation influences the type of soil eventually formed from a given parent material (calcareous glacial till, in this example). The grassland vegetation is likely to occur in a somewhat less humid climate than the deciduous forest. The amount, and especially the vertical distribution, of organic matter that accumulates in the upper part of the profile differs markedly between the vegetation types. The forested soil exhibits surface layers (O horizons) of leaves and twigs in various stages of decomposition, along with a thin, mineral A horizon, into which some of the surface litter has been mixed. In contrast, most of the organic matter in the grassland is added as fine roots distributed throughout the upper 1 m or so, creating a thick, mineral A horizon. Also note that calcium carbonate has been solubilized and has moved down to the lower horizons (Ck) in the grassland soils, while it has been completely removed from the profile in the more acidic, leached forested soil. Under both types of vegetation, clay and iron oxides move downward from the A horizon and accumulate in the B horizon, encouraging the formation of characteristic soil structure. In the forested soil, the zone above the B horizon usually becomes a distinctly bleached E horizon, partly because most of the organic matter is restricted to the near-surface layers, and partly because decomposition of the forest litter generates organic acids that remove the brownish iron oxide coatings. Compare these mature profiles to the changes over time discussed in Sections 2.7 and 2.8. (Diagrams courtesy of R. Weil)

Cation Cycling by Trees.[4] The ability of natural vegetation to accelerate the release of nutrient elements from minerals by biogeochemical weathering, and to take up these elements from the soil, strongly influences the characteristics of the soils that develop. Soil acidity is especially affected. Differences occur not only between grassland and forest vegetation, but also between different species of forest trees. Litter falling from coniferous trees (e.g., pines, firs, spruces, and hemlocks) will recycle only small quantities of calcium, magnesium, and potassium compared to those recycled by litter from some deciduous trees (e.g., beech, oaks, and maples) that take up and store much larger amounts of these cations (Figure 2.17).

Heterogeneity in Rangelands. In arid and semiarid rangelands, competition for limited soil water does not permit vegetation dense enough to completely cover the soil surface. Scattered shrubs or bunch grasses are interspersed with openings in the plant canopy where the soil is bare or partially covered with plant litter. The widely scattered vegetation alters soil properties in several ways. Plant canopies trap windblown dust that is often relatively rich in silt and clay. Roots scavenge nutrients such as nitrogen, phosphorus, potassium, and sulfur from the interplant areas. These nutrients are then deposited with the leaf litter under the plant canopies. The decaying litter adds organic acids, which lower the soil pH and stimulate mineral weathering. As time goes on, the relatively bare soil areas between plants decline in fertility and may increase in size as they become impoverished and even less inviting for the establishment of plants. Simultaneously, the vegetation creates "islands" of enhanced fertility, thicker A horizons, and often more deeply leached calcium carbonate (see Figure 2.18).

[4]For an intriguing paper on the ecological impacts and competing theories of how trees affect nutrient cycling and soil formation, see Binkley and Giardina (1998).

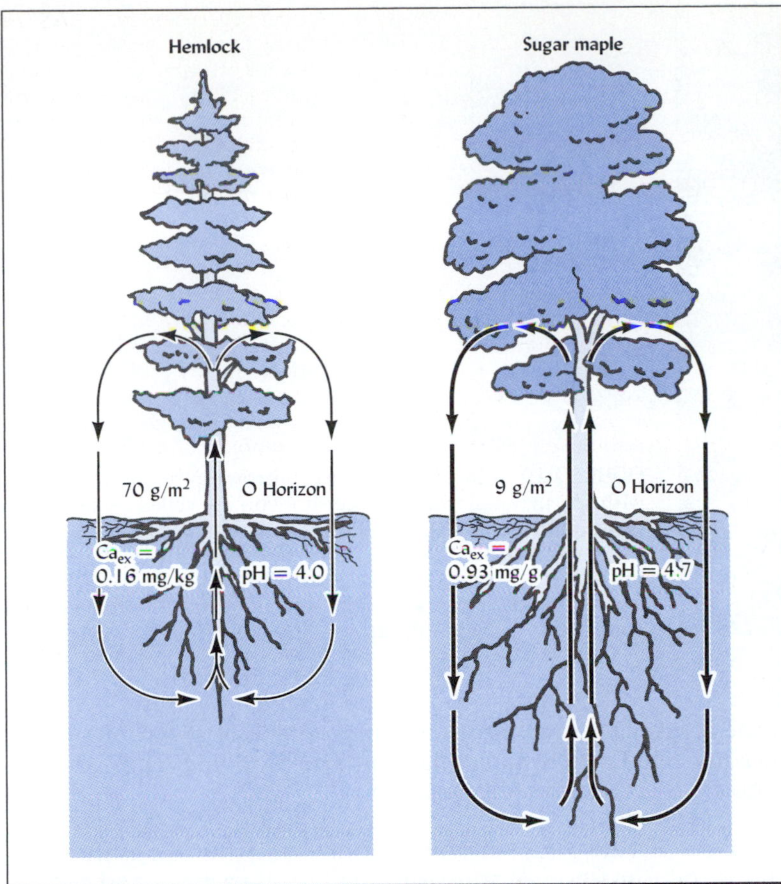

FIGURE 2.17 Nutrient cycling is an important process by which plants affect the soil in which they grow, altering the course of soil development and the suitability of the soil environment for future generations of vegetation. For example, hemlock (a conifer) and sugar maple (a deciduous hardwood) differ markedly in their ability to accelerate mineral weathering, mobilize nutrient cations, and recycle them to the upper soil horizons. Sugar maple roots are efficient at taking up Ca from soil minerals, and the maple leaves produced contain high concentrations of Ca. When these leaves fall to the ground, they decompose rapidly and release large amounts of Ca^{2+} ions that become adsorbed as exchangeable Ca^{2+} on humus and clay in the O and A horizons. This influx of Ca^{2+} ions may somewhat retard acidification of the surface layers. However, the maple roots' efficient extraction of Ca from minerals in the parent material may accelerate acidification and weathering in deeper soil horizons. In contrast, hemlock needles are Ca-poor, much slower to decompose, and therefore result in a thicker O horizon, greater acidity in the O and upper mineral horizons, but possibly less rapid weathering of minerals in the underlying parent material. [Data for a Connecticut forest reported by van Breeman and Finzi (1998).]

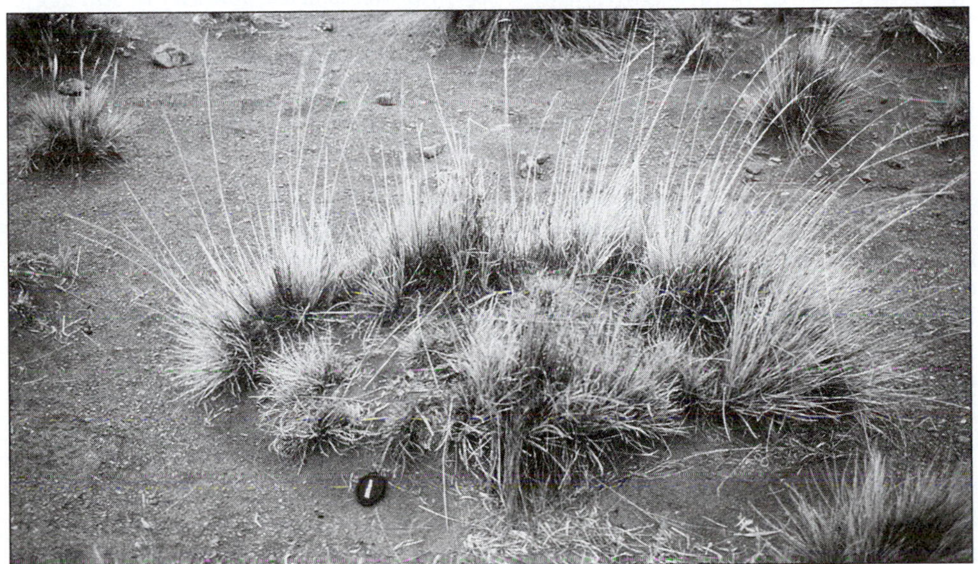

FIGURE 2.18 The scattered bunch grasses of this semiarid rangeland in the Patagonia region of Argentina have created "islands" of soil with enhanced fertility and thicker A horizons. A lens cap placed at the edge of one of these islands provides scale and highlights the increased soil thickness under the plant canopies. Such small-scale, plant-associated soil heterogeneity is common where soil water limitations prevent complete plant ground cover. (Photo courtesy of Ingrid C. Burke, Short-Grass Steppe Long-Term Ecological Research Program, Colorado State University)

Role of Animals

The role of animals in soil-formation processes must not be overlooked. Large animals such as gophers, moles, and prairie dogs bore into the lower soil horizons, bringing materials to the surface. Their tunnels are often open to the surface, encouraging movement of water and air into the subsurface layers. In localized areas, they enhance mixing of the lower and upper horizons by creating, and later refilling, underground tunnels. For example, dense populations of prairie dogs may completely turn over the

FIGURE 2.19 Abandoned animal burrows in one horizon filled with soil material from another horizon are called *crotovinas*. In this Illinois prairie soil, dark, organic-matter-rich material from the A horizon has filled in old prairie dog burrows that extend into the B horizon. The dark circular shapes in the subsoil mark where the pit excavation cut through these burrows. Scale marked every 10 cm. (Photo courtesy of R. Weil)

upper meter of soil in the course of several thousand years. Old animal burrows in the lower horizons often become filled with soil material from the overlying A horizon, creating profile features known as **crotovinas** (Figure 2.19).

EARTHWORMS, ANTS, AND TERMITES. Earthworms, ants, and termites mix the soil as they burrow, significantly affecting soil formation. Earthworms ingest soil particles and organic residues, enhancing the availability of plant nutrients in the material that passes through their bodies. They aerate and stir the soil and increase the stability of soil aggregates, thereby assuring ready infiltration of water. Ants and termites, as they build mounds, also transport soil materials from one horizon to another. In general, the mixing activities of animals, sometimes called **pedoturbation**, tends to undo or counteract the tendency of other soil-forming processes to accentuate the differences among soil horizons.

HUMAN INFLUENCES. Human activities also influence soil formation. For example, it is believed that Native Americans regularly set fires to maintain several large areas of prairie grasslands in Indiana and Michigan. In more recent times, human destruction of natural vegetation (trees and grass) and subsequent tillage of the soil for crop production has abruptly modified soil formation. Likewise, irrigating an arid region soil drastically influences the soil-forming factors, as does adding fertilizer and lime to soils of low fertility. In surface mining and urbanizing areas today, bulldozers may have an effect on soils almost akin to that of the ancient glaciers; they level and mix soil horizons and set the clock of soil formation back to zero.

2.6 TOPOGRAPHY

Topography relates to the configuration of the land surface and is described in terms of differences in elevation, slope, and landscape position—in other words, the lay of the land. The topographical setting may either hasten or retard the work of climatic forces (as shown in Box 2.1). Steep slopes generally encourage rapid soil loss by erosion and allow less rainfall to enter the soil before running off. In semiarid regions, the lower effective rainfall on steeper slopes also results in less complete vegetative cover, so there is less plant contribution to soil formation. For all of these reasons, steep slopes prevent the formation of soil from getting very far ahead of soil destruction. Therefore, soils on steep terrain tend to have rather shallow, poorly developed profiles in comparison to soils on nearby, more level sites (Figure 2.20).

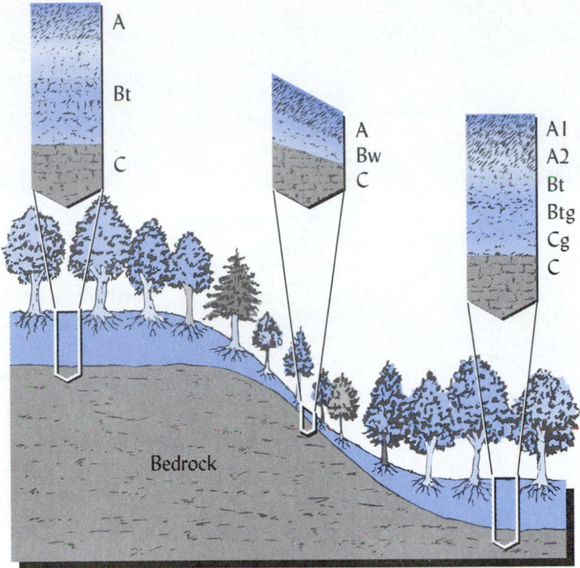

FIGURE 2.20 Topography influences soil properties, including soil depth. The diagram on the left shows the effect of slope on the profile characteristics and the depth of a soil on which forest trees are the natural vegetation. The photo on the right illustrates the same principle under grassland vegetation. Often a relatively small change in slope can have a great effect on soil development. See Section 2.9 for explanation of horizon symbols. (Photo courtesy of R. Weil)

Relief or topography as a factor in soil formation:
http://grunwald.ifas.ufl.edu/Nat_resources/soil_forming_factors/formation.htm#relief

In swales and depressions where runoff water tends to concentrate, the regolith is usually more deeply weathered and soil profile development is more advanced. However, in the lowest landscape positions, water may saturate the regolith to such a degree that drainage and aeration are restricted. Here the weathering of some minerals and the decomposition of organic matter are retarded, while the loss of iron and manganese is accelerated. In such low-lying topography, special profile features characteristic of wetland soils may develop (see Section 7.7 on the soils of wetlands).

Soils commonly occur together in the landscape in sequence called a **catena** (from the Latin meaning *chain*), with each member of the catena occupying a characteristic topographic position. Soils in a catena generally exhibit properties that reflect the influence of topography on water movement and drainage. A **toposequence** is a type of catena, in which the differences among the soils result predominantly from the influence of topography because the soils in the sequence all share similar conditions regarding the other four soil forming factors: climate, vegetation, parent material, and time (see, for example, Figure 2.20 and Plate 16 after page 114).

Interaction with Vegetation. Topography often interacts with vegetation to influence soil formation. In grassland–forest transition zones, trees are commonly confined to the slight depressions where soil is generally wetter than in upland positions. As would be expected, the nature of the soil in the depressions is quite different from that in the uplands. If water stands for part or all of the year in a depression in the landscape, climate is less influential in regulating soil development. Low-lying areas may give rise to peat bogs and, in turn, to organic soils (see Figure 2.14).

Slope Aspect. Topography affects the absorbance of solar energy in a given landscape. In the northern hemisphere, south-facing slopes are more perpendicular to the sun's rays and are generally warmer and thereby commonly lower in moisture than their north-facing counterparts (see also Figure 7.19). Consequently, soils on the south slopes tend to be lower in organic matter and are not so deeply weathered.

Salt Buildup. In arid and semiarid regions, topography influences the buildup of soluble salts. Dissolved salts from surrounding upland soils move on the surface and through the underground water table to the lower-lying areas (see Section 9.13). There they rise to the surface as the water evaporates, often accumulating to plant-toxic levels.

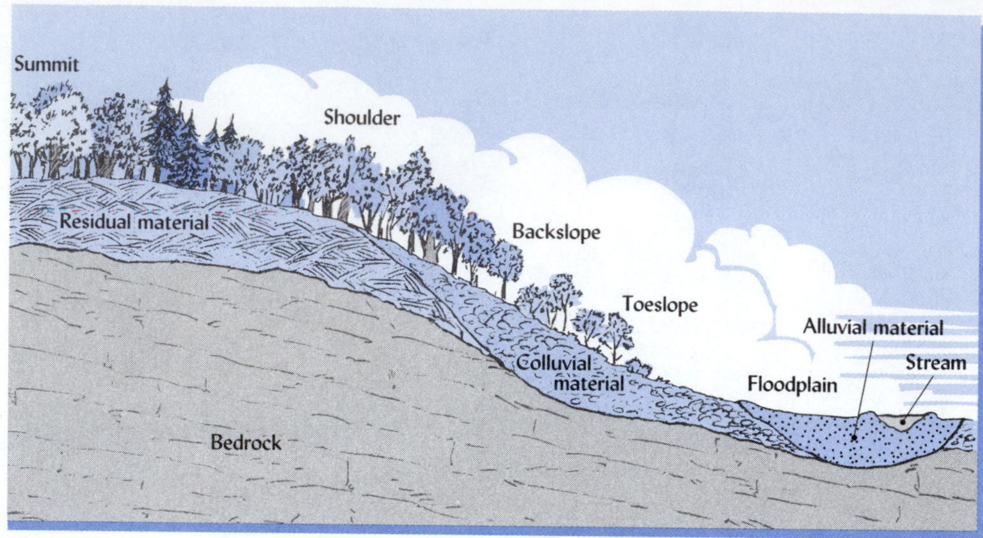

FIGURE 2.21 An interaction of topography and parent material as factors of soil formation. The soils on the summit, toe-slope, and floodplain in this idealized landscape have formed from residual, colluvial, and alluvial parent materials, respectively.

Parent Material Interactions. Topography can also interact with parent material. For example, in areas of tilted beds of sedimentary rock, the ridges often consist of resistant sandstone, while the valleys are underlaid by more weatherable limestone. In many landscapes, topography reflects the distribution of residual, colluvial, and alluvial parent materials, with residual materials on the upper slopes, colluvium covering the lower slopes, and alluvium filling the valley bottom (Figure 2.21).

2.7 TIME

Soil-forming processes take time to show their effects. The clock of soil formation starts ticking when a landslide exposes new rock to the weathering environment at the surface, when a flooding river deposits a new layer of sediment on its floodplain, when a glacier melts and dumps its load of mineral debris, or when a bulldozer cuts and fills a landscape to level a construction or mine-reclamation site.

Rates of Weathering. When we speak of a "young" or a "mature" soil, we are not so much referring to the age of the soil in years, as to the degree of weathering and profile development. Time interacts with the other factors of soil formation. For example, on a level site in a warm climate, with much rain falling on permeable parent material rich in reactive minerals, weathering and soil profile differentiation will proceed far more rapidly than on a site with steep slopes and resistant parent material in a cold, dry climate.

In a few instances soils form so rapidly that the effect of time on the process can be measured in a human life span. For example, dramatic mineralogical, structural, and color changes occur within a few months to a few years when certain sulfide-containing materials are first exposed to air by excavation, wetland drainage, or sediment dredging (see Sections 9.7 and 12.22). Under favorable conditions, organic matter may accumulate to form a darkened A horizon in freshly deposited, fertile alluvium in a mere decade or two. In some cases, incipient B horizons become discernible on humid-region mine spoils in as few as 40 years. Structural alteration and coloring by accumulated iron may form a simple B horizon within a few centuries if the parent material is sandy and the climate is humid. The same degree of B-horizon development would take much longer under conditions less favorable for weathering and leaching. The accumulation of silicate clays and the formation of blocky structure in B horizons usually become noticeable only after several thousand years. Developing in resistant rock, a mature, deeply weathered soil may be hundreds of thousands of years in the making (Figure 2.22).

Example of Soil Genesis over Time. Figure 2.22 illustrates changes that typically take place during soil development on residual rock in a warm, humid climate. During the first

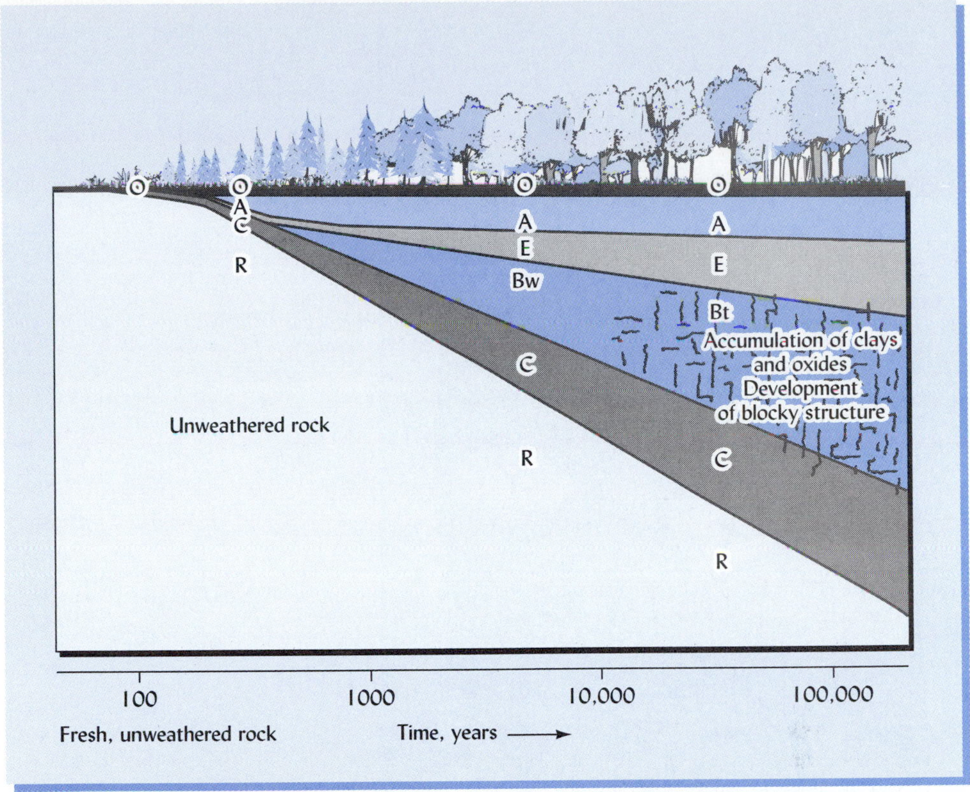

FIGURE 2.22 Progressive stages of soil profile development over time for a residual igneous rock, in a warm, humid climate that is conducive to forest vegetation. The time scale increases logarithmically from left to right, covering more than 100,000 years. Note that the mature profile (right side of this figure) expresses the full influence of the forest vegetation as illustrated in Figure 2.16. This mature soil might be classified as an Ultisol (see Section 3.14).

100 years, lichens and mosses establish themselves on the bare exposed rock and begin to accelerate its breakdown and the accumulation of dust and organic matter. Within a few hundred years, grasses, shrubs, and stunted trees have taken root in a deepening layer of disintegrated rock and soil, adding greatly to the accumulation of organic materials and to the formation of the A and C horizons. During the next 10,000 years or so, successions of forest trees establish themselves and the activities of a multitude of tiny soil organisms transform the surface plant litter into a distinct O horizon. The A horizon thickens somewhat, becomes darker in color, and develops a stable granular structure. Soon, a bleached zone appears just below the A horizon as soluble weathering products, iron oxides, and clays are moved down with the water and organic acids percolating down from the litter layer. These transported materials begin to accumulate in a deeper layer, forming a B horizon. The process continues, with more silicate clay accumulating and blocky structure forming as the B horizon thickens and becomes more distinct. Eventually, the silicate clays themselves break down, some silica is leached away, and new clays containing less silica form in the B horizon. These clays often become mixed or coated with oxides of iron and aluminum, causing the B horizon to take on a reddish hue. As the entire profile continues to deepen over time, the zone of weathering below the B horizon may become many meters thick.

2.8 FOUR BASIC PROCESSES OF SOIL FORMATION[5]

The accumulation of regolith from the breakdown of bedrock, or the deposition (by wind, water, ice, etc.) of unconsolidated geologic materials may precede or, more commonly, occur simultaneously with the development of the distinctive horizons of a soil

[5]For the classic presentation of the processes of soil formation see Simonson (1958). In addition, detailed discussion of these basic processes and their specific manifestations can be found in Birkeland (1999), Fanning and Fanning (1989), and Buol et al. (1997).

FIGURE 2.23 A schematic illustration of additions, losses, translocations, and transformations as the fundamental processes driving soil-profile development. (Diagram courtesy of R. Weil)

profile. During the formation (**genesis**) of a soil from parent material, the regolith undergoes many profound changes. These changes are brought about by variations in the four broad soil-forming processes (Figure 2.23) considered next. These four basic processes—often referred to as the soil-forming, or **pedogenic**, processes—help distinguish soils from layers of sediment deposited by geologic processes.

Transformations

Transformations occur when soil constituents are chemically or physically modified or destroyed and others are synthesized from the precursor materials. Many transformations involve weathering of primary minerals, disintegrating and altering some to form various kinds of silicate clays. As other primary minerals decompose, the decomposition products recombine into new minerals that include additional types of silicate clays and hydrous oxides of iron and aluminum (see Figure 2.3). Other important transformations involve the decomposition of organic residues and the synthesis of organic acids, humus, and other products. Still other transformations change the size (e.g., physical weathering to smaller particles) or arrangement (e.g., aggregation) of mineral particles.

Translocations

Translocations involve the movement of inorganic and organic materials laterally within a horizon or vertically from one horizon up or down to another. Water, either percolating down with gravity or rising up by capillary action (see Section 5.2), is the most common translocation agent. The materials moved within the profile include dispersed fine clay particles, dissolved salts, and dissolved organic substances. Translocations of materials by soil organisms also have a major influence on soil genesis. Important examples include incorporation of surface organic litter into the A and B horizons by certain earthworms, transport of B and C horizon material to the surface by mound-building termites, and the widespread burrowing actions of rodents.

Additions

Inputs of materials to the developing soil profile from outside sources are considered *additions*. A very common example is the input of organic matter from fallen plant leaves and sloughed-off roots (the carbon having originated in the atmosphere). Another ubiquitous addition is dust particles falling on the surface of the soil (wind may have blown these particles from a source just a few meters away, or across the ocean). Still another example, common in arid regions, is the addition of salts or silica dissolved in the groundwater and deposited near or at the soil surface when the rising water evaporates.

Losses

Materials are lost from the soil profile by leaching to groundwater, erosion of surface materials, or other forms of removal. Evaporation and plant use cause losses of water. Leaching and drainage cause the loss of water, dissolved substances such as salts or silica weathered from parent minerals, or organic acids produced by microorganisms or plant roots. Erosion, a major loss agent, often removes the finer particles (humus, clay, and silt), leaving the surface horizon relatively sandier and less rich in organic matter than before. Organic matter is also lost by microbial decomposition (see Section 11.2). Grazing by animals or harvest by people can remove large amounts of both organic matter and nutrient elements. Of course, animals and people can also contribute additions, such as manure and fertilizers.

These processes of soil genesis, operating under the influence of the environmental factors discussed previously, give us a logical framework for understanding the relationships between particular soils and the landscapes and ecosystems in which they function. In analyzing these relationships for a given site, ask yourself: What are the materials being added to this soil? What transformations and translocations are taking place in this profile? What materials are being removed? And how have the climate, organisms, topography, and parent material at this site affected these processes over time?

Soil-Forming Processes in Action: A Simplified Example

More specific processes of soil formation: http://grunwald.ifas.ufl.edu/Nat_resources/processes_soil_formation/processes.htm

Consider the changes that might take place as a soil develops from a thick layer of relatively uniform loess parent material in a climate conducive to grass vegetation (Figure 2.24). Although some physical weathering and leaching of carbonates and salts may be necessary to allow plants to grow in certain parent materials, soil formation really gets started when plants become established and begin to provide *additions* of litter and root residues on and in the surface layers of the parent material. The plant residues are *transformed* by soil organisms into humus and other new organic substances. The accumulation of humus enhances the capacity of the soil to hold water and nutrients, providing a positive feedback for accelerated plant growth and further humus buildup. Earthworms, ants, termites, and a host of smaller animals come to live in the soil and feed on the newly accumulating organic resources. In so doing, they accelerate the organic *transformations,* as well causing the *translocation* of plant residues, loosening the mineral material as they burrow into the soil.

A-Horizon Development. The resulting organic-mineral mixture near the soil surface, which comes into being rather quickly, is commonly the first soil horizon developed, the A horizon. It is darker in color and its chemical and physical properties differ from those of the original parent material. Individual soil particles in this horizon commonly clump together under the influence of organic substances to form granules, differentiating this layer from the deeper layers and from the original parent material. On sloping land, erosion may remove some materials from the newly forming upper horizon, retarding, somewhat, the progress of horizon development.

Formation of B and C Horizons. Carbonic and other organic acids are carried by percolating waters into the soil, where they stimulate weathering reactions. The acid-charged

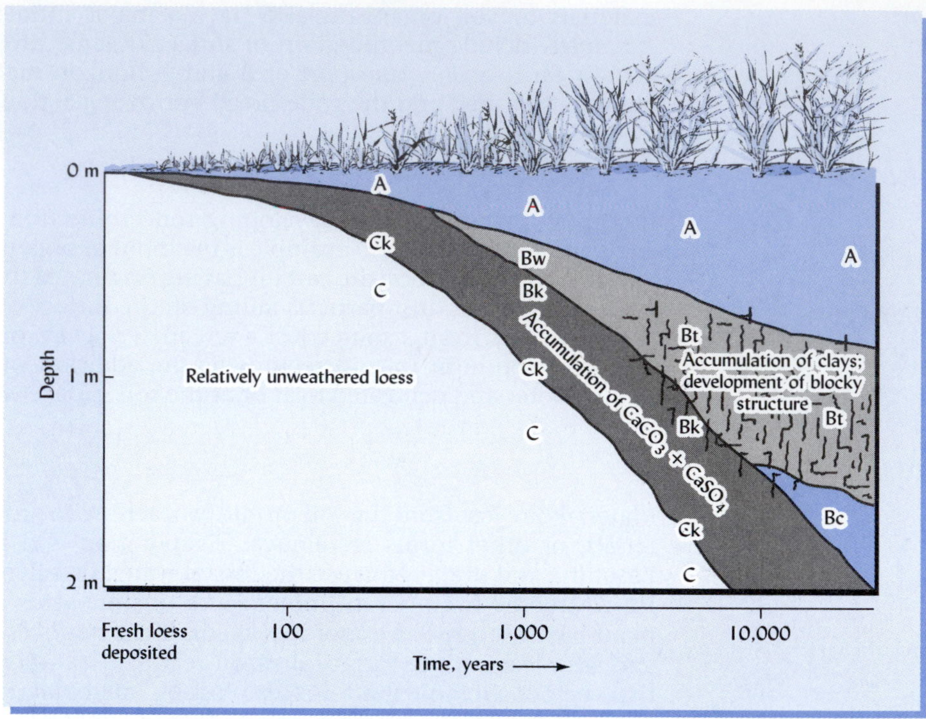

FIGURE 2.24 Development of a hypothetical soil in about 2 m of uniform calcareous loess deposits where the warm subhumid climate is conducive to tall grass prairie vegetation. The time scale increases logarithmically from left to right, covering about 12,000 years. In the initial stages, rainwater, charged with organic acids from microbial respiration, dissolves carbonates from the loess and moves them downward to a zone of accumulation (Bk horizon). As this happens, plant roots take hold in the upper layer and add the organic matter necessary to create an A horizon. Ants, beetles, earthworms, and a host of smaller creatures take up residence and actively mix in surface litter and speed the release of nutrients from the minerals and plant residues. Over time the carbonate concentration zone moves deeper, the A horizon thickens, and noncalcareous B horizons develop as changes in color and structure occur in the weathering loess above the zone of carbonate accumulation. Eventually, silicate clay accumulates in the B horizon (giving it the designation Bt), both by stationary weathering of primary minerals and by traveling there with water percolating from the upper horizons. Compare the stages of development and rates of change to those illustrated in Figure 2.23 for a different parent material, vegetation, and climate. Note that the mature profile (right side of this figure) expresses the full influence of the grassland vegetation as illustrated in Figure 2.16. This mature soil would be classified as a Mollisol (see Section 3.12).

percolating water dissolves various minerals (a *transformation*) and leaches the soluble products (a *translocation*) from upper to lower horizons, where they may precipitate. This combination of transformation and translocation creates zones of depletion in the upper layers and zones of accumulation in the lower. The dissolved substances include both positively charged ions (cations; e.g., Ca^{2+}) and negatively charged ions (anions; e.g., CO_3^- and SO_4^{2-}) released from the breakdown of minerals and organic matter. In semiarid and arid regions, precipitation of such ions produces horizons enriched in calcite ($CaCO_3$), designated as a Bk layer in Figure 2.24.

Over time, the leached surface layer thickens, and the zone of Ca accumulation is moved downward to the maximum depth of water penetration. Where rainfall is great enough to cause significant drainage to the groundwater, some of the dissolved materials may be completely removed from the developing soil profile (*losses*), and the zone of accumulation may move below the reach of plant roots, or be dissipated altogether. On the other hand, deep-growing plant roots may intercept some of these soluble weathering products and return them, through leaf and litter fall, to the soil surface, thus retarding somewhat the processes of acid weathering and horizon differentiation.

The weathering of primary minerals into clay minerals becomes evident only long after the dissolution and movement of Ca is well underway. The newly formed clay minerals may accumulate where they are formed, or they may move downward and accumulate deeper in the profile. As clay is removed from one layer and accumulates in another, adjacent layers become more distinct from each other, and a Bt horizon (one enriched in silicate clay) is formed. When the accumulated clay in the Bt horizon periodically dries out and cracks, blocklike or prismatic units of soil structure begin to develop (see Figure 2.24 and Section 4.7). As the soil matures, the various horizons within the profile generally become more numerous and more distinctly different from each other.

2.9 THE SOIL PROFILE

At each location on the land, the earth's surface has experienced a particular combination of influences from the five soil-forming factors, causing a different set of layers (horizons) to form in each part of the landscape, thus slowly giving rise to the natural bodies we call **soils.** Each soil is characterized by a given sequence of these horizons. A vertical exposure of this sequence is termed a **soil profile.** We will now consider the major horizons making up soil profiles and the terminology used to describe them.

The Master Horizons and Layers

Five **master** soil horizons are recognized and are designated using the capital letters O, A, E, B, and C (Figure 2.25). Subordinate horizons may occur within some master horizons and these are designated by lowercase letters following the capital master horizon letter (e.g., Bt, Ap, or Oi).

O HORIZONS. The O group is comprised of organic horizons that generally form above the mineral soil or occur in an organic soil profile. They derive from dead plant and animal residues. Generally absent in grassland regions, O horizons usually occur in forested areas and are commonly referred to as the forest floor (see Plate 10). Often three subordinate O horizons can be distinguished, Oi, Oe, and Oa (Figure 2.25).

A HORIZONS. The topmost mineral horizons, designated A horizons, generally contain enough partially decomposed (humified) organic matter to give the soil a color darker than that of the lower horizons. The A horizons are often coarser in texture, having lost some of the finer materials by translocation to lower horizons and by erosion.

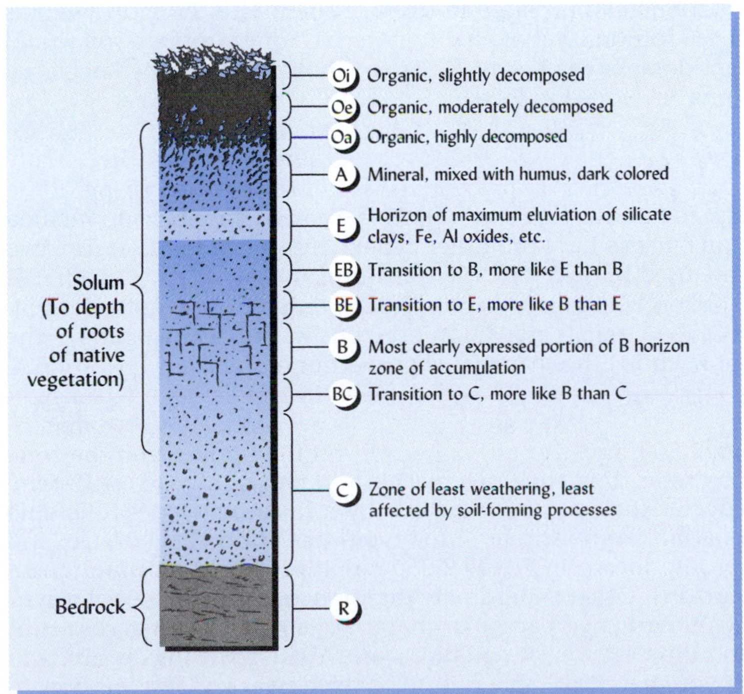

FIGURE 2.25 Hypothetical mineral soil profile showing the major horizons that may be present in a well-drained soil in the temperate humid region. Any particular profile may exhibit only some of these horizons, and the relative depths vary. In addition, however, a soil profile may exhibit more detailed subhorizons than indicated here. The solum includes the A, E, and B horizons plus some cemented layers of the C horizon.

E Horizons. These are zones of maximum leaching or **eluviation** (from Latin *ex* or *e,* out, and *lavere,* to wash) of clay, iron, and aluminum oxides, which leaves a concentration of resistant minerals, such as quartz, in the sand and silt sizes. An E horizon is usually found underneath the A horizon and is generally lighter in color than either the A horizon above it or the horizon below. Such E horizons are quite common in soils developed under forests, but they rarely occur in soils developed under grassland. Distinct E horizons can be seen in the soils in Plates 10 and 19.

B Horizons. B horizons form below an O, A, or E horizon and have undergone sufficient changes during soil genesis so that the original parent material structure is no longer discernable. In many B horizons, materials have accumulated, typically by washing in from horizons above, a process termed **illuviation** (from the Latin *il,* in, and *lavere,* to wash). In humid regions, B horizons are the layers of maximum accumulation of materials such as iron and aluminum oxides (Bo or Bs horizons) and silicate clays (Bt horizons), some of which may have illuviated from upper horizons and some of which may have formed in place. Such B horizons can be clearly seen in the middle depths of the profiles shown in Plates 1, 10, and 11. In arid and semiarid regions, calcium carbonate or calcium sulfate may accumulate in the B horizons (giving Bk or By horizons).

C Horizon. The C horizon is the unconsolidated material underlying the solum (A and B horizons). It may or may not be the same as the parent material from which the solum formed. The C horizon is below the zones of greatest biological activity and has not been sufficiently altered by soil genesis to qualify as a B horizon. Accumulations of calcium carbonate and gypsum may occur. Although loose enough to be dug with a shovel, C horizon material often retains some of the structural features of the parent rock or geologic deposits from which it formed (see, for example, the lower third of the profiles shown in Plates 7 and 11). Its upper layers may in time become a part of the solum as weathering and erosion continue.

R Layers. The R layers are consolidated rock, with little evidence of weathering.

Subdivisions within Master Horizons

Often distinctive layers exist within a given master horizon, and these are indicated by a numeral following the letter designation. For example, if three different combinations of structure and colors can be seen in the B horizon, then the profile may include a B1–B2–B3 sequence.

If two different geologic parent materials (e.g., loess over glacial till) are present within the soil profile, the numeral 2 is placed in front of the master horizon symbols for horizons developed in the second layer of parent material. For example, a soil would have a sequence of horizons designated O–A–B–2C if the C horizon developed in glacial till while the upper horizons developed in loess.

Transition Horizons

Transitional layers between the master horizons (O, A, E, B, and C) may be dominated by properties of one horizon but also have prominent characteristics of another. The two applicable capital letters are used to designate the transition horizons (e.g., AE, EB, BE, and BC), the dominant horizon being listed before the subordinate one. Letter combinations such as E/B are used to designate transition horizons where distinct parts of the horizon have properties of E while other parts have properties of B.

Subordinate Distinctions

Since the capital letter designates the nature of a master horizon in only a very general way, specific horizon characteristics may be indicated by a lowercase letter following the master horizon designation. For example, three types of O horizons (Oi, Oe, and Oa) are indicated in the profile shown in Figure 2.25, which presents a commonly encountered sequence of horizons. Other subordinate distinctions include special physical properties and the accumulation of particular materials, such as clays and salts. A list of the recognized subordinate letter designations and their meanings is given in Table 2.2. We suggest that you mark this table for future reference, and study it now to

TABLE 2.2 Lowercase Letter Symbols to Designate Subordinate Distinctions Within Master Horizons

Letter	*Distinction*	*Letter*	*Distinction*
a	Organic matter, highly decomposed	n	Accumulation of sodium
b	Buried soil horizon	o	Accumulation of Fe and Al oxides
c	Concretions or nodules	p	Plowing or other disturbance
d	Dense unconsolidated materials	q	Accumulation of silica
e	Organic matter, intermediate decomposition	r	Weathered or soft bedrock
f	Frozen soil	s	Illuvial accumulation of O.M.[a] and Fe and Al oxides
g	Strong gleying (mottling)	ss	Slickensides (shiny clay wedges)
h	Illuvial accumulation of organic matter	t	Accumulation of silicate clays
i	Organic matter, slightly decomposed	v	Plinthite (high iron, red material)
j	Jarosite (yellow sulfur mineral)	w	Distinctive color or structure
jj	Cryoturbation (frost churning)	x	Fragipan (high bulk density, brittle)
k	Accumulation of carbonates	y	Accumulation of gypsum
m	Cementation or induration	z	Accumulation of soluble salts

[a]O.M. = organic matter

get an idea of the distinctive soil properties that can be indicated by horizon designations. By way of illustration, a Bt horizon is a B horizon characterized by clay accumulation (*t* from the German *ton,* meaning "clay"); likewise, in a Bk horizon, carbonates (k) have accumulated.

Horizons in a Given Profile

It is not likely that the profile of any one soil will show all of the horizons that collectively are shown in Figure 2.25. The ones most commonly found in well-drained soils are Oi and Oe (or Oa) if the land is forested; A or E (or both, depending on circumstances); Bt or Bw; and C. Conditions of soil genesis will determine which others are present and their clarity of definition.

When a virgin (never cultivated) soil is plowed and cultivated, the upper 15 to 20 cm becomes the plow layer or Ap horizon. Cultivation, of course, obliterates the original layered condition of the upper portion of the profile, and the Ap horizon becomes more or less homogeneous. An example of an Ap horizon can be seen in the upper 20 cm of the soil shown in Plate 1. In some soils, the A and E horizons are deeper than the plow layer (Figure 2.26). In other cases, where the upper horizons are quite thin, the plow line is just at the top of, or even down in, the B horizon.

In some cultivated land, serious erosion produces a **truncated** profile. As the surface soil is swept away over the years, the plow reaches deeper and deeper into the profile.

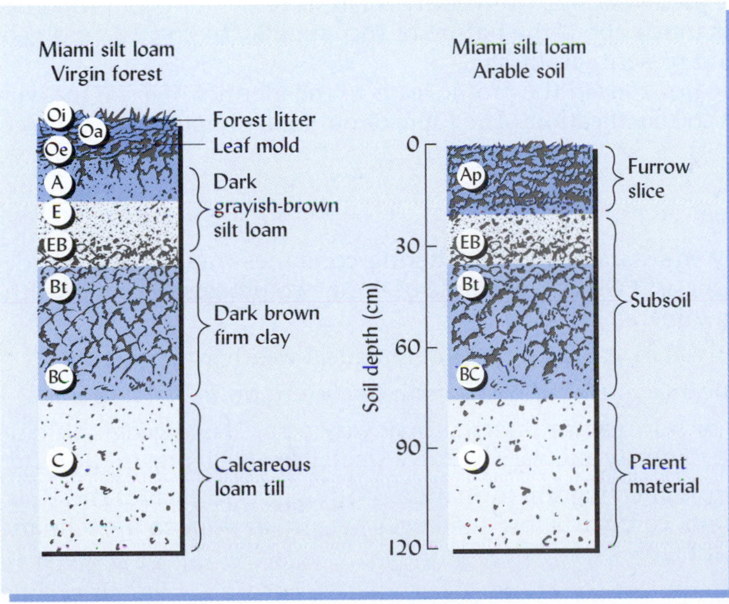

FIGURE 2.26 Generalized profile of the Miami silt loam, one of the Alfisols of the eastern United States, before and after land is plowed and cultivated. The surface layers (O, A, and E) are mixed by tillage and are termed the *Ap* (plowed) *horizon.* If erosion occurs, they may disappear, at least in part, and some of the B horizon will be included in the furrow slice.

Hence, the plowed zone in many cases consists almost entirely of former B-horizon material, and the C horizon is correspondingly near the surface. Comparison to a nearby noneroded site can show how much erosion has occurred. Another, sometimes perplexing, profile feature is the presence of a buried soil resulting from natural or human action. In profile study and description, such a situation requires careful analysis.

Soil Genesis in Nature

Not every contrasting layer of material found in soil profiles is a **genetic horizon** that developed as a result of the processes of soil genesis, such as those just described. The parent materials from which many soils develop contained contrasting layers *before* soil genesis started. For example, such parent materials as glacial outwash, marine deposits, or recent alluvium may consist of various layers of fine and coarse particles laid down by separate episodes of sedimentation. Consequently, in characterizing soils, we must recognize not only the genetic horizons and properties that come into being during soil genesis, but also those layers or properties that may have been inherited from the parent material.

2.10 CONCLUSION

The parent materials from which soils develop vary widely around the world and from one location to another only a few meters apart. Knowledge of these materials, their sources or origins, mechanisms for their weathering, and means of transport and deposition are essential to understanding soil genesis.

Soil formation is stimulated by *climate* and living *organisms* acting on *parent materials* over periods of *time* and under the modifying influence of *topography.* These five major factors of soil formation determine the kinds of soil that will develop at a given site. When all of these factors are the same at two locations, the kind of soil at these locations should be the same.

Soil genesis starts when layers or horizons not present in the parent material begin to appear in the soil profile. Organic matter accumulation in the upper horizons, the downward movement of soluble ions, the synthesis and downward movement of clays, and the development of specific soil particle groupings (structure) in both the upper and lower horizons are signs that the process of soil formation is underway. As we have learned, soil bodies are dynamic in nature. Their genetic horizons continue to develop and change. Consequently, in some soils the process of horizon differentiation has only begun, while in others it is well advanced.

The four general processes of soil formation (gains, losses, transformations, and translocations) and the five major factors influencing these processes provide us with an invaluable logical framework in site selection and in predicting the nature of soil bodies likely to be found on a particular site. Conversely, analysis of the horizon properties of a soil profile can tell us much about the nature of the climatic, biological, and geological conditions (past and present) at the site.

Characterization of the horizons in the profile leads to the identity of a soil individual, which is then subject to classification—the topic of our next chapter.

STUDY QUESTIONS

1. What is meant by the statement, "Weathering combines the processes of destruction and synthesis"? Give an example of these two processes in the weathering of a primary mineral.
2. How is water involved in the main types of chemical weathering reactions?
3. Explain the significance of the ratio of silicon to aluminum in soil minerals.
4. Give an example of how parent material may vary across large geographic regions on one hand, but may also vary within a small parcel of land, on the other.
5. Name the five factors affecting soil formation. With regard to each of these factors of soil formation, compare a forested Rocky Mountain slope to the semiarid grassland plains far below.

6. How do *colluvium, glacial till,* and *alluvium* differ in appearance and agency of transport?
7. What is *loess,* and what are some of its properties as a parent material?
8. Give two specific examples for each of the four broad processes of soil formation.
9. Assuming a level area of granite rock was the parent material in both cases, describe in general terms how you would expect two soil profiles to differ—one in a warm, semiarid grassland and the other in a cool, humid pine forest.
10. For the two soils described in question 5, make a profile sketch using master horizon symbols and subordinate suffixes to show the approximate depths, sequence, and nature of the horizons you would expect to find in each soil.
11. Visualize a slope in the landscape near where you live. Discuss how specific soil properties (such as colors, horizon thickness, types of horizons present, etc.) would likely change along the toposequence of soils on this slope.

REFERENCES

Binkley, D., and C. Giardina. 1998. "Why do tree species affect soils? The warp and woof of tree–soil interactions," *Biogeochemistry* **42**:89–106.

Birkeland, P. W. 1999. *Soils and Geomorphology,* 3d ed. (New York: Oxford University Press).

Buol, S. W., F. D. Hole, R. J. McCraken, and R. J. Southard. 1997. *Soil Genesis and Classification,* 4th ed. (Ames, Iowa: Iowa State University Press).

Fanning, D. S., and C. B. Fanning. 1989. *Soil: Morphology, Genesis, and Classification.* (New York: John Wiley and Sons).

Hilgard, E. W. 1921. *Soils: Their Formation, Properties, Composition, and Plant Growth in the Humid and Arid Regions.* (London: Macmillan).

Jenny, Hans. 1941. *Factors of Soil Formation: A System of Quantitative Pedology.* (originally published by McGraw-Hill; Mineola, N.Y.: Dover).

Jenny, Hans. 1980. *The Soil Resource—Origins and Behavior.* Ecological Studies, vol. 37. (New York: Springer-Verlag).

Likens, G. E., and F. H. Bormann. 1995. *Biogeochemistry of a Forested Ecosystem,* 2d ed. (New York: Springer-Verlag).

Logan, William Bryant. 1995. *Dirt: The Ecstatic Skin of the Earth.* (New York: Riverhead Books).

Richter, D. D., and D. Markewitz. 2001. *Understanding Soil Change.* (Cambridge, U.K.: Cambridge University Press).

Rieder, R., T. Economou, H. Wanke, A. Turkevich, J. Crisp, J. Bruckner, G. Dreibus, and H. Y. McSween Jr. 1997. "The chemical composition of Martian soil and rock returned by the mobile alpha proton x-ray spectrophotometer: Preliminary results from the x-ray mode," *Science* **278**:1771–1774.

Simonson, R. W. 1958. "Outline of a generalized theory of soil genesis," *Soil Sci. Soc. Amer. Proc.,* **22**:152–156.

Soil Survey Division Staff. 1993. *Soil Survey Manual.* Agricultural Handbook 18. (Washington, D.C.: U.S. Government Printing Office).

van Breemen, N., and A. C. Finzi. 1998. "Plant–soil interactions: Ecological aspects and evolutionary implications," *Biogeochemistry* **42**:1–19.

3

Soil Classification

Roy Simonson investigates a soil. (R. Weil)

It is embarrassing not to be able to agree on what soil is. In this the pedologists are not alone. Biologists cannot agree on a definition of life and philosophers on philosophy.
—*Hans Jenny,* The Soil Resource: Origin and Behavior

We classify things in order to make sense of our world. When we call things by group names, we are classifying. Imagine a world without classifications. Imagine surviving in the woods knowing only that each plant was a plant, not which are edible by people, which attract wildlife, or which are poisonous to touch. So, too, our understanding and management of soils and terrestrial systems would be hobbled if we knew only that a soil was a soil. How could we organize our information about soils? How could we learn from others' experiences or communicate our knowledge to clients, colleagues, or workers?

Throughout history people have used various systems to name and classify soils, grouping them according to their suitability for different uses by giving them descriptive names such as *black cotton soils, rice soils,* or *olive soils*. Other soil names still in common use today have geological connotations, suggesting the parent materials from which the soils formed: *limestone soils, piedmont soils,* and *alluvial soils*. As farmers passed their experiences from one generation to the next, they summarized their knowledge about soils by developing unique systems of soil classification. In many parts of the world, local languages have a large vocabulary for the different types of soils, reflecting a sophisticated and detailed knowledge of how soils differ from one another. However, such terms are inadequate for helping us to organize our scientific knowledge of soils, or for defining the relationships among the soils of the world.

In this chapter we will learn how soils are classified as natural bodies, on the basis of their profile characteristics, not merely on their suitability for a particular use. Such a soil classification system is essential to foster global communications about soils among soil scientists and all people concerned with the management of land and the conservation of the soil resource. Through such systems we can take advantage of research and experience at one location to predict the behavior of similarly classified soils at another location. Soil names such as *Mollisols* or *Oxisols* conjure up similar mental images in the minds of soil scientists everywhere, whether they live in the United States, Europe, Japan, developing countries, or elsewhere. Through the classification system, we can create a universal language of soils that enhances communication among users of soils around the world.

3.1 CONCEPT OF INDIVIDUAL SOILS

The natural body concept of soils recognizes the existence of individual entities, each of which we call *a soil*. Just as human individuals differ from one another, soil individuals have characteristics distinguishing each from the others. The gradation in soil properties from one soil individual to an adjacent one can be compared to the gradation in the wavelengths of light from one color to another in a rainbow. The change is gradual, and yet we identify a boundary differentiating between what we call green and what we call blue.

Pedon, Polypedon and Series

Soils in the field are heterogenous; that is, the profile characteristics are not exactly the same in any two points within the soil individual you may choose to examine. Consequently, it is necessary to characterize a soil individual in terms of an imaginary three-dimensional unit called a **pedon** (rhymes with "head on," from the Greek *pedon*, ground; see Figure 3.1). It is the smallest sampling unit that displays the full range of properties characteristic of a particular soil.

Pedons occupy from about 1 to 10 m^2 of land area. The pedon is actually examined during field investigation of soils and serves as the fundamental unit of soil classification. However, a soil unit in a landscape usually consists of a group of very similar pedons, closely associated together in the field. Such a group of similar pedons, or **polypedon,** is of sufficient size to be recognized as the landscape component termed a *soil individual*.

All the soil individuals in the world that have in common a suite of soil-profile properties and horizons that fall within a particular range are said to belong to the same **soil series.** A soil series, then, is class of soils, not a soil individual, in the same way that *Pinus sylvestrus* is a species of tree, not a particular individual tree.

Groupings of Soil Individuals

We now identify the most specific and most general extremes in the concept of soils. One extreme is that of a natural body called *a soil,* characterized by a three-dimensional sampling unit (pedon), related groups of which (polypedons) are included in a soil individual.

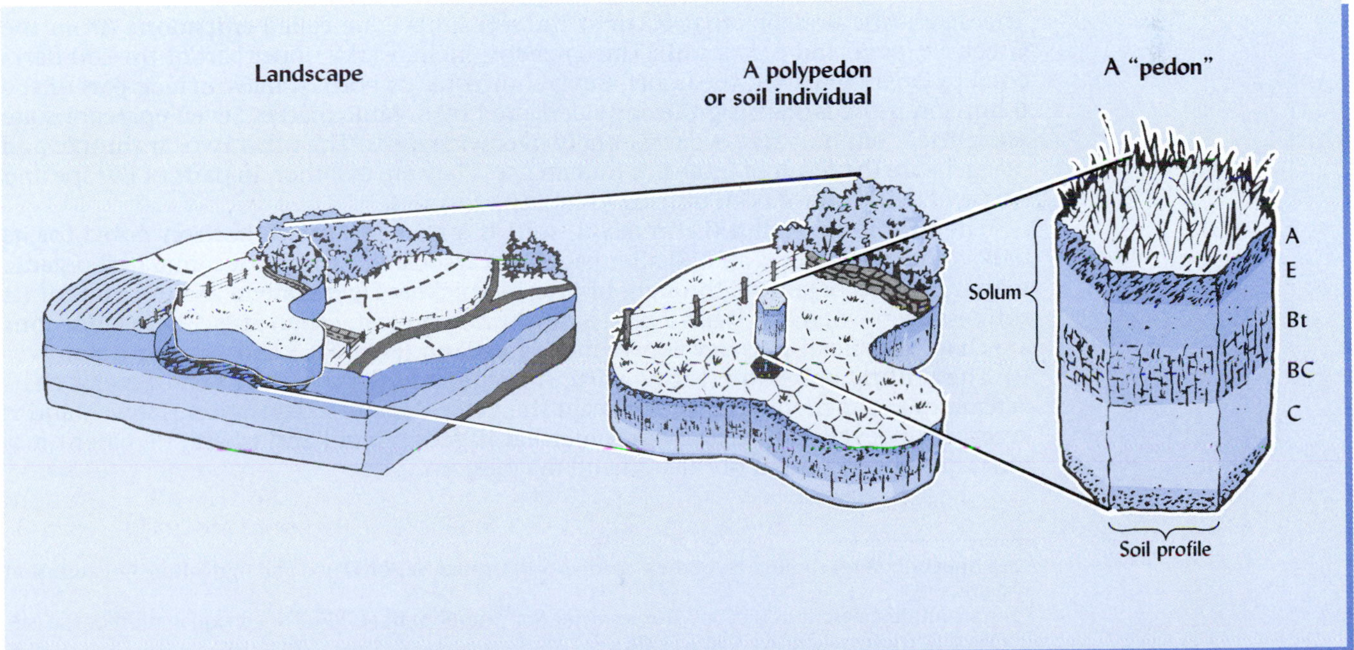

FIGURE 3.1 A schematic diagram to illustrate the concept of pedon and of the soil profile that characterizes it. Note that several contiguous pedons with similar characteristics are grouped together in a larger area (outlined by broken lines) called a *polypedon* or soil *individual*. Several soil individuals are present in this landscape.

World reference base for soil resources (FAO/UN):
http://www.fao.org/docrep/W8594E/W8594E00.htm

At the other extreme is *the soil,* a collection of all these natural bodies that is distinct from water, solid rock, and other natural parts of the earth's crust. Hierarchical soil classification schemes generally group soils into classes at increasing levels of generality between these two extremes.

There are a number of soil classification systems in use in different parts of the world.[1] In the United States, U.S. Department of Agriculture in cooperation with soil scientists in other countries developed a comprehensive system of soil classification. This system, with modifications, is used in the United States and in over 50 other countries.

3.2 COMPREHENSIVE CLASSIFICATION SYSTEM: SOIL TAXONOMY[2]

The comprehensive soil classification system called *Soil Taxonomy,*[3] provides a hierarchical grouping of natural soil bodies. The system is based on *soil properties* that can be objectively observed or measured rather than on presumed mechanisms of soil formation. The system employs a unique nomenclature that gives a definite connotation of the major characteristics of the soils in question. It is truly international because it is not based on any one national language.

Bases of Soil Classification

Most of the chemical, physical, and biological properties presented in this text are used as criteria for *Soil Taxonomy.* A few examples are moisture and temperature status throughout the year, as well as color, texture, and structure of the soil. Chemical and mineralogical properties, such as the contents of organic matter, clay, iron and aluminum oxides, silicate clays, salts, the pH, the percentage *base saturation,*[4] and soil depth are other important criteria for classification. While many of the properties used may be observed in the field, others require precise measurements on samples taken to a sophisticated laboratory. This precision makes the system more objective, but in some cases may make the proper classification of a soil quite expensive and time-consuming. Precise measurements are also used to define certain **diagnostic soil horizons,** the presence or absence of which help determine the place of a soil in the classification system.

Diagnostic Surface Horizons

The diagnostic horizons that occur at the soil surface are called **epipedons** (from the Greek *epi,* over, and *pedon,* soil). The epipedon includes the upper part of the soil darkened by organic matter, the upper eluvial horizons, or both. It may include part of the B horizon if the latter is significantly darkened by organic matter. Seven epipedons are recognized, but only five occur naturally over wide areas. The other two, anthropic and plaggen, are the result of intensive human use. They are common in parts of Europe and Asia where soils have been utilized for many centuries.

The **mollic epipedon** (Latin *mollis,* soft) is a mineral surface horizon noted for its dark color (see Plates 8 and 20 after page 114) associated with its accumulated organic matter (> 0.6% organic C throughout), for its thickness (generally > 25 cm), and for its softness even when dry. It has a high base saturation greater than 50%. These epipedons are characteristic of soils developed under grassland (Figure 3.2 and Plate 8).

The **umbric epipedon** (Latin *umbra,* shade; hence, dark) has the same general characteristics as the mollic epipedon except the percentage base saturation is lower, and it commonly develops in areas with somewhat higher rainfall and where the parent material has lower content of calcium and magnesium.

---

[1] See Appendix A for summaries of the Canadian and United Nations Food and Agriculture Organization systems.

[2] For a complete description of *Soil Taxonomy* see Soil Survey Staff (1999). For an explanation of the earlier classification system, see USDA (1938).

[3] Taxonomy is the science of the principles of classification. For a review of the achievements and challenges of *Soil Taxonomy,* see SSSA (1984).

[4] The percentage base saturation is the percentage of the soil's negatively charged sites (cation exchange capacity) that are satisfied by attracting nonacid (or *base*) cations (such as Ca^{2+}, Mg^{2+}, and K^{+}) (see Section 9.4).

FIGURE 3.2 The mollic epipedon (a diagnostic horizon) in this soil includes genetic horizons designated Ap, an A2, and Bt1, all darkened by the accumulation of organic matter. A subsurface diagnostic horizon, the argillic horizon, overlaps the mollic. The argillic horizon is the zone of illuvial clay accumulation (Bt1 and Bt2 horizons in this profile). Scale marked every 10 cm. (Photo courtesy of R. Weil)

The **ochric epipedon** (Greek *ochros,* pale) is a mineral horizon that is either too thin, too light in color, or too low in organic matter to be either a mollic or an umbric horizon. It may be hard and massive when dry (see Plates 1, 7, and 11).

The **melanic epipedon** (Greek *melas,* melan, black) is a black, organic-matter-rich mineral horizon characteristic of soils developed from volcanic ash. It is more than 30 cm thick, and is extremely light in weight and fluffy for a mineral soil (see Plate 2).

The **histic epipedon** (Greek *histos,* tissue) is a 20- to 60-cm-thick layer of **organic soil materials** overlying a mineral soil. Formed in wet areas, the histic epipedon is a layer of black to dark brown peat or muck with a very low density.

Diagnostic Subsurface Horizons

Some 18 subsurface diagnostic horizons (Figure 3.3) are used to characterize different soils in *Soil Taxonomy.* Each diagnostic horizon provides a characteristic that helps place a soil in its proper class in the system. We will briefly discuss a few of the more commonly encountered subsurface diagnostic horizons.

The **argillic horizon** is a subsurface accumulation of high-activity silicate clays that have moved downward from the upper horizons or have formed in place. Examples are shown in Figure 3.2 and in Plate 1. Clay often coats pore walls (as shown in Figure 4.2) and surfaces of structural groupings. The coatings usually appear as shiny surfaces or as clay bridges between sand grains. Termed **argillans** or *clay skins,* they are concentrations of clay translocated from upper horizons (see Plates 24 and 27, after page 114).

The **natric horizon** likewise has silicate clay accumulation (with clay skins), but the clays are accompanied by more than 15% exchangeable sodium on the colloidal complex and by columnar or prismatic soil structural units. The natric horizon is found mostly in arid and semiarid areas. Examples are shown in Figures 4.10*e* and 9.24.

The **kandic horizon** has an accumulation of Fe and Al oxides as well as low-activity silicate clays (e.g., kaolinite), but clay skins need not be evident. The clays are low in

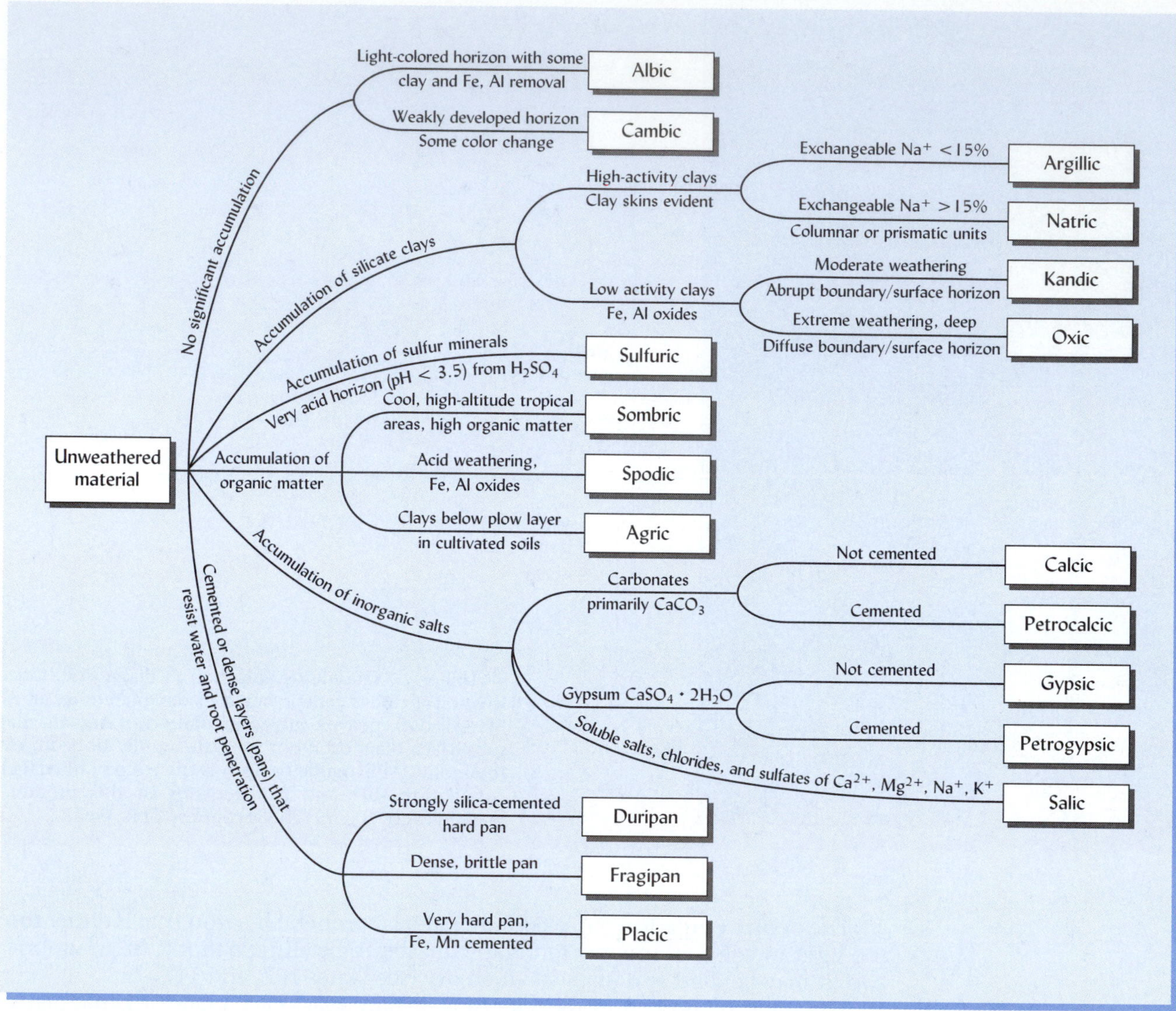

FIGURE 3.3 Names and major distinguishing characteristics of subsurface diagnostic horizons. Note that the focus is primarily on soil properties, not on how the properties have presumably evolved. Among the characteristics emphasized is the accumulation of silicate clays, organic matter, Fe and Al oxides, calcium compounds, and soluble salts, as well as materials that become cemented or highly acidified, thereby constraining root growth. The presence or absence of these horizons plays a major role in determining in which class a soil falls in *Soil Taxonomy.* See Chapter 8 for discussion of low- and high-activity clays.

activity as shown by their low cation-holding capacities (<16 $cmol_c$/kg clay). The epipedon that overlies a kandic horizon has commonly lost much of its clay content (see Figure 3.4).

The **oxic horizon** is a highly weathered subsurface horizon that is very high in Fe and Al, oxides, and in low-activity silicate clays (e.g., kaolinite). The cation-holding capacity is <16 $cmol_c$/kg clay. The horizon is at least 30 cm deep and has <10% weatherable minerals in the fine fraction. It is generally physically stable, crumbly, and not very sticky, despite its high clay content. It is found mostly in humid tropical and subtropical regions (see Plate 9, between about 1 and 3 feet on the scale).

The **spodic horizon** is an illuvial horizon that is characterized by the accumulation of colloidal organic matter and aluminum oxide (with or without iron oxide). It is commonly found in highly leached forest soils of cool humid climates, typically on sandy-textured parent materials (see Plate 10, reddish brown and black layers below the whitish layer).

The **sombric horizon** is an illuvial horizon, dark in color because of high organic matter accumulation. It has a low degree of base saturation and is found mostly in the cool, moist soils of high plateaus and mountains in tropical and subtropical regions (Plate 23).

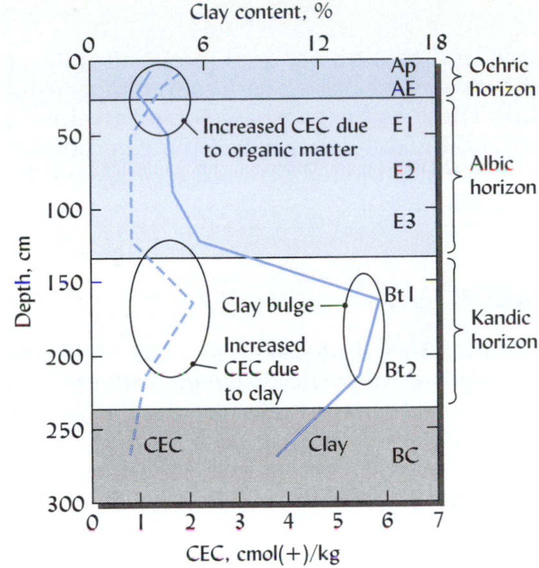

FIGURE 3.4 Vertical variation in clay content and cation exchange capacity (CEC) in a soil with thick albic and kandic horizons (E1–E2–E3 and Bt1–Bt2, respectively). Clays and iron and aluminum oxides have been translated downward from the albic horizon, leaving it lighter in color and with a low CEC. Note the well-expressed "clay bulge" that marks the kandic horizon. Similar clay enrichment (plus clay skins or other visual evidence of clay illuviation) characterizes an argillic horizon. This is a kandic rather than an argillic horizon because the accumulated clay is of *low-activity* types. The sharp increase in clay at the upper boundary of the kandic horizon, the considerable thickness (more than 100 cm) of the clay-rich layer, and the low CEC per kg of clay are all indications that this is a very old, highly mature soil. It formed under humid, subtropical conditions in sandy sediments in the upper coastal plain of Georgia. It is classified in the Kandiudults great group in *Soil Taxonomy.* (Data from Shaw et al., 2000)

The **albic horizon** is a light-colored eluvial horizon that is low in clay and oxides of Fe and Al. These materials have largely been moved downward from this horizon (see Plate 10, at about 10 cm depth).

A number of horizons have accumulations of saltlike chemicals that have leached from upper horizons in the profile. **Calcic horizons** contain an accumulation of carbonates (mostly $CaCO_3$) that often appear as white chalklike nodules (see the Bk horizon in the lower part of Figure 3.2). **Gypsic horizons** have an accumulation of gypsum ($CaSO_4 \cdot 2H_2O$), and **salic horizons** an accumulation of soluble salts. These are found mostly in soils of arid and semiarid regions.

In some subsurface diagnostic horizons, the materials are cemented or densely packed, resulting in relatively impermeable layers called **pans** (**duripan, fragipan,** and **placic horizons**). These can resist water movement and the penetration of plant roots. Such pans constrain plant growth and may encourage water runoff and erosion because rainwater cannot move readily downward through the soil. Figure 3.3 explains the genesis of these and the other subsurface diagnostic horizons.

Soil Moisture Regimes

A soil moisture regime (SMR) refers to the presence or absence of either water-saturated conditions (usually groundwater) or plant-available soil water.

Aquic. Soil is saturated with water and virtually free of gaseous oxygen for sufficient periods of time for evidence of poor aeration (gleying and mottling) to occur.

Udic. Soil moisture is sufficiently high year-round in most years to meet plant needs. This regime is common for soils in humid climatic regions and characterizes about one-third of the worldwide land area. An extremely wet moisture regime with excess moisture for leaching throughout the year is termed **perudic.**

Ustic. Soil moisture is intermediate between udic and aridic regimes—generally there is some plant-available moisture during the growing season, although significant periods of drought may occur.

Aridic. The soil is dry for at least half of the growing season and moist for less than 90 consecutive days. This regime is characteristic of arid regions. The term *torric* is used to indicate the same moisture condition in certain soils that are both hot and dry in summer, though they may not be hot in winter.

Xeric. This soil moisture regime is found in typical Mediterranean-type climates, with cool, moist winters and warm, dry summers. Like the ustic regime, it is characterized by having long periods of drought in the summer.

Soil Temperature Regimes

Soil temperature regimes, such as *frigid, mesic,* and *thermic,* are used to classify soils at some of the lower levels in *Soil Taxonomy.* The cryic (Greek *kryos,* very cold) temperature regime distinguishes some higher-level groups. These regimes are based on mean annual soil temperature, mean summer temperature, and the difference between mean summer and winter temperatures, all at 50 cm depth. The specific temperature regimes will be described in the discussion of soil families (Section 3.17).

3.3 CATEGORIES AND NOMENCLATURE OF SOIL TAXONOMY

Updated summary of the USDA Soil Taxonomy: http://soils.usda.gov/education/facts/formation.htm

There are six hierarchical categories of classification in *Soil Taxonomy:* (1) *order,* the highest (broadest) category, (2) *suborder,* (3) *great group,* (4) *subgroup,* (5) *family,* and (6) *series* (the most specific category). Each order has several suborders, each suborder has several subgroups, and so forth. This system may be compared with those used for the classification of plants or animals, as shown in Table 3.1. Just as *Trifolium repens* identifies a specific kind of plant, the Miami series identifies a specific kind of soil.

Nomenclature of Soil Taxonomy

Soil Taxonomy (2nd Ed) is available here in PDF format for printing or viewing: http://soils.usda.gov/classification/taxonomy/main.htm

Although unfamiliar at first sight, the nomenclature system has a logical construction and conveys much information about the nature of the soils named. The system is easy to learn after a bit of study. The nomenclature is used throughout this book, especially to identify the kinds of soils shown in illustrations. When reading, if you make a conscious effort to identify the parts of each soil class mentioned in the text and figure captions, and recognize the level of category indicated, the system will become second nature.

The names of the classification units are combinations of syllables, most of which are derived from Latin or Greek and are root words in several modern languages. Since each part of a soil name conveys a concept of soil character or genesis, the name automatically describes the general kind of soil being classified. For example, soils of the order **Aridisols** (from the Latin *aridus,* dry, and *solum,* soil) are characteristically dry soils in arid regions. Those of the order **Inceptisols** (from the Latin *inceptum,* beginning, and *solum,* soil) are soils with only the beginnings or inception of profile development. Thus, the names of orders are combinations of (1) formative elements, which generally define the characteristics of the soils, and (2) the ending *sols.*

The names of suborders automatically identify the order of which they are a part. For example, soils of the suborder **Aquolls** are the wetter soils (from the Latin *aqua,* water) of the Mollisols order. Likewise, the name of the great group identifies the suborder and order of which it is a part. **Argiaquolls** are Aquolls with clay or argillic (Latin *argilla,* white clay) horizons. In the following illustration, note that the three letters *oll* identify each of the lower categories as being in the Mollisols order.

M**oll**isols	Order
Aqu**oll**s	Suborder
Argiaqu**oll**s	Great group
Typic Argiaqu**oll**s	Subgroup

TABLE 3.1 Comparison of the Classification of a Common Cultivated Plant, White Clover (*Trifolium repens*), and a Soil, Miami Series

Plant classification			*Soil classification*	
Phylum	Pterophyta	Increase Specificity ↓	Order	Alfisols
Class	Angiospermae		Suborder	Udalfs
Subclass	Dicotyledoneae		Great group	Hapludalfs
Order	Rosales		Subgroup	Oxyaquic Hapludalfs
Family	Leguminosae		Family	Fine loamy, mixed, mesic, active
Genus	*Trifolium*		Series	Miami
Species	*repens*		Phase[a]	Miami silt loam

[a]Technically not a category in *Soil Taxonomy* but used in field surveying. *Silt loam* refers to the texture of the A horizon.

If one is given only the subgroup name, the great group, suborder, and order to which the soil belongs are automatically known.

Family names in general identify subsets of a subgroup that are similar in texture, mineral composition, and mean soil temperature at a depth of 50 cm. Thus, the name "Typic Argiaquolls, fine, mixed, mesic, active" identifies a family in the Typic Argiaquolls subgroup with a fine texture, mixed clay mineral content, mesic (8 to 15°C) soil temperature, and clays active in cation exchange.

Soil series are named after a geographic feature (town, river, etc.) near where they were first recognized. Thus, names such as *Fort Collins, Cecil, Miami, Norfolk,* and *Ontario* identify soil series first described near the town or geographic feature named. Approximately 19,000 soil series have been classified in the United States alone.

With this brief explanation of the nomenclature of *Soil Taxonomy,* we will now consider the general nature of soils in each of the soil orders.

3.4 SOIL ORDERS

Each of the world's soils is assigned to one of 12 **orders**, largely on the basis of soil properties that reflect a major course of development, with considerable emphasis placed on the presence or absence of major diagnostic horizons (Table 3.2). As an example, many soils that developed under grassland vegetation have the same general sequence of horizons and are characterized by a *mollic* epipedon—a thick, dark, surface horizon that is high in base-forming cations. These soils belong to the order, *Mollisols*. Note that all order names have a common ending, *sols* (from the Latin *solum,* soil).

Maps, photos and information about each soil order:
http://soils.ag.uidaho.edu/soilorders/orders.htm

The general conditions that enhance the formation of soils in the different orders are shown in Figure 3.5. From soil profile characteristics, soil scientists can ascertain the relative degree of soil development in the different orders, as shown in this figure. Note that soils with essentially no profile layering (Entisols) have the least development, while the deeply weathered soils of the humid tropics (Oxisols and Ultisols) show the greatest soil development. The effect of climate (temperature and moisture) and of vegetation (forests or grasslands) on the kinds of soils that develop is also indicated in Figure 3.5.

To some degree, most of the soil orders occur in climatic regions that can be described by moisture and temperature regimes. Figure 3.6 illustrates some of the relationships among the soil orders with regard to these climatic factors. While only the Gelisols and Aridisols orders are defined directly in relation to climate, Figure 3.6 indicates that orders with the most highly weathered soils tend to be associated with the warmer and wetter climates. Profiles typical of each soil order are shown in color on Plates 1 through 12 (after page 114). A color-coded soil order map for the United States can be found on the back endpaper of this book; a world map of the 12 soil orders is printed on the front endpaper. In studying these maps, try to confirm that the

TABLE 3.2 Names of Soil Orders in Soil Taxonomy with Their Derivation and Major Characteristics

The bold letters in the order names indicate the formative element used as the ending for suborders and lower taxa within that order.

Name	*Formative element*	*Derivation*	*Pronunciation*	*Major characteristics*
Alfisols	alf	Nonsense symbol	Ped**alf**er	Argillic, natric, or kandic horizon; high to medium base saturation
Andisols	and	Jap. *ando,* black soil	**And**esite	From volcanic ejecta, dominated by allophane or Al-humic complexes
Ar**id**isols	id	L. *aridus,* dry	Ar**id**	Dry soil, ochric epipedon, sometimes argillic or natric horizon
Entisols	ent	Nonsense symbol	Rec**ent**	Little profile development, ochric epipedon common
G**el**isols	el	Gk. *gelid,* very cold	J**el**ly	Permafrost, often with cryoturbation (frost churning)
H**ist**osols	ist	Gk. *histos,* tissue	H**ist**ology	Peat or bog; >20% organic matter
Inc**ept**isols	ept	L. *inceptum,* beginning	Inc**ept**ion	Embryonic soils with few diagnostic features, ochric or umbric epipedon, cambic horizon
M**oll**isols	oll	L. *mollis,* soft	M**oll**ify	Mollic epipedon, high base saturation, dark soils, some with argillic or natric horizons
Oxisols	ox	Fr. *oxide,* oxide	**Ox**ide	Oxic horizon, no argillic horizon, highly weathered
Sp**od**osols	od	Gk. *spodos,* wood ash	P**od**zol; odd	Spodic horizon commonly with Fe, Al oxides and humus accumulation
Ultisols	ult	L. *ultimus,* last	**Ult**imate	Argillic or kandic horizon, low base saturation
V**ert**isols	ert	L. *verto,* turn	Inv**ert**	High in swelling clays; deep cracks when soil dry

Histosols (organic)
Gelisols (permafrost)
Aridisols (dry)
Alfisols (mildly acid clays)
Ultisols (strongly acid clays)
Entisols (recent)
Oxisols (oxides)
Spodosols (spodic horizon)
Mollisols (soft, dark)
Vertisols (swelling clays)
Inceptisols (beginning B)
Andisols (volcanic ejecta)
Deep accumulation of organic materials, wet conditions
Very cold permafrost
Desert shrubs, grasses, dry
Moist, mildly acid clay accumulation
Wet tropical and subtropical forests, acid silicate and Fe, Al oxides
Wet, tropical forest, extreme weathering, low activity clays, Fe, Al oxides
Cool, wet, sandy acid, coniferous forest
Semiarid to moist grasslands, mollic epipedon
High base status, high-activity clays, dry season
Mild weathering, various conditions
Mild weathering on volcanic ejecta
Degree of weathering and soil development
Slight
Intermediate
Strong

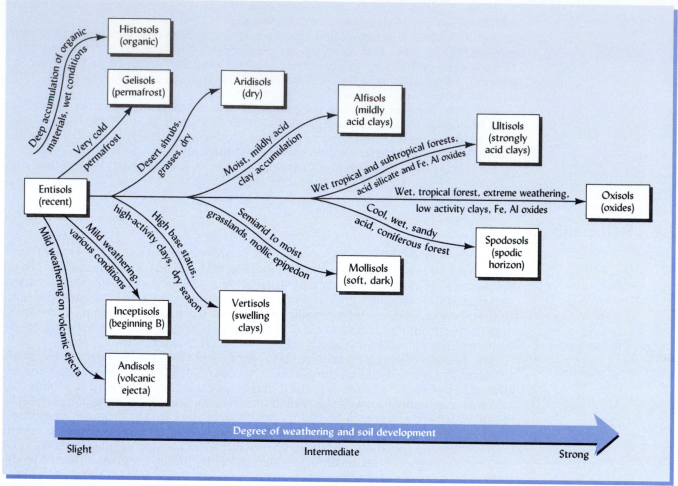

FIGURE 3.5 Diagram showing general degree of weathering and soil development in the different soil orders classified in *Soil Taxonomy.* Also shown are the general climatic and vegetative conditions under which soils in each order are formed.

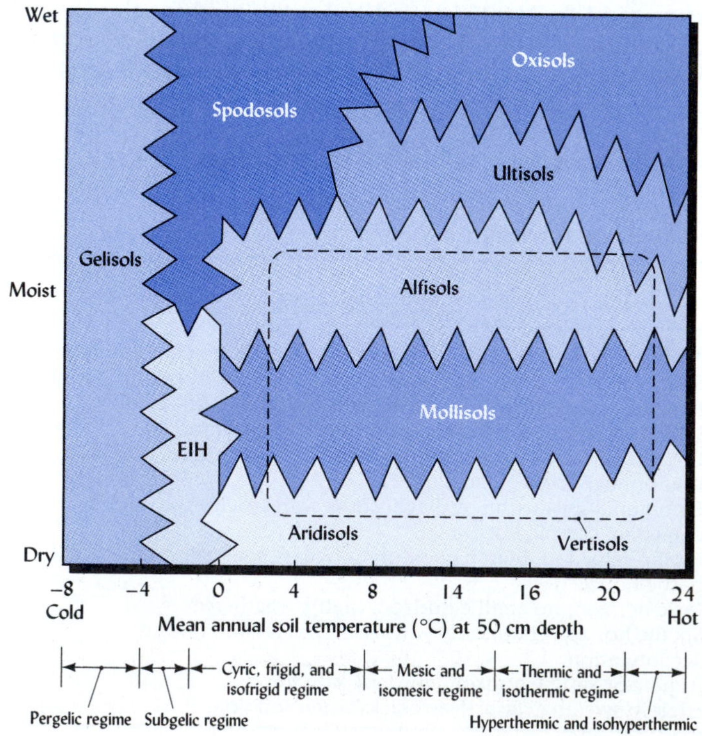

FIGURE 3.6 Diagram showing the general soil moisture and soil temperature regimes that characterize the most extensive soils in each of eight soil orders. Soils of the other four orders (Andisols, Entisols, Inceptisols, and Histosols) may be found under any of the soil moisture and temperature conditions (including the area marked EIH). Major areas of Vertisols are found only where clayey materials are in abundance and are most extensive where the soil moisture and temperature conditions approximate those shown inside the box with broken lines. Note that these relationships are only approximate and that less extensive areas of soils in each order may be found outside the indicated ranges. For example, some Ultisols (Ustults) and Oxisols (Ustox) have soil moisture levels for at least part of the year that are much lower than this graph would indicate. (The terms used at the bottom to describe the soil temperature regimes are those used in helping to identify soil families.)

See a brief description, US map and a typical profile for each of the 12 soil orders: http://soils.usda.gov/classification/orders/main.htm

distribution of the soil orders is in accordance with what you know about the climate in various regions of the world.

Although a detailed description of all the lower levels of soil categories is far beyond the scope of this (or any other) book, a general knowledge of the 12 soil orders is essential for understanding the nature and function of soils in different environments. The simplified key given in Figure 3.7 helps illustrate how *Soil Taxonomy* can be used to key

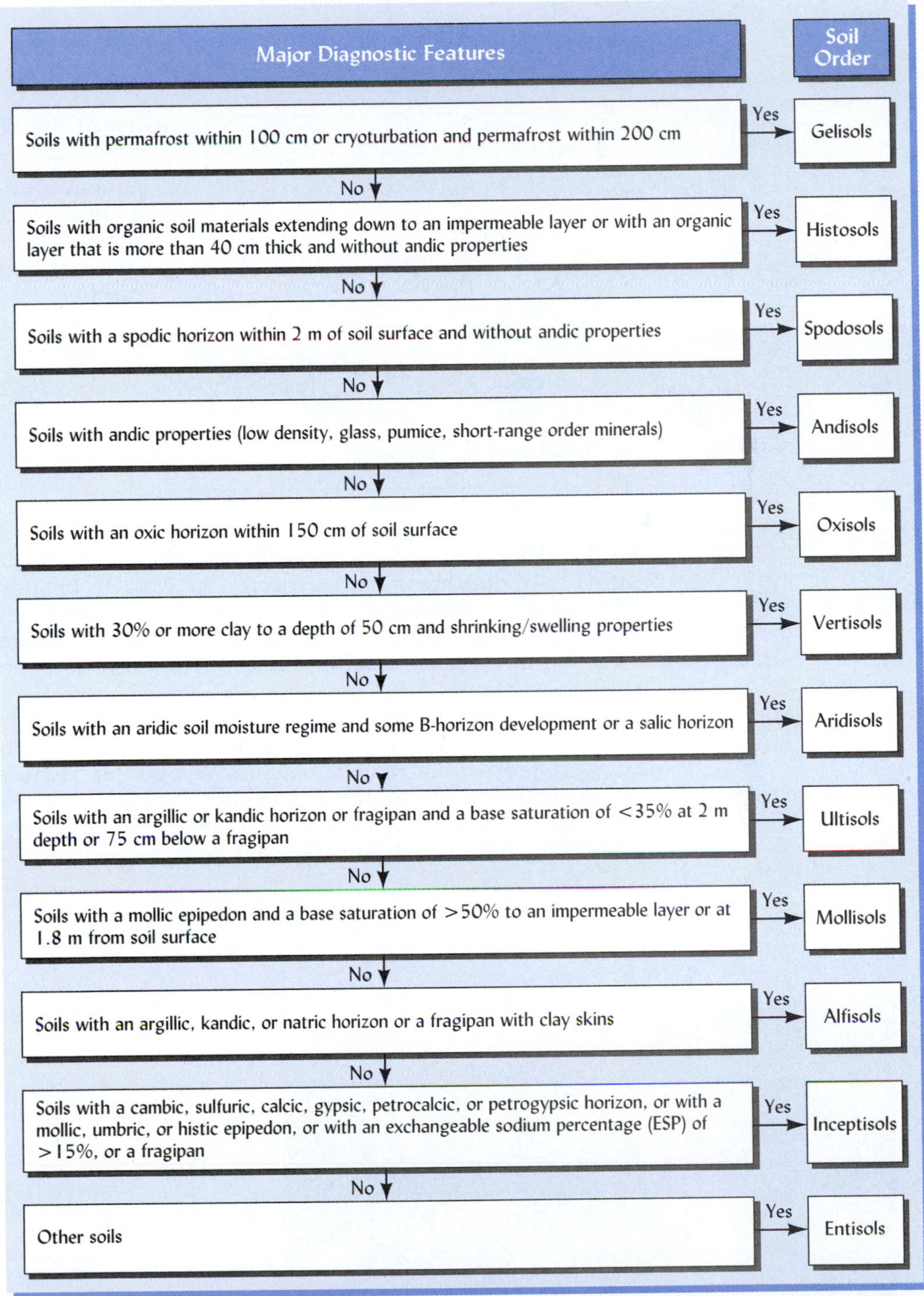

FIGURE 3.7 A simplified key to the 12 soil orders in *Soil Taxonomy*. In using the key, always begin at the top. Note how diagnostic horizons and other profile features are used to distinguish each soil order from the remaining orders. Entisols, having no such special diagnostic features, key out last. Also note that the sequence of soil orders in this key bears no relationship to the degree of profile development, and adjacent soil orders may not be more similar than nonadjacent ones. See Section 3.2 for explanations of the diagnostic horizons. (Diagram courtesy of R. Weil)

out the order of any soil based on observable and measurable properties of the soil profile. Because certain diagnostic properties take precedence over others, the key must always be used starting at the top, and working down. It will be useful to review this key and the color soil order maps after reading about the general characteristics, nature, and occurrence of each soil order.

We will now consider each of the soil orders, beginning with those characterized by little profile development and progressing to those with the most highly weathered profiles (as represented from left to right in Figure 3.5).

3.5 ENTISOLS (RECENT: LITTLE IF ANY PROFILE DEVELOPMENT)

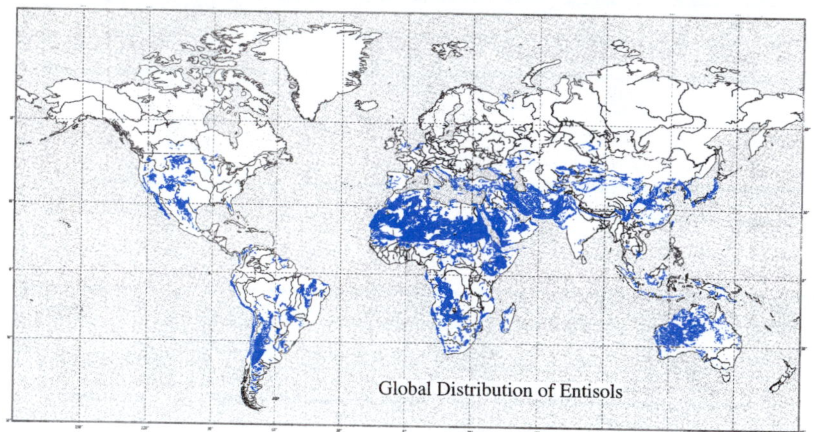

16.3% of global ice-free land.

Suborders are:

- Aquent (wet)
- Arents (mixed horizons)
- Fluvents (alluvial deposits)
- Orthents (typical)
- Psamments (sandy)

Weakly developed mineral soils without natural genetic (subsurface) horizons, or with only the beginnings of such horizons (see Plate 4), belong to the Entisols order. Most have an ochric epipedon and a few have human-made anthropic or agric epipedons. Some have albic subsurface horizons. Soil productivity ranges from very high for certain Entisols formed in recent alluvium to very low for those forming in shifting sand or on steep rocky slopes.

This is an extremely diverse group of soils with little in common, other than the lack of evidence for all but the earliest stages of soil formation (Figure 3.8). Entisols are either young in years or their parent materials have not reacted to soil-forming factors. On parent materials such as fresh lava flows or recent alluvium (Fluvents), there has been too little time for much soil formation. In extremely dry areas, scarcity of water and vegetation may inhibit soil formation. Likewise, frequent saturation with water (Aquents)

FIGURE 3.8 Profile of an Entisol (a Psamment) formed on sandy alluvium in Virginia. Note the accumulation of organic matter in the A horizon but no other evidence of profile development. The A horizon is 30 cm thick. (Photo courtesy of R. Weil)

may delay soil formation. Some Entisols occur on steep slopes, where the rates of erosion may exceed the rates of soil formation, preventing horizon development. Others occur on construction sites where bulldozers destroy or mix together the soil horizons, causing the existing soils to become Entisols (some have suggested that these be called *urbents,* or urban Entisols).

The agricultural productivity of the Entisols varies greatly depending on their location and properties. Entisols developed on alluvial floodplains (Fluvents) are among the world's most productive soils. However, the productivity of most Entisols is restricted by inadequate soil depth, clay content, or water availability.

3.6 INCEPTISOLS (FEW DIAGNOSTIC FEATURES: INCEPTION OF B HORIZON)

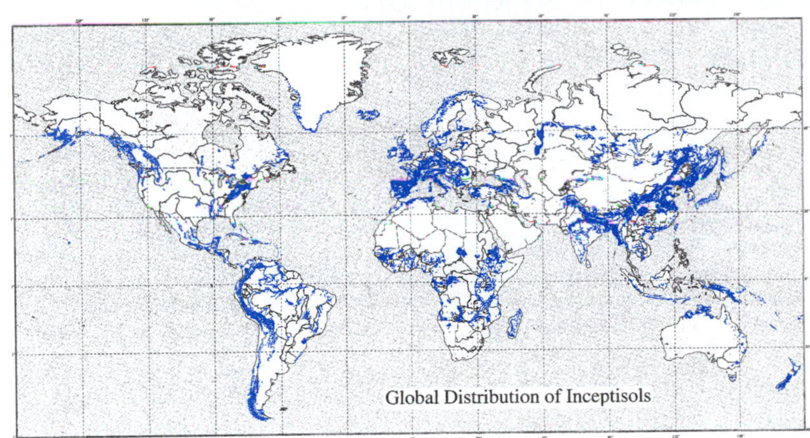

9.9% of global ice-free land.

Suborders are:

- Anthrepts (dark, human-made, high phosphorus)
- Aquepts (wet)
- Cryepts (very cold)
- Udepts (humid climate)
- Ustepts (semiarid)
- Xerepts (dry summers, wet winters)

In Inceptisols the beginning or *inception* of profile development is evident, and some diagnostic features are present. However, the well-defined profile characteristics of soils thought to be more mature have not yet developed. For example, a cambic horizon showing some color or structural change is common in Inceptisols, but a more mature illuvial B horizon such as an argillic cannot be present. Other subsurface diagnostic horizons that may be present in Inceptisols include duripans, fragipans, and calcic, gypsic, and sulfidic horizons. The epipedon in most Inceptisols is an ochric, although a plaggen or weakly expressed mollic or umbric epipedon may be present. Inceptisols show more significant profile development than Entisols, but are defined to exclude soils with diagnostic horizons or properties that characterize certain other soil orders. Thus, soils with only slight profile development occurring in arid regions, or containing permafrost or andic properties, are excluded from the Inceptisols. They fall, instead, in the soil orders Aridisols, Gelisols, or Andisols, as discussed in later sections. Inceptisols are widely distributed throughout the world and their natural productivity varies greatly.

3.7 ANDISOLS (VOLCANIC ASH SOILS)

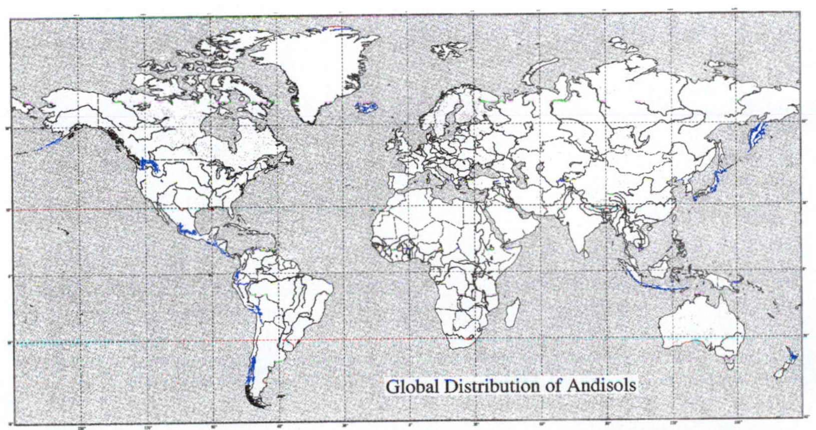

0.7% of global ice-free land.

Suborders are:

- Aquands (wet)
- Cryands (cold)
- Torrands (hot, dry)
- Udands (humid)
- Ustands (moist/dry)
- Vitrands (volcanic glass)
- Xerands (dry summers, moist winters)

Andisols are usually formed on volcanic ash and cinders deposited in recent geological times. They are commonly found near the volcano source or in areas downwind from the volcano, where a sufficiently thick layer of ash has been deposited during eruptions. Andisols have not had time to become highly weathered. The principal soil-forming process has been the rapid weathering (transformation) of volcanic ash to produce amorphous or poorly crystallized silicate minerals such as **allophane** and **imogolite** and the iron oxy-hydroxide, **ferrihydrite.** Some Andisols have a melanic epipedon, a surface diagnostic horizon that has a high organic matter content and dark color (see Plate 2). The accumulation of organic matter is quite rapid due largely to its protection in aluminum–humus complexes. Some Andisols have a dark melanic epipedon (Figure 3.9). Little downward translocation of the colloids, or other profile development, has taken place. Like the Entisols and Inceptisols, Andisols are young soils, usually having developed for only 5000 to 10,000 years.

Unlike the previous two orders of immature soils, Andisols have a unique set of **andic properties** in at least 35 cm of the upper 60 cm of soil due to common types of parent materials. Materials with andic properties are characterized by a high content of volcanic glass and/or a high content of amorphous or poorly crystalline iron and aluminum minerals. The combination of these minerals and the high organic matter results in light, fluffy soils that are easily tilled, but have a high water-holding capacity and resist erosion by water. They are mostly found in regions where rainfall keeps them from being susceptible to erosion by wind. Andisols are usually of high natural fertility, except that phosphorus availability is severely limited by the extremely high phosphorus retention capacity of the andic materials (see Section 13.8). Fortunately, proper management of plant residues and fertilizers can usually overcome this difficulty.

FIGURE 3.9 An Andisol developed on layers of volcanic ash and pumice in central Africa. (Photo courtesy of R. Weil)

3.8 GELISOLS (PERMAFROST AND FROST CHURNING)

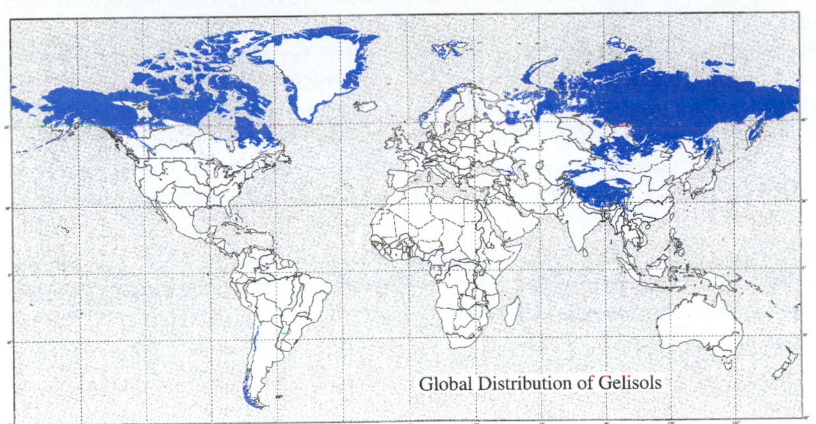

8.6% of global ice-free land.

Suborders are:
- Histels (organic)
- Orthels (no special features)
- Turbels (cryoturbation)

Gelisols are young soils with little profile development. Cold temperatures and frozen conditions for much of the year slow the process of soil formation. The principal defining feature of these soils is the presence of a **permafrost** layer (see Plates 5 and 14 after page 114). Permafrost is a layer of material that remains at temperatures below 0°C for more than 2 consecutive years. It may be a hard, ice-cemented layer of soil material (e.g., designated Cfm in profile descriptions), or, if dry, it may be uncemented (e.g., designated Cf). In Gelisols, the permafrost layer lies within 100 cm of the soil surface, unless **cryoturbation** is evident within the upper 100 cm, in which case the permafrost may begin as deep as 200 cm from the soil surface.

Cryoturbation, or *frost churning,* moves soil material to form broken, convoluted horizons (e.g., designated Cjj), and/or patches of organic matter at the top of the permafrost. The frost churning also may form patterns on the ground surface, such as hummocks and ice-rich polygons that may be several meters across. In some cases rocks forced to the surface form rings or netlike patterns.

Gelisols showing evidence of cryoturbation are called *Turbels*. Other Gelisols, often found in wet environments, have developed in accumulations of mainly organic materials, making them *Histels* (Greek *histos,* tissue; Figure 3.10). Most of the soil-forming

FIGURE 3.10 Gelisol with a histic epipedon and permafrost. This Histel was photographed in Alaska in July. Scale in centimeters. (Photo courtesy of James G. Bockheim, University of Wisconsin)

processes that occur take place above the permafrost in the **active layer** that thaws every year or two. Various types of diagnostic horizons may have developed in different Gelisols, including mollic, histic, umbric, calcic, and, occasionally, argillic horizons.

Blanketed under snow and ice for much of the year, most Gelisols support tundra or bog vegetation of lichens, grasses, and low shrubs. Plant productivity is low because of the extremely short potential growing season in the far northern latitudes, the low levels of solar radiation (except during the fleeting summer), and the waterlogged condition of many Gelisols in which permafrost inhibits internal drainage during the summer thaw (Figure 3.10). Millions of caribou, reindeer, and muskox survive on this vegetation during the summer and the many bogs and pools serve as nesting sites for migratory birds.

Gelisols present unique problems for construction projects. Many of the Histels are very wet and have little bearing strength to support roads or foundations. The permafrost in some Gelisols is unstable once disturbed.

3.9 HISTOSOLS (ORGANIC SOILS WITHOUT PERMAFROST)

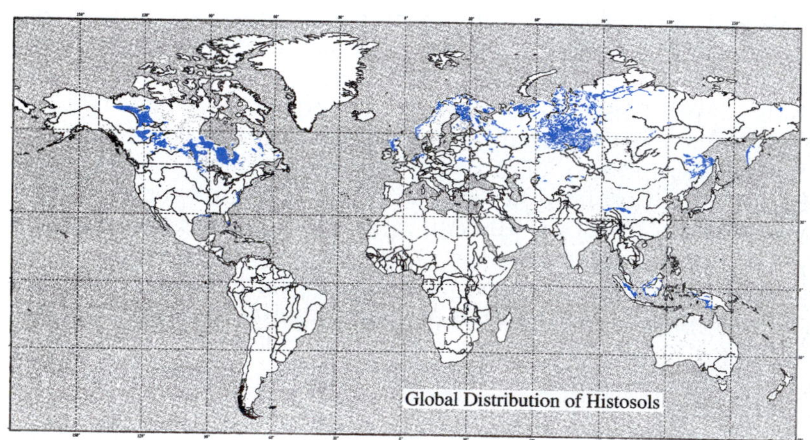

1.2% of global ice-free land.

Suborders are:

Fibrists (fibers of plants obvious)
Folists (leaf mat accumulations)
Hemists (fibers partly decomposed)
Saprists (fibers not recognizable)

Histosols are soils that have undergone little profile development because of the anaerobic environment in which they form. The main process of soil formation evident in Histosols is the accumulation of partially decomposed organic parent material without permafrost (which would cause the soil to be classified in the Histels suborder of Gelisols). Histosols consist of one or more thick layers of *organic soil material*. Generally, Histosols have organic soil materials in more than half of the upper 80 cm of soil (Plate 6, after page 114), or in two-thirds of the soil overlying shallow rock.

While not all wetlands contain Histosols, all Histosols (except Folists) occur in wetland environments. They can form in almost any moist climate in which plants can grow, from equatorial to arctic regions, but they are most prevalent in cold climates, up to the limit of permafrost. Horizons are differentiated by the type of vegetation contributing the residues, rather than by translocations and accumulations within the profile.

Organic deposits accumulate in marshes, bogs, and swamps, which are habitats for water-loving plants such as pond-weeds, cattails, sedges, reeds, mosses, shrubs, and even some trees. Generation after generation, the residues of these plants sink into the water, which inhibits their oxidation by reducing oxygen availability and, consequently, acts as a partial preservative (see Figure 2.14).

Because of their high content of organic matter, Histosols are generally black to dark brown in color. They are extremely lightweight when dry, being only about 10 to 20% as dense as most mineral soils. Histosols also have high water-holding capacities on a mass basis. While a mineral soil will absorb and hold from 20 to 40% of its weight of water, a cultivated Histosol may hold a mass of water equal to 200 to 400% of its dry weight. However, because of the low density of the organic material, most Histosols will hold only about the same amount of plant-available water as will a good mineral soil on a per-hectare or per-unit-volume basis.

FIGURE 3.11 Profile of a drained Histosol on which onions are being produced in New York State. The organic soil rests on lake-laid (lacustrine) mineral material. The thickness of this Histosol has been much reduced by subsidence during nearly a century of drainage and cultivation. The surface of this black soil appears light-colored because it was dry when the photograph was taken. The knife is 24 cm long. (Photo courtesy of R. Weil)

The ecological roles of natural wetland environments have not always been appreciated (or protected by law), and more than 50% of the original wetland area in the United States has been drained for agricultural or other uses, especially for horticultural production (Figure 3.11). Some Histosols make very productive farmlands, but the organic nature of the materials requires liming, fertilization, tillage, and drainage practices quite different from those applied to soils in the other 11 orders. In some places, Histosols are also mined for their peat, which is sold for use in potting media, as a mulch, and to make peat-fiber pots. Peat deposits are also used for fuel in some countries, especially in Russia, where several power stations are fueled by this material.

3.10 ARIDISOLS (DRY SOILS)

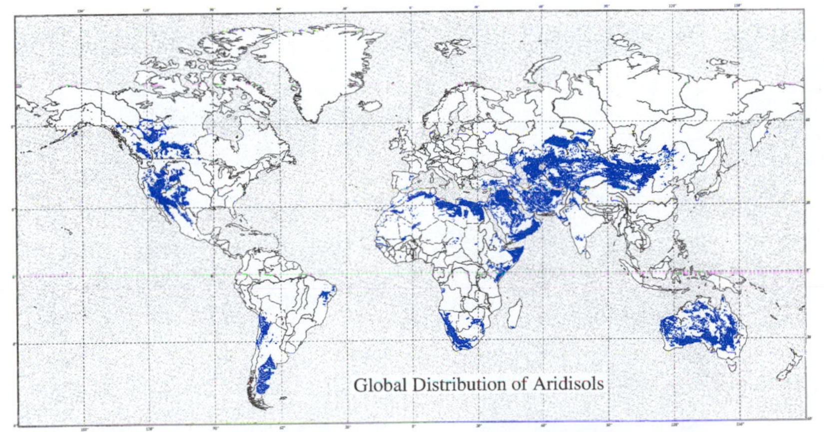

12.1% of global ice-free land.

Suborders are:

- Argids (clay)
- Calcids (carbonate)
- Cambids (typical)
- Cryids (cold)
- Durids (duripan)
- Gypsids (gypsum)
- Salids (salty)

Aridisols occupy a larger area globally than any other soil order (more than 12%) except Entisols. Water deficiency is a major characteristic of these soils. The soil moisture level is sufficiently high to support plant growth for no longer than 90 consecutive days. The natural vegetation consists mainly of scattered desert shrubs and short bunchgrasses. Soil properties, especially in the surface horizons, may differ substantially between interspersed bare and vegetated areas (see Section 2.5).

Aridisols are characterized by an ochric epipedon that is generally light in color and low in organic matter (see Plate 3). The processes of soil formation have brought about a redistribution of soluble materials, but there is generally not enough water to leach these materials completely out of the profile. Therefore, they often accumulate at a lower level in the profile. These soils may have a horizon of accumulation of calcium carbonate (calcic), gypsum (gypsic), soluble salts (salic), or exchangeable sodium (natric). Under certain circumstances, carbonates may cement together the soil particles and coarse fragments in the layer of accumulation, producing hard layers known as **petrocalcic horizons** (Figure 3.12). These hard layers act as impediments to plant root growth and also greatly increase the cost of excavations for buildings.

Some Aridisols (the *Argids*) have an argillic horizon, most probably formed under a wetter climate that long ago prevailed in many areas that are deserts today. With time, and the addition of carbonates from calcareous dust and other sources, many argillic horizons become engulfed by carbonates (*Calcids*). On steeper land surfaces subject to erosion, argillic horizons do not get a chance to form, and the dominant soils are often *Cambids* (Aridisols with only weakly differentiated cambic subsurface B horizons).

In rocky soils, erosion may remove all the fine particles from the surface layers, leaving behind a layer of wind-rounded pebbles called **desert pavement** (see Figure 3.12). The surfaces of the pebbles in desert pavement often have a shiny coating called *desert varnish*. This coating is thought to be produced by algae that extract iron and manganese from the minerals and leave an oxide coating on the pebbles.

Except where there is shallow groundwater or irrigation, the soil layers are moist only for short periods during the year. These short moist periods may be sufficient for native desert shrubs and annual plants, but not for conventional crop production. If groundwater is present near the soil surface, soluble salts may accumulate in the upper horizons to levels that most crop plants cannot tolerate. However, if irrigated valleys are carefully managed to prevent salt accumulation, their Aridisols can be among the most productive of soils.

Some Aridisols are used for low-intensity grazing, but the production per unit area is low. The overgrazing of Aridisols leads to increased heterogeneity of both soils and vegetation. The animals graze the relatively even cover of palatable grasses, giving a competitive advantage to less palatable shrubs not eaten. The scattered shrubs compete against the struggling grasses for water and nutrients. The once-grassy areas become increasingly bare, and the soils between the scattered shrubs succumb to erosion by the desert winds and occasional thunderstorms. The desertification of areas of sub-Saharan Africa and the western United States is partly due to such degradation.

FIGURE 3.12 Two features characteristic of some Aridisols. (Left) Wind-rounded pebbles have given rise to a desert pavement. (Right) A petrocalcic horizon of cemented calcium carbonate. (Photos courtesy of R. Weil)

3.11 VERTISOLS (DARK, SWELLING, AND CRACKING CLAYS)[5]

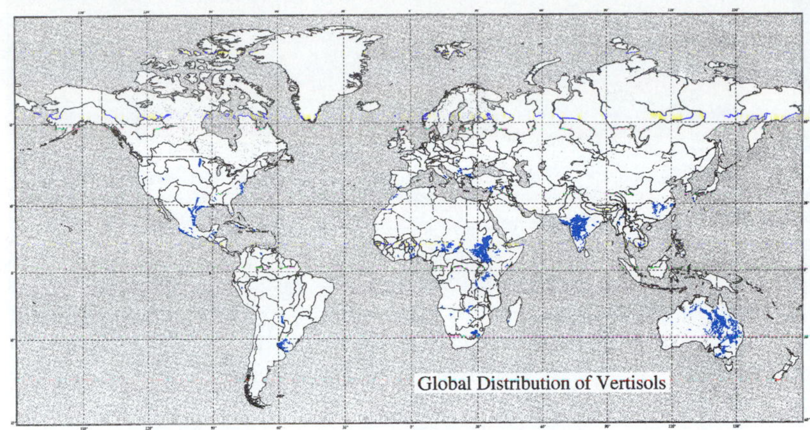

2.4% of global ice-free land.

Suborders are:

- Aquerts (wet)
- Cryerts (cold)
- Torrerts (hot summer, very dry)
- Uderts (humid)
- Usterts (moist/dry)
- Xererts (dry summers, moist winters)

The main soil-forming process affecting Vertisols is the shrinking and swelling of clay as these soils go through periods of drying and wetting. To a depth of 1 m or more, Vertisols have a high content (> 30%) of sticky, swelling, and shrinking-type clays. Most Vertisols are dark, even blackish in color (Plate 12). However, unlike for most other soils, the dark color of Vertisols is not necessarily indicative of a high organic matter content. The organic matter content of dark Vertisols typically ranges from as much as 5 or 6% to as little as 1%.

Vertisols typically develop from limestone, basalt or other calcium, and magnesium-rich parent materials. The presence of these cations encourages the formation of swelling-type clays (see Section 8.5).

Guy Smith Memorial slides on Vertisols:
http://www.nrcs.usda.gov/technical/worldsoils/vertisols/vert-start.html

Vertisols are found mostly in subhumid to semiarid environments in warm regions, but a few (Cryerts) occur where the average soil temperatures are as low as 0°C (see Figure 3.6). The native vegetation is usually grassland. Vertisols generally occur where the climate features dry periods of several months. In dry seasons, the clay shrinks, causing the soils to develop deep, wide cracks that are diagnostic for this order (Figure 3.13*a*). The surface soil generally forms granules, of which a significant number may slough off into the cracks, giving rise to a partial inversion of the soil (Figure 3.14). This accounts for the association with the term *invert,* from which this order derives its name.

When the rains come, water entering the large cracks moistens the clay in the subsoils, causing it to swell. The repeated shrinking and swelling of the subsoil clay results in a kind of imperceptively slow "rocking" movement of great masses of soil. As the subsoil swells, blocks of soil shear off from the mass and rub past each other under pressure, giving rise in the subsoil to shiny, grooved, tilted surfaces called **slickensides** (Figure 3.14d). Eventually, this back-and-forth motion may form bowl-shaped depressions with relatively deep profiles surrounded by slightly raised areas in which little soil development has occurred and in which the parent material remains close to the surface (see Figure 3.14). The resulting pattern of microhighs and microlows on the land surface, called **gilgai**, is usually discernable only where the soil is untilled (Figure 3.13*b*).

The high shrink-swell potential of Vertisols makes them extremely problematic for any kind of highway or building construction (Figure 8.20 and Plate 33). This property can also make agricultural management extremely difficult. Because they are very sticky and plastic when wet and become very hard when dry, the timing of tillage operations is critical. Some farmers refer to Vertisols as *24-hour soils,* because they are said to be too wet to plow one day and too dry the next.

[5]Knowledge of the properties and mode of formation of Vertisols has increased greatly in recent years. See Coulombe et al. (1996) for a detailed review.

(a)

(b)

FIGURE 3.13 (*a*) Wide cracks formed during the dry season in the surface layers of this Vertisol in India. Surface debris can slough off into these cracks and move to subsoil. When the rains come, water can move quickly to the lower horizons, but the cracks are soon sealed, making the soils relatively impervious to the water. Once the cracks have sealed, water may collect in the microlows, making the gilgai relief easily visible as in the Texas vertisol (*b*). (Photo (*a*) courtesy of N. C. Brady; (*b*) courtesy of K. N. Potter, USDA/ARS, Temple, Texas)

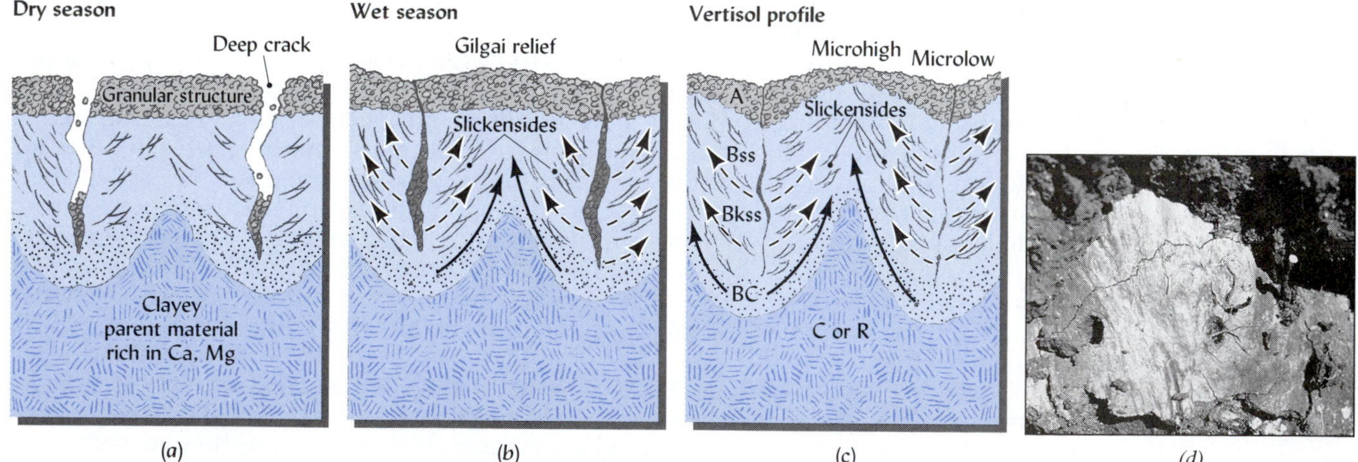

FIGURE 3.14 Vertisols typically are high in swelling-type clay and have developed wedgelike structures in the subsoil horizons. (*a*) During the dry season, large cracks appear as the clay shrinks upon drying. Some of the surface soil granules fall into cracks under the influence of wind and animals. This action causes a partial mixing, or *inversion,* of the horizons. (*b*) During the wet season, rainwater pours down the cracks, wetting the soil near the bottom of the cracks first, and then the entire profile. As the clay absorbs water, it swells the cracks shut, entrapping the collected granular soil. The increased soil volume causes lateral and upward movement of the soil mass. The soil is pushed up between the cracked areas. (*c*) As the subsoil mass shears from the strain, smooth surfaces or *slickensides* form. These processes result in a Vertisol profile that typically exhibits *gilgai,* cracks more than 1 m deep, and slickensides in a Bss horizon. Calcium carbonate concretions often accumulate in a Bkss horizon. (*d*) An example of slickenside showing the shining appearance. Also note the white concretions of carbonate minerals. (Diagram and photo courtesy of R. Weil)

3.12 MOLLISOLS (DARK, SOFT SOILS OF GRASSLANDS)

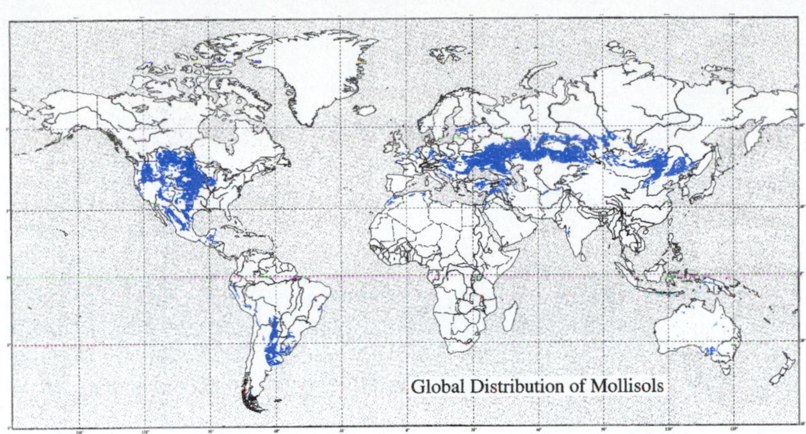

6.9% of global ice-free land.

Suborders are:

Albolls (albic horizon)
Aquolls (wet)
Cryolls (very cold)
Rendolls (calcareous)
Udolls (humid)
Ustolls (moist/dry)
Xerolls (dry summers, moist winters)

The principal process in the formation of Mollisols is the accumulation of calcium-rich organic matter, largely from the dense root systems of prairie grasses, to form the thick, soft Mollic epipedon that characterizes soils in this order (Plate 8). This humus-rich surface horizon is often 60 to 80 cm in depth and high in calcium and magnesium. Its cation exchange capacity is more than 50% saturated with non-acid cations (Ca^{2+}, Mg^{2+}, etc.). Mollisols in humid regions generally have higher organic matter and darker, thicker mollic epipedons than their lower-moisture-regime counterparts (see Section 11.10).

The surface horizon generally has granular or crumb structures, largely resulting from an abundance of organic matter and swelling-type clays. The aggregates are not hard when the soils are dry, hence the name *Mollisol,* implying softness (Table 3.2). In addition to the mollic epipedon, Mollisols may have an argillic (clay), natric, albic, or cambic subsurface horizon, but not an oxic or spodic horizon.

Most Mollisols have developed under grass vegetation (Figure 3.15). Grassland soils of the central part of the United States, lying between Aridisols on the west and the

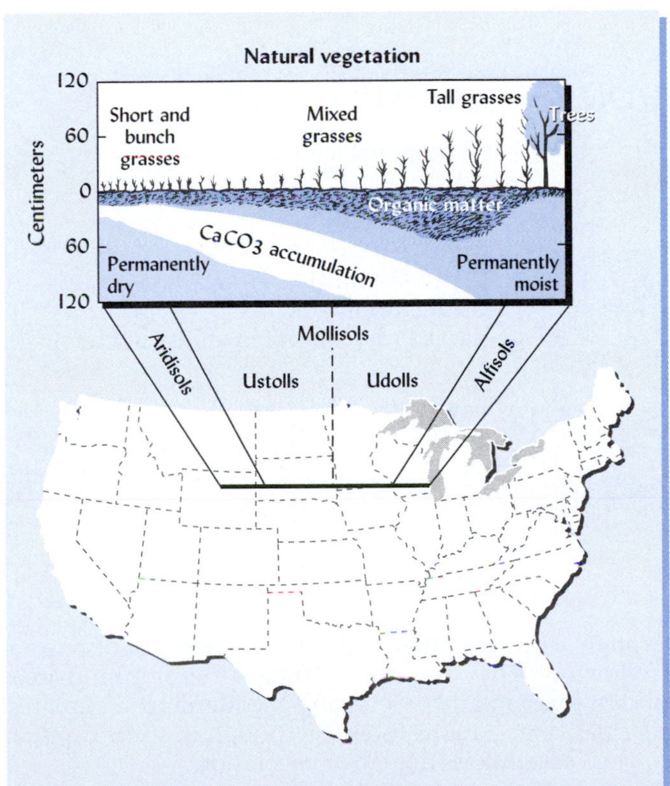

FIGURE 3.15 Correlation between natural grassland vegetation and certain soil orders is graphically shown for a transect across north central United States. The controlling factor, of course, is climate. Note the deeper organic matter and deeper zone of calcium accumulation, sometimes underlain by gypsum, as one proceeds from the drier areas in the west toward the more humid region where prairie soils are found. Alfisols may develop under grassland vegetation, but more commonly occur under forests and have lighter-colored surface horizons.

FIGURE 3.16 Typical landscape dominated by Ustolls. These productive soils produce much of the food and feed in the United States. (Photos courtesy of R. Weil)

Alfisols on the east, make up the central concept of this order. However, a few soils developed under forest vegetation (primarily in depressions) have a mollic epipedon and are included among the Mollisols.

Mollisols cover a larger land area in the United States than any other soil order (about 22%). Mollisols are dominant in the Great Plains of North America (Figure 3.16), as well as in Illinois (see end papers). The largest area of Mollisols in the world stretches from east to west across the heartlands of Kazakhstan, Ukraine, and Russia (see front papers). Because the high native fertility of Mollisols makes them among the world's most productive soils, few Mollisols have been left uncultivated in regions with sufficient rainfall for crop production. In the United States, efforts are underway to preserve the few remnants of the once vast and diverse prairie ecosystem.

Global perspectives on Mollisols and other soils: http://www.nrcs.usda.gov/technical/worldsoils/gsr/

3.13 ALFISOLS (ARGILLIC OR NATRIC HORIZON, MODERATELY LEACHED)

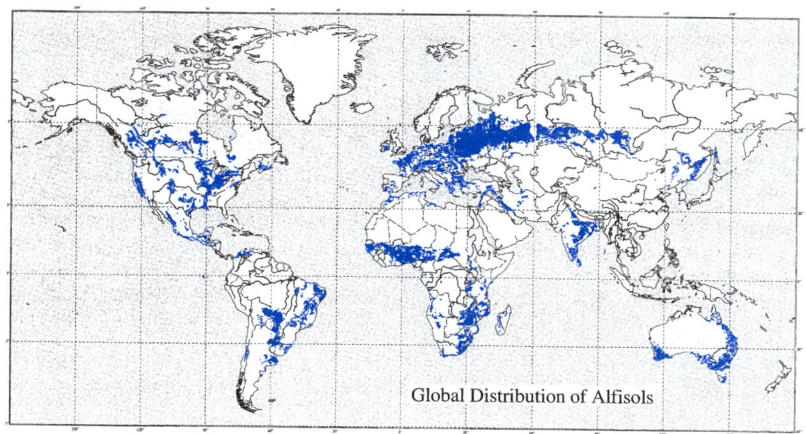

9.7% of global ice-free land.

Suborders are:

- Aqualfs (wet)
- Cryalf (cold)
- Udalfs (humid)
- Ustalfs (moist/dry)
- Xeralfs (dry summers, moist winters)

The Alfisols appear to be more strongly weathered than soils in the orders just discussed, but less so than Spodosols (see following). They are formed in cool to hot humid areas (see Figure 3.6), but also are found in the semiarid tropics and Mediterranean climates. Most often, Alfisols develop under native deciduous forests, although in some cases, as in California and parts of Africa, grass savanna is the native vegetation.

Alfisols are characterized by a subsurface diagnostic horizon in which silicate clay has accumulated by illuviation (see Plate 1). Clay skins or other signs of clay movement are present in such a B horizon (see Plates 24 and 27). In Alfisols, this clay-rich horizon is only moderately leached, and its cation exchange capacity is more than 35% "base saturated" (i.e. occupied with non-acid cations such as Ca^{2+}). In most Alfisols this horizon is termed *argillic* because of its accumulation of silicate clays. The horizon is termed *natric* if, in addition to having an accumulation of clay, it is more than 15% saturated with sodium and has prismatic or columnar structure (see Figure 4.9). In some Alfisols in subhumid tropical regions, the accumulation is termed a *kandic* (from the mineral kandite) horizon because the clays have a low cation exchange capacity.

Alfisols rarely have a mollic epipedon, for such soils would be classified in the Argiudolls or other suborder of Mollisols with an argillic horizon. Instead, Alfisols typically have a relatively thin, gray to brown ochric epipedon (Plate 1 shows an example) or an umbric epipedon. Those formed under deciduous temperate forests commonly have a light-colored, leached *albic* E horizon immediately under the A horizon (see Plate 21 and Figure 3.17).

In general, Alfisols are productive soils. Good hardwood forest growth and crop yields are favored by their medium to high "base-saturation" status, generally favorable texture, and location (except for some Xeralfs) in regions with sufficient rainfall for plants for at least part of the year. In the United States these soils rank favorably with the Mollisols and Ultisols in their productive capacity. Many, especially the sandier ones, are quite susceptible to erosion by heavy rains if deprived of their natural surface litter. Alfisols in udic moisture regimes are sufficiently acidic to require amendment with limestone for many kind of plants (see Chapter 9).

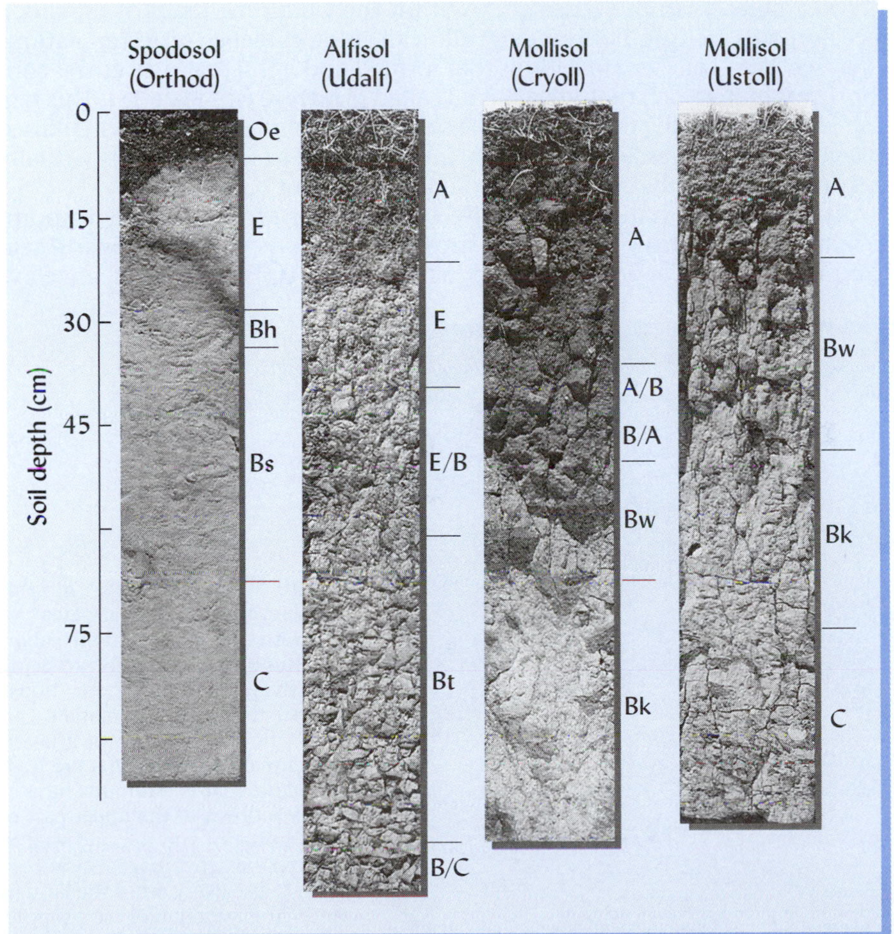

FIGURE 3.17 Monoliths of profiles representing three soil orders. The suborder names are in parentheses. Genetic (not diagnostic) horizon designations are also shown. Note the spodic horizons in the Spodosol characterized by humus (Bh) and iron (Bs) accumulation. In the Alfisol is found the illuvial clay horizon (Bt), and the structural B horizon (Bw) is indicated in the Mollisols. The thick dark surface horizon (mollic epipedon) characterizes both Mollisols. Note that the zone of calcium carbonate accumulation (Bk) is near the surface in the Ustoll, which has developed in a dry climate. The E/B horizon in the Alfisol has characteristics of both E and B horizons.

3.14 ULTISOLS (ARGILLIC HORIZON, LOW BASES)

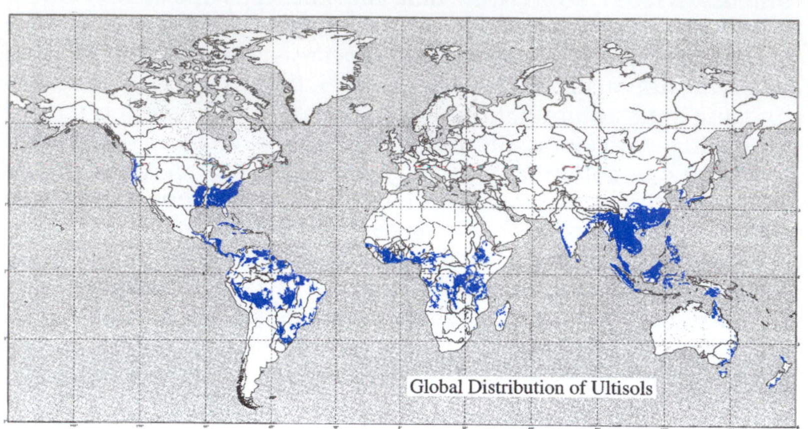

8.5% of global ice-free land.

Suborders are:

Aquults (wet)
Humults (high humus)
Udults (humid)
Ustults (moist/dry)
Xerults (dry summers, moist winters)

The principal processes involved in forming Ultisols are clay mineral weathering, translocation of clays to accumulate in an argillic or kandic horizon, and leaching of base-forming cations from the profile. Most Ultisols have developed under moist conditions in warm to tropical climates. Ultisols are formed on old land surfaces, usually under forest vegetation, although savanna or even swamp vegetation is also common. They often have an ochric or umbric epipedon, but are characterized by a relatively acidic B horizon that has less than 35% of the exchange capacity satisfied with non-acid cations. The clay accumulation may be either an argillic horizon or, if the clay is of low activity, a kandic horizon. Ultisols commonly have both an epipedon and a subsoil that is quite acid and low in plant nutrients.

Ultisols are more highly weathered and acidic than Alfisols, but less acid than Spodosols and less highly weathered than the Oxisols. Except for the wetter members of the order, their subsurface horizons are commonly red or yellow in color, evidence of accumulations of oxides of iron (see Plate 11). Certain Ultisols that formed under fluctuating wetness conditions have horizons of iron-rich mottled material called **plinthite** (see Plate 15). This material is soft and can be easily dug from the profile so long as it remains moist. When dried in the air, however, plinthite hardens irreversibly into a kind of ironstone that is virtually useless for cultivation, but can be used to make durable bricks for building.

Although Ultisols are not naturally as fertile as Alfisols or Mollisols, they respond well to good management. They are located mostly in regions of long growing seasons and of ample moisture for good crop production (Figure 3.18). The silicate clays

FIGURE 3.18 The soils in this high-elevation, tropical area of south Asia are Ultisols in the suborder Humults. These soils are being intensively used both for house construction and for market gardens. The combination of a favorable climate and soils that are high in organic matter (Humults have at least 9% down to the upper part of the B horizon) and respond well to fertilizer has encouraged local residents to use every bit of the land in producing vegetables to supplement their incomes. (Photo courtesy of R. Weil)

of Ultisols are usually of the nonsticky type, which, along with the presence of iron oxides and aluminum, assures ready workability. Where adequate levels of fertilizers and lime are applied, Ultisols are quite productive. In the United States, well-managed Ultisols compete well with Mollisols and the Alfisols as first-class agricultural soils. They also support the most productive commercial softwood and hardwood forests in the country.

3.15 SPODOSOLS (ACID, SANDY, FOREST SOILS, LOW BASES)

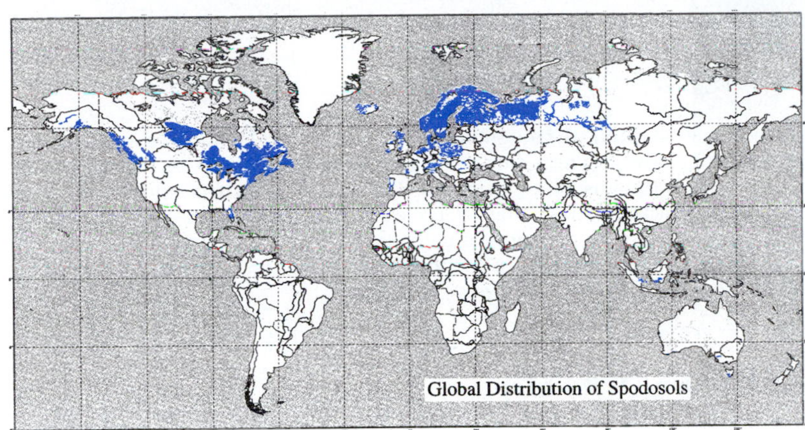

2.6% of global ice-free land.

Suborders are:
Aquods (wet)
Cryods (icy cold)
Humods (humus)
Orthods (typical).

Official Michigan State Soil—a Spodosol named Kalkaska:
http://www.michigan.gov/mda/0,1607,7-125-2961_2971_2972-11120—,00.html

Spodosols occur mostly on coarse-textured, acid parent materials subject to ready leaching. They occur only in moist to wet areas, commonly where it is cold or temperate (see Figure 3.6), but also in some tropical and subtropical areas. Intensive acid leaching is the principal soil-forming process. They are mineral soils with a *spodic* horizon, a subsurface accumulation of illuviated organic matter, and an accumulation of aluminum oxides with or without iron oxides (see Plate 10 and Figure 3.17). This usually thin, dark, illuvial horizon typically underlies a light, ash-colored, eluvial *albic* horizon.

Spodosols form under forest vegetation, especially under coniferous species whose needles are low in non-acid cations such as calcium and high in acid resins. As this acid litter decomposes, strongly acid organic compounds are released and carried down into the permeable profile by percolating waters. These acids bind with iron and aluminum and carry them downward until they precipitate in the spodic horizon. Similarly, soluble organic compounds move downward and are likewise precipitated above or intermixed with the iron and aluminum compounds. Most minerals except quartz are removed by this acid leaching, generally removing the coloring agents from the E horizon. Deeper in the profile, precipitated oxides of aluminum and iron form a Bs spodic horizon that is often reddish brown in color. The leaching organic compounds may precipitate and form a black Bh spodic horizon. The precipitation often occurs along wavy wetting fronts, thus yielding striking Spodosol profiles (Figure 3.19). The depth of these weathering and leaching processes, hence the spodic horizons, can vary from less than 20 cm to several meters.

Spodosols are not naturally fertile. When properly fertilized, however, these soils can become quite productive. For example, most potato-producing soils of northern Maine are Spodosols, as are some of the vegetable- and fruit-producing soils of Florida, Michigan, and Wisconsin. Because of their sandy nature and occurrence in regions of high rainfall, groundwater contamination by leaching of soluble fertilizers and pesticides has proved to be a serious problem where these soils are used in crop production. They are now covered mostly with forests, the vegetation under which they originally developed. Most Spodosols should remain as forest habitats. Because they are already quite acid and poorly buffered, many Spodosols and the lakes in watersheds dominated by soils of this order are susceptible to damage from acid rain (see Section 9.7).

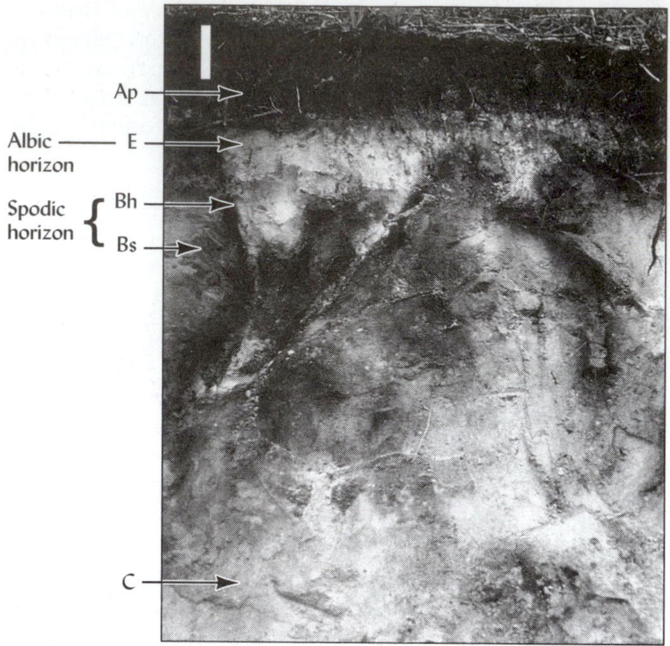

FIGURE 3.19 Spodosol in the Upper Penninsula of Michigan shows a relatively deep, very wavy spodic horizon under a discontinuous albic horizon. It has a uniform ochric Ap horizon, indicating that it has been cleared and plowed. Spodosols are sandy, very acidic, and usually best suited to supporting coniferous forests such as influenced their formation. The white bar is 10 cm long. (Photo courtesy of R. Weil)

3.16 OXISOLS (OXIC HORIZON, HIGHLY WEATHERED)

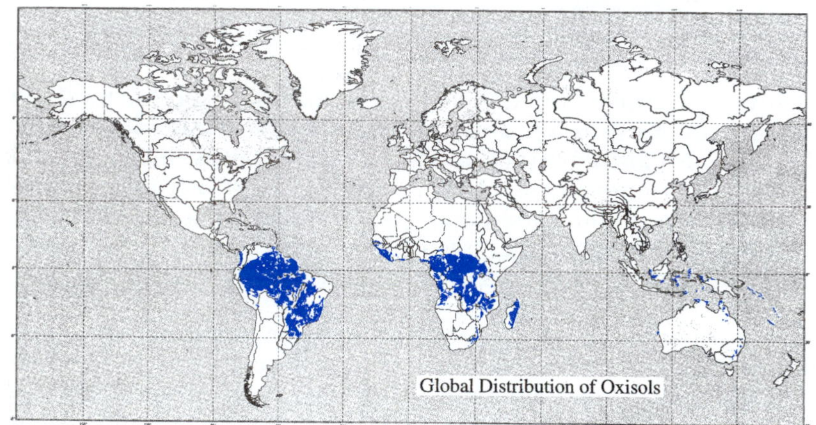

7.6% of global ice-free land.

Suborders are:

- Aquox (wet)
- Perox (very humid)
- Torrox (hot, dry)
- Udox (humid)
- Ustox (moist/dry).

Guy Smith Memorial slide on Oxisols. Large PDF file with links:
http://www.ncrs.usda.gov/technical/worldsoils/oxisols/Oxisols.pdf

The Oxisols are the most highly weathered soils in the classification system (see Figure 3.5). They form in hot climates with nearly year-round moist conditions; hence, the native vegetation is generally thought to be tropical rain forest. However, some Oxisols (Ustox) are found in areas which are today much drier than was the case when the soils were forming their oxic characteristics. Their most important diagnostic feature is a deep oxic subsurface horizon. This horizon is generally very high in clay-size particles dominated by hydrous oxides of iron and aluminum. Weathering and intense leaching have removed a large part of the silica from the silicate materials in this horizon. Some quartz and 1:1-type silicate clay minerals remain, but the hydrous oxides are dominant (see Chapter 8 for information on the various clay minerals). The epipedon in most Oxisols is either ochric or umbric. Usually the boundaries between subsurface horizons are indistinct, giving the subsoil a relatively uniform appearance with depth.

The clay content of Oxisols is generally high, but the clays are of the low-activity, nonsticky type. Consequently, when the clay dries out it is not hard and cloddy, but is easily worked. Also, Oxisols are resistant to compaction, so water moves freely through the profile. The depth of weathering in Oxisols is typically much greater than for most of the other soils, 20 m or more having been observed. The low-activity clays have a very limited capacity to hold nutrient cations such as Ca^{2+}, Mg^{2+}, and K^{+}, so these soils are typically of low natural fertility and moderately acid. The high concentration of iron and aluminum oxides also gives these soils a capacity to bind so tightly with what little phosphorus is present, that phosphorus deficiency often limits plant growth once the natural vegetation is disturbed.

Oxisols are found on old land surfaces that have not been disturbed by glaciation or erosion. Although nearly all Oxisols occur in the tropics, most tropical soils are *not* Oxisols. Because of their low natural fertility, most Oxisols have been left under forest vegetation or are farmed by shifting cultivation methods. Nutrient cycling by deep-rooted trees is especially important to the productivity of these soils. Road and building construction is relatively easily accomplished on most Oxisols because these soils are easily excavated, do not shrink and swell, and are physically very stable on slopes. The very stable aggregation of the clays, stimulated largely by iron compounds, makes these soils quite resistant to erosion.

3.17 LOWER-LEVEL CATEGORIES IN SOIL TAXONOMY

Suborders

Within each soil order just described, soils are grouped into suborders (see list under each order heading in Sections 3.5 to 3.16) on the basis of soil properties that reflect major environmental controls on current soil-forming processes. Many suborders are indicative of the moisture regime or, less frequently, the temperature regime under which the soils are found. Thus, soils formed under wet conditions generally are identified under separate suborders (e.g., Aquents, Aquerts, and Aquepts) as being wet soils.

To determine the relationship between suborder names and soil characteristics, refer to Table 3.3. Here the formative elements for suborder names are identified and their connotations given. Thus, the Ustolls are dry Mollisols. Likewise, soils in the Udults suborder (from the Latin *udus,* humid) are moist Ultisols.

TABLE 3.3 Formative Elements in Names of Suborders in *Soil Taxonomy*

Formative element	*Derivation*	*Connotation of formative element*
alb	L. *albus,* white	Presence of albic horizon (a bleached eluvial horizon)
anthr	Gk. *anthropos,* human	Presence of anthropic or plaggen epipedon
aqu	L. *aqua,* water	Characteristics associated with wetness
ar	L. *arare,* to plow	Mixed horizons
arg	L. *argilla,* white clay	Presence of argillic horizon (a horizon with illuvial clay)
calc	L. *calcis,* lime	Presence of calcic horizon
camb	L. *cambriare,* to change	Presence of cambric horizon
cry	Gk. *kryos,* icy cold	Cold
dur	L. *durus,* hard	Presence of a duripan
fibr	L. *fibra,* fiber	Least decomposed stage
fluv	L. *fluvius,* river	Floodplains
fol	L. *folia,* leaf	Mass of leaves
gyps	L. *gypsum,* gypsum	Presence of gypsic horizon
hem	Gk. *hemi,* half	Intermediate stage of decomposition
hist	Gk. *histos,* tissue	Presence of histic epipedon
hum	L. *humus,* earth	Presence of organic matter
orth	Gk. *orthos,* true	The common ones
per	L. *per,* throughout time	Of year-round humid climates, perudic moisture regime
psamm	Gk. *psammos,* sand	Sand textures
rend	Modified from Rendzina	Rendzinalike—high in carbonates
sal	L. *sal,* salt	Presence of salic (saline) horizon
sapr	Gk. *sapros,* rotten	Most decomposed stage
torr	L. *torridus,* hot and dry	Usually dry
turb	L. *turbidus,* disturbed	Cryoturbation
ud	L. *udus,* humid	Of humid climates
ust	L. *ustus,* burnt	Of dry climates, usually hot in summer
vitr	L. *vitreus,* glass	Resembling glass
xer	Gk. *xeros,* dry	Dry summers, moist winters

Great Groups

The **great groups** are subdivisions of suborders. More than 300 great groups are recognized. They are defined largely by the presence or absence of diagnostic horizons and the arrangements of those horizons. These horizon designations are included in the list of formative elements for the names of great groups shown in Table 3.4. Note that these formative elements refer to epipedons such as umbric and ochric (see Table 3.1), to subsurface horizons such as argillic and natric, and to certain diagnostic impervious layers such as duripans and fragipans (see Figure 3.20).

Remember that the great group names are made up of these formative elements attached as prefixes to the names of suborders in which the great groups occur. Thus, Ustolls with a natric horizon (high in sodium) belong to the Natrustolls great group. As can be seen in the example discussed in Box 3.1, soil descriptions at the great group level can

TABLE 3.4 Formative Elements for Names of Great Groups and Their Connotation

These formative elements combined with the appropriate suborder names give the great group names.

Formative element	*Connotation*	*Formative element*	*Connotation*	*Formative element*	*Connotation*
acr	Extreme weathering	fol	Mass of leaves	petr	Cemented horizon
agr	Agric horizon	fragi	Fragipan	plac	Thin pan
al	High aluminum, low iron	fragloss	See *fragi* and *gloss*	plagg	Plaggen horizon
alb	Albic horizon	fulv	Light-colored melanic horizon	plinth	Plinthite
and	Ando-like	gyps	Gypsic horizon	psamm	Sand texture
anhy	Anhydrous	gloss	Tongued	quartz	High quartz
aqu	Water saturated	hal	Salty	rhod	Dark red colors
argi	Argillic horizon	hapl	Minimum horizon	sal	Salic horizon
calc, calci	Calcic horizon	hem	Intermediate decomposition	sapr	Most decomposed
camb	Cambic horizon	hist	Presence of organic materials	somb	Dark horizon
chrom	High chroma	hum	Humus	sphagn	Sphagnum moss
cry	Cold	hydr	Water	sulf	Sulfur
dur	Duripan	kand	Low-activity 1:1 silicate clay	torr	Usually dry and hot
dystr, dys	Low base saturation	lithic	Near stone	ud	Humid climates
endo	Fully water saturated	luv, lu	Illuvial	umbr	Umbric epipedon
epi	Perched water table	melan	Melanic epipedon	ust	Dry climate, usually hot in summer
eutr	High base saturation	molli	With a mollic epipedon	verm	Wormy, or mixed by animals
ferr	Iron	natr	Presence of a natric horizon	vitr	Glass
fibr	Least decomposed	pale	Old development	xer	Dry summers, moist winters
fluv	Floodplain				

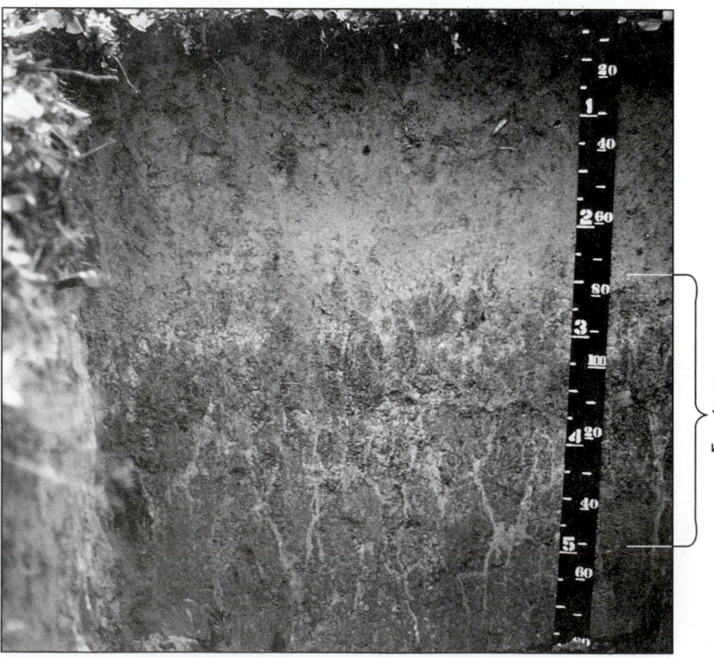

FIGURE 3.20 A forested Fragiudalf in Missouri containing a typical well-developed fragipan with coarse prismatic structure (outlined by gray, iron-depleted coatings). Fragipans (usually Bx or Cx horizons) are extremely dense and brittle. They consist mainly of silt, often with considerable sand, but not very much clay. One sign of encountering a fragipan in the field is the ringing noise that your shovel will make when you attempt to excavate it. Digging through a fragipan is almost like digging concrete. Plant roots cannot penetrate this layer. Yet, once a piece of a fragipan is broken loose, it fairly easily crushes with hand pressure. But it does not squash or act in a plastic manner as a claypan would; instead, it bursts in a *brittle* manner. (Photo courtesy of Fred Rhoton, Agricultural Research Service, U.S. Department of Agriculture)

BOX 3.1 GREAT GROUPS, FRAGIPANS, AND ARCHAEOLOGIC DIGS

Soil Taxonomy is a communications tool that helps scientists and land managers share information. In this box we will see how *misclassification*, even at a lower level in *Soil Taxonomy* such as the great group, can have costly ramifications.

To preserve our national heritage, both federal and state laws require that an archaeological impact statement be prepared prior to starting major construction works on the land. The archaeological impact is usually assessed in three phases. Selected sites are then studied by archaeologists, with the hope that at least some of the artifacts can be preserved and interpreted before construction activities obliterate them forever. Only a few relatively small sites can be subjected to actual archaeological digs because of the expensive skilled hand labor involved (Figure 3.21).

Such an archaeological impact study was ordered as a precursor to construction of a new highway in a mid-Atlantic state. In the first phase, a consulting company gathered soils and other information from maps, aerial photographs, and field investigations to determine where neolithic people may have occupied sites. Then the consultants identified about 12 ha of land where artifacts indicated significant neolithic activities. The soils in one area were mapped mainly as Typic Dystrudepts. These soils formed in old colluvial and alluvial materials that, many thousands of years ago, had been along a river bank. Several representative soil profiles were examined by digging pits with a backhoe. The different horizons were described, and it was determined in which horizons artifacts were most likely to be found. What was not noted was the presence in these soils of a **fragipan**, a dense, brittle layer that is extremely difficult to excavate using hand tools.

A fragipan is a subsurface diagnostic horizon used to classify soils, usually at the great group or subgroup level (see Figure 3.20). Its presence would distinguish Fragiudepts from Dystrudepts.

When it came time for the actual hand excavation of sites to recover artifacts, a second consulting company was awarded the contract. Unfortunately, their bid on the contract was based on soil descriptions that did not specifically classify the soils as Fragiudepts—soils with very dense, brittle, hard fragipans in the layer that would need to be excavated by hand. So difficult was this layer to excavate and sift through by hand, that it nearly doubled the cost of the excavation—an additional expense of about $1 million. Needless to say, there ensued a controversy as to whether this cost would be borne by the consulting firm that bid with faulty soils data, the original consulting firm that failed to adequately describe the presence of the fragipan, or the highway construction company that was paying for the survey.

This episode gives us an example of the practical importance of soil classification. The formative element *Fragi* in a soil great group name warns of the presence of a dense, impermeable layer that will be very difficult to excavate, will restrict root growth (often causing trees to topple in the wind or become severely stunted), may cause a perched water table (**epiaquic** conditions), and will interfere with proper percolation in a septic drain field.

FIGURE 3.21 *An archaeological dig. (Photo courtesy of Antonio Segovia, University of Maryland)*

provide important information not indicated at the higher, more general levels of classification. The names identify the suborder and order in which the great groups are found. Thus, Argiudolls are Mollisols of the Udolls suborder characterized by an argillic horizon.

Subgroups

Subgroups are subdivisions of the great groups. More than 2000 subgroups are recognized. The central concept of a great group makes up one subgroup, termed *Typic*. Thus, the Typic Hapludolls subgroup typifies the Hapludolls great group. Other subgroups may have characteristics that intergrade between those of the central concept and soils of other orders, suborders, or great groups. A Hapludoll with restricted drainage would be classified as an Aquic Hapludoll. One with evidence of intense earthworm activity would fall in the Vermic Hapludolls subgroup. Some intergrades may have properties in common with other orders or with other great groups. Thus, soils in the Entic Hapludolls subgroup are weakly developed Mollisols, close to being in the Entisols order. The subgroup concept illustrates very well the flexibility of this classification system.

Families

Within a subgroup, soils fall into a particular family if, at a specified depth, they have similar physical and chemical properties affecting the growth of plant roots. About 8000 families have been identified. The criteria used include broad classes of particle size, mineralogy, cation exchange activity of the clay, temperature, and depth of the soil penetrable by roots. Table 3.5 gives examples of the classes used. Terms such as *loamy, sandy,* and *clayey* are used to identify the broad particle size classes. Terms used to describe the mineralogical classes include *smectitic, kaolinitic, siliceous, carbonatic,* and *mixed.* The clays are described as *superactive, active, semiactive,* or *subactive* with regard to their capacity to hold cations. For temperature classes, terms such as *cryic, mesic,* and *thermic* are used. The terms *shallow* and *micro* are sometimes used at the family level to indicate unusual soil depths.

Thus, a Typic Argiudoll from Iowa, loamy in texture, having a mixture of moderately active clay minerals and with annual soil temperatures (at 50 cm depth) between 8 and 15°C, is classed in the *Typic Argiudolls loamy, mixed, active, mesic* family. In contrast, a

TABLE 3.5 Some Commonly Used Particle-Size, Mineralogy, Cation Exchange Activity, and Temperature Classes Used to Differentiate Soil Families

The characteristics generally apply to the subsoil or 50 cm depth. Other criteria used to differentiate soil families (but not shown here) include the presence of calcareous or highly aluminum toxic (allic) properties, extremely shallow depth (shallow or micro), degree of cementation, coatings on sand grains, and the presence of permanent cracks.

				Soil temperature regime class		
		Cation exchange activity class[b]				
Particle-size class	Mineralogy class	Term	CEC/% clay	Mean annual temperature, °C	>6°C difference between summer and winter	<6°C difference between summer and winter
Ashy	Mixed	Superactive	0.60	<−10	Hypergelic[c]	—
Fragmental	Micaceous	Active	0.4 to 0.6	−4 to −10	Pergelic[c]	—
Sandy-skeletal[a]	Siliceous	Semiactive	0.24 to 0.4	+1 to −4	Subgelic[c]	—
Sandy	Kaolinitic	Subactive	< 0.24	<+8	Cryic	—
Loamy	Smectitic			<+8	Frigid[d]	Isofrigid
Clayey	Gibbsitic			<+ 8 to +15	Mesic	Isomesic
Fine-silty	Gypsic			+15 to +22	Thermic	Isothermic
Fine-loamy	Carbonic			>+22	Hyperthermic	Isohyperthermic
Etc.	Etc.					

[a] *Skeletal* refers to presence of up to 35% rock fragments by volume.
[b] Cation exchange activity class is not used for taxa already defined by low CEC (e.g., kandic or oxic groups).
[c] Permafrost present.
[d] Frigid is warmer in summer than Cryic.

sandy-textured Typic Haplorthod, high in quartz and located in a cold area in eastern Canada, is classed in the *Typic Haplorthods sandy, siliceous, frigid* family (note that clay activity classes are not used for soils in sandy textural classes).

Series

The series category is the most specific unit of the classification system. It is a subdivision of the family, and each series is defined by a specific range of soil properties involving primarily the kind, thickness, and arrangement of horizons. Features such as a hardpan within a certain distance below the surface, a distinct zone of calcium carbonate accumulation at a certain depth, or striking color characteristics greatly aid in series identification. In the United States, each series is named after a geographic feature such as a town or river in the locality where the soil was first described. Fargo, Muscatine, Cecil, Mohave, and Ontario soils are examples. There are about 19,000 soil series in the United States. For practical reasons, soil series are sometimes subdivided into **phases** based on certain properties of importance to land use and management (such as texture of the surface horizon, stoniness, slope, degree of erosion or soluble salt content). Thus, a "Pinole loam, 2 to 9 percent slopes" and a "Hagerstown silt loam, stony phase" are examples of phases of soil series. Although not technically a category in *Soil Taxonomy,* soil phases are commonly represented on soil maps and used as management units.

Official USDA Soil Series Names and Descriptions. Click on Soil Series Name Search: http://ortho.ftw.nrsc.usda.gov/osd/osd.html

The complete classification of a Mollisol, the Kokomo series, is given in Figure 3.22. This figure illustrates how *Soil Taxonomy* can be used to show the relationship between *the soil,* a comprehensive term covering all soils, and a specific soil series. The figure deserves study because it reveals much about the structure and use of *Soil Taxonomy.* If a soil series name is known, the complete *Soil Taxonomy* classification of the soil may be found on the World Wide Web (see URL in margin). Box 3.2 illustrates how soil taxonomic information can assist in understanding the nature of a landscape such as shown in Figure 3.23.

Learn all about your State Soil here. Click on the state of your choice: http://soils.usda.gov/gallery/state/main.htm

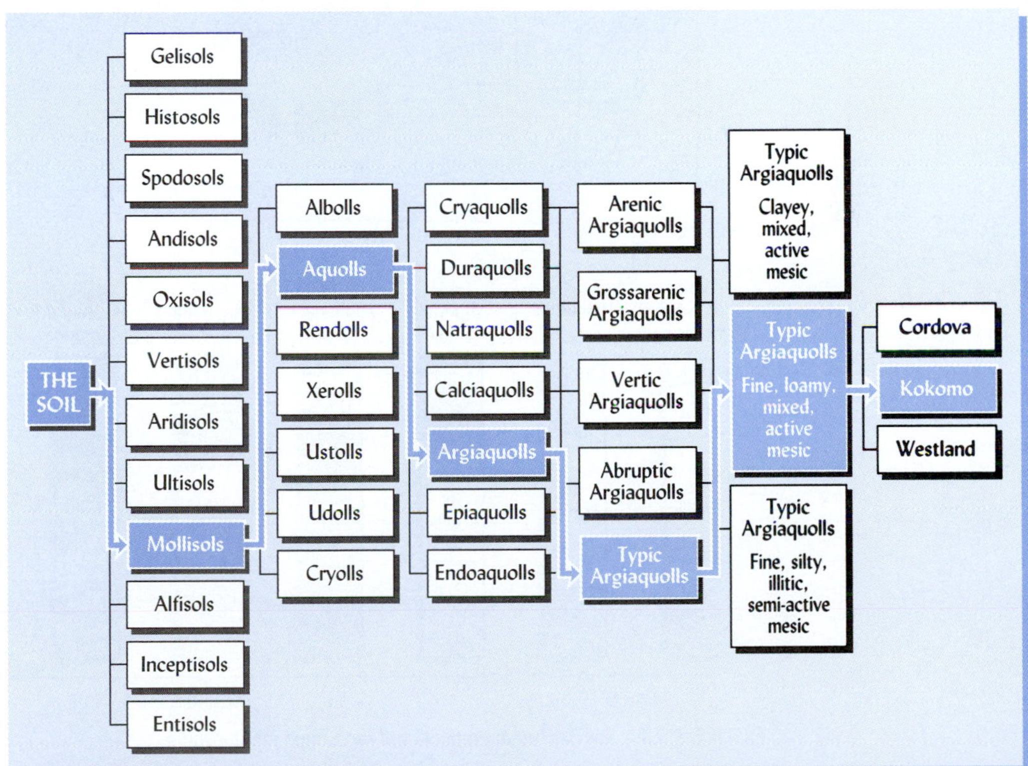

FIGURE 3.22 Diagram illustrating how one soil (Kokomo) keys out in the overall classification scheme. The shaded boxes show that this soil is in the Mollisols order, Aquolls suborder, Argiaquolls great group, and so on. In each category, other classification units are shown in the order in which they key out in *Soil Taxonomy.*

BOX 3.2 SOIL TAXONOMY HELPS IN UNDERSTANDING THE LANDSCAPE

In reviewing different categories in *Soil Taxonomy*, we have made little reference to the fact that soils exist alongside each other, often in complex patterns. Figure 3.23 depicts a landscape in a humid temperate region (Iowa) where 2 to 7 m of loess overlies leached glacial till and the native vegetation was principally tall grass prairie interspersed with small areas of trees. Seven soil map units are shown in the block landscape diagram, along with a profile diagram for the dominant series in each map unit. The soils include two Alfisols (Fayette and Downs) and five Mollisols (Tama, Wabash, Dinsdale, Muscatine, and Garwin).

The particular set of soil horizons present in each profile relates to the (1) parent material, (2) vegetation, and (3) topography and drainage. These factors are reflected in the *Soil Taxonomy* names. The Dinsdale soil differs from the Tama because two *parent materials*, loess and glacial till, contributed to its profile. The Fayette soil has a thin Ochric epipedon, a bleached albic horizon, and an argillic horizon because of the influence of forest *vegetation*. The soils on the concave or level *topographic* positions are wetter and less permeable than those on the steeper slopes (Downs 2 to 15% means this soil is found on slopes that vary from 2 to 15%). Restricted drainage has retarded the development of an argillic horizon in the wetter soils (those with it *aqu* in their names) so that only a gleyed (waterlogged) cambic B horizon (Bg) is present in these soils.

Careful study of this landscape diagram will reveal the practical manner in which diagnostic horizons and other features of *Soil Taxonomy* are used to organize our soils information and to highlight the relationships among soils. Recognition of such relationships in the field is important in making soil surveys and in interpreting geographic soils information.

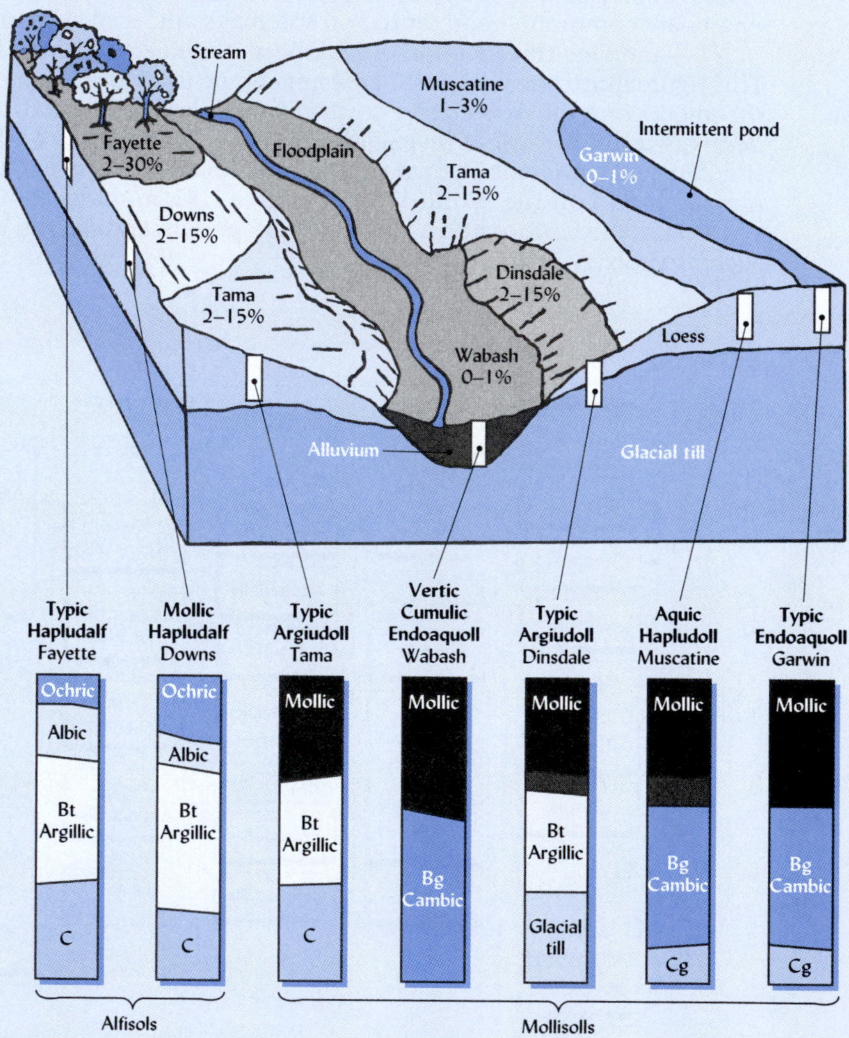

FIGURE 3.23 *Soil Taxonomy reflects soil-landscape relationships.*

3.18 SOIL SURVEYS AND MAPS

List of online U.S. soil survey manuscripts:
http://soils.usda.gov/soil_survey/surveys/main.htm

Even a small tract of land such as a suburban home lot or farm field is likely to contain several soil series that may be quite different from one another. They may even be members of different soil orders. A map showing where these different soils occur would provide valuable information for making the best use of the land. In fact, **soil maps** are in great demand as tools that communicate geographic information about soils for practical land planning and management. Many soil scientists therefore specialize in mapping soils. Their basic task is threefold: (1) to define each soil unit to be mapped, (2) to compile information about the nature of each soil, and (3) to delineate the boundaries where each soil unit occurs in the landscape.

While soil mappers may use such high-technology tools as computers and geopositioning satellites, they mainly rely on simple field tools, hands, and eyes to study soils in the field. The thorough study of a soil requires that a soil pit be dug by hand or with a backhoe. One or more soil scientists enter the pit and study a typical pedon (see Section 3.1) as exposed on the pit face. They write a soil description in a standard format that includes the texture, colors, structure, and other observable properties for each horizon.

Example of HTML soil survey from California:
http://www.ca.nrcs.usda.gov/mlra/wmendo/wmendofrm.html

Of course, in making a soil map, soil scientists cannot dig pits all over the landscape. Instead, they bring up soil material from small boreholes made with a hand-turned **soil auger.** The nature of the soils and the locations of the boundary lines surrounding them are inferred from information obtained by auger borings at numerous locations across a landscape. The soil mappers usually use an aerial photo as a base on which to draw the soil boundaries, coordinating their field observations with the landscape features visible on the aerial photo (see Figure 3.24).

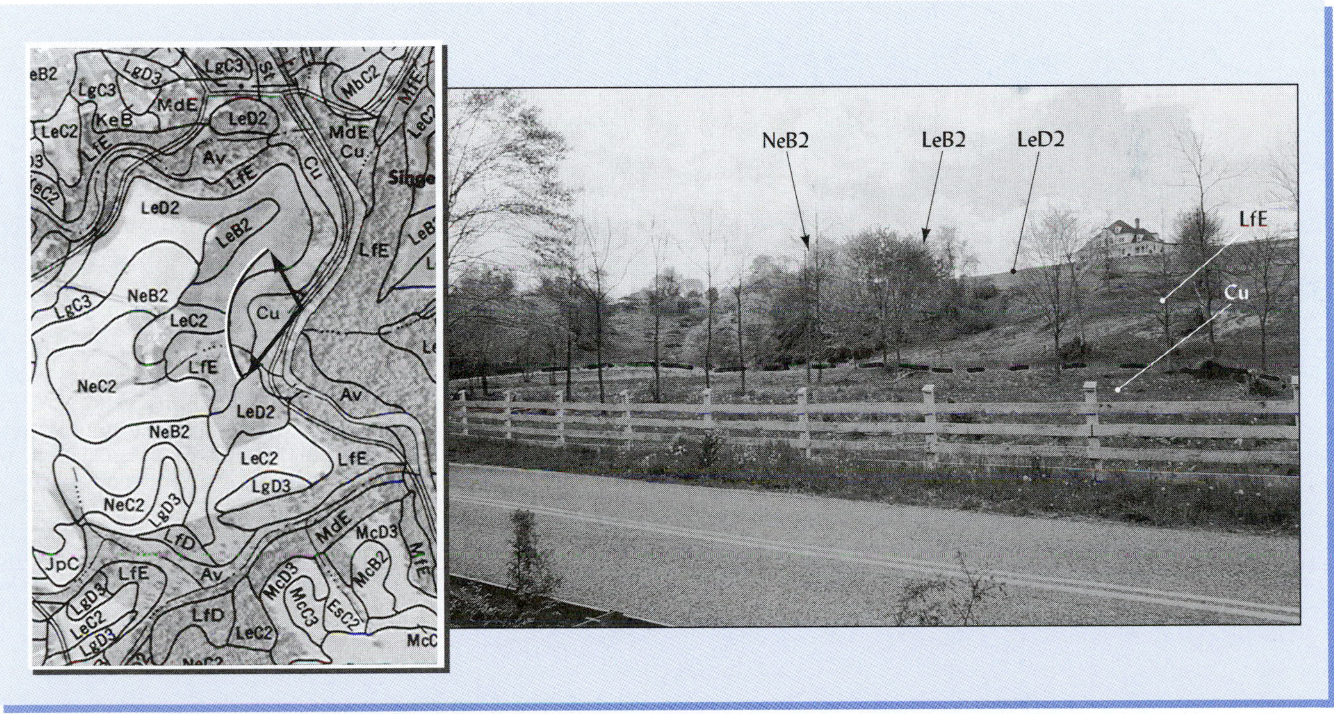

FIGURE 3.24 A small section of detailed soil survey map of Harford County, Maryland (left), and a ground-level view of the part of the landscape represented on the map by the view arrows (right). The map is reproduced here at the original scale (1:15,840) and represents an area of about 1.1 km^2. The map symbols represent soil consociations such as LeD2, which is named for the soil phase "Legore silt loam, 12 to 15% moderately eroded," a soil in the Alfisols order. The level land in the foreground is an alluvial soil in the Inceptisols order, the Codorus silt loam. [Map from Smith and Matthews (1975); photo courtesy of R. Weil]

Mapping Units

Soil Taxonomy (or some other classification system) is usually the basis for preparing a soil survey. Local features, planned land use, and map scale[6] (level of detail) will dictate the specific soil units that are mapped. The field **mapping units** may be somewhat different from the classification units of *Soil Taxonomy.* The mapping units may represent some further differentiation below the soil series level—namely, phases of soil series; or the soil mappers may choose to group similar or associated soils into conglomerate mapping units. Examples of such soil mapping units follow.

Soil Consociation. A **soil consociation** is the smallest practical mapping unit for most detailed soil surveys. It is an area that contains primarily one phase of a soil series. For example, a mapping unit may be labeled as the consociation "Legore silt loam, 12 to 15% slopes, moderately eroded." Although small areas of other soils may exist in this mapping unit, these so-called impurities should be so similar to the named soil that the differences do not affect land management.

Soil Complex. Sometimes contrasting rather than similar soils occur adjacent to each other in a pattern so intricate that it cannot be drawn on a soil map. In such cases, a **soil complex** is delineated on the soil map to indicate an area containing two or three distinctly different soil series.

Soil Association. A **soil association** is a more general grouping of individual soils that occur together in a landscape. It is used mainly in small-scale, regional soil maps. Soil associations are named after the two or three dominant soils in the group, but may contain several additional, less extensive soils. The soils may have formed in the same or in different parent materials and may be from the same or different soil orders (see Figure 3.25). The only requirement is that the soils occur together in the same area. A given soil association represents a defined range of soil properties and landscape relationships, even though the range of conditions included may be quite large.

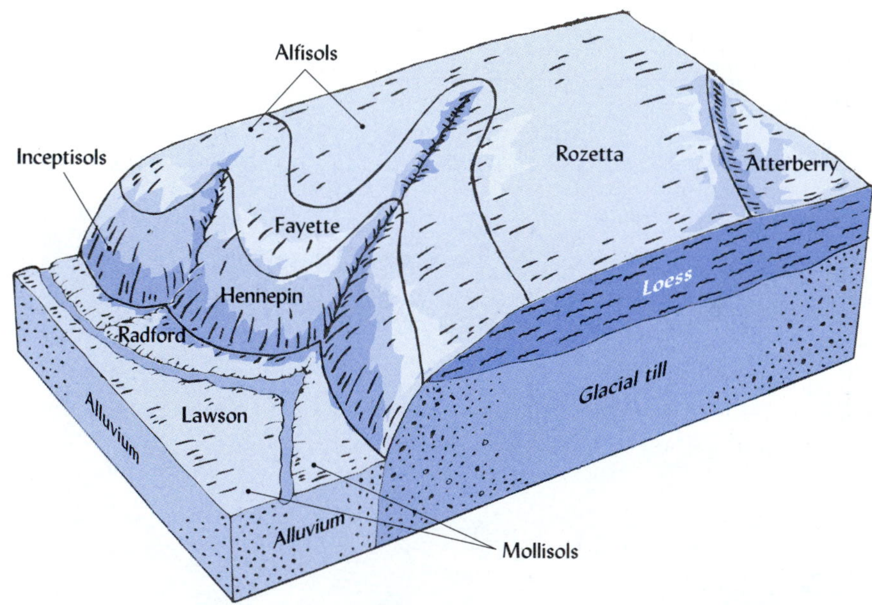

FIGURE 3.25 Diagram of the Rozetta-Fayette-Hennepin soil association in Bureau County, Illinois. This soil association includes three types of parent material (loess, glacial till, and alluvium) and a range of topography and vegetation types which give rise to five main soil series belonging to three different soil orders. [Based on Zwicker (1992)]

[6] The scale tells us the ratio of distance on a map to distance on the ground. On a relatively small-scale map, 1 cm may represent 100 km (scale = 1:10,000,000). On a relatively large-scale map, 1 cm may represent only 100 m (scale = 1:10,000). The map scale determines the level of detail that can be represented. Detailed (large-scale) soil maps in the United States are commonly made at scales of 1:12,000 (1 in. = 1000 ft) and 1:15,840 (4 in. = 1 mile).

Soil Survey Interpretations

In addition to the soil map, a modern **soil survey** report contains descriptive information on the soil mapping units and their suitability for various land uses. In the United States, the USDA Natural Resource Conservation Service produces most soil survey reports on a county by county basis in collaboration with state and local governments.

To be most useful for land use or site planning, the geographic soils information on the map should be integrated with the descriptive information in the rest of the report. The report usually contains detailed soil profile descriptions for all of the mapping units, as well as tables that provide soil characterization data and interpretive rankings for the mapping units. For each mapping unit, the soil survey offers interpretive information on agricultural (e.g., yield potentials, suitability for irrigation, drainage requirements, and land capability classification) and nonagricultural (e.g., wildlife habitat, forestry, landscaping, waste disposal, building construction, and source of roadbed materials) land uses. Table 3.6 lists a few of the many types of interpretive ratings given in the soil survey report for the soils mapped in Figure 3.24. If you were planning to develop a park on this site, the information in Table 3.6 would suggest that the Cu mapping unit would be suitable for a nature trail, but not for a visitor center with restrooms.

For complex land-planning analyses the enormous quantity of information stored in a county soil survey report can be used to best advantage with the help of a computerized **geographic information system (GIS)**. In many land-planning projects, soil properties comprise just one of several types of geographic information that must be integrated by a GIS in order to take best advantage of a given site or to find the site best suited for the proposed land use. Planning the best use of different farm fields, finding a site suitable for a sanitary landfill, and determining where ponds can be constructed for a wildlife refuge are some examples of projects that could make good use of both soils and nonsoils geographic information. Nonsoils information to be considered might include topography, streams, vegetation, present land use, and legal zoning categories.

Download SSURGO soil map files at:
http://www.ftw.nrcs.usda.gov/ssur_data.html

Canadian soils maps:
http://res.agr.ca/CANSIS/SYSTEMS/onlinemaps.htp
Choose "map" and construct a map of "soil development."

Efforts are underway to digitize soil survey maps so they can be easily entered into a GIS to help in planning rational use of the land. Electronic versions of some U.S. county-level soil survey manuscripts are available online (listed at URL in margin), but at this writing many contain only the textual information (profile descriptions and soil attributes) and not the actual soil maps. Although many county soil survey maps have yet to be digitized, detailed geo-referenced soils information for areas in the U.S. can be accessed by downloading files from the soil survey geographic database named SSURGO (URL in margin). Using appropriate GIS software, one can integrate SSURGO data with

TABLE 3.6 A Sampling of Some of the Interpretive Information Provided by the Soil Survey of Harford County Area, Maryland

The four mapping units described are those in the site shown in Figure 3.24. The interpretations included are some of those that would be of interest for developing the site as a county park.

Mapping unit code	*Name of mapping unit*	*Land-use Capability Classification*[a]	*Limitations for use of soil*			*Suitability for open land wildlife*
			Camp areas	*Septic filter fields*	*Paths and trails*	
Cu	Codorus silt loam	IIw-7	Severe—flood hazard	Severe—high water table, flood hazard	Moderate—flood hazard	Good
LeB2	Legore silt loam, 3 to 8% slopes, moderately eroded	IIe-10	Slight	Slight	Slight	Good
LeD2	Legore silt loam, 15 to 25% slopes, moderately eroded	IVe-10	Severe—slope	Severe—slope	Moderate—slope	Fair
LfE	Legore very stony silt loam, 25 to 45% slopes	VIIs-3	Severe—slope	Severe—slope	Severe—slope	Poor
NeB2	Neshaminy silt loam, 3 to 8% slopes, moderately eroded	IIe-4	Slight	Moderate—mod. permeability	Slight	Good

[a]See Section 15.10 for an explanation of the USDA Land Capability Classification System.
Data abstracted from Smith and Matthews (1975).

non-soils information to construct maps showing the suitability of land areas for various uses. For areas in Canada, one can directly construct generalized, small scale soils maps of regions in Canada at an Agriculture Canada website (URL in margin).

3.19 CONCLUSION

International Soil Reference and Information Center. Includes geo-referenced soil profile data sets and alternatives to Soil Taxonomy for classification: http://lime.isric.nl

The soil which covers the earth is actually comprised of many individual soils, each with distinctive properties. Among the most important of these properties are those associated with the horizontal layers, or *horizons,* found in a soil profile. These horizons reflect the physical, chemical, and biological processes soils have undergone during their development. Horizon properties greatly influence how soils can and should be used.

Knowledge of the kinds and properties of soils around the world is critical to humanity's struggle for survival and well-being. A soil classification system based on these properties is equally critical if we expect to use knowledge gained at one location to solve problems at other locations where similarly classed soils are found. *Soil Taxonomy,* a classification system based on measurable soil properties, helps fill this need in some 55 countries. Scientists constantly update the system as they learn more about the nature and properties of the world's soils and the relationships among them. In the remaining chapters of this book we will use taxonomic names when appropriate to indicate the kinds of soils to which a concept or illustration may apply.

Making soil surveys is both a science and an art by which many soil scientists apply their understanding of soil classification to landscapes in the real world. Mapping soils is not simply a profession; many would say that it is a way of life. Soil survey reports are used in countless practical ways by soil scientists and nonscientists alike. The soil survey, combined with powerful geographic information systems, enables planners to make rational decisions about *what* should go *where* on the land. The challenge for soil scientists and concerned citizens is to develop the foresight and fortitude to use the planning process to preserve our most valuable soils, rather than hasten their destruction under shopping malls and landfills.

STUDY QUESTIONS

1. Diagnostic horizons are used to classify soils in *Soil Taxonomy.* Explain the difference between a diagnostic horizon (such as an argillic horizon) and a genetic horizon designation (such as a Bt1 horizon). Give a field example of a diagnostic horizon that contains several genetic horizon designations.
2. Explain the relationships among a *soil individual,* a *polypedon,* a *pedon,* and a *landscape.*
3. Rearrange the following soil orders from the *least* to the *most* highly weathered: Oxisols, Alfisols, Mollisolls, Entisols, and Inceptisols.
4. What is the principal soil property by which Ultisols differ from Alfisols? Inceptisols from Entisols?
5. Use the key given in Figure 3.7 to determine the soil order of a soil with the following characteristics: a spodic horizon at 30 cm depth, permafrost at 80 cm depth. Explain your choice of soil order.
6. Of the five soil-forming factors discussed in Chapter 2 (parent material, climate, organisms, topography, and time), choose *two* that have had the dominant influence on developing soil properties characterizing each of the following soil orders: Vertisols, Mollisols, Spodosols, and Oxisols.
7. To which soil order does each of the following belong: Psamments, Udolls, Argids, Udepts, Fragiudalfs, Haplustox, and Calciusterts?
8. What's in a name? Write a hypothetical soil profile description and land-use suitability interpretation for a hypothetical soil that is classified in the Aquic Argixerolls subgroup.
9. Explain why *Soil Taxonomy* is said to be a hierarchical classification system.
10. Name the soil taxonomy category and discuss the engineering implications of these soil taxonomy classes: Aquic Paleudults, Fragiudults, Haplusterts, Saprists, and Turbels.

11. A soil mapper drew a boundary around an area in which he made six randomly located auger borings, two in soil A, with an argillic horizon more than 60 cm thick and strong brown in color, and the other four in soil B, with a somewhat lighter brown argillic horizon between 50 and 70 cm thick. Other soil properties, as well as management considerations, were similar for the two types of soils. Would the map unit delineated likely be a *soil association,* a *soil consociation,* or a *soil complex?* Explain.
12. Assume you are planning to buy a 4 ha site on which to start a small orchard. Explain how you could use the county soil survey to help determine if the prospective sight was suitable for your intended use.
13. If you were hired by a state government to produce a GIS-based map showing where investments should be made to protect farmland from suburban development, what "layers" of information would you want to include in the GIS?

REFERENCES

Coulombe, C. E., L. P. Wilding, and J. B. Dixon. 1996. "Overview of Vertisols: Characteristics and impacts on society," *Advances in Agronomy* **17**:289–375.

Eswaran, H. 1993. "Assessment of global resources: Current status and future needs," *Pedologie* **43**(1):19–39.

McCracken, R. J., et al. 1985. "An appraisal of soil resources in the USA," in R. F. Follet and B. A. Stewart (eds.), *Soil Erosion and Crop Productivity.* (Madison, Wis.: Amer. Soc. Agron.).

Shaw, J. N., L. T. West, D. E. Radcliffe, and D. D. Bosch. 2000. "Preferential flow and pedotransfer functions for transport properties in sandy Kandiustults," *Soil Sci. Soc. Amer. J.* **64**:670–678.

Smith, H., and E. Matthews, 1975. *Soil Survey of Harford County Area, Maryland.* (Washington, D.C.: U.S. Soil Conservation Service).

SSSA. 1984. *Soil Taxonomy, Achievements and Challenges.* SSSA Special Publication no. 14. (Madison, Wis.: Soil Sci. Soc. Amer.).

Soil Survey Staff. 1998. *Keys to Soil Taxonomy.* (Washington, D.C.: USDA Natural Resources Conservation Service).

Soil Survey Staff. 1999. *Soil Taxonomy: A Basic System of Soil Classification for Making and Interpreting Soil Surveys,* 2d ed. (Washington, D.C.: USDA Natural Resources Conservation Service).

Star, J., and J. Estes. 1990. *Geographic Information Systems: An Introduction.* (Englewood Cliffs, N.J.: Prentice Hall).

U.S. Department of Agriculture. 1938. *Soils and Men.* USDA Yearbook. (Washington, D.C.: U.S. Government Printing Office).

Zwicker, S. E. 1992. *Soil Survey of Bureau County, Illinois.* (Washington, D.C.: USDA Natural Resources Conservation Service).

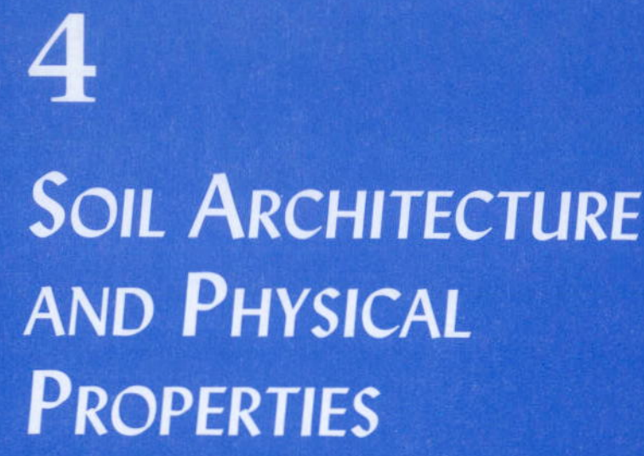

4 Soil Architecture and Physical Properties

Structure and texture of a Mollisol. (Ray Weit)

And when that crop grew, and was harvested, no man had crumbled a hot clod in his fingers and let the earth sift past his fingertips.
—JOHN STEINBECK, THE GRAPES OF WRATH

Soil physical properties profoundly influence how soils function in an ecosystem and how they can best be managed. Success or failure of both agricultural and engineering projects often hinges on the physical properties of the soil used. The occurrence and growth of many plant species are closely related to soil physical properties, as is the movement over and through soil of water and its dissolved nutrients and chemical pollutants.

Soil scientists use the color, texture, and other physical properties of soil horizons in classifying soil profiles and in making determinations about soil suitability for agricultural and environmental projects. Knowledge of the basic soil physical properties is not only of great practical value in itself, but will also help in understanding many aspects of soils considered in later chapters.

The physical properties discussed in this chapter relate to the solid particles of the soil and the manner in which they are aggregated. If we think of the soil as a house, the solid particles in soil are the building blocks from which the house is constructed. **Soil texture** describes the size of the soil particles. The larger mineral fragments usually are embedded in, and coated with, clay and other colloidal materials. Where the larger mineral particles predominate, the soil is gravelly or sandy; where the mineral colloids are dominant, the soil is claylike. All gradations between these extremes are found in nature.

In building a house, the manner in which the building blocks are put together determines the nature of the walls, rooms, and passageways. Organic matter and other substances act as cement between individual particles, encouraging the formation of clumps or aggregates of soil. **Soil structure** describes the manner in which soil particles are aggregated. This property, therefore, defines the nature of the system of pores and channels in a soil.

The physical properties considered in this chapter focus on soil solids and on the pore spaces between the solid particles. Together, soil texture and structure help determine the ability of the soil to hold and conduct the water and air necessary for sustaining life. These factors also determine how soils behave when used for highways and building construction and foundations, or when manipulated by tillage. In fact, through their influence on water movement through and off soils, soil physical properties also exert considerable control over the destruction of the soil itself by erosion.

4.1 SOIL COLOR

Click on *color order system*, then *color notation:* http://www.munsell.com/

Soil colors have little effect on the behavior and use of soils; however, they provide valuable clues to the nature of other soil properties and conditions. Because of the importance of color in soil classification and interpretation, a standard system for accurate color description has been developed: the **Munsell color system** (carefully study Plate 22 after page 114). In this system, a small piece of soil is compared to standard color chips in a soil color book. Each color chip is described by the three components of color: the **hue** (in soils, usually redness or yellowness), the **chroma** (intensity or brightness, a chroma of 0 being neutral gray), and the **value** (lightness or darkness, a value of 0 being black).

Soils display a wide range of reds, browns, yellows, and even greens (see Plates 15 and 20, following page 114). Some soils are nearly black, others are nearly white. Some soil colors are very bright, others are dull grays. Soil colors may vary from place to place in the landscape, as when adjacent soils have differing surface horizon colors (e.g., Plate 16). Colors also typically change with depth through the various layers (horizons) within a soil profile. In many soils, the horizons in a given profile have colors that are similar in hue, but vary with respect to chroma and value. Even within a single horizon or clod of soil, colors may vary from spot to spot (see Plates 26 and 29).

Causes of Soil Colors

The color of a soil is influenced primarily by its content of organic matter and water, and by the presence and oxidation state of iron and manganese oxides. Organic matter (as in peat) is dark brown to black, and tends to darken and mask the brighter colors of mineral compounds such as iron oxides (see Plate 8 following page 114). Wet soils are generally darker in color than dry ones, but water levels have a more significant indirect effect on soil color: They influence the level of oxygen in the soil, and thereby the oxidation state of several elements, especially iron and manganese (see Plates 9 and 15). In well-drained uplands, especially in warm climates, well oxidized iron compounds impart bright (high chroma) reds and browns to the soil. These colors are in contrast to the gray and bluish colors (low chroma) that dominate where wet, low-oxygen conditions chemically reduce the iron in more poorly drained soil profiles (Plates 21 and 26). Other minerals that influence soil color are manganese oxide (black), glauconite (green), and calcite (imparts a whitish color to subsoils on semiarid regions) (see Plate 13).

4.2 SOIL TEXTURE (SIZE DISTRIBUTION OF SOIL PARTICLES)

The size of mineral particles in soil may seem too mundane a subject to warrant much attention, yet knowledge of the proportions of different-sized particles in a soil (i.e., the soil texture) is critical for understanding soil behavior and management. When investigating soils on a site, the texture of various soil horizons is often the first and most important property to determine, for a soil scientist can draw many conclusions from this information. Furthermore, the texture of a soil in the field is not readily subject to change, so it is considered a basic property of a soil.

Nature of Soil Separates

Soil texture and its assessment: www.soils.umn.edu/academics/classes/soil3125/doc/lab3txt.htm

Diameters of individual soil particles range over six orders of magnitude, from boulders (1 m) to submicroscopic clays ($<10^{-6}$ m). Scientists group these particles into **soil separates** according to several classification systems, as shown in Figure 4.1. The classification established by the U.S. Department of Agriculture is used in this textbook. The size ranges for these separates are not purely arbitrary, but reflect major changes in how the particles behave and in the physical properties they impart to soils.

Gravels, cobbles, boulders, and other **coarse fragments** greater than 2 mm in diameter may affect the behavior of a soil, but they are not considered to be part of the **fine earth fraction** to which the term *soil texture* properly applies.

SAND. Sand particles are those smaller than 2 mm but larger than 0.05 mm. They may be rounded or angular (Figure 4.2), depending on the extent to which they have been

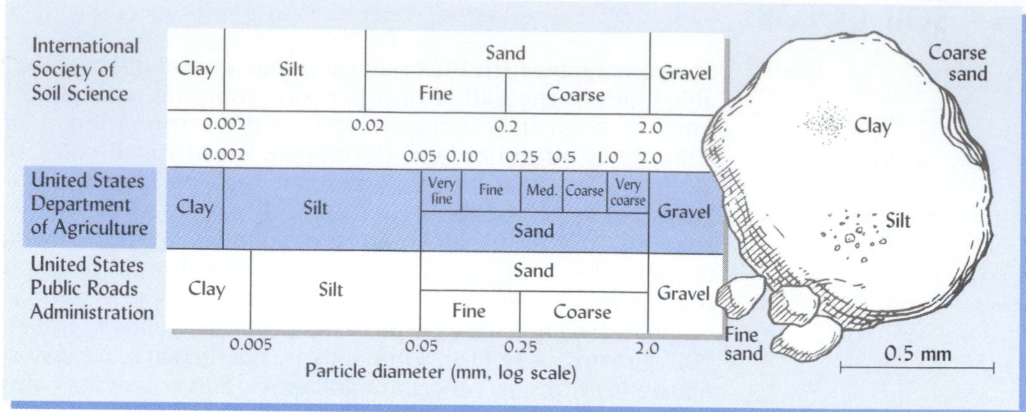

FIGURE 4.1 Classification of soil particles according to their size. The shaded scale in the center and the names on the drawings of particles follow the U.S. Department of Agriculture system, which is widely used throughout the world and in this book. The other two systems shown are also widely used by soil scientists and by highway construction engineers. The drawing illustrates the size of soil separates (note scale).

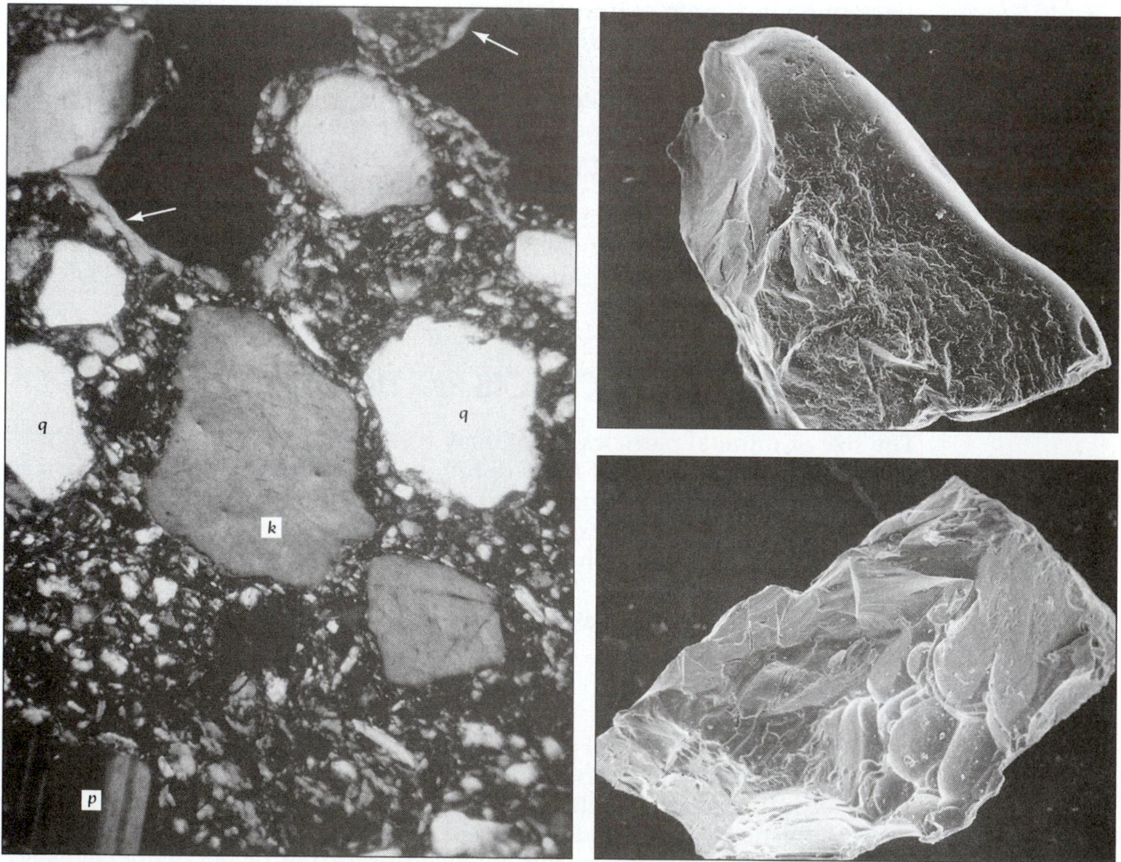

FIGURE 4.2 (Left) A thin section of a loamy soil as seen through a microscope using polarized light (empty pores appear black). Both the sand and silt particles shown are irregular in size and shape, the silt being only smaller. Although quartz (*q*) dominates the sand and silt fractions in this soil, several other silicate minerals can be seen (*p* = plagioclase, *k* = feldspar). Clay films can be seen to coat the walls of the large pores (arrow). Scanning electron micrographs of sand grains show quartz sand (bottom right) and a feldspar grain (upper right) magnified about 40 times. (Left photo courtesy of Martin Rabenhorst, University of Maryland; right photos courtesy of J. Reed Glasmann, Union Oil Research)

worn down by abrasive processes during soil formation. The coarsest sand particles may be rock fragments containing several minerals, but most sand grains consist of a single mineral, usually quartz (SiO_2) or other primary silicate (Figure 4.3). The dominance of quartz means that the sand separate generally has a far smaller total content of plant nutrients than do the finer separates.

Sand feels gritty between the fingers and the particles are generally visible to the naked eye. As sand particles are relatively large, so, too, the voids between them are relatively large and promote free drainage of water and entry of air into the soil. The relationship between particle size and **specific surface area** (the surface area for a given volume or mass of particles) is illustrated in Figure 4.4. Because of their large size, particles of sand have relatively low specific surface area. Therefore, sand particles can hold little water, and soils dominated by sand are prone to drought. Sand particles are considered noncohesive; that is, they do not tend to stick together in a mass (see Section 4.9).

SILT. Particles smaller than 0.05 mm but larger than 0.002 mm in diameter are classified as silt. Individual silt particles are not visible to the unaided eye (see Figure 4.2), nor do they feel gritty when rubbed between the fingers. These are essentially microsand particles, with quartz generally the dominant mineral. Where silt is composed of weatherable minerals, the smaller size of the particles allows weathering to proceed rapidly enough to release significant amounts of plant nutrients.

Although silt consists of particles similar in shape to sand, it feels smooth or silky, like flour. The pores between silt particles are much smaller (and much more numerous) than those in sand, so silt retains more water and lets less drain through. However, even when wet, silt itself does not exhibit much **stickiness** or **plasticity** (malleability). What little plasticity, **cohesion** (stickiness), and adsorptive capacity some silt fractions exhibit is largely due to a film of adhering clay.

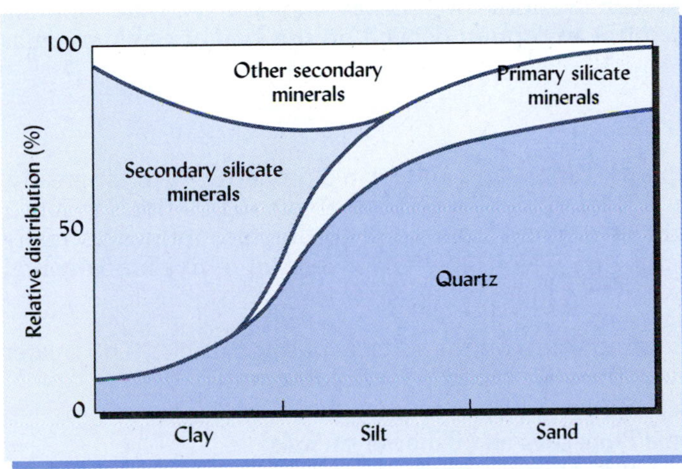

FIGURE 4.3 General relationship between particle size and kinds of minerals present. Quartz dominates the sand and coarse silt fractions. Primary silicates such as the feldspars, hornblende, and micas are present in the sands and, in decreasing amounts, in the silt fraction. Secondary silicates dominate the fine clay. Other secondary minerals, such as the oxides of iron and aluminum, are prominent in the fine silt and coarse clay fractions.

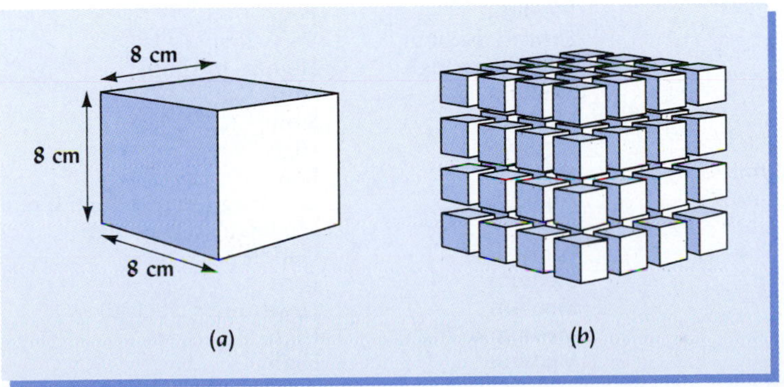

FIGURE 4.4 The relationship between the surface area of a given mass of material and the size of its particles. In the single large cube (*a*) each face has 64 cm^2 of surface area. The cube has six faces, so the cube has a total of 384 cm^2 surface area (6 faces $\times$ 64 cm^2 per face). If the same cube of material was cut into smaller cubes (*b*) so that each cube was only 2 cm on each side, then the same mass of material would now be present as 64 smaller cubes ($4 \times 4 \times 4$). Each face of each small cube would have 4 cm^2 (2×2 cm) of surface area, giving 24 cm^2 of surface area for each cube (6 faces $\times$ 4 cm^2 per face). The total mass would therefore have 1536 cm^2 (24 cm^2 per cube $\times$ 64 cubes) of surface area. This is four times as much surface area as the single large cube. Since clay particles are very, very small and usually platelike in shape, their surface area is thousands of times greater than that of the same mass of sand particles.

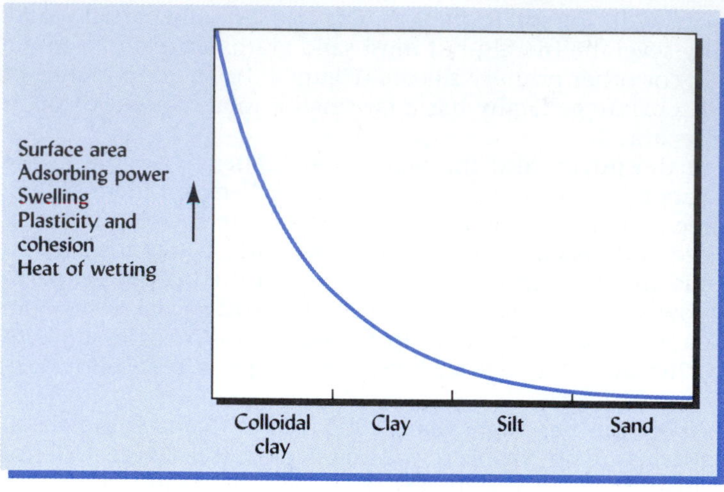

FIGURE 4.5 The finer the texture of a soil, the greater is the effective surface exposed by its particles. Note that adsorption, swelling, and the other physical properties cited follow the same general trend and that their intensities go up rapidly as the colloidal size is approached.

CLAY. Particles smaller than 0.002 mm are classified as clay and have a very large specific surface area, giving them a tremendous capacity to adsorb water and other substances. A spoonful of clay may have a surface area the size of a football field (see Section 8.1). This large adsorptive surface causes clay particles to cohere in a hard mass after drying. When wet, clay is sticky and can be easily molded.

Clay size particles are so small that they behave as **colloids**—if suspended in water they do not readily settle out. Unlike most sand and silt particles, clay particles tend to be shaped like tiny flakes or flat platelets. The pores between clay particles are very small and convoluted, so movement of both water and air is very slow. Each unique clay mineral (see Chapter 8) imparts very different properties to the soils in which it is prominent. Therefore, soil properties such as shrink-swell behavior, plasticity, water-holding capacity, soil strength, and chemical adsorption depend on the *kind* of clay present as well as the *amount.*

Influence of Surface Area on Other Soil Properties

When particle size decreases, specific surface area and related properties increase greatly, as shown graphically in Figure 4.5. Fine colloidal clay has about 10,000 times as much surface area as the same weight of medium-sized sand. Soil texture influences many other soil properties in far-reaching ways (see Table 4.1) as a result of five fundamental surface phenomena:

1. Water is retained in soils as thin films on the surfaces of soil particles. The greater the surface area, the greater the soil's capacity for holding water.

TABLE 4.1 Generalized Influence of Soil Separates on Some Properties and Behavior of Soils[a]

	Rating associated with soil separates		
Property/behavior	*Sand*	*Silt*	*Clay*
Water-holding capacity	Low	Medium to high	High
Aeration	Good	Medium	Poor
Drainage rate	High	Slow to medium	Very slow
Soil organic matter level	Low	Medium to high	High to medium
Decomposition of organic matter	Rapid	Medium	Slow
Warm-up in spring	Rapid	Moderate	Slow
Compactability	Low	Medium	High
Susceptibility to wind erosion	Moderate (high if fine sand)	High	Low
Susceptibility to water erosion	Low (unless fine sand)	High	Low if aggregated, high if not
Shrink-swell potential	Very Low	Low	Moderate to very high
Sealing of ponds, dams, and landfills	Poor	Poor	Good
Suitability for tillage after rain	Good	Medium	Poor
Pollutant leaching potential	High	Medium	Low (unless cracked)
Ability to store plant nutrients	Poor	Medium to high	High
Resistance to pH change	Low	Medium	High

[a]Exceptions to these generalizations do occur, especially as a result of soil structure and clay mineralogy.

2. Both gases and dissolved chemicals are attracted to and adsorbed by mineral particle surfaces. The greater the surface area, the greater the soil's capacity to retain nutrients and other chemicals.
3. Weathering takes place at the surface of mineral particles, releasing constituent elements into the soil solution.
4. The surfaces of mineral particles often carry both negative and some positive electromagnetic charges so that particle surfaces and the water films between them tend to attract each other (see Section 4.7).
5. Microorganisms tend to grow on and colonize particle surfaces.

4.3 SOIL TEXTURAL CLASSES

Textural classes of Florida soils:
http://edis.ifas.ufl.edu/BODY_SS169

Textural class names convey an idea of the size distribution of particles in a soil, and indicate the general physical properties of that soil. The range in percentages of sand, silt, and clay for the 12 major textural classes are shown in the triangle in Figure 4.6. A study of this figure shows a graduated sequence from the sands, which are coarse in texture and easy to move about, to the clays, which are very fine and difficult to handle physically. Note that sands and loamy sands are dominated by the properties of sand, for the sand separate comprises at least 70% of the material by weight (less than 15% of the material is clay). Characteristics of the clay separate are distinctly dominant in clays, sandy clays, and silty clays.

Loams. The loam group contains many subdivisions. An ideal **loam** may be defined as a mixture of sand, silt, and clay particles that exhibits the *properties* of those separates in about equal proportions. This definition does not mean that the three separates are present in equal *amounts* (as will be revealed by careful study of Figure 4.6). This anomaly exists because a relatively small percentage of clay is required to engender clayey properties in a soil, whereas small amounts of sand and silt have a lesser influence on how a soil behaves.

Most soils are some type of loam. They may possess the ideal makeup of equal proportions previously described and be classed simply as loam. However, a loam in which sand is dominant is classified as a *sandy loam.* In the same way, there may occur silt loams, silty clay loams, sandy clay loams, and clay loams.

Coarse Fragment Modifiers. For some soils, qualifying factors such as stone, gravel, and the various grades of sand become part of the textural class name. Fragments that range from 2 to 75 mm along their greatest diameter are termed *gravel* or *pebbles,* those ranging from 75 to 250 mm are called *cobbles* (if round) or *flags* (if flat), and those more than 250 mm across are called *stones* or *boulders*. A *cobbly, fine sandy loam* is an example of such a modified textural class.

Alteration of Soil Textural Class

Over very long periods of time, pedologic processes (see Chapter 2) such as erosion, deposition, illuviation, and weathering can alter the textures of various soil horizons. However, management practices generally do not alter the textural class of a soil on a field scale. The texture of a given soil can be changed only by mixing it with another soil of a different textural class. For example, the incorporation of large quantities of sand to change the physical properties of a clayey soil for use in greenhouse pots or for turf would bring about such a change. However, where specifications (as for a landscape design) call for soil materials of a certain textural class, it is advisable to find a naturally occurring soil that meets the specification, rather than attempt to alter the textural class by mixing in sand or clay.

Note also that adding peat or compost to a soil while mixing a potting medium does not constitute a change in texture, as this property refers only to the mineral particles. The term *soil texture* is not relevant to artificial media that contain mainly perlite, peat, styrofoam, or other nonsoil materials.

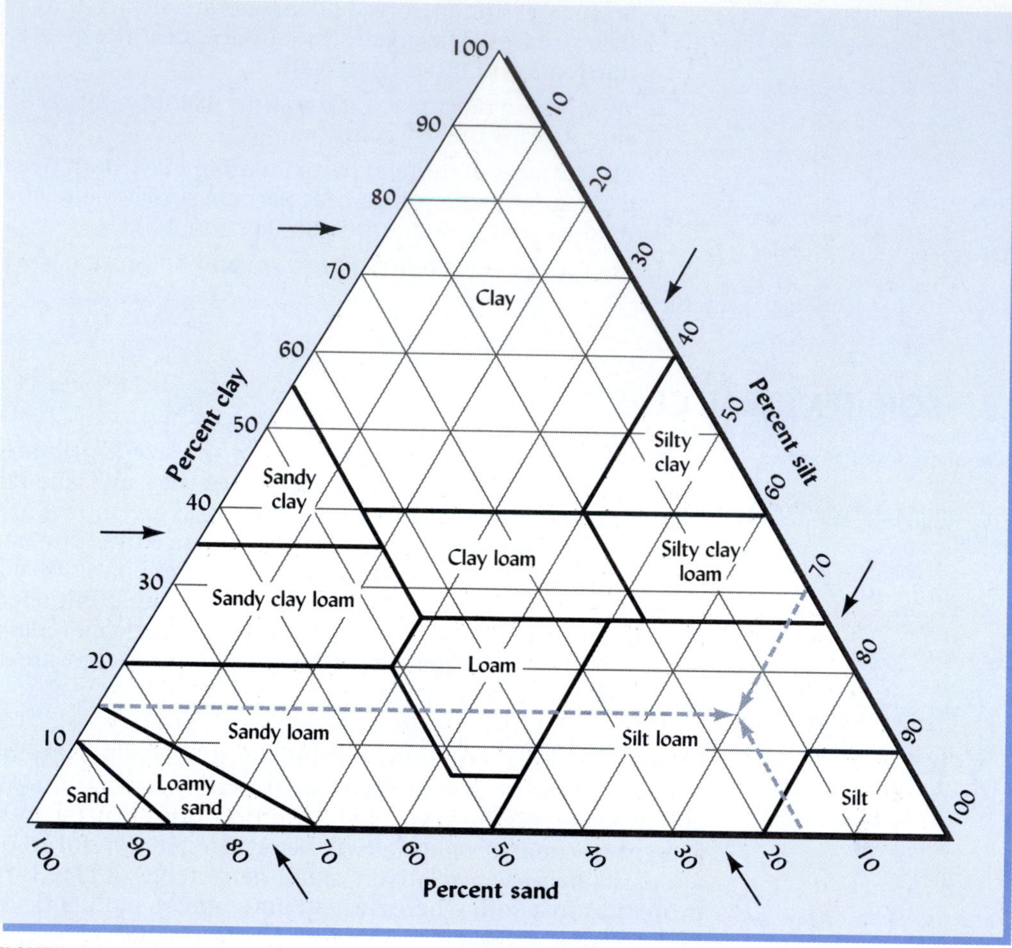

FIGURE 4.6 The major soil textural classes are defined by the percentages of sand, silt, and clay according to the heavy boundary lines shown on the textural triangle. If these percentages have been determined for a soil sample by particle size analysis, then the triangle can be used to determine the soil textural class name that applies to that soil sample. To use the graph, first find the appropriate clay percentage along the left side of the triangle, then draw a line from that location across the graph going parallel to the base of the triangle. Next find the sand percentage along the base of the triangle, then draw a line inward going parallel to the triangle side labeled "Percent silt." The small arrows indicate the proper direction in which to draw the lines. The name of the compartment in which these two lines intersect indicates the textural class of the soil sample. Percentages for any two of the three soil separates is all that is required. Because the percentages for sand, silt, and clay add up to 100%, the third percentage can easily be calculated if the other two are known. If all three percentages are used, the three lines will all intersect at the same point. Consider, as an example, a soil that has been determined to contain 15% sand, 15% clay, and 70% silt. This example is indicated by the light dashed lines that intersect in the compartment labeled "Silt loam." What is the textural class of another soil sample that has 33% sand, 33% silt, and 33% clay? The lines (not shown) for this second example would intersect in the center of the "Clay loam" compartment.

Determination of Textural Class by the "Feel" Method

The textural class of a soil is commonly determined in the field by the *feel* method. This involves rubbing a sample of the soil, usually in a moist or wet condition, between the thumb and fingers as illustrated in Box 4.1 and Figure 4.7. The feel method for determining textural class is one of the first field skills a soil scientist should develop. This skill is of great practical value in soil survey, land classification, and any investigation in which soil texture may play a role. Accuracy depends largely on experience, so practice whenever you can, beginning with soils of known texture to calibrate your fingers. Keep in mind the textural triangle in Figure 4.6 as you use the feel method to ascertain the textural class.

BOX 4.1 A METHOD FOR DETERMINING TEXTURE BY FEEL

The first, and most critical, step in the texture-by-feel method is to knead a walnut-sized sample of moist soil into a uniform puttylike consistency, slowly adding water if necessary. This step may take a few minutes, but a premature determination is likely to be in error as hard clumps of clay and silt may feel like sand grains. The soil should be moist, but not quite glistening. Try to do this with only one hand so as to keep your other hand clean for writing in a field notebook (and shaking hands with your client).

While squeezing and kneading the sample, note its maleability, stickiness, and stiffness, all properties associated with the clay content. A high silt content makes a sample feel smooth and silky, with little stickiness or resistance to deformation. A soil with a significant content of sand feels rough and gritty, and makes a grinding noise when rubbed near one's ear.

Get a feel for the amount of clay by attempting to squeeze a ball of properly moistened soil between your thumb and the side of your forefinger, making a ribbon of soil. Make the ribbon as long as possible until it breaks from its own weight (see Figure 4.7).

Interpret your observations as follows:

1. Soil will not cohere into a ball, falls apart: **sand**
2. Soil forms a ball, but will not form a ribbon: **loamy sand**
3. Soil ribbon is dull and breaks off when less than 2.5 cm long and
   a. Grinding noise is audible; grittiness is prominent feel: **sandy loam**
   b. Smooth, floury feel prominent; no grinding audible: **silt loam**
   c. Only slight grittiness and smoothness; grinding not clearly audible: **loam**
4. Soil exhibits moderate stickiness and firmness, forms ribbons 2.5 to 5 cm long, and
   a. Grinding noise is audible; grittiness is prominent feel: **sandy clay loam**
   b. Smooth, floury feel prominent; no grinding audible: **silty clay loam**
   c. Only slight grittiness and smoothness; grinding not clearly audible: **clay loam**
5. Soil exhibits dominant stickiness and firmness, forms shiny ribbons longer than 5 cm, and
   a. Grinding noise is audible; grittiness is dominant feel: **sandy clay**
   b. Smooth, floury feel prominent; no grinding audible: **silty clay**
   c. Only slight grittiness and smoothness; grinding not clearly audible: **clay**

A more precise estimate of sand content (and hence more accurate placement in the horizontal dimension of the textural class triangle) can be made by wetting a pea-sized clump of soil in the palm of your hand and smearing it around with your finger until your palm becomes coated with a souplike suspension of soil. The sand grains will stand out visibly and their volume as compared to the original "pea" can be estimated, as can their relative size (fine, medium, coarse, etc.).

It is best to learn the method using samples of known textural class. With practice, accurate textural class determinations can be made on the spot.

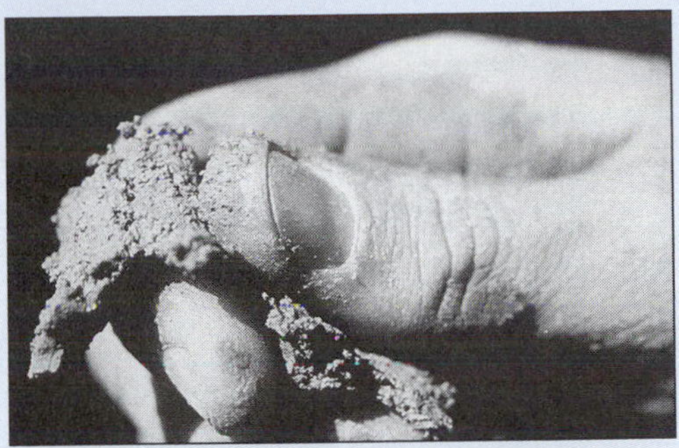

FIGURE 4.7 *The "feel" method for determining soil textural class. A moist soil sample is rubbed between the thumb and forefingers and squeezed out to make a "ribbon." (Top) The gritty, noncohesive appearance and short ribbon of a sandy loam with about 15% clay. (Middle) The smooth, dull appearance and crumbly ribbon characteristic of a silt loam. (Bottom) The smooth, shiny appearance and long, flexible ribbon of a clay. (Photos courtesy of R. Weil)*

Laboratory Particle-Size Analyses

Particle analyses are also made in the laboratory. A sample of soil is completely dispersed in water and passed through a set of sieves to separate out the sand fractions. A sedimentation procedure is then used to ascertain the quantities of silt and clay that had passed through the sieves. The principle involved is simple. Because soil particles are more dense than water, they tend to sink, settling at a velocity that is proportional to their size. In other words: "The bigger they are, the faster they fall." The equation that

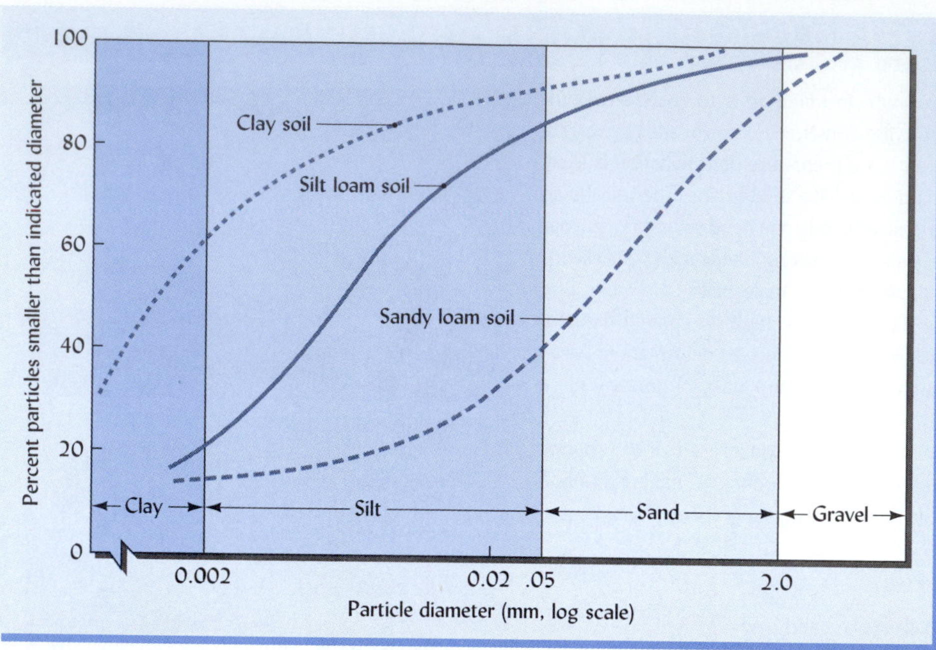

FIGURE 4.8 Particle-size distribution in three soils varying widely in their textures. Note that there is a gradual transition in the particle-size distribution in each of these soils.

Soil texture and mechanical analysis illustrated: http://www.soils.umn.edu/academics/classes/soil2125/doc/s6chap1.htm

describes this relationship is referred to as *Stokes' law.* In its simplest form it tells us that the velocity of settling V is proportional to the square of a particle's diameter d

$$V = kd^2$$

where k is a constant related to the acceleration due to gravity and the density and viscosity of water. By measuring the amount of soil still in suspension after various amounts of settling time, the percentages of each size fraction can be determined so as to identify the soil textural class and generate particle-size distribution curves such as those shown in Figure 4.8.

Figure 4.8 presents particle-size distribution curves for soils representative of three textural classes. The fact that these curves are smooth emphasizes that there is no sharp line of demarcation in the distribution of sand, silt, and clay fractions, and suggests a gradual change of properties with change in particle size.

It is important to note that soils are assigned to textural classes *solely* on the basis of the mineral particles of sand size and smaller; therefore, the percentages of sand, silt, and clay always add up to 100%. The amounts of stone and gravel are rated separately. Organic matter is usually removed from a soil sample by oxidation before the mechanical separation.

The relationship between such analyses and textural class names is commonly shown diagrammatically as a triangular graph (see Figure 4.6). This *textural triangle* also enables us to use laboratory particle-size analysis data to check the accuracy of field textural determinations by feel.

4.4 STRUCTURE OF MINERAL SOILS

Photos and drawing of structural types: http://ltpwww.gsfc.nasa.gov/globe/pvg/prop1.htm

The term **structure** relates to the arrangement of primary soil particles into groupings called **aggregates** or **peds.** The pattern of pores and peds defined by soil structure greatly influences water movement, heat transfer, aeration, and porosity in soils. Activities such as timber harvesting, grazing, tillage, trafficking, drainage, liming, and manuring impact soils largely through their effect on soil structure, especially in the surface horizons.

The processes involved in the formation, stabilization, and management of soil structure will be discussed in Sections 4.7 and 4.8. Here we will examine the nature of the various types of structural peds.

Many types of structural peds occur in soils, often within different horizons of a particular soil profile. Soil structure is characterized in terms of the shape (or *type*), size, and distinctness (or *grade*) of the peds. The four principal shapes of soil structure are *spheroidal, platy, prismlike,* and *blocklike.* A study of Figure 4.9 will help you visualize these structural types (and some subtypes).

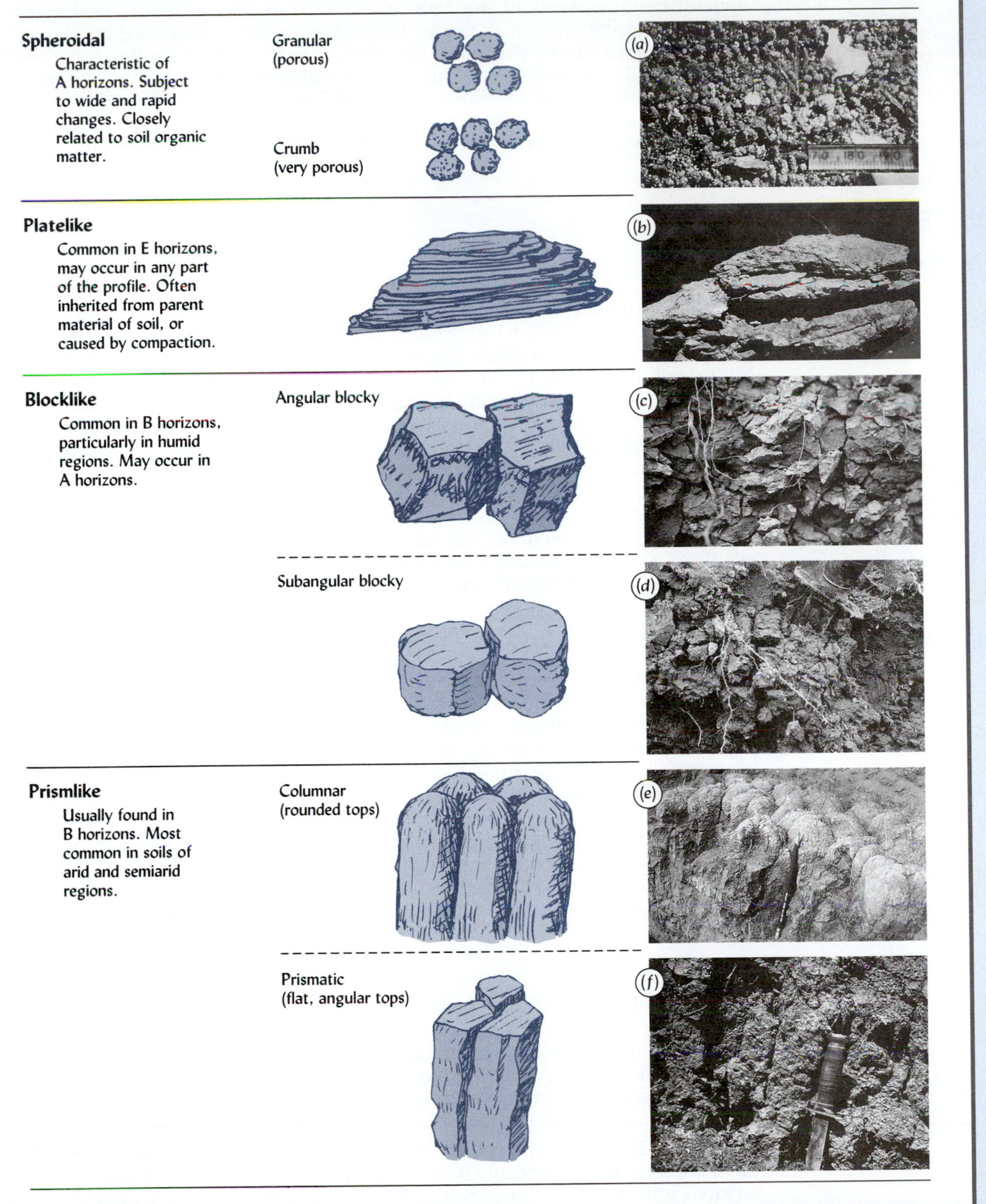

Spheroidal
Characteristic of A horizons. Subject to wide and rapid changes. Closely related to soil organic matter.
Granular (porous)
Crumb (very porous)
(a)
Platelike
Common in E horizons, may occur in any part of the profile. Often inherited from parent material of soil, or caused by compaction.
(b)
Blocklike
Common in B horizons, particularly in humid regions. May occur in A horizons.
Angular blocky
(c)
Subangular blocky
(d)
Prismlike
Usually found in B horizons. Most common in soils of arid and semiarid regions.
Columnar (rounded tops)
(e)
Prismatic (flat, angular tops)
(f)

FIGURE 4.9 The various structure types (shapes) found in mineral soils. Their typical location is suggested. The drawings illustrate their essential features and the photos indicate how they look in situ. For scale, note the 15-cm-long pencil in (*e*) and the 3-cm-wide knife blade in (*d*) and (*f*). [Photo (*e*) courtesy of J. L. Arndt, now with Petersen Environmental Consulting; North Dakota State University. Others courtesy of R. Weil]

Soil Under a Microscope—micromorphology of soil structure:
http://www.nrcs.usda.gov/technical/worldsoils/microscope

In describing soil structure, soil scientists note not only the *type* (shape) of the structural peds present, but also the relative *size* (fine, medium, coarse) and degree of development or distinctness of the peds (*grades* such as strong, moderate, or weak). For example, the soil horizon shown in Figure 4.9*d* might be described as having "weak, fine, subangular blocky structure." Generally, the structure of a soil is easier to observe when the soil is relatively dry. When wet, structural peds may swell and press closer together, making the individual peds less well defined. In any case, the structural arrangement of soil particles and the pore spaces between structural peds greatly influence soil density, an aspect of soil architecture that we will now examine in detail.

4.5 SOIL DENSITY

Particle Density

Soil **particle density** D_p is defined as the mass per unit volume of soil *solids* (in contrast to the volume of the *soil*, which would also include spaces between particles). Thus, if 1 cubic meter (m^3) of soil solids weighs 2.6 megagrams (Mg), the particle density is 2.6 Mg/m^3 (which can also be expressed as 2.6 grams per cubic centimeter).

Particle density is essentially the same as the **specific gravity** of a solid substance. The chemical composition and crystal structure of a mineral determines its particle density. Particle density is *not* affected by pore space, and therefore is not related to particle size or to the arrangement of particles (soil structure).

Particle densities for most mineral soils vary between the narrow limits of 2.60 to 2.75 Mg/m^3 because quartz, feldspar, micas, and the colloidal silicates that usually comprise the major portion of mineral soils have densities within this range. For general calculations concerning arable mineral surface soils (1 to 5% organic matter), a particle density of about 2.65 Mg/m^3 may be assumed if the actual particle density is not known.

Bulk Density

A second important mass measurement of soils is **bulk density** D_b, which is defined as the mass of a unit volume of dry soil. This volume includes both solids and pores. A careful study of Figure 4.10 should make clear the distinction between particle and bulk density. Both expressions of density use only the mass of the solids in a soil; therefore, any water present is excluded from consideration.

There are several methods of determining soil bulk density by obtaining a known volume of soil, drying it to remove the water, and weighing the dry mass. A special coring instrument (Figure 4.11) can obtain a sample of known volume without disturbing the natural soil structure. For surface soils, perhaps the simplest method is to dig a small hole, dry and weigh all the excavated soil, and then determine the soil volume by lining the hole with plastic film and filling it completely with a measured volume of water.

Factors Affecting Bulk Density

Soils with a high proportion of pore space to solids have lower bulk densities than those that are more compact and have less pore space. Consequently, any factor that influences soil pore space will affect bulk density. Typical ranges of bulk density for various soil materials and conditions are illustrated in Figure 4.12. It would be worthwhile to study this figure until you have a good feel for these ranges of bulk density.

EFFECT OF SOIL TEXTURE. As illustrated in Figure 4.12, fine-textured soils such as silt loams, clays, and clay loams generally have lower bulk densities than do sandy soils.[1] In fine-textured soils, the solid particles tend to be organized in porous granules, especially if

[1] This fact may seem counterintuitive at first because sandy soils are commonly referred to as "light" soils, while clays and clay loams are referred to as "heavy" soils. The terms *heavy* and *light* in this context refer not to the mass per unit volume of the soils, but to the amount of effort that must be exerted to manipulate these soils with tillage implements—the sticky clays being much more difficult to till.

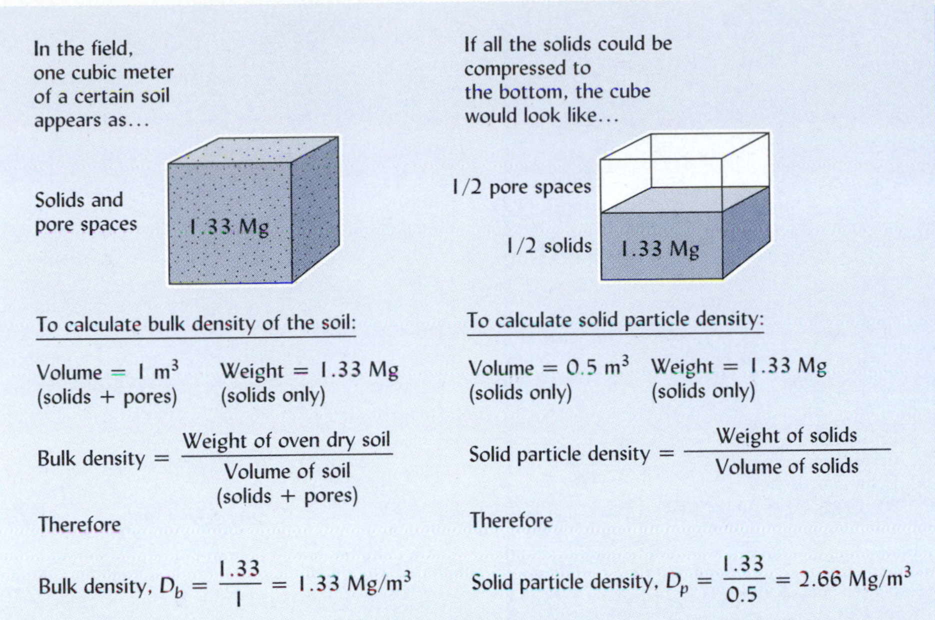

In the field, one cubic meter of a certain soil appears as...
If all the solids could be compressed to the bottom, the cube would look like...
Solids and pore spaces
1.33 Mg
1/2 pore spaces
1/2 solids
1.33 Mg
To calculate bulk density of the soil:
Volume = 1 m³ (solids + pores)
Weight = 1.33 Mg (solids only)
Bulk density = Weight of oven dry soil / Volume of soil (solids + pores)
Therefore
Bulk density, Db = 1.33 / 1 = 1.33 Mg/m³
To calculate solid particle density:
Volume = 0.5 m³ (solids only)
Weight = 1.33 Mg (solids only)
Solid particle density = Weight of solids / Volume of solids
Therefore
Solid particle density, Dp = 1.33 / 0.5 = 2.66 Mg/m³

FIGURE 4.10 Bulk density D_b and particle density D_p of soil. Bulk density is the weight of the solid particles in a standard volume of field soil (solids plus pore space occupied by air and water). Particle density is the weight of solid particles in a standard volume of those solid particles. Follow the calculations through carefully and the terminology should be clear. In this particular case, the bulk density is one-half the particle density, and the percent pore space is 50.

adequate organic matter is present. As can be seen in Figure 4.13, there are pore spaces both within and between the granules of these soils. Consequently, they have higher total pore space and lower bulk density than their sandy counterparts.

The bulk density of different sandy soils varies somewhat, as it is affected by the size and packing arrangement of the sand grains (see Figure 4.14). Where the grains are fairly uniform in size they are more apt to be loosely packed, resulting in higher total pore space and lower bulk density. If the grains vary greatly in size, however, the small grains can fit in pores around the larger grains, thereby permitting tighter packing, lower pore space, and higher bulk density.

Depth in Soil Profile. Deeper in the soil profile, bulk densities are generally higher, probably as a result of lower organic matter contents, less aggregation, fewer roots and other soil-dwelling organisms, and compaction caused by the weight of the overlying layers. Very compact subsoils may have bulk densities of 2.0 Mg/m^3 or even greater. Many soils formed from glacial till (see Section 2.3) have extremely dense subsoils as a result of past compaction by the enormous mass of ice.

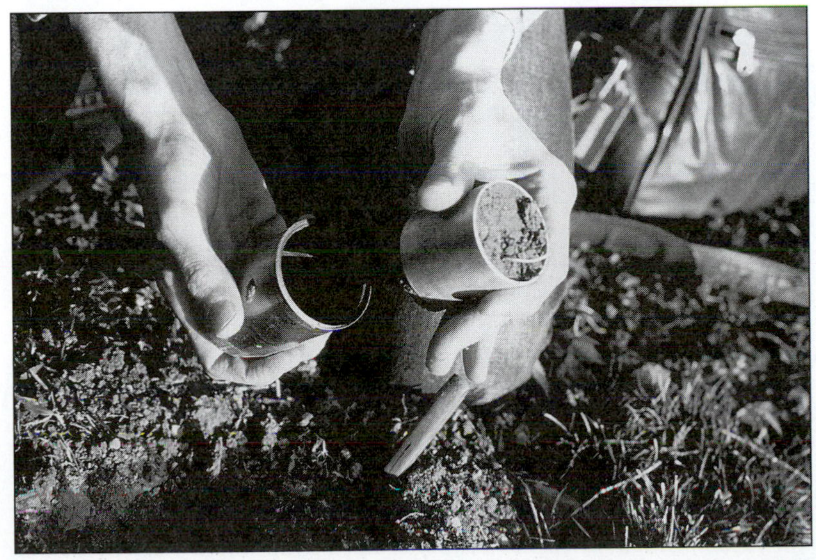

FIGURE 4.11 A special sampler designed to remove a cylindrical core of soil without causing disturbance or compaction. The sampler head contains an inner cylinder and is driven into the soil with blows from a drop hammer. The inner cylinder containing an undisturbed soil core is then removed and trimmed on the end with a knife to yield a core whose volume can easily be calculated from its length and diameter. The weight of this soil core is then determined after drying in an oven. (Photo courtesy of R. Weil)

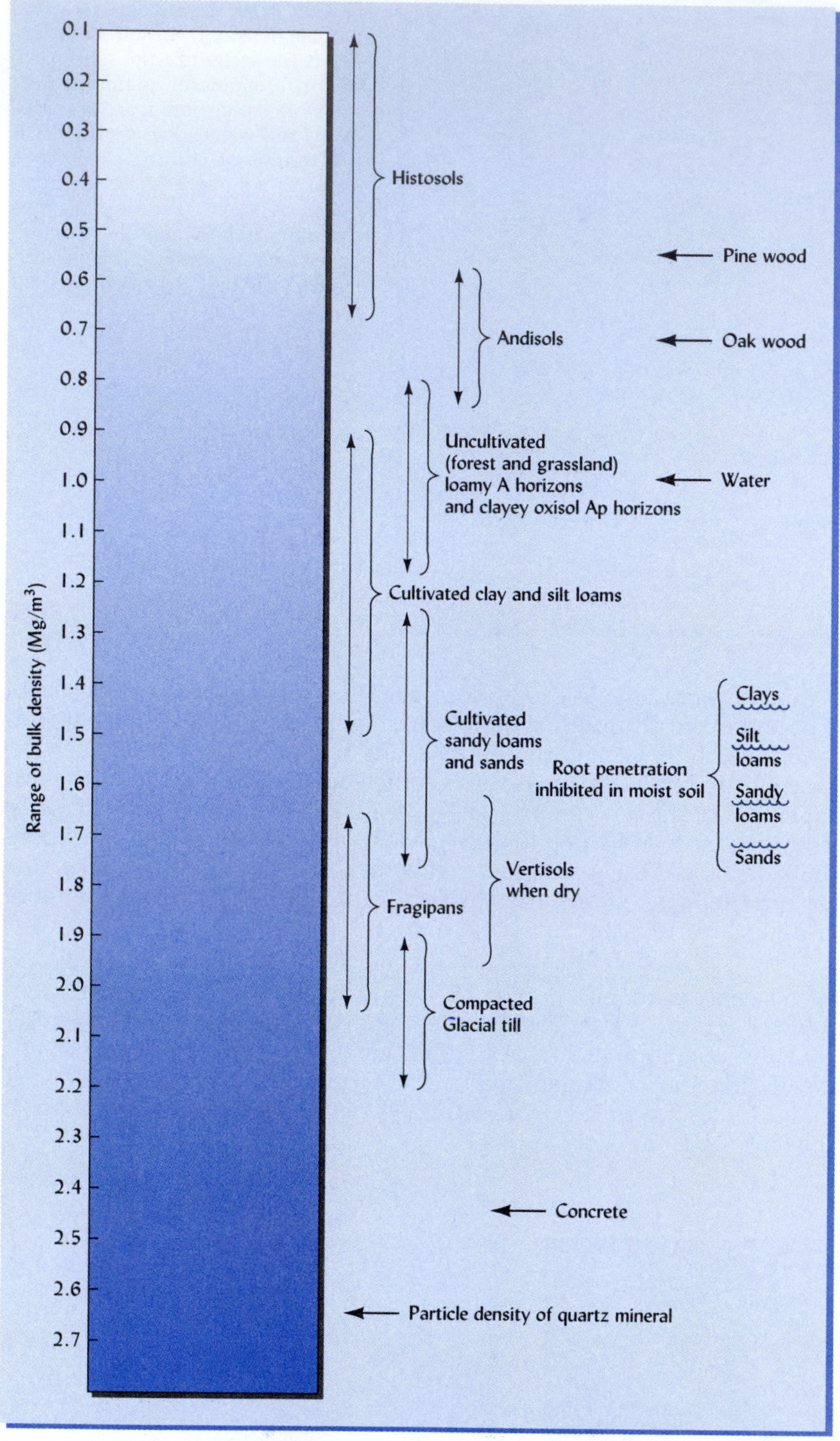

FIGURE 4.12 Bulk densities typical of a variety of soils and soil materials.

Useful Density Figures

For engineers involved with moving soil during construction, or for landscapers bringing in topsoil by the truckload, a knowledge of the bulk density of various soils is useful in estimating the weight of soil to be moved. A typical medium-textured mineral soil might have a bulk density of 1.25 Mg/m^3, or 1250 kg in a cubic meter.[2] People are often

[2] Most commercial landscapers and engineers in the United States still use English units. To convert values of bulk density given in units of Mg/m^3 into values of lb/yd^3, multiply by 1686. Therefore, 1 yd^3 of a dry medium-textured mineral soil with a bulk density of 1.25 Mg/m^3 would weigh over a ton ($1686 \times 1.25 = 2108$ lb/yd^3).

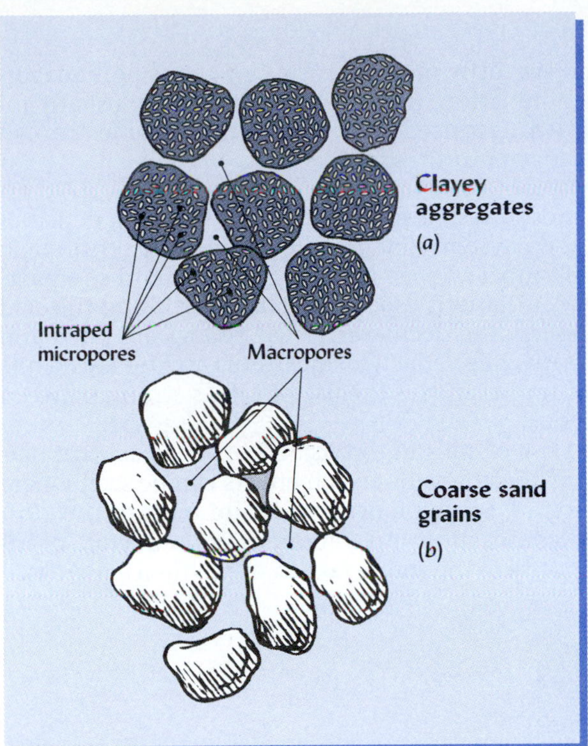

FIGURE 4.13 A schematic comparison of sandy and clayey soils showing the relative amounts of large (macro-) pores and small (micro-) pores in each. There is less total pore space in the sandy soils than in the clayey soils because the clayey soil contains a large number of fine pores within each aggregate (*a*), but the sand particles (*b*), while similar in size to the clayey aggregates, are solid and contain no pore spaces within them. This is the reason why, among surface soils, those with coarse texture are usually more dense than those with fine textures.

surprised by how heavy soil is. Imagine driving your pickup truck to a nursery where natural topsoil is sold in bulk and filling your truck bed with a nice, rounded load. Of course, you would not really want to do this, as the load might break your rear axle; you certainly would not be able to drive away with the load. A typical "half-ton" (1000 lb or 454 kg) load capacity pickup truck could carry less than 0.4 m^3 of this soil, even though the truck bed has room for about five times this volume of material.

The mass of soil in 1 ha to a depth of normal plowing (15 cm) can be calculated from soil bulk density. If we assume a bulk density of 1.3 Mg/m^3 for a typical arable surface soil, such a hectare–furrow slice 15 cm deep weighs about 2 million kg.[3] This estimate of the mass of surface soil in a hectare of land is very useful in calculating lime and fertilizer application rates and organic matter mineralization rates (see Boxes 8.1, 9.2, and 12.1 for detailed examples). However, this estimated mass must be adjusted if the bulk density is other than 1.3 Mg/cm^3 or the depth of the layer under consideration is more or less than 15 cm.

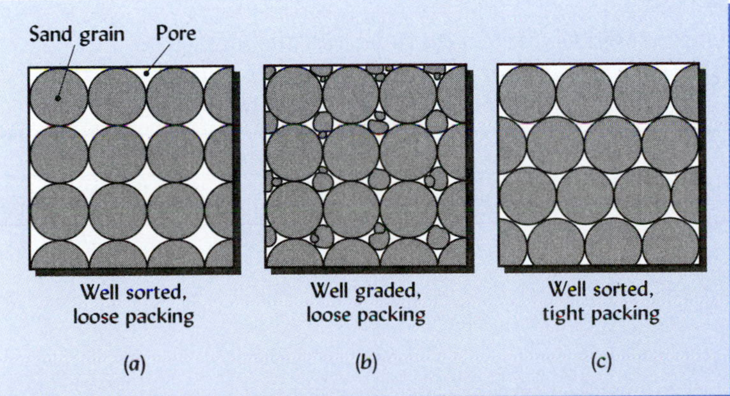

FIGURE 4.14 The uniformity of grain size and the type of packing arrangement significantly affect the bulk density of sandy materials. Materials consisting of all similar-sized grains are termed *well sorted* (or poorly graded). Those with a variety of grain sizes are *well graded* (or poorly sorted). In either case, compaction of the particles into a tight packing arrangement markedly increases the bulk density of the material and decreases its porosity. Note that the size distribution of sand and gravel particles can be described as either graded or sorted, the two terms having essentially opposite meanings. Geologists usually speak of rivers having sorted the sand grains by size as deposits were laid down. Engineers usually are concerned as to whether or not sand consists of a gradation of sizes (i.e., is well graded or not).

[3] 10,000 m^2/ha × 1.3 Mg/m^3 × 0.15 m = 1950 Mg/ha, or about 2 million kg per ha 15 cm deep. A comparable figure in the English system is 2 million lb per acre–furrow slice 6 to 7 in. deep.

Management Practices Affecting Bulk Density

Changes in bulk density for a given soil are easily measured and can alert soil managers to changes in soil quality and ecosystem function. Increases in bulk density usually indicate a poorer environment for root growth, reduced aeration, and undesirable changes in hydrologic function, such as reduced water infiltration.

Forest Lands. The surface horizons of most forested soils have rather low bulk densities (see Figure 4.12). Tree growth and forest ecosystem function are particularly sensitive to increases in bulk density. Conventional timber harvest generally disturbs and compacts 20% to 40% of the forest floor (Figure 4.15) and is especially damaging along the skid trails where logs are dragged and at the landing decks—areas where logs are piled and loaded onto trucks (Table 4.2). An expensive, but effective, means of moving logs while minimizing compactive degradation of forest lands is the use of cables strung between towers or hung from large balloons.

Intensive recreational and transport use of soils in forests and other areas with natural vegetation can also lead to increased bulk densities. Such effects can be seen where access roads, trails, and campsites are found. Some pioneer roads and trails of the 19th century still show the ill effects of soil compaction. An important consequence of increased bulk density is a diminished capacity of the soil to take in water, hence increased

FIGURE 4.15 Timber harvest with a conventional rubber-tired skidder in a boreal forest in western Alberta, Canada results in disturbance to the surface soil horizons (forest floor). Such practices cause significant soil compaction that can impair soil ecosystem functions for many years. Timber harvest practices that can reduce such damage to forest soils include selective cutting, use of flexible-track vehicles and overhead cable transport of logs, and abstaining from harvest during wet conditions. (Photos courtesy of Andrei Startsev, Alberta Environmental Center)

TABLE 4.2 Effects of Timber Harvest on Bulk Density at Different Depths in Two Forested Ultisols in Georgia

Rubber-wheeled skidders were used to harvest the logs. Note the generally higher bulk densities of the sandy loam soil compared to the clay loam, and the greater effect of timber harvest on the skidder trails.

	Bulk density, Mg/m³		
Soil Depth, cm	*Preharvest*	*Postharvest, off trails*	*Postharvest, skidder trails*
Upper coastal plain, sandy loam			
0–8	1.25	1.50	1.47
8–15	1.40	1.55	1.71
15–23	1.54	1.61	1.81
23–30	1.58	1.62	1.77
Piedmont, clay loam			
0–8	1.16	1.36	1.52
8–15	1.39	1.49	1.67
15–23	1.51	1.51	1.66
23–30	1.49	1.46	1.61

Data from Gent et al. (1984, 1986).

losses by surface runoff. Damage from hikers can be minimized by restricting foot traffic to well-designed, established trails that may include a thick layer of wood chips, or even a raised boardwalk in the case of heavily traveled paths over very fragile soils, such as in wetlands.

Soil compaction problems in cities:
http://www.stormwatercenter.net/Practice/36-The%20Compaction%20of%20urban%20Soils.pdf

Urban Areas. Trees planted for landscaping purposes must often contend with severely compacted soils. While it is usually not practical to modify the entire root zone of a tree, several practices can help. First, making the planting hole as large as possible will provide a zone of loose soil for early root growth. Second, a thick layer of mulch spread out to the drip line (but not too near the trunk) will enhance root growth, at least near the surface. Third, the tree roots may be given paths for expansion by digging a series of narrow trenches radiating out from the planting hole and backfilled with loose, enriched soil. It may be desirable to create an "artificial soil" that includes a skeleton of coarse angular gravel to provide stability, and a mixture of loam-textured topsoil and organic matter to provide nutrient- and water-holding capacities.

Agricultural Land. Although tillage may temporarily loosen the surface soil, in the long term, intense tillage increases soil bulk density because it depletes soil organic matter and weakens soil structure (see Section 4.8). The effect of cultivation can be minimized by adding crop residues or farm manure, and by rotating cultivated crops with a grass sod or tap-rooted cover crops.

Heavy machines used on wet soils to pull implements, apply amendments, or harvest crops can create yield-limiting soil compaction. Certain tillage implements, such as the moldboard plow, compact the soil below their working depth even as they lift and loosen the soil above. This results in **plow pans** (or *traffic pans*), dense zones immediately below the plowed layer (Figure 4.16). Other tillage implements, such as the chisel plow and the spring-tooth harrow, do not press down upon the soil beneath them, and so are useful in breaking up plow pans and stirring the soil with a minimum of compaction. Large chisel-type plows (Figure 4.17) can be used in **subsoiling** to break up dense subsoil layers, thereby permitting root penetration (Figure 4.18). However, in

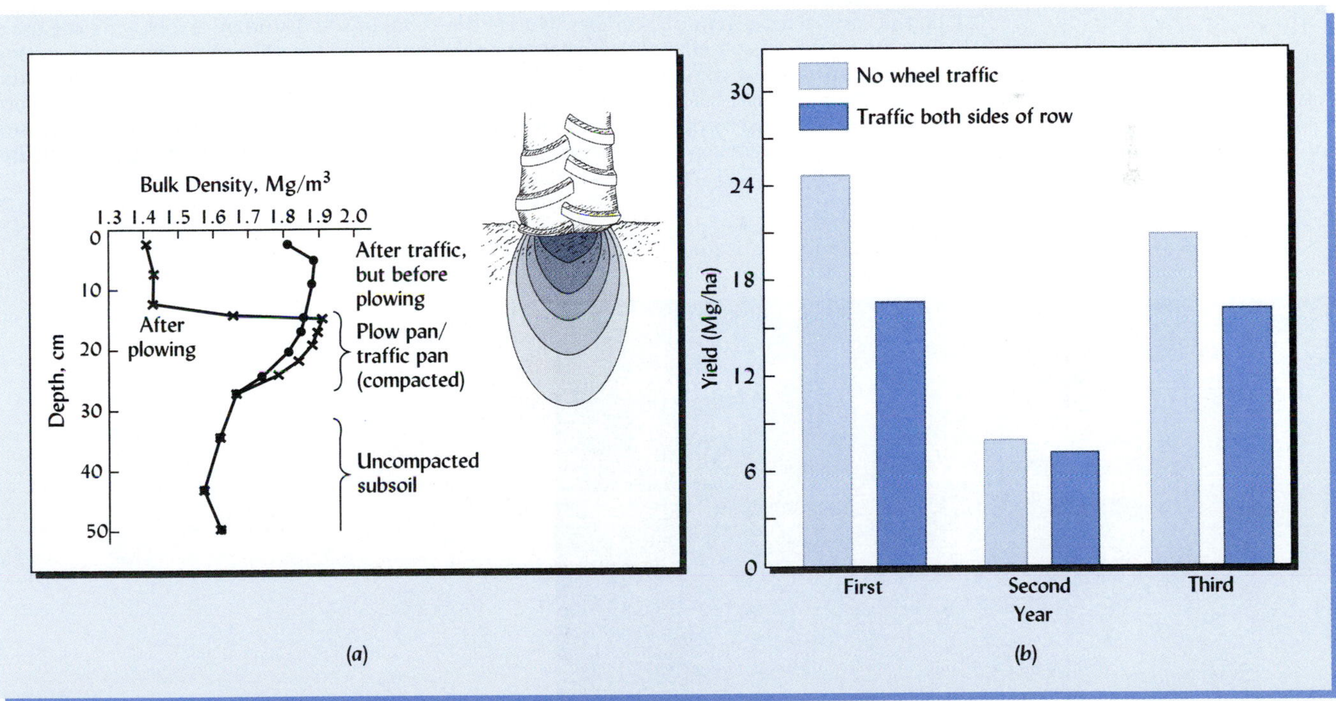

FIGURE 4.16 Tractors and other heavy equipment compact the soil to considerable depths, increasing bulk density and reducing plant growth and crop yields. The effects are especially damaging if the soil is wet when trafficked. (*a*) The tires of a heavy vehicle compact a sandy loam soil to about 30 cm, creating a traffic pan. Plowing temporarily loosens the compacted surface soil (plow layer), but increases compaction just below the plowed layer, creating a combined traffic pan and plow pan. Bulk densities in excess of 1.8 Mg/m^3 prevented the penetration of cotton roots in this case. (*b*) The yield of potatoes was reduced in two out of three years in this test on a clay loam in Minnesota. Yield reductions are often most pronounced in relatively dry years when plants have the greatest need for subsoil moisture. [Based on data from Camp and Lund (1964) and Voorhees (1984)]

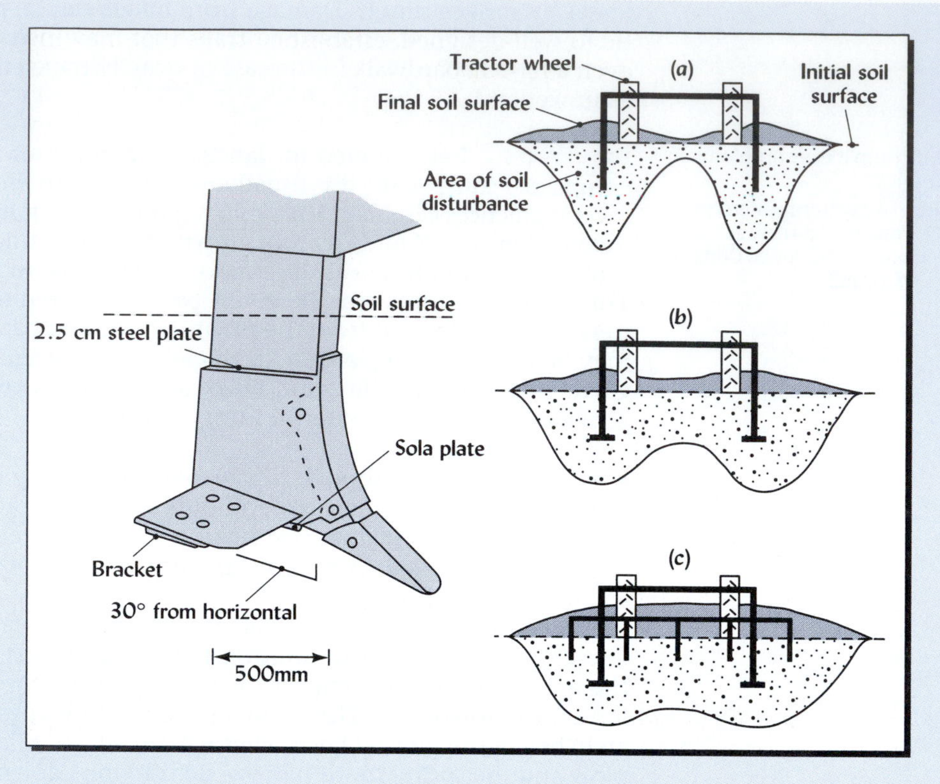

FIGURE 4.17 In the reclamation of severely compacted soil sites such as those created by heavy vehicles during the construction and use of landfills, it is sometimes necessary to mechanically break up the soil to permit the establishment of trees or other natural vegetation. This is commonly accomplished by the use of a subsoiler with penetrating tines such as the one shown above left. As these tines are pulled through the ground, they loosen and lift the soil, thereby breaking up the compacted condition. On the right are the zones of soil disturbance following the use of (*a*) a conventional subsoiler with two vertical tines but no wings, (*b*) equipment with two tines and associated lifting wings (as shown at left), and (*c*) a winged subsoiler plus shallower tines that loosen the upper soil before the main lines pass through. Note the increased quantity of compacted soil disturbed by the more extensive treatments. Subsoilers are also used to loosen the subsoil of some very tight clay soils used for agricultural purposes. [Modified from McRae (1999); used with permission of CRC Press, Boca Raton, Fla.]

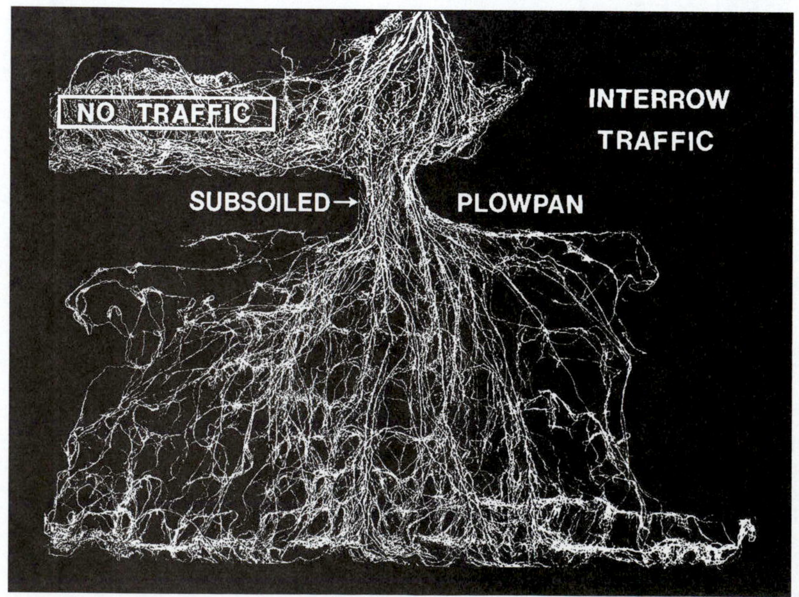

FIGURE 4.18 Root distribution of a cotton plant. On the right, interrow tractor traffic and plowing have caused a plowpan that restricts root growth. Roots are more prolific on the left where there had been no recent tractor traffic. The roots are seen to enter the subsoil through a loosened zone created by a subsoiling chisel-type implement. (Courtesy USDA National Tillage Machinery Laboratory)

More on how to manage compaction on farm soils: http://www.extension.umn.edu/distribution/cropsystems/DC7400.html

some soils the effects of subsoiling are quite temporary. Any tillage tends to reduce soil strength, thus making the soil less resistant to subsequent compaction.

Another approach for minimizing compaction is to carefully restrict all wheel traffic to specific lanes, leaving the rest of the field (usually 90% or more of the area) free from compaction. Such **controlled traffic** systems are widely used in Europe, especially on clayey soils. Gardeners can practice controlled traffic by establishing permanent foot paths between planting beds. The paths may be enhanced by covering with a thick mulch, planting to sod grass, or paving with flat stones.

Some managers attempt to reduce compaction using an opposite strategy in which special wide tires are fitted to heavy equipment so as to spread the weight over more soil surface, thus reducing the force applied per unit area (Figure 4.19*a*). Home gardeners can avoid concentrating their body weight on just the few square centimeters of their footprints by standing on wooden boards when preparing seedbeds in relatively wet soil (Figure 4.19*b*).

Influence of Bulk Density on Soil Strength and Root Growth

Excessively dense soils inhibit root penetration, especially if the soil is relatively dry. Roots penetrate the soil by pushing their way into pores; but in soils with high bulk densities, many of the pores are too small to accommodate the root cap, and the root must push the soil particles aside and enlarge the pore. Such root penetration, however, is further limited by **soil strength**, the property of the soil that causes it to resist deformation (see Section 4.9). One way to quantify soil strength is to measure the force needed to push a standard cone-tipped rod (a **penetrometer**) into the soil. Compaction generally increases both bulk density and soil strength, thereby restricting root penetration.

Soil strength is also affected by soil water content, being higher in a relatively dry soil than in that same soil after it is wetted (see Figure 4.20). Consequently, the effect of bulk density on root growth is most pronounced in a relatively dry soil, a higher bulk density being necessary to prevent root penetration when the soil is moist. For example, a traffic pan having a bulk density of 1.6 Mg/m^3 may completely prevent the penetration of roots when the soil is rather dry, yet roots may readily penetrate this same layer when it is in a moist condition.

The more clay present in a soil, the smaller the average pore size, and the greater the resistance to penetration at a given bulk density. Therefore, if the bulk density is the

(a)

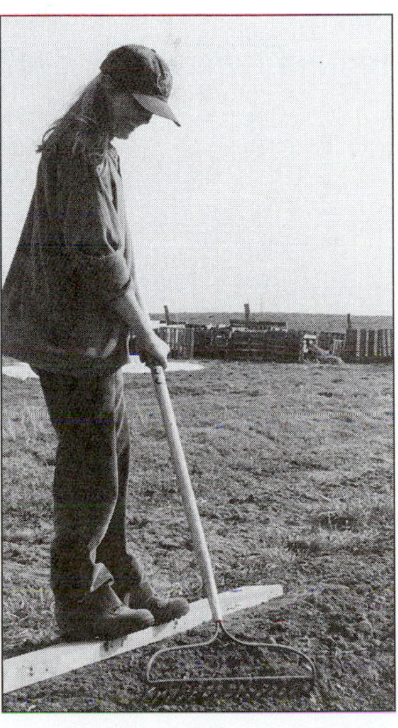

(b)

FIGURE 4.19 One approach to reducing soil compaction is to spread the applied weight over a larger area of the soil surface. Examples are extra-wide wheels on heavy vehicles used to apply soil amendments (a) and standing on a wooden board while preparing a garden seedbed in early spring (b). (Photos courtesy of R. Weil)

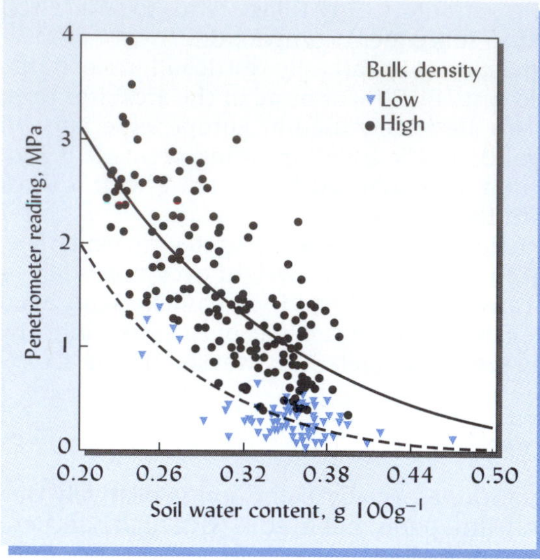

FIGURE 4.20 Both water content and bulk density affect soil strength as measured by penetrometer resistance. The data are for the clay textured Bt horizon of a Tatum soil in Virginia (Hapludults), which was either severely compacted (bulk density 1.7 Mg/m^3) or not compacted (bulk density 1.3 Mg/m^3). Note that soil strength decreases as water content increases and is very low regardless of bulk density when the soil is nearly saturated with water. [Unpublished data of R. Gilker, R. Weil, and D. Krizek, University of Maryland and USDA/ARS]

same, roots more easily penetrate a moist sandy soil than a moist clayey one. The growth of roots into moist soil is generally limited by bulk densities ranging from 1.45 Mg/m^3 in clays to 1.85 Mg/m^3 in loamy sands (see Figure 4.12).

4.6 PORE SPACE OF MINERAL SOILS

One of the main reasons for measuring soil bulk density is that this value can be used to calculate pore space. For soils with the same particle density, the lower the bulk density, the higher the percent pore space (**total porosity**). See Box 4.2 for derivation of the formula expressing this relationship.

Factors Influencing Total Pore Space

In Chapter 1 (Figure 1.12) we noted that for an "ideal" medium-textured, well-granulated surface soil in good condition for plant growth, approximately 50% of the soil volume would consist of pore space, and that the pore space would be about half filled with air and half filled with water. Actually, total porosity varies widely among soils for the same reasons that bulk density varies. Values range from as low as 25% in compacted subsoils to more than 60% in well-aggregated, high-organic-matter surface soils. As is the case for bulk density, management can exert a decided influence on the pore space of soils (see Table 4.3). Data from a wide range of soils show that cultivation tends to lower the total pore space compared to that of uncultivated soils. This reduction usually is associated with a decrease in organic matter content and a consequent lowering of **granulation**.

Size of Pores

Bulk density values help us predict only *total* porosity. However, soil pores occur in a wide variety of sizes and shapes (see Figure 4.21). The size of a pore largely determines what role the pore can play in the soil. We will simplify our discussion at this point by referring only to **macropores** (larger than about 0.08 mm) and **micropores** (smaller than about 0.08 mm).

MACROPORES. The macropores characteristically allow the ready movement of air and the drainage of water. They also are large enough to accommodate plant roots and the wide range of tiny animals that inhabit the soil (see Chapter 11). Several types of macropores are illustrated in Figure 4.22.

BOX 4.2 CALCULATION OF PERCENT PORE SPACE IN SOILS

The bulk density of a soil can be easily measured and particle density can usually be assumed to be 2.65 Mg/m^3 for most silicate-dominated mineral soils. Direct measurement of the pore space in soil requires the use of much more tedious and expensive techniques. Therefore, when information on the percent pore space is needed, it is often desirable to calculate the pore space from data on bulk and particle densities.

The derivation of the formula used to calculate the percentage of total pore space in soil follows:

Let D_b = bulk density, Mg/m^3 $\quad$ V_s = volume of solids, m^3 $\quad$ $V_s + V_p$ = total soil volume V_s, m^3

D_p = particle density, Mg/m^3 $\quad$ V_p = volume of pores, m^3 $\quad$ W_s = Weight of soil (solids), Mg

By definition, $\frac{W_s}{V_s} = D_p$ and $\frac{W_s}{V_s + V_p} = D_b$

Solving for W_s gives $W_s = D_p \times V_s$ and $W_s = D_b(V_s + V_p)$

Therefore $D_p \times V_s = D_b(V_s + V_p)$ and $\frac{V_s}{V_s + V_p} = \frac{D_b}{D_p}$

Since $\frac{V_s}{V_s + V_p} \times 100 = \%$ solid space then $\%\text{ solid space} = \frac{D_b}{D_p} \times 100$

Since % pore space + % solid space = 100, and % pore space = 100 − % solid space, then $\%\text{ pore space} = 100 - \left(\frac{D_b}{D_p} \times 100\right)$

EXAMPLE

Consider the cultivated clay soil from Canada in Table 4.3 (the Cambid). The bulk density was determined to be 1.28 Mg/m^3. Since we have no information on the particle density, we assume that the particle density is approximately that of the common silicate minerals (i.e., 2.65 Mg/m^3). We calculate the percent pore space using the formula derived above:

$$\%\text{ pore space} = 100 - \left(\frac{1.28\text{ Mg/m}^3}{2.65\text{ Mg/m}^3} \times 100\right) = 100 - 48.3 = 51.7$$

This value of pore space, 51.7%, is quite close to the typical percentage of air and water space described in Figure 1.12 for a well-granulated, medium- to fine-textured soil in good condition for plant growth. This simple calculation tells us nothing about the relative amounts of large and small pores, however, and so must be interpreted with caution.

For certain soils it is inaccurate to assume that the soil particle density is 2.65 Mg/m^3. For instance, a soil with a high organic matter content can be expected to have a particle density somewhat lower than 2.65. Similarly, a soil rich in iron oxide minerals will have a particle density greater than 2.65, because these minerals have particle densities as high as 3.5. As an example of the latter type of soil, let us consider the uncultivated clay soils from Zimbabwe (Ustalfs) described in Table 4.3. These are red clays, high in iron oxides. The particle density for these soils was determined to be 3.21 Mg/m^3 (not shown in Table 4.3). Using this value and the bulk density value from Table 4.3, we calculate the pore space as follows:

$$\%\text{ pore space} = 100 - \left(\frac{1.20}{3.21} \times 100\right) = 100 - 37.4 = 62.6$$

Such a high percentage pore space is an indication that this soil is in an uncompacted, very well granulated condition typical for soils found under undisturbed natural vegetation.

TABLE 4.3 Bulk Density and Pore Space of Some Surface Soils from Cultivated and Nearby Uncultivated Areas

Cultivation increased the bulk density and proportionately decreased the pore space in every case.

			Bulk density, Mg/m^3		*Pore space, %*	
Soil	*Texture*	*Years cropped*	*Cultivated*	*Uncultivated*	*Cultivated*	*Uncultivated*
Ustoll (Canada)	Silt loam	90	1.30	1.04	50.9	60.8
Cambid (Canada)	Clay	70	1.28	0.98	51.7	63.0
Mean of 3 Ustalfs (Zimbabwe)	Clay	20–50	1.44	1.20	54.1	62.6
Mean of 3 Ustalfs (Zimbabwe)	Sandy loam	20–50	1.54	1.43	42.9	47.2

Data for Canadian soils from Tiessen, et al. (1982), for Zimbabwe soils from Weil (unpublished).

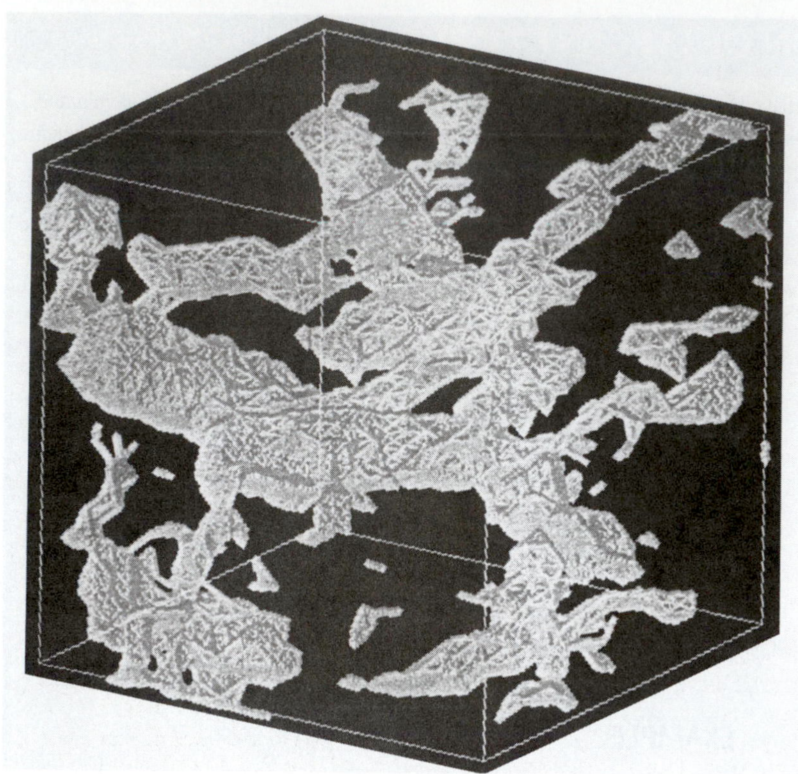

FIGURE 4.21 A three-dimensional representation of the network of pores in a small block of undisturbed soil in France (edges about 2 mm in length). The pores (light in color) exhibit great variability in size and cross-sectional area. Note that the pore channels are tortuous and that not all pores are connected to each other, some being isolated from channels that could transport water and air into and out of the soil. This suggests that tiny volumes of water and air may be trapped in localized pockets, thereby preventing their ready movement downward or upward in the soil. (Image courtesy of Dr. Isabelle Cousin, INRA Unité de Science du Sol—SESCPF Centre de Recherche d'Orléans Domaine de Limere, Ardon, France)

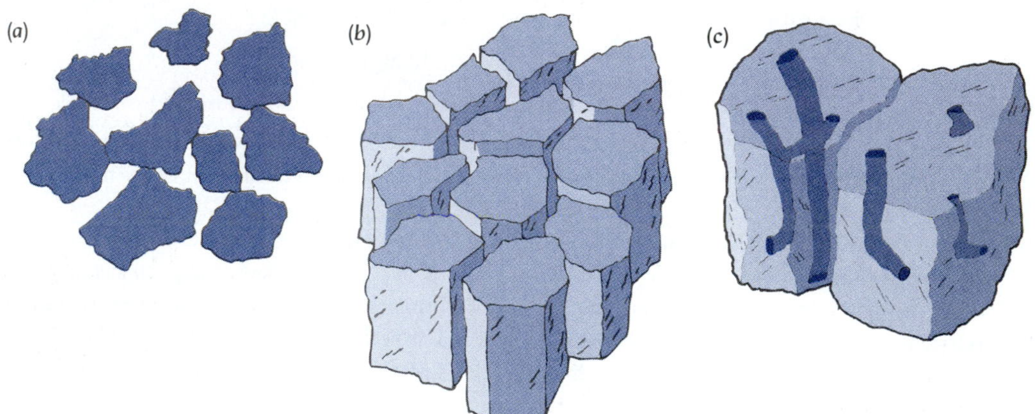

FIGURE 4.22 Various types of soil pores. (*a*) Many soil pores occur as *packing pores,* spaces left between primary soil particles. The size and shape of these spaces is largely dependent on the size and shape of the primary sand, silt, and clay particles and their packing arrangement. (*b*) In soils with structural peds, the spaces between the peds form *interped pores*. These may be rather planar in shape, as with the cracks between prismatic peds, or they may be more irregular, like those between loosely packed granular aggregates. (*c*) *Biopores* are formed by organisms such as earthworms, insects, and plant roots. Most of these are long, sometimes branched channels, but some are round cavities left by insect nests and the like.

In well-structured soils, the macropores are generally found between peds. These **interped pores** may occur as spaces between loosely packed granules or as the planar cracks between tight-fitting blocky and prismatic peds (see Plate 39, after page 498). Figure 4.23 shows that the decrease in organic matter and increase in clay that occur with depth in many profiles are associated with a shift from macropores to micropores.

Macropores created by roots, earthworms, and other organisms constitute a very important type of pores termed **biopores.** (See tubular pores in Plate 29, after page 114). In some clayey soils, biopores are the principal form of macropores, greatly facilitating the growth of plant roots (Table 4.4, Plates 38 and 40, after page 498).

PLATE 1 Alfisols—an Aeric Epiaqualf from western New York. Argillic horizon between 20–80 cm. Scale in centimeters.

PLATE 2 Andisols—a Typic Melanudand from western Tanzania. Scale in 10 cm.

PLATE 3 Aridisols—a Typic Haplocambid from western Nevada. Scale in feet.

PLATE 4 Entisols—a Typic Quartzipsamment from eastern Texas. Scale in feet.

Photo credits follow plate 67, preceding page 499

PLATE 5 Gelisols—a Typic Aquaturbel from Alaska. Permafrost below 32 cm on scale.

PLATE 6 Histosols—a Limnic Haplosaprist from southern Michigan. Buried mineral soil at bottom of scale. Scale in feet.

PLATE 7 Inceptisols—a Typic Dystrudept from West Virginia. Knife is in cambic horizon.

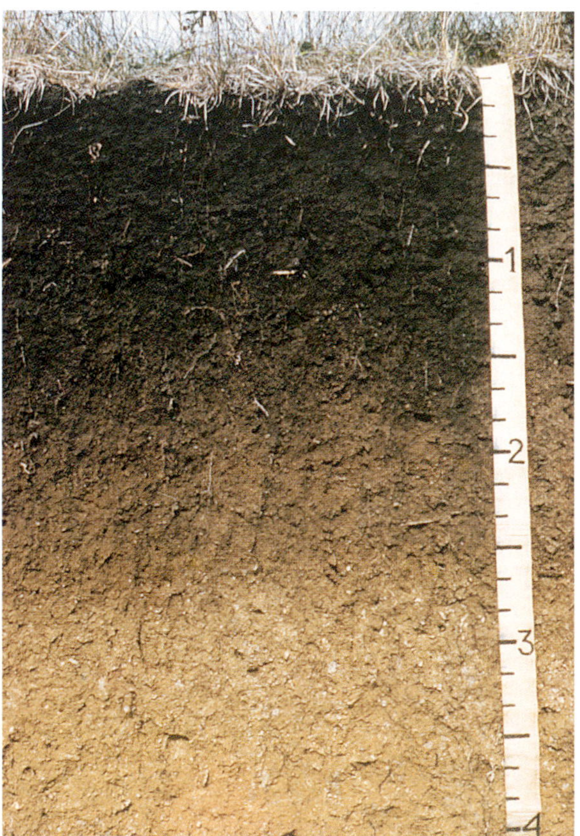

PLATE 8 Mollisols—a Typic Hapludoll from central Iowa. Mollic epipedon to 1.8 ft. Scale in feet.

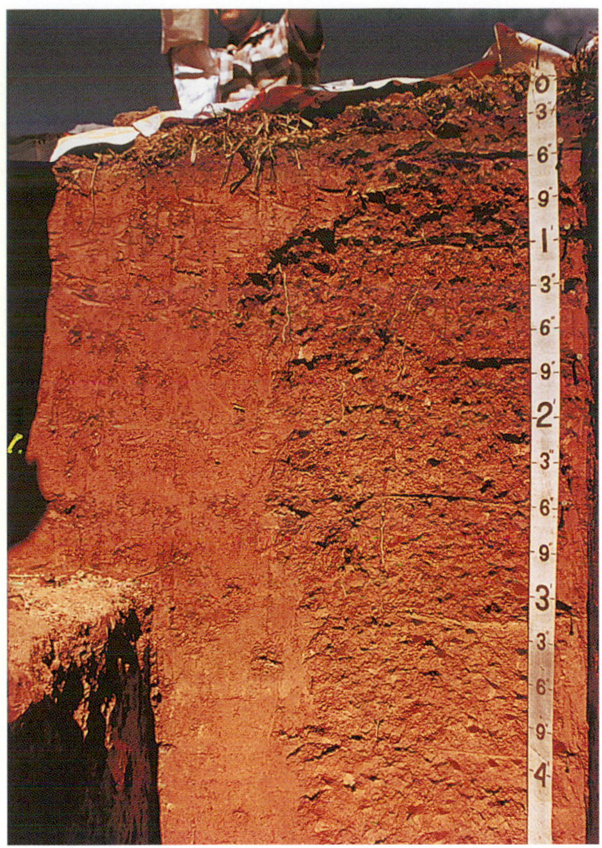

PLATE 9 Oxisols—a Udeptic Hapludox from central Puerto Rico. Scale in feet and inches.

PLATE 10 Spodosols—a Humic Cryorthod from southern Quebec. Albic horizon at about 10 cm. Bar = 10 cm.

PLATE 11 Ultisols—a Typic Hapludult from central Virginia showing metamorphic rock structure in the saprolite below the 60-cm-long shovel.

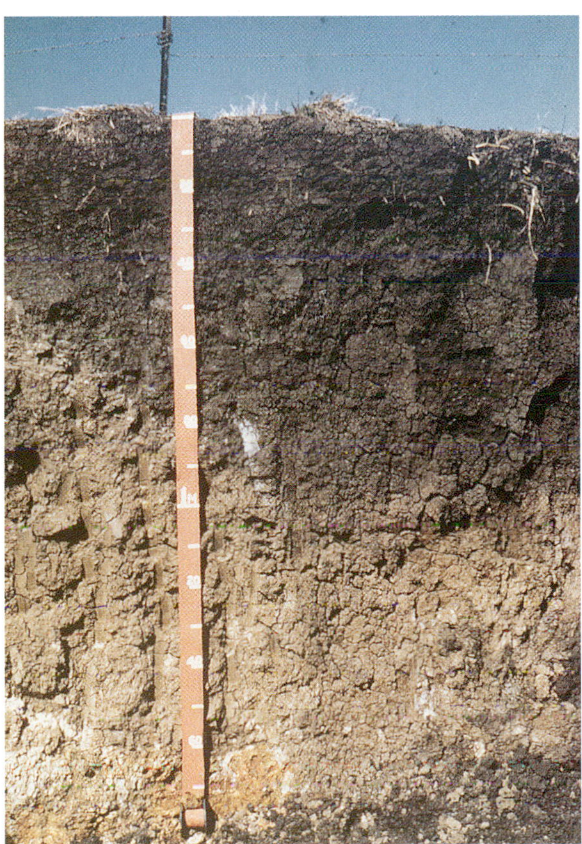

PLATE 12 Vertisols—a Typic Haplustert from Queensland, Australia, during wet season. Scale in meters.

PLATE 13 Typic Argiustolls in eastern Montana with a chalky white calcic horizon (Bk and Ck) overlain by a Mollic epipedon (Ap, A2, and Bt).

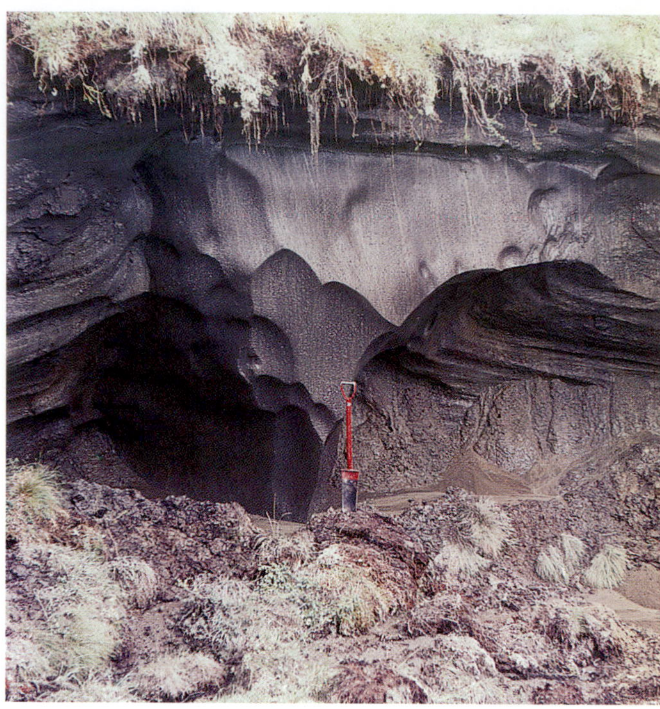

PLATE 14 Ice wedge and permafrost underlying Gelisols in the Seward Peninsula of Alaska. Shovel is 1 m long.

PLATE 15 A Typic Plinthudult in central Sri Lanka. Mottled zone is plinthite, in which ferric iron concentrations will harden irreversibly if allowed to dry.

PLATE 16 A soil catena or toposequence in central Zimbabwe. Redder colors indicate better internal drainage. Inset: B-horizon clods from each soil in the catena.

PLATE 17 Uneven water infiltration and movement in a sandy soil due to hydrophobic organic coatings.

PLATE 18 The darker surface soil was brushed aside to expose a hydrophobic layer caused by burning the chaparral vegetation. Water beads up rather than soaking into this layer. See page 222.

PLATE 19 The boundary between the Oe and the E horizons of a forested Ultisol.

PLATE 20 The effect of moisture on soil color. Right side of this Mollisol profile was sprayed with water.

PLATE 21 Effect of poor drainage on soil color. Gray colors and red redox concentrations in the B horizons of a Plinthaquic Paleudalf.

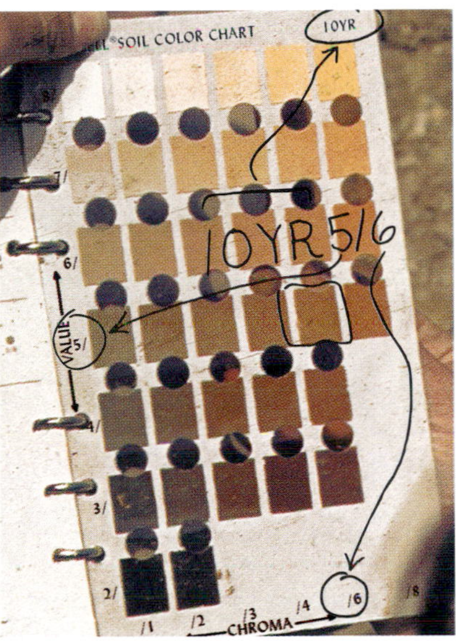

PLATE 22 The 10YR hue page of a Munsell color book. The standard notation is handwritten for the color with hue 10YR, value 5, and chroma 6.

PLATE 23 Road cut in southern Brazil exposing the profile of an Udult with a sombric horizon. Humid, high-altitude tropical and subtropical mountains are the typical environments for the formation of this dark, humus-rich subsurface horizon.

PLATE 24 Thick clay skins or argillans in an argillic Bt horizon. Image made from a very thin, polished slice of soil, magnified with a petrographic microscope using plain polarized light (*left*) and cross-polarized light (*right*). Note the thin layers of illuvial clay.

PLATE 25 Connecticut River valley in western Massachusetts. Note variable alluvial soils and presence of riparian forest buffer along the river bank.

PLATE 26 Redox concentrations (red) and depletions (gray) in a Btg horizon from an Aquic Paleudalf.

PLATE 27 The waxlike ped surfaces in this Bt horizon from an Ultisol are clay skins (argillans). Bar = 1 cm.

PLATE 28 Erosion of convex sites by tillage and water has exposed red B-horizon material.

PLATE 29 Oxidized (red) root zones in the A and E horizons indicate a hydric soil. They result from oxygen diffusion out from roots of wetland plants having aerenchyma tissues (air passages).

PLATE 30 Dark (black) humic accumulation and gray humus depletion spots in the A horizon are indicators of a hydric soil. Water table is 30 cm below the soil surface.

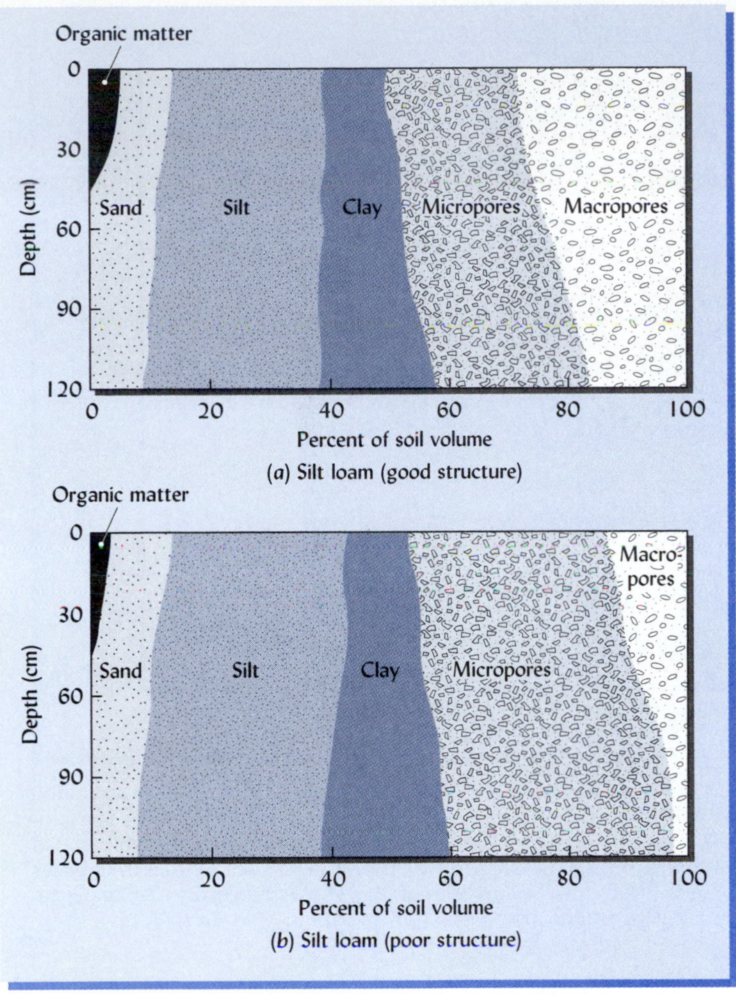

FIGURE 4.23 Volume distribution of organic matter, sand, silt, clay, and pores of macro- and microsizes in two representative silt loams, one with good soil structure (*a*) and the other with poor soil structure (*b*). The silt loam with poor structure has a smaller volume of larger (macro) pores than the other soil. Note that at the lowest depths in both silt loams, about one-third of the mineral matter is clay, giving the lower horizons enough clay to be classified as silty clay loams.

TABLE 4.4 Distribution of Different-Sized Loblolly Pine Roots in the Soil Matrix and in Old Root Channels in the Uppermost Meter of an Ultisol in South Carolina

The root channels (1 to 5 cm in diameter) filled with loose surface soil and decaying organic matter were easy for roots to penetrate.

	Number of roots counted per 1 m² of the soil profile		
Root size, diameter	*Soil matrix*	*Old root channels*	*Comparative increase in root density in the old channels, %*
Fine roots, <4 mm	211	3617	94
Medium roots, 4–20 mm	20	361	95
Coarse roots, >20 mm	3	155	98

Calculated from Parker and Van Lear (1996).

Micropores. In contrast to macropores, micropores are usually filled with water in field soils. Even when not water filled, they are too small to permit much air movement. Water movement in micropores is slow, and much of the water retained in these pores is not available to plants (see Chapter 5). Fine-textured soils, especially those without a stable granular structure, may have a preponderance of micropores, thus allowing relatively slow gas and water movement, despite the relatively large volume of total pore space. Aeration, especially in the subsoil, may be inadequate for satisfactory root development and desirable microbial activity. While the larger micropores

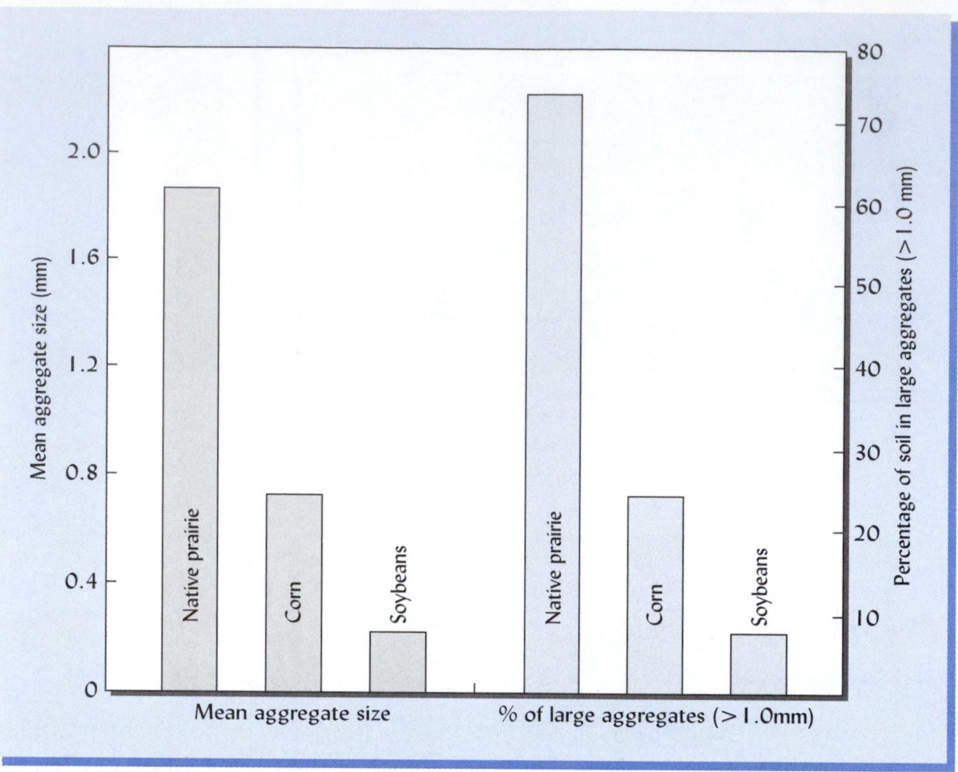

FIGURE 4.24 Soil aggregates in a Mollisol in Iowa are larger and more stable under native prairie vegetation than where cultivated crops had been grown for some 90 years. In this study, soil samples were taken from a prairie area and from two nearby fields, where either corn or soybeans had been grown the previous year. Differences in past management may in part account for differences between the corn and soybean fields, but the soil in both of these fields shows distinct aggregate breakdown compared to the native grassland area. [Drawn from data in Martens (2000)]

accommodate plant root hairs and microorganisms, the smaller micropores (sometimes termed *ultramicropores* and *cryptopores*) are so tiny that their radii are measured in nanometers (10^{-9} meters), giving rise to the term *nanopores*. Such pores are too small to permit the entrance of even the smallest bacteria or some decay-stimulating enzymes produced by the bacteria. Thus, these pores can act as hiding places for some adsorbed organic compounds (both naturally occurring and pollutants), thereby protecting them from breakdown for long periods of time, perhaps for centuries (see Section 11.8).

When native prairie lands are plowed and planted to row crops such as corn or soybeans, soil organic matter contents and total pore space are reduced. The most striking effect of such cropping, however, is the drastic reduction in the amount of large aggregates and associated macropore space that is so critical for ready air movement (see Figure 4.24).

In recent years, conservation tillage practices, which minimize plowing and associated soil manipulations, have been widely adopted in the United States (see Sections 4.8 and 15.5). Because of increased accumulation of organic matter near the soil surface and the development of a long-lived network of macropores (especially biopores), some conservation tillage systems lead to greater macroporosity of the surface layers. These benefits are particularly likely to accrue in soils with extensive production of earthworm burrows, which may remain undisturbed in the absence of tillage. Unfortunately, such improvements in porosity do not always occur in soils with poor internal drainage.

4.7 FORMATION AND STABILIZATION OF SOIL AGGREGATES

The formation and maintenance of a high degree of aggregation are among the most difficult tasks of soil management, yet they are also among the most important, since they are a potent means of influencing ecosystem function. The organization of surface soils into relatively large structural aggregates provides for the low bulk density and a high proportion of macropores so desirable for most soil uses.

Some aggregates readily succumb to the beating of rain and the rough-and-tumble of plowing and tilling the land. Others resist disintegration, thus making the maintenance of a suitable soil structure comparatively easy (see Figures 4.29 and 4.30 later in this chapter). Generally, the smaller aggregates are more stable than the larger ones, so maintaining the much-prized larger aggregates requires much care.

Hierarchical Organization of Soil Aggregates[4]

The large aggregates (>1 mm) so desirable for most soil uses are typically composed of smaller aggregates, which in turn are composed of still smaller units, down to clusters of clay and humus less than 0.001 mm in size. You may easily demonstrate the existence of this *hierarchy of aggregation* by selecting a few of the largest aggregates in a soil and gently crushing or picking them apart to separate them into many smaller-sized aggregates. Then try rubbing the tiniest of these granules between your thumb and forefinger. You will find that most of these break down into a smear of still smaller aggregates composed of silt, clay, and humus. The hierarchical organization of aggregates (Figure 4.25) seems to be characteristic of most soils, with the exception of certain Oxisols and some very young Entisols. At each level in the hierarchy of aggregates, different factors are responsible for binding together the subunits.

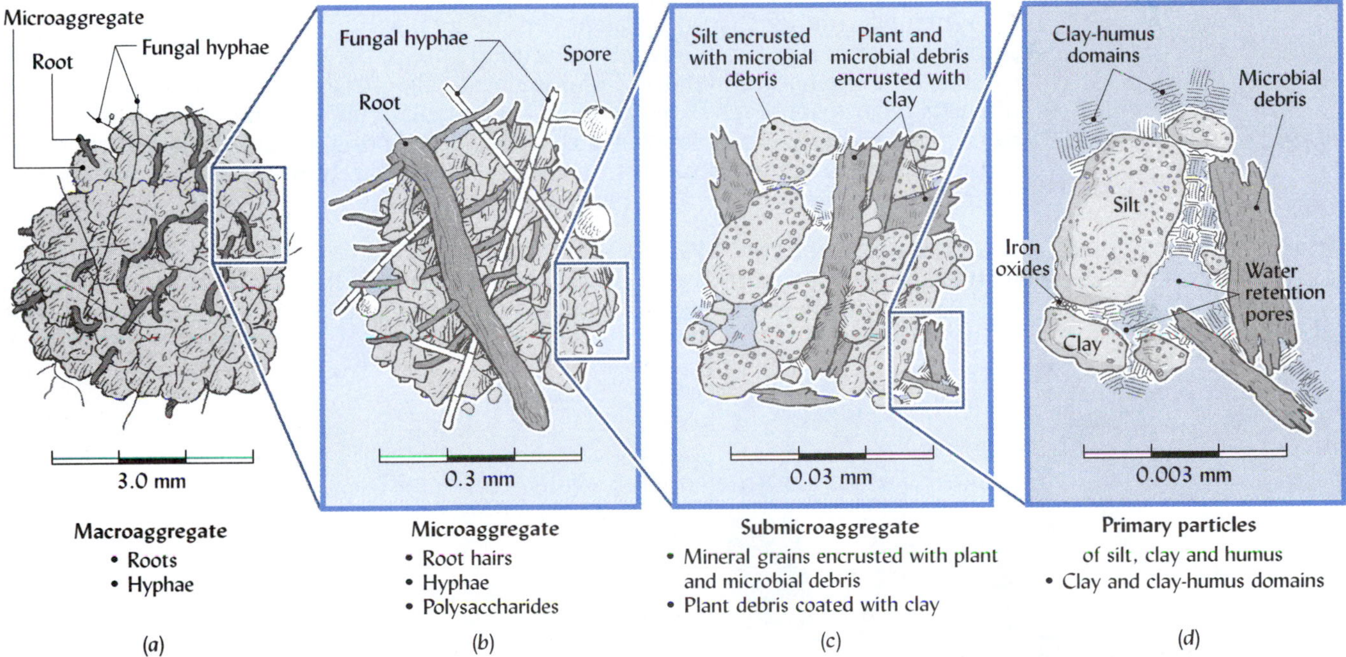

FIGURE 4.25 Larger aggregates are often composed of an agglomeration of smaller aggregates. This illustration shows four levels in this hierarchy of soil aggregates. The different factors important for aggregation at each level are indicated. (*a*) A *macroaggregate* composed of many microaggregates bound together mainly by a kind of sticky network formed from fungal hyphae and fine roots. (*b*) A *microaggregate* consisting mainly of fine sand grains and smaller clumps of silt grains, clay, and organic debris bound together by root hairs, fungal hyphae, and microbial gums. (*c*) A very small submicroaggregate consisting of fine silt particles encrusted with organic debris and tiny bits of plant and microbial debris (called *particulate organic matter*) encrusted with even smaller packets of clay, humus, and Fe or Al oxides. (*d*) Clusters of parallel and random clay platelets interacting with Fe or Al oxides and organic polymers at the smallest scale. These organoclay clusters or *domains* bind to the surfaces of humus particles and the smallest of mineral grains. (Diagram courtesy of R. Weil)

---

[4]The role of organic matter and biological processes in the hierarchical organization of soil aggregates was originally put forward by Tisdall and Oades (1982) and elaborated on by Oades (1993) and Tisdall (1994).

Factors Influencing Aggregate Formation and Stability in Soils

Both biological and physical-chemical (abiotic) processes are involved in the formation of soil aggregates. The physical-chemical processes tend to be most important at the smaller end of the scale, biological processes at the larger end. Also, the physical-chemical processes of aggregate formation are associated mainly with clays and, hence, tend to be of greater importance in finer-textured soils. In sandy soils that have little clay, aggregation is almost entirely dependent on biological processes.

Physical-Chemical Processes

Most important among the physical-chemical processes are (1) the mutual attraction among clay particles and (2) the swelling and shrinking of clay masses.

Flocculation of Clays and the Role of Adsorbed Cations. Except in very sandy soils that are almost devoid of clay, aggregation begins with the **flocculation** of clay particles into microscopic clumps or *floccules* (Figure 4.26). If two clay platelets come close enough to each other, the cations compressed in a layer between them will attract the negative charges on both platelets, thus serving as bridges to hold the platelets together. This process is repeated until a small "stack" of parallel clay platelets, termed a *clay domain,* is formed (Figure 4.25). Other types of clay domains are more random in orientation, resembling a house of cards. These form when the positive charges on the edges of the clay platelets attract the negative charges on the planar surfaces (Figure 4.26). Clay floccules or domains, along with charged organic colloids (humus), provide much of the long-term stability for the smaller (<0.03 mm) **microaggregates.** The cementing action of inorganic compounds, such as iron oxides, produces very stable small aggregates sometimes called **pseudosand** in certain clayey soils (Ultisols and Oxisols) of hot, humid regions.

When Na^+ (rather than polyvalent cations such as Ca^{2+} or Al^{3+}) is a prominent adsorbed ion, as in some soils of arid and semiarid areas, the attractive forces are not able to overcome the natural repulsion of one negatively charged clay by another. The clay platelets cannot approach closely enough to flocculate, so remain dispersed as far apart from one another as possible (see Figure 4.26). Clay in this dispersed, gel-like condition causes the soil to become almost structureless, impervious to water and air, and very undesirable from the standpoint of plant growth (see Sections 9.15–9.16).

Volume Changes in Clayey Materials. As a soil dries out and water is withdrawn, the platelets in clay domains move closer together, causing the domains and, hence, the soil mass to shrink in volume. As a soil mass shrinks, cracks will open up along planes

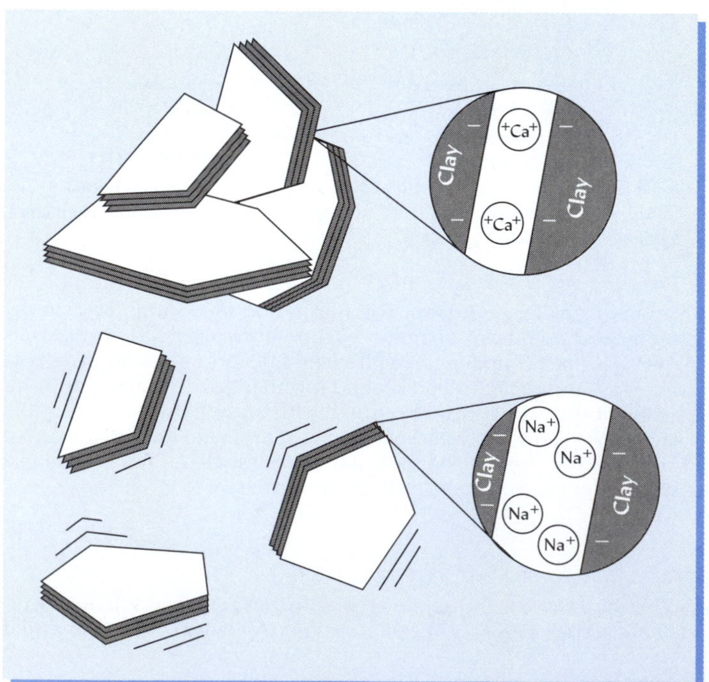

FIGURE 4.26 The role of cations in the flocculation of soil clays. The di- and trivalent cations, such as Ca^{2+} and Al^{3+}, are tightly absorbed and can effectively neutralize the negative surface charge on clay particles. These cations can also form bridges that bring clay particles close together. Monovalent ions, especially Na^+, with relatively large hydrated radii, can cause clay particles to repel each other and create a dispersed condition. Two things contribute to the dispersion: (1) the large hydrated sodium ion does not get close enough to the clay to effectively neutralize the negative charges, and (2) the single charge on sodium is not effective in forming a bridge between clay particles.

of weakness. Over the course of many cycles (as occur between rain or irrigation events in the field) the network of cracks becomes more extensive and the aggregates between the cracks better defined. Water uptake, especially by fibrous-rooted perennial grasses, accentuates the physical aggregation processes associated with wetting and drying. This effect is but one of many examples of ways in which physical and biological soil processes interact. Freezing and thawing cycles have a similar effect, because the formation of ice crystals is a drying process that also draws water out of clay domains.

Biological Processes

Activities of Soil Organisms. Among the biological processes of aggregation, the most prominent are (1) the burrowing and molding activities of earthworms, (2) the enmeshment of particles by sticky networks of roots and fungal hyphae, and (3) the production of organic glues by microorganisms, especially bacteria and fungi.

Plant roots (particularly root hairs) and fungal hyphae exude sugarlike polysaccharides and other organic compounds, forming sticky networks that bind together individual soil particles and tiny microaggregates into larger agglomerations called **macroaggregates** (see Figure 4.25*a*). The threadlike fungi that associate with plant roots (called *mycorrhizae;* see Section 10.9) are especially effective in providing this type of relatively short-term stabilization of large aggregates, because they secrete a gooey sugar-protein called **glomalin**, which is very effective as a cementing agent (Box 4.3 and Figure 4.27).

BOX 4.3 SOIL AGGREGATES AND MYCORRHIZAL FUNGI

The favorable influences of soil microorganisms, and especially fungi, on the formation and stability of soil aggregates have long been known. However, we are still discovering the specific mechanisms by which these favorable effects come into being. One such discovery relates to the aggregating and stabilizing effect of glomalin, a glycoprotein that is sloughed off the hyphae of mycorrhizal fungi. These organisms live in a symbiotic state with the roots of many plants, including many tree species. The following graph shows the relationship between glomalin content and aggregate stability.

Note the very high aggregate stability and glomalin content in the grassland plots and the very low levels of each in the plots tilled conventionally (plowed annually). The soil in this experiment was an Ultisol, although similar effects of glomalin have been noted on other soils of temperate regions.

The level of carbon dioxide in the atmosphere has been found to affect both glomalin levels and aggregate formation. The results of one study in southern California illustrate this relationship.

Apparently, higher atmospheric CO_2 stimulated the growth of plant roots, which in turn, increased the growth of the associated mycorrhizal fungi that produce the glomalin. Increases in carbon dioxide content of the atmosphere in the last few decades may have had some influence on aggregate formation in soils.

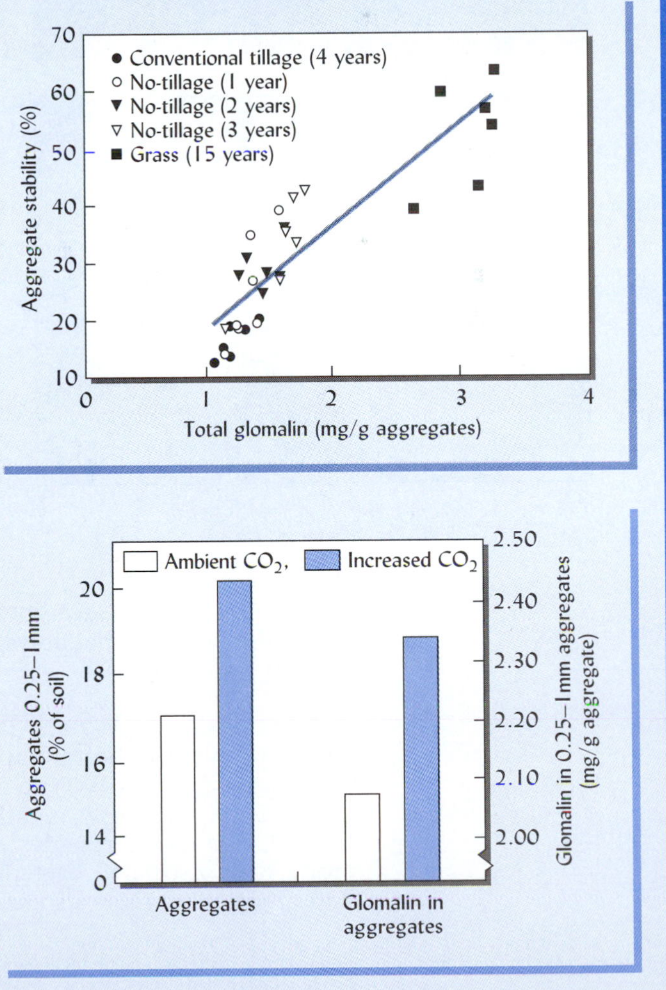

Graph sources: *Upper*, Wright, et al., (1999). *Lower*, drawn from data in Rillig, et al. (1999).

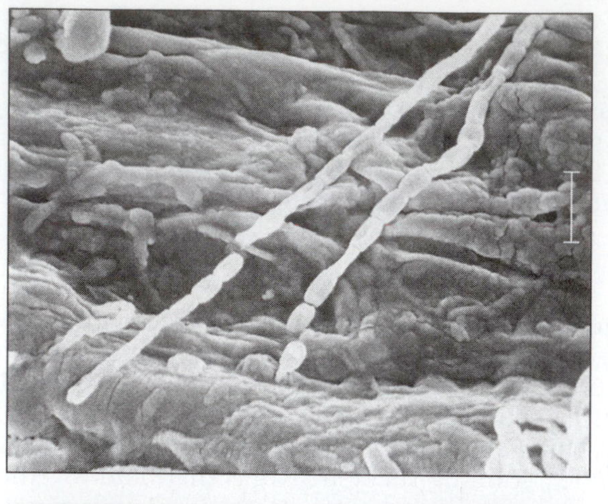

(a)

(b)

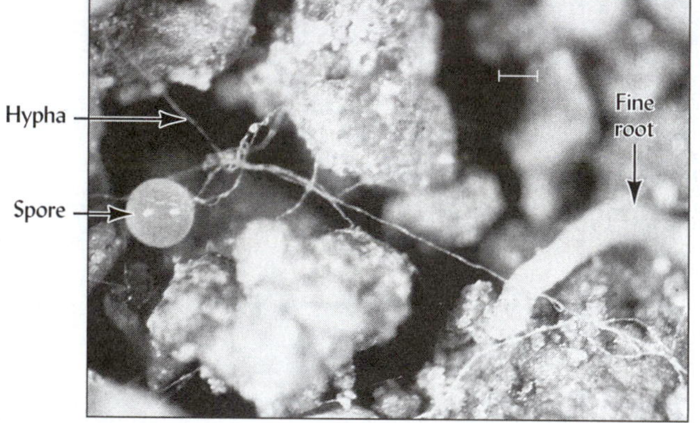

(c)

FIGURE 4.27 Fungal hyphae binding soil particles into aggregates. (*a*) Close-up of a hyphae growing over the surface of a mineral grain encrusted with microbial cells and debris. Bar = 10 μm. (*b*) An advanced stage of aggregation during the formation of soil from dune sands. Note the net of fungal hyphae and the encrustation of the mineral grains with organic debris. Bar = 50 μm. (*c*) Hyphae of the root-associated fungus from the genus *Gigaspora* interconnecting particles in a sandy loam from Oregon. Note also the fungal spore and the plant root. Bar = 320 μm. [Photos (*a*) and (*b*) courtesy of Sharon L. Rose, Willamette University; photo (*c*) courtesy of R. P. Schreiner, USDA-ARS, Corvallis, Ore.]

Bacteria also produce polysaccharides and other organic glues as they decompose plant residues. Bacterial polysaccharides are shown intermixed with clay at a very small scale in Figure 4.28.

Influence of Organic Matter. In most temperate zone soils, organic matter is *the* major agent stimulating the formation and stabilization of granular and crumb-type aggregates (see Figure 4.29). First, organic matter provides the energy substrate that makes possible the previously mentioned activities of the fungi, bacteria, and soil animals. Second, as organic residues decompose, gels and other viscous microbial products, along with associated bacteria and fungi, encourage crumb formation. Organic exudates from plant roots also participate in this aggregating action.

Influence of Tillage. Tillage can have both favorable and unfavorable effects on aggregation. If the soil is not too wet or too dry when the tillage is performed, the short-term effect of tillage is generally favorable. Tillage implements break up large clods, incorpo-

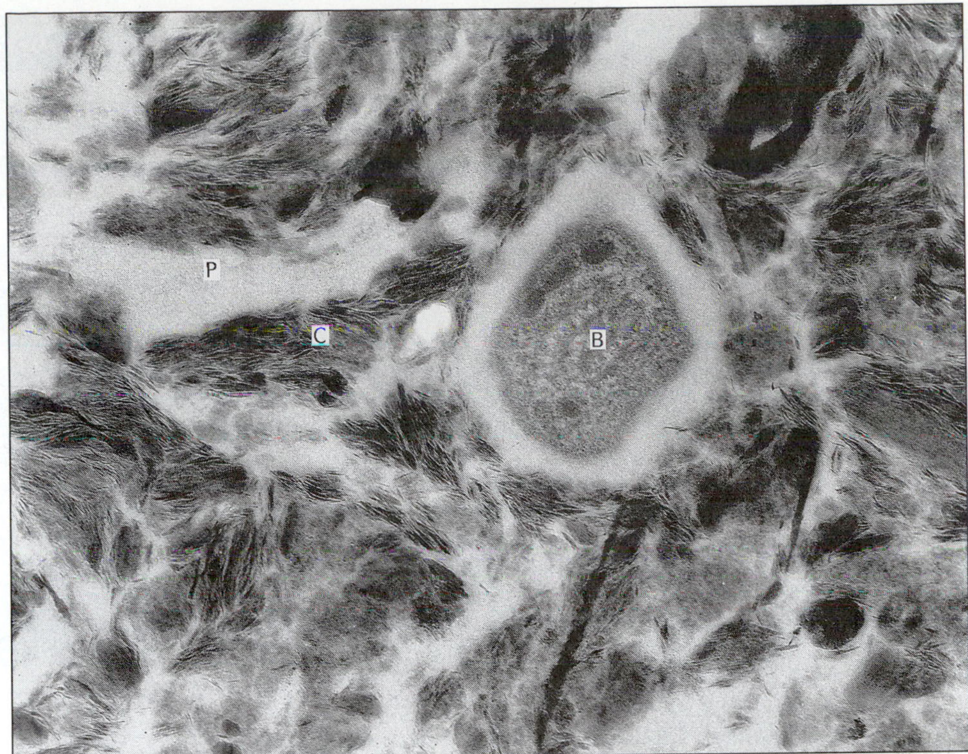

FIGURE 4.28 An ultrathin section illustrating the interaction among organic materials and silicate clays in a water-stable aggregate. The dark-colored materials (C) are groups of clay particles that are interacting with organic polysaccharides (P). A bacterial cell (B) is also surrounded by polysaccharides. Note the generally horizontal orientation of the clay particles, an orientation encouraged by the organic materials. [From Emerson, et al. (1986); photograph provided by R. C. Foster, CSIRO, Glen Osmond, Australia]

rate organic matter into the soil, kill weeds, and generally create a more favorable seedbed (see Box 4.4). Immediately after plowing, the surface soil is loosened (its cohesive strength is decreased) and total porosity is increased.

Over longer periods, however, tillage greatly hastens the oxidation of soil organic matter, thus reducing the aggregating effects of this soil component. Tillage operations, especially if carried out when the soil is wet, also tend to crush or smear stable soil aggregates, resulting in loss of macroporosity and the creation of a *puddled* condition (see Figure 4.30).

Influence of Iron/Aluminum Oxides. Scientists have noted that some well-weathered soils of the tropics have more stable aggregation than soils with comparable or even higher organic matter levels in temperate regions. This is thought to be due primarily to the binding effects of the iron and aluminum oxides that are plentiful in many tropical soils. Films of such compounds coat and cement soil aggregates, thereby preventing their ready breakdown when the soil is tilled.

FIGURE 4.29 The aggregates of soils high in organic matter are much more stable than are those low in this constituent. The low-organic-matter soil aggregates fall apart when they are wetted; those high in organic matter maintain their stability.

BOX 4.4 PREPARING A GOOD SEEDBED

Early in the growing season, one of the main activities of a farmer or gardener is the preparation of a good seedbed to ensure that the sowing operation goes smoothly, and that the plants come up quickly, evenly, and well spaced.

A good seedbed consists of soil loose enough to allow easy root elongation and seedling emergence [see photo (*a*)]. At the same time, a seedbed should be packed firmly enough to ensure good contact between the seed and moist soil so that the seed can easily imbibe water to begin the germination process. The seedbed should also be relatively free of large clods. Seeds could fall between such clods and become lodged too deeply for proper emergence and without sufficient soil contact.

(a) A bean seed emerges from a seedbed. (Photo courtesy R. Weil)

Tillage may be needed to loosen compacted soil, help control weeds, and, in cool climates, help the soil dry out so it will warm more rapidly. On the other hand, the objectives of seedbed preparation may be achieved with little or no tillage if soil and climatic conditions are favorable and a mulch or herbicide is used to control early weeds.

Mechanical planters can assist in maintaining a good seedbed. Most planters are equipped with coulters (sharp steel disks) designed to cut a path through plant residues on the soil surface. No-till planters usually follow the coulter with a pair of sharp cutting wheels called a *double disk opener* that opens a groove in the soil into which the seeds can be dropped [see photos (*b*) and (*c*)]. Most planters also have a press wheel that follows behind the seed dropper and packs the loosened soil just enough to ensure that the groove is closed and the seed is pressed into contact with moist soil.

Ideally, only a narrow strip in the seed row is packed down to create a seed *germination zone*, while the soil between the crop rows is left as loose as possible to provide a *good rooting zone*. The surface of the interrow rooting zone may be left in a rough condition to encourage water infiltration and discourage erosion. The above principles also apply to the home gardener who may be sowing seeds by hand.

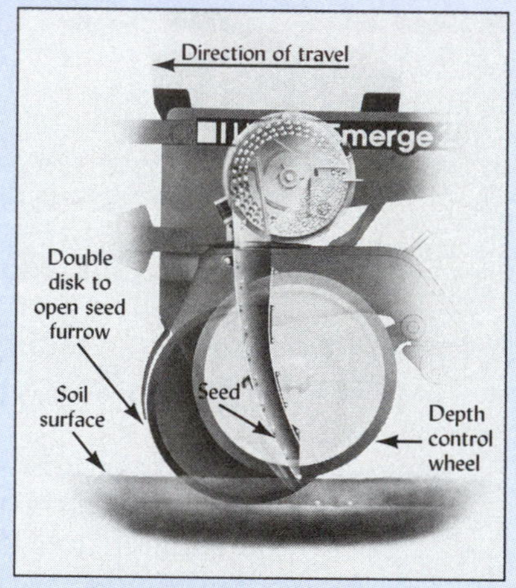

(b) A no-till planter in action and (c) a diagram showing how it works. [Photo (b) and diagram (c) courtesy of Deere & Company, Moline, Ill.]

4.8 TILLAGE AND STRUCTURAL MANAGEMENT OF SOILS

Conservation tillage—pros and cons:
http://www.cals.ncsu.edu/sustainable/peet/tillage/c03tilla.html

When protected under dense vegetation and undisturbed by tillage, most soils (except perhaps some sparsely vegetated soils in arid regions) possess a surface structure sufficiently stable to allow rapid infiltration of water and to prevent crusting. However, for the manager of cultivated soils, the development and maintenance of stable surface soil structure is a major challenge. Many studies have shown that aggregation and associated desirable soil properties such as water infiltration rate decline under long periods of tilled row-crop cultivation (Table 4.5).

FIGURE 4.30 Puddled soil (left) and well-granulated soil (right). Plant roots and especially humus play the major role in soil granulation. Thus a sod tends to encourage development of a granular structure in the surface horizon of cultivated land. (Courtesy USDA Natural Resources Conservation Service)

TABLE 4.5 Effect of Period of Corn Cultivation on Soil Organic Matter, Aggregate Stability, and Water Infiltration[a] in Five Silt Loam Inceptisols from Southwest France

In soils with depleted organic-matter levels, aggregates easily broke down under the influence of water, forming smaller aggregates and dispersed materials that sealed the soil surface and inhibited infiltration. A level of 3% soil organic matter seems to be sufficient for good structural stability in these temperate region silt loam soils.

			Infiltration	
Period of cultivation, years	*Organic matter, %*	*Aggregate stability, mm MWD[b]*	*Of total rain, %*	*Prior to ponding, mm*
100	0.7	0.35	25	6
47	1.6	0.61	34	9
32	2.6	0.76	38	15
27	3.1	1.38	47	25
15	4.2	1.52	44	23

[a]Infiltration is the amount of water that entered the soil when 64 mm of "rain" was applied during a 2-hour period.
[b]MWD = mean weighted diameter or average size of the aggregates that remained intact after sieving under water.
[Data from Le Bissonnais and Arrouays (1997)]

Tillage and Soil Tilth

Simply defined, **tilth** refers to the physical condition of the soil in relation to plant growth. Tilth depends not only on aggregate formation and stability, but also on such factors as bulk density, soil moisture content, degree of aeration, rate of water infiltration, drainage, and capillary water capacity. As might be expected, tilth often changes rapidly and markedly. For instance, the workability of fine-textured soils may be altered abruptly by a slight change in moisture.

Clayey soils are especially prone to puddling and compaction because of their high plasticity and cohesion. When puddled clayey soils dry, they usually become dense and hard. Proper timing of trafficking is more difficult for clayey than for sandy soils, because the former take much longer to dry to a suitable moisture content and may also become too dry to work easily.

Some clayey soils of humid tropical regions are much more easily managed than those just described. The clay fraction of these soils is dominated by hydrous oxides of iron and aluminum, which are not as sticky, plastic, and difficult to work. These soils may have very favorable physical properties, since they hold large amounts of water but have such stable aggregates that they respond to tillage after rainfall much like sandy soils.

Farmers in temperate regions typically find their soils too wet for tillage just prior to planting time (early spring), while farmers in tropical regions may face the opposite problem of soils too dry for easy tillage just prior to planting (end of dry season). In tropical and subtropical regions with a long dry season, soil often must be tilled in a very dry state to prepare the land for planting with the onset of the first rains. Tillage under such dry conditions can be very difficult and can result in hard clods if the soils contain much sticky-type silicate clay.

Conventional Tillage and Crop Production

Since the Middle Ages, the moldboard plow has been the primary tillage implement most used in the Western world.[5] Its purpose is to lift, twist, and invert the soil while incorporating crop residues and animal wastes into the plow layer (Figure 4.31). The moldboard plow is often supplemented by the disk plow, which is used to cut up residues and partially incorporate them into the soil. In conventional practice, such primary tillage has been followed by a number of secondary tillage operations, such as harrowing to kill weeds and to break up clods, thereby preparing a suitable seedbed. After the crop is planted, the soil may receive further secondary tillage to control weeds and to break up crusting of the immediate soil surface.

Conservation Tillage and Soil Tilth

Pictures and words to help you understand the differences among tillage implements, what they do to residues and soil, and their consequences for erosion and productivity: http://www.WTAMU.EDU/~crobinson/TILLAGE/tillage.htm

In recent years, land management systems have been developed that minimize the need for soil tillage. Since these systems also leave considerable plant residues on or near the soil surface, they protect the soil from erosion (see Section 17.6 for a detailed discussion). For this reason, the tillage practices followed in these systems are called *conservation tillage*.[6] Box 4.4 [photo (*b*)] illustrates a no-till operation, where one crop is planted in the residue of another, with virtually no tillage. Other minimum-tillage systems permit some stirring of the soil, but still leave a high proportion of the crop residues on the surface. These organic residues protect the soil from the beating action of raindrops and the abrasive action of the wind, thereby reducing water and wind erosion and maintaining soil structure.

FIGURE 4.31 While the action of the moldboard plow lifts, turns, and loosens the upper 15 to 20 cm of soil (the furrow slice), the counterbalancing downward force compacts the next lower layer of soil. This compacted zone can develop into a *plowpan*. Compactive action can be understood by imagining that you are lifting a heavy weight—as you lift the weight your feet press down on the floor below. (Photo courtesy of R. Weil)

[5] For an early but still valuable critique of the moldboard plow, see Faulkner (1943).

[6] The U.S. Department of Agriculture defines *conservation tillage* as any system that leaves at least 30% of the soil surface covered by residues.

Soil Crusting

Falling drops of water during heavy rains or sprinkler irrigation beat apart the aggregates exposed at the soil surface. In some soils the dilution of salts by this water stimulates the dispersion of clays. Once the aggregates become dispersed, small particles and dispersed clay tend to wash into and clog the soil pores. Soon the soil surface is covered with a thin layer of fine, structureless material called a **surface seal.** The surface seal inhibits water infiltration and increases erosion losses.

As the surface seal dries, it forms a hard **crust.** Seedlings, if they emerge at all, can do so only through cracks in the crust. A crust-forming soil is compared to one with stable aggregates in Figure 4.32. Formation of a crust soon after a crop is sown may allow so few seeds to emerge that the crop has to be replanted. In arid and semiarid regions, soil sealing and crusting can have disastrous consequences because high runoff losses leave little water available to support plant growth.

Crusting can be minimized by keeping some vegetative or mulch cover on the land to reduce the impact of raindrops. Improved management of soil organic matter and use of certain soil amendments can "condition" the soil and help prevent clay dispersion and crust formation (see also Section 9.19).

Soil Conditioners

GYPSUM. Gypsum (calcium sulfate) can improve the physical condition of many types of soils, from some highly weathered acid soils to some low-salinity, high-sodium soils of semiarid regions (see Chapter 9). The more soluble gypsum products provide enough electrolytes (cations and anions) to promote flocculation and inhibit the dispersion of aggregates, thus preventing surface crusting. Field trials have shown that gypsum-treated soils permit greater water infiltration and are less subject to erosion than untreated soils. Similarly, gypsum can reduce the strength of hard subsurface layers, thereby allowing greater root penetration and subsequent plant uptake of water from the subsoil.

ORGANIC POLYMERS. Certain synthetic organic polymers can stabilize soil structure in much the same way as do natural organic polymers such as polysaccharides. Figure 4.33 shows the dramatic stabilizing effect of synthetic polyacrylamides used in irrigation water. A number of research reports indicate that the best results can be obtained by combining the use of PAM and gypsum products.

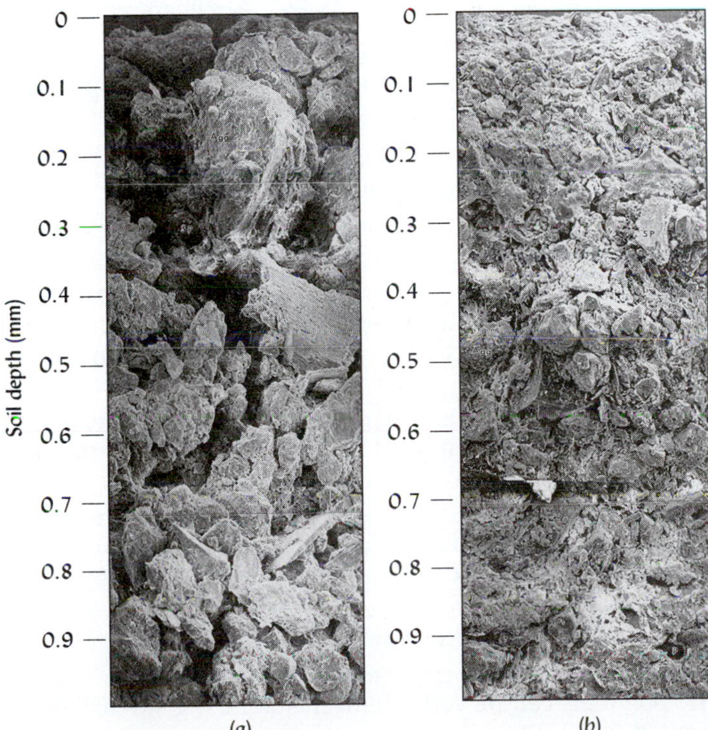

FIGURE 4.32 Scanning electron micrographs of the upper 1 mm of a soil with stable aggregation (*a*) compared to one with unstable aggregates (*b*). Note that the aggregates in the immediate surface have been destroyed and a surface crust has formed. The bean seedling (*c*) must break the soil crust as it emerges from the seedbed. [Photos (*a*) and (*b*) from O'Nofiok and Singer (1984), used with permission of Soil Science Society of America; photo (*c*) courtesy of R. Weil]

FIGURE 4.33 The remarkable stabilizing effect of a synthetic polyacrylamide is seen in the furrow on the right compared to the left, untreated, furrow. Irrigation water broke down much of the structure of the untreated soil but had no effect on the treated row. The two furrows are about 1 meter apart. [From Mitchell (1986)]

OTHER SOIL CONDITIONERS. Several species of algae that live near the soil surface are known to produce quite effective aggregate-stabilizing compounds. Application of small quantities of commercial preparations containing such algae may bring about a significant improvement in surface soil structure.

Various humic materials are marketed for their soil conditioning effects when incorporated at low rates (<500 kg/ha). However, carefully conducted research at many universities has failed to show that these materials have significantly affected aggregate stability or crop yield, as claimed.

General Guidelines for Managing Soil Tilth

Although each soil presents unique problems and opportunities, the following principles are generally relevant to managing soil tilth:

1. Minimizing tillage, especially moldboard plowing, disk harrowing, or rototilling, reduces the loss of aggregate-stabilizing organic matter.
2. Timing traffic activities to occur when the soil is as dry as possible and restricting tillage to periods of optimum soil moisture conditions will minimize destruction of soil structure.
3. Mulching the soil surface with crop residues or plant litter adds organic matter, encourages earthworm activity, and protects aggregates from beating rain and direct solar radiation.
4. Adding crop residues, composts, and animal manures to the soil is effective in stimulating microbial supply of the decomposition products that help stabilize soil aggregates.
5. Including sod crops in the rotation favors stable aggregation by helping to maintain soil organic matter, providing maximal aggregating influence of fine plant roots, and assuring a period without tillage.
6. Using cover crops and green manure crops, where practical, provides another good source of root action and organic matter for structural management.
7. Applying gypsum (or calcareous limestone if the soil is acidic) by itself or in combination with synthetic polymers can be very useful in stabilizing surface aggregates, especially in irrigated soils.

4.9 SOIL PROPERTIES RELEVANT TO ENGINEERING USES

Field Rating of Soil Consistence and Consistency

CONSISTENCE. Soil **consistence** is a term used by soil scientists to describe the resistance of a soil to mechanical stresses or manipulations at various moisture contents. Soils are rated for consistence as part of describing a soil profile and for estimating suitability for traffic and tillage. This property is a composite expression of those forces of mutual attraction among soil particles, and between particles and pore water, that determine the ease with which a soil can be reshaped or ruptured.

Moisture content greatly influences how a soil responds to stress; hence, moist and dry soils are given separate consistence ratings (Table 4.6). A dry, clayey soil that cannot be crushed between the thumb and forefinger but can be crushed easily underfoot would be designated as *hard*. If a clod of moist soil crumbles into aggregates when crushed with only light pressure, it is said to be friable. Friable soils are easily tilled or excavated.

The degrees of *stickiness* and *plasticity* (malleability) of soil in the wet condition are often included in describing soil consistence (although not shown in Table 4.6).

CONSISTENCY. The term **consistency** is used in a similar manner by soil engineers to describe the degree to which a soil resists deformation when a force is applied. However, consistency is determined by the soil's resistence to *penetration* by an object, while the soil scientist's consistence describes resistence to *rupture*. Instead of crushing a clod of soil, the engineer attempts to penetrate it with either the blunt end of a pencil (some use their thumbs) or a thumbnail. For example, if the blunt end of a pencil makes only a slight indentation, but the thumbnail penetrates easily, the soil is rated as *very firm* (Table 4.5).

Field observations of both consistence and consistency provide valuable information to guide decisions about loading and manipulating soils. For construction purposes, however, soil engineers usually must make more precise measurements of a number of related soil properties that help predict how a soil will respond to applied stress.

TABLE 4.6 Field Tests and Terms Used to Describe the Consistence and Consistency of Soils

The consistency of cohesive materials is closely related to, but not exactly the same as, their consistence. Conditions of least coherence are represented by terms at the top of each column, those of greater coherence near the bottom.

Soil consistence[a]				*Soil consistency*[b]	
Dry soil	*Moist to wet soil*	*Soil dried then submerged in water*	*Field rupture (crushing) test*	*Soil at in situ moisture*	*Field penetration test*
Loose	Loose	Not applicable	Specimen not obtainable	Soft	Blunt end of pencil penetrates deeply with ease
Soft	Very friable	Noncemented	Crumbles under very slight force between thumb and forefinger	Medium firm	Blunt end of pencil can penetrate about 1.25 cm with moderate effort
Slightly hard	Friable	Extremely weakly cemented	Crumbles under slight force between thumb and forefinger	Firm	Blunt end of pencil can penetrate about 0.5 cm
Hard	Firm	Weakly cemented	Crushes with difficulty between thumb and forefinger	Very firm	Blunt end of pencil makes slight indentation; thumbnail easily penetrates
Very hard	Extremely firm	Moderately cemented	Cannot be crushed between thumb and forefinger, but can be crushed slowly underfoot	Hard	Blunt end of pencil makes no indentation; thumbnail barely penetrates
Extremely hard	Slightly rigid	Strongly cemented	Cannot be crushed by full body weight underfoot		

[a]Abstracted from Soil Survey Division Staff (1993).
[b]Modified from McCarthy (1993).

Soil Strength and Sudden Failure

Soil liquefaction:
www.ce.washington.edu/~liquefaction/html/what/what1.html

Perhaps the most important property of a soil for engineering uses is its **strength.** This is a measure of the capacity of a soil mass to withstand stresses without giving way to those stresses by rupturing or becoming deformed. The failure of a soil to withstand stress can be seen where a structure topples as its weight exceeds the soil's bearing strength, where earthen dams give way under the pressure of impounded water, or where pavements and structures slide down unstable hillsides (Figure 1.6).

Cohesive Soils. The strength of cohesive soils (essentially soils with a clay content of more than about 15%) declines dramatically if the material is very wet and the pores are nearly filled with water. Then the particles are forced apart so that neither the cohesive nor the frictional component is very strong, making the soil prone to failure, often with catastrophic results such as the mudslide shown in Figure 4.34. On the other hand, if cohesive soils become dry or more compacted, their strength increases as particles are forced into closer contact with one another—a result that has implications for plant root growth as well as for engineering (see Section 4.5).

Noncohesive Soils. The strength of dry, noncohesive soil materials such as sand depends entirely on frictional forces, including the interlocking of rough particle surfaces. One reflection of the strength of a noncohesive material is its **angle of repose,** the steepest angle to which it can be piled without slumping. Smooth, rounded sands grains cannot be piled as steeply as can rough, interlocking sands. If a small amount of water bridges the gaps between particles, electrostatic attraction of the water for the mineral surfaces will increase the soil strength (see Figure 4.35 for an example). Interparticle water bridges explain why cars can drive along the edge of the beach where the sand is moist, but their tires sink in and lose traction on loose, dry sand or in saturated quicksand (see Figure 5.20 for an example of the latter).

Collapsible Soils. Certain soils that exhibit considerable strength at low in situ water contents lose their strength suddenly if they become wet. Such soils may collapse without warning under a roadway or building foundation. A special case of soil collapse is **thixotropy,** the sudden assumption of liquid properties (liquification) of a wet soil mass when subjected to vibrations, such as those accompanying earthquakes and blasting.

Most **collapsible soils** are noncohesive materials in which loosely packed sand grains are cemented at their contact points by small amounts of gypsum, clay, or water under tension. These soils usually occur in arid and semiarid regions, where such ce-

FIGURE 4.34 Houses damaged by a mudslide that occurred when the soils of a steep hillside in Oregon became saturated with water after a period of heavy rains. The weight of the wet soil exceeded its shear strength, causing the slope to fail. Excavations for roads and houses near the foot of a slope can contribute to the lack of slope stability, as can removal of tree roots by large-scale clear-cutting on the slope itself. (Photo courtesy of John Griffith, Coos Bay, Ore.)

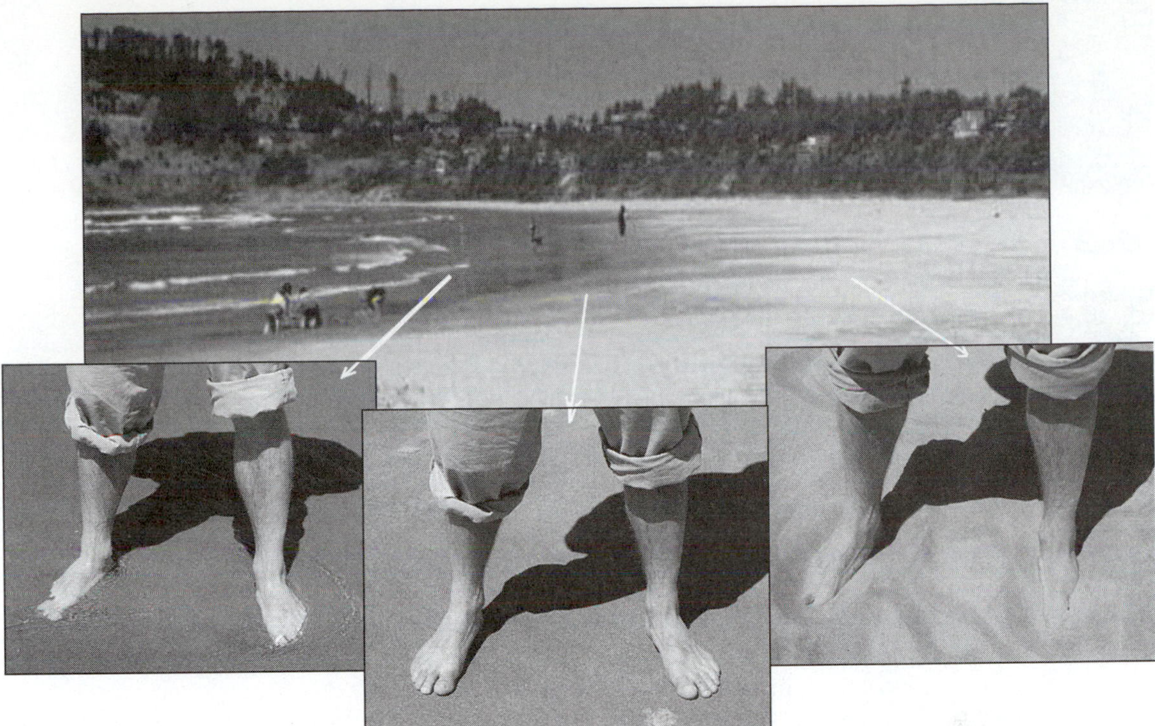

FIGURE 4.35 A walk along a beach such as this one in Oregon illustrates the concept of soil strength for sandy materials. The dry sand (lower right) has little strength and your feet easily mire into it as you walk along. There is nothing to hold the individual sand particles together to permit them to serve as a firm base. As you move toward the ocean where the soil has been thoroughly wetted by the incoming waves, but where there is no standing water (lower center), you find firm footing, indicating considerably higher soil strength. Thin water films act as bridges between sand particles, holding them together and thereby resisting penetration by the feet. If you stand in shallow water along the edge of the ocean (lower left), once again your feet penetrate the surface sand, indicating that soil strength has been reduced. Each sand particle is completely surrounded by water, which acts more as a lubricant than as a binding force. If you were to drive an automobile over the same areas, only the wetted but not submerged sand would provide a firm base. Soil strength is a property of great concern to engineers, but also must be reckoned with by plant roots as they try to penetrate soils with high bulk densities. (Photos courtesy of R. Weil)

menting agents are relatively stable. When these soils are wetted, excess water may dissolve cements such as gypsum or disperse clays that form bridges between particles, causing a sudden loss of strength.

Settlement—Gradual Compression

While embankments and hill slopes commonly fail due to stresses that exceed the soil's strength, most buildings and roads are unlikely to provide loads that cause the soil to rupture. Instead, most foundation problems result from slow, often uneven, vertical subsidence or **settlement** of the soil.

COMPACTION CONTROL. For the purposes of growing plants, soil compaction is to be avoided; however, soils are purposely compacted prior to being used as a roadbed or for a building foundation. Compaction occurring later under the heavy load would result in uneven settlement and a cracked pavement or foundation. Compaction to an optimum density is usually achieved on clayey soils by heavy sheepsfoot rollers that knead the soil like a baker kneads bread dough (Figure 4.36). For sandy soils, vibrating rollers or impact hammers do a better job because they can jar the particles into a tight packing arrangement (see Figure 4.14). Some colloidal silicate clays, and micas of all sizes, can be compressed when a load is placed upon them, but regain their original shape when the load is removed. Consequently, these materials make poor road bases and foundations.

The **Proctor test** is the most common method used to obtain data that can guide efforts at compacting soil materials before construction. Samples of a given soil at different moisture levels are compacted by a drop hammer and the resulting bulk densities are measured. This shows the maximum that the soil can be compacted and the moisture

FIGURE 4.36 Compaction of soils used as foundations and roadbeds is accomplished by heavy equipment such as this sheepsfoot roller. The knobs ("sheepsfeet") concentrate the mass on a small impact area, punching and kneading the loose, freshly graded soil to optimum density. (Photo courtesy of R. Weil)

content at which the greatest compaction occurs. On the construction site, water tank trucks will be used, if needed, to bring the water content of the soil to the determined optimum level, and compaction equipment such as that shown in Figure 4.36 will be used to achieve the desired density.

COMPRESSIBILITY. A **consolidation test** may be conducted on a soil specimen to determine its **compressibility**—how much its volume will be reduced by a given applied force. Because of the relatively low porosity and equidimensional shape of the individual mineral grains, very sandy soils resist compression once the particles have settled into a tight packing arrangement. They make excellent soils to bear foundations. The high porosity of clay floccules and the flakelike shape of clay particles gives clayey soils much greater compressibility. Soils consisting mainly of organic matter (peats) have the highest compressibilities and generally are unsuitable for foundations.

Expansive Soils

Some clays, particularly the smectites, swell when wet and shrink when dry (see Section 8.13). Expansive soils are rich in these types of clay (see Figure 3.13). Damage caused by such expansive soils in the United States rarely makes the evening news programs, although the total cost annually exceeds that caused by tornados, floods, earthquakes, or any other type of natural disaster. Expansive clays occur on about 20% of the land area in the United States and cause upwards of $4 billion in damages annually to pavements, foundations, and utility lines. The damages can be severe in certain sites in all parts of the country, but are most extensive in regions that have long dry periods alternating with periods of rain (e.g., California, Texas, Wyoming, and Colorado).

Atterberg Limits

As a dry soil containing expansive clays takes on increasing amounts of water, it eventually begins to increase in volume, and changes from a hard, rigid solid to a crumbly (friable) semisolid. The soil moisture level at which this change takes place is known as the **shrinkage limit.** As more water is added, the soil becomes plastic, is subject to molding, and the water content is said to be at the **plastic limit.** If still more water is added, the soil-water system reaches a semiliquid state, and the water content is referred to as having reached the **liquid limit.** These critical water contents (measured in units of percent) are termed the **Atterberg limits**, after the Swedish engineer who developed the

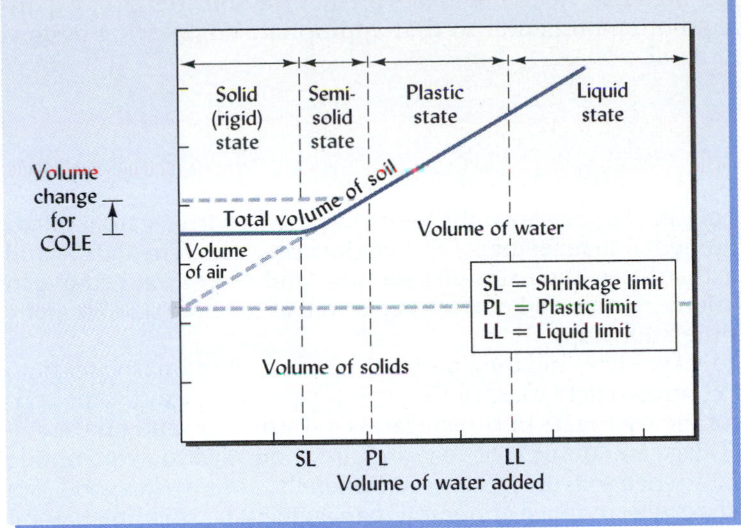

FIGURE 4.37 A common depiction of the Atterberg limits, which mark major shifts in the behavior of a cohesive soil as its water content changes (from left to right). As water is added to a certain volume of dry solids, first air is displaced; then, as more water is added, the total volume of the soil increases (if the soil has some expansive properties). When the shrinkage limit (SL) is reached, the once rigid, hard solid becomes a crumbly semisolid. With more water, the plastic limit (PL) is reached, after which the soil becomes plastic and can be molded. It remains in a plastic stage over a range of water contents until the liquid limit (LL) is exceeded, at which point the soil begins to behave as a viscous liquid that will flow when jarred. The volume change for calculating the coefficient of linear extensibility (COLE) is shown at the left.

system. Soil engineers determine the Atterberg limits as part of their investigation of soil consistency, to help predict the behavior of particular soils and the suitability of these materials for different construction purposes.

Plasticity Index. The plasticity index (PI) is the difference between the plastic limit (PL) and liquid limit (LL) and indicates the water-content range over which the soil has plastic properties:

$$\text{PI} = \text{LL} - \text{PL}$$

Soils with a high plasticity index (greater than about 25) are usually expansive clays that make poor roadbeds or foundations. Figure 4.37 shows the relationship among the Atterberg limits and the changes in soil volume associated with increasing water contents for a hypothetical soil. Smectite clays (see Section 8.13) generally have high liquid limits and plasticity indices, especially if saturated with sodium. Kaolinite and other nonexpansive clays have low liquid limit values.

Coefficient of Linear Extensibility. The expansiveness of a soil (and therefore the hazard of its destroying foundations and pavements) can be quantified as the *coefficient of linear extensibility* (COLE). Suppose a sample of soil is moistened to its plastic limit and molded into the shape of a bar with length LM. If the bar of soil is allowed to air dry, it will shrink to length LD. The COLE is the percent reduction in length of the soil bar upon shrinking:

$$\text{COLE} = \frac{\text{LM} - \text{LD}}{\text{LM}} \times 100$$

Figure 4.37 indicates how the volume change used to calculate the COLE relates to the Atterberg limits.

Unified Classification System for Soil Materials

The U.S. Army Corps of Engineers and the U.S. Bureau of Reclamation have established a widely used system of classifying soil materials to aid in predicting the engineering behavior of different soils. Each type of soil is given a two-letter designation based primarily on its particle-size distribution (texture), Atterberg limits, and organic-matter content.

The system first groups soils into coarse-grained soils (more than 50% retained on a 0.075 mm sieve) and fine-grained soils (at least half smaller than 0.075 mm). The coarse materials are further divided on the basis of grain size (gravels and sands), amount of fines present, and uniformity of grain size (well or poorly graded). The fine-grained soils are divided into silts, clays, and organic materials. These classes are further subdivided on the basis of their liquid limit (above or below 50) and their plasticity index.

This classification of soil materials helps engineers predict the soil strength, expansiveness, compressibility, and other properties so that appropriate engineering designs can be made for the soil at hand.

4.10 CONCLUSION

Physical properties exert a marked influence on the behavior of soils with regard to plant growth, hydrology, environmental management, and engineering uses. The nature and properties of the individual particles, their size distribution, and their arrangement in soils determine the total volume of nonsolid pore space, as well as the pore sizes, thereby impacting on water and air relationships.

The properties of individual particles and their proportionate distribution (soil texture) are subject to little human control in field soils. However, it is possible to exert some control over the arrangement of these particles into aggregates (soil structure) and on the stability of these aggregates. Tillage and traffic must be carefully controlled to avoid undue damage to soil tilth, especially when soils are rather wet. Generally, nature takes good care of soil structure, and humans can learn much about soil management by studying natural systems. Vigorous and diverse plant growth, generous return of organic residues, and minimal physical disturbance are attributes of natural systems worthy of emulation. Proper plant species selection, crop rotation, and management of chemical, physical, and biological factors can help ensure maintenance of soil physical quality. In recent years, these management goals have been made more practical by the advent of conservation tillage systems that minimize soil manipulations while decreasing soil erosion and water runoff.

Particle size, moisture content, and plasticity of the colloidal fraction all help determine the stability of soil in response to loading forces from traffic, tillage, or building foundations. The physical properties presented in this chapter greatly influence nearly all other soil properties and uses, as discussed throughout this book.

STUDY QUESTIONS

1. If you were investigating a site for a proposed housing development, how could you use soil colors to help predict where problems might be encountered?
2. You are considering the purchase of some farmland in a region with variable soil textures. The soils on one farm are mostly sandy loams and loamy sands, while those on a second farm are mostly clay loams and clays. List the potential advantages and disadvantages of each farm as suggested by the texture of its soils.
3. Revisit your answer to question 2. Explain how soil structure in both the surface and subsurface horizons might modify your opinion of the merits of each farm.
4. Two different timber-harvest methods are being tested on adjacent forest plots with clay loam surface soils. Initially, the bulk density of the surface soil in both plots was 1.1 Mg/m^3. One year after the harvest operations, plot A soil had a bulk density of 1.48 Mg/m^3, while that in plot B was 1.29 Mg/m^3. Interpret these values with regard to the relative merits of systems A and B, and the likely effects on the soil's function in the forest ecosystem.
5. What are the textural classes of two soils, the first with 15% clay and 45% silt, and the second with 80% sand and 10% clay? (Hint: Use Figure 4.6)
6. For the forest plot B in question 4, what was the change in percent pore space of the surface soil caused by timber harvest? Would you expect that most of this change was in the micropores or in the macropores? Explain.
7. Discuss the positive and negative impacts of tillage on soil structure. What is another physical consideration that you would have to take into account in deciding whether to change from a conventional to a conservation tillage system?
8. What would you, as a home gardener, consider to be the three best and three worst things that you could do with regard to managing the soil structure in your home garden?

9. What does the Proctor test tell an engineer about a soil, and why would this information be important?
10. In a humid region characterized by expansive soils, a homeowner experienced burst water pipes, doors that no longer closed properly, and large vertical cracks in the brick walls. The house had had no problems for over 20 years, and a consulting soil scientist blamed the problems on a large tree that was planted near the house some 10 years before the problems began to occur. Explain.

REFERENCES

Camp, C. R., and J. F. Lund. 1964. "Effects of soil compaction on cotton roots," *Crops and Soils* **17**:13–14.

Emerson, W. W., R. C. Foster, and J. M. Oades. 1986. "Organomineral complexes in relation to soil aggregation and structure," in P. M. Huang and M. Schnitzer (eds.), *Interaction of Soil Minerals with Natural Organics and Microbes*. SSSA Special Publication no. 17. (Madison, Wis.: Soil Sci. Soc. Amer.).

Faulkner, E. H. 1943. *Plowman's Folly*. (Norman, Okla.: University of Oklahoma Press).

Gent, J. A., Jr., R. Ballard, A. E. Hassan, and D. K. Cassel. 1984. "Impact of harvesting and site preparation on physical properties of Piedmont forest soils," *Soil Sci. Soc. Amer. J.* **48**:173–177.

Gent, J. A., Jr., and L. A. Morris. 1986. "Soil compaction from harvesting and site preparation in the Upper Gulf Coastal Plain," *Soil Sci. Soc. Amer. J.* **50**:443–446.

Le Bissonnais, Y., and D. Arrouays. 1997. "Aggregate stability and assessment of soil crustability and erodibility: II. Application to humic loamy soils with various organic carbon contents," *European J. Soil Sci.* **48**:39–48.

Martens, D. A. 2000. "Management and crop residue influence soil aggregate stability," *J. Environ. Qual.* **29**:723–727.

McCarthy, D. F. 1993. *Essentials of Soil Mechanics and Foundations,* 4th ed. (Englewood Cliffs, NJ: Prentice Hall).

McRae, S. G. 1999. "Land reclamation after open-pit mineral extraction in Britain," *Remediation and Management of Degraded Lands*. (Boca Raton, FL: CRC Press).

Mitchell, A. R. 1986. "Polyacrylamide application in irrigation water to increase infiltration," *Soil Sci.* **141**:353–358.

Oades, J. M. 1993. "The role of biology in the formation, stabilization, and degradation of soil structure," *Geoderma* **56**:377–400.

O'Nofiok, O., and M. J. Singer. 1984. "Scanning electron microscope studies of surface crusts formed by simulated rainfall," *Soil Sci. Soc. Amer. J.* **48**:1137–1143.

Parker, M. M., and D. H. Van Lear. 1996. "Soil heterogeneity and root distribution of mature loblolly pine stands in Piedmont soils," *Soil Sci. Soc. Amer. J.* **60**:1920–1925.

Rillig, M. C., S. F. Wright, M. F. Allen, and C. B. Field. 1999. "Rise in carbon dioxide changes soil structure," *Nature* **400**:628.

Six, J., E. T. Elliott, and K. Paustian. 2000. "Soil structure and soil organic matter II. A normalized stability index and the effect of mineralogy," *Soil Sci. Soc. Amer. J.* **64**:1042–1049.

Soil Survey Division Staff. 1993. *Soil Survey Manual*. USDA-NRSC Agricultural Handbook 18. (Washington, D.C.: U.S. Government Printing Office), pp. 174–175.

Tiessen, H., J. W. B. Stewart, and J. R. Bottany. 1982. "Cultivation effects on the amounts and concentration of carbon, nitrogen, and phosphorus in grassland soils," *Agron. J.* **74**:831–835.

Tisdall, J. M. 1994. "Possible role of soil microorganisms in aggregation in soils," *Plant and Soil* **159**:115–121.

Tisdall, J. M., and J. M. Oades. 1982. "Organic matter and water-stable aggregates in soils," *J. Soil Sci.* **33**:141–163.

Voorhees, W. B. 1984. "Soil compaction, a curse or a cure?" *Solutions* **28**: 42–47 (Peoria, Ill.: Solutions Magazine Inc.)

Wright, S. F., J. L. Starr, and I. C. Paltineanu. 1999. "Changes in aggregate stability and concentration of glomalin during tillage management transition," *Soil Sci. Soc. Amer. J.* **63**: 1825–1829.

5

Soil Water: Characteristics and Behavior

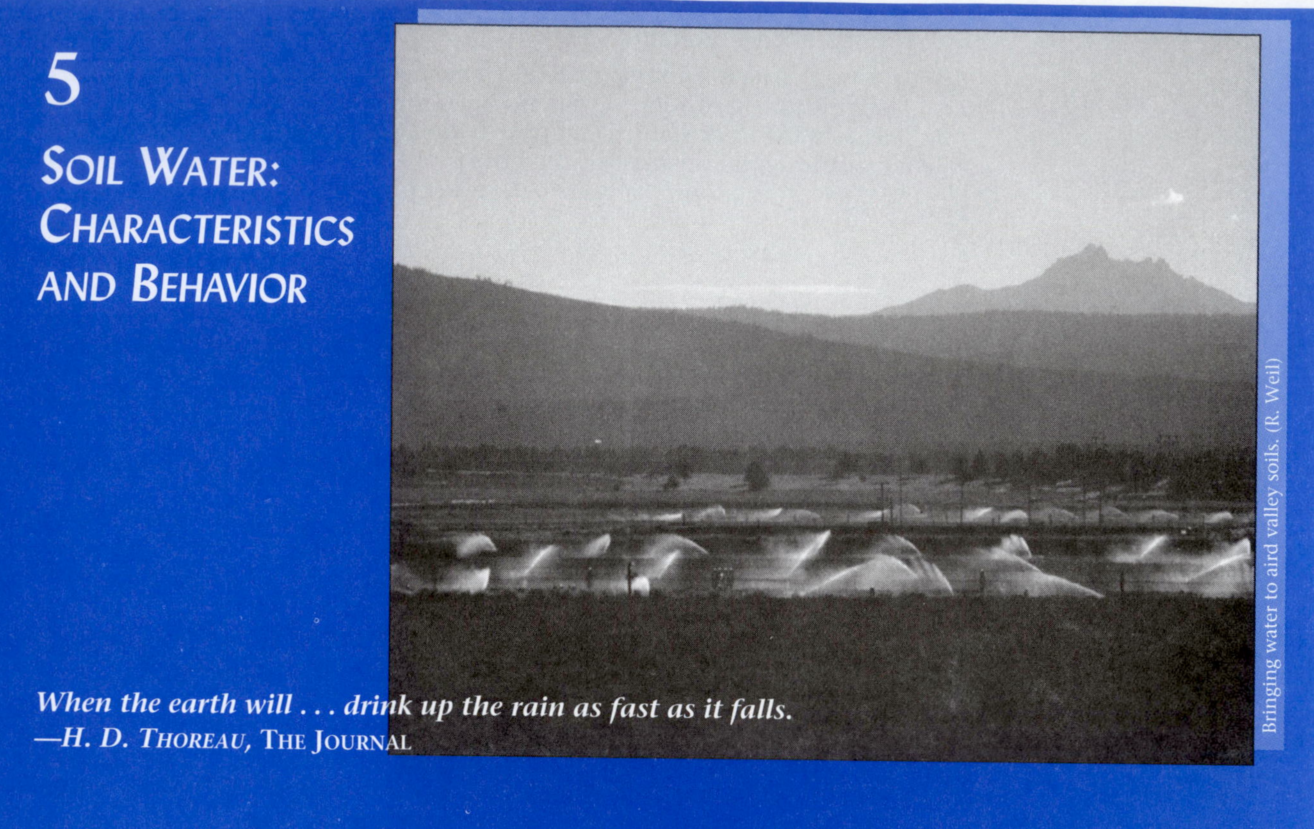

Bringing water to aird valley soils. (R. Weil)

When the earth will . . . drink up the rain as fast as it falls.
—H. D. Thoreau, The Journal

Water is a vital component of every living thing. Although it is one of nature's simplest chemicals, water has unique properties that promote a wide variety of physical, chemical, and biological processes. These processes greatly influence almost every aspect of soil development and behavior, from the weathering of minerals to the decomposition of organic matter, from the growth of plants to the pollution of groundwater.

We are all familiar with water. We drink it, wash with it, swim in it, and irrigate our crops with it. But water in the soil is something quite different from water in a drinking glass. In the soil, water is intimately associated with solid particles, particularly those that are colloidal in size. The interaction between water and soil solids changes the behavior of both.

Water causes soil particles to swell and shrink, to adhere to each other, and to form structural aggregates. Water participates in innumerable chemical reactions that release or tie up nutrients, create acidity, and wear down minerals so that their constituent elements eventually contribute to the saltiness of the oceans.

Attraction to solid surfaces restricts some of the free movement of water molecules, making it less liquid and more solidlike in its behavior. In the soil, water can flow up as well as down. Plants may wilt and die in a soil whose profile contains a million kilograms of water in a hectare. A layer of sand and gravel in a soil profile may actually inhibit drainage, causing the upper horizons to become muddy and saturated with water during much of the year. These and other soil water phenomena seem to contradict our intuition about how water ought to behave.

Soil–water interactions influence many of the ecological functions of soils and practices of soil management. These interactions determine how much rainwater runs into and through the soil and how much runs off the surface. Control of these processes in turn determines the movement of chemicals to the groundwater, and of both chemicals and eroded soil particles to streams and lakes. The interactions affect the rate of water loss through leaching and evapotranspiration, the balance between air and water in soil pores, the rate of change in soil temperature, the rate and kind of metabolism of soil organisms, and the capacity of soil to store and provide water for plant growth.

The characteristics and behavior of water in the soil is a common thread that interrelates nearly every chapter in this book. The principles contained in this chapter will help us understand why mudslides occur in water-saturated soils (Chapter 4), why earthworms improve soil quality (Chapter 10), why rice paddies contribute to global ozone depletion (Chapter 12); and why famine stalks humanity in certain regions of the world. Mastery of the principles presented in this chapter are fundamental to a working knowledge of the soil system.

5.1 STRUCTURE AND RELATED PROPERTIES OF WATER[1]

More on basic properties of water: http://www.geog.ouc.bc.ca/physgeog/contents/8a.html

The ability of water to influence so many soil processes is determined primarily by the structure of the water molecule. This structure also is responsible for the fact that water is a liquid, not a gas, at temperatures found on earth. Water is, with the exception of mercury, the *only* inorganic (not carbon-based) liquid found on Earth. Water is a simple compound, its individual molecules containing one oxygen atom and two much smaller hydrogen atoms. The elements are bonded together covalently, each hydrogen atom sharing its single electron with the oxygen.

Polarity

The arrangement of the three atoms in a water molecule is not symmetrical, as one might expect. Instead of the atoms being arranged linearly (H-O-H), the hydrogen atoms are attached to the oxygen in a V-shaped arrangement at an angle of only 105 degrees. As shown in Figure 5.1, this results in an asymmetrical molecule with the shared electrons spending most of the time nearer to the oxygen than to the hydrogen. Consequently, the water molecule exhibits *polarity;* that is, the charges are not evenly distributed. Rather, the side on which the hydrogen atoms are located tends to be electropositive and the opposite side electronegative.

This imbalance of charge allows water to play unique roles in the soil environment. For example, the attraction of the *negative* (oxygen) end of water molecules to cations such as H^+, Na^+, K^+, and Ca^{2+} accounts for the hydration of these cations. Simultaneously, the *positive* (hydrogen) end of water molecules is attracted to the negatively charged clay surfaces. Many physical and chemical properties of soils are affected by such relationships.

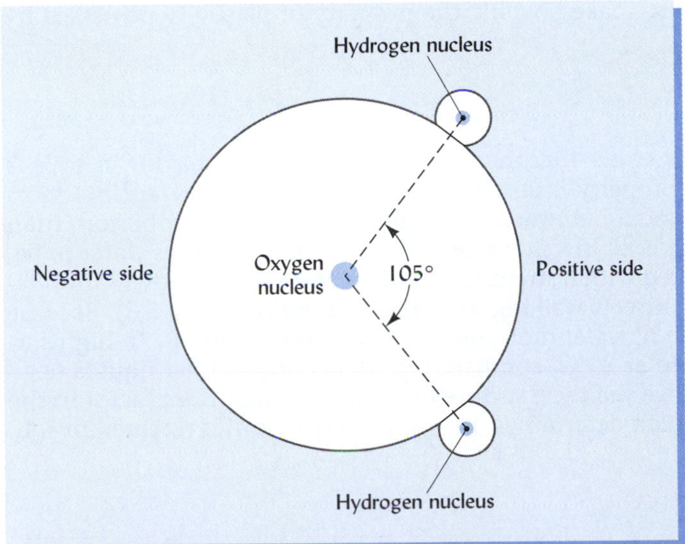

FIGURE 5.1 Two-dimensional representation of a water molecule showing a large oxygen atom and two much smaller hydrogen atoms. The HOH angle of 105° results in an asymmetrical arrangement. One side of the water molecule (that with the two hydrogens) is electropositive; the other is electronegative. This accounts for the polarity of water.

[1] For a comprehensive treatise on these and other physical soil concepts, see Hillel (1998).

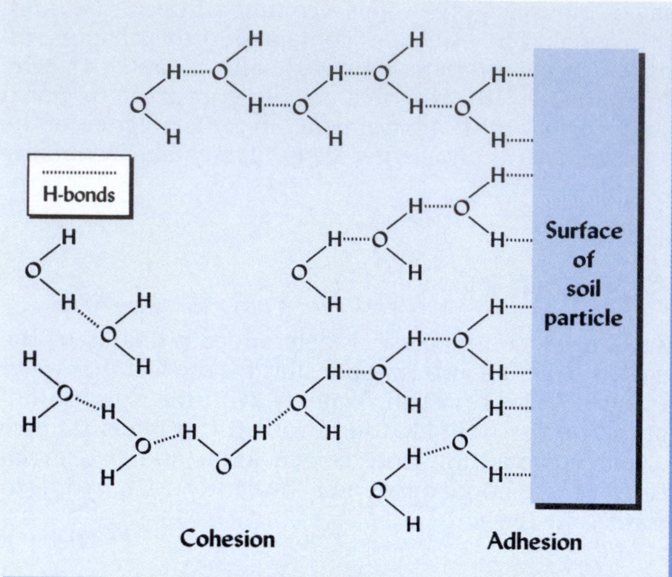

FIGURE 5.2 The forces of cohesion (between water molecules) and adhesion (between water and solid surface) in a soil–water system. The forces are largely a result of H-bonding, shown as broken lines. The adhesive or adsorptive force diminishes rapidly with distance from the solid surface. The cohesion of one water molecule to another results in water molecules forming temporary clusters that are constantly changing in size and shape as individual water molecules break free or join up with others. The cohesion between water molecules also allows the solid to indirectly restrict the freedom of water for some distance beyond the solid–liquid interface.

Hydrogen Bonding

Through a phenomenon called **hydrogen bonding,** a hydrogen atom may be shared between two electronegative atoms forming a relatively low-energy link. Because of its high electronegativity, an O atom in one water molecule exerts some attraction for the H atom in a neighboring water molecule. This type of bonding accounts for the polymerization of water.

Cohesion versus Adhesion

Hydrogen bonding accounts for two basic forces responsible for water retention and movement in soils: the attraction of water molecules for each other (**cohesion**) and the attraction of water molecules for solid surfaces (**adhesion**). By adhesion (also called *adsorption*), some water molecules are held rigidly at the surfaces of soil solids. In turn, these tightly bound water molecules hold by cohesion other water molecules farther removed from the solid surfaces (Figure 5.2). Together, the forces of adhesion and cohesion make it possible for the soil solids to retain water and control its movement and use. Adhesion and cohesion also make possible the property of plasticity possessed by clays (see Section 4.9).

Surface Tension

Another important property of water that markedly influences its behavior in soils is that of **surface tension.** This property is commonly evidenced at liquid–air interfaces and results from the greater attraction of water molecules for each other (cohesion) than for the air above. The net effect is an inward force at the surface that causes water to behave as if its surface were covered with a stretched elastic membrane, an observation familiar to those who have seen insects walking on water in a pond (Figure 5.3). Because of the relatively high attraction of water molecules for each other, water has a high surface tension (72.8 newtons/mm at 20°C) compared to that of most other liquids (e.g., ethyl alcohol, 22.4 N/mm). As we shall see, surface tension is an important factor in the phenomenon of capillarity, which determines how water moves and is retained in soil.

5.2 CAPILLARY FUNDAMENTALS AND SOIL WATER

The movement of water up a wick typifies the phenomenon of capillarity. Two forces cause capillarity: (1) the attraction of water for the solid (adhesion or adsorption), and (2) the surface tension of water, which is due largely to the attraction of water molecules for each other (cohesion). Capillarity can be demonstrated by placing one end of a fine,

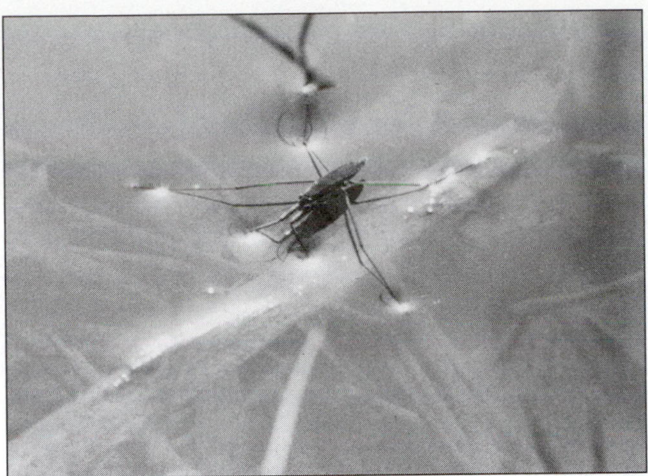

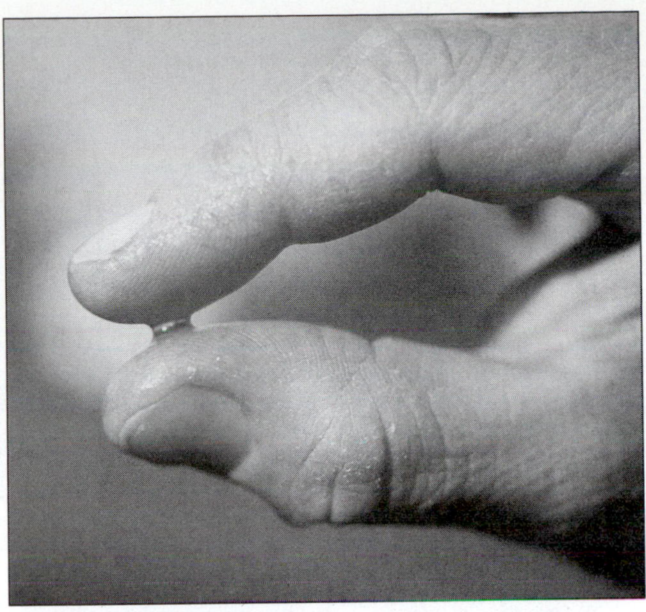

FIGURE 5.3 Everyday evidences of water's surface tension (left) as insects walk on water and do not sink, and of forces of cohesion and adhesion (right) as a drop of water is held between the fingers. (Photos courtesy of R. Weil)

clean glass tube in water. The water rises in the tube: The smaller the tube bore, the higher the water rises. The water molecules are attracted to the sides of the tube (adhesion) and start to spread out along the glass in response to this attraction. At the same time, the cohesive forces hold the water molecules together and create surface tension, causing a curved surface (called a *meniscus*) to form at the interface between water and air in the tube (Figure 5.4). Lower pressure under the meniscus in the glass tube (P2) allows the higher pressure (P1) on the free water to push water up the tube. The process continues until the water in the tube has risen high enough that its weight just balances the pressure differential across the meniscus.

The height of rise in a capillary tube is inversely proportional to the tube radius *r*. Capillary rise is also inversely proportional to the density of the liquid, and is directly

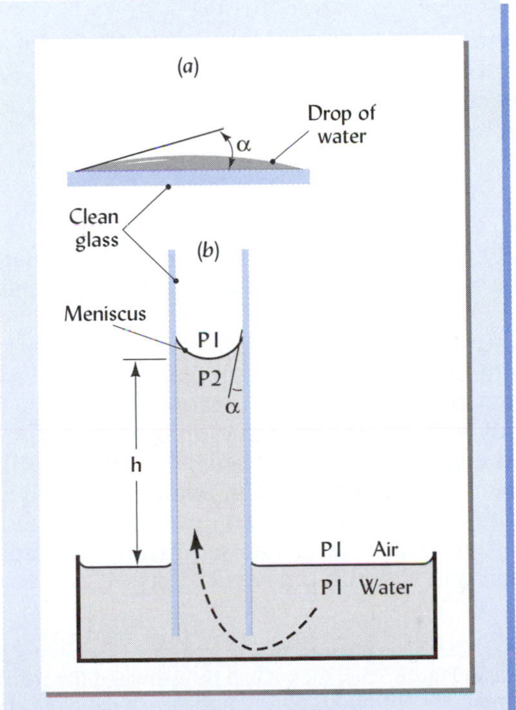

FIGURE 5.4 The interaction of water with a hydrophillic surface (*a*) results in a characteristic *contact angle* (α). If the solid surface surrounds the water as in a tube, a curved water–air interface termed the *meniscus* forms because of adhesive and cohesive forces. When air and water meet in a curved meniscus, pressure on the convex side of the curve is lower than on the concave side. (*b*) Capillary rise occurs in a fine hydrophillic (e.g., glass) tube because pressure under the meniscus (P2) is less than pressure on the free water.

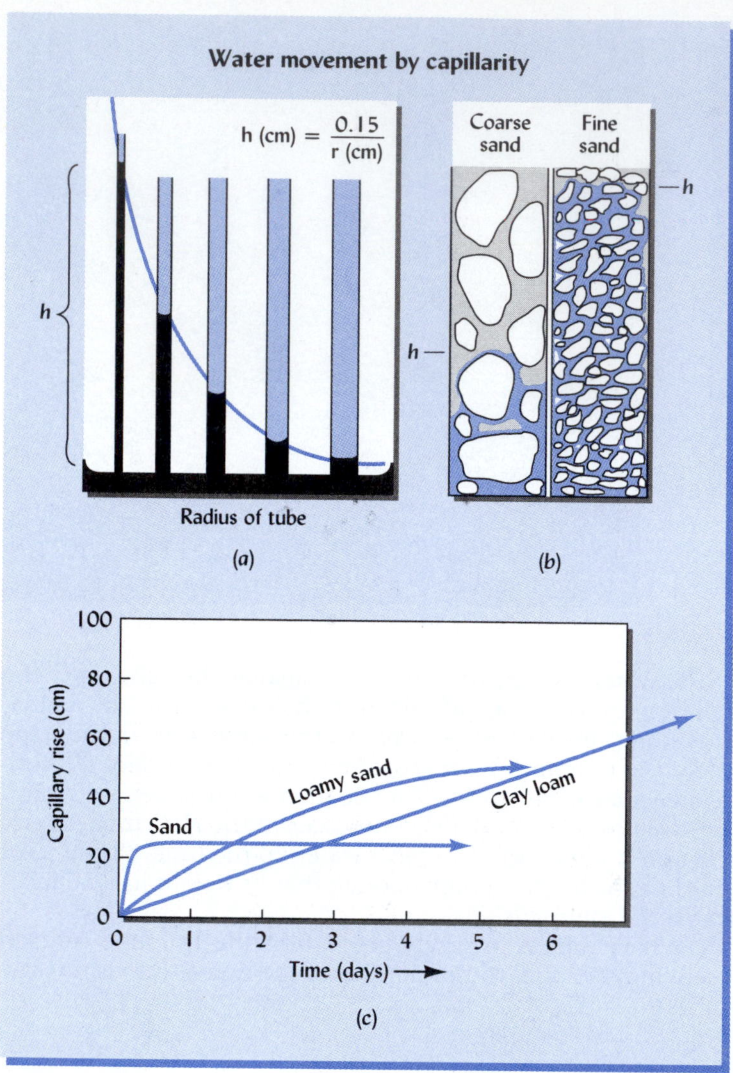

FIGURE 5.5 Upward capillary movement of water through tubes of different bore and soils with different pore sizes. (*a*) The capillary equation can be graphed to show that the height of rise *h* doubles when the tube inside radius is halved. The same relationship can be demonstrated using glass tubes of different bore size. (*b*) The same principle also relates to pore sizes in a soil and height of capillary rise, but the rise of water in a soil is rather jerky and irregular because of the tortuous shape and variability in size of the soil pores (as well as because of pockets of trapped air). (*c*) The finer the soil texture, the greater the proportion of small-sized pores and, hence, the higher the ultimate rise of water above a free water table. However, because of the much greater frictional forces in the smaller pores, the capillary rise is much slower in the finer-textured soil than in the sand.

Capillary soil water demonstration:
http://www.wtamu.edu/~crobinson/SoilWater/capact.html#1

proportional to the liquid's surface tension and the degree of its adhesive attraction to the soil surface. If we limit our consideration to water at a given temperature (e.g., 20°C), then these factors can be combined into a single constant, and we can use a simple capillary equation to calculate the height of rise *h:*

$$h = \frac{0.15}{r} \tag{5.1}$$

where both *h* and *r* are expressed in centimeters. This equation tells us that the smaller the tube bore, the greater the capillary force and the higher the water rise in the tube (Figure 5.5*a*).

The upward movement due to capillarity in soils is illustrated in Figure 5.5. Usually the eventual height of rise resulting by capillarity is greater with fine-textured soils, but the rate of flow may be extremely slow because of frictional forces in the tiny pores. The large pores in sandy soils present little frictional resistance to rapid capillary water movement. However, as expected from our discussion of the capillary equation, the large radii of the pores between sand grains result in relatively little height of capillary rise.[2]

Although traditionally illustrated as an upward adjustment, capillary movement takes place in any direction (Figure 5.6). The significance of capillarity in controlling water movement in small pores will become evident as we turn to soil water energy concepts.

[2] For example, if water rises by capillarity to a height of 25 cm above a free-water surface in a sand (as shown in the example in Figure 5.5*c*), then it can be estimated (by rearranging the capillary equation to $r = 0.15/h$) that the smallest continuous pores must have a radius of about 0.006 cm (0.15/25 = 0.006). This calculation gives an approximation of the minimum effective pore radius in a soil.

FIGURE 5.6 As this field irrigation scene in Arizona shows (left), water has moved up by capillarity from the irrigation furrow toward the top of the ridge. The photo on the right illustrates some horizontal movement to both sides and away from the irrigation water.

5.3 SOIL WATER ENERGY CONCEPTS

As we consider energy, we should keep in mind that all substances, including water, tend to move or change from a higher to a lower energy state. Therefore, if we know the pertinent energy levels at various points in a soil, we can predict the direction of water movement. It is the *differences* in energy levels from one contiguous site to another that influence this water movement.

Forces Affecting Potential Energy

The discussion of the structure and properties of water in the previous section suggests three important forces affecting the energy level of soil water. First, adhesion, or the attraction of water to the soil solids (matrix), provides a **matric** force (responsible for adsorption and capillarity) that markedly reduces the energy state of water near particle surfaces. Second, the attraction of water to ions and other solutes, resulting in **osmotic** forces, tends to reduce the energy state of water in the soil solution. Osmotic movement of pure water across a semipermeable membrane into a solution (osmosis) is evidence of the lower energy state of water in the solution. The third major force acting on soil water is **gravity**, which always pulls the water downward. The energy level of soil water at a given elevation in the profile is thus higher than that of water at some lower elevation. This difference in energy level causes water to flow downward.

Soil Water Potential

The *difference* in energy level of water from one site or one condition (e.g., in wet soil) to another (e.g., in dry soil) determines the direction and rate of water movement in soils and in plants. In a wet soil, most of the water is retained in large pores or thick water films around particles. Therefore, most of the water molecules in a wet soil are not very close to a particle surface and so are not held very tightly by the soil solids (the matrix). In this condition, the water molecules have considerable freedom of movement, so their energy level is near that of water molecules in a pool of pure water outside the soil. In a drier soil, however, the water that remains is located in small pores and thin water films, and is therefore held tightly by the soil solids. Thus, the water molecules in a drier soil have little freedom of movement, and their energy level is much lower than that of the water molecules in wet soil. If wet and dry soil samples are brought in touch with each other, water will move from the wet soil (higher energy state) to the drier soil (lower energy).

To predict how water will move in soils, the energy status of soil water in a particular location in the profile is compared with that of pure water at standard pressure and temperature, unaffected by the soil and located at some reference elevation. The *difference* in energy levels between this pure water in the reference state and that of the soil water is termed **soil water potential** (Figure 5.7), the term *potential,* like the term *pressure,* implying a difference in energy status. Water will move from a soil zone having a high soil water potential to one having a lower soil water potential (Figure 5.8). Keep this fact in mind when considering the behavior of water in soils.

The soil water potential is due to several forces, each of which is a component of the **total soil water potential** ψ_t. These components are due to differences in energy levels resulting from gravitational, matric, submerged hydrostatic, and osmotic forces and are termed **gravitational potential** ψ_g, **matric potential** ψ_m, **submergence potential,** and

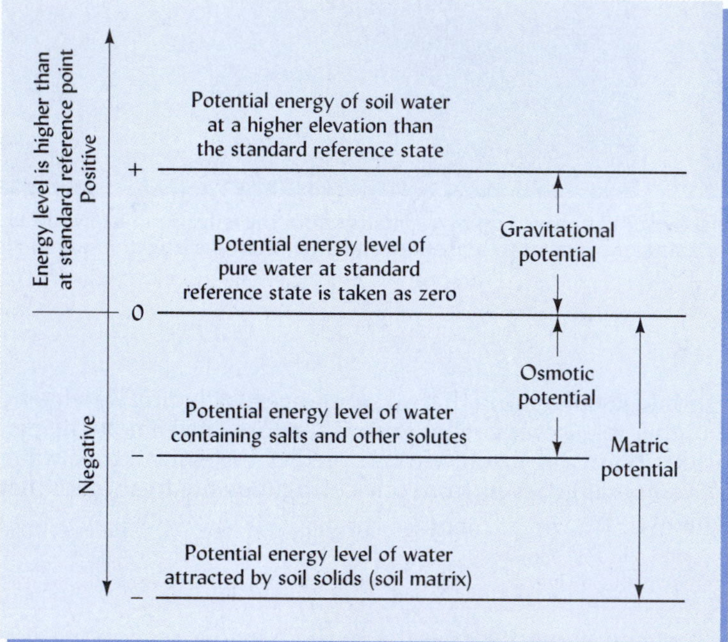

FIGURE 5.7 Relationship between the potential energy of pure water at a standard reference state (pressure, temperature, and elevation) and that of soil water. If the soil water contains salts and other solutes, the mutual attraction between water molecules and these chemicals reduces the potential energy of the water, the degree of the reduction being termed *osmotic potential*. Similarly, the mutual attraction between soil solids (soil matrix) and soil water molecules also reduces the water's potential energy. In this case the reduction is called *matric potential*. Since both of these interactions reduce the water's potential energy level compared to that of pure water, the changes in energy level (osmotic potential and matric potential) are both considered to be negative. In contrast, differences in energy due to gravity (gravitational potential) are always positive. This is because the reference elevation of the pure water is purposely designated at a site in the soil profile below that of the soil water. A plant root attempting to remove water from a moist soil would have to overcome all three forces simultaneously.

FIGURE 5.8 Whether concerning matric potential, osmotic potential, or gravitational potential (as shown here), water always moves to where its energy state will be lower. In this case the energy lost by the water is used to turn the historic Mabry Mills waterwheel and grind flour. (Photo courtesy of R. Weil)

osmotic potential ψ_o, respectively. All of these components act simultaneously to influence water behavior in soils. The general relationship of soil water potential to potential energy levels is shown in Figure 5.7 and can be expressed as:

$$\psi_t = \psi_g + \psi_m + \psi_o + \ldots \tag{5.2}$$

where the ellipses (. . .) indicate the possible contribution of additional potentials not yet mentioned.

Gravitational Potential

The force of gravity acts on soil water the same as it does on any other body (Figure 5.8), the attraction being toward the earth's center. The gravitational potential ψ_g of soil water may be expressed mathematically as:

$$\psi_g = gh \tag{5.3}$$

where g is the acceleration due to gravity and h is the height of the soil water above a reference elevation. The reference elevation is usually chosen within the soil profile or at its lower boundary to ensure that the gravitational potential of soil water above the reference point will always be positive.

Following heavy precipitation or irrigation, gravity plays an important role in removing excess water from the upper horizons and in recharging groundwater below the soil profile. It will be given further attention when the movement of soil water is discussed (see Section 5.5).

Pressure Potential (Including Submergence and Matric Potentials)

Pressure potential accounts for the effects on soil water potential of all factors other than gravity and solute levels. It most commonly includes (1) the positive hydrostatic pressure due to the weight of water in saturated soils and aquifers, and (2) the negative pressure due to the attractive forces between the water and the soil solids or the soil matrix.

The hydrostatic pressures give rise to what is often termed the submergence potential ψ_s, a component that is operational only for water in saturated zones below the water table. Anyone who has dived to the bottom of a swimming pool has felt hydrostatic pressure on the eardrums.

The attraction of water to solid surfaces gives rise to the matric potential ψ_m, which is always negative because the water attracted by the soil matrix has an energy state lower than that of pure water. (These negative pressures are sometimes referred to as *suction* or *tension*.) The matric potential is operational in unsaturated soil above the water table, while the submergence potential applies to water in saturated soil or below the water table (Figure 5.9).

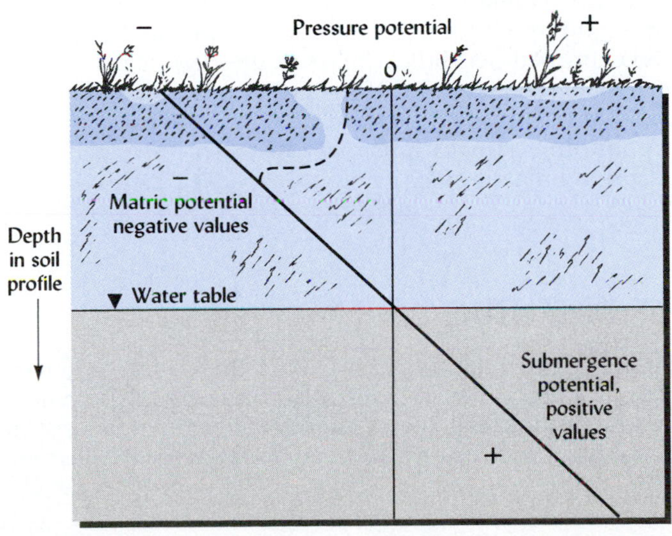

FIGURE 5.9 The matric potential and submergence potential are both pressure potentials that may contribute to total water potential. The matric potential is always negative and the submergence potential is positive. When water is in unsaturated soil above the water table (top of the saturated zone) it is subject to the influence of matric potentials. Water below the water table in saturated soil is subject to submergence potentials. In the example shown here, the matric potential decreases linearly with elevation above the water table, signifying that water rising by capillary attraction up from the water table is the only source of water in this profile. Rainfall or irrigation (see dotted line) would alter or curve the straight line, but would not change the fundamental relationship described.

While each of these pressures is significant in specific field situations, the matric potential is important in all unsaturated soils because there are omnipresent interactions between soil solids and water. The movement of soil water, the availability of water to plants, and the solutions to many civil engineering problems are determined to a considerable extent by matric potential. Consequently, matric potential will receive primary attention in this textbook, along with gravitational and osmotic potential.

Matric potential ψ_m, which results from the phenomena of adhesion (or adsorption) and of capillarity, influences soil moisture retention as well as soil water movement. Differences between the ψ_m of two adjoining zones of a soil encourage the movement of water from moist zones (high energy state) to dry zones (low energy state) or from large pores to small pores. Although this movement may be slow, it is extremely important, especially in supplying water to plant roots.

Osmotic Potential

Osmotic potential and plant cells:
http://www.biologie.uni-hamburg.de/b-online/e22/22c.htm

The osmotic potential ψ_o is attributable to the presence of solutes in the soil solution. The solutes may be inorganic salts or organic compounds. Their presence reduces the potential energy of water, primarily because of the reduced freedom of movement of the water molecules that cluster around each solute ion or molecule. The greater the concentration of solutes, the more the osmotic potential is lowered. As always, water will tend to move to where its energy level will be lower, in this case to the zone of higher solute concentration. However, liquid water will move in response to differences in osmotic potential (the process termed *osmosis*) only if a *semipermeable membrane* exists between the zones of high and low osmotic potential, allowing water through but *preventing the movement of the solute*. If no membrane is present, the solute, rather than the water, generally moves to equalize concentrations.

Because soil zones are *not* generally separated by membranes, the osmotic potential ψ_o has little effect on the mass movement of water in soils. Its major effect is on the uptake of water by plant root cells that *are* isolated from the soil solution by their semipermeable cell membranes. In soils high in soluble salts, ψ_o may be lower (have a greater negative value) in the soil solution than in the plant root cells. This leads to constraints in the uptake of water by the plants. In very salty soil, the soil water osmotic potential may be low enough to cause cells in young seedlings to collapse (plasmolyze) as water moves from the cells to the lower osmotic potential zone in the soil.

Methods of Expressing Energy Levels

Several units can be used to express differences in energy levels of soil water. One is the *height of a water column* (usually in centimeters) whose weight just equals the potential under consideration. We have already encountered this means of expression since the h in the capillary equation (Section 5.2) tells us the matric potential of the water in a capillary pore. A second unit is the standard *atmosphere* pressure at sea level, which is 760 mm Hg or 1020 cm of water. The unit termed *bar* approximates the pressure of a standard atmosphere. Energy may be expressed per unit of mass (joules/kg) or per unit of volume (newtons/m^2). In the International System of Units (SI), 1 pascal (Pa) equals 1 newton (N) acting over an area of 1 m^2. In this textbook we use Pa or kilopascals (kPa) to express soil water potential. Since other publications may use other units, Table 5.1 shows the equivalency among common means of expressing soil water potential.

5.4 SOIL WATER CONTENT AND SOIL WATER POTENTIAL

Types of soil water sensors, how they work, how to buy them:
http.//www.sowacs.com/sensors/index.html

The previous discussions suggest an inverse relationship between the water content of soils and the tenacity with which the water is held in soils. Water is more likely to flow out of a wet soil than from one low in moisture. Many factors affect the relationship between soil water potential ψ and moisture content θ. A few examples will illustrate this point.

TABLE 5.1 Approximate Equivalents Among Expressions of Soil Water Potential

Height of unit column of water, cm	*Soil water potential, bars*	*Soil water potential, kPa*[a]
0	0	0
10.2	−0.01	−1
102	−0.1	−10
306	−0.3	−30
1,020	−1.0	−100
15,300	−15	−1,500
31,700	−31	−3,100
102,000	−100	−10,000

[a]The SI unit kilopascal (kPa) is equivalent to 0.01 bar.

Soil Water Versus Energy Curves

The relationship between soil water potential ψ and moisture content θ of three soils of different textures is shown in Figure 5.10. Such curves are sometimes termed *water release characteristic curves,* or simply *water characteristic curves.* The clay soil holds much more water at a given potential than does the loam or sand. Likewise, at a given moisture content, the water is held much more tenaciously in the clay than in the other two soils (note that soil water potential is plotted on a log scale).

Soil structure also influences soil water content–energy relationships. A well-granulated soil has more total pore space and greater overall water-holding capacity than one with poor granulation or one that has been compacted. The compacted soil will hold less water, most of which will be held tenaciously in the small and midsize pores that characterize such a soil.

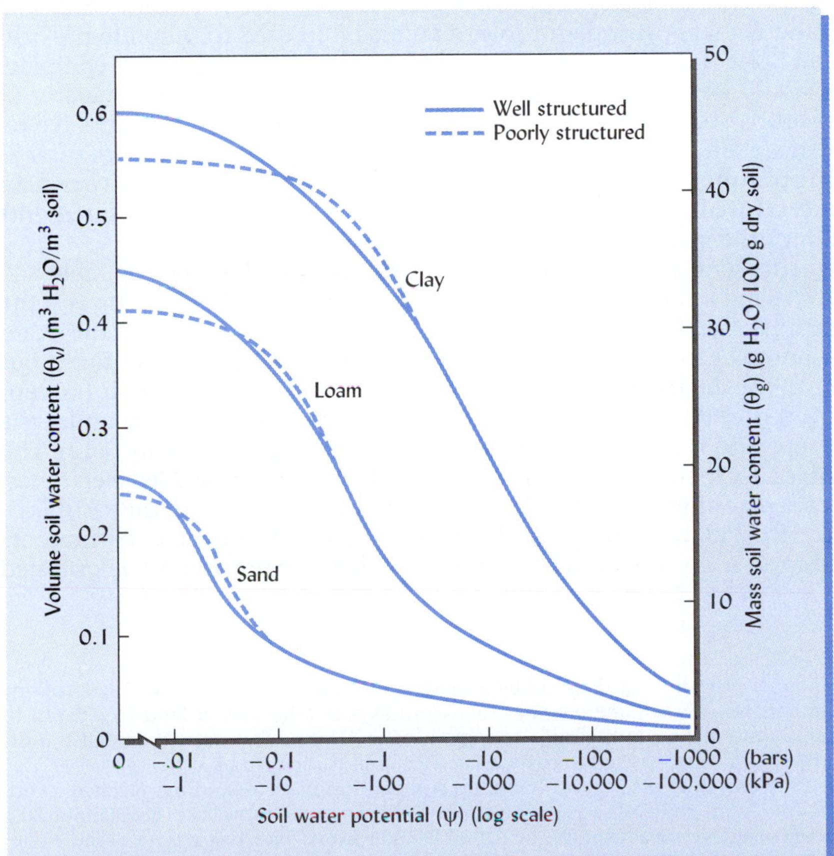

FIGURE 5.10 Soil water potential curves for three representative mineral soils. The curves show the relationship obtained by slowly drying completely saturated soils. The dashed lines show the effect of compaction or poor structure. The soil water potential ψ (which is negative) is expressed in terms of bars (upper scale) and kilopascals (kPa) (lower scale). Note that the soil water potential is plotted on a log scale.

Measurement of Soil Water Status

The soil water characteristic curves (Figure 5.10) highlight the importance of making two general kinds of soil water measurements: the *amount* of water present (water content) and the *energy status* of the water (soil water potential). To understand or manage water supply and movement in soils it is essential to have information (directly measured or inferred) on *both* types of measurements. For example, a soil water potential measurement might tell us whether water will move toward the groundwater, but without a corresponding measurement of the soil water content, we would not know the possible significance of the contribution to groundwater.

Generally, the behavior of soil water is most closely related to the energy status of the water, not to the amount of water in a soil. Thus, a clay loam soil and a loamy sand will both feel moist and will easily supply water to plants when the ψ_m is, say, -10 kPa. However, the amount of water held by the clay loam, and thus the length of time it could supply water to plants, would be far greater at this potential than would be the case for the loamy sand.

Measuring Water Content

The **volumetric water content** θ is defined as the volume of water associated with a given volume (usually 1 m^3) of dry soil (see Figure 5.10). A comparable expression is the **mass water content** θ_m, or the mass of water associated with a given mass (usually 1 kg) of dry soil. Both of these expressions have advantages for different uses. In most cases we shall use the volumetric water content θ in this text.

Because in the field we think of plant root systems as exploring a certain depth of soil, and because we express precipitation (and sometimes irrigation) as a depth of water (e.g., mm of rain), it is often convenient to express the volumetric water content as a *depth ratio* (depth of water per unit depth of soil). Conveniently, the numerical values for these two expression are the same. For example, for a soil containing 0.1 m^3 of water per m^3 of soil (10% by volume), the depth ratio of water is 0.1 m of water per m of soil depth.[3]

Of the three methods of measuring soil water *content* listed in Table 5.2, the **gravimetric method** is the most straightforward and is often used to calibrate the others. As illustrated in Box 5.1, a sample of moist soil is weighed before and after complete drying, and the mass of water lost is divided by the dry soil mass to calculate grams of water per g of dry soil.[4] Although inexpensive and simple, this method does involve removing a sample of soil and bringing it back to the lab. Instrumental methods, such as **neutron scattering** or **time domain reflectrometry** (TDR) (Figure 5.11), involve certain limitations and great initial expense, but are capable of making repeated measurements on the same soil once the instrument probe has been installed.

Several instruments for measuring soil water *potential* are also described in Table 5.2, each with limitations and advantages. The most straightforward of these methods is the **tensiometer,** which is essentially a water-filled tube with one end sealed and the other end capped by a porous ceramic tip. It is inserted into the soil, and water flows out through the tip into the adjacent soil until the water potential is the same in the tensiometer tube and in the soil. As the water is drawn out, a vacuum develops under the sealed end of the tube and is measured by an attached vacuum gauge (Figure 5.12). The **electrical resistance block** method uses small porous blocks with embedded electrodes. When the blocks are placed in a moist soil they absorb water, decreasing the resistance to the flow of electricity between the embedded electrodes. Both of these instruments can be automated to turn on irrigation systems when the soil reaches a predetermined water potential.

---

[3]When measuring amounts of water added to soil by irrigation, it is customary to use units of volume such as m^3 and hectare-meter (the volume of water that would cover a hectare of land to a depth of 1 m). Generally, farmers and ranchers in the irrigated regions of the United States use the English units ft^3 or acre-foot (the volume of water needed to cover an acre of land to a depth of 1 ft).

[4]Enough drying time must be allowed so that the soil has stopped losing water and has reached a constant weight. To save time, a microwave oven may be used. A dozen small samples of soil (about 20 g each) in glass beakers may be dried on a turntable in a 1000-W microwave oven using three or more consecutive 3-minute periods, stirring the soil between periods.

TABLE 5.2 Some Methods of Measuring Soil Water

Note that more than one method may be needed to cover the entire range of soil moisture conditions.

Method	Measures soil water: Content	Measures soil water: Potential	Useful range, kPa	Used mainly in: Field	Used mainly in: Lab	Comments
1. Gravimetric	×		0 to <−10,000		×	Destructive sampling; slow (1 to 2 days) unless microwave used. The standard for calibration.
2. Resistance blocks		×	−100 to <−1,500	×		Can be automated; not sensitive near optimum plant water contents.
3. Neutron scattering	×		0 to <−1,500	×		Radiation permit needed; expensive equipment; not good in high-organic-matter soils; requires access tube.
4. Time domain reflectometry (TDR)	×		0 to <−10,000	×	×	May be automated; accurate to 1 kPa; requires wave guides; expensive instrument.
5. Tensiometer		×	0 to −85	×		Accurate to 0.1 to 1 kPa; limited range; inexpensive; can be automated; needs periodic servicing.
6. Thermocouple psychrometer		×	50 to <−10,000	×	×	Moderately expensive; wide range; accurate only to ±50 kPa.
7. Pressure membrane apparatus		×	50 to <−10,000		×	Used in conjunction with gravimetric method to construct water characteristic curve.

BOX 5.1 GRAVIMETRIC DETERMINATION OF SOIL WATER CONTENT

The gravimetric procedures for determining mass soil water content θ_m are relatively simple. Assume that you want to determine the water content of a 100 g sample of moist soil. You dry the sample in an oven kept at 105°C and then weigh it again. Assume that the dried soil now weighs 70 g, which indicates that 30 g of water has been removed from the moist soil. Expressed in kilograms, this is 30 kg water associated with 70 kg dry soil.

Since the mass soil water content θ_m is commonly expressed in terms of kg water associated with 1 kg dry soil (*not* 1 kg of wet soil), it can be calculated as follows:

$$\frac{30 \text{ kg water}}{70 \text{ kg dry soil}} = \frac{X \text{ kg water}}{1 \text{ kg dry soil}}$$

$$X = \frac{30}{70} = 0.428 \text{ kg water/kg dry soil} = \theta_m$$

To calculate the volume soil water content θ, we need to know the bulk density of the dried soil, which in this case we shall assume to be 1.3 Mg/m^3. In other words, a cubic meter of this soil has a mass of 1300 kg. From the above calculations we know that the mass of water associated with this 1300 kg is 0.428 × 1300 or 556 kg.

Since 1 m^3 of water has a mass of 1000 kg, the 556 kg of water will occupy 556/1000 or 0.556 m^3.

Thus, the volume water content is 0.556 m^3/m^3 of dry soil:

$$\frac{1300 \text{ kg soil}}{\text{m}^3 \text{ soil}} * \frac{\text{m}^3 \text{ water}}{1000 \text{ kg water}} * \frac{0.428 \text{ kg water}}{\text{kg soil}} = \frac{0.556 \text{ m}^3 \text{ water}}{\text{m}^3 \text{ soil}}$$

The relationship between the mass and volume water contents for a soil can be summarized as:

$$\theta = D_b \times \theta_m \qquad (5.4)$$

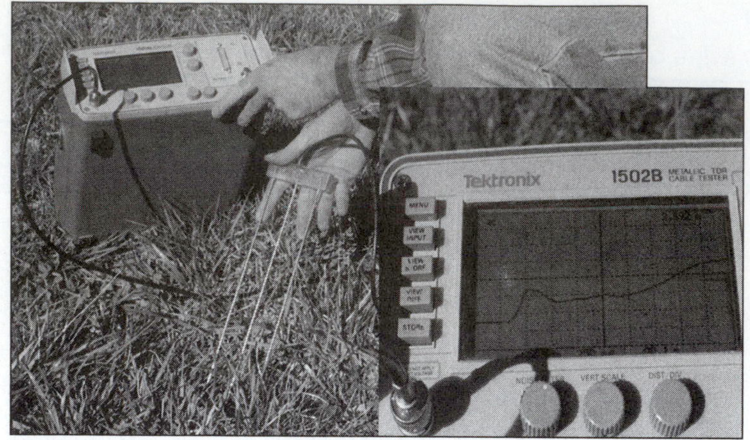

FIGURE 5.11 Instrumental measurement of soil water content using time domain reflectrometry (TDR). The electronic instrument sends a pulse of electromagnetic energy down the three parallel metal rods of a waveguide that the soil scientist is pushing into the soil. The TDR instrument makes precise picosecond measurements of the speed at which the pulse travels down the rods, a speed influenced by the nature of the surrounding soil. Microprocessors in the instrument analyze the wave patterns generated and calculate the apparent dielectric constant of the soil. Since the dielectric constant of a soil is mainly influenced by its water content, the instrument can accurately convert its measurements into volumetric water content of the soil. (Photos courtesy of R. Weil)

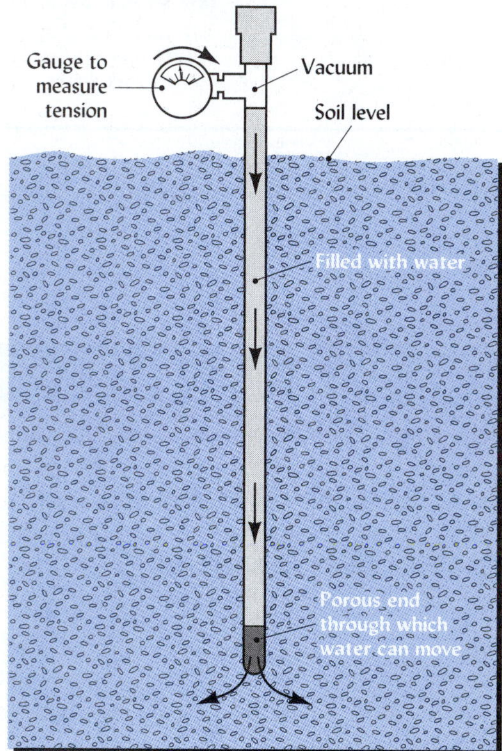

FIGURE 5.12 Tensiometer method of determining water potential in the field. Cross section showing the essential components of a tensiometer. Water moves through the porous end of the instrument in response to the pull (matric potential) of the soil. The vacuum so created is measured by a gauge that reads in kPa of tension (−kPa water potential).

5.5 THE FLOW OF LIQUID WATER IN SOIL

Advanced concepts in saturated flow:
http://fbe.uwe.ac.uk/public/geocal/soilmech/water/water.htm

Three types of water movement within the soil are recognized: (1) saturated flow, (2) unsaturated flow, and (3) vapor movement. In all cases water flows in response to energy gradients, with water moving from a zone of higher to one of lower water potential. *Saturated flow* takes place when the soil pores are completely filled (or saturated) with water. *Unsaturated flow* occurs when the larger pores in the soil are filled with air, leaving only the smaller pores to hold and transmit water. *Vapor movement* occurs as vapor pressure differences develop in relatively dry soils.

Saturated Flow Through Soils

Under some conditions, at least part of a soil profile may be completely saturated; that is, all pores, large and small, are filled with water. The lower horizons of poorly drained soils are often saturated, as are portions of well-drained soils above stratified layers of clay. During and immediately following a heavy rain or irrigation, pores in the upper soil zones are often filled entirely with water.

The quantity of water per unit of time Q/t that flows through a column of saturated soil can be expressed by Darcy's law, as follows:

$$\frac{Q}{t} = AK_{sat}\frac{\Delta\psi}{L} \tag{5.5}$$

where A is the cross-sectional area of the column through which the water flows, K_{sat} is the **saturated hydraulic conductivity,** $\Delta\psi$ is the change in water potential between the ends of the column (for example, $\psi_1 - \psi_2$), and L is the length of the column. For a given column, the rate of flow is determined by the ease with which the soil transmits water (K_{sat}) and the amount of force driving the water, namely the **water potential gradient** $\Delta\psi/L$. For saturated flow this force may also be called the **hydraulic gradient.** By analogy, think of pumping water through a garden hose, with K_{sat} representing the size of the hose (water flows more readily through a larger hose) and $\Delta\psi/L$ representing the size of the pump that drives the water though the hose.

The units in which K_{sat} is measured are length/time, typically cm/s or cm/h. The K_{sat} is an important property that helps determine how well a soil or soil material will perform in such uses as irrigated cropland, sanitary landfill cover material, wastewater storage lagoon lining, and septic tank drain field (Table 5.3). Figure 5.13 illustrates how K_{sat} can be determined by measuring the rate of downward vertical saturated flow, but keep in mind that saturated flow in other situations may be horizontal or even upwards (as when groundwater under pressure wells up under a stream).

Factors Influencing the Hydraulic Conductivity of Saturated Soils

Any factor affecting the size and configuration of soil pores will influence hydraulic conductivity. The total flow rate in soil pores is proportional to the fourth power of the radius. Thus, flow through a pore 1 mm in radius (say, a small earthworm channel) is equivalent to that in 10,000 pores with a radius of 0.1 mm, even though it takes only 100 pores of radius 0.1 mm to give the same cross-sectional area as a 1 mm pore. As a result, macropores (radius >0.05 mm) account for most water movement in saturated soils. The presence of biopores, such as root channels and earthworm burrows (typically >1 mm in radius), may have a marked influence on the saturated hydraulic conductivity of different soil horizons (Table 5.4). Because they usually have more macropore space, sandy soils generally have higher saturated conductivities than finer-textured soils. Likewise, soils with stable granular structure conduct water much more rapidly than do those with unstable structural units, which break down upon being wetted. Saturated conductivity of soils under natural vegetation is commonly much higher than where cultivated crops have been grown (Figure 5.14).

Preferential Flow. In natural field soil, water may flow rapidly through certain pathways in preference to slower, more uniform movement through the bulk of the soil. These

TABLE 5.3 Some Approximate Values of Saturated Hydraulic Conductivity (in Various Units) and Interpretations for Soil Uses

K_{sat}, *cm/s*	K_{sat}, *cm/h*	K_{sat}, *in./h*	*Comments*
1×10^{-2}	36	14	Typical of beach sand.
5×10^{-3}	18	7	Typical of very sandy soil, too rapid to effectively filter pollutants in wastewater.
5×10^{-4}	1.8	0.7	Typical of moderately permeable soils, K_{sat} between 1.0 and 15 cm/h considered suitable for most agricultural, recreational, and urban uses calling for good drainage.
5×10^{-5}	0.18	0.07	Typical of fine-textured, compacted or poorly structured soils. Too slow for proper operation of septic tank drain fields, most types of irrigation, and many recreational uses such as playgrounds.
$<1 \times 10^{-8}$	$<3.6 \times 10^{-5}$	$<1.4 \times 10^{-5}$	Extremely slow; typical of compacted clay. K_{sat} of 10^{-5} to 10^{-8} cm/h may be required where nearly impermeable material is needed, as for wastewater lagoon lining or landfill cover material.

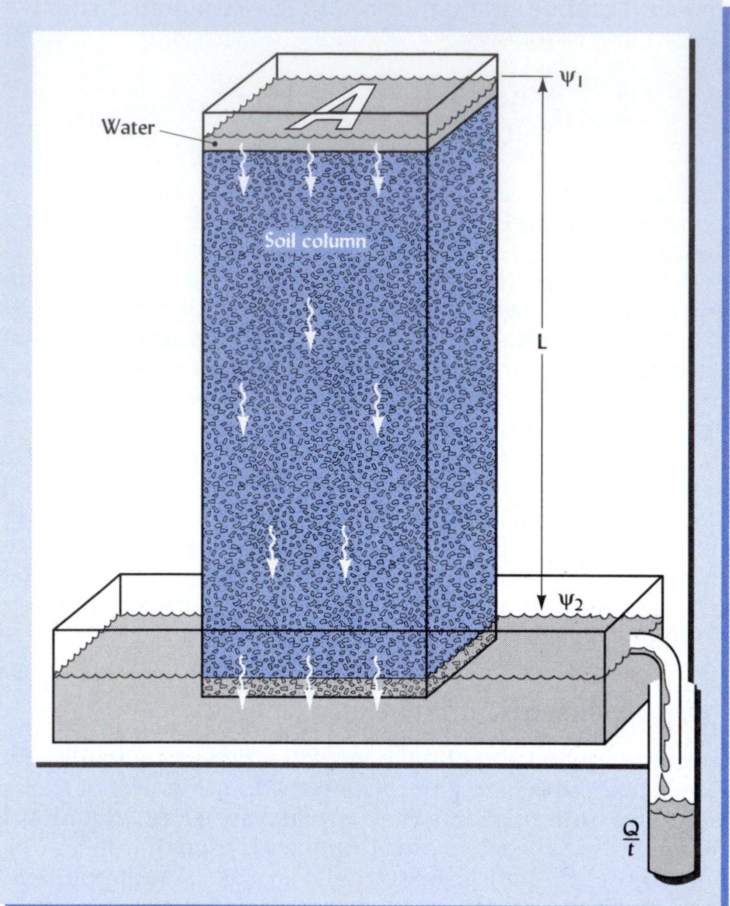

FIGURE 5.13 Saturated flow (percolation) in a column of soil with cross-sectional area A, cm^2. All soil pores are filled with water. At lower right, water is shown running off into a container to indicate that water is actually moving down the column. The force driving the water through the soil is the water potential gradient, $\psi_1 - \psi_2/L$, where both water potentials and length are expressed in cm (see Table 5.1). If we measure the quantity of water flowing out Q/t as cm^3/s we can rearrange Darcy's law (eq. 5.5) to calculate the saturated hydraulic conductivity of the soil K_{sat} in cm/s as:

$$K_{sat} = \frac{Q}{A} \times \frac{L}{\psi_1 - \psi_2} \quad (5.6)$$

Remember that the same principles apply where the water potential gradient moves the water in a horizontal direction.

preferential flow pathways may allow for rapid movement of chemical-laden water deep into the soil profile, increasing the likelihood of groundwater contamination. Deep clay shrinkage cracks deep earthworm burrows, and old root channels may serve as preferential flow paths. Organic and inorganic chemicals, and fecal bacteria from surface-applied animal manures, can move rapidly down these preferential channels (see Table 5.5).

TABLE 5.4 The Saturated Hydraulic Conductivity K_{sat} and Related Properties of Various Horizons in a Typic Hapludult Profile

The upper horizons had many biopores (mainly earthworm burrows), which resulted in high values of K_{sat} as well as extreme variability from sample to sample. The presence of a clay-enriched argillic horizon resulted in reduced K_{sat} values. Apparently most large biopores in this soil did not extend below 30 cm.

Horizon	*Depth, cm*	*Clay, %*	*Bulk density, Mg/m³*	*Mean K_{sat}, cm/h*	*Range of K_{sat} values[a], cm/h*
Ap	0–15	12.6	1.42	22.4	0.80–70
E	15–30	11.1	1.44	7.9	0.50–24
E/B	30–45	14.5	1.47	0.93	0.53–1.33
Bt	45–60	22.2	1.40	0.49	0.19–0.79
Bt	60–75	27.2	1.38	0.17	0.07–0.27
Bt	75–90	24.1	1.28	0.04	0.01–0.07

[a]For each soil layer K_{sat} was determined on five soil cores of 7.5 cm diameter.
Data from Waddell and Weil (1996).

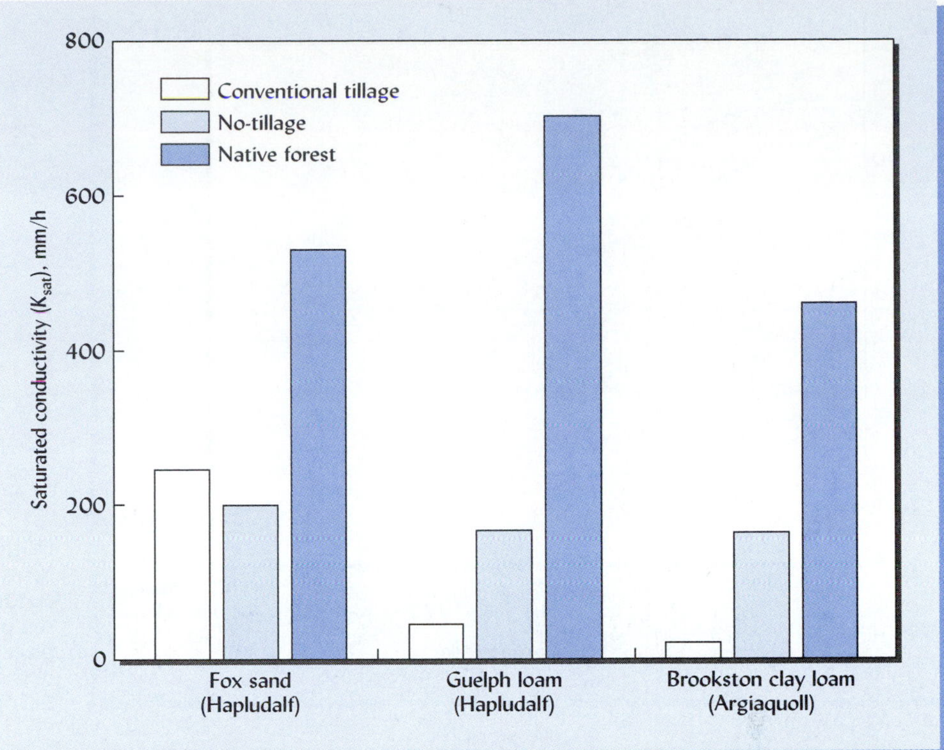

FIGURE 5.14 The effect of land management and soil texture on saturated conductivity (K_{sat}) of three soils in Canada. Soils under native woodlots had higher K_{sat} values, apparently due to higher organic matter contents and to preferential flow channels provided by decayed root channels and burrowing animals. Tillage practices had little effect on conductivity in sand, but in loam and clay loam soils conductivity was higher where no-tillage systems had been used, suggesting that no-till had increased the proportion of larger, water-conducting pores. [Drawn from averages of three methods of measuring K_{sat} in Reynolds et al. (2000)].

Unsaturated Flow in Soils

In unsaturated soils, the macropores are filled with air, leaving only the finer pores to accommodate water movement. The driving force for unsaturated water flow is mainly the **matric potential gradient**, the difference in the matric potential of the moist soil areas and nearby drier areas into which the water is moving. Movement will be from a zone of thick moisture films (high matric potential, e.g., −1 kPa) to one of thin films (lower matric potential, e.g., −100 kPa).

TABLE 5.5 Leaching of Pesticides by Preferential Flow

Most of the spring leaching of three widely used pesticides from a slowly permeable Alfisol in Indiana took place following the first major storm of the year. This leaching was mainly by preferential flow through macropores rather than through the general body of the soil. Preferential flow accounts for much of the contamination of groundwater from pesticide-treated soils.

	Leaching of pesticide applied, % (3-year average)		
Chemical	*First storm*	*Spring season*	*First storm/spring season total, %*
Carbofuran	0.22	0.25	88
Atrazine	0.037	0.053	68
Cyanazine	0.02	0.02	100

Calculated from Kladivco et al. (1999).

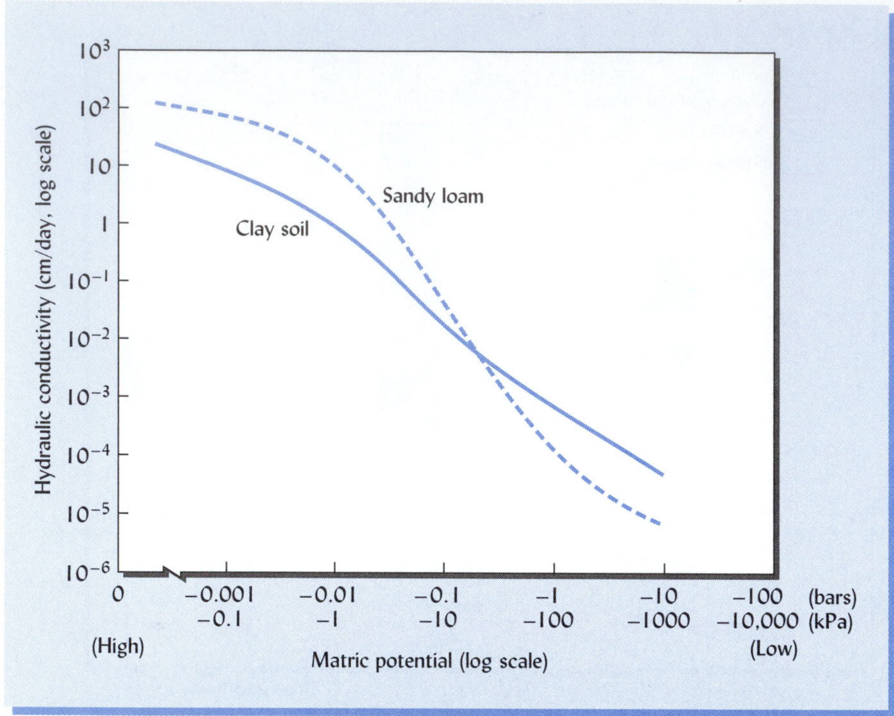

FIGURE 5.15 Generalized relationship between matric potential and hydraulic conductivity for a sandy soil and a clay soil (note log scales). Saturation flow takes place at or near zero potential, while much of the unsaturated flow occurs at a potential of −0.1 bar (−10 kPa) or below.

Influence of Texture. Figure 5.15 shows the general relationship between matric potential ψ_m (and, in turn, water content) and hydraulic conductivity of a sandy loam and clay soil. Note that at or near zero potential (which characterizes the saturated flow region), the hydraulic conductivity is thousands of times greater than at potentials that characterize typical unsaturated flow (−10 kPa and below).

At high potential levels (high moisture contents), hydraulic conductivity is higher in the sand than in the clay. The opposite is true at low potential values (low moisture contents). This relationship is to be expected because the sandy soil contains many large pores which are water-filled when the soil water potential is high (and the soil is quite wet), but most of these have been emptied by the time the soil water potential becomes lower than about −10 kPa. The clay soil has many more micropores which are still water-filled at lower soil water potentials (drier soil conditions) and can participate in unsaturated flow.

5.6 INFILTRATION AND PERCOLATION

Improving infiltration in rangeland soils:
http://www.ag.ndsu.nodak.edu/streeter/98report/chad98.htm

The process by which water enters the soil pore spaces and becomes soil water is termed **infiltration.** The rate at which water enters the soil is called the **infiltration capacity,** which is commonly measured in terms of mm/s or cm/h. This pivotal process greatly influences the moisture regime for plants, and the potential for soil degradation, chemical runoff, and down-valley flooding. The infiltration capacity is highest when rain or irrigation water is first applied, and decreases with time as the soil becomes saturated (see Figure 5.16).

Infiltration

Like hydraulic conductivity, infiltration capacity is a characteristic of each particular soil and depends mainly upon the texture and structure at the soil surface, but also upon the presence of any layers in the soil profile that might restrict the downward movement of water. The infiltration capacity of a soil may be easily measured using a simple device known as a **double ring infiltrometer** (see Figure 5.16).

Once the water has infiltrated the soil, the water moves downward into the profile by the process termed **percolation.** Both saturated and unsaturated flow are involved in percolation of water down the profile, and rate of percolation is related to the soil's hy-

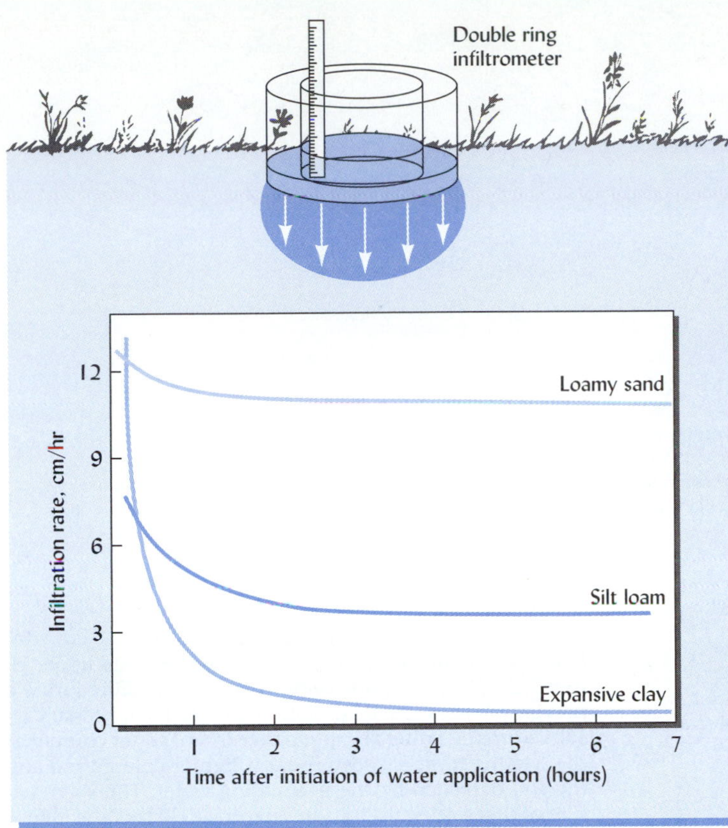

FIGURE 5.16 The potential rate of water entry into the soil, or infiltration capacity, can be measured by recording the drop in water level in a double ring infiltrometer (top). Changes in the infiltration rate of several soils during a period of water application by rainfall or irrigation are shown (bottom). Generally, water enters a dry soil rapidly at first, but its infiltration rate slows as the soil becomes saturated. The decline is least for very sandy soils with macropores that do not depend on stable structure or clay shrinkage. In contrast, a soil high in expansive clays may have a very high initial infiltration rate when large cracks are open, but a very low infiltration rate once the clays swell with water and close the cracks. Most soils fall between these extremes, exhibiting a pattern similar to that shown for the silt loam soil.

draulic conductivity. In the case of water that has infiltrated a relatively dry soil, the progress of water movement can be observed by the darkened color of the soil as it becomes wet (Figure 5.17). There usually appears to be a sharp boundary, termed a **wetting front**, between the dry underlying soil and the soil already wetted.

During an intense rain or heavy irrigation, water movement near the soil surface occurs mainly by saturated flow in response to gravity. At the wetting front, however, water is moving into the underlying drier soil in response to matric potential gradients as well as gravity. The relative influence of these forces on wetting fronts in a sandy loam and a clay loam is illustrated in Figure 5.18.

Water Movement in Stratified Soils

The fact that, at the wetting front, water is moving by unsaturated flow has important ramifications for how percolating water behaves when it encounters an abrupt change in pore sizes due to such layers as fragipans or claypans, or sand and gravel lenses. In some cases, such pore-size stratification may be created by soil managers, as when coarse plant residues are plowed under in a layer, or a layer of gravel is placed under finer soil in a planting container. In all cases, the effect on water percolation is similar—that is, the downward movement is impeded—even though the causal mechanism may vary. It is not surprising that percolating water should slow markedly when it reaches a layer with finer pores, which therefore has a lower hydraulic conductivity. However, the fact that a layer of *coarser* pores will temporarily stop the movement of water may not be obvious. It has this effect because the macropores of the sand offer less attraction for the water than do the finer pores of the overlying material (Figure 5.19). Because it retards downward movement, stratification markedly increases the amount of water the upper part of the soil holds in the field (Figure 5.20). Interestingly, a coarse sand layer in an otherwise fine-textured soil profile would also inhibit the *rise* of water from moist subsoil layers up to the surface soil, a situation that could be illustrated by turning Figure 5.19*b* upside down. This principle also allows a layer of gravel to act as a capillary barrier under a concrete slab foundation to prevent water from soaking up from the soil and through the concrete floor of a home basement.

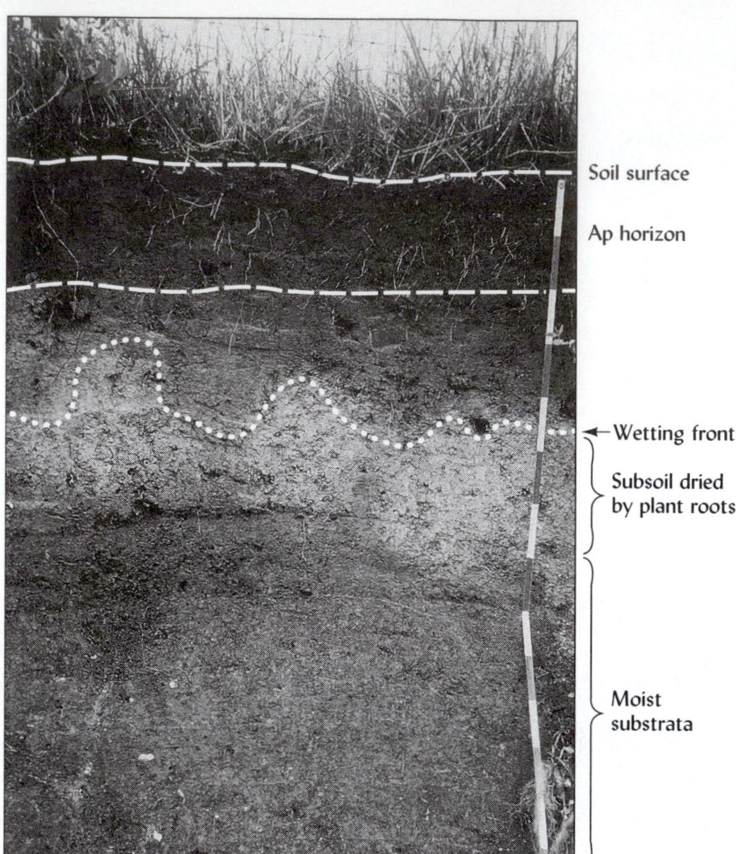

FIGURE 5.17 The wetting front 24 hours after a 5 cm rainfall. Water removal by plant roots had dried the upper 70 to 80 cm of this humid-region (Alabama) profile during a previous 3-week dry spell. The clearly visible boundary results from the rather abrupt change in soil water content at the wetting front between the dry, lighter-colored soil and the soil darkened by the percolating water. The wavy nature of the wetting front in this natural field soil is evidence of the heterogeneity of pore sizes. Scale in 10 cm intervals. (Photo courtesy of R. Weil)

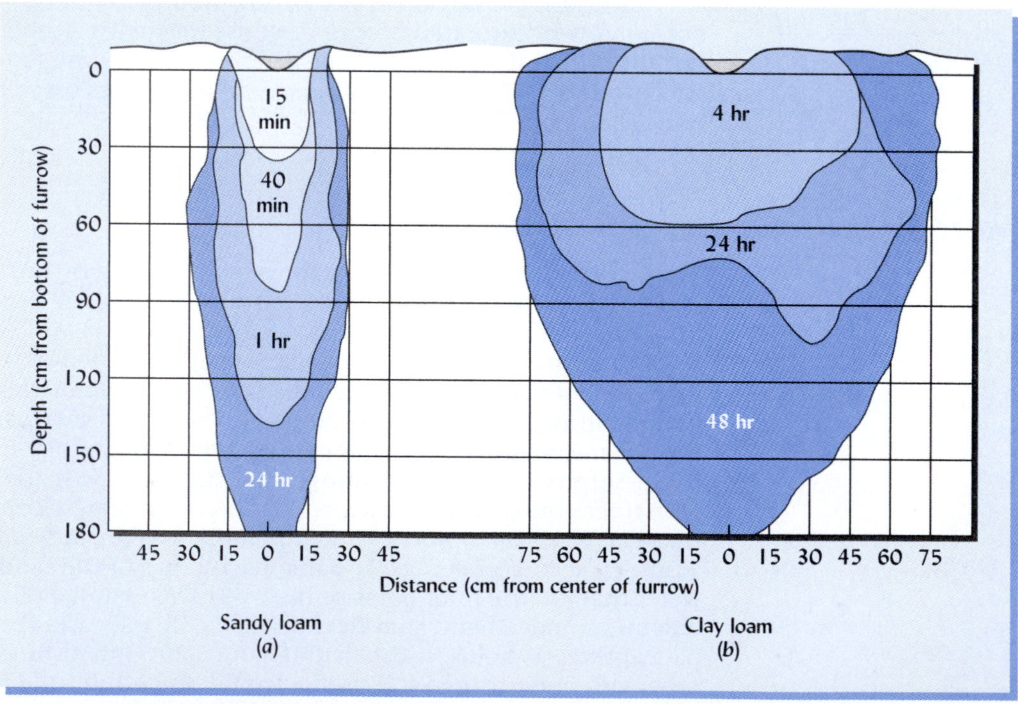

FIGURE 5.18 Comparative rates of irrigation water movement into a sandy loam and a clay loam. Note the much more rapid rate of movement in the sandy loam, especially in a downward direction. [Redrawn from Cooney and Peterson (1955)]

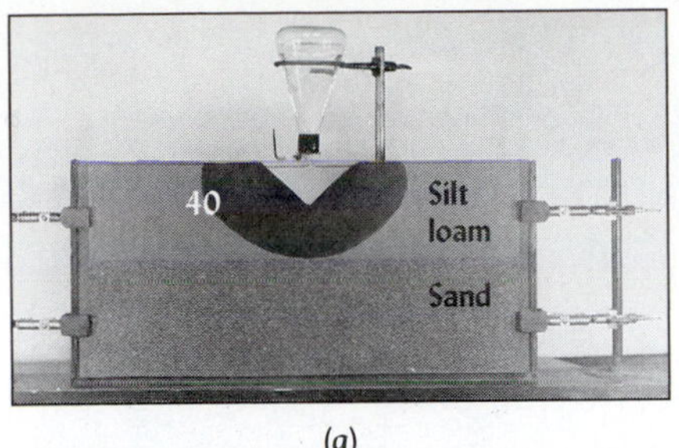

(a)

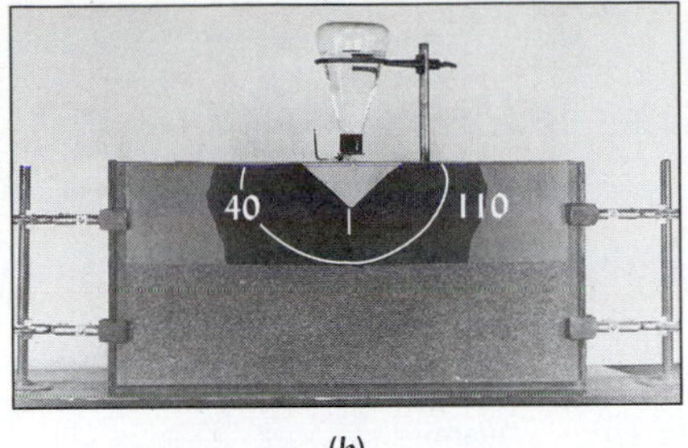

(b)

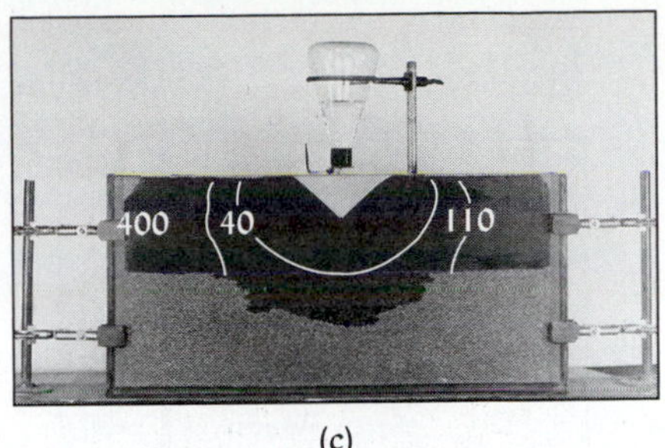

(c)

FIGURE 5.19 Downward water movement in soils with a stratified layer of coarse material. (*a*) Water is applied to the surface of a medium-textured topsoil. Note that after 40 min, downward movement is no greater than movement to the sides, indicating that in this case the gravitational force is insignificant compared to the matric potential gradient between dry and wet soil. (*b*) The *downward* movement stops when a coarse-textured layer is encountered. After 110 min, no movement into the sandy layer has occurred. The macropores of the sand provide less attraction for water than the finer-textured soil above. Only when the water content (and in turn the matric potential gradient) is raised sufficiently will the water move into the sand. (*c*) After 400 min, the water content of the overlying layer becomes sufficiently high to give a water potential of about −1 kPa or more, and downward movement into the coarse material takes place. (Courtesy W. H. Gardner, Washington State University)

FIGURE 5.20 One result of soil layers with contrasting texture. This North Carolina soil has about 50 cm of loamy sand coastal plain material atop deeper layers of silty clay-loam-textured material derived from the piedmont. Rainwater rapidly infiltrates the sandy surface horizons, but its downward movement is arrested at the finer-textured layer, resulting in saturated conditions near the surface and a quick-sandlike behavior. (Photo courtesy of R. Weil)

5.7 WATER VAPOR MOVEMENT IN SOILS

Advanced physics of water movement in soils: http://fbe.uwe.ac.uk/public/geocal/soilmech/water/water.htm#

Water vapor moves from one point to another within the soil in response to differences in vapor pressure. Thus, water vapor will move from a moist soil where the soil air is nearly 100% saturated with water vapor (high vapor pressure) to a drier soil where the vapor pressure is somewhat lower. Also, water vapor will move from a zone of low salt content to one with a higher salt content (e.g., around a fertilizer granule). The salt lowers the vapor pressure of the water and encourages water movement from the surrounding soil.

If the temperature of one part of a uniformly moist soil is lowered, the vapor pressure will decrease and water vapor will tend to move toward this cooler part. Heating will have the opposite effect in that heating will increase the vapor pressure and the water vapor will move away from the heated area. Figure 5.21 illustrates these relationships.

Because the amount of water vapor is small, its movement in soils is of limited practical importance if the soil moisture is kept near optimum for plant growth. However, enough water can move in the vapor form to stimulate the germination of some seeds (see Figure 5.22). Such movement also may be of considerable significance to drought-resistant desert plants (*xerophytes*).

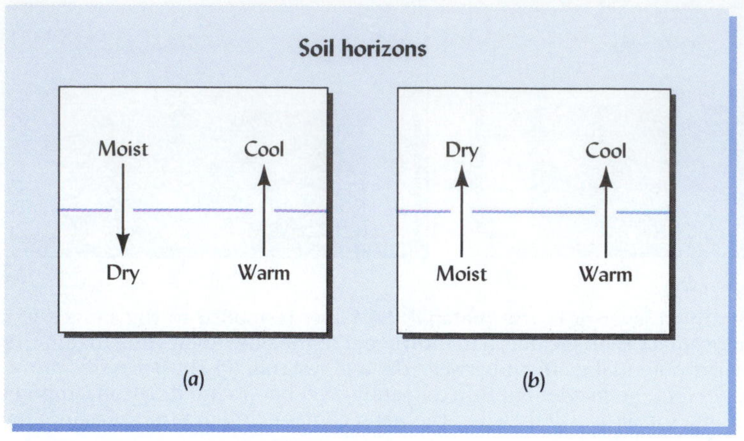

FIGURE 5.21 Vapor movement tendencies that may be expected between soil horizons differing in temperature and moisture. In (*a*) the tendencies more or less negate each other, but in (*b*) they are coordinated and considerable vapor transfer might be possible if the liquid water in the soil capillaries does not interfere.

5.8 QUALITATIVE DESCRIPTION OF SOIL WETNESS

The measurement of soil water potential and the observable behavior of soil water always depend on that portion of the soil water that is farthest from a particle surface and therefore has the highest potential. As an initially water-saturated soil dries down, both the soil as a whole and the soil water it contains undergo a series of gradual changes in physical behavior and in their relationships with plants. These changes are due mainly to the fact that the water remaining in the drying soil is found in smaller pores and thinner films where the water potential is lowered principally by the action of matric forces. Matric potential therefore accounts for an increasing proportion of the total soil water potential, while the proportion attributable to gravitational potential decreases.

Maximum Retentive Capacity

When all soil pores are filled with water from rainfall or irrigation, the soil is said to be **saturated** with respect to water (Figure 5.23) and at its **maximum retentive capacity.** The matric potential is close to zero, nearly the same as that of pure water. The volumetric water content is essentially the same as the total porosity. The soil will remain at maximum retentive capacity only so long as water continues to infiltrate, for the water in the largest pores (sometimes termed **gravitational water**) will percolate downward, mainly under the influence of gravitational forces.

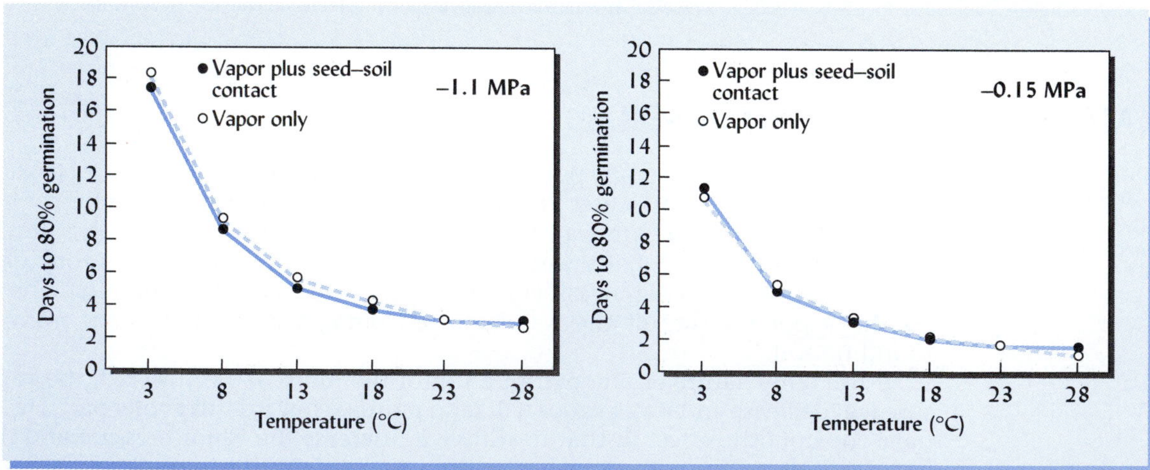

FIGURE 5.22 The rate of water vapor movement from nearby soil particles is sufficient to germinate the seeds of wheat as demonstrated in these graphs. In this experiment, either soil is firmly packed around the planted seed (vapor plus seed–soil contact), or the seed is kept from such contact by wrapping it in fiberglass cloth, permitting the movement of only water vapor to the seed (vapor only). Note that where water moves only in the vapor form, seed germination at different temperatures and at two soil water potentials is essentially as high as where there is soil–seed contact. Apparently water vapor movement is of significant practical importance in enhancing seed germination. [From Wuest, Albrecht, and Skirvin (1999)]

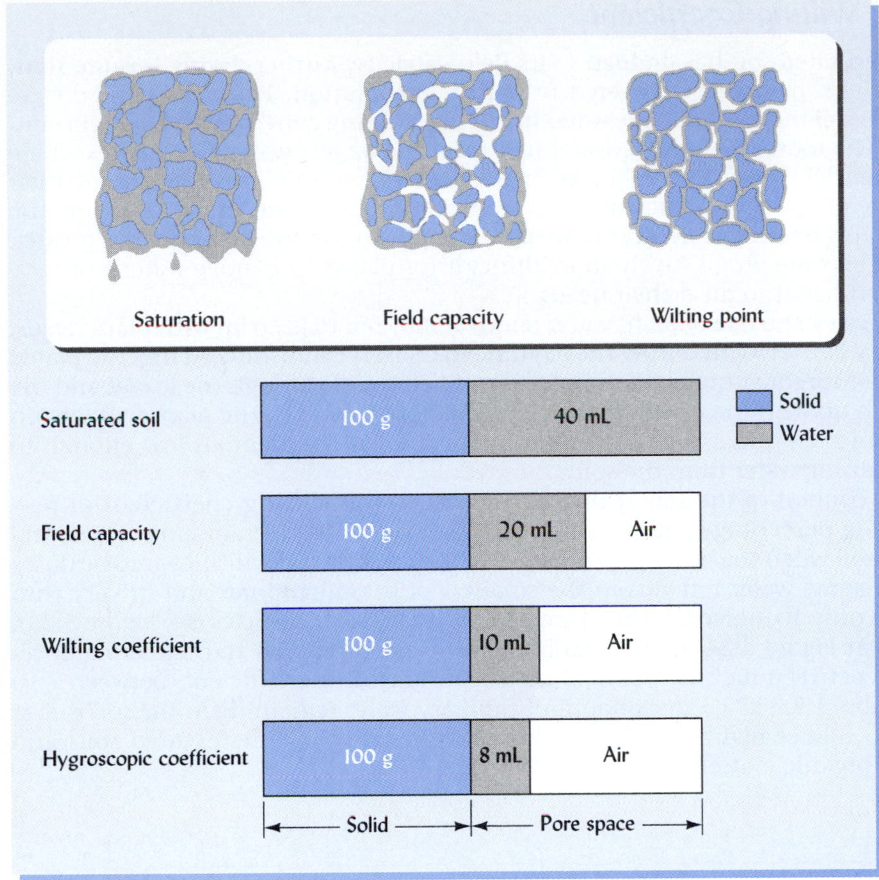

FIGURE 5.23 Volumes of water and air associated with a 100 g slice of soil solids in a well-granulated silt loam at different moisture levels. The top bar shows the situation when a representative soil is completely saturated with water. This situation will usually occur for short periods of time during a rain or when the soil is being irrigated. Water will soon drain out of the larger pores (*macropores*). The soil is then said to be at the *field capacity*. Plants will remove water from the soil quite rapidly until they begin to wilt. When permanent wilting of the plants occurs, the soil water content is said to be at the *wilting coefficient*. There is still considerable water in the soil, but it is held too tightly to permit its absorption by plant roots. A further reduction in water content to the *hygroscopic coefficient* is illustrated in the bottom bar. At this point the water is held very tightly, mostly by the soil colloids. (Top drawings modified from *Irrigation on Western Farms*, published by the U.S. Departments of Agriculture and Interior)

Field Capacity

Once the rain or irrigation has ceased, water in the largest soil pores will drain downward quite rapidly in response to the hydraulic gradient (mostly gravity). After 1 to 3 days, this rapid downward movement will become negligible as matric forces play a greater role in the movement of the remaining water. The soil then is said to be at its **field capacity**. In this condition, water has moved out of the macropores and air has moved in to take its place. The micropores or capillary pores are still filled with water and can supply plants with needed water. The matric potential will vary slightly from soil to soil but is generally in the range of -10 to -30 kPa, assuming drainage into a less-moist zone of similar porosity.[5] Water movement will continue to take place by unsaturated flow, but the rate of movement is very slow since it now is due primarily to capillary forces, which are effective only in micropores (Figure 5.23). The water found in pores small enough to retain it against rapid gravitational drainage, but large enough to allow capillary flow in response to matric potential gradients, is sometimes termed **capillary water.**

Field capacity is a very useful term because it refers to an approximate degree of soil wetness at which several important soil properties are in transition. At field capacity, a soil is holding the maximal amount of water useful to plants. Additional water would remain in the soil for only a short time before draining, and would reduce soil aeration. At field capacity, the soil is near its lower plastic limit; that is, the soil behaves as a crumbly semisolid at water contents below field capacity, and as a plastic puttylike material that easily turns to mud at water contents above field capacity (see Section 4.9). Finally, at field capacity, sufficient pore space is filled with air to allow optimal aeration for most aerobic microbial activity and for the growth of most plants (see Section 7.6).

[5]Note that because of the relationships pertaining to water movement in stratified soils (see Section 5.6), soil in a flower pot will cease drainage while much wetter than field capacity.

Permanent Wilting Percentage or Wilting Coefficient

Once an unvegetated soil has drained to its field capacity, further drying is quite slow, especially if the soil surface is covered to reduce evaporation. However, if plants are growing in the soil they will remove water from their rooting zone, and the soil will continue to dry. The roots will remove water first from the largest water-filled pores where the water potential is relatively high. As these pores are emptied, roots will draw their water from the progressively smaller pores and thinner water films in which the matric water potential is lower and the forces attracting water to the solid surfaces are greater. Hence, it will become progressively more difficult for plants to remove water from the soil at a rate sufficient to meet their needs.

As the soil dries, the rate of plant water removal may fail to keep up with plant needs, and plants may begin to wilt during the daytime to conserve moisture. At first the plants will regain their turgor at night when water is not being lost through the leaves and the roots can catch up with the plant's demand. Ultimately, however, the plant will remain wilted night and day when the roots cannot generate water potentials low enough to coax the remaining water from the soil.

The water content of the soil at this stage is called the **wilting coefficient** or **permanent wilting percentage** and by convention is taken to be that amount of water retained by the soil when the water potential is −1500 kPa. The soil will appear to be dusty dry, although some water remains in the smallest of the micropores and in very thin films (perhaps only 10 molecules thick) around individual soil particles (see Figure 5.23). As illustrated in Figure 5.24, **plant-available water** is considered to be that water retained in soils between the states of field capacity and wilting coefficient (between −10 to −30 kPa and −1500 kPa). The amount of capillary water remaining in the soil that is unavailable to higher plants can be substantial, especially in fine-textured soils and those high in organic matter.

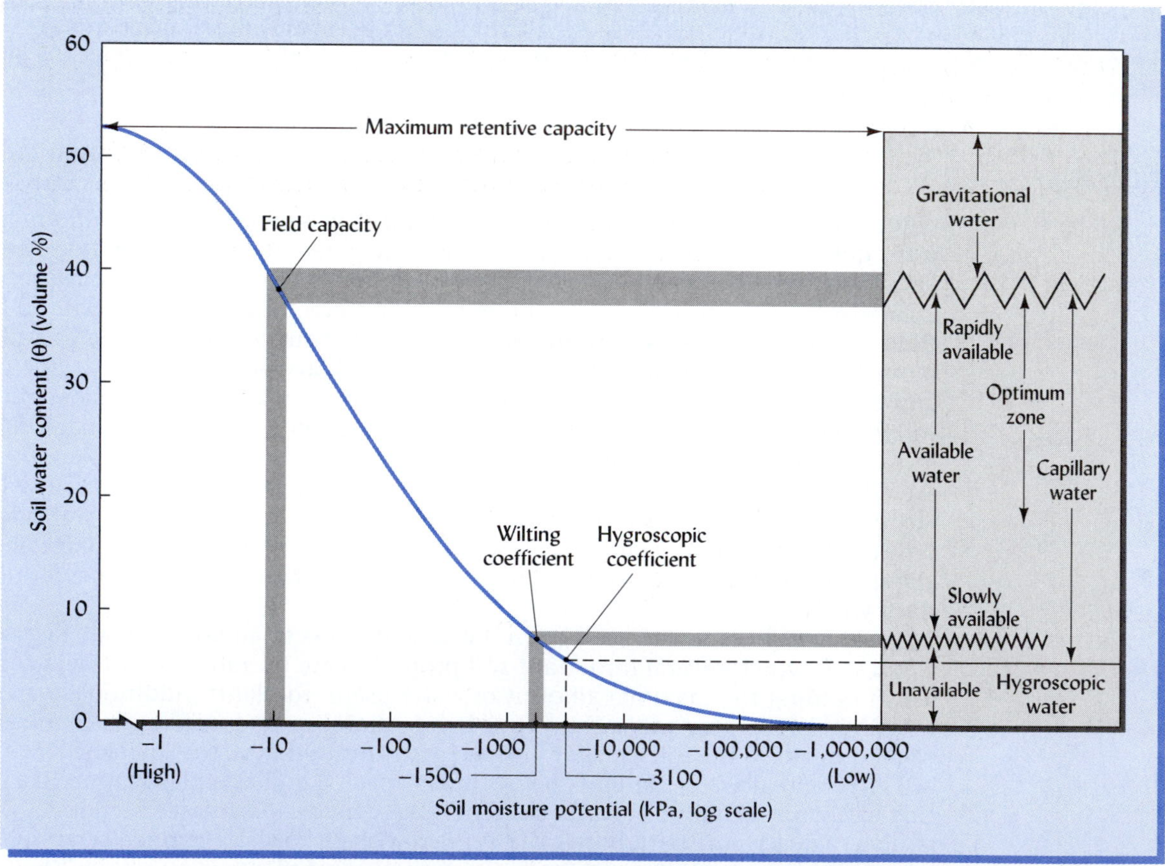

FIGURE 5.24 Water content–matric potential curve of a loam soil as related to different terms used to describe water in soils. The wavy lines in the diagram to the right suggest that measurements such as field capacity are only approximations. The gradual change in potential with soil moisture change discourages the concept of different "forms" of water in soils. At the same time, such terms as *gravitational* and *available* assist in the qualitative description of moisture utilization in soils.

Hygroscopic Coefficient

Although plant roots do not generally dry the soil beyond the permanent wilting percentage, if the soil is exposed to the air, water will continue to be lost by evaporation. When soil moisture is lowered below the wilting point, the water molecules that remain are very tightly held, mostly being adsorbed by colloidal soil surfaces. This state is approximated when the atmosphere above a soil sample is essentially saturated with water vapor (98% relative humidity) and equilibrium is established at a water potential of −3100 kPa. The water is thought to be in films only 4 or 5 molecules thick and is held so rigidly that much of it is considered nonliquid and can move only in the vapor phase. The moisture content of the soil at this point is termed the **hygroscopic coefficient.** Soils high in colloidal materials (clay and humus) will hold more water under these conditions than will sandy soils that are low in clay and humus. Soil water considered to be *unavailable* to plants includes the hygroscopic water and that portion of capillary water retained at potentials below −1500 kPa (see Figure 5.24).

Availability of Soil Water to Plants

Downloadable program estimates soil water tension, conductivity and water holding capability based on the soil texture, organic matter, gravel content, and salinity.
http://www.bsyse.wsu.edu/saxton/soilwater/

Many factors determine the amount of plant-available water in a soil, including soil texture, organic matter content, soil compaction, soil and rooting depths, and soil stratification. Each will be discussed briefly.

Effect of Texture and Organic Matter. For soils with greater contents of silt and clay (finer textures), the amount of water held at field capacity tends to be greater. The water held at potentials below the wilting coefficient (water unavailable to plants) also increases with finer soil texture, but not as sharply. Therefore, the amount of water held at potentials between those of field capacity and wilting point (i.e., plant-available water) tends to be greatest in medium-textured soils such as silt loams. For most temperate-region soils, higher clay contents increase mainly the water held too tightly for plants to use (Figure 5.25). However, with increasing contents of low-activity-type clays (typical of humid tropical regions) and/or organic matter, soils tend to hold more plant-available water because the greater amount of water held at field capacity more than compensates for that held at potentials below wilting point (Figure 5.26). Increased soil organic matter can also enhance water availability to plants by improving soil structure and infiltration capacity.

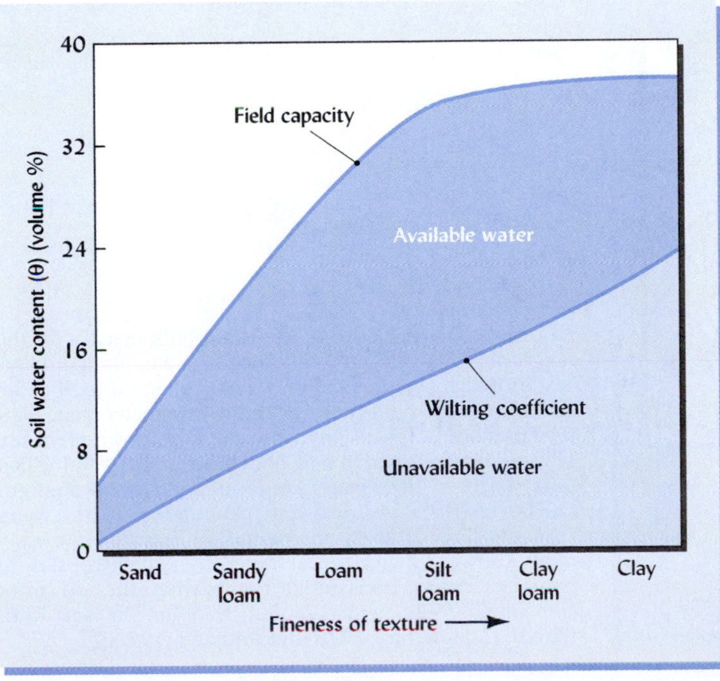

FIGURE 5.25 General relationship between soil water characteristics and soil texture. Note that the wilting coefficient increases as the texture becomes finer. The field capacity increases until we reach the silt loams, then levels off. Remember these are representative curves; individual soils would probably have values different from those shown.

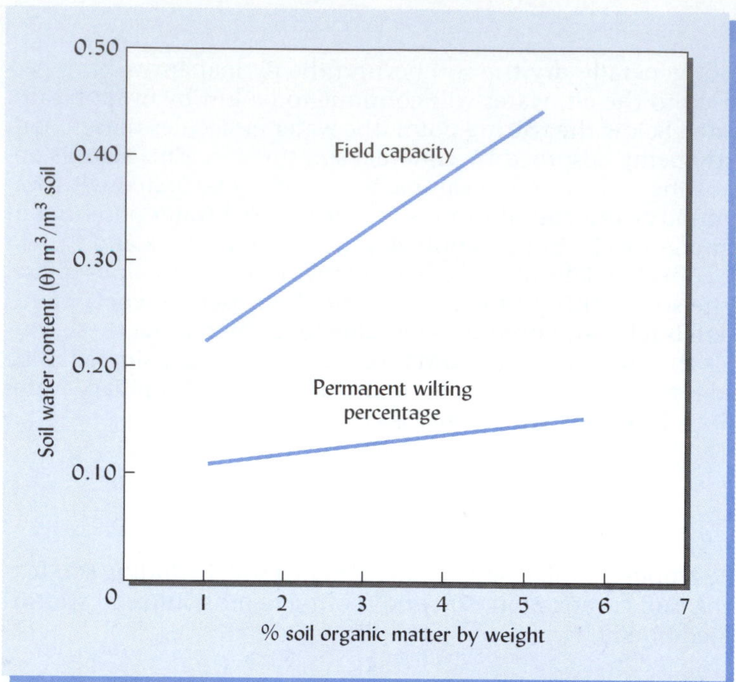

FIGURE 5.26 The effects of organic matter content on the field capacity and permanent wilting percentage of a number of silt loam soils. The difference between the two lines shown is the available soil moisture content, which was obviously greater in the soils with higher organic matter levels. [Redrawn from Hudson (1994); used with permission of the Soil & Water Conservation Society]

Effect of Compaction. Soil compaction (increased bulk density) and weakening of soil structure can reduce the amount of water available to plants for several reasons. The compression of macropores into micropores results in less water being held at field capacity, more water being held at the wilting coefficient, fewer air-filled pores when the soil is wet, and more cohesion of clay particles when the soil is relatively dry. The latter effect may increase soil strength enough to prevent roots from being able to penetrate soil layers which contain otherwise "available" water (see Figure 5.27).

Effect of Soil Depth and Layering. We have already seen (Section 5.6) how soil profiles stratified with layers of abruptly differing texture can retard downward water movement and increase the amount of water held. In any field soil, each layer or horizon will

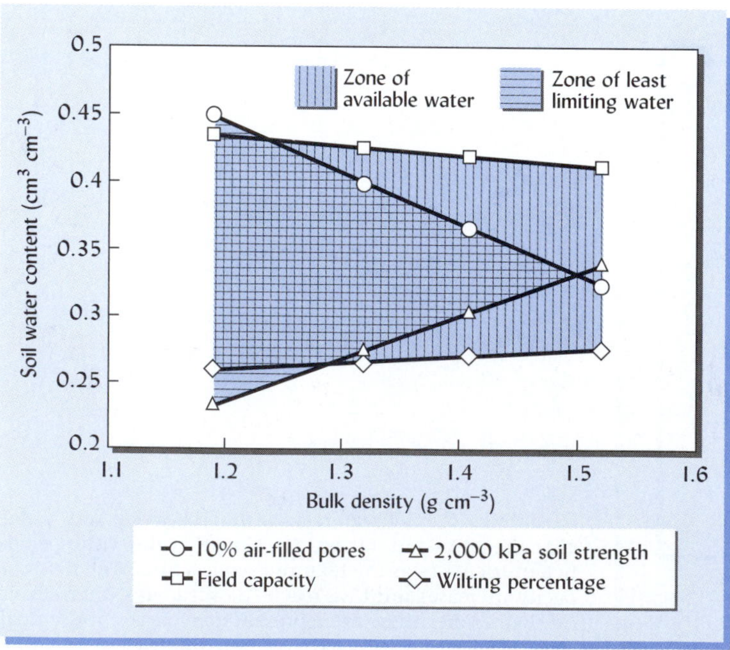

FIGURE 5.27 Influence of increased bulk density on the range of soil water contents available for plant uptake. Traditionally, plant-available water is defined as that retained between field capacity and wilting percentage (vertical hatching). If soils are compacted, however, plant use of water may be restricted by poor aeration (<10% air-filled pore space) at high water contents and by soil strength (>2000 kPa) that restricts root penetration at low water contents. The latter criteria define the *least limiting water range* shown by horizontal hatching. The two sets of boundary criteria give similar results when the soil is not compacted (bulk density about 1.25 for the soil illustrated). [Adapted from Da Silva and Kay (1997)]

BOX 5.2 CALCULATION OF THE TOTAL AVAILABLE WATER HOLDING CAPACITY OF A SOIL PROFILE

The total amount of water available to a plant growing in a field soil is a function of the rooting depth of the plant and the sum of the water held between field capacity and wilting percentage in each of the soil horizons explored by the roots. For each soil horizon, then, the mass available water-holding capacity is estimated as the difference between the mass water contents θ_m (Mg water per 100 Mg soil) at field capacity and permanent wilting percentage. This value can be converted into a volume water content θ by multiplying by the ratio of the bulk density of the soil to the density of water. Finally, this volume ratio is multiplied by the thickness of the horizon to give the total centimeters of available water capacity (AWC) in that horizon. For the first horizon described in Table 5.6 the calculation (with units) is:

$$\left(\frac{22\ \text{g}}{100\ \text{g}} - \frac{8\ \text{g}}{100\ \text{g}}\right) * \frac{1.2\ \text{Mg}}{\text{m}^3} * \frac{1\ \text{m}^3}{1\ \text{Mg}} * 20\ \text{cm} = 3.36\ \text{cm AWC}$$

Note that all units cancel out except cm, resulting in the depth of available water (cm) held by the horizon. In Table 5.6, the available water-holding capacity of all horizons within the rooting zone are summed to give a total AWC for the soil–plant system. Since no roots penetrated to the last horizon (1.0 to 1.25 m), this horizon was not included in the calculation. We can conclude that for the soil–plant system illustrated, 14.13 cm of water could be stored for plant use. At a typical summertime water-use rate of 0.5 cm of water per day, this soil could hold about a 4-week supply.

TABLE 5.6 Calculation of Estimated Soil Profile Available Water-Holding Capacity

Soil depth, cm	*Relative root length*	*Soil bulk density, Mg/m³*	*Field capacity (FC), g/100g*	*Wilting percentage (WP), g/100g*	*Available water-holding capacity (AWC), cm*
0–20	xxxxxxxxx	1.2	22	8	$20 * 1.2\left(\frac{22}{100} - \frac{8}{100}\right) = 3.36$ cm
20–40	xxxx	1.4	16	7	$20 * 1.4\left(\frac{16}{100} - \frac{7}{100}\right) = 2.52$ cm
40–75	xx	1.5	20	10	$35 * 1.5\left(\frac{20}{100} - \frac{10}{100}\right) = 5.25$ cm
75–100	xx	1.5	18	10	$25 * 1.5\left(\frac{18}{100} - \frac{10}{100}\right) = 3.00$ cm
100–125	—	1.6	15	11	No roots
Total					3.36 + 2.52 + 5.25 + 3.00 = 14.13 cm

have different characteristics that affect the amount of water held and the availability of that water to growing plants. The total depth of soil into which roots can explore also helps determine how much water will be available for plant use (see Plate 36 following page 498). A quantitative example involving some of these considerations is given in Box 5.2 and Table 5.6.

How Plant Roots are Supplied with Water. At any one time, only a small proportion of the soil water is adjacent to the absorptive root surfaces. How then do roots get access to the immense amount of water needed by growing plants? One answer is that water slowly moves to the roots by capillary flow in response to the matric potential gradient set up as roots dry out the soil in their immediate vicinity. At the same time, the root tips of growing plants are continually extending into wetter soil areas. The roots of most plants proliferate profusely in the upper 20 to 30 cm of soil (Table 5.7), but roots that explore pores and crevices 1 or 2 or more meters deep are critical in obtaining water, especially during droughty periods when the upper soil layers are quite dry. Roots that spread widely near the surface and roots that penetrate many meters deep are two rooting patterns used by native plants in dry regions to obtain sufficient water under adverse conditions.

TABLE 5.7 Percentage of Root Mass of Three Crops and Two Trees Found in the Upper 30 cm Compared with Deeper Depths (30–180 cm)

Plant species	*Percentage of roots*	
	Upper 30 cm	*30–180 cm*
Soybeans	71	29
Corn	64	36
Sorghum	86	14
Pinus radiata	82	18
Eucalyptus marginata	86	14

Data for crops from Mayaki et al. (1976), for trees, estimated from Bowen (1985).

5.9 CONCLUSION

Water impacts all life. The interactions and movement of this simple compound in soils help determine whether these impacts are positive or negative. An understanding of the principles that govern the attraction of water for soil solids and for dissolved ions can help maximize the positive impacts while minimizing the less-desirable ones.

The water molecule has a polar structure that results in electrostatic attraction of water to both soluble cations and soil solids. These attractive forces tend to reduce the potential energy level of soil water below that of pure water. The extent of this reduction, called soil water potential ψ, has a profound influence on a number of soil properties, but especially on the movement of soil water and its uptake by plants.

The water potential due to the attraction between soil solids and water (the matric potential ψ_m) combines with the force of gravity ψ_g to largely control water movement. This movement is relatively rapid in soils high in moisture and with an abundance of macropores. In drier soils, however, the adsorption of water on the soil solids is so strong that its movement in the soil and its uptake by plants are greatly reduced. As a consequence, plants die for lack of water—even though there are still significant quantities of water in the soil—because that water is unavailable to plants.

Water is supplied to plants by capillary movement toward the root surfaces and by growth of the roots into moist soil areas. In addition, vapor movement may be of significance in supplying water for drought-resistant desert species (xerophytes). The osmotic potential ψ_o becomes important in soils with high soluble salt levels that can impede plant uptake of water from the soil. Such conditions occur most often in soils with restricted drainage in areas of low rainfall and in potted indoor plants.

The characteristics and behavior of soil water are very complex. As we have gained more knowledge, however, it has become apparent that soil water is governed by relatively simple, basic physical principles. Furthermore, researchers are discovering the similarity between these principles and those governing the movement of groundwater and the uptake and use of soil moisture by plants—the subject of the next chapter.

STUDY QUESTIONS

1. What is the role of the *reference state of water* in defining soil water potential? Describe the properties of this reference state of water.
2. Imagine a root of a cotton plant growing in the upper horizon of an irrigated soil in California's Imperial Valley. As the root attempts to draw water molecules from this soil, what forces (potentials) must it overcome? If this soil were compacted by a heavy vehicle, which of these forces would be most affected? Explain.
3. Using the terms *adhesion, cohesion, meniscus, surface tension, atmospheric pressure,* and *hydrophilic surface,* write a brief essay to explain why water rises from the water table in a mineral soil.
4. Suppose you were hired to design an automatic irrigating system for a wealthy homeowner's garden. You determine that the flower beds should be kept at a water potential above −60 kPa, but not wetter than −10 kPa, as the annual flow-

ers here are sensitive to both drought and lack of good aeration. The rough turf areas, however, can do well if the soil dries to as low as −300 kPa. Your budget allows either tensiometers or electrical resistance blocks to be hooked to electronic switching valves. Which instruments would you use and where? Explain.

5. Suppose the homeowner referred to in question 4 increased your budget and asked to use the TDR method to measure soil water contents. What additional information about the soils, not necessary for using the tensiometer, would you have to obtain to use the TDR instrument? Explain.

6. A greenhouse operator was growing ornamental woody plants in 15-cm-tall plastic containers filled with a loamy sand. He watered the containers daily with a sprinkler system. His first batch of 1000 plants yellowed and died from too much water and not enough air. As an employee of the greenhouse, you suggest that he use 30-cm-tall pots for the next batch of plants. Explain your reasoning.

7. Suppose you measured the following data for a soil:

		θ_m *at different water tensions, kg water/kg dry soil*		
Horizon	*Bulk density, Mg/m*^{3}	*−10kPa*	*−100kPa*	*−1500kPa*
A (0–30 cm)	1.28	0.28	0.20	0.08
B_t (30–70 cm)	1.40	0.30	0.25	0.15
B_x (70–120 cm)	1.95	0.20	0.15	0.05

Estimate the total available water-holding capacity (AWC) of this soil in centimeters of water.

8. A forester obtained a cylindrical core (L = 15 cm, r = 3.25 cm) of soil from a field site. She placed all the soil in a metal can with a tight-fitting lid. The empty metal can weighed 300 g and when filled with the field-moist soil weighed 972 g. Back in the lab, she placed the can of soil, with lid removed, in an oven for several days until it ceased to lose weight. The weight of the dried can with soil (including the lid) was 870 g. Calculate both θ_m and θ.

9. Give four reasons why compacting a soil is likely to reduce the amount of water available to growing plants.

10. Since even, rapidly growing, finely branched root systems rarely contact more than 1 or 2% of the soil particle surfaces, how is it that the roots can utilize much more than 1 or 2% of the water held on these surfaces?

REFERENCES

Bowen, G. D. 1985. "Roots as components of tree productivity," in M. G. R. Cannell and J. E. Jackson (eds.), *Attributes of Trees as Crop Plants.* (Midlothian, Scotland: Institute of Terrestrial Ecology).

Cooney, J. J., and J. E. Peterson. 1955. *Avocado Irrigation.* Leaflet 50. California Agricultural Extension Service.

Da Silva, A. P., and B. D. Kay. 1997. "Estimating the least limiting water range of soil from properties and management," *Soil Sci. Soc. Amer. J.* **61**:877–883.

Hillel, D. 1998. *Environmental Soil Physics.* (Orlando, Fla.: Academic Press).

Hudson, B. D. 1994. "Soil organic matter and available water capacity," *J. Soil and Water Cons.* **49**:189–194.

Kladivko, E. J., et al. 1999. "Pesticide and nitrate transport into subsurface tile drains of different spacing," *J. Environ. Qual.* **28**:997–1004.

Mayaki, W. C., L. R. Stone, and I. D. Teare. 1976. "Irrigated and nonirrigated soybean, corn, and grain sorghum root systems," *Agron. J.* **68**:532–534.

Reynolds, W. D., et al. 2000. "Comparison of tension infiltrometer, pressure infiltrometer and soil core estimates of saturated conductivity," *Soil Sci. Soc. Am. J.* **64**:478–484.

Waddell, J. T., and R. R. Weil. 1996. "Water distribution in soil under ridge-till and no-till corn," *Soil Sci. Soc. Amer. J.* **60**:230–237.

Wuest, S. B., S. L. Albrecht, and K. W. Skirvin. 1999. "Vapor transport vs seed–soil contact in wheat germination." *Agron. J.* **91**:783–787.

6

Soil and the Hydrologic Cycle

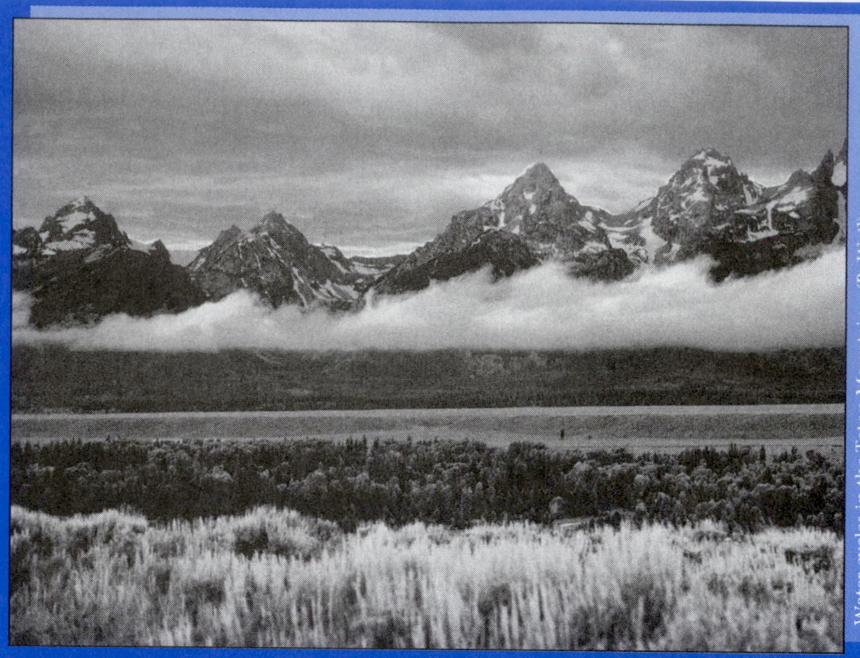

Water cycles in the Teton Mountains. (R. Weil)

Both soil and water belong to the biosphere, to the order of nature, and—as one species among many, as one generation among many yet to come—we have no right to destroy them.
—Daniel Hillel, *Out of Earth*

One of the most striking—and troubling—features of human society is the yawning gap in wealth between the world's rich and poor. The life experience of the one group is quite incomprehensible to the other. So it is with the distribution of the world's water resources. The rain forests of the Amazon and Congo basins are drenched by more than 2000 mm of rain each year, while the deserts of North Africa and central Asia get by with less than 200. Nor is the supply of water distributed evenly throughout the year. Rather, periods of high rainfall and flooding alternate with dry spells or periods of drought.

Yet one could say that everywhere the supply of water is adequate to meet the needs of the plants and animals native to the natural communities of the area. Of course, this is so only because the plants and animals have adapted to the local availability of water. Early human populations, too, adapted to the local water supplies by settling where water was plentiful from rain or rivers, by learning which underground gourds and plant stems could quench one's thirst, by developing techniques to harvest water for agriculture and store it in underground cisterns, and by adopting nomadic lifestyles that allowed them and their herds to follow the rains and the grass supply.

But "civilized" humans have not been willing to adapt their needs and cultures to their environment. Rather, they have joined in organized efforts to adapt their environment to their desires. Hence, the ancients tamed the flows of the Tigris and Euphrates. We moderns dig wells in the Sahel, bottle up the mighty Nile at Aswan, channel the waters of the Colorado to the chaparral region of southern California, pump out the aquifers under farms and suburbs, and create sprawling cities (with swimming pools and bluegrass lawns!) in the deserts of the American Southwest or on the sands of Arabia. Truly, cities like Las Vegas are gambling in more ways than one.

There is plenty of room for improvement in managing water resources, and many of the improvements are likely to come as a result of better management of soils. The soil plays a central role in the cycling and use of water. For instance, by serving as a massive reservoir, soil helps moderate the adverse effects of excesses and deficiencies of water. The soil can help us treat and reuse wastewaters from animal, domestic, and industrial sources. The flow of water through the soil in these and other circumstances connects the chemical pollution of soils to the possible contamination of groundwater.

In Chapter 5 we considered the nature and movement of water in soils. In this chapter we will see how those characteristics apply to the cycling of water between the soil, the atmosphere, and vegetation. We will then examine the unique role of the soil in water resource management.

6.1 THE GLOBAL HYDROLOGIC CYCLE

Global Stocks of Water

Hydrologic cycle processes in motion: http://observe.arc.nasa.gov/nasa/earth/hydrocycle/hydro3.html

There are nearly 1400 million km^3 of water on the earth—enough (if it were all aboveground at a uniform depth) to cover the earth's surface to a depth of some 3 km. Most of this water, however, is relatively inaccessible and is not active in the annual cycling of water that supplies rivers, lakes, and living things.

The more actively cycling water is in the surface layer of the oceans, in shallow groundwater, in lakes and rivers, in the atmosphere, and in the soil (see Figure 6.1). Although the combined volume is a tiny fraction of the water on earth, these pools of water are accessible for movement in and out of the atmosphere and from one place on the earth's surface area to another.

The Hydrologic Cycle

Solar energy drives the cycling of water from the earth's surface to the atmosphere and back again in what is termed the **hydrologic cycle** (Figure 6.2). About one-third of this energy stimulates *evaporation*—the conversion of liquid water into water vapor. The water vapor moves up into the atmosphere, eventually forming clouds that can move from one region of the globe to another. Within an average of about 10 days, pressure and temperature differences in the atmosphere cause the water vapor to condense into liquid droplets or solid particles, which return to the earth as rain or snow.

About 500,000 km^3 of water are evaporated from the earth's surfaces and vegetation each year, some 110,000 km^3 of which falls as rain or snowfall on the continents. Some of the water falling on land runs off the surface of the soil, and some infiltrates the soil and drains into the groundwater. Both the surface runoff and groundwater seepage enter streams and rivers that, in turn, flow downstream into the oceans. The volume of water returned in this way is about 40,000 km^3, which balances the same quantity of water that is transferred annually in clouds from the oceans to the continents.

Water Balance Equation

It is often useful to consider the components of the hydrologic cycle as they apply to a given **watershed**, an area of land drained by a single system of streams and bounded by ridges that separate it from adjacent watersheds. All the precipitation falling on a watershed is either stored in the soil, returned to the atmosphere (see Section 6.2), or discharged from the watershed as surface or subsurface flow (runoff). Water is returned to the atmosphere either by **evaporation** from the land surface (vaporization of soil water)

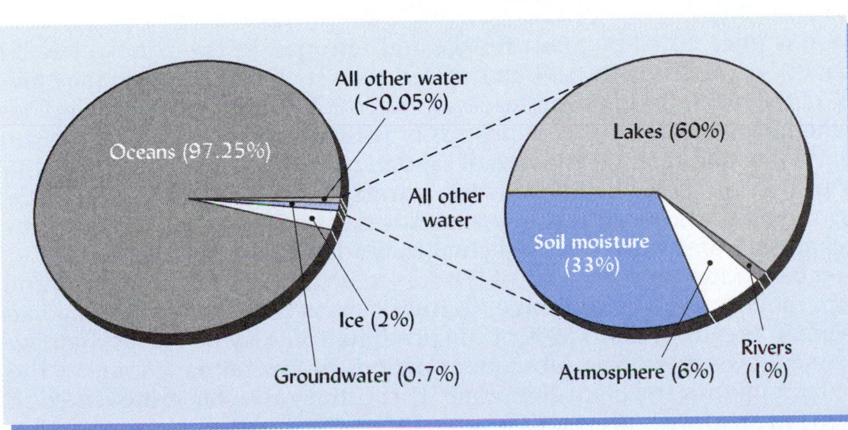

FIGURE 6.1 The sources of the earth's water. The preponderance of water is found in the oceans, glaciers and ice caps, and deep groundwater (left) but most of these waters are inaccessible for rapid exchange with the atmosphere and the land. The sources on the right, though much smaller in quantity, are actively involved in water movement through the hydrologic cycle. (Data from several sources)

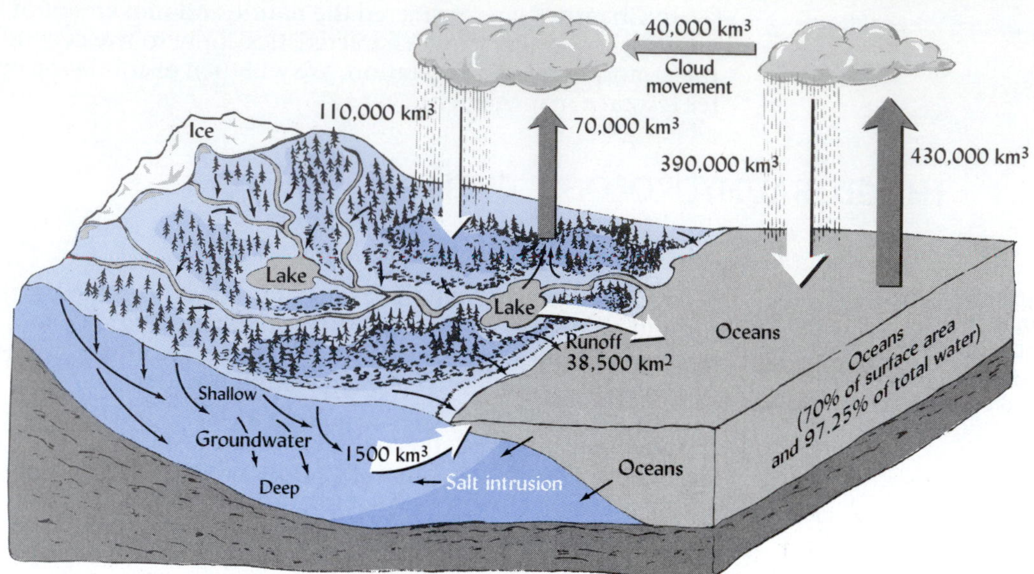

FIGURE 6.2 The hydrologic cycle upon which all life depends is very simple in principle. Water evaporates from the earth's surface, both the oceans and continents, and returns in the form of rain or snowfall. The net movement of clouds brings some 40,000 km^3 of water to the continents and an equal amount of water is returned through runoff and groundwater seepage that is channeled through rivers to the ocean. About 86% of the evaporation and 78% of the precipitation occurs in the ocean areas. However, the processes occurring on land areas where the soils are influential have impacts not only on humans but on all other forms of life, including those residing in the sea.

or, after plant uptake and use, by vaporization from the stomata on the surfaces of leaves (a process termed *transpiration*). Together, these two pathways of evaporative loss to the atmosphere are called **evapotranspiration.**

The disposition of water in a watershed is often expressed by the **water-balance equation,** which in its simplest form is:

$$P = ET + SS + D \tag{6.1}$$

where P = precipitation, ET = evapotranspiration, SS = soil storage, and D = discharge.

For a forested watershed (sometimes termed a *catchment*), management may aim to maximize D so as to provide more water to downstream users. The equation makes it clear that discharge can be increased only if ET and/or SS are decreased, changes that may or may not be desirable. In the case of an irrigated field, water applied in irrigation would be included on the left side of the equation. Irrigation managers may want to save water applied by minimizing unnecessary losses in D and allowing negative values for soil storage (withdrawals of soil water) during parts of the year.

6.2 FATE OF PRECIPITATION AND IRRIGATION WATER

Some precipitation is intercepted by plant foliage and returned to the atmosphere by evaporation without ever reaching the soil. In forested areas such **interception** may prevent 30% to 50% of the precipitation from reaching the soil. Most of the water that does reach the soil penetrates downward by the process of **infiltration,** especially if the soil surface structure is loose and open. If the rate of rainfall or snowmelt exceeds the infiltration capacity of the soil, considerable **surface runoff** and erosion may take place, thereby reducing the proportion of the water that moves into the soil, and causing the pollution of lakes and streams with dissolved chemicals and detached particles.

Once the water penetrates the soil, some of it is subject to downward percolation and eventual loss from the root zone by **drainage.** In humid areas and in some irrigated areas of arid and semiarid regions, up to 50% of the precipitation may be lost as drainage below the root zone; however, during subsequent periods of low rainfall, some of this water may move back up into the plant-root zone by **capillary rise.** Such movement is important to plants in areas with deep soils, especially in dry climates.

The water retained by the soil is referred to as **soil storage** water, some of which eventually moves upward by capillarity and is lost by evaporation from the soil surface. Much of the remainder is absorbed by plants and then moves through the roots and stems to the leaves, where it is lost by transpiration. The water thus lost to the atmosphere by evapotranspiration may later return to the soil as precipitation or irrigation water, and the cycle starts again.

Factors Affecting Water Partitioning

Timing of Precipitation. The amount of water moving through each of the channels just discussed is influenced greatly by the timing, rate, and form of the precipitation. Heavy rainfall, even if of short duration, can supply water faster than most soils can absorb it. This accounts for the fact that in some arid regions with only 200 mm of annual rainfall, a cloudburst that brings 20 to 50 mm of water in a few minutes can result in flash flooding and gully erosion. A larger amount of precipitation spread over several days of gentle rain could move slowly into the soil, thereby increasing the stored water available for plant absorption, as well as replenishing the underlying groundwater. As illustrated in Figure 6.3, the timing of snowfall can affect the partitioning between surface runoff and infiltration.

Type of Vegetation. Plants help determine the proportion of water that runs off and that which penetrates the soil. The vegetation and surface residues of perennial grasslands and dense forests protect the porous soil structure from the beating action of raindrops. Therefore, they encourage water infiltration and reduce the likelihood that soil will be carried off by any runoff that does occur. In general, very little runoff occurs from land under undisturbed forests or well-managed turf grass.

Stem Flow. Many plant canopies direct rainfall toward the plant stem, thus altering the spatial distribution of rain reaching the soil. Under a forest canopy, more than half of the rainfall may trickle down the leaves, twigs, and branches to the tree trunk, there to

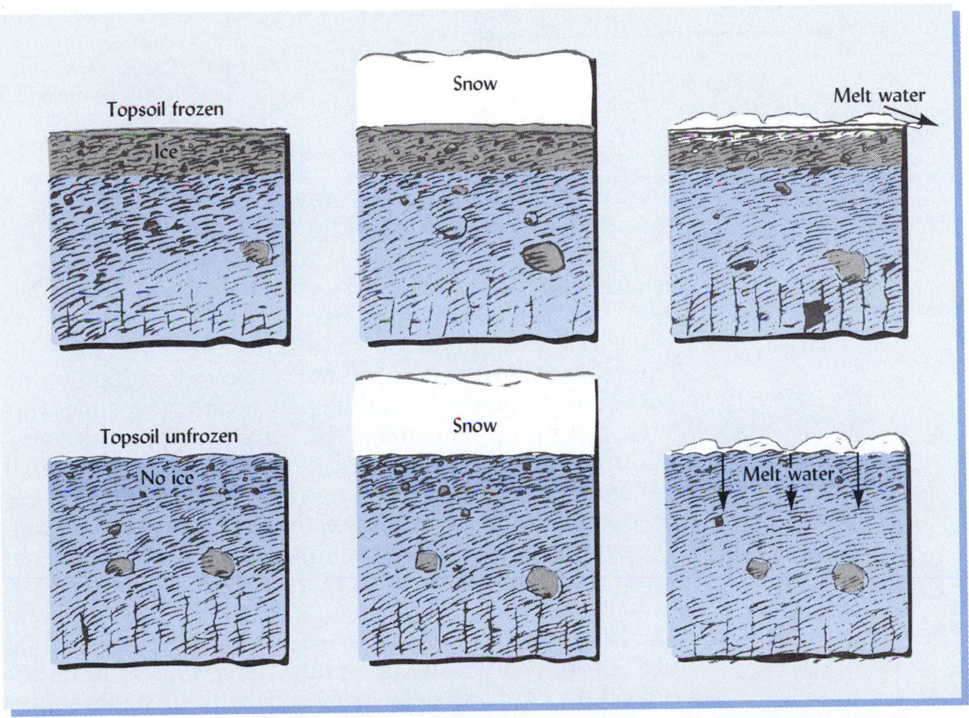

FIGURE 6.3 The relative timing of freezing temperatures and snowfall in the fall in some temperate regions drastically influences water runoff and infiltration into soils in the spring. The upper three diagrams illustrate what happens when the surface soil freezes before the first heavy snowfall. The snow insulates the soil so that it is still frozen and impermeable as the snow melts in the spring. The lower sequence of diagrams illustrates the situation when the soil is unfrozen in the fall when it is covered by the first deep snowfall.

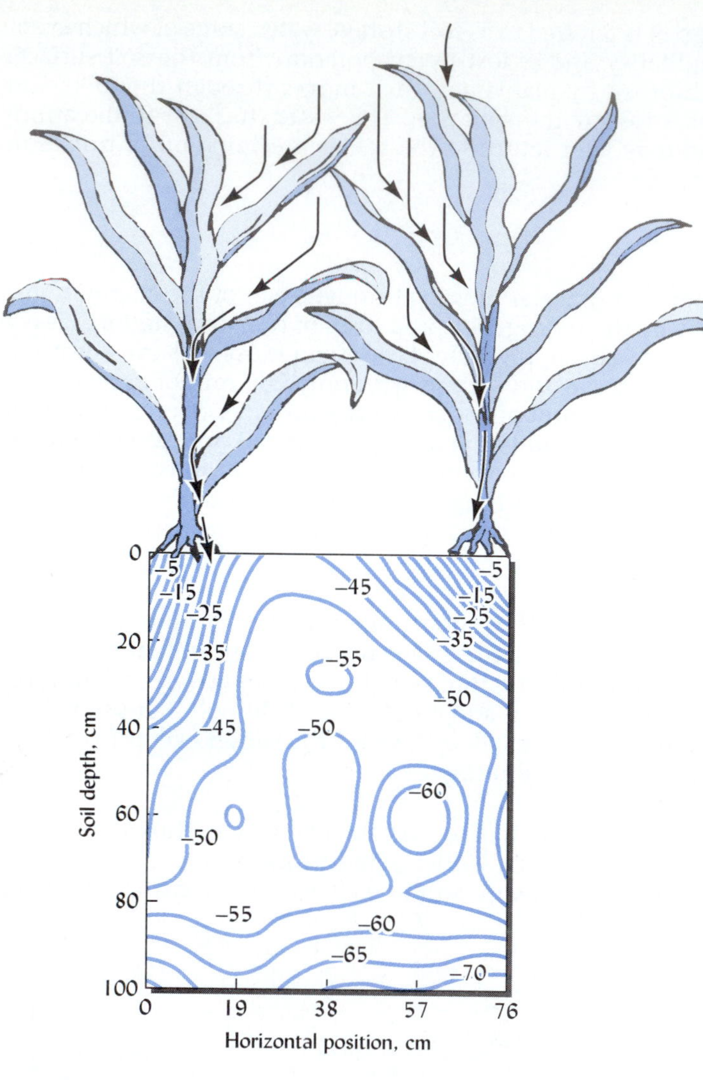

FIGURE 6.4 Vertical and horizontal distribution of soil water resulting from stem flow. The contours indicate the soil water potential in kPa between two corn rows in a sandy loam soil. During the previous two days, 26 mm of rain fell on this field. Many plant canopies, including that of the corn crop shown, direct a large proportion of rainfall toward the plant stem. Stem flow results in uneven spatial distribution of water. In cropland this may have ramifications for the leaching of chemicals such as fertilizers, depending on whether they are applied in or away from the zone of highest wetting near the plant stems. The concentration of water by stem flow may also increase the likelihood of macropore flow in soils after only moderate rainfall. [Data from Waddell and Weil (1996); used by permission of the Soil Science Society of America]

progress downward as **stem flow.** Likewise, certain crop canopies, such as that of corn, funnel a large proportion of the rainfall to the soil in the crop row (Figure 6.4).

Soil Management. A major objective of soil and water management systems is to encourage water infiltration rather than runoff. This may be achieved by enhancing soil surface storage to allow more time for infiltration to take place (Figure 6.5). Another approach is to establish dense vegetation during periods of high rainfall. This may include the use of **cover crops,** plants established between the principle crop-growing seasons (Figure 6.5*b*). Such cover crops encourage the activities of earthworms and protect the soil structure, leading to greatly enhanced water infiltration. However, remember that cover crops also transpire water. If the following crop will be dependent on soil storage water, care may be needed to kill the cover crop before it can dry out the soil profile.

Compaction by heavy equipment used in farming and forestry can sharply reduce water infiltration rates, thus increasing runoff losses.

Soil Properties. If the soil is loose and open (e.g., sands and well-granulated soils), a high proportion of the incoming water will infiltrate the soil, and relatively little will run off. In contrast, heavy clay soils with unstable soil structures resist infiltration and encourage runoff. These differences, attributable to soil properties as well as vegetation, are illustrated in Figure 6.6.

Having considered the overall hydrologic cycle, we will now focus our attention on the component of the cycle for which soils and plants play the most prominent roles.

(a)

(b)

FIGURE 6.5 Managing soils to increase infiltration of rainwater. (*a*) Small furrow dikes on the right side of this field in Texas retain rainwater long enough for it to infiltrate rather than run off. (*b*) This saturated soil under a winter cover crop of hairy vetch is riddled with earthworm burrows that greatly increased the infiltration of water from a recent heavy rain. Scale in centimeters. [Photo (*a*) courtesy of O. R. Jones, USDA Agricultural Research Service, Bushland, Texas; photo (*b*) courtesy of R. Weil]

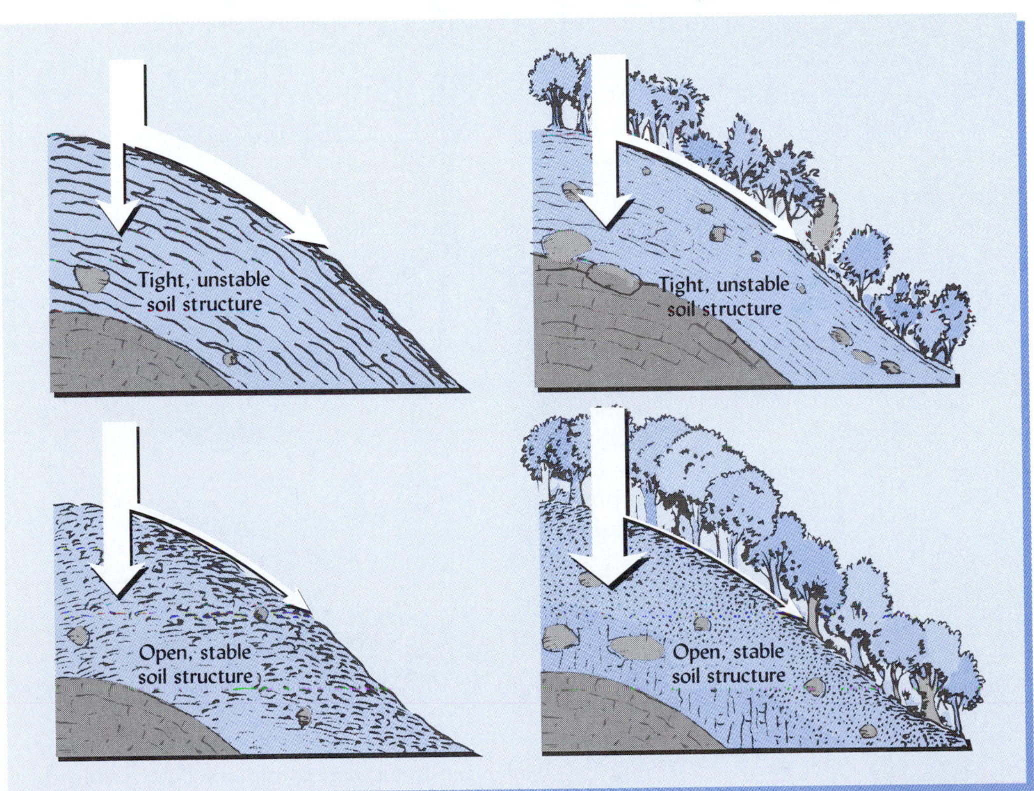

FIGURE 6.6 Influence of soil structure and vegetation on the partitioning of rainfall into infiltration and runoff. The upper two diagrams show soils with tight, unstable structure that resists infiltration and percolation. The bare soil is especially prone to surface sealing and resulting high losses by runoff. Even with forest cover, the low-permeability soils cannot accept all the rain in an intense storm. The two lower diagrams show much greater infiltration into soils that have open, stable structures with significant macropore space. The more open soil structure combined with the protective effects of the forest floor and canopy nearly eliminate surface runoff.

6.3 THE SOIL–PLANT–ATMOSPHERE CONTINUUM

Water Movement in the Plant: http://www.hort.purdue.edu/hort/courses/HORT301/MikesLectures/WATERMovement.html

The flow of water through the soil–plant–atmosphere continuum (SPAC) is a major component of the overall hydrologic cycle. Figure 6.7 ties together many of the processes we have just discussed: *interception, surface runoff, percolation, drainage, evaporation, plant water uptake, water movement to plant leaves,* and *transpiration* of water from the leaves back into the atmosphere.

It further suggests that the same basic principles govern the retention and movement of water whether it is in soil, in plants, or in the atmosphere. Consequently, if a plant is to absorb water from the soil, the water potential must be lower in the plant root than in the soil adjacent to the root. Likewise, movement up the stem to the leaf cells is in response to differences in water potential, as is the movement from leaf surfaces to the atmosphere.

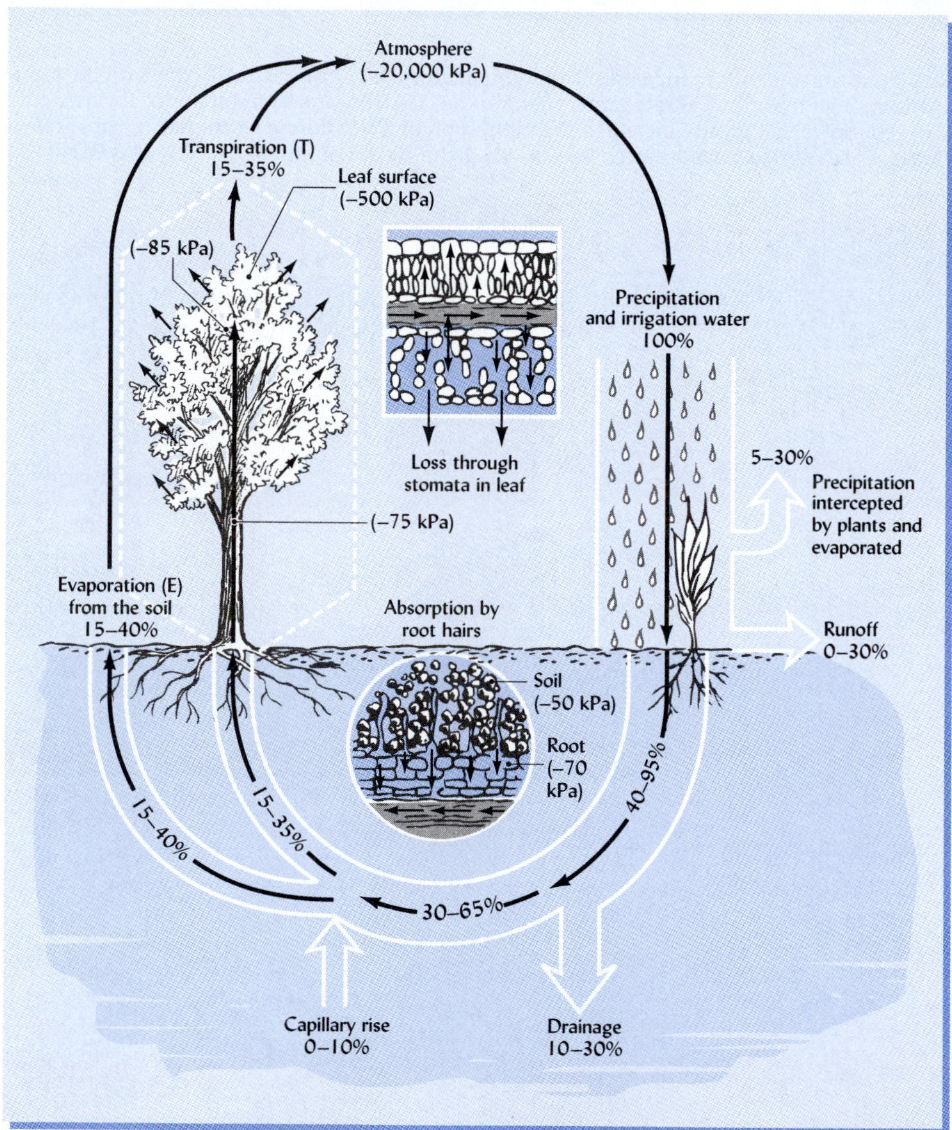

FIGURE 6.7 Soil–plant–atmosphere continuum (SPAC) showing water movement from soil to plants to the atmosphere and back to the soil in a humid to subhumid region. Water behavior through the continuum is subject to the same energy relations covering soil water that were discussed in Chapter 5. Note that the moisture potential in the soil is −50 kPa, dropping to −70 kPa in the root, declining still further as it moves upward in the stem and into the leaf, and is very low (−500 kPa) at the leaf–atmosphere interface, from whence it moves into the atmosphere, where the moisture potential is −20,000 kPa. Moisture moves from a higher to a lower moisture potential. Note the suggested ranges for partitioning of the precipitation and irrigation water as it moves through the continuum. Over 98% of the water absorbed by the roots of irrigated crops is transpired as water vapor over the course of the season.

Two Points of Resistance

Two primary factors determine whether plants are well supplied with water: (1) the rate at which water is supplied by the soil to the absorbing roots, and (2) the rate at which water is evaporated from the plant leaves. Since factors affecting the soil's ability to supply water were discussed in Section 5.8, we will now address the loss of water by evaporation from soil–plant systems.

Evapotranspiration

It is quite difficult to determine just how much water loss occurred directly from the soil (by evaporation, E) and how much occurred from the leaf surfaces after plant uptake (by transpiration, T). Therefore, information is most commonly available on evapotranspiration (ET), the combined loss resulting from these two processes. The evaporation component of ET may be viewed as a "waste" of water from the standpoint of plant productivity. However, at least some of the transpiration component is essential for plant growth, providing the water that plants need for cooling, nutrient transport, photosynthesis and turgor maintenance.

POTENTIAL ET. The **potential evapotranspiration** (PET) rate tells us how fast water vapor *would* be lost from a densely vegetated plant–soil system *if* soil water content were continuously maintained at an optimal level. The PET is largely determined by climatic variables, such as temperature, relative humidity, cloud cover, and wind speed, that influence the *vapor pressure gradient* between a wet soil, leaf, or body of water and the atmosphere.

A number of mathematical models have been devised to estimate PET from climatological data, but in practice, PET can be most easily estimated by applying a correction factor to the amount of water evaporated from an open pan of water of standard design (a class-A evaporation pan—see Figure 6.8). Dense, well-watered vegetation typically transpires water about 65% as rapidly as water evaporates from an open pan; hence, the correction factor for dense vegetation such as a lawn is typically 0.65 (it is lower for less-dense vegetation):

$$\text{PET} = 0.65 * \text{pan evaporation} \tag{6.2}$$

Values of PET range from more than 1500 mm per year in hot, arid areas to less than 40 mm in very cold regions. During the winter in temperate regions, PET may be less than 1 mm per day. By contrast, hot, dry wind will continually sweep away water vapor from a wet surface, creating a particularly steep vapor pressure gradient and PET levels as high as 10 to 12 mm per day.

EFFECT OF SOIL MOISTURE. Evaporation from the soil surface E at a given temperature is determined to a large extent by soil surface wetness and by the ability of the soil to replenish this surface water as it evaporates. In most cases, the upper 15 to 25 cm of soil provides most of the water for surface evaporation E. Unless a shallow water table exists, the upward capillary movement of water is very limited and the surface soil soon dries out, greatly reducing further evaporation loss.

However, because plant roots penetrate deep into the profile, a significant portion of the water lost by evapotranspiration ET comes from the subsoil layers (see Figure 6.9). Water stored deep in the profile is especially important to vegetation in regions having alternating moist and dry seasons (such as Ustic or Xeric moisture regimes). Water stored in the subsoil during rainy periods is available for evapotranspiration during dry periods. Plate 42 following page 498 illustrates the death of a rooftop lawn because of the inability of the shallow soil to hold sufficient water during a prolonged summer drought.

PLANT WATER STRESS. For dense vegetation growing in soil that is well supplied with water, ET will nearly equal PET. When soil water content is less than optimal, the plant will not be able to withdraw water from the soil fast enough to satisfy the PET. If water evaporates from the leaves faster than it enters the roots, the plant will lose turgor pressure and wilt. Under these conditions, actual evapotranspiration is less than potential evapotranspiration and the plant experiences **water stress**. The difference between PET and actual ET is termed the **water deficit**. A large deficit is indicative of high water stress and aridity.

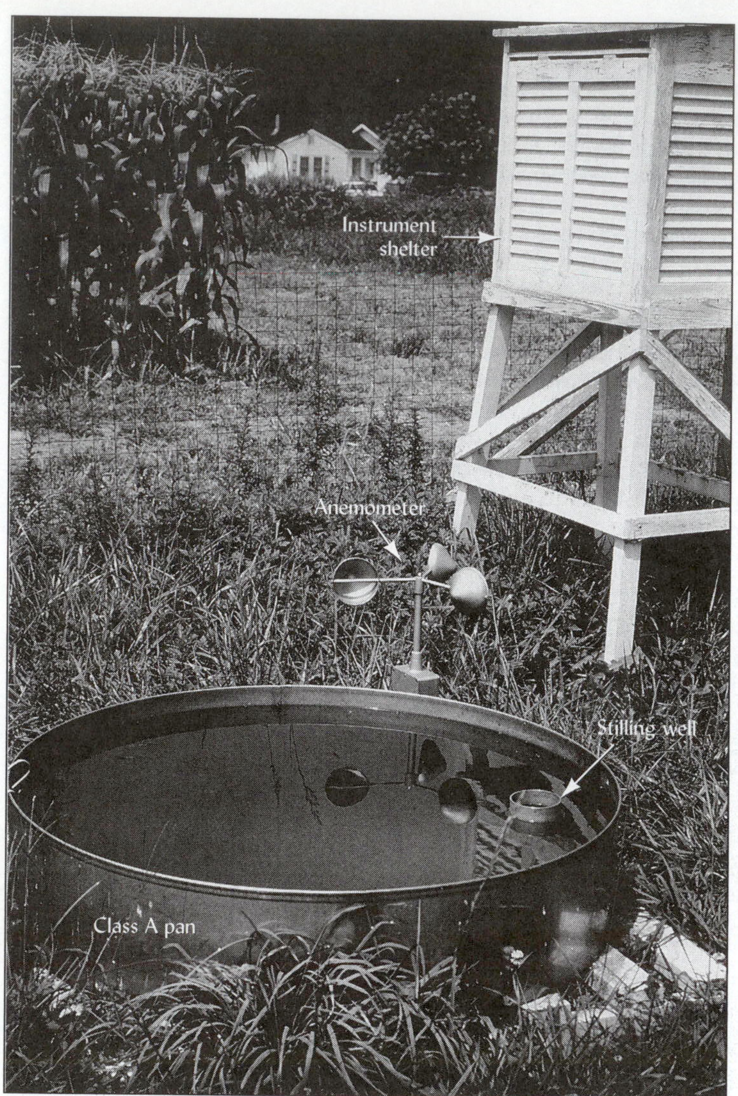

FIGURE 6.8 A class-A evaporation pan used to help estimate potential evapotranspiration (PET). Once a day, the water level is determined in the stilling well (small cylinder) and a measured amount of water is added to bring the level back up to the original mark. Evaporation from the pan integrates the effects of relative humidity, temperature, wind speed, and other climatic variables related to the vapor pressure gradient. Also shown are an anemometer to measure wind speed and a shelter containing instruments to measure temperature and relative humidity. (Photo courtesy of R. Weil)

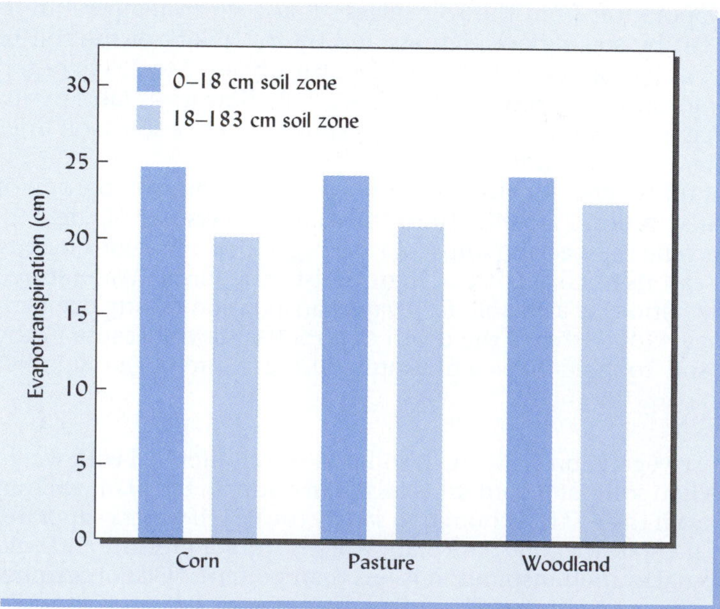

FIGURE 6.9 Evapotranspiration loss from the surface layer (0 to 18 cm) compared with loss from the subsoil (18 to 183 cm). Note that more than half the water loss came from the upper 18 cm and only half from the *next 165 cm* of depth. Periods of measurement: corn, May 23 to September 25; pasture, April 15 to August 23; woodland, May 25 to September 28. [Calculated from Dreibelbis and Amerman (1965)]

Influence of Plant Characteristics. In vegetated areas, incoming solar radiation is either absorbed by plant leaves on its way through the plant canopy or it reaches the ground and is absorbed by the soil. Therefore, for annual plants, as the leaf area per unit land area (a ratio termed the **leaf area index** [LAI]) increases, more radiation will be absorbed by the foliage to stimulate transpiration (T), and less will reach the soil to promote evaporation (E).

In contrast, perennial vegetation, such as pastures and forests, have very high leaf area indices both early and late in the growing season. Where the forest floor (leaf litter) is undisturbed, even less direct sunlight strikes the soil and E is very low throughout the year.

In summary, water losses from the soil surface and from transpiration are determined by (1) climatic conditions, (2) plant cover in relation to soil surface (leaf area index), (3) efficiency of water use by different plants, and (4) length and season of the plant growing period.

6.4 EFFICIENCY OF WATER USE

Crop residues, tillage and irrigation water management:
http://www.ianr.unl.edu/pubs/irrigation/g1154.htm

The vegetative dry matter produced while using a given amount of water is an important measure of efficiency, especially in areas where moisture is scarce. This efficiency may be expressed in terms of dry matter yield per unit of water transpired (*T efficiency*) or the dry matter yield per unit of water lost by evapotranspiration (*ET efficiency*).

Transpiration Ratio. Another expression of water use efficiency is the **transpiration ratio,** which is the inverse of T efficiency and is expressed as kilograms of water transpired to produce 1 kg of dry matter. The transpiration ratio for a given plant is markedly affected by climatic conditions, ranging from 200 to 500 in humid regions and almost twice as much in arid regions (Table 6.1).

At a given location there are considerable differences in transpiration ratio among different plant species. For example, the data in Table 6.1 show that in semiarid Akron, Colorado, the transpiration ratios for millet and maize (corn) are only half as large as those for clover or peas (Table 6.1).

We can see the significance of these figures by considering a wheat crop growing in a semiarid region and having a transpiration ratio of 500. For every kilogram of aboveground dry matter produced, about 500 kg (or liters) of water were transpired. If we assume that only about 40% of the dry matter produced by the time of harvest is in the grain (the rest is in the straw), we see that it takes about 1250 L of water to produce 1 kg of wheat grain. When we take into account the water used to grow fruits, vegetables, and feed for cattle, as well as grains, no wonder it is estimated that almost 7000 L (1700 gal) of water are required to grow a single day's food supply for one adult in the United States.

Indeed, the amount of water necessary to bring a crop to maturity is very large. For example, a representative crop of wheat containing 5000 kg/ha of dry matter and having a transpiration ratio of 500 will withdraw water from the soil during the growing season equivalent to about 250 mm of rain. This amount of water, *in addition* to that

TABLE 6.1 Transpiration Ratios of Various Crops as Determined at Different Locations

Kilograms of water used in production of 1 kg of dry matter.

Crop	*Harpenden, England*	*Dahme, Germany*	*Madison, Wisconsin*	*Pusa, India*	*Akron, Colorado*
Beans	209	282	—	—	736
Clover	269	310	576	—	797
Maize (corn)	—	—	271	337	368
Millet	—	—	—	—	310
Oats	—	376	503	469	597
Peas	259	273	477	563	788
Potatoes	—	—	385	—	636
Wheat	247	338	—	544	513

Data compiled by Lyon et al. (1952).

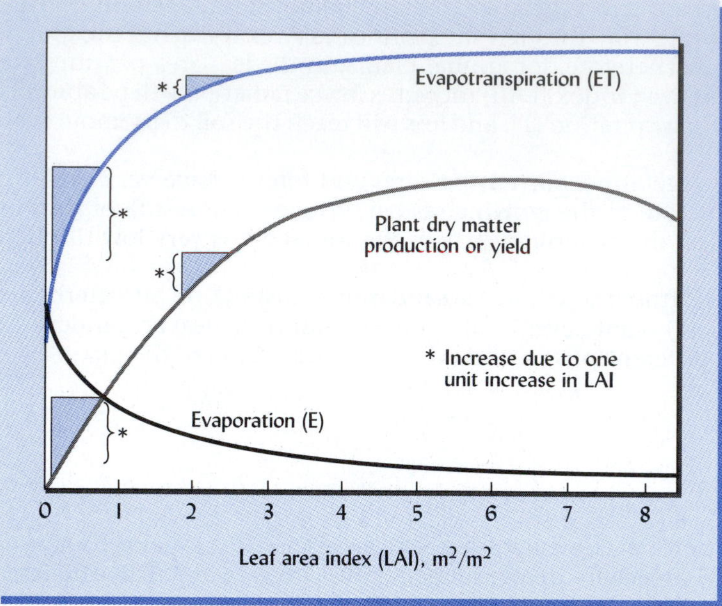

FIGURE 6.10 Generalized effect of leaf area index (LAI) on plant productivity, evapotranspiration (ET), and evaporation (E). The shaded areas show the effects of increasing LAI by 0.6 unit. Note that ET increases with increasing LAI up to a point and then levels off when the foliage intercepts nearly all the incoming solar radiation. Simultaneously, E decreases and levels off at a very low value. Except where ground cover is sparse (LAI < 1) or the plant canopy is overly dense (LAI > 7), practices (such as fertilization or closer plant spacing) that increase LAI are likely to increase plant productivity by a greater percentage than the corresponding increase in ET, resulting in improved water-use efficiency. (Diagram courtesy of R. Weil)

evaporated from the surface, must be supplied during the growing season. It is not surprising that water—and the ability of the soil to store it—is often the most critical factor in plant growth.

ET Efficiency. Since evapotranspiration (ET) includes both transpiration (T) from plants and evaporation (E) from the soil surface, ET efficiency is more subject to management than T efficiency. Highest ET efficiency is attained where plant density and other growth factors are optimum for plant growth. Increases in leaf area index (LAI) lead to increases in plant production, up to the point at which virtually all available solar radiation is absorbed by leaf surfaces. Furthermore, as shown in Figure 6.10, when the LAI is increased over the range of 1 to about 7, plant production increases by a greater amount than does ET. The plant production per unit of water vaporized (ET efficiency) is therefore increased.

Effect of Crop Management. Figure 6.11 illustrates the effect on ET efficiency of increased yields from improved crop management. The increased production per unit of water vaporized results from more vigorous plants (1) sending their roots deeper into the soil profile, and (2) producing a greater leaf area, which intercepts a larger portion of the

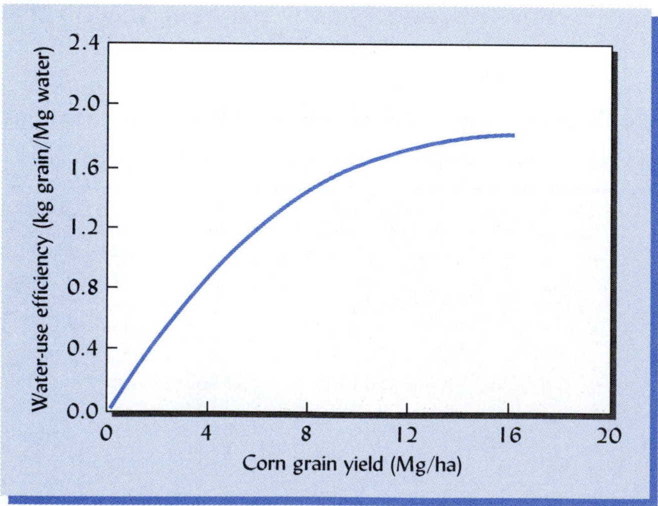

FIGURE 6.11 Water-use efficiency generally increases with increased plant yields, more plant biomass being produced with a given amount of water transpired at the higher yield levels. This graph shows this relationship for corn grown on a number of research plots on a Mollisol in the U.S. Southern High Plains. Efforts to increase the very low yields pay off in greatly increased water-use efficiency, but the rate of increase levels out at the higher yield levels. [Modified from Howell and Tolk (1998)]

solar radiation, leaving less radiation to evaporate water from the soil (see also Figure 6.10). Therefore, as plant production is increased by improved nutrition, T is increased substantially, but E is decreased, so that ET is increased only a little. We can conclude that so long as the supply of water is not too limited, maintaining optimum conditions for plant growth (by closer plant spacing, fertilization, or selection of more vigorous varieties) increases the efficiency of water use by plants.

In harsh semiarid environments, one should apply this principle with caution, however, if irrigation water is not available and the period of rainfall is very short. The modest increase in total water use (ET) by the more vigorously growing plants may deplete the stored soil water before the life cycle of the plant is complete, resulting in serious water stress, or even plant death before any harvestable yield has been produced.

6.5 CONTROL OF EVAPOTRANSPIRATION (ET)

Evapotranspiration and water use (click on evaporation):
http://www.engineering.usu.edu/uwrl/atlas/ch3/index.html

As we have just seen, water loss by ET is closely related to the total leaf area exposed to solar radiation. To some degree, adjusting the total leaf area exposed can bring ET into balance with PET (and thus lessen water stress). Five approaches to accomplish this are the following:

1. If water can be added by irrigation (an option that will be discussed in Section 6.11), both ET and plant production are likely to increase—often dramatically. Of course, for reasons of cost or lack of a water source, this practice is often not feasible.
2. Where rapid growth might prematurely deplete the available water, it may be wise to limit plant-growth factors such as nutrient supply to only moderate levels, thus keeping LAI in check.
3. The LAI of desired plants can be limited by sowing fewer seeds per unit land area or by spacing plants farther apart. It should be noted, however, that plants growing farther apart tend to individually produce a greater leaf area, compensating somewhat for the lower density. Also, wider spacing allows greater evaporation losses from the soil.
4. Elimination of undesired plants (i.e., weeds) can remove a major cause of T losses that could greatly reduce the soil water supplies available for the desired species.
5. In semiarid regions, elimination of all vegetation for a period of time may allow rainwater to accumulate in the soil profile for use by subsequent plantings.

Farming systems that alternate summer **fallow** (unvegetated period) one year with traditional cropping the next have been used to conserve soil moisture in some low-rainfall environments. By controlling all vegetative growth during the fallow year, T loss of water is minimized, thereby leaving more water in the profile for the next year's crop. However, the growing use of conservation tillage practices (see Section 6.6) that leave most of the crop residues on the soil surface has led to sufficient saving of soil moisture as to make fallow cropping unnecessary in some environments. The data from several no-till trials in eastern Colorado (Figure 6.12) confirm this point.

6.6 CONTROL OF SURFACE EVAPORATION (E)

More than half the precipitation in semiarid and subhumid areas is usually returned to the atmosphere by evaporation (E) directly from the soil surface. In natural rangeland systems, E is a large part of ET because plant communities tend to self-regulate toward low plant densities and leaf area configurations that minimize the deficit between PET and ET. In addition, plant residues on the soil surface are sparse. Evaporation losses are also high in arid-region irrigated soil, especially if inefficient practices are used (see Section 6.11). Even in humid-region rain-fed areas, E losses are significant during hot rainless periods. Such moisture losses rob the plant community of much of its growth

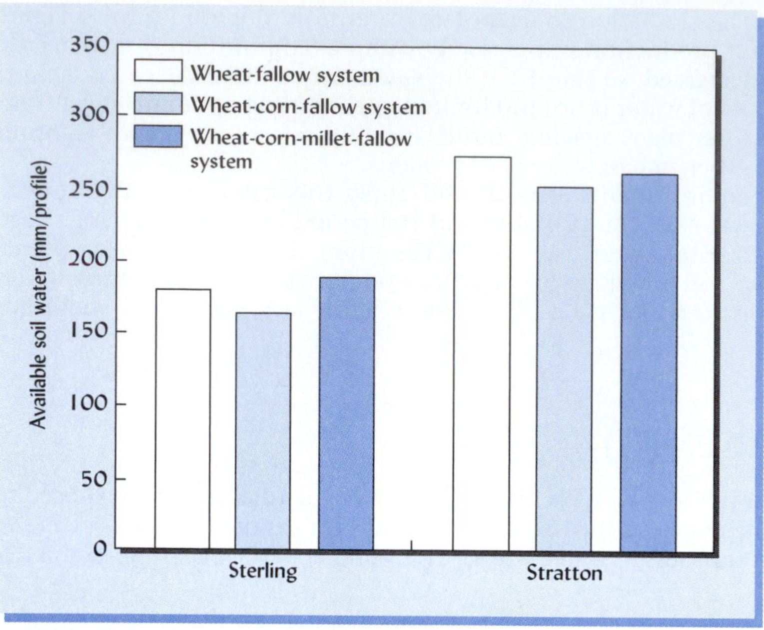

FIGURE 6.12 Available water (mm/profile) in the soil profile at wheat planting time in systems with fallow every second, third, or fourth year at two semiarid locations in eastern Colorado where conservation tillage practices had been adopted. Note that the available soil water level was about the same whether the soil was fallowed every other year or every third or fourth year. Apparently soil evaporation losses were so high during the fallow years that little additional water was conserved by leaving the land fallow. [Redrawn from Farahani et al. (1998)]

potential and reduce the water available for discharge to streams. Careful study of Figure 6.13 will clarify these relationships and the principles discussed in Sections 6.4 and 6.5.

For arable soils, the most effective practices aimed at controlling E are those that provide some cover for the soil. This cover can best be provided by mulches and by selected conservation tillage practices that leave plant residues on the soil surface, mimicking the soil cover of natural ecosystems.

Vegetative Mulches. A mulch is any material placed on the soil surface primarily for the purpose of reducing evaporation or controlling weeds. Examples include sawdust, manure, straw, leaves, and crop residues. Mulches can be highly effective in checking evaporation, but they may be expensive and labor-intensive. Mulching is therefore most practical for small areas (gardens and landscaping beds) and for high-value crops such as cut flowers, berries, fruit trees, and certain vegetables. In addition to reducing evaporation, vegetative mulches may provide these benefits: (1) reducing the spread of soilborne diseases; (2) providing a clean path for foot traffic; (3) reducing weed growth (if applied thickly); (4) moderating soil temperatures, especially preventing overheating in summer months (see Section 7.11); (5) increasing water

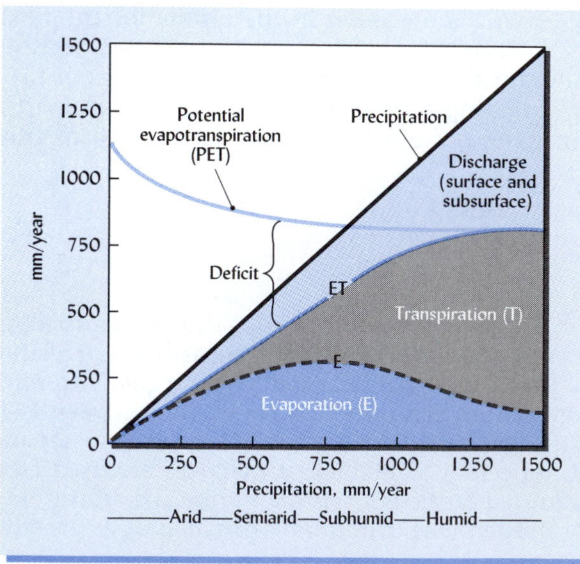

FIGURE 6.13 Partitioning of liquid water losses (discharge) and vapor losses (evaporation and transpiration) in regions varying from low (arid) to high (humid) levels of annual precipitation. The example shown assumes that temperatures are constant across the regions of differing rainfall. Potential evapotranspiration (PET) is somewhat higher in the low-rainfall zones because the lower relative humidity there increases the vapor pressure gradient at a given temperature. Evaporation (E) represents a much greater proportion of total vapor losses (ET) in the drier regions due to sparse plant cover caused by interplant competition for water. The greater the gap between PET and ET, the greater the deficit and the more serious the water stress to which plants are subject. (Diagram courtesy of R. Weil)

FIGURE 6.14 For crops with high cash value, plastic mulches are commonly used. The plastic is installed by machine (left) and at the same time the plants are transplanted (right). Plastic mulches help control weeds, conserve moisture, encourage rapid early growth, and eliminate need for cultivation. The high cost of plastic makes it practical only with the highest-value crops. (Courtesy K. Q. Stephenson, Pennsylvania State University)

infiltration; (6) providing organic matter and, possibly, plant nutrients to the soil; (7) encouraging earthworm populations; and (8) reducing soil erosion. Most of these ancillary benefits do not accrue from the use of plastic or paper mulches, discussed next.

Paper and Plastic Mulches. Specially prepared paper and plastics are also used as mulches to control evaporative water losses. In the case of black plastic and dark paper, they also effectively control weeds. The mulch is often applied by machine and plants grow through slits or holes in the sheeting (Figure 6.14). These mulches are widely used for vegetable and small fruit crops and in landscaping beds, where they are often covered with a layer of tree-bark mulch or gravel for a more pleasing appearance. As long as the ground is covered, evaporation is checked, and in some cases remarkable increases in plant growth have been reported. Unless rainfall is very heavy, the mulch does not seriously interfere with the infiltration of rainwater into the soil.

The temperature of the soil under plastic mulch is commonly 8° to 10°C higher than under a straw mulch. The higher temperature may be helpful for crops established early in spring, but it can reduce the yields of heat-sensitive plants in summer (see Section 7.11). A common problem with plastic mulches is that they can rarely be completely removed at the end of the growing season, so after a number of years scraps of plastic accumulate in the soil, causing an unsightly mess and interfering with water movement and cultivation.

Managing Residue to Store Precipitation in Semiarid Colorado:
http://www.akron.ars.usda.gov/fs_managing.html

Crop Residue and Conservation Tillage. Plant residues left on the soil surface are effective in reducing E and, in turn, in conserving soil moisture. **Conservation tillage** practices leave a high percentage of the residues from the previous crop on or near the surface (Figure 6.15). A conservation tillage practice widely used in subhumid and semiarid regions is **stubble mulch** tillage. With this method, residues such as wheat stubble or cornstalks from the previous crop are uniformly spread on the soil surface. The land is then tilled with special implements that permit much of the plant residue to remain on or near the surface. Conservation tillage planters are capable of planting through the stubble and allow much of it to remain on the surface during the establishment of the next crop (see Figure 6.15). Combining summer fallow with stubble mulching saves soil moisture in dry areas, as shown in Table 6.2. Unfortunately, plant growth in dry regions is usually insufficient to produce residue mulches at the levels needed to reduce E.

Other conservation tillage systems that leave residues on the soil surface include no-tillage (see Figure 6.15), where the new crop is planted directly into the sod or residues of the previous crop, with no plowing or disking. The effect of such tillage practices on soil moisture conservation is shown in Box 6.1. Conservation tillage systems will receive further attention in Section 15.5.

FIGURE 6.15 Conservation tillage leaves plant residues on the soil surface, reducing both evaporation losses and erosion. (Left) In a semiarid region (South Dakota) the straw from the previous year's wheat crop was only partially buried to anchor it against the wind while still allowing it to cover much of the soil surface. In the next year the left half of the field, now shown growing wheat, will be thusly *stubble mulched* and the right half sown to wheat. (Right) *No-till* planted corn in a more humid region grows up through the straw left on the surface by a previous wheat crop. Note that with no-till, almost no soil is directly exposed to solar radiation, rain, or wind. (Photos courtesy of R. Weil, left, and USDA Natural Resources Conservation Service, right)

6.7 PERCOLATION–EVAPORATION BALANCE

In our discussion of the hydrologic cycle we noted two types of liquid losses of water from soils: (1) percolation or drainage water, and (2) surface runoff water (see Figure 6.2). *Percolation water* recharges the groundwater and moves chemicals out of the soil. *Runoff water* often carries appreciable amounts of soil (erosion) as well as dissolved chemicals.

When the amount of rainfall entering a soil exceeds the water-holding capacity of the soil, losses by percolation will occur. Percolation losses are influenced by (1) the amount of rainfall and its distribution, (2) runoff from the soil, (3) evaporation, (4) the character of the soil, and (5) the nature of the vegetation. Figure 6.16 illustrates the relationships among precipitation, runoff, soil storage, soil water depletion, and percolation for representative humid and semiarid temperate regions and for an irrigated arid region.

Humid Region. In the *humid temperate* region, the rate of water infiltration into the soil (precipitation minus runoff) is greater, at least during certain seasons, than the rate of evapotranspiration. As soon as the soil field capacity is reached, percolation into the substrata occurs. Maximum percolation occurs during the winter and early spring, when evaporation is lowest. This period is when most leaching and chemical pollution of groundwater takes place. During the summer, little percolation occurs. In fact, evapotranspiration exceeds precipitation, resulting in a depletion of soil water. Normal plant

TABLE 6.2 Gains in Soil Water from Different Rates of Straw Mulch During Fallow Periods at Four Great Plains Locations

Location	Av. annual precipitation, mm	Soil water gain, mm, at each mulch rate			
		0 Mg/ha	2.2 Mg/ha	4.4 Mg/ha	6.6 Mg/ha
Bushland, Tex.	508	71	99	99	107
Akron, Colo.	476	134	150	165	185
North Platte, Nebr.	462	165	193	216	234
Sidney, Mont.	379	53	69	94	102
Average		107	127	145	157
Average gain by mulching			20	38	50

From Greb (1983).

BOX 6.1 WATER CONSERVATION PAYS OFF

During the past few decades, the world's farmers dramatically increased their yields of grain, enabling food production to more than keep up with the doubling of the human population. The increased productivity stemmed largely from (1) improved crop varieties, (2) greater availability of plant nutrients, and (3) greater amounts of available water (from irrigation and water conservation). Research on grain production in semiarid areas of the United States shows the important role of water conservation in the nearly quadrupling of grain yields in Texas over the 40-year period since 1938 (left figure).

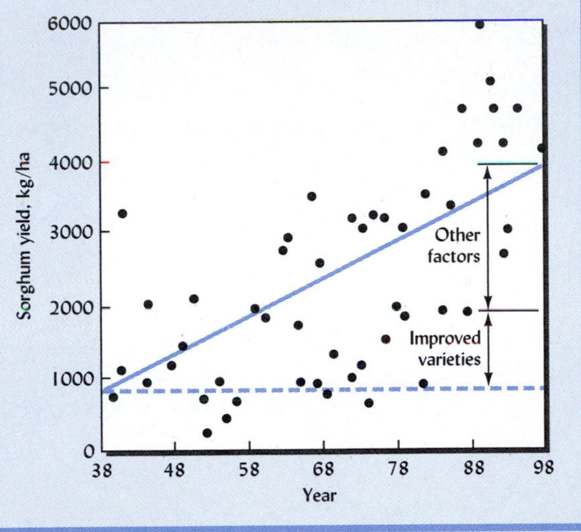

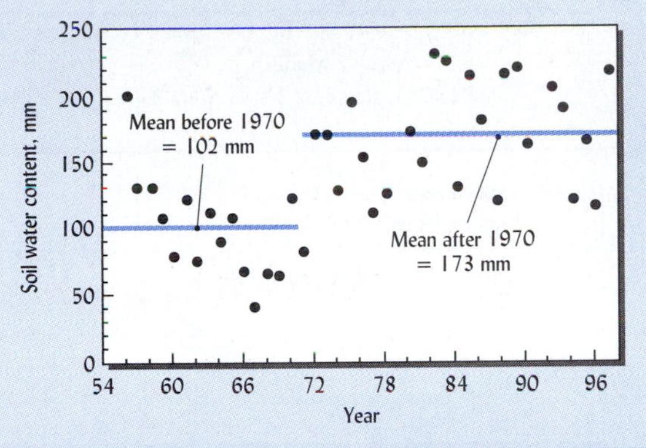

Research suggested that about a third of the productivity increase could be ascribed to improved crop genetics, with nearly two-third of the increase due to other factors. Because no nutrients were applied, researchers ruled out increased nutrient availability, leaving increased water availability as the most likely yield-enhancing factor. Analysis of precipitation records showed no increase in rain or snow that might account for the upward yield trend. However, measurements of water present in the soil profile at planting time were much higher in the 25-year period after 1972 than in the previous 20 years. Records showed that, prior to 1972, the soil was tilled between crops to control weeds and consequently, little residue cover was left on the soil. After 1972, crops were managed with no-tillage techniques that maintained crop residues on the soil surface and controlled weeds with herbicide sprays rather than cultivation. The good weed control reduced transpiration losses while the surface residue mulch greatly reduced evaporation losses, leaving considerable extra water in the soil to stimulate the increased sorghum yields.

From Unger and Baumhardt (1999)

growth is possible only because of water stored in the soil from the previous winter and early spring.

Semiarid Region. In the *semiarid region,* as in the humid region, water is stored in the soil during the winter months and is used to meet the moisture deficit in the summer. But because of the low rainfall, little runoff and essentially no percolation out of the profile occurs. Water may move to the lower horizons, but it is absorbed by plant roots and ultimately is lost by transpiration.

Irrigated Soils. The irrigated soil in the arid region shows a unique pattern. Irrigation in the early spring, along with a little rainfall, provides more water than is being lost by evapotranspiration. The soil is charged with water, and some percolation may occur. As we shall see in Section 10.8, irrigation systems must provide enough water for some percolation, in order to remove excess soluble salts. During the summer, fall, and winter months, the stored water is depleted because the amount being added is less than the very high evapotranspiration that takes place in response to large vapor pressure gradients.

The comparative losses of water by evapotranspiration and percolation through soils found in different climatic regions are shown in Figure 6.17. These differences should be kept in mind while reading the following section on percolation and groundwaters.

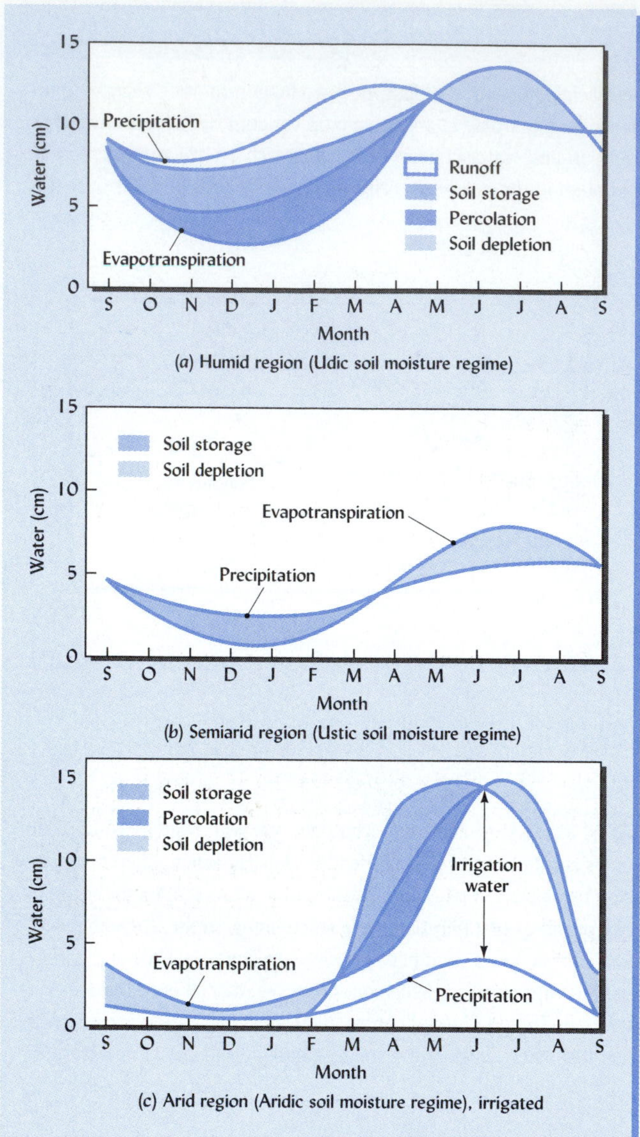

FIGURE 6.16 Generalized curves for precipitation and evapotranspiration for three temperate zone regions: (*a*) a humid region, (*b*) a semiarid region, and (*c*) an irrigated arid region. Note the absence of percolation through the soil in the semiarid region. In each case water is stored in the soil. This water is released later when evapotranspiration demands exceed the precipitation. In the semiarid region evapotranspiration would likely be much higher if ample soil moisture were available. In the irrigated arid region soil, the very high evapotranspiration needs are supplied by irrigation. Soil moisture stored in the spring is utilized by later summer growth and lost through evaporation during the late fall and winter.

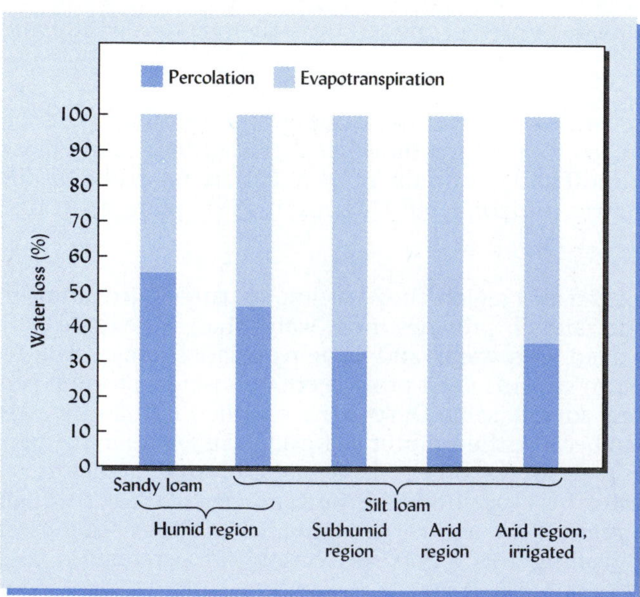

FIGURE 6.17 Percentage of the water entering the soil that is lost by downward percolation and by evapotranspiration. Representative figures are shown for different climatic regions.

6.8 PERCOLATION AND GROUNDWATERS

When drainage water moves downward and out of the soil, it eventually encounters a zone in which the pores are all saturated with water. Often this saturated zone lies above an impervious soil horizon (Figure 6.18) or a layer of impermeable rock or clay. The upper surface of this zone of saturation is known as the **water table**, and the water within the saturated zone is termed **groundwater.** The water table (Figure 6.19) is commonly only 1 to 10 m below the soil surface in humid regions but may be several hundred meters deep in arid regions. In swamps it is essentially at the land surface.

The unsaturated zone above the water table is termed the **vadose zone** (see Figures 6.19 and 6.20). The vadose zone may include unsaturated materials underlying the soil profile, and so may be considerably deeper than the soil itself. In some cases, however, the saturated zone may be sufficiently high to include the lower soil horizons, with the vadose zone confined to the upper soil horizons.

Shallow groundwater receives downward-percolating drainage water. Most of the groundwater, in turn, seeps laterally through porous geological materials (termed **aquifers**) until it is discharged into springs and streams. Groundwater may also be removed by pumping for domestic and irrigation uses. The water table will move up or down in response to the balance between the amount of drainage water coming in through the soil and the amount lost through pumped wells and natural seepage to springs and streams. In humid temperate regions, the water table is usually highest in early spring following winter rains and snow melt, but before evapotranspiration begins to withdraw the water stored in the soil above.

Groundwater that is near the surface can serve as a reciprocal water reservoir for the soil. As plants remove water from the soil, it may be replaced by upward capillary movement from a shallow water table. The zone of wetting by capillary movement is known as the **capillary fringe** (see Figure 6.20). Such movement can provide a steady and significant supply of water that enables plants to survive during periods of low rainfall.[1]

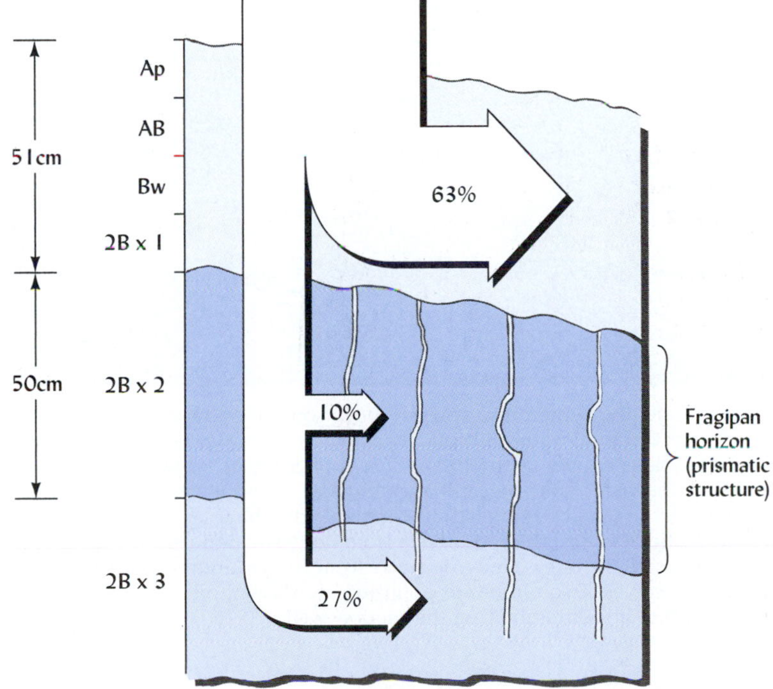

FIGURE 6.18 Soil profile characteristics largely determine the lateral and vertical movement of water. This graph shows the effect of a fragipan layer on the horizontal and vertical movement of water in the profile. The soil is a Typic Fragiochrept on a 12% hill slope in Pennsylvania. Note that most of the downward-moving water (63%) did not enter the fragipan but flowed downslope above the fragipan. Of the 37% that entered the fragipan, only 10% moved laterally, the remaining 27% moving downward along the faces of prismatic structural units of the fragipan. The mottled appearance of the soil from 39 cm downward suggests poor aeration conditions that, along with the very dense fragipan, restrict the growth of plant roots. [Drawn from data in Day et al. (1998); used with permission of Lippincott, Williams, and Wilkins, Baltimore]

[1]Capillary rise from shallow groundwater may also bring a steady supply of salts to the surface if the groundwater is brackish (see Section 9.13 for details on this soil-degrading process).

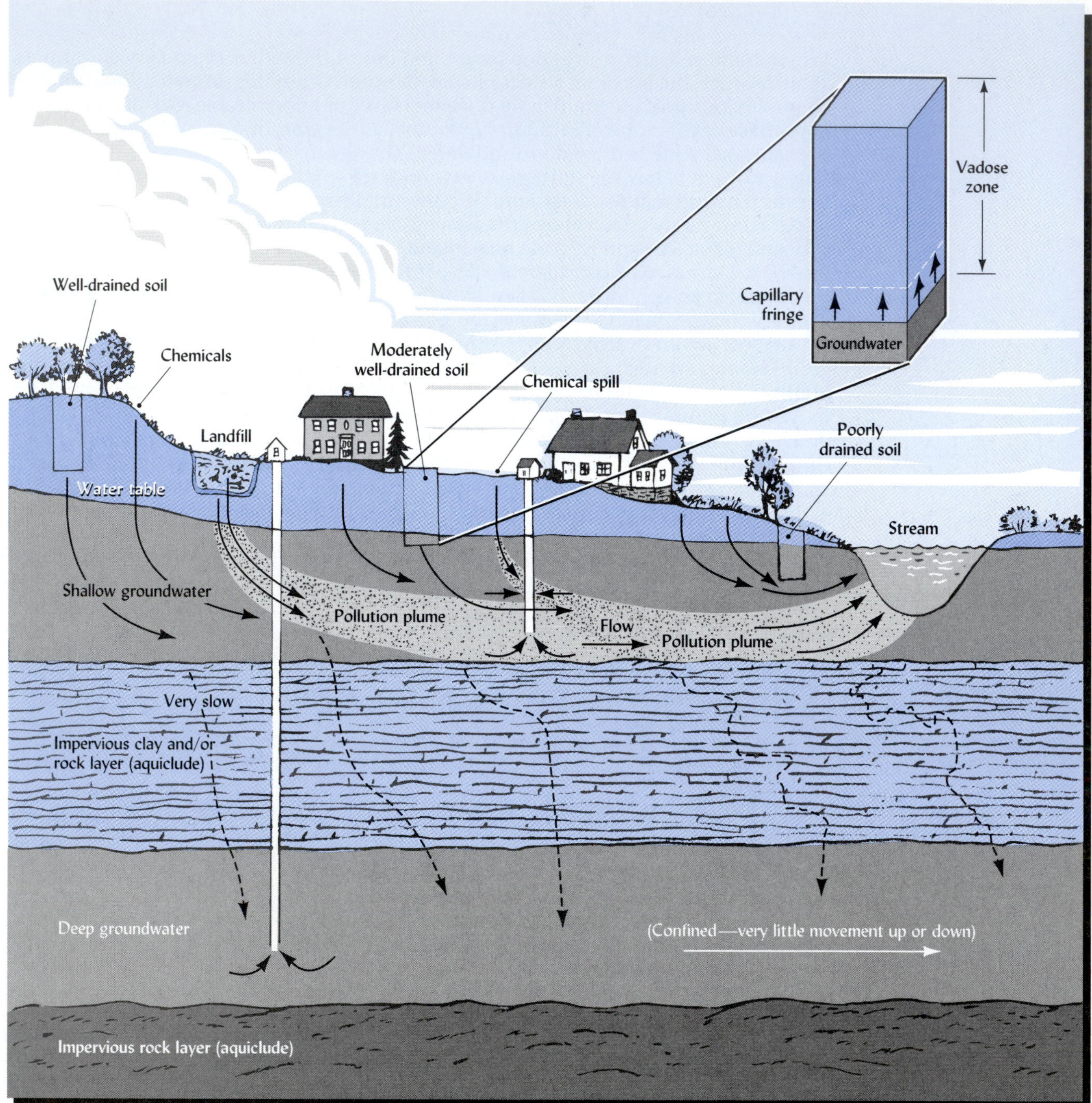

FIGURE 6.19 Relationship of the water table and groundwater to water movement into and out of the soil. Precipitation and irrigation water move down the soil profile under the influence of gravity (gravitational water), ultimately reaching the water table and underlying shallow groundwater. The unsaturated zone above the water table is known as the *vadose zone* (upper right). As water is removed from the soil by evapotranspiration, groundwater moves up from the water table by capillarity in what is termed the *capillary fringe*. Groundwater also moves horizontally down the slope toward a nearby stream, carrying with it chemicals that have leached through the soil, including essential plant nutrients (N, P, Ca, etc.) as well as pesticides and other pollutants from domestic, industrial, and agricultural wastes. Groundwaters are major sources of water for wells, the shallow ones removing water from the groundwater near the surface, while deep wells exploit deep and usually large groundwater reserves. Two plumes of pollution are shown, one originating from landfill leachate, the other from a chemical spill. The former appears to be contaminating the shallow well.

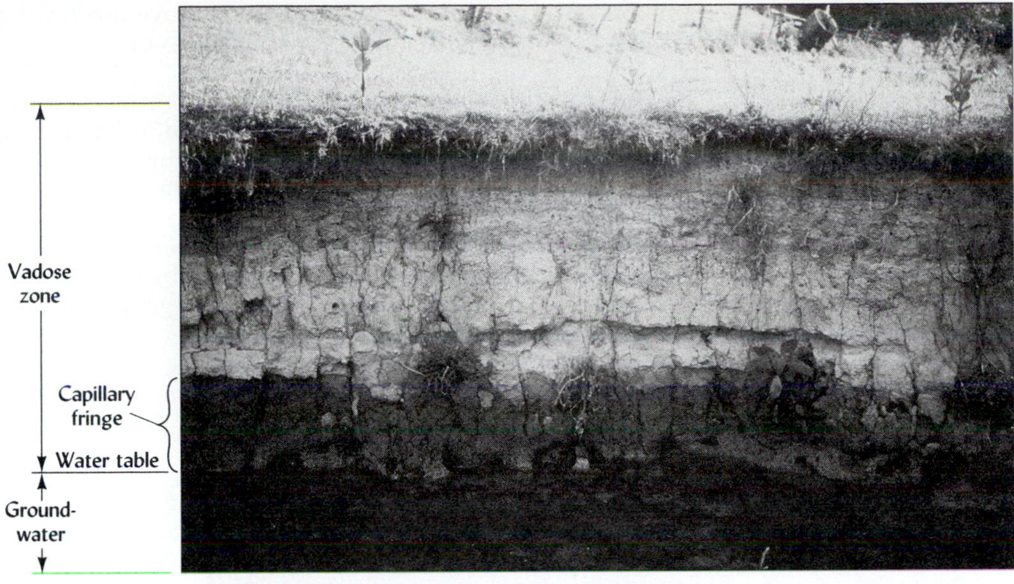

FIGURE 6.20 The water table, capillary fringe, zone of unsaturated material above the water table (vadose zone), and groundwater are visible. The groundwater can provide significant quantities of water for plant uptake. (Photo courtesy of R. Weil)

Basic groundwater flow: http://www.earthdrx.org/poresizegwflow.html

LEACHING.[2] Percolation of water through the soil to the water table not only replenishes the groundwater, it also dissolves and carries downward a variety of inorganic and organic chemicals found in the soil or on the land surface. Chemicals **leached** from the soil to the groundwater (and eventually to streams and rivers) in this manner include elements weathered from minerals, natural organic compounds resulting from the decay of plant residues, plant nutrients derived from natural and human sources, and various synthetic chemicals applied intentionally or inadvertently to soils.

In the case of plant nutrients, especially nitrogen, downward movement through the soil and into underlying groundwaters has two serious implications. First, the leaching of these chemicals represents a depletion of plant nutrients from the root zone (see Section 14.2). Second, accumulation of these chemical nutrients in ponds, lakes, reservoirs, and groundwater downstream may stimulate a process called **eutrophication**, which ultimately depletes the oxygen content of the water, with disastrous effects on fish and other aquatic life (see also Section 13.2). Also, in some areas underground sources of drinking water may become contaminated with excess nitrates to levels unsafe for human consumption (see Section 12.9).

Of even more concern is the contamination of groundwater with human pathogens and various highly toxic synthetic compounds, such as pesticides and their breakdown products or chemicals leached out of waste disposal sites. Figure 6.19 illustrates how the groundwater can be charged with these pathogens or chemicals, and how a plume of contamination spreads to downstream wells and bodies of water.

Chemical Movement Through Macropores

Large macropores play a critical role in determining field hydraulic conductivity (see Section 5.5). Old root channels, earthworm burrows, and clay shrinkage cracks commonly contribute large macropores that may provide channels for rapid water flow from the soil surface to depths of 1 m or more.

Once chemicals are carried below the zone of greatest root and microbial activity, they are less likely to be removed or degraded before being carried further down to the groundwater. This means that chemicals that are normally broken down in the soil by

[2] For a review of chemical transport through field soils, see Jury and Fluhler (1992).

microorganisms within a few weeks may move down through large macropores to the groundwater before their degradation can occur.

Investigating preferential flow using colored water: http://observe.ivv.nasa.gov/nasa/earth/hydrocycle/hydro1.html

Bypass Flow. In some cases, leaching of chemicals is most serious if the chemicals are merely applied on the soil surface. As shown in Figure 6.21, chemicals may be washed from the soil surface into large pores, through which they can quickly move downward. Research suggests that most of the water flowing through large macropores does not come into contact with the bulk of the soil. Such flow is sometimes termed **preferential** or **bypass flow**, as it tends to move rapidly around, rather than through, the soil matrix. As a result, if chemicals have been well mixed with the upper few centimeters of soil, their movement into the larger pores is reduced, and downward leaching is greatly curtailed.

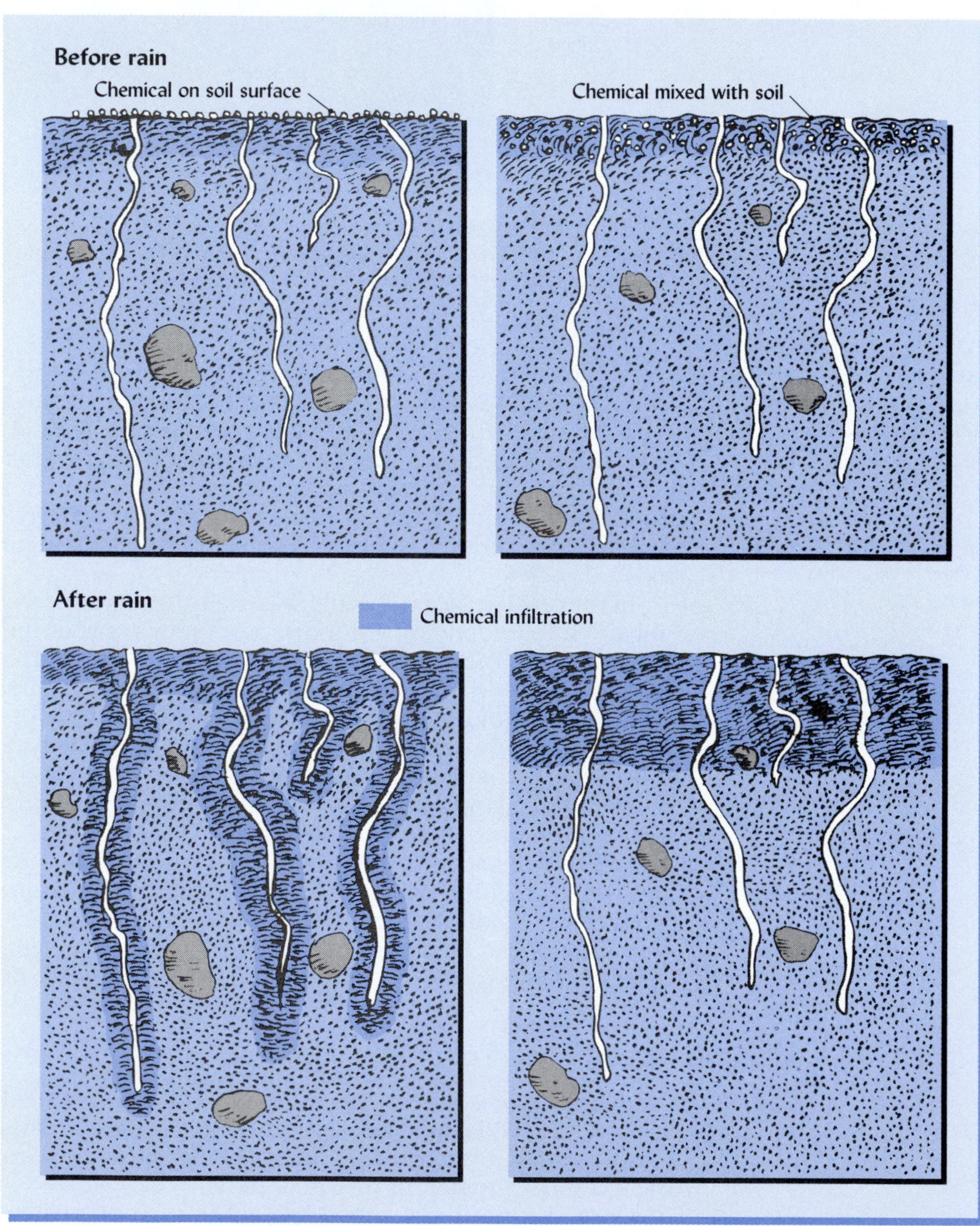

FIGURE 6.21 Preferential or bypass flow in macropores transports soluble chemicals downward through a soil profile. Where the chemical is on the soil surface (left), and can dissolve in surface-ponded water when it rains, it may be transported rapidly down cracks, earthworm channels, and other macropores. Where the chemical is dispersed within the soil matrix in the upper horizon (right), most of the water moving down through the macropores will bypass the chemical, and thus little of the chemical will be carried downward.

TABLE 6.3 Influence of Water Application Intensity on the Leaching of Pesticides Through the Upper 50 cm of a Grass Sod Covered Mollisol

The tests were conducted on 20-cm-diameter soil columns with the natural structure, including worm channels, undisturbed. All columns received 2.54 cm of simulated rain per week for 4 weeks, but the heavy rain was applied in four doses of 2.54 cm each, while the light rain was applied as 16 evenly spaced doses of 0.64 cm each. Metalaxyl is much more highly water soluble than Isazofos, but in each case the heavy rain stimulated much more pesticide leaching through the macropores.

	Pesticide leached as percentage of that applied to surface	
Pesticide	*Heavy rains*	*Light rains*
Isazofos	8.8	3.4
Metalaxyl	23.8	13.9

Data from Starrett et al. (1996).

Intensity of Rain or Irrigation. Bypass flow occurs most often during high-intensity rainfall or irrigation events that follow a dry period during which cracks and other such macropores have formed. The chemical-bearing water quickly flows down through the macropores before the mass of the soil is wetted and before the swelling clays can shut the cracks and greatly reduce other macropore sizes. In contrast, a gentle rain that in time may provide as much water as the more intense event, would likely thoroughly wet the entire upper soil mass, thereby reducing the size of macropores and preventing the rapid downward movement of water along with the chemicals it carries. Table 6.3 illustrates the effects of rainfall intensity on the leaching of pesticides applied to turf grass.

Manipulations of soil macropores and irrigation intensity provide us with two examples of how environmental quality can be impacted by human activities that influence the hydrologic cycle. Likewise, we have considered steps that can be taken to influence the infiltration of water into soils and to maximize plant biomass production from the water stored in soils. The picture would not be complete if we did not consider briefly three other anthropogenic modifications to the hydrologic cycle: (1) the use of artificial drainage to enhance the downward percolation of excess water from some soils; conversely, (2) the application of additional water to soils as a means of wastewater disposal; and (3) the use of irrigation to supplement water available for plant growth. We shall consider artificial drainage first.

6.9 ENHANCING SOIL DRAINAGE[3]

Soil waterlogging and drainage: www.maf.govt.nz/MAFnet/schools/activities/swi/swi-05.htm

Some soils tend to be water saturated in the upper part of their profile for extended periods during part or all of the year. The prolonged saturation may be due to the low-lying landscape position in which the soil is found, such that the regional water table is at or near the soil surface for extended periods (**endoaquic**). In other soils, water may accumulate above an impermeable layer in the soil profile (**epiaquic**), creating a **perched water table** (see Figure 6.22). Soils with either type of saturation may be components of wetlands (see Section 7.7), transitional ecosystems between land and water.

Reasons for Enhancing Soil Drainage

Water-saturated, poorly aerated soil conditions are essential to the normal functioning of wetland ecosystems and to the survival of many wetland plant species. However, for most other land uses, these conditions are a distinct detriment.

[3] For a review of all aspects of artificial drainage, see Skaggs and van Schilfgaarde (1999).

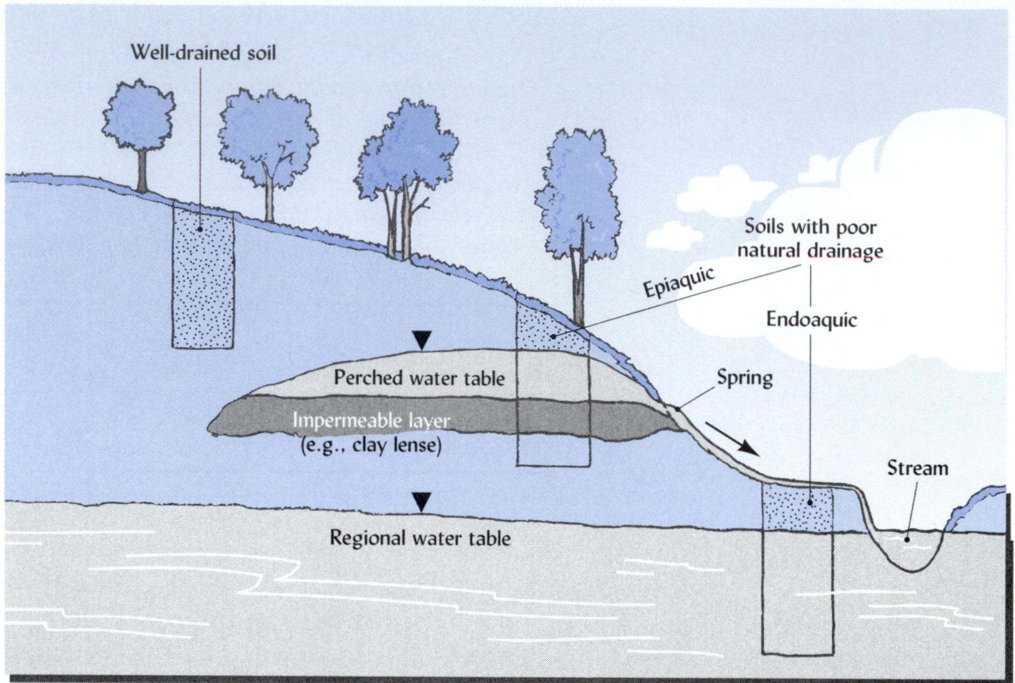

FIGURE 6.22 Cross section of a landscape showing the regional and perched water tables in relation to three soils, one well-drained and two with poor internal drainage. By convention, an inverted triangle (▼) identifies the level of the water table. The soil containing the perched water table is wet in the upper part, but unsaturated below the impermeable layer, and therefore is said to be *epiaquic* (Greek *epi,* upper), while the soil saturated by the regional water table is said to be *endoaquic* (Greek *endo,* under). Artificial drainage can help to lower both types of water tables.

How subsurface drainage affects soil water and the water balance:
http://www.extension.umn.edu/distribution/cropsystems/DC7644.html

Engineering Problems. During construction, the muddy, low-bearing-strength conditions of saturated soils make it very difficult to operate machinery (see Section 4.4). Houses built on poorly drained soils may suffer from uneven settlement and flooded basements during wet periods. Similarly, a high water table will result in capillary rise of water into roadbeds and around foundations, lowering the soil strength and leading to damage from frost-heaving (see Section 7.8) if the water freezes in winter. Heavy trucks traveling over a paved road underlaid by a high water table create potholes and eventually destroy the pavement.

Plant Production. Water-saturated soils make the production of most upland crops and forest species difficult, if not impossible. In wet soil, farm equipment used for planting, tillage, or harvest operations may bog down. Except for a few specially adapted plant species (bald cypress trees, rice, cattails, etc.), most crop and forest species grow best in well-drained soils since their roots require adequate oxygen for respiration (see Section 7.1). Furthermore, a high water table early in the growing season will confine the plant roots to a shallow layer of partially aerated soil; the resulting restricted root system can lead to water stress later in the year, when the weather turns dry and the water table drops rapidly.

For these and other reasons, artificial drainage systems have been widely used to remove excess (gravitational) water and lower the water table in poorly drained soils. Land drainage is practiced in select areas in almost every climatic region, but is most widely used to enhance the agricultural productivity of clayey alluvial and lacustrine soils. Drainage systems are also a vital, if sometimes neglected, component of arid region irrigation systems, where they are needed to remove excess salts and prevent waterlogging.

Artificial drainage is a major alteration of the soil system, and the following list of potential beneficial and detrimental effects should be carefully considered. Also take note that in many instances laws designed to protect wetlands require that a special permit be obtained for the installation of a new artificial drainage system.

Benefits of Artificial Drainage

1. Increased bearing strength and improved soil workability, which allow more timely field operations and greater access to vehicular or foot traffic (e.g., in recreational facilities).
2. Reduced frost-heaving of foundations, pavements, and crop plants (e.g., see Figure 7.15).
3. Enhanced rooting depth, growth, and productivity of most upland plants due to improved oxygen supply and, in acid soils, lessened toxicity of manganese and iron (see Sections 7.3 and 7.5).
4. Reduced levels of fungal disease infestation in seeds and on young plants.
5. More rapid soil warming in spring, resulting in earlier maturing crops (see Section 7.11).
6. Less production of methane and nitrogen gases that cause global environmental damages (see Sections 11.2 and 12.10).
7. Removal of excess salts from irrigated soils, and prevention of salt accumulation by capillary rise in areas of salty groundwater (see Section 9.13).

Detrimental Effects of Artificial Drainage

1. Loss of wildlife habitat, especially waterfowl breeding and overwintering sites.
2. Reduction in nutrient assimilation and other biochemical functions of wetlands (see Section 7.7).
3. Increased leaching of nitrates and other contaminants to groundwater.
4. Accelerated loss of soil organic matter, leading to subsidence of certain soils (see Sections 11.10 and 11.12).
5. Increased frequency and severity of flooding due to loss of runoff water retention capacity.
6. Greater cost of damages when flooding occurs on alluvial lands developed after drainage.

Artificial drainage systems are designed to promote two general types of drainage: (1) *surface drainage* and (2) internal or *subsurface drainage*. Each will be discussed briefly.

Surface Drainage Systems

Types of subsurface drainage, including mole drains: http://www.maf.govt.nz./MAFnet/schools/activities/swi/swi-05.htm#P862_36683

Surface drainage systems are designed to remove water from the land before it infiltrates the soil. They are extensively used, primarily on nearly level lands with fine-textured soils that have slow internal drainage.

Surface Drainage Ditches. The removal of surface water may be hastened by constructing shallow ditches with gentle side slopes that do not interfere with equipment traffic. If there is some slope on the land, the shallow ditches are usually oriented across the slope and across the direction of planting and cultivating, thereby permitting the interception of water as it runs off down the slope. For removing surface water from landscaped lawns, this system of drainage can be modified by constructing gently sloping swales rather than ditches.

Land Forming. Often, surface drainage ditches are combined with **land forming** or smoothing to eliminate the ponding of water and facilitate its removal from the land (Figure 6.23). Small ridges are cut down and depressions are filled in using precision, laser-guided field-leveling equipment. The resulting land configuration permits excess water to move at a controlled rate over the soil surface to the outlet ditch and then on to a natural drainage channel. Land smoothing is also commonly used to prepare a field for flood irrigation (see Section 6.11).

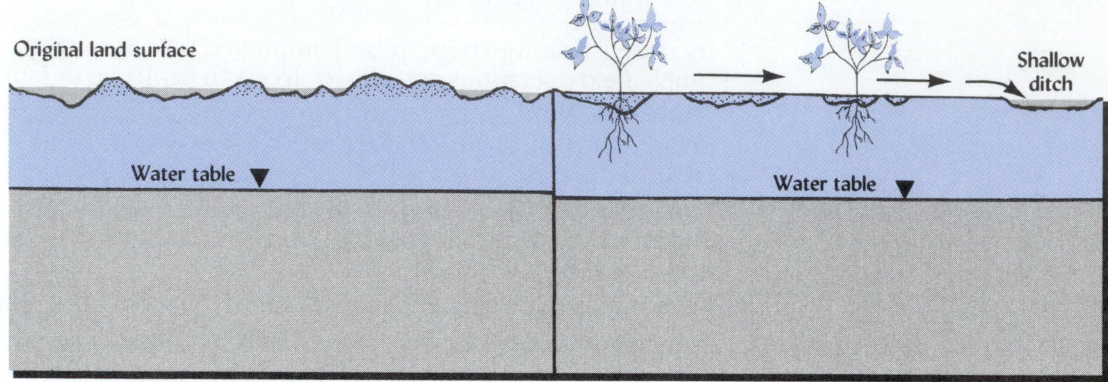

FIGURE 6.23 Land surface before (left) and after (right) land forming or smoothing. The original soil surface (left) is uneven and water ponds in depressions. After land smoothing (right), the land has a uniform slope and water runs off the surface at a controlled rate, to be carried away by a shallow ditch. Note that the soil from the ridges (stippled) is used to fill in the depressions.

Subsurface (Internal) Drainage

Subsurface drainage systems are set up to remove the groundwater from within the soil and to subsequently lower the water table. They require channels such as deep ditches, underground pipes, or installed "moles" into which excess water can flow in response to positive gravitational and submergence potentials. Internal drainage occurs only when the pathway for drainage is located below the level of the water table (Figure 6.24). Box 6.2 provides an example of how knowledge of basic soil properties and water movement principles can be applied in designing a systems to alleviate drainage problems.

Buried Perforated Pipes (Drain Tiles). A network of perforated plastic pipe can be laid underground using specialized equipment (Figure 6.25). Water moves through the perforations into the pipe which has sufficient slope to permit rapid flow of the water to an outlet ditch or stream. The pipe outlet should be covered by a gate or wire mesh, to prevent the entrance of rodents in dry weather but allow the free flow of water. A properly installed subsurface drainage system may continue to operate for many decades.

Building Foundation Drains. Surplus water around foundations and underneath basement floors of houses and other buildings can cause serious damage. The removal of this excess water is commonly accomplished using buried perforated pipe, placed alongside and slightly below the foundation or underneath the floor (Figure 6.24*c*). The perforated pipe must be sloped to allow water to move rapidly to an outlet ditch or sewer. If the drain successfully prevents the water table from rising above the floor level, water will not seep into the basement for the same reason that it will not seep into a drainpipe placed above the water table.

6.10 SEPTIC TANK DRAIN FIELDS

Our discussion of practical water management in soil systems would be incomplete without consideration of the role of soil and, in particular, water movement through soil in treating wastewater in areas not served by a centralized sewage system.

Thousands of ordinary suburbanites get their first exposure to the importance of water movement through soils when they choose a beautiful home site "in the country," make plans to build a dream home, and apply for the required local government building permit. The local authorities will usually not allow a home to be built until arrangements are made for wastewater treatment. Typically, a soil scientist will come out to inspect the soils at the homesite and judge their suitability for use as a **septic tank drain field.** If the soils fail to perc, or fall short in some other attribute described later in this section, the landowner may be denied the permit to build.

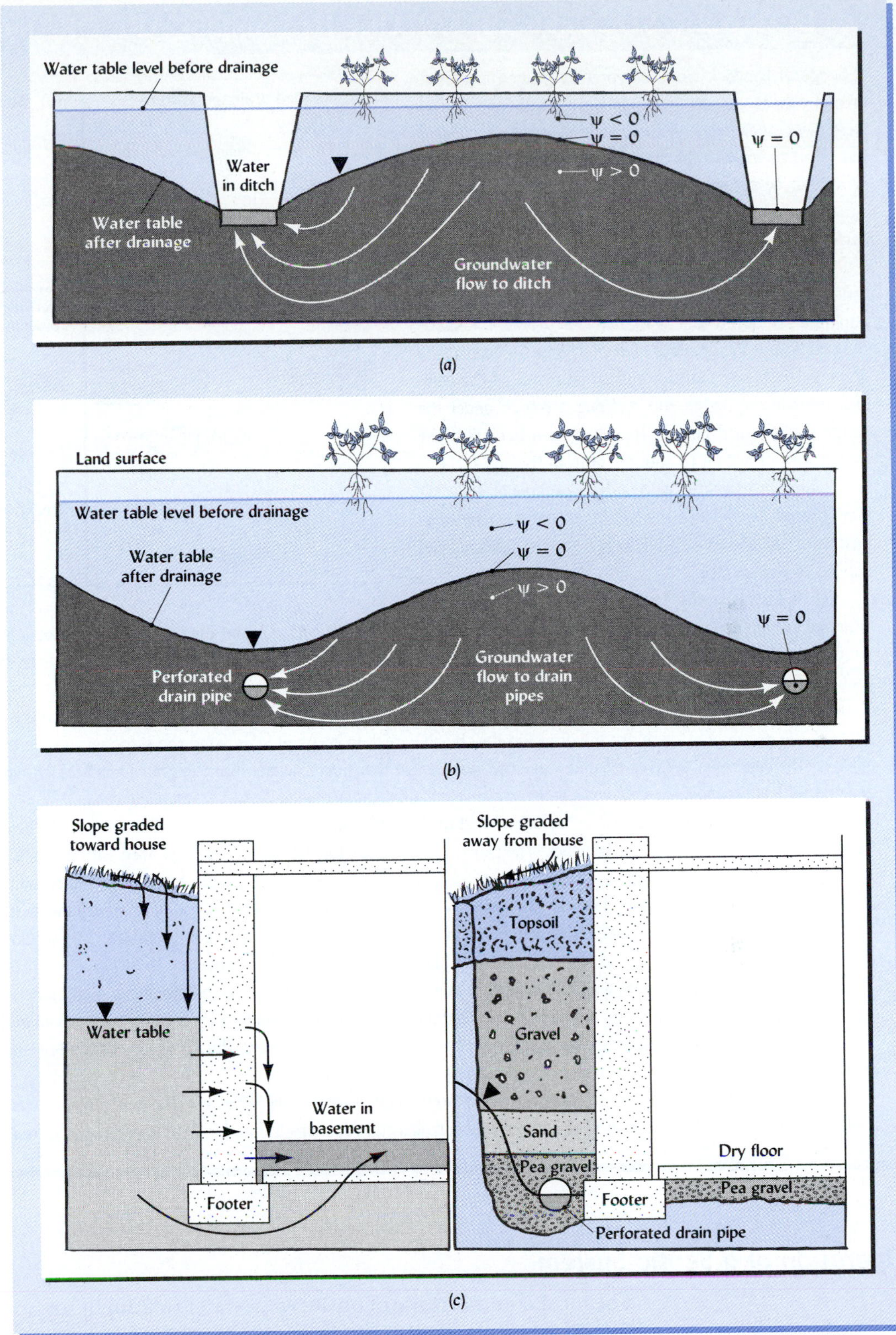

FIGURE 6.24 Three types of subsurface drainage systems. (*a*) Open ditches are used to lower the water table in a poorly drained soil. The wet season levels of the water table before and after ditch installation are shown. The water table is deepest next to the ditch, and the drainage effect diminishes with distance from the ditch. (*b*) Buried "tile lines" made of perforated plastic pipe act very much as the ditches in (*a*), but have two advantages: they are not visible after installation and they do not present any obstacle for surface equipment. Note the flow lines indicating the paths taken by water moving to the drainage ditches or pipes in response to the water potential gradients between the submerged water ($\psi > 0$) and free water in the drainage ditch and pipes ($\psi = 0$). (*c*) The water table around a building foundation before (left) and after (right) installation of a *footer drain* and correction of surface grading. The principles of water movement in soils are applied to keep the basement dry.

BOX 6.2 SUCCESS OR FAILURE IN LANDSCAPE DRAINAGE DESIGN

A hedge of hemlock trees, all carefully pruned into ornamental shapes as part of an intricate landscape design, were dying again because of poor drainage. A very expensive effort to improve the drainage under the hemlocks had proved to be a failure and the replanted hemlocks were again causing an unsightly blemish in an otherwise picture-perfect garden.

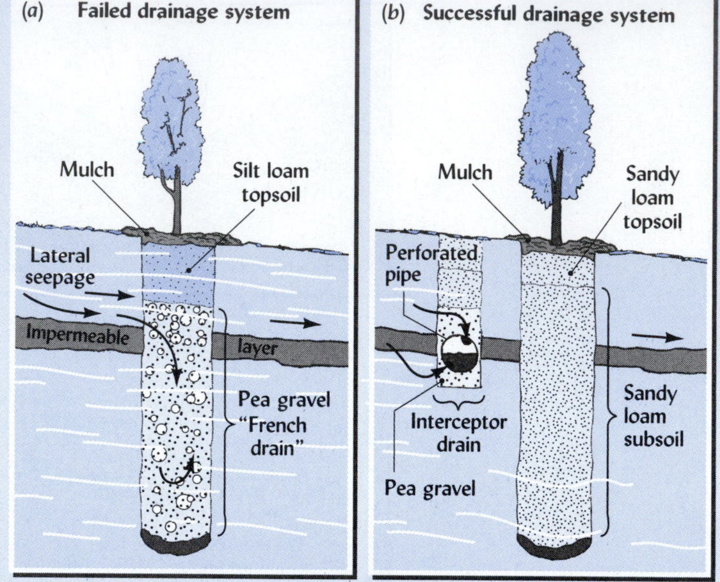

Finally, the landscape architect for the world-famous ornamental garden called in a soil scientist to assist her in finding a solution to the dying hemlock problem. Records showed that in the previous, failed attempt to correct the drainage problem, contractors had removed all the hemlock trees in the hedge and had dug a trench under the hedge some 3 m deep (*a*). They had then backfilled the trench with gravel up to about 1 m from the soil surface, completing the backfill with a high-organic-matter, silt loam topsoil. It was into this silt loam that the new hemlock trees had been planted. Finally a bark mulch had been applied to the surface.

The landscaper who had designed and installed the drainage system apparently had little understanding of the various soil horizons and their relation to the local hydrology. When the soil scientist examined the problem site, he found an impermeable claypan that was causing water from upslope areas to move laterally into the hemlock root zone.

Basic principles of soil water movement also told him that water would not drain from the fine pores in the silt loam topsoil into the large pores of the gravel, and therefore the gravel in the trench would do no good in draining the silt loam topsoil. In fact, the water moving laterally over the impermeable layer created a perched water table that poured water into the gravel-filled trench, soon saturating both the gravel and the silt loam topsoil.

To cure the problem, the previous "solution" had to be undone (*b*). The dead hemlocks were removed, the ditches were reexcavated, and the gravel was removed from the trenches. Then the trench, except for the upper ½ m, was filled with a sandy loam subsoil to provide a suitable rooting medium for the replacement evergreen trees (a different species was chosen for reasons unrelated to drainage). The upper ½ m of the trench was filled with a sandy loam topsoil, which was also acid but contained a higher level of organic matter. The interface between the subsoil and surface soil was mixed so that there would be no abrupt change in pore configuration. This would allow an unsaturated wetting front to move down from the upper to the lower layers, drawing down any excess water.

About 1 m uphill from the trench an *interceptor drain* was installed (*b*). This involved digging a small trench through the impermeable clay layer that was guiding water to the area. A perforated drainage pipe surrounded by a layer of gravel was laid in the bottom of this trench with about a 1% slope to allow water to flow away from the area to a suitable outlet. The interceptor drain prevented the water moving laterally over the impermeable soil layer from reaching the evergreen hedge root zone.

Even though the replanting of the hedge was followed by an exceptionally rainy year, the new drainage system kept the soil well aerated, and the trees thrived. The principles of water movement explained in Sections 5.6 and 6.9 of this textbook were applied successfully in the field.

Operation of a Septic System

The most common type of onsite wastewater treatment for homes not connected to municipal sewage systems is the *septic tank* and associated *drain field* (sometimes called *filter field* or *absorption field*). Over 25 million homes in the United States use septic tank drain fields to treat sewage and wastewater. This method of sewage and wastewater treatment depends on several soil processes, of which the most fundamental is the movement of water through the soil.

In essence, a septic drain field operates like artificial soil drainage in reverse. A network of perforated underground pipes is laid in trenches, very much like the network of drainage pipes used to lower the water table in a poorly drained soil; but instead of draining water away from the soil, the pipes in a septic drain field carry wastewater *to*

(a)

(b)

FIGURE 6.25 (*a*) An open-ditch drainage system designed to lower the water table during the wet season. The water flowing in the ditch has seeped in from the saturated soil. Such ditches also speed the removal of surface runoff water. The capillary fringe above the water table can be seen as a dark band of soil. In order to protect water quality from chemicals and sediments that might reach the ditch via surface runoff, buffer strips of grass or forest vegetation should be established on both sides. Note in the background the piles of *spoil,* subsoil excavated to make the ditches. This material is usually spread thinly and mixed by tillage with the surface soil in the field, but it should first be tested for extreme acidity or other detrimental properties that could impair the quality of the surface soil in which it is to be mixed. (*b*) Specialized, laser-guided equipment laying a drain line made of corrugated plastic pipe. The pipe has perforations on the underside that allow water to seep in from a saturated soil. The trench will be backfilled with the soil shown piled on the left. The drain line must be laid deeper than the seasonal high water table if it is to remove any water. [Photo (*a*) courtesy of R. Well; photo (*b*) courtesy of USDA Natural Resources Conservation Service]

the soil, the water entering the soil via slits or perforations in the pipes. In a properly functioning septic drain field, the wastewater will enter the soil and percolate downward, undergoing several purifying processes before it reaches the groundwater. One of the advantages of this method of sewage treatment is that it has the potential to replenish groundwater supplies for other uses.

Pollutants that potentially come from on-site septic systems: http://www.deh.enr.state.nc.us/oww/nonpointsource/NPS.htm

The Septic Tank. The basic design of a standard septic tank and drain field is shown in Figure 6.26*a*. Water carrying wastes from toilets, sinks, and bathtubs flows by gravity through sealed pipes to a large underground concrete box, called the septic tank. Baffles in the tank cause the inflowing wastewater to slow down and drop most (70%) of its load of suspended solid materials. As these organic solids partially decompose by microbial action in the septic tank, their volume is reduced so that many years usually pass before the septic tank becomes too full and the accumulated **sludge** (also called *septage*) has to be pumped out.

On-site wastewater treatment and good homeowner practices for septic systems: http://www.groundwater.org/GWBasics/septics.htm

The Drain Field. The water exiting the septic tank via a pipe near the top is termed the septic tank **effluent.** Although its load of suspended solids has been much reduced, it still carries organic particles, dissolved chemicals (including nitrogen), and microorganisms (including pathogens). The flow is directed to one or more buried pipes that constitute the **drain field.** Blanketed in gravel and buried in trenches about 0.6 to 1 m under the soil surface, these pipes are perforated on the bottom to allow the wastewater to seep out and enter the soil.

It is at this point that soil properties play a crucial role. Septic systems depend on the soil in the drain field to (1) keep the effluent out of sight and out of contact with people, (2) treat or purify the effluent, and (3) conduct the purified effluent to the groundwater.

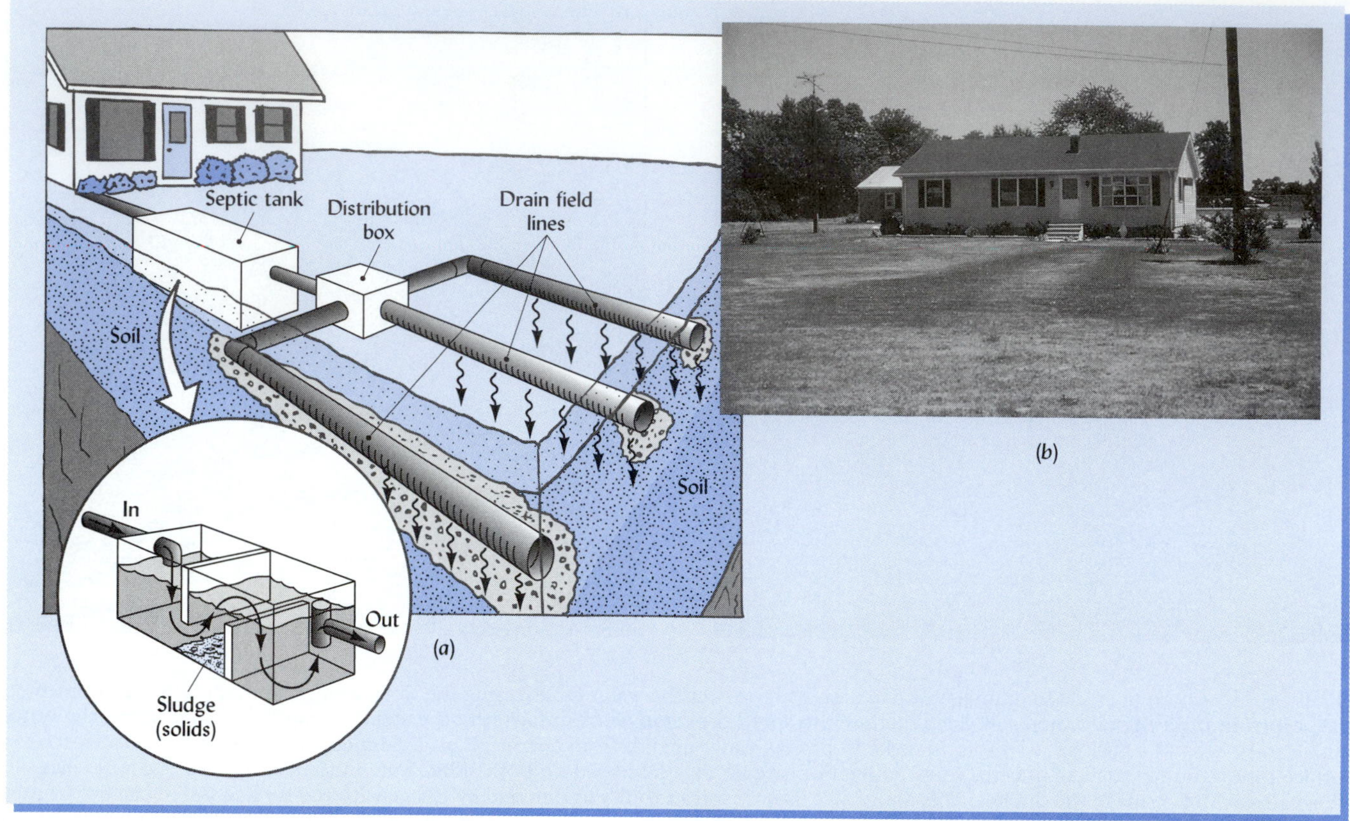

FIGURE 6.26 (*a*) A septic tank and drain field constituting a standard system for onsite wastewater treatment. Most of the solids suspended in the household wastewater settle out in the concrete septic tank. Effluent from the tank flows to the drain field, where it seeps out of the perforated pipes and into the soil. In the soil, the effluent is purified by microbial, chemical, and physical processes as it percolates toward the groundwater. (*b*) Photo shows the telltale dark strips of lawn where poorly functioning septic tank drain lines are stimulating the grass with wastewater and nitrogen. (Photo courtesy of R. Weil)

Soil Properties Influencing Suitability for a Septic Drain Field

Pathogens reduction by septic drainfield soil: http://www.deh.enr.state.nc.us/oww/nonpointsource/NPSseptic/npspathog.htm

The soil should have a *saturated hydraulic conductivity* (see Section 5.5) that will allow the wastewater to enter and pass through the soil profile rapidly enough to avoid backups that might saturate the surface soil with effluent, but slowly enough to allow the soil to purify the effluent before it reaches the groundwater. The soil should be sufficiently *well aerated* to encourage *microbial breakdown* of the wastes and *destruction of pathogens*. The soil should have some fine pores and clay or organic matter to adsorb and filter contaminants from the wastewater.

Properties that can be observed in the soil profile that indicate the presence of a high water table at some time of the year, such as mottled and gleyed colors (see Plates 26 and 29 after page 114), may disqualify a site for use as a septic drain field. Likewise, impermeable layers such as a fragipan or a heavy claypan could eliminate a building site from consideration.

Septic tank drain fields installed where soil properties are not appropriate may result in extensive pollution of groundwater and in health hazards caused by seepage of untreated wastewater (Figure 6.26*b*).

SUITABILITY RATING. The suitability of a site for septic drain field installation depends largely on soil properties that affect water movement and the ease of installation (see Table 6.4). For example, a septic drain field laid out on a slope greater than 15% may allow considerable lateral movement of the percolating water such that at some point downslope, the wastewater will seep to the surface and present a potential health hazard.

PERC TEST. One of the most important tests conducted to determine the suitability of a soil for the installation of a septic tank is called the **perc test.** This is a test that deter-

TABLE 6.4 Soil Properties Influencing Suitability for a Septic Tank Drain Field

Note that most of these soil properties pertain to the movement of water through the soil profile.

	Limitations		
Soil property[a]	*Slight*	*Moderate*	*Severe*
Flooding	—	—	Floods frequent to occasional
Depth to bedrock or impermeable pan, cm	>183	102–183	<102
Ponding of water	No	No	Yes
Depth to seasonal high water table, cm	>183	122–183	<122
Permeability (perc test) at 60 to 152 cm soil depth, mm/h	50–150	15–50	<15 or >150[b]
Slope of land, %	<8	8–15	>15
Stones > 7.6 cm, % of dry soil by weight	<25	25–50	>50

[a]Assumes soil does not contain permafrost and has not subsided more than 60 cm.
[b]Soil permeability (as determined by a perc test) greater than 150 mm/h is considered too fast to allow for sufficient filtering and treatment of wastes.
Adapted from Soil Survey Staff (1993), Table 620-17.

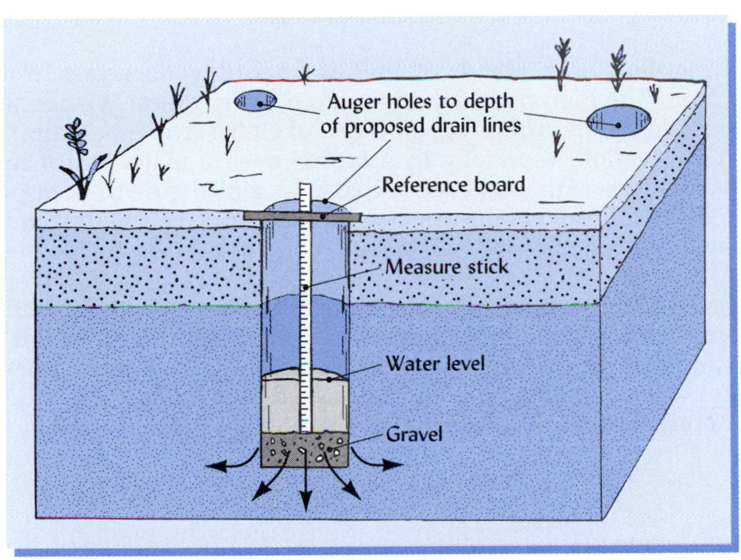

FIGURE 6.27 The perc test used to help determine soil suitability for a septic tank drain field. On the site proposed for the drain field, a number of holes are drilled to the depth where the perforated pipes are to be laid. The bottom of each hole is lined with a few centimeters of gravel, and the holes are filled with water as a pretreatment to ensure that the soil is wet when the test is conducted. After the water has drained, the hole is refilled with water and a measuring rod is used to determine how rapidly the water level drops during a period of several hours or a day.

mines the *percolation rate* (which is related to the saturated hydraulic conductivity described in Section 5.5). The percolation rate is expressed in millimeters (or other unit of depth) of water entering the soil per hour and indicates whether the soil can accept wastewater rapidly enough to provide a practical disposal medium (Table 6.4). The test is simple to conduct (Figure 6.27) and should be carried out during the wettest part of the year.

6.11 IRRIGATION PRINCIPLES AND PRACTICES[4]

Drainage and irrigation methods:
www.soils.umn.edu/academics/classes/soil3125/doc/6chap4.htm

In most regions of the world, insufficient water is the prime limitation to agricultural productivity. In semiarid and arid regions, intensive crop production is all but impossible without supplementing the meager rainfall provided by nature. However, if given supplemental water through irrigation, the sunny skies and fertile soils of some arid regions stimulate extremely high crop yields. It is no wonder, then, that many of the earliest civilizations and city-states depended on irrigated agriculture (and vice versa). During the 20th century, the area of irrigated cropland expanded greatly in many parts of the world, including 17 semiarid states of the western United States. The total area

[4]For a fascinating account of water resources and irrigation management in the Middle East, see Hillel (1995). For a practical manual on small-scale irrigation with simple but efficient microtechnology, see Hillel (1997). CAST (1988, 1996) provides overviews of methods and prospects for commercial irrigated agriculture in the Western United States.

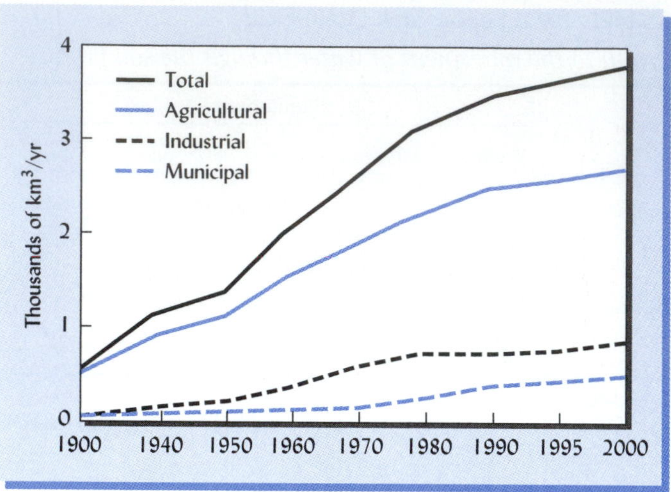

FIGURE 6.28 Trends in global consumptive water use by major categories of users. Note that agriculture (mainly irrigation) uses by far the greatest amount of water, although its share has declined in recent years. [Redrawn from Kuylenstierna, et al., (1997)]

under irrigation seems to be leveling off at about 250 million ha worldwide, about 16% of the world's agricultural land. The high productivity induced by irrigation is evident in the fact that this irrigated land is responsible for some 40% of global crop production.

Expanded and improved irrigation, especially in Asia, has been a major factor in helping global food supplies keep up with, and even surpass, the global growth in population. As a result, irrigated agriculture remains the largest **consumptive use** of water resources, accounting for about 80% of all water consumed, both worldwide and in the United States (Figure 6.28).

Water resources for irrigation are most rapidly becoming scarce and/or expensive in arid regions where they are most needed. Reducing waste and achieving greater efficiency of water use in irrigation are increasingly important aspects that will be emphasized in this section. One other major problem associated with irrigation, the salinization of soils and drainage waters, is considered in Chapter 9.

Water-Use Efficiency

Efficient use of water in gardens and landscapes—the view from Texas:
http://aggie-horticulture.tamu.edu/extension/homelandscape/water/water.html

Various measures of water-use efficiency are used to compare the relative benefits of different irrigation practices and systems. The most meaningful overall measure of efficiency would compare the output of a system (crop biomass or value of marketable product) to the amount of water allocated as an input into the system. There are many factors to consider (type of plants grown, reuse of "wasted" water by others downstream, etc.), so such comparisons must be made with caution.

Application Efficiency. A simpler measure of water-use efficiency, sometimes termed the *water application efficiency,* compares the amount of water allocated to irrigate a field to the amount of water actually used by the irrigated plants. In this regard, most irrigation systems are very inefficient, with as little as 30% to 50% of the water that is taken from the source ever reaching the plant roots. Much water is lost from unlined canal and ditch delivery systems, either by evaporation from the water surfaces or by moving outward and downward into the groundwater (however, some of the latter may be of use to irrigators downstream). These water distribution losses can be reduced by lining the ditches with concrete or plastic (see Figure 6.29).

Field Water Efficiency. Water-use efficiency *in the field* may be expressed as

$$\text{Field water efficiency} = \frac{\text{Water transpired by the crop}}{\text{Water applied to the field}} \times 100 \qquad (6.3)$$

Values are usually quite low (50 to 60%), especially if traditional irrigation systems are used. Commonly, more than half of the water actually delivered to the field is not transpired by the crop, but is lost as surface runoff, deep percolation below the root zone,

FIGURE 6.29 Concrete-lined irrigation ditches (center) and standard-sized siphon pipes (right) can increase the efficiency of water delivery to the field. Unlined ditches (left) lose much of the water to adjacent soil areas or to the groundwater. Note the evidence of capillary movement above the water level in the unlined ditch.

and/or evaporation from the soil surface. Achieving a high level of field water efficiency (or a low level of water wastage) is very dependent on the skill of the irrigation manager and on the methods of irrigation used (Table 6.5).

Surface Irrigation

In these systems water is applied to the upper end of a field and allowed to distribute itself by gravity flow. Usually the land must be leveled and shaped so that the water will flow uniformly across the field. The water may be distributed in **furrows** graded to a slight slope so that water applied to the upper end of the field will flow down the furrows at a controlled rate (Figure 6.30). In **border irrigation** systems, the land is shaped into broad strips 10 to 30 m wide, bordered by low dikes.

WATER CONTROL. Water is usually brought to surface-irrigated fields in supply ditches or gated pipes (such as are shown in Figures 6.29 and 6.30). The amount of water that enters the soil is determined by the permeability of the soil and by the length of time a given spot in the field is inundated with water. If the soil is highly permeable (e.g., some loamy sands), too much water may infiltrate and leach the soil near the upper end of the field, and too little may reach the lower end (Figure 6.31). On the other hand, for a fine-textured soil, infiltration may be so slow that water flows across the field and ponds up or runs off the lower end without a sufficient amount soaking into the soil.

The **level basin** technique of surface irrigation, as is used for paddy rice and certain tree crops, alleviates these problems because each basin has no slope and is completely surrounded by dikes that allow water to stand on the area until infiltration is complete. This method is not practical for highly permeable soils into which water would infiltrate so fast that the basin would never fill. On sloping land, terraces can be built in a modification of the level-basin method (Figure 6.32).

TABLE 6.5 Some Characteristics of the Three Principal Methods of Irrigation

Methods and specific examples	*Direct costs of installation, 1996 dollars*[a]	*Labor requirements*	*Field water efficiency, %*[b]	*Suitable soils*
Surface: basin, flood, furrow	400–700	High to low, depending on system	40–50	Nearly level land; not too sandy or rocky
Sprinkler: center pivot, movable pipe, solid set	600–1200	Medium to low	60–70	Level to moderately sloping; not too clayey
Microirrigation: drip, porous pipe, spitter, bubbler	700–1500	Low	80–90	Steep to level slopes; any texture, including rocky or gravelly soils

[a]Average ranges from many sources. Costs of required drainage systems not included.
[b]Values from Hillel (1997). Field water efficiency = 100 × (water transpired by crop/water applied to field).

FIGURE 6.30 Typical irrigation scene early in the spring in arid regions of the western United States. The water is delivered to the field in concrete ditches, and then (left) siphoned from the ditch into miniponds that can supply water to every row or alternate rows as shown (right). These small ponds also reduce furrow head erosion since the siphoned water moves into a pool of water rather than onto wet soil. Also, any suspended soil particles settle out before the slow-moving water starts flowing down the furrows. Note the upward capillary movement of the water along the sides of the rows. (Photos courtesy of N. C. Brady)

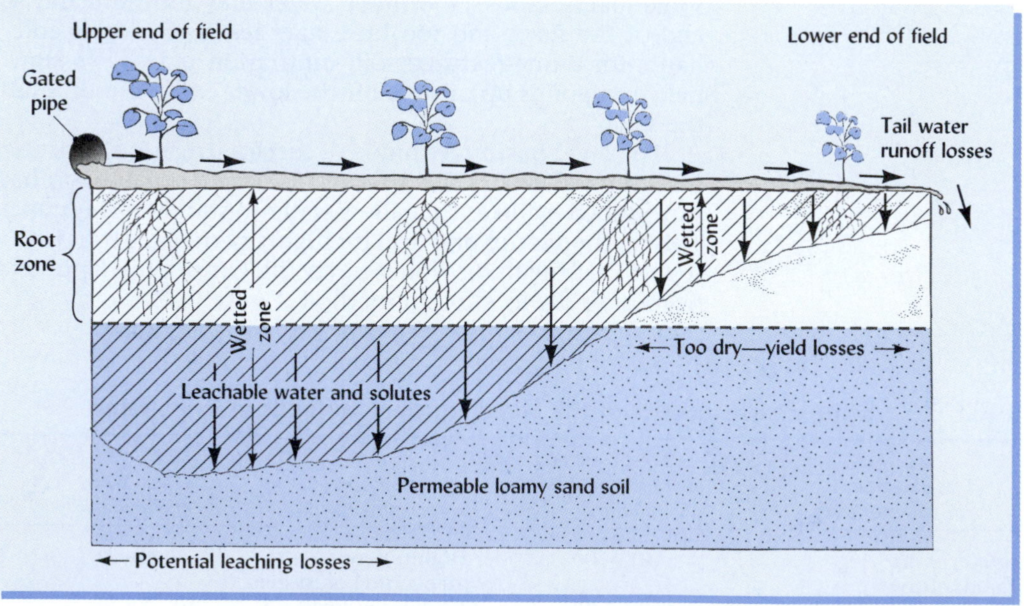

FIGURE 6.31 Penetration of water into a coarse-textured soil under surface irrigation. The high infiltration rate causes most of the water to soak in near the gated pipe at the upper end of the field. The uneven penetration of water results in the potential for leaching losses of water and dissolved chemicals at the upper end of the field, while plants at the lower end may receive insufficient water to moisten the entire potential root zone: On a less-permeable soil, or on a field with a steep slope, the tail water runoff losses would likely be greater at the lower end and leaching potential less at the upper end. (Diagram courtesy of R. Weil)

FIGURE 6.32 The level-basin type of surface irrigation modified for sloping land in South Asia by construction of terraces. Paddy rice is growing in the flooded terrace basins. This type of irrigation is practical only on soils of low permeability. (Upper photo courtesy of R. Weil)

Sprinkler Systems

In sprinkler irrigation, water is sprayed through the air onto a field, simulating rainfall. Thus, the entire soil surface, as well as plant foliage (if present), is wetted. This leads to evaporative losses similar to those described for surface systems. Furthermore, an additional 5% to 20% of the applied water may be lost by evaporation or windblown mist as the drops fly through the air. One advantage is that plants often respond positively to the cooler, better-aerated sprinkler water. A disadvantage is that wet leaves may increase the incidence of fungal diseases in some plants, such as grapes, fruit trees, and roses, so sprinkler systems are not often used for these plants.

Water Control. A sprinkler system should be designed to deliver water at a rate that is less than the infiltration capacity of the soil, so that runoff or excessive percolation will not occur. In practice, runoff and erosion may be problems if the soil infiltration capacity is low, the land is relatively steep, or too much water is applied in one place. Because of better control over application rates, the field water-use efficiency is generally higher for sprinkler systems than for surface systems, especially on coarse-textured soils.

Suitable Soils. Sprinkler irrigation is practical on a wider range of soil conditions than is the case for the surface systems. Various types of sprinkler systems are adapted to moderately sloping as well as level land (Figure 6.33). They can be used on soils with a wide range of textures, even those too sandy for surface irrigation systems.

FIGURE 6.33 A center-pivot irrigation system in eastern Colorado. The system is rotating toward the left (note the dry soil at the extreme left), and the center pivot of the system is located approximately 0.5 km in the distance. Each self-propelled tower moves in coordination with all the others. The shed at the right houses a diesel-powered water pump that taps a deep aquifer. The tank next to the shed holds liquid fertilizer that can be metered into the irrigation water. (Photo courtesy of R. Weil)

Microirrigation

The most efficient irrigation systems in use today are those using microirrigation, whereby only a small portion of the soil is wetted in contrast to the complete wetting accomplished by most surface and sprinkler systems. Microirrigation may also refer to the tiny amounts of water applied at any one time and the miniature size of the equipment involved.

Perhaps the best-established microirrigation system is *drip* (or *trickle*) *irrigation,* in which tiny emitters attached to plastic tubing apply water to the soil surface alongside individual plants. In some cases the tubing and emitters are buried 20 to 50 cm deep so the water soaks directly into the root zone. In either case, water is applied at a low rate (sometimes drop by drop) but at a high frequency, with the objective of maintaining optimal soil water availability in the immediate root zone while leaving most of the soil volume dry (Figure 6.34). Table 6.6 illustrates the high field-water efficiency of drip irrigation.

Other forms of microirrigation that are especially well adapted for irrigating individual trees include *spitters* (microsprayers) and *bubblers* (small vertical standpipes)

FIGURE 6.34 Two examples of microirrigation. (Left) *Drip* or *trickle* irrigation with a single emitter for each seedling in a cabbage field in Africa. (Right) A *microsprayer* or *spitter* irrigating an individual tree in a home garden in Arizona. In both cases, irrigation wets only the small portion of the soil in the immediate root zone. Small quantities of water applied at high frequency (such as once or twice a day) ensures that the root zone is kept almost continuously at an optimal moisture content. (Left photo courtesy of R. Weil)

TABLE 6.6 **Efficiencies of Selected Irrigation Methods in the High Plains of Texas**

Note that improved irrigation methods show 20–35% higher efficiency and considerably lower water requirements.

Irrigation method	*Typical efficiency (percent)*	*Water application needed to add 100 mm to root zone (millimeters)*	*Water savings over conventional furrow (percent)*
Conventional furrow	60	167	—
Furrow with surge valve	80	125	25
Low-pressure sprinkler	80	125	25
LEPA sprinkler[a]	90–95	105	37
Drip	95	105	37

[a]LEPA refers to low-energy precision application that delivers the water close to the plants.
From Postel (1999) based on data from the High Plains Underground Water Conservation District (Lubbock, Texas).

(Figure 6.34). The bubblers (and usually the spitters) require that a small level basin be formed in the soil under each tree.

WATER CONTROL. Water is normally carried to the field in pipes, run through special filters to remove any grit or chemicals that might clog the tiny holes in the emitters (filtering is not necessary for bubblers), and then distributed throughout the field by means of a network of plastic pipes. Soluble fertilizers may be added to the water as needed.

If properly maintained and managed, microirrigation allows much more control over water application rates and spatial distribution than do either surface or sprinkler systems. Losses by supply-ditch seepage, sprinkler-drop evaporation, runoff, drainage (in excess of that needed to remove salts), soil evaporation, and weed transpiration can be greatly reduced or eliminated. Once in place, the labor required for operation is modest.

Microirrigation often produces healthier plants and higher crop yields because the plant is never stressed by low water potentials or low aeration conditions that are associated with the feast-or-famine regime of infrequent, heavy water applications made by all surface irrigation systems and most sprinkler systems. A disadvantage or risk is that there is very little water stored in the soil at any time, so even a brief breakdown of the system could be disastrous in hot, dry weather.

EQUIPMENT. The capital costs for microirrigation tend to be higher than for other systems (see Table 6.5), but the differences are not so great if the cost of drainage systems for control of salinity and waterlogging in surface systems, the costs of high-pressure pumping in sprinkler systems, and the real value of wasted water are taken into account. Because of its high water-use efficiency, microirrigation is most profitable where water supplies are scarce and expensive, and where high-valued plants such as fruit trees are being grown.

Irrigation Water Management

Irrigation scheduling for water managers: http://www.wateright.org/

The two most serious irrigation management problems relate to the quality of the water being applied and to the efficiency of the irrigation water in stimulating plant production.

SALINITY BUILDUP. Most irrigation systems are located in semiarid and arid regions where the levels of soluble salts in the drainage water and, in turn, in the streams and rivers are relatively high. When this water is added to the soil and percolation takes place, still more salts are dissolved from the soil itself, making the drainage water even more saline than the originally added water. As the drainage water is repeatedly reused downstream, the salt buildup in the water can become very damaging to both the physical and chemical properties of the soils to which the water is applied (see Section 9.17).

EFFICIENT WATER USE. The microirrigation systems are clearly the most efficient in water use, but whether they are the most economical to use for a given situation depends on many factors. The use of crop residues or mulches to reduce evaporation from the soil surface while simultaneously reducing the soil temperature (see Section 7.11) can also enhance overall productivity and water-use efficiency. It is also wise to concentrate irrigation on those crops that produce high-value products with relatively low levels of water use. Much water is currently squandered in producing low-value, high-transpiration-ratio crops (such as forages, cereals, and cotton) that would be more economically imported from rain-fed areas if all the costs of obtaining irrigation water were actually paid for by the irrigators (as opposed to being subsidized by governments).

Likewise, arid-region homeowners' attempts to sustain humid-region landscaping (e.g., green lawns and high-water-use shrubs) around their homes rather than using xerophytic (desert) plants, mulches, and stones is equally wasteful.

Finally, to maintain or increase plant production in the face of dwindling water supplies, the application of water must be scheduled according to plant needs, which change with the weather, the LAI, and other factors. The basic principles of water retention and movement in soils and water use by plants, as outlined in this and the previous chapter, must be applied in developing efficient irrigation schedules.

6.12 CONCLUSION

The hydrologic cycle encompasses all movements of water on or near the earth's surface. It is driven by solar energy, which evaporates water from the ocean, the soil, and vegetation. The water cycles into the atmosphere, returning elsewhere to the soil and the oceans in rain and snow.

The soil is an essential component of the hydrologic cycle. It receives precipitation from the atmosphere, rejecting some of it, which is then forced to run off into streams and rivers, and absorbing the remainder, which then moves downward, to be either transmitted to the groundwater, taken up and later transpired by plants, or evaporated directly from soil surfaces and returned to the atmosphere.

The behavior and movement of water in soils and plants are governed by the same set of principles: Water moves in response to differences in energy levels, moving from higher to lower water potential. These principles can be used to manage water more effectively and to increase the efficiency of its use.

Management practices should encourage movement of water into well-drained soils while minimizing evaporative (E) losses from the soil surface. These two objectives will provide as much water as possible for plant uptake and groundwater recharge. Water from the soil must satisfy the transpiration (T) requirements of healthy leaf surfaces; otherwise, plant growth will be limited by water stress. Practices that leave plant residues on the soil surface and that maximize plant shading of this surface will help achieve high efficiency of water use.

Extreme soil wetness, characterized by surface ponding and saturated conditions, is a natural and necessary condition for wetland ecosystems. However, for most other land uses extreme wetness is detrimental. Drainage systems have therefore been developed to hasten the removal of excess water from soil and lower the water table so that upland plants can grow without aeration stress, and so the soil can better bear the weight of vehicular and foot traffic.

A septic tank drain field operates as a drainage system in reverse. Septic wastewaters can be disposed of and treated by soils if the soils are freely draining. Soils with low permeability or high water tables may indicate good conditions for wetland creation or appropriate sites for installation of artificial drainage for agricultural use, but they are not generally suited for septic tank drain fields.

Irrigation waters from streams or wells greatly enhance plant growth, especially in regions with scarce precipitation. With increasing competition for limited water resources, it is essential that irrigators manage water with maximal efficiency so that the greatest production can be achieved with the least waste of water resources. Such efficiency is encouraged by practices that favor transpiration over evaporation, such as mulching and the use of microirrigation.

As the operation of the hydrologic cycle causes constant changes in soil water content, other soil properties are also affected, most notably soil aeration and temperature, the subjects of the next chapter.

STUDY QUESTIONS

1. You know that the forest vegetation that covers a 120 km^2 wildland watershed uses an average of 4 mm of water per day during the summer. You also know that the soil averages 150 cm in depth and at field capacity can store 0.2 mm of water per mm of soil depth. However, at the beginning of the season the soil was quite dry, holding an average of only 0.1 mm/mm. As the watershed manager, you are asked to predict how much water will be carried by the streams draining the watershed during the 90-day summer period when 450 mm of precipitation falls on the area. Use the water balance equation to make a rough prediction of the stream discharge as a percentage of the precipitation and in cubic meters of water.
2. Draw a simple diagram of the hydrologic cycle using a separate arrow to represent these processes: *evaporation, transpiration, infiltration, interception, percolation, surface runoff,* and *soil storage.*
3. Describe and give an example of the *indirect* effects of plants on the hydrologic balance through their effects on the soil.
4. State the basic principle that governs how water moves through the SPAC. Give two examples, one at the soil–root interface and one at the leaf–atmosphere interface.
5. Define *potential evapotranspiration* and explain its significance to water management.
6. What is the role of evaporation from the soil (E) in determining water-use efficiency, and how does it affect ET? List three practices that can be used to control losses by E.
7. Weed control should reduce water losses by what process?
8. Comment on the relative advantages and disadvantages of organic versus plastic mulches.
9. What does conservation tillage conserve? How does it do it?
10. Explain under what circumstances earthworm channels might increase downward saturated water flow, but not have much effect on the leaching of soluble chemicals applied to the soil.
11. What will be the effect of placing a perforated drainage pipe in the capillary fringe zone just above the water table in a wet soil? Explain in terms of water potentials.
12. What soil features may limit the use of a site for a septic tank drain field?
13. Which irrigation systems are likely to be used where: (a) water is expensive and the market value of crops produced per hectare is high, and (b) the cost of irrigation water is subsidized and the value of crop products that can be produced per hectare is low? Explain.

REFERENCES

CAST. 1988. *Effective Use of Water in Irrigated Agriculture.* Task Force Report No. 113. (Ames, Iowa: Council for Agricultural Science and Technology).

CAST. 1996. *Future of Irrigated Agriculture.* Task Force Report No. 127. (Ames, Iowa: Council for Agricultural Science and Technology).

Day, R. L., et al. 1998. "Water balance and flow patterns in a fragipan using in situ soil block," *Soil Sci.,* **163**:517–528.

Dreibelbis, F. R., and C. R. Amerman. 1965. "How much topsoil moisture is available to your crops?" *Crops and Soils* **17**:8–9.

Farahani, H. J., et al. 1998. "Soil water storage in dryland cropping systems: The significance of cropping intensification," *Soil Sci. Soc. Am. J.* **62**:984–991.

Greb, B. W. 1983. "Water conservation: Central Great Plains," in H. E. Dregue and W. O. Willis (eds.), *Dryland Agriculture.* Agronomy Series No. 23. (Madison, Wis.: Amer. Soc. of Agron.).

Hillel, D. 1995. *The Rivers of Eden.* (New York: Oxford University Press).

Hillel, D. 1997. *Small-Scale Irrigation for Arid Zones.* FAO Development Series 2. (Rome: U.N. Food and Agriculture Organization).

Howell, T. A., and J. A. Tolk. 1998. "Water use efficiency of corn in the U. S. Southern High Plains," *Agron Abstract* (Madison, Wis.: Amer. Soc. Agron.), p. 14.

Jury, W. A., and H. Fluhler. 1992. "Transport of chemicals through soil: Mechanisms, models, and field applications," *Advances in Agronomy* **47**:141–201.

Kuylenstierna, J. L., G. Björklund, and P. Najlis. 1997. "Future sustainable water use: Challenges and constraints," *J. Soil Water Conserv.* **52**:151–156.

Lyon, T. L., H. O. Buckman, and N. C. Brady. 1952. *The Nature and Properties of Soils,* 5th ed. (New York: Macmillan).

Postel, S. 1999. *Pillar of Sand: Can the Irrigation Miracle Last?* (Washington, D.C.: Worldwatch Institute).

Skaggs, R. W., and J. van Schilfgaarde (eds.). 1999. *Agricultural Drainage.* Agronomy Series No. 38. (Madison, Wis.: Amer. Soc. Agron., Crop Sci. Soc. Amer., Soil Sci. Soc. Amer.).

Soil Survey Staff. 1993. *National Soil Survey Handbook.* Title 430-VI. (Washington, D.C.: USDA Natural Resources Conservation Service).

Starrett, S. K., N. E. Christians, and T. Al Austin. 1996. "Movement of pesticides under two irrigation regimes applied to turfgrass," *J. Environ. Qual.* **25**:566–571.

Unger, P. W., and R. L. Baumhardt. 1999. "Factors related to dryland grain sorghum yield increases: 1939 through 1997," *Agron J.* **91**:870–875.

Waddell, J., and R. Weil. 1996. "Water distribution in soil under ridge-till and no-till corn," *Soil Sci. Soc. Amer. J.* **60**:230–237.

7

Soil Aeration and Temperature

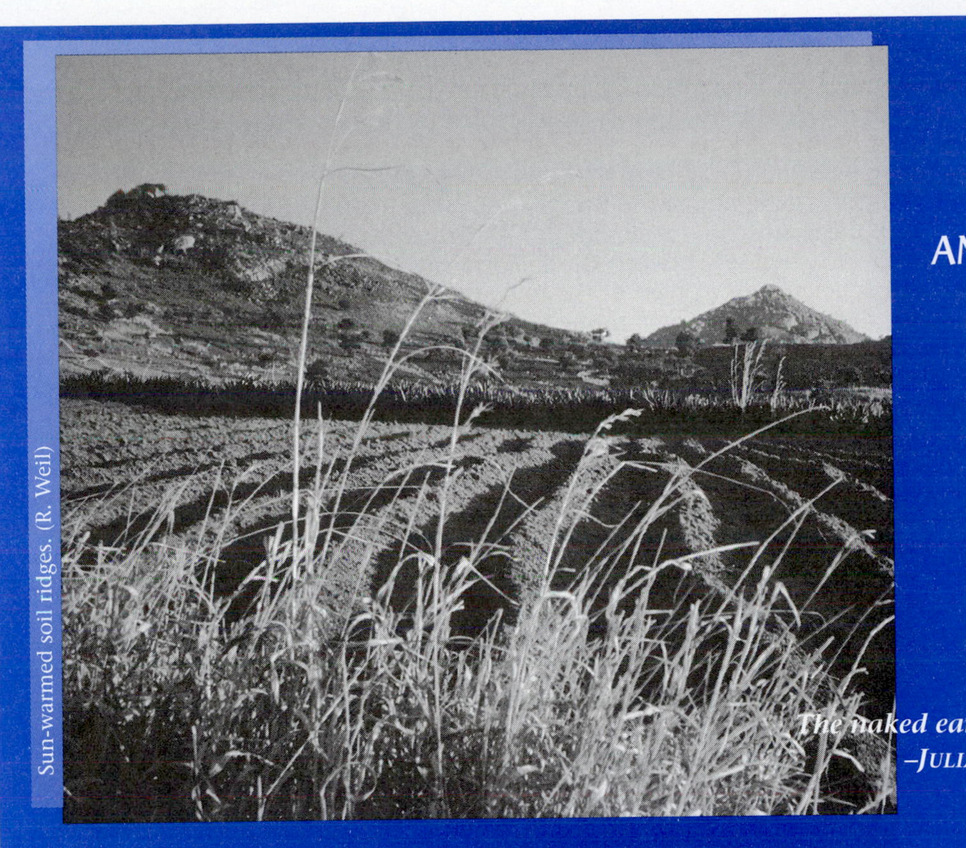

Sun-warmed soil ridges. (R. Weil)

The naked earth is warm with Spring.
—Julian Grenfell, Into Battle

It is a central maxim of ecology that "everything is connected to everything else." This interconnectedness is one reason why soils are such fascinating (and challenging) objects of study. In this chapter we explore two aspects of the soil environment, aeration and temperature, that are not only closely connected to each other but are both also intimately influenced by many of the soil properties discussed in other chapters.

Since air and water share the pore space of soils, it is not surprising that much of what we learned about the texture, structure, and porosity of soils (Chapter 4) and the retention and movement of water in soils (Chapters 5 and 6) will have direct bearing on soil aeration. These are some of the physical parameters affecting aeration status, but chemical and biological processes also affect, and are affected by, soil aeration.

For the growth of plants and the activity of microorganisms, soil aeration status can be just as important as soil moisture status, and can sometimes be even more difficult to manage. In most forest, range, agricultural, and ornamental applications, a major management objective is to maintain a high level of oxygen in the soil for root respiration. Yet it is also vital that we understand the chemical and biological changes that take place when the oxygen supply in the soil is depleted.

Soil temperatures affect plant and microorganism growth and also influence soil drying by evaporation. The movement and retention of heat energy in soils are often ignored but hold the key to understanding many important soil phenomena, from frost-damaged pipelines and pavements to the spring awakening of biological activity in soils. The unusually high soil temperatures that result from fires on forest-, range-, or croplands can markedly change critical physical and chemical soil properties.

We will see that increasing soil temperatures influence soil aeration largely through their stimulating effects on the growth of plants and soil organisms and on the rates of biochemical reactions. Nowhere are these interrelationships more critical than in the water-saturated soils of wetlands, ecosystems that will therefore receive special attention in this chapter.

7.1 SOIL AERATION—THE PROCESS

Aeration involves the ventilation of the soil, with gases moving both into and out of the soil. Aeration determines the rate of gas exchange with the atmosphere, the proportion of pore spaces filled with air, the composition of that soil air, and the resulting chemical oxidation or reduction potential in the soil environment.

For respiration to be carried out by soil organisms, including plant roots, oxygen (O_2) must be supplied and carbon dioxide (CO_2) removed. In a well-aerated soil, the exchange of these two gases between the soil and the atmosphere is sufficiently rapid to prevent the deficiency of oxygen or the toxicity of excess carbon dioxide. For most upland plants, the supply of oxygen in the soil air must be kept above 0.1 L/L (as compared to 0.2 L/L in the atmosphere). In turn, the concentrations of CO_2 and other potentially toxic gases, such as methane or ethylene, must not be allowed to build up excessively.

Soil Aeration in the Field

Soil aeration and plant growth:
http://www.uoguelph.ca/~mgoss/five/410_NO6.html

Oxygen availability in field soils is regulated by three principal factors: (1) *soil macroporosity* (as affected by texture and structure), (2) *soil water content* (as it affects the proportion of porosity that is filled with air), and (3) *O_2 consumption* by respiring organisms (including plant roots and microorganisms). The term *poor soil aeration* refers to a condition in which the availability of O_2 in the root zone is insufficient to support optimal growth of most plants and aerobic microorganisms. Typically, poor aeration becomes a serious impediment to plant growth when more than 80 to 90% of the soil pore space is filled with water (leaving less than 10 to 20% of the pore space filled with air). The high soil water content not only leaves little pore space for air storage, but, more important, the water blocks the pathways by which gases could exchange with the atmosphere. Compaction can also cut off gas exchange, even if the soil is not very wet and has a large percentage of air-filled pores.

Excess Moisture

The extreme case of excess water occurs when all or nearly all of the soil pores are filled with water. The soil is then said to be **water saturated** or **waterlogged.** Waterlogged soil conditions are typical of wetlands, and may also occur for short periods of time in depressions and flat areas on upland sites. In well-drained soils, saturated conditions may occur temporarily when excess water is applied.

Certain plants adapted to life in waterlogged soils are termed ***hydrophytes.*** For example, a number of grass species, including rice, eastern gama grass, and spartina marsh grasses transport oxygen for respiration down to their roots via hollow structures in their stems and roots known as ***aerenchyma*** tissues. Mangroves and other hydrophytic trees produce aerial roots and other structures that allow their roots to obtain O_2 while growing in water-saturated soils.

Most plants, however, are dependent on a supply of oxygen from the soil, and so suffer dramatically if good soil aeration is not maintained by drainage or other means (Figure 7.1). Some plants succumb to O_2 deficiency or toxicity of other gases within hours after the soil is saturated.

Gaseous Interchange

The more rapidly roots and microbes use up oxygen and release carbon dioxide, the greater is the need for the exchange of gases between the soil and the atmosphere. This exchange is facilitated by two mechanisms, ***mass flow*** and ***diffusion.*** Mass flow of air is much less important than diffusion in determining the total exchange that occurs. It is enhanced, however, by fluctuations in soil moisture content that force air in or out of the soil or by wind and changes in barometric pressure.

The great bulk of the gaseous interchange in soils occurs by *diffusion.* Through this process, each gas moves in a direction determined by its own partial pressure. The *partial pressure* of a gas in a mixture is simply the pressure this gas would exert if it alone were present in the volume occupied by the mixture. Thus, if the pressure of air is 1 atmo-

FIGURE 7.1 Most plants depend on the soil to supply oxygen for root respiration, and therefore are disastrously affected by even relatively brief periods of soil saturation during which oxygen becomes depleted. (Left) Sugar beets on a clay loam soil dying where the soil has become water saturated in a compacted area. (Right) Pine trees dying in a sandy soil area that has become saturated as a result of flooding by beavers. A new community of plants better adapted to poorly aerated soil conditions is taking over the site. (Photos courtesy of R. Weil)

sphere (~100 kPa), the partial pressure of oxygen, which makes up about 21% (0.21 L/L) of the air by volume, is approximately 21 kPa.

Diffusion allows extensive gas movement from one area to another even though there is no overall pressure gradient for the total mixture of gases. There is, however, a concentration gradient for each individual gas, which may be expressed as a *partial pressure gradient.* As a consequence, the higher concentration of oxygen in the atmosphere will result in a net movement of this particular gas into the soil. Carbon dioxide and water vapor normally move in the opposite direction, since the partial pressures of these two gases are generally higher in the soil air than in the atmosphere. A representation of the principles involved in diffusion is given in Figure 7.2.

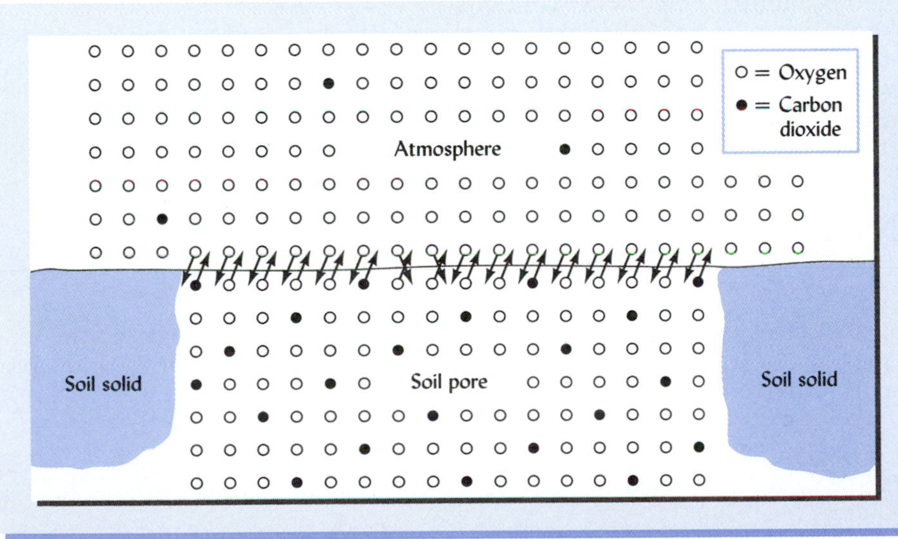

FIGURE 7.2 The process of diffusion between gases in a soil pore and in the atmosphere. The total gas pressure is the same on both sides of the boundary. The partial pressure of oxygen is greater, however, in the atmosphere. Therefore, oxygen tends to diffuse into the soil pore where fewer oxygen molecules per unit volume are found. The carbon dioxide molecules, on the other hand, move in the opposite direction owing to the higher partial pressure of this gas in the soil pore. This diffusion of O_2 into the soil pore and of CO_2 into the atmosphere will continue as long as the respiration of root cells and microorganisms consumes O_2 and releases CO_2.

7.2 MEANS OF CHARACTERIZING SOIL AERATION

The aeration status of a soil can be characterized in several ways, including (1) the content of oxygen and other gases in the soil atmosphere, (2) the air-filled soil porosity, and (3) the chemical oxidation-reduction (redox) potential.

Gaseous Composition of the Soil Air

OXYGEN. The atmosphere above the soil contains nearly 21% O_2, 0.035% CO_2, and more than 78% N_2. In comparison, soil air has about the same level of N_2 but is consistently lower in O_2 and higher in CO_2. The O_2 content may be only slightly below 20% in the upper layers of a soil with a stable structure and an abundance of macropores. It may drop to less than 5% or even to near zero in the lower horizons of a poorly drained soil with few macropores. Once the supply of O_2 is virtually exhausted, the soil environment is said to be **anaerobic.**

CARBON DIOXIDE. Because the N_2 content of soil air is relatively constant, there is a general inverse relationship between the contents of the other two major components of soil air—O_2 and CO_2—with O_2 decreasing as CO_2 increases. Although the actual differences in CO_2 amounts may not be impressive, they are significant, comparatively speaking. Thus, when the soil air contains only 0.35% CO_2, this gas is about 10 times as concentrated as it is in the atmosphere. In cases where the CO_2 content becomes as high as 10%, it may be toxic to some plant processes.

OTHER GASES. Soil air usually is much higher in water vapor than is the atmosphere, being essentially saturated except at or very near the surface of the soil (see Section 5.7). Also, under waterlogged conditions, the concentrations of gases such as methane (CH_4) and hydrogen sulfide (H_2S), which are formed as organic matter decomposes, are notably higher in soil air. Another gas produced by anaerobic microbial metabolism is ethylene (C_2H_4). This gas is particularly toxic to plant roots, even in concentrations lower than 1 μ L/L (0.0001%). Root growth of a number of plants has been shown to be inhibited by ethylene that accumulates when gas exchange rates between the atmosphere and the soil are too slow.

Air-Filled Porosity

In Chapter 1 (Figure 1.12) we noted that the ideal soil composition for plant growth would include close to a 50:50 mix of air and water in the soil pore space, or about 25% air in the soil, by volume (assuming a total porosity of 50%). Many researchers believe that, in most soils, microbiological activity and plant growth become severely inhibited when air-filled porosity falls below 20% of the pore space or 10% of the total soil volume (with correspondingly high water contents).

One of the principal reasons that high water contents cause oxygen deficiencies for roots is that water-filled pores block the diffusion of oxygen into the soil to replace that used by respiration. In fact, oxygen diffuses 10,000 times faster through a pore filled with air than through a similar pore filled with water.

7.3 OXIDATION-REDUCTION (REDOX) POTENTIAL[1]

Redox Potential (E_h)

The reaction that takes place as the reduced state of an element is changed to the oxidized state may be illustrated by the oxidation of two-valent iron [Fe^{2+} or Fe(II)] in FeO to the trivalent form [Fe^{3+} or Fe(III)] in FeOOH.

$$\underset{\text{Fe(II)}}{\overset{(2+)}{2\text{FeO}}} + 2\text{H}_2\text{O} \rightleftharpoons \underset{\text{Fe(III)}}{\overset{(3+)}{2\text{FeOOH}}} + 2\text{H}^+ + 2e^- \qquad (7.1)$$

[1]For a review of redox reactions in soils, see Bartlett and James (1993).

Iron and sulfur redox transformations:
http://ucs.byu.edu/bioag/aghort/514pres/fe/sld019.htm

Note that Fe(II) loses an electron e^- as it changes to Fe(III), and that H^+ ions are formed in the process. The loss of an electron suggests that there are potentials for the transfer of electrons from one substance to another. This **redox potential** can be measured using a platinum electrode.

The redox potential E_h provides a measure of the tendency of a substance to accept or donate electrons. It is usually measured in volts or millivolts. As is the case for water potential (Section 5.3), redox potential is related to a reference state, in this case the hydrogen couple $\frac{1}{2} H_2 \rightleftharpoons H^+ + e^-$, whose redox potential is arbitrarily taken as zero. If a substance will accept electrons easily, it is known as an *oxidizing agent;* if a substance supplies electrons easily, it is a *reducing agent.*

Role of Oxygen (O_2)

Oxidation-reduction reactions:
http://www.shodor.org/UNChem/advanced/redox/

Oxygen gas (O_2) is an important example of a strong oxidizing agent, since it rapidly accepts electrons from many other elements. All aerobic respiration requires O_2 to serve as the electron acceptor as living organisms oxidize organic carbon to release energy for life.

Oxygen can oxidize both organic and inorganic substances. Keep in mind, however, that as it oxidizes another substance, O_2 is in turn reduced. This reduction process can be seen in the following reaction.

$$\overset{(0)}{1/2O_2} + 2H^+ + 2e^- \rightleftharpoons \overset{(2-)}{H_2O} \tag{7.2}$$

Note that the oxygen atom having zero charge in O_2 accepts two electrons, taking on a charge of -2 when it becomes part of the water molecule. These electrons could have been donated by two molecules of FeO undergoing oxidation, as shown in equation 7.1. If we combine equations 7.1 and 7.2 we can see the overall effect of oxidation and reduction.

$$\begin{array}{l} 2FeO + 2H_2O \rightleftharpoons 2FeOOH + 2H^+ + 2e^- \\ 1/2O_2 + 2H^+ + 2e^- \rightleftharpoons H_2O \\ \hline 2FeO + 1/2O_2 + H_2O \rightleftharpoons 2FeOOH \end{array} \tag{7.3}$$

The donation and acceptance of electrons (e^-) and H^+ ions on each side of the equation have balanced each other and therefore do not appear in the combined reaction, but for the specific reduction and oxidation reactions they are both very important.

The redox potential E_h of a soil is dependent on both the presence of electron acceptors (oxygen or other oxidizing agents) and pH. There is generally a positive correlation between O_2 content of soil air and E_h (redox potential). In a well-drained soil, the E_h is in the 0.4 to 0.7 volt (V) range. As aeration is reduced, the E_h declines to a level of about 0.3 to 0.35 V when gaseous oxygen is depleted. Under flooded conditions, in warm, organic-matter-rich soils, E_h values as low as -0.3 V can be found.

Other Electron Acceptors

Other elements in addition to oxygen can act as terminal electron acceptors (oxidizers). For example, N(V) in nitrate accepts two electrons when it is reduced to N(III) in nitrite:

$$\underset{N(V)}{\overset{(5+)}{NO_3}} + 2e^- + 2H^+ \rightleftharpoons \underset{N(III)}{\overset{(3+)}{NO_2^-}} + H_2O \tag{7.4}$$

Similar reactions involve the reduction or oxidation of Fe, Mn, and S (see Figure 7.3).

The effect of pH on redox potentials relating to several important reactions taking place in soils is shown in Figure 7.3. Note that in all cases the E_h decreases as the pH rises from 2 to 8. Since both pH and E_h are easily measured, it is not too difficult to ascertain whether a specified reaction would be likely to occur in a given soil. For example, at pH 6 the E_h would need to be somewhat lower than +0.5 V to encourage the reduction of nitrates to nitrites, and about +0.2 volts to stimulate the reduction of FeOOH to the Fe^{2+} ion. For methane to form in a waterlogged soil at the same pH (6), an E_h of about -0.2 V would be required.

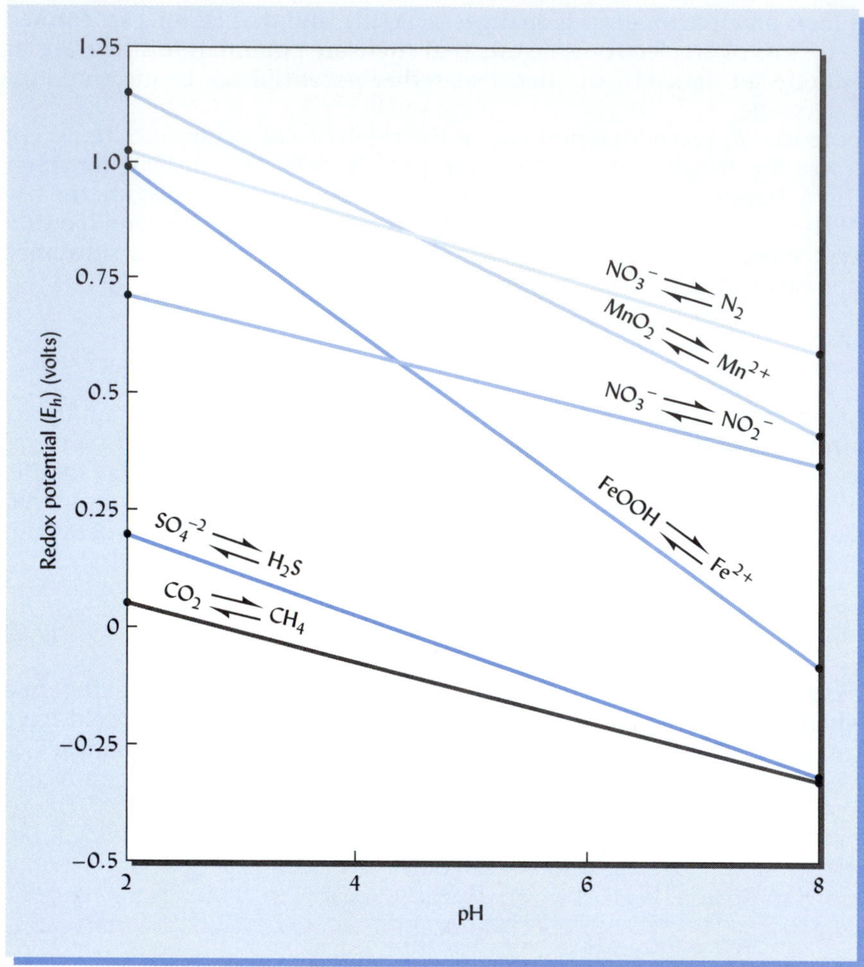

FIGURE 7.3 The effect of pH on the redox potential E_h at which several important reduction-oxidation reactions take place in soil. [From McBride (1994); used with permission of Oxford University Press]

The E_h value at which oxidation-reduction reactions occur varies with the specific chemical to be oxidized or reduced. Table 7.1 lists oxidized and reduced forms of several elements important in soils, along with the approximate redox potentials at which the oxidation-reduction reactions occur. The E_h values explain the sequence of reactions that are known to occur when a well-aerated soil becomes saturated with water.

At first, respiration will reduce the concentrations of O_2 in the soil air and dissolved in the soil water. As the concentration of this electron acceptor is lowered, so, too, the redox potential of the soil is lowered. Since O_2 is reduced to water at E_h levels of 0.38 to

TABLE 7.1 Oxidized and Reduced Forms of Certain Elements in Soils and the Redox Potentials E_h at Which Changes in Form Occur in a Soil at pH 6.5

Note that gaseous oxygen is depleted at E_h levels of 0.38 to 0.32 V. At lower E_h levels microorganisms utilize elements other than oxygen as the electron acceptor in their metabolism. By donating electrons they transform these elements into their reduced valence state.

Oxidized form	*Reduced form*	*E_h at which change of form occurs, V*
O_2	H_2O	0.38 to 0.32
NO_3^-	N_2	0.28 to 0.22
Mn^{4+}	Mn^{2+}	0.22 to 0.18
Fe^{3+}	Fe^{2+}	0.11 to 0.08
SO_4^{2-}	S^{2-}	−0.14 to −0.17
CO_2	CH_4	−0.20 to −0.28

From Patrick and Jugsujinda (1992).

0.32 V, once the soil redox potential falls below this level the soil is essentially devoid of O_2. At lower E_h values, the only microorganisms that can function are those able to use elements other than oxygen as their metabolic electron acceptors.

The next most easily reduced element present is usually N(V) (in nitrate, NO_3^-). If the soil contains much nitrate, the E_h will remain near 0.28 to 0.22 V as the nitrate is reduced. Once nearly all the nitrate has disappeared [N(V) has been transformed into N(III) and other N species], the E_h will drop further. At this point, organisms capable of reducing Mn will become active, and so on. Thus, as E_h values fall, the elements N, Mn, Fe, and S (in SO_4^{-2}) and C (in CO_2) accept electrons and become reduced, predominantly in the order listed.

In other words, the soil E_h must be lowered to zero or less before methane is produced, but an E_h value of 0.28 to 0.22 is low enough to result in the reduction of nitrate-N. Thus, soil aeration helps determine the specific chemical species present and, in turn, the availability, mobility, and possible toxicity of various elements in soils.

7.4 FACTORS AFFECTING SOIL AERATION

Drainage of Excess Water

Drainage of gravitational water out of the profile and concomitant diffusion of air into the soil takes place most readily in macropores. The most important factors influencing the aeration of well-drained soils are therefore those that determine the volume of the soil macropores. Macropore content has a major influence on the total air space as well as on gaseous exchange and biochemical reactions. Soil texture, bulk density, aggregate stability, organic matter content, and biopore formation are among the soil properties that help determine macropore content and, in turn, soil aeration (see Section 4.6).

Rates of Respiration in the Soil

A study of respiration in an orchard soil: http://www.ppsystems.com/apple_orchard.html

The concentrations of both oxygen and carbon dioxide are largely dependent on microbial activity, which in turn depends on the availability of organic carbon compounds as food. Incorporation of large quantities of manure, crop residues, or sewage sludge may alter the soil air composition appreciably. Likewise, the cycling of plant residues by leaf fall, root mass decay, and root excretion in natural ecosystems provides the substrate for microbial activity. Respiration by plant roots and enhanced respiration by soil organisms near the roots are also significant processes. All these processes are very much enhanced as soil temperature increases (see Section 7.8).

Soil Heterogeneity

Subsoils are usually more deficient in oxygen than are topsoils. Not only is the water content usually higher (in humid climates), but also the total pore space, as well as the macropore space, is generally much lower in the deeper horizons. In addition, the pathway for diffusion of gases into and out of the soil is longer for deeper horizons. However, if organic substrates are in low supply in the subsoil, it may still be aerobic. For this reason, certain recently flooded soils are anaerobic in the upper 50 to 100 cm and are aerobic below.

Profile. As seen in Figure 7.4, the aeration status varies greatly in different locations in a soil profile. In well-drained, relatively uniform soils the trend is a general reduction in O_2 and increase in CO_2 as one moves down the profile. However, poorly aerated zones or pockets may be found in any horizon of an otherwise well-drained and well-aerated soil.

Tillage. One cause of soil heterogeneity is tillage, which has both short-term and long-term effects on soil aeration. In the short term, stirring the soil often allows it to dry out faster and mixes in large quantities of air. These effects are especially evident on somewhat compacted, fine-textured soils, on which plant growth often responds immediately after a cultivation to control weeds or "knife in" fertilizer. In the long term, however, tillage may reduce macroporosity (see Section 4.6).

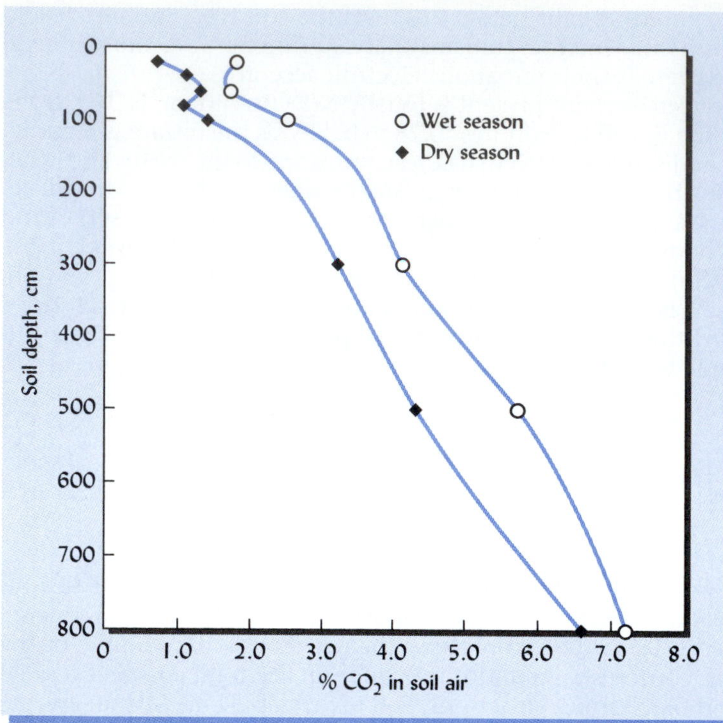

FIGURE 7.4 Changes in the concentration of CO_2 in soil air with depth into the profile of a Haplustox soil under a tropical rainforest in the Brazilian Amazon region. The source of the CO_2 was likely a combination of root and microbial respiration. Although by far the highest rates of CO_2 production were in the upper soil layers, gas produced there had little distance to travel to reach the atmosphere and many large pores to travel through. The concentration of CO_2 increased with depth because the increasing travel distance to the surface and the much lower macroporosity at depth greatly slowed the movement of gases, causing CO_2 to accumulate. [Data from Davidson and Trumbore (1995)]

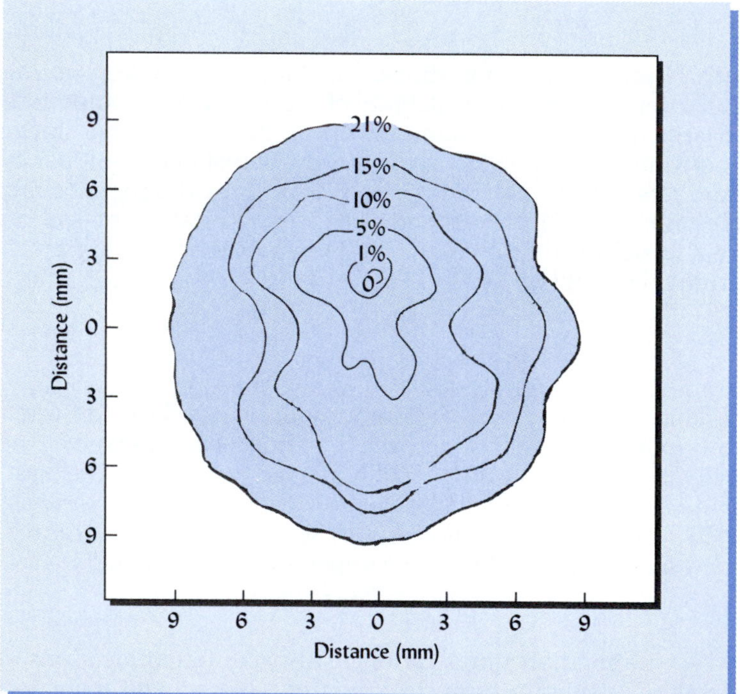

FIGURE 7.5 The oxygen content of soil air in a wet aggregate from an Aquic Hapludoll (Muscatine silty clay loam) from Iowa. The measurements were made with a unique microelectrode. Note that the oxygen content near the aggregate center was zero, while that near the edge of the aggregate was 21%. Thus, pockets of oxygen deficiency can be found in a soil whose overall oxygen content may not be low. [From Sexstone et al. (1985)]

LARGE MACROPORES. Poorly aerated zones may result from a heavy-textured or compacted soil layer, or it may be merely the inside of a soil ped (structural unit) where the smallness of pores may limit ready air exchange (Figure 7.5). In well-drained soils, the large pores (cracks) between peds, and old root channels in the subsoil, may periodically fill with water, causing localized zones of poor aeration. In saturated soil, such large pores may cause the opposite effect, as they facilitate O_2 diffusion into the soil during periods of drying.

Plant Roots. Likewise, the roots of growing plants may either reduce or increase the O_2 concentration in their immediate vicinity. In somewhat poorly drained soils, respiration by roots of upland plants may deplete the O_2 in nearby soil. In contrast, hydrophytic plants with aerenchyma tissues may transport surplus O_2 into their roots, allowing some to diffuse into the soil and produce an oxidized zone in an otherwise anaerobic soil (see, for example, Plate 29, after page 114).

For these reasons, oxidation reactions may be occurring within a few centimeters or millimeters from another location where reducing conditions exist. This heterogeneity of soil aeration should be kept in mind when attempting to understand the role that soil plays in elemental cycling and ecosystem function.

Seasonal Differences

There is marked seasonal variation in the composition of soil air. In the springtime in temperate humid regions, the soils are often wet and opportunities for ready gas exchange are poor. But due to low soil temperatures, the respiration of plant roots and soil microorganisms is restricted, so the utilization of oxygen and release of CO_2 are also restrained. In the summer months, the soils are commonly lower in moisture content, and the opportunity for gaseous exchange is increased. However, more favorable temperatures stimulate vigorous respiration by plant roots and microorganisms, releasing copious quantities of carbon dioxide.

Effects of Vegetation

In addition to the root respiration effects mentioned previously, vegetation may affect soil aeration by removing large quantities of water via transpiration, enough to lower the water table in some poorly drained soils. The data in Table 7.2 illustrate this effect for a pine forest. The effects of soil depth and season are also evident.

7.5 ECOLOGICAL EFFECTS OF SOIL AERATION

Effects on Organic Residue Degradation

Identify soil compaction and determine if compaction is limiting crop yield: http://www.ianr.unl.edu/pubs/soil/g831.htm

Soil aeration influences many soil reactions and, in turn, many soil properties. The most obvious of these reactions are associated with microbial activity, especially the breakdown of organic residues and other microbial reactions. Poor aeration slows the rate of decay, as evidenced by the relatively high levels of organic matter that accumulate in poorly drained soils.

TABLE 7.2 Effect of Timber Harvest with Minimal Compaction on Soil Aeration and Temperature Regimes in a Subtropical Pine Forest

Once the 55-year-old loblolly pine trees were cut, evapotranspiration and shading were reduced, resulting in a higher water table, lower redox potentials, and warmer spring temperatures. The latter stimulated microbial use of oxygen and further lowered redox potentials in this Vertic Ochraqualf. Note that the redox potentials were lower with the warmer temperatures in spring, even though the soil was not as wet in spring as in winter.

		Soil temperature, °C		Soil redox potentials E_h, V	
Site treatment	Time soil is saturated, %	Winter	Spring	Winter	Spring
		Measured at 50-cm depth			
Undisturbed pine stand	31	11.8	18.3	0.83	0.65
Trees cut, not compacted	64	11.7	20.5	0.51	0.11
		Measured at 100-cm depth			
Undisturbed pine stand	54	13.3	17.3	0.83	0.49
Trees cut, not compacted	46	13.2	18.7	0.54	0.22

Data from Tiarks et al. (1996).

The nature and the rate of microbial activity are determined by the O_2 content of the soil. Where O_2 is present, aerobic organisms are active, and oxidation reactions such as shown in equation 11.1 occur rapidly. In the absence of gaseous oxygen, anaerobic organisms take over and much slower breakdown takes place. Poorly aerated soils therefore tend to contain a wide variety of only partially oxidized products such as ethylene gas (C_2H_4), alcohols, and organic acids, many of which can be toxic to higher plants and to many decomposer organisms. The latter effect helps form Histosols in wet areas where inhibition of decomposition allows thick layers of organic matter to accumulate.

Oxidation-Reduction of Elements

Nutrients. Through its effects on the redox potential, the level of soil oxygen largely determines the forms of several inorganic elements, as shown in Table 7.1. The oxidized states of nitrogen and sulfur are readily utilizable by higher plants. In alkaline soils reduced forms of iron and manganese may be desirable since they are more soluble and can thereby alleviate plant deficiencies of these elements. But in acid soils, the reduced forms of these elements are so soluble that toxicities may occur. However, some reduction of iron in acid soils may be beneficial as it will release phosphorus from insoluble iron-phosphate compounds. Such phosphorus release has implications for eutrophication (see Section 13.2) when it occurs in saturated soils or in underwater sediments.

Such differences illustrate the interaction of aeration and soil pH in supplying available nutrients to plants (see Chapters 9 and 13).

Other Elements. Redox potential determines the species of such toxic elements as chromium, arsenic, and selenium, markedly affecting their impact on the environment and food chain. Reduced forms of arsenic are most mobile and toxic, giving rise to toxic levels of this element in drinking water, a serious human health problem in many places around the world, but especially in Bangladesh. In contrast, oxidized hexavalent Cr^{6+} compounds are mobile and are very toxic to humans. In neutral to acid soils, easily decomposed organic materials can be used to reduce the chromium to the Cr^{3+} form which is not subject to ready oxidation (Figure 7.6).

Soil Colors. As was discussed in Section 4.1, soil color is influenced markedly by the oxidation status of iron and manganese. Colors such as red, yellow, and reddish brown are characteristic of well-oxidized conditions. More subdued shades such as grays and blues predominate if insufficient oxygen is present. Soil color can be used in field methods for determining the status of soil drainage. Imperfectly drained soils are characterized by contrasting streaks of oxidized and reduced materials (see Plates 15 and 29, after page 114). Such a mottled condition indicates a zone of alternating good and poor aeration, a condition not conducive to the optimum growth of most plants.

Methane Production. Methane gas is produced by the reduction of CO_2. Its formation occurs when the E_h is reduced to about -0.2 V, a condition common in natural wetlands and in rice paddies. It is estimated that wetlands in the United States emit about 100 million metric tons of methane annually. Because of the biological productivity and diversity of these environments (see Section 7.7), soil scientists are seeking means of managing methane release without resorting to drainage of the wetlands.

Effects on Activities of Higher Plants

Plants are adversely affected in at least three ways by conditions of poor aeration: (1) The growth of the plant, particularly the roots, is curtailed; (2) the absorption of nutrients and water is decreased; and (3), as discussed in the previous section, the formation of certain inorganic compounds toxic to plant growth is favored.

Different plant species vary in their ability to tolerate poor aeration (Table 7.3). Sugar beets and barley are examples of crop species that require high air porosities for best growth (see Figure 7.1, left photo). In contrast, ladino clover and reed canary grass can grow with very low air porosity. Rice and cranberries are crops that can grow with their roots submerged in water. Furthermore, the tolerance of a given plant to low porosity may be different for seedlings than for rapidly growing plants. A case in point

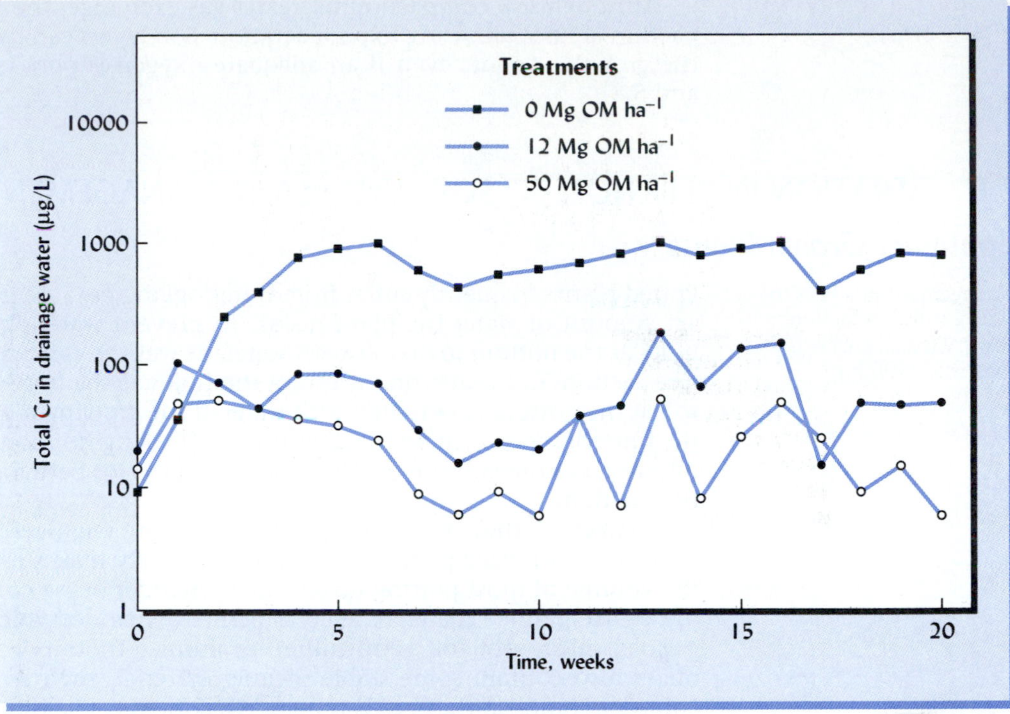

FIGURE 7.6 Effect of adding decomposable organic matter (OM) on the concentration of chromium in water draining from a chromium-contaminated soil. Here dried cattle manure was added as the decomposable OM. As the manure oxidized, it caused the reduction of the toxic, mobile Cr^{6+} to the relatively immobile, nontoxic Cr^{3+}. Note the log scale for the Cr in the water, indicating that the high level of manure addition caused the Cr level to be lowered approximately 100-fold. The coarse textured soil was a Typic Torripsamment in California. [Data from Losi et al. (1994)]

is the tolerance of red pine to restricted drainage during its early development and its poor growth or even death on the same site at later stages (see Figure 7.1, right photo).

Knowledge of plant tolerance to poor aeration is useful in choosing appropriate species to revegetate wet sites. The occurrence of plants specially adapted to anaerobic conditions is diagnostic for identifying wetland sites (see Section 7.7).

TABLE 7.3 Examples of Plants with Varying Degrees of Tolerance to a High Water Table and Accompanying Restricted Aeration

The plants in the leftmost column commonly thrive in wetlands. Those in the rightmost column are very sensitive to poor aeration.

Plants adapted to grow well with a water table at the stated depth				
<10 cm	*15 to 30 cm*	*40 to 60 cm*	*75 to 90 cm*	*>100 cm*
Bald cypress	Alsike clover	Birdsfoot trefoil	Beech	Arborvitae
Black spruce	Black willow	Black locust	Birch	Barley
Common cattail	Cottonwood	Bluegrass	Cabbage	Beans
Cranberries	Deer tongue	Linden	Corn	Cherry
Duckgrass	Eastern gama grass	Mulberry	Hairy vetch	Hemlock
Fragmites grass	Ladino clover	Mustard	Millet	Oats
Mangrove	Loblolly pine	Red maple	Peas	Peach
Pitcher plant	Orchard grass	Sorghum	Red oak	Sand lovegrass
Reed canarygrass	Redtop grass	Sycamore		Sugar beets
Rice	Tall fescue	Weeping lovegrass		Walnut
Skunk cabbage		Willow oak		Wheat
Spartina grass				White pine
Swamp white oak				
Swamp rosemalow				

Although soil compaction decreases gas exchange, the negative effects of soil compaction are not all owing to poor aeration. Soil layers can become so dense as to impede the growth of roots even if an adequate oxygen supply is available (see Sections 4.5 and 5.8).

7.6 AERATION IN RELATION TO SOIL AND PLANT MANAGEMENT

Container-Grown Plants

Management practices to reduce soil compaction: http://www.ianr.unl.edu/pubs/soil/g896.htm

Potted plants frequently suffer from waterlogging because it is difficult to supply the exact amount of water the plant needs. To prevent waterlogging, most containers have holes at the bottom to drain excess water. As was the case with stratified soils in the field (see Section 5.6), water drains out of the holes at the bottom of the pot *only* when the soil at the bottom is saturated with water. If the growing medium is mostly mineral soil, the fine soil pores remain filled with water, leaving no room for air, and anaerobic conditions soon prevail. Use of taller pots will allow for better aeration in the upper part of the medium.

To manage these problems, potting mixes are engineered to meet the requirements of the containerized plants. Mineral soil generally makes up no more than one-third of the volume of most potting mixes, the remainder being composed of inert lightweight but coarse-grained materials such as perlite (expanded volcanic glass), vermiculite (expanded mica—not soil vermiculite), or pumice (porous volcanic rock). Most modern mixes also contain some stable organic material, such as peat, shredded bark, wood chips, or compost, that holds water as well as adds macroporosity.

Tree and Lawn Management

In transplanting a young tree seedling or any woody species, special caution must be taken to prevent poor aeration or waterlogging immediately around the young roots. Figure 7.7 illustrates the right and the wrong way to manage the transplanting of trees in compacted soils.

The aeration of well-established, mature trees must also be safeguarded. If operators push surplus excavated soil around the base of a tree during landscape grading, serious consequences are soon noticed (Figure 7.8). The tree's feeder roots near the original soil surface are soon deficient in oxygen even if the overburden is no more than 5 to 10 cm in depth. Building a protective wall (a *dry well*) around the base of a valuable tree before grading operations begin will preserve enough of the original surface to allow the tree roots access to the O_2 they need, thereby saving the tree.

Management systems for heavily trafficked lawns commonly have components relating to soil aeration. For example, one means of increasing the aeration in compacted lawn areas is to use *core cultivation* that actually removes small cores of soil from the surface horizon, thereby permitting gas exchange to take place more easily (Figure 7.9).

7.7 WETLANDS AND THEIR POORLY AERATED SOILS[2]

Areas of poorly aerated soils called **wetlands** cover approximately 14% of the world's ice-free land, with the greatest areas occurring in the cold regions of Canada, Alaska, and Russia. In the continental United States, about 400,000 km^2 exist today, less than half of the area that is estimated to have existed when European settlement of the nation began. Most wetland losses occurred as farmers used artificial drainage (see Section 6.9) to convert them into cropland. In recent decades, filling and drainage for urban development has also taken its toll of wetland areas. Since environmental consciousness has become a force in modern societies, wetland preservation has become a major issue, and the loss of wetlands has been slowed from about 114,000 ha/yr during the 1970s and 1980s to 23,000 ha/yr during the 1990s.

---

[2]Two well-illustrated, nontechnical, yet informative publications on wetlands are Welsh et al. (1995) and CAST (1994). For a compilation of technical papers on hydric soils and wetlands, see Richardson and Vepraskas (2001).

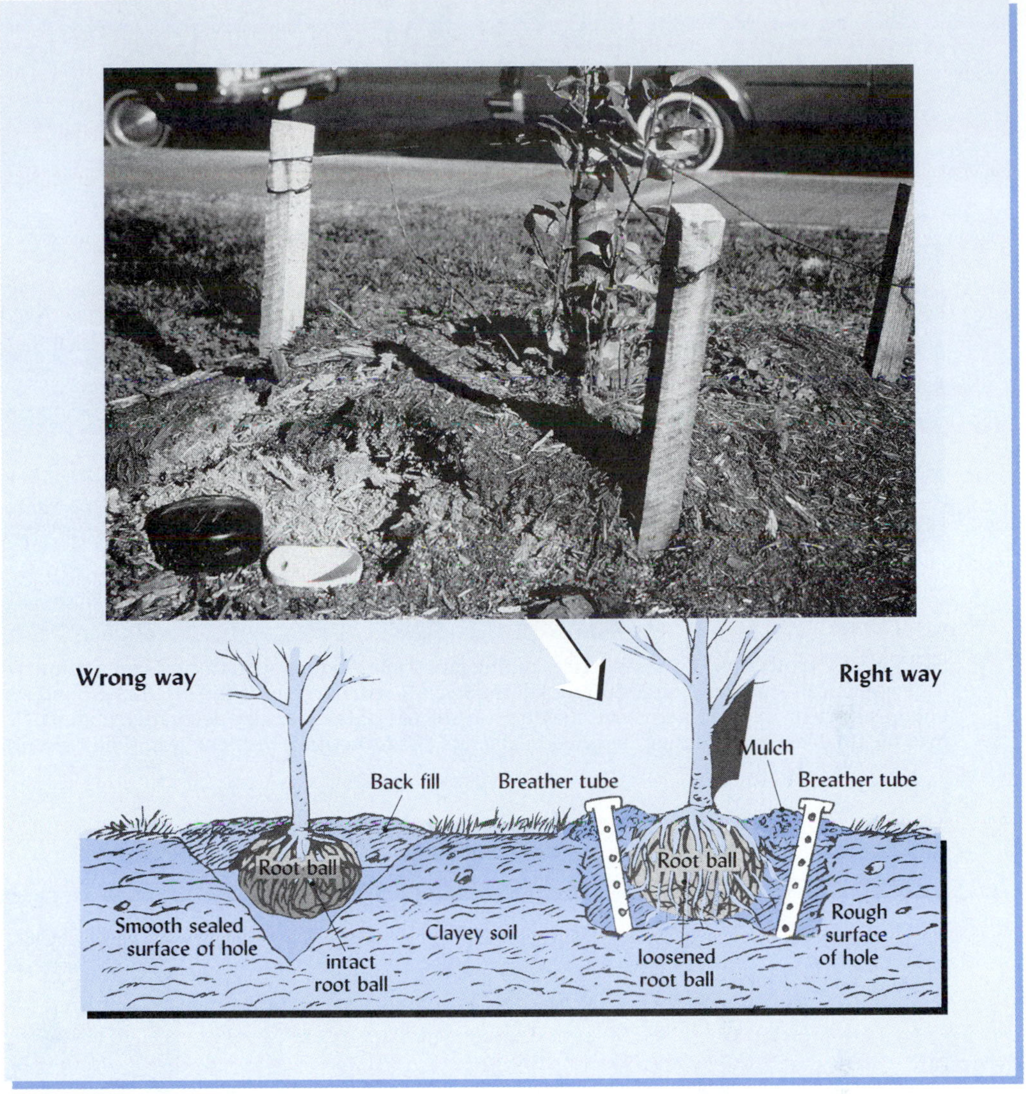

FIGURE 7.7 Providing a good supply of air to tree roots can be a problem, especially when trees are planted in fine-textured, compacted soils of urban areas. A machine-dug hole with smooth sides will act as a "tea cup" and fill with water, suffocating tree roots. Breather tubes, a larger rough-surfaced hole, and a layer of surface mulch in which some fine tree roots can grow are all measures that can improve the aeration status of the root zone. (Photo courtesy of R. Weil)

Defining a Wetland[3]

Wetland is a scientific term for ecosystems that are transitional between land and water. These systems are neither strictly terrestrial (land based) nor aquatic (water based). While there are many different types of wetlands, they all share a key feature, namely *soils that are water-saturated near the surface for prolonged periods when soil temperatures and other conditions are such that plants and microbes can grow and remove the soil oxygen, thereby assuring anaerobic conditions.* It is largely the prevalence of anaerobic conditions that determines the kinds of plants, animals, and soils found in these areas. There is widespread agreement that the wetter end of a wetland occurs where the water is too deep for rooted, emergent vegetation to take hold. The difficulty is in precisely defining the so-called *drier end* of the wetland, the boundary beyond which exist nonwetland, upland

[3]In 1987, the U.S. Army Corps of Engineers and the Environmental Protection Agency agreed on the following definition to be used in enforcing the Clean Water Act: "The term wetlands means those areas that are inundated or saturated by surface or ground water at a frequency and duration sufficient to support, and that under normal circumstances do support, a prevalence of vegetation typically adapted for life in saturated soil conditions."

FIGURE 7.8 Protection of valuable trees during landscape grading operations. Even a thin layer of soil spread over a large tree's root system can suffocate the roots and kill the tree. (Inset) To preserve the original ground surface so that tree feeder roots can obtain sufficient oxygen, a dry well may be constructed of brick or any decorative material. The dry well may be incorporated into the final landscape design, or filled in at a rate of a few centimeters per year. (Photos courtesy of R. Weil)

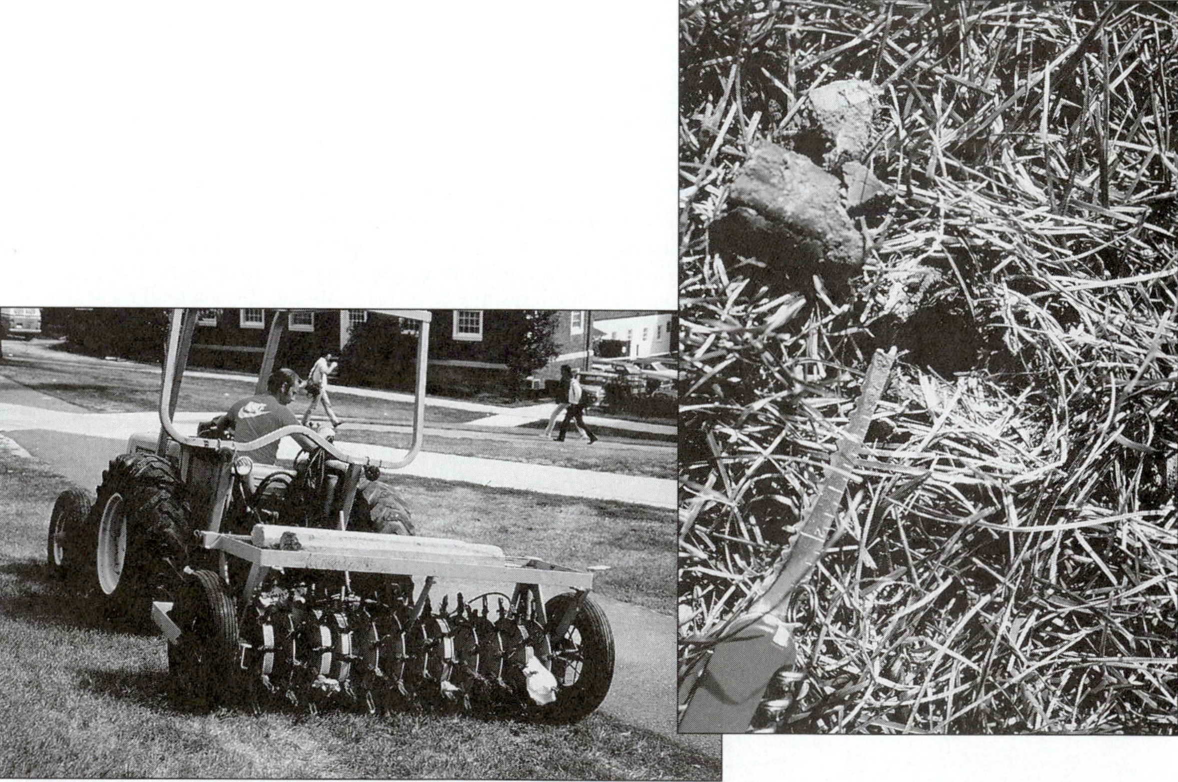

FIGURE 7.9 One means of increasing the aeration of compacted soil is by core cultivation. The machine removes small cores of soil, leaving holes about 2 cm in diameter and 5 to 8 cm deep. This method is commonly used on high-traffic turf areas. Note that the machine *removes* the cores and does not simply punch holes in the soil, a process that would increase compaction around the hole and impede air diffusion into the soil. (Photos courtesy of R. Weil)

systems in which the plant–soil–animal community is no longer predominantly influenced by the presence of water-saturated soils.

What are the characteristics scientists look for to indicate the existence of a wetland system? Most authorities agree that three characteristics can be found in any wetland:

1. A wetland hydrology or water regime
2. Hydric soils
3. Hydrophytic plants

We shall briefly examine each.

Wetland Hydrology

Wetlands characterized, the view from South Africa: http://psybergate.com/wetfix/ShareNet/Sharenet2/Share.htm

Guide to recognizing wetlands in the field: http://www.nao.usace.army.mil/Regulatory/RBwetlands.html

Water Balance. Water flows into wetlands from surface runoff (e.g., bogs and marshes), from groundwater seepage, and from direct precipitation. It flows out by surface and subsurface flows, as well as by evaporation and transpiration (see Section 6.3 and Table 7.2). The balance between inflows and outflows, as well as the water storage capacity of the wetland itself, determines how wet it will be and for how long.

Hydroperiod. Tidal marshes in coastal areas may be flooded for part of each day (Figure 7.10). Other wetlands may be flooded for only a month or so each year, while some may never be flooded, although they are saturated within the upper soil horizons. If the period of saturation occurs when the soil is too cold for microbial or plant-root activity to take place, anaerobic conditions may not develop, even in flooded soils. This temperature requirement is sometimes referred to as the *growing season* in reference to the time of year during which plants are actively growing. As we shall see in Section 7.8, microbial activity continues down to about 5°C, so O_2 may be depleted when the soil is warmer than this (even if plants are dormant). Remember, it is the anaerobic condition, not just saturation, that makes a wetland a wetland.

Indicators. In the field, even during dry periods, there are many signs one can look for to indicate where saturated conditions frequently occur. Past periods of flooding will leave water stains on trees and rocks, and a coating of sediment on the plant leaves and

FIGURE 7.10 An example of a wetland with a daily hydroperiod that follows the rise and fall of the slightly brackish estuary tides. This tidal marsh with a beaver house is seen at low tide when some saturated soils and emergent plants are exposed. The drier-end boundary of the wetland is probably just beyond the treeline in the background. (Photo courtesy of R. Weil).

litter. Drift lines of once-floating branches, twigs, and other debris also suggest previous flooding. Trees with extensive root masses above ground indicate an adaptation to saturated conditions. But perhaps the best indicator of saturated conditions is the presence of **hydric soils.**

Hydric Soils[4]

U.S. national hydric soils list:
http://soils.usda.gov/soil_use/hydric/national.htm

To assist in delineating wetlands, soil scientists developed the concept of hydric soils. In *Soil Taxonomy* (see Chapter 3) these soils are mostly (but not exclusively) classified in the order Histosols, in Aquic suborders such as Aquents, Aquepts, and Aqualfs, or in Aquic subgroups. These soils generally have an aquic or peraquic moisture regime (see Section 3.2). Three properties help define hydric soils. First, they are subject to *periods of saturation* that inhibit the diffusion of O_2 into the soil. Second, for substantial periods of time they undergo *reduced conditions* (see Section 7.3); that is, electron acceptors other than O_2 are reduced. Third, they exhibit certain features termed *hydric soil indicators* (Box 7.1).

Hydrophytic Vegetation

What exactly does it take to be a hydric soil?
http://soils.usda.gov/soil_use/hydric/intro.htm

Although the vast majority of plant species cannot survive the conditions characteristic of wetlands, there do exist varied and diverse communities of plants that have evolved special mechanisms to adapt to life in saturated, anaerobic soils. These plants comprise the **hydrophytic vegetation** that distinguishes wetlands from other systems.

Typical adaptive features include hollow aerenchyma tissues that allow plants like spartina grass to transport O_2 down to their roots. Certain trees (such as bald cypress) produce adventitious roots, buttress roots, or knees. Other species spread their roots in a shallow mass on or just under the soil surface, where some O_2 can diffuse even under a layer of ponded water. The leftmost column in Table 7.3 lists a few common hydrophytes. Not all the plants in a wetland are likely to be hydrophytes, but the majority usually are.

Wetland Chemistry

Field indicators of hydric soils in the United States:
http://soils.usda.gov/soil_use/hydric/field_ind.pdf

Even in a flooded wetland, O_2 will be able to diffuse from the atmosphere or from oxygenated water into the upper 1 or 2 cm of soil, creating a thin *oxidizing zone* (see Figure 7.12). The diffusion of O_2 within the saturated soil is extremely limited, so that a few centimeters deeper into the profile, O_2 is eliminated and the redox potential becomes low enough for reactions such as nitrate reduction to take place. The close proximity of the oxidized and anaerobic zones allows water passing through wetlands to be stripped of N by the sequential oxidation of ammonium N to nitrate N, and then the reduction of the nitrate to various nitrogen gases that escape into the atmosphere (see Section 12.10).

REDOX. Redox potentials may become low enough for iron reduction to produce redoximorphic features, and for sulfate reduction to produce rotten-egg-smelling hydrogen sulfide (H_2S) gas. The anaerobic zone may extend downward, or in some cases may be limited to the upper horizons where microbial activity is high. The anaerobic carbon reactions discussed in Section 7.5 are characteristic of this zone, including methane (swamp gas) production. These and other chemical reactions involving the cycling of C, N, and S are explained in Sections 11.2, 12.10, and 12.22. Toxic elements such as chromium and selenium undergo redox reactions that may help remove them from the water before it leaves the wetland. Acids from industry or mine drainage may also be neutralized by reactions in hydric soils.

---

[4] The U.S. Department of Agriculture Natural Resources Conservation Service defines a hydric soil as one "that formed under conditions of saturation, flooding or ponding long enough during the growing season to develop anaerobic conditions in the upper part." For an illustrated field guide to features that indicate hydric soils, see Hurt et al. (1996).

BOX 7.1 HYDRIC SOIL INDICATORS

Hydric soil indicators are features associated (sometimes only in specific geographic regions) with the occurrence of saturation and reduction. Most of the indicators can be observed in the field by digging a small pit to a depth of about 50 cm. They principally involve the loss or accumulation of various forms of Fe, Mn, S, or C. The carbon (organic matter) accumulations are most evident in Histosols, but thick, *dark surface layers* in other soils can also be indicators of hydric conditions in which organic matter decomposition has been inhibited (see, for example, Plates 6 and 30).

Iron, when reduced to Fe(II), becomes sufficiently soluble that it migrates away from reduced zones and may precipitate as Fe(III) compounds in more aerobic zones. Zones where reduction has removed or depleted the iron coatings from mineral grains are termed **redox depletions.** They commonly exhibit the gray, low-chroma colors of the bare, underlying minerals (see Section 4.1 for an explanation of chroma). Also iron itself turns gray to blue-green when reduced. The contrasting colors of redox depletions or reduced iron and zones of reddish oxidized iron result in unique mottled **redoximorphic features** (see, for example, Plates 15 and 26 following page 114). Other redoximorphic features involve reduced Mn. These include the presence of hard black *nodules* that sometimes resemble shotgun pellets. Under severely reduced conditions the entire soil matrix may exhibit *low-chroma colors,* termed *gley.* Colors with a chroma of 1 or less quite reliably indicate reduced conditions (Figure 7.11).

Always keep in mind that redoximorphic features are indicative of hydric soils only when they occur in the upper horizons. Many soils of upland areas exhibit redoximorphic features only in their deeper horizons, due to the presence of a fluctuating water table at depth. Upland soils that are saturated or even flooded for short periods, especially if during cold weather, are *not* wetland (hydric) soils.

A unique redoximorphic feature associated with certain wetland plants is the presence, in an otherwise gray matrix, of reddish oxidized iron around root channels where O_2 diffused out from the aerenchyma-fed roots of a hydrophyte (see Plate 29). These *oxidized root zones* exemplify the close relationship between hydric soils and *hydrophytic vegetation.*

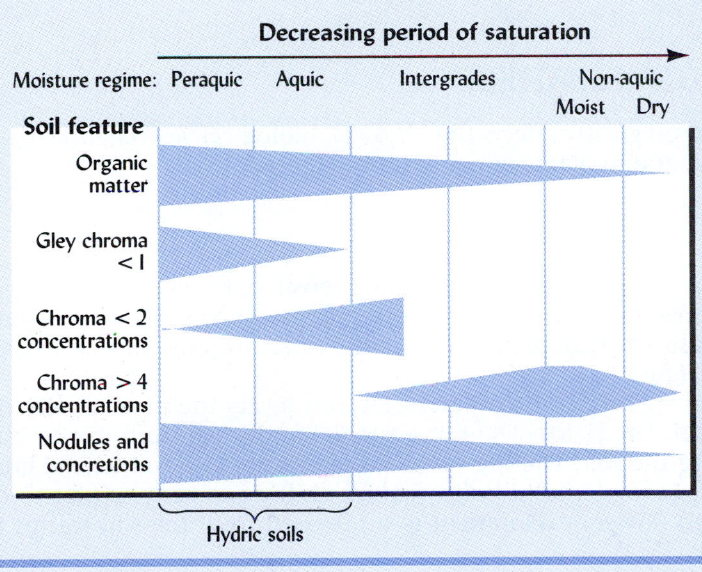

FIGURE 7.11 The relationship between the occurrence of some soil features and the annual duration of water-saturated conditions. The absence of iron concentrations (mottles) with colors of chroma >4, and the presence of strong expressions of the other features are indications that a soil may be hydric. [Adapted from Veneman et al. (1999)]

Constructed Wetlands

Realizing all the beneficial functions of wetlands, scientists and engineers have begun not only to find ways to preserve natural wetlands, but to construct artificial ones for specific purposes, such as wastewater treatment (see, for example, Box 13.2).

Another reason for attempting to construct wetlands is the provision in several regulations that allows for the destruction of certain natural wetland areas, provided that an equal or larger area of new wetlands is constructed or that previously degraded wetlands are restored. This process, termed ***wetland mitigation,*** has been only partially successful, as scientists cannot be expected to create what they do not fully understand.

We have seen the influence of soil water on soil aeration. We now turn to another soil physical property, soil temperature, that is also greatly influenced by the content of water in a soil.

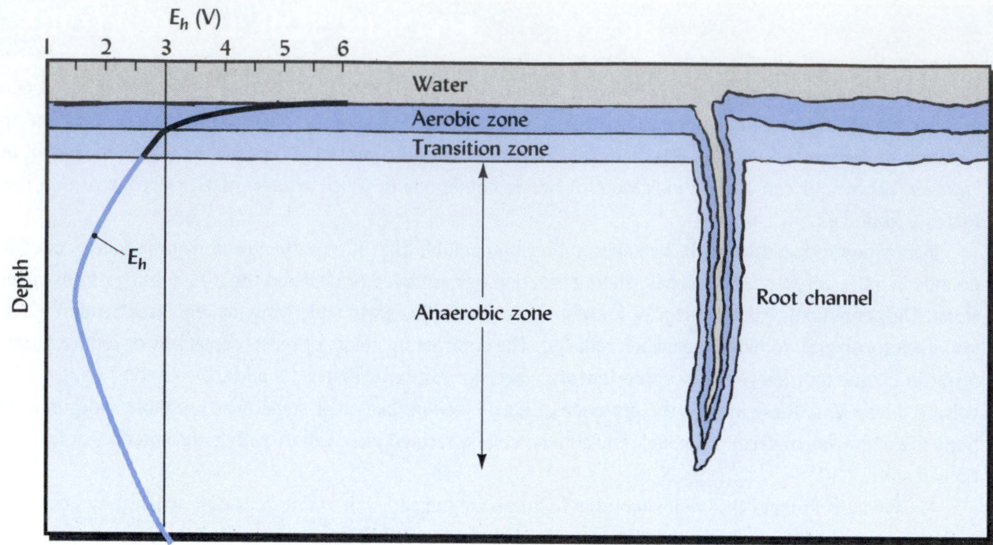

FIGURE 7.12 Representative redox potentials within the profile of an inundated hydric soil. Many of the biological and chemical functions of wetlands depend on the close proximity of reduced and oxidized zones in the soil. The changes in redox potential at the lower depths depend largely on the vertical distribution of organic matter. In some cases, low subsoil organic matter results in a second oxidized zone beneath the reduced zone. (Diagram courtesy of R. Weil)

7.8 PROCESSES AFFECTED BY SOIL TEMPERATURE

Temperature effects on no-till crops:
http://www.agric.gov.ab.ca/agdex/500/9000002.html

The temperature of a soil greatly affects the physical, biological, and chemical processes occurring in that soil, and in plants growing on it (Figure 7.13).

Plant Processes

The growth rates of most plants are actually more sensitive to *soil* temperature than to aboveground *air* temperature, but this is not often appreciated since air temperature is more commonly measured. Most plants have a rather narrow range of soil temperatures for optimal growth (Figure 7.13).

In cool temperate regions, soil temperature often limits the productivity of crops and natural vegetation. The yields of some vegetables and small fruits can be markedly increased by warming the soil. The life cycles of plants are also influenced greatly by soil temperature. For example, tulip bulbs require chilling in early winter to develop flower buds, although flower development is suppressed until the soil warms the following spring.

In warm regions, and in the summer in temperate regions, soil temperatures may be too high for optimal plant growth, especially in the upper few centimeters of soil. Even plants of tropical origin, such as corn, are adversely affected by soil temperatures higher than 35°C. Root growth near the surface may be encouraged by shading the soil with either live vegetation, plant residues, or the use of a mulch.

SEED GERMINATION. Different plant processes have different optimal temperatures. One of the processes most sensitive to soil temperature is seed germination. For example, farmers know that if they plant corn seed into soils cooler than 7 to 10°C, germination will not occur and the seed will likely rot. Many herbaceous annual plants require specific soil temperatures to trigger seed germination, accounting for much of the difference in species between early- and late-season weeds in cultivated land. Likewise, the seeds of certain plants adapted to open gaps in a forest stand are stimulated to germinate by the higher daily maximum soil temperatures and greater fluctuations in soil temperature that occur when the forest canopy is disturbed by timber harvest or wind-thrown trees. The seeds of certain prairie grasses require a period of cold soil temperatures (2 to 4°C) to enable them to germinate the following spring, a process termed *vernalization.*

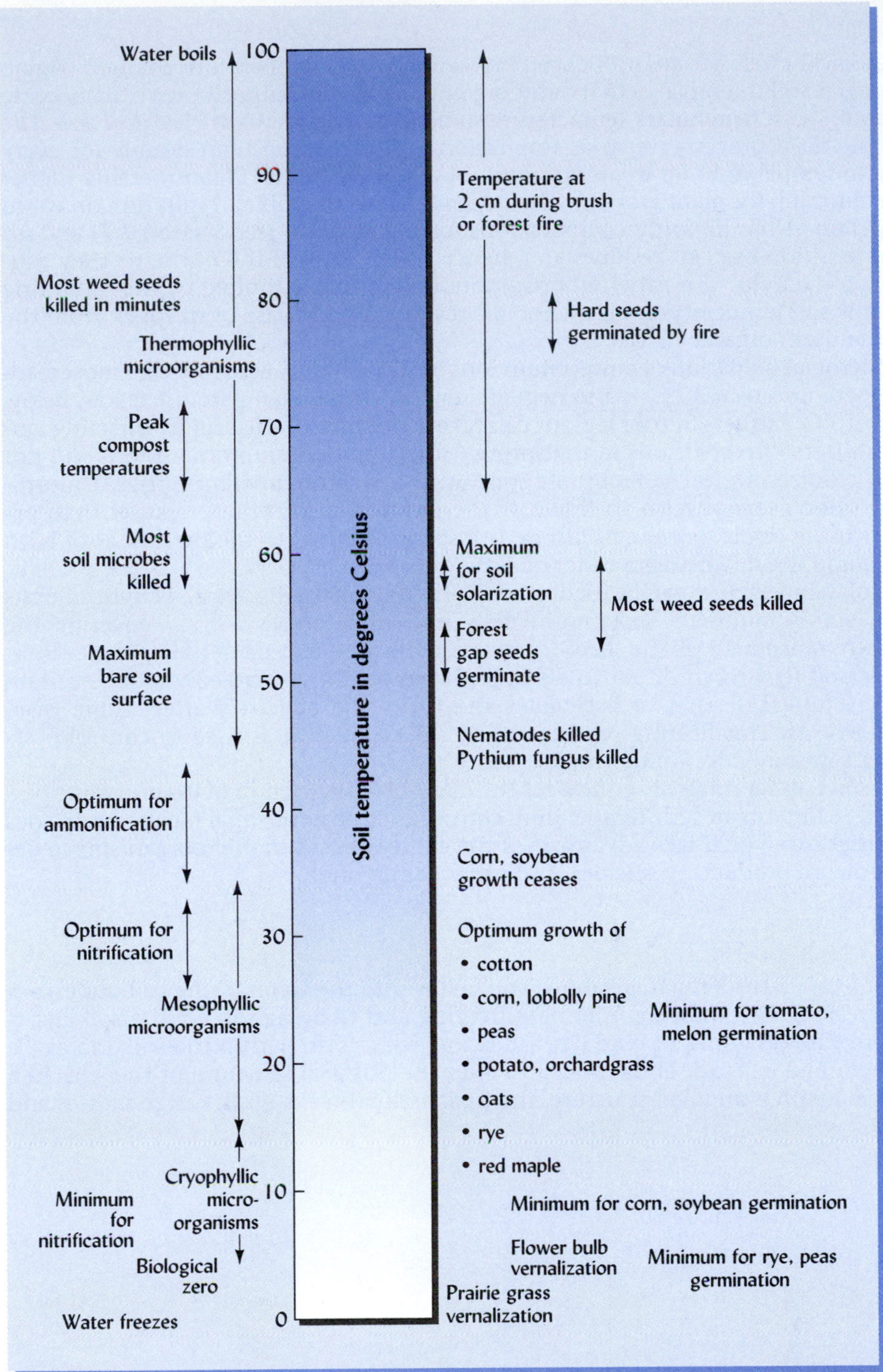

FIGURE 7.13 Soil temperature ranges associated with a variety of soil processes.

Root Functions. Root functions such as nutrient uptake and water uptake are sluggish in cool soils with temperatures below the optimum for the particular species. One result is that nutrient deficiencies, especially of phosphorus, often occur in young plants in early spring, only to disappear when the soil warms later in the season. The phenomenon of *winter burn* of plant foliage is another consequence of low soil temperature that particularly affects evergreen shrubs. On bright sunny days in winter and early spring when the soil is still cold, evergreen plants may become desiccated and even die because the slow water uptake by roots in the cold soils cannot keep up with the high evaporative demand of bright sun on the foliage. The problem can be prevented by covering the shrubs with a shade cloth.

Microbial Processes

Microbiological processes are influenced markedly by soil temperature changes (Figure 7.14). General soil microbial activity and organic matter decomposition virtually cease below about 5°C, a benchmark temperature sometimes referred to as *biological zero*. The rates of microbial processes such as respiration typically more than double for every 10°C rise in temperature up to an optimum of about 35 to 40°C (considerably higher than the optimum for plant growth). The dependence of microbial respiration on warm soil temperatures has important implications for soil aeration (see Section 7.7) and for the decomposition of plant residues and, hence, the cycling of the nutrients they contain. The productivity of northern (boreal) forests is probably limited by the inhibiting effect of low soil temperatures on microbial recycling and release of nitrogen from the tree litter and soil organic matter.

The microbial oxidation of ammonium ions to nitrate ions, which occurs most readily at temperatures near 30°C, is also negligible when the soil temperature is low, below about 8 to 10°C. Farmers in cool regions can take advantage of this fact by injecting ammonia fertilizers into cold soils in the spring, expecting that ammonium ions will not be readily oxidized to leachable nitrate ions until the soil temperature rises. Unfortunately, in some years a warm spell allows the production of nitrates earlier than expected, with the result that much nitrate is lost by leaching, to the detriment of both the farmer and the downstream water quality.

High soil temperatures can be used to control certain plant diseases. In environments with hot, sunny summers—maximum daily air temperatures >35°C—covering the ground with transparent plastic sheeting can raise the temperature of the upper few centimeters of soil to as high as 50° to 60°C. Such temperatures markedly reduce certain wilt-causing fungal diseases of vegetables and fruits and adversely affect some weed seeds and insects. This heating process, called *soil solarization,* is used to control pests and diseases in some high-value crops.

Warm soil temperatures are critical for the microbial destruction of toxic organic pesticides and pollutants in soil. Temperature control is critical for some new technologies (**bioremediation**) which take advantage of the ability of certain microorganisms to degrade petroleum products, pesticides, and other compounds.

Freezing and Thawing

When soil temperatures fluctuate above and below 0°C, the water in the soil undergoes cycles of freezing and thawing. Alternate freezing and thawing subject the soil aggregates to pressures as zones of pure ice, called *ice lenses,* form within the soil and as ice crystals form and expand. These pressures alter the physical structure of the soil. In a saturated soil with a puddled structure, the frost action breaks up the large masses and

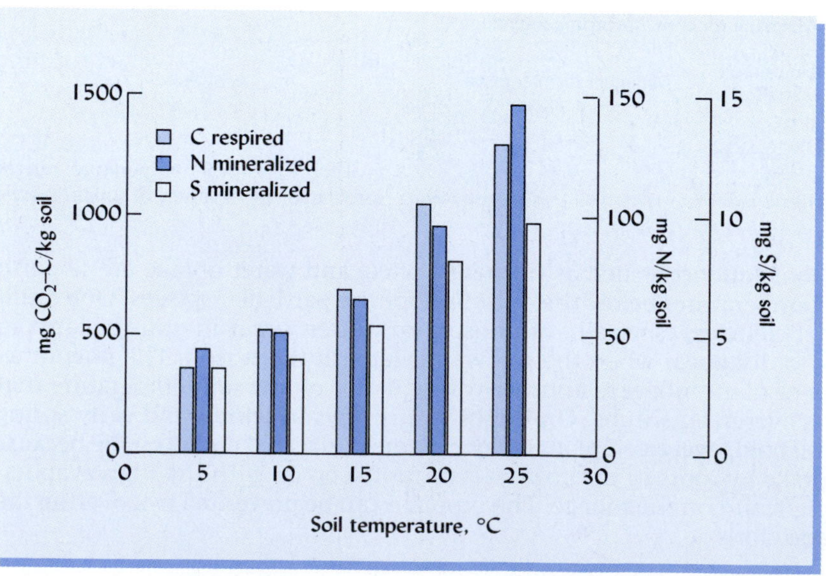

FIGURE 7.14 Effect of soil temperature on the cumulative microbial respiration (CO_2 release) and net nitrogen and sulfur mineralization in surface soils under hardwood forests over a period of 32 weeks during which the soil water content was adequate for microbial activity. Note the near doubling of activity with each 10°C increase in temperature. Data are averages of four sites in Michigan. From MacDonald et al. (1995).

greatly improves granulation. In contrast, for soils with good aggregation to begin with, freeze-thaw action when the soil is very wet can lead to structural deterioration.

Alternate freezing and thawing can force objects upward in the soil, a process termed *frost heaving* (Figure 7.15). Objects subject to heaving include stones, fence posts, and perennial taprooted plants. This action, which is most severe where the soil is silty in texture, wet, and lacking a covering of snow or dense vegetation, can drastically reduce stands of alfalfa, some clovers, and trefoil.

Freezing can also heave shallow foundations, roads, and runways which have fine material as a base. Gravels and pure sands are normally resistant to frost damage, but silts and sandy soils with modest amounts of finer particles are particularly susceptible. Very clay-rich soils do not usually exhibit much frost heave, but ice-lens segregation can still occur and can lead to severe loss of strength when thawing occurs. To avoid damage by freezing soil temperatures, foundation footings (as well as water pipelines) should be set into the soil below the maximum depth to which the soil freezes—a depth that

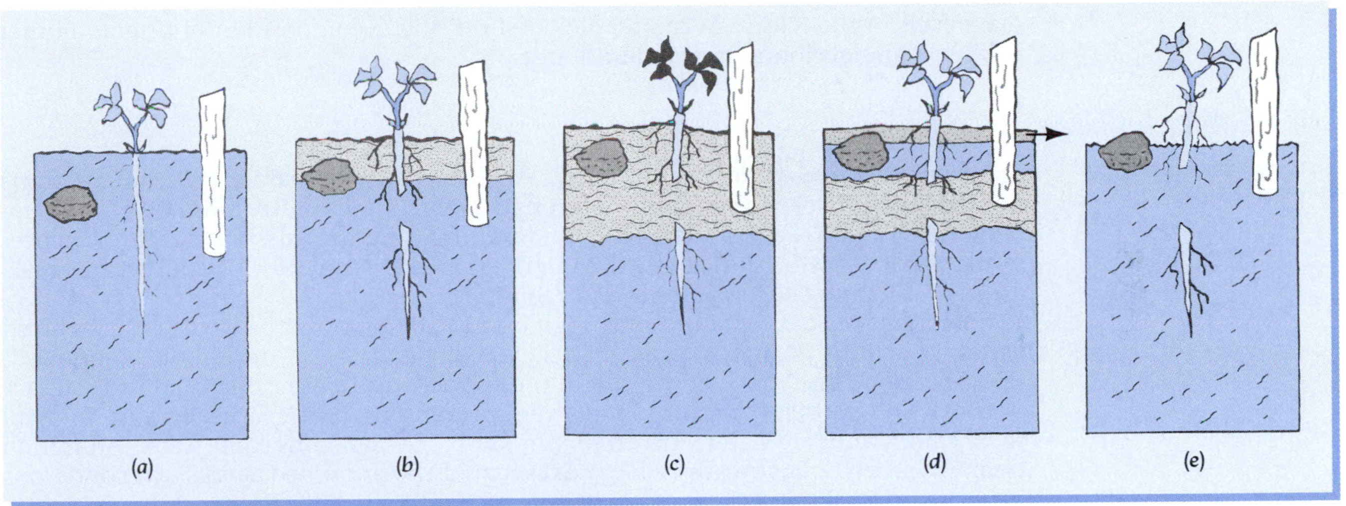

(f)

(g)

FIGURE 7.15 How frost heaving moves objects upward. (*a*) Position of the object (stone, plant, or fence post) before the soil freezes. (*b*) As lenses of pure ice form in the freezing soil by attraction of water from the unfrozen soil below, the frozen soil tightens around the upper part of the object, lifting it somewhat—enough to break the root in the case of the plant. (*c*) The objects are lifted upwards as ice-lens formation continues with deeper penetration of the freezing front. (*d*) As for freezing, thawing commences from the surface downward. Water from thawing ice lenses escapes to the surface because it cannot drain downward through the frozen soil. The soil surface subsides while the heaved objects are held in the "jacked-up" position by the still-frozen soil around their lower parts. (*e*) After complete thaw, the stone is closer to the surface than previously (although rarely at the surface unless erosion of the thawed soil has occurred), and the upper part of the broken plant's root is exposed, so that is likely to die. (*f*) Alfalfa plants lifted out of the ground by frost action. (*g*) Fence posts encased in concrete that have been progressively "jacked out" of the ground by frost action over several years. [Photo (*f*) courtesy of R. Weil; photo (*g*) courtesy of R. L. Berg, Corps of Engineers, Cold Regions Research and Engineering Laboratory, Hanover, N.H.]

ranges from less than 10 cm in subtropical zones, such as South Texas and Florida, to more than 200 cm in very cold climates.

Permafrost[5]

Perhaps the most significant global event involving soil temperatures is the thawing in recent years of some of the permafrost (permanently frozen ground) in northern areas of Canada, Russia, China, Mongolia, and Alaska. Nearly 25% of the land areas of the earth are underlain by permafrost. Rising temperatures since the late 1980s have caused some of the upper layers of permafrost to thaw. In parts of Alaska, for example, temperatures in top layers of permafrost have risen about 3.5°C since the late 1980s, resulting in melting rates of about a meter in a decade. Such melting has serious implications, as it can drastically affect the physical foundation of buildings and roads, as well as the stability of root zones of forests and other such vegetation in the region. Some trees have fallen, and homes and other buildings have collapsed as the frozen layers melt. Thawing also may have implications for future global warming, since the increasing temperatures may result in the release of carbon dioxide from the decomposition of peat and other organic materials found in permafrost areas.

Soil Heating by Fire

Interactive data on soil temperatures and prairie fires:
http://www.konza.ksu.edu/keep/temperature/soil_tp.asp

Fire is one of the most far-reaching ecosystem disturbances in nature. It definitely affects soil temperature, even though the effect may be brief and limited primarily to the upper few centimeters of soil. Temperatures resulting from the "slash and burn" practices in the tropics may be sufficiently high in the upper few mm of soil to cause the breakdown of minerals such as gibbsite and kaolinite.

The heat may also be sufficiently high (>125°C being common) to essentially distill various fractions of the organic matter, yielding volatile hydrocarbon compounds (Figure 7.16). These move downward in the soil and condense (solidify) on the surface of cooler soil particles and fill some of the surrounding pore spaces. Because some of these compounds are water repellent (hydrophobic), when rains come water infiltration even in a sandy soil is greatly reduced over that of the unburned areas. Such conditions in burned-over sloping chaparral areas may account for disastrous mudslides that occur when the layer of soil above the hydrophobic zone becomes saturated with rainwater.

Fires also affect the germination of certain seeds, which have hard coatings that prevent them from germinating until they are heated above 70° to 80°C. On the other

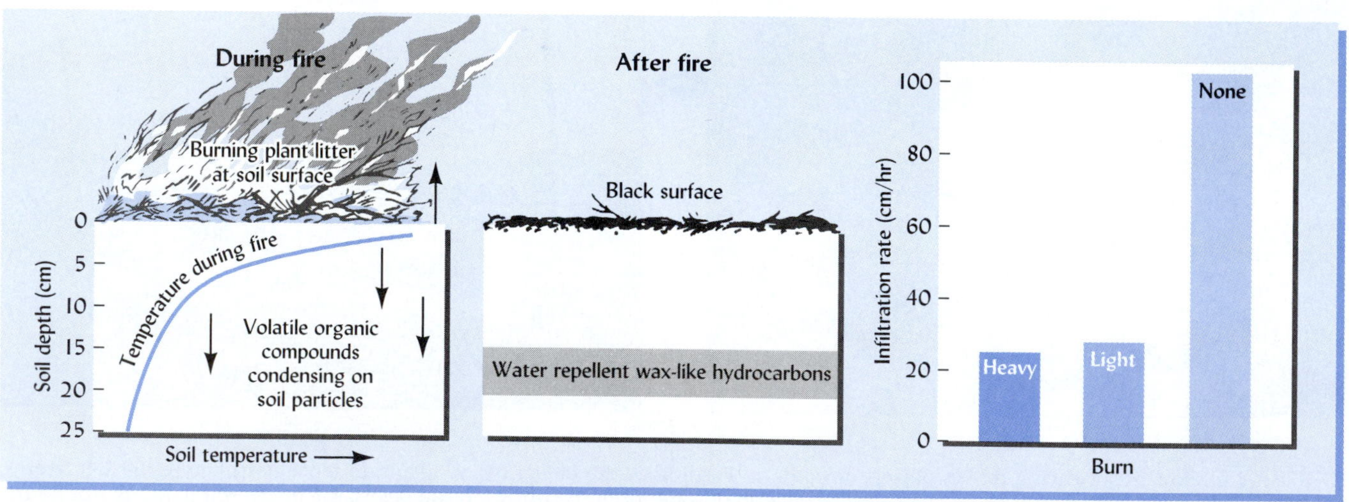

FIGURE 7.16 (Left) Wildfires of a lodgepole pine stand heat up the surface layers of this sandy soil (an Inceptisol) in Oregon. (Center) Note that the soil temperature is increased sufficiently near the surface to volatilize organic compounds, some of which then move down into the soil and condense (solidify) on the surface of cooler soil particles. These condensed compounds are waxlike hydrocarbons that are water repellent. As a consequence (right) the infiltration of water into the soil is drastically reduced and remains so for a period of at least 6 years. [From Dryness (1976)]

[5]For a recent article on permafrost in Alaska, see Wuethrich (2000).

Effects of forest fires on soils: http://www.fire.r9.fws.gov/ifcc/monitor/EFGuide/soils.htm

hand, burning of straw in wheat fields generates similar soil temperatures, but with the effect of killing most of the weed seeds near the surface and thus greatly reducing subsequent weed infestation. The heat and ash may also hasten the cycling of nutrient plant nutrients (see Chapter 14). Fires set to clear land of timber slash may burn long and hot enough to seriously deplete soil organic matter and kill so many soil organisms that forest regrowth is inhibited.

Contaminant Removal

The removal of certain organic pollutants from contaminated soils can be accomplished by raising the soil temperature. But the process may be prohibitively expensive if the soil has to be excavated and hauled to and from an extraction bin. Techniques are under development to warm the soil in place using electromagnetic radiation. The resulting temperatures are sufficiently high to vaporize some contaminants which can then be flushed from the soil by air. Figure 7.17 shows the results of one such operation set up to remove diesel fuel from soil under an air base.

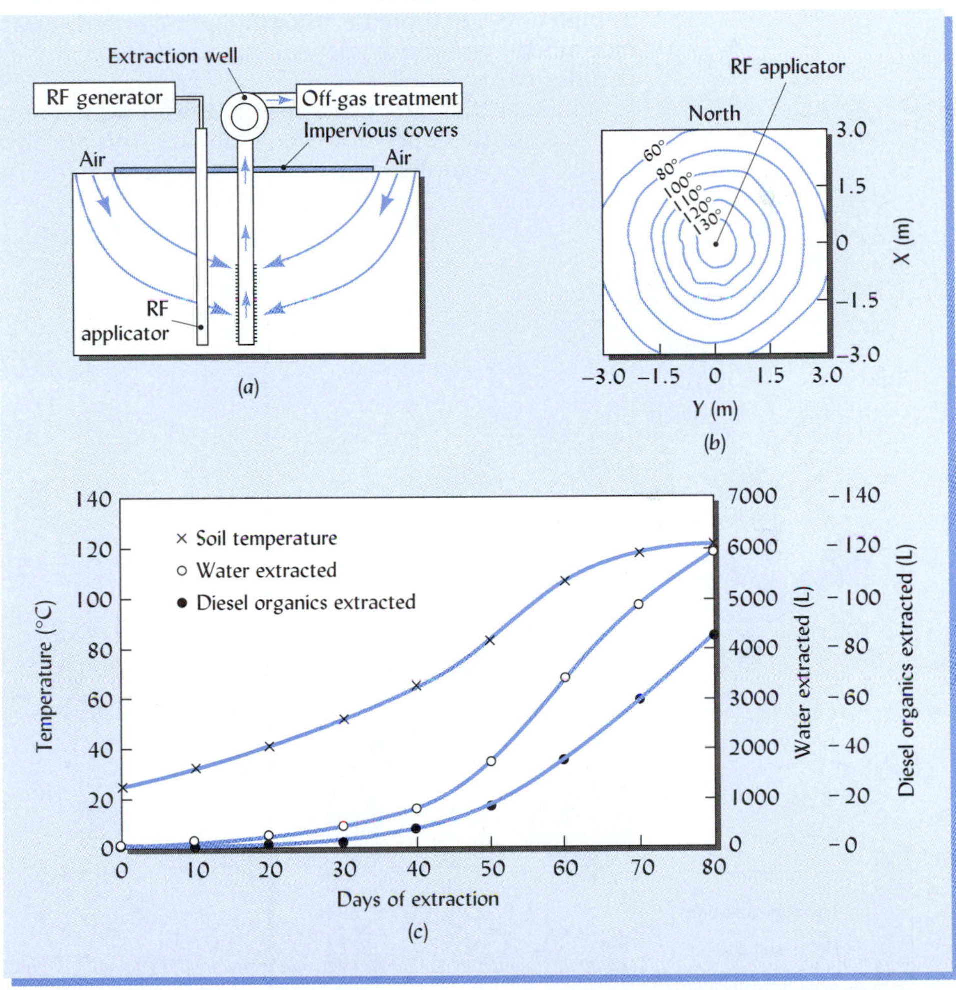

FIGURE 7.17 By increasing soil temperatures, environmental engineers can extract some organic pollutants from soils. (*a*) Electromagnetic radiation from radio frequency (RF) was used to supply the energy to gradually increase temperatures in a block of soil containing diesel fuel at the Kirkland Air Force Base near Albuquerque, New Mexico. At the higher temperatures the organic hydrocarbons were vaporized (along with water). The vapors were then extracted from the soil using an extraction well, and were subsequently removed from the air (off-gas treatment). (*b*) The effect of 80 days' radiation on soil temperatures outward from the RF applicator in a small block of polluted soil at the Kirkland AFB site. Note that extremely high temperatures did not prevail. (*c*) Soil temperature increase with time near the RF applicator, along with the quantity of organic compounds and water extracted from the soil. While this procedure is rather expensive, it permits remediation of the soil without having to remove it from its natural setting, and does not result in extremely high temperatures as the soil is heated. [Modified from figures in Lowe et al. (2000)]

7.9 ABSORPTION AND LOSS OF SOLAR ENERGY[6]

Urban heat islands: http://www.energy.ca.gov/coolcommunity/strplan.html

The temperature of soils in the field is directly or indirectly dependent on at least three factors: (1) the net amount of heat energy the soil absorbs; (2) the heat energy required to bring about a given change in the temperature of a soil; and (3) the energy required for processes such as evaporation, which are constantly occurring at or near the surface of soils.

Solar radiation is the primary source of energy to heat soils. But clouds and dust particles intercept the sun's rays and absorb, scatter, or reflect most of the energy (Figure 7.18). Only about 35 to 40% of the solar radiation actually reaches the earth in cloudy humid regions, and 75% in cloud-free arid areas. The global average is about 50%.

Little of the solar energy reaching the earth actually results in soil warming. The energy is used primarily to evaporate water from the soil or leaf surfaces, or is radiated or reflected back to the sky. Only about 10% is absorbed by the soil and can be used to warm it. Even so, this energy is of critical importance to soil processes and to plants growing on the soils.

ALBEDO. The fraction of incident radiation that is reflected by the land surface is termed the **albedo**, and ranges from as low as 0.1 to 0.2 for dark-colored, rough soil surfaces to as high as 0.5 or more for smooth, light-colored surfaces. Vegetation may affect the surface albedo either way, depending on whether it is dark green and growing or yellow and dormant.

The fact that dark-colored soils absorb more energy than lighter-colored soils does not necessarily imply, however, that dark soils are always warmer. In fact, in most landscapes, the darkest soils are usually the wettest and therefore the slowest to warm.

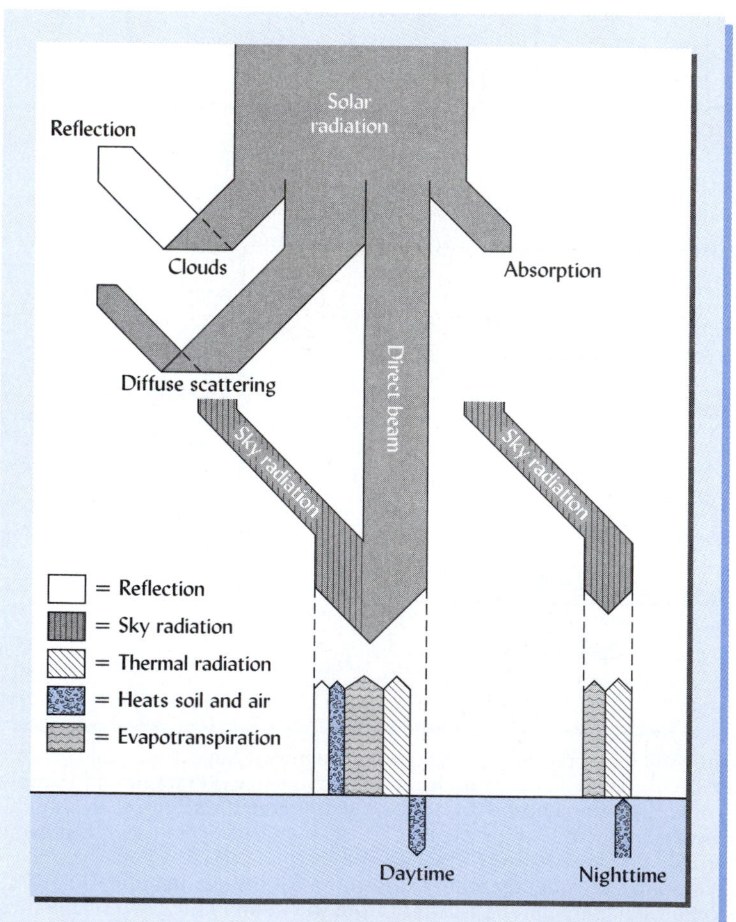

FIGURE 7.18 Schematic representation of the radiation balance in daytime and nighttime in the spring or early summer in a temperate region. About half the solar radiation reaches the earth, either directly or indirectly, from sky radiation. Most radiation that strikes the earth in the daytime is used as energy for evapotranspiration or is radiated back to the atmosphere. Only a small portion, perhaps 10%, actually heats the soil. At night the soil loses some heat, and some evaporation and thermal radiation occur.

[6]For a review of models that describe these processes on a global scale, see Sellers et al. (1997).

ASPECT. The angle at which the sun's rays strike the soil also influences soil temperature. If the sun is directly overhead, the incoming path of the rays is perpendicular to the soil surface, and energy absorption (and soil temperature increase) is greatest. The effect of the direction of slope, or **aspect**, on soil temperature and forest species is illustrated in Figure 7.19.

SOIL COVER. Whether the soil is bare or is covered with vegetation, mulch, or snow is another factor markedly influencing the amount of solar radiation reaching the soil. Bare soils warm more quickly and cool more rapidly than covered soils. Frost penetration during the winter is considerably greater in bare, noninsulated land.

Even low-growing vegetation such as turf grass has a noticeable influence on soil temperature and on the temperature of the surroundings (Table 7.4). Much of the cooling effect is due to heat dissipated by transpiration of water. To experience this effect, on a blistering hot day, try having a picnic on an asphalt parking lot instead of on a growing green lawn!

The effect of a dense forest is universally recognized. Timber-harvest practices that leave sufficient canopy to provide about 50% shade will likely prevent undue soil warming that could hasten the loss of soil organic matter or the onset of anaerobic conditions

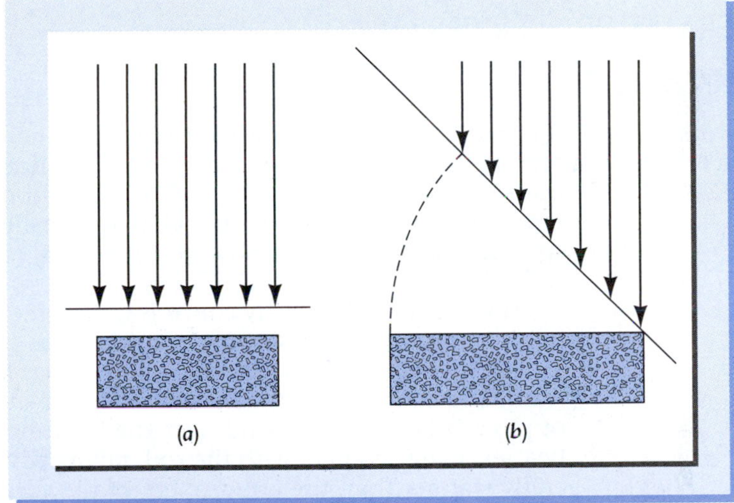

FIGURE 7.19 (Upper) Effect of the angle at which the sun's rays strike the soil on the area of soil that is warmed. (*a*) If a given amount of radiation from the sun strikes the soil at right angles, the radiation is concentrated in a relatively small area, and the soil warms quite rapidly. (*b*) If the same amount of radiation strikes the soil at a 45° angle, the area affected is larger by about 40%, the radiation is not so concentrated, and the soil warms up more slowly. This is one of the reasons why north slopes tend to have cooler soils than south slopes. It also accounts for the colder soils in winter than in summer. (Lower) A view looking eastward of a forested area in Virginia illustrates the temperature effect. The main ridge (left to right) is running north and south and the smaller side ridges east and west (up and down). The dark patches are pine trees in this predominantly hardwood deciduous forest. The pines dominate the southern (warmer and drier) slopes on each east-west ridge. (Photo courtesy of R. Weil)

TABLE 7.4 **Maximum Surface Temperatures for Four Types of Surfaces on a Sunny August Day in College Station, Texas**

Type of surface	*Maximum temperature, °C*	
	Day	*Night*
Green, growing turf grass	31	24
Dry, bare soil	39	26
Brown, summer-dormant grass	52	27
Dry synthetic sports turf	70	29

Data from Beard and Green (1994).

in wet soils. The effect of timber harvest on soil temperature as deep as 50 cm is seen in the case presented in Table 7.2, where tree removal warmed the soil in spring, even though it also raised the soil water content. However, as we shall see in the next section, a higher water content normally slows the warming of soils in spring.

7.10 THERMAL PROPERTIES OF SOILS

Specific Heat of Soils

Heat and energy concepts: http://hyperphysics.phy-astr.gsu.edu/ hbase/heacon.html#heacon

A dry soil is more easily heated than a wet one, because the amount of energy required to raise the temperature of water by 1°C (its heat capacity) is much higher than that required to warm soil solids by 1°C. When heat capacity is expressed per unit mass—for example, in calories per gram (cal/g)—it is called **specific heat** or heat capacity c. The specific heat of pure water is about 1.00 cal/g (or 4.18 joules per gram, J/g); that of dry soil is about 0.2 cal/g (0.8 J/g).

The specific heat largely controls the degree to which soils warm in the spring, wetter soils warming more slowly than drier ones (see Box 7.2).

Temperature control systems referred to as *heat pumps,* which are designed to both warm and cool buildings, take advantage of the high specific heat of soils. A network of pipes is laid underground near the building to be heated and cooled to maximize heat-exchange contact with the soil, taking advantage of the fact that subsoils are generally warmer than the atmosphere in the winter and cooler than the atmosphere in the summer. Water circulating through the network of pipes absorbs heat from the soil during the winter and releases it to the soil in the summer. The high specific heat of soils permits a large exchange of energy to take place without greatly modifying the soil temperature.

Heat of Vaporization

The evaporation of water from soil surfaces requires a large amount of energy, 540 kilocalories (kcal) or 2.257 mega joules (mJ) for every kilogram of water vaporized. This energy must be provided by solar radiation or it must come from the surrounding soil. In either case, evaporation has the potential of cooling the soil, much the way it chills a person who comes out from swimming on a windy day. The low temperature of wet soils in spring is due partially to evaporation and partially to high specific heat.

Thermal Conductivity of Soils

As shown in Section 7.9, some of the solar radiation that reaches the earth slowly penetrates the profile largely by conduction; this is the same process by which heat moves along an iron pipe when one end is placed in a fire. The movement of heat in soil is analogous to the movement of water (see Section 5.5), the rate of flow being determined by

BOX 7.2 CALCULATING THE SPECIFIC HEAT OR HEAT CAPACITY OF MOIST SOILS

Soil water content markedly impacts soil temperature changes through its effect on the specific heat or heat capacity c of a soil. For example, consider two soils with comparable characteristics, soil A, a relatively dry soil with 10 g water/100 g soil solids, and soil B, a wetter soil with 30 g water/100 g soil solids.

We can assume the following values for specific heat:

$$\text{Water} = 1.0 \text{ cal/g and dry mineral soil} = 0.2 \text{ cal/g}$$

For soil A with 10 g water/100 g dry soil, or 0.1 g water/g dry soil, the number of calories required to raise the temperature of 0.1 g of water by 1°C is

$$0.1 \text{ g} \times 1.0 \text{ cal/g} = 0.1 \text{ cal}$$

The corresponding figure for the 1.0 g of soil solids is

$$1 \text{ g} \times 0.2 \text{ cal/g} = 0.2 \text{ cal}$$

Thus, a total of 0.3 cal (0.1 + 0.2) is required to raise the temperature of 1.1 g (1.0 + 0.1) of the moist soil by 1°C. Since the specific heat is the number of calories required to raise the temperature of 1 g of moist soil by 1°C, we can calculate the specific heat of soil A as follows:

$$c_{\text{soil A}} = \frac{0.3}{1.1} = 0.273 \text{ cal/g}$$

These calculations can be expressed as a simple equation to calculate the weighted average specific heat of a mixture of substances:

$$c_{\text{moist soil}} = \frac{c_1 m_1 + c_2 m_2}{m_1 + m_2} \tag{7.5}$$

where c_1 and m_1 are the specific heat and mass of substance 1 (the dry mineral soil, in this case), and c_2 and m_2 are the specific heat of substance 2 (the water, in this case).

Applying this equation to soil A, we again calculate that $c_{\text{soil A}}$ is 0.273 cal/g, as follows:

$$c_{\text{soil A}} = \frac{0.2 \text{ cal/g} * 1.0 \text{ g} + 1.0 \text{ cal/g} * 0.10 \text{ g}}{1.0 \text{ g} + 0.10 \text{ g}} = \frac{0.30 \text{ cal/g}}{1.1 \text{ g}} = 0.273 \text{ cal/g}$$

In the same manner, we calculate the specific heat of the wetter soil B:

$$c_{\text{soil B}} = \frac{0.2 \text{ cal/g} * 1.0 \text{ g} + 1.0 \text{ cal/g} * 0.30 \text{ g}}{1.0 \text{ g} + 0.30 \text{ g}} = \frac{0.50 \text{ cal/g}}{1.3 \text{ g}} = 0.385 \text{ cal/g}$$

The wetter soil B has a specific heat c_B of 0.385 cal/g, whereas the drier soil A has a specific heat c_A of 0.273 cal/g. Because it must absorb an additional 0.112 cal (0.385 − 0.273) of solar radiation for every degree of temperature rise, the wetter soil will warm up much more slowly than the drier soil.

a driving force and by the ease with which heat flows through the soil. This can be expressed as Fourier's law:

$$Q_h = K \times \frac{\Delta T}{x} \tag{7.6}$$

where Q_h is the *thermal flux,* the quantity of heat transferred across a unit cross-sectional area in a unit time; K is the **thermal conductivity** of the soil; and $\Delta T/x$ is the temperature gradient over distance x that serves as the driving force for the conduction of heat.

The thermal conductivity K of soil is influenced by a number of factors, the most important being the moisture content of the soil and the degree of compaction. Heat

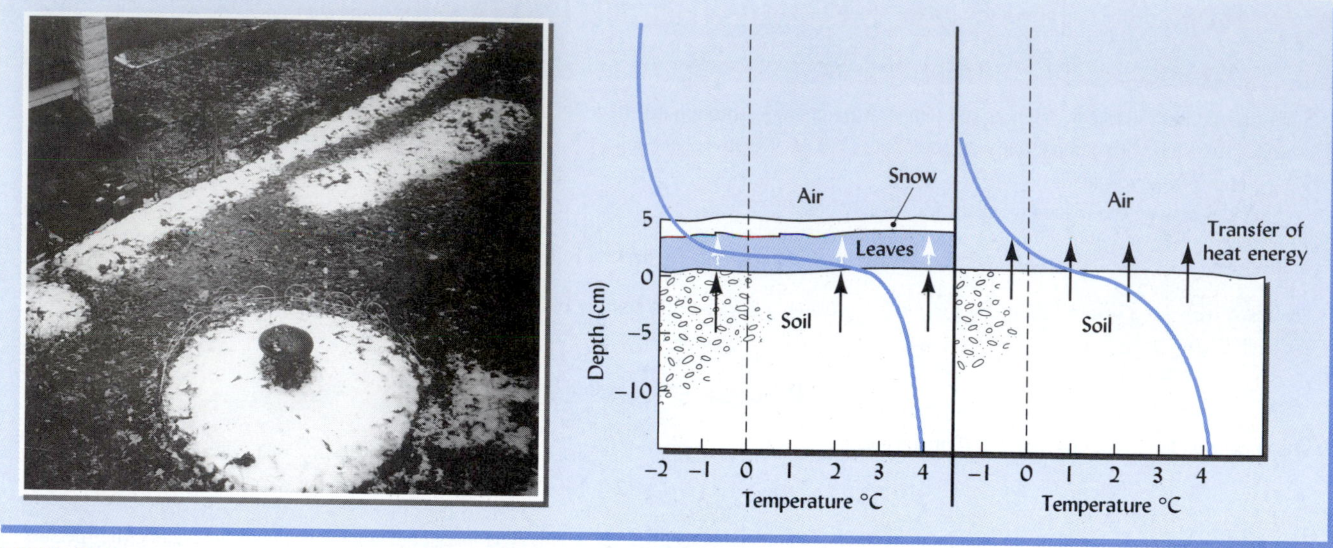

FIGURE 7.20 Transfer of heat energy from soil to air. The scene, looking down on a garden after an early fall snow storm, shows snow on the leaf-mulched flower beds, but not on areas where the soil is bare or covered with thin turf. The reason for this uneven accumulation of snow can be seen in the temperature profiles. Having stored heat from the sun, the soil layers are often warmer than the air as temperatures drop in fall (this is also true at night during other seasons). On bare soil, heat energy is transferred rapidly from the deeper layers to the surface, the rate of transfer being enhanced by high moisture content or compaction, which increase the *thermal conductivity* of the soil. As a result, the soil surface and the air above it are warmed to above freezing, so snow melts and does not accumulate. The leaf mulch, which has a low thermal conductivity, acts as an insulating blanket that slows the transfer of stored heat energy from the soil to the air. The upper surface of the mulch is therefore hardly warmed by the soil, and the snow remains frozen and accumulates. A heavy covering of snow can itself act as an insulating blanket. (Photo and diagram courtesy of R. Weil)

passes through water many times faster than through air. As the water content increases in a soil, the air content decreases, and the transfer resistance is decidedly lowered. When sufficient water is present to form a bridge between most of the soil particles, further additions will have little effect on heat conduction. Heat moves through mineral particles even faster than through water, so when particle-to-particle contact is increased by soil compaction, heat-transfer rates are also increased. Therefore, a wet, compacted soil would be the poorest insulator or the best conductor of heat. Here again the interconnectedness of soil properties is demonstrated.

Relatively dry soil makes a good insulating material. Buildings built mostly underground can take advantage of both the low thermal conductivity and relatively high heat capacity of large volumes of soil. Soil thermal conductivity can also affect air temperatures above the soil, as shown in Figure 7.20.

Vertical and Seasonal Temperature Changes

It is apparent from Figures 7.21 and 7.22 that considerable seasonal and monthly variations of soil temperature occur, even at the lower depths. The surface layer temperatures vary more or less according to the temperature of the air, although these layers are generally warmer than the air for most of the year.

In the subsoil, the seasonal temperature increases and decreases lag behind changes registered in the surface soil and in the air. Accordingly, the temperature data for March at College Station suggest that the surface soil temperatures have already begun to respond to the warming of the spring, while temperatures of the deep subsoil seem to still be responding to the cold winter weather.

Daily Variations

With a clear sky, the air temperature in temperate regions rises from lowest in the morning to a maximum at about 2 P.M. The surface soil, however, does not reach its maximum until later in the afternoon because of the usual lag. This retardation is greater and the temperature change is less as the depth increases. The lower subsoil shows little daily or weekly fluctuation; the variation there, as already emphasized, is a slow monthly or seasonal change.

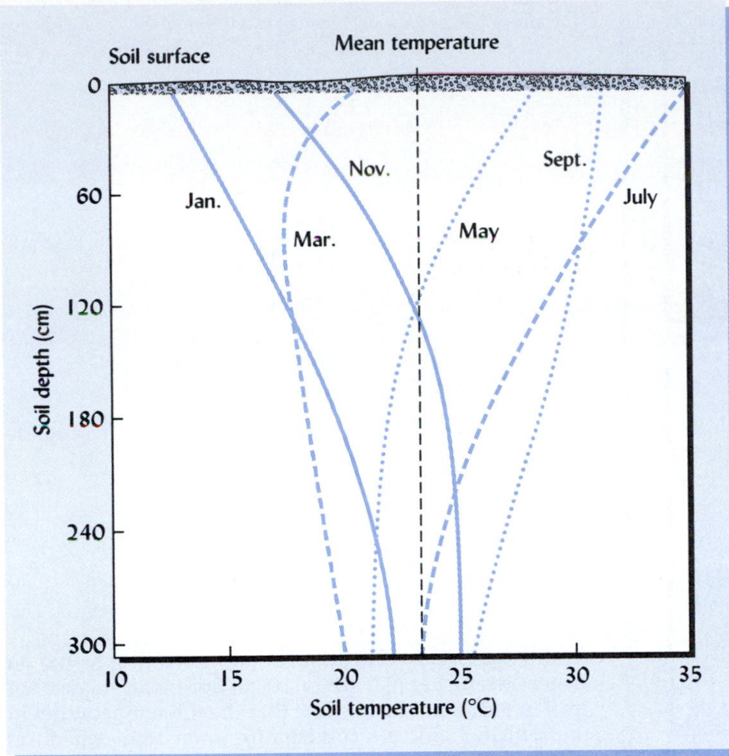

FIGURE 7.21 Average monthly soil temperatures for 6 of the 12 months of the year at different soil depths at College Station, Texas (1951–1955). Note the lag in soil temperature change at the lower depths. [From Fluker (1958)]

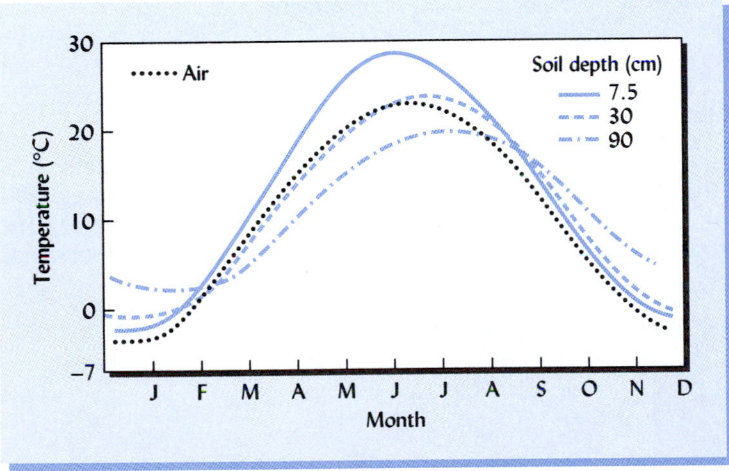

FIGURE 7.22 Average monthly air and soil temperatures at Lincoln, Nebraska (12 years). Note that the 7.5 cm soil layer is consistently warmer than the air above and that the 90 cm soil horizon is cooler in spring and summer, but warmer in the fall and winter, than surface soil.

7.11 SOIL TEMPERATURE CONTROL

Mulching effects: http://www.ianr.unl.edu/pubs/horticulture/g1257.htm

The temperature of field soils is not subject to radical human regulation. However, two kinds of management practice have significant effects on soil temperature: those that affect the cover or mulch on the soil, and those that reduce excess soil moisture. These effects have meaningful biological implications.

Organic Mulches and Plant-Residue Management

Soil temperatures are influenced by soil cover and especially by organic residues or other types of mulch on the soil surface. Figure 7.23 shows that mulches effectively buffer extremes in soil temperatures. In periods of hot weather, they keep the surface soil cooler

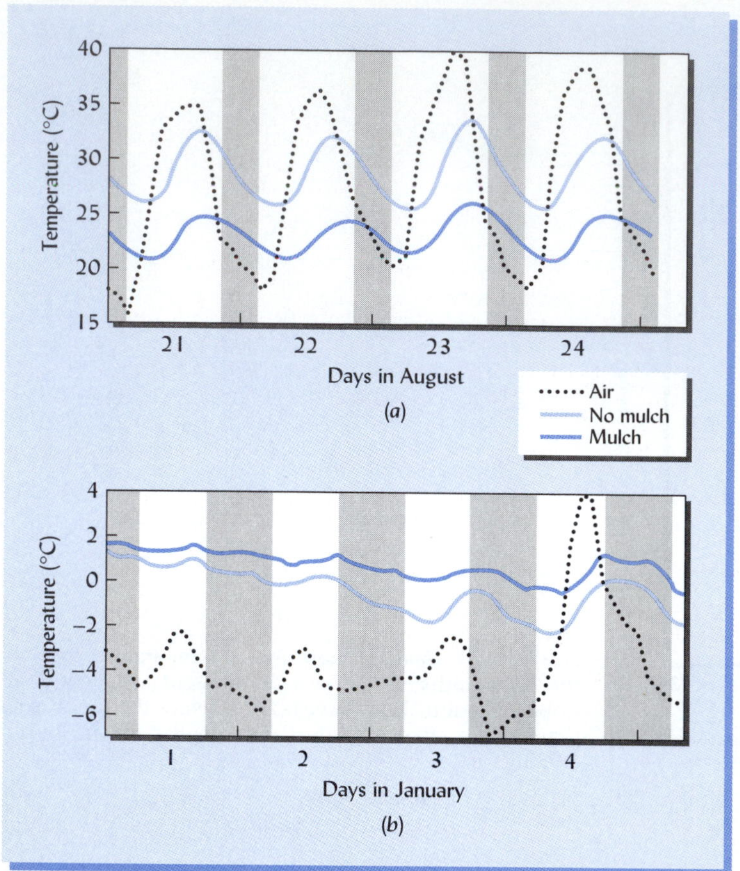

FIGURE 7.23 (*a*) Influence of straw mulch (8 tons/ha) on air temperature at a depth of 10 cm during an August hot spell in Bushland, Texas. Note that the soil temperatures in the mulched area are consistently lower than where no mulch was applied. (*b*) During a cold period in January the soil temperature was higher in the mulched than in the unmulched area. The shaded bars represent nighttime. [Redrawn from Unger (1978); used with permission of American Society of Agronomy]

than where no cover is used; in contrast, during cold weather they keep the soil warmer than it would be if bare.

The forest floor is a prime example of a natural temperature-modifying mulch. It is not surprising, therefore, that timber harvest practices can markedly affect forest soil temperature regimes (Figure 7.24). Disturbance of the leaf mulch, changes in water content due to reduced evapotranspiration, and compaction by machinery are factors that

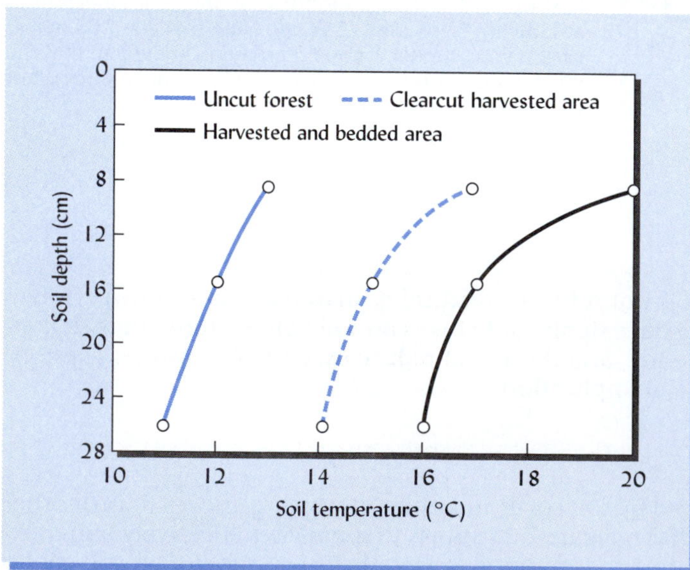

FIGURE 7.24 Influence of timber harvest and site preparation on soil temperature profiles in June on a forested wetland in Michigan. Clear-cut timber harvest greatly disturbed the site and raised soil temperatures by 4 to 5°C. Tillage to form beds for tree replanting eliminated the remaining forest floor mulch and dried the soil, thus causing temperatures to rise another 2 to 3°C. The higher temperatures increased the rate of soil organic matter decomposition. [Redrawn from Trettin et al. (1996)]

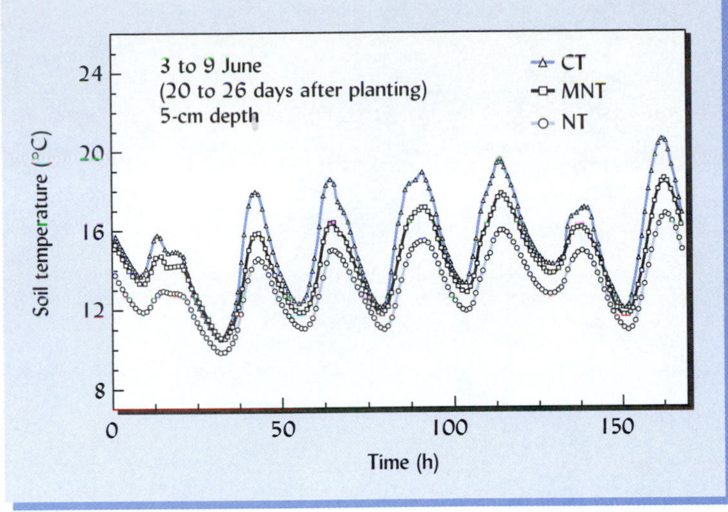

FIGURE 7.25 Tillage effects on hourly temperature change near the surface of a cold Alfisol in northern British Columbia. The soil had been managed to grow barley under no-tillage (NT) and conventional (clean surface without residues) tillage (CT) systems for the previous 14 years. In the clean-tilled soil, midafternoon temperatures peaked at 4°C higher than those in the residue mulch covered no-tillage soil. A modification of the no-tillage system (MNT) that pushed aside the residues in a narrow (7.5-cm-wide) band over the seeding row eliminated much of the temperature depression while keeping most of surface covered by the soil- and water-conserving mulch. Note the daily temperature changes and the general warming trend during the seven days shown. [From Arshad and Azooz (1996)]

influence soil temperatures through thermal conductivity. Reduced shading after tree removal also lets in more solar radiation.

Concerns in Cool Climates. While mulch provides great control over erosion (see Section 15.6), the soil-temperature-depressing effects of some mulch practices have a serious negative impact on the production of crops such as corn in cold climates. The lower temperatures in May and early June resulting from these practices inhibit seed germination, seedling performance, and, often, the yields of corn. The effect of the residue mulch is most pronounced in lowering the midday maximum temperature, and has much less effect on the minimum temperature reached at night. This effect is well illustrated by the data in Figure 7.25, which also features an innovative way to alleviate this problem by pushing aside the residues in just a narrow band over the seed row in the no-tillage system. Another approach to solving this problem is to ridge the soil, permit water to drain out of the ridge, and then plant on the drier, warmer ridgetop (or on the south side of the ridge—see Section 7.9).

Advantages in Warm Climates. In warm regions, delayed planting is not a problem. In fact, the cooler near-surface soil temperatures under a mulch may reduce heat stress on roots during summer. Plant-residue mulches also conserve soil moisture by decreasing evaporation. The resulting cooler, moist surface layer of soil is an important part of no-tillage systems because it allows roots to proliferate in this zone, where nutrient and aeration conditions are optimal.

Plastic Mulches

One of the reasons for the popularity of plastic mulches for gardens and high-value specialty crops is their effect on soil temperature. In contrast to the organic mulches, plastic mulches generally increase soil temperature, clear plastic having a greater heating effect than black plastic. In temperate regions, this effect can be used to extend the growing season or to hasten production to take advantage of the higher prices offered by early-season markets (Figure 7.26).

In warmer climates, and during the summer months, the soil-heating effect of plastic mulches may be quite detrimental, inhibiting root growth in the upper soil layers and sometimes seriously decreasing crop yields (see Table 7.5, for example). Obviously, the weed-control and moisture-conservation effects of plastic mulches may be outweighed in some circumstances by their excessive heating effects.

FIGURE 7.26 These winter-grown southern California strawberries will come to market when prices are still high because of the effect of the clear plastic mulch on soil temperature. (Photo courtesy of R. Weil)

TABLE 7.5 **Soil Temperature and Tomato Yield with Straw or Black Plastic Mulch**

The data are averages for 2 years of tomato production on a sandy loam Ultisol near Griffin, Georgia. The straw kept the surface soil from rising to detrimentally high temperatures, while it also increased infiltration of rainwater and reduced soil compaction. Daily drip irrigation supplied plenty of water, but could not overcome the temperature effects of the black plastic mulch.

	Not irrigated		*Irrigated daily*	
	Straw mulch	*Plastic mulch*	*Straw mulch*	*Plastic mulch*
Average soil temperature, °C[a]	24	37	24	35
Tomato yield, Mg/ha	68	30	70	24

[a]Soil temperature measured at 5 cm below the soil surface, average of weeks 2–10 of the growing season.
Data calculated from Tindall et al. (1991).

Moisture Control

Another means of exercising some control over soil temperature is by controlling soil moisture. Poorly drained soils in temperate regions that are wet in the spring have temperatures 3° to 6°C lower than comparable well-drained soils. Removing the excess water by installing drainage systems (see Section 6.9) or using ridging systems of tillage can cause the soil to warm quickly in the spring.

7.12 CONCLUSION

Soil aeration and soil temperature critically affect the quality of soils as habitats for plants and other organisms. Most plants have definite requirements for soil oxygen along with limited tolerance for carbon dioxide, methane, and other such gases found in poorly aerated soils. Some microbes, such as the nitrifiers and general-purpose decay organisms, are also constrained by low levels of soil oxygen.

Soils with extremely wet moisture regimes are unique with respect to their morphology and chemistry and to the plant communities they support. Such hydric soils are characteristic of wetlands and help these ecosystems perform a myriad of valuable functions.

Plants as well as microbes are also quite sensitive to differences in soil temperature, particularly in temperate climates where low temperatures can limit essential biological processes. Soil temperature also impacts on the use of soils for engineering purposes, again primarily in the cooler climates. Frost action, which can move perennial plants such as alfalfa out of the ground, can do likewise to building foundations, fence posts, sidewalks, and highways.

Soil water exerts a major influence over both soil aeration and soil temperature. It competes with soil air for the occupancy of soil pores and interferes with the diffusion of gases into and out of the soil. Soil water also resists changes in soil temperature by virtue of its high specific heat and its high energy requirement for evaporation.

The tendency of a soil to retain water is largely regulated by the content of colloidal particles of clay and humus, soil components on which the following chapter is focused.

STUDY QUESTIONS

1. What are the two principal gases involved with soil aeration, and how do their relative amounts change as one samples deeper into a soil profile?
2. It is sometimes said that organisms in anaerobic environments will use the combined oxygen in nitrate or sulfate instead of the oxygen in O_2. Why is this statement incorrect? What actually happens when organisms reduce sulfate or nitrate?
3. If an alluvial forest soil were flooded for 10 days and you sampled the gases evolving from the wet soil, what gases would you expect to find (other than oxygen and carbon dioxide)? In what order of appearance? Explain.
4. Explain why warm weather during periods of saturation is required to form a hydric soil.
5. If you were in the field trying to delineate the so-called drier end of a wetland area, what are three soil properties and three other indicators that you might look for?
6. What are the three major components that define a wetland?
7. Discuss four plant processes that are influenced by soil temperature.
8. Explain how a brush fire might lead to subsequent mudslides, as often occurs in California.
9. If you were to build a house below ground, to save heating and cooling costs, would you firmly compact the soil around the house? Explain your answer.
10. If you measured a daily maximum air temperature of 28°C at 1 P.M., what might you expect the daily maximum temperature to be at a 15 cm depth in the soil? At about what time of day would the maximum temperature occur at this depth? Explain.
11. In relation to soil temperature, explain why conservation tillage has been more popular in Missouri than in Minnesota.

REFERENCES

Arshad, A., and R. H. Azooz. 1996. "Tillage effects on soil thermal properties in a semi-arid cold region," *Soil Sci. Soc. Am. J.* **60**:561–567.

Bartlett, R. J., and B. R. James. 1993. "Redox chemistry of soils," *Advances in Agronomy* **50**:151–208.

Beard, J. B., and R. L. Green. 1994. "The role of turfgrasses in environmental protection and their benefits to humans," *J. Environ. Qual.* **23**:452–460.

CAST. 1994. *Wetland Policy Issues*. Publication No. CC1994-1. (Ames, Iowa: Council for Agricultural Science and Technology).

Davidson, E. A., and S. E. Trumbore. 1995. "Gas diffusivity and production of CO_2 in deep soils of the eastern Amazon," *Tellus* **47B**:550–565.

Dryness, C. T. 1976. "Effects of wildfire on soil wetability in the high cascades of Oregon," USDA Forest Service Research Paper PNW-202. (Washington, D.C.: USDA).

Fluker, B. J. 1958. "Soil temperature," *Soil Sci.* **86**:35–46.

Hurt, G. W., P. M. Whited, and R. F. Pringle (eds.). 1996. *Field Indicators of Hydric Soils in the United States.* (Fort Worth, Tex.: USDA Natural Resources Conservation Service).

Losi, M. E., C. Amrheim, and W. T. Frankenberger Jr. 1994. "Bioremediation of chromatic-contaminated groundwater by reduction and precipitation in surface soils," *J. Environ. Qual.* **23**:1141–1150.

Lowe, D. F., C. L. Oubre, and C. H. Ward (eds.). 2000. *Soil Vapor Extraction Using Radio Frequency Heating: Resource Manual and Technology Demonstration.* (New York: Lewis Publishers).

MacDonald, N. W., D. R. Zac, and K. S. Pregitzer. 1995. "Temperature effects on kinetics of microbial respiration and net nitrogen and sulfur mineralization," *Soil Sci. Soc. Amer. J.* **59**:233–240.

McBride, M. B. 1994. *Environmental Chemistry of Soils.* (New York: Oxford University Press).

Patrick, W. H., Jr., and A. Jugsujinda. 1992. "Sequential reduction and oxidation of inorganic nitrogen, manganese, and iron in flooded soil," *Soil Sci. Soc. Amer. J.* **56**:1071–1073.

Richardson, J. L., and M. J. Vepraskas. 2001. *Wetland Soils—Genesis, Hydrology, Landscapes, and Classification.* (Boca Raton, Fla.: Lewis).

Sellers, P., et al. 1997. "Modeling the exchange of energy, water, and carbon between continents and the atmosphere," *Science* **275**:502–509.

Sexstone, A. J., et al. 1985. "Direct measurement of oxygen profiles and denitrification rates in soil aggregates," *Soil Sci. Soc. Amer. J.* **49**:645–651.

Tiarks, A. E., W. H. Hudnall, J. F. Ragus, and W. B. Patterson. 1996. "Effect of pine plantation harvesting and soil compaction on soil water and temperature regimes in a semi-tropical environment," in A. Schulte and D. Ruhiyat (eds.), *Proceedings of International Congress on Soils of Tropical Forest Ecosystems 3rd Conference on Forest Soils: Vol. 3, Soil and Water Relationships.* (Samarinda, Indonesia: Mulawarmon University Press).

Tindall, J. A., R. B. Beverly, and D. E. Radcliff. 1991. "Mulch effect on soil properties and tomato growth using micro-irrigation," *Agron. J.* **83**:1028–1034.

Trettin, C. C., M. Davidian, M. F. Jurgensen, and R. Lea. 1996. "Organic matter decomposition following harvesting and site preparation of a forested wetland," *Soil Sci. Soc. Am. J.* **60**:1994–2003.

Unger, P. W. 1978. "Straw mulch effects on soil temperatures and sorghum germination and growth," *Agron. J.* **70**:858–864.

Veneman, P. L. M., D. L. Lindbo, and L. A. Spokas. 1999. "Soil moisture and redoximorphic features: A historical perspective," in M. J. Rabenhorst, J. C. Bell, and P. A. McDaniel (eds.), *Quantifying Soil Hydromorphology.* Special Publication No. 54. (Madison, Wis.: Soil Sci. Soc. Amer.).

Welsh, D., D. Smart, J. Boyer, P. Minkin, H. Smith, and T. McCandless (eds.). 1995. *Forested Wetlands: Functions, Benefits, and Use of Best Management Practices.* (Radnor, Pa.: USDA Forest Service).

Wuethrich, B. (2000). "When permafrost isn't," *Smithsonian,* Comments and Notes, February 2000.

Yavitt, J. B., T. J. Fahley, and J. A. Simmons. 1995. "Methane and carbon dioxide dynamics in a northern hardwood ecosystem," *Soil. Sci. Soc. Am. J.* **59**:796–804.

Mica weathers to clay.(Serge Jolicoeur, Université de Moncton)

8 The Colloidal Fraction: Seat of Soil Chemical and Physical Activity

The landscape of the clays is like—the intricate folds of the womb—whose activity is to receive, contain, enfold, and give birth.
—William Bryant Logan

In the aftermath of a nuclear power plant accident, why would milk, vegetables, and grains produced on some soils contain dangerous levels of radioactivity, while the same foods produced on other soils would be safe to eat? How does the practice of irrigating certain soils with polluted sewage water lead to clean water for the recharge of groundwater aquifers? The answers to these and many other environmental mysteries lie in the nature of the smallest of soil particles, the clay and humus **colloids.** These particles are not just extra-small fragments of rock and organic matter. They are highly reactive materials with electrically charged surfaces. Because of their size and shape, they give the soil an enormous amount of reactive **surface area.** It is the colloids, then, that allow the soil to serve as nature's great electrostatic chemical reactor.

Each tiny colloid particle carries a swarm of positively and negatively charged ions (cations and anions) attracted to electrostatic charges on its surface. The ions are held tightly enough by the **soil colloids** to greatly reduce their loss in drainage waters, but loosely enough to allow plant roots access to the nutrients among them. Other modes of adsorption bind ions more tightly, so that they are no longer available for plant uptake, reaction with the soil solution, or leaching loss to the environment. In addition to plant nutrient ions, soil colloids also bind with water molecules, biomolecules, viruses, toxic metals, pesticides, and a host of other mineral and organic substances. Hence, soil colloids greatly influence nearly all ecosystem functions in the soil.

We shall see that different soils are endowed with different types of clays that, along with humus, elicit very different types of physical and chemical behaviors. Certain clay minerals are much more reactive than others are, and some are more dramatically influenced by the acidity of the soil and other environmental factors. Studying the soil colloids in some depth will deepen your understanding of soil architecture (Chapter 4) and soil water (Chapters 5 and 6). Knowledge of structure, origin, and behavior of the different types of soil colloids will also help you understand soil chemical and biological processes, and make appropriate decisions regarding the use of soil resources.

8.1 GENERAL PROPERTIES AND TYPES OF SOIL COLLOIDS

Size

The clay and humus particles in soils are referred to collectively as the colloidal fraction because of their extremely small size and colloid-like behavior. Too small to be seen with an ordinary light microscope, they can be made visible only with an electron microscope. Particles behave as colloids if they are less than about 1 μm (0.000001 meter) in diameter, although some soil scientists consider 2 μm to mark the upper boundary of the colloidal fraction to coincide with the definition of clay (see Section 4.2).

Surface Area

As discussed in Section 4.2, the smaller the size of the particles in a given mass of soil, the greater the surface area exposed for adsorption, catalysis, precipitation, microbial colonization, and other surface phenomena. Because of their small size, all soil colloids expose a large **external surface** area per unit mass, more than 1000 times the surface area of the same mass of sand particles. Some silicate clays also possess extensive **internal surface** area between the layers of their platelike crystal units.

The total surface area of soil colloids ranges from 10 m^2/g for clays with only external surfaces, to more than 800 m^2/g for clays with extensive internal surfaces. To put this in perspective, we can calculate that the surface area exposed within 1 ha (about the size of a football field) of a 1.5-m-deep fine-textured soil (45% clay) might be as great as 8,700,000 km^2 (the land area of the entire United States).

Surface Charges

The internal and external surfaces of soil colloids carry positive and/or negative electrostatic charges. For most soil colloids, electronegative charges predominate, although some mineral colloids in very acid soils have a net electropositive charge. As we shall see in Sections 8.3 to 8.7, the amount and origin of surface charge differs greatly among the different types of soil colloids and, in some cases, is influenced by changes in chemical conditions, such as soil pH. The charges on the colloid surfaces attract or repulse substances in the soil solution as well as neighboring colloid particles. These reactions, in turn, greatly influence soil chemical and physical behavior.

Adsorption of Cations and Anions

Of particular significance is the attraction of positively charged ions (**cations**) to the surfaces of negatively charged soil colloids. Each colloid particle (sometimes referred to as a **micelle** or microcell) attracts thousands of Al^{3+}, Ca^{2+}, Mg^{2+}, K^+, H^+, and Na^+ ions and lesser numbers of other cations. The cations in moist soils exist in the *hydrated* state (surrounded by a shell of water molecules), but for simplicity in this textbook, we will show just the cations (e.g., Ca^{2+} or H^+) rather than the hydrated forms (e.g., $Ca(H_2O)_6{}^{2+}$ or the hydronium ion, H_3O^+). These hydrated cations constantly vibrate about in a swarm near the colloid surface, held there by electrostatic attraction to the colloid's negative charges. Frequently, an individual cation will break away from the swarm and move out to the soil solution. When this happens, another cation of equal charge will simultaneously move in from the soil solution and take its place. This process of **cation exchange** will be discussed in detail (Section 8.8) because of its fundamental importance in nutrient cycling and other environmental processes. The cations swarming about near the colloidal surface are said to be **adsorbed** (loosely held) on the colloid surface. Because these cations can *exchange places* with those moving freely about in the soil solution, the term **exchangeable ions** is also used to refer to the ions in this adsorbed state.

The colloid with its adsorbed cations is sometimes described as an **ionic double layer** in which the negatively charged colloid micelle acts as a huge anion constituting the inner ionic layer, and the swarm of adsorbed cations constitutes the outer ionic layer (Figure 8.1). Because cations from the soil solution are constantly trading places with those that are adsorbed to the colloid, the ionic composition of the soil solution reflects that of the adsorbed swarm.

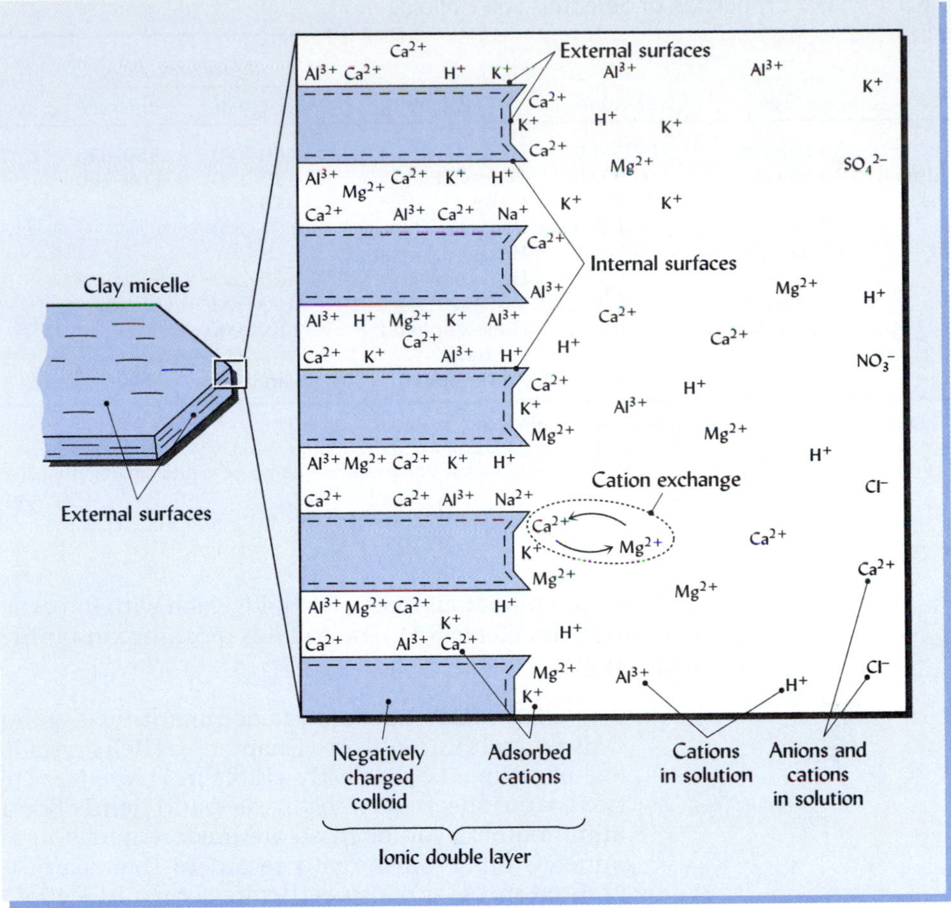

FIGURE 8.1 Simplified representation of a silicate clay crystal (micelle), its complement of adsorbed cations, and ions in the surrounding soil solution. The enlarged view (right) shows that the clay comprises sheetlike layers with both external and internal negatively charged surfaces. The negatively charged micelle acts as a huge anion and a swarm of positively charged cations is adsorbed to the micelle because of attraction between charges of opposite sign. Cation concentration decreases with distance from the clay. Anions (such as Cl^-, NO_3^-, and SO_4^{2-}), which are repulsed by the negative charges, can be found in the bulk soil solution farthest from the clay (far right). Some clays (not shown) also exhibit positive charges that can attract anions.

Anions such as Cl^-, NO_3^-, and SO_4^{2-} (also surrounded by water molecules, though, again, we do not show these water shells) may also be attracted to certain soil colloids that have *positive* charges on their surfaces. While adsorption of **exchangeable anions** is not as extensive as that for exchangeable cations, we shall see (Section 8.11) that it is an important mechanism for holding negatively charged constituents, especially in acid subsoils. When thinking about the colloids in soil, we should always keep in mind that they carry with them a complement of exchangeable cations and anions, along with certain other more tightly bound ions and molecules.

Adsorption of Water

In addition to adsorbing cations and anions, soil colloids attract and hold a large number of water molecules. The charges on the internal and external colloid surfaces attract the oppositely charged end of the polar water molecule. Some water molecules are attracted to the exchangeable cations, each of which is hydrated with a shell of water molecules. Water adsorbed between the crystal layers can cause the layers to move apart, making the clay more plastic and swelling its volume. Colloids that adsorb a great deal of water may make soil unsuitable for construction purposes (see Sections 4.9 and 8.13). As a soil colloid dries, any water in the internal surfaces is removed, and the layers are brought closer together. Generally, the greater the external surface area of the soil colloids, the greater the amount of water held when the soil is air-dry.

TABLE 8.1 Major Properties of Selected Soil Colloids

Colloid	Type	Size, μm	Shape	Surface area, m^2/g External	Surface area, m^2/g Internal	Interlayer spacing[a], nm	Net charge[b], $cmol_c/kg$
Smectite	2:1 silicate	0.01–1.0	Flakes	80–150	550–650	1.0–2.0	−80 to −150
Vermiculite	2:1 silicate	0.1–0.5	Plates, flakes	70–120	600–700	1.0–1.5	−100 to −200
Fine mica	2:1 silicate	0.2–2.0	Flakes	70–175	—	1.0	−10 to −40
Chlorite	2:1 silicate	0.1–2.0	Variable	70–100	—	1.41	−10 to −40
Kaolinite	1:1 silicate	0.1–5.0	Hexagonal crystals	5–30	—	0.72	−1 to −15
Gibbsite	Al-oxide	<0.1	Hexagonal crystals	80–200	—	0.48	+10 to −5
Goethite	Fe-oxide	<0.1	Variable	100–300	—	0.42	+20 to −5
Allophane & Imogolite	Noncrystalline silicates	<0.1	Hollow spheres or tubes	100–1000	—	—	+20 to −150
Humus	Organic	0.1–1.0	Amorphous	Variable[c]	—	—	−100 to −500

[a]From the top of one layer to the next similar layer, 1 nm = 10^{-9} m = 10 Å.
[b]Centimoles of charge per kilogram of colloid ($cmol_c/kg$), a measure of ion exchange capacity (see Section 8.9).
[c]It is very difficult to determine the surface area of organic matter. Different procedures give values ranging from 20 to 800 m^2/g.

Types of Soil Colloids[1]

Soils contain numerous types of colloids, each with its particular composition, structure, and properties (Table 8.1). The colloids most important in soils can be grouped in four major types:

1. *Crystalline silicate clays* are the dominant type in most soils (except in Andisols, Oxisols, and Histosols—see Chapter 3). Their crystalline structure is layered much like pages in a book (clearly visible in Figure 8.2*a*). Each layer (page) consists of two to four sheets of closely packed and tightly bonded oxygen, silicon, and aluminum atoms. Although all are predominately negatively charged, silicate clay minerals differ widely with regard to their particle shapes (**kaolinite,** a **fine-grained mica,** and a **smectite** are shown in Figure 8.2*a–c*), intensity of charge, stickiness, plasticity, and swelling behavior.
2. *Noncrystalline silicate clays* also consist mainly of tightly bonded silicon, aluminum, and oxygen atoms, but they do not exhibit ordered, crystalline sheets. The two principal clays of this type, **allophane** and **imogolite,** usually form from

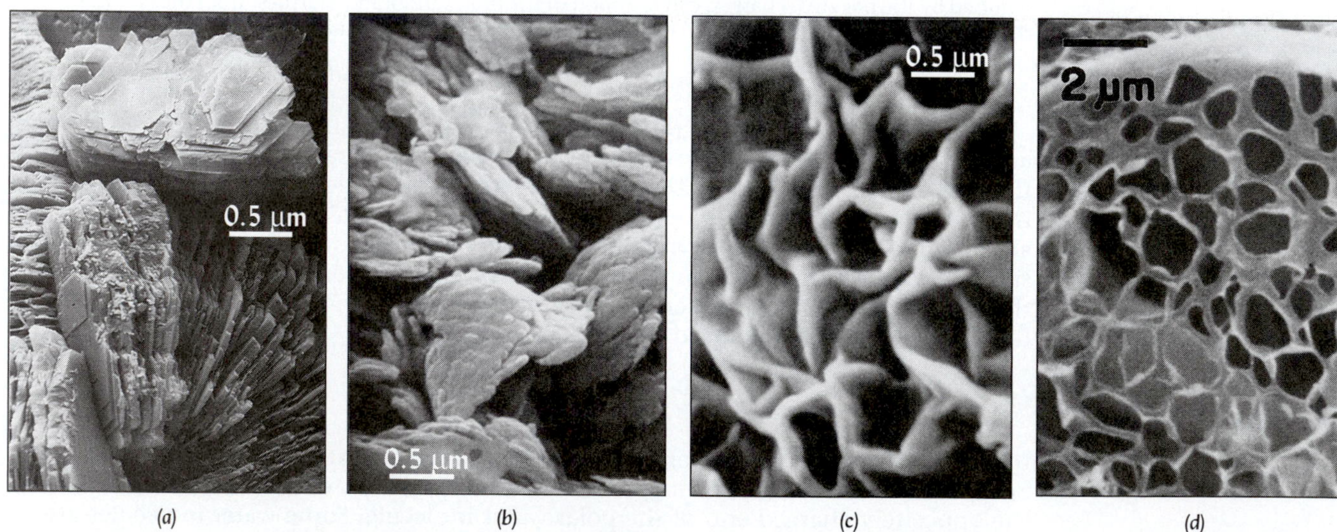

FIGURE 8.2 Crystals of three silicate clay minerals and a photomicrograph of humic acid found in soils. (*a*) Kaolinite from Illinois (note hexagonal crystal at upper right). (*b*) A fine-grained mica from Wisconsin. (*c*) Montmorillonite (a smectite group mineral) from Wyoming. (*d*) Fulvic acid (a humic acid) from Georgia. [(*a*)–(*c*) Courtesy Dr. Bruce F. Bohor, Illinois State Geological Survey; (*d*) from Dr. Kim H. Tan, University of Georgia; used with permission of Soil Science Society of America]

[1] For a review of the composition and properties of clays and humus, see Dixon and Schulze (2002).

volcanic ash and are characteristic of Andisols (Section 3.7). They have high amounts of both positive and negative charge, and high water-holding capacities. Although malleable (plastic) when wet, they exhibit a very low degree of stickiness. Allophane and imogolite are also known for their extremely high capacities to strongly adsorb phosphate and other anions, especially under acid conditions.

3. *Iron and aluminum oxide* clays are found in many soils, but are especially important in the more highly weathered soils of warm, humid regions (e.g., Ultisols and Oxisols). They consist mainly of either iron or aluminum atoms coordinated with oxygen atoms (the latter are often associated with hydrogen ions to make hydroxyl groups). Some, like **gibbsite** (an Al-oxide) and **goethite** (an Fe-oxide) consist of crystalline sheets. Other oxide minerals are noncrystalline, often occurring as amorphous coatings on soil particles. The oxide colloids are relatively low in plasticity and stickiness. Their net charge ranges from slightly negative to moderately positive.
4. *Organic (humus)* colloids are important in nearly all soils, especially in the upper parts of the soil profile. Humus colloids are not minerals, nor are they crystalline (Figure 8.2*d*). Instead, they consist of convoluted chains and rings of carbon atoms bonded to hydrogen, oxygen, and nitrogen. Humus particles are often among the smallest of soil colloids and exhibit very high capacities to adsorb water, but almost no plasticity or stickiness. Because humus is noncohesive, soils composed mainly of humus (Histosols) have very little bearing strength and are unsuitable for making building or road foundations. Humus has high amounts of both negative and positive charge per unit mass, but the net charge is always negative and varies with soil pH.

8.2 FUNDAMENTALS OF LAYER SILICATE CLAY STRUCTURE[2]

Slide set showing silicate mineral structures:
http://www.cs.odu.edu/~manch_v/fall2001/ch8/sld007.htm

To see why soils rich in one silicate clay mineral, say kaolinite, behave so very differently from soils dominated by another silicate clay, say montmorillonite, it is necessary to understand the main structural features of the silicate clay minerals. We will begin by examining the main building blocks from which the layer silicates are constructed, then consider the particular arrangements that give rise to the all-important surface charges.

Silicon Tetrahedral and Aluminum-Magnesium Octahedral Sheets

The most important silicate clays are known as **phyllosilicates** (Greek *phyllon,* leaf) because of their leaflike or planar structure. As shown in Figure 8.3, they are composed of two kinds of horizontal **sheets.**

TETRAHEDRAL SHEETS. This kind of sheet consists of two **planes** of oxygens with mainly silicon in the spaces between the oxygens. The basic building block for the tetrahedral sheet is a unit composed of one silicon atom surrounded by four oxygen atoms. It is called a **tetrahedron** because (as shown in Figure 8.3, top left) the oxygens define the apices of a *four*-sided geometric solid that resembles a pyramid (having three "sides" and the base). An interlocking array of such tetrahedra, each sharing its basal oxygens with its neighbor, give rise to a **tetrahedral sheet.**

Three-dimensional rotatable models of silicate clay minerals and their building blocks:
http://soils1.cses.vt.edu/MJE/CSES3124/VR_exports/index.html

OCTAHEDRAL SHEETS. The building block for the second kind of sheet is composed of six oxygen atoms coordinating with a central aluminum or magnesium atom, forming the shape of an *eight*-sided geometric solid, or **octahedron.** Numerous octahedra linked together horizontally constitute the **octahedral sheet.**

The tetrahedral and octahedral sheets are the fundamental structural units of silicate clays. Two to four of these sheets may be stacked together in sandwich-like arrangements, with adjacent sheets strongly bound together by sharing some of the same oxygen atoms (see Figure 8.3). The specific nature and combination of sheets in these layers vary from one type of clay to another and largely control the physical and chemical properties exhibited. The relationship between *planes, sheets,* and *layers* shown in Figure 8.3 should be carefully studied.

[2] The authors are indebted to Dr. Darrel G. Schultze of Purdue University for kindly providing the structural models for the silicate clay minerals.

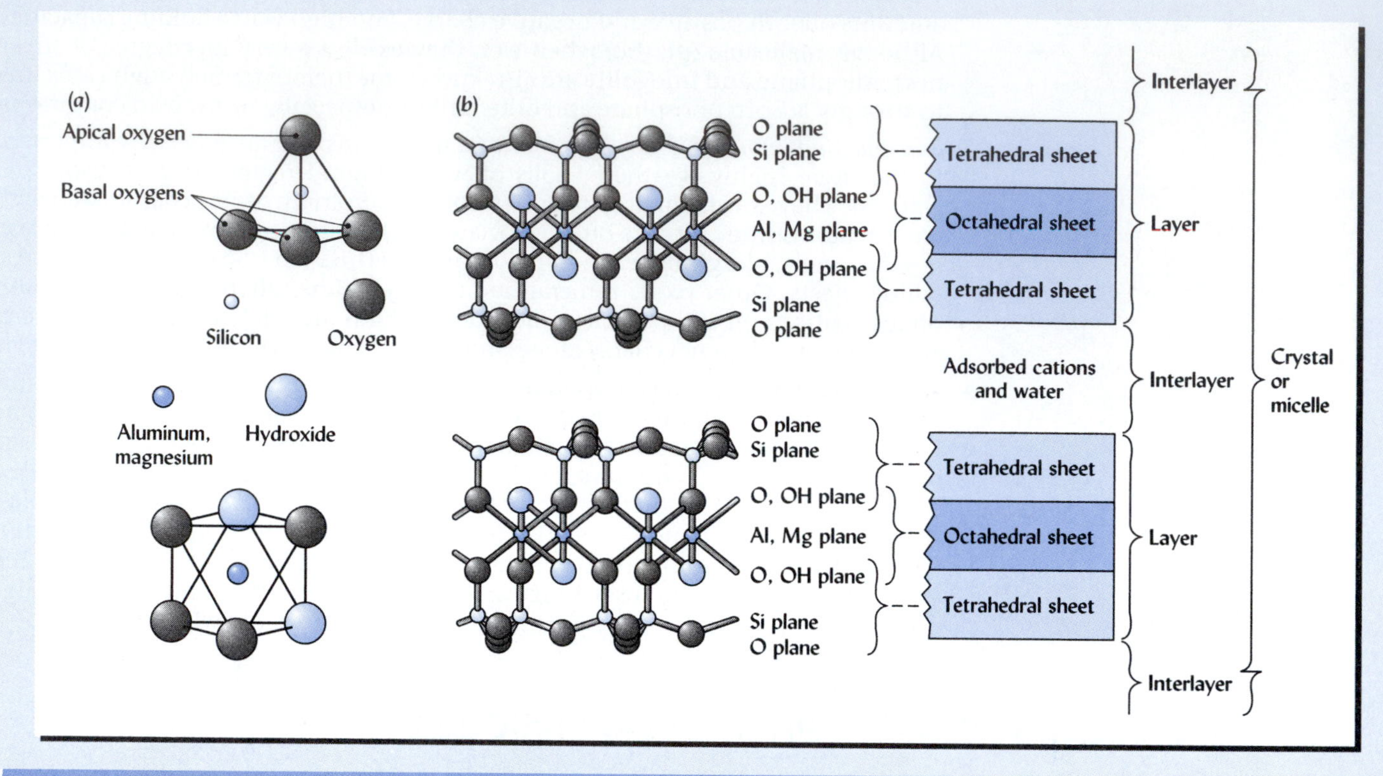

FIGURE 8.3 The basic molecular and structural components of silicate clays. (*a*) A single *tetrahedron*, a four-sided building block composed of a silicon ion surrounded by four oxygen atoms; and a single eight-sided *octahedron*, in which an aluminum (or magnesium) ion is surrounded by six hydroxy groups or oxygen atoms (the number of each varying for different minerals). (*b*) In clay crystals thousands of these tetrahedral and octahedral building blocks are connected to give planes of silicon and aluminum (or magnesium) ions. These planes alternate with planes of oxygen atoms and hydroxy groups. Note that apical oxygen atoms are common to adjoining tetrahedral and octahedral sheets. The silicon plane and associated oxygen-hydroxy planes make up a *tetrahedral sheet*. Similarly, the aluminum-magnesium plane and associated oxygen-hydroxy planes constitute the *octahedral sheet*. Different combinations of tetrahedral and octahedral sheets are termed *layers*. In some silicate clays these layers are separated by *interlayers* in which water and adsorbed cations are found. Many layers are found in each *crystal* or *micelle* (microcell).

Isomorphous Substitution

The structural arrangements just described suggest a very simple relationship among the elements making up silicate clays. In nature, however, clays have formulas that are more complex. During the weathering of rocks and minerals, many different elements are present in the weathering solution. As the clay minerals crystallize, cations of comparable size (see Table 8.2) may substitute for silicon, aluminum, and magnesium ions in the respective tetrahedral and octahedral sheets.

Note from Table 8.2 that aluminum is only slightly larger than silicon. Consequently, aluminum can fit into the center of the tetrahedron in the place of the silicon without much change in the basic structure of the crystal. This process by which one element fills a position usually filled by another of similar size is called **isomorphous substitution.** This phenomenon is responsible for much of the variability in the nature of silicate clays.

Isomorphous substitution can also occur in the octahedral sheets. For example, iron and zinc ions are not much different in size from aluminum and magnesium ions (Table 8.2). Any of these ions can fit in the central ion in the octahedra of the octahedral sheet. Thus, isomorphous substitution can occur in tetrahedral and in octahedral sheets. In some layer silicates, it occurs in both.

Source of Charges

Isomorphous substitution is of vital importance because it is the primary source of both negative and positive charges of silicate clays. For example, the Mg^{2+} ion is only slightly larger than the Al^{3+} ion, but it has one less positive charge. If a Mg^{2+} ion substitutes for an Al^{3+} ion in an octahedral sheet, there will be insufficient positive charges to balance

TABLE 8.2 Ionic Radii and Location of Elements Found in Silicate Clays

Ion	*Radius, nm*[a]	*Found in*
Si^{4+}	0.042	Tetrahedral sheet (Si^{4+}, Al^{3+}, Fe^{3+})
Al^{3+}	0.051	Octahedral sheet (Al^{3+}, Fe^{3+}, Mg^{2+}, Fe^{2+})
Fe^{3+}	0.064	
Mg^{2+}	0.066	Exchange sites (Al^{3+} to Ca^{2+})
Fe^{2+}	0.070	
Zn^{2+}	0.074	
Na^{+}	0.097	
Ca^{2+}	0.099	
K^{+}	0.133	
O^{2-}	0.140	Both sheets (O^{2-}, OH^{-})
OH^{-}	0.155	

[a]1 nm = 10^{-9}m.

the negative charges from the oxygens; hence, the lattice is left with a -1 net charge (see Figure 8.4, right). Similarly, every Al^{3+} that substitutes for a Si^{4+} in a tetrahedral sheet creates a net negative charge at that site because the negative charges from the four oxygens will be only partially balanced. As we shall see (Sections 8.3 and 8.6), additional, more temporary charges can also develop on the $\gtrapprox$ Si—O or Al—OH groups exposed at the micelle edges of the tetrahedral and octahedral surfaces, respectively.

8.3 MINERALOGICAL ORGANIZATION OF SILICATE CLAYS

Based on the number and arrangement of tetrahedral (Si) and octahedral (Al, Mg, Fe) sheets contained in the crystal units or layers, crystalline clays may be classed into two main groups: **1:1 silicate clays,** in which each layer contains *one* tetrahedral and *one* octahedral sheet, and **2:1 silicate clays,** in which each layer has *one* octahedral sheet sandwiched between *two* tetrahedral sheets.

1:1-Type Silicate Clays

The 1:1 silicate clays include **kaolinite,** *halloysite, nacrite,* and *dickite.* To illustrate the properties of 1:1 silicate clays we will focus on kaolinite, which is by far the most common in soils.

As implied by the term *1:1 silicate clay,* each kaolinite layer consists of one silicon tetrahedral sheet and one aluminum octahedral sheet. The two types of sheets are tightly held together because the apical oxygen atom (the oxygen atom that forms the

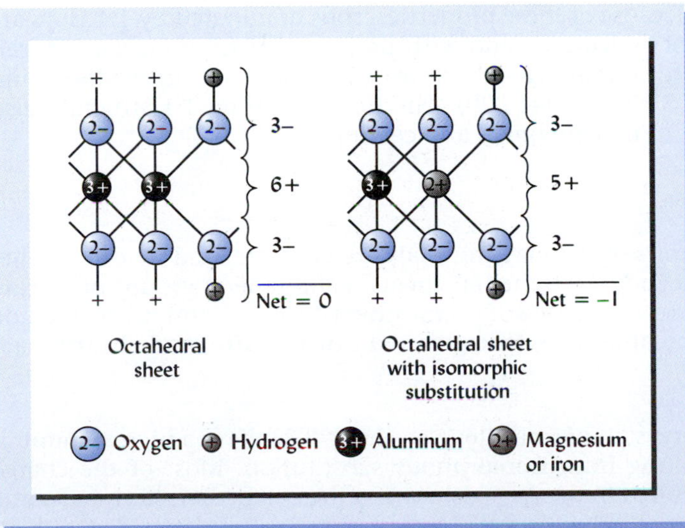

FIGURE 8.4 Simplified diagrams of octahedral sheets in a silicate mineral illustrating the effect of isomorphous substitution on the net charge. Note that for each oxygen atom, one of the two − charges is balanced by a + charge from either a H^{+} (making a hydroxyl group) or a Si atom in the tetrahedral sheet (not shown, but represented by a +). During the crystallization of the octahedral sheet shown on the right, a Mg^{2+} atom occupied one of the positions normally occupied by an Al^{3+} atom, thus leaving a -1 net charge on the sheet. Such net negative charges in the crystal can be balanced by cations from the soil solution adsorbed to the crystal surface.

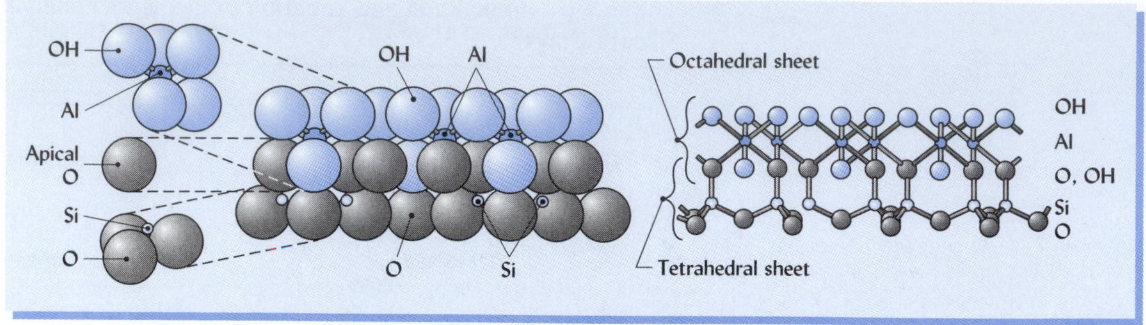

FIGURE 8.5 Models of the 1:1-type clay kaolinite. The primary elements of the octahedral (upper left) and tetrahedral (lower left) sheets are depicted as they might appear separately. In the crystal structure, however, these sheets are held together by mutually shared *apical* oxygen atoms. Note that a layer consists of one octahedral and one tetrahedral sheet—hence, the designation 1:1. The result is a layer with hydroxyls on one surface and oxygens on the other. A schematic drawing of the ionic arrangement (right) shows a cross-sectional view of a crystal layer. Note the common apical oxygen ions that hold the sheets together. The kaolinite mineral is comprised of a stack of these flat layers tightly held together (with no interlayer spaces) by the attraction of the hydroxyl plane on the top of one layer to the oxygen plane on the bottom of the next layer.

apex or tip of the "pyramid") in each tetrahedron also forms a bottom corner of one or more of the octahedra in the adjoining sheet (Figure 8.5). Note that because a kaolinite crystal layer consists of these two sheets, it exposes a plane of oxygen atoms on the bottom surface, but a plane of hydroxyls on the upper surface.

Learn how clay structures are identified by downloading X-ray diffraction program: http://neon.mems.cmu.edu/degraef/xray/

This arrangement has two very important consequences. First, as will be discussed in Section 8.6, where the hydroxyl plane is exposed on the clay particle surface, removal or addition of hydrogen ions can produce either positive or negative charges, depending on the pH of the soil. The exposed **hydroxylated surface** can also react with and strongly bind specific anions. Second, when the layers consisting of alternating tetrahedral and octahedral sheets are stacked on top of one another, the hydroxyls of the octahedral sheet in one layer are adjacent to the oxygens of the tetrahedral sheet of the next layer, permitting adjacent layers to be bound together by **hydrogen bonding** (see Section 5.1).

Because of the interlayer hydrogen bonding, the structure of kaolinite is fixed, and no expansion can occur between the layers when the clay is wetted. Cations and water do not enter between the structural layers of a 1:1 mineral particle. The effective surface of kaolinite is thus restricted to its outer faces or external surface area. This fact and the lack of significant isomorphous substitution in this mineral account for the relatively small capacity of kaolinite to adsorb exchangeable cations (see Table 8.1).

3-D, movable model of kaolinite with menu to highlight parts: http://www.soils.umn.edu/virtual_museum/kaolinite/index.html

Kaolinite crystals are usually hexagonal in shape (see Figure 8.2*a*) and larger than most other clays (Table 8.1). In contrast to some 2:1 silicate clays, 1:1 clays like kaolinite exhibit less plasticity, stickiness, cohesion, shrinkage, and swelling and can also hold less water than other clays. Because of these properties, soils dominated by 1:1 clays are relatively easy to cultivate for agriculture, and with proper nutrient management, can be quite productive. Kaolinite-containing soils are well suited for use in roadbeds and building foundations (Plate 33 after page 000). The nonexpanding 1:1 structure also makes kaolinite clays useful for making bricks and ceramics.

Expanding 2:1-Type Silicate Clays

As mentioned, the crystal units in 2:1 clay minerals are characterized by *one* octahedral sheet sandwiched between *two* tetrahedral sheets. Four general groups of silicate clays have this basic crystal structure. Two of them, **smectite** and **vermiculite**, include expanding-type minerals; the other two, **fine-grained micas** (**illite**) and **chlorite**, are relatively nonexpanding.

SMECTITE GROUP. The flakelike crystals of smectites (see Figure 8.2*c*) have a high amount of mostly negative charge resulting from isomorphous substitution. Most of the charge derives from Mg^{2+} ions substituted in the Al^{3+} positions of the octahedral sheet, but some also derives from substitution of Al^{3+} ions for Si^{4+} in the tetrahedral sheets (Figure 8.6).

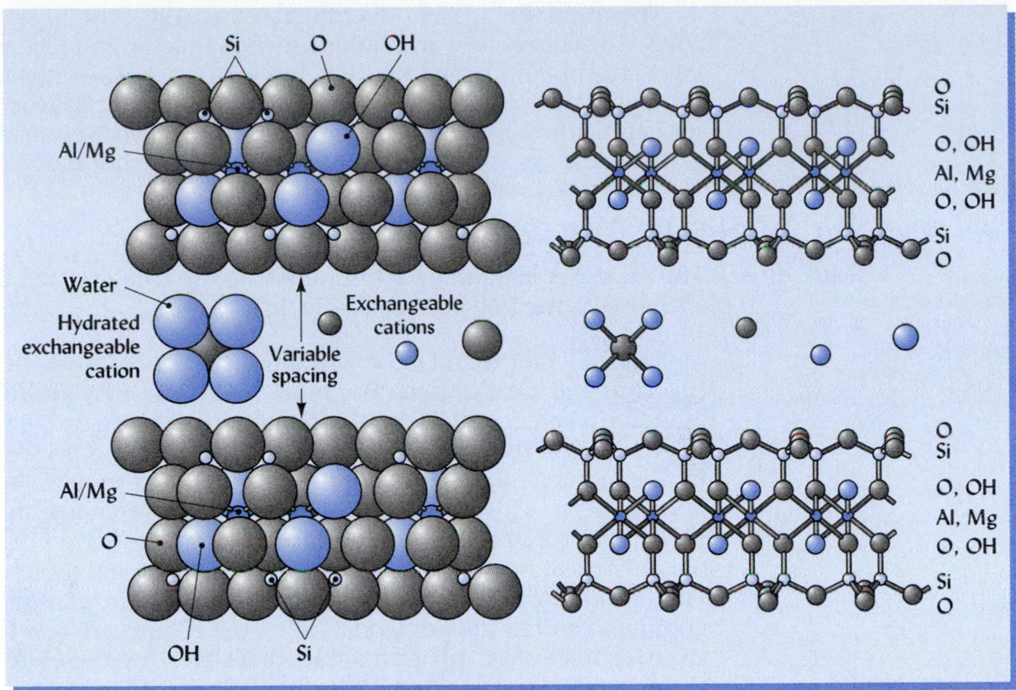

FIGURE 8.6 Model of two crystal layers and an interlayer characteristic of montmorillonite, a smectite expanding-lattice 2:1-type clay mineral. Each layer is made up of an octahedral sheet sandwiched between two tetrahedral sheets with shared apical oxygen ions that hold the sheets together. There is little attraction between oxygen atoms in the bottom tetrahedral sheet of one unit and those in the top tetrahedral sheet of another. This permits a variable space between layers, which is occupied by water and exchangeable cations. The internal surface area thus exposed far exceeds the surface around the outside of the crystal. Note that magnesium has replaced aluminum in some sites of the octahedral sheet. Likewise, some silicon atoms in the tetrahedral sheet may be replaced by aluminum (not shown). These substitutions give rise to a negative charge, which accounts for the high cation exchange capacity of this clay mineral. A ball-and-stick model of the atoms and chemical bonds is at the right.

Because of these substitutions the capacity to adsorb cations is very high—about 20 to 40 times that of kaolinite.

In contrast to the situation just described for kaolinite, in the 2:1 structure of smectites both the top and bottom planes of each layer consist of oxygen atoms. Therefore, adjacent layers are only loosely bound to each other by very weak oxygen-to-oxygen and cation-to-oxygen linkages (Figure 8.6). Exchangeable cations and associated water molecules are strongly attracted to the spaces between the layers (interlayer spaces). The crystal expands as the water pushes apart the layers. The internal surface area thus exposed by far exceeds the external surface area of these minerals and contributes to the very high total **specific surface area** (Table 8.1). This expansion with interlayer water contributes to the very high degree of plasticity, stickiness, and cohesion for which smectites are well known. These properties make smectitic soils difficult to cultivate or excavate, as they are very sticky when wet and hard when dry. **Montmorillonite** is the most prominent of the smectites in soils.

The adsorption of interlayer water also leads to severe swelling when soils high in smectite are wetted and shrinkage when they are dried. Wide cracks commonly appear during the drying of smectite-dominated soils (such as Vertisols, Figure 3.13). The change in soil volume can crack pavements and foundations and misalign pipelines and utility poles. Therefore, smectitic soils are quite undesirable for most construction activities, but are especially well suited for a number of applications that require a high adsorptive capacity and the ability to form seals of very low permeability (see Section 8.13).

VERMICULITE GROUP. The most common **vermiculites** are 2:1-type minerals in which the octahedral sheet is aluminum dominated, but some magnesium-dominated vermiculites also exist. The tetrahedral sheets of most vermiculites have considerable substitution of aluminum in the silicon positions, giving rise to most of the very large quantity of negative charge associated with these clays. The cation exchange capacity of vermiculites usually exceeds that of all other silicate clays, including smectites (Table 8.1).

The interlayer spaces of vermiculites usually contain strongly adsorbed water molecules, Al-hydroxy ions and cations such as magnesium (Figure 8.7). However, these interlayer constituents act primarily as bridges to hold the units together, rather than wedges driving them apart. The degree of swelling and shrinkage is, therefore, considerably less for vermiculites than for smectites. For this reason, vermiculites are considered limited-expansion clays, expanding more than kaolinite, but much less than the smectites.

Nonexpanding 2:1 Silicate Minerals

More on clay minerals, their properties and uses:
http://galleries.com/minerals/silicate/clays.htm

The main nonexpanding 2:1 minerals are the **fine-grained micas** and the **chlorites.** We will discuss the fine-grained micas first.

MICA GROUP. Biotite and muscovite are examples of unweathered micas typically found in the sand and silt fractions. The more weathered **fine-grained micas,** such as **illite** and **glauconite,** are found in the clay fraction of soils. Their 2:1-type structures are quite similar to those of their unweathered cousins. Unlike in smectites, the main source of charge in fine-grained micas is the substitution of Al^{3+} in about 20% of the Si^{4+} sites in the tetrahedral sheets. This results in a high net negative charge in the tetrahedral sheet, even higher than that found in vermiculites. The negative charge attracts cations, among which potassium (K^+) is just the right size to fit snugly into certain spaces (hexagonal "holes") that exist between the tetrahedral oxygen groups (Figure 8.7). Because of the goodness of fit in the hexagonal holes, the K^+ ions are able to get very close to the negatively charged sites. Adjacent tetrahedral sheets are therefore strongly bound together by their mutual attraction to the K^+ ions in between. The layers in fine-grained micas are thus strongly bound together, preventing the type of expansion that characterizes smectite clays. Hence, the fine-grained micas are quite *nonexpansive.* Because of their nonexpansive character, the fine-grained micas are more like kaolinite than smectites with regard to their capacity to adsorb water and cations and their degree of plasticity and stickiness.

CHLORITES. In the 2:1 layers of soil **chlorites,** iron or magnesium, rather than aluminum, occupy most of the octahedral sites. In most chlorite clays, a magnesium-dominated octahedral hydroxide sheet is sandwiched in between adjacent 2:1 layers (Figure 8.7). Thus, chlorite is sometimes said to have a 2:1:1 structure. Chlorites are nonexpansive

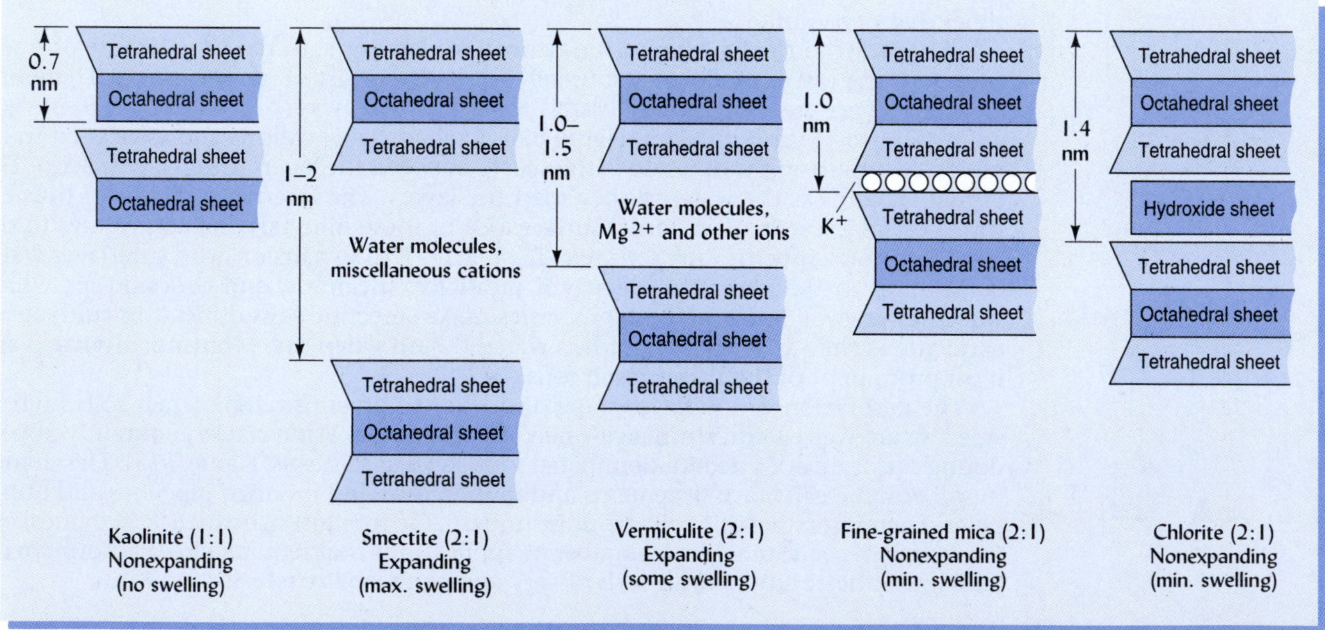

FIGURE 8.7 Schematic drawing illustrating the organization of tetrahedral and octahedral sheets in one 1:1-type mineral (kaolinite) and four 2:1-type minerals. The octahedral sheets in each of the 2:1-type clays can be either aluminum dominated or magnesium dominated. Note that kaolinite is nonexpanding, the layers being held together by hydrogen bonds. Maximum interlayer expansion is found in smectite, with somewhat less expansion in vermiculite because of the moderate binding power of numerous Mg^{2+} ions. Fine-grained mica and chlorite do not expand because K^+ ions (fine-grained mica) or an octahedral-like sheet of hydroxides of Al, Mg, Fe, and so forth (chlorite) tightly bind the 2:1 layers together.

because the hydroxylated surfaces of an intervening Mg-octahedral sheet are hydrogen-bonded to the oxygen atoms of the two adjacent tetrahedral sheets, binding the layers tightly together. The colloidal properties of the chlorites are therefore quite similar to those of the fine-grained micas (Table 8.1).

8.4 STRUCTURAL CHARACTERISTICS OF NONSILICATE COLLOIDS

Iron and Aluminum Oxides

3-D, moveable model of gibbsite with menu to highlight parts: http://www.soils.umn.edu/virtual_museum/gibbsite/index.html

These clays consist of modified octahedral sheets with either iron (e.g., goethite, FeOOH) or aluminum (e.g., gibbsite, $Al(OH)_3$) in the cation positions. They have neither tetrahedral sheets nor silicon in their structures. Little or no isomorphous substitution occurs, so these clays do not have a large negative charge. The small amount of net charge these clays possess (positive and negative) is caused by the removal of hydrogen ions from, or the addition of hydrogen ions to, the surface oxy-hydroxyl groups. The structure of gibbsite is shown in Figure 8.8. The presence of these covalently bound oxygen and hydroxyl groups enables the surfaces of these clays to strongly adsorb and combine with anions such as phosphate or arsenate. The oxide clays are nonexpansive and generally exhibit relatively little stickiness, plasticity, and cation adsorption. They make quite stable materials for construction purposes.

In many soils, iron and aluminum oxide clays are mixed with silicate clays. The oxide clays may form coatings on the external surfaces of the silicate clays, or they may occur as "islands" in the interlayer spaces of such 2:1 clays as vermiculites and smectites. In either case, the presence of iron and aluminum oxides can substantially alter the colloidal behavior of the associated silicate clays by masking charge sites, interfering with shrinkage and swelling, and providing anion-retentive surfaces.

Humus

As mentioned in Section 8.1, humus is a noncrystalline organic substance. It consists of very large organic molecules whose chemical composition varies considerably, but generally contains 40 to 60% C, 30 to 50% O, 3 to 7% H, and 1 to 5% N. The molecular weights of humic acids, a major type of colloidal humus, range from 10,000 to 100,000

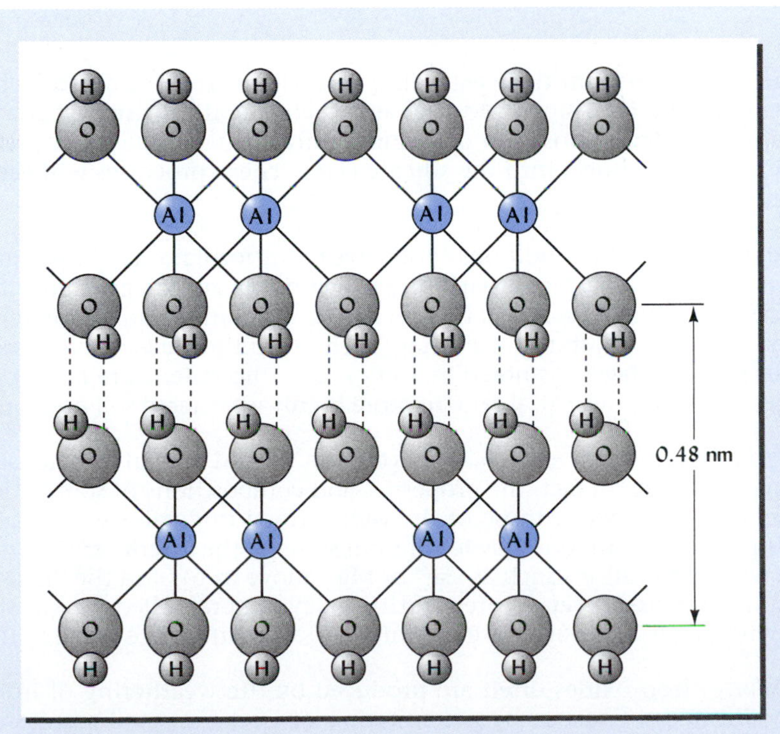

FIGURE 8.8 A simplified diagram showing the structure of gibbsite, an aluminum oxide clay common in highly weathered soils. This clay consists of octahedral sheets (two are shown) that are hydrogen-bonded together. Other oxide-type clays have iron instead of aluminum in the octahedral positions, and their structures are somewhat less regular and crystalline than that shown for gibbsite. The surface plane of covalently bonded hydroxyls gives this, and similar clays, the capacity to strongly adsorb certain anions (see Section 8.8).

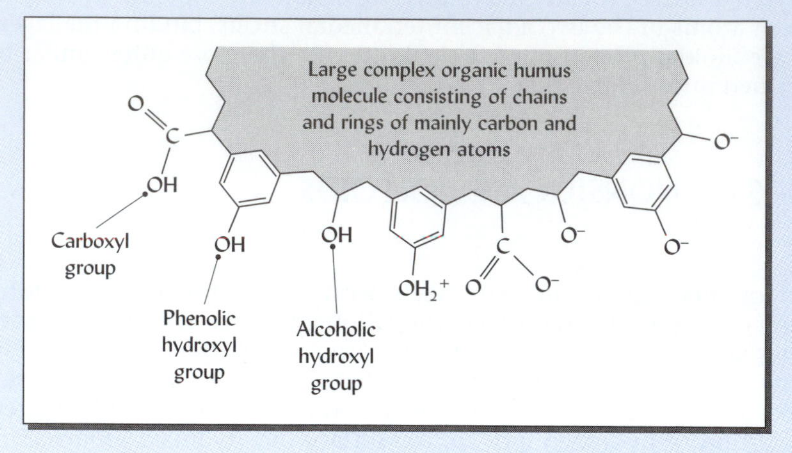

FIGURE 8.9 A simplified diagram showing the principal chemical groups responsible for the high amount of negative charge on humus colloids. The three groups highlighted all include—OH that can lose its hydrogen ion by dissociation and thus become negatively charged. Note that the **alcoholic**, **phenolic**, and **carboxylic** groups on the right side of the diagram are shown in their disassociated state, while those on the left side still have their associated hydrogen ions. Note also that association with a second hydrogen ion causes a site to exhibit a net positive charge.

g/mol. Identification of the actual structure of humus colloids is very difficult. Figure 8.9 provides a simplified diagram to illustrate the three main types of −OH groups thought to be responsible for the high amount of charge associated with these colloids. Negative or positive charges on the humus colloid develop as H^+ ions are either lost or gained by these groups. Both cations and anions are therefore attracted to and adsorbed by the humus colloid. The negative sites always outnumber the positive ones, and a very large *net* negative charge is associated with humus (Table 8.1). Because of its great surface area and many hydrophilic (water-loving) groups, humus can adsorb very large amounts of water per unit mass. However, humus also contains many hydrophobic sites and therefore can strongly adsorb a wide range of hydrophobic, nonpolar organic compounds (see Section 8.12). Because of its extraordinary influence on soil properties and behavior, we will delve much more deeply into the nature and function of soil humus in Chapter 11.

Highlight the protein, sugar, and other parts of a model humic acid:
http://www.soils.umn.edu/virtual_museum/som/index.html

8.5 GENESIS AND GEOGRAPHIC DISTRIBUTION OF SOIL COLLOIDS

Genesis of Clays and Humus

The silicate clays are developed from the weathering of a wide variety of minerals by at least two distinct processes: (1) a slight physical and chemical **alteration** of certain primary minerals, and (2) a **decomposition** of primary minerals with the subsequent **recrystallization** of certain products into the silicate clays. These processes will each be given brief consideration.

RELATIVE STAGES OF WEATHERING. Specific conditions conducive to the formation of important clay types are shown in Figure 8.10. Note that fine-grained micas and magnesium-rich chlorites represent earlier weathering stages of the silicates, and kaolinite and (ultimately) iron and aluminum oxides the most advanced stages. The smectites (e.g., montmorillonite) represent intermediate stages. As noted in Section 2.1, silicon tends to be lost as weathering progresses, leaving a lower Si:Al ratio in more highly weathered soil horizons.

MIXED AND INTERSTRATIFIED LAYERS. In a given soil, it is common to find several silicate clay minerals in an intimate mixture. In fact, the properties and compositions of some mineral colloids are intermediate between those of the well-defined minerals described in Section 8.3. For example, potassium ions may leach from some of the interlayers in fine-grained mica, while water and cations such as Ca^{2+} or Mg^{2+} move in to bind the 2:1 layers together. The result is a **mixed layer** or **interstratified** clay mineral called *fine-grained mica-vermiculite,* in which some layers are more like mica and some more like vermiculite.

IRON AND ALUMINUM OXIDES. Iron oxides often are produced by the weathering of iron-containing primary silicate minerals or by precipitation of iron from soil solutions. Under aerated weathering conditions, divalent iron (Fe^{2+}) oxides rapidly to trivalent

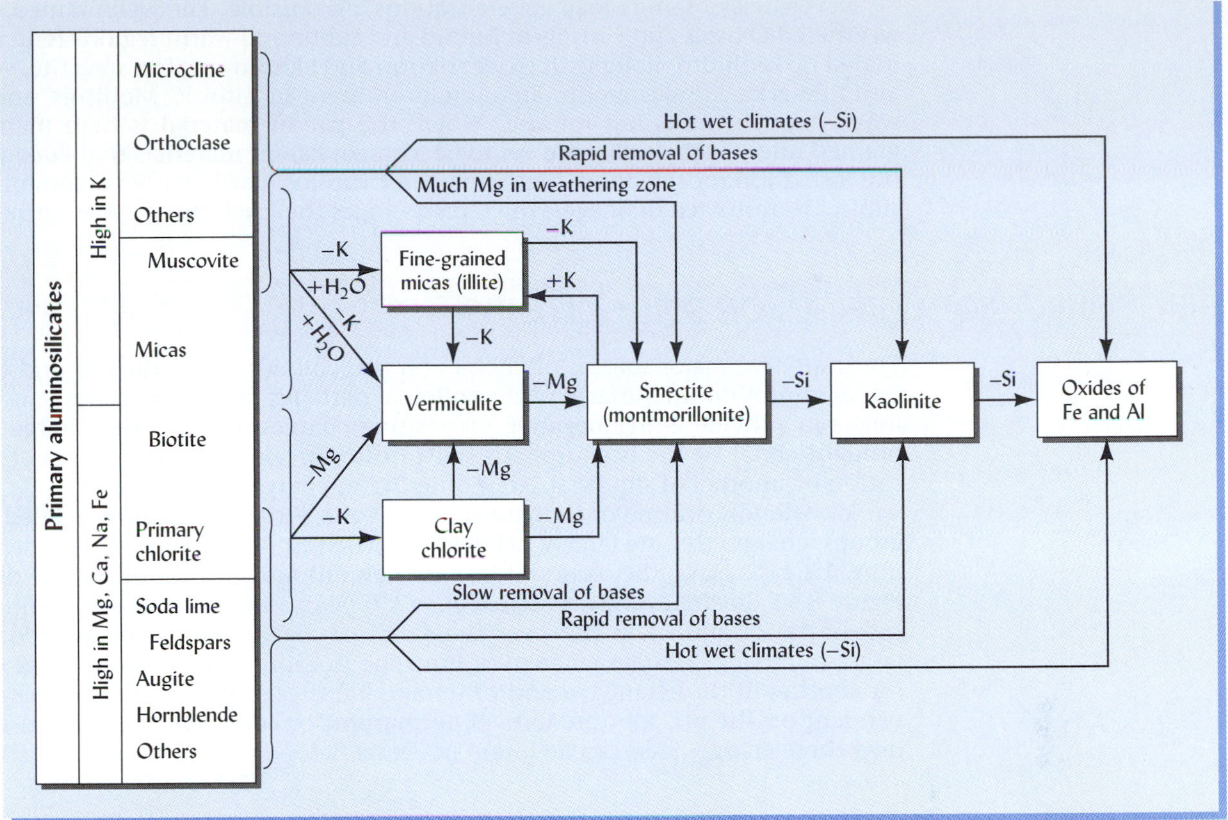

FIGURE 8.10 General conditions for the formation of the various layer silicate clays and oxides of iron and aluminum. Fine-grained micas, chlorite, and vermiculite are formed through rather mild weathering of primary aluminosilicate minerals, whereas kaolinite and oxides of iron and aluminum are products of much more intense weathering. Conditions of intermediate weathering intensity encourage the formation of smectite. In each case silicate clay genesis is accompanied by the removal in solution of such elements as K, Na, Ca, and Mg.

iron (Fe^{3+}), either while still within the structure of primary minerals or after its release into the soil solution. The Fe^{3+} forms stable oxides and hydroxides by reacting with oxygen atoms and water. The yellow-brown colored **goethite** (FeOOH) is the most stable iron oxide under most conditions and therefore tends to be the dominant iron oxide in most soils, especially in temperate, humid regions.

Aluminum oxides, mainly **gibbsite** [$Al(OH)_3$], are produced by strong weathering environments in which acid leaching rapidly removes the Si released from the breakdown of primary and secondary silicate minerals. During weathering, hydrogen ions replace the K^+, Mg^{2+}, and other such ions in the crystal, causing the framework to break down and releasing the silicon and aluminum. Gibbsite is extremely stable in soils and typically represents the most advanced stage of weathering in soils.

Humus. The breakdown and alteration of plant residues by microorganisms and the concurrent synthesis of new, more stable, organic compounds results in the formation of the dark-colored colloidal organic material called *humus* (see Section 11.4 for details). The various organic structural units associated with the decay and synthesis provide charged sites for the attraction of both cations and anions.

Distribution of Clays by Geography and Soil Orders

The clay of any particular soil is generally made up of a mixture of different colloidal minerals. In a given soil, the mixture may vary from horizon to horizon, because the kind of clay that develops depends not only on climatic influences and profile conditions but also on the nature of the parent material. The situation may be further complicated by the presence in the parent material itself of clays that were formed under a preceding and perhaps an entirely different type of climatic regime.

Nevertheless, some broad generalizations are possible. The well-drained and highly weathered Oxisols and Ultisols of humid and subhumid warm regions tend to be dominated by kaolinite, along with oxides of iron and aluminum. The smectite, vermiculite, and fine-grained mica groups are more prominent in Alfisols, Mollisols, and Vertisols, where weathering is less intense. Where the parent material is high in micas, fine-grained micas such as illite are apt to be formed. Parent materials that encourage smectite formation include those high in metallic cations (particularly magnesium) or those subject to restricted drainage, which discourages the leaching of these cations.

8.6 SOURCES OF CHARGES ON SOIL COLLOIDS

There are two major sources of charges on soil colloids: (1) hydroxyls and other functional groups on the surfaces of the colloidal particles that by releasing or accepting H^+ ions can provide either negative or positive charges, and (2) the charge imbalance brought about by the isomorphous substitution in some clay crystal structures of one cation by another of similar size but differing in charge.

All colloids, organic or inorganic, exhibit the surface charges associated with OH^- groups, charges that are largely **pH dependent.** Most of the charges associated with humus, 1:1-type clays, the oxides of iron and aluminum, and allophane are of this type. Figure 8.11 illustrates the occurence of this type of *variable* charge in kaolinite. In the case of the 2:1-type clays, however, these surface charges are complemented by a much larger number of charges emanating from the isomorphous substitution of one cation for another in the octahedral and/or tetrahedral sheets. Since these charges are not dependent on the pH, they are termed **permanent** or **constant charges.** An example of how these charges arise can be found in Figure 8.4.

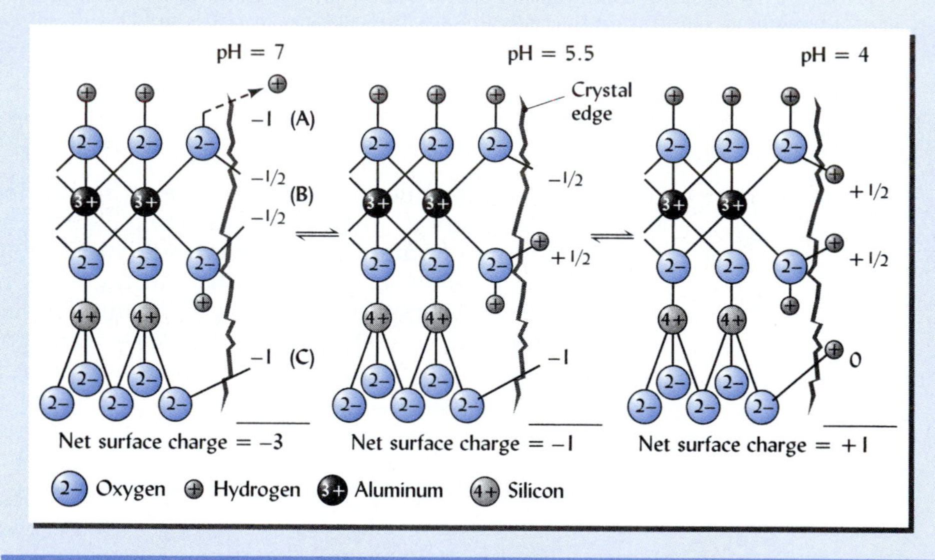

FIGURE 8.11 How pH-dependent charges develop at the broken edge of a kaolinite crystal. Three sources of net negative surface charge at a high pH are illustrated (left): (A) One (−1) charge from octahedral oxygen that has lost its H^+ ion by dissociation (the H broke away from the surface hydroxyl group and escaped into the soil solution). *Note that such dissociation can generate negative charges all along the surface hydroxyl plane, not just at a broken edge.* (B) One half (−½) charge from each octahedral oxygen that would normally be sharing its electrons with a second aluminum. (C) One (−1) charge from a tetrahedral oxygen atom that would normally be balanced by bonding to another silicon if it were not at the broken edge. The middle and right diagrams show the effect of acidification (lowering the pH), which increases the activity of H^+ ions in the soil solution. As more H^+ ions bond to the oxygen atoms at the clay surface, the 1+ charge of each H^+ ion either just balances a −1 charge on a tetrahedral oxygen or *more* than balances a −½ charge on an octahedral oxygen (middle). At the lowest pH shown (right), all of the edge oxygens have an associated H^+ ion, giving rise to a net positive charge on the crystal. These mechanisms of charge generation are similar to those illustrated for humus in Figure 8.9.

TABLE 8.3 **Charge Characteristics of Representative Colloids Showing Comparative Levels of Permanent (Constant) and pH-Dependent Charges**

		Negative charge		
Colloid type	*Total at pH 7, $cmol_c/kg$*	*Constant, %*	*pH dependent, %*	*pH dependent positive charge, $cmol_c/kg$*
Organic	200	10	90	0
Smectite	100	95	5	0
Vermiculite	150	95	5	0
Fine-grained micas	30	80	20	0
Chlorite	30	80	20	0
Kaolinite	8	5	95	2
Gibbsite (Al)	4	0	100	5
Goethite (Fe)	4	0	100	5
Allophane	30	10	90	15

Positive and negative charges may be exhibited at the same time. In most soils of temperate regions, the negative charges far exceed the positive ones (Table 8.3). However, in some acid soils high in Fe, Al oxides or allophane, the overall net charge may be positive.

The charge characteristics of selected soil colloids are shown in Table 8.3. Note the high percentage of constant negative charges in some 2:1-type clays (e.g., smectites and vermiculites). Humus, kaolinite, allophane and Fe, Al oxides have mostly variable (pH-dependent) negative charges and exhibit modest positive charges at low pH values. The negative and positive charges on soil colloids are of vital importance to the behavior of soil colloids in nature, especially as they absorb oppositely charged ions from the soil solution. This subject will be taken up next.

8.7 ADSORPTION OF CATIONS AND ANIONS

In soil, the negative and positive surface charges on the colloids attract and hold a complex swarm of cations and anions. In Figure 8.1 this concept was illustrated in a simplified manner, showing positive cations held on the negatively charged surfaces of a soil colloid. Actually, both cations and anions are usually attracted to the same colloid. In temperate-region soils, anions are commonly adsorbed in much smaller quantities than cations because these soils generally contain predominately 2:1-type silicate clays on which negative charges predominate. In the tropics, where soils are more highly weathered, acid, and rich in 1:1 clays and Fe, Al oxides, the amount of negative charge on the colloids is not so high and positive charges are more abundant. Therefore, the adsorption of anions is more prominent in these soils. Table 8.4 lists some important cations and anions. The adsorption of these ions by soil colloids greatly affects their biological availability and mobility, thereby influencing both soil fertility and environmental quality.

Figure 8.12 shows how both cations and anions may be attracted to the same colloid if it has both positively and negatively charged sites. This figure also illustrates that adsorption of ions on colloidal surfaces occurs by the formation of two quite different general types of colloid-ion complexes referred to as *outer-sphere* and *inner-sphere* complexes.

Outer- and Inner-Sphere Complexes

Remembering that water molecules surround (hydrate) the cations and anions in the soil solution, we can visualize that in an **outer-sphere complex** water molecules form a bridge between the adsorbed ion and the charged colloid surface. Sometimes several layers of water molecules are involved. Thus, the ion itself never comes close enough to the colloid surface to form a bond with a specific charged site. Instead, the ion is only weakly held by electrostatic attraction, the charge on the oscillating hydrated ion balancing, in a general way, an excess charge of opposite sign on the colloid surface. Ions in an outer-sphere complex are therefore easily replaced by other similarly charged ions.

In contrast, adsorption via formation of an **inner-sphere complex** does *not* involve any intervening water molecules. Therefore, one or more direct bonds are

TABLE 8.4 Selected Cations and Anions Commonly Adsorbed to Soil Colloids and Important in Plant Nutrition and Environmental Quality

The listed ions form inner- and/or outer-sphere complexes with soil colloids. Ions marked by an asterisk (*) are among those that predominate in most soil solutions. Many other ions may be important in certain situations.

Cation	Formula	Comments	Anion	Formula	Comments
Ammonium	NH_4^+	Plant nutrient	Arsenate	AsO_4^{3-}	Toxic to animals
Aluminum	Al^{3+} etc.[a]	Toxic to many plants	Borate	$B(OH)_4^-$	Plant nutrient, can be toxic
Calcium*	Ca^{2+}	Plant nutrient	Bicarbonate	HCO_3^-	Toxic in high-pH soils
Cadmium	Cd^{2+}	Toxic pollutant	Carbonate*	CO_3^{2-}	Forms weak acid
Cesium	Cs^+	Radioactive contaminant	Chromate	CrO_4^{2-}	Toxic pollutant
Copper	Cu^{2+}	Plant nutrient, toxic pollutant	Chloride*	Cl^-	Plant nutrient, toxic in large amounts
Hydrogen*	H^+	Causes acidity	Fluoride	Fl^-	Toxic, natural and pollutant
Iron	Fe^{2+}	Plant nutrient	Hydroxyl*	OH^-	Alkalinity factor
Lead	Pb^{2+}	Toxic to animals, plants	Nitrate*	NO_3^-	Plant nutrient, pollutant in water
Magnesium*	Mg^{2+}	Plant nutrient	Molybdate	MoO_4^{2-}	Plant nutrient, can be toxic
Manganese	Mn^{2+}	Plant nutrient	Phosphate	HPO_4^{2-}	Plant nutrient, water pollutant
Nickel	Ni^{2+}	Plant nutrient, toxic pollutant	Selenate	SeO_4^{2-}	Animal nutrient and toxic pollutant
Potassium*	K^+	Plant nutrient	Selenite	SeO_3^{2-}	Animal nutrient and toxic pollutant
Sodium*	Na^+	Used by animals, some plants, can damage soil	Silicate*	SiO_4^{4-}	Mineral weathering product, used by plants
Strontium	Sr^{2+}	Radioactive contaminant	Sulfate*	SO_4^{2-}	Plant nutrient
Zinc	Zn^{2+}	Plant nutrient, toxic pollutant			

[a]Important aluminum cations include Al^{3+}, $AlOH^{2-}$, and $Al(OH)_2^-$.

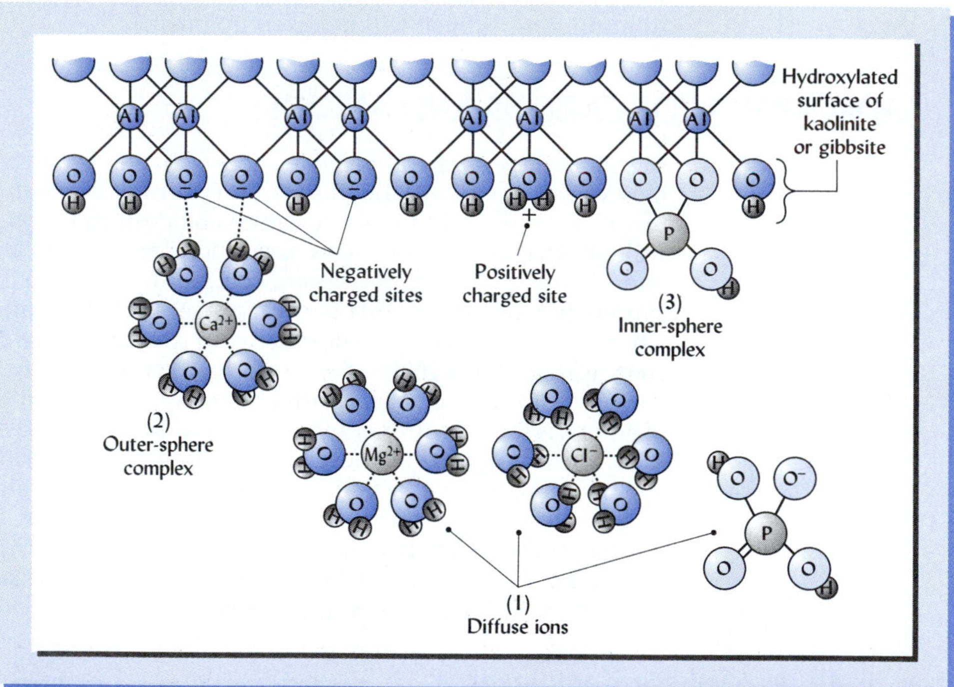

FIGURE 8.12 Adsorption of ions on a colloid by the formation of outer-sphere and inner-sphere complexes. (1) Water molecules surround diffuse cations and anions (such as the Mg^{2+}, Cl^-, and HPO_4^- shown) in the soil solution. (2) In an **outer-sphere complex** (such as the adsorbed Ca^{2+} ion shown), water molecules form a bridge between the adsorbed cation and the charged colloid surface. (3) In the case of an **inner-sphere complex** (such as the adsorbed $H_2PO_4^-$ anion shown), no water molecules intervene and the cation or anion binds directly with the metal atom (aluminum in this case) in the colloid structure. Outer-sphere complexes typify easily exchangeable ions that satisfy, in a general way, the net charge on the colloid surface. Inner-sphere complexes, on the other hand, are not easily replaced from the colloid surface, as they represent strong bonding of specific ions to specific sites on the colloid. In this example, all the charges originate with the dissociation of H^+ ions from surface hydroxyl groups. Although not shown in this example, permanent charges from isomorphous substitution in the interior structure of a colloid could also cause adsorption of outer-sphere complexes.

formed between the adsorbed ion and the atoms in the colloid surface. One example already discussed is the case of the K^+ ions that fit so snugly into the spaces between silicon tetrahedra in a mica crystal. Since there are no intervening water molecules, the K^+ ions are directly bonded by sharing electrons with the negatively charged tetrahedral oxygen atoms. Similarly, strong inner-sphere complexes may be formed by reactions of Cu^{2+} or Ni^{2+} with the oxygen atoms in silica tetrahedral.

Another important example, this time involving an anion, occurs when a $H_2PO_4^-$ ion is directly bonded by shared electrons with the octahedral aluminum in the colloid structure (Figure 8.12). Other ions cannot easily replace an ion held in an inner-sphere complex because this type of adsorption involves relatively strong bonds that are dependent on the compatible nature of specific ions and specific sites on the colloid.

8.8 CATION EXCHANGE REACTIONS

Because of its enormous ecological significance, the process of **cation exchange** requires our close attention. Consider the case of an outer-sphere complex between a negatively charged colloid surface and a hydrated cation (such as the Ca^{2+} ion shown in Figure 8.12). Such an outer-sphere complex is only loosely held together by electrostatic attraction, and the adsorbed ion remains in constant motion near the colloid surface. There are moments (microseconds) when the adsorbed cation is located a bit farther than average from the colloid surface. This moment provides an opportunity for another hydrated cation from the soil solution (say the Mg^{2+} ion shown in Figure 8.12) to diffuse into a position a bit closer to the negative site on the colloid. The instant this occurs, the second ion would replace the first ion, freeing the formerly adsorbed ion to diffuse out into the soil solution. In this manner, an exchange of cations between the adsorbed and diffuse state takes place. If a hydrated anion similarly replaces another hydrated anion at a positively charged colloid site, the process is called **anion exchange.** The ions held in outer-sphere complexes from which they can be replaced by exchange reactions are said to be **exchangeable** cations or anions. As a group, all the colloids in a soil, inorganic and organic, capable of holding exchanging cations or anions are termed the cation or anion **exchange complex.**

Principles Governing Cation Exchange Reactions

Reversibility. We can illustrate the process of cation exchange using a simple reaction in which a hydrogen ion (perhaps generated by organic matter decomposition—see Section 9.1) displaces a sodium ion from its adsorbed state on a colloid surface:

$$\underset{\text{(soil colloid)}}{\boxed{\text{Micelle}}\,Na^+} + \underset{\text{(soil solution)}}{H^+} \rightleftharpoons \underset{\text{(soil colloid)}}{\boxed{\text{Micelle}}\,H^+} + \underset{\text{(soil solution)}}{Na^+} \qquad (8.1)$$

The reaction takes place rapidly and, as shown by the double arrows, the reaction is reversible. It will go to the left if sodium is added to the system. This reversibility is a fundamental principle of cation exchange.

Animated cation exchange reaction. Keep track of the charges!
http://www.une.edu.au/~agronomy/SSCATXCH.dcr

Charge Equivalence. Another basic principle of cation exchange reactions is that the exchange is chemically equivalent; that is, it takes place on a *charge-for-charge* basis. Therefore, although one H^+ ion exchanged with *one* Na^+ ion in the reaction just shown, it would require *two* singly charged H^+ ions to exchange with or replace *one* divalent Ca^{2+} ion. If the reaction is reversed, one Ca^{2+} ion will displace two H^+ ions. In other words, two charges from one cation species replace two charges from the other:

$$\underset{\text{(soil colloid)}}{\boxed{\text{Micelle}}\,Ca^{2+}} + \underset{\text{(soil solution)}}{2H^+} \rightleftharpoons \underset{\text{(soil colloid)}}{\boxed{\text{Micelle}}\,{}^{H^+}_{H^+}} + \underset{\text{(soil solution)}}{Ca^{2+}} \qquad (8.2)$$

Note that by this principle, it would require three Na^+ ions to replace a single Al^{3+} ion, and so on.

RATIO LAW. Consider an exchange reaction between two similar cations, say Ca^{2+} and Mg^{2+}. If there are a large number of Ca^{2+} ions adsorbed on a colloid, and some Mg^{2+} is added to the soil solution, the added Mg^{2+} ions will begin displacing the Ca^{2+} from the colloid. This will bring more Ca^{2+} into the soil solution and these Ca^{2+} ions will, in turn, displace some of the Mg^{2+} back off the colloid. The *ratio law* tells us that, at equilibrium, the ratio of Ca^{2+} to Mg^{2+} on the colloid will be the same as the ratio of Ca^{2+} to Mg^{2+} in the solution and both will be the same as the ratio in the overall system. To illustrate this concept, assume that 20 Ca^{2+} ions are initially adsorbed on a soil colloid and 5 Mg^{2+} ions are added to the system:

$$\underset{\text{(colloid)}}{\boxed{\text{Micelle}}\,20\,Ca^{2+}} + \underset{\text{(soil solution)}}{5\,Mg^{2+}} \rightleftharpoons \underset{\text{(colloid)}}{\boxed{\text{Micelle}}\begin{matrix}16\,Ca^{2+}\\4\,Mg^{2+}\end{matrix}} + \underset{\text{(soil solution)}}{1\,Mg^{2+} + 4\,Ca^{2+}} \qquad \text{Ratio: 4Ca:1Mg} \quad (8.3)$$

If the two exchanging ions are not of the same charge (e.g., K^+ exchanging with Mg^{2+}), the reaction becomes somewhat more complicated and a modified version of the ratio law would apply.

Up to this point, our discussion of exchange reactions has assumed that both ionic species (elements) exchanging places take part in the exchange reaction in exactly the same way. This assumption must be modified to take into account three additional factors if we are to understand how exchange reactions actually proceed in nature.

ANION EFFECTS ON MASS ACTION. The laws of **mass action** tell us that an exchange reaction will be more likely to proceed to the right if the released ion is prevented from reacting in the reverse direction. This may be accomplished if the product formed with the released cation on the right side of the reaction either *precipitates, volatilizes,* or *strongly associates* with an anion. To illustrate this concept, consider the displacement of H^+ ions on an acid colloid by Ca^{2+} ions added to the soil solution as calcium carbonate:

$$\boxed{\text{Micelle}}\begin{matrix}H^+\\H^+\end{matrix} + \underset{\text{(added)}}{CaCO_3} \rightleftharpoons \boxed{\text{Micelle}}\,Ca^{2+} + \underset{\text{Water}}{H_2O} + \underset{\text{(gas)}}{CO_2\uparrow} \qquad (8.4)$$

Where $CaCO_3$ is added, when a hydrogen ion is displaced off the colloid, it combines with an oxygen atom from the $CaCO_3$ to form water. Furthermore, the CO_2 produced is a gas, which can volatilize out of the solution and leave the system. The removal of these products pulls the reaction to the right. This principle explains why $CaCO_3$ (in the form of limestone) is effective in neutralizing an acid soil, while calcium chloride is not (see Section 9.9).

CATION SELECTIVITY. Until now we have assumed that both cation species taking part in the exchange reaction are held with equal tenacity by the colloid and therefore have an equal chance of displacing each other. In reality, some cations are held much more tightly than others, and so are less likely to be displaced from the colloid. In general, the higher the charge and the smaller the hydrated radius of the cation, the more strongly it will adsorb to the colloid. The order of strength of adsorption for the most common cations is:

$$Al^{3+} > Ca^{2+} > Mg^{2+} > K^+ = NH_4^+ > Na^+$$

The less tightly held cations oscillate farther from the colloid surface and therefore are the most likely to be displaced into the soil solution and carried away by leaching. This series therefore explains why the soil colloids are dominated by Al^{3+} (and other aluminum ions) and Ca^{2+} in humid regions and by Ca^{2+} in drier regions, even though the weathering of minerals in many parent materials provides relatively larger amounts of K^+, Mg^{2+}, and Na^+ (see Section 8.10).

The relative strengths of adsorption order may be altered on certain colloids whose properties favor adsorption of particular cations. An important example of such colloidal "preference" for specific cations is the very high affinity for K^+ ions (and the similarly sized NH_4^+ and Cs^+ ions) exhibited by vermiculite and fine-grained micas (Section 8.3), which attracts these ions to intertetrahedral spaces exposed at weathered crystal edges. The influence of different colloids on the adsorption of specific cations impacts the availability of cations for leaching or plant uptake (see Section 13.16). Certain metals such as copper, mercury, and lead have very high selective affinities for sites on humus and iron oxide colloids, making most soils quite efficient at removing these potential pollutants from water leaching through the profile.

COMPLEMENTARY CATIONS. In soils, colloids are always surrounded by many different adsorbed cation species. The likelihood that a given adsorbed cation will be displaced from a colloid is influenced by how strongly its neighboring cations are adsorbed to the colloid surface. For example, consider an adsorbed Mg^{2+} ion. An ion diffusing in from the soil solution is more likely to displace one of the neighboring ions rather than the Mg^{2+} ion, if the neighboring adsorbed ions (sometimes called **complementary ions**) are loosely held. If they are tightly held, then the chances are greater that the Mg^{2+} ion will be displaced. In Section 8.10, we shall discuss the influence of complementary ions on the availability of nutrient cations for plant uptake.

8.9 CATION EXCHANGE CAPACITY

Previous sections have dealt qualitatively with exchange reactions. We now turn to a consideration of the quantitative **cation exchange capacity (CEC).** This property is defined simply as the sum total of the exchangeable cations that a soil can absorb.

Means of Expression

The CEC is expressed as the number of moles of positive charge adsorbed per unit mass. To deal with whole numbers of a convenient size, many publications, including this textbook, report CEC values in centimoles of charge per kilogram ($cmol_c/kg$). Some publications still use the older unit, milliequivalents per 100 grams (me/100 g), which gives the same value as $cmol_c/kg$ (1 me/100 g = 1 $cmol_c/kg$). A particular soil may have a CEC of 15 $cmol_c/kg$, indicating that 1 kg of the soil can hold 15 $cmol_c$ of H^+ ions, for example, and can exchange this number of charges from H^+ ions for the same number of charges from any other cation. This means of expression emphasizes that exchange reactions take place on a charge-for-charge (not an ion-for-ion) basis. The concept of a mole of charges and its use in CEC calculations are reviewed in Box 8.1.

Methods of Determining CEC

The CEC is an important soil chemical property that is used for classifying soils in *Soil Taxonomy* (e.g., in defining Oxic, Mollic, and Kandic diagnostic horizons, Section 3.2) and for assessing their fertility and environmental behavior. Several different standard methods can be used to determine the CEC of a soil. In general, a concentrated solution of a particular exchanger cation (for example, Ba^{2+}, NH_4^+, or Sr^{2+}) is used to leach the soil sample. This provides an overwhelming number of the exchanger cations that can completely replace all the exchangeable cations initially in the soil. Then, the CEC can be determined by measuring either (1) the number of exchanger cations adsorbed or (2) the amounts of each of the elements originally held on the exchange complex (usually Ca^{2+}, Al^{3+}, Mg^{2+}, K^+, and Na^+).

BUFFER CEC METHODS. The CEC procedure often calls for use of a solution buffered to maintain a certain pH (usually either pH 7.0 using ammonium as the exchanger cation or pH 8.2 using barium as the exchanger cation). If the native soil pH is less than the pH of the buffered solution, then these methods measure not only the cation exchange sites active at the pH of the particular soil, but also any pH-dependent exchange sites (see Section 8.6) that would become negatively charged at pH 7.0 or 8.2.

EFFECTIVE CEC (UNBUFFERED). Alternatively, the CEC procedure may use unbuffered solutions to allow the exchange to take place at the actual pH of the soil. The buffered methods (NH_4^+ at pH 7.0 or Ba^{2+} at pH 8.2) measure the *potential* or *maximum* cation exchange capacity of a soil. The unbuffered method measures only the **effective cation exchange capacity** (ECEC), which can hold exchangeable cations at the pH of the soil

BOX 8.1 CHEMICAL EXPRESSION OF CATION EXCHANGE

One mole of any atom, molecule, or charge is defined as 6.02×10^{23} (Avogadro's number) of atoms, molecules, or charges. Thus, 6.02×10^{23} negative charges associated with the soil colloidal complex would attract 1 mole of positive charge from adsorbed cations such as Ca^{2+}, Mg^{2+}, and H^+. The number of moles of the positive charge provided by the adsorbed cations in any soil gives us a measure of the *cation exchange capacity* (CEC) of that soil. The CEC of soils commonly varies from 0.03 to 0.5 mole of positive charge per kilogram.

Using the mole concept, it is easy to relate the mole charges to the mass of ions or compounds involved in cation or anion exchange. Consider, for example, the exchange that takes place when adsorbed sodium ions in an alkaline arid-region soil are replaced by hydrogen ions:

$$\boxed{\text{Micelle}}\ Na^+ + H^+ \rightleftharpoons \boxed{\text{Micelle}}\ H^+ + Na^+$$

If 2 $cmol_c$ of adsorbed Na^+ ions per kilogram of soil were replaced by H^+ ions in this reaction, how many grams of Na^+ ions would be replaced?

Since the Na^+ ion is singly charged, the mass of Na^+ needed to provide 1 mole of charge (1 mol_c) is the gram atomic weight of sodium, or 23 g (see periodic table in Appendix C). The mass providing 1 *centimole* of charge ($cmol_c$) is 1/100 of this amount, or 0.23 g. Thus, the mass of the 2 *$cmol_c$* Na^+ replaced from 1 kg of soil is:

$$2\ cmol_c\ Na^+/kg \times 0.23\ g\ Na^+/cmol_c = 0.46g\ Na^+/kg\ soil$$

The 0.46 g Na^+ would be replaced by only 0.02 g H, which is the mass of 2 $cmol_c$ of this much lighter element.

Another example is the replacement of H^+ ions when lime [$Ca(OH)_2$] is added to an acid soil. This time assume that 4 $cmol_c$ H^+/kg soil is replaced by the $Ca(OH)_2$, which reacts with the acid soil as follows:

$$\boxed{\text{Micelle}}\begin{matrix}H^+\\H^+\end{matrix} + Ca(OH)_2 \rightleftharpoons \boxed{\text{Micelle}}\ Ca^{2+} + 2H_2O$$

Since the Ca^{2+} ion in each molecule of $Ca(OH)_2$ has two positive charges, the mass of $Ca(OH)_2$ needed to replace 1 mole of charge from the H^+ ions is only one-half of the gram molecular weight of this compound, or 74/2 = 37 g. A comparable figure for 1 *centimole* is 37/100, or 0.37 grams.

Note that the number of charges provided by the replacing ion is *equivalent* to the number associated with the ion being replaced. Thus, 1 mole of negative charges attracts 1 mole of positive charges whether the charges come from H^+, K^+, Na^+, NH_4^+, Ca^{2+}, Mg^{2+}, Al^{3+}, or any other cation. Keep in mind, however, that only one-half the atomic weights of divalent cations, such as Ca^{2+} or Mg^{2+}, and only one-third the atomic weight of trivalent Al^{3+} are needed to provide 1 mole of charge. This *chemical equivalency* principle applies to both cation and ion exchange.

as sampled. For colloids with substantial pH-dependent charge (see Table 8.3) the different methods may give quite different results. Therefore it is important that the method used be known when comparing soils based on their CEC.

Cation Exchange Capacities of Soils

The CEC of a given soil horizon is determined by the relative amounts of different colloids in that soil and by the CEC of each of these colloids. Figure 8.13 illustrates the common range in CEC among different soils and other organic and inorganic exchange materials. Note that sandy soils, which are generally low in all colloidal material, have low CECs compared to those exhibited by silt loams and clay loams. Also note the very high CECs associated with humus compared to those exhibited by the inorganic clays, especially kaolinite and Fe, Al oxides. The CEC coming from humus generally plays a prominent role, and sometimes a dominant one, in cation exchange reactions in A horizons. For example, in a clayey Ultisol (pH = 5.5) containing 2.5% humus and 30% kaolinite, about 75% of the CEC is associated with humus.

Using the CEC range from Figure 8.13 it is possible to estimate the CEC of a soil if the quantities of the different soil colloids in the soil are known. Box 8.2 illustrates how this can be done.

Cation Exchange Capacity explained with 3-D Models and Animations:
http://soils1.cses.vt.edu/MJE/CSES3124/cec_demo/version1.1/cec.html

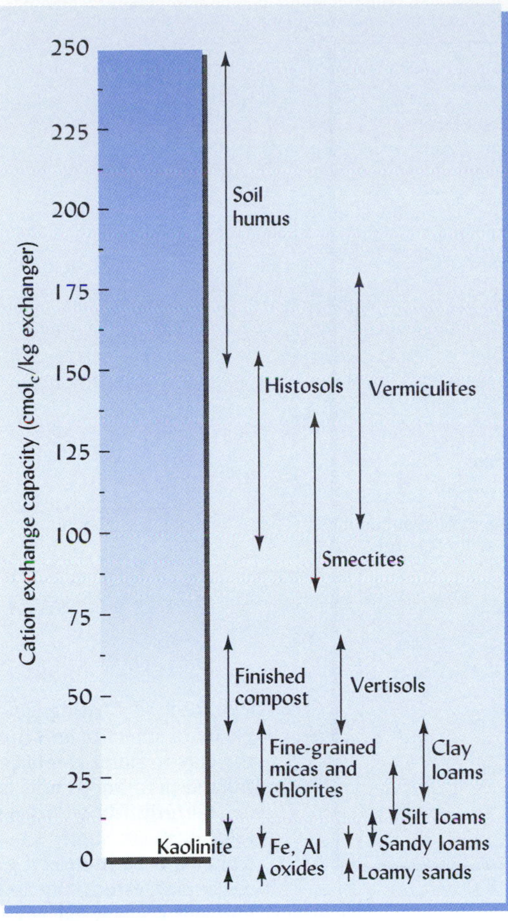

FIGURE 8.13 Ranges in the cation exchange capacities (at pH 7) that are typical of a variety of soils and soil materials. The high CEC of humus shows why this colloid plays such a prominent role in most soils, and especially those high in kaolinite and Fe, Al oxides, clays that have low CECs.

pH and Cation Exchange Capacity

The previous sections explained that the cation exchange capacity of most soils increases with pH. At very low pH values only the permanent charges of the 2:1-type clays (see Section 8.8) and a small portion of the potential pH-dependent charges hold exchangeable ions (Figure 8.14). As the pH is raised, the negative charge on some 1:1-type silicate clays, allophane, humus, and even Fe, Al oxides increases, thereby increasing the cation exchange capacity. As noted earlier, to obtain a measure of this maximum retentive capacity, the CEC is commonly determined at a pH of 7.0 or 8.2. At neutral or slightly alkaline pH, the CEC reflects most of the potential pH-dependent charges as well as the permanent ones.

BOX 8.2 ESTIMATING CATION EXCHANGE CAPACITIES

Assume you want to estimate the cation exchange capacity (CEC) of the A horizon of a cultivated Mollisol from Iowa (pH = 7.0, 20% clay, and 4% organic matter). The most dominant clays in the Mollisol are likely the 2:1 types, and their average CEC would be about 80 cmol$_c$/kg. At pH 7.0, the CEC of the organic matter is about 200 cmol$_c$/kg (see Table 8.3, page 000). Since 1 kg of the soil has 0.2 kg (20%) of clay and 0.04 kg (4%) of organic matter, we can calculate the CEC associated with each of these sources.

From the clays: $0.2 \text{ kg} \times 80 \text{ cmol}_c/\text{kg} = 16 \text{ cmol}_c$

From the O.M.: $0.04 \text{ kg} \times 200 \text{ cmol}_c/\text{kg} = 8 \text{ cmol}_c$

The total CEC of the Mollisol is: $16 + 8 = 24 \text{ cmol}_c/\text{kg soil}$

In a very general way, the type of clay mineralogy can be estimated by reversing the preceding steps when CEC and contents of clay and organic matter are known but the type of clay is uncertain.

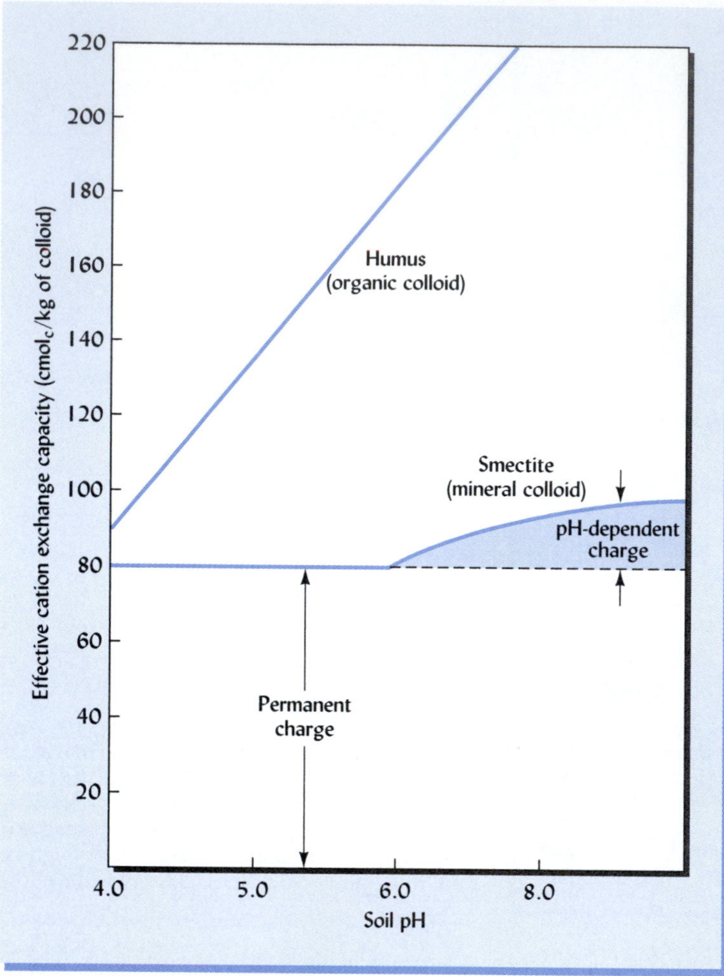

FIGURE 8.14 Influence of pH on the cation exchange capacity of smectite and humus. Below pH 6.0 the charge for the clay mineral is relatively constant. This charge is considered permanent and is due to ionic substitution in the crystal unit. Above pH 6.0 the charge on the mineral colloid increases slightly because of ionization of hydrogen from exposed hydroxyl groups at crystal edges. In contrast to the clay, essentially all of the charges on the organic colloid are considered pH dependent. [Smectite data from Coleman and Mehlich (1957); organic colloid data from Helling et al. (1964)]

8.10 EXCHANGEABLE CATIONS IN FIELD SOILS

The specific exchangeable cations associated with soil colloids differ from one climatic region to another—Ca^{2+}, Al^{3+}, complex aluminum hydroxy ions, and H^+ being most prominent in humid regions, and Ca^{2+}, Mg^{2+}, and Na^+ dominating in low-rainfall areas (Table 8.5). The cations that dominate the exchange complex have a marked influence on soil properties.

In a given soil, the proportion of the cation exchange capacity satisfied by a particular cation is termed the ***percentage saturation*** for that cation. Thus, if 50% of the CEC is satisfied by Ca^{2+} ions, the exchange complex is said to have a ***percentage calcium saturation*** of 50.

This terminology is especially useful in identifying the relative proportions of sources of acidity and alkalinity in the soil solution. Thus, the percentage saturation with Al^{3+} and H^+ ions gives an indication of the acid conditions, while increases in the percentage nonacid cation saturation (sometimes referred to as the **percentage base saturation**[5]) indicate the tendency toward neutrality and alkalinity. These relationships will be discussed further in Chapter 9.

Cation Saturation and Nutrient Availability

Exchangeable cations generally are available to both higher plants and microorganisms. By cation exchange, hydrogen ions from the root hairs and microorganisms replace nutrient cations from the exchange complex. The nutrient cations are forced into the soil

[5]Technically speaking, nonacid cations such as Ca^{2+}, Mg^{2+}, K^+, and Na^+ are not bases. When adsorbed by soil colloids in the place of H^+ ions, however, they reduce acidity and increase the soil pH. For that reason, they are traditionally referred to as *bases* and the portion of the CEC that they satisfy is often termed *percentage base saturation*.

TABLE 8.5 Cation Exchange Properties Typical for Unamended Clay Loam Surface Soils in Different Climatic Regions

Note that soils with coarser textures would have less clay and organic matter and therefore lower amounts of exchangeable cations and lower CEC values.

Property	*Warm, humid region (Ultisols)*[a]	*Cool, humid region (Alfisols)*	*Semiarid region (Ustolls)*	*Arid region (Natrargids)*[b]
Exchangeable H^+ and Al^{3+}, $cmol_c$/kg (% of CEC)	7.5 (75%)	5 (28%)	0 (0%)	0 (0%)
Exchangeable Ca^{2+}, $cmol_c$/kg (% of CEC)	2.0 (20%)	9 (50%)	17 (65%)	13 (50%)
Exchangeable Mg^{2+}, $cmol_c$/kg (% of CEC)	0.4 (4%)	3 (17%)	6 (23%)	5 (19%)
Exchangeable K^+, $cmol_c$/kg (% of CEC)	0.1 (1%)	1 (5%)	2 (8%)	3 (12%)
Exchangeable Na^+, $cmol_c$/kg (% of CEC)	Tr	0.02 (0.1%)	1 (4%)	5 (19%)
Cation exchange capacity (CEC)[c], $cmol_c$/kg	10	18	26	26
Probable pH	4.5–5.0	5.0–5.5	7.0–8.0	8–10
Nonacid cations (% of CEC)[d]	25%	68%	100%	100%

[a]See Chapter 3 for explanation of soil group names.
[b]Natrargids are Aridisols with natric horizons. They are sodic soils, high in exchangeable sodium, as explained in Section 9.15.
[c]The sum of all the exchangeable cations measured at the pH of the soil. This is termed the effective CEC or ECEC (see Section 8.9).
[d]Traditionally referred to as "base" saturation.

solution, where they can be assimilated by the adsorptive surfaces of roots and soil organisms, or they may be removed by drainage water. Cation exchange reactions affecting the mobility of organic and inorganic pollutants in soils will be discussed in Section 8.12. Here we focus on the plant nutrition aspects.

The percentage saturation of essential nutrient cations such as calcium and potassium also greatly influences the uptake of these elements by growing plants. For example, if the percentage calcium saturation of a soil is high, the displacement of this cation is comparatively easy and rapid. Thus, 6 cmol/kg of exchangeable calcium in a soil whose exchange capacity is 8 cmol/kg (75% calcium saturation) probably would mean ready availability, but 6 cmol/kg when the total exchange capacity of a soil is 30 cmol/kg (20% calcium saturation) would produce lower availability. This is one reason that, for calcium-loving plants such as alfalfa, the calcium saturation of at least part of the soil should approach or even exceed 80%.

Influence of Complementary Cations

A second factor influencing plant uptake of a given cation is the effect of the complementary ions held on the colloids (Section 8.8). Consider, for example, the loosely held ion, K^+. If the complementary ions surrounding a K^+ ion are held tightly (that is, they oscillate very close to the colloid surface), then a H^+ ion from a root is less likely to "find" a complementary ion and more likely, instead, to "bump into" and replace a K^+ ion (see Figure 8.15).

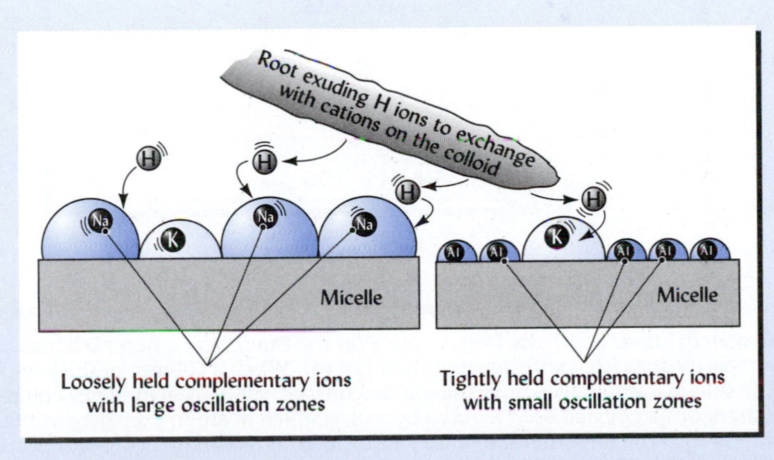

FIGURE 8.15 Effect of complementary ions on the availability of a particular exchangeable nutrient cation. The half spheres represent the zones in which the ion oscillates, the more loosely held ions moving within larger zones of oscillation. For simplicity, the water molecules that hydrate each ion are not shown. (Left) H^+ ions from the root are more likely to encounter and exchange with loosely held Na^+ ions rather than the more tightly held K^+ ion. (Right) The likelihood that H^+ ions from the root will encounter and exchange with a K^+ ion is increased by the inaccessibility of the neighboring tightly held Al^{3+} ions. The K^+ ion on the right colloid is comparatively more vulnerable to being replaced and sent into the soil solution, and is therefore more available for plant uptake or leaching than the K^+ ion on the left colloid.

Plant uptake of one cation may also be reduced by excessive uptake of another cation. For instance, potassium uptake by plants is limited by high levels of calcium in some soils. Likewise, high potassium levels are known to limit the uptake of magnesium even when significant quantities of magnesium are present in the soil.

8.11 ANION EXCHANGE

Anions are held by soil colloids in two major ways. First, they are held by anion adsorption mechanisms similar to those responsible for cation adsorption. Second, they may actually react with surface oxides or hydroxides, forming more definitive *inner-sphere complexes*. We shall consider anion adsorption first.

The basic principles of **anion exchange** are similar to those of cation exchange, except that the charges on the colloids are positive and the exchange is among negatively charged anions. The positive charges on colloid surfaces attract anions such as SO_4^{2-} and NO_3^-. A simple example of an anion exchange reaction is as follows:

$$\underset{\text{(positively charged soil solid)}}{\boxed{\text{Micelle}}\ NO_3^-} + \underset{\text{(soil solution)}}{Cl^-} \rightleftharpoons \underset{\text{(positively charged soil solid)}}{\boxed{\text{Micelle}}\ Cl^-} + \underset{\text{(soil solution)}}{NO_3^-} \qquad (8.5)$$

Just as in cation exchange, *equivalent* quantities of NO_3^- and Cl^- are exchanged; the reaction can be reversed; and plant nutrients so released can be absorbed by plants.

In contrast to cation exchange capacities, anion exchange capacities of soils generally *decrease* with increasing pH. Figure 8.16 illustrates this fact for an Ultisol in Georgia. In some very acid tropical soils that are high in kaolinite and iron and aluminum oxides, the anion exchange capacity may actually exceed the cation exchange capacity.

Anion exchange is very important in making anions available for plant growth while at the same time retarding the leaching of such anions from some soils. For example, anion exchange restricts the loss of sulfates from subsoils in many Ultisols and Oxisols (see Figure 8.16 and Section 12.23). Even the leaching of nitrate may be retarded by an-

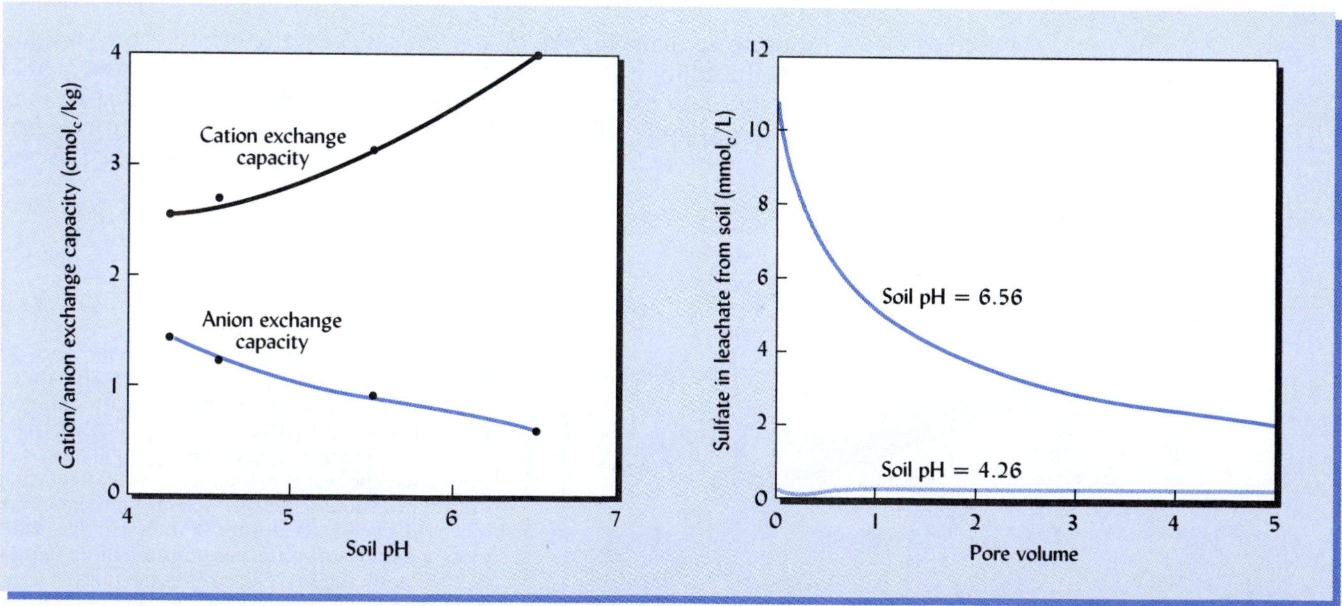

FIGURE 8.16 (Left) Effect of increasing the pH of subsoil material from an Ultisol from Georgia on the cation and anion exchange capacities. Note the significant decrease in anion exchange capacity associated with the increased soil pH. When a column of the low-pH material (pH = 4.6) was leached with $Ca(NO_3)_2$ (right), little sulfate was removed from the soil. In contrast, similar leaching of a column of the soil with the highest pH (6.56), where the anion exchange capacity had been reduced by half, resulted in anion exchange of NO_3^- ions for SO_4^- ions and significant leaching of sulfate from the soil. The importance of anion adsorption in retarding movement of specific anions or other negatively charged substances is illustrated. [Data from Bellini et al. (1996)]

ion exchange in the B horizons of certain highly weathered soils of the humid tropics. Similarly, the downward movement into groundwater of some charged organic pollutants can be retarded by such anion and/or cation exchange reactions.

Inner-Sphere Complexes

Some anions, such as phosphate, arsenate, molybdate, and sulfate, can react with particle surfaces, forming *inner-sphere complexes* (see Figure 8.12). For example, the $H_2PO_4^-$ ion may react with the protonated hydroxyl group rather than remain as an easily exchanged anion.

$$\underset{\text{(soil solid)}}{>\text{Al—OH}_2^+} + \underset{\text{(soil solution)}}{\text{H}_2\text{PO}_4^-} \longrightarrow \underset{\text{(soil solid)}}{>\text{Al—H}_2\text{PO}_4} + \underset{\text{(soil solution)}}{\text{H}_2\text{O}} \qquad (8.6)$$

This reaction actually reduces the net positive charge on the soil colloid. Also, the $H_2PO_4^-$ is held very tightly and is not readily available for plant uptake.

Anion adsorption and exchange reactions regulate the mobility and availability of many important ions. Together with cation exchange they largely determine the ability of soils to hold nutrients in a form that is accessible to plants, and to retard movement of pollutants in the environment.

Weathering and CEC/AEC Levels

The range of CEC levels of different clay minerals shown in Figure 8.17 shows that clays developed under mild weathering conditions (e.g., smectites and vermiculites) have much higher CEC levels than those developed under more extreme weathering. In contrast, the AEC levels tend to be much higher in clays developed under strong weathering conditions (e.g., kaolinite) than in those found under more mild weathering. This generalized figure is helpful in obtaining a first approximation of CEC and AEC levels in soils of different climatic regions. It must be used with caution, however, since in some soils the type of clay present may reflect past rather than current weathering conditions.

8.12 SORPTION OF ORGANIC COMPOUNDS

Soil colloids help control the movement of pesticides and other organic compounds into groundwater. The retention of these chemicals by soil colloids can prevent their downward movement through the soil or can delay that movement until the compounds are broken down by soil microbes.

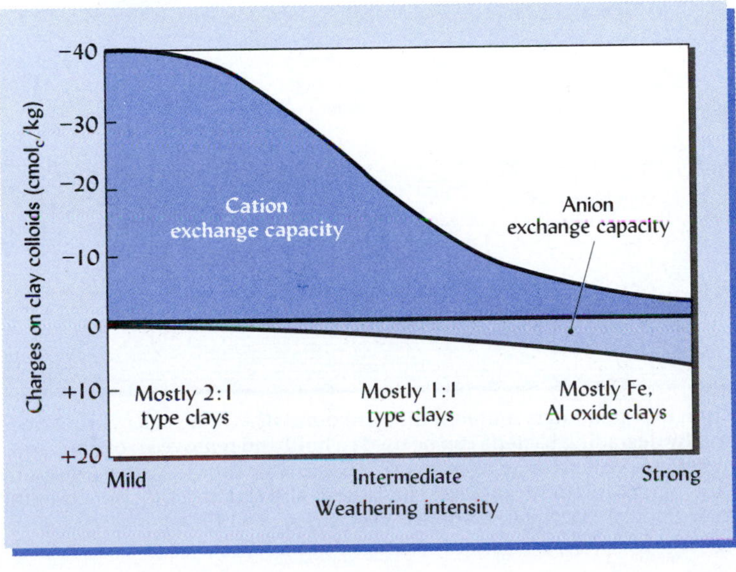

FIGURE 8.17 The effect of weathering intensity on the charges on clay minerals and, in turn, on their cation and anion exchange capacities (CECs and AECs). Note the high CEC and very low AEC associated with mild weathering, which has encouraged the formation of 2:1-type clays such as fine-grained micas, vermiculites, and smectites. More intense weathering destroys the 2:1-type clays and leads to the formation of first kaolinite and then oxides of Fe and Al. These have much lower CECs and considerably higher AECs. Such changes in clay dominance account for the curves shown.

By accepting or releasing protons (H^+ ions), groups such as —OH, —NH_2, and —COOH in the chemical structure of some organic compounds provide positive or negative charges that participate in anion or cation exchange reactions. Other organic compounds participate in inner-sphere complexation and adsorption reactions as do the inorganic ions we have discussed; however, it is more common for organic colloids to be *absorbed* within the soil organic colloids by a process termed **partitioning.** The soil organic colloids tend to act as a solvent for organic chemicals, thereby partitioning their concentrations between those held on the soil colloids and those left in the soil solution.

Because we seldom know the exact involvement of adsorption, complexation, or partitioning processes, we use the general term **sorption** to describe the retention of organic compounds by soil colloids. Many organic compounds are **hydrophobic,** meaning they are repelled by water. In moist soil, clay particles are coated with a layer of water and, therefore, cannot effectively sorb hydrophobic organic compounds. However, it is possible to replace some of the hydrated metal cations (e.g., Ca^{2+}) on such clays with large organic cations, giving rise to **organoclays** that *can* effectively retain hydrophobic organic compounds. Environmental soil scientists take advantage of this phenomenon by injecting quaternary ammonium compounds into contaminated clay soils to create organoclays that can stop the movement of organic contaminants and thus protect groundwater (see Figure 8.18).

Distribution Coefficients

The tendency of an organic contaminant to leach into the groundwater is determined by the solubility of the compound and by the ratio of the amount of chemical sorbed by the soil to that remaining in solution. This ratio is known as the soil **distribution coefficient** K_d.

$$K_d = \frac{\text{mg chemical sorbed/kg soil}}{\text{mg chemical/L solution}} \tag{8.7}$$

The K_d therefore has units of L/kg or mL/g. The K_d for a given compound may vary widely depending on the nature of the soil in which the compound is distributed. The variation is related mainly to the amount of organic matter (organic carbon) in the soils.

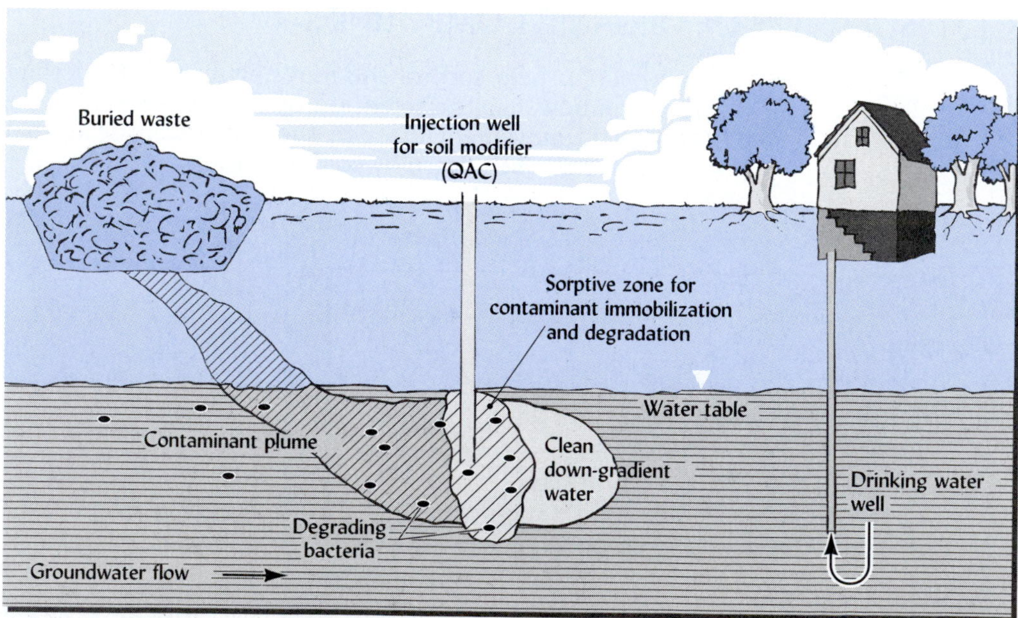

FIGURE 8.18 How a combination of a quaternary ammonium compound (QAC), hexadecyltrimethylammonium, and bioremediation by degrading bacteria can be used to hold and remove an organic contaminant. The pollutant is moving into groundwater from a buried waste site. The QAC reacts with soil clays to form organoclays and soil organic matter complexes that adsorb and stabilize the contaminant, giving microorganisms time to degrade or destroy it. [Redrawn from Xu et al. (1997)]

Therefore, most scientists prefer to use a similar ratio that focuses on sorption by organic matter. This ratio is termed the organic carbon distribution coefficient K_{oc}:

$$K_{oc} = \frac{\text{mg chemical sorbed/kg organic carbon}}{\text{mg chemical/L solution}} \tag{8.8}$$

The K_{oc} can be calculated by dividing the K_d by the fraction of organic C (g/g) in the soil. Therefore, K_{oc} value is often 100 times as large as K_d values. Higher K_d or K_{oc} values indicate the chemical is more tightly sorbed by the soil and therefore less susceptible to leaching and movement to the groundwater. On the other hand, if the management objective is to wash the chemical out of a soil, this will be more easily accomplished for chemicals with lower coefficients. Equations 8.7 and 8.8 emphasize the importance of the sorbing power of the soil colloidal complex, and especially of humus, in the management of organic compounds added to soils.

Binding of Biomolecules to Clay and Humus

The enormous surface area and charged sites on the clay and humus in soils attract and bind many types of biologically active substances, including DNA (genetic code material), enzymes, toxins, and even viruses. Adsorption of biomolecules takes place rapidly (in a matter of minutes) and the amount adsorbed is related to the type of clay mineral (Figure 8.19). The bond between the biomolecule and the colloid is often quite strong, so that the biomolecule cannot be easily removed by washing or exchange reactions. In most cases, biomolecules bound to clays do not enter the interlayer spaces, but are attached to the outer planar surfaces and edges of the clay crystals.

The binding of biomolecules to soil colloids in this manner has important environmental implications for two reasons. First, such binding usually protects the biomolecules from enzymatic attack, meaning that the molecules will persist in the soil much longer than studies of unbound biomolecules might suggest. Second, it has been shown that many biomolecules retain their biological activity in the bound state. Toxins remain toxic to susceptible organisms, enzymes continue to catalyze reactions, viruses can lyse (break open) cells or transfer genetic information to host cells, and DNA strands retain the ability to transform the genetic code of living cells, even while bound to colloidal surfaces and protected from decay. The presence of such DNA is not detectable by the usual chemical tests since it is not in a living cell and therefore not expressing its genes.

When genetically modified plants or bacteria are introduced to the soil environment (see Section 10.16), these cryptic (hidden) genes may represent a serious, undetected

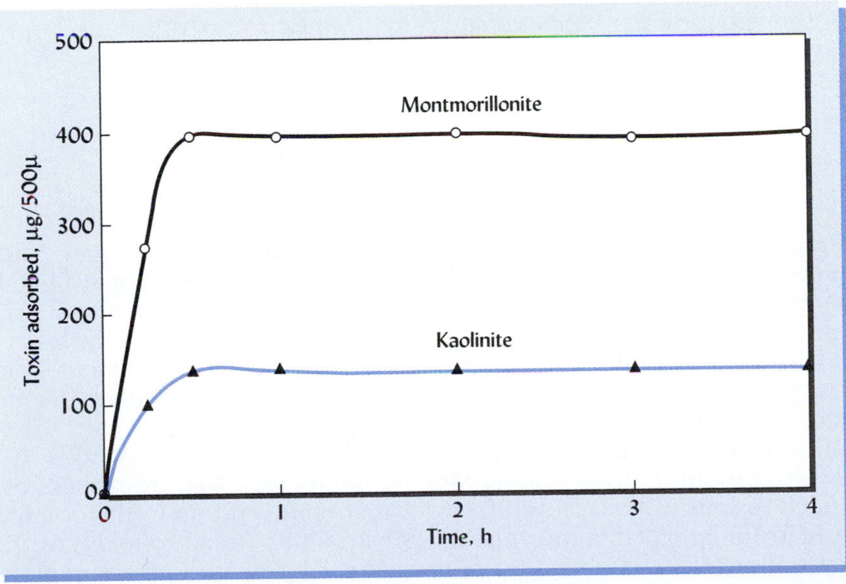

FIGURE 8.19 Adsorption of a toxin called *Bt*, the insecticidal protein produced by the soil bacteria, *Bacillus thuringiensis*, and used to protect some crop plants. Highly active clay minerals such as montmorillonite (a member of the smectite group of 2:1 clays) adsorb and bind much larger amounts of these biomolecules than do low-activity clays such as kaolinite (a 1:1 mineral). In both cases, the adsorption reaction was completed in 30 minutes or less. Since only 500 µg of either clay mineral was used in the experiment, it appears that the clays adsorbed an amount of the toxin equal to 30 to 80% of their mass. [Redrawn from Stotzky (2000)]

hazard for the potential transfer of genetic information to organisms for which they were not intended. A similar concern exists regarding plants genetically modified to produce certain compounds, such as crops given a bacterial gene that codes for the production of the insecticidal toxin Bt. The Bt toxin is released into the soil by root excretion and by the decomposition of crop residues containing the toxin. Over time, large amounts of the toxin may accumulate in soils in the colloid-bound state, threatening to upset the ecological balance by killing susceptible nontarget organisms that play a critical role in soil processes (see Sections 10.2 to 10.5).

8.13 PHYSICAL IMPLICATIONS OF SWELLING-TYPE CLAYS

Engineering Hazards

Map and discussion of swelling clays in the U.S.: http://www.surevoid.com/surevoid_web/soil_maps/intro1.html

Soil colloids differ widely in their physical properties, including plasticity, cohesion, swelling, shrinkage, dispersion, and flocculation. These properties greatly influence the usefulness of soils for both engineering and agricultural purposes. As discussed in Section 4.9, the tendency of certain clays to swell in volume when wetted (their coefficient of linear extensibility [COLE]) is a major concern for the construction of roads and foundations. Values for the COLE and plasticity index are both much higher for smectitic soils than kaolinitic ones.

Figure 8.20 gives an example of steps that must be taken in constructing buildings to overcome the adverse effects of the shrinking and swelling properties of smectite on some sites. If preventative measures are not taken before houses are constructed on smectite clays, homeowners will likely pay dearly in the future. This is just one example of critical role soil colloids play in determining the usefulness of our soils.

Environmental Uses of Swelling-Type Clays

The physical and chemical properties of swelling-type clays, and particularly the smectites, make them useful tools in preventing unwanted movement of water and water contaminants. Such clays are widely used in the construction of liners to seal ponds, sewage lagoons, industrial waste lagoons, and landfills. A layer of the smectite placed on the bottom and sides of the pond or lagoon expands when wetted and forms a relatively impenetrable barrier to the movement not only of water but of organic and inorganic contaminants contained in the water. These contaminants are held in the containment area and are prevented from moving downward into the groundwater.

Smectites are also used to prevent the undesirable upward or downward movement of chemicals through boreholes drilled to monitor the presence of contaminants in groundwater. The wetted smectite (bentonite) forms an impenetrable plug between the sampling tube and the larger borehole, thereby preventing the upward movement of chemicals from the groundwater. Increasingly, environmental scientists are using swelling-type clays for such environmental monitoring, as well as for the removal of organic chemicals from water by partitioning.

8.14 CONCLUSION

The complex structures, enormous surface area (both internal and external), and tremendous numbers of charges of soil colloids combine to make these tiniest of soil particles the seats of chemical and physical activity in soils. The chemical activity of colloids results largely from the charges on or near their surfaces. These charged sites attract ions and molecules of opposite charge from the soil solution. The negative sites on the colloids attract positive ions (cations) such as Ca^{2+}, Cu^{2+}, K^+, or Al^{3+}. Positive sites on the colloids attract negative anions such as Cl^-, SO_4^{2-}, NO_3^-, or HPO_4^{2-}.

Although both positive and negative charges occur on colloids, in most soils the negative charges far outnumber the positive. Most elements dissolved from rocks by weathering or added to soils in lime or fertilizer will eventually end up in the oceans, but it is very fortunate for land plants and animals that attraction to soil colloids greatly slows the journey. Colloidal attraction is a major mechanism by which soils accumulate the stocks of nutrients necessary to support forests, crops, and, ultimately, civilizations.

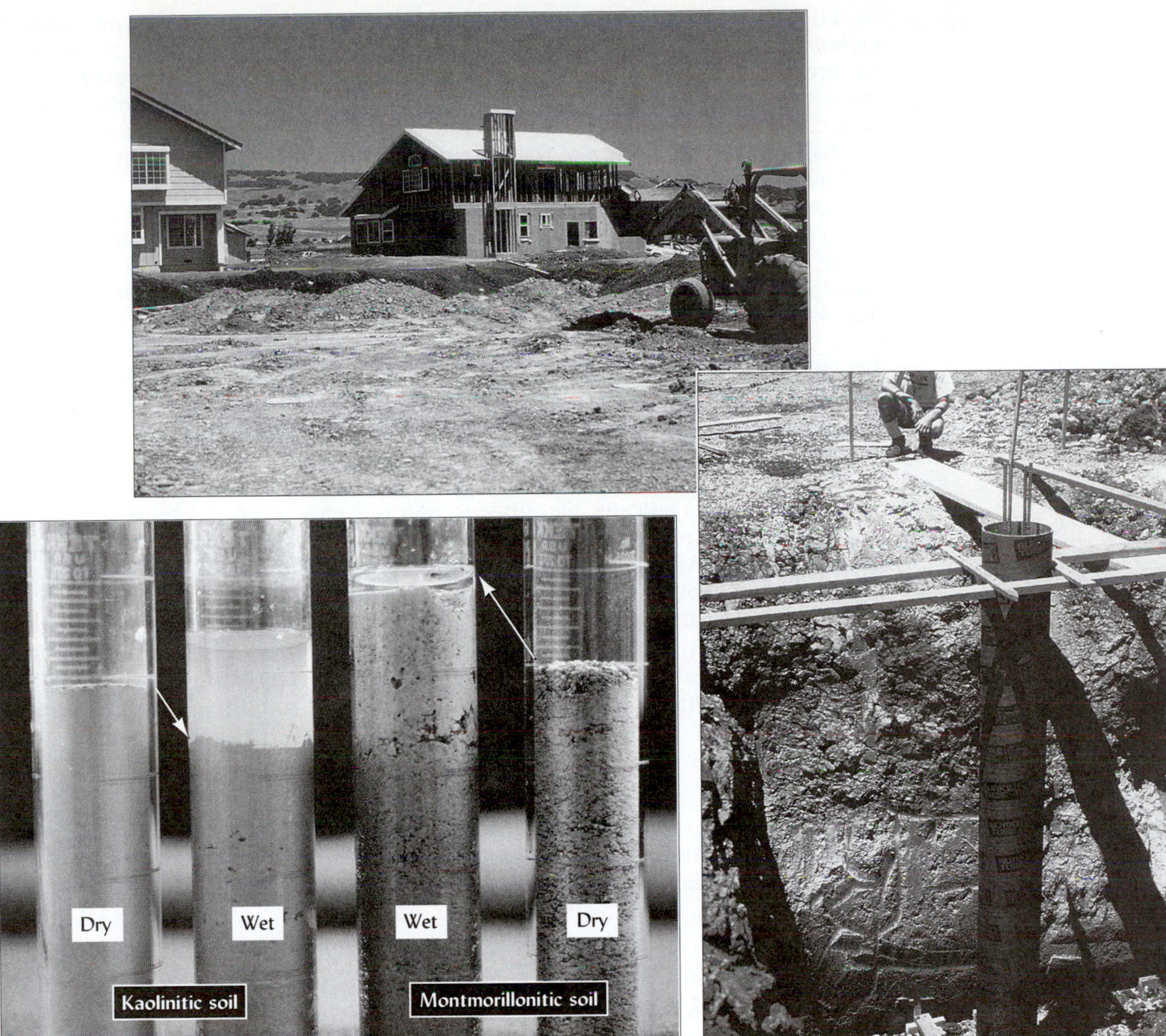

FIGURE 8.20 The different swelling tendencies of two types of clay are illustrated in the lower left. All four cylinders initially contained dry, sieved clay soil, the two on the left from the B horizon of a soil high in kaolinite, the two on the right from one of a soil high in montmorillonite. An equal amount of water was added to the two center cylinders. The kaolinitic soil settled a bit and was not able to absorb all the water. The montmorillonitic soil swelled about 25% in volume and absorbed nearly all the added water. The scenes to the right and above show a practical application of knowledge about these clay properties. (Upper) Soils containing large quantities of smectite undergo pronounced volume changes as the clay swells and shrinks with wetting and drying. Such soils (e.g., the California Vertisol shown here) make very poor building sites. The normal-appearing homes (upper) are actually built on deep, reinforced-concrete pilings (lower right) that rest on nonexpansive substrata. Construction of the 15 to 25 such pilings needed for each home more than doubles the cost of construction. (Photo courtesy of R. Weil)

When ions are attracted to a colloid, they may enter into two general types of relationships with the colloid surface. If the ion bonds directly to atoms of colloidal structure with no water molecules intervening, the relationship is termed an *inner-sphere complex*. This type of reaction is quite specific, and once created, is not easily reversed. In contrast, ions that keep their hydration shell of water molecules around them cannot come close enough to be bound chemically to the colloidal surfaces. Instead, they are held only loosely by electrostatic forces, and their interaction is less specific and more easily reversed.

The latter type of adsorption is termed *outer-sphere complexation*. Ions in such a state of dynamic adsorption are termed *exchangeable ions* because they break away from the colloid when another ion from the solution moves in closer and takes over neutralizing the colloid's charges. Plant roots can exchange H^+ for nutrient cations or OH^- ions for nutrient anions.

The replacement of one ion for another in the outer-sphere complex is termed *ion exchange.* Except in certain highly weathered subsurface horizons, cation exchange is far greater than anion exchange. Cation exchange reactions are reversible and balanced charge for charge (rather than ion for ion). The extent of the reaction is influenced by mass action, the relative charge and size of the hydrated ions, the nature of the colloid, and the nature of the other (complementary) ions already adsorbed on the colloid.

The colloids in soils are both organic in nature (humus) and mineral (clays). In most surface soils, half or more of the charges are contributed by organic matter colloids, while in most subsurface horizons, clays provide the majority of charges. The total number of negative colloid charges per unit mass is termed the *cation exchange capacity (CEC).* Expressed in moles of charge per kilogram, the CEC of different colloids varies from about 1 to over 200, and that of whole mineral soils commonly varies from about 1 to 40. The CEC of a soil, as well as its capacity to strongly adsorb particular ions (such as K^+ or HPO_4^{2-}) depends on the amount of humus in the soils and on both the amount and type of clays present. Low-activity clays (iron and aluminum oxides and 1:1-type silicate clays such as kaolinite) tend to dominate highly weathered soils of warm, humid regions. High-activity clays (expanding 2:1 silicates like smectite and vermiculite, and nonexpanding 2:1 silicates like fine-grained mica and chlorite) tend to dominate soils in cooler or drier regions where weathering is less advanced. Most of the charge on humus and low-activity clays is pH-dependent (becomes more negative as pH rises), while most of the charge on high-activity clays is permanent.

The differing ability of soil colloids to adsorb ions and molecules is key to managing soils for both plant production and to understanding how CEC may regulate movement of both nutrients and toxins in the environment. Among the important properties influenced by the colloids is the acidity or alkalinity of the soil, the topic of the next chapter.

STUDY QUESTIONS

1. Describe the *soil colloidal complex,* indicate its various components, and explain how it tends to serve as a "bank" for plant nutrients.
2. How do you account for the difference in surface area associated with a grain of kaolinite clay compared to that of montmorillonite, a smectite?
3. Contrast the difference in crystalline structure among *kaolinite, smectites, fine-grained micas, vermiculites,* and *chlorites.*
4. There are two basic processes by which silicate clays are formed by weathering of primary minerals. Which of these would likely be responsible for the formation of (a) fine-grained mica, and (b) kaolinite from muscovite mica? Explain.
5. If you wanted to find a soil high in kaolinite, where would you go? The same for (a) smecitite and (b) vermiculite?
6. Which of the silicate clay minerals would be *most* and *least* desired if one were interested in (a) a good foundation for a building, (b) a high cation exchange capacity, (c) an adequate source of potassium, and (d) a soil on which hard clods form after plowing?
7. Which of the following would you expect to be *most* and *least* sticky and plastic when wet: (a) a soil with significant sodium saturation in a semiarid area, (b) a soil high in exchangeable calcium in a subhumid temperate area, or (c) a well-weathered acid soil in the tropics? Explain your answer.
8. A soil contains 4% humus, 10% montmorillonite, 10% vermiculite, and 10% Fe, Al oxides. What is its approximate cation exchange capacity?
9. Calculate the number of grams of Al^{3+} ions needed to replace 10 $cmol_c$ of Ca^{2+} ion from the exchange complex of 1 kg of soil.
10. Explain the importance of K_d and K_{oc} in assessing the potential pollution of drainage water. Which of these expressions is likely to be most consistently characteristic of the organic compounds in question regardless of the type of soil involved? Explain.

REFERENCES

Bellini, G., M. E. Sumner, D. E. Radcliffe, and N. P. Qafoku. 1996. "Anion transport through columns of highly weathered acid soil: Adsorption and retardation," *Soil Sci. Soc. Amer. J.* **60**:132–137.

Coleman, N. T., and A. Mehlich. 1957. "The chemistry of soil pH," in *The Yearbook of Agriculture (Soil)*. (Washington, D.C.: U.S. Department of Agriculture).

Dixon, J. B., and D. J. Schulze. 2002. *Soil Mineralogy with Environmental Applications*. (Madison, Wis.: Soil Sci. Soc. Amer.).

Helling, C. S., et al. 1964. "Contribution of organic matter and clay to soil cation exchange capacity as affected by the pH of the saturated solution," *Soil Sci. Soc. Amer. Proc.* **28**:517–520.

Reid, D. A. and A. L. Ulery. 1998. "Environmental applications of smectites," in J. Dixon, D. Schultze, W. Bleam, and J. Amonette (eds.), *Environmental Soil Mineralogy*. (Madison, Wis.: Soil Sci. Soc. Amer.).

Stotzky, G. 2000. "Persistence and biological activity in soil of insecticidal proteins from *Bacillus thuringiensis* and of bacterial DNA bound on clays and humic acids," *J. Environ. Quality* **29**:691–705.

Xu, S., G. Sheng, and S. A. Boyd. 1997. "The use of organoclays in pollution abatement," *Advances in Agronomy* **59**:25–62.

9

Soil Acidity, Alkalinity, and Salinity

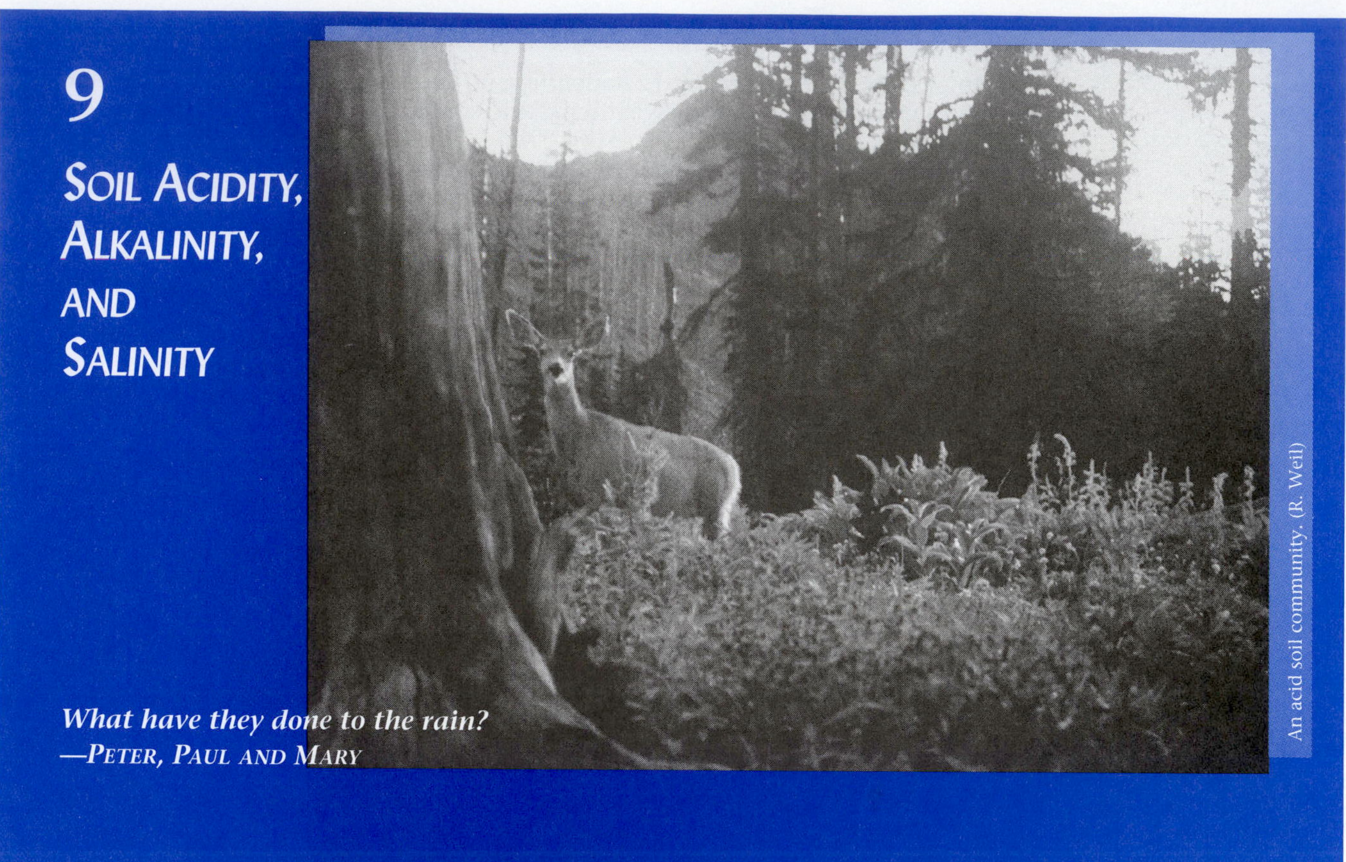

An acid soil community. (R. Weil)

What have they done to the rain?
—Peter, Paul and Mary

The degree of soil acidity or alkalinity, expressed as soil pH, is a *master variable* that affects a wide range of soil properties—chemical, biological, and, indirectly, even physical. This chemical variable greatly influences the availability for root uptake of many elements, including both nutrients and toxins. The activity of soil microorganisms is also affected. The mix of plant species that dominate a landscape under natural conditions often reflects the pH of the soil. For people attempting to produce crops or ornamental plants, soil pH is a major determinant of which species will grow well or even grow at all in a given site.

Soil pH affects the *mobility* of many pollutants in soil by influencing the rate of their biochemical breakdown, their solubility, and their adsorption to colloids. Thus, soil pH is a critical factor in predicting the likelihood that a given pollutant will contaminate groundwater, surface water, and food chains. Furthermore, there are certain situations in which so much acidity is generated that the acid itself becomes a significant environmental pollutant. For example, in this chapter we will learn how soils developing in waste materials from certain mines or other disturbed land generate extremely acid drainage water that can cause massive fish kills when it reaches a lake or stream.

Many complex factors affect soil pH, but none more than two simple balances: the balance between acid and nonacid cations on colloid surfaces and the balance between H^+ and OH^- ions in the soil solution. These balances, in turn, are largely controlled by the nature of the soil colloids. Therefore, to understand and manage soil acidity, it is essential to have a good grasp of the concepts of charged colloidal surfaces and cation exchange as presented in Chapter 8. Also inextricably tied to soil acidity (and therefore considered in this chapter) are the supply of the plant macronutrients, **calcium** and **magnesium**, and the toxicity of the nonnutrient element, **aluminum.**

Acidification is a natural process involved in soil formation. It reaches its greatest expression in humid regions where rainfall is sufficient to thoroughly leach the soil profile. Not surprisingly, acidity is one of the main constraints to crop production in these areas. Forests, which comprise the natural vegetation in most humid regions, are better adapted to acid soils than are most crops. Nonetheless, increasing soil acidity seems to be a problem for forest health in certain regions, especially where soils are incapable of resisting the effects of human-induced acidification.

By contrast, in drier regions leaching is much less extensive, allowing soils to retain enough Ca^{2+}, Mg^{2+}, K^+, and Na^+ to prevent a buildup of acid cations. Soils in semiarid and arid regions, therefore, tend to have **alkaline** pH levels (that is, pH >7). Many soils of dry regions also accumulate detrimental levels of soluble salts (**saline soils**) or exchangeable sodium ions (**sodic soils**), or both. The chemical conditions associated with **alkalinity, salinity**, and **sodicity** can lead to severe problems in the physical condition and fertility of soils in these areas. More than 2100 years ago, the Roman armies spread salt (sodium chloride) on the lands of their vanquished enemies in the city-state of Carthage, to ensure that they would never have to fight them again. In this chapter, we learn why sodium and salts are so damaging to soils and how they can be managed. Indeed, though only one of many problems unique to alkaline soils, we will learn that salt accumulation is perhaps the most vexing problem for long-term sustainable use of arid lands.

9.1 THE PROCESS OF SOIL ACIDIFICATION

Economic evaluation of soil acidification, the view from Australia:
http://www.regional.org.au/au/roc/1986/roc198663.htm

Acidity and alkalinity are usually quantified using the pH scale, which expresses the activity or concentration of H^+ ions present in a solution. Therefore, before we begin our discussion of the chemical nature and causes of acidity in soils, it will be useful to review the fundamentals about pH in Box 9.1 and Figure 9.1. It will also be of value to develop a feel for the range of pH values found in soils and to compare these to the pH values of other common substances (see Figure 9.2).

In order to understand soil acidification and its management, we need to know where the H^+ ions come from, how they enter the soil system, and how they may be lost. We will begin by describing a variety of processes that add H^+ ions to the soil system and therefore contribute to acidification. Although most of these acid-producing processes occur naturally, human activities have a major impact on some of them, as we shall see in Section 9.7.

Sources of Hydrogen Ions

Fundamentals of acid-base chemistry:
http://www.shodor.org/UNChem/basic/ab/index.html

CARBONIC AND OTHER ORGANIC ACIDS. Perhaps the most ubiquitous contributor to soil acidity is the formation and subsequent dissociation of H^+ ions from **carbonic acid.** This weak acid is formed when carbon dioxide gas from soil air dissolves in water. Because H_2CO_3 is a weak acid (has a relatively high pK_a),[1] its contribution of H^+ ions is negligible when the pH is much below 5.0. Root respiration and the decomposition of soil organic matter by microorganisms produce high levels of CO_2 in soil air, pushing the following reaction to the right.

$$CO_2 + H_2O \longrightarrow H_2CO_3 \rightleftharpoons HCO_3^- + H^+ \qquad pK_a = 6.35 \tag{9.1}$$

Many other organic acids are generated as microbes break down soil organic matter. Some of these are low molecular weight organic acids, such as citric or malic acids, that only weakly dissociate. Others are more complex and stronger acids, such as the carboxylic and phenolic acid groups in humic substances produced by litter breakdown. The pK_a of carboxylic groups ranges from 3 to 5, depending on the nature of the associated organic structures. Other organic groups, such as phenolic (R-OH), dissociate at higher pH levels.

Soil acidification—the causes:
http://regional.org.au/au/roc/1981/roc198131.htm

ACCUMULATION OF ORGANIC MATTER. The accumulation of organic matter tends to acidify the soil for two reasons. First, organic matter forms soluble complexes with nonacid nutrient cations such as Ca^{2+} and Mg^{2+}, thus facilitating the loss of these cations by

[1] The pK_a is the negative logarithm (analogous to the pH) of the equilibrium constant for a dissociation reaction—for example, $[H^+] \times [HCO_3^-]/H_2CO_3 = K_a$. It indicates the pH at which one half of the acid will be dissociated. Dissociation of the acid can usually be considered negligible when the pH is more than 1.0 to 2.0 units lower than the pK_a.

BOX 9.1 SOIL pH, SOIL ACIDITY, AND ALKALINITY

Whether a soil is acid, neutral, or alkaline is determined by the comparative concentrations of H^+ and OH^- ions.[a] Pure water provides these ions in equal concentrations:

$$H_2O \rightleftharpoons H^+ + OH^-$$

The equilibrium for this reaction is far to the left, only about 1 of every 10 million water molecules is dissociated into H^+ and OH^- ions. The *ion product* of the concentrations of the H^+ and OH^- ions is a constant (K_w), which at 25°C is known to be 1×10^{-14}:

$$[H^+] \times [OH^-] = K_w = 10^{-14}$$

Since in pure water the concentration of H ions $[H^+]$ must be equal to that of OH^- ions $[OH^-]$, this equation shows that the concentration of each is 10^{-7} ($10^{-7} \times 10^{-7} = 10^{-14}$). It also shows the inverse relationship between the concentrations of these two ions. As one increases, the other must decrease proportionately. Thus, if we were to increase the H^+ ion concentration $[H^+]$ by 10 times (from 10^{-7} to 10^{-6}), the $[OH^-]$ would be decreased by 10 times (from 10^{-7} to 10^{-8}) since the product of these two concentrations must equal 10^{-14}:

$$10^{-6} \times 10^{-8} = 10^{-14}$$

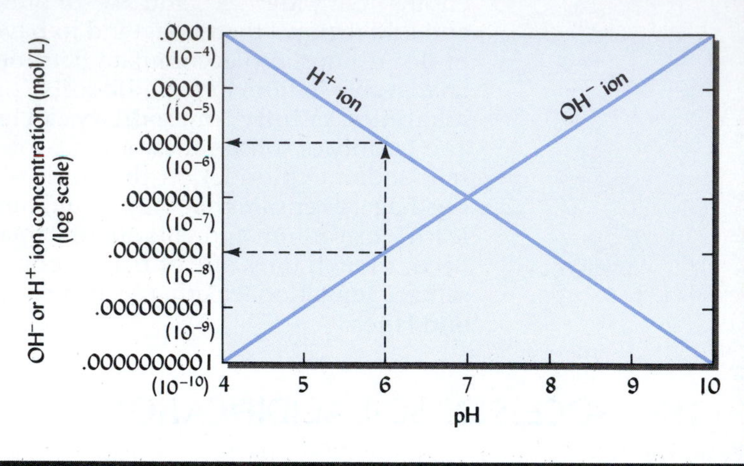

FIGURE 9.1 *The relationship between pH, pOH, and the concentrations of hydrogen and hydroxyl ions in water solution.*

Scientists have simplified the means of expressing the very small concentrations of H^+ and OH^- ions by using the *negative logarithm of the H^+ ion concentration*, termed the *pH*. Thus, if the H^+ concentration in an acid medium is 10^{-5}, the pH is 5; if it is 10^{-9} in an alkaline medium, the pH is 9.

Note that the pH also gives us an indirect measure of the OH^- ion concentration since the product of $[H^+] \times [OH^-]$ must always equal 10^{-14}. Thus, at pH 5 the $[OH^-]$ is 10^{-9} ($10^{-5} \times 10^{-9} = 10^{-14}$); at pH 8 it is 10^{-6} ($10^{-8} \times 10^{-6} = 10^{-14}$).

Figure 9.1 above shows the relationship between pH and the concentrations of H^+ and OH^- ions. Note that as one goes down, the other goes up, and vice versa. The dotted line illustrates the concentrations of these two ions at pH = 6.0: $[H^+] = 10^{-6}$; $[OH^-] = 10^{-8}$. This reciprocal relationship between H^+ and OH^- ions should always be kept in mind in studying soil acidity and alkalinity.

[a]Technically speaking, chemical reactions are influenced by the *activity* of an ion rather than by its *concentration*, because of the electrostatic effect of one ion on the activity of its nearby neighbors. Ionic activities are thus essentially effective concentrations. Since the difference between ionic activities and concentrations in soils are not so great, we will use the term *concentration* in this text.

leaching. Second, organic matter is a source of H^+ ions because it contains numerous acid **functional groups** from which these ions can dissociate (see Figure 8.9).

OXIDATION OF NITROGEN (NITRIFICATION). Oxidation reactions generally produce H^+ ions as one of their products. Reduction reactions, on the other hand, tend to consume H^+ ions and raise soil pH. Ammonium ions (NH_4^+) from organic matter or from most fertilizers react with oxygen, in a process termed nitrification, releasing two H^+ ions for each NH_4^+ ion oxidized. Because the nitrate (NO_3) produced is the anion of a **strong acid** (nitric acid, HNO_3), it does not tend to recombine with the H^+ ion to make the reaction go to the left.

$$NH_4^+ + 2O_2 \rightleftharpoons H_2O + \underbrace{H^+ + H^+ + NO_3^-}_{\text{Dissociated nitric acid}} \quad (9.2)$$

OXIDATION OF SULFUR. The decomposition of plant residues commonly involves the oxidation of organic –SH groups to yield sulfuric acid (H_2SO_4). Another important source of this strong acid is the oxidation of reduced sulfur in minerals such as pyrite. This and

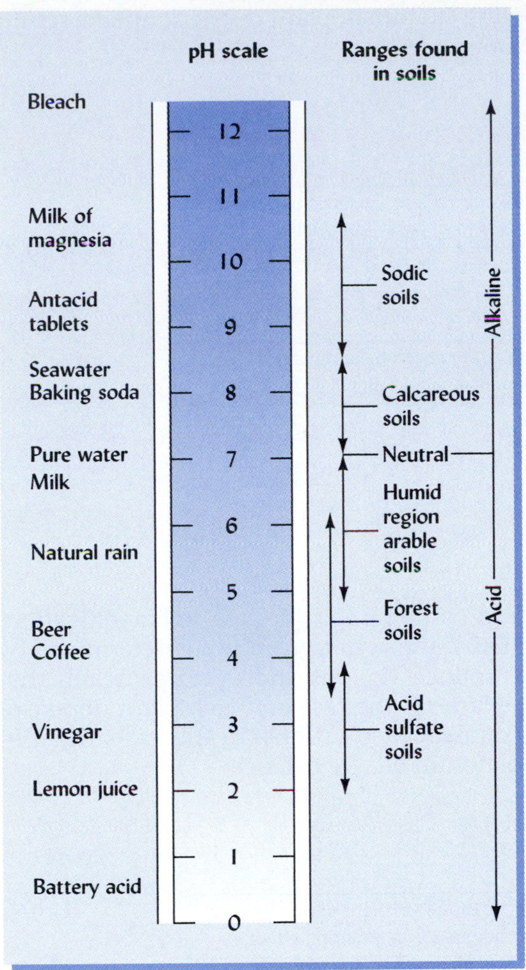

FIGURE 9.2 Some pH values for familiar substances (left) compared to ranges of pH typical for various types of soils (right).

related reactions are responsible for producing large amounts of acidity in certain soils in which reduced sulfur is plentiful and oxygen levels are increased by drainage or excavation (see Section 9.7).

$$\underset{\text{Pyrite}}{FeS_2} + 3\tfrac{1}{2}O_2 + H_2O \rightleftharpoons \underset{\text{Ferrous sulfate}}{FeSO_4} + \underset{\text{Dissociated sulfuric acid}}{2H^+ + SO_4^{2-}} \tag{9.3}$$

Acids in Precipitation. Precipitation (rain, snow, fog, and dust) contains a variety of acids that contribute H^+ ions to the soil receiving the precipitation. As raindrops fall through unpolluted air, they dissolve CO_2 and form enough carbonic acid to lower the pH of the water from 7.0 (the pH of pure water) to about 5.6. Varying amounts of sulfuric and nitric acids form in precipitation from certain nitrogen and sulfur gases produced by lightning, volcanic eruptions, forest fires, and the combustion of fossil fuels. Unlike carbonic acid, these strong acids completely dissociate to form H^+ ions and sulfate or nitrate anions. In recent decades, combustion of coal and petroleum products has significantly increased the amounts of these strong acids found in precipitation (see Section 9.7).

Plant Uptake of Cations. Plants must maintain a balance between the positive and negative charges on the ions they take up from the soil solution. For every positive charge taken in on a cation, a root can maintain charge balance either by taking up a negative charge as an anion or by exuding a positive charge as a different cation. When they take up far more of certain cations (e.g., K^+, NH_4^+, and Ca^{2+}) than they do of anions (e.g., NO_3^-, SO_4^{2-}), plants usually exude H^+ ions into the soil solution to maintain charge balance. In the first two of the following examples, plant nutrient uptake results in the addition of H^+ ions to the soil solution:

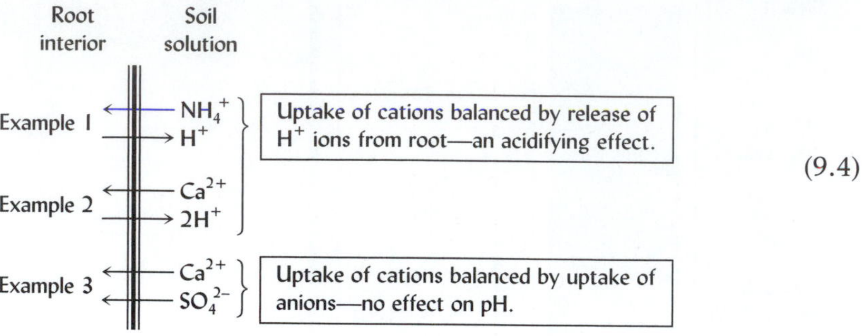

(9.4)

Balance Between Production and Consumption of H^+ Ions

The H^+ ion-producing processes just discussed combine to stimulate soil acidification. However, the degree of acidification that actually occurs in a given soil is determined by the balance between those processes that produce H^+ ions and other processes that *consume* them (Table 9.1). An example of the latter is the case in which plant uptake of an anion such as NO_3^- exceeds the uptake of associated cations. In this case, the root exudes the anion, bicarbonate (HCO_3^-) to maintain charge balance:

Root interior | Soil solution

NO_3^- (taken up by root) / HCO_3^- (released to soil solution) — Uptake of anion balanced by release of bicarbonate ion—alkalizing effect. (9.5)

The increased concentration of bicarbonate ions tends to reverse the dissociation of carbonic acid (see Equation 9.1), thereby *consuming* H^+ ions, and raising the pH of the soil solution. Another H^+ ion–consuming process involving nitrogen is the reduction of nitrate to nitrogen gases under anaerobic conditions (see denitrification, Sections 7.3 and 12.10).

TABLE 9.1 The Main Processes that Produce or Consume Hydrogen Ions (H^+) in Soil Systems

Production of H^+ ions increases soil acidity, while consumption of H^+ ions delays acidification and leads to alkalinity. The pH level of a soil reflects the long-term balance between these two types of processes.

Acidifying (H^+ ion–producing) processes	*Alkalinizing (H^+ ion–consuming) processes*
Formation of carbonic acid from CO_2	Input of bicarbonates or carbonates
Acid dissociation such as:	Anion protonation such as:
$RCOOH \rightarrow RCOO^- + \mathbf{H^+}$	$RCOO^- + \mathbf{H^+} \rightarrow RCOOH$
Oxidation of N, S, and Fe compounds	Reduction of N, S, and Fe compounds
Atmospheric H_2SO_4 and HNO_3 deposition	Atmospheric Ca, Mg deposition
Cation uptake by plants	Anion uptake by plants
Accumulation of acidic organic matter (e.g., fulvic acids)	Specific (inner sphere) adsorption of anions (especially SO_4^{2-})
Cation precipitation such as:	Cation weathering from minerals such as:
$Al^{3+} + 3H_2O \rightarrow \mathbf{3H^+} + Al(OH)_3^0$	$\mathbf{3H^+} + Al(OH)_3^0 \rightarrow Al^{3+} + 3H_2O$
$SiO_2 + 2Al(OH)_3 + Ca^{2+} \rightarrow CaAl_2SiO_6 + 2H_2O + \mathbf{2H^+}$	$CaAl_2SiO_6 + 2H_2O + \mathbf{2H^+} \rightarrow SiO_2 + 2Al(OH)_3 + Ca^{2+}$
Deprotonation of pH-dependent charges	Protonation of pH-dependent charges

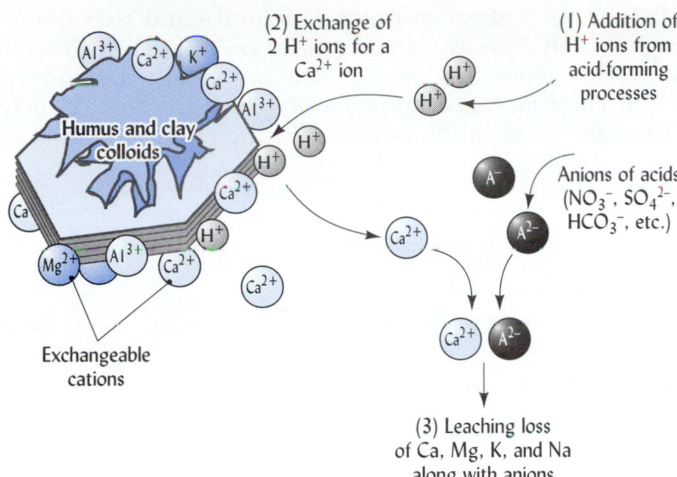

FIGURE 9.3 Soils become acid when H^+ ions added to the soil solution exchange with nonacid Ca^{2+}, Mg^{2+}, K^+, and Na^+ ions held on humus and clay colloids. The nonacid cations can then be exported in leaching water along with accompanying anions. As a result, the exchange complex (and therefore also the soil solution) becomes increasingly dominated by acid cations (H^+ and Al^{3+}). Because of this sequence of events, H^+ ion–producing processes acidify soils in humid regions where leaching is extensive, but cause little long-term soil acidification in arid regions where the Ca^{2+}, Mg^{2+}, K^+, and Na^+ are mostly *not* removed by leaching. In the latter case, the Ca^{2+}, Mg^{2+}, K^+, and Na^+ remain in the soil and re-exchange with the acid cations, preventing a drop in pH level.

The weathering of nonacid cations from minerals (Section 2.1) is a slow but very important H^+ ion–consuming process that counteracts acidification. An example is the weathering of calcium from a silicate mineral:

$$\text{Ca-silicate} + 2H^+ \longrightarrow H_4SiO_4 + Ca^{2+} \tag{9.6}$$

Some of the nonacid cations (Ca^{2+}, Mg^{2+}, K^+, and Na^+) so released become exchangeable cations on the soil colloids. Hydrogen ions added to the soil solution from acids in rain (and other sources just discussed) may replace these cations on the cation exchange sites of humus and clay. The displaced nonacid cations are then subject to loss by leaching along with the anions of the added acids (Figure 9.3). The soil slowly becomes more acid if the leaching of Ca^{2+}, Mg^{2+}, K^+, and Na^+ proceeds faster than the release of these cations from weathering minerals. Thus, the formation of acid soils is favored by high rainfall, parent materials low in Ca, Mg, K, and Na, and a high degree of biological activity (favoring formation of H_2CO_3). The positive relationship between annual rainfall and soil acidity can be seen along climosequences on every continent.

9.2 CAUSES OF ALKALINITY: HIGH SOIL pH

In regions where precipitation is less than potential evapotranspiration (see Section 6.3)—generally temperate-zone regions with less than about 500 mm (20 in) of rain per year—the cations released by mineral weathering accumulate because there is not enough rain to thoroughly leach them away. The pH of soils in these arid and semiarid environments is generally in the alkaline range, that is, 7 or above.

There seems to be considerable confusion about the terms **alkaline** and **alkalinity**, and people often use these terms *wrongly* to describe soils characterized by detrimental levels of soluble salts or sodium. Alkaline soils are simply those with a pH above 7.0. *Alkalinity* refers to the concentration of OH^- ions, much as *acidity* refers to that of H^+ ions. *Alkaline* soils should not be confused with *alkali* soils, an obsolete term for what are now called **sodic** or **saline-sodic soils**, those with levels of sodium high enough to be detrimental to plant growth (see Section 9.15).

Sources of Alkalinity

Minimal leaching in dry environments means that soil acidification is minimized. The cations in the soil solution and on the exchange complex are mainly Ca^{2+}, Mg^{2+}, K^+, and Na^+. As explained in Section 9.3, these cations are nonhydrolyzing and so do not produce acid (H^+) upon reacting with water as Al^{3+} or Fe^{3+} do. However, they generally do not produce OH^- ions either. Rather, their effect in water is neutral,[2] and soils dominated by them have a pH no higher than 7 unless certain *anions* are present in the soil solution. The basic, hydroxyl (OH^-)–generating anions are principally **carbonate** (CO_3^{2-}) and **bicarbonate** (HCO_3^-). These anions originate from the dissolution of such minerals as calcite ($CaCO_3$) or from the dissociation of carbonic acid (H_2CO_3).

$$\underset{\text{Calcite (solid)}}{CaCO_3} \rightleftharpoons \underset{\text{(dissolved in water)}}{Ca^{2+}} + \underset{\text{(dissolved in water)}}{CO_3^{2-}} \quad (9.7)$$

$$CO_3^{2-} + H_2O \rightleftharpoons HCO_3^- + OH^- \quad (9.8)$$

$$HCO_3^- + H_2O \rightleftharpoons H_2CO_3 + OH^- \quad (9.9)$$

$$\underset{\text{Carbonic acid}}{H_2CO_3} \rightleftharpoons H_2O + \underset{\text{(gas)}}{CO_2\uparrow} \quad (9.10)$$

In this series of linked equilibrium reactions, carbonate and bicarbonate act as bases because they react with water to form hydroxyl ions and thus raise the pH.

Influence of Carbon Dioxide and Carbonates

The direction of the overall reaction (Equations 9.7 to 9.10) determines whether OH^- ions are consumed (proceeding to the left) or produced (proceeding to the right). The reaction is controlled mainly by the precipitation or dissolution of calcite on the one end, and by the production (by respiration) or loss (by volatilization to the atmosphere) of carbon dioxide at the other end. The concentration of CO_2 in the atmosphere is about 0.0035%, but may be as high as 0.5% in soil air due to respiration by roots and microorganisms (see Section 7.2). Therefore, biological activity in soils tends to lower the pH by driving the reaction series to the left.

The other process that limits the rise in pH is the precipitation of $CaCO_3$ that occurs when the soil solution becomes saturated with respect to Ca^{2+} ions. Such precipitation removes Ca from the solution, again driving the reaction series to the left (lowering pH). Because of the limited solubility of $CaCO_3$, the pH of the solution cannot rise above 8.4 when the CO_2 in solution is in equilibrium with that in the atmosphere. The pH at which $CaCO_3$ precipitates in soil is typically only about 7.0 to 8.0, depending on how much the CO_2 concentration is enhanced by biological activity. This is an important point to remember, because it suggests that if other carbonate minerals more soluble than $CaCO_3$ (for example, Na_2CO_3) were present, the pH would rise considerably higher (see Section 9.15). Indeed this is the case; calcite-laden **calcareous soil** horizons range in pH from 7 to 8.4 (tolerable by most plants) while sodic (sodium carbonate–laden) horizons may range in pH from 8.5 to as high as 10.5 (levels toxic to many plants).

Because sodium carbonate (and sodium bicarbonate) is much more water soluble than calcium carbonate, producing high concentrations of CO_3^{2-} ions in solution, and a pH of 10 or higher, it is fortunate for plants that Ca^{2+}, not Na^+, ions are dominant in most soils.

Calcium-Rich Layers

A common characteristic of many soils in low-rainfall areas is the accumulation of calcium carbonate at some depth in the soil profile (see Plate 13 after page 114). Soil materials with free calcium carbonates are termed *calcareous* and can be distinguished in

[2]The cations Ca^{2+}, Mg^{2+}, K^+, Na^+, and NH_4^+ have been traditionally called *base* or *base-forming* cations as a convenient way of distinguishing them from the acid cation, H^+, and the H^+-forming cations Al^{3+} and Fe^{3+}. However, since Ca^{2+} and associated cations do not accept protons, they are technically not bases. Consequently, it is less misleading to refer simply to acid cations (H^+, Al^{3+}, and Fe^{3+}) and nonacid cations (most other cations). Likewise, the term *nonacid saturation* should be used rather than *base saturation* to refer to the percentage of the exchange capacity satisfied by nonacid cations (usually Ca^{2+}, Mg^{2+}, K^+, and Na^+) on the exchange complex (see also Section 9.4).

Colonies of bacteria called "desert varnish," and colonies of symbiotic algae and fungi called lichens: http://www.desertusa.com/magdec97/varnish/dec_varnish.html

the field by the effervescence (fizzing) that occurs if a drop of acid (10% HCl or strong vinegar) is applied. The high carbonate concentrations in these *calcic* horizons can inhibit root growth for some plants. Carbonate concentrations are often found at or near the soil surface in eroded spots or in regions of very low rainfall (<25 cm/yr). In these soils, serious micronutrient and phosphorus deficiencies can result for plants that are not adapted to the calcareous conditions (Plate 55 after page 498).

In other alkaline soils, one or more subsoil layers may be cemented into hard, concretelike horizons such as petrocalcic layers or duripans. Many alkaline soils also contain layers rich in calcium sulfate (gypsum), a mineral much more soluble than calcium carbonate. The depth of these *gypsic* horizons (and also of calcic horizons) is largely determined by the age of the soil and by the amount of rainfall available to leach these minerals downward (see Figure 3.15).

9.3 ROLE OF ALUMINUM IN SOIL ACIDITY

Aluminum toxicity in rice and its control: http://www.knowledgebank.irri.org/ricedoctor_mx/Fact_Sheets/DeficienciesToxicities/Aluminum.html

Although low pH is defined as a high concentration of H^+ ions, **aluminum** also plays a central role in soil acidity. Aluminum is a major constituent of most soil minerals (**aluminosilicates** and aluminum oxides), including clays. When H^+ ions are adsorbed on a clay surface, they usually do not remain as exchangeable cations for long, but instead they attack the structure of the minerals, releasing Al^{3+} ions in the process. The Al^{3+} ions then become adsorbed on the colloid's cation-exchange sites. These exchangeable Al^{3+} ions, in turn, are in equilibrium with dissolved Al^{3+} in the soil solution.

The exchangeable and soluble Al^{3+} ions play two critical roles in the soil acidity story. First, aluminum is *highly toxic* to most organisms and is responsible for much of the deleterious impact of soil acidity on plants and aquatic animals. We will discuss this role in Section 9.8. Second, Al^{3+} ions have a strong tendency to hydrolyze, splitting the water molecules into H^+ and OH^- ions. The aluminum combines with the OH^- ions, leaving the H^+ to lower the pH of the soil solution. For this reason, Al^{3+} and H^+ together are considered **acid cations.** A single Al^{3+} ion can thus release up to three H^+ ions as the following reversible reaction series proceeds to the right in stepwise fashion:

$$Al^{3+} \underset{H_2O \;\; H^+}{\overset{H_2O \;\; H^+}{\rightleftharpoons}} AlOH^{2+} \underset{H_2O \;\; H^+}{\overset{H_2O \;\; H^+}{\rightleftharpoons}} Al(OH)_2^{+} \underset{H_2O \;\; H^+}{\overset{H_2O \;\; H^+}{\rightleftharpoons}} \underset{\text{Gibbsite or amorphous (solid)}}{Al(OH)_3^0} \quad (9.11)$$

$$pK_a = 5.0 \qquad pK_a = 5.1 \qquad pK_a = 6.7$$

Most of the hydroxy aluminum ions $[Al(OH)_x^{y-}]$ formed as the pH increases are strongly adsorbed to clay surfaces or complexed with organic matter. Often the hydroxy aluminum ions join together, forming large polymers with many positive charges. When tightly bound to the colloid's negative charge sites, these polymers are not exchangeable and so mask much of the colloid's potential cation exchange capacity. As the pH is raised and more of the hydroxyl aluminum ions precipitate as uncharged $Al(OH)_3^0$, the negative sites on the colloids become available for cation exchange. This is one reason for the increase in soil cation exchange capacity (CEC) as the pH is raised from pH 4.5 to pH 7.0 [above which virtually all the aluminum cations have precipitated as $Al(OH)_3^0$].

9.4 POOLS OF SOIL ACIDITY

Principal Pools of Soil Acidity

Research suggests that three major pools of acidity are common in soils: (1) **active acidity** due to the H^+ and Al^{3+} ions in the soil solution; (2) **salt-replaceable (exchangeable) acidity**, involving the aluminum and hydrogen that are *easily exchangeable* by other cations in a simple unbuffered salt solution, such as KCl; and (3) **residual acidity,** which can be neutralized by limestone or other alkaline materials but cannot be detected by the salt-replaceable technique. These types of acidity comprise the **total acidity** of a soil. In addition, a much less common, but sometimes very important fourth pool, namely **potential acidity**, arises from the oxidation of sulfur compounds in certain acid sulfate soils (see Section 9.7).

Fundamentals of acid-base chemistry: www.shodor.org/UNChem/basic/ab/index.html#char

ACTIVE ACIDITY. The active acidity pool is defined by the H^+ ion activity in the soil solution. This pool is very small compared to the acidity in the exchangeable and residual pools. Even so, the active acidity is extremely important, as it determines the solubility of many substances and provides the soil solution environment to which plant roots and microbes are exposed.

EXCHANGEABLE (SALT-REPLACEABLE) ACIDITY. Salt-replaceable acidity is primarily associated with exchangeable aluminum and hydrogen ions that are present in large quantities in very acid soils (see Figure 9.3). These ions can be released into the soil solution by cation exchange with an unbuffered salt, such as KCl.

Once released to the soil solution, the aluminum hydrolyzes to form additional H^+, as explained in Section 9.3. The chemical equivalent of salt-replaceable acidity in strongly acid soils is commonly thousands of times that of active acidity in the soil solution. Even in moderately acid soils, the limestone needed to neutralize this type of acidity is commonly more than 100 times that needed to neutralize the soil solution (active acidity). At a given pH value, exchangeable acidity is generally highest for smectites, intermediate for vermiculites, and lowest for kaolinite.

RESIDUAL ACIDITY. Together, exchangeable (salt-replaceable) and active acidity account for only a fraction of the total soil acidity. The remaining *residual acidity* (or *reserve acidity*) is generally associated with hydrogen and aluminum ions (including the aluminum hydroxy ions) that are bound in nonexchangeable forms by organic matter and clays (see Figure 9.4). As the pH increases, the bound hydrogen dissociates and the bound aluminum ions are released and precipitate as amorphous $Al(OH)_3^0$. These changes free up negative cation exchange sites and increase the cation exchange capacity.

The residual acidity is commonly far greater than either the active or salt-replaceable acidity. It may be 1000 times greater than the soil solution or active acidity in a sandy soil and 50,000 or even 100,000 times greater in a clayey soil high in organic matter. The amount of ground limestone recommended to at least partly neutralize residual acidity in the upper 15 cm of soil is commonly 5 to 10 metric tons (Mg) per hectare (2.25 to 4.5 tons per acre).

POTENTIAL ACIDITY FROM REDUCED SULFUR. A fourth pool of soil acidity is found in potential acid sulfate soils that contain reduced sulfur compounds such as the mineral pyrite ($Fe^{II}S_2$). If oxygen is introduced into these normally anaerobic soils, acidifying reactions may occur (see Section 9.7). The term *potential acidity* refers to the acidity that could thus be produced by such reactions. Potential acidity is estimated from the total sulfur content of the soil material in the reduced state. The potential acidity of sulfide-rich soils can require even greater amounts of liming material for neutralization than does the residual acidity found in most acid soils. These soils are found only under limited circumstances and will be discussed in Section 9.7.

TOTAL ACIDITY. For most soils (not potential acid-sulfate soils) the total acidity that must be overcome to raise the soil pH to a desired value can be defined as:

$$\text{Total acidity} = \text{active acidity} + \text{salt replaceable acidity} + \text{residual acidity} \tag{9.12}$$

We can conclude that the pH of the soil solution is only the tip of the iceberg in determining how much lime may be needed to overcome the ill effects of soil acidity.

Soil pH and Cation Associations

EXCHANGEABLE AND BOUND CATIONS. Figure 9.4 illustrates the relationship between soil pH and the prevalence of various exchangeable and tightly bound cations in a mineral and an organic soil. Two forms of hydrogen and aluminum are shown in Figure 9.4: (1) that

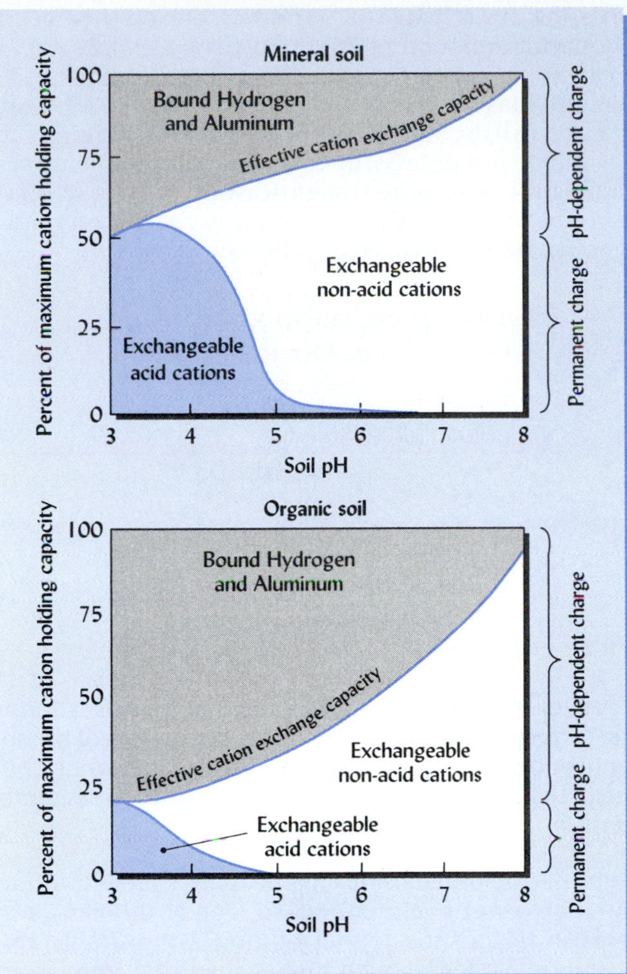

FIGURE 9.4 General relationship between soil pH and cations held in exchangeable form or tightly bound to colloids in two representative soils. Note that any particular soil would give somewhat different distributions. (*Upper*) A mineral soil with mixed mineralogy and a moderate organic matter level exhibits a moderate decrease in effective cation exchange capacity as pH is lowered, suggesting that *pH-dependent charges* and *permanent charges* (see Section 8.6 for explanation of these terms) each account for about half of the maximum CEC. At pH values above 5.5, the concentrations of exchangeable aluminum and H^+ ions are too low to show in the diagram, and the effective CEC is essentially 100% saturated with exchangeable nonacid cations (Ca^{2+}, Mg^{2+}, K^+, and Na^+, the so-called base cations). As pH drops from 7.0 to about 5.5, the effective CEC is reduced because H^+ ions and $Al(OH)_x^{y-}$ ions [which may include $AlOH_2^+$, $Al(OH)_2^+$, etc.] are tightly bound to some of the pH-dependent charge sites. As pH is further reduced from 5.5 to 4.0, aluminum ions (especially Al^{3+}), along with some H^+ ions, occupy an increasing portion of the remaining exchange sites. Exchangeable H^+ ions occupy a major portion of the exchange complex only at pH levels below 4.0. (*Lower*) The CEC of an organic soil is dominated by pH-dependent (variable) charges with only a small amount of permanent charge. Therefore, as pH is lowered, the effective CEC of the organic soil declines more dramatically than the effective CEC of the mineral soil. At low pH levels, exchangeable H^+ ions are more prominent and Al^{3+} less prominent on the organic soil than on the mineral soil.

tightly held by the pH-dependent sites (*bound*), and (2) that associated with negative charges on the colloids (*exchangeable*). The bound forms contribute to the residual acidity pool, but only the exchangeable ions have an immediate effect on soil pH. As we shall see in Section 9.9, both forms are very much involved in determining how much lime or sulfur is needed to change soil pH.

EFFECTIVE CEC AND pH. Note that in both soils illustrated in Figure 9.4, the effective CEC increases as the pH level rises. This change in effective CEC results mainly from two factors: (1) the binding and release of H^+ ions on pH-dependent charge sites (as explained in Sections 8.6 and 8.9), and (2) the hydrolysis reactions of aluminum species (as explained in Section 9.3). The change in effective CEC will be most dramatic for organic soils (Figure 9.4, lower) and highly weathered mineral soils dominated by iron and aluminum oxide clays. However, effective CEC changes with pH even in surface soils dominated by 2:1 clays, which carry mainly permanent charges, because a substantial amount of variable charge is usually supplied by the organic matter and the weathered edges of clay micelles.

Cation Saturation Percentages

The proportion of the CEC occupied by a given ion is termed its **saturation percentage.** Consider a soil with a CEC of 20 $cmol_c/kg$ holding these amounts of exchangeable cations (in $cmol_c/kg$): 10 of Ca^{2+}, 3 of Mg^{2+}, 1 of K^+, 1 of Na^+, 1 of H^+, and 4 of Al^{3+}. This soil with 10 $cmol_c$ Ca^{2+}/kg and a CEC of 20 $cmol_c/kg$ is said to be 50% calcium saturated. Likewise, the aluminum saturation of this soil is 20% (4/20 = 0.20 or 20%). Together, the 4 $cmol_c/kg$ of exchangeable Al^{3+} and 1 $cmol_c/kg$ of exchangeable H^+ ions give this soil an **acid saturation** of 25% [(4 + 1)/20 = 0.25]. Similarly, the term **nonacid saturation** can be used to refer to the proportion of Ca^{2+}, Mg^{2+}, K^+, and Na^+, etc. on the CEC. Thus, the soil in our example has a nonacid saturation of 75% [(10 + 3 + 1 + 1)/20 = 0.75].

Traditionally, the nonacid cations have been referred to as "base" cations and their proportion on the CEC the **percent "base" saturation.** Cations such as Ca^{2+}, Mg^{2+}, K^+, and Na^+ do not hydrolyze as Al^{3+} and Fe^{3+} do, and therefore are not acid-forming cations. However, they are also *not* bases and do not necessarily form bases in the chemical sense of the word.[3] Because of this ambiguity, it is more straightforward to refer to *acid saturation* when describing the degree of acidity on the soil cation exchange complex. The relationships among these terms can be summarized as follows:

$$\text{Percent acid saturation} = \frac{\text{cmol}_c \text{ of exchangeable } Al^{3+} + H^+}{\text{cmol}_c \text{ of CEC}} \tag{9.13}$$

$$\begin{matrix}\text{Percent}\\ \text{nonacid}\\ \text{saturation}\end{matrix} = \begin{matrix}\text{percent}\\ \text{"base"}\\ \text{saturation}\end{matrix} = \frac{\text{cmol}_c \text{ of exchangeable } Ca^{2+} + Mg^{2+} + K^+ + Na^+}{\text{cmol}_c \text{ of CEC}}$$

$$= 100 - \begin{matrix}\text{percent acid}\\ \text{saturation}\end{matrix} \tag{9.14}$$

Acid (or Nonacid) Cation Saturation and pH

The percentage saturation of a particular cation (e.g., Al^{3+} or Ca^{2+}) or class of cations (e.g., nonacid cations or acid cations) is often more closely related to the nature of the soil solution than is the absolute amount of these cations present. Generally, when the acid cation percentage increases, the pH of the soil solution decreases. However, a number of factors can modify this relationship.

Effect of Type of Colloid. The type of clay minerals or organic matter present influences the pH of different soils at the same percent acid saturation, due to differences in the ability of various colloids to furnish H^+ ions to the soil solution. For example, the dissociation of adsorbed H^+ ions from smectites is much higher than that from Fe and Al oxide clays. Consequently, the pH of soils dominated by smectites is appreciably lower than that of the oxides at the same percent acid saturation.

Effect of Kind of Adsorbed Nonacid Cations. The relative amounts of each of the nonacid cations present on the colloidal complex is another factor influencing soil pH. For example, soils with high sodium saturation (usually highly alkaline soils of arid and semiarid regions) have much higher pH values than those dominated by calcium and magnesium (as explained in Section 9.2).

Effect of Method of Measuring CEC. An unfortunate ambiguity in the cation saturation percentage concept is that the actual percentage calculated depends on whether the effective CEC (which itself changes with pH) or the maximum potential CEC (which is a constant for a given soil) is used in the denominator. The different methods of measuring CEC are explained in Section 8.9.

When the concept of cation saturation was first developed, the percent nonacid saturation (then termed "base saturation") was calculated by dividing the level of these exchangeable cations by the *potential* cation exchange capacity that is measured at high pH values (7.0 or 8.2). Thus, if a representative mineral soil such as shown in Figure 9.4 has a potential CEC of 20 cmol_c/kg, and at pH 6 has a nonacid exchangeable cation level of 15 cmol_c/kg, the percent nonacid cation saturation would be calculated as 15 $\text{cmol}_c/20\ \text{cmol}_c \times 100 = 75\%$. The percent "base" saturation determined by this method is still used as a soil classification criterion in *Soil Taxonomy.*

A second method relates the exchangeable cation levels to the *effective* CEC at the pH of the soil. As Figure 9.4 shows, the effective CEC of the representative soil at pH 6

[3] A base is a substance that combines with H^+ ions, while an acid is a substance that releases H^+ ions. The anions OH^- and HCO_3^- are strong bases because they react with H^+ to form the weak acids, H_2O and H_2CO_3, respectively.

would be only about 15 $cmol_c/kg$. At this pH level essentially all the exchangeable sites are occupied by nonacid cations (15 $cmol_c/kg$). Using the effective CEC as our base, we find that the nonacid cation saturation is 15 $cmol_c$/15 $cmol_c \times 100 = 100\%$. Thus, this soil at pH 6 is either 75% or 100% saturated with nonacid cations, depending on whether we use the potential CEC or the effective CEC in our calculations.

Uses of Cation Saturation Percentages. Which nonacid saturation percentage just described is the correct one? It depends on the purpose at hand. The first percentage (75% of the potential CEC) indicates that significant acidification has occurred and is used in soil classification (for example by definition Ultisols must have a nonacid or "base" saturation of less than 35%). The second percentage (100% of the effective CEC) is more relevant to soil fertility and the availability of nutrients. It indicates what proportion of the total exchangeable cations at a given soil pH is accounted for by nonacid cations. For example, when the effective CEC is less than 80% saturated with nonacid cations (that is, more than 20% acid saturated), aluminum toxicity is likely to be a problem in many soils.

While the factors responsible for soil acidity are far from simple, the described relationships indicate that two dominant groups of elements are in control. The different aluminum-containing ions and H^+ ions generate acidity, and most of the other cations do not. This simple statement is worth remembering.

9.5 BUFFERING OF pH IN SOILS

Soils tend to resist change in the pH of the soil solution when either acid or base is added. This resistance to change is called **buffering** and can be demonstrated by comparing the *titration curves* for pure water with those for various soils (Figure 9.5).

Titration Curves

A titration curve is obtained by monitoring the pH of a solution as an acid or base is added in small increments. For example, consider the addition of 0.1 $cmol_c$ of a strong acid like HCl to a liter of water initially at pH 6 (in which the H^+ ion concentration = 10^{-6} = 0.000001 mol/L). The acid supplies 0.001 mol (= 0.1 cmol) of H^+ ions. Because the water is unbuffered, the new H^+ concentration is 0.001001, which is approximately 10^{-3} mol/L or pH 3. The pH has dropped about three units (from 6 to about pH 3) in response to this tiny addition of acid (curve A in Figure 9.5). If the same amount of acid were added to soil, the change in pH would be almost too small to measure (curves B and C in Figure 9.5). By comparing the slopes of these titration curves, we can conclude that the better buffered the soil, the smaller the change in pH caused by a given addition of acid (or base).

The titration curves shown in Figure 9.5 suggest that the soils are most highly buffered when aluminum compounds (low pH) and carbonates (high pH) are controlling the buffer reactions. The soil is least well buffered at intermediate pH levels where

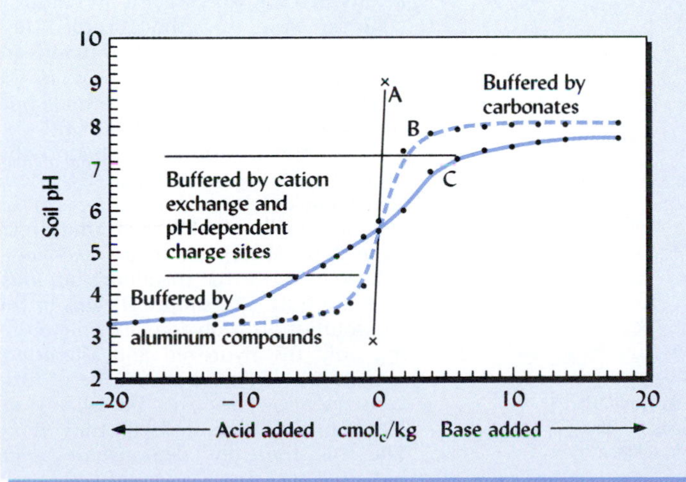

FIGURE 9.5 Buffering of soils against changes in pH when acid (H_2SO_4) or base ($CaCO_3$) is added. A moderately buffered soil (B) and a strongly buffered soil (C) are compared to unbuffered water (A). Most soils are strongly buffered at low pH by the **hydrolysis and precipitation of aluminum** compounds and at high pH by the precipitation and dissolution of **calcium carbonate.** Most of the buffering at intermediate pH levels (pH 4.5 to 7.5) is provided by **cation exchange** and **protonation or deprotonation** (gain or loss of H^+ ions) of pH-dependent exchange sites on clay and humus colloids. The well-buffered soil (C) would have a higher amount of organic matter and/or highly charged clay than the moderately buffered soil (B). [Curves based on data from Magdoff and Bartlett (1985) and Lumbanraja and Evangelou (1991)]

H^+ ion dissociation and cation exchange are the primary buffer mechanisms. However, considerable variability exists in the titration curves for various soils. This may be due to differences among soils with regard to the amounts and types of dominant colloids and contents of bound Al hydroxy complexes that can absorb OH^- ions as the pH rises.

Mechanisms of Buffering

For soils with intermediate pH levels (5 to 7), buffering can be explained in terms of the equilibrium that exists among the three principal pools of soil acidity: active, salt replaceable, and residual (Figure 9.6). If just enough base (lime, for example) is applied to neutralize the H^+ ions in the soil solution, they are largely replenished as the reactions move to the right, thereby minimizing the change in soil solution pH (Figure 9.6). Likewise, if the H^+ ion concentration of the soil solution is increased (for example, by organic decay or fertilizer applications) the reactions in Figure 9.6 are forced to the left, consuming H^+ and again minimizing changes in soil solution pH. Because of the involvement of residual and exchangeable acidity, we can see that soils with higher clay and organic matter contents are likely to be better buffered in this pH range.

Throughout the entire pH range, reactions that either consume or produce H^+ ions provide mechanisms to buffer the soil solution and prevent rapid changes in soil pH. We will now briefly consider some of the specific mechanisms involved.

Aluminum Hydrolysis. In very acid soils (pH 4 to 5.5), hydrolysis, dissolution, or precipitation of gibbsite [$Al(OH)_3$] and other aluminum and iron hydroxyoxide clay minerals provide a major mechanism for pH buffering. These reactions were discussed in Sections 9.1 and 9.3. They include the associated hydrolysis reactions by which the released Al and Fe ions react with water. For example:

$$Al(OH)_2{}^+ + H_2O \rightleftharpoons Al(OH)_3 + H^+ \qquad (9.15)$$

The addition of H^+ ions would drive these reactions to the left, so relatively few of the H^+ ions would accumulate in the soil solution and the pH would be reduced only modestly. Likewise, if OH^- ions were added, they would remove H^+ from the solution by forming water and therefore force the reaction to the right. In both cases, changes in soil solution pH would be buffered.

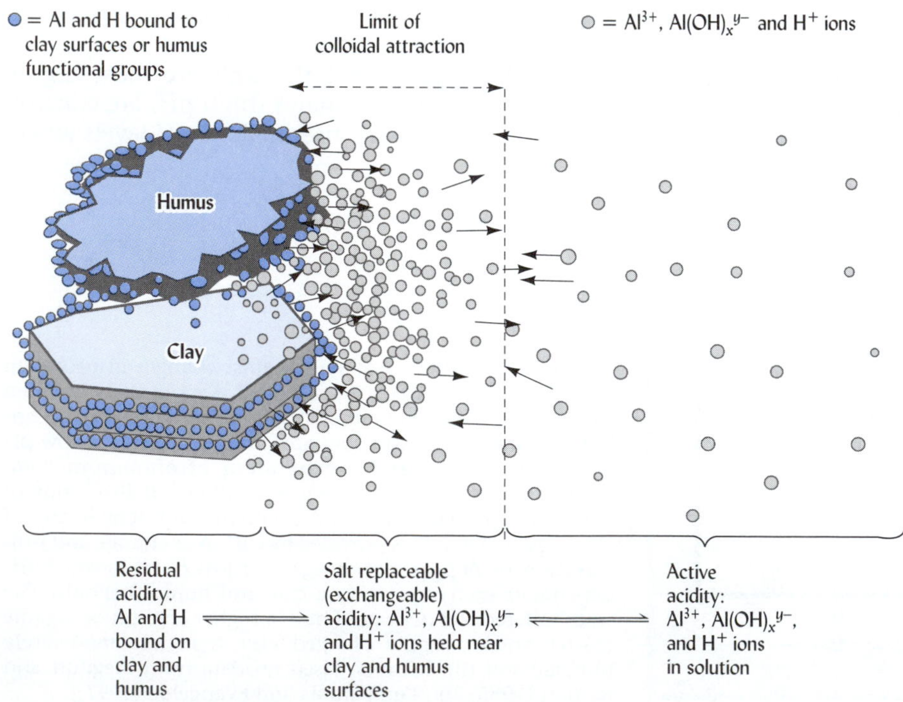

FIGURE 9.6 Equilibrium relationship among residual, salt-replaceable (exchangeable), and soil solution (active) acidity in a soil with organic and mineral colloids. Note that the adsorbed (exchangeable) and residual (bound) ions are much more numerous than those in the soil solution, even when only a small portion of the ions associated with the colloids is shown. Most of the bound aluminum is in the form of $Al(OH)_x{}^{y-}$ ions that are held tightly on the surfaces of the clay or complexed with the humus; relatively few $Al(OH)_x{}^{y-}$ ions are exchangeable. Remember that the aluminum ions, by hydrolysis, also supply H^+ ions in the soil solution. It is obvious that neutralizing only the hydrogen and aluminum ions in the soil solution will be of little consequence. They will be quickly replaced by ions associated with the colloid. The soil, therefore, demonstrates high buffering capacity.

ORGANIC MATTER REACTIONS. The protonation or deprotonation of organic matter functional (R-OH) groups provides much of the pH buffering in soils. As discussed in Section 9.1, the tenacity with which H^+ ions are held by organic matter is different for different functional groups and is also influenced by the nature of the nearby C to C bonds. Therefore, soil organic matter, with its complex structure, has many types of sites to which H^+ ions can bond with various degrees of strength. At low pH, H^+ ions will dissociate from the most acid sites (that is, from sites with a low pK_a). If a base is added to such an acid soil, these sites will dissociate, providing H^+ ions to replace those neutralized by the base. The sites from which the H^+ ions dissociated then become negatively charged, adding to the effective CEC. The reverse occurs if an acid is added to the soil. Organic matter buffers the soil over a wide range of pH levels as different sites on organic matter dissociate or associate.

Organic matter also buffers soil pH by the release of aluminum ions from organic complexes in which they are held with varying degrees of strength. If a base is added to raise pH of the solution, more aluminum escapes from these complexes and subsequently replaces the neutralized H^+ ions by hydrolysis reactions. Again, the pH is stabilized.

PH-DEPENDENT CHARGE SITES ON CLAY. As discussed in Section 8.6, aluminum and silicon atoms at the surface of clay minerals are bonded to hydroxyl groups and oxygen atoms that can associate with or dissociate with H^+ ions. Adding a base to raise the solution OH^- concentration would induce the dissociation of H^+ ions from these sites, especially on the weathered edges of clays. This release of H^+ ions would buffer the pH by neutralizing most of the added OH^- ions. At the same time, the sites from which the H^+ ions dissociated would become negatively charged and would increase the soil's CEC.

CATION EXCHANGE. Exchangeable ions on clay and humus are in equilibrium with the ions in solution. Therefore, if additional H^+ is added to the soil, most of these ions will be attracted to the cation exchange complex where they will exchange with other ions such as Ca^{2+}. The H^+ concentration in the solution will therefore not change very much. If H^+ ions in solution are neutralized by addition of a base, they will be immediately replaced by H^+ ions and various aluminum ions released from exchange sites on clay and humus. Because of the relatively small amount of exchangeable ions present in soils, these reactions contribute most significantly to pH buffering when the amount of H^+ ions in solution is relatively low (that is, when the pH is above 6).

CARBONATE DISSOLUTION AND PRECIPITATION. Buffer reactions involving carbonates, bicarbonates, carbonic acid, and water are most important in alkaline soils (pH 7 and above).

$$CaCO_3 + H^+ + HCO_3 \rightleftharpoons Ca(HCO_3)_2$$
$$Ca(HCO_3)_2 + 2H_2O \rightleftharpoons Ca^{2+} + 2OH^- + 2H_2CO_3 \quad (9.16)$$

If acid were added, the H^+ ions would be consumed as the reactions shifted to the right. If a base were added, the OH^- ions would be consumed and H^+ ions produced as the reactions shifted to the left. In both cases, the pH change would be buffered.

Importance of Soil Buffering Capacity

Soil buffering is important for two primary reasons. First, buffering tends to ensure some stability in the soil pH, preventing drastic fluctuations that might be detrimental to plants, soil microorganisms, and aquatic ecosystems. For example, well-buffered soils resist the acidifying effect of acid rain, preventing the acidification of both the soil and the drainage water. Second, buffering influences the amount of amendments, such as lime or sulfur, required to bring about a desired change in soil pH.

Soils vary greatly in their buffering capacity. Other things being equal, the higher the cation exchange capacity (CEC) of a soil, the greater its buffering capacity. This relationship exists because in a soil with a high CEC, more reserve and exchangeable acidity must be neutralized or increased to affect a given change in soil pH.

9.6 DETERMINATION OF SOIL pH

Readable, illustrated details about measuring pH: http://www.ph-measurement.com/background.pdf

Soil pH is measured routinely, and the determination is easy and rapid to make. In one method, a few drops of a pH-sensitive organic dye solution are placed in contact with the soil, usually on a white spot plate (see Plate 48). The color of the dye is then compared to a color chart that indicates the approximate pH. Such dye methods are accurate within about 0.2 to 0.5 pH unit.

The most accurate method of determining soil pH is with a pH-sensitive glass electrode and a standard reference electrode (or a probe that combines both electrodes in one). The electrodes are inserted into a soil-water suspension that simulates the soil solution. The difference between the H^+ ion activities in the soil suspension and in the glass electrode gives rise to an electrometric potential that is read on a special millivolt meter (called a *pH meter*).

Most soil testing laboratories (see Section 14.11) in the United States measure the pH of a suspension of soil in water (usually at a ratio of 1:1 or 1:2.5). This is designated the pH_{water}. Other labs may measure the pH using a soil suspended in a neutral salt solution instead of in pure water. This modification of the pH measurement procedure generally gives lower (but more repeatable) readings. It is important to be aware of which method is used before comparing results from different labs. In this textbook (as in most U.S. publications), we will report values for pH_{water}, unless specified otherwise.

Variability in the Field

More may be inferred regarding the chemical and biological conditions in a soil from the pH value than from any other single measurement. However, the interpretation of this measurement must take into consideration the substantial variation that occurs in the field.

Spatial Variation. Soil pH may vary spatially over very small distances (millimeter or smaller). For example, plant roots taking up NH_4^+ ions release H^+ ions, resulting in a lower pH in the rhizosphere soil immediately around the root than in the bulk soil a few mm away. Such variability of the soil solution is important in many respects. For example, organisms unfavorably influenced by a given hydrogen ion concentration may find, at an infinitesimally short distance away, a different environment that is more satisfactory.

Concentrations of fertilizers or ashes from forest fires may give sizable pH variations within the space of a few centimeters to a few meters. Other factors, such as erosion or drainage, may cause pH to vary considerably over larger distances (to 100s of meters), often ranging over two or more pH units within a few hectares. A carefully planned sampling procedure may minimize errors due to such variability (see Section 14.11). In addition to differences from place to place, a sampling scheme should also recognize variations with depth and time (see, for example, Figure 9.7).

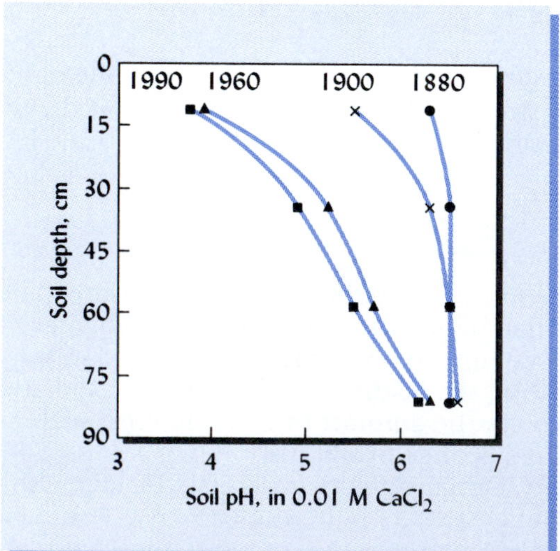

FIGURE 9.7 The change in soil pH (measured in 0.01 *M* $CaCl_2$) during a 110-year period in which a former agricultural field was allowed to revert to natural vegetation (eventually a mature oak forest). The fine-textured (clay loam to clay) Alfisol at the famous Rothamstead experiment station in England, was untilled, unfertilized, and unlimed. Note that in the first 20 years acidification was most pronounced near the soil surface. In the ensuing years acidity continued to increase most dramatically at the surface, but eventually increased throughout the profile. By 1960, the surface horizon had reached the pH range in which strong buffering by aluminum compounds probably slowed acidification. [Drawn from data in Blake et al. (1999); used with permission of Blackwell Science, Ltd.]

9.7 HUMAN-INFLUENCED SOIL ACIDIFICATION

In certain situations, the natural processes of soil acidification are greatly (and usually inadvertently) accelerated by human activities. We will consider three major types of human-influenced soil acidification: (1) nitrogen amendments, (2) acid precipitation, and (3) exposure of acid sulfate soils.

Nitrogen Fertilization

CHEMICAL FERTILIZERS. The widely used ammonium-based fertilizers, such as ammonium sulfate $(NH_4)_2 SO_2$ and diammonium phosphate $(NH_4)_2 HPO_4$, are oxidized in the soil by microbes to produce strong inorganic acids by reactions such as the following:

$$(NH_4)_2 SO_4 + 4O_2 \rightleftharpoons 2HNO_3 + H_2SO_4 + 2H_2O \quad (9.17)$$

However, because plant uptake of nitrate (and sulfate) anions consumes H^+ ions (as shown in equation 9.5), soil acidification is related only to that proportion of oxidized nitrogen not taken up by plants. Excessive nitrogen fertilizer rates popularized since the 1960s have ensured that soil acidification from this cause is not trival—in the United States, neutralization of this source of acidity on agricultural lands alone would require some 20 million Mg of limestone each year.

ACID-FORMING ORGANIC MATERIALS. Application of organic materials such as sewage sludge or animal manures to agricultural and forested lands can decrease soil pH, both by oxidation of the nitrogen released and by organic and inorganic acids formed during decomposition. Therefore, a program of regular organic matter additions should also include regular additions of liming materials to counteract this acidification. In many countries, control of soil pH after amendment with sewage sludge is regulated, with the objective of minimizing the mobility of toxic metals found in this type of waste material. It should be noted that certain types of sewage sludge (often referred to as *lime-stabilized sludge*) have been treated with large quantities of lime to control pathogens and odors. Lime-stabilized sludge may contain 20 to 30% lime by weight and have a pH of 7.5 to 8.5. Rather than acidifying the soil, application of lime-stabilized sewage sludge may result in *overliming* if the lime content is not taken into account when determining application rates.

Acid Deposition from the Atmosphere

ORIGINS OF ACID RAIN. Industrial activities such as the combustion of fuel and the smelting of sulfur-contain metal ores, emit enormous quantities of nitrogen and sulfur-containing gases into the atmosphere (Figure 9.8). The gases react with water and other substances in the atmosphere to form HNO_3 and H_2SO_4. These strong acids are then returned to the earth in **acid rain** (as well as in snow, fog, and dry deposition). Normal rainfall that is in equilibrium with atmospheric carbon dioxide has a pH of about 5.5. The pH of acid rain commonly is between 4.0 and 4.5, but may be as low as 2.0.

Overview of environmental damages from acid rain:
http://www.epa.gov/airmarkets/acidrain/effects/index.html

EFFECTS OF ACID RAIN. Acid rain causes expensive damage to buildings and car finishes, but the principal environmental reasons for concern about acid rain are its effects on (1) fish and (2) forests. Since the 1970s, scientists have documented the loss of normal fish populations in thousands of lakes and streams. More recently, studies have suggested that the health of certain forest ecosystems is suffering because of acid rain. Furthermore, scientists have learned that the health of both the lakes and the forests is not usually affected directly by the rain, but rather by the interaction of the acid rain with the soils in the watershed (Figure 9.8).

SOIL ACIDIFICATION. The incoming strong acids mobilize aluminum in the soil minerals and the aluminum displaces Ca and other nonacid cations from the exchange complex. The presence of the strong acid anions (SO_4^{2-} and NO_3^-) facilitates the leaching of the displaced Ca ions (as explained in Figure 9.3). Soon Al^{3+} and H^+ ions, rather than Ca^{2+} ions, become dominant on the exchange complex, as well as in the soil solution and drainage waters. Figure 9.9 illustrates the increase in acid saturation of the cation

FIGURE 9.8 Simplified diagram showing the formation of acid rain in urban areas and its impact on distant watersheds. Combustion of fossil fuels in electric power plants and in vehicles accounts for the largest portions of the nitrogen and sulfur emissions. About half of the acidity is due to sulfur gases and about half is due to nitrogen gases. The gases are carried hundreds of kilometers by the wind and are oxidized to form sulfuric and nitric acid in the clouds. These acids then return to earth in precipitation and in dry deposition. The H^+ cations and NO_3^- and SO_4^{2-} anions cause acidification to occur in soils, soil aluminum to mobilize, and the loss of calcium and magnesium to accelerate. The mobilized aluminum percolates through the soil mantle, eventually reaching lakes and streams. The principal ecological effects of concern in sensitive watersheds are (1) possible decline in forest health and (2) decline or even death of aquatic ecosystems.

exchange complex over a 28-year period. However, in this and other studies, it is not easy to sort out how much acidification is due to natural processes internal to the soil ecosystem (see left side of Table 9.1) and how much is due to acid rain.

Effects on Forests. Some scientists are concerned that trees, which have a high requirement for calcium to synthesize wood, may eventually suffer from insufficient supplies of this and other nutrient cations in acidified soils. The leaching of calcium and the mobilization of aluminum may result in Ca/Al ratios (mol_c/mol_c) of less than 1.0 in both the soil solution and on the exchange complex. A Ca/Al ratio of 1.0 is widely considered a threshold for aluminum toxicity, root growth inhibition, reduced calcium uptake, and reduced survival for forest vegetation. While there is little doubt that aluminum in acidified soils is toxic to many forest species (see Section 9.8), the scientific evidence for forest calcium deficiencies is less clear. The calcium supply in most forested

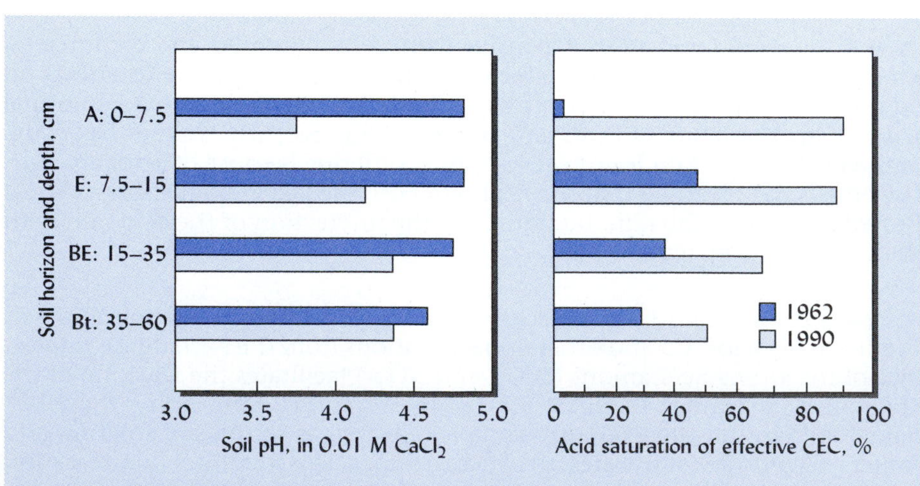

FIGURE 9.9 Changes in two indicators of soil acidity in a South Carolina Ultisol profile during 28 years under a loblolly pine forest. Note that the greatest acidification took place in the uppermost horizon (data for the O horizon were not reported). The soils at this site were limed regularly for agricultural crops prior to 1952 and therefore were less acid at the beginning of the study than they would be in their natural state. The acid saturation percentage = ($cmol_c$/Al + H)/($cmol_c$ effective CEC). The effective CEC for these soil horizons ranged from 0.8 to 3.3 $cmol_c$/kg (data not shown). [Calculated and drawn from data in Markewitz et al. (1998); with permission of the Soil Science Society of America]

soils in the humid eastern United States is being depleted as the rate of calcium loss by leaching, tree uptake, and harvest exceeds the rate of calcium deposition. However, it seems that even in very acid soils low in exchangeable Ca^{2+}, the weathering of soil minerals often releases sufficient calcium for good tree growth—at least in the short term.

Effects on Aquatic Ecosystems. Acid soil water, containing elevated levels of aluminum, and often of sulfate and nitrate, eventually drains into streams and lakes. The water in the lakes and streams becomes lower in calcium, less well buffered, more acid, and higher in aluminum. The aluminum is directly toxic to fish, partly because it damages the gill tissues. As the lake water pH drops to about 6.0, acid-sensitive organisms in the aquatic food web die off and reproductive performance of such fish as trout and salmon declines. With a further drop in water pH to about 5.0, virtually all fish are killed. Although the acidified water may be crystal clear (in part due to the flocculating influence of aluminum), the lake or stream is considered to be "dead" except for a few algae, mosses, and other acid tolerant organisms.

Ecological damage from acid rain is most likely to occur where the rain is most acid and the soils are most susceptible to acidification. Areas of the world are most sensitive to acid rain damage where soils have low CEC and high acid saturation percentage, and low reserves of weatherable minerals. Ecosystems in much of northeastern United States and eastern Canada, northern Europe, and southeastern China are among those most likely to be damaged.[4]

Tightening air-quality standards in industrialized countries should continue to reduce emissions of N- and S-gases. The resulting decrease in acid inputs into sensitive ecosystems is expected to eventually restore a suitable chemical balance in the soils of these areas (and therefore in the lakes and streams). Meanwhile, national and international agencies monitoring the effects of acid rain on a global basis warn that rapid industrialization without strict environmental controls in such countries as China, Russia, and Brazil may shift the problem focus to other susceptible regions.

Exposure of Potential Acid Sulfate Materials

Drainage of Certain Coastal Wetlands. Due to the microbial reduction of sulfates originally in seawater, certain coastal sediments contain significant quantities of pyrite (FeS_2), iron monosulfides (FeS), and elemental sulfur (S). Coastal wetland areas in the southeastern United States, Southeast Asia, northern Australia, and West Africa commonly contain soils formed in such sediments. As long as waterlogged conditions prevail, the *potential* **acid sulfate soils** retain the sulfur and iron in their reduced forms. However, if these soils are drained for agriculture, forestry, or other development, air enters the soil pores and both the sulfur (S^0, S^- or S^{2-}) and the iron (Fe^{2+}) are oxidized, changing the potential acid sulfate soils into *active* **acid sulfate soils.** Ultimately such soils earn their name by producing prodigious quantities of sulfuric acid, resulting in soil pH values below 3.5 and in some cases as low as 2.0. The principal reactions involved are:

$$\underset{\text{Pyrite}}{Fe^{II}S_2^{-I}} + 3\tfrac{1}{2}O_2 + H_2O \rightleftharpoons \underset{\text{Ferrous sulfate}}{Fe^{II}S^{VI}O_4} + \underset{\text{Sulfuric acid}}{H_2S^{VI}O_4}$$

$$\underset{\text{Ferrous sulfate}}{Fe^{II}SO_4} + \tfrac{1}{4}O_2 + 1\tfrac{1}{2}H_2O \rightleftharpoons \underset{\text{Iron oxyhydroxides}}{Fe^{III}OOH} + \underset{\text{Sulfuric acid}}{H_2SO_4} \qquad (9.18)$$

$$\underset{\text{Elemental S}}{S^0} + 1\tfrac{1}{2}O_2 + H_2O \rightleftharpoons \underset{\text{Sulfuric acid}}{H_2SO_4} \qquad (9.19)$$

The iron sulfide compounds in the potential acid sulfate soils often give these soils a black color. The black color has sometimes led, with disastrous results, to the use of

[4]For an assessment of the recovery of lakes and streams from acidification in North America and Europe, see Stoddard et al. (1999).

such soils by those seeking black, organic-matter-rich "topsoil" material for landscaping installations. The pH of the potential acid sulfate soils is in the neutral range (typically near 7.0) while they are still reduced, but drops precipitously within days or weeks of being exposed to air. When in doubt, the pH of the soil should be monitored for a week or two while a sample is incubated in a moist, well-aerated, warm condition.

Excavation of Pyrite-Containing Materials. Coastal marsh soils are not the only places where problems with acid sulfate soils occur. The sediments dredged up to deepen shipping lanes in coastal harbors may also contain high concentrations of reduced sulfur compounds (see Plate 65 before page 499). Furthermore, many geological sulfide-rich sediments long ago became sedimentary rock. When these once deeply buried materials are exposed to air and water by the mining excavations, the reduced sulfur and iron compounds in them begin to oxidize and hydrolyze by the types of reactions already shown.

Research on acid mine drainage with pH less than 0:
http://www.pnas.org/cgi/content/full/96/7/3455

As water percolates through the oxidizing materials it becomes an extremely acid and toxic brew known as **acid mine drainage.** Typical acid mine drainage has a pH in the range of 0.5 to 2.0, but pH values below zero have been measured. When this drainage water reaches a stream (as in Plate 64), iron sulfides dissolved in the drainage water continue to produce acid by oxidation and hydrolysis. The aquatic community can be devastated by the pH shock and the iron and aluminum that is mobilized.

Avoidance as the Best Solution. The amount of sulfur present in such soil material is considered an indication of its potential acidity and of how much liming material ($CaCO_3$) would eventually be required to neutralize the sulfuric acid that could be produced. Often the volume of soil that would need to be treated and the amounts of $CaCO_3$ required are so enormous as to make neutralization an impractical solution to the problem.

Usually the best approach to solving this environmental challenge is to *prevent* the oxidation that produces the acidity in the first place. This means that sulfide-bearing wetland soils are best left undisturbed or returned to their undrained, wetland conditions. In the case of mining and excavation, any sulfide-bearing materials exposed must be identified and eventually deeply reburied to prevent their oxidation. If some acid drainage is unavoidable (as from abandoned, poorly designed mines), one of the most effective and least expensive treatments is to route the drainage through a wetland area, even if this requires the construction of a wetland (see Section 7.7). The reducing conditions in a saturated wetland will re-reduce the iron and sulfur, causing much of it to precipitate as iron sulfides, while simultaneously raising the pH of the water.

9.8 BIOLOGICAL EFFECTS OF SOIL pH

The acid soil headache: a view from tropical Hawaii:
http://www2.ctahr.hawaii.edu/tpss/research_extension/rxsoil/acid.html

The pH of the soil solution is a critical environmental factor for the growth of all organisms that live in the soil, including plants, animals, and microbes. Although we have already mentioned some ways that soil acidity or alkalinity affects plant growth, we will now take a more detailed look at the impacts that various pH conditions have on soil organisms.

To nonadapted plants, strongly acid soil presents a host of problems. These include toxicities of aluminum, manganese, and hydrogen, as well as deficiencies of calcium, magnesium, molybdenum, and phosphorus. As it is difficult to separate one problem from another, the situation is sometimes simply referred to as the acid soil "headache."

Aluminum Toxicity[5]

Aluminum toxicity stands out as the most common and severe problem associated with acid soils. Not only plants are affected; many bacteria, such as those that carry out certain transformations in the nitrogen cycle, are also adversely impacted by the high levels of aluminum associated with low soil pH.

[5]For a review of aluminum toxicity and the development of plant tolerance to aluminum, see de la Fuente-Martinez and Herrera-Estralla (2000).

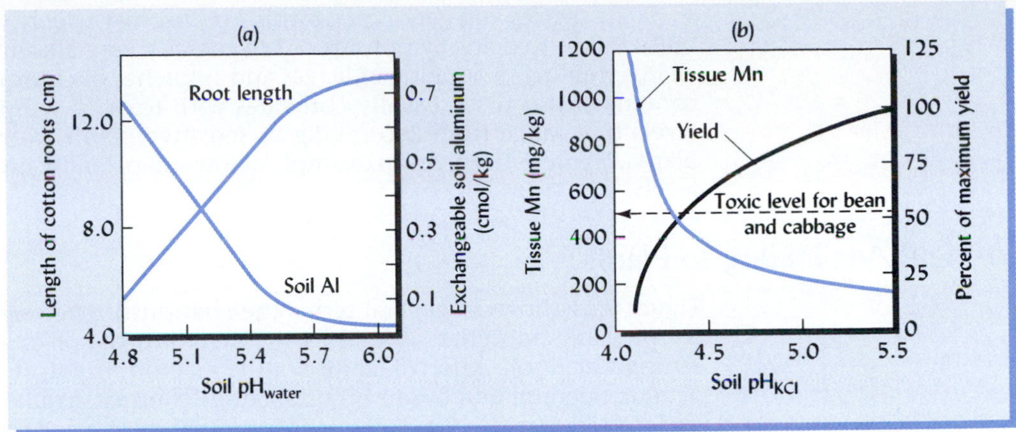

FIGURE 9.10 Plant responses to toxicity of aluminum (*a*) and manganese (*b*) at low soil pH. As soil pH_{water} drops below 5.2, exchangeable Al increases and cotton root length is severely restricted in an Ultisol (*a*). Plant shoot growth (the average of bean and cabbage) declines and Mn content of foliage increases at low pH levels in manganese-rich soils from East Africa (average data for an Andisol and an Alfisol). [(*a*) From Adams and Lund (1966); (*b*) Redrawn from data in Weil (2000)]

As can be deduced from Figure 9.4, aluminum toxicity is rarely a problem when the soil pH is above about 5.2 (above pH_{CaCl} 4.8), because little Al^{3+} or $AlOH^{2+}$ exists in the solution or exchangeable pools above this pH level. Aluminum toxicity to plants is often greatest near pH 4.5 because the $AlOH^{2+}$ species, which is even more toxic than the Al^{3+} species, is most soluble at that pH. Figure 9.10*a* shows an example of the effect of soil acidity (and aluminum toxicity) on root growth in a mineral soil. Aluminum toxicity is much less of a problem in most organic soils (or in organic soil horizons) at comparable pH levels. This is because there is far less total aluminum in these soils—and what aluminum exists in them is likely to be complexed in nonexchangeable form by the organic matter.

Effects on Plants. Aluminum causes several types of damage in plants. On the membranes of young root tips, aluminum can block the sites where calcium is normally taken in. Aluminum interferes with the metabolism of phosphorus containing compounds essential for energy transfers (ATP) and genetic coding (DNA). Aluminum also restricts cell wall expansion.

The most common symptom of aluminum toxicity is a stunted root system with short, thick, stubby roots that show little branching or growth of laterals. Because of the restricted root system, plants suffering from aluminum toxicity often show drought stress.

Generally, plant species that originated in areas dominated by acid soils (such as most humid regions) tend to be less sensitive than species originating in areas of neutral to alkaline soils (such as the Mediterranean region). Fortunately, plant breeders have been able to find genes that confer tolerance to aluminum even in species that are typically sensitive to this toxicity.

Manganese Toxicity to Plants

Although not as widespread as aluminum toxicity, manganese (Mn) toxicity is a serious problem for plants in certain acid soils with a high content of manganese-containing minerals. Unlike aluminum, manganese is an essential plant nutrient (see Section 13.21), but is toxic when taken up in excessive quantities. Like aluminum, manganese becomes increasingly soluble as pH drops, but in the case of manganese, toxicity is common at pH_{water} levels as high as 5.6 (about 0.5 unit higher than for aluminum).

Since the reduced form [Mn(II)] is far more soluble than the oxidized form [Mn(IV)], toxicity is greatly increased by low oxygen conditions associated with a combination of oxygen-demanding decomposable organic matter and overly wet conditions. Flooding, including that of acid rice paddy soils, can induce toxicity. Manganese toxicity is also common in certain high-organic-matter surface horizons of volcanic soils (e.g., Melanudands).

Plant species and genotypes within species vary widely with regard to their susceptibility to Mn toxicity. Symptoms of Mn toxicity vary among plant species, but often include crinkling or cupping of leaves and splotches of chlorotic tissue. Unlike Al, the Mn content of leaf tissue usually correlates with toxicity symptoms, toxicity beginning at levels that range from 200 mg/kg in sensitive plants to over 5000 mg/kg in tolerant plants. Figure 9.10*b* gives an example of low soil pH inducing plant uptake of Mn to toxic levels.

Nutrient Availability to Plants

Figure 9.11 shows in general terms the relationship between soil pH and the availability of plant nutrients, as well as the activities of soil microorganisms. Note that in strongly acid soils, the availability of the macronutrients (Ca, Mg, K, P, N, and S) as well as molybdenum and boron is curtailed. In contrast, availability of most micronutrient cations (Fe, Mn, Zn, Cu, and Co) is increased by low soil pH, even to the extent of toxicity to higher plants and microorganisms.

In slightly to moderately alkaline soils, molybdenum and all of the macronutrients (except phosphorus) are amply available, but levels of available Fe, Mn, Zn, Cu, and Co

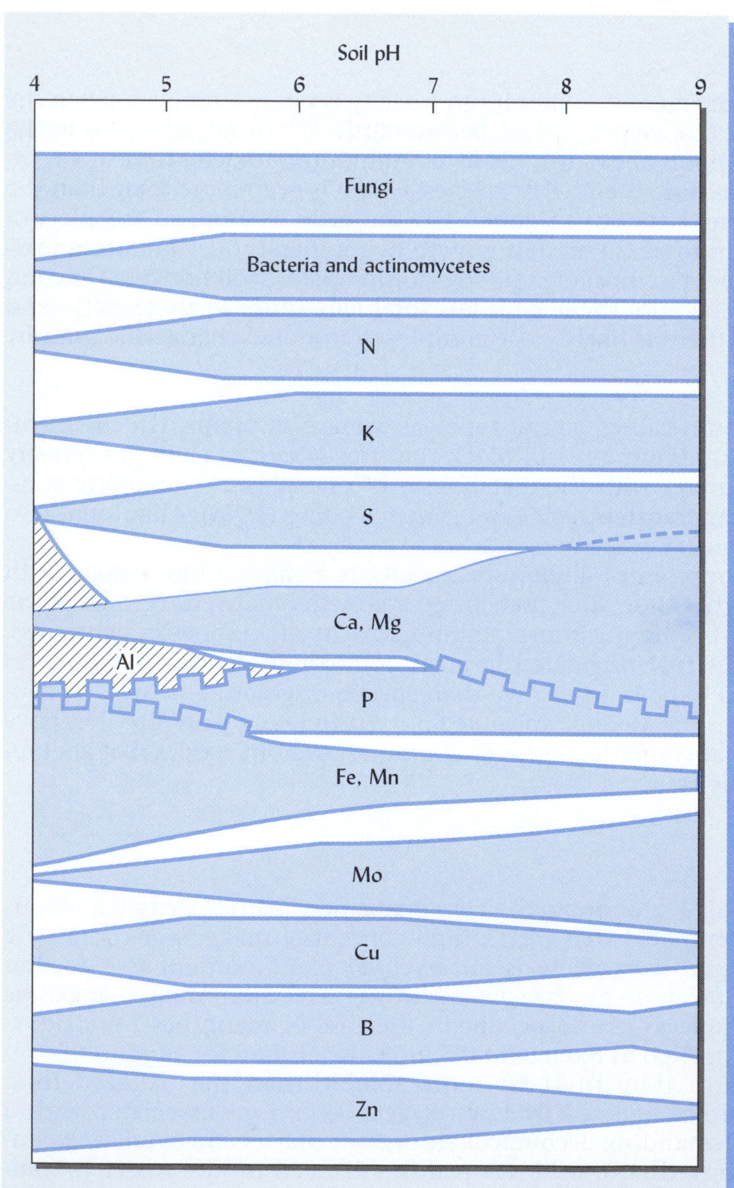

FIGURE 9.11 Relationships existing in mineral soils between pH and the availability of plant nutrients. The relationship with activity of certain microorganisms is also indicated. The width of the bands indicates the relative microbial activity or nutrient availability. The jagged lines between the P band and the bands for Ca, Al, and Fe represent the effect of these metals in restraining the availability of P. When the correlations are considered as a whole, a pH range of about 5.5 to perhaps 7.0 seems to be best to promote the availability of plant nutrients. In short, if the soil pH is suitably adjusted for phosphorus, the other plant nutrients, if present in adequate amounts, will be satisfactorily available in most cases.

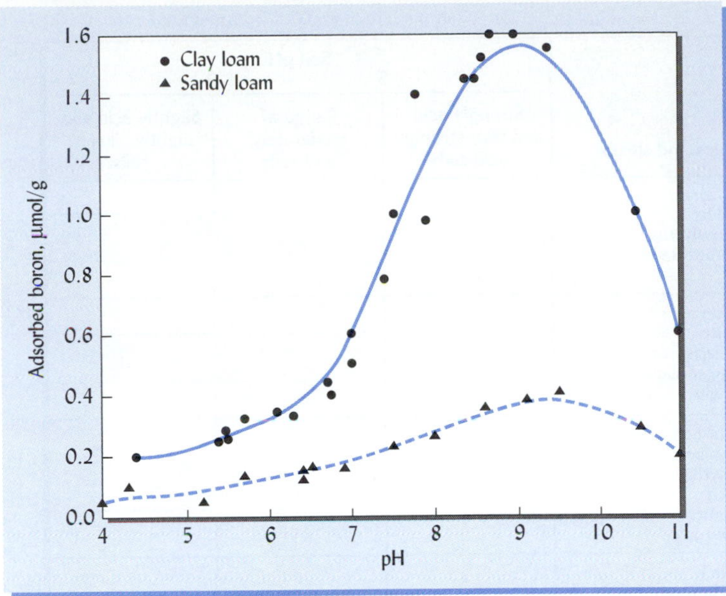

FIGURE 9.12 Adsorption of boron increases as soil pH increases up to about 9. Boron is thought to form difficult-to-reverse (inner-sphere) complexes with hydroxyls on the surfaces of iron and aluminum oxides and silicate clays. Because of their greater surface area, fine-textured soils can tightly adsorb much more boron than can sandy soils, which may lose most of their boron by leaching. Therefore, boron deficiency is common at high pH in both sandy soils (because of low boron content), and in fine textured clayey soils (because the boron is tightly held by the clay). [Redrawn from Goldberg et al. (2000) and used with permission of the Soil Science Society of America]

are so low that plant growth is constrained. Phosphorus and boron availability is likewise reduced in alkaline soil—commonly to a deficiency level. The P reacts with free Ca to form insoluble compounds. The B is strongly sorbed to colloids at high pH (Figure 9.12).

It is difficult to generalize about nutrient-pH relationships. However, it appears from Figure 9.11 that the pH range of 5.5 to 6.5 may provide the most satisfactory plant nutrient levels overall. However, this generalization may not be valid for all soil and plant combinations. For example, certain micronutrient deficiencies are common on some sandy soils when the soils are limed to pH values of only 6.0 to 6.5. Plants vary considerably in their tolerance to acid and/or alkaline conditions (Figure 9.13).

Microbial Effects

Figure 9.11 suggests that most generalist bacteria (including actinomycetes) function well at intermediate and high pH values. Fungi seem to be particularly versatile, flourishing satisfactorily over a wide pH, range. Therefore, fungal activity tends to predominate in acid soils. However, at intermediate and higher pH, bacteria are strong competitors and tend to dominate the microbial activity. We may take advantage of these relationships to control certain soil-borne plant diseases (see Section 10.14).

Soil pH and Organic Molecules

Soil pH influences environmental quality in many ways, but we will discuss only one example here—the influence of pH on the mobility of ionic organic molecules in soils. The molecular structure of certain ionic herbicides includes such chemical groups as —NH_2 and —COO^-. If soil pH is low, the excess H^+ ions (protons) present in solution are attracted to and bond with these chemical groups, neutralizing negative charges and creating positively charged sites on the molecule. This process is called **protonation**.

Atrazine, a chemical that is widely used to control weeds in corn, is an example of an herbicide whose mobility is greatly influenced by soil pH. In a low pH environment, the atrazine molecule protonates, developing a positive charge. The positively charged molecule is then adsorbed on the negatively charged soil colloids, where it is held until it can be decomposed by soil organisms. The adsorbed pesticide is less likely to move downward and into the groundwater. At pH values above 5.7, however, the adsorption is greatly reduced and the tendency for the herbicide to move downward in the soil is increased. Of course, the adsorption in acidic soils also reduces the availability of atrazine to weed roots, thus reducing its effectiveness as a weed killer.

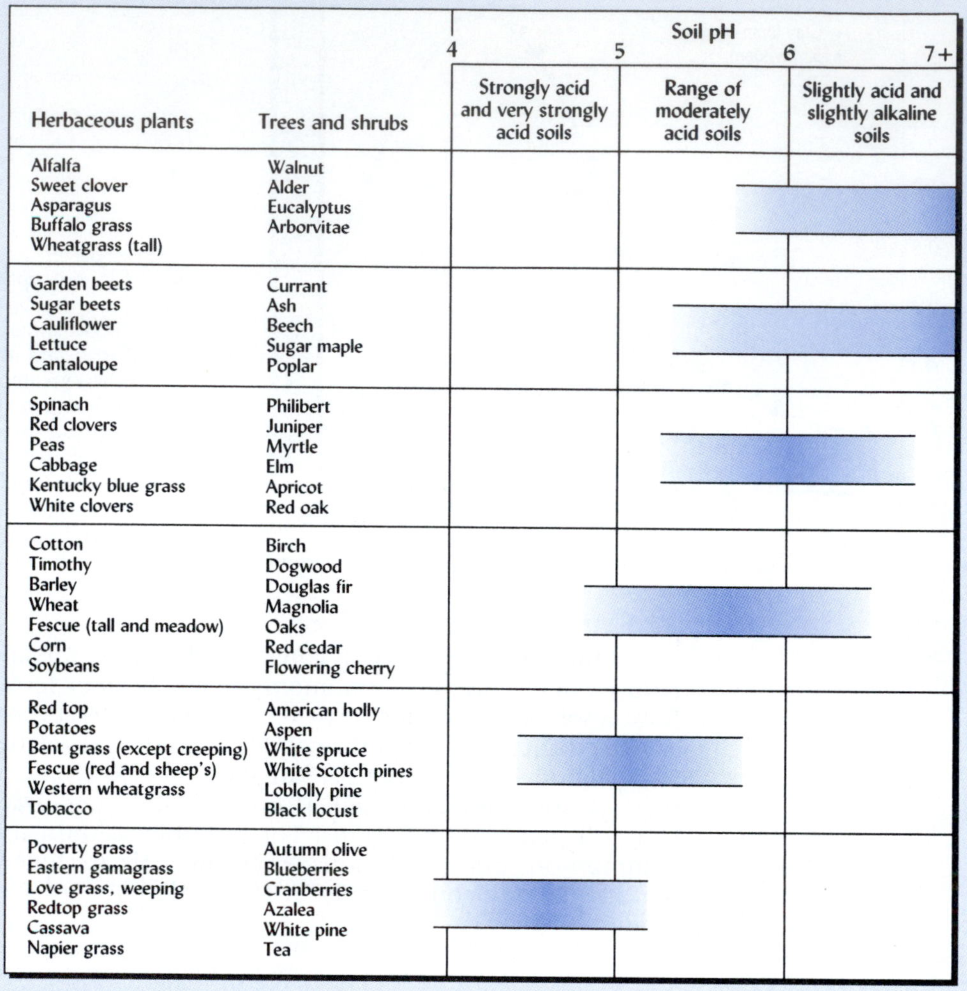

Herbaceous plants	Trees and shrubs	Soil pH 4–5: Strongly acid and very strongly acid soils	Soil pH 5–6: Range of moderately acid soils	Soil pH 6–7+: Slightly acid and slightly alkaline soils
Alfalfa Sweet clover Asparagus Buffalo grass Wheatgrass (tall)	Walnut Alder Eucalyptus Arborvitae			
Garden beets Sugar beets Cauliflower Lettuce Cantaloupe	Currant Ash Beech Sugar maple Poplar			
Spinach Red clovers Peas Cabbage Kentucky blue grass White clovers	Philibert Juniper Myrtle Elm Apricot Red oak			
Cotton Timothy Barley Wheat Fescue (tall and meadow) Corn Soybeans	Birch Dogwood Douglas fir Magnolia Oaks Red cedar Flowering cherry			
Red top Potatoes Bent grass (except creeping) Fescue (red and sheep's) Western wheatgrass Tobacco	American holly Aspen White spruce White Scotch pines Loblolly pine Black locust			
Poverty grass Eastern gamagrass Love grass, weeping Redtop grass Cassava Napier grass	Autumn olive Blueberries Cranberries Azalea White pine Tea			

FIGURE 9.13 Ranges of pH in mineral soils optimal for growth of selected plants.

9.9 RAISING SOIL pH BY LIMING

Agricultural Liming Materials

To decrease soil acidity (raise the pH), the soil is usually amended with alkaline materials that provide conjugate bases of weak acids. Examples of such conjugate bases include carbonate (CO_3^-), hydroxide (OH^-), and silicate (SiO_3^{2-}). These conjugate bases are anions that are capable of consuming (reacting with) H^+ ions to form weak acids (such as water). For example:

$$CO_3^- + 2H^+ \longrightarrow CO_2 + H_2O \qquad (9.20)$$

Most commonly, these bases are supplied in their calcium or magnesium forms ($CaCO_3$, etc.) and are referred to as **agricultural limes.** Some liming materials contain oxides or hydroxides of alkaline earth metals (e.g., CaO or MgO), which form hydroxide ions in water. Unlike fertilizers, which are used to supply plant nutrients in relatively small amounts that are determined by plant nutrient uptake needs, *the purpose of liming is usually to change the chemical makeup of a substantial part of the root zone.* Therefore, lime must be added in large enough quantities to chemically react with a large volume of soil. This requirement dictates that inexpensive, plentiful materials are normally used for liming soils—most commonly finely ground limestone, or materials derived from it (see Table 9.2). Dolomitic limestone products should be used if magnesium levels are low.

TABLE 9.2 Common Liming Materials: Their Composition and Use

Common name of liming material	*Chemical formula (of pure materials)*	*% $CaCO_3$ equivalent*	*Comments on manufacture and use*
Calcitic limestone	$CaCO_3$	100	Natural rock ground to a fine powder. Low solubility; may be stored outdoors uncovered. Noncaustic, slow to react.
Dolomitic limestone	$CaMg(CO_3)_2$	95–108	Natural rock ground to a fine powder; somewhat slower reacting than calcitic limestone. Supplies Mg to plants.
Burned lime (oxide of lime)	CaO (+ MgO)[a]	178	Caustic, difficult to handle, fast acting, can burn foliage, expensive. Made by heating limestone. Protect from moisture.
Hydrated lime (hydroxide of lime)	$Ca(OH)_2$ (+ $Mg(OH)_2$)[a]	134	Even more caustic and more difficult to handle than CaO. Fast acting, can burn foliage, expensive. Protect from moisture.
Basic slag	$CaSiO_3$	70	By-product of pig-iron industry. Must be finely ground. Also contains 1–7%P.
Marl	$CaCO_3$	40–70	Usually mined from shallow coastal beds, dried, and ground before use.
Wood ashes	CaO, MgO, K_2O, K(OH), etc.	40	Caustic, largely water-soluble, must be protected from water.
Misc. lime-containing by-products	Usually $CaCO_3$ with various impurities	20–100	Variable composition; test for toxic impurities.

[a]If made from dolomitic limestone.

How Liming Materials React to Raise Soil pH

Most liming materials—whether they be oxide, hydroxide, or carbonate—react with carbon dioxide and water to yield bicarbonate when applied to an acid soil. The carbon dioxide partial pressure in the soil, usually several hundred times greater than that in atmospheric air, is generally high enough to drive such reactions to the right. For example:

$$\underset{\text{Dolomitic limestone}}{CaMg(CO_3)_2} + 2H_2O + 2CO_2 \rightleftharpoons Ca + \underset{\text{Bicarbonate}}{2HCO_3^-} + Mg + \underset{\text{Bicarbonate}}{2HCO_3^-} \qquad (9.21)$$

The Ca and Mg bicarbonates are much more soluble than are the carbonates, so the bicarbonate formed is quite reactive with the exchangeable and residual soil acidity. The Ca^{2+} and Mg^{2+} replace H^+ and Al^{3+} on the colloidal complex:

$$\boxed{\text{Clay or humus}}\begin{matrix}H^+\\ Al^{3+}\end{matrix} + 2Ca^{2+} + \underset{\text{Bicarbonate}}{4HCO_3^-} \rightleftharpoons \boxed{\text{Clay or humus}}\begin{matrix}Ca^{2+}\\ Ca^{2+}\end{matrix} + \underset{\text{(solid)}}{Al(OH)_3} + H_2O + 4CO_2\uparrow \qquad (9.22)$$

The insolubility of $Al(OH)_3$, the weak dissociation of water, and the release of CO_2 gas to the atmosphere all pull these reactions to the right. In addition, the adsorption of the calcium and magnesium ions lowers the percentage acid saturation of the colloidal complex, and the pH of the soil solution increases correspondingly.

The amount of liming material required to ameliorate acid soil conditions is determined by several factors, including (1) the change required in the pH or exchangeable Al saturation, (2) the buffer capacity of the soil, (3) the chemical composition of the liming materials to be used, and (4) the fineness of the liming material. The limestone requirements of soils with several different textures (and therefore likely to have different buffering capacities) are estimated in Figure 9.14. Because of greater buffering capacity, the lime requirement for a clay loam is much higher than that of a sandy loam with the same pH value (see Section 9.5).

Within the pH range of 4.5 to 7.0, the degree of change in pH brought about by additions of base to an acid soil is determined by the buffering capacity of the particular soil (Figure 9.5). In turn, the capacity to buffer pH is closely related to organic matter and clay contents of a soil as reflected in the CEC. Given the titration curve for a given type of soil (such as the two examples shown in Figure 9.5), the pH of the unlimed soil,

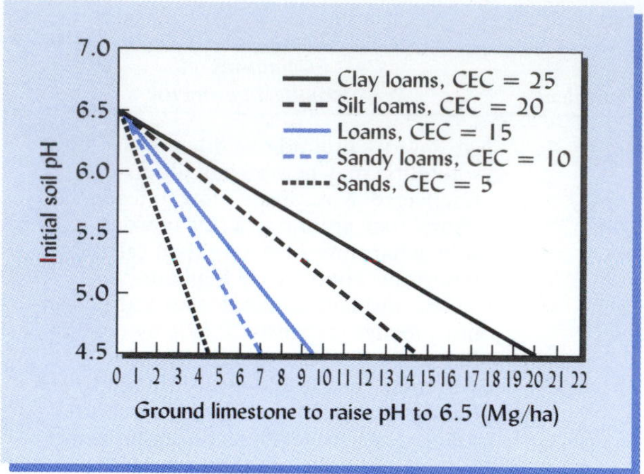

FIGURE 9.14 Effect of soil textural class on the amount of limestone required to raise the pH of soils from their initial level to pH 6.5. Note the very high amounts of lime needed for fine-textured soils that are strongly buffered by their high levels of clay, organic matter, and CEC. The chart is most applicable to soils in cool, humid regions where 2:1 clays predominate. In warmer regions where organic matter levels are lower and clays provide less CEC, the target pH would likely be closer to 5.8 and the amounts of lime required would be one-half to one-third of those indicated here. In any case, however, it is unwise to apply more than 7 to 9 Mg/ha (3 to 4 tons/acre) of liming materials in a single application. If more is needed, subsequent applications can be made at 2- to 3-year intervals until the desired pH is achieved.

and the desired pH level, it is possible to calculate the approximate amount of liming material needed (see Box 9.2).

Buffer pH Methods for Lime Requirement. Another widely used approach to determining lime requirements is to equilibrate a sample of soil with a special buffered solution that contains a small quantity of a neutral salt. The greater the total acidity of the soil, the more the solution buffering is overcome and the solution pH reduced, the degree of the reduction being proportional to the acidity released from the soil.

Exchangeable Aluminum. For most purposes, amelioration of adverse soil acidity will be achieved if the pH is raised just enough to eliminate exchangeable aluminum. This approach to liming, rather than attempting to achieve a certain soil pH, has been found especially appropriate for highly weathered soils such as Ultisols and Oxisols. By this approach, the required amount of lime can be calculated using values for the initial CEC and the percent Al saturation.

For example, if a soil has a CEC of 10 $cmol_c/kg$ and is 50% Al-saturated, then 5 $cmol_c/kg$ of Al^{3+} ions must be displaced (and their acidity from Al hydrolysis neutralized). This would require 5 $cmol_c/kg$ from $CaCO_3$:

$$5\ cmol_c/kg \times (100\ g/mol\ CaCO_3) \times (1\ mol\ CaCO_3/2\ mol_c) \times (0.01\ mol_c/cmol_c) = 2.5\ g\ CaCO_3/kg\ soil$$

This amount is equivalent to 5000 kg/ha (2.5 g/kg $\times$ 2 $\times$ 10^6 kg/ha). Experience suggests that to ensure a complete reaction in the field, the amount of limestone so calculated must be multiplied by a factor of 1.5 or 2.0 to give the actual amount of lime to apply.

How Lime is Applied

Practical tutorial on agricultural liming practices: http://hubcap.clemson.edu/~blpprt/acidity.html

The action of limestone is not instantaneous. Rather, the material slowly reacts with the soil acidity, gradually raising the pH to the desired level over a period ranging from a few weeks in the case of hydrated lime to a year or so with finely ground limestone.

Because of its gradual effects, lime should be spread about 6 to 12 months ahead of the crop that has the highest pH and calcium requirements. Thus, in a rotation with corn, wheat, and two years of alfalfa, the lime may be applied after the corn harvest to favorably influence the growth of the alfalfa crop that follows. However, because most lime is bulk-spread using heavy trucks, applications are most feasible on the sod or hay crop. This prevents adverse soil compaction by the truck tires, which might occur if the lime were applied on recently tilled land.

Liming will be most beneficial to acid sensitive plants if as much as possible of the root environment is altered. However, in most instances it is economically and physically feasible to mix lime into only the upper 15 to 20 cm of soil. The Ca^{2+} and Mg^{2+} ions provided by limestone replace acid cations on the exchange complex and do not move readily down the profile. Therefore, the effects of limestone are mainly limited to the soil layer into which the material is incorporated.

BOX 9.2 CALCULATING LIME NEEDS BASED ON pH BUFFERING

Your client wants to grow a high-value crop of asparagus in a 2 ha field. The soil is a sandy loam with a current pH of 5.0. Asparagus is a calcium-loving crop that requires a high pH for best production (see Figure 9.13), so you recommend that the soil pH be raised to 6.8. Since the soil texture is a sandy loam, we will assume that its buffer curve is similar to that of the moderately buffered soil B in Figure 9.5. In actual practice, a soil test laboratory using this method to calculate lime requirement should have buffer curves for the major types of soils in its service area.

1. Extrapolating from curve B in Figure 9.5, we estimate that it will require about 2.5 $cmol_c$ of lime/kg of soil to change the soil pH from 5.0 to 6.8. (Draw a horizontal line from 5.0 on the *y*-axis of Figure 9.5 to curve B, then draw a vertical line from this intersection down to the x-axis. Repeat this procedure beginning at 6.8 on the *y*-axis. Then use the x-axis scale to compare the distance between where your two vertical lines intersect the x-axis.)
2. Each molecule of $CaCO_3$ neutralizes 2 H^+ ions.

$$CaCO_3 + 2H^+ \rightarrow \rightarrow Ca^{2+} + CO_2 + H_2O$$

3. The mass of 2.5 $cmol_c$ of pure $CaCO_3$ can be calculated using the molecular weight of $CaCO_3$ = 100 g/mol.

$$(2.5 \text{ cmol}_c/\text{kg soil}) \times (100 \text{ g/mol CaCO}_3)\ (1 \text{ mol CaCO}_3/2 \text{ mol}_c) \times (0.01 \text{ mol}_c/\text{cmol}_c)$$

$$= 1.25 \text{ g CaCO}_3/\text{kg soil}$$

4. Using the conversion factor of 2×10^6 kg/ha of surface soil (see footnote, page 107), we calculate the amount of pure $CaCO_3$ needed per hectare:

$$(1.25 \text{ g CaCO}_3/\text{kg soil}) \times (2 \times 10^6 \text{ kg soil/ha}) = 2{,}500{,}000 \text{ g CaCO}_3/\text{ha}$$

$$(2{,}500{,}000 \text{ g CaCO}_3/\text{ha}) \times (1 \text{ kg CaCO}_3/1000 \text{ g CaCO}_3) = 2500 \text{ kg CaCO}_3/\text{ha}$$

$$2500 \text{ kg CaCO}_3/\text{ha} = 2.5 \text{ Mg/ha (or about 1.1 tons/acre).}$$

5. Since our limestone has a $CaCO_3$ equivalence of 90%, 100 kg of our limestone would be the equivalent of 90 kg of pure $CaCO_3$. Consequently, we must adjust the amount of our limestone needed by a factor of 100/90:

$$2.5 \text{ Mg pure CaCO}_3 \times 100/90 = 2.8 \text{ Mg limestone/ha}$$

6. Finally, because not all the $CaCO_3$ in the limestone will completely react with the soil, the amount calculated from the laboratory buffer curve is usually increased by a factor of 2.

$$(2.8 \text{ Mg limestone/ha}) \times 2 = 5.6 \text{ Mg limestone/ha}$$

(using Appendix B, this value can be converted to about 2.5 tons/acre)

Note that this result is very similar to the amount of lime indicated by the chart in Figure 9.14 for this degree of pH change in a sandy loam.

OVERLIMING. Another practical consideration is the danger of *overliming*—the application of so much lime that the resultant pH values are too high for optimal plant growth. Overliming is not very common on fine-textured soils with high buffer capacities, but it can occur easily on coarse-textured soils that are low in organic matter. The detrimental results of excess lime include deficiencies of iron, manganese, copper, and zinc; reduced availability of phosphate; and constraints on the plant absorption of boron from the soil solution. It is an easy matter to add a little more lime later, but quite difficult to counteract the results of applying too much. Therefore, liming materials should be added conservatively to poorly buffered soils. For some Ultisols and Oxisols, overliming may occur if the pH is raised even to 6.0.

9.10 ALTERNATIVE WAYS TO AMELIORATE THE ILL EFFECTS OF SOIL ACIDITY

Where the principal acidity problem is in the surface horizon, and where sources of limestone are readily accessible, traditional liming procedures are quite effective and economical. However, where subsoil acidity is a problem, or where liming materials are not accessible or affordable, several other approaches to combating the negative effects of soil acidity may be appropriate for use with or without traditional liming. Particularly deserving of attention are the use of gypsum and organic materials to reduce aluminum toxicity and the use of plant species or genotypes that tolerate acid conditions.

Gypsum Applications

Gypsum ($CaSO_4 \cdot 2H_2O$) is a widely available material found in natural deposits or as industrial by-products. Gypsum can ameliorate aluminum toxicity despite the fact that it does not increase soil pH. In fact, gypsum has been found more effective than lime in reducing exchangeable aluminum in subsoils, and thereby in improving root growth and crop yields (see Figure 9.15).

One reason for the effectiveness of gypsum is that the calcium from surface-applied gypsum moves down the soil profile more readily than that from lime. As lime dissolves, its reactions raise the pH, thus increasing the pH-dependent charges on the soil colloids, which in turn retain the released Ca^{2+}, preventing its downward leaching. In addition, the anion released by lime is either OH^- or CO_3^-, both of which are largely removed by the lime reactions (either by forming water or carbon dioxide gas), thus depriving the Ca^{2+} cations of surplus anions that could accompany them in the leaching process. By contrast, gypsum as a neutral salt does not raise the soil pH, and so does not increase the CEC. Furthermore, the SO_4^{2-} anion released by the dissolution of gypsum is available to accompany Ca^{2+} cations in leaching.

Using Organic Matter

Practices such as the application of organic wastes, production of cover crops (see Section 14.2), and mulching or return of crop residues all increase the organic matter in soil. In so doing they can ameliorate the effects of soil acidity in at least three ways:

1. Humified organic matter can bind tightly with aluminum ions and prevent them from reaching toxic concentrations in the soil solution (see Figure 9.16).
2. Low-molecular-weight organic acids produced by microbial decomposition or root exudation can combine with aluminum, forming soluble complexes that are nontoxic to plants and microbes.
3. Many organic amendments contain substantial amounts of calcium held in organic complexes that leach quite readily down the soil profile. Therefore, if such amendments as legume residues, animal manure, or sewage sludge are high in Ca, they can effectively combat aluminum toxicity and raise Ca and pH levels, not only in the surface soil where they are incorporated, but also quite deep into the subsoil (see Figure 9.15).

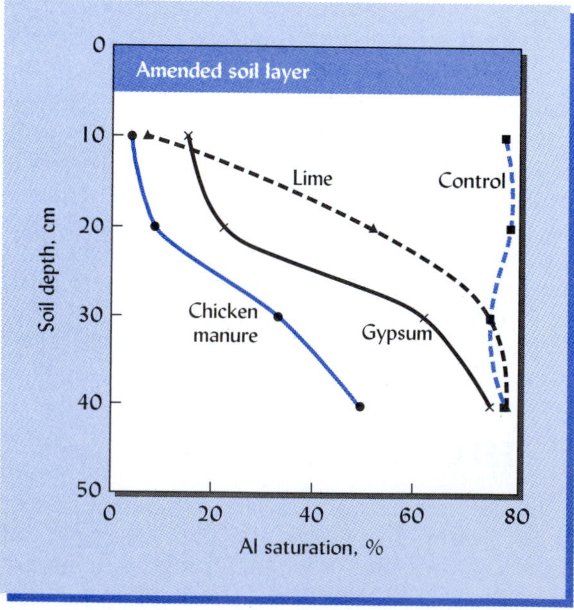

FIGURE 9.15 Aluminum saturation percentage in the subsoil of a fine-textured Hawaiian Ultisol after treatment of the surface soil with chicken manure, limestone, or gypsum. The soil was slowly leached with the 380 mm of water. The lime raised the pH and thereby effectively reduced Al saturation, though only in the upper 10 to 20 cm. The effect of gypsum extended somewhat more deeply. Gypsum does not raise the pH or increase the CEC, but is more soluble than lime and provides the SO_4^{2-} anions to accompany Ca^{2+} cations as they leach downward in solution. The greatest and deepest reduction in Al saturation resulted from the chicken manure. Another Ca-rich organic amendment, sewage sludge, gave similar results (not shown). The manure probably formed soluble organic complexes with Ca^{2+} ions, which then move down the profile where Ca exchanged with Al to form nontoxic organic Al complexes. [Redrawn from Hue and Licudine (1999); used with permission of the American Society of Agronomy]

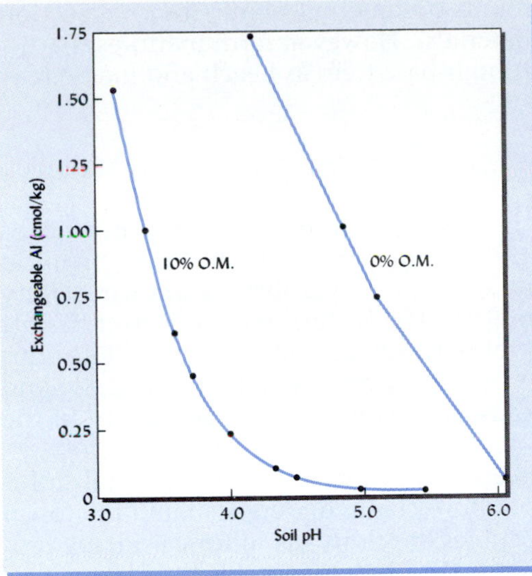

FIGURE 9.16 Effect of humified organic matter (peat) on the relationship between exchangeable aluminum and soil pH. Where the soil was mixed with peat to give a 10% organic matter content, exchangeable aluminum was much reduced between pH 3.5 and 5.5. The peat apparently binds the aluminum in nonexchangeable (and nontoxic) form. This effect helps to explain why a much lower pH can be tolerated by roots growing in high-organic-matter soils such as Histosols or forest O horizons. [Modified from Hargrove and Thomas (1981); used with permission of the American Society of Agronomy]

Enhancing these organic matter reactions may be much more practical than standard liming practices for resource-poor farmers or those in areas far from limestone deposits. **Green manure** crops (vegetation grown specifically for the purpose of adding organic matter to the soil) and mulches can provide the organic matter needed to stimulate such interactions and thereby reduce the level of Al^{3+} ions in the soil solution. Aluminum-sensitive crops can then be grown following the green manure crop. One caution regarding use of organic amendments to ameliorate soil acidity is that the amounts of these materials effective for this purpose may exceed amounts suggested by nutrient management guidelines designed to prevent polluting water from excessive leaching and runoff losses of nitrogen and phosphorus.

Selecting Adapted Plants

It is often more judicious to solve soil acidity problems by changing the plant to be grown rather than by trying to change the soil pH. The choice of grass and tree species should consider soil pH adaptation, whether revegetating a former coal mine site on acidic mining debris or landscaping a suburban front yard on alkaline desert soils.

Considerable natural variability exists in the acid tolerance of different strains within a species. Plant breeders and biotechnologists have taken advantage of this variability to develop cultivars of widely grown food crops that are quite tolerant of very acid conditions. These varieties are especially valuable in some areas of the tropics where even modest liming applications are economically impractical. These advances show how collaboration between plant and soil scientists can enhance the ability of low-income families to grow their traditional food crops in acid, degraded soils.

9.11 LOWERING SOIL pH

It is often desirable to reduce the pH of highly alkaline soils. Furthermore, some acid-loving plants cannot even tolerate near-neutral pH values. For example, rhododendrons and azaleas, favorites of gardeners around the world, grow best on soils having pH values of 5.0 and below. Blueberry is an acid-loving crop that requires similar low pH levels for good production. To accommodate these plants, it is sometimes desirable to increase the acidity of even acid soils. This is done by adding acid-forming organic and inorganic materials.

Acid Organic Matter

As organic residues decompose, organic and inorganic acids are formed. These can reduce the soil pH if the organic material is low in calcium and other nonacid cations. Leaf mold from coniferous trees, pine needles, tanbark, pine sawdust, and acid peat moss are quite satisfactory organic materials to add around ornamental plants (but see Section 11.3 for nitrogen considerations with these materials). However, farm manures (particularly poultry manures) and leaf mold of such high-base trees as beech and maple may be alkaline, and may increase the soil pH.

Inorganic Chemicals

When the addition of acid organic matter is not feasible, inorganic chemicals such as aluminum sulfate ("alum") or ferrous sulfate ($Fe^{II}SO_4$) may be used. The latter chemical provides available iron (Fe^{2+} ions) for the plant and, upon hydrolysis, enhances acidity by reactions similar to that shown in Equation 9.18. The H^+ ions released lower the pH locally and liberate some of the iron already present in the soil. Ferrous sulfate thus serves a double purpose for iron-loving plants by supplying available iron directly and by reducing the soil pH—a process that may cause a release of fixed iron present in the soil (see Plates 43 and 45, after page 498).

Other materials that are often used to increase soil acidity include elemental sulfur and, in some irrigation systems, sulfuric acid. Sulfur usually undergoes rapid microbial oxidation in the soil (see Section 13.22), and 2 moles of acidity (as sulfuric acid) are produced for every mole of S oxidized (see Equation 9.19). Under favorable conditions, sulfur is 4 to 5 times more effective, kilogram for kilogram, in developing acidity than is ferrous sulfate. Although ferrous sulfate brings about more rapid plant response, sulfur is less expensive, is easy to obtain, and is often used for other purposes. The quantities of ferrous sulfate or sulfur that should be applied will depend upon the buffering capacity of the soil and its original pH level. Figure 9.5 suggests that for each unit drop in pH desired, a well-buffered soil (e.g., a silty clay loam with 4% organic matter) will require about 4 $cmol_c$ of sulfur per kilogram of soil. This is about 1200 kg S/ha (since $cmol_c$ of $S = 0.32/2 = 0.16$ g, the 2 mol_c/mol being based on the 2 mol of H^+ ion produced by each mole of S).

9.12 CALCIUM AND MAGNESIUM AS PLANT NUTRIENTS

Calcium and magnesium are both macronutrients essential for all plants. The ability of a soil to supply these two elements to plants is intimately tied to soil alkalinity, acidity and liming. Therefore, we will consider their roles and dynamics here, rather than in later chapters that are devoted to the other essential nutrients. It should be noted that the amount of liming materials needed to supply Ca and Mg as nutrients is generally less than 10% of that needed to ameliorate soil acidity.

Calcium

Plants generally use Ca in amounts second only to N and K. The amounts of calcium taken up are greater for dicots (0.5% to >2% of dry matter) than for monocots (0.15% to 0.5% of dry matter). Much calcium is stored in woody tissues, the net calcium uptake by many trees being close to that of nitrogen.

Deficiency Symptoms. Calcium is the most plentiful cation on the exchange complex of nearly all soils that are not so acidic as to have a high aluminum saturation. For this reason, deficiencies of calcium are quite rare for most plants, except in very acid soils. When calcium deficiency does occur, it is usually associated with growing points (meristems) such as buds, fruits, and root tips. It is difficult to distinguish calcium deficiency from the Al or Mn toxicity that so often accompanies it in very acid soils. When the calcium concentration in the soil solution falls below a critical value, many metals (including those like magnesium, manganese, or zinc that are normally beneficial as nutrients) become highly toxic and the roots become stunted and gelatinous (Figure 9.17).

Soil calcium is found mainly in three pools: (1) calcium-containing minerals (such as calcite or plagioclase), (2) calcium complexed with soil humus, and (3) calcium on the clay and humus cation exchange complex. The cycling of calcium among these and other soil pools, and the gains and losses of calcium by such mechanisms as plant uptake, liming, and leaching comprise the calcium cycle as illustrated in Figure 9.18. In most soils, the principal sources of Ca for plant uptake are (1) exchangeable Ca and (2) Ca in readily

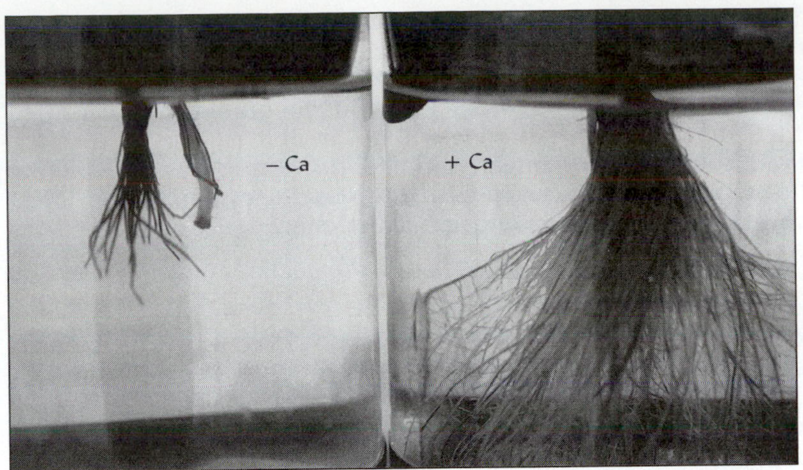

FIGURE 9.17 Root growth was almost completely inhibited by low calcium in the nutrient solution (left) compared to healthy roots in the same nutrient solution but with calcium added (right). If the ratio of calcium to all other cations in solution drops below 5:1, the integrity of root membranes is lost, causing many other elements to become toxic to the plants. (Photo courtesy of R. Weil)

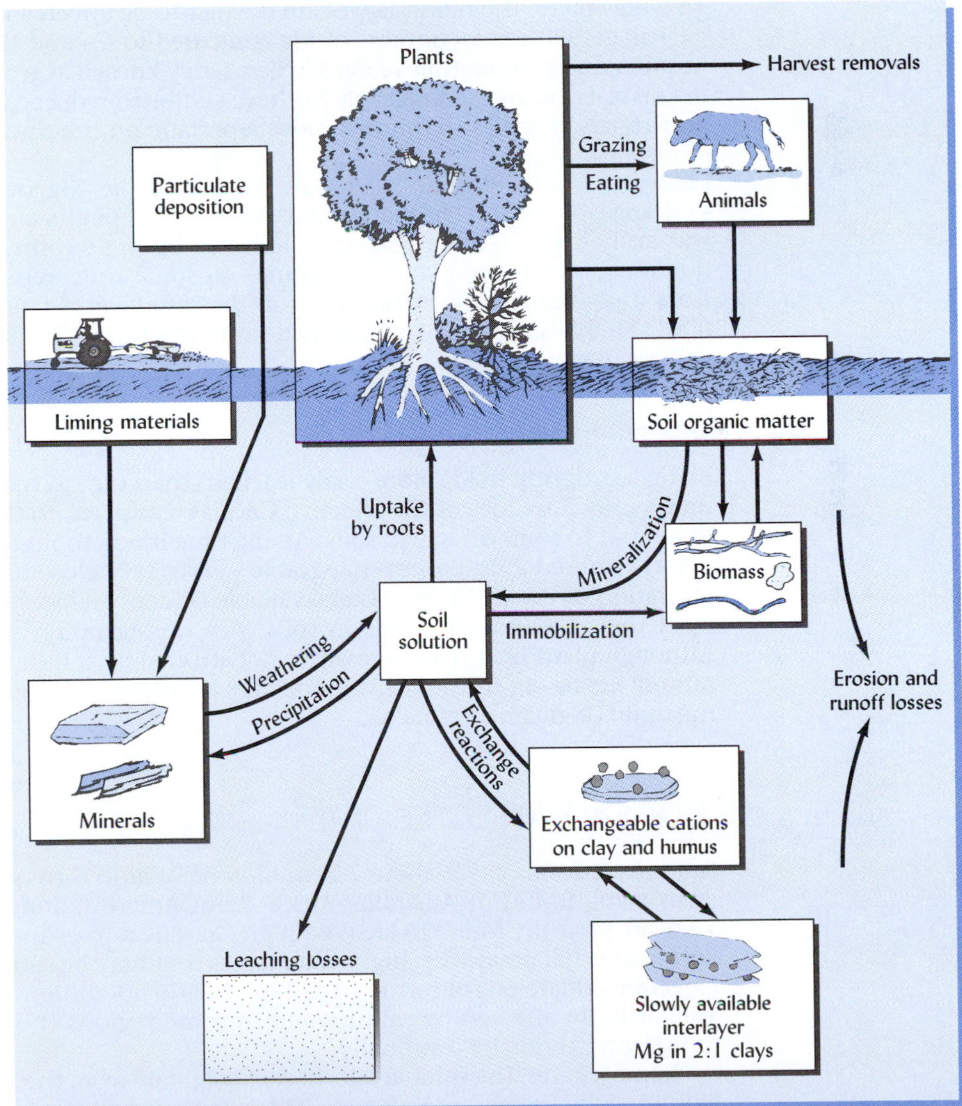

FIGURE 9.18 Simplified diagram of the cycling of calcium and magnesium in soils. The rectangular compartments represent pools of these elements in various forms, while the arrows represent processes by which the elements are transformed or transported from one pool to another.

weathered minerals (such as carbonates). Note that in arid and semiarid regions, the high pH, high carbonate nature of the soil solution greatly diminishes the solubility of calcium carbonate minerals. In humid regions, leaching losses of Ca often exceed plant uptake.

Scientists are concerned that for some forested soils in the humid eastern United States, the release of Ca from mineral weathering will not be able to keep up with the losses, and that acid rain combined with intensive timber harvesting may be depleting the Ca reserves in the more poorly buffered watersheds. Nonetheless, up to now forest growth has rarely showed a positive response to applied calcium.

Magnesium

Plants generally take up Mg in somewhat smaller amounts (0.15 to 0.75% of dry matter) than Ca. Magnesium is a component of the chlorophyll molecule, and so is intimately involved with photosynthesis in plants. It also plays critical roles in the synthesis of oils and proteins and in the activation of enzymes involved in energy metabolism.

The deficiency of Mg is much more common than that of Ca, at least when the soil pH is at an appropriate level. The most common symptom of Mg deficiency is interveinal chlorosis on the older leaves, which appears as a mottled green and yellow coloring in dicots (Plate 46 following page 498) and a striping in monocots. Because Mg (unlike Ca) is readily translocated in most plants from the older to the younger, still-growing leaves, the oldest leaves are the first to be affected by low Mg supplies.

Forages with low contents of Mg compared to Ca and K can cause grazing animals to suffer from a sometimes-fatal Mg deficiency known as *grass tetany*. Spruce trees growing on soils low in exchangeable Mg have exhibited reduced Mg in needle tissue, stunted growth, and needles that turn yellow, especially on the tips.

FORMS IN SOIL. The main source of plant-available Mg in most soils is the pool of exchangeable Mg on the clay-humus complex. As plants and leaching remove this Mg, the easily exchangeable pool is replenished by Mg weathered from minerals (such as dolomite, hornblende, and serpentine). In some soils, replenishment also takes place from a pool of slowly available Mg in the interlayers of certain 2:1 clays (see Section 8.3). Variable amounts of Mg are made available by the breakdown of plant residues and soil organic matter.

Ratio of Calcium to Magnesium

Being less tightly held (more easily leached) than Ca, exchangeable Mg commonly saturates only 5 to 20% of the effective CEC, as compared to the 60 to 90% typical for Ca in neutral to somewhat acid soils. At one time it was thought that for optimum growth plants required a ratio of exchangeable Ca:Mg very close to 6:1 (65% Ca and 10% Mg saturation of the CEC). However, available evidence suggests that plants grow well and meet their Ca and Mg needs in soils with Ca:Mg ratios anywhere from 1:1 to 15:1. Although plant health and growth is not affected by soil Ca:Mg ratios in this range, the ratio of Ca to Mg in the plant tissue may be affected enough to influence the mineral nutrition of grazing animals.

9.13 DEVELOPMENT OF SALT-AFFECTED SOILS[6]

Salt-affected soils cover about 320 million ha of land throughout the world, the largest areas being found in Australia, Africa, Latin America, and the Near and Middle East. They are typically found in areas with precipitation-to-evaporation ratios of 0.75 or less, and in low, flat areas with high water tables that may be subject to seepage from higher elevations (Plate 60 following page 498). Nearly 50 million ha of cropland and pasture are currently affected by salinity and in some regions the area of land so affected is growing by about 10% annually.

In most cases, the soluble salts in soils originate from the weathering of primary minerals in rocks and parent materials. The salts are usually transported to a developing salt-

[6]For a discussion of these soils, which are also referred to as *halomorphic soils*, see Abrol et al. (1988) and Szabolcs (1989).

affected soil as ions dissolved in water. The salt-containing water moves through a landscape from areas of higher to lower elevations, and from soil zones that are wetter to those that are drier. The water eventually evaporates, but the dissolved salts are left behind to accumulate in the soil. This is true in both irrigated and unirrigated landscapes.

Many salt-affected soils develop because changes in the local water balance, usually brought about by human activities, increase the input of salt-bearing water more than they increase the output of drainage water. Increased evaporation, waterlogging, and rising water tables usually result. It is worth remembering the irony that salts usually become a problem when too much water is supplied, not too little.

Accumulation of Salts in Nonirrigated Soils

Chloride and sulfate salts of calcium, magnesium, sodium, and potassium accumulate naturally in some surface soils because there is insufficient rainfall to flush them from the upper soil layers. In coastal areas, sea spray (Plate 61) and inundation with seawater can be locally important sources of salt in soils, even in humid regions.

Other localized but important sources are fossil deposits of salts laid down during geological time. These fossil salts can dissolve in underground waters and move horizontally over impervious geological layers and ultimately rise to the surface of the soil in the low-lying parts of the landscape. The low-lying areas where the saline groundwater emerges and salts accumulate are termed **saline seeps.**

Saline seeps occur naturally in some locations, but their formation is often greatly increased when the water balance in a semiarid landscape is disturbed by bringing land under cultivation (Figure 9.19). Replacement of native, deep-rooted perennial vegetation

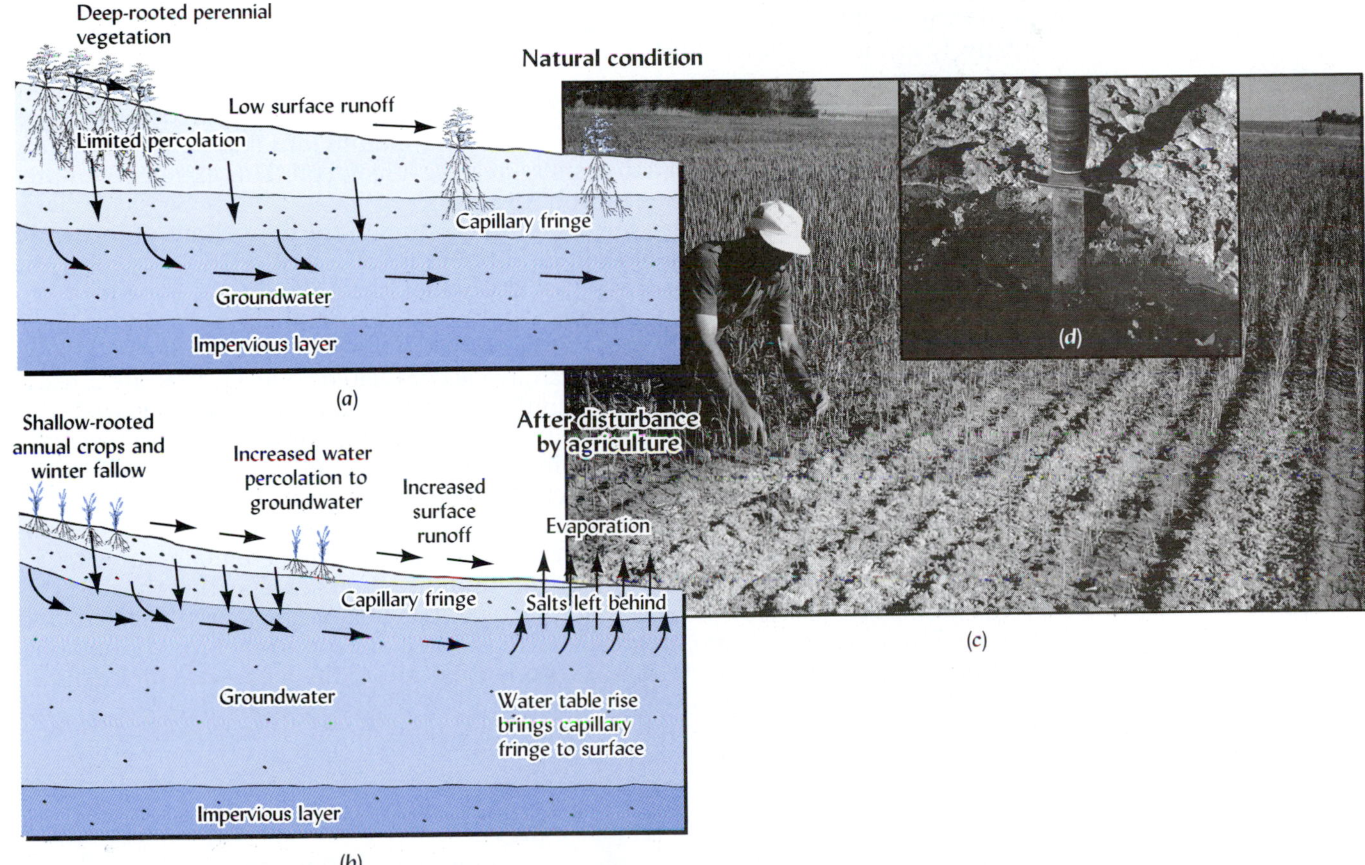

FIGURE 9.19 Saline seep formation in a semiarid area where the salt-rich substrata are underlain by an impermeable layer. (*a*) Under deep-rooted perennial vegetation, transpiration is high and the water table is kept low. (*b*) After conversion to agriculture, shallow-rooted annual crops take up much less water, especially if fallow is practiced, allowing more water to percolate through the salt-bearing substrata. Consequently, in lower-elevation landscape positions, the wet-season water table rises close to the soil surface. This allows the salt-laden groundwater to rise by capillary flow to the surface, from which it evaporates, leaving behind an increasing accumulation of salts. Note that the diagrams greatly exaggerate vertical distances. (*c*) A spreading saline seep area in eastern Montana where wheat fallow cropping has replaced natural deep-rooted prairie vegetation. (*d*) Close-up of salt crust over moist soil. (Photos and drawings courtesy of R. Weil)

with annual crop species greatly reduces the annual evapotranspiration, especially if the cropping system includes periods of fallow during which the soil is unvegetated. The decreased evapotranspiration allows more rainwater to percolate through the soil, thus raising the water table and increasing the flow of groundwater to lower elevations. In dry regions, soils and substrata may contain substantial amounts of soluble salts that can be picked up by the percolating water. If a shallow layer restricts percolation, it will further encourage the flow of the salt-laden groundwater across the landscape toward the lowest elevations. Eventually, water table may rise to within 1 m or less of the soil surface, and capillary rise will begin to contribute a continuous stream of salt-laden water to replace the water lost at the surface by evaporation. Year by year the evaporation zone will creep up the slope, and the barren, salinized area will become larger and more saline. Millions of hectares of land in North America, Australia, and other semiarid regions have been degraded in this fashion.

Irrigation-Induced Salinity and Alkalinity

Irrigation not only alters the water balance by bringing in more water, it also brings in more salts. Whether taken from a river or pumped from the groundwater, even the best quality freshwater contains some dissolved salts (see Section 9.17). The amount of salt brought in with the water may seem negligible, but the amounts of water applied over the course of time are huge. Again, pure water is lost by evaporation, but the salt stays and accumulates. The effect is accentuated in arid regions for two reasons: (1) The water available from rivers or from underground is relatively high in salts because it has flowed through dry-region soils which typically contain large amounts of easily weatherable minerals; and (2) the dry climate creates a relatively high evaporative demand, so large amounts of water are needed for irrigation. An arid-region farmer may need to apply 90 cm of water to grow an annual crop. Even if this is good-quality water relatively low in salts, it will likely dump more than 6 Mg/ha (3 tons/acre) of salt on the land every year.

Nature and management of salt affected soils with special reference to Canada: http://sis.agr.gc.ca/cansis/publications/health/chapter08.html

If irrigation water carries a significant quantity of Na^+ ions compared to Ca^{2+} and Mg^{2+} ions, and especially if the HCO_3^- ion is present, sodium ions may come to saturate a major part of the colloidal exchange sites, creating a *sodic* soil. This extremely unproductive type of soil is discussed in Section 9.15.

During the past three decades, low-income countries in the dry regions of the world have greatly expanded the area of their land under irrigation in order to produce the food needed by their rapidly growing human populations. Initially, the expanded irrigation stimulated phenomenal increases in food-crop production. Unfortunately, many irrigation projects failed to provide for adequate drainage. As a result, the process of **salinization** has accelerated, and salts have accumulated to levels that are already adversely affecting crop production. In some areas, sodic soils have been created.

These events remind the world of the flat wastelands of southeastern Iraq that have been barren since the 12th century. In biblical times, these areas were irrigated by the Euphrates and Tigris Rivers and so productive that the overall region was called the Fertile Crescent. Unfortunately, salinization eventually set in because many of the soils were not naturally well drained and the societies sometimes failed to maintain what artificial drains were created. Today, societies around the world are repeating the mistakes of the past. Some observers believe that each year, the area of previously irrigated land degraded by severe salinization is greater than the area of land newly brought under irrigation. Truly, the world needs to give serious attention to the large-scale problems associated with salt-affected soils.

9.14 MEASURING SALINITY AND SODICITY[7]

Salt-affected soils adversely affect plants because of the total concentration of salts (*salinity*) in the soil solution and because of concentrations of specific ions, especially sodium (*sodicity*).

[7]For an informative discussion of these methods, see Rhoades et al. (1999) or the Web site of the U.S. Salinity Laboratory at *www.ussl.ars.usda.gov.*

Salinity

Total Dissolved Solids. In concept, the simplest way to determine the total amount of dissolved salt in a sample of water is to heat the solution in a container until all of the water has evaporated and only a dry residue remains. The residue can then be weighed and the total dissolved solids (TDS) expressed as milligrams of solid residue per liter of water (mg/L). In water to be used for irrigation, TDS typically ranges from about 5 to 1000 mg/L, while in the solution extracted from a soil sample, TDS may range from about 500 to 12,000 mg/L.

Home page of U.S. Soil Salinity Lab. Try choosing "assessment": http://www.ussl.ars.usda.gov/

Electrical Conductivity. Pure water is a poor conductor of electricity, but conductivity increases as more and more salt is dissolved in the water. Thus, the **electrical conductivity (EC)** of the soil solution gives us an indirect measurement of the salt content. The EC can be measured both on samples of soil or on the bulk soil in situ (Table 9.3). It is expressed in terms of decisiemens per meter (dS/m).[8] In the most common procedure, a soil sample is saturated with distilled water and mixed into a paste; then the water is extracted and the EC of the **saturated paste extract** is measured (Figure 9.20).

Mapping EC in the Field. Advances in instrumentation now allow rapid, continuous field measurement of bulk soil conductivity, which, in turn, is directly related to soil salinity (see Table 9.3). One such method involves inserting a **four-electrode conductivity apparatus** into moist soil to make direct measurements of apparent EC in the field (EC_a). If the apparatus is fitted with a geopositioning system receiver and driven across a field, a map of salinity variation can be produced (Figure 9.21).

Kit to determine—in the field—whether soils contain too much salt or sodium: http://www.ussl.ars.usda.gov/hachkit.htm

Electromagnetic Induction. A second rapid field method employs **electromagnetic induction** (EM) of electrical current in the body of the soil, the level of which is related to electrical conductivity and, in turn, to soil salinity. A small transmitter coil located in one end of the EM instrument generates a magnetic field within the soil. This magnetic field, in turn, induces small electric currents within the soil whose values are related to the soil's conductivity. These small currents generate their own secondary magnetic fields, which can be measured by a small receiving cell in the opposite end of the EM instrument. The EM instrument thus can measure ground EC (designated EC_a^*) to considerable depths in the soil profile without mechanically probing the soil.

Sodium Status

Two expressions are commonly used to characterize the sodium status of soils. The **exchangeable sodium percentage (ESP)** identifies the degree to which the exchange complex is saturated with sodium:

$$\text{ESP} = \frac{\text{Exchangeable sodium, cmol}_c\text{/kg}}{\text{Cation exchange capacity, cmol}_c\text{/kg}} \times 100 \quad (9.23)$$

TABLE 9.3 Different Measurements for Estimating Soil Salinity

The methods are well intercorrelated, so each can be converted to any other. The EC_e is the most common standard for comparison.

	Measured on soil sample
EC_e	Conductivity of the solution extracted from a water-saturated soil paste
EC_p	Conductivity of the water-saturated soil paste itself
EC_w	Conductivity of the solution extracted from a 1:2 soil–water mixture
TDS	Total dissolved solids in water or the solution extracted from a water-saturated soil paste[a]
	Measured on bulk soil in place
EC_a	Apparent conductivity of bulk soil sensed by metal electrodes in soil
EC_a^*	Electromagnetic induction of an electric current using surface transmitter and receiving coils

[a]Note that TDS can be converted to EC_w using these relationships: for Na salts, TDS = 640 × EC_w; for Ca salts, TDS = 800 × EC_w.

[8]Formerly expressed as millimhos per centimeter (mmho/cm). Since 1 S = 1 mho, 1 dS/m = 1 mmho/cm.

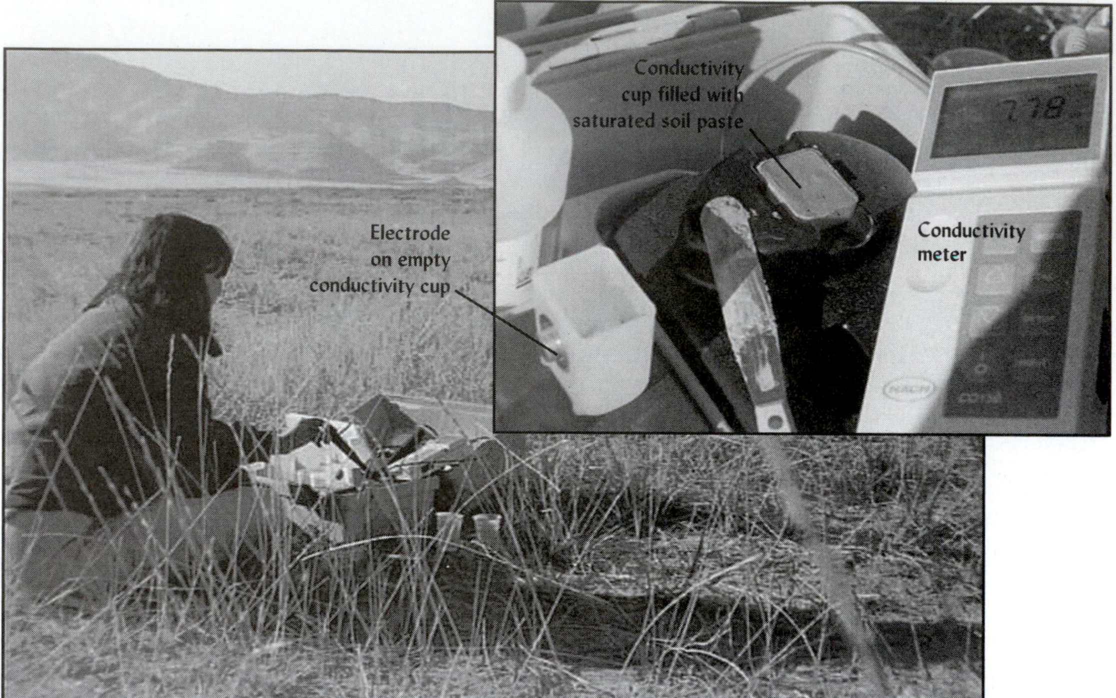

FIGURE 9.20 Measuring the electrical conductivity (EC) of a soil sample in a field of wheatgrass to determine the level of salinity. A sample of the soil is stirred with pure water until a saturated paste is made. The paste is then transferred into a special conductivity cup that has a flat, circular electrode on either side (inset). This is then inserted into a stand that connects the electrodes to a conductivity meter. Note readout of 7.78 dS/m on the conductivity meter. This level of EC_p indicates a highly saline soil that would inhibit the growth of many crops. (Photos courtesy of R. Weil)

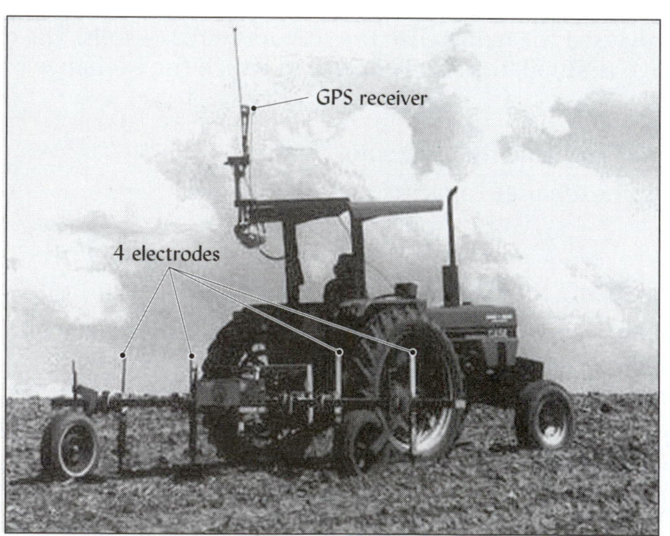

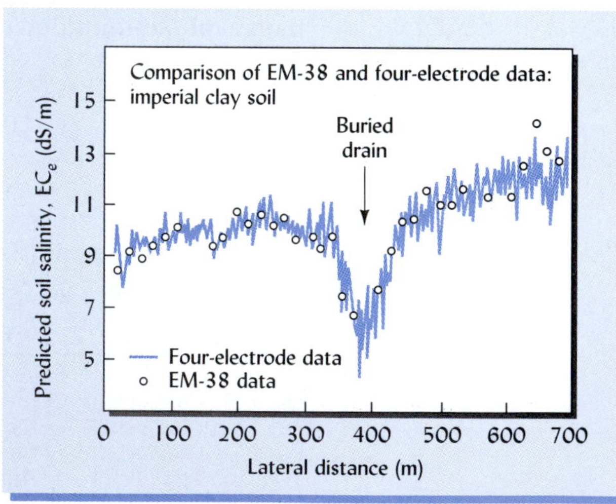

FIGURE 9.21 Surveying electrical conductivity in the field. (*Left*) A mobile four-electrode apparatus with integrated geopositioning system receiver generates a continuous readout of soil conductivity across a field. (*Right*) Soil electrical conductivity data from a transect across an irrigated field show that the EC (measured in the upper meter of soil) tends to increase down the length of an irrigation run across the field, indicating that the lower end of the field is more saline. This end of the field probably gets less irrigation water, and the salts in it are therefore less thoroughly leached out. The dip in the middle corresponds to the location of a buried drain that has reduced the salinity by encouraging drainage. The data points (small circles) in the graph were measured using an EM conductivity sensor. (Data and photo courtesy of J. D. Rhoades and S. Lesch, U.S. Salinity Lab)

ESP levels greater than 15 are associated with severely deteriorated soil physical properties and pH values of 8.5 and above.

The **sodium adsorption ratio (SAR)** is a second more easily measured property that is becoming even more widely used than ESP. The SAR gives information on the comparative concentrations of Na^+, Ca^{2+}, and Mg^{2+} in soil solutions. It is calculated as follows:

$$\mathrm{SAR} = \frac{[\mathrm{Na}^+]}{(0.5[\mathrm{Ca}^{2+}] + 0.5[\mathrm{Mg}^{2+}])^{1/2}} \tag{9.24}$$

where $[Na^+]$, $[Ca^{2+}]$, and $[Mg^{2+}]$ are the concentrations (in mmol of charge per liter) of the sodium, calcium, and magnesium ions in the soil solution. An SAR value of 13 for the solution extracted from a saturated soil paste is approximately equivalent to an ESP value of 15. The SAR of a soil extract takes into consideration that the adverse effect of sodium is moderated by the presence of calcium and magnesium ions. The SAR also is used to characterize irrigation water applied to soils (see Section 9.17).

9.15 CLASSES OF SALT-AFFECTED SOILS

Saline Soils

Using EC, ESP (or SAR), and soil pH, salt-affected soils are classified as **saline, saline-sodic,** and **sodic** (Figure 9.22). Soils that are not greatly salt-affected are classed as **normal.** The processes that result in the accumulation of neutral soluble salts are referred to as **salinization.** The concentration of these salts sufficient to interfere with plant growth (see Section 9.16) is generally defined as that which produces an electrical conductivity in the saturation extract (EC_e) greater than 4 dS/m. However, some sensitive plants are adversely affected when the EC_e is only about 2 dS/m.

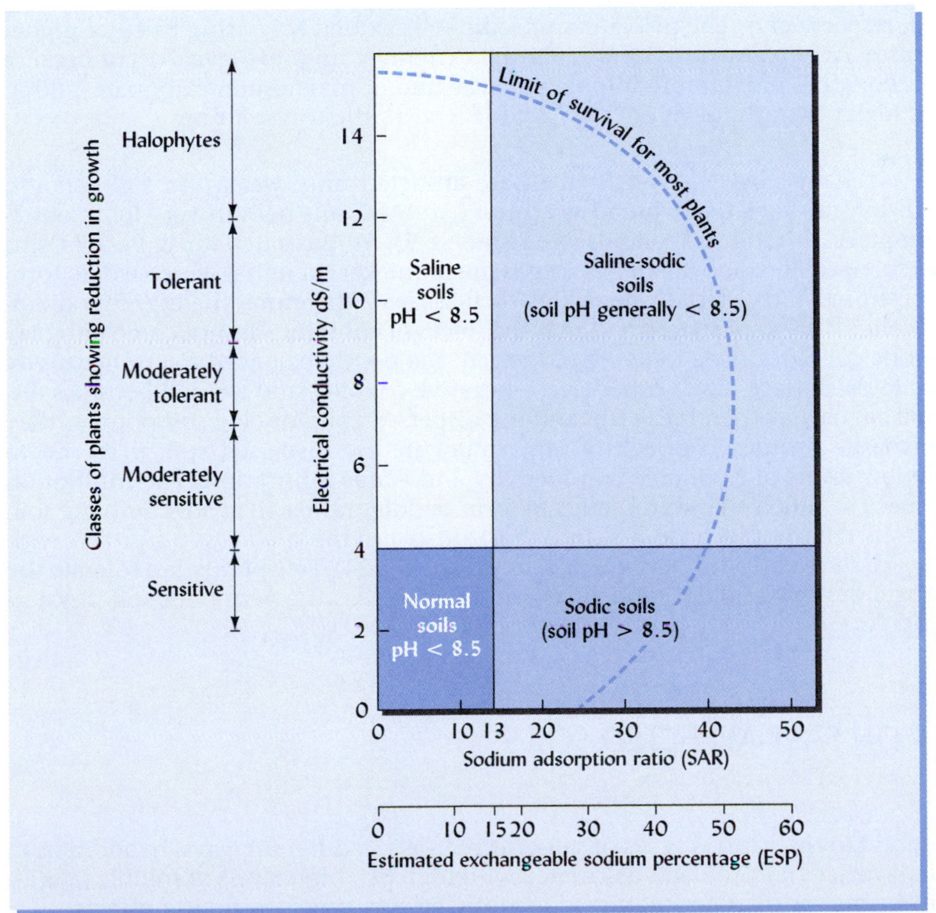

FIGURE 9.22 Diagram illustrating the classification of normal, saline, saline-sodic, and sodic soils in relation to soil pH, electrical conductivity, sodium adsorption ratio (SAR), and exchangeable sodium percentage (ESP). Also shown are the ranges for different degrees of sensitivity of plants to salinity.

Saline soils contain sufficient salinity to give EC_e values greater than 4 dS/m, but have an ESP less than 15 (or an SAR less than 13) in the saturation extract. Thus, the exchange complex of saline soils is dominated by calcium and magnesium, not sodium. The pH of saline soils is usually below 8.5. Because soluble salts help prevent dispersion of soil colloids, plant growth on saline soils is not generally constrained by poor infiltration, aggregate stability, or aeration. In many cases, the evaporation of water creates a white salt crust on the soil surface (see Figure 9.19*d* and Plate 60 following page 498), which accounts for the name *white alkali* that was previously used to designate saline soils.

Saline-Sodic Soils

Soils that have both detrimental levels of neutral soluble salts (EC_e greater than 4 dS/m) *and* a high proportion of sodium ions (ESP greater than 15 or SAR greater than 13) are classified as **saline-sodic soils** (see Figure 9.22). Plant growth in these soils can be adversely affected by both excess salts and excess sodium levels. However, the high concentration of neutral salts moderates the dispersing influence of the sodium. The salts provide excess cations that move in close to the negatively charged colloidal particles, thereby reducing their tendency to repel each other, or to disperse (Figure 9.23).

Unfortunately, this situation is subject to rather rapid change if the soluble salts are leached from the soil, especially if the SAR of the leaching waters is high. In such a case, salinity will drop, but the exchangeable sodium percentage will increase, and the saline-sodic soil will become a sodic soil.

Sodic Soils

Definition. Sodic soils are, perhaps, the most troublesome of the salt-affected soils. While their levels of neutral soluble salts are low (EC_e less than 4.0 dS/m), they have relatively high levels of sodium on the exchange complex (ESP and SAR values are above 15 and 13, respectively). The pH values of sodic soils exceed 8.5, rising to 10 or higher in some cases. As explained in Section 9.2 the extremely high pH levels occur because sodium carbonate is much more soluble than calcium or magnesium carbonate, and so maintains high concentrations of CO_3^{2-} and HCO_3^- in the soil solution.

Dispersion. Exchangeable Na^+ ions, which are attracted only weakly to soil colloids, spread out to form a relatively broad swarm of hydrated ions held in very loose outer-sphere complexes around the colloids (see Section 8.7). As illustrated in Figure 9.23, this layer of exchangeable monovalent Na^+ ions is much thicker than that which would form with more strongly attracted divalent ions such as Ca^{2+}. Therefore, the highly sodium-saturated colloids are kept so far apart that the forces of cohesion cannot come into play to attract one colloid surface to another. Instead, the poorly balanced electronegativity of each colloidal surface repels other electronegative colloids, and the soil becomes dispersed. Soil aggregates then break up, and the dispersed colloids clog soil pores as they move down the profile. The lack of large pores in the dispersed soil gives rise to extremely low levels of hydraulic conductivity and water infiltration. The infiltration rate is reduced so much that water tends to form puddles rather than soak into the soil.

Sodic soils commonly have a thin A horizon overlying a clayey B horizon with columnar structure and a rubbery consistency (Figure 9.24). Few plants can tolerate the specific toxicities of Na^+, OH^-, and HCO_3^- ions, as well as the very poor soil physical conditions.

9.16 GROWTH OF PLANTS ON SALT-AFFECTED SOILS

How Salts Affect Plants

Plants respond to the various types of salt-affected soils in different ways. In addition to the nutrient-deficiency problems associated with high pH, high levels of soluble salts affect plants by two primary mechanisms: **osmotic effects** and **specific ion effects.**

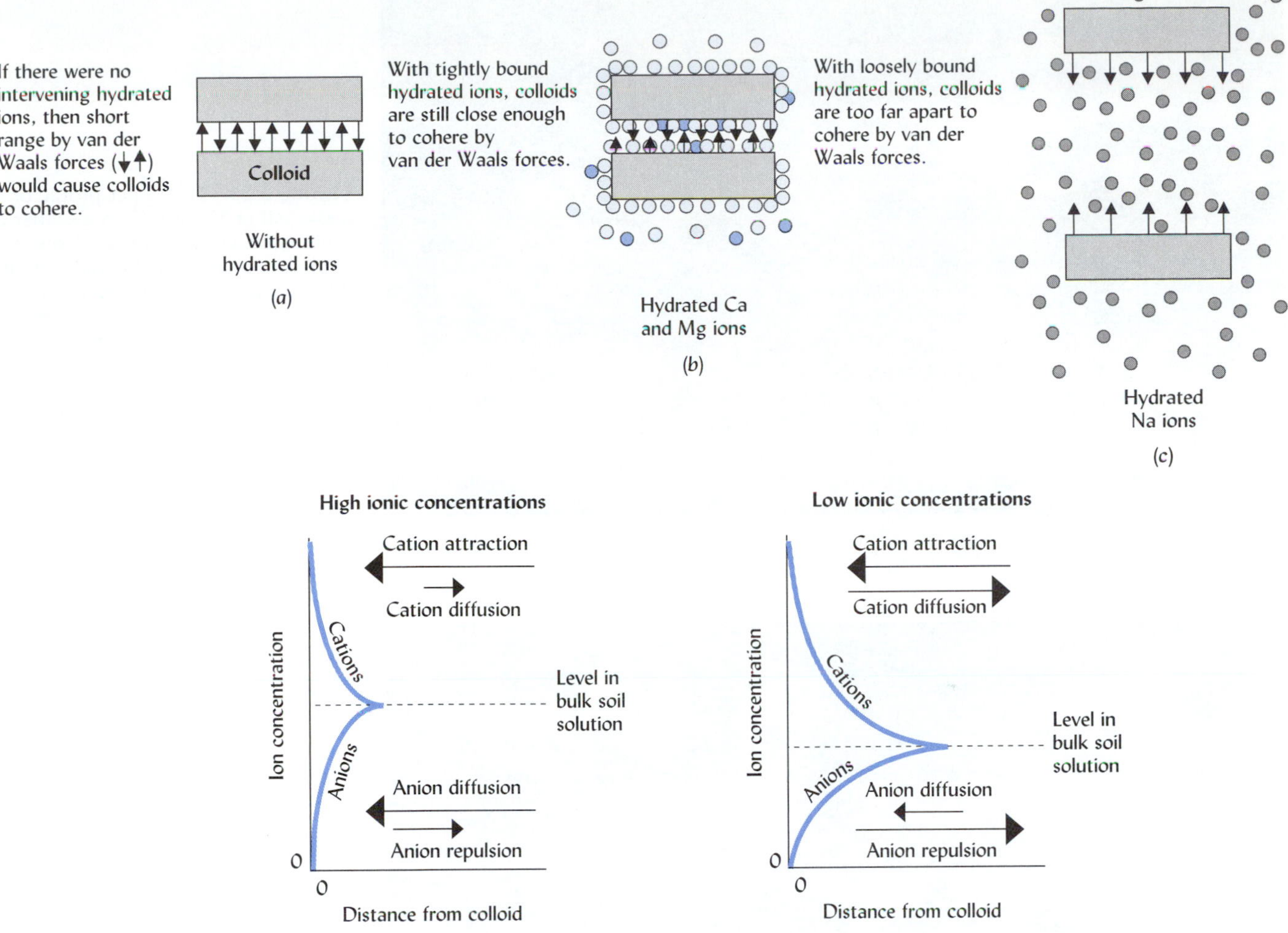

FIGURE 9.23 Soil dispersion and flocculation are influenced by both the type of salts (ions) present (upper) and their concentrations (lower). If colloids could approach closely (*a*), they would be held together (cohere) by short-range van der Waals forces. In soil, the colloids are surrounded by a swarm of hydrated exchangeable ions, which prevent the colloids from approaching so closely. If these are strongly attracted calcium and magnesium ions (*b*), they do not keep the colloids very far apart, so cohesive forces still have some effect. However, if they are loosely attracted sodium ions (*c*), the more spread out ionic swarm keeps the colloids too far apart for cohesive forces to come into play. The graphs show that when the salt concentration (both anions and cations) in the bulk soil solution is high (*d*), the ionic swarm is compressed, allowing the colloids to approach more closely. High salt concentration causes the anions to diffuse more strongly toward the colloid surface (because their concentration gradient is greater) and the cations to diffuse away less strongly (because the cation concentration gradient is diminished). When the ionic concentration is low (*e*), the swarm is more diffuse and the colloids are dispersed. Low salt (ion) concentrations and weakly attracted ions (e.g., sodium) encourage soil dispersion, while high salt (ion) concentrations and strongly attracted ions (e.g., calcium) promote flocculation. (Diagrams courtesy of R. Weil)

Osmotic Effects. Soluble salts lower the osmotic potential of the soil water (see Section 5.3), making it more difficult for roots to remove water from the soil. Established plants expend more energy accumulating organic and inorganic solutes to lower the osmotic potential *inside* their cells to counteract the low osmotic potential of the soil solution outside.

Plants are most susceptible to salt damage in the early stages of growth. Salinity may delay, or even prevent, the germination of seeds (Figure 9.25). Young seedlings may be killed by saline conditions that older plants of the same species could tolerate. As young plant cells encounter a soil solution high in salts, they may lose water by osmosis to the more concentrated soil solution. The cells then collapse.

Specific Ion Effects. The kind of salt can make a big difference in how plants respond to salinity. Certain ions, including Na^+, Cl^-, $H_4BO_4^-$, and HCO_3^-, are quite toxic to many plants. In addition to specific toxic effects, high levels of Na^+ can cause imbalances in the

FIGURE 9.24 The upper profile of a sodic soil (a Natrustalf) in a semiarid region of western Canada. Note the thin A horizon (knife handle is about 12 cm long) underlain by columnar structure in the natric (Btn) horizon. The white, rounded "caps" of the columns are comprised of soil dispersed because of the high sodium saturation. The dispersed clays give the soil an almost rubbery consistency when wet. [Photo courtesy of Agriculture Canada, Canadian Soils Information System (CANSIS)]

FIGURE 9.25 Foliar symptoms (*inset*), reduced germination, and stunted growth of soybean plants with increasing levels of soil salinity due to additions of NaCl to a sandy soil. The large numbers written on the pots indicate the electrical conductivity (EC_e) of the soil. Note that serious growth reductions occurred for this sensitive cultivar even at EC_e levels considered normal. The soybean cultivar used (Jackson) is more sensitive to salinity than most other cultivars of soybean. (Photos courtesy of R. Weil)

uptake and utilization of other cations such as K^+ or Ca^{2+}. Deterioration of physical properties in sodic soils may harm plants in at least two ways: (1) Oxygen becomes deficient due to the breakdown of soil structure and the very limited air movement that results, and (2) water relations are poor due largely to the very slow infiltration and percolation rates.

Plant Symptoms. In response to excessive soil salinity, many plants become severely stunted and exhibit small dark bluish green leaves with dull surfaces. High levels of sodium or chloride typically produce scorching or necrosis of the leaf margins and tips (see Figure 9.25). Salt-stressed plants may also lose their leaves prematurely.

Selective Tolerance of Higher Plants to Saline and Sodic Soils[9]

Satisfactory plant growth on salty soils depends on a number of interrelated factors, including the physiological constitution of the plant, its stage of growth, and its rooting habits. For example, old alfalfa plants are more tolerant to salt-affected soils than young ones, and deep-rooted legumes show a greater resistance to such soils than those that are shallow-rooted.

Plant Sensitivity. While it is difficult to forecast precisely the tolerance of plant species to salty soils, numerous tests have made it possible to classify many plants into four general salt-tolerance groups. Table 9.4 groups selected plants according to this classification. Note that trees, shrubs, grasses, fruits, vegetables, and field crops are included in the different categories.

A number of wild *halophytes* (salt-loving plants) have been found that are quite tolerant to salts and that possess qualities that could make them useful for human and/or animal consumption. They are also valuable for restoring disturbed or degraded land under saline conditions.

Plant breeders have developed new plant strains with improved salt tolerance. Recently, a major advance was made by the discovery of a single gene that enables halophytes to sequester high amounts of Na^+ in their cellular vacuoles (large, membrane-enclosed storage structures inside individual cells). In the vacuole, the Na^+ ions can act to the plants' advantage by contributing to low internal osmotic potential while remaining isolated from cellular systems that are susceptible to Na^+ toxicity. Work is

TABLE 9.4 Relative Salt Tolerance of Selected Plants

Approximate EC_e resulting in a 10% reduction in plant growth for the most sensitive species in each column

Tolerant, 12 dS/m	*Moderately tolerant, 8 dS/m*	*Moderately sensitive, 4 dS/m*	*Sensitive, 2 dS/m*
Alkali grass, Nutall	Ash (white)	Alfalfa	Almond
Alkali sacaton	Asparagus	Arborvitae	Alders
Barley (grain)	Aspen	Boxwood	Apple
Bermuda grass	Black cherry	Broad bean	Azalea
Bougainvillea	Beet (garden)	Cabbage	Beech
Canola (rapeseed)	Broccoli	Celery	Bean
Cotton	Brome grass	Clover	Birch
Date	Cedar (red)	Corn	Blackberry
Guayule	Cowpea	Cucumber	Carrot
Jojoba	Fescue (tall)	Grape	Dogwood
Kallar grass	Honeysuckle	Hickory	Elm (American)
Kenaf	Hydrangea	Juniper	Hemlock
Natal plum	Juniper	Lettuce	Hibiscus
Oak (red and white)	Locust (honey)	Locust (black)	Larch
Olive	Mandarin	Pea	Linden
Prostrate kochia	Orchard grass	Peanut	Maple (sugar and red)
Redwort	Oak (red and white)	Radish	Onion
Rescue grass	Oats	Rice (paddy)	Orange
Rosemary	Privet	Soybean	Peach
Rugosa	Ryegrass (perennial)	Squash	Pine (red and white)
Rye (grain)	Safflower	Sugar cane	Pineapple
Salt grass (desert)	Sorghum	Sweet clover	Plum
Sugar beet	Soybean	Sweet potato	Potato
Tamarix	Sudan grass	Timothy	Raspberry
Wheat grass (crested)	Trefoil (birdsfoot)	Tomato	Rose
Wheat grass (tall)	Wheat	Vetch	Strawberry
Wild rye (Russian)	Wheat grass (western)	Viburnum	Tomato

Crop plant ratings from Maas and Grattan (1999); other plants from many sources.

[9] For a review of natural halophytes, see NAS (1990). For a perspective on the potential of plant biotechnology to deal with problems of salt-affected soils, see Frommer et al. (1999).

underway to use this gene to produce crops that can tolerate saline soils and salty irrigation water. Plant selection and improvement will almost certainly make important contributions to food production on salt-affected soils in the future. However, improved plant tolerance must not be viewed as a substitute for proper salinity control, as discussed in Section 9.18.

Salt Problems Not Related to Arid Climates

DEICING SALTS. In areas where deicing salts are used to keep roads and sidewalks free of snow and ice during winter months, these salts may impact roadside soils and plants. Repeated application of deicing salts can result in salinity levels sufficiently high as to adversely affect plants and soil organisms living alongside highways or sidewalks. In humid regions, such salt contamination is usually temporary, as the abundant rainfall leaches out the salts in a matter of weeks or months. To avoid the specific chemical and physical problems associated with sodium salts, many municipalities have switched from NaCl to KCl for deicing purposes. Sand can be used to improve traction, thereby reducing the need for deicing salt.

CONTAINERIZED PLANTS. Salinity can also be a serious problem for potted plants, particularly perennials that remain in the same pot for long periods. Greenhouse operators producing containerized plants must carefully monitor the quality of the water used for irrigation. Salts in the water, as well as those applied in fertilizers, can build up if care is not taken to flush them out occasionally with excess water. Chlorinated urban tap water used for indoor plants should be left overnight in an open container to allow some of the dissolved chlorine to escape, thus reducing the load of Cl^- ions added to the potting soil.

9.17 WATER-QUALITY CONSIDERATIONS FOR IRRIGATION[10]

Whether in a single field or in a large regional watershed, understanding the **salt balance** is a basic prerequisite for wise management of salt-affected soils. To achieve salt balance, the amount of salt coming in must be matched by the amount being removed. Meeting this condition is a fundamental challenge to the long-term sustainability of irrigated agriculture. In irrigated areas, this principally means managing the quality and amount of the irrigation water brought in and the quality and amount of soil drainage water removed.

IRRIGATION WATER QUALITY. Table 9.5 provides some guidelines on water quality for irrigation. If the salt content of irrigation water is high, salt balance will be difficult to achieve. However, even very salty water can be used successfully if soils are sufficiently well drained to allow careful management of salt inputs and outputs. Where irrigation water is low in salts but has a high SAR, the formation of sodic soils is likely to accelerate. In addition, irrigation water high in carbonates or bicarbonates can reduce Ca^{2+} and Mg^{2+} concentrations in the soil solution by precipitating these ions as insoluble carbonates. This leaves a higher proportion of Na^+ in the soil solution and can increase its SAR, moving the soil toward the sodic class.

DRAINAGE WATER SALINITY. Since some portion of added water must be drained away to combat salt buildup, the quality and disposition of *waste irrigation waters* must also be carefully monitored and controlled to minimize potential harm to downstream users and habitats. In any irrigation system, the drainage water leaving a field will be considerably more concentrated in salts than the irrigation water applied to the same field (Figure 9.26 explains why). What to do with the increasingly saline drainage water presents a major challenge to the sustainability of irrigated agriculture.

---

[10]For an overview of water-quality problems facing irrigated agriculture in California, see Letey (2000).

TABLE 9.5 Some Water-Quality Guidelines for Irrigation

Note that with regard to effects on physical structure of soils, higher total salinity (EC_w) in the irrigation water compensates, somewhat, for increasing sodium hazard (SAR). In addition, note that while water low in salts (low EC_w) avoids problems of restricted water availability to plants, it may worsen soil physical properties, especially if the SAR is high.

Potential irrigation problem	*Units*	*Degree of restriction on use*		
		None	*Slight to moderate*	*Severe*
Salinity (affects crop water availability)				
EC_w	dS/m	<0.7	0.7–3.0	>3.0
TDS	mg/l	<450	450–2000	>2000
Physical structure and water infiltration (Evaluate using EC_w and SAR together)				
SAR = 0–3 and EC_w =	dS/m	>0.7	0.7–0.2	<0.2
SAR = 3–6 and EC_w =	dS/m	>1.2	1.2–0.3	<0.3
SAR = 6–12 and EC_w =	dS/m	>1.9	1.9–0.5	<0.5
SAR = 12–20 and EC_w =	dS/m	>2.9	2.9–1.3	<1.3
SAR = 20–40 and EC_w =	dS/m	>5.0	5.0–2.9	<2.9
Boron (B) specific ion toxicity (affects sensitive crops)				
	mg/l	<0.7	<0.7–3.0	>3.0

Modified from Abrol et al. (1988) with permission of the Food and Agriculture Organization of the United Nations.

Different irrigation projects take different approaches, but rarely is the problem solved without some downstream environmental damage. Though not yet in common use, perhaps the most efficient approach is to collect the drainage water, keep it isolated from the relatively high-quality canal water, and reuse it to irrigate a more salt-tolerant crop in a lower field. This approach provides both high-quality canal water and lower-quality drainage water for use on appropriate crops. After several cycles of reuse, the drainage water must be disposed of, as it will have become too saline for irrigating even the most salt-tolerant species.

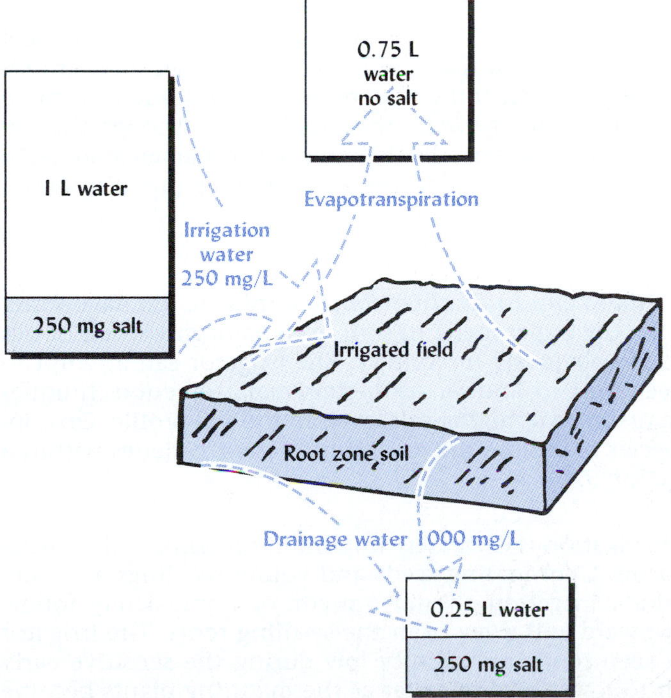

FIGURE 9.26 Evapotranspiration and salt balance together ensure that the drainage water from irrigated fields is much saltier than the irrigation water applied. In this example, the irrigation water contains 250 mg salts per liter. Some 75% of the applied water is lost to the atmosphere by evapotranspiration. About 25% of the water applied is used for drainage, which is necessary to maintain the salt balance (prevent the buildup of salts) in the field. The added salts are leached away with the drainage water, which then contains the same amount of salt as was added, but in only 25% of the added amount of water. The concentration of salt in the drainage water is thereby four times as great (1000 mg/L) as in the irrigation water. Disposal and/or reuse of the highly saline drainage water present challenges for any irrigation project. (Diagram courtesy of R. Weil)

Toxic Elements in Drainage Water. If either the irrigation water or the soil of the irrigated fields contains significant quantities of certain toxic trace elements, these too will become increasingly concentrated in the drainage water. The elements of concern include molybdenum (Mo), arsenic (As), boron (B), and selenium (Se). Molybdenum and selenium are necessary nutrients for animals and humans in trace amounts, but all four elements can be toxic to cattle, wildlife, or people if they become concentrated in water or food. At some locations in the western United States, such trace elements have accumulated to toxic levels in downstream wetlands or evaporative ponds. Plants growing on these affected areas can accumulate levels of these elements that are unsafe for livestock and/or wildlife.

9.18 RECLAMATION OF SALINE SOILS

The restoration of soil chemical and physical properties conducive to high productivity is referred to as soil ***reclamation.*** Reclamation of saline soils is largely dependent on the provision of effective drainage and the availability of good-quality irrigation water (see Table 9.5) so that salts can be leached from the soil. In areas where irrigation water is not available, such as in saline seeps in the Northern Plains states, the leaching of salts is not practical. In these areas, deep-rooted vegetation may be used to lower the water table and reduce the upward movement of salts.

If the natural soil drainage is inadequate to accommodate the leaching water, an artificial drainage network must be installed. Intermittent applications of excess irrigation water may be required to effectively reduce the salt content to a desired level. The process can be monitored by measuring the soil's EC, using either the saturation extract procedure or one of the field instruments described in Section 9.15.

Leaching Requirement

The amount of water needed to remove the excess salts from saline soils, called the **leaching requirement (LR)**, is determined by the characteristics of the crop to be grown, the irrigation water, and the soil. As demonstrated in Box 9.3, an approximation of the *LR* is given for relatively uniform salinity conditions by the ratio of the salinity of the irrigation water (expressed as its EC_{iw}) to the maximum acceptable salinity of the soil solution for the crop to be grown (expressed as EC_{dw}, the EC of the drainage water).

$$LR = \frac{EC_{iw}}{EC_{dw}} \tag{9.25}$$

The *LR* indicates water added in excess of that needed to thoroughly wet the soil and meet the crop's evapotranspiration needs. Note that if EC_{iw} is high and a salt-sensitive crop is chosen (dictating a low EC_{dw}), a very large leaching requirement *LR* will result. As mentioned in Section 9.17, disposal of the drainage water that has leached through the soil can present a major problem. Therefore, it is generally desirable to use management techniques that minimize the *LR* and the amount of drainage water that requires disposal.

Management of Soil Salinity

Management of irrigated soils should aim to simultaneously minimize drainage water and protect the root zone (usually the upper meter of soil) from damaging levels of salt accumulation. These two goals are obviously in conflict. The irrigator can attempt to find the best compromise between the two, and can use certain management techniques that allow plants to tolerate the presence of higher salt levels in the soil profile. One option is to plant salt-tolerant species or choose the most salt-tolerant varieties within a crop species (as discussed in Section 9.16).

Irrigation Timing. The timing of irrigation is extremely important on saline soils, particularly early in the growing season. Germinating seeds and young seedlings are especially sensitive to salts. Therefore, irrigation should precede or immediately follow planting to move the salts downward and away from the seedling roots. The irrigator can use high-quality water to keep root-zone salinity low during the sensitive early growth stages, and then switch to lower quality water as the maturing plants become more salt tolerant.

BOX 9.3 LEACHING REQUIREMENT FOR SALINE SOILS

The salt balance in a field can be described by equating the salt inputs and outputs:

$$\underbrace{S_{iw} + S_p + S_f + S_m}_{\text{Salt inputs}} = \underbrace{S_{dw} + S_c + S_{ppt}}_{\text{Salt outputs}} \tag{9.26}$$

The salt inputs include those from irrigation water (S_{iw}), atmospheric deposition (S_p), fertilizers (S_f), and the weathering or dissolution of existing soil minerals (S_m). The salt outputs are due to drainage water (S_{dw}), crop removal (S_c), and chemical precipitation (S_{ppt}) of carbonates and sulfates. Usually S_{iw} and S_{dw} are far larger than the other terms in Equation 9.26, so the main concern is to balance the salt coming in with the irrigation water and that leaving with the drainage water:

$$\underbrace{S_{iw}}_{\text{Salt in with irrigation water}} = \underbrace{S_{dw}}_{\text{Salt out with drainage water}} \tag{9.27}$$

We can estimate the quantity of salt S carried in drainage or irrigation as the product of the volume of the water (expressed as cm depth applied to an area of land) and the concentration of salt in that water (as approximated by its electrical conductivity, EC). Therefore, we can rewrite Equation 9.27 as follows (using the same subscripts as before):

$$D_{iw} \times EC_{iw} = D_{dw} \times EC_{dw} \tag{9.28}$$

Rearranging the terms, we obtain the following expression that defines the **leaching requirement *LR*** or the ratio of drainage water depth to irrigation water depth (D_{dw}/D_{iw}) needed to maintain salt balance:

$$\frac{EC_{iw}}{EC_{dw}} = \frac{D_{dw}}{D_{iw}} = LR \tag{9.29}$$

The *LR* tells farmers how much irrigation water (in excess of that required to wet the soil) they should apply for sufficient leaching. The goal is usually to assure that the upper two-thirds of the root zone does not accumulate salts beyond the level acceptable for a particular crop.

From Equation 9.29, we see that *LR* also equals the ratio of the irrigation water EC to the drainage water EC:

$$LR = \frac{EC_{iw}}{EC_{dw}} \tag{9.25}$$

where EC_{dw} is an acceptable level for the crop being grown. What is considered an acceptable EC_{dw} is open to interpretation. An acceptable EC_{dw} might be interpreted to mean the EC_e that allows 90% of maximum crop yield. A more conservative interpretation of acceptable EC_{dw} is the threshold EC_e at which the growth of the particular crop just begins to decline (usually 1 to 2 dS/m lower than that which gives the 90% yield level).

As an example, consider the situation where the irrigation water has an EC_{iw} of 2.5 dS/m and a moderately tolerant crop (e.g., broccoli) is to be grown. If information that is more specific is unavailable, the acceptable EC_{dw} for the crop can be (roughly) estimated from the column heading in Table 9.4. For a moderately tolerant crop, we can use 8 dS/m as the acceptable EC_{dw} to produce 90% of the maximum yield. Then,

$$LR = \frac{2.5 \text{ ds/m}}{8 \text{ ds/m}} = 0.31 \tag{9.30}$$

If this *LR* (0.31) is multiplied by the amount of water needed to wet the root zone—let us suppose it is 12 cm of water—the amount of water to be leached would be 3.7 cm (12 cm × 0.31). This is the minimum amount of water that must be leached through a water-saturated soil to maintain the root zone salinity at the acceptable level. A more sensitive crop could be grown in this soil, but it would have a lower acceptable EC_{dw} and therefore would require the application of a greater amount of water for leaching.

Location of Salts in the Root Zone. Tillage and planting practices can influence the location and accumulation of salts in arid-region soils. Tillage or surface-residue management practices (such as mulches or conservation tillage) that reduce evaporation from the soil surface should also reduce the upward transport of soluble salts. Likewise, specific techniques for applying irrigation water that direct salt concentrations away from young plant roots can allow higher levels of salt to accumulate without damage to the crop (Figure 9.27).

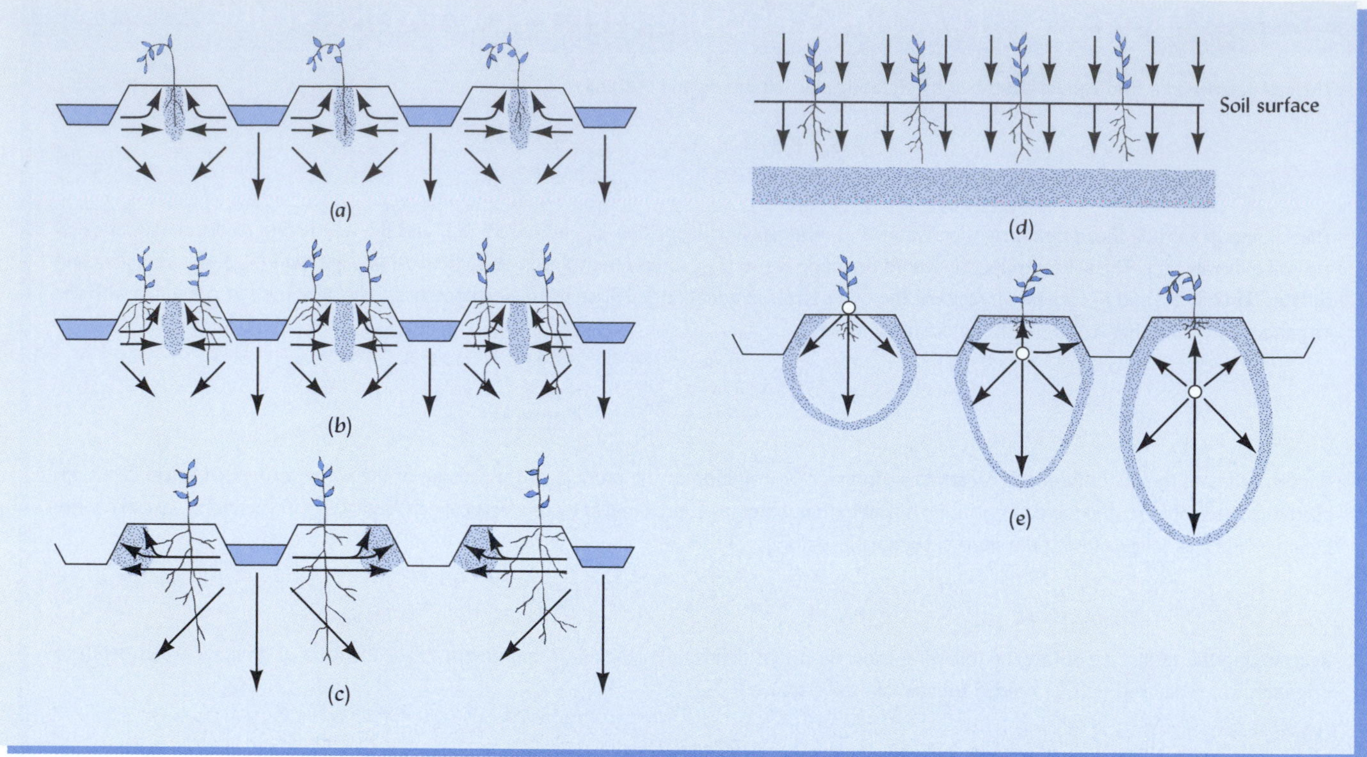

FIGURE 9.27 Effect of irrigation techniques on salt movement and plant growth in saline soils. (*a*) With irrigation water applied to furrows on both sides of the row, salts move to the center of the ridge and damage young plants. (*b*) Placing plants on the edges of the bed rather than in the center helps them avoid the most concentrated salts. (*c*) Application of water to every other furrow and placement of plants on the side of the bed nearest the water helps plants avoid the highest salt concentrations. (*d*) Sprinkler irrigation or uniform flooding temporarily moves salts downward out of the root zone, but the salts will return afterward as the soil surface dries out and water moves up by capillary flow. (*e*) Drip irrigation at low rates provides a nearly continuous flow of water, creating a low-salt soil zone with the salts concentrated at the wetting front. The placement of the drip emitters largely determines whether the salts are moved toward or away from the plant roots. (Courtesy of Wesley M. Jarrell)

Some Limitations of the Leaching Requirement Approach

The leaching requirement approach to managing irrigated soils is only an approximation and has several inherent weaknesses. First, additional leaching may be needed, in some cases, to reduce the excess concentration of specific elements, such as boron. Second, the *LR* by itself does not take into account the rise in the water table that is likely to result from increased leaching, and so may lead to waterlogging and, eventually, increased salinization. Third, irrigation using a simple *LR* approach usually overapplies water because an entire field is treated to avoid salt damage in its most saline spots. Fourth, the *LR* method does not consider salts that may be picked up from fossil salt deposits already in the soil and substrata. Fifth, it assumes that the EC of the drainage water is known, but in fact this may be largely unknown, since it may take years or even decades for the water applied in irrigation to reach the main drains where it can be easily sampled. In other words, the drainage water sampled today may represent the leaching conditions of several months or years ago.

An alternative approach would be to closely monitor the salinity in the soil profile by taking repeated measurements across the field, using the EM sensor or four-electrode methods discussed in Section 9.14. This more complex approach, combined with site-specific management techniques, seems to hold promise for future management and reclamation of salt-affected soils under irrigation.

9.19 RECLAMATION OF SALINE-SODIC AND SODIC SOILS

Saline-sodic soils have some of the adverse properties of both saline and sodic soils. If attempts are made to leach out the soluble salts in saline-sodic soils, as was discussed for saline soils, the exchangeable Na^+ level as well as the pH would likely increase and the soil would take on adverse characteristics of sodic soils. Consequently, for both saline-sodic and sodic soils, attention must first be given to reducing the level of exchangeable Na^+ ions and then to the problem of excess soluble salts.

Gypsum

Removing Na^+ ions from the exchange complex is most effectively accomplished by replacing them with either the Ca^{2+} or the H^+ ion. Providing Ca^{2+} in the form of gypsum ($CaSO_4 \cdot 2H_2O$) is the most practical way to bring about this exchange. When gypsum is added, the replaced sodium forms the soluble salt Na_2SO_4, which can be easily leached from the soil as was done in the case of the saline soils.

Several tons of gypsum per hectare are usually necessary to achieve reclamation. In Box 9.4, calculations are made to approximate the amount of gypsum that is theoretically needed to remove an acceptable portion of the Na^+ ion from the exchange complex. The

BOX 9.4 CALCULATING THE THEORETICAL GYPSUM REQUIREMENT

PROBLEM

How much gypsum is needed to reclaim a sodic soil with an exchangeable sodium percentage (ESP) of 25% and a cation exchange capacity of 18 $cmol_c/kg$? Assume that you want to reduce the ESP of the upper 30 cm of soil to about 5% so that a crop like alfalfa could be grown.

SOLUTION

First, determine the amount of Na^+ ions to be replaced by multiplying the CEC (18) by the change in Na^+ saturation desired (25 − 5 = 20%).

$$18\ cmol_c/kg \times 0.20 = 3.6\ cmol_c/kg$$

From the reaction that occurs when the gypsum ($CaSO_4 \cdot 2H_2O$) is applied,

$$\boxed{\text{Micelle}}\begin{matrix}Na^+\\Na^+\end{matrix} + CaSO_4 \cdot 2H_2O \rightleftharpoons \boxed{\text{Micelle}}\ Ca^{2+} + Na_2SO_4 + 2H_2O \qquad (9.31)$$

We know that the Na^+ is replaced by a *chemically equivalent* amount of Ca^{2+} in the gypsum ($CaSO_4 \cdot 2H_2O$). In other words, 3.6 $cmol_c$ of $CaSO_4 \cdot 2H_2O$ will be needed to replace 3.6 $cmol_c$ of Na^+.

Second, calculate the weight in grams of gypsum needed to provide the 3.6 $cmol_c/kg$ soil. This can be done by first dividing the molecular weight of $CaSO_4 \cdot 2H_2O$ (172) by 2 (since Ca^{2+} has two charges and Na^+ only one) and then by 100 since we are dealing with *centimole*$_c$ rather than $mole_c$.

$$\frac{172}{2} = 86\ g\ CaSO_4 \cdot 2H_2O/mol_c$$

$$\text{and } \frac{86}{100} = 0.86\ g\ CaSO_4 \cdot 2H_2O/cmol \text{ required to replace 1 } cmol_c\ Na$$

The 3.6 $cmol_c$ Na^+/kg would require

$$3.6\ cmol_c/kg \times 0.86\ g/cmol_c = 3.1\ g\ CaSO_4 \cdot 2H_2O/kg \text{ of soil}$$

Last, to express this in terms of the amount of gypsum needed to treat 1 ha of soil to a depth of 30 cm, multiply by 4×10^6, which is twice the weight in kg of a 15-cm-deep hectare–furrow slice (see Section 4.5).

$$3.1\ g/kg \times 4 \times 10^6\ k/ha = 12{,}400{,}000\ g\ gypsum/ha$$

This is 12,400 kg/ha, 12.4 Mg/ha, or about 5.5 tons/acre.

Because of impurities in the gypsum and incomplete exchange of Ca^{2+} for Na^+ (see Section 8.8), these amounts would likely be adjusted upward by 20 to 30% in actual field practice. Finally, it is advisable to apply only half of the recommended amount at one time, with the remaining half applied about 6 months later, if necessary.

soil must be kept moist to hasten the reaction, and the gypsum should be thoroughly mixed into the surface by cultivation—not simply plowed under. The treatment must be supplemented later by a thorough leaching of the soil with irrigation water to leach out most of the sodium sulfate.

Sulfur and Sulfuric Acid

Why *not* to use gypsum on calcium carbonate rich soils: http://www.colostate.edu/Depts/CoopExt/TRA/PLANTS/alksoil.html

Elemental sulfur and sulfuric acid can be used to advantage on sodic soils, especially where sodium bicarbonate abounds. The sulfur, upon biological oxidation (see Sections 9.7 and 13.21), yields sulfuric acid, which not only changes the sodium bicarbonate to the less harmful and more leachable sodium sulfate but also decreases the pH.

In research trials, sulfur and even sulfuric acid have proven to be very effective in the reclamation of sodic soils, especially if large amounts of $CaCO_3$ are present. In practice, however, gypsum is much more widely used than the acid-forming materials. Gypsum is less expensive, is widely available in both natural and in industrial by-product forms, and is more easily handled.

Physical Condition

The effects of gypsum and sulfur on the physical condition of sodic soils is perhaps more spectacular than are the chemical effects. Sodic soils are almost impermeable to water, since the soil colloids are largely dispersed and the soil is essentially void of stable aggregates. When the exchangeable Na^+ ions are replaced by Ca^{2+} or H^+, soil aggregation and improved water infiltration results (Figure 9.28).

The reclamation effects of gypsum or sulfur are greatly accelerated by plants growing on the soil. Crops that have some degree of tolerance to saline and sodic soils, such as sugar beets, cotton, barley, sorghum, berseem clover, or rye, can be grown initially. Their roots help provide channels through which gypsum can move downward into the soil. Deep-rooted crops, such as alfalfa, are especially effective in improving the water conductivity of gypsum-treated sodic soils. Figure 9.28 illustrates the ameliorating effects of the combination of gypsum and deep-rooted crops.

Management of Reclaimed Soils

Once salt-affected soils have been reclaimed, prudent management steps must be taken to be certain that the soils remain productive. For example, surveillance of the EC and SAR and trace element composition of the irrigation water is essential. The number and timing of irrigation episodes should be adjusted to maintain the balance of salts entering and leaving the soil. Likewise, the maintenance of good internal drainage is essential for the removal of excess salts.

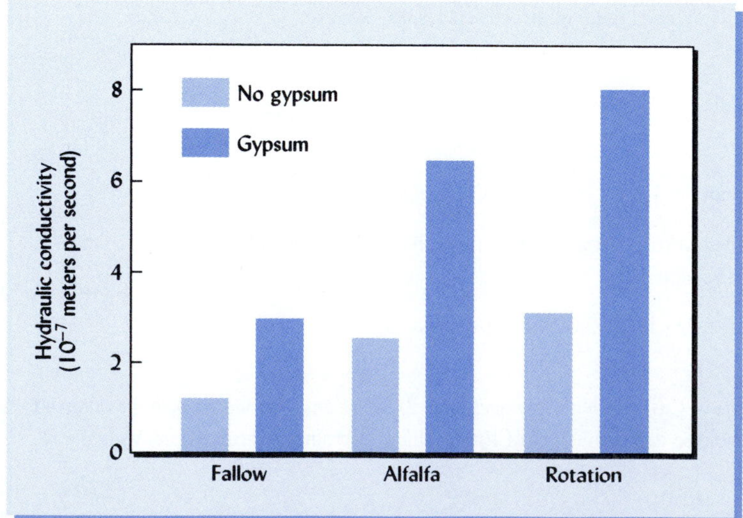

FIGURE 9.28 The influence of gypsum and growing crops on the hydraulic conductivity (readiness of water movement) of the upper 20 cm of a saline-sodic soil (Natrustalf) in Pakistan. The use of gypsum to increase water conductivity in all plots was more effective when deep-rooted alfalfa and rotation of sesbania-wheat were grown. [Drawn from selected data from Ilyas et al. (1993); used with permission of the Soil Science Society of America]

Steps should also be taken to monitor appropriate chemical characteristics of the soils, such as pH, EC, and SAR, as well as specific levels of such elements as boron, chlorine, molybdenum, and selenium that could lead to chemical toxicities.

Crop and soil fertility management to maintain satisfactory yield levels is essential to maintain the overall quality of salt-affected soils. The crop residues (roots and aboveground stalks) will help maintain organic matter levels and good physical condition of the soil.

9.20 CONCLUSION

No other chemical soil characteristic is more important in determining the chemical environment of higher plants and soil microbes than the pH. There are few reactions involving the soil or its biological inhabitants that are not sensitive to soil pH. This sensitivity must be recognized in any soil-management system.

Acidification is a natural process in soil formation that is accentuated in humid regions where processes that produce H^+ ions outpace those that consume them. Natural acidification is largely driven by the production of organic acids (including carbonic acid) and the leaching away of the nonacid cations (Ca^{2+}, Mg^{2+}, K^+, and Na^+) that the H^+ ions from the acids displace from the exchange complex. Emissions from power plants and vehicles, as well as inputs of nitrogen into agricultural systems are the principal means by which human activities have accelerated acidification during recent decades.

Aluminum is the other principal acid cation besides hydrogen. Its hydrolysis reactions produce H^+ ions and its toxicity comprises one of the main detrimental effects of soil acidity. The total acidity in soils is the sum of the active, exchangeable (salt replaceable), and residual pools of soil acidity. Changes in the soil solution pH (the active acidity) are buffered by the presence of the other two pools. In certain anaerobic soils and sediments, the presence of reduced sulfur provides the potential for enormous acid production if the material is exposed to air by drainage or excavation.

Alkaline and salt-affected soils dominate arid and semiarid regions. They cover vast regions and support important desert and grassland ecosystems. These soils typically exhibit above-neutral pH values throughout their profiles and often have calcic (calcium carbonate–rich) or gypsic (gypsum-rich) horizons at some depth. People use these soils for rangeland and dryland farming.

High-pH, calcium-rich conditions of alkaline soils commonly lead to deficiencies of certain essential micronutrients (especially iron and zinc) and macronutrients (especially phosphorus). In localized areas boron, molybdenum, and selenium may be so readily available as to accumulate in plants to levels that can harm grazing animals.

Soil pH is largely controlled by the humus and clay fractions and their associated exchangeable cations. The maintenance of satisfactory soil fertility levels in humid regions depends considerably on the judicious use of lime to balance the losses of calcium and magnesium from the soil. Gypsum and organic matter represent other tools that can be used to ameliorate soil acidity instead of, or in addition to, liming. On the other hand, it is sometimes most judicious to use acid-tolerant plants, rather than attempt to change the chemistry of the soil.

Knowing how pH is controlled, how it influences the supply and availability of essential plant nutrients as well as toxic elements, how it affects higher plants and human beings, and how it can be ameliorated is essential for the conservation and sustainable management of soils throughout the world.

Scientists have grouped salt-affected soils into three classes based on their total salt content [indicated by electrical conductivity (EC)] and the proportion of sodium among the cations [indicated by either the sodium adsorption ration (SAR) or the exchangeable sodium percentage (ESP)]. The physical conditions of saline and saline-sodic soils are satisfactory for plant growth, but the colloids in sodic soils are largely dispersed, the soil is puddled and poorly aerated, and water infiltration rate is extremely slow.

If irrigators in arid regions apply only enough water to meet plant evapotranspiration needs, salts will build up relentlessly in the soil until the land becomes too salinized to grow crops. Therefore, just as the humid-region farmer must periodically use lime to restore a favorable balance between H^+ ion consumption and production, the irrigator must use periodic leaching to restore the balance between the import and export of salts.

Leaching and drainage are both essential components of any successful irrigation scheme. However, the leaching of salts is not a simple matter, and it inevitably leads to additional problems both in the field being leached and in sites further downstream. At best, salinity management in irrigation agriculture provides a compromise between unavoidable evils.

In order to make saline-sodic and sodic soils permeable enough for leaching to take place, the excess exchangeable sodium ions must first be removed from the exchange complex. This is accomplished by replacing the Na^+ ions with either Ca^{2+} or H^+ ions.

The world increasingly depends on irrigated agriculture in dry regions for the production of food for its growing population. About 17% of the world's farmland is irrigated, but this land produces some 40% of our food supply. However, irrigated agriculture is in inherent disharmony with the nature of arid-region ecology. We have seen that it is therefore fraught with difficulties that require constant and careful management if human and environmental tragedies are to be avoided.

STUDY QUESTIONS

1. Soil pH gives a measure of the concentration of H^+ ions in the soil solution. What, if anything, does it tell you about the concentration of OH^+ ions? Explain.
2. Describe the role of aluminum and its associated ions in soil acidity. Identify the ionic species involved and the effect of these species on the CEC of soils.
3. What are the primary sources of alkalinity in soils? Explain.
4. If you could somehow extract the soil solution from the upper 16 cm of 1 ha of moist acid soil (pH = 5), how many kg of pure $CaCO_3$ would be needed to neutralize the soil solution (bring its pH to 7.0)? Under field conditions, up to 6 Mg of limestone may be required to bring the pH of this soil layer to a pH of 6.5. How do you explain the difference in the amounts of $CaCO_3$ involved?
5. What is meant by *buffering?* Why is it so important in soils, and what are the mechanisms by which it occurs?
6. What is acid rain, and why does it seem to have greater impact on forests than on commercial agriculture?
7. Compare the availability of the following essential elements in alkaline soils with that in acid soils: (1) iron, (2) nitrogen, (3) molybdenum, and (4) phosphorus.
8. How much limestone with a $CaCO_3$ equivalent of 90 would you need to apply to eliminate exchangeable aluminum in an Ultisol with CEC = 8 $cmol_c$/kg and an aluminum saturation of 60%?
9. A landscape contractor purchased 10 dump-truck loads of "topsoil" excavated from a black, rich-appearing soil in a coastal wetland. Samples of the soil were immediately sent to a lab to be sure they met the specified properties (silt loam texture, pH 6 to 6.5) for the topsoil to be used in a landscaping job. The lab reported the texture and pH in the specified range, so the topsoil was installed and an expensive landscape of beautiful plants established. Unfortunately, within a few months all the plants began to die. Replacement plants also died. The topsoil was again tested. It was still a silt loam, but now its pH was 3.5. Explain why this pH change likely occurred and suggest appropriate management solutions.
10. A neighbor complained when his azaleas were adversely affected by a generous application of limestone to the lawn immediately surrounding the azaleas. To what do you ascribe this difficulty? How would you remedy it?
11. An arid-region soil, when it was first cleared for cropping, had a pH of about 8.0. After several years of irrigation, the crop yield began to decline, the soil aggregation tended to break down, and the pH had risen to 10. What is the likely explanation for this situation?
12. What physical and chemical treatments would you suggest to bring the soil described in question 11 back to its original state of productivity?
13. Calculate the leaching requirement to prevent the buildup of salts in the upper 45 cm of a soil if the EC_{dw} of the drainage water is 6 dS/m and the EC_{iw} of the irrigation water is 1.2 dS/m.
14. Calculate the quantity of gypsum needed to reclaim a sodic soil (ESP = 30%) when CEC = 25 $cmol_c$/kg and pH is 10.2. Assume you want an ESP no higher than 4%.

REFERENCES

Abrol, I. P., J. S. P. Yadov, and F. I. Massoud. 1988. "Salt-affected soils and their management," *FAO Soils Bulletin* **39**. Rome: Food and Agriculture Organization of the United Nations. Complete text available at *www.fao.org/docrep/x5871e/x5871e00.htm*.

Adams, F., and Z. F. Lund. 1966. "Effect of chemical activity of soil solution aluminum on cotton root penetration of acid subsoils," *Soil Sci.* **101**:193–198.

Beyrouty, C. A., J. K. Keino, E. E. Gbur, and M. G. Hanson. 1999. "Phytotoxic concentrations of subsoil aluminum as influenced by soils and landscape position," *Soil Sci.* **165**:135–143.

Blake, L., W. T. Goulding, C. J. B. Mott, and A. E. Johnston. 1999. "Changes in soil chemistry accompanying acidification over more than 100 years under woodland and grass at Rothamsted Experiment Station, UK," *European J. Soil Sci.* **50**:401–412.

de la Fuente-Martinez, J. M., and L. Herrera-Estralla. 2000. "Advances in the understanding of aluminum toxicity and the development of aluminum-tolerant transgenic plants," *Advances in Agronomy* **66**:103–120.

Frommer, W. B., U. Ludewig, and D. Rentsch. 1999. "Taking transgenic plants with a pinch of salt," *Science* **285**:1222–1223.

Goldberg, S., S. M. Lesch, and D. L. Suarez. 2000. "Predicting boron adsorption by soils using soil chemical parameters in the constant capacitance model," *Soil Sci. Soc. Amer. J.* **64**:1356–1363.

Hargrove, W. L., and G. W. Thomas. 1981. "Effect of organic matter on exchangeable aluminum and plant growth in acid soils," in *Chemistry in the Soil Environment.* ASA Special Publication no. 40. (Madison, Wis.: Amer. Soc. Agron. and Soil Sci. Soc. Amer.).

Hue, N. V., and D. L. Licudine. 1999. "Amelioration of subsoil acidity through surface applications of organic manures," *J. Environ. Qual.* **28**:623–632.

Ilyas, M., R. W. Miller, and R. H. Qureski. 1993. "Hydraulic conductivity of saline-sodic soil after gypsum application," *Soil Sci. Soc. Amer. J.* **57**:1580–1585.

Letey, J. 2000. "Soil salinity poses challenges for sustainable agriculture and wildlife," *California Agriculture* **54**(2):43–48.

Likens, G. E., C. T. Driscoll, and D. C. Buso. 1996. "Long-term effects of acid rain: Response and recovery of a forest ecosystem," *Science* **272**:244–246.

Lumbanraja, J., and V. P. Evangelou. 1991. "Acidification and liming influence on surface charge behavior of Kentucky subsoils," *Soil Sci. Soc. Amer. J.* **54**:26–34.

Maas, E. V., and S. R. Grattan. 1999. "Crop yields as affected by salinity," Chap. 3 in R. W. Skaggs and J. van Schilfgaarde (eds.), *Agricultural Drainage,* Agronomy Monograph no. 38. (Madison, Wis.: ASA, CSSA, SSSA).

Magdoff, F. R., and R. J. Bartlett. 1985. "Soil pH buffering revisited," *Soil Sci. Soc. Amer. J.* **49**:145–148.

Markewitz, D., D. D. Richter, H. L. Allen, and J. B. Urrego. 1998. "Three decades of observed soil acidification in the Calhoun experimental forest: Has acid rain made a difference?" *Soil Sci. Soc. Amer. J.* **62**:1428–1439.

National Academy of Sciences. 1990. *Saline Agriculture: Salt Tolerant Plants for Developing Countries.* (Washington, D.C.: National Academy Press).

Okusmi, T. A., H. Rust, and A. S. R. Juo. 1987. "Reactive characteristics of certain soils from south Nigeria," *Soil Sci. Soc. Amer. J.* **51**:1256–1262.

Rhoades, J. D., F. Chanduvi, and S. Lesch. 1999. *Soil Salinity Assessment Methods and Interpretation of Electrical Conductivity Measurements.* FAO Irrigation and Drainage Paper no. 57. (Rome: Food and Agriculture Organization of the United Nations).

Robarge, W. P., and D. W. Johnson. 1992. "The effects of acidic deposition on forested soils," *Advances in Agronomy* **47**:1–84.

Shoemaker, H. E., E. O. McLean, and P. F. Pratt. 1961. "Buffer methods for determining lime requirements of soils with appreciable amounts of extractable aluminum," *Soil Sci. Soc. Amer. Proc.* **25**:274–277.

Stoddard, J. L., D. S. Jeffries, A. Lükewille, et al. 1999. "Regional trends in recovery from acidification in North America and Europe," *Nature* **401**:575–578.

Sumner, M. E. 1993. "Gypsum and acid soils: The world scene," *Advances in Agronomy* **51**:1–32.

Szabolcs, I. 1989. *Salt-Affected Soils.* (Boca Raton, Fla.: CRC Press).

Weil, R. R. 2000. "Soil and plant influences on crop response to two African phosphate rocks," *Agron. J.* **92**:1167–1175.

10

Organisms and Ecology of the Soil[1]

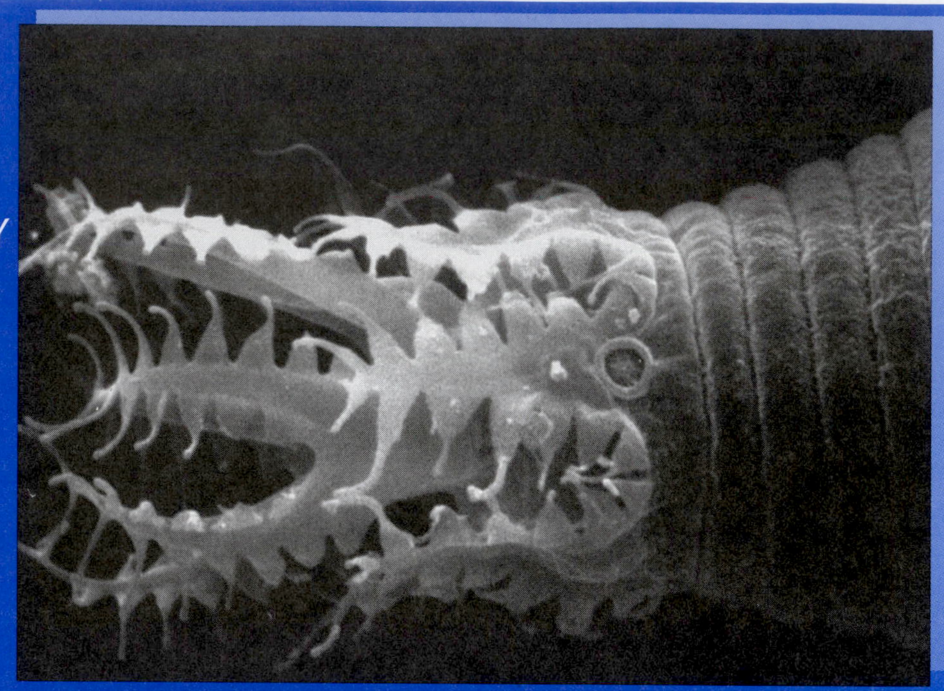

Head of a bacteria-feeding nematode. (Sven Boström, Swedish Museum of Natural History)

Under the silent, relentless chemical jaws of the fungi, the debris of the forest floor quickly disappears. . . .
—A. Forsyth and K. Miyata, Tropical Nature

The terms *ecosystem* and *ecology* usually call to mind scenes of lions stalking vast herds of wildebeest on the grassy savannas of East Africa, or the interplay of phytoplankton, fish, and fishermen in some great estuary. Like a savanna or an estuary, a soil is an ecosystem in which thousands of different creatures interact and contribute to the global cycles that make all life possible. This chapter will introduce some of the organisms in the living drama that goes on largely unseen in the soil beneath our feet. We will explore a world populated by a wild array of creatures all fiercely competing for every leaf, root, fecal pellet, dead body that reaches the soil. We will also find predators, some with fearsome jaws to snatch unwary victims, others whose jellylike bodies simply engulf and digest their hapless prey. We will emphasize the activities rather than the scientific classification of these organisms. Consequently, we will consider only very broad, simple taxonomic categories.

Most of the work of the soil community is carried out by creatures whose "jaws" are chemical enzymes that eat away at all manner of organic substances left in the soil by their coinhabitants. Most of these enzymes are found in the soil solution, having been excreted by microorganisms to promote biochemical reactions in their immediate environment. The diversity of substrates and environmental conditions found in every handful of soil spawns a diversity of adapted organisms that staggers the imagination. The collective vitality, diversity, and balance among these organisms engender a healthy ecosystem and make possible the functions of a high-quality soil.

We will learn how these organisms, both flora and fauna, interact with one another, what they eat, how they affect the soil, and how soil conditions affect them. The central theme will be how this community of organisms assimilates plant and animal materials, creating soil humus, recycling carbon and mineral nutrients, and supporting plant growth. A subtheme will be how people can manage soils to encourage a healthy, diverse soil community that efficiently makes nutrients available to higher plants, protects plant roots from pests and disease, and helps protect the global environment from some of the excesses of the human species.

---

[1] For a review of soil biology see Coyne (1999), Paul and Clark (1996), or Sylvia et al. (1997).

10.1 THE DIVERSITY OF ORGANISMS IN THE SOIL

Soil ecology studies in the Baltimore Ecosystem Study: http://www.beslter.org/frame4-page_3a_02.html

Soil organisms are creatures that spend all or part of their lives in the soil environment. Every handful of soil is likely to contain billions of organisms, with representatives of nearly every phylum of living things. A simplified, general classification of soil organisms is shown in Table 10.1.[2]

SIZES OF ORGANISMS. The animals (**fauna**) of the soil range in size from **macrofauna** (such as moles, prairie dogs, earthworms, and millipedes), through **mesofauna** (such as tiny springtails and mites), to **microfauna** (such as nematodes and single-celled protozoans). Plants (**flora**) include the roots of higher plants, as well as microscopic algae and diatoms. Other microorganisms (too small to be seen without the aid of a microscope) include fungi, and bacteria which tend to predominate in terms of numbers, mass, and metabolic capacity.

---

[2] Most biologists classify living organisms into five kingdoms rather than two, since it has proven difficult to fit many microorganisms into either the plant or animal kingdom. More recently, classifications based on similarities in genetic material place all living things into three primary groups: *Eucarya* (which includes all plants, animals, and fungi), *Bacteria*, and *Archaea*.

TABLE 10.1 General Classification of Some Important Groups of Soil Organisms

Some organisms subsist on living plants (herbivores), others on dead plant debris (detritivores). Some consume animals (predators), some devour fungi (fungivore) or bacteria (bacterivores), and some live off of, but do not consume, other organisms (parasites). Heterotrophs rely on organic compounds for their carbon and energy needs while autotrophs obtain their carbon mainly from carbon dioxide and their energy from photosynthesis or oxidation of various elements.

Generalized grouping (width in mm)	*Major specific groups*	*Examples*
Macrofauna (>2 mm)		
All heterotrophs, largely herbivores and detritivores	Vertebrates	Gophers, mice, moles
	Arthropods	Ants, beetles and their larvae, centipedes, grubs, maggots, millipedes, spiders, termites, woodlice
	Annelids	Earthworms
	Mollusks	Snails, slugs
Macroflora		
Largely autotrophs	Vascular plants	Feeder roots
	Bryophytes	Mosses
Mesofauna (0.1–2 mm)		
All heterotrophs, largely detritivores	Arthropods	Mites, collembola (springtails)
	Annelids	Enchytraeid (pot) worms
All heterotrophs, largely predators	Arthropods	Mites, protura
Microfauna (<0.1 mm)		
Detritivores, predators, fungivores, bacterivores	Nematodes	Nematodes
	Rotifera[a]	Rotifers
	Protozoa[a]	Amoebae, ciliates, flagellates
Microflora (<0.1 mm)		
Largely autotrophs	Vascular plants	Root hairs
	Algae	Greens, yellow-greens, diatoms
Largely heterotrophs	Fungi	Yeasts, mildews, molds, rusts, mushrooms
	Actinomycetes	Many kinds of actinomycetes
Heterotrophs and autotrophs	Bacteria[b] (and Archaea[b])	Aerobes, anaerobes
	Cyanobacteria	Blue-green algae

[a]Generally classified in the kingdom *Protista*.
[b]Traditionally classified together in the kingdom *Monera*, these organisms have prokaryotic cells but are classed in the domains *Bacteria* or *Archaea* based on differences in RNA.

The Soil Biodiversity Program in Scotland: http://www.nmw.ac.uk/soilbio/

Note that we have classified the organisms according to their size (*macro* being larger than 2 mm in width, *meso* being between 0.1 and 2 mm, and *micro* being less than 0.1 mm), as well as by their ecological functions (what they eat). A typical, healthy soil might contain several species of vertebrate animals (mice, gophers, snakes, etc.), a half dozen species of earthworms, 20 to 30 species of mites, 50 to 100 species of insects (collembola, beetles, ants, etc.), dozens of species of nematodes, hundreds of species of fungi, and perhaps thousands of species of bacteria (including archaeans and actinomycetes).

Types of Diversity. Great diversity is possible because of the nearly limitless variety of foods and the wide range of habitat conditions found in soils. Within a handful of soil there may be areas of good and poor aeration, high and low acidity, cool and warm temperatures, moist and dry conditions, and localized concentrations of dissolved nutrients, organic substrates, and competing organisms. The populations of soil organisms tend to be concentrated in zones of favorable conditions, rather than evenly distributed throughout the soil.

Soil scientists use the concept of biological diversity as an indicator of soil quality. A high **species diversity** indicates that the organisms present are fairly evenly distributed among a large number of species. Most ecologists believe that such complexity and species diversity are usually paralleled by a high degree of **functional diversity**—the capacity to utilize a wide variety of substrates and carry out a wide array of processes.

Ecosystem Dynamics. In most healthy soil ecosystems **functional redundancy**—the presence of several organisms to carry out each task—leads both to **ecosystem stability** and **resilience.** *Stability* describes the ability of soils, even in the face of wide variations in environmental conditions and inputs, to continue to perform such functions as the cycling of nutrients, assimilation of organic wastes, and maintainence of soil structure. *Resilience* describes the ability of the soil to "bounce back" to functional health after a severe disturbance has disrupted normal processes.

Given a high degree of diversity, no single organism is likely to become completely dominant. By the same token, the loss of any one species is unlikely to cripple the entire system. Nonetheless, for certain soil processes, such as ammonium oxidation (see Section 12.7), methane oxidation (see Section 11.11), or the creation of aeration macropores (see Section 4.6), primary responsibility may fall to only one or two species. The activity and abundance of these **keystone species** (for example, certain nitrifying bacteria or burrowing earthworms) merit special attention, for their populations may indicate the health of the entire soil ecosystem.

Genetic Resources. Soils make an enormous contribution to **global biodiversity.** Many scientists believe that there are more species in existence below the surface of the earth than above it. The soil is therefore a major storehouse of the genetic innovations that nature has written into the DNA code over hundreds of millions of years. Humans have always found ways to make use of some of the genetic material in soil organisms (beer, yogurt, and antibiotics are examples). However, with recent developments in biological engineering that allow the transfer of genetic material from one type of organism to another, the soil DNA bank has taken on a much larger practical importance for human welfare. The genes from soil organisms may now be used to produce plants and animals of superior utility to the human community.

10.2 ORGANISMS IN ACTION[3]

The activities of soil flora and fauna are intimately related in what ecologists call a *food chain* or, more accurately, a *food web.* Some of these relationships are shown in Figure 10.1, which illustrates how various soil organisms are involved in the degradation of

[3] For a succinct and well-illustrated summary of soil communities and ecological dynamics, see Tugel and Lewandowski (1999). For a more detailed discussion, see Coleman and Crossley (1996) and Killham (1994).

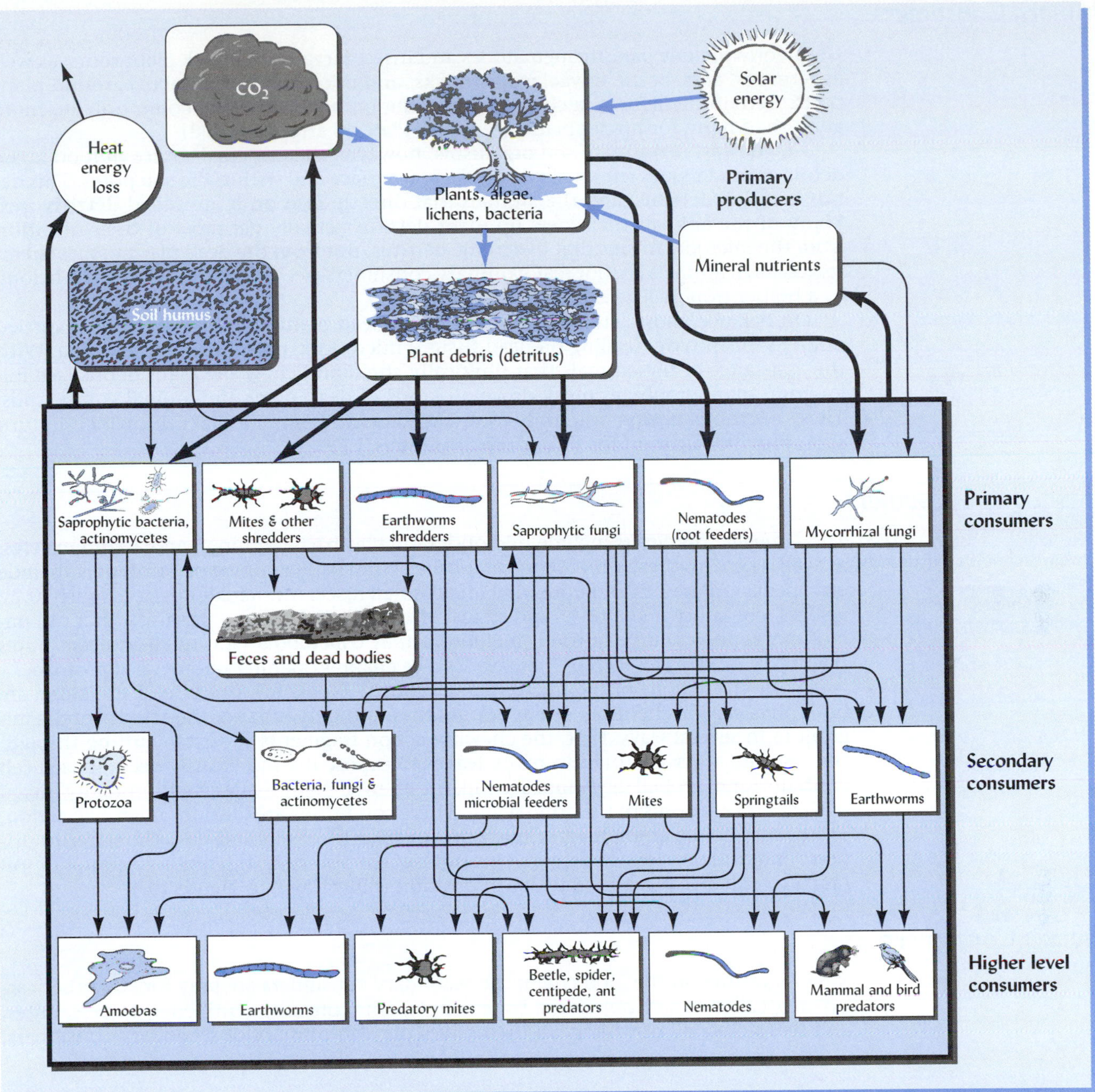

FIGURE 10.1 Greatly simplified and generalized diagram of the soil food web involved in the breakdown of plant tissue, the formation of humus, and the cycling of carbon and nutrients. The large shaded compartment represents the community of soil organisms. The rectangular boxes represent various groups of organisms; the arrows represent the transfer of carbon from one group to the next as predator eats prey. The thick arrows entering the top of the shaded compartment represent primary consumption of carbon originating from the tissue of the producers—higher plants, algae, and cyanobacteria. The rounded boxes represent other inputs that support, and outputs that result from, the soil food web. Although all groups shown play important roles in the process, some 80 to 90% of the total metabolic activity in the food web can be ascribed to the fungi and the bacteria (including actinomycetes). As a result of this metabolism, soil humus is synthesized, and carbon dioxide, heat energy, and mineral nutrients are released into the soil environment.

residues from higher plants. As one organism eats another, nutrients and energy are said to be passed from one **trophic level** to a higher one. The first trophic level is that of the **primary producers** (mainly vascular plants and algae). The second trophic level consists of the **primary consumers** (including **herbivores** that eat living plants). The third trophic level would be those **predators** that eat the primary consumers, the fourth level would be predators that eat predators, and so on.

Primary Consumers

Herbivores include parasitic nematodes and insect larvae that attack plant roots, as well as termites, ants, beetle larvae, woodchucks, and mice that devour aboveground plant parts. Certain herbivores, such as soil-dwelling immature cicadas, commonly do more good than harm for higher plants (see color Plate 40 after page 498).

For the vast majority of soil organisms, however, the principal source of food is the debris of dead tissues left by plants on the soil surface and within the soil pores. This debris is called **detritus**, and the animals that directly feed on it are called **detritivores.** Many of the animals that chew up plant detritus actually get most of their nutrition from the microorganisms that live in the detritus, not from the dead plant tissues themselves. Such animals are not really primary consumers (second trophic level), but belong to a higher trophic level.

On balance, most actual decomposition of dead plant and animal debris is carried out by **saprophytic** (feeding on dead tissues) microflora, both fungi and bacteria. With the assistance of the animals that physically shred and chew the plant debris (see following), the saprophytes break down all kinds of dead plant and animal compounds. These microorganisms commonly form large colonies of microbial cells, which in turn soon provide nutrition for secondary consumers.

Secondary Consumers

Image gallery of mites, including many that live in soils:
http://www.uq.edu.au/entomology/mite/mitetxt.html

The *secondary consumers* include microflora, such as bacteria, fungi, and actinomycetes, as well as **carnivores**, which consume other animals. Examples of carnivores include centipedes (Figure 1.5) and mites that attack small insects or nematodes (see Figure 10.2), spiders, predatory nematodes, and snails. **Microphytic feeders**, organisms that use microflora as their source of food, include certain collembola (springtail insects), mites (Figure 10.3), termites, certain nematodes (see chapter opening photo), and protozoa.

While the actions of the microflora are mostly biochemical, those of the fauna are both physical and chemical. The actions of these animals enhance the activity of the microflora in several ways. First, the chewing action fragments the litter, cutting through the resistant waxy coatings on many leaves to expose the more easily decomposed cell contents for microbial digestion. Second, the chewed plant tissues are thoroughly mixed with microorganisms in the animal gut, where conditions are ideal for microbial action. Third, the mobile animals carry microorganisms with them and help the latter to disperse and find new food sources to decompose. The ability of different size groups of soil fauna to enhance plant residue decomposition is illustrated in Figure 10.4.

Tertiary Consumers

Moving farther up the food chain, the secondary consumers are prey for still other carnivores, called *tertiary consumers*. For example, ants consume centipedes, spiders, mites, and scorpions—all of which can themselves prey on primary or secondary consumers.

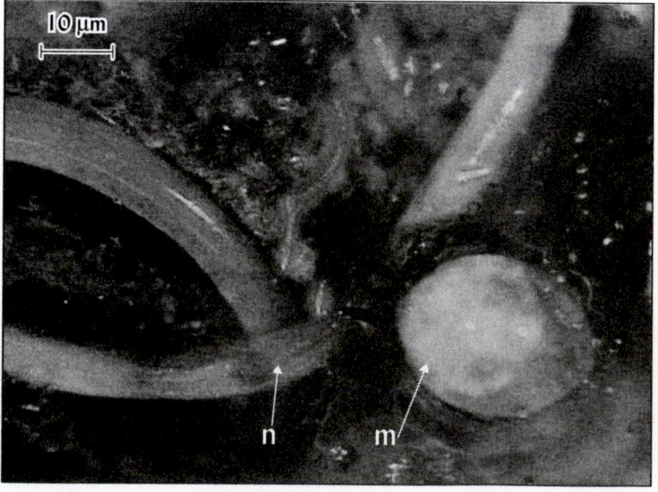

FIGURE 10.2 A predatory mite (an Astigmatid, *m*) dining on its prey, a microscopic roundworm (a nematode, *n*). Predation of this type keeps the populations of various groups of organisms in balance and releases nutrients previously tied up in the bodies of the prey. (Photo courtesy of Marie Newman, North Carolina State University)

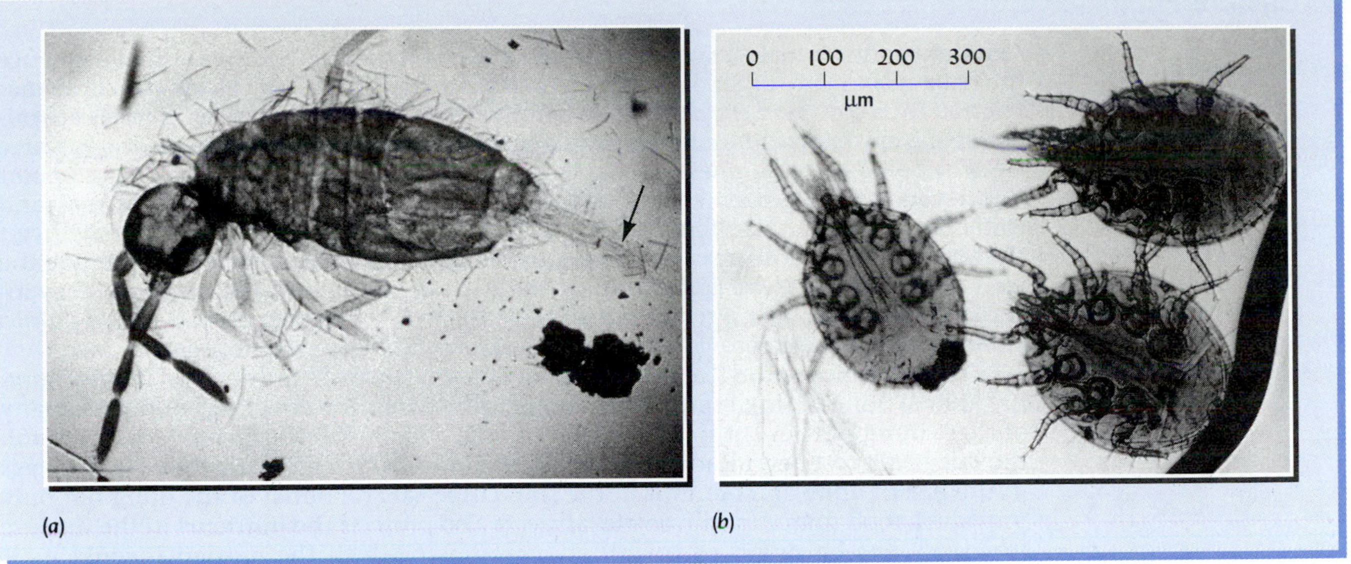

FIGURE 10.3 Springtails (collembola) (*a*) and mites (*b*) are mesofauna detritivores commonly found in soils. The springtail gets its name from the springlike furcula or "tail" that it uses to hop about (see arrow). (Photos courtesy of R. Weil)

Many species of birds specialize in eating soil animals such as beetles and earthworms. Whether the prey is a nematode or an earthworm, predation serves important ecological functions, including the release of nutrients tied up in the living cells.

The microflora are intimately involved in every level of the process. In addition to their direct attack on plant tissue (as primary consumers), microflora are active within the digestive tracts of many soil animals, helping these animals digest more resistant organic materials. The microflora also attack the finely shredded organic material in animal feces, and they decompose the bodies of dead animals. For this reason they are referred to as the ultimate decomposers.

As indicated in Figure 10.1, broad groups of organisms, such as the fungi or mites, include some species that are primary consumers, some that are secondary consumers, and others that are tertiary consumers. For example, some mites attack detritus directly, while others eat mainly fungi or bacteria that grow on the detritus. Still others attack and devour the mites that eat fungi.

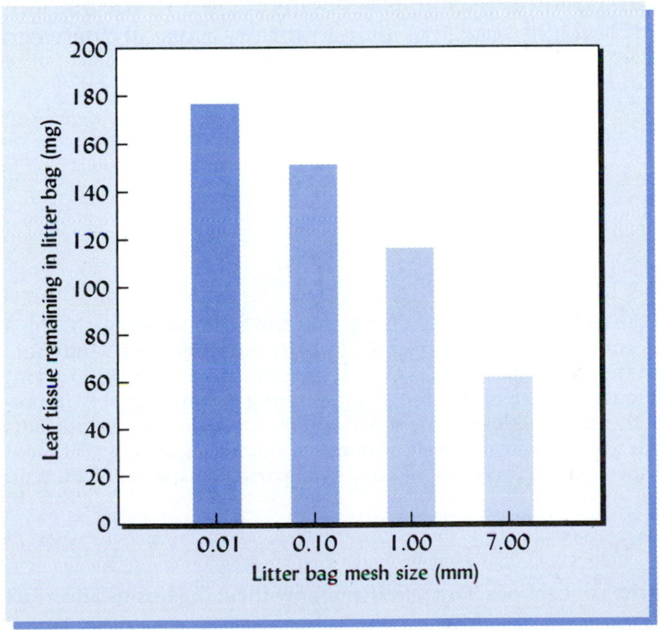

FIGURE 10.4 Influence of various sizes of soil organisms on the decomposition of corn leaf tissue buried in soil. Small bags made of nylon material with four different-size openings (mesh size) were filled with 558 mg (dry weight) of corn leaf tissue and buried in the soil for 10 weeks. The amount of corn leaf tissue remaining in the bags was considerably greater (less decomposition had taken place) where the meso- and macrofauna were excluded by the smaller mesh sizes. [Data from Weil and Kroontje (1979)]

Ecosystem Engineers[4]

Certain organisms make major alterations to their physical environment that influence the habitats of many other organisms in the ecosystem. These organisms are sometimes referred to as *ecosystem engineers*. For example, some of these "engineer" species are microorganisms that create an impermeable surface crust that spatially concentrates scarce water supplies in certain desert soils. Others are burrowing animals that create opportunities and challenges for other organisms by digging channels that greatly alter air and water movement in soils. Termites (see Section 10.5) and ants, for example, may invert the soil profile in local areas (bringing subsoil material to the surface) and denude the soil surface in wide areas. Earthworms literally eat their way through the soil as they incorporate plant residues into the mineral soil. Burrows encourage more water and air to enter the soil and also provide passages that plant roots can easily follow.

Certain beetles of the *Scarabaeidae* family greatly enhance nutrient cycling by burying animal dung in tiny tunnels that they excavate into the upper soil horizons. Many of these **dung beetles** cut round balls from large mammal feces, enabling them to roll the dung balls to a new location (Figure 10.5). The female dung beetle then lays her eggs in the ball of dung and buries it in the soil. Dispersal and burial of the dung not only provides a food source for the beetle larvae, it also protects the nutrients in the manure from easy loss by runoff or volatilization—fates to which the nutrients would most likely succumb if left on the soil surface. Dung beetles therefore play important roles in nutrient cycling and conservation in many grazed ecosystems.

10.3 ORGANISM ABUNDANCE, BIOMASS, AND METABOLIC ACTIVITY

Soil organism numbers are influenced primarily by the amount and quality of food available. Other factors affecting their numbers include physical factors (e.g., moisture and temperature), biotic factors (e.g., predation and competition), and chemical characteristics of the soil (e.g., acidity, dissolved nutrients, and salinity). The species that inhabit the soil in a desert will certainly be different from those in a humid forest, which, in turn, will be quite different from those in a cultivated field. Acid soils are populated by species different from those in alkaline soils. Likewise, species diversification and abundance in a tropical rain forest are different from those in a cool temperate area.

Despite these variations, a few generalizations can be made. For example, forested areas usually support a more diverse soil fauna and more fungal-dominated microflora than do grasslands, although the total faunal mass per hectare and level of faunal activity are generally higher in grasslands. Cultivated fields are generally lower than undisturbed native lands in numbers and biomass of soil organisms, especially the fauna, partly because tillage destroys much of the soil habitat.

Total **soil biomass,** the living fraction of the soil, is generally related to the amount of organic matter present. On a dry-weight basis, the living portion is usually between

FIGURE 10.5 Two dung beetles (*Scarabaeidae*), one on top and the other underneath, roll a ball that they have fashioned out of buffalo dung over the leaf-strewn surface of a sandy soil. The female will lay her eggs in the ball of dung and bury it in the soil (some coarse sand grains from which are seen sticking to the surface of the dung ball). Burying the dung is very important for making the nutrients therein available to the soil food web. Dung burial also prevents the reproduction of carnivorous flies and other pests of dung-producing mammals. Different dung beetles have evolved to specialize in the burial of dung from particular species of animals. (Photo courtesy of R. Weil)

[4] See Lavelle (1997) for a discussion of the concept of *soil engineers* and how these organisms affect soil properties.

1 and 5% of the total soil organic matter. In addition, scientists commonly observe that the ratios of soil organic matter to detritus to microbial biomass to faunal biomass are approximately 1000:100:10:1.

Comparative Organism Activity

The importance of specific groups of soil organisms is commonly identified by (1) the numbers of individuals in the soil, (2) their weight (biomass) per unit volume or area of soil, and (3) their metabolic activity (often measured as the amount of carbon dioxide given off in respiration). Representative measurements of soil biological activity are not easily made because the soil is such a heterogeneous medium and the organisms living in it are very unevenly distributed. Concentrations of microbial activity (hotspots) occur in the immediate vicinity of living plant roots and decaying bits of detritus, in the organic material lining earthworm burrows, in fecal pellets of soil fauna, and in other favored soil environments.

As might be expected, the microorganisms are the most numerous and have the highest biomass. Together with earthworms (or termites, in the case of some tropical soils), the microflora also dominate the biological activity in most soils (Table 10.2). It is estimated that 80-90% of the total soil metabolism is due to the microflora, although, as previously mentioned, their activity is enhanced by the actions of soil fauna. Despite their relatively small total biomass, such microfauna as nematodes and protozoa play important roles in nutrient cycling by preying on bacteria and fungi.

Even those animals that account for only very small fractions of the total metabolism in the soil can play important roles in soil formation (see Chapter 2) and management. Rodents pulverize, mix, and granulate soil, as well as incorporate surface organic residues into lower horizons. They provide large channels through which water and air can move freely. Ants are important in localized areas, especially in warm-region grasslands and in boreal forests, for their exceptional ability to break down woody materials and turn over soil materials as they build their nests. Mesofauna detritivores (mostly mites and collembola; see Figure 10.3) translocate and partially digest organic residues and leave their excrement for microfloral degradation (see Plate 37, after page 498, noting the center of the left image, where mites feeding on soft inner root tissue have left a pile of fecal pellets). By their movements in the soil, many animals rearrange soil particles to form biopores, thus favorably affecting the soil's physical condition (also visible in Plate 37). Others animals use the soil as a habitat for destructive action against higher plants.

TABLE 10.2 Relative Numbers and Biomass of Fauna and Flora Commonly Found in the Surface 15 cm of Soil[a]

Since metabolic activity is generally related to biomass, microflora and earthworms dominate the life of most soils.

Organisms	Number		Biomass[b]	
	Per m²	*Per gram*	*kg/ha*	*g/m²*
Microflora				
Bacteria	10^{13}–10^{14}	10^{8}–10^{9}	400–5000	40–500
Actinomycetes	10^{12}–10^{13}	10^{7}–10^{8}	400–5000	40–500
Fungi	10^{10}–10^{11}	10^{5}–10^{6}	1,000–15,000	100–1500
Algae	10^{9}–10^{10}	10^{4}–10^{5}	10–500	1–50
Fauna				
Protozoa	10^{9}–10^{10}	10^{4}–10^{5}	20–200	2–20
Nematodes	10^{6}–10^{7}	10–10^{2}	10–150	1–15
Mites	10^{3}–10^{6}	1–10	5–150	.5–1.5
Collembola	10^{3}–10^{6}	1–10	5–150	.5–1.5
Earthworms	10–10^{3}		100–1500	10–150
Other fauna	10^{2}–10^{4}		10–100	1–10

[a]A greater depth is used for earthworms.
[b]Biomass values are on a liveweight basis. Dry weights are about 20 to 25% of these values.

TABLE 10.3 Metabolic Grouping of Soil Organisms According to Their Source of Metabolic Energy and Their Source of Carbon for Biochemical Synthesis

Source of carbon	Source of energy: Biochemical oxidation	Source of energy: Solar radiation
Combined organic carbon	**Chemoheterotrophs**: All animals, fungi, actinomycetes, and most bacteria Examples: *Earthworms* *Aspergillus* *Azotobacter* *Pseudomonas*	**Photoheterotrophs**: A few algae
Carbon dioxide	**Chemoautotrophs** Examples: Ammonia oxidizers—*Nitrosomonas* Sulfur oxidizers—*Thiobacillus denitrificans*	**Photoautotrophs**: Algae and cyanobacteria Examples: *Chorella* *Nostoc*

Source of Energy and Carbon

Soil organisms may be classified as either **autotrophic** or **heterotrophic** based on where they obtain the *carbon* needed to build their cell constituents (Table 10.3). The heterotrophic soil organisms obtain their carbon from the breakdown of organic materials previously produced by other organisms. Nearly all heterotrophs also obtain their *energy* from the oxidation of the carbon in organic compounds. They are responsible for organic decay. These organisms, which include the soil fauna, the fungi, actinomycetes, and most bacteria, are far more numerous than the autotrophs.

The autotrophs obtain their carbon from simple carbon dioxide gas (CO_2) or carbonate minerals, rather than from carbon already fixed in organic materials. Autotrophs can be further classified based on how they obtain energy. Some use solar energy (photoautotrophs), while others use energy released by the oxidation of inorganic elements such as nitrogen, sulfur, and iron (chemoautotrophs). While the autotrophs are in the distinct minority, their carbon fixation and inorganic oxidation reactions allow them to play crucial roles in the soil system.

10.4 EARTHWORMS[5]

Information and images for common earthworm species: http://www.sarep.ucdavis.edu/worms/

Earthworms are probably the most important macroanimals in soils (Figure 10.6). They (along with their much smaller mesofauna cousins, the enchytraeid worms) are egg-laying *hermaphrodites* (organisms without separate male and female genders) that eat detritus, soil organic matter, and microorganisms found on these materials. They do not eat living plants or their roots, and so do not act as pests to crops.[6]

Epigeic, Endogeic, and Anecic Earthworms. The 7000 or so species of earthworms reported worldwide can be grouped according to their burrowing habits and habitat. The relatively small **epigeic** earthworms live in the litter layer or in the organic-rich soil very near the surface. This group includes the common compost worm, *Eisenia foetida*. **Endogeic** earthworms move about, making shallow, largely horizontal burrows as they eat their way through the upper 10 to 30 cm of soil. The pale, pink *Allolobophora caliginosa* (known as "red worm") is an example common to Europe and the United States that lives mainly head-down in the soil, and therefore expels its casts on the soil surface as well as in its burrows. Finally, the relatively large **anecic** earthworms make mainly vertical burrows as much as several meters deep. They emerge from their relatively permanent burrows to forage on the surface for pieces of litter to drag back into their homes, often covering the entrance to their burrow with a midden of leaves (see Figure 10.6).

---

[5] For a readable overview of earthworm biology and roles in the soil, see Lee (1985). For a more technical but wide-ranging review of earthworm ecology, see Hendrix (1995). Edwards (1998) provides information on earthworm distribution and ecology.

[6] Earthworms should not be confused, in this regard, with wireworms, rootworms, or other soil-dwelling insect larvae that are commonly referred to as "worms," and which may be plant pests.

FIGURE 10.6 Earthworms are perhaps the most significant macroorganism in soils of humid temperate regions, particularly in relation to their effects on the physical conditions of soils. Surface feeding species such as *Lumbricus terrestris* incorporate large amounts of plant litter into the soil and often gather plant debris into piles called *middens* near their burrow entrances. This species, known in North America as *night crawler,* was introduced from Europe in the root balls of settlers' fruit trees and is now the most common anecic earthworm both in North America and in its native Europe. (Photo courtesy of R. Weil)

Influence on Soil Fertility and Productivity

Earthworms can eat their way through the soil, ingesting a weight of soil equal to 2 to 30 times their own weight in a single day. In so doing, they create extensive systems of **burrows.** The burrows may be mainly vertical or horizontal, depending on the species, some providing continuous macropores to 1 m depth. Plant roots commonly grow down earthworm channels, sometimes through otherwise impenetrable soil layers (see Plate 38 following page 498).

After passing through the earthworm gut, ingested soil is expelled as globular soil aggregates called **casts,** some of which can be seen in Figure 10.6. During the passage through the earthworm's gut, organic materials are thoroughly shredded and mixed with mineral soil materials. Probably because of enhanced bacterial activity, earthworm casts are usually high in polysaccharides, which are credited with stabilizing the granular structure (see Section 4.7). Compared to the bulk soil, the casts are significantly higher in bacteria and organic matter and available plant nutrients (see Table 10.4).

The activities of earthworms greatly enhance soil fertility and productivity by altering both physical and chemical conditions in the soil, especially in the upper 15 to 35 cm of soil. Earthworms increase the availability of mineral nutrients to plants in two ways. First, as soil and organic materials pass through an earthworm's body, they are ground up physically as well as attacked chemically by the digestive enzymes of the earthworm and its gut microflora. Second, as earthworms ingest detritus and soil organic matter of relatively low nitrogen, phosphorus, and sulphur concentrations, they assimilate part of this material into their own body tissues. Consequently, earthworm tissues contain high concentrations of these nutrients, which are then readily released into plant-available form when the earthworms die and decay.

TABLE 10.4 Comparative Characteristics of Earthworm Casts and Bulk Soil at Six Sites in Nigeria

Characteristic	*Earthworm casts*	*Soils*
Silt and clay, %	38.8	22.2
Bulk density, Mg/m^3	1.11	1.28
Structural stability[a]	849	65
Cation exchange capacity, cmol/kg	13.8	3.5
Exchangeable Ca^{2+}, cmol/kg	8.9	2.0
Exchangeable K^+, cmol/kg	0.6	0.2
Soluble P, ppm	17.8	6.1
Total N, %	0.33	0.12

[a]Numbers of raindrops required to destroy structural aggregates.
From de Vleeschauwer and Lal (1981).

TABLE 10.5 Improvements in Soil Ecosystem Quality Nine Years After Inoculating Clay Loam Soil with Earthworm on a Restored Coal Mine in Wales (U.K.)

The soil had lost all its earthworms while stored in a huge stock pile for four months during the coal-mining operation. Earthworms of several indigenous species were added (70/m²) to experimental plots, but not to control plots. Without inoculation, normal populations of earthworms may take 20 to 30 years to recolonize such soils.

	Earthworm population[a] *number/m²*	*Carbohydrate content in soil,*[b] *g/100 g soil*	*Clay dispersed by wetting soil,*[c] *g/100 g soil*	*Metabolic quotient,*[d] *mg $CO_2-C\ g^{-1}\ h^{-1}$*
Control (no earthworms added)	8	0.90	23.7	15.3
Earthworms added	106	1.28	16.9	10.9
Percent change	+ 1200	+ 42	−29	− 29

[a]Earthworm populations were measured 6 years after restoration began.
[b]Carbohydrates were mainly polysaccharides, which may have been produced by the earthworms themselves or by bacteria stimulated by the earthworms.
[c]Clay dispersed when dry soil aggregates are shaken in water is a measure of the stability of the aggregate structure and the soil's resistance to erosion. More dispersed clay indicates weaker structure.
[d]Metabolic quotient is the amount of respiration occurring per gram of microbial biomass. A higher value indicates a stressed ecosystem in which microorganisms must devote increased energy to survival rather than growth. The units are mg of C respired in the form of CO_2 per gram of microbial biomass per hour. See also Table 11.4. Data from Scullen and Malik (2000).

Physical incorporation of surface residues into the soil by earthworms and other fauna reduces the loss of nutrients, especially nitrogen, by erosion and volatilization. This physical action, which is particularly important in untilled soils (including grasslands and no-till croplands), has earned the earthworm the title of "nature's tiller". Table 10.5 illustrates the value of using earthworms to assist in the restoration of soil after a strip-mining operation.

Beneficial Physical Effects. Earthworms are important in other ways. The holes left in the soil (see, for example, Figure 6.5) serve to increase aeration and drainage, an important consideration in plant productivity and soil development. In turf grass, the mixing activity of earthworms can alleviate or reduce compaction problems and nearly eliminate the formation of an undesirable thatch layer. Under conditions of heavy rainfall, earthworm burrows may greatly increase the infiltration of water into the soils, conserving water and reducing erosion. One aspect of concern is that water rapidly percolating down vertical earthworm burrows may carry potential pollutants toward the groundwater (see Figure 6.21).

Factors Affecting Earthworm Activity

Earthworms prefer a moist, but well-aerated environment with an abundant supply of calcium (which is an important component of their slime excretions). Most earthworms are quite sensitive to excess salinity. Enchytraeid worms are much more tolerant of acid conditions and are more active than earthworms in some forested Spodosols.

Maximum earthworm activity occurs in spring and autumn in temperate regions. In soils not insulated by residue mulches, a sudden heavy frost in the fall may decimate earthworm populations before they can move deeper into the profile.

Other factors that depress earthworm populations include predators (moles, mice, and certain mites and millipedes), very sandy soils (partly because of the abrasive effect of sharp sand grains), direct contact with ammonia fertilizer, application of certain insecticides (especially carbamates), and most importantly, tillage. The last factor is often the overriding deterrent to earthworm populations in agricultural soils. Minimum tillage, with plenty of crop residues left as a mulch on the soil surface, is ideal for encouraging earthworms.

A simple method of assessing earthworm populations in grassland or arable soils is to dig into the surface soil to a depth of about 20 cm. During relatively cool, moist conditions, an average of 5 to 10 earthworms in every shovelful of upturned soil would suggest a healthy soil ecosystem.

10.5 ANTS AND TERMITES

Ants

Nearly 9000 species of soil-inhabiting ants have been identified. They occur in many climates, being most diverse in the humid tropics, but are perhaps most functionally prominent in temperate semiarid grasslands. Different ant species typify different trophic levels in soil food webs, some acting as detritivores, other as herbivores, and still others as predators. Many species build mounds and underground tunnels somewhat similar to (though generally smaller than) those of termites described in the following paragraphs. Ant nest-building activity can improve soil aeration, increase water infiltration, and modify soil pH. Some ants feed on the sugary fluid produced by aphids as these sucking insects attack higher plants. The ants may "raise" the aphids in their nest or "herd" them while they feed on plant sap. Although ants occur more widely than termites and may play ecological roles that are more varied, the termites have been much more intensively studied.

Termites

DIVERSITY AND DISTRIBUTION. Termites are sometimes called *white ants,* although they are quite distinct from ants (one can distinguish an ant by its narrow "waist" between the abdomen and thorax). The world's 2000 species of termites are major contributors to the breakdown of organic material in or at the surface of soils. Perhaps the most infamous of the termites are those that invade (and subsequently destroy) the woodwork of houses built without protective metal termite shields. The species build protective soil tunnels, which enable them to return to the soil for their daily water supply.

Termites are found in about two-thirds of the land areas of the world but are most prominent in the grasslands (savannas) and forests of tropical and subtropical areas (both humid and semiarid). Their activity is on a scale comparable to that of earthworms. In the drier tropics (less than 800 mm annual rainfall) termites surpass earthworms in dominating the soil fauna.

MOUND-BUILDING ACTIVITIES. Termites are social animals that live in complex labyrinths of nests, passages, and chambers that they build both below and above the soil surface (Figure 10.7). In building their mounds, termites transport soil from lower layers to the surface, thereby extensively mixing the soil and incorporating into it the plant residues they use as food. Scavenging a large area around each mound, these insects remove up to 4000 kg/ha of leaf and woody material annually, a substantial portion of the plant litter produced in many tropical ecosystems. They also annually move 300 to 1200 kg/ha of soil in their mound-building activities. These activities have significant impacts on soil formation, as well as on current soil fertility and productivity.

(*a*)

(*b*)

FIGURE 10.7 Termite mounds in cultivated fields in Africa. (*a*) A termite mound in a cornfield. (*b*) Individual worker termites of another species drag cut pieces of leaves into their underground nest as soldier termites stand guard. (Photos courtesy of R. Weil)

Environmental Effects of Termites. Termites mix plant residues into the soil only in the localized areas of their nests, while denuding the remaining land area of surface residues. This termite behavior can reduce soil quality over large areas and make it extremely difficult to provide cropland with the benefits of a protective crop residue mulch.

Because termites build their mounds mainly with subsoil, termite mound material often has a lower organic matter and nutrient content than the surrounding undisturbed topsoil. Crop growth in soil in areas where these mounds have existed is often poor. However, on soils with a high water table, termite mounds may provide islands of better drainage and aeration that produce much better plant growth. In certain semiarid and savanna regions, the stable macrochannels constructed by termites greatly increase water infiltration into soils that otherwise tend to form impermeable surface crusts.

Before leaving the subject of termites, it should be mentioned that the bacterial metabolism in the guts of these widespread soil animals accounts for a substantial fraction of the global production of methane (CH_4), an important greenhouse gas (see Section 11.11).

10.6 SOIL MICROANIMALS

From the viewpoint of microscopic animals, soils present many habitats that are essentially aquatic, at least intermittently so. For this reason the soil microfauna are closely related to the microfauna found in lakes and streams. The two groups exerting the greatest influence on soil processes are the nematodes and protozoa.

Nematodes

Nematodes are found in almost all soils, often in surprisingly large numbers (see Table 10.2). They are unsegmented roundworms, about 4 to 100 μm in cross section and up to several millimeters in length (Figure 10.8). Therefore, they can seldom be seen with the naked eye. Moist, sandy soils typically have especially high nematode populations, as they contain abundant pores large enough to accommodate the swimming activities of these highly mobile creatures. When the large soil pores dry up, the nematodes survive by coiling up into a **cryptobiotic** or resting state, in which they seem to be nearly impervious to environmental conditions and use no detectable oxygen for respiration. In semiarid rangelands, nematode activity has been shown to be largely restricted to the first few days after each rainfall that awakens the nematodes from their cryptobiotic state.

Closer than you want to get-scanning electron micrographs of nematodes: http://www.nrm.se/ev/dok/nemgal.html.en

Predation. The nematodes comprise a highly diverse group. Most nematodes are predatory on other nematodes, fungi, bacteria, algae, protozoa, and insect larvae. As a result, certain predatory nematodes are sold for use as biological control agents for soilborne insect pests, such as the corn rootworm (a far more environmentally friendly method than the use of toxic pesticides). The head of a bacteria-eating nematode is illustrated in Figure 10.9, along with other soil microorganisms of various sizes. Nematode grazing can have a marked effect on the growth and activities of fungal and bacterial populations. Since bacterial cells contain more nitrogen than the nematodes can use, nematode activity often stimulates the cycling and release of plant-available nitrogen in the soil, accounting for 30 to 40% of the nitrogen released in some ecosystems.

Plant Parasites. Some nematodes, especially those of the genus *Heterodera,* can infest the roots of practically all plant species by piercing the plant cells with a sharp spearlike mouth part (Figure 10.8). These wounds often allow infection by secondary pathogens and cause

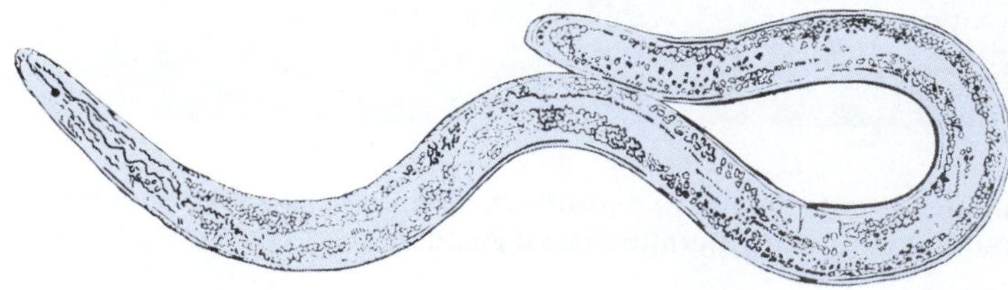

FIGURE 10.8 A nematode commonly found in soil (magnified about 120 times). More than 1000 species of soil nematodes are known. Most nematodes feed on bacteria and fungi, but plant parasitic species, such as this nematode, can be very detrimental to plant growth. (Courtesy William F. Mai, Cornell University)

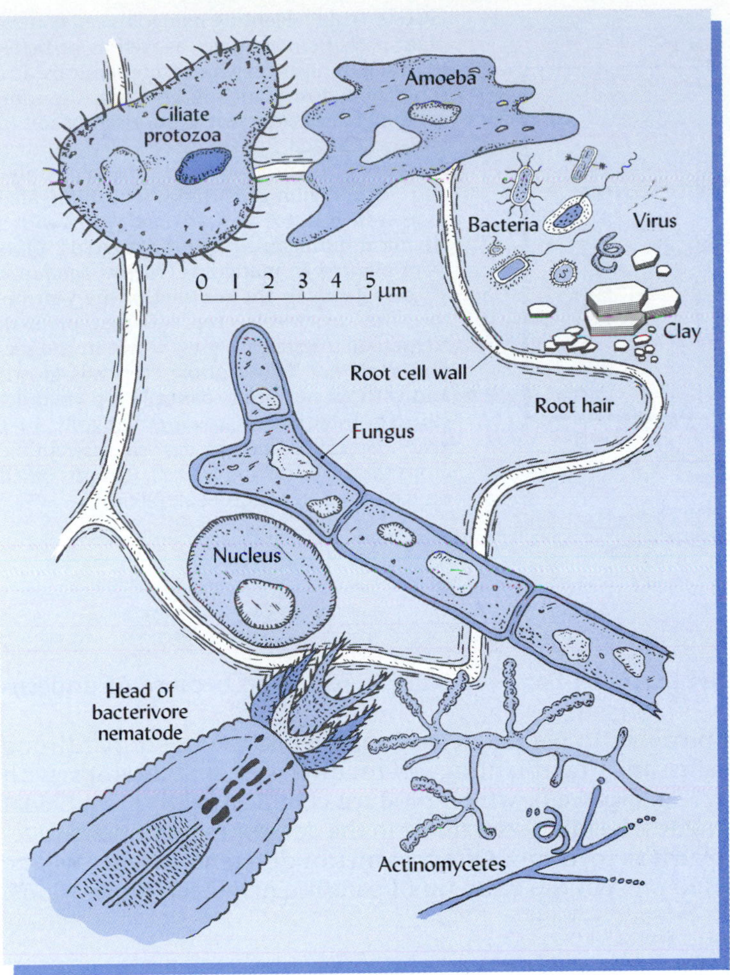

FIGURE 10.9 A depiction of representative groups of soil microorganisms, showing their relative sizes. (Drawing courtesy of R. Weil)

the formation of knotlike growths on the roots. Minor nematode infestations are nearly ubiquitous and often have little observable effect on the host plant. However, infestations beyond a certain threshold level result in serious stunting of the plant. Cyst- (egg-sac-) forming nematodes are major pests of sugar beets and soybeans, while root-knot-forming nematodes cause widespread damage to fruit trees and solanaceous crops (Figure 10.10).

Nematode Control. Until recently the principal methods of controlling plant parasitic nematodes were long rotations with nonhost crops (often 5 years is required for the parasitic nematode populations to sufficiently dwindle), use of genetically resistant crop varieties, and soil fumigation with toxic chemicals (**nematicides**). The use of soil

FIGURE 10.10 The root system of a tobacco plant that was heavily infested with root-knot nematodes, which stunted the roots and produced knotlike deformities. The aboveground portion of the infested plant was severely stunted. (Photo courtesy of R. Weil)

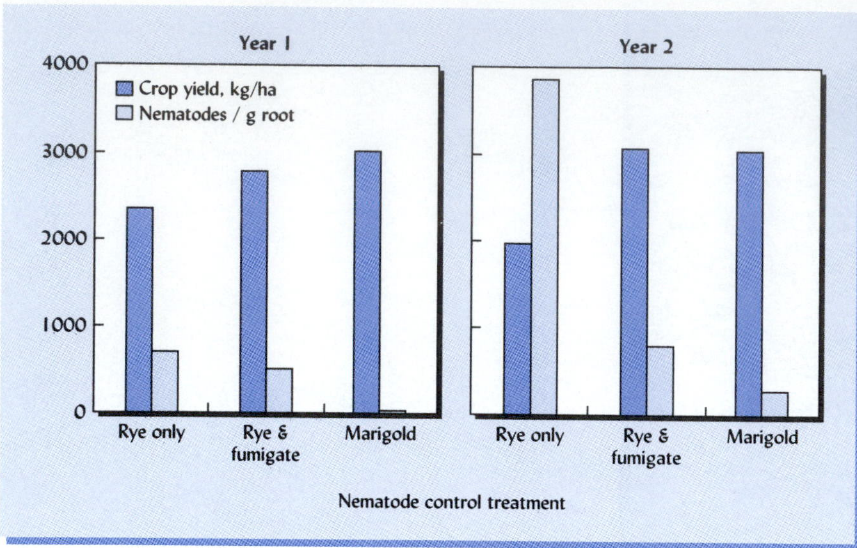

FIGURE 10.11 Planting marigolds can control plant-parasitic nematodes as well as or better than soil fumigation with toxic chemicals. The susceptible host plant was grown in the summers of year 1 and year 2 on a sandy soil in Ontario, Canada. Rye (*Secale cereale*) was grown as a cover crop each winter in the "Rye only" and "Rye + Fumigate" plots. The fumigated plots were injected in year 1 *and* year 2 with a chemical fumigant. The "Marigold" plots were planted to marigolds (*Tagetes patula* cv. "Creole") *only* in the summer of the year before the susceptible crop was first planted. Nematode infestations grew worse in the second year that a susceptible crop was grown and without nematode control crop yield declined substantially. Growing marigolds in 1 year controlled nematodes in susceptible crops for the following 2 years. [Graphs based on data from Reynolds et al. (2000)]

This page links to movies of the nematode, C. elegans, in action:
http://www.bio.unc.edu/faculty/goldstein/lab/movies.html

nematicides, such as methyl bromide, has been sharply restricted because of undesirable environmental effects.

New, less dangerous approaches to nematode control include the use of hardwood bark for containerized plants and interplanting or rotating susceptible crops with marigolds or brasicca crops (e.g. rapeseed), which produce chemicals with nematicidal properties (Figure 10.11). Progress has also been made in the development of nematode-resistant varieties of such plants as soybeans. Often a series of different varieties will be planted in succeeding years to prevent the build up of parasitic nematode populations.

Protozoa

Protozoa are mobile, single-celled creatures that capture and engulf their food. They are the most varied and numerous of the soil microfauna (see Figure 10.12 and Table 10.2). They are considerably larger than bacteria (see Figure 10.9), having a diameter range of 6 to 100 μm. Their cells do not have true cell walls, and have a distinctly more complex organization than bacterial cells. Soil protozoa include amoebas (which move by extending and contracting pseudopodia), ciliates (which move by waving hairlike structures; see Figure 10.12), and flagellates (which move by waving a whiplike appendage called a *flagellum*). Like nematodes, they swim about in the water-filled pores and water films in the soil and can form resistant resting stages (called *cysts*) when the soil dries out or food becomes scarce.

More than 350 species of protozoa have been isolated in soils; sometimes as many as 40 or 50 of species occur in a single sample of soil. A considerable number of serious animal and human diseases are attributed to infection by protozoa, but mainly by those that are waterborne, rather than soilborne. Most soil-inhabiting protozoa prey upon soil bacteria, exerting a significant influence on the populations of these microflora in soils.

Protozoa generally thrive best in moist, well-drained soils and are most numerous in surface horizons. In pursuit of their bacterial prey, soil-dwelling protozoa may squeeze into soil pores with openings as small as 10 μm. Protozoa are especially active in the area immediately around plant roots. Their main influence on organic matter decay and nutrient release is through their effects on bacterial populations.

10.7 ROOTS OF HIGHER PLANTS

Higher plants store the sun's energy and are the primary producers of organic matter (see Figure 10.1). Their roots live and die in the soil and are classified as soil organisms in this textbook. They typically occupy about 1% of the soil volume and may be responsible for a quarter to a third of the respiration occurring in a soil. Roots usually compete

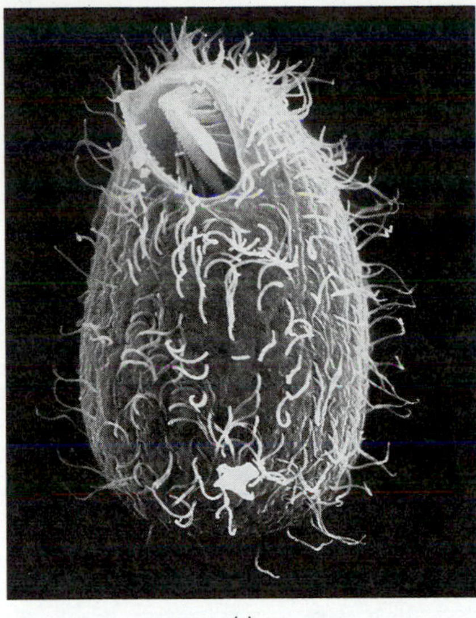

(a)

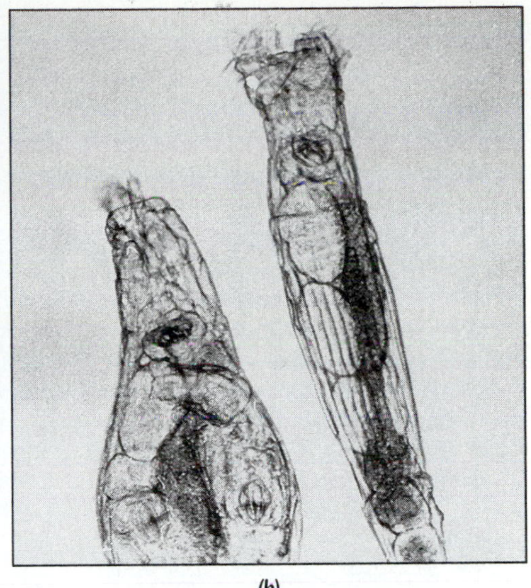

(b)

FIGURE 10.12 Two microanimals typical of those found in soils. (*a*) Scanning electron micrograph of a ciliated protozoan, *Glaucoma scintillaus*. (*b*) Photomicrograph of two species of rotifer, *Rotaria rotatoria* (stubby one at left) and *Philodina acuticornus* (slender one at right). [(*a*) Courtesy of J. O. Corliss, University of Maryland; (*b*) courtesy of F. A. Lewis and M. A. Stirewalt, Biomedical Research Institute, Rockville, Md.]

More on the chemistry and microbiology of the root-soil interface:
http://www.uoguelph.ca/~mgoss/seven/410_n08.html

for oxygen, but they also supply much of the carbon and energy needed by the soil community of fauna and microflora. Depending on their size, roots may be considered to be either meso- or microorganisms. Fine feeder roots range in diameter from 100 to 400 μm, while root hairs are only 10 to 50 μm in diameter—similar in size to the strands of microscopic fungi (see Figure 10.9). Root hairs are elongated protuberances of single cells of the outer (epidermal) layer (Figure 10.13). One function of root hairs is to anchor the root as it pushes its way through the soil. Another function is to increase the amount of root surface area available to absorb water and nutrients from the soil solution. Root-hair formation is stimulated by contact with soil particles and by low nutrient supply.

Root Effects on the Soil

Living roots physically modify the soil as they push through existing cracks and make new openings of their own. Once extended into a pore, the root matures and expands, exerting lateral forces that enlarge the pore. By removing moisture from the soil, plant roots stabilize organic–mineral bonds and encourage soil shrinkage and cracking, which, in turn, increase stable soil aggregation. Root exudates also support a myriad of microorganisms, which help to further stabilize soil aggregates. In addition, when roots die and decompose, they provide building materials for humus, not only in the top few centimeters, but also to greater soil depths.

The importance of root residues in helping to maintain soil organic matter is often overlooked. In natural grasslands and forests, roots may account for about 50% of the net primary biomass production (total plant biomass). For annual crops a figure of 25% is typical. Good crops of oats, corn, and sugarcane would be expected to leave about 2500, 4500, and 8500 kg/ha of root residues, respectively. The mass of organic compounds contributed by roots is seen to be even greater when the rhizosphere effects discussed in the following subsections are considered.

Rhizosphere[7]

The zone of soil significantly influenced by living roots is termed the **rhizosphere** and usually extends about 2 mm out from the root surface. Soil acidity may be 10 times higher (or lower) in the rhizosphere than in the bulk soil. Roots greatly affect the nutrient

[7] For a detailed review of the rhizosphere and its management, see Bowen and Rovira (1999).

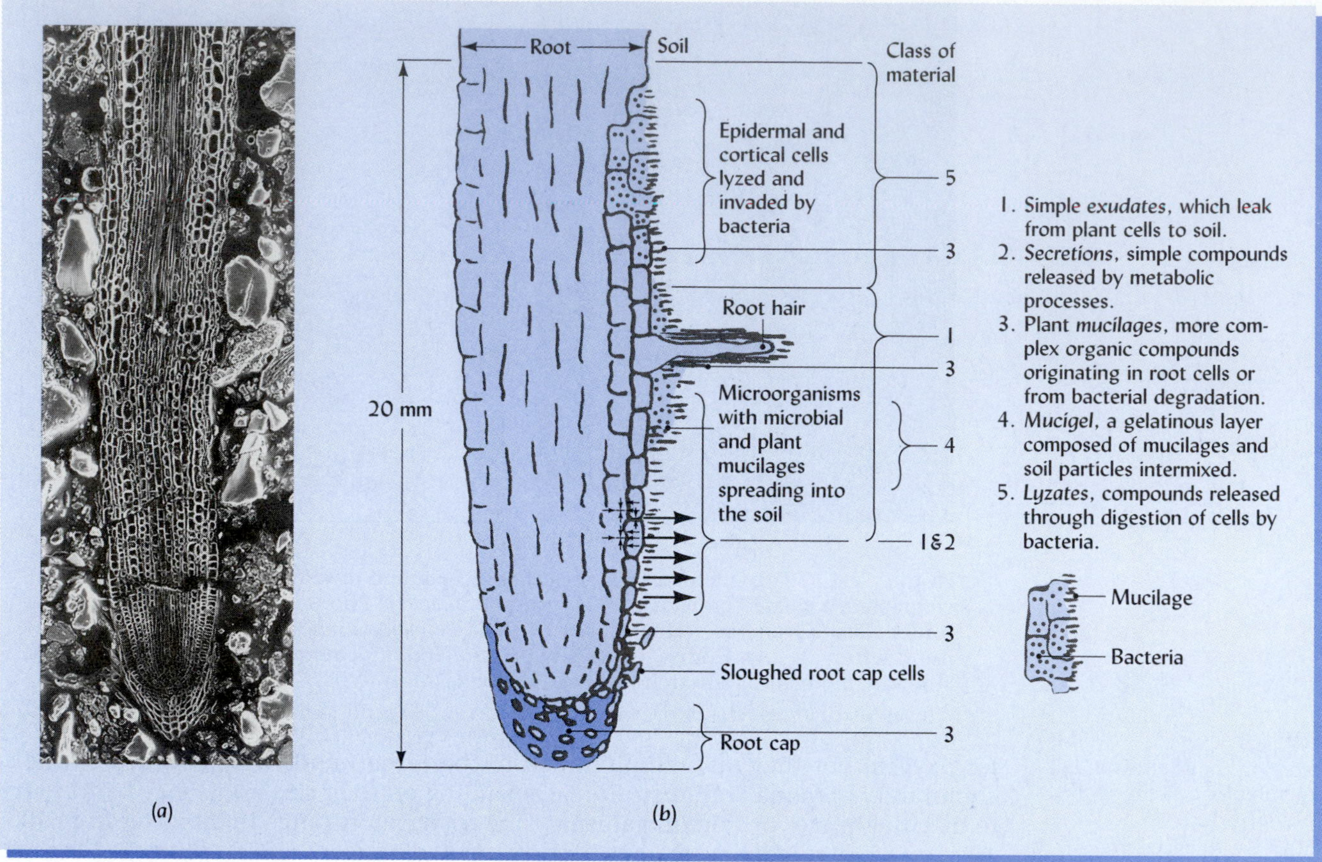

FIGURE 10.13 (*a*) Photograph of a root tip illustrating how roots penetrate soil and emphasizing the root cells through which nutrients and water move into and up the plant. (*b*) Diagram of a root showing the origins of organic materials in the rhizosphere. [(*a*) From Chino (1976), used with permission of Japanese Society of Soil Science and Plant Nutrition, Tokyo; (*b*) redrawn from Rovira et al. (1979), used with permission of Academic Press, London]

Images of plant roots in their natural environment: http://ic.ucsc.edu/~wxcheng/wewu/index.html

supply in this zone by withdrawing dissolved nutrients on one hand and by solubilizing nutrients from soil minerals on the other. By these and other means, roots affect the mineral nutrition of soil microbes, just as the microbes affect the nutrients available to the plant roots.

RHIZODEPOSITION. Significant quantities of at least three broad types of organic compounds are released at the surface of young roots (see Figure 10.13). First, low-molecular-weight organic compounds are exuded by root cells, including organic acids, sugars, amino acids, and phenolic compounds. Some of these root exudates, especially the phenolics, exert growth-regulating influences on other plants and soil microorganisms. Second, high-molecular-weight mucilages secreted near root tips form a substance called **mucigel** when mixed with microbial cells and clay particles. Third, cells from the root cap and epidermis continually slough off as the root grows and enrich the rhizosphere with a wide variety of cell contents.

Taken together, these types of **rhizodeposition** typically account for 2 to 30% of total dry-matter production in young plants. Sometimes when a plant is carefully uprooted from a loose soil, these root exudates cause a thin layer of soil, approximating the rhizosphere, to form a sheath around the actively growing root tips (Figure 10.14). The amount of organic material contributed to the rhizosphere during the growing season of annual plants may be more than twice the amount remaining in the root system at the end of the growing season. Rhizodeposition decreases with plant age but increases with soil stresses, such as compaction and low nutrient supply. Because of carbon substrates and specific growth factors released by roots, microbial numbers in the rhizosphere are typically 2 to 10 times as great as in the bulk soil.

The processes just described explain why plant roots are among the most important organisms in the soil ecosystem.

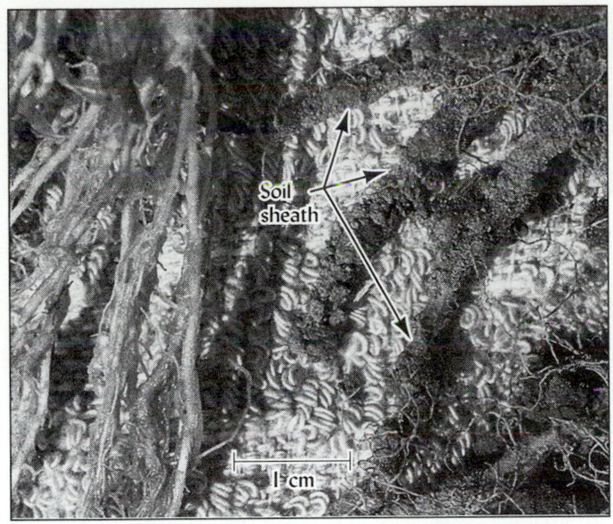

FIGURE 10.14 The zone of soil within 1 to 2 mm of living plant roots is termed the *rhizosphere*. This zone is greatly enriched in organic compounds excreted by the roots. These exudates and the microorganisms they support have caused the rhizosphere soil to adhere to some of these wheat roots as a sheath (roots on the right, see arrows). The roots on the left have been washed in water and the soil of the rhizosphere soil has been removed. The background is a terrycloth towel. (Photo courtesy of R. Weil)

10.8 SOIL ALGAE

Like higher plants, algae consist of eukaryotic cells, those with nuclei organized inside a nuclear membrane. (Organisms formerly called *blue-green algae* are prokaryotes and therefore will be considered with the bacteria.) Also like higher plants, algae are equipped with chlorophyll, enabling them to carry out photosynthesis. As photoautotrophs, (Table 10.3) most algae need light, and are therefore mostly found very near the surface of the soil.

Common soil algae range in size from 2 to 20 μm. Many algal species are motile and swim about in soil pore water, some by means of flagella (whiplike "tails"). Most grow best under moist to wet conditions, but some are also very important in hot or cold desert environments. Sometimes the growth of algae may be so great that the soil surface is covered with a green or orange algal mat. Some algae (as well as certain cyanobacteria) form *lichens,* symbiotic associations with fungi. These are important in colonizing bare rock and other low-organic-matter environments. In unvegetated patches in deserts, algae commonly contribute to the formation of **microbiotic crusts** that reduce water evaporation and soil erosion, but are very sensitive to disruption by trampling or off-road vehicle traffic (see Section 10.15).

In addition to producing a substantial amount of organic matter in some fertile soils, certain algae excrete polysaccharides that have very favorable effects on soil aggregation (see Section 4.8).

10.9 SOIL FUNGI

Soil fungi comprise an extremely diverse group of microorganisms. Tens of thousands of species have been identified in soils, representing some 170 genera. As many as 2500 species have been reported to occur in a single location. Scientists believe there are at least 1 million fungal species in the soil still awaiting discovery. Although the numbers of individuals are usually somewhat smaller than those of bacteria, their relatively large size results in their dominating the biomass and metabolic activity in many soils (Table 10.2).

Fungi are eukaryotes with a nuclear membrane and cell walls. As heterotrophs, they depend on living or dead organic materials for both their carbon and their energy. Fungi are aerobic organisms, although some can tolerate the rather low oxygen concentrations and high levels of carbon dioxide found in wet or compacted soils. Strictly speaking, fungi are not entirely microscopic, since some of these organisms, such as mushrooms, form macroscopic structures that can easily be seen without magnification.

For convenience of discussion, fungi may be divided into three groups: (1) yeasts, (2) molds, and (3) mushroom fungi. *Yeasts,* which are single-celled organisms, live principally in waterlogged, anaerobic soils. *Molds* and *mushrooms* are both considered to be filamentous fungi, because they are characterized by long, threadlike, branching chains of cells. Individual fungal filaments, called **hyphae** (Figure 10.20*a* on page 339), are often twisted together to form **mycelia** that appear somewhat like woven ropes. Fungal mycelia are often visible as thin, white or colored strands running through decaying plant litter. Filamentous fungi reproduce by means of spores, often formed on fruiting bodies, which may be microscopic (e.g., molds) or macroscopic (such as that shown in Figure 10.15).

Molds

The molds are distinctly filamentous, microscopic or semimacroscopic fungi that play a much more important role in soil organic matter breakdown than the mushroom fungi. Molds develop vigorously in acid, neutral, or alkaline soils. Some are favored, rather than harmed, by lowered pH (see Figure 9.11). Consequently, they may dominate the microflora in acid surface soils, where bacteria and actinomycetes offer only mild competition. The ability of molds to tolerate low pH is especially important in decomposing organic residues in acid forest soils. Many genera of molds are found in soils. Four of the most common are *Penicillium, Mucor, Fusarium,* and *Aspergillus.*

Mushroom Fungi

These fungi are associated with forest and grass vegetation where soil moisture and organic residues are ample. Although the mushrooms of many species are extremely poisonous to humans, some are edible—and a few have been domesticated. The aboveground fruiting body of most mushrooms is only a small part of the total organism. An extensive network of hyphae permeates the underlying soil or organic residue. In fact, the largest living organism known is thought to be a fungus growing in the soil under a temperate rain forest in the Pacific Northwest. While mushrooms are not as widely distributed as the molds, these fungi are very important, especially in the breakdown of woody tissue, and because some species form a symbiotic relationship with plant roots (see "Mycorrhizae," following).

Activities of Fungi

As decomposers of organic materials in soil, fungi are the most versatile and persistent of any group. Cellulose, starch, gums, and lignin, as well as the more easily metabolized proteins and sugars, succumb to their attack. Fungi play major roles in the processes of humus formation (see Section 11.4) and aggregate stabilization (see Section 4.7). They usually dominate in the upper horizons of forested soils, as well as in very acid or sandy soils.

Fungi function more efficiently than bacteria in that they assimilate into their tissues a larger proportion of the organic materials they metabolize. Up to 50% of the substances decomposed by fungi may become fungal tissue, compared to about 20% for bacteria. Soil *fertility* depends in no small degree on nutrient cycling by fungi, since they continue to decompose complex organic materials after bacteria and actinomycetes have essentially ceased to function. Some of the nutrient cycling and ecological activi-

FIGURE 10.15 Fruiting bodies of a Bird's Nest Fungus (*Cyathus olla*) contain disc-shaped spores ("eggs") that are scattered forcefully when the cone is hit by a raindrop. Like other members of the Basidiomycota, this fungus produces enzymes that break down cellulose, hemicellulose, and lignin in materials such as woody forest litter or corncobs left on agricultural fields after harvest. See tip of ballpoint pen for scale. (Photo courtesy of R. Weil)

FIGURE 10.16 A "fairy ring" of fungal growth and the fungi's fruiting bodies (mushrooms, inset). As the fairy ring fungi (most commonly *Marasmius* spp.) metabolize accumulated grass thatch and residues, they release excess nitrogen that stimulates the lush green growth of the grass. Later, bacteria decompose the aging and dead fungi, producing a second release of nitrogen. The fungus produces a chemical (thought to be hydrogen cyanide) that is toxic to itself. Therefore, each generation must grow into uncolonized soil, producing an ever-expanding circle of fungi and decay marked by an ever-larger ring of dark green grass. The grass in the center of the ring is often brown, stunted, and water-stressed, most likely because the fungi render the upper soil layers somewhat hydrophobic. (Photos courtesy of R. Weil)

ties of soil fungi are made easily visible in the case of "fairy rings," commonly seen on lawns in early spring (Figure 10.16).

In addition to the breakdown of organic residues and the formation of humus, numerous other fungal activities have significant impact on soil ecology. Some fungi are even predators of soil animals (e.g., see Figure 10.17). Soil fungi can synthesize a wide range of complex organic compounds in addition to those associated with soil humus. It was from a soil fungus, a *Penicillium* species, that the first modern antibiotic drug, penicillin, was obtained. By killing bacteria, such compounds probably help fungi to outcompete rival microorganisms in the soil.

Unfortunately, not all the compounds produced by soil fungi benefit humans or higher plants. A few fungi produce chemicals (**mycotoxins**) that are highly toxic to plants or animals (including humans). An important example of the latter is the production of highly carcinogenic aflatoxin by the fungus *Aspergillus flavus* growing on grains such as corn or peanuts, especially when grain is exposed to soil and moisture.

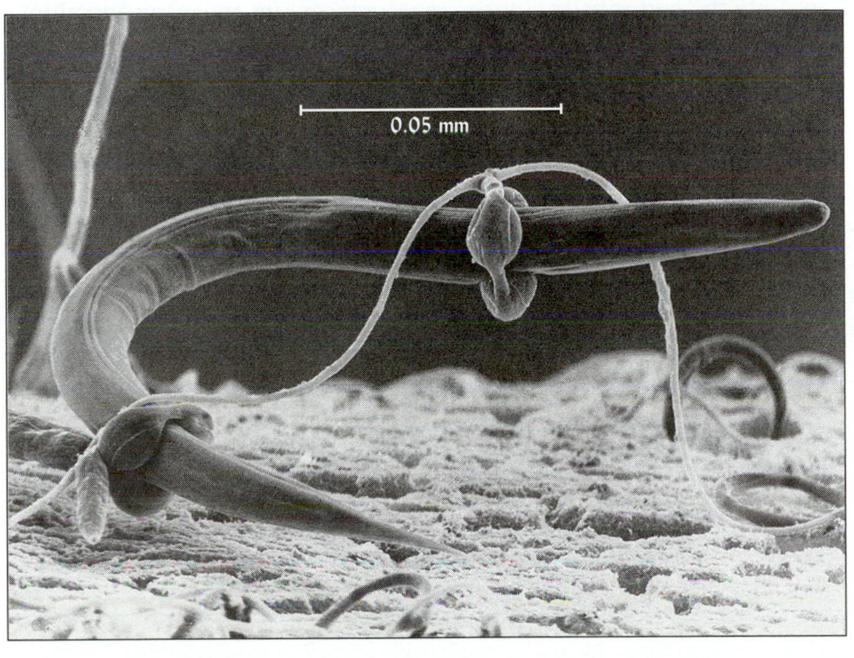

FIGURE 10.17 Several species of fungi prey on soil nematodes—often on those nematodes that parasitize higher plants. Some species of nematode-killing fungi attach themselves to, and slowly digest, the nematodes. Others, like this *Arthrobotrys anchonia,* make loops with their hyphae and wait for a nematode to swim through these lassolike structures. The loop is then constricted and the nematode is trapped. The nematode shown here is being crushed by two such fungal loops. Additional loops can be seen in their nonconstricted configuration. (Photo courtesy of George L. Barron, University of Guelph)

Other fungi produce compounds that allow them to invade the tissues of higher plants (see Section 10.14), causing such serious plant diseases as wilts (e.g., *Verticillium*) and root rots (e.g., *Rhizoctonia*).

On the other hand, efforts are now under way to develop the potential of certain fungi (such as *Beauvaria*) as biological control agents against some insects and mites that damage higher plants. These examples merely hint at the impact of the complex array of fungal activities in the soil.

Mycorrhizae[8]

Excellent overview of mycorrhizal symbioses: http://dmsylvia.ifas.ufl.edu/mycorrhiza.htm

One of the most ecologically and economically important activities of soil fungi is the mutually beneficial association **(symbiosis)** between certain fungi and the roots of higher plants. This association is called **mycorrhizae,** a term meaning "fungus root." In natural ecosystems many plants are quite dependent on mycorrhizal relationships and cannot survive without them. Mycorrhizae are the rule, not the exception, for most plant species, including the majority of economically important plants.

Mycorrhizal fungi derive an enormous survival advantage from teaming up with plants. Instead of having to compete with all the other soil heterotrophs for decaying organic matter, the mycorrhizal fungi obtain sugars directly from the plant's root cells. In return, plants receive some extremely valuable benefits from the fungi. The fungal hyphae grow out into the soil some 5 to 15 cm from the infected root, reaching farther and into smaller pores than could the plant's own root hairs. This extension of the plant root system increases its efficiency, providing perhaps 10 times as much absorptive surface as the root system of an uninfected plant.

Mycorrhizae greatly enhance the ability of plants to take up phosphorus and other nutrients that are relatively immobile and present in low concentrations in the soil solution (Table 10.6). Water uptake may also be improved by mycorrhizae, making plants more resistant to drought. In soils contaminated with high levels of metals or salts, mycorrhizae protect the plants from excessive uptake of these potential toxins. There is evidence that mycorrhizae also protect plants from certain soilborne diseases and parasitic nematodes by producing antibiotics, altering the root epidermis, and competing with fungal pathogens for infection sites. For all these reasons, the use of mycorrhizae can be a powerful tool in land restoration projects (see Box 10.1).

ECTOMYCORRHIZA. Two types of mycorrhizal associations are of considerable practical importance: **ectomycorrhiza** and **endomycorrhiza.** The ectomycorrhiza group includes hundreds of different fungal species associated primarily with temperate- or semiarid-region trees and shrubs, such as pine, birch, hemlock, beech, oak, spruce, and fir. These fungi, stimulated by root exudates, cover the surface of feeder roots with a fungal mantle. Their hyphae develop in the free space around the cells of the root cortex, but do not penetrate the cortex cell walls (hence the term *ecto,* meaning "outside").

[8]For two excellent reviews of mycorrhizae and their effects on the soil–plant system, see Bethlenfalvay and Linderman (1992) and Read et al. (1992).

TABLE 10.6 Effect of Inoculation with Mycorrhizae and of Added Phosphorus on the Content (μg/plant) of Different Elements in the Shoots of Corn

	No phosphorus		*25 mg/kg phosphorus added*	
Element in plant	*No mycorrhizae*	*Mycorrhizae*	*No mycorrhizae*	*Mycorrhizae*
P	750	1,340	2,970	5,910
K	6,000	9,700	17,500	19,900
Ca	1,200	1,600	2,700	3,500
Mg	430	630	990	1,750
Zn	28	95	48	169
Cu	7	14	12	30
Mn	72	101	159	238
Fe	80	147	161	277

From Lambert et al. (1979).

BOX 10.1 PRACTICAL MYCORRHIZAE TECHNOLOGIES

Although definite differences exist in mycorrhizal efficiency among various species and strains of fungi, the ubiquitous distribution of native strains ensures natural infection in plants under many circumstances. Thus, gains from inoculation with efficient mycorrhizal fungi are usually quite small. Exceptions occur where native populations of fungi are low due to soil fumigation or elimination of host plants, such as by surface mining or other drastic soil disturbance.

Soil tillage disrupts hyphal networks, and therefore minimum tillage increases the effectiveness of native mycorrhizae. Use of effective host species as cover crops or in crop rotations (as opposed to the use of monocropping, nonhost species, or long periods of bare fallow) favors the buildup of effective mycorrhizae in soils.

Likely Problem Areas

Surface soil stockpiled during mining and construction activities (as shown in Plate 35 following page 498) may lose its mycorrhizal inoculum potential because of the long period without any host plants. Satisfactory survival and growth of introduced plants may depend on inoculation with mycorrhizal fungi where such stockpiled soil is used for revegetation, on other sites in which the native fungal populations have been drastically reduced, or on sites which have not previously been vegetated with the intended plant species. Small areas of disturbed soils often become reinoculated by airborne ectomycorrhizal spores from adjacent undisturbed areas within a year or two. Sometimes soil from a mature forest containing the desired tree species can be used as inoculum to bring the necessary mycorrhizal fungi to forest plantings in such highly disturbed sites as reclaimed mines and landslides.

Along the west coast of the United States and in Canada, Douglas fir regrowth may be greatly hampered by the common practice of clear-cutting, followed by suppression of hardwood trees, which are viewed as weed species. On some sites without hardwood trees, young Douglas fir trees either fail to survive or grow much more slowly than those planted where some hardwood saplings remain to provide host continuity and serve as an inoculum source for AM fungi.

Commercial Inoculants

Because many ectomycorrhizal fungi are facultative symbionts that can also live independently in the soil they can be cultured in large quantities on artificial media. Effective and pathogen-free commercial inoculants have been developed for such trees as conifers, oaks, hickory, birch, willow, poplar, pecan, and eucalyptus. Inoculants for use in planting such trees in warm climates usually contain the ectomycorrhizal fungus *Pisolithus tinctorus*, while those used to inoculate soils in cold climates typically contain *Rhizopogon* species. On infertile soils, it is not uncommon for these ectomycorrhizal inoculants to increase tree survival and growth by 50 to 500%.

Although AM fungi cannot be grown in pure culture, commercial inoculants are made from infected root fragments, hyphal fragments, and spores, mainly of fungi in the *Glomus* genus. Again, little response to inoculation is likely on healthy, biologically active soils, but the results can be dramatic on disturbed or infertile sites.

Ectomycorrhizae cause the infected root system to consist primarily of stubby, white rootlets with a characteristic Y shape (Figure 10.18*a*).

Endomycorrhiza. The most important members of the endomycorrhiza group are called **arbuscular mycorrhizae (AM)** (formerly called *vesicular-arbuscular mycorrhizae, VAM)*. When forming AM, fungal hyphae actually penetrate the cortical root cell walls and, once inside the plant cell, form small, highly branched structures known as **arbuscules.** These structures serve to transfer mineral nutrients from the fungi to the host plants and sugars from the plant to the fungus. Other structures, called **vesicles,** are usually also formed and serve as storage organs for the mycorrhizae (see Figure 10.18*b*).

The AM fungi are the most common and widespread group. Nearly 100 identified species of fungi form these endomycorrhizal associations in soils from the tropics to the arctic. The roots of most grain, fiber, fruit, and vegetable crops form AM. Many forest trees, including maple, yellow poplar, and redwood, as well as such important tree crops as cacao, coffee, and rubber, also form AM. Two important groups of crop plants that do *not* form mycorrhizae are the *Cruciferae* (cabbage, mustard, canola, and broccoli) and the *Chenopodiaceae* (sugar beet, red beet, and spinach). AM fungi are agriculturally most important where soils are low in nutrients (especially phosphorus) and fertilizer inputs are limited.

Research on AM continues to reveal the ecological and practical significance of this symbiosis. The importance of mycorrhizal hyphae in stabilizing soil aggregate

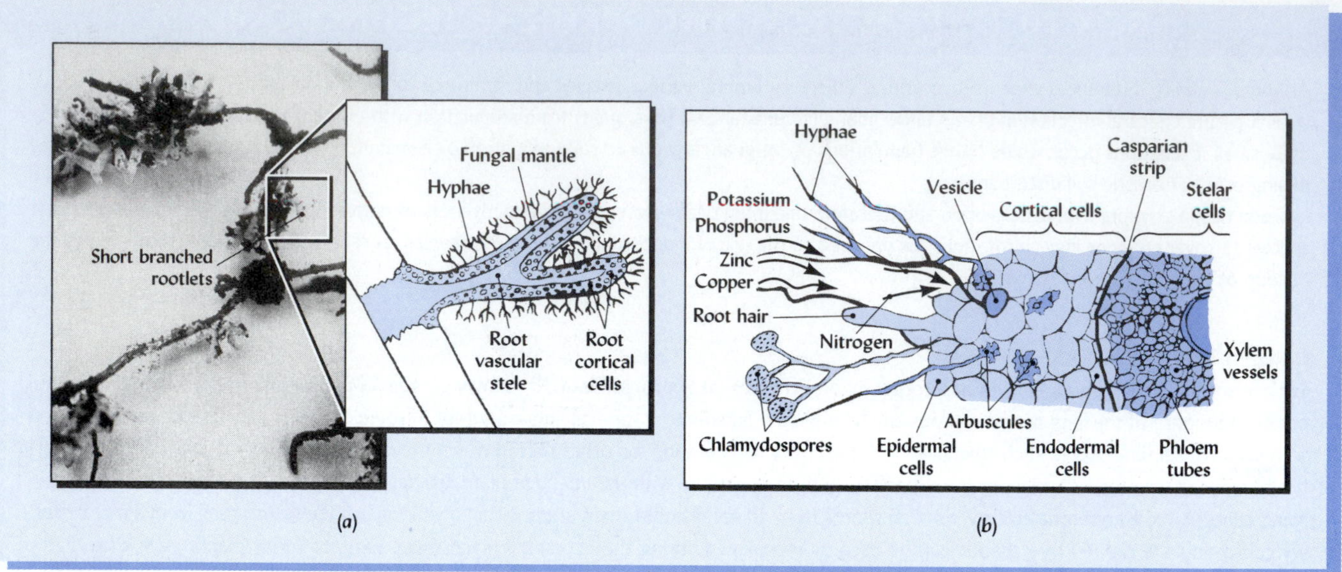

FIGURE 10.18 Diagram of ectomycorrhiza and arbuscular mycorrhiza (AM) associations with plant roots. (*a*) The ectomycorrhiza association produces short branched rootlets that are covered with a fungal mantle, the hyphae of which extend out into the soil and between the plant cells but do not penetrate the cells. (*b*) In contrast, the AM fungi penetrate not only between cells but into certain cells as well. Within these cells, the fungi form structures known as *arbuscules* and *vesicles*. The former transfer nutrients to the plant, and the latter store these nutrients. In both types of association, the host plant provides sugars and other food for the fungi and receives in return essential mineral nutrients that the fungi absorb from the soil. [Redrawn from Menge (1981); photo courtesy of R. Weil]

structure is becoming increasingly clear (see Figure 4.27). The presence of mycorrhizae is known to enhance nodulation and nitrogen fixation by legumes. Also, AM have been observed to transfer nutrients from one plant to another via hyphal connections, sometimes resulting in a complex four-way symbiotic relationship (Figure 10.19).

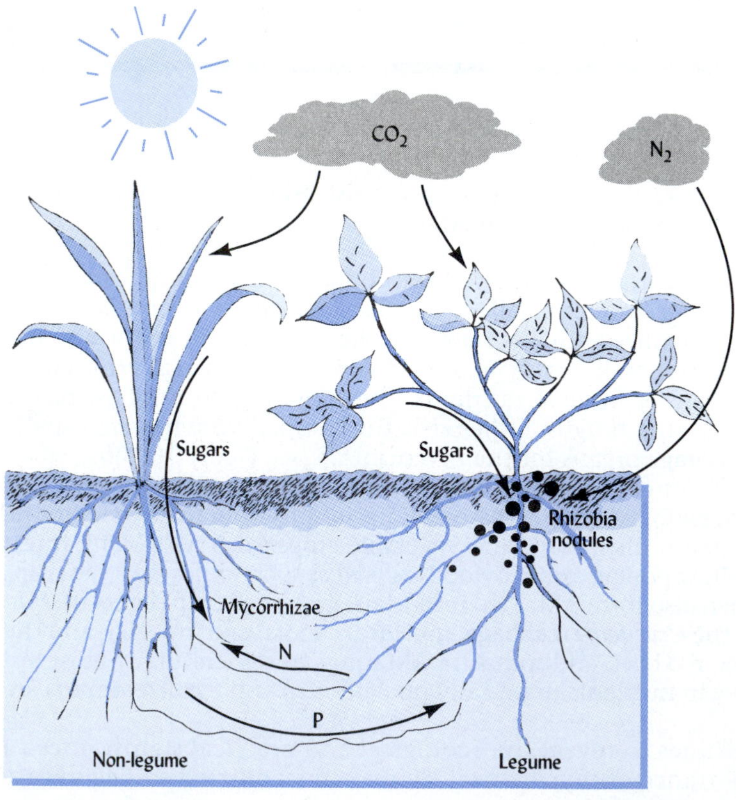

FIGURE 10.19 Mycorrhizal fungi, rhizobia bacteria, legumes, and nonlegume plants can all interact in a four-way, mutually beneficial relationship. Both the fungi and the bacteria obtain their energy from sugars supplied through photosynthesis by the plants. The rhizobia form nodules on the legume roots and enzymatically capture atmospheric nitrogen, providing the legume with nitrogen to make amino acids and proteins. The mycorrhizal fungi infect both types of plants and form hyphal interconnections between them. The mycorrhizae then not only assist in the uptake of phosphorus from the soil, but can also directly transfer nutrients from one plant to the other. Isotope tracer studies have shown that, by this mechanism, nitrogen is transferred from the nitrogen-fixing legume to the nonlegume (e.g., grass) plant, and phosphorus is mostly transferred to the legume from the nonlegume. The nonlegume grass plant has a fibrous root system and an extensive mycorrhizal network, which is relatively more efficient in extracting P from soils than the root system of the legume. Research indicates that some direct transfer of nutrients via mycorrhizal connections occurs in many mixed plant communities, such as in forest understories, grass–legume pastures, and mixed cropping systems.

10.10 SOIL BACTERIA

Although analysis of their genetic material suggests that organisms classed in the domain *Archaea* are evolutionarily distinct from bacteria, under the microscope they are indistinguishable and our discussion of bacteria will include both groups.

Bacteria are very small, single celled, and prokaryotic—meaning they have no distinct nucleus. They range in size from 0.5 to 5 μm, the smaller ones approaching the size of the average clay particle (see Figure 10.9). Bacteria are found in various shapes, but in the soil, rod-shaped organisms seem to predominate (see Figure 10.20*a* and *b*). Many bacteria are motile, swimming about in the soil water by means of hairlike cilia or whiplike flagella.

Bacteria are perhaps the most diverse group of soil organisms; a gram of soil typically contains 20,000 different *species*. They have evolved mechanisms to adapt to life in the most extreme of environments, from the Antarctic to the Amazon, from anaerobic wetlands to desiccated salt flats, from forest litter to deep groundwaters, from sodic soils at pH 10 to acid sulfate soils at pH 2.

Bacterial Populations in Soils

As with other soil organisms, the numbers of bacteria are extremely variable but high, ranging from a few billion to more than a trillion in each gram of soil. A biomass of a 400- to 5000-kg/ha liveweight is commonly found in fertile soils (see Table 10.2).

Their ability to form extremely resistant resting stages that survive dispersal by winds, sediments, ocean currents, and animal digestive tracts has allowed bacteria to spread to almost all soil environments. Their extremely rapid reproductive potential (generation times of only a few hours) enables bacteria to increase their populations quickly in response to favorable changes in soil environment and food availability. As the early soil microbiologist Beijerinck expressed it, "everything is everywhere" when it comes to bacteria. Change the soil environment, add a new substrate—even an industrial waste—and soon you'll find populations of bacteria feeding on it.

Importance of Bacteria

Bacteria participate vigorously in virtually all of the organic transactions that characterize a healthy soil system. Most soil bacteria are heterotrophs, but photoautotrophs and chemautotrophs also exist (Table 10.3). Because soil bacteria as a group possess such a broad range of enzymatic capabilities, scientists are now finding ways of harnessing, even improving, the metabolic activities of bacteria to help with the remediation of soils polluted by crude oil, pesticides, and various other organic toxins. Bacteria are usually the most important group in the breakdown of hydrocarbon compounds, such as gasoline and diesel fuel.

(*a*)

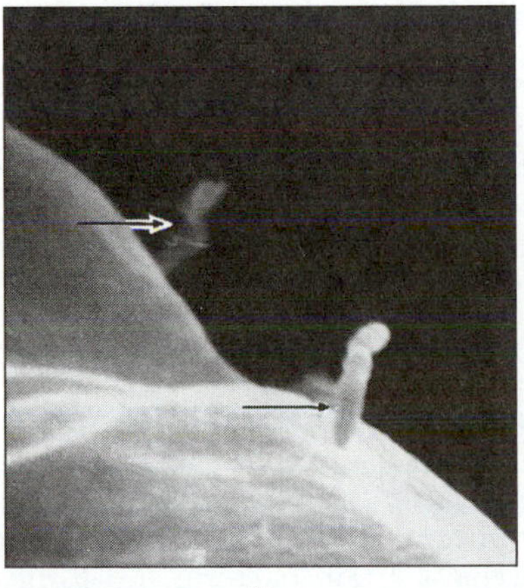

(*b*)

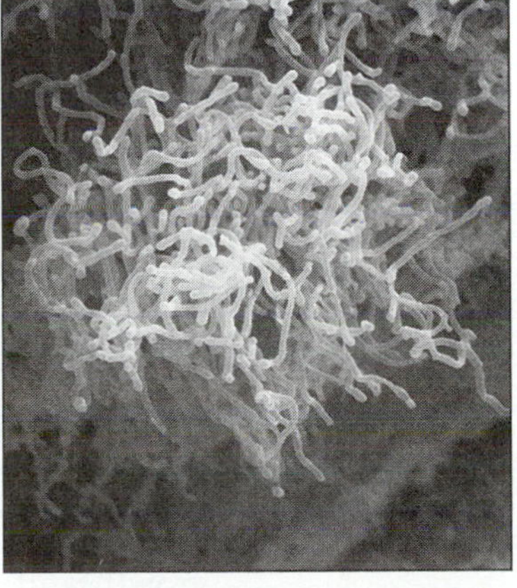

(*c*)

FIGURE 10.20 Scanning electron micrographs of fungi, bacteria, and actinomycetes. (*a*) Fungal hyphae associated with much smaller rod-shaped bacteria. (*b*) Rod-shaped bacteria attached to a plant root hair. (*c*) Actinomycete threads. [(*a*) Courtesy of R. Campbell, University of Bristol, used with permission of American Phytopathological Society; (*b*, *c*) courtesy of Maureen Petersen, University of Florida]

In addition, bacteria hold near monopolies in several basic enzyme-mediated transformations. One such process is the oxidation or reduction of selected chemical elements in soils (see Sections 7.3 and 7.5). Many of these biochemical oxidation and reduction reactions have significant implications for environmental quality as well as for plant nutrition. A second critical process in which bacteria are prominent is nitrogen fixation—the biochemical combining of atmospheric nitrogen with hydrogen to form organic nitrogen compounds usable by higher plants (see Section 12.11).

Cyanobacteria

Previously classified as blue-green algae, **cyanobacteria** contain chlorophyll, allowing them to photosynthesize like plants. Cyanobacteria are especially numerous in rice paddies and other wetland soils, and fix appreciable amounts of atmospheric nitrogen when such lands are flooded. Certain cyanobacteria, growing within the leaves of the aquatic fern *azolla,* are able to fix nitrogen in quantities great enough to be of practical significance for rice production (see Section 12.13). These organisms also exhibit considerable tolerance to saline environments and are important in forming microbiological crusts on desert soils (see Section 10.15).

10.11 SOIL ACTINOMYCETES

Like the fungi, **actinomycetes** are filamentous and often profusely branched (see Figure 10.20*c*), though their mycelial threads are much smaller than those of fungi. Although historically they have often been classified with the fungi, actinomycetes are bacterialike in that they are unicellular, of about the same diameter, have no nuclear membrane (prokaryote), and often break up into spores that closely resemble bacterial cells. They are now commonly classified as bacteria; however, we will consider them separately because of the unique roles they play in the soil.

Generally aerobic heterotrophs, actinomycetes live on decaying organic matter in the soil or on compounds supplied by plants with which certain species form parasitic or symbiotic relationships. Many actinomycete species produce antibiotic compounds that kill other microorganisms. Actinomycin, neomycin, and streptomycin are examples of familiar antibiotic drugs produced by growing soil actinomycetes in pure culture. In forest ecosystems, much of the nitrogen supply depends on certain actinomycetes that are capable of fixing atmospheric nitrogen gas into ammonium nitrogen (see Section 12.13).

Actinomycetes develop best in moist, warm, well-aerated soil. However, in times of drought, they remain active to a degree not usually exhibited by either bacteria or fungi. Actinomycetes are tolerant of low osmotic potential and are important in arid-region, salt-affected soils. They are generally rather sensitive to acid soil conditions, with optimum development occurring at pH values between 6.0 and 7.5.

Numbers and Activities of Actinomycetes

Actinomycete numbers in soil exceed those of all other organisms except bacteria (see Table 10.2). These organisms are especially numerous in soils high in humus, such as old meadows or pastures, where the acidity is not too great. The earthy aroma of organic-rich soils and freshly plowed land is mainly due to actinomycete-produced *geosmins,* volatile derivatives of terpene. Actinomycetes often become dominant in the later stages of decay when the easily metabolized substrates have been used up.

10.12 CONDITIONS AFFECTING THE GROWTH OF SOIL MICROORGANISMS

Organic Matter Requirements. The addition of almost any energy-rich organic substance, including the compounds excreted by plant roots, stimulates heterotrophic microbial growth and activity. Certain bacteria and fungi are stimulated by specific amino acids and other growth factors found in the rhizosphere or produced by other organisms. Bacteria tend to respond most rapidly to additions of simple compounds such as starch and sugars, while fungi and actinomycetes overshadow the bacteria if cellulose and more resistant compounds dominate the added organic materials. If organic materials are left on the soil surface (as in coniferous forest litter), fungi dominate the microbial activity. Bacteria commonly play a larger role if the substrates are mixed into the soil, as by earthworms, root distribution, or tillage.

Oxygen Requirements. While most soil microorganisms are *aerobic* and use O_2 as the electron acceptor in their metabolism, some bacteria are *anaerobic* and use substances other than O_2 (e.g., NO_3^-, SO_4^{2-}), as electron acceptors. *Facultative* bacteria can use either aerobic or anaerobic forms of metabolism. All three of these types of metabolism are usually carried out simultaneously in different habitats within a soil.

Moisture and Temperature. Optimum moisture level for higher plants (moisture potential of −10 to −70 kPa) is usually best for most aerobic microbes. Too high a water content will limit the oxygen supply. Microbial activity is generally greatest when temperatures are 20° to 40°C. The warmer end of this range tends to favor actinomycetes. Ordinary soil temperature extremes seldom kill bacteria, and commonly only temporarily suppress their activity. However, except for certain **cryophilic** species, most microorganisms cease metabolic activity below about 5°C, a temperature sometimes referred to as *biological zero* (see Section 7.8).

Exchangeable Calcium and pH. Levels of exchangeable calcium and pH help determine which specific organisms thrive in a particular soil. Although in any chemical condition found in soils some bacterial species will thrive, high calcium and near-neutral pH generally result in the largest, most diverse bacterial populations. Low pH allows fungi to become dominant. The effect of pH and calcium helps explain why fungi tend to dominate in forested soils, while bacterial biomass generally exceeds fungal biomass in most subhumid to semiarid prairie and rangeland soils.

10.13 BENEFICIAL EFFECTS OF SOIL ORGANISMS

Organic Material Decomposition. Perhaps the most significant contribution of the soil fauna and flora to higher plants is that of plant residue decomposition. By this process, dead leaves, roots, and other plant tissues are broken down, converting organically held nutrients into mineral forms available for renewed plant uptake. As a by-product of their metabolism, microbes synthesize new compounds, some of which help to stabilize soil structure and others of which contribute to humus formation.

Breakdown of Toxic Compounds. Many organic compounds toxic to plants or animals find their way into the soil. Some of these toxins are produced by soil organisms as metabolic by-products. Some toxins are **xenobiotic** (artificial) compounds foreign to biological systems, applied purposefully by humans as agrochemicals to kill pests, and some are deposited in the soil because of unintentional environmental contamination. If these compounds accumulated unchanged, they would do enormous ecological damage. Fortunately, soil ecosystems include organisms that produce enzymes that allow them to use most toxins as food (Box 10.2). The detoxifying activity of these microorganisms is by far the greatest in the surface layers of soil, where microbial numbers are concentrated in response to the greater availability of organic matter and oxygen (Figure 10.21).

Inorganic Transformations. Nitrates, sulfates, and, to a lesser degree, phosphate ions are present in soils primarily due to the action of microorganisms. Bacteria and fungi assimilate some of the N, P, and S in the organic materials they digest. Excess amounts of these nutrients may be excreted into the soil solution in inorganic form either by the bacteria and fungi themselves or by the nematodes and protozoa that feed on them. In this manner, the soil food web converts organically bound forms of nitrogen, phosphorus, and sulfur into mineral forms that can be taken up by higher plants.

Likewise, the availabilities of other essential elements, such as iron and manganese, are determined largely by microbial action, especially that which enhances oxidation and reduction. Microbial oxidation also controls the potential for toxicity in soil contaminated with selenium or chromium.

BOX 10.2 HARNESSING SOIL BACTERIA TO FIGHT POLLUTION

Organic pollutants from industrial processes, leaks and spills have caused *acutely* toxic contamination of soils, producing barren areas sometimes referred to as **brownfields.** Industry and government are spending billions of dollars annually to *remediate* (clean up) these contaminated soils. Expensive, but widespread methods of soil remediation involve physical and/or chemical treatment of the soil, either in situ (in place) or by excavating and moving the soil to treatment facilities. The polluted soil is incinerated washed with solvents, or heated and subjected to vacuum extraction to remove or destroy toxic contaminants.

For many contaminated soils there is a more economical *biological* alternative to incineration, soil washing, or landfilling—namely **bioremediation.** Simply put, this technology uses microbes—usually soil bacteria—to degrade organic toxins into harmless metabolic products. Certain soil bacteria can even break down such toxic petroleum constituents as polyacrylic aromatic hydrocarbons (PAHs), as well as such synthetic toxins as pentachlorophenol and trichloroethylene. The job may be carried out by organisms currently in the soil, or microbes specifically selected for their ability to destroy the contaminants may be introduced into the soil. For example, bacteria isolated from sites with a long history of the specific contamination (or grown in laboratory culture on a diet rich in the pollutant) tend to evolve an unusually high efficiency at metabolizing the target chemical.

Inoculating a polluted soil with such specialized exotic organisms is called **bioaugmentation.** For example, a bacterium has been identified that can detoxify perchloroethene (PCE) a common, highly toxic groundwater pollutant suspected of causing cancer. The bacteria's enzymes remove the four chlorines from the PCE producing ethylene, a naturally occurring and relatively harmless gas.

$$\underset{\text{Toxic PCE}}{Cl_2C{=}CCl_2} \xrightarrow{2H \quad HCl} Cl_2C{=}CHCl \xrightarrow{2H \quad HCl} ClHC{=}CHCl \xrightarrow{2H \quad HCl} ClHC{=}CH_2 \xrightarrow{2H \quad HCl} \underset{\text{Harmless ethylene gas}}{H_2C{=}CH_2}$$

In some cases indigenous microorganisms are capable of degrading the contaminant, but their growth and metabolic rates may be far too slow because of insufficient mineral nutrients, especially nitrogen and phosphorus (see Section 10.3 for a discussion of the C/N ratio in organic decomposition). To alleviate such nutritional limitations, special fertilizers have been formulated and used with innovative application techniques to stimulate and greatly speed up microbial degradation processes in soils. This **biostimulation** approach was pioneered in 1989 when an oil-soluble fertilizer was used to stimulate the breakdown of oil spilled on Alaska's pristine beaches (Entisols) by the tanker *Exxon Valdez.*

Some situations call for the use of bioremediation techniques deep in the soil where low soil porosity causes oxygen deficiency that limits microbial activity. For example, deep-soil biostimulation can be used to clean up toxic organic solvents in the subsoil and groundwater under leaking storage tanks. Typically, special horizontal drilling rigs are used to pipe in a mixture of air (for oxygen), methane (to act as a carbon source to stimulate specific methanotrophic bacteria), and phosphorus (a nutrient needed for bacteria growth). Soil scientists in the young field of bioremediation are constantly developing new ways to harness the abilities of soil microorganisms to clean up our environment.

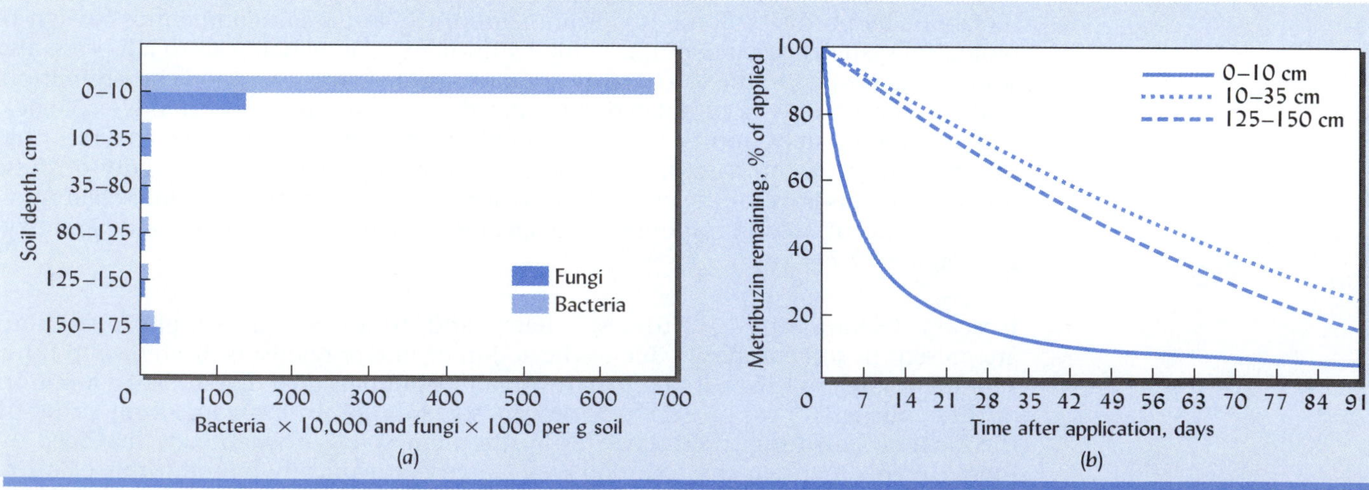

FIGURE 10.21 (*a*) Populations of aerobic bacteria and fungi at various depths in a Dundee soil (Aqualfs). (*b*) Breakdown of the herbicide metribuzin in the same soil at various depths. Soil was sampled from each depth and incubated with the herbicide. Note that the fungi and bacteria were concentrated in the upper 10 cm, and that the breakdown of the herbicide was far more rapid in soil from this layer than in soil from the deeper layers. [From Mooreman and Harper (1989)]

NITROGEN FIXATION. The fixation of elemental nitrogen gas, which cannot be used directly by higher plants, into compounds usable by plants is one of the most important microbial processes in soils. Actinomycetes in the genus *Frankia* fix major amounts of nitrogen in forest ecosystems; cyanobacteria are important in flooded rice paddies, wetlands, and deserts; and rhizobia bacteria are the most important group for the capture of gaseous nitrogen in agricultural soils. By far the greatest amount of nitrogen fixation by these organisms occurs in root nodules or in other associations with plants. Worldwide, enormous quantities of atmospheric nitrogen are fixed annually into forms usable by higher plants (see Section 12.10).

RHIZOBACTERIA.[9] As pointed out in Section 10.7, the rhizosphere soil supports a dense population of microorganisms. Bacteria especially adapted to living in this zone are termed **rhizobacteria,** many of which are beneficial to higher plants (the so-called **plant growth-promoting rhizobacteria**). Because in nature root surfaces are almost completely encrusted with bacterial cells, little interaction between the soil and root can take place without some intervening microbial influence. The world of rhizobacteria is still largely uncharted, but research is beginning to uncover useful ways to take advantage of interactions that can benefit higher plants. In addition to those that ward off plant diseases (see Section 10.14), certain rhizobacteria promote plant growth in other ways, such as enhanced nutrient uptake or hormonal stimulation (for an example, see Table 10.7).

Some bacteria that live in the rhizosphere damage plants by invading the root cells and living as parasites. Others, called **deleterious rhizobacteria,** inhibit root growth and function by various noninvasive chemical interactions. These nonparasitic bacteria can cause stunting, wilting, foliar discoloration, nutrient deficiency, and even death of affected plants, but often the effects are subtle and difficult to detect. The discussion of deleterious rhizobacteria is warranted in this section on beneficial effects mainly because of their potential as a weed management strategy. By managing soils and cropping systems in ways that favor the deleterious rhizobacteria associated with certain weeds, scientists hope to be able to reduce weed seed germination and seedling growth and thereby reduce the use of herbicide (weed-killer) sprays on cropland and rangeland.

PLANT PROTECTION. Certain soil organisms protect plant roots from invasion by soil parasites and pathogens. Plant diseases and the protective action of the soil microflora will be discussed in the next section.

TABLE 10.7 Rice Plants Respond to Inoculation with Growth-Promoting Rhizobacteria

Rice plants were grown in pots with clay soil that was puddled and flooded with water. All pots were fertilized with adequate N fertilizer, but only some were inoculated with various strains of Rhizobia or Bradyrhizobia bacteria. The bacteria colonized the rice rhizosphere (hence the term rhizobacteria) *and changed the physiology of the rice plant, partly by producing the plant growth hormone IAA. Inoculated roots were more efficient at nutrient uptake. Analysis of N isotope tracers showed that the rhizobacteria did* not *cause significant amounts of N fixation.*

		Uptake of nutrients by rice plants, mg/pot				
Treatment	*Grain yield, g/pot*	*N*	*P*	*K*	*Fe*	*IAA[a] in the rhizosphere, mg/L*
Control—no inoculation	36.7	488	111	902	18.9	1.0
Inoculated with rhizobacteria	44.3	612	134	1020	23.6	2.1
Percent change	+21	+25	+21	+13	+25	+110

[a]IAA = indole-3-acetic acid, a plant growth hormone.
Data calculated from Biswas et al. (2000)

[9] For a general review of rhizobacteria ecology with an emphasis on deleterious effects on plants, see Nehl et al. (1996); for a review of weed control using deleterious rhizobacteria, see Kremer (1998).

10.14 SOIL ORGANISMS AND DAMAGE TO HIGHER PLANTS

Although most of the activities of soil organisms are vital to a healthy soil ecosystem and economic plant production, some soil organisms affect plants in detrimental ways that cannot be overlooked.

SOIL FAUNA. It already has been suggested that certain of the soil fauna are injurious to higher plants. For instance, some rodents may severely damage crops. Snails and slugs in some climates are dreaded pests, especially of vegetables. Ants transfer aphids onto certain plants, and so contribute to plant damage. However, the greatest damage to plants by soil fauna is caused by the feeding of nematodes and insect larvae. In nature, plants commonly sustain a low level of nematode and insect infestation, but under certain circumstances, infestation may be so great that the successful growth of some crops is both difficult and expensive. Such infestations, which are often associated with a lack of proper crop rotation and organic matter addition, account for a major proportion of nematicide and insecticide chemicals used in agriculture.

MICROFLORA AND PLANT DISEASE. Disease infestations occur in great variety and are induced by many different organisms. Among microorganisms, bacteria and actinomycetes contribute their quota of plant diseases, but fungi are responsible for most of the common soilborne diseases of crop plants. Included are wilts, damping-off, root rots, clubroot of cabbage, and blight of potatoes. Soils are easily infested with disease organisms, which are transferred from soil to soil by many means, including tillage or planting implements, transplant material, manure from animals that were fed infected plants, soil erosion, and windborne fungal spores and bacteria. Once a soil is infested, it is apt to remain so for a long time.

Plant Disease Control by Soil Management

PREVENTION. Prevention is the best defense against soilborne diseases. Strict quarantine systems will restrict the transfer of soilborne pathogens from one area to another. Elimination of the host crop from the infested field will help in cases where no alternate hosts are available. Crop rotation and tillage practices can be used to help control a disease by growing nonsusceptible plants for several years between susceptible crops, and by burying plant residues on which fungal spores might overwinter. By preventing the splashing of soil onto foliage during rainstorms, mulches can reduce the transmission of soilborne diseases.

SOIL PH. Regulation of soil pH is effective in controlling some diseases. For example, keeping the pH low (<5.2) can control both the actinomycete-caused *potato scab* and the fungal disease of turf grass known as *spring dead spot.* Raising soil pH to about 7.0 can control *clubroot* disease in the cabbage family, because the spores of the fungal pathogen germinate poorly, if at all, under alkaline conditions.

AIR AND TEMPERATURE. Wet, cold soils favor some seed rots and seedling diseases, known as *damping-off.* Good drainage and planting on ridges can help control these diseases. High levels of ammonium (as compared to nitrate) nitrogen tend to increase wilt diseases caused by *Fusarium* fungi. Steam or chemical sterilization is a practical method of treating greenhouse soils for a number of pathogens. **Solarization**, the heating of soil under clear plastic sheeting, is a practical way to partially sterilize the upper few cm of soil in some field situations. It should be remembered, however, that sterilization kills beneficial microorganisms, such as mycorrhizal fungi, as well as pathogens, and so may do more harm than good.

DISEASE-SUPPRESSIVE SOILS. Research on plant diseases ranging from *take-all* of wheat to *Fusarium wilt* of bananas has documented the existence of ***disease-suppressive soils***, in which a particular disease fails to develop *even though both the virulent pathogen and a susceptible host are present.* The reason that certain soils become disease suppressive is not entirely understood, but much evidence suggests that the pathogenic organisms are inhibited by **antagonism** from certain beneficial bacteria and fungi.

In some cases, disease suppressiveness has developed through long-term crop monoculture in which the buildup of the pathogen during the first few years is eventually overshadowed by a subsequent buildup of specific organisms antagonistic to the

List of pests and biological products that help to manage them:
www.agrobiologicals.com

pathogen (Figure 10.22). In other cases, the addition of large amounts and varied types of decomposable organic materials, sometimes combined with lime, encourages a high level of microbial activity and diversity—properties that, in turn, provide a more general suppression of soilborne pathogens.

The effects are sometimes much greater for certain types of organic additions than for others. For example, scientists in California have discovered that rotating cauliflower with broccoli can provide a practical means of controlling verticillium wilt, a serious disease of cauliflower, even in fields where the disease-causing fungus is present. The leaf residues left after broccoli harvest break down in the soil to release glucosinolates and other volatile compounds specifically toxic to the *Verticillium dahliae* fungus, which causes the disease in cauliflower. The level of disease control was equal to that achieved using two conventional toxic fumigants.

Similarly, horticulturalists have been able to control *Fusarium* diseases in containerized plants by replacing traditional potting mixes with growing media made mainly from certain well-aged **composts.** Apparently, large numbers of antagonistic organisms colonize the organic material during the final stages of composting (see Section 11.5). These examples merely hint at the potential that exists for controlling plant diseases and pests through ecological management rather than applications of toxic chemicals.

Induced Systemic Resistance. Some beneficial rhizobacteria have an intriguing mode of action called **induced systemic resistance**, which helps plants ward off infection by diseases or insect pests both above and below ground. The process begins when a plant root system is colonized by beneficial rhizobacteria that cause the accumulation of a signaling chemical. The chemical signal is translocated up to the shoot, where it induces leaf cells to mount a chemical defense against a specific pathogen, even before the pathogen has arrived on the scene.

Competition for Nutrients and Oxygen. Soil organisms may harm higher plants, at least temporarily, by competition for available nutrients. Nitrogen and iron are the elements for which competition is usually greatest, although similar competition occurs for phosphorus, potassium, calcium, and most micronutrients. Under conditions of poor drainage, active soil microflora may suffocate roots by depleting the already limited soil oxygen supply. This oxygen depletion limits root growth and nutrient uptake. It also affects the availability of several plant nutrients such as iron and manganese whose oxidation state and, in turn, solubility are determined by the soil oxidation level.

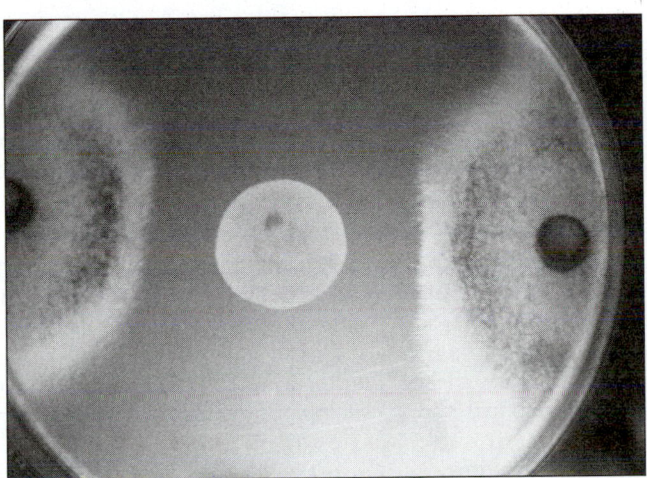

FIGURE 10.22 The biological basis of disease-suppressive soils that control take-all disease of wheat. (Left) In the center of the petri dish grows a colony of certain *Pseudomonas* bacteria isolated from the rhizosphere of wheat that grew in soils suppressive of take-all disease. The *Pseudomonas* colony is producing an antibiotic that is toxic to *Gaeumannomyces graminis* (the fungal pathogen that causes take-all disease), preventing the pathogen colonies from growing to the center of the plate. (Right) These plots in eastern Washington have grown monoculture wheat for 15 years and have developed high populations of organisms antagonistic to the take-all pathogen. In the 15th year of the study, the entire field was inoculated with *G. graminis* for experimental purposes, but the disease developed (seen as light-colored, prematurely ripened plots) only where the soil was fumigated prior to the inoculation. The fumigation killed most of the antagonistic organisms, leaving the pathogenic fungi free to infect the wheat plants. (Photos by R. J. Cook; courtesy of the American Phytopathological Society)

10.15 ECOLOGICAL RELATIONSHIPS AMONG SOIL ORGANISMS

Mutualistic Associations

"Desert varnish," and symbiotic algae and fungi called lichens:
http://www.desertusa.com/magdec97/varnish/dec_varnish.html

We have already mentioned a number of mutually beneficial associations between plant roots and other soil organisms (e.g., mycorrhizae and nitrogen-fixing nodules) and between several microorganisms (e.g., lichens). Other examples of such associations abound in soils. For example, photosynthetic algae reside within the cells of certain protozoans. Several types of associations, among them algal-fungal associations on or in soils and rocks, are very important cyclers of nutrients and producers of biomass in desert ecosystems. Next, we will briefly consider the nature of such associations.

Microbiotic Crusts[10]

In relatively undisturbed arid- and semiarid-region ecosystems, where the vegetation cover is quite patchy, it is common to find an irregular, usually dark-colored crust covering the soil in the areas between clumps of grasses and shrubs. In many cases this forms a sort of miniature landscape of tiny, jagged pinnacles only a few centimeters tall (see Figure 10.23). This crust is not at all like the physical-chemical crusts associated with degraded soils that have a hard, smooth surface seal (see Section 4.8). Rather, the **microbiotic crusts** of arid rangelands are intricate living systems that greatly benefit the associated natural vegetation communities. They consist of mutualistic associations that usually include algae or cyanobacteria along with fungi, mosses, bacteria, and/or liverworts. An intact microbiotic crust is considered a sign of a healthy ecosystem.

Soil biological crusts—by U.S. Geol Serv.:
http://www.soilcrust.org/

Such crusts improve arid-region ecosystem productivity by (1) protecting the soil from erosion, (2) helping to conserve and cycle nutrients, (3) increasing nitrogen supplies via the nitrogen-fixing activities of the cyanobacteria, (4) enhancing water supplies in some cases by increasing infiltration and reducing evaporation, and (5) contributing to net organic matter production by crust photosynthesis, which may continue during environmental conditions that inhibit photosynthesis by higher plants in the ecosystem. Unfortunately, the crusts can be easily destroyed by trampling or burial under windblown soil and are very slow to reestablish. Off-road vehicles, off-trail hiking, and overgrazing are especially damaging to these important ecosystem components.

Competitive Interactions

COMPETITION FOR FOOD. The population of soil microbes is generally limited by the food supply, increasing rapidly whenever more organic materials become available. As a result, the soil presents an intensely competitive environment for its microscopic inhabitants. When fresh organic matter is added, heterotrophic soil organisms (bacteria, fungi, and actinomycetes) compete with each other for this source of food. If sugars, starches, and amino acids are available in the organic material, the bacteria dominate initially because they reproduce most rapidly and prefer such simple compounds. As these simple compounds are used up, fungi and actinomycetes become more competitive.

ANTIBIOTICS AND OTHER COMPETITIVE MECHANISMS. The soil microbes use many different tactics in their battle for resources. Certain organisms alter the acidity level in their vicinity to the disadvantage of competitors. Others produce substances (**siderophores**) that so strongly bind to iron that other organisms in their immediate vicinity cannot grow for lack of this essential element. Certain soil bacteria, fungi, and actinomycetes produce antibiotics, compounds that kill specific types of other organisms on contact (Figure 10.24). Many antibiotics for the treatment of human and animal diseases, such as penicillin, streptomycin, and aureomycin, are produced by organisms found in soils. When these and other antagonistic activities of soil microorganisms are turned against potentially plant-pathogenic organisms, signif-

[10]Other names used to refer to these microbiotic crusts include *biological crusts, cryptograms, cryptobiotic crusts, microfloral crusts,* and *microphytic crusts.* All refer to the same thing. For a technical review of scientific knowledge on the ecology of microbial crusts, see Belnap et al. (2001).

FIGURE 10.23 Tiny pinnacles of a microbiotic crust in Arches National Monument, Utah, seem to reflect the larger pinnacles of an arid landscape. These crusts consist of algae, cyanobacteria, fungi, and other organisms living together in a mutualistic relationship. The inset shows a scanning electron micrograph of cyanobacteria filaments that make up the backbone of many crusts. Microbiotic crusts typically cover the soil surface in the unvegetated patches between clumps of desert shrubs and grasses. The crusts provide considerable protection against erosion by wind and water. They also help conserve and cycle nutrients, add nitrogen, enhance water supplies, and improve desert productivity. However, the fragile crusts can be easily destroyed by wheels, feet, and hooves. [Large photo courtesy of Ben Waterman; inset courtesy of Jayne Belnap (U.S. Geologic Service, Moab, Utah) and John Gardner (Brigham Young University, Provo, Utah)]

icant suppression of plant diseases can occur. Microbial ecologists are attempting to harness some of these competitive mechanisms to favor desirable microorganisms in soils (Box 10.3).

Effects of Management Practices on Soil Organisms

Practical advice on soil biology and soil management: http://www.extension.umn.edu/distribution/cropsystems/DC7403.html

Monocultures or even common crop rotations greatly reduce the number of plant species and so provide a much narrower range of plant materials and rhizosphere environments than nature provides in forests or grasslands.

While agricultural practices have different effects on different organisms, a few generalizations can be made (Table 10.8). For example, some agricultural practices (e.g., extensive tillage and monoculture) generally reduce the diversity of soil organisms as well as the number of individuals. However, monoculture may *increase* the population of a few species.

Adding lime and fertilizers (either organic or inorganic) to an infertile soil generally will increase microbial and fauna activity, largely due to the increase in the plant biomass that is likely to be returned to the soil as roots, root exudates, and shoot residues. Tillage, on the other hand, is a drastic disturbance of the soil ecosystem. Tillage tends to increase the role of bacteria at the expense of the fungi and usually decreases overall organism numbers as organic matter is depleted (Table 10.9). Addition of animal manure or compost stimulates higher microbial and faunal (especially earthworm) activity.

Pesticides are highly variable in their effects on soil ecology. Soil fumigants and nematicides can sharply reduce organism numbers, especially for fauna, at least on a temporary basis. On the other hand, application of a particular pesticide often stimulates the population of a specific microorganism because the predators of that organism have been killed. Figure 10.26 illustrates how a practice that affects one group of organisms will likely affect other groups as well and will eventually impact the productivity and functioning of the whole soil ecosystem. It is wise to remember that the interrelationships among soil organisms are intricate and the effects of any perturbation of the system are difficult to predict.

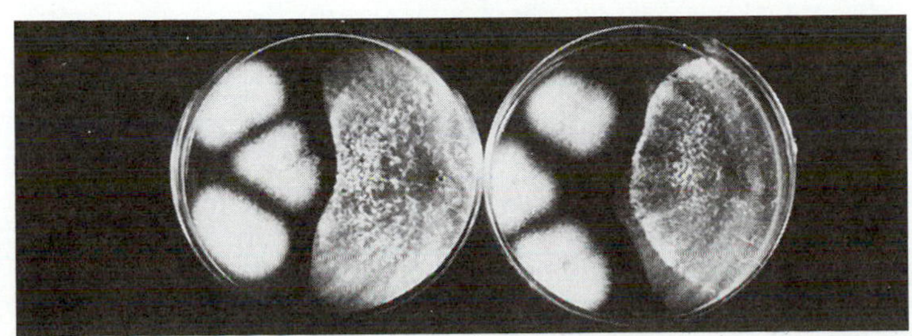

FIGURE 10.24 Fungal antagonism. The three smaller colonies in each petri dish are exuding an antibiotic compound that prevents the larger fungal colony from spreading over their half of the plate. Note the clear zone around each of the antibiotic-producing colonies. (Photo courtesy of American Society of Agronomy)

BOX 10.3 CHOOSING SIDES IN THE WAR AMONG THE MICROBES

Recent advances in our understanding of microbial genetics, combined with new techniques for transferring genetic material among different species of microorganisms, have opened many interesting and promising opportunities for scientists to join in the microbial wars in the soil. Here we summarize but two examples of innovative approaches scientists are taking to harness the antagonisms long observed among microorganisms.

Biological-control seed treatments are now commercially available. The seed treatments use antagonistic soil bacteria instead of chemical fungicide to protect seed from rot-causing soil pathogens. Strains of *Pseudomonas fluorescens* have been specially selected and enhanced to inhabit rhizosphere soil and help protect plant roots from soilborne diseases. When a preparation of these bacteria is applied in the furrow as wheat seeds are sown, the seeds germinate safely, even in soil known to be infested with seed-rotting fungi. Seeds sown without the treatment decay before they can germinate (Figure 10.25).

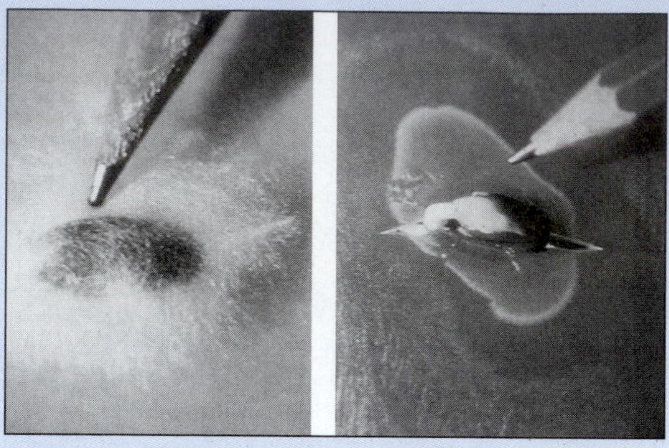

FIGURE 10.25 *(Right) The wheat seeds protected with antagonistic bacteria remain healthy, while unprotected seed (left) becomes heavily infected with pathogenic fungi. (Photo courtesy of Ecogen, Inc.)*

Scientists have found strains of rhizobia and bradyrhizobia bacteria that have superior nitrogen-fixing ability on legume crop roots. However, to take advantage of the improved strains it is necessary to give them a competitive edge over the many native strains already in most soils. These native strains compete strongly for a place on the crop roots, but are relatively inefficient at nitrogen fixation. USDA scientists in Wisconsin [see Breil et al. (1996)] discovered that the strain *Rhizobium leguminosarum* bv. *trifolii* produced a substance they called *trifolitoxin* that inhibited other rhizobia strains but did not inhibit the strain that produced it. The scientists were able to transfer the gene for trifolitoxin production to other strains of rhizobia. Their hope is that, armed with these genes, improved rhizobia strains that are highly efficient in nitrogen fixation will be able to outcompete the less efficient indigenous strains for nodule formation sites on legume roots. If widely successful, this approach should allow farmers to use less nitrogen fertilizer in their crop rotations.

TABLE 10.8 The Generalized Effects of Major Soil-Management Practices on the Overall Diversity and Abundance of Soil Organisms

Note that the practices that tend to enhance biological diversity and activity in soils are also those associated with efforts to make agricultural systems more sustainable.

Decreases biodiversity and populations	*Increases biodiversity and populations*
Fumigants	Balanced fertilizer use
Nematicides	Lime on acid soils
Some insecticides	Proper irrigation
Compaction	Improved drainage and aeration
Soil erosion	Animal manures and composts
Industrial wastes and heavy metals	Domestic (clean) sewage sludge
Moldboard plow-harrow tillage	Reduced or zero tillage
Monocropping	Crop rotations
Row crops	Grass-legume pastures
Bare fallows	Cover crops or mulch fallows
Residue burning or removal	Residue return to soil surface
Plastic mulches	Organic mulches

10.16 GENETICALLY ENGINEERED MICROORGANISMS

Advances in molecular biology have allowed scientists to identify the genes or pieces of genetic material (DNA or RNA) responsible for a particular trait in an organism. Scientists have also developed techniques for transferring specific genetic material to another not-necessarily-related organism. These technologies enable the engineering of new genetic combinations—**recombinant organisms**—not seen in nature.

TABLE 10.9 Tillage Systems Effects on Biomass Carbon (kg/ha) of Microbial and Faunal Groups in Soil

Researchers in Georgia grew grain sorghum in summer and a rye cover crop each winter on a sandy loam (Kanhapludults). With plow tillage, plant residues were mixed into the soil, but with no-tillage residues were left as surface mulch. Decomposer, microphytic feeder, and detritivore functions were dominated by larger organisms (fungi, microarthropods, and earthworms) in the no-till system, while smaller organisms (bacteria, protozoa, nematodes, and enchytraeid worms) were more prominent in the plowed system.

				Nematodes[b]				
Tillage	*Fungi*[a]	*Bacteria*[a]	*Protozoa*[b]	*Fungivores*	*Bacterivores*	*Microarthropods*[a]	*Enchytraeids*[c]	*Earthworms*[c]
No-till	360	260	24	0.14	0.82	1.31	5.55	60
Plowed	240	270	39	0.47	1.27	0.49	4.79	21

[a]0 to 5 cm depth.
[b]0 to 21 cm depth.
[c]0 to 15 cm depth.
Calculated from data of various sources presented by Beare (1997).

Potential Benefits. There are many potential uses for **genetically engineered microorganisms (GEMs)** in the soil environment. Already, soils contaminated by chemical spills are being cleaned by soil bacteria that have been endowed with an enhanced ability to degrade certain organic pollutants. Other GEMs produce toxins that kill plant pathogens or pests. An example is the transfer of genes from the soil bacterium *Bacillus thuringiensis* (**Bt**) to a strain of the bacterium *Pseudomonas fluorescens* that is adapted to life in the rhizosphere of corn plants. The Bt producing rhizobacteria can then protect the corn from the corn rootworm, without making any genetic modification to the corn plant itself. Other potential uses of GEMs in soil are increased nitrogen fixation in roots or rhizospheres, improved mycorrhizal efficiency, enhanced antagonism against plant pathogens, and the production of life-saving drugs.

Environmental and Safety Concerns. The use of GEMs in soils brings the risk of undesirable side effects, which are difficult to predict. As with any alteration to the soil ecosystem, a change in one component will lead to changes in many other components. In

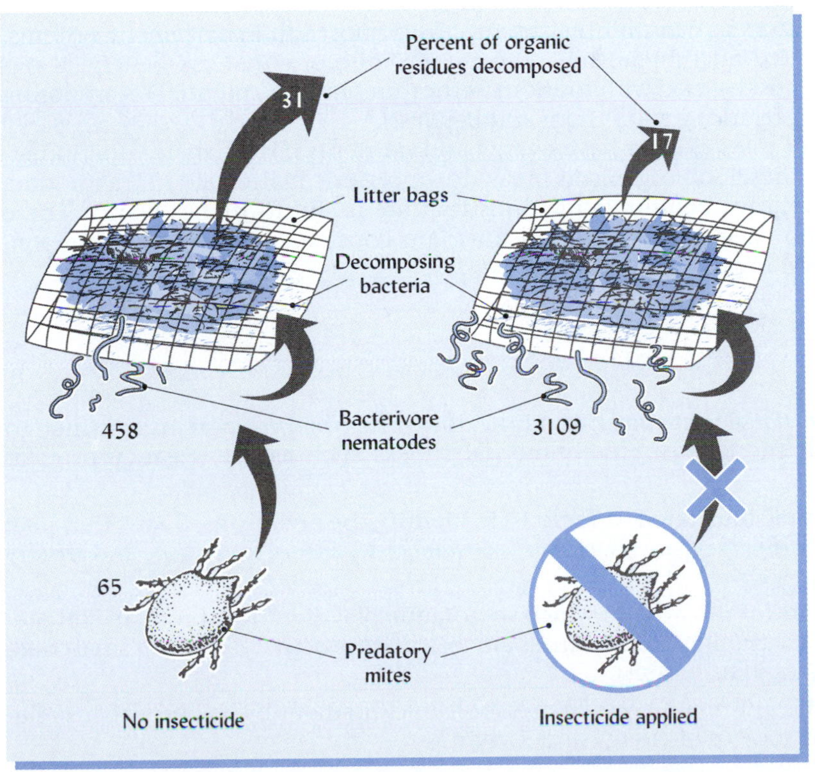

FIGURE 10.26 The indirect effects of insecticide treatment on the decomposition of creosote bush litter in desert ecosystems. Litter bags filled with creosote bush leaves and twigs were buried in desert soils in Arizona, Nevada, and California, either with or without an insecticide (chlordane) treatment. The insecticide killed virtually all the insects and mites. Without predatory mites to hold them in check, bacterivore nematodes multiplied rapidly and devoured a large portion of the bacterial colonies responsible for litter decomposition and nutrient cycling. Thus the insecticide reduced the rate of litter decomposition nearly in half, not by any direct effect on the bacteria, but by the indirect effect of killing the predators of their predators. [Data calculated from Whitford et al. (1982)]

Leakage of transgenic proteins into soil—a balanced overview:
http://www.colostate.edu/programs/lifesciences/TransgenicCrops/soilleak.html

addition, nontarget organisms may be killed. Some GEMs are given so-called suicide genes that cause the organism to die under certain conditions when its job is done. However, even if the GEM itself quickly dies out, recent studies (see Section 8.12) suggest that its DNA and/or the toxic protein may remain in the soil in an active, but difficult to detect, form. Empowering a microbe (or plant) to produce an insecticidal toxin is essentially a novel method for delivery of a single chemical approach and carries with it all the dangers of inducing pest resistance and nontarget toxicity that are inherent in the pesticide approach to pest control.

Of broader concern is the possibility that the new genetic material might be transferred (by virus infection or by transfer between bacteria of pieces of chromosome-free DNA) to indigenous organisms well adapted to survival and dispersal in the soil, possibly increasing the virulence of existing pathogens.

Finally, genetic engineers use antibiotic-resistance genes as markers to help identify and track the genetic material. Resistance to antibiotics might be transferred from the introduced GEM to another microbe possibly even to human pathogens, rendering our life-saving drugs useless. Such concerns continue to require extensive testing and caution in releasing GEMs into the soil environment.

10.17 CONCLUSION

The soil is a complex ecosystem with a diverse community of organisms. Soil organisms are vital to the cycle of life on earth. They incorporate plant and animal residues into the soil and digest them, returning carbon dioxide to the atmosphere, where it can be recycled through higher plants. Simultaneously, they create humus, the organic constituent so vital to good physical and chemical soil conditions. During digestion of organic substrates, they release essential plant nutrients in inorganic forms that can be absorbed by plant roots or be leached from the soil. They also mediate the redox reactions that influence soil color and nutrient cycling in wetlands.

Animals, particularly earthworms, mechanically incorporate residues into the soil and leave open channels through which water and air can flow. Microorganisms such as fungi, actinomycetes, and bacteria are responsible for most organic decay, although their activity is greatly influenced by the soil fauna. Certain of the microorganisms form symbiotic associations with higher plants, playing special roles in plant nutrition and nutrient cycling. Competition among soil microbes, and between these organisms and higher plants, for mineral nutrients can result in plant nutrient deficiencies. Microbial requirements are factors in determining the success of most soil management systems. Genetically engineered microorganisms offer the promise of enhanced beneficial microbial action, but must be used with caution, as they present the potential for troubling unintended side effects in the soil system and beyond.

The organisms of the soil must have energy and nutrients if they are to function efficiently. To obtain these, soil organisms break down organic matter, aid in the production of humus, and leave behind compounds that are useful to higher plants. These decay processes and their practical significance are considered in the next chapter, which deals with soil organic matter.

STUDY QUESTIONS

1. What is *functional redundancy,* and how does it help soil ecosystems continue to function in the face of environmental shocks such as fire, clear-cutting, or tillage?
2. In the example illustrated in Table 10.9, identify the organisms, if any, that play the roles of *primary producers, primary consumers, secondary consumers,* and *tertiary consumers.*
3. Describe some of the ways in which mesofauna play significant roles in soil metabolism even though their biomass and respiratory activity is only a small fraction of the total in the soil.
4. What are the four main types of metabolism carried out by soil organisms relative to their sources of energy and carbon?

5. What role does O_2 play in aerobic metabolism? What elements take its place under anaerobic conditions?
6. A *mycorrhiza* is said to be a symbiotic association. What are the two parties in this symbiosis, and what are the benefits derived by each party?
7. In what ways is soil improved as a result of earthworm activity? Are there possible detrimental effects as well?
8. What is the *rhizosphere,* and in what ways does the soil in the rhizosphere differ from the rest of the soil?
9. Explain and compare the effects of tillage and manure application on the abundance and diversity of soil organisms.
10. What are some of the potential benefits and potential risks involved with the release of *genetically engineered organisms (GEMs)* into the soil?

REFERENCES

Beare, M. H. 1997. "Fungal and bacterial pathways of organic matter decomposition and nitrogen mineralization in arable soils," in L. Brussaard and R. Ferrera-Cerrato (eds.), *Soil Ecology in Sustainable Agricultural Systems.* (Boca Raton, Fla: Lewis Publishers).

Belnap, J., J. Kaltenecker, R. Rosentreter, J. Williams, S. Leonard, and D. Eldridge. 2001. *Biological Soil Crusts: Ecology and Management.* U.S. Department of the Interior, Bureau of Land Management, Report no. BLM/ID/ST-01/001+ 1730. PDF file available at *www.soilcrust.org/advbody.htm.*

Bethlenfalvay, G. J., and R. G. Linderman (eds.). 1992. *Mycorrhizae in Sustainable Agriculture.* ASA Special Publication no. 54. (Madison, Wis.: Amer. Soc. of Agron.).

Biswas, J. C., J. K. Ladha, and F. B. Dazzo. 2000. "Rhizobia inoculation improves nutrient uptake and growth of lowland rice," *Soil Sci. Soc. Amer. J.* **64**:1644–1650.

Bowen, G. D., and A. D. Rovira. 1999. "The rhizosphere and its management to improve plant growth," *Advances in Agronomy* **66**:1–102.

Breil, B. T., J. Borneman, and E. W. Triplett. 1996. "A newly discovered gene, tfuA, involved in the production of the ribosomally synthesized peptide antibiotic trifolitoxin," *J. Bacteriol.* **178**:4150–4156.

Chino, M. 1976. "Electron microprobe analysis of zinc and other elements within and around rice root growth in flooded soils," *Soil Sci. and Plant Nut. J.* **22**:449.

Coleman, D. C., and D. A. Crossley, Jr. 1996. *Fundamentals of Soil Ecology.* (San Diego, Calif.: Academic Press, Inc.).

Coyne, M. 1999. *Soil Microbiology: An Exploratory Approach.* (Albany, NY: Delmar Publishers), 462 pp.

de Vleeschauwer, D., and R. Lal. 1981. "Properties of worm casts under secondary tropical forest regrowth," *Soil Sci.* **132**:175–181.

Edwards, C. (ed.). 1998. *Earthworm Ecology.* (Delray Beach, Fla: St. Lucie Press).

Hendrix, P. F. (ed.). 1995. *Earthworm Ecology and Biogeography in North America.* (Boca Raton, Fla.: Lewis Publishers).

Killham, K. 1994. *Soil Ecology.* (New York: Cambridge University Press).

Kremer, R. J. 1998. "Microbial interaction with weed seeds and seedlings and its potential for weed management," in J. L. Hatfield, D. D. Buhler, and B. A. Stewart (eds.), *Integrated Weed and Soil Management* (Chelsea, Mich.: Ann Arbor Press), pp. 161–179.

Lambert, D. H., D. E. Baker, and H. Cole, Jr. 1979. "The role of Mycorrhizae in the interactions of phosphorus with zinc, copper, and other elements," *Soil Sci. Soc. Amer. J.* **43**:976–980.

Lavelle, P. 1997. "Faunal activities and soil processes: Adaptive strategies that determine ecosystem function," *Advances in Ecological Research* **27**:93–132.

Lee, K. E. 1985. *Earthworms, Their Ecology and Relationships with Soils and Land Use.* (New York: Academic Press).

Menge, J. A. 1981. "Mycorrhizae agriculture technologies," in *Background Papers for Innovative Biological Technologies for Lesser Developed Countries,* Paper no. 9., Office of Technology Assessment Workshop, Nov. 24–25, 1980. (Washington, D.C.: U.S. Government Printing Office), pp. 383–424.

Mooreman, T. B., and S. S. Harper. 1989. "Transformation and mineralization of Metribuzin in surface and subsurface horizons of a Mississippi Delta soil," *J. Environ. Qual.* **18**:302–306.

Nehl, D. B., S. J. Allen, and J. F. Brown. 1996. "Deleterious rhizobacteria: An integrating perspective," *Applied Soil Ecology* **5**:1–20.

Paul, E. A., and F. E. Clark. 1996. *Soil Microbiology and Biochemistry,* 2nd ed. (New York: Academic Press).

Read, D. J., D. Lewis, A. Fitter, and I. Alexander. 1992. *Mycorrhizas in Ecosystems.* (Wallingford, U.K.: CAB International).

Reynolds, L. B., J. W. Potter, and B. R. Ball-Coelho. 2000. "Crop rotation with *Tagetes* sp. is an alternative to chemical fumigation for control of root-lesion nematodes," *Agronomy J.* **92**:957–966.

Rovira, A. D., R. C. Foster, and J. K. Martin. 1979. "Origin, nature and nomenclature of the organic materials in the rhizosphere," in J. L. Harley and R. S. Russell (eds.), *The Soil–Root Interface.* (New York: Academic Press).

Scullen, J., and A. Malik. 2000. "Earthworm activity affecting organic matter, aggregation and microbial activity in soils restored after opencast mining for coal," *Soil Biology and Biochemistry* **32**:119–126.

Sylvia, D., J. Fuhrmann, P. Hartel, and D. Zuberer. 1997. *Principles and Applications of Soil Microbiology.* (Upper Saddle River, N.J.: Prentice Hall).

Tugel, A. J., and A. M. Lewandowski (eds.). 1999. *Soil Biology Primer.* (Ames, Iowa: Natural Resource Conservation Service Soil Quality Institute), *www.statlab.iastate.edu/survey/SQI/SoilBiologyPrimer/index.htm.*

Weil, R. R., and W. Kroontje. 1979. "Organic matter decomposition in a soil heavily amended with poultry manure," *J. Environ. Qual.* **8**:584–588.

Whitford, W. G., D. W. Freckman, P. F. Santos, N. Z. Elkins, and L. W. Parker. 1982. "The role of nematodes in decomposition in desert ecosystems," in Diana Freckman (ed.), *Nematodes in Soil Ecosystems.* (Austin, Tex.: University of Texas Press), pp. 98–116.

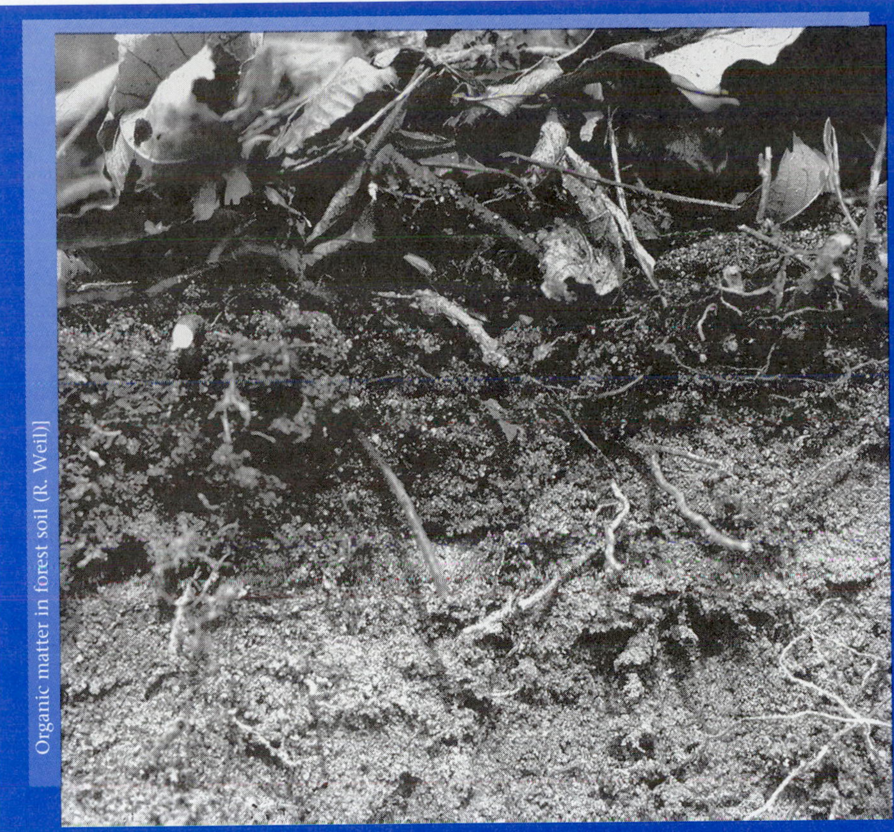

Organic matter in forest soil (R. Weil)]

11
Soil Organic Matter

The earth lay rich and dark and fell apart lightly under the points of their hoes.
—*P. S. Buck,* The Good Earth

The element *carbon* is the foundation of all life. From cellulose to chlorophyll, the compounds that comprise living tissues are made of carbon atoms arranged in chains or rings and associated with many other elements. The cycle of carbon on earth is the story of life on this planet.

All organic substances, by definition, contain carbon. Organic matter in the world's soils contains 3 to 4 times as much carbon as is found in all the world's living vegetation. Soil organic matter,[1] therefore, plays a critical role in the global carbon balance that is thought to be the major factor affecting global warming, or the **greenhouse effect.** Although organic matter comprises only a small fraction of the total mass of most soils, this dynamic soil component exerts a dominant influence on many soil physical, chemical, and biological properties.

Soil organic matter is a complex and varied mixture of organic substances. It provides much of the cation exchange and water-holding capacities of surface soils. Certain components of soil organic matter are largely responsible for the formation and stabilization of soil aggregates. Soil organic matter also contains large quantities of plant nutrients and acts as a slow-release nutrient storehouse, especially for nitrogen. Furthermore, organic matter supplies energy and body-building constituents for most of the microorganisms whose general activities were discussed in Chapter 10. In addition to enhancing plant growth through the just-mentioned effects, certain organic compounds found in soils have direct growth-stimulating effects on plants. For all these reasons, the quantity and quality of soil organic matter is a central factor in determining **soil quality.**

We will first examine the role of soil organic matter in the **global carbon cycle** and the process of decomposition of organic residues. Next, we will focus on inputs and losses with regard to soil carbon in specific ecosystems. Finally, we will study the processes and consequences involved in soil organic matter management.

[1] For a detailed review of organic matter in soils see Tate (1987). For a practical review of organic matter management in agricultural soils, see Magdoff and Van Es (2000).

11.1 THE GLOBAL CARBON CYCLE

Pathways

The C cycle simplified and made fun:
http://library.thinkquest.org/11226/why.htm

The basic pathways involved in the global carbon cycle are shown in Figure 11.1. Plants take in carbon dioxide from the atmosphere. Then, through the process of photosynthesis, the energy of sunlight is trapped in the carbon-to-carbon bonds of organic molecules (such as those described in Section 11.2). Some of these organic molecules are used as a source of energy (via respiration) by the plants themselves (especially by the plant roots), with the carbon being returned to the atmosphere as carbon dioxide. The remaining organic materials are stored temporarily as constituents of the standing vegetation, most of which is eventually added to the soil as plant litter (including crop residues) or root deposition (see Section 10.7). Some plant material may be eaten by animals (including humans), in which case about half of the carbon eaten is exhaled into the atmosphere as carbon dioxide. The carbon not returned to the atmosphere is eventually returned to the soil as bodily wastes or body tissues. Once deposited on or in the soil, these plant or animal tissues are metabolized (digested) by soil organisms, which gradually return this carbon to the atmosphere as carbon dioxide.

Much smaller amounts of carbon dioxide react in the soil to produce carbonic acid (H_2CO_3) and (bi)carbonates. The bicarbonates are readily soluble and may be removed in drainage. Eventually, much of the carbon in the carbonates and bicarbonates is also returned to the atmosphere as carbon dioxide.

Microbial metabolism and interaction with clay in the soil produces some organic materials of such stability that decades or even centuries may pass before the carbon in them is returned to the atmosphere as carbon dioxide. Such resistance to decay allows organic matter to accumulate in soils.

Carbon Sources

Globally, at any one time, approximately 2400 petagrams (Pg, 10^{15}) of carbon are stored in soil profiles as soil organic matter (including O horizons), about one-third of that at depths below 1 m. An additional 700 Pg are stored as soil carbonates, which can release CO_2 upon weathering. Altogether, about twice as much carbon is stored in the soil than

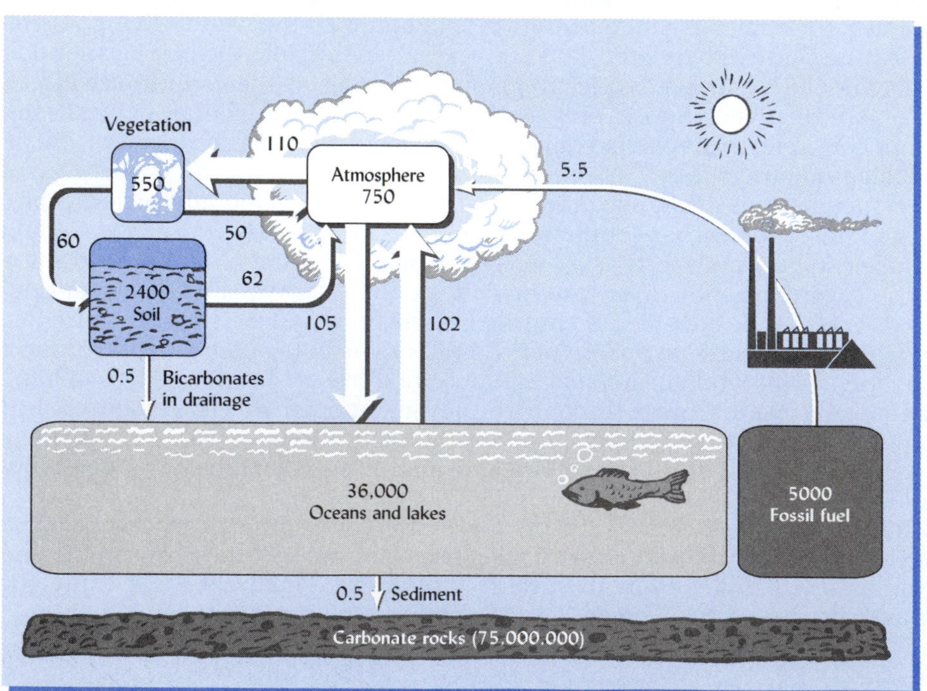

FIGURE 11.1 A simplified representation of the global carbon cycle emphasizing those pools of carbon which interact with the atmosphere. The numbers in boxes indicate the petagrams (Pg = 10^{15}/g) of carbon stored in the major pools. The numbers by the arrows show the amount of carbon annually flowing (Pg/yr) by various pathways between the pools. Note that the soil contains almost twice as much carbon as the vegetation and the atmosphere combined. Imbalances caused by human activities can be seen in the flow of carbon to the atmosphere from fossil fuel burning (5.5) and in the fact that more carbon is leaving (62 + 0.5) than entering (60) the soil. These imbalances are only partially offset by increased absorption of carbon by the oceans. The end result is that a total of 219.5 Pg/yr enters the atmosphere while only 215 Pg/yr of carbon is removed. It is easy to see why carbon dioxide levels in the atmosphere are rising. [Data from several sources; soil C estimate from Batjes (1996)]

TABLE 11.1 Mass of Organic Carbon in the World's Soils

Values for the upper 1 m represent most of the carbon in the soil profile. The upper 15 cm is most readily influenced by land use and soil management.

		Organic carbon[a] in upper 15 cm			Organic carbon[a] in upper 100 cm	
Soil order	*Global area, 10^3 km^2*	*Mg/ha*	*Global Pg[b]*	*% of global*	*Range,[c] %*	*Typical,[c] %*
Entisols	14,921	99	148	9	0.06–6.0	—[d]
Inceptisols[e]	21,580	163	352	22	0.06–6.0	—[d]
Histosols[e]	1,745	2,045	357	23	12–57	47
Andisols	2,552	306	78	5	1.2–10	6
Vertisols	3,287	58	19	1	0.5–1.8	0.9
Aridisols	31,743	35	110	7	0.1–1.0	0.6
Mollisols	5,480	131	73	5	0.9–4.0	2.4
Spodosols	4,878	146	71	5	1.5–5.0	2.0
Alfisols	18,283	69	127	8	0.5–3.8	1.4
Ultisols	11,330	93	105	7	0.9–3.3	1.4
Oxisols	11,772	101	119	8	0.9–3.0	2.0
Misc. land	7,644	24	18	1	—	—
Total	135,215		1576	100		

[a] Values do not include surface litter horizons. Organic matter may be roughly estimated as 1.7 to 2.0 times this value. The value traditionally used is 1.72. Organic nitrogen may also be estimated from organic carbon by dividing by 12 for most soils, but see Section 11.3.
[b] Petagram = 10^{15} g = 1 billion metric tons.
[c] Percent on mass basis (i.e., g/100 g).
[d] These soils are too variable to suggest a typical value.
[e] Carbon stored in Gelisols is included with these soils.
Data calculated from Eswaran et al. (1993) and Brady (1990).

Introductory Carbon Balance Model attempts to show the world's soil carbon dynamics. Download it and try it out:
http://www.mv.slu.se/vaxtnaring/olle/ICBM.html

in the world's vegetation and atmosphere combined (see Figure 11.1). Of course, this carbon is not equally distributed among all types of soils (Table 11.1). About 45% of the total is contained in soils of just three orders, Histosols, Inceptisols, and Gelisols. Histosols (and Histels in the order Gellisols) are of limited extent but contain very large amounts of organic matter per unit land area. Inceptisols (and nonhistic Gellisols) contain only moderate concentrations of carbon, but cover vast areas of the globe. The reasons for the varying amounts of organic carbon in different soils will be detailed in Section 11.10.

Figure 11.1 shows that, globally, the release of carbon from soils into the atmosphere is about 62 Pg/yr, while only about 60 Pg/yr enter the soils from the atmosphere via plant residues. This imbalance of about 2 Pg/yr, along with about 5 Pg/yr of carbon released by the burning of fossil fuels (in which carbon was sequestered from the atmosphere millions of years ago) is only partially offset by increased absorption of atmospheric carbon dioxide by the ocean. Fossil fuel burning and degrading land-use practices have increased the concentration of carbon dioxide in the atmosphere at an accelerating rate since the beginning of the Industrial Revolution, some 400 years ago. The levels have increased from 290 to 370 ppm during the past century alone. The implications of carbon dioxide imbalances and of other gaseous emissions on the greenhouse effect will be discussed in Section 11.11, after we consider the processes involved in the carbon cycle.

11.2 THE PROCESS OF DECOMPOSITION IN SOILS

Composition of Plant Residues

Genetic diversity associated with chitin degradation in soil:
http://www.nmw.ac.uk/soilbio/Download/newsletter5.PDF

Plant residues are the principal material undergoing decomposition in soils and, hence, are the primary source of soil organic matter. Green plant tissues contain from 60 to 90% water by weight (Figure 11.2). If plant tissues are dried to remove all water, the *dry matter* remaining consists mostly (at least 90 to 95%) of carbon, oxygen, and hydrogen. During photosynthesis plants obtain these elements from carbon dioxide and water. If plant dry matter is burned (oxidized), these elements become carbon dioxide and water once more. Of course, some ash and smoke will also be formed upon burning, accounting for the remaining 5 to 10% of the dry matter. In the ash and smoke can be found the many nutrient elements originally taken up by the plants from the soil. The

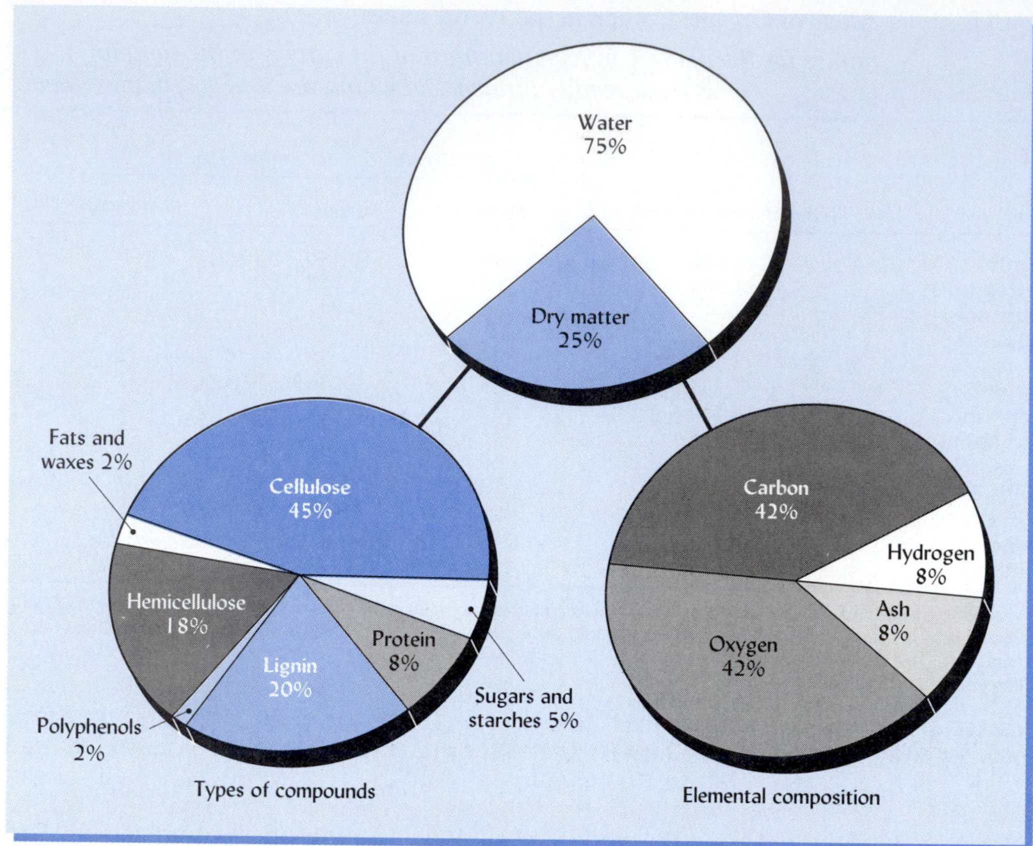

FIGURE 11.2 Typical composition of representative green-plant materials. The major types of organic compounds are indicated at left and the elemental composition at right. The *ash* is considered to include all the constituent elements other than carbon, oxygen, and hydrogen (nitrogen, sulfur, calcium, etc.).

essential nutrient elements in the ash, such as nitrogen, sulfur, phosphorus, potassium, and micronutrients, will be given detailed consideration in later chapters.

Organic Compounds in Plant Residues. The organic compounds in plant tissue can be grouped into broad classes. Although representative percentages of these classes are shown in Figure 11.2, tissues from different plant species, as well as from different parts (leaves, roots, stems, etc.) of a given plant, differ considerably in their makeup. Carbohydrates, which range in complexity from simple sugars and starches to cellulose, are usually the most plentiful of plant organic compounds.

Lignins, which are complex compounds with multiple ring-type or *phenol* structures, are components of plant cell walls. The content of lignin increases as plants mature and is especially high in woody tissues. Other **polyphenols,** such as tannins, may comprise as much as 6 or 7% of the leaves and bark of certain plants (for example, the brown color of steeped tea is due to tannins). Certain plant parts, especially seed and leaf coatings, contain significant amounts of fats, waxes, and oils, which are more complex than carbohydrates but less so than lignins. Proteins contain about 16% nitrogen and smaller amounts of other essential elements, such as sulfur, manganese, copper, and iron.

Rate of Decomposition. Organic compounds may be listed in terms of ease of decomposition as follows:

1. Sugars, starches, and simple proteins — Rapid decomposition ↑
2. Crude proteins
3. Hemicellulose
4. Cellulose
5. Fats, waxes, and so forth
6. Lignins and phenolic compounds — ↓ Very slow decomposition

Decomposition of Organic Compounds in Aerobic Soils

When organic tissue is added to an aerobic soil, three general reactions take place. Carbon compounds are enzymatically oxidized to produce carbon dioxide, water, energy, and decomposer biomass. The essential nutrient elements, such as nitrogen, phosphorus, and sulfur, are released and/or immobilized by a series of specific reactions that are relatively unique for each element. Compounds very resistant to microbial action are formed, either through modification of compounds in the original tissue or by microbial synthesis.

Decomposition: An Oxidation Process. In a well-aerated soil, all of the organic compounds found in plant residues are subject to oxidation.

$$\underset{\text{Carbon- and hydrogen-containing compounds}}{R\text{—}(C, 4H)} + 2O_2 \xrightarrow[\text{oxidation}]{\text{Enzymatic}} CO_2\uparrow + 2H_2O + \text{energy } (478 \text{ kJ mol}^{-1} \text{ C}) \quad (11.1)$$

Many intermediate steps are involved in this overall reaction, and it is accompanied by important side reactions that involve elements other than carbon and hydrogen. Even so, this basic reaction accounts for most of the organic matter decomposition in the soil, as well as for the oxygen consumption and CO_2 release.

Breakdown of Proteins. When plant proteins decay, they yield not only carbon dioxide and water, but also amino acids such as glycine (CH_2NH_2COOH) and cysteine ($CH_2HSCHNH_2COOH$). In turn, these nitrogen and sulfur compounds further break down, eventually yielding such simple inorganic ions as ammonium (NH_4^+), nitrate (NO_3^-), and sulfate (SO_4^{2-}) forms available for plant nutrition.

Breakdown of Lignin. Lignin molecules are very large and complex, consisting of hundreds of interlinked phenolic ring subunits. Because the linkages among these structures are so varied and strong, only a few microorganisms (mainly *white rot fungi*) can break them down. Decomposition proceeds very slowly at first, and is generally assisted by the physical activities of soil fauna. Once the lignin subunits are separated, many types of microorganisms participate in their breakdown. It is thought that microorganisms use some of the ring structures from lignin in the synthesis of stable soil organic matter.

Example of Organic Decay

The process of organic decay in time sequence is illustrated in Figure 11.3. Assume the soil has not been disturbed or amended with plant residues for some time. Initially, no readily decomposable materials are present. The principal microorganisms actively metabolizing are small populations of **autochthonous organisms**, which survive by slowly and steadily digesting the very resistant, stable soil organic matter. Competition for food is severe and microbial activity is relatively low, as reflected in the low **soil respiration** rate or level of CO_2 production in the soil. The supply of soil carbon is slowly but steadily being depleted.

Then, suddenly, an abundance of fresh, decomposable tissue is added to the soil. Maybe deciduous trees are losing their leaves in fall, or a farmer has plowed in the residue of a harvested crop. In any case, the appearance of easily decomposable and often water-soluble compounds, such as sugars, starches, and amino acids, stimulates an almost immediate increase in metabolic activity among the soil microbes. Soon the slower-acting autochthonous populations are overtaken by rapidly multiplying populations of **zymogenous (opportunist) organisms** that have been awakened from their dormant state by the presence of new food supplies. Cellulose-digesting organisms rapidly join in the attack. Microbial numbers and carbon dioxide evolution from microbial respiration both increase exponentially in response to the new food resource.

Soon microbial activity is at its peak, energy is being rapidly liberated, and carbon dioxide is being formed in large quantities. As they multiply and increase their biomass, the microbes are also synthesizing new organic compounds. The microbial biomass at this point may account for as much as one-sixth of the organic matter in a soil. The intense microbial activity may even stimulate the breakdown of some resistant soil organic matter, a phenomenon known as the **priming effect.**

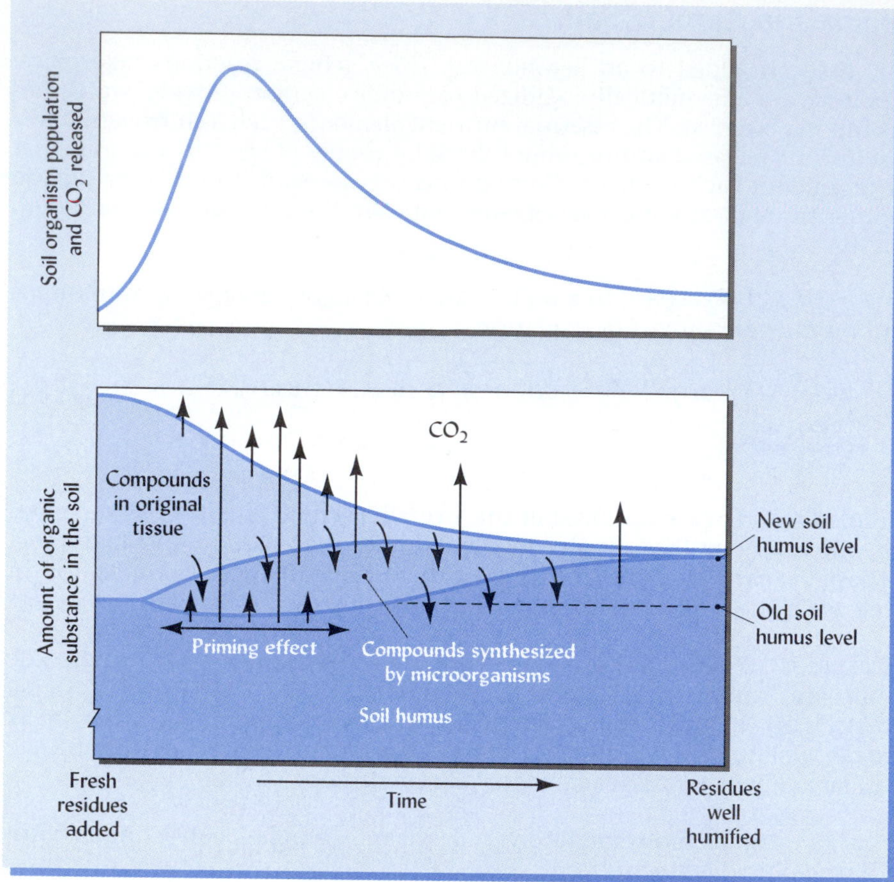

FIGURE 11.3 Diagram of the general changes that take place when fresh plant residues are added to a soil. The arrows indicate transfers of carbon among compartments. The time required for the process will depend on the nature of the residues and the soil. Most of the carbon released during the initial rapid breakdown of the residues is converted to carbon dioxide, but the smaller amounts of carbon converted into microbially synthesized compounds (biomass) and, eventually, into soil humus should not be overlooked. Although the peak of microbial activity appears to accelerate the decay of the original humus, a phenomenon known as the *priming effect*, the humus level is increased by the end of the process. Where vegetation, environment, and management remain stable for a long time, the soil humus content will reach an *equilibrium level* in which, the carbon added to the humus pool through the decomposition of plant residues each year is balanced by carbon lost through the decomposition of existing soil humus.

With all this frenetic microbial activity, the easily decomposed compounds are soon exhausted. While the cellulose and lignin decomposers continue their slow work, most of the zymogenous microorganisms begin to die of starvation. As microbial populations plummet, the dead cells provide a readily digestible food source for the survivors, which continue to evolve carbon dioxide and water. The decomposition of the dead microbial cells is also associated with the **mineralization** or release of simple inorganic products, such as nitrates and sulfates.

As food supplies are further reduced, microbial activity continues to decline, and the general-purpose zymogenous soil organisms again sink back into comparative quiescence. Little of the original residue material persists, mainly tiny particles that have become ***physically*** **protected** from decay by lodging inside soil pores too tight to allow access by most organisms. Some of the remaining carbon has been ***chemically*** **protected** by conversion into **soil humus**, the dark-colored, heterogeneous, mostly colloidal mixture of modified lignin and newly synthesized organic compounds that strongly resists further decay. Some of the fine humus is further protected by binding strongly to clay particles. Thus, a small percentage of the carbon in the added residues has been retained, increasing slightly the pool of stable soil organic matter. In a mature ecosystem, this increase will likely be offset during each annual cycle by slow authochthonous decomposition, resulting in little net change in the level of soil organic matter from year to year.

Decomposition in Anaerobic Soils

Microbial decomposition proceeds most rapidly in the presence of plentiful supplies of O_2, which acts as the electron acceptor during aerobic oxidation of organic compounds. Oxygen supplies may become depleted when soil pores filled with water prevent the diffusion of O_2 into the soil from the atmosphere. Without sufficient oxygen present, aerobic organisms cannot function, so anaerobic or facultative organisms become dominant. Under low-oxygen or anaerobic conditions, decomposition takes place much more slowly than when oxygen is plentiful. Hence, wet, anaerobic soils tend to accumulate large amounts of organic matter in a partially decomposed condition.

The products of anaerobic decomposition include a wide variety of partially oxidized organic compounds, such as organic acids, alcohols, and methane gas. Anaerobic decomposition releases relatively little energy for the organisms involved; therefore, the end products still contain much energy. (For this reason, alcohol and methane can serve as fuel.) Some of the products of anaerobic decomposition are of concern because they produce foul odors or inhibit plant growth. The methane gas produced in wet soils by **methanogenic bacteria** is a major contributor to the greenhouse effect (Section 11.11).

Production of Simple Inorganic Products

As proteins are attacked by microbes, the amide (—R—NH_2) and sulfide (—R—S) groups of the amino acids, are broken off to produce, first, ammonium (NH_4^+) and sulfide (S^{2-}) compounds and, finally, nitrates (NO_3^-) and sulfates (SO_4^{2-}). Similar decomposition of other organic compounds releases these and other inorganic nutrient ions. The term *mineralization* applies to the overall process that releases elements from organic compounds to produce inorganic (mineral) forms. Most of the inorganic ions released by mineralization are readily available to higher plants and to microorganisms. The decay of organic tissues is an important source of nitrogen, sulfur, phosphorus, and other essential elements for plants.

11.3 FACTORS CONTROLLING RATES OF DECOMPOSITION AND MINERALIZATION[2]

The time needed to complete the processes of decomposition and mineralization may range from days to years, depending mainly on two broad factors: (1) the environmental conditions in the soil, and (2) the quality of the added residues as a food source for soil organisms.

The environmental conditions conducive to rapid decomposition and mineralization include a near-neutral pH, sufficient soil moisture, and good aeration (about 60% of the soil pore space filled with water), and warm temperatures (25 to 35°C).

Physical Factors Influencing Residue Quality

The location of residues in or on the soil is a physical factor with a critical impact on decomposition rates. Surface placement of plant residues, as in forest litter or conservation tillage mulch, usually results in slower, more variable rates of decomposition than where similar residues are incorporated into the soil by root deposition, faunal action, or tillage. Surface residues are subject to drying, as well as extremes of temperature. Nutrient elements mineralized from surface-applied residues are also more susceptible to loss in runoff or by volatilization than are those from incorporated residues. Surface residues are physically out of reach for most soil organisms. Compared to surface residue, incorporated residues are in intimate contact with soil moisture and soil organisms, decompose more quickly, and may lose nutrients more easily by leaching.

Residue particle size is another important physical factor—the smaller the particles, the more rapid the decomposition. Small particle size may result from the nature of the residues (e.g., twigs versus branches), from mechanical treatment (grinding, chopping, tillage, etc.), or from the chewing action of soil fauna. Diminution of residues into smaller particles physically exposes more surface area to decomposition, and also breaks up lignacious cell walls and waxy outer coatings on leaves so as to expose the more readily decomposed tissues and cell contents.

[2]For an excellent collection of papers dealing with litter decomposition in soils, see Cadisch and Giller (1997). Many of our current ideas about decomposition were first put forward in a classic book by Swift et al. (1979).

Carbon/Nitrogen Ratio of Organic Materials and Soils

The carbon content of typical plant dry matter is about 42% (see Figure 11.2); that of soil organic matter ranges from 40 to 58%. In contrast, the nitrogen content of plant residues is much lower and varies widely (from <1% to >6%). The ratio of carbon to nitrogen C/N in organic residues applied to soils is important for two reasons: (1) Intense competition among microorganisms for available soil nitrogen occurs when residues having a high C/N ratio are added to soils, and (2) the C/N ratio in residues helps determine their rate of decay and the rate at which nitrogen is made available to plants.

C/N Ratio in Plants and Microbes. The C/N ratio in plant residues ranges from between 10:1 to 30:1 in legumes and young green leaves to as high as 600:1 in some kinds of sawdust (Table 11.2). Generally, as plants mature, the proportion of protein in their tissues declines, while the proportion of lignin and cellulose, and the C/N ratio, increase. These differences in composition have pronounced effects on the rate of decay when plant residues are added to the soil. In the bodies and cells of microorganisms, the C/N ratio is not only less variable than in plant tissues, but also much lower, ordinarily falling between 5:1 and 10:1.

C/N Ratio in Soils. The C/N ratio in the organic matter of arable (cultivated) surface (Ap) horizons commonly ranges from 8:1 to 15:1, the median being near 12:1. The ratio is generally lower for subsoils than for surface layers in a soil profile. In a given climatic region, little variation occurs in the C/N ratio for similarly managed soils. For instance, in calcium-rich soils of semiarid grasslands (e.g., Mollisols and tropical Alfisols), the C/N ratio is relatively narrow. In more severely leached and acidic A horizons in humid regions, the C/N is relatively wide; C/N ratios as high as 30:1 are not uncommon. Forest O horizons commonly have C/N ratios of 30 to 50. When such soils are brought under cultivation and limed to increase their pH and calcium content, the enhanced decomposition tends to lower the C/N ratio to near 12:1.

Influence of Carbon/Nitrogen Ratio on Decomposition

Soil microbes, like other organisms, require a balance of nutrients from which to build their cells and extract energy. Soil organisms need carbon for building essential organic compounds and to obtain energy for life processes. Organisms must also obtain sufficient nitrogen to synthesize nitrogen-containing cellular components, such as amino

TABLE 11.2 Typical Carbon and Nitrogen Contents and C/N Ratios of Some Organic Materials

Organic material	*% C*	*% N*	*C/N*
Spruce sawdust	50	0.05	600
Newspaper	39	0.3	120
Wheat straw	38	0.5	80
Corn stover	40	0.7	57
Sugarcane trash	40	0.8	50
Bluegrass from fertilized lawn	37	1.2	31
Rye cover crop, vegetative stage	40	1.5	26
Rotted barnyard manure	41	2.1	20
Broccoli residues	35	1.9	18
Finished household compost	30	2.0	15
Young alfalfa hay	40	3.0	13
Hairy vetch cover crop	40	3.5	11
Digested municipal sewage sludge	31	4.5	7
Soil microorganisms			
Bacteria	50	10.0	5
Actinomycetes, nematodes	50	8.5	6
Fungi	50	5.0	10
Soil organic matter			
Average forest O horizons	50	1.3	45
Average forest A horizons	50	2.8	20
Mollisol Ap horizon	56	4.9	11
Average B horizon	46	5.1	9

Data calculated from many sources.

acids, enzymes, and DNA. On the average, soil microbes must incorporate into their cells about eight parts of carbon for every one part of nitrogen (i.e., C/N ratio of 8:1). Because only about one-third of the carbon metabolized by microbes is incorporated into their cells (the remainder is respired and lost as CO_2), the microbes need to find about 1 g of N for every 24 g of C in their "food."

This requirement results in two extremely important practical consequences. First, if the C/N ratio of organic material added to soil exceeds about 25:1, soil microbes will have to scavenge the soil solution to obtain enough nitrogen, a process called nitrogen **immobilization.** Thus, the incorporation of high C/N residues will deplete the soil's supply of soluble nitrogen, causing higher plants to suffer from nitrogen deficiency. Second, the decay of organic materials can be delayed if sufficient nitrogen to support microbial growth is neither present in the material undergoing decomposition nor available in the soil solution.

Examples of Inorganic Nitrogen Release During Decay

The practical significance of the C/N ratio becomes apparent if we compare the changes that take place in the soil when residues of either high or low C/N ratio are added (Figure 11.4). Consider a soil with a moderate level of soluble nitrogen (mostly nitrates). General-purpose decay organisms are at a low level of activity in this soil, as evidenced by low carbon dioxide production. If no nitrogen were lost or taken up by plants, the level of nitrates would very slowly increase as the native soil organic matter decays.

Low Nitrogen Material. Now consider what happens when a large quantity of readily decomposable organic material is added to this soil. If this material has a C/N ratio greater than 25, changes will occur according to the pattern shown in Figure 11.4*a*. In the example shown, the initial C/N ratio of the residues is about 55, typical for cornstalks or many kinds of leaf litter. As soon as the residues contact the soil, the microbial community responds to the new food supply (see Section 11.2). Because of the microbial demand for nitrogen, little or no mineral nitrogen (NH_4^+ or NO_3^-) is available to higher plants during this period.

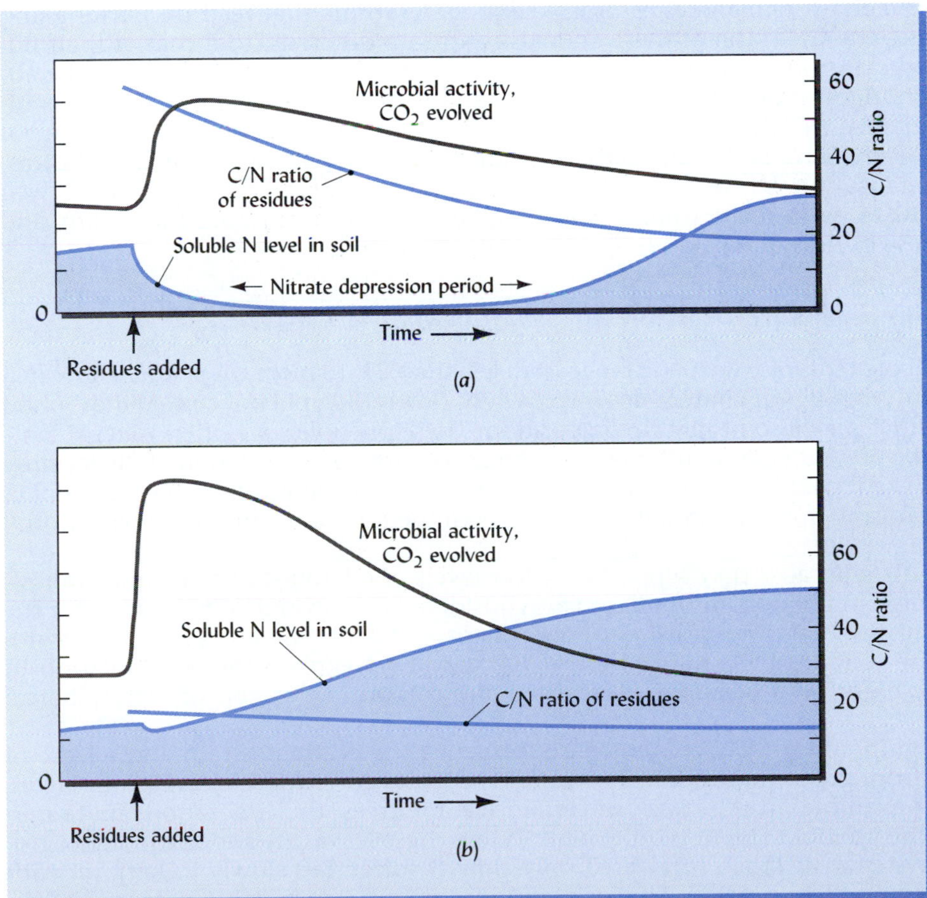

FIGURE 11.4 Changes in microbial activity, in soluble nitrogen level, and in residual C/N ratio following the addition of either high (*a*) or low (*b*) C/N ratio organic materials. Where the C/N ratio of added residues is above 25, microbes digesting the residues must supplement the nitrogen contained in the residues with soluble nitrogen from the soil. During the resulting nitrate depression period, competition between higher plants and microbes would be severe enough to cause nitrogen deficiency in the plants. Note that in both cases soluble N in the soil ultimately increases from its original level once the decomposition process has run its course. The trends shown are for soils without growing plants, which, if present, would continually remove a portion of the soluble nitrogen as soon as it is released.

Nitrate Depression. This condition, often called the **nitrate depression period**, persists until the activities of the decay organisms gradually subside due to lack of easily oxidizable carbon. As decay proceeds, the C/N ratio of the remaining plant material decreases because carbon is being lost (by respiration) and nitrogen is being conserved (by incorporation into microbial cells). Generally, one can expect mineral nitrogen to begin to be released when the C/N ratio of the remaining material drops below about 20. Then, nitrates appear again in quantity, and the original conditions prevail, except that the soil is somewhat richer in both nitrogen and humus.

The nitrate depression period may last for a few days, a few weeks, or even several months. To avoid producing seedlings that are stunted, chlorotic, and nitrogen-starved, planting should be delayed until after the nitrate depression period, or additional sources of nitrogen can be applied to satisfy the nutritional requirements of both the microbes and the plants.

High-Nitrogen Material. The effects on soil nitrate level will be quite different if the residues added have a C/N ratio lower than 20, as in the case represented by Figure 11.4*b*. With organic materials of low C/N ratio, more than enough nitrogen is present to meet the needs of the decomposing organisms. Therefore, soon after decomposition begins, some of the nitrogen from organic compounds is released into the soil solution, augmenting the level of soluble nitrogen available for plant uptake. Generally, nitrogen-rich materials decompose quite rapidly, resulting in a period of intense microbial growth and activity, but no nitrate depression period.

Influence of Soil Ecology

In nature, the process of nitrogen mineralization involves the entire food web (see Section 10.2), not just the saprophytic bacteria and fungi. For example, when organic residues are added to soil, bacteria and fungi grow rapidly on this food source, producing a large biomass of bacterial and fungal cells that contain much of the nitrogen originally in the residues. Until the microbial biomass begins to die off, this nitrogen is immobilized and not available to plants. However, a healthy soil ecosystem is likely to contain certain nematodes, protozoa, and earthworms that feed on bacteria and fungi. Since the C/N ratio of these animals is not too different from that of their microbial food, and since most of the carbon is converted to CO_2 by respiration, the animals soon ingest more nitrogen than they can use. They then excrete the excess nitrogen, mainly as NH_4^+, into the soil solution, providing plant-available mineral nitrogen. The microbial feeding activity of soil animals may increase the rate of nitrogen mineralization by 100%. Soil management that favors a complex food web (Section 10.15) with many trophic levels can be expected to enhance the cycling and efficient use of nutrients.

Influence of Lignin and Polyphenol Content of Organic Materials

The lignin contents of plant litter range from less than 2% to more than 50%. Those materials with high lignin content decompose very slowly. Polyphenol compounds found in plant litter may also inhibit decomposition. These phenolics are often water soluble and may be present in concentrations as high as 5 to 10% of the dry weight. By forming highly resistant complexes with proteins during residue decomposition, these phenolics can dramatically slow the rates of both nitrogen mineralization and carbon oxidation.

Litter Quality. Because they support only low levels of microbial activity and biomass, residues high in phenols and/or lignin are considered to be *poor quality resources* for the soil organisms that cycle carbon and nutrients. The production of such slow-to-decompose residues by certain forest plants may help explain the accumulation of extremely high levels of humified nitrogen and carbon in the soils of mature boreal forests.

The lignin and phenol contents also influence the decomposition and release of nitrogen from **green manures**—plant residues used to enrich agricultural soils (Figure 11.5). For example, in the leaves of certain legume trees, the C/N ratio is quite narrow, but the phenol content is quite high. When these leaves are added to soil in agroforestry systems, nitrogen is released only slowly—often too slowly to keep up with the needs of a growing crop. Similarly, residues with a lignin content of more than

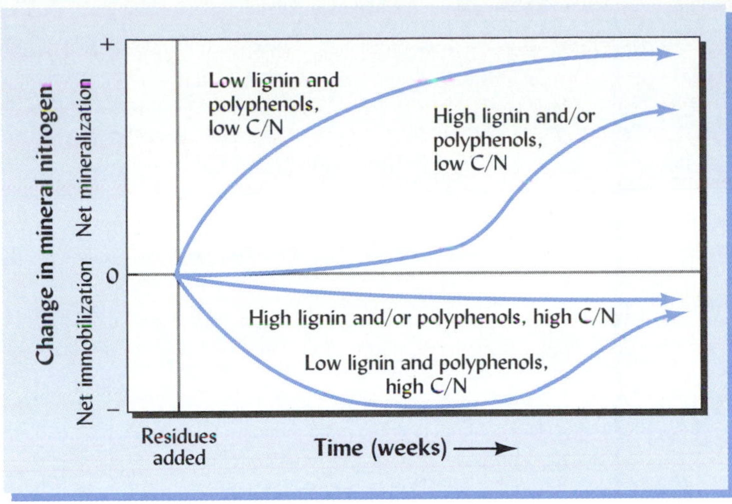

FIGURE 11.5 Temporal patterns of nitrogen mineralization or immobilization with organic residues differing in C/N ratios and contents of lignin and polyphenols. Lignin contents greater than 20%, polyphenol contents greater than 3%, and C/N ratios greater than 30 would all be considered high in the context of this diagram, the combination of these properties characterizing litter of poor quality—that is, litter that has a limited potential for microbial decomposition and mineralization of plant nutrients.

20 to 25% will decompose too slowly to be effective as green manure for rapidly growing annual crops. However, for perennial crops, or forests, the slow release of nitrogen from such residues may be advantageous in the long run, as the nitrogen may be less subject to losses. By the same token, the slow decomposition of phenol- or lignin-rich materials means that even if their C/N ratio is very high, the nitrate depression will not be pronounced.

11.4 HUMUS: GENESIS AND NATURE

Everything you ever wanted to know about humus—the view from Poland: http://www.ar.wroc.pl/~weber/humic.htm

The general term **soil organic matter (SOM)** encompasses all the organic components of a soil: (1) living biomass (intact plant and animal tissues and microorganisms), (2) dead roots and other recognizable plant residues, and (3) a largely amorphous and colloidal mixture of complex organic substances no longer identifiable as tissues. Only the third category of organic material is properly referred to as **soil humus** (Figure 11.6).

Microbial Transformations

As decomposition of plant residues proceeds, microbes polymerize (link together) some of the simpler new compounds with each other and with the complex residual products into long, complex chains that resist further decomposition. These high-molecular-weight compounds interact with nitrogen-containing amino compounds, giving rise to a significant component of resistant humus. The presence of colloidal clays stimulates

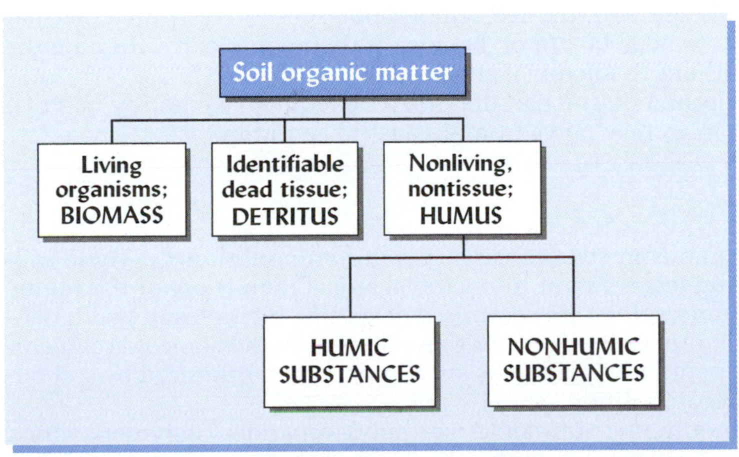

FIGURE 11.6 Classification of soil organic matter components separable by chemical and physical criteria. Although surface residues (litter) are not universally considered to be part of the soil organic matter, we include them because they are the principal component of the O horizons in soil profiles. Some soil scientists use the term *soil organic matter* synonymously with *humus*. However, we exclude soil biomass and dead but recognizable tissues from the latter, while including them in the former. As a practical matter, particles of plant residue that do not pass a 2-mm sieve opening are often excluded from consideration as soil organic matter.

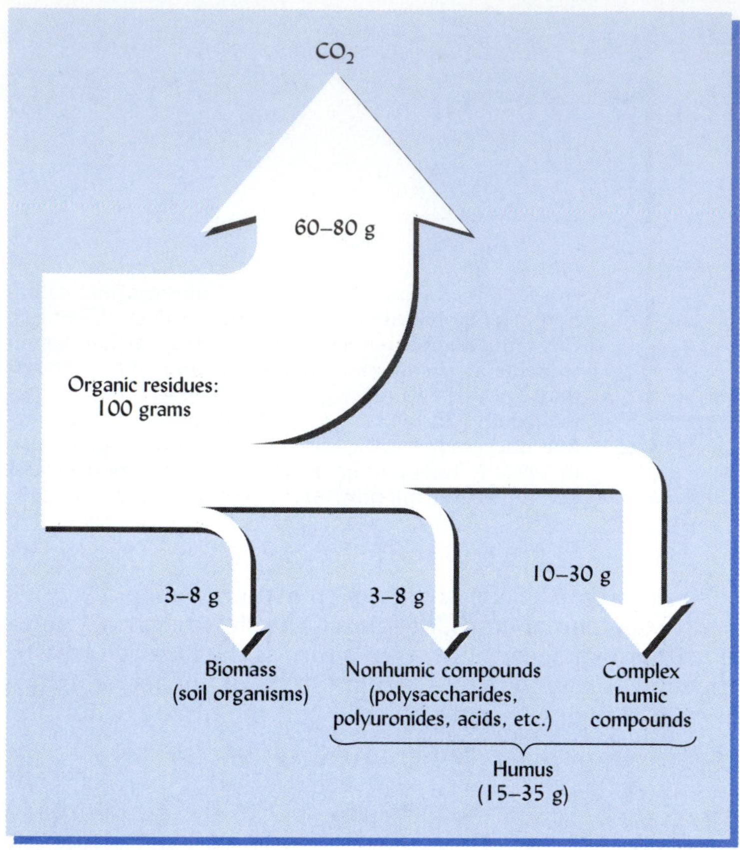

FIGURE 11.7 Disposition of 100 g of organic residues 1 year after they were incorporated into the soil. More than two-thirds of the carbon has been oxidized to CO_2, and less than one-third remains in the soil—some in the cells of soil organisms, but a larger component as soil humus. The amount converted to CO_2 is generally greater for aboveground residues than for belowground (root) residues.

3-D models of amino acids: http://www.bio.cmu.edu/Courses/BiochemMols/AAViewer/AAVFrameset.htm

the complex polymerization. These ill-defined, complex, resistant, polymeric compounds are called **humic substances.** The term **nonhumic substances** refers to the group of identifiable biomolecules that are mainly produced by microbial action and are generally less resistant to breakdown.

One year after plant residues are added to the soil, most of the carbon has returned to the atmosphere as CO_2, but one-fifth to one-third is likely to remain in the soil either as live biomass (≃ 5%) or as the humic (≃ 20%) and nonhumic (≃5%) fractions of soil humus (Figure 11.7). The proportion remaining from root residues tends to be somewhat higher than that remaining from incorporated leaf litter.

Humic Substances

Highlight the protein, sugar and other parts of a model humic acid: http://www.soils.umn.edu/virtual_museum/som/index.html

Humic substances comprise 60 to 80% of the soil organic matter. They are comprised of huge molecules with variable, complex structures characterized by many aromatic rings. Humic substances generally are dark-colored, amorphous substances with molecular weights varying from 2000 to 300,000 g/mol. Because of their complexity, they are the organic materials most resistant to microbial attack.

Depending on the environment, the half-life (the time required to destroy half of a substance) of humic substances may range from decades to centuries.

Nonhumic Substances

About 20 to 30% of the humus in soils consists of nonhumic substances. These substances are less complex and less resistant to microbial attack than those of the humic group. Unlike humic substances, they are comprised of specific biomolecules with definite physical and chemical properties. Some of these nonhumic substances are microbially modified plant compounds, while others are compounds synthesized by soil microbes as by-products of decomposition.

Included among the nonhumic substances are polysaccharides, polymers which have sugarlike structures and a general formula of $C_n(H_2O)_m$, where n and m are variable.

Polysaccharides are especially important in enhancing soil aggregate stability (see Section 4.7). Also included are polyuronides, which are not found in plants, but are synthesized by soil microbes.

Some even simpler compounds (such as low-molecular-weight organic acids and some proteinlike materials) are part of the nonhumic group. Although none of these simpler materials are present in large quantities, they may influence the availability of plant nutrients, such as nitrogen and iron, and may also directly affect plant growth.

Clay–Humus Combinations

Interaction with clay minerals often stabilizes soil nitrogen and organic matter. High-activity clays especially attract and hold such substances as amino acids, peptides, and proteins, forming complexes that protect the nitrogen-containing compounds from microbial degradation. Organic matter that is entrapped in the very small (<1 μm) pores formed by clay particles is physically inaccessible to decomposing organisms. It is also possible that layer silicate clays, such as vermiculite, may bind organic matter in their interlayers in forms that strongly resist decomposition. Iron and aluminum oxide coatings in highly weathered soils and allophane in volcanic soils also can bind with nitrogenous organic molecules and protect them from decay.

Thus, clay, along with humic and polysaccharide polymers, can protect relatively simple nitrogen compounds from microbial attack. In many soils more than half the organic matter is associated with clay and other inorganic constituents. Although the extent and mechanisms are not yet fully understood, clay–humus interactions partially account for the high organic matter content of clay soils.

Colloid Characteristics of Humus

The colloidal nature of humus was emphasized in Section 8.1. The surface area of humus colloids per unit mass is very high, generally exceeding that of silicate clays. The colloidal surfaces of humus are negatively charged, as a result of H^+ dissociation from carboxylic (—COOH) or phenolic (—OH) groups. The extent of the negative charge is pH-dependent. At high pH values, the cation exchange capacity of humus on a mass basis (150 to 300 cmol/kg) far exceeds that of most silicate clays. However, because humus is much less dense than clay, the cation exchange capacity of humus on a volume basis (40 to 80 cmol/L) is comparable to that of silicate clays.

The water-holding capacity of humus on a mass basis (but not on a volume basis) is 4 to 5 times that of the silicate clays. Humus plays a role in aggregate formation and stability. The highly complex humus molecules contain chemical structures that absorb nearly all wavelengths of visible light, giving the substance a characteristic black color (see, for example, the humus-rich A horizons in Plates 2 and 8, following page 114).

11.5 COMPOSTS AND COMPOSTING

Compost heated greenhouses:
http://attra.ncat.org/attra-pub/compostheatedgh.html

Composting is the practice of creating humuslike organic materials by aerobic decomposition outside of the soil. The finished product, **compost**, is popular as a mulching material, as an ingredient for potting mixes, as an organic soil conditioner, and as a slow-release fertilizer.

High-quality compost can be made at ambient temperatures by a slow decomposition process, or more quickly by a process called **vermicomposting**, in which certain litter-dwelling (epigeic) earthworms are added to help transform the material. However, we will focus on the most commonly practiced type of composting, in which intense decomposition activity occurs within large, well-aerated piles. This approach is called **thermophilic composting** because the large mass of rapidly decomposing material, combined with the insulating properties of the pile, results in considerable buildup of heat.

Composting Process. When a sufficient mass of suitable organic materials is kept in a moist, well-aerated, state, it goes through a three-stage process of decomposition. (1) During a brief initial **mesophilic** stage sugars and readily available microbial food sources are rapidly metabolized, causing the temperature in the compost pile to gradually rise from ambient to over 40°C. (2) A **thermophilic** stage occurs during the next week or two, during which temperatures rise to 50 to 75°C while thermophilic organisms decompose cellulose and other more resistant materials. Frequent mixing during this stage is essential to maintain oxygen supplies and assure even heating of all the material (Figure 11.8). The easily decomposed compounds are used up, and humuslike compounds are formed during this stage. (3) A second mesophilic or *curing stage* follows, during which the temperature falls back to near ambient, and the material is recolonized by mesophilic organisms, including certain beneficial organisms that produce plant-growth-stimulating compounds or are antagonistic to plant pathogenic fungi. This stage may last from several weeks to several months. Longer curing tends to encourage a more fungal-dominated community of microorganisms and a compost that may be of special value in amending bacteria-dominated soils.

Nature of the Compost Produced. As raw organic materials are humified in a compost pile, the content of nonhumic substances declines, and the content of humic acid increases markedly. During the composting process, the C/N ratio of organic materials in the pile decreases until a fairly stable ratio, in the range of 14:1 to 20:1, is achieved. The CEC of the organic matter increases to about 50 to 70 $cmol_c$/kg of compost.

Although 50% or more of the carbon in the initial material is typically lost during composting, mineral nutrients are mostly conserved. Finished compost is therefore generally more concentrated in nutrients than the initial combination of raw materials used and can serve as an effective means of building soil fertility. Weed seeds and pathogenic organisms are generally killed during the thermophilic phase; however, inorganic contaminants such as heavy metals are *not* destroyed by composting.

Proper management of compost is essential if the finished product is to be desirable for use as a potting media or soil amendment (see Box 11.1).

The Don't Bag It Compost Plan to reduce landscape and kitchen waste entering landfills:
http://aggie-horticulture.tamu.edu/earthknd/compost/compost.html

Benefits of Composting. Although making compost may involve more work and expense than applying uncomposted organic materials directly to the soil, the process offers many advantages. These include safe storage and easy handling of some otherwise rather sloppy materials like manure. For residues with a high initial C/N ratio, proper composting ensures that any nitrate depression period will occur in the compost pile, not in the soil, thereby avoiding induced plant nitrogen deficiency. When applied to the soil, composted materials generally decompose and mineralize much more slowly than uncomposted organic materials. **Cocomposting** low-C/N-ratio materials (such as livestock manure and sewage sludge) with high-C/N-ratio materials (such as sawdust, wood chips, senescent tree leaves, or municipal solid waste) provides sufficient carbon for microbes to immobilize the excess nitrogen and minimize any nitrate leaching hazard from the low-C/N materials.

Many weed seeds, pathogens and toxic compounds (pesticides, natural phytotoxic chemicals, etc.) are destroyed by the high temperatures achieved in well-managed compost piles. Composting is therefore often used as a method of biological treatment of polluted soils and wastes. Under less than ideal conditions, temperatures in parts of the pile may not exceed 40 to 50°C, so weeks or months may be required to achieve the same results. Some composts can effectively suppress soilborne plant diseases by encouraging microbial antagonisms (see Section 10.15).

11.6 DIRECT INFLUENCES OF ORGANIC MATTER ON PLANT GROWTH

Long ago, the observation that plants generally grow better on organic-matter-rich soils led people to think that plants derive much of their nutrition by absorbing humus from the soil. We now know that higher plants derive their carbon from carbon dioxide and that most of their nutrients come from inorganic ions dissolved in the soil solution. In fact, plants can complete their life cycles growing totally without humus, or even without soil (as in soilless or **hydroponic** production systems using only aerated nutrient solutions). This is not to say that soil organic matter is less important to plants than was once supposed, but rather that most of the benefits accrue to plants indirectly through the many influences of organic matter on soil properties. These will be discussed in Section 11.7, after we consider two types of direct organic matter effects on plants.

BOX 11.1 MANAGEMENT OF A COMPOST PILE

MATERIALS TO USE. Materials good for home composting include tree leaves (best if shredded), grass clippings, weeds (best before going to seed), kitchen scraps, wood shavings, gutter cleanings, pine needles, spoiled hay, straw, and even vacuum cleaner dust. Large-scale commercial compost is often made from materials such as municipal garbage, sewage sludge, wood chips, animal manures, municipal leaves, and food-processing wastes.

MATERIALS TO AVOID. Some materials to avoid include meat scraps (odors, rodents), cat droppings (carry microbes harmful to infants and pregnant women), sawdust from pressure-treated lumber and plywood (heavy metals and arsenic), and plastics and glass (nonbiodegradable; nuisance or dangerous in final product).

BALANCING NUTRIENTS. Although highly carbonaceous materials can be made to compost satisfactorily if they are turned frequently and kept moist, best results are obtained if high-C/N-ratio materials (e.g., brown leaves, straw, or paper) are mixed with low-C/N-ratio materials (e.g., green grass clippings, legume hay, blood meal, sewage sludge, or livestock manure) so as to achieve an overall C/N ratio between 20 and 30. Nitrogen fertilizer can also be added to lower the C/N ratio.

Other materials commonly added to improve nutrient balance and content include mixed fertilizers (N, P, and K), wood ashes (K, Ca, and Mg), bone meal or phosphate rock powder (P and Ca), and seaweed (K, Mg, Ca, and micronutrients). Some of these materials contain enough soluble salts to necessitate leaching of the finished compost before using it on salt-sensitive plants.

COMPOSTING METHODS. Provide good aeration throughout the pile, but build the pile large enough to provide mass sufficient to generate heat and prevent excessive drying. Backyard compost piles should be at least 1 m square and 1 m high. Compost bins are available to make turning the compost easier. Large-scale composting is usually carried out in *windrows*, about 2 to 3 m wide, 1 to 2 m tall, and many meters long (Figure 11.8*b*). In dry climates, compost may be made in pits dug about 1 m deep to protect the material from drying. The various materials to be composted can be mixed together or applied in thin layers. Often, a small amount of garden soil or finished compost is added to ensure that plenty of decomposer organisms will be immediately available. Compost activators containing microbial inoculum or herbal extracts are available, but while some may speed the initial heating of the pile, scientific tests rarely show any other advantage to using these preparations.

(*a*)

(*b*)

FIGURE 11.8 *(a) An efficient and easily managed method of composting suitable for homeowners is the three-bin method in which materials are turned with a pitchfork from one bin to the next. The perforated white plastic pipes enhance aeration. The leftmost bin contains relatively fresh materials, while the one at the right contains finished compost. (b) A special machine turns large-scale compost windrows (direction of travel is away from the reader) to mix the material and maintain well-aerated conditions at a facility in North Carolina where university dining hall food scraps are processed into compost for campus landscaping. [Photos courtesy of R. Weil]*

OXYGEN AND MOISTURE CONTROL. Low oxygen levels, usually due to inadequate turning combined with excessive moisture, can produce putrid odors as anaerobic decomposition takes over. Monitoring temperature and oxygen levels in the pile can help avoid this situation. To promote good aeration it is best to mix in a bulking agent such as wood chips, avoid excessive packing, and either turn the pile or pull a stream of air through it (Figure 11.8). If turning is used, this should be done during the thermophilic stage whenever the temperature begins to drop. The compost water content should be maintained at 50 to 70%. Properly moist compost will feel damp—but not dripping wet—when squeezed. Turning the pile during dry weather can help reduce excess moisture, while turning it during rain can help moisten a too-dry pile.

TABLE 11.3 Some Direct Effects of Humic Substances on Plant Growth

Effect on plant growth	*Humic substance*	*Concentration range, mg/L*
Accelerated water uptake and enhanced germination of seeds	Humic acid	1–100
Stimulated root initiation and elongation	Humic and fulvic acids	50–300
Enhanced root cell elongation	Humic acid	5–25
Enhanced growth of plant shoots and roots	Humic and fulvic acids	50–300

From Chen and Aviad (1990).

Direct Influence of Humus on Plant Growth

Forage brassicas are useful for control of nematodes in soil, and for allelopathic prevention of weed seeds germination: http://www.abc.net.au/gardening/stories/s124457.htm

Plants can absorb a very small portion of their nitrogen and phosphorus needs as soluble organic compounds. Vitamins, amino acids, auxins, and gibberellins formed as organic matter decays may at times stimulate growth in both higher plants and microorganisms. Small quantities of humic substances in the soil solution are known to enhance certain aspects of plant growth (Table 11.3). Components of these humic substances probably act as regulators of specific plant-growth functions, such as cell elongation and lateral root initiation. The concentrations of humic substances commonly present in the soil solution (50 to 100 mg/L or ppm) are effective in stimulating plant growth. Commercial humate products have been marketed with claims that small amounts enhance plant growth, but scientific tests of many of these products have failed to show any benefit from their use. Perhaps this is because effective levels of humic substances are naturally present in most soils.

ALLELOCHEMICAL EFFECTS.[3] **Allelopathy** is the process by which one plant infuses the soil with a chemical that affects the growth of other plants. The plant may do this by root exudation of **allelochemicals,** or the compounds may be leached out of the plant foliage by through-fall rainwater. In other cases, microbial metabolism of dead plant tissues (residues) forms the allelochemicals (Figure 11.9). Occasionally, the term *allelochemical* is also applied to plant chemicals that inhibit microorganisms. In principal, the interactions are much like the antagonistic relationships among certain microorganisms discussed in Section 10.15.

Allelochemicals present in the soil are responsible for many of the effects observed when various plants grow in association with one another. Because they produce such chemicals, certain weeds (e.g., johnsongrass and giant foxtail) damage crops far out of proportion to the size and number of weeds present. Crop residues left on the soil surface may inhibit the germination and growth of weeds or of the next crop planted (e.g., wheat residues often inhibit sorghum plants). Other allelopathic interactions influence the succession of species as forest and grassland ecosystems mature.

Allelopathic interactions are usually very specific, involving only certain species, or even varieties, on both the producing and receiving ends. The effects of allelopathic chemicals are many and varied. Although the term *allelopathy* most commonly refers to

FIGURE 11.9 Positive and negative allelopathic effects of winged beans on grain amaranth plants. In the pot on the left (T4) amaranth is growing in fresh soil (no association with winged beans). In the center pot (T20) amaranth is growing in soil previously used to grow winged beans (positive effect). In the pot on the right (T28) the amaranth is growing in fresh soil, but the plant was watered three times with a water extract of winged bean tissue (negative effect). The average dry weight of the amaranth plants for each treatment is shown. All pots were watered with a complete nutrient solution. [Data from Weil and Belmont (1987); photo courtesy of R. Weil]

[3]For a comprehensive review of allelopathy, see Inderjit et al. (1999).

negative effects, allelochemical effects can also be positive (as in certain **companion plantings**). While they vary in chemical composition, most allelochemicals are relatively simple phenolic or organic acid compounds that could be included among the nonhumic substances found in soils. Because most of these compounds can be rapidly destroyed by soil microorganisms or easily leached out of the root zone, effects are usually relatively short-lived once the source is removed.

11.7 INFLUENCE OF ORGANIC MATTER ON SOIL PROPERTIES AND THE ENVIRONMENT

Soil organic matter affects so many soil properties and processes that a complete discussion of the topic is beyond the scope of this chapter. Indeed, in almost every chapter in this book there is mention of the roles of soil organic matter. Figure 11.10 summarizes some of the more important effects of organic matter on soil properties and on soil–environment interactions. Often one effect leads to another, so that a complex chain of multiple benefits results from the addition of organic matter to soils. For example (beginning at the upper left in Figure 11.10), adding organic mulch to the soil surface encourages earthworm activity, which in turn leads to the production of burrows and other biopores, which in turn increases the infiltration of water and decreases its loss as runoff, a result that finally leads to less pollution of streams and lakes.

Influence on Soil Physical Properties. Humus tends to give surface horizons dark brown to black colors. Granulation and aggregate stability are encouraged, especially by the nonhumic substances produced by fungi and bacteria during decomposition. The humic fractions help reduce the plasticity, cohesion, and stickiness of clayey soils, making these soils easier to manipulate. Soil water retention is also improved, since organic matter increases both infiltration rate and water-holding capacity.

Influence on Soil Chemical Properties. Humus generally accounts for 50 to 90% of the cation-adsorbing power of mineral surface soils. Like clays, humus colloids hold nutrient cations (potassium, calcium, magnesium, etc.) in easily exchangeable form, wherein they can be used by plants but are not too readily leached out of the profile by percolating waters. Through its cation exchange capacity and acid and base functional groups, organic matter also provides much of the pH buffering capacity in soils (see Section 9.5). In addition, nitrogen, phosphorus, sulfur, and micronutrients are stored as constituents of soil organic matter, from which they are slowly released by mineralization.

Humic acids also attack soil minerals and accelerate their decomposition, thereby releasing essential nutrients as exchangeable cations. Organic acids and polysaccharides can attract such cations as Fe^{3+}, Cu^{2+}, Zn^{2+}, and Mn^{2+} from the edges of mineral structures and **chelate** or bind them in stable organomineral complexes. Some of these metals are made more available to plants as micronutrients because they are kept in soluble, chelated form (see Chapter 13). In very acid soils, organic matter alleviates aluminum toxicity by binding the aluminum ions in nontoxic complexes (see Sections 9.3 and 9.10).

Biological Effects. Soil organic matter greatly influences the biology of the soil, because it provides most of the food for the community of heterotrophic soil organisms described in Chapter 10. In Section 11.3 it was shown that the quality of plant litter and soil organic matter markedly affects decomposition rates and, therefore, the amount of organic matter accumulating in soils. The type and diversity of organic residues added to a soil can influence the type and diversity of organisms that make up the soil community.

11.8 MANAGEMENT OF AMOUNT AND QUALITY OF SOIL ORGANIC MATTER

Perhaps the most useful approach to defining soil organic matter quality is to recognize different pools of organic carbon that vary in their susceptibility to microbial metabolism. The model in Figure 11.11 denotes metabolic (sugars, proteins) and structural (lignin, cellulose) pools within the plant residues. The model shows the total organic matter in soil as the sum of several pools of soil organic matter, namely, **active**, **slow**, and **passive**.

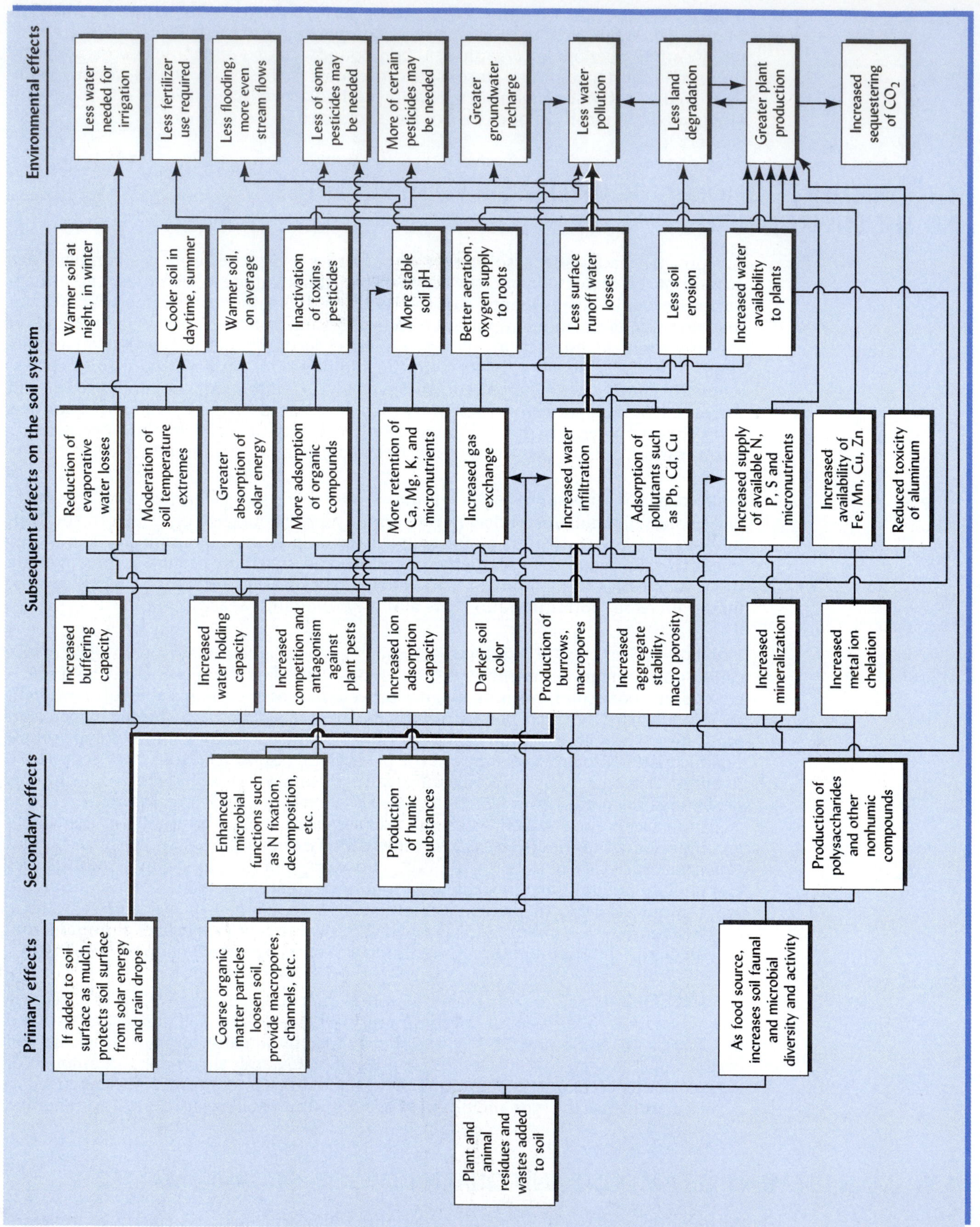

FIGURE 11.10 Some of the ways in which soil organic matter influences soil properties, plant productivity, and environmental quality. Many of the effects are indirect, the arrows indicating the cause-and-effect relationships. The influences of soil organic matter are far out of proportion to the relatively small amounts present in most soils. Many of these influences are discussed in this and other chapters in this book. The thicker line shows the sequence of effects referred to in the text in Section 11.7. (Diagram courtesy of R. Weil)

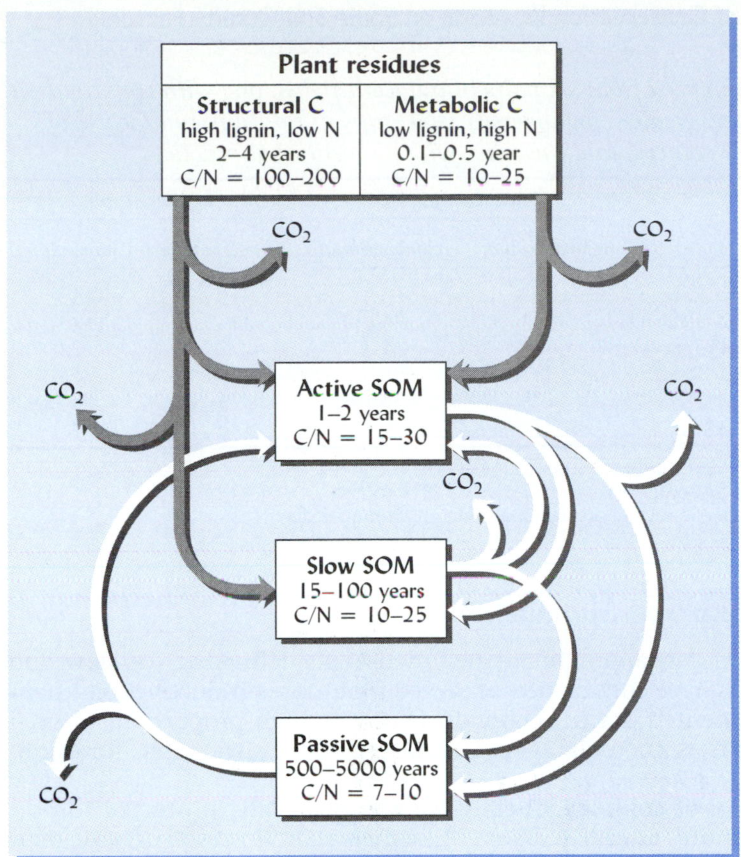

FIGURE 11.11 A conceptual model that recognizes various pools of soil organic matter (SOM) differing by their susceptibility to microbial metabolism. Models that incorporate *active, slow,* and *passive* fractions of soil organic matter have proven very useful in explaining and predicting real changes in soil organic matter levels and in attendant soil properties. Note that microbial action can transfer organic carbon from one pool to another. For example, when the nonhumic substances and other components of the active fraction are rapidly broken down, some resistant, complex by-products may be formed, adding to the slow and passive fractions. Note that all these metabolic changes result in some loss of carbon from the soil as CO_2. [Adapted from Paustian et al. (1992)]

Active Fraction of Soil Organic Matter

The active fraction of soil organic matter consists of materials with a relatively high average C/N ratio (about 15 to 30) and short half-lives (half of these materials can be metabolized in a matter of a few months to a few years). Components include the living biomass, various carbohydrates, some of the fine particles of detritus (referred to as **particulate organic matter,** or **POM**), most of the polysaccharides, and other nonhumic substances described in Section 11.4, as well as some of the more labile (easily decomposed) fulvic acids. This active fraction provides most of the readily accessible food for the soil organisms and most of the readily mineralizable nitrogen. It is responsible for most of the beneficial effects on structural stability that lead to enhanced infiltration of water, resistance to erosion, and ease of tillage. The active fraction can be readily increased by root exudates and the addition of fresh plant and animal residues, but it is also very readily lost when such additions are reduced or tillage is intensified. This fraction rarely comprises more than 10 to 20% of the total soil organic matter.

Passive and Slow Fractions of Organic Matter

The passive fraction of soil organic matter consists of very stable materials remaining in the soil for hundreds or even thousands of years. This fraction includes most of the humus physically protected in clay–humus complexes, most of the humin, and much of the humic acids. The passive fraction accounts for 60 to 90% of the organic matter in most soils, and its quantity is increased or diminished only slowly. The passive fraction is most closely associated with the colloidal properties of soil humus, and it is responsible for most of the CEC and water-holding capacity contributed to the soil by organic matter.

Intermediate in properties between the active and passive fractions is the *slow fraction* of soil organic matter. This fraction probably includes very finely divided plant tissues, high in lignin and other slowly decomposable and chemically resistant components. The half-lives of these materials are typically measured in decades. The slow fraction is an important source of mineralizable nitrogen and other plant nutrients, and it provides the underlying food source for the steady metabolism of the autochthonous soil microbes (see Section 11.2). The slow fraction also probably makes some contribution to the effects associated primarily with the active and passive fractions.

TABLE 11.4 The Effect of Conservation Practices on Some Soil-Quality Factors Related to Organic Matter

In each region, soil was analyzed from six pairs of adjacent fields, one with conservation practices (reduced tillage, greater crop diversity, sod crops in rotation, and/or use of organic nutrient sources) and the other with conventional practices.

	Coastal plain soils		*Piedmont soils*	
Properties	*Conservation*	*Conventional*	*Conservation*	*Conventional*
Total organic C, g/kg	12.5	8.3	19.6	15.5
Active organic C,[a] mg/kg	121	75	134	112
Microbial biomass C, % of total organic	2.4	1.3	2.6	2.3
Nitrogen mineralization rate constant[b]	38	33	42	36
Aggregate stability, %	73	58	74	66
Specific maintenance respiration[c]	41	72	18	32

[a]Mainly sugars extracted from soils after disruption with microwaves.
[b]The rate constant k, day^{-1}, in the first-order decay equation $N_p = N_o e^{kt}$
[c]Higher numbers indicate ecosystem stress. Units are mg CO_2/g microbial biomass/day.
Data from Islam and Weil (2000).

Changes in Active and Passive Fractions with Soil Management

Soil organic matter losses in long term cropping systems experiments in the Canadian prairies: http://www.agric.gov.ab.ca/agdex/500/536-1.html#effects

Soil analytical methods can only approximate the functionally defined active, slow, and passive fractions. Scientists have consistently observed that more productive soils managed with conservation-oriented practices contain relatively high proportions of such active-fraction components as microbial biomass and oxidizable sugars (see, for example, Table 11.4).

The existence of a pool of complex, chemically and physically protected soil organic matter (passive fraction), as well as a pool of easily metabolized soil organic matter (active fraction), explains why conversion of forests or grassland into cultivated cropland results in a very rapid decline in soil organic matter during the first few years, followed by a much slower decline thereafter (see, for example, Figure 11.12).

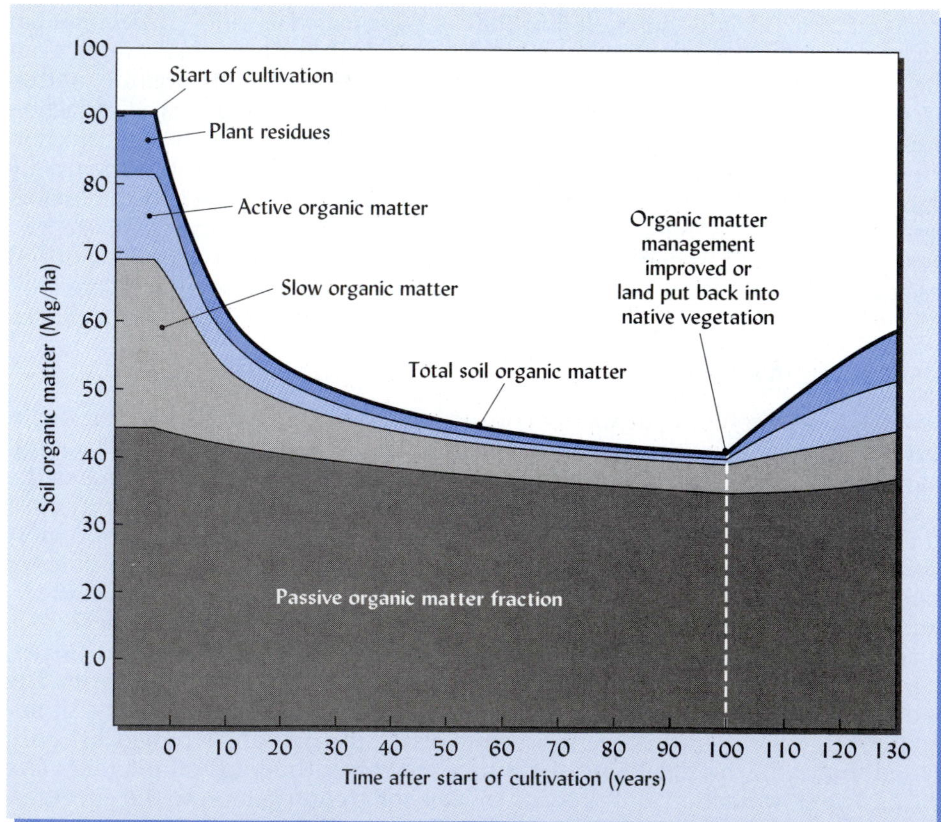

FIGURE 11.12 Changes in various fractions of organic matter in the upper 25 cm of a representative soil after bringing virgin land under cultivation. Initially, under natural vegetation, this soil contained about 91 Mg/ha of total organic matter. The resistant *passive fraction* accounted for about 44 Mg/ha, or about half of the total soil organic matter. The rapidly decomposing *active fraction* accounted for about 14 Mg, or about 16% of the total soil organic matter. After about 40 years of cultivation, the passive fraction had declined by about 11% to about 39 Mg/ha, while the active fraction had lost 90% of its mass, declining to only 1.4 Mg/ha. Note that much of the organic loss due to the change in land management came at the expense of the active fraction. This was also the fraction that most quickly increased when improved organic matter management was adopted after the 100th year. The susceptibility of the active fraction to rapid change explains why even relatively small changes in total soil organic matter can produce dramatic changes in important soil properties, such as aggregate stability and nitrogen mineralization, associated with this soil organic matter fraction. (Diagram courtesy of R. Weil)

Soil management practices that cause only very small changes in total soil organic matter often cause rather pronounced alterations in aggregate stability, nitrogen mineralization rate, or other soil properties attributed to organic matter. This occurs because the relatively small pool of active organic matter may undergo a large percentage increase or decrease without having a major effect on the much larger pool of total organic matter (Figure 11.12). As a result, the soil organic matter remaining after some years is far less effective in promoting structural stability and nutrient cycling than the original organic matter in the virgin soil. If a favorable change in environmental conditions or management regime occurs, the plant litter and active fractions of soil organic matter are also the first to positively respond (see right-hand portion of Figure 11.12). Therefore, some scientists advocate monitoring active-fraction soil C as an early warning of changes in total soil organic matter and soil quality.

11.9 CARBON BALANCE IN THE SOIL–PLANT–ATMOSPHERE SYSTEM

Management of organic inputs in soils of the tropics homepage:
http://ppathw3.cals.cornell.edu/mba_project/moist/home2.html/

Whether the goal is to reduce greenhouse gas emissions or to enhance soil quality and plant production, proper management of soil organic matter requires an understanding of the cycling and balance of carbon. Although each type of ecosystem, whether a deciduous forest, a prairie, or a wheat field, will emphasize particular compartments and pathways in the carbon cycle, consideration of a specific example, such as that described in Box 11.2, will be helpful.

The rate at which soil organic matter either increases or decreases is determined by the balance between *gains* and *losses* of carbon. The gains come primarily from plant residues grown in place and from applied organic materials. The losses are due mainly to respiration (CO_2 losses), plant removals, and erosion (Table 11.5).

Conservation of Soil Carbon in Agroecosystems. To halt or reverse the net carbon loss shown in Figure 11.13, land managers would have to either *increase the additions* or *decrease the losses* of carbon. Since all crop residues and animal manures in the example are already being returned to the soil, additional carbon inputs could most practically be achieved by increasing crop production or growing cover crops during the winter. Using a no-till production system would leave crop residues as mulch on the soil surface where they would decompose much more slowly. Refraining from tillage might also reduce the annual respiration losses from the original 2.5% to perhaps 1.5%. A combination of these changes in management would convert the system in our example from one in which soil organic matter is degrading (declining) to one in which it is aggrading (increasing).

Forest Ecosystems. If forest soil fertility were not too low, the total annual biomass production might be similar to that shown in Figure 11.13. The standing biomass, on the other hand, would be much greater in the forest since the tree crop is not removed each year. While some litter would fall to the soil surface, much of the annual biomass production would remain stored in the trees.

Humus oxidation in the undisturbed forest would be considerably slower in the absence of tillage. The litter from certain tree species may also be rich in phenolics and lignin, factors that greatly slow decomposition (see Figure 11.6). In forest soils, decomposition of leaf litter produces copious quantities of dissolved organic carbon (DOC) compounds such as fulvic acids. Consequently, 5 to 40% of the total C losses may occur by leaching—a much greater proportion than from cropland. However, losses of organic matter through soil erosion would be much smaller on the forested site. Taken together, these factors allow annual net gains in soil organic matter in a young forest and maintenance of high soil organic matter levels in mature forests.

Grasslands. Similar trends occur in natural grasslands, although the total biomass production is likely to be considerably less, depending mainly on the annual rainfall. Among the principles illustrated in Box 11.2, and applicable to most ecosystems, is the dominant role that plant root biomass plays in maintaining soil organic matter levels. In a grassland, the contribution from the plant roots is relatively more important than in a forest. Therefore, a greater proportion of the total biomass produced tends to accumulate as soil organic matter, and this soil organic C is distributed more uniformly with depth.

BOX 11.2 CARBON BALANCE—AN AGROECOSYSTEM EXAMPLE

The principal carbon pools and annual flows in a terrestrial ecosystem are illustrated in Figure 11.13 using a hypothetical cornfield in a warm temperate region. During a growing season the corn plants produce (by photosynthesis) 17,500 kg/ha of dry matter containing 7500 kg/ha of carbon (C). This C is equally distributed (2500 kg/ha each) among the roots, grain, and unharvested aboveground residues. In this example, the harvested grain is fed to cattle, which oxidize and release as CO_2 about 50% of this C (1250 kg/ha), assimilate a small portion as weight gain, and void the remainder (1100 kg/ha) as manure. The corn stover and roots are left in the field and, along with the manure from the cattle, are incorporated into the soil by tillage or by earthworms.

The soil microbes decompose the crop residues (including the roots) and manure, releasing as CO_2 some 75% of the manure C, 67% of the root C, and 85% of the C in the surface residues. The remaining C in these pools is assimilated into the soil as humus. Thus, during the course of one year, some 1475 kg/ha of C enters the humus pool (825 kg from roots, plus 375 from stover, plus 275 from manure). These values are in general agreement with Figure 11.7, but they will vary widely among different soil conditions and ecosystems.

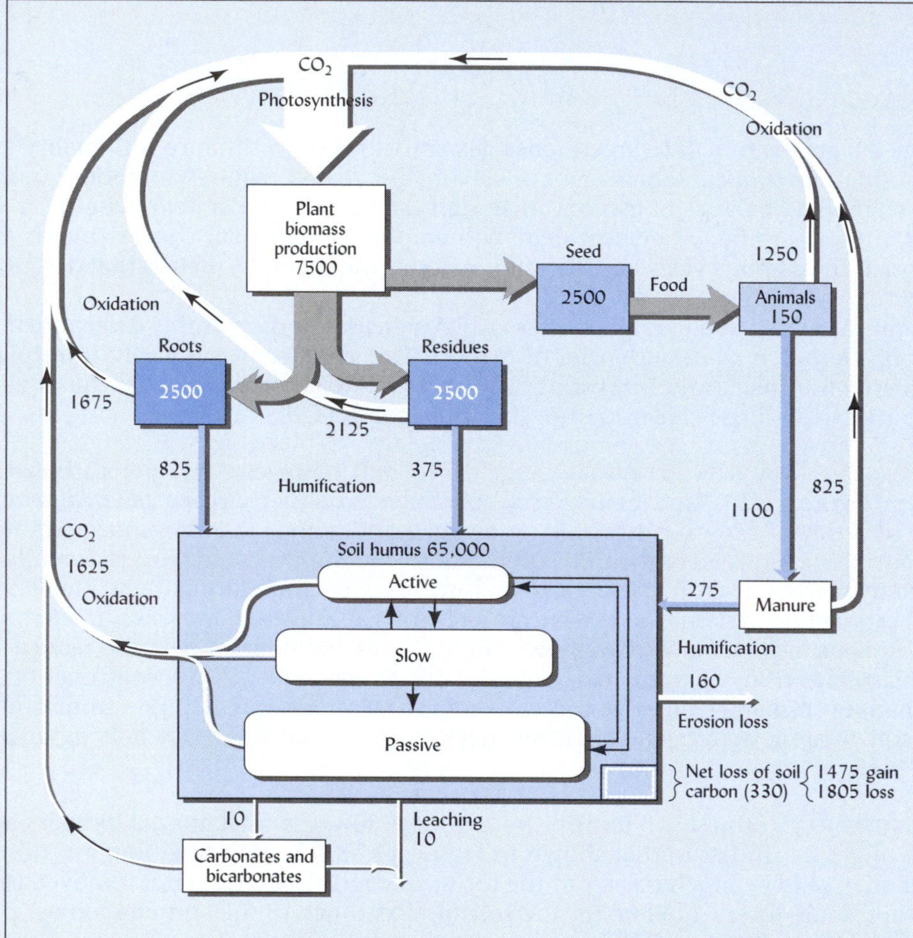

FIGURE 11.13 *Carbon cycling in an agroecosystem.*

At the beginning of the year, the upper 30 cm of soil in our example contained 65,000 kg/ha organic C in humus. Such a soil cultivated for row crops would typically lose about 2.5% of its organic C by soil respiration each year. In our example this loss amounts to some 1625 kg/ha of C. Smaller losses of soil organic C occur by soil erosion (160 kg/ha), leaching (10 kg/ha), and formation of carbonates and bicarbonates (10 kg/ha).

Comparing total losses (1805 kg/ha) with the total gains (1475 kg/ha) for the pool of soil humus, we see that the soil in our example suffered a *net annual loss* of 330 kg/ha of C, or 0.5% of the total C stored in the soil humus. If this rate of loss were to continue, degradation of soil quality and productivity would surely result.

TABLE 11.5 Factors Affecting the Balance between Gains and Losses of Organic Matter in Soils

Factors promoting gains	*Factors promoting losses*
Green manures or cover crops	Erosion
Conservation tillage	Intensive tillage
Return of plant residues	Whole plant removal
Low temperatures and shading	High temperatures and exposure to sun
Controlled grazing	Overgrazing
High soil moisture	Low soil moisture
Surface mulches	Fire
Application of compost and manures	Application of only inorganic materials
Appropriate nitrogen levels	Excessive mineral nitrogen
High plant productivity	Low plant productivity
High plant root:shoot ratio	Low plant root:shoot ratio

11.10 FACTORS AND PRACTICES INFLUENCING SOIL ORGANIC MATTER LEVELS

More on the nature and management of soil organic matter:
http://www.extension.umn.edu/distribution/cropsystems/components/7402_02.html

Soil organic matter content is usually estimated from analysis of soil organic carbon content, because the latter can be determined more precisely. Therefore, for quantitative discussions scientists usually refer to soil organic carbon. We will now consider factors that influence the rates of soil organic carbon gain and loss (see Table 11.5 for some of the management-oriented factors).

Climate

Temperature. Mean annual temperature influences soil organic matter levels because of the different manner in which the processes of organic matter production (plant growth) and organic matter destruction (microbial decomposition) respond to increases in this climatic variable. At low temperatures plant growth outstrips decomposition, but that the opposite is true above approximately 25°C. Therefore, as one moves from a warmer to a cooler climate, the organic matter and associated nitrogen content of comparable soils tend to increase. Some of the most rapid rates of organic matter decomposition occur in irrigated soils of hot desert regions.

Within zones of uniform moisture conditions and comparable vegetation, the average total amounts of organic matter and nitrogen in soils increase from 2 to 3 times for each 10°C decline in mean annual temperature. This temperature effect can be readily observed by noting the darkening color of well-drained surface soils as one travels from south (Texas) to north (Minnesota) in the humid grasslands of the North American Great Plains region (Figure 11.14). Similar changes in soil organic matter are evident as one climbs from warm lowlands to cooler highlands in mountainous regions.

Moisture. Soil moisture also exerts a major influence on the accumulation of organic matter and nitrogen in soils. Under comparable conditions, the nitrogen and organic matter content of soils increase as the effective moisture becomes greater. These relationships are illustrated by the darker and thicker A horizons encountered as one travels across the North American Great Plains region (within a belt of similar mean annual temperature), from the drier zones in the west (Colorado) to the higher rainfall east (Illinois). The explanation lies mostly in the sparser vegetation of the drier regions.

The lowest natural levels of soil organic matter and the greatest difficulty in maintaining those levels are found where annual mean temperature is high and rainfall is low. These relationships are extremely important to the productivity and conservation of soils and to the relative difficulty of sustainable natural resource management.

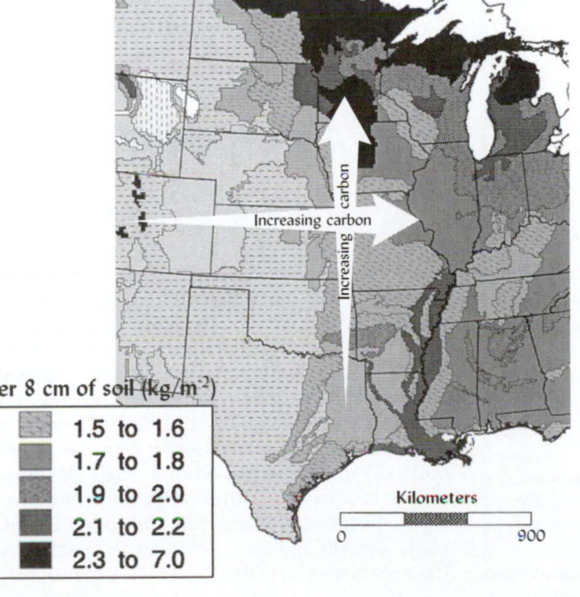

FIGURE 11.14 Influence of mean annual temperature and precipitation on organic matter levels in soils and on the difficulty of sustaining the soil resource base. The large white arrows on the map indicate that in the North American Great Plains region, soil organic matter increases with cooler temperatures going north, and with higher rainfall going east, provided that the soils compared are similar in texture, type of vegetation, drainage, and all other aspects except temperature and rainfall. These trends can be further generalized for global environments. [Kern (1994); Map courtesy of J. Kern, U.S. Environmental Protection Agency]

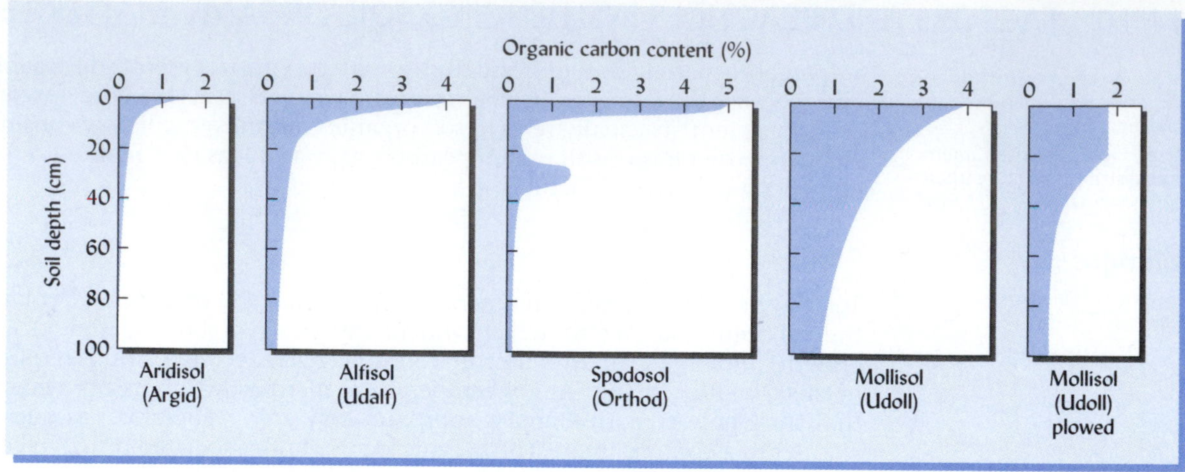

FIGURE 11.15 Vertical distribution of organic carbon in well-drained soils of four soil orders. Note the higher content and deeper distribution of organic carbon in the soils formed under grassland (Mollisols) compared to the Alfisol and Spodosol, which formed under forests. The bulge of organic carbon in the Spodosol subsoil is due to illuvial humus in the spodic horizon (see Chapter 3). The Aridisol has little organic carbon in the profile, as is typical of dry-region soils.

Influence of Natural Vegetation

Storing Carbon in Soil: Why and How? A brief overview by a soil microbiologist: http://www.agiweb.org/geotimes/jan02/feature_carbon.html

Climate and vegetation usually act together to influence the soil contents of organic carbon and nitrogen. The greater plant productivity engendered by a well-watered environment leads to greater additions to the pool of soil organic matter. Grasslands generally dominate the subhumid and semiarid areas, while trees are dominant in humid regions. In climatic zones where the natural vegetation includes both forests and grasslands, the total organic matter is higher in soils developed under grasslands than under forests (see Figure 11.15). With grassland vegetation, a relatively high proportion of the plant residues consist of root matter, which decomposes more slowly and contributes more efficiently to soil humus formation than does forest leaf litter.

Effects of Texture and Drainage

While climate and natural vegetation affect soil organic matter over broad geographic areas, soil texture and drainage are often responsible for marked differences in soil organic matter within a local landscape. All else being equal, soils high in clay and silt are generally higher in organic matter than are sandy soils (Figure 11.16). The finer-textured

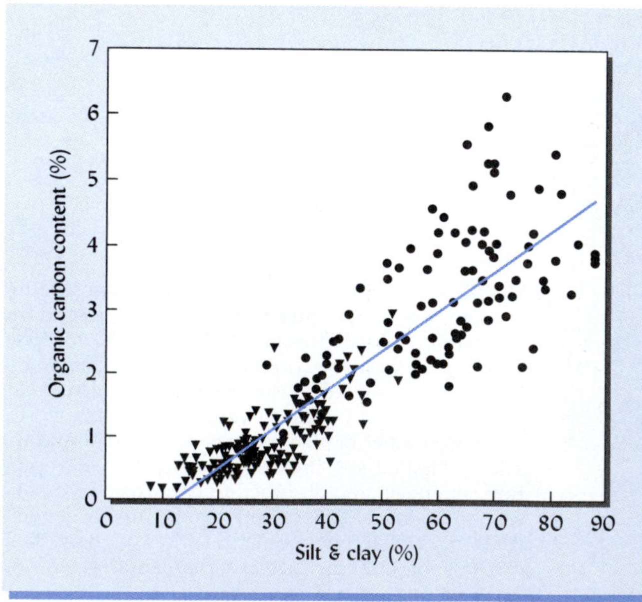

FIGURE 11.16 Soils high in silt and clay tend to contain high levels of organic carbon. The data shown are for surface soils in 279 tilled maize fields in subhumid regions of Malawi (▼) and Honduras (●). All soils were moderately well to well drained and tilled. Variability (scatter of data points) among soils with the same silt + clay content is probably due to differences in (1) the type of clay minerals present (2:1 silicates tend to stabilize more organic carbon), (2) site elevation (cooler, high elevation locations being conducive to greater organic carbon accumulation), and (3) years since cultivation began (longer history of cultivation leading to lower organic carbon levels). (Data courtesy of R. Weil and M. A. Stine, University of Maryland, and S. K. Mughogho, University of Malawi)

soils accumulate more organic matter for several reasons: (1) They produce more plant biomass, (2) they lose less organic matter because they are less well aerated, and (3) more of the organic material is protected from decomposition by being bound in clay–humus complexes (see Section 11.4) or sequestered inside soil aggregates.

DRAINAGE EFFECTS. In poorly drained soils, the high moisture supply promotes plant dry-matter production and relatively poor aeration inhibits organic matter decomposition. Poorly drained soils therefore generally accumulate much higher levels of organic carbon and nitrogen than similar but better-aerated soils (Figure 11.17). Histosols represent an extreme example of this principle.

Influence of Agricultural Management and Tillage

The most productive cultivated soils will likely be considerably lower in organic matter than similar soils carrying undisturbed natural vegetation, except in the case of irrigated soil in desert areas, where natural vegetation is sparse. This is not surprising; under natural conditions erosion is low, all the organic matter produced by the vegetation is returned to the soil, and the soil is not disturbed by tillage. In contrast, in cultivated areas much of the plant material is removed for human or animal food and relatively less finds its way back to the land. Also, soil tillage aerates the soil and breaks up the organic residues, making them more accessible to microbial decomposition. The rapid decline in organic matter upon conversion of natural vegetation to cultivation was illustrated by Figure 11.12. Many long-term experimental plots suggest that cultivated soils kept highly productive by supplemental applications of nutrients, lime, and manure and by the choice of high-yielding crop varieties are likely to have more organic matter than comparable, less productive soils. With higher productivity comes larger amounts of root and shoot residues returned to the soil.

Recommendations for Managing Soil Organic Matter

While the total carbon stabilized in the soil is important in relation to the global greenhouse effect (see Section 11.11), in terms of productivity and other ecological functions, achieving a particular level of total soil organic matter is far less important than maintaining a substantial amount in the active fraction. It is the active fraction C that fuels the food web and allows biological metabolism to constantly enhance soil tilth and nutrient cycling.

A continuous supply of organic materials must be added to the soil to maintain an appropriate level of soil organic matter, especially in the active fraction. Plant residues (roots and tops), animal manures, composts, and organic wastes are the primary sources of these organic materials. Cover crops can provide protective cover and additional organic material for the soil. Moderate applications of lime and nutrients may be needed to help free plant growth from the constraints imposed by chemical toxicities and nutrient deficiencies. It is almost always preferable to keep the soil vegetated than to keep

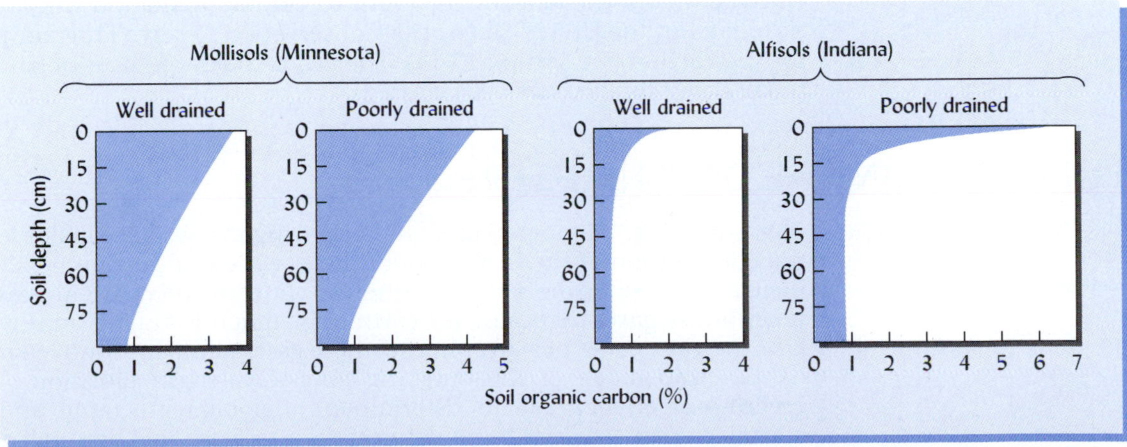

FIGURE 11.17 Distribution of organic carbon in four soil profiles, two well drained and two poorly drained. Poor drainage results in higher organic carbon content, particularly in the surface horizon.

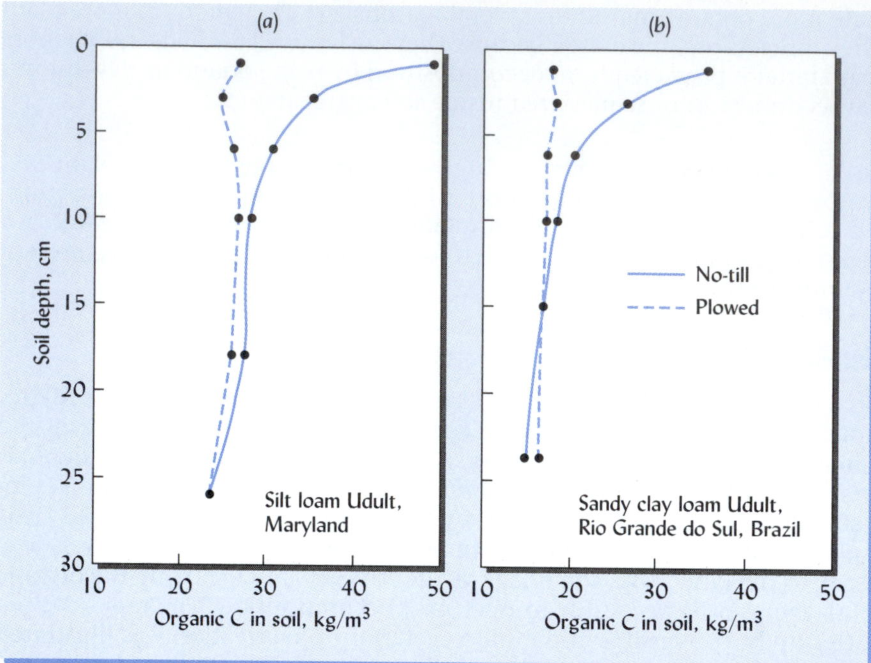

FIGURE 11.18 Less tillage means more soil organic carbon. In both cases, the no-till system had been used on the experimental plots for 8 to 10 years when the data were collected. In the plowed plots, the soil was disturbed annually by tillage to about 20 cm deep. The soils in Maryland (*a*) and Brazil (*b*) were well-drained Ultisols and the climate was temperate (Maryland) to subtropical (Brazil). In Maryland, corn was grown every year with a rye cover crop. In Brazil, oats were rotated with corn using legume cover crops in between. In both cases, no-till encouraged the accumulation of organic C, but only in the upper 5 to 10 cm of the soil. [Data from Weil et al. (1988) and Bayer et al. (2000)]

it in bare fallow. It is generally not practical to try to maintain higher soil organic matter levels than the soil–plant–climate control mechanisms dictate. For example, 1.5% organic matter might be an excellent level for a sandy soil in a warm climate, but would be indicative of a very poor condition for a finer-textured soil in a cool climate.

Because of the linkage between soil nitrogen and organic matter, adequate nitrogen inputs are requisite for adequate organic matter levels. Accordingly, the inclusion of legumes in the crop rotation and the judicious use of nitrogen-containing fertilizers to enhance high soil productivity are two desirable practices. At the same time, steps must be taken to minimize the loss of nitrogen by leaching, erosion, or volatilization (see Chapter 12).

Because tillage accelerates organic matter losses both by increased oxidation of soil organic matter and by erosion, it should be limited to that needed to control weeds and to maintain adequate soil aeration. Conservation practices that minimize tillage leave much of the plant residues on or near the soil surface, and thereby slow down the rate of residue decay and reduce erosion losses (see Sections 4.8 and 15.6). In time, conservation tillage can lead to higher organic matter levels (see Figure 11.18), especially in surface horizons.

Perennial vegetation, especially natural ecosystems, should be encouraged and maintained where feasible. Improved agricultural production on existing farmlands should be pursued to allow land currently supporting natural ecosystems to be left relatively undisturbed. In addition, there should be no hesitation about taking land out of cultivation and encouraging its return to natural vegetation where such a move is appropriate. In the United States, the Conservation Reserve Program provides incentives for such action (see Section 15.14). The fact is that large areas of land under cultivation today never should have been cleared.

11.11 SOILS AND THE GREENHOUSE EFFECT

Intergovernmental Panel on Climate Change: www.ipcc.ch/

GLOBAL WARMING. Biological processes occurring in soils have major long-term effects on the composition of the Earth's atmosphere, which in turn influences all living things, including those in the soil. Of particular concern today are increases in the levels of **greenhouse gases** that cause the earth to be much warmer than it would otherwise be. Like the glass panes of a greenhouse, these gases allow short-wavelength solar radiation in, but trap much of the outgoing long-wavelength radiation. This heat-trapping **greenhouse effect** is a major determinant of global temperature and, hence, global climates. Human activities, from deforestation to fossil fuel burning, are major contributors to global warming. Gases produced by biological processes, such as those occurring in the soil, account for approximately half of the greenhouse effect (Figure 11.19).

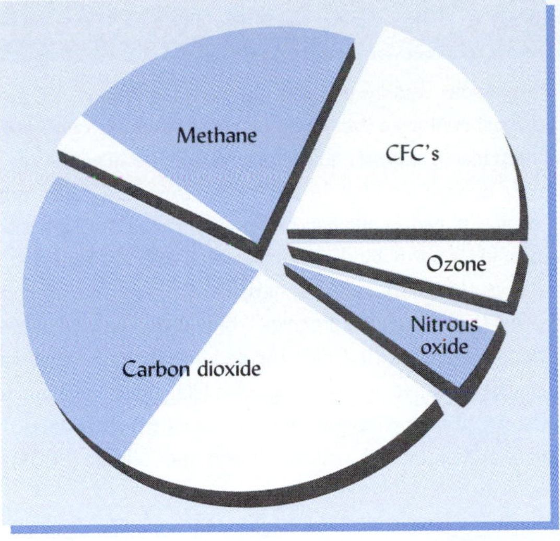

FIGURE 11.19 Relative contribution of different gases to the global greenhouse effect. The shaded portions indicate the emissions related to biological systems (in which soils play a role), the white portions being industrial contributions. Of the five primary greenhouse gases, only chlorofluorocarbons (CFCs) are exclusively of industrial origin. [Modified from Dale et al. (1993)]

Beautiful scientific photo-documentation of global climate change. Includes links under "references": http://www.worldviewofglobalwarming.org/

Most scientists believe that the average global temperature has increased by 0.5 to 1.0°C during the past century, and predict that it is likely to increase by another 2 to 3° C in this century. Major changes in the earth's climate are already taking place, including changes in rainfall distribution and growing season length, increases in sea level, and greater frequency and severity of storms. Much effort and expense will have to be directed at reducing the anthropogenic (human-caused) contributions to climate change. Fortunately, soil science has the potential to contribute greatly to our ability to deal with global warming and the increasing levels of greenhouse gases.

CARBON DIOXIDE. In 2000, the atmosphere contained about 370 ppm CO_2, as compared to about 280 ppm before the Industrial Revolution. Levels of this greenhouse gas are increasing at about 0.5% per year. Although the burning of fossil fuels is a major contributor, much of the increase comes from a net loss of organic matter in the world's soils. Box 11.3 illustrates the importance of soil organic matter in regulating atmospheric CO_2 levels. Research indicates a two-way feedback between the soil and atmosphere—changes in the levels of gases beneficial or harmful to plants influence the rate at which carbon accumulates in soil organic matter (Figure 11.22).

We have already discussed many ways that land managers can increase levels of soil organic matter (Section 11.10) by changing the balance between gains and losses. Gains in soil organic matter occur first in the active fractions, but eventually some of the carbon moves into the stabile passive fraction, where it may be *sequestered* for hundreds or thousands of years. The opportunities for sequestering carbon are greatest for degraded soils that currently contain only a small portion of the organic matter levels they contained originally under natural conditions. Reforestation of denuded areas is one such opportunity. Others include switching cropland from conventional tillage to no-tillage practices or converting it to perennial vegetation.

Such management changes could significantly enhance society's efforts to stem the rise in atmospheric CO_2, and at the same time improve soil quality and plant productivity. Some estimates suggest that during a 50-year period, improved management of agricultural lands could provide about 15% of the CO_2 emission reductions that the United States may be obligated to make. However, in accordance with the factors discussed in Section 11.10, soils have only a finite capacity to assimilate carbon into stable soil organic matter. Therefore carbon sequestration in soils can only buy time before other kinds of actions (shifts to renewable energy sources, increased fuel efficiency, etc.) are fully implemented to reduce carbon emission to levels that will not threaten climate stability. Soils may continue to combat increased CO_2 by supporting biomass fuels and increased timber production.

METHANE. Methane (CH_4) occurs in the atmosphere in far smaller amounts than CO_2. However, methane's contribution to the greenhouse effect is nearly half as great as that from CO_2 because each molecule of CH_4 is 30 to 50 times as effective as CO_2 in trapping outgoing radiation. The level of CH_4 is rising at about 0.6% per year. Soils both add CH_4 and remove it from the atmosphere.

BOX 11.3 CARBON CYCLE IN MINIATURE

Soil organic matter definitely impacted the eight biospherians who lived in Biosphere 2, a 1.3-ha sealed glass building in the Arizona desert (Figure 11.20). This structure was designed to hold a self-contained ecosystem in which scientists could live and study as part of the ecosystem. Biosphere 2 contains a miniature ocean, coral reef, marsh, forests, and farms designed to maintain the ecological balance. However, not only did the biospherians have a very hard time growing enough food for themselves (Figure 11.21), their atmosphere soon ran so low on oxygen and became so enriched in carbon dioxide that engineers had to pump in oxygen and install a scrubber to cleanse carbon dioxide from the air. It turned out that the ecosystem was thrown out of kilter partly by the organic-matter-rich soil hauled in for the Biosphere farms. The soil, made from a mixture of pond sediment (1.8% C), compost (22% C), and peat moss (40% C) was installed uniformly about 1 m deep. This artificial soil contained about 2.5% organic C at all depths. In contrast, the organic C content of a natural soil from the desert environment would probably be only 1% in the surface soil, and would decline sharply with depth. The designers had underestimated the rate at which soil organic matter would decompose when placed in the warm environment and aerated by garden tillage (see Section 11.2). Decomposition of the soil organic matter was carried out by aerobic soil microbes, which use up oxygen and give off carbon dioxide as they respire. For more information on Biosphere 2 in general, see *www.bio2.com/research*, and on the soil used, see Torbert and Johnson (2001).

FIGURE 11.20 *The Biosphere 2 structure, a huge, sealed, ecological laboratory. (Photo by C. Allen Morgan, © 1995 by Decisions Investments Corp. Reprinted with permission.)*

FIGURE 11.21 *Biospherians at work growing their own food supply in the intensive agriculture biome with compost-amended soils rich in organic matter. (Photo by Pascale Maslin, © 1995 by Decisions Investments Corp. Reprinted with permission.)*

Climate change impacts in the Gulf Coast region of the United States:
http://www.ucsusa.org/gulf

Biological soil processes account for much of the methane emitted into the atmosphere. When soils are strongly anaerobic, as in wetlands and rice paddies, bacteria produce CH_4, rather than CO_2, as they decompose organic matter (see Sections 7.4 and 11.2). Among the factors influencing the amount of CH_4 released to the atmosphere from wet soils are: (1) the maintenance of a redox potential (Eh) near 0 mV, (2) the availability of easily oxidizable carbon, and (3) the nature and management of the plants growing on these soils (70% to 80% of the CH_4 released from flooded soils escapes to the atmosphere through the hollow stems of wetland plants). Periodically draining rice paddies prevents the development of extremely anaerobic conditions, and therefore can substantially decrease CH_4 emissions. Such management practices should be given serious consideration, as rice paddies are thought to be responsible for up to 25% of global CH_4 production. Significant quantities of methane are also produced by the anaerobic decomposition of cellulose in the guts of termites living in well-aerated soils (see Section 10.5), and of garbage buried deep in landfills.

USEPA's Global Warming site:
http://yosemite.epa.gov/oar/globalwarming.nsf/content/index.html

In well-aerated soils, certain **methanotrophic bacteria** produce the enzyme *methane monooxygenase,* which allows them to oxidize methane to methanol as an energy source. This reaction, which is largely carried out in soils, reduces the global greenhouse gas burden by about 1 billion Mg of methane annually. Unfortunately, the long-term use of inorganic (especially ammonium) nitrogen fertilizer on cropland, pastures, and forests has been shown to reduce the capacity of the soil to oxidize methane. The rapid availability of ammonium from fertilizer may stimulate ammonium-oxidizing bacteria at the expense of methane-oxidizing bacteria. In contrast, supplying nitrogen in organic form (as manure) may enhance the soil's capacity for methane oxidation (Table 11.6).

No-till tactic—farmers may turn to carbon storage for cash:
http://www.washingtonpost.com/wp-dyn/articles/A55389-2002Aug23.html

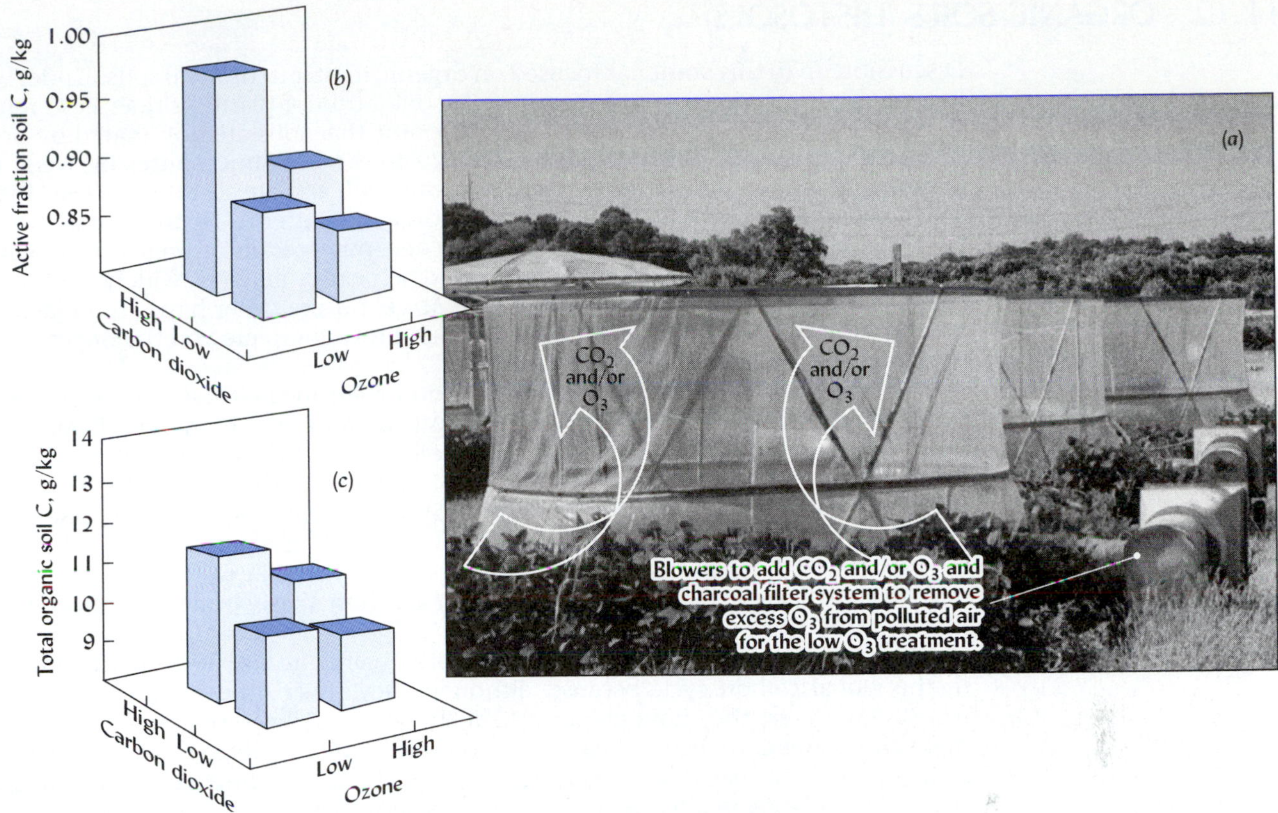

FIGURE 11.22 Atmospheric composition in these open-top field chambers (*a*) altered plant growth and physiology, and thereby also affected the amounts and forms of soil organic carbon. Increasing atmospheric CO_2 from low (360 mg/L, the ambient level) to high (500 mg/L, the level expected by 2050) enhanced photosynthesis in the plant, and thus increased the amount of fixed carbon available for translocation to the roots and eventually to the soil. (*b*) Increased root growth and exudation of carbon compounds contributed first to the active fraction of soil carbon as suggested by the pronounced effect after only 5 years of elevated atmospheric CO_2. (*c*) The level of total soil organic carbon, most of which is stable humus, was also beginning to increase. Ozone, a pollutant at ground level, injures plants, reducing photosynthesis and therefore impacting the soil in a manner opposite that of CO_2. The data suggest an interaction between the two gases, by which the full effect of CO_2 is seen only when ozone is kept low. [Data from Weil et al. (2000); photo courtesy of R. Weil]

Nitrous oxide (N_2O) is another greenhouse gas produced by microorganisms in poorly aerated soils, but since it is not directly involved in the carbon cycle, it will be discussed in the next chapter (see Section 12.10). Because the soil can act as a major source or sink for carbon dioxide, methane, and nitrous oxide, it is clear that, together with steps to modify industrial outputs, soil management has a major role to play in controlling the atmospheric levels of greenhouse gases.

TABLE 11.6 Effect of Nitrogen Fertility Management Systems on Methane Oxidation by an Arable Soil (Mollisol) in Germany

The four nitrogen treatments were applied annually for 92 years to a rotation of sugar beets, spring barley, potatoes, and winter wheat. The measurements were made on soil sampled in spring, just before the annual nitrogen applications were made. Note that farmyard manure increased methane oxidation, while inorganic nitrogen fertilizer (NH_4NO_3) reduced methane oxidation from the levels in the control and the manure-only treatment.

Soil treatment	*Soil pH*	*Soil NO_3^--N, kg/ha*	*Soil NH_4^+-N, kg/ha*	*Methane oxidation rate, nL CH_4 L^{-1}/hr^{-1}*
1. Control—no N added in any form	6.8	0.83	0.20	4.60
2. Fertilizer N—40 to 130 kg/ha N as NH_4NO_3 to meet crop needs	6.9	15.36	3.1	1.34
3. Farmyard manure applied at 20 Mg/ha	7.0	1.98	0.22	11.2
4. Farmyard manure plus N fertilizer as in #3	7.2	5.01	0.71	3.76

Data from the Static Fertilization Experiment begun in 1902 at Bad Lauchstädt, Germany, and reported in Willison et al. (1996).

11.12 ORGANIC SOILS (HISTOSOLS)

Discussion up to this point has focused on organic matter in mineral soils. Under waterlogged conditions in bogs and marshes, the oxidation of plant residues is so retarded that organic matter may accumulate to the point that soils are dominated by organic matter. If these soils contain greater than 20 to 30% organic matter by weight, depending on clay content, they are termed *organic soils* and are classified in the Histosols order in *Soil Taxonomy* (see Chapter 3). The organic matter in Histosols may be either peat or muck. *Peat* is brownish, only partially decomposed, fibrous remains of plant tissues. Some of these soils are mined and sold as peat, a material widely used in containerized plant production (see Box 11.4). *Muck,* on the other hand, is a black, powdery material in which decomposition is much more complete and the organic matter is highly humified.

Whether artificially drained for cultivation or left in their natural water-saturated state, Histosols possess unique properties resulting from their high organic matter contents. These include dark brown to black colors, very low bulk densities (0.2 to 0.4 Mg/m^3), very high water-holding capacities (often 2 to 3 times their dry weight), and high cation exchange capacities (typically 150 to 300 $cmol_c/kg$) that increase with increasing soil pH. The water- and cation-holding capacities are much higher than those of even clay-rich mineral soils on a weight basis, but similar to those of mineral soils rich in 2:1 silicate clays when considered on a volume basis (water or cation held per liter of soil).

Histosols (and Histels—permafrost soils with organic surface horizons) are important in the global carbon cycle because, although they cover little more than 1% of the world's land area, they hold about 20% of the global soil carbon. While drainage can make these soils extremely valuable for vegetable and floriculture crops (see Chapter 3), it also speeds up oxidation of organic matter, which over time leads to Histosol destruction and increased release of CO_2 to the atmosphere.

As the organic matter oxidizes in Histosols, the land surface is actually lowered, a process termed **subsidence** (see Figure 11.23). To preserve these valuable soils and reduce land subsidence, the water table should be kept as near to the surface as possible. Except for the production of such crops as rice and cranberries under flooded conditions, agriculture on Histosols is not sustainable in the long term.

BOX 11.4 ORGANIC MATERIALS FOR POTTING MEDIA

The economic importance of plants grown in containers (containerized production) has greatly increased in recent years. Container-grown plants are used in forest nurseries, for bedding-plant production, for indoor ornamental plantings, as architectural landscape elements, and in scientific greenhouse experiments. As noted in Chapters 4, 5, 7, and 9, growing plants in small pots or other containers presents special problems not encountered with field-grown plants. Most of the problems of containerized rooting media are physical in nature, principally involving water, air, and temperature. Peat and similar organic materials are lightweight, have great water-holding capacity, and provide large air-filled pores. The organic materials also provide considerable cation exchange capacity for holding available plant nutrients.

Organic wastes stabilized by thorough composting are increasingly used as the main ingredient in soilless potting mixes. In addition to providing most of the same benefits just described for peat, composted materials may also release nitrogen and other nutrients as they slowly decompose. This use of composts made from such organic wastes as sewage sludge, poultry manure, or municipal garbage not only provides a low-cost renewable source of organic potting media for the greenhouse industry, but solves some very vexing waste disposal problems, as well. Nonetheless, mined peat continues to be used in large quantities for containerized plant production.

The peats used in potting mixes are mined from deep Histosols, mainly in cool climates. Three general types of peats are recognized: *fibrous, woody,* and *sedimentary*. Fibrous peats are most desirable for potting mixes. These peats have unusually high water-holding capacities; they may hold 10 to 20 times their dry weight (2 to 4 times their volume) in water. Some fibrous peat (e.g., moss peats) may be quite acid because of their low ash content.

FIGURE 11.23 Soil subsidence due to rapid organic matter decomposition after artificial drainage of Histosols in the Florida Everglades. The house was built at ground level, with the septic tank buried about 1 m below the soil surface. Over a period of about 60 years, more than 1.2 m of the organic soil has "disappeared." The loss has been especially rapid because of Florida's warm climate, but artificial drainage that lowers the water table and continually dries out the upper horizons is an unsustainable practice on any Histosol. (Photo courtesy of George H. Snyder, Everglades Research and Education Center, Belle Glade, Fla.)

11.13 CONCLUSION

Organic matter is a complex and dynamic soil component that exerts a major influence on soil behavior, properties, and functions in the ecosystem. Because of the enormous amount of carbon stored in soil organic matter and the dynamic nature of this soil component, soil management may be an important tool for moderating the global greenhouse effect.

Organic residue decay, nutrient release and humus formation are controlled by environmental factors and by the quality of the organic materials. High contents of lignin and polyphenols, along with high C/N ratios, markedly slow the decomposition process, causing organic matter to accumulate while reducing the availability of nutrients.

For some purposes it is advantageous to manage the decomposition of organic matter outside of the soil in a process known as *composting*. Composting transforms various organic waste materials into a humuslike product that can be used as a soil amendment or a component of potting mixes. The aerobic decomposition in a compost pile can conserve nutrients while avoiding certain problems, such as noxious odors and the presence of either excessive or deficient quantities of soluble nitrogen, which can occur if fresh organic wastes are applied directly to soils.

Soil organic matter comprises three major fractions of organic compounds. The *active fraction* consists of microbial biomass and relatively easily decomposed compounds, such as polysaccharides and other nonhumic substances. Although only a small percentage of the total carbon, the active pool plays a major role in nutrient cycling, micronutrient chelation, maintenance of structural stability and soil tilth, and as a food source underpinning biological diversity and activity in soils.

Most of the organic matter is in the *passive fraction,* which contains very stable materials that may persist for centuries. This fraction provides cation exchange and water-holding capacities, but is relatively inert biologically. The so-called *slow fraction* is intermediate in stability and resistance to decomposition. It provides sources of food and energy for the steady metabolism of the authochthonous soil organisms that exist in the soil between times of residue additions. When soil is cleared of natural vegetation and brought under cultivation, the initial decline in soil organic matter is principally at the expense of the active fraction. The passive fraction is depleted only slowly and over very long periods of time.

The carbon-to-nitrogen (C/N) ratio of most soils is relatively constant, generally near 12:1. This means that the level of organic matter will be partially determined by the level of nitrogen available for assimilation into humus. Soil management for enhancing organic matter levels must, therefore, include some means of supplying nitrogen, for example, by the inclusion of leguminous plants.

The level of soil organic matter is influenced by climate (being higher in cool, moist regions), drainage (being higher in poorly drained soils), and vegetation type (being generally higher where root biomass is greatest, as under grasses). Maintenance of soil organic matter, especially the active fractions, is one of the great challenges in natural resource management. By encouraging vigorous growth of crops or other vegetation, abundant residues (which contain both carbon and nitrogen) can be returned to the soil directly or through feed-consuming animals. Also, the rate of destruction of soil organic matter can be minimized by restricting soil tillage, controlling erosion, and keeping most of the plant residues at or near the soil surface.

Organic soils (Histosols) contain a large share of the world's soil organic carbon. Histosols are far more permeable to air and water, and generally can hold somewhat more water and cations per hectare, than most mineral soils. While drainage of these soils allows them to be used for production of high-value vegetable and floriculture crops, it also leads to accelerated oxidation of the organic matter they contain and increased CO_2 release.

The decay and mineralization of soil organic matter is one of the main processes governing the economy of nitrogen and sulfur in soils—the subject of the next chapter.

STUDY QUESTIONS

1. Compare the amounts of carbon in Earth's standing vegetation, soils, and atmosphere.
2. If you wanted to apply an organic material that would make a long-lasting mulch on the soil surface, you would choose an organic material with what chemical and physical characteristics?
3. Describe how the addition of certain types of organic materials to soil can cause a nitrate depression period. What are the ramifications of this phenomenon for plant growth?
4. In addition to humic substances, what other categories of organic materials are found in soils?
5. Some scientists include plant litter (surface residues) in their definition of soil organic matter, while others do not. Write two brief paragraphs, one justifying the inclusion of litter as soil organic matter and one justifying its exclusion.
6. What soil properties are mainly influenced by the active and passive fractions, respectively, of organic matter?
7. In this textbook and elsewhere, the terms *soil organic carbon* and *soil organic matter* are used to mean almost the same thing. How are these terms related, conceptually and quantitatively? Why is the term *organic carbon* generally more appropriate for quantitative scientific discussions?
8. Explain, in terms of the balance between gains and losses, why agricultural soils generally contain much lower levels of organic carbon than similar soils under natural vegetation.
9. In what ways are soils involved in the greenhouse effect that is thought to be warming up the Earth? What are some common soil management practices that could be changed to reduce the negative effects and increase the beneficial effects of soils on the greenhouse effect?
10. What causes subsidence of Histosols? How is this phenomenon related to the sustainable use of Histosols?

REFERENCES

Batjes, N. H. 1996. "Total carbon and nitrogen in the soils of the world," *European J. Soil Sci.* **47**:151–163.

Bayer, C., J. Mielniczuk, T. Amado, L. Martin-Neto, and S. Fernandes. 2000. "Organic matter storage in a sandy clay loam Acrisol affected by tillage and cropping system in southern Brazil," *Soil and Tillage Research* **54**:101–109.

Bouwman, A. F. (ed.). 1990. *Soils and the Greenhouse Effect.* (Chichester, U.K.: Wiley).

Brady, N. C. 1990. *The Nature and Properties of Soils,* 10th ed. (New York: MacMillan).

Cadisch, G., and K. E. Giller (eds.). 1997. *Driven by Nature—Plant Litter Quality and Decomposition.* (Walingford, U.K.: CAB International).

Chen, Y., and T. Aviad. 1990. "Effects of humic substances in plant growth," in P. MacCarthy, C. E. Clapp, R. L. Malcolm, and P. R. Bloom (eds.), *Humic Substances in Soil and Crop Sciences: Selected Readings.* (Madison, Wis.: ASA Special Publications), pp. 161–186.

Dale, V. H., R. A. Houghton, A. Grainger, A. E. Lugo, and S. Brown. 1993. "Emissions of greenhouse gases from tropical deforestation and subsequent uses of the land," in National Research Council, *Sustainable Agriculture and the Environment in the Humid Tropics.* (Washington, D.C.: National Academy Press), pp. 215–260.

Eswaran, H., E. Van Den Berg, and P. Reich. 1993. "Organic carbon in soils of the world," *Soil Sci. Soc. Amer. J.* **57**:192–194.

Inderjit, K. M., M. Dakshini, and C. L. Foy (eds.). 1999. *Principles and Practices in Plant Ecology: Allelochemical Interactions.* (Boca Raton, Fla.: CRC Press).

Islam, K. R., and R. R. Weil. 2000. "Soil quality indicator properties in Mid-Atlantic soils as influenced by conservation management," *J. Soil Water Cons.* **55**:69–78.

Kern, J. S. 1994. "Spatial patterns of soil organic carbon in the contiguous United States," *Soil Sci. Soc. Amer. J.* **58**:439–455.

Magdoff, F., and H. Van Es. 2000. *Building Soils for Better Crops,* 2nd ed. Sustainable Agriculture Network Handbook Series no. 4. (Burlington, Vt.: Sustainable Agriculture Publications).

Paustian, K., W. J. Parton, and J. Persson. 1992. "Modeling soil organic matter-amended and nitrogen-fertilized long-term plots," *Soil Sci. Soc. Amer. J.* **56**:476–488.

Stevenson, F. J. 1994. *Humus Chemistry—Genesis, Composition Reactions,* 2nd ed. (New York: Wiley).

Swift, M. J., O. W. Heal, and J. M. Anderson. 1979. *Decomposition in Terrestrial Ecosystems.* Studies in Ecology, vol. 5. (Berkeley: University of California Press).

Tate, R. L. 1987. *Soil Organic Matter: Biological and Ecological Effects.* (New York: Wiley).

Torbert, H., and H. Johnson. 2001. "Soil of the intensive agriculture biome of Biosphere 2," *J. Soil Water Cons.* **56**:4–11.

Weil, R. R., K. R. Islam, and C. L. Mulchi. 2000. "Impact of elevated CO_2 and ozone on C cycling processes in soil," in *Agronomy Abstracts.* (Madison, Wis.: American Society of Agronomy), p. 47.

Weil, R. R., P. W. Benedetto, L. J. Sikora, and V. A. Bandel. 1988. "Influence of tillage practices on phosphorus distribution and forms in three Ultisols," *Agron. J.* **80**:503–509.

Weil, R. R., and G. S. Belmont. 1987. "Interactions between winged bean and grain amaranth," *Amaranth Newsletter* **3**(1):3–6.

Willison, T., R. Cook, A. Müller, and D. Powlson. 1996. "CH_4 oxidation in soils fertilized with organic and inorganic N: Differential effects," *Soil Biol. and Biochem.* **28**:135–136.

12

NITROGEN AND SULFUR ECONOMY OF SOILS

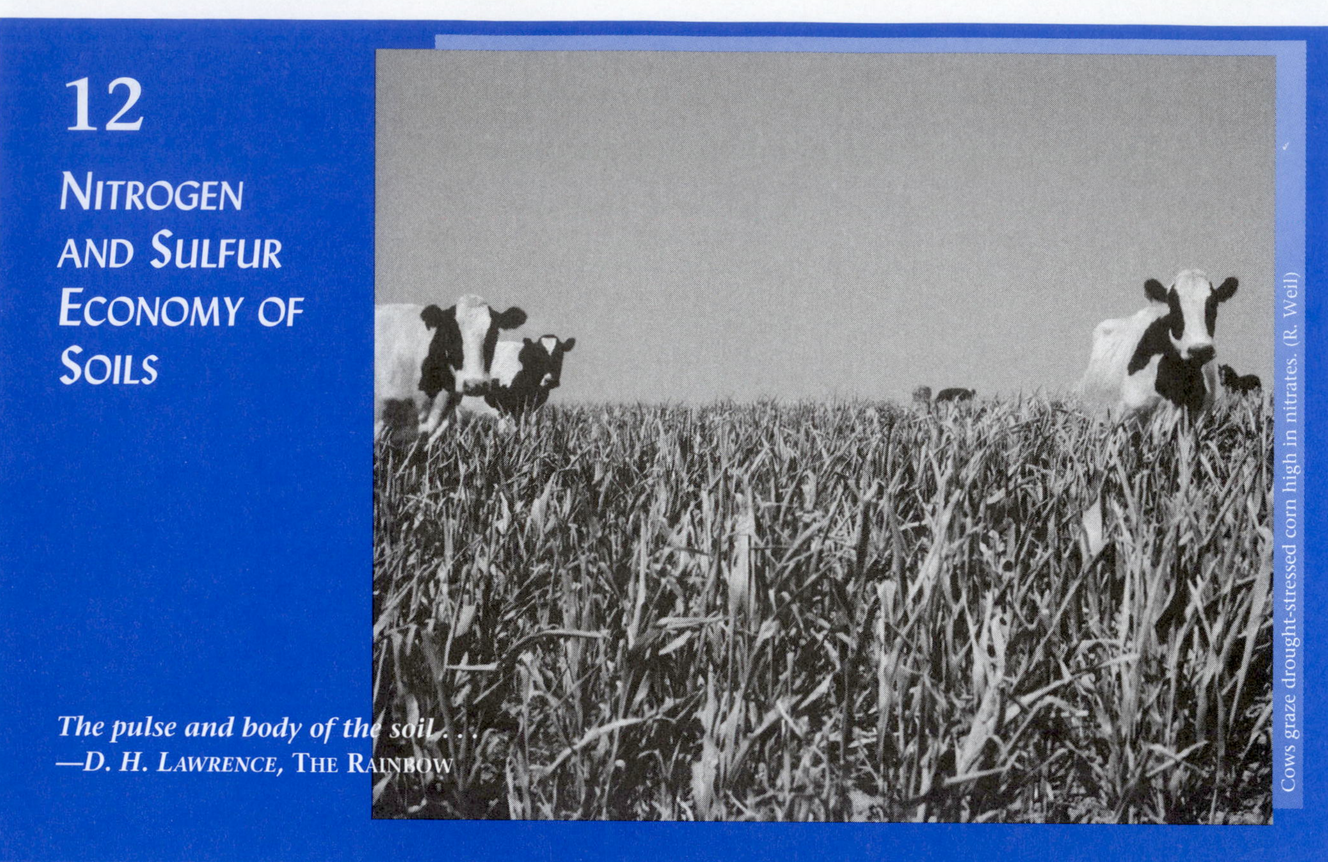

Cows graze drought-stressed corn high in nitrates. (R. Weil)

The pulse and body of the soil . . .
—D. H. LAWRENCE, THE RAINBOW

Nitrogen and sulfur share some important characteristics as essential plant nutrients. Both are found primarily in organic forms in soils, both move in soils and plants mostly in the anionic form, and both are responsible for serious environmental problems. Consequently, they will be considered together, starting with nitrogen.

More money and effort have been, and are being, spent on the management of nitrogen[1] than on any other mineral element. And for good reason: The world's ecosystems are probably influenced more by deficiencies or excesses of nitrogen than by those of any other essential element. The pale yellowish-green foliage of nitrogen-starved crops forebodes crop failure, financial ruin, and hunger for people in all corners of the world. Likewise, deficiencies of nitrogen are widespread among plants growing in the wild. Were it not for the biological fixation of nitrogen from the atmosphere by certain microorganisms, and for the recycling back to the soil of much of the nitrogen taken up in natural ecosystems, deficiencies of nitrogen would be even more widespread and ecosystem and farm productivity much reduced.

Excesses of some nitrogen compounds in soils can adversely affect human and animal health and can denigrate the quality of the environment. High nitrate levels in soils can lead to sufficiently high nitrates in drinking water as to endanger the health of human infants and ruminant animals. For this reason, nitrate levels are monitored in wells, reservoirs, and other drinking supplies. Likewise, the movement of soluble nitrogen compounds from soils to aquatic systems can distrupt the balance of those systems, leading to eutrophication, decline in oxygen content of the water, and the subsequent death of fish and other aquatic species.

Nitrogen is an essential component of protein. Because of its nutritional importance and relative scarcity, protein is highly sought after by most animals, humans included. Supplying sufficient nitrogen often represents a major expense in agricultural production. The manufacture of nitrogen fertilizer also accounts for a large part of the fossil

[1]Nitrogen in soils and in crop production is covered extensively in the review by Smil (1999). The human influence on the global nitrogen cycle is discussed by Galloway and Cowling (2002).

fuel energy used by the agricultural sector. Yet another way in which nitrogen links the soil to the wider environment is the ozone-destroying action of soil-generated nitrous oxide gas. Clearly, for the management of the soil nitrogen cycle, the ecological, financial, and environmental stakes are very high.

12.1 INFLUENCE OF NITROGEN ON PLANT GROWTH AND DEVELOPMENT

Visually rating nitrogen sufficiency: http://www.ipm.iastate.edu/ipm/icm/1997/7-28-1997/visualN.html

Roles in the Plant. Nitrogen is an integral component of many essential plant compounds. It is a major part of all amino acids, which are the building blocks of all proteins—including the enzymes, which control virtually all biological processes. Other critical nitrogenous plant components include the nucleic acids, in which hereditary control is vested, and chlorophyll, which is at the heart of photosynthesis. Nitrogen is also essential for carbohydrate use within plants. A good supply of nitrogen stimulates root growth and development, as well as the uptake of other nutrients.

Deficiency. Plants deficient in nitrogen tend to have a pale yellowish-green color (**chlorosis**), have a stunted appearance, and develop thin, spindly stems (see Plates 44, 50, and 58 after page 498). Nitrogen is quite mobile (easily translocated) within the plant and when plant uptake is inadequate, supplies are transferred to the newest foliage, causing the older leaves to show pronounced chlorosis first.

Oversupply. When too much nitrogen is applied, excessive vegetative growth occurs; the cells of the plant stems become enlarged but relatively weak, and the top-heavy plants are prone to falling over (lodging) with heavy rain or wind. High-nitrogen applications may delay plant maturity and cause the plants to be more susceptible to disease (especially fungal disease) and to insect pests. These problems are especially noticeable if other nutrients, such as potassium, are in relatively low supply.

An oversupply of nitrogen degrades crop quality. The color and flavor of fruits may be poor, as may the sugar and vitamin levels of certain vegetables and root crops. Flower production in ornamentals is reduced in favor of abundant foliage. Under certain conditions, an oversupply may result in high nitrates that are harmful to livestock in the case of foliage and to human babies in the case of leafy vegetables.

Forms of Nitrogen Taken Up by Plants. Plant roots take up nitrogen from the soil solution principally as nitrate (NO_3^-) and ammonium (NH_4^+) ions.[2] Although certain plants grow best when provided mainly one or the other of these forms, a relatively equal mixture of the two ions gives the best results with most plants. These two ions differ in their effect on the pH of the rhizosphere. Nitrate *anions* (negatively charged ions) move easily to the root with the flow of soil water and exchange at the root surface with HCO_3^- or OH^- ions that, in turn, increase the pH of the soil solution immediately around the root. In contrast, ammonium *cations* (positively charged ions) exchange at the root surface with hydrogen ions, thereby lowering the pH of the solution around the roots (Section 9.6).

12.2 ORIGIN AND DISTRIBUTION OF NITROGEN

Some 75,000 Mg of nitrogen is found in the air above 1 ha of soil,[3] but the very strong triple bond between two nitrogen atoms makes this gas quite inert and not directly usable by plants or animals. Were it not for the ability of certain microorganisms to break this triple bond and to form nitrogen compounds (see Section 12.11), vegetation in terrestrial ecosystems around the world would be rather sparse, and little nitrogen would be found in soils.

[2] Nitrite (NO_2^-) can also be taken up, but this ion is toxic to plants. Fortunately, it rarely occurs in greater then trace quantities in soils.

[3] About 98% of the earth's nitrogen is contained in the igneous rocks deep under the planet's crust, where it is effectively out of contact with the soil–plant–air–water environment in which we live. We will therefore focus our attention on the remaining 2% that cycles in the biosphere.

The nitrogen content of surface mineral soils normally ranges from 0.02 to 0.5%, a value of about 0.15% being representative for cultivated soils. A hectare of such a soil would contain about 3.5 Mg nitrogen in the A horizon and perhaps an additional 3.5 Mg in the deeper layers. In forest soils, the litter layer (O horizon) might contain another 1 to 2 Mg of nitrogen. Although these figures are low compared to those for the atmosphere, the soil contains 10 to 20 times as much nitrogen as does the standing vegetation (including roots) of either forested or cultivated areas. Most of the nitrogen in terrestrial systems is found in the soil.

Most soil nitrogen occurs as part of organic molecules. Soil organic matter typically contains about 5% nitrogen; therefore, the distribution of soil nitrogen closely parallels that of soil organic matter (see Section 11.10). Except where large amounts of chemical fertilizers have been applied, inorganic (i.e., mineral) nitrogen seldom accounts for more than 1 to 2% of the total nitrogen in the soil. Unlike most of the organic nitrogen, the mineral forms of nitrogen are mostly quite soluble in water, and may be easily lost from soils through leaching and volatilization.

12.3 THE NITROGEN CYCLE

Overview of the nitrogen cycle: http://users.rcn.com/jkimball.ma.ultranet/BiologyPages/N/NitrogenCycle.html

As it moves through the **nitrogen cycle**, an atom of nitrogen may appear in many different chemical forms, each with its own properties, behaviors, and consequences for the ecosystem. This cycle explains why vegetation (and, indirectly, animals) can continue to remove nitrogen from a soil for centuries without depleting the soil of this essential nutrient. The **biosphere** does not run out of nitrogen, because it uses the same nitrogen over and over again. The principal pools and forms of nitrogen, and the processes by which they interact in the cycle, are illustrated in Figure 12.1. This figure deserves careful study; we will refer to it frequently as we discuss each of the major divisions of the nitrogen cycle.

12.4 IMMOBILIZATION AND MINERALIZATION

The great bulk (95 to 99%) of the soil nitrogen is in organic compounds that protect it from loss but leave most of it unavailable to higher plants. Much of this nitrogen is present as amine groups ($R{-}NH_2$), largely in proteins or as part of humic compounds. When soil microbes attack these compounds, simple amino compounds[4] ($R{-}NH_2$) are formed. Then the amine groups are hydrolyzed, and the nitrogen is released as ammonium ions (NH_4^+), which can be oxidized to the nitrate form. This enzymatic process, termed ***mineralization*** (Figure 12.1), may be indicated as follows, using an amino compound as an example of the organic nitrogen source:

$$\xrightarrow{\text{Mineralization}}$$

$$R{-}NH_2 \underset{-2H_2O}{\overset{+2H_2O}{\rightleftharpoons}} OH^- + R{-}OH + NH_4^+ \underset{-O_2}{\overset{+O_2}{\rightleftharpoons}} 4H^+ + \text{energy} + NO_2^- \underset{-\frac{1}{2}O_2}{\overset{+\frac{1}{2}O_2}{\rightleftharpoons}} \text{energy} + NO_3^- \quad (12.1)$$

$$\xleftarrow{\text{Immobilization}}$$

Many studies have shown that only about 1.5 to 3.5% of the organic nitrogen of a soil mineralizes annually. Even so, this rate of mineralization provides sufficient mineral nitrogen for normal growth of natural vegetation in most soils. Research shows that mineralized soil nitrogen constitutes a major part of the nitrogen taken up by a crop. If the organic matter content of a soil is known, one can make a rough estimate of the amount of nitrogen likely to be released by mineralization during a typical growing season (Box 12.1).

The opposite of mineralization is ***immobilization,*** the conversion of inorganic nitrogen ions (NO_3^- and NH_4^+) into organic forms (see Figure 12.1). As microorganisms decompose carbonaceous organic residues in the soil, they may require more nitrogen than is contained in the residues themselves. The microorganisms then incorporate

[4] Amino acids such as lysine (CH_2NH_2COOH) and alanine (CH_3CHNH_2COOH) are examples of these simpler compounds. The R in the generalized formula represents the part of the organic molecule with which the amino group (NH_2) is associated. For example, for lysine, the R is CH_2COOH.

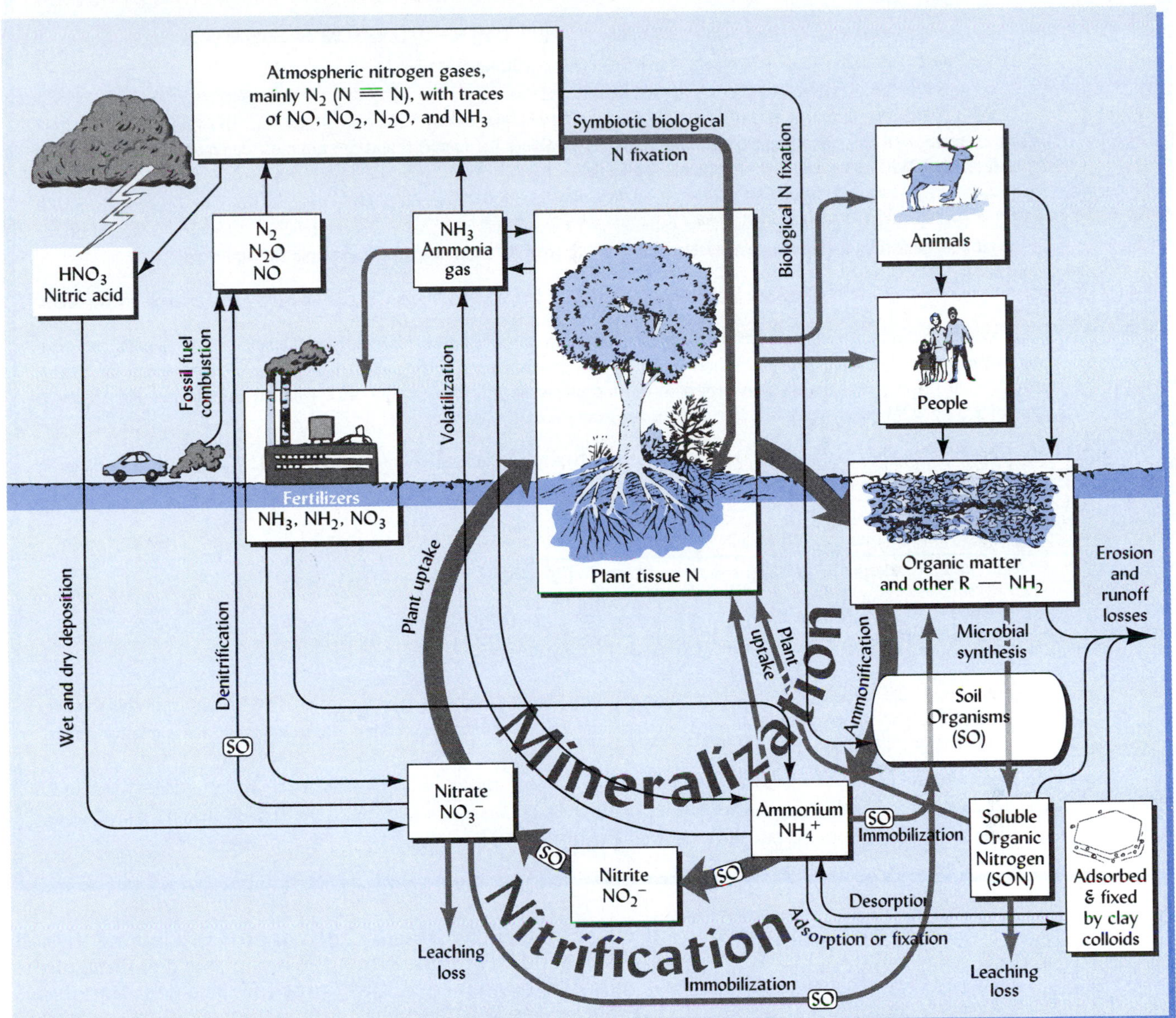

FIGURE 12.1 The nitrogen cycle, emphasizing the primary cycle (heavy, dark arrows) in which organic nitrogen is mineralized, plants take up the mineral nitrogen, and eventually organic nitrogen is returned to the soil as plant residues. Note also the pathways by which nitrogen is lost from the soil and the means by which it is replenished. The compartments represent various forms of nitrogen; the arrows represent processes by which one form is transformed into another. Soil organisms (rounded boxes labeled "SO") are the driving force for most of the reactions in the cycle.They create enzymes that catalyze the different biochemical reactions, either within their microbial bodies or in adjacent sites where the enzymes may have been excreted.

Ammonium nitrogen is subject to five possible fates in the nitrogen cycle: (1) *immobilization* by microorganisms; (2) removal by *plant uptake;* (3) *fixation* in the interlayers of certain 2:1 clay minerals; (4) transformation into ammonia gas and loss to the atmosphere by *volatilization;* and (5) oxidation to nitrite and subsequently to nitrate by a microbial process called *nitrification.*

Nitrogen in the nitrate form is highly mobile in the soil and in the environment. Whether added as fertilizer or produced in the soil by nitrification, nitrate may take any of four paths in the nitrogen cycle: (1) *immobilization* by microorganisms; (2) removal by *plant uptake;* (3) loss by *leaching* in drainage water; or (4) *volatilization* to the atmosphere as several nitrogen-containing gases.

BOX 12.1 CALCULATION OF NITROGEN MINERALIZATION

If the organic matter content of a soil, soil management practices, climate, and soil texture are known, it is possible to make a rough estimate of the amount of N likely to be mineralized each year. The following equation may be used:

$$\frac{\text{kg N mineralized}}{\text{ha 15 cm deep}} = \left(\frac{\text{A kg SOM}}{\text{100 kg soil}}\right)\left(\frac{\text{B kg soil}}{\text{ha 15 cm deep}}\right)\left(\frac{\text{C kg N}}{\text{100 kg SOM}}\right)\left(\frac{\text{D kg SOM mineralized}}{\text{100 kg SOM}}\right) \qquad (12.2)$$

where A = The amount of soil organic matter (SOM) in the upper 15 cm of soil, given in kg SOM per 100 kg soil. This value may range from close to zero to over 75% (in a Histosol) (see Section 11.12). Values between 0.5 and 5% are most common. ☞ Use a value of 2.5% (2.5 kg SOM/100 kg soil) for the example shown below.

B = The weight of soil per hectare to the depth of 15 cm. Most nitrogen used by plants is likely to come from this upper horizon. If it is 15 cm deep, 2×10^6 kg/ha is a reasonable estimate of its weight per hectare. See Section 4.5 to calculate the weight of this horizon if bulk density of a soil is known. ☞ Use 2×10^6 kg soil/ha 15 cm deep in the example shown below.

C = The amount of nitrogen in the SOM (see Section 11.3). ☞ Use the typical figure of 5 kg N/100 kg SOM in the example shown below.

D = The amount of SOM likely to be mineralized in 1 year for a given soil. This figure depends upon the soil texture, climate, and management practices. Values of around 2% are typical for a fine-textured soil, while values of around 3.5% are typical for coarse-textured soils. Slightly higher values are typical in warm climates; slightly lower values are typical in cool climates. ☞ Assume a value of 2.5 kg SOM mineralized/100 kg SOM for the example shown below.

The amount of nitrogen likely to be released by mineralization during a typical growing season may be calculated by substituting the example values indicated above into equation 12.2 as follows:

$$\frac{\text{kg N mineralized}}{\text{ha 15 cm deep}} = \left(\frac{\text{2.5 kg SOM}}{\text{100 kg soil}}\right)\left(\frac{2 \times 10^6\text{ kg soil}}{\text{ha 15 cm deep}}\right)\left(\frac{\text{5 kg N}}{\text{100 kg SOM}}\right)\left(\frac{\text{2.5 kg SOM mineralized}}{\text{100 kg SOM}}\right)$$

$$\frac{\text{kg N mineralized}}{\text{ha}} = \left(\frac{2.5}{100}\right)\left(\frac{2 \times 10^6}{1}\right)\left(\frac{5}{100}\right)\left(\frac{2.5}{100}\right) = 62.5 \text{ kg N/ha}$$

Contributions from the deeper layers of this soil might be expected to bring total nitrogen mineralized in the root zone of this soil during a growing season to over 120 kg N/ha. Most nitrogen mineralization occurs during the growing season when the soil is relatively moist and warm.

These calculations estimate the nitrogen mineralized annually from a soil that has not had large amounts of organic residues added to it. Animal manures, legume residues, or other nitrogen-rich organic soil amendments would mineralize much more rapidly than the native soil organic matter, and thus would substantially increase the amount of nitrogen available in the soil.

mineral nitrogen ions into their cellular components, such as proteins, leaving the soil solution essentially void of NO_3^- and NH_4^+ ions. When the organisms die, some of the organic nitrogen in their cells may be converted into forms that make up the humus complex, and some may be released as NO_3^- and NH_4^+ ions. Mineralization and immobilization occur simultaneously in the soil; whether the *net* effect is an increase or a decrease in the mineral nitrogen available depends primarily on the ratio of carbon to nitrogen in the organic residues undergoing decomposition (see Section 11.3).

12.5 SOLUBLE ORGANIC NITROGEN[5]

Until recently it has been uncertain as to whether any organic nitrogen compounds were sufficiently soluble to be taken up by plants and to be leached from the soil. We now know that soluble organic nitrogen (SON) compounds exist in both arable and natural soil systems. They account for about 0.3 to 1.5% of the total organic nitrogen in soils, a pool size similar to that of mineral nitrogen (NH_4^+ and NO_3^-). In fact, where organic manures have been applied to arable soils, or where permanent grassland has been

[5] For a review of what is known of the nature, significance, and analyses of soluble organic nitrogen in agricultural soils, see Murphy et al., 2000. Perakis and Hedin (2002) report on dissolved organic nitrogen in unpolluted forests.

grown for many years, SON contents are often considerably higher than those of mineral nitrogen. Also the nitrogen mobilized from the litter of some forest species may have SON to inorganic N ratios of 10:1 or higher.

Plants can absorb organic nitrogen. In fact, in some nitrogen-limited ecosystems SON is the primary source of this nutrient. The SON may be taken up directly by the plant roots, or it may be assimilated through mycorrhizal associations. Most root uptake of SON occurs, however, after the readily-decomposable organic N compounds are mineralized to NH_4^+ and NO_3^- ions.

Soluble organic nitrogen can move downward in leaching waters and into streams and rivers, thereby contributing to environmental problems downstream. For example, nearly one-fourth of the nitrogen the Mississippi River carries into the Gulf of Mexico is in the SON form. The chemical constituents of SON have not been fully identified, although it is known that some may interact with inorganic colloids while others react primarily with the soil organic matter.

12.6 AMMONIUM FIXATION BY CLAY MINERALS[6]

Like other positively charged ions, ammonium ions are held in exchangeable form, available for plant uptake, but partially protected from leaching. However, because of the particular size of the ammonium and potassium ions, they can become entrapped within cavities in the crystal structure of certain clays (see Figure 12.1). Several 2:1-type clay minerals, especially vermiculites, have the capacity to *fix* both ammonium and potassium ions in this manner (see Figures 8.7 and 13.24). Ammonium and potassium ions fixed in the rigid part of a crystal structure are held in a nonexchangeable form, from which they are released only slowly to higher plants and microorganisms. Ammonium fixation by clay minerals is generally greater in subsoil than in topsoil, due to the higher clay content of subsoils.

12.7 AMMONIA VOLATILIZATION

Ammonia gas (NH_3) that is produced in the soil–plant system from the breakdown of organic residues and farm manures, and from chemical fertilizers such as anhydrous ammonia, is in equilibrium with ammonium ions according to the following reversible reaction:

$$\underset{\text{Dissolved ions}}{NH_4^+ + OH^-} \rightleftharpoons H_2O + \underset{\text{Gas}}{NH_3\uparrow} \qquad (12.3)$$

From this reaction we can draw three conclusions. First, ammonia volatilization will be more pronounced at high pH levels (i.e., OH^- ions drive the reaction to the right); second, ammonia-gas-producing amendments will drive the reaction to the left, raising the pH of the solution in which they are dissolved; and, finally, as a moist soil dries, the water is removed from the right side of the equation, pulling the equation to the right.

Incorporation of manure and fertilizers into the top few centimeters of soil can reduce ammonia losses by 25 to 75% compared to leaving these materials on the soil surface. In natural grasslands and pastures, incorporation of animal wastes by earthworms and dung beetles is critical in maintaining a favorable nitrogen balance and a high animal-carrying capacity in these ecosystems.

Volatilization from Wetlands. Gaseous ammonia loss from nitrogen fertilizers applied to the surface of fishponds and flooded rice paddies can also be appreciable, even on slightly acid soils. As algae in the water photosynthesize, they extract CO_2 from the water and reduce the amount of carbonic acid formed. As a result, the pH of the water increases markedly, especially during daylight hours, to levels commonly above 9.0. At these pH levels, ammonia is released from ammonium compounds and goes directly into the atmosphere. Natural wetlands lose ammonium by a similar daily cycle.

[6] This chemical fixation of ammonia is caused by the entrapment or other strong binding of NH_4^+ ions by certain silicate clays. This type of fixation (by which K ions are similarly bound) is not to be confused with the very beneficial biological fixation of atmospheric nitrogen gas into compounds usable by plants (see Section 12.10).

Ammonia Absorption. By the reverse of the ammonium loss mechanisms just described, both soils and plants can absorb ammonia from the atmosphere; thus, the soil–plant system can help cleanse ammonia from the air, while deriving usable nitrogen for plants and soil microbes. Forests may receive a significant proportion of their nitrogen requirements as ammonia carried by wind from fertilized cropland and cattle feedlots located many kilometers away.

12.8 NITRIFICATION

Ammonium ions in the soil may be enzymatically oxidized by certain soil bacteria, yielding first nitrites and then nitrates. These bacteria are classed as **autotrophs,** because they obtain their energy from oxidizing the ammonium ions rather than organic matter. The process termed **nitrification** (see Figure 12.1) consists of two main sequential steps. The first results in the conversion of ammonium to nitrite by a specific group of autotrophic bacteria (***Nitrosomonas***). The nitrite so formed is then immediately acted upon by a second group of autotrophs, ***Nitrobacter.*** The enzymatic oxidation releases energy:

Step 1

$$\underset{\text{Ammonium}}{NH_4^+} + 1\tfrac{1}{2}O_2 \xrightarrow[\text{bacteria}]{\textit{Nitrosomonas}} \underset{\text{Nitrite}}{NO_2^-} + 2H^+ + H_2O + 275 \text{ kJ energy} \qquad (12.4)$$

Step 2

$$\underset{\text{Nitrite}}{NO_2^-} + \tfrac{1}{2}O_2 \xrightarrow[\text{bacteria}]{\textit{Nitrobacter}} \underset{\text{Nitrate}}{NO_3^-} + 76 \text{ kJ energy} \qquad (12.5)$$

So long as conditions are favorable for both reactions, the second transformation is thought to follow the first closely enough to prevent accumulation of nitrite. This is fortunate, because even at concentrations of just a few parts per million, nitrite is quite toxic to most plants and to mammals.

Regardless of the source of ammonium (i.e., ammonia-forming fertilizer, sewage sludge, animal manure, or any other organic nitrogen source), nitrification can significantly increase soil acidity by producing H^+ ions, as shown in the preceding reaction (see also Section 9.7). These H^+ ions and associated Al^{3+} replace Ca^{2+} and Mg^{2+} ions on the exchange complex, adversely affecting some forest species (see Figure 12.2). In humid regions, liming materials are commonly added to counteract acidity in cultivated soils (see Chapter 9).

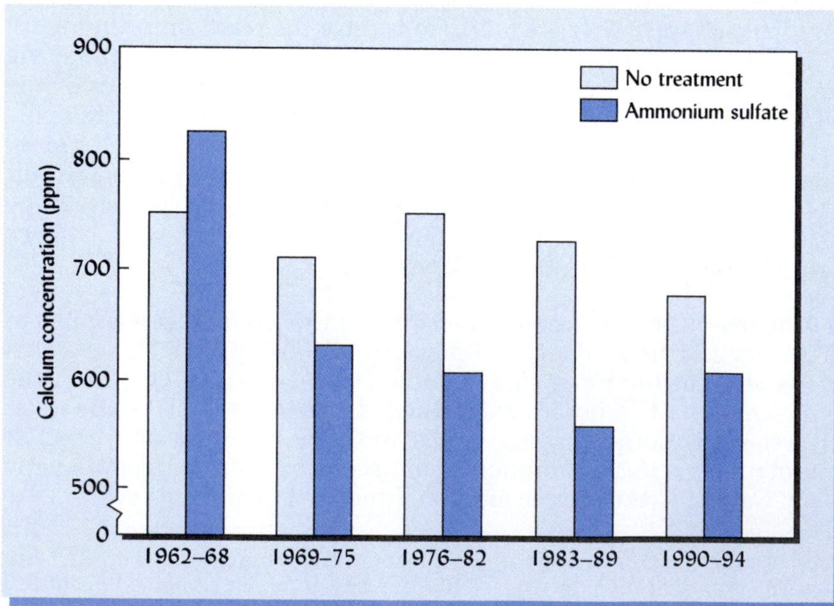

FIGURE 12.2 Depletion of calcium in sugar maple tree rings in response to applications of ammonium sulfate equivalent to twice the normal rates of bulk N and S deposition from the atmosphere. The hydrogen ions from nitric and sulfuric acids formed by microbial oxidation of ammonium sulfate replace calcium from the exchange complex. This element is leached from the tree root zone and is unavailable for plant uptake. [Redrawn from De Walle et al. (1999)]

Soil Conditions Affecting Nitrification

The nitrifying bacteria are much more sensitive to environmental conditions than are the broad groups of heterotrophic organisms responsible for the release of ammonium from organic nitrogen compounds (**ammonification**). Nitrification requires a supply of ammonium ions, but excess NH_4^+ is toxic to *Nitrobacter* and must be avoided. The nitrifying organisms, being aerobic, require oxygen to make NO_2^- and NO_3^- ions, and are therefore favored in well-drained soils. The optimum moisture for these organisms is about the same as that for most land plants (about 60% of the pore space filled with water). Since they are autotrophs, their carbon sources are bicarbonates and CO_2. They perform best if the temperature is kept between 20 and 30°C, and perform very slowly if the soil is cold (below 5°C).

Nitrification proceeds most rapidly where there is an abundance of exchangeable Ca^{2+} and Mg^{2+} and where nutrient levels are optimum for the growth of higher plants. Nitrifying organisms are quite sensitive to some pesticides, being constrained if these pesticides are applied at high rates. Studies suggest, however that at ordinary field rates, most pesticides have only a minimal effect on nitrification.

Use of nitrification inhibitors:
http://ohioline.osu.edu/agf-fact/0201.html

In recent years, chemicals have been found that can inhibit or slow the nitrification process, thereby reducing the nitrate leaching potential. Such compounds, as well as others that retard the dissolution of urea, are discussed in Box 12.2.

12.9 THE NITRATE LEACHING PROBLEM

Nitrogen management and water quality:
http://www.soil.ncsu.edu/publications/Soilfacts/AG-439-02/

The negatively charged nitrate ions are repelled rather than adsorbed by the negatively charged colloids that dominate most soils. Consequently, they move downward freely with drainage water and are thus readily leached from the soil. The loss of nitrogen in this manner is of concern for two basic reasons: (1) Such loss represents an impoverishment of the ecosystem regardless if cultivated crops are grown, and (2) leaching of nitrate causes several serious environmental problems.

Productivity Loss

Productivity of the ecosystem suffers not only because of the loss of nitrogen, but also because the leaching of nitrate from acidic sources (nitrification or acid rain) facilitates the loss of calcium and other nutrient cations. Both types of nutrient losses will generally reduce the ecosystem productivity (see also Section 12.17). In the case of managed land, there is also an economic loss equal to the value of the lost nitrogen.

Environmental Impact

Nitrate and other chemical entry into groundwater:
http://sis.agr.gc.ca/cansis/publications/health/chapter10.html

Environmental problems caused by nitrogen are mainly associated with the movement of nitrate through drainage waters to the groundwater. It may reach domestic wells, and may also eventually flow underground to surface waters, such as streams, lakes, and estuaries. The nitrate may contaminate drinking water and cause eutrophication and associated life-threatening consequences to fish and other aquatic species (Box 12.3).

Nitrate losses are determined by the amount of water leaching through the soil, and the nitrate content of that water. They would be highest in a heavily fertilized sandy soil in a high-rainfall area, or a similar soil under irrigation. Under some circumstances leaching may also be accentuated in conservation tillage systems which increase water infiltration and thereby increase water available for leaching.

Some mature forest ecosystems achieve a very close balance between plant uptake of nitrogen and the return of nitrogen in litter fall and dead trees, the groundwater commonly containing less than 1 mg/L nitrate. However, inputs of nitrogen from the atmosphere (e.g., nitrate in acid rain or from nitrogen fixation) and disturbances of the system (e.g., timber harvest) can overload the forest ecosystem, resulting in annual leaching losses of up to 25 or 30 kg N/ha.

Even greater losses may come from agricultural systems in which inputs of nitrogen regularly exceed the amounts removed by plant uptake and harvest. Heavy

BOX 12.2 REGULATION OF SOLUBLE NITROGEN WITH FERTILIZER TECHNOLOGY

Perhaps the most challenging aspect of nitrogen control is the regulation of the soluble forms of this element after it enters the soil. Availability at the proper time and in suitable amounts, with a minimum of loss, is the ideal. Even where commercial fertilizers are used to supply much of the nitrogen, maintaining an adequate but not excessive quantity of available nitrogen is not an easy task. Split applications of nitrogen fertilizer is one way to accomplish this. This technique involves splitting nitrogen application into several small doses applied as the growing crop develops, rather than applying the entire amount at or before planting time. In sandy soils or where rainfall or irrigation is high early in the season, applying all the nitrogen at planting time may result in much of the nitrogen leaching below the root zone before the crop has had a chance to use it. In regions with high rainfall during the winter, it is important to avoid leaving a large amount of soluble nitrogen in the soil profile after the crop has completed its growth.

Nitrification Inhibitors.[a] Because ammonium is much less susceptible than nitrate to losses by leaching or denitrification, nitrogen fertilizers might be used more efficiently if nitrification of ammonium from fertilizers could be slowed down until the crop is ready to make use of the nitrogen. To this end chemical companies have developed compounds called ***nitrification inhibitors***, which inhibit the activity of the *Nitrosomonas* bacteria that convert ammonium to nitrate in the first step of nitrification. (Note that a chemical that inhibited *Nitrobacter* would ***not*** be useful, as this would cause toxic nitrite to accumulate.) Three commercially available nitrification inhibitors are dicyandiamide (DCD), nitrapyrin (N-Serve®) and etridiazol (Dwell®). When mixed with nitrogen fertilizers, these compounds can temporarily prevent nitrate formation. The key word here is *temporarily*, for when conditions are favorable for nitrification, the inhibition usually lasts only a few weeks (less if soils are above 20°C). Research has shown that these materials may pay for themselves with improved fertilizer efficiency when conditions are very conducive to nitrogen loss by leaching or denitrification. Thus, nitrification inhibitors have been of value primarily in relatively wet years on sandy soils (reduced leaching) and on imperfectly drained soils (reduced denitrification).

Slow-Release Nitrogen Fertilizers. The use of slow-release fertilizer materials provides another means of reducing nitrogen losses from fertilized soils. Unlike most inorganic fertilizers, these materials release soluble nitrogen slowly in the soil so that nitrogen availability is better sychronized with plant uptake. Some stabilized organic materials, such as compost and digested sewage sludge (e.g., Milorganite®), are used for this purpose. However, most slow-release nitrogen fertilizers are made by treating urea with materials that slow its dissolution or inhibit its hydrolysis to ammonium. Urea-formaldehyde, isobutylidene diurea (IBDU), resin-coated fertilizers (e.g., Osmocote®), and sulfur-coated urea are all examples of slow-release nitrogen fertilizers. In the case of the latter, the sulfur content (10 to 20%) may be a benefit where sulfur is in low supply (Section 12.18), or a problem where the extra acidity generated by the sulfur (Section 12.22) would be undesirable. Because of additional manufacturing costs and lower concentrations of nitrogen, slow-release fertilizers cost from 1.5 (for sulfur-coated ureas) to 4 times as much as plain urea per unit of nitrogen. Nonetheless, slow-release materials are practical for high-value plant production where the cost of fertilizers is not critical (certain vegetables, turf, and ornamentals) and for certain situations in which nitrogen loss from urea would be very high (e.g., some rice paddy soils[b]). They are widely applied to turf grass because they are not likely to cause fertilizer burn, and because they provide the convenience of fewer applications. Turf fertilized by a single annual application of soluble nitrogen may lose 10 to 50% of the applied nitrogen by leaching, while 3% or less is lost if slow-release fertilizers are used.

[a]For a review of this subject, see Prasad and Power (1995).

[b]Sulfur-coated urea should not be applied in flooded rice paddies, as the reduced iron in the floodwater will combine with the sulfur coating to form insoluble FeS, which locks up the nitrogen in the fertilizer.

---

nitrogen fertilization (especially common for vegetables, corn, and other cash crops) exceeds what the plants are able to utilize, and can be a major cause of excessive nitrate leaching (Figure 12.3).

How does nitrate poisoning happen?
http://muextension.missouri.edu/explore/envqual/wq0258.htm

Ineffective management of manure from concentrated livestock production facilities is another common cause of nitrate contamination of ground and surface waters. Especially when combined with fertilizer nitrogen applications, manure can provide this element in quantities far in excess of plant uptake and can result in the pollution of both water and the atmosphere. Data from three European countries illustrate this point (Table 12.1). It is no wonder that shallow groundwater under some farm conditions contains nitrate in excess of 45 mg/L (10 mg N/L),[7] the legal limit for drinking water in the United States and the European Union.

[7] Nitrate concentrations are sometimes reported as *N in nitrate form*. To convert mg nitrate (NO_3^-)/L to mg (N in NO_3^- form)/L, divide by 4.4.

BOX 12.3 NITROGEN AND THE HEALTH OF HUMANS AND AQUATIC ANIMALS

High nitrates in groundwater or surface water that is used as drinking water can cause toxic effects. **Methemoglobinemia,** also called *blue baby syndrome,* occurs when nitrates are converted to nitrites in the guts of human infants and ruminant animals. The nitrites decrease the blood's ability to carry oxygen to the body cells. Since inadequately oxygenated blood lacks the red color, infants with the condition take on a bluish skin color. Although death from methemoglobinemia is quite rare, regulatory agencies in most countries set a limit on the amount of nitrates in drinking water. In the United States and European Union this limit is 45 mg/L nitrate (or 10 mg/L N in the nitrate form).

A much more widespread nitrogen induced water-quality problem is the degradation of aquatic ecosystems. A fishing trip off the coast of Louisiana would alert you to a "dead zone" in the Gulf of Mexico that is largely devoid of fish. As shown in the map, this zone reaches westward from the mouth of the Mississippi river nearly 300 miles, to near the Texas border. It varies in size from year to year, but typically covers an area as large as the state of Connecticut. In depth it varies from 4 to 5 m near the shore to as deep as 60 m.

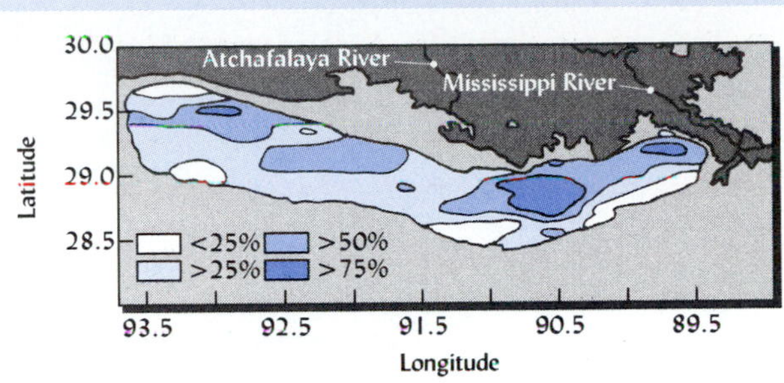

Frequency of midsummer hypoxia occurrence, 1985 to 1997

Increased human input of nutrients, particularly nitrogen, into these coastal waters is the cause of this environmental malady. Nutrient-rich freshwater carried by the Mississippi River glides over the gulf waters that are cooler, more salty, and heavier. There is little mixing of the fresh and gulf waters. The nitrogen and other nutrients stimulate explosive growth of algae and other organisms, which sink to the bottom when they die. In decomposing this dead tissue, bacteria and other organisms utilize the dissolved oxygen in the saltwater layer, leaving the water too low in oxygen to sustain animal life.

The fish and other aquatic species either die or move to other areas more adequately supplied with oxygen. This state of low oxygen in the water (less than 2 to 3 mg O_2/L) is known as ***hypoxia,*** and the process that brings it about is called ***eutrophication*** (see also Box 13.1). While the Mississippi-related dead zone is the largest hypoxic area in the western hemisphere, even larger dead zones can be found in the Baltic Sea and the northwestern shelf of the Black Sea.

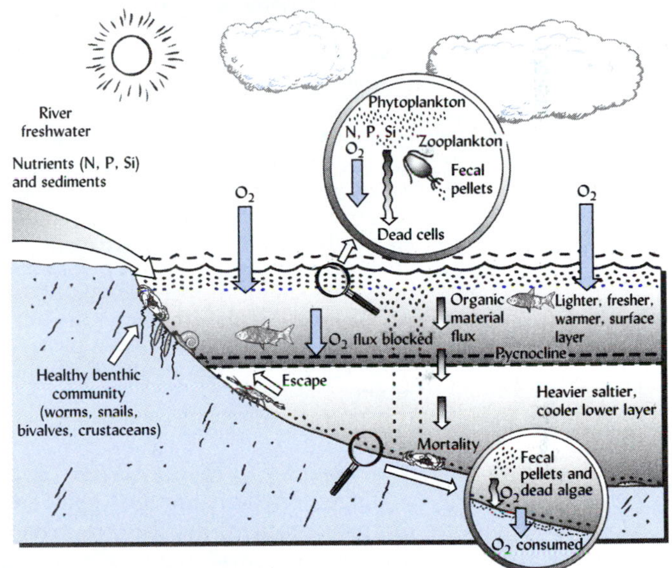

Eutrophication process leading to hypoxia

Historical records show that concentrations of nitrogen in the Mississippi River and its tributaries have tripled in the past 30 years. While some of this increase (about 30%) may be attributed to increased precipitation in the watershed, the bulk is due to human activities, especially those in agriculture. For example, fertilizer usage in the watershed has increased some sevenfold since the 1950s, while legume and pasture hectarage has increased about 50% over the same period. Critical assessments suggest that about half of the 1.5 million Mg of nitrogen the Mississippi carries to the gulf annually is from fertilizer or the mineralization of organic nitrogen within the soil (see chart). Manure accounts for about 15% of this nitrogen, while about 11% comes from point sources from industrial municipal activities. The remainder (24%) is from runoff, erosion, atmospheric deposition, and other such sources. In the coming decades, we surely must make organized efforts to constrain such transfers of nitrogen and other nutrients from upstream watersheds to oceans and other downstream water bodies.

11% Point sources
15% Manure
50% Fertilizer and soil N pool
24% Runoff/erosion deposition and others

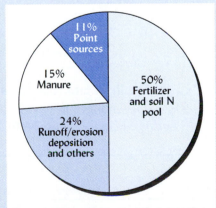

Sources of N carried by the Mississippi River

Sources: Upper map and center figure. CAST (1999): Lower chart drawn from data in Goolsby et al. (1999). For worldwide examples of hypoxia, see Diaz (2001).

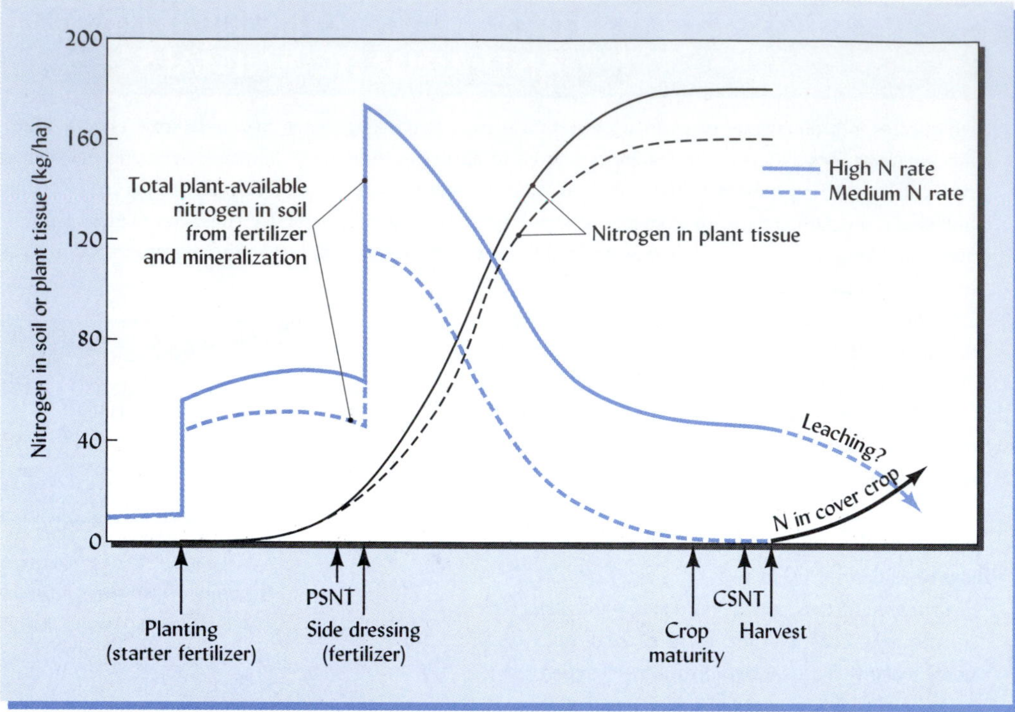

FIGURE 12.3 Two nitrogen fertilizer systems for corn production in the Midwest. The system that has been dominant in the past (solid lines) involves very high nitrogen applications and continuous corn culture. At least 150 kg N/ha is applied—part as starter fertilizer at or before planting time, the remainder as a side-dressing just before the most rapid growth stage. Unfortunately, when the crop matures and is harvested, much soluble nitrogen remains in the soil, probably in the form of nitrates. If not captured by a cover crop planted in the corn at harvest time, the nitrogen is subject to leaching during the fall and winter months. This leads to contamination of groundwater and, eventually, surface water. More environmentally sound systems involve crop rotation that include legumes, or if corn is grown continuously, the rate of nitrogen fertilizer is greatly reduced (broken lines). A preside-dress nitrate soil test (PSNT) is used to determine the amounts of nitrogen to apply. At the end of the season, a cornstalk nitrate test (CSNT) is used to assess plant nitrogen status. Crop yields and economic returns are about the same, but the nitrogen remaining in the soil at crop harvest is low and nitrate contamination is minimized.

Management Can Reduce Losses. Even in regions of high leaching potential, careful soil management can prevent excessive nitrate losses. The timing of modest fertilizer and manure applications should provide nitrogen when the plants need it, not much before or after the period of active plant uptake. If this is not economically possible, other crops in the rotation, including cover crops, should be planted immediately following the cash crop to take up the unused nitrates (see Figure 12.3). If such guidelines are followed, nitrogen leaching may be kept less than 5 or 10% of the nitrogen applied.

Nitrate Recycling in the Humid Tropics. Much of the nitrate mineralized in certain highly weathered, acid, tropical Oxisols and Ultisols leaches below the root zone before annual

TABLE 12.1 The Input into Soils of Nitrogen from Animal Manures and Fertilizer in Three European Countries

The uptake of nitrogen by crops and residual nitrogen not assimilated are shown. Some of the residual nitrogen is incorporated into soil organic matter, but much moves into the environment through leaching, runoff, and volatilization into the atmosphere.

	Nitrogen supply, 1000 Mg				*Residual nitrogen*	
Country	*Manure*	*Fertilizer*	*Total*	*N uptake by crops, 1000 Mg*	*Total, 1000 Mg*	*Per hectare, kg*
Belgium/Luxembourg	380	199	580	211	369	240
Denmark	434	381	816	287	529	187
The Netherlands	752	504	1,255	285	970	480

Data from Leuch et al. (1995).

crops such as corn can take it up. Recently it has been found that some of this leached nitrate is not lost to groundwater, but is stored several meters deep in the profile on the highly weathered positively charged clays. Deep-rooted trees are capable of taking up this deep subsoil nitrate and subsequently using it to enrich the surface soil when they shed their leaves. Trees such as *Sesbania,* grown in rotation with annual food crops, can make this pool of leached nitrogen available for food production and prevent its further movement to groundwater. Agroforestry practices such as this have the potential to make a significant contribution to both crop protection and environmental quality in the humid tropics.

12.10 GASEOUS LOSSES BY DENITRIFICATION

Nitrogen may be lost to the atmosphere when nitrate ions are converted to gaseous forms of nitrogen by a series of widely occurring biochemical reduction reactions termed **denitrification.**[8] The organisms that carry out this process are mostly facultative anaerobic bacteria in genera such as *Pseudomonas, Bacillus, Micrococcus,* and *Achromobacter.* These organisms are heterotrophs, which obtain their energy and carbon from the oxidation of organic compounds. Other denitrifying bacteria are autotrophs, such as *Thiobacillus denitrificans,* which obtain their energy from the oxidation of sulfide. The exact mechanisms vary depending on the conditions and organisms involved. In the reaction, NO_3^- [N(V)] is reduced in a series of steps to NO_2^- [N(III)], and then to nitrogen gases that include NO [N(II)], N_2O [N(I)], and eventually N_2 [N(0)]:

$$\underset{\substack{\text{Nitrate ions}\\(+5)}}{2NO_3^-} \xrightarrow{-2[O]} \underset{\substack{\text{Nitrite ions}\\(+3)}}{2NO_2^-} \xrightarrow{-2[O]} \underset{\substack{\text{Nitric oxide gas}\\(+2)}}{2NO\uparrow} \xrightarrow{-[O]} \underset{\substack{\text{Nitrous oxide gas}\\(+1)}}{N_2O\uparrow} \xrightarrow{-[O]} \underset{\substack{\text{Dinitrogen gas}\\(0)\ \longleftarrow\ \text{Valence state of nitrogen}}}{N_2\uparrow} \quad (12.6)$$

Although not shown in this simplified reaction, the oxygen released at each step would be used to oxidize organic carbon (or sulfide, if *Thiobacillus* is the nitrifying organism) and CO_2 would be released.

For these reactions to take place, organic carbon (or sulfide) should be available to provide the energy the denitrifiers need. The soil air in the microsites where denitrification occurs should contain no more than 10% oxygen, and lower levels of oxygen are preferred. Optimum temperatures for denitrification are from 25 to 35°C, but the process will occur between 2 and 50°C. Very strong acidity (pH <5.0) inhibits rapid denitrification and favors the formation of N_2O.

Generally, when oxygen levels are very low, the end product released from the overall denitrification process is dinitrogen gas (N_2); however, NO and N_2O are also released during denitrification under the fluctuating aeration conditions that often occur in the field (Figure 12.4). The proportion of the three main gaseous products seems to be dependent on the prevalent pH, temperature, degree of oxygen depletion, and concentration of nitrate and nitrite ions available. For example, the release of nitrous oxide (N_2O) is favored if the concentrations of nitrite and nitrate are high and the supply of oxygen is not too low. Under very acid conditions, almost all of the loss occurs in the form of N_2O. Nitric oxide (NO) loss is generally small and apparently occurs most readily under acid conditions.

Atmospheric Pollution

The question of how much of each nitrogen-containing gas is produced is not merely of academic interest. Dinitrogen gas is quite inert and environmentally harmless, but the oxides of nitrogen are very reactive gases and have the potential to do serious environmental damage in at least four ways. First, NO and N_2O can contribute to the formation of nitric acid, one of the principal components of acid rain. Second, the nitrogen oxide gases can react with volatile organic pollutants to form ground-level ozone, a major air pollutant in the photochemical smog that plagues many urban areas. Third, when NO rises into the upper atmosphere it contributes to the greenhouse effect (as much as 300 times that of an equal amount of CO_2) by absorbing infrared radiation that would otherwise escape into space (see Section 11.11).

[8] Nitrate can also be reduced to nitrite and to nitrous oxide gas by nonbiological chemical reactions, but these reactions are quite minor in comparison with biological denitrification.

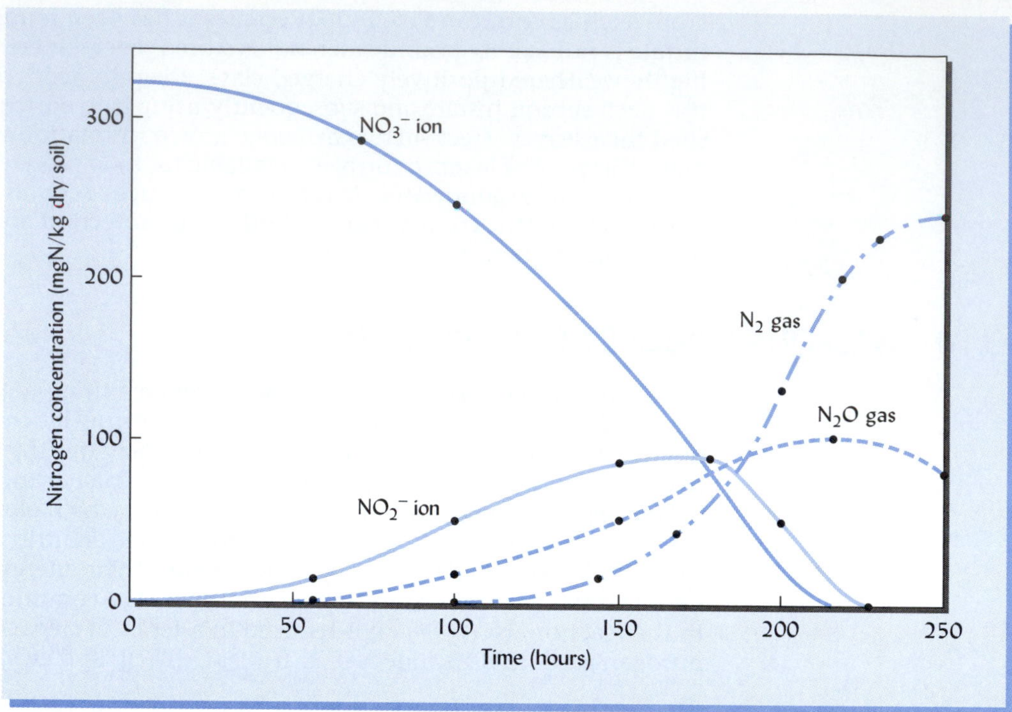

FIGURE 12.4 Changes in various forms of nitrogen during the process of denitrification in a moist soil incubated in the absence of atmospheric oxygen. [From Leffelaar and Wessel (1988)]

Finally, and perhaps most significantly, as N_2O moves up into the stratosphere, it may participate in reactions that result in the destruction of ozone (O_3), a gas that helps shield Earth from harmful ultraviolet solar radiation. In recent years this protective ozone layer has been measurably depleted, in part by reaction with N_2O and other gases. As this protective layer is further degraded, thousands of additional cases of skin cancer are likely to occur annually. While there are other important sources of N_2O, such as automobile exhaust fumes, denitrification in soils makes a major contribution, especially in rice paddies, wetlands, and heavily fertilized or manured agricultural soils.

Quantity of Nitrogen Lost Through Denitrification

The exact magnitude of denitrification loss depends on management practices and soil conditions. Studies of forest ecosystems have shown that during periods of adequate soil moisture, denitrification results in a slow, but relatively steady, loss of nitrogen from these undisturbed natural systems. In contrast, most field losses from agricultural soils are highly variable in both time and space. The greater part of the annual nitrogen loss often occurs during just a few days in summer, when heavy rain has temporarily caused the warm soils to become poorly aerated (Figure 12.5).

Low-lying, organic-rich areas and other hot spots may lose nitrogen 10 times as fast as the average rate for a typical upland field which is rarely more than 5 to 15 kg N/ha annually. But where drainage is restricted and where large amounts of fertilizer nitrogen are applied, annual losses as high as 30 to 60 kg N/ha/yr have been observed.

Much of the nitrogen that moves from watersheds into shallow streams and slow-moving rivers is lost by denitrification. As the water moves through these shallow transport channels there is opportunity for the nitrogen to come in contact with the river bottoms where the denitrification occurs. Recent research suggests that 5 to 20% of the nitrogen in some streams or rivers may be lost by denitrification.

Denitrification in Flooded Soils

In flooded soils, such as those found in natural wetlands or rice paddies (Figure 12.6), losses by denitrification may be very high. The soils of rice paddies are commonly subject to alternate periods of wetting and drying. Nitrates that are produced

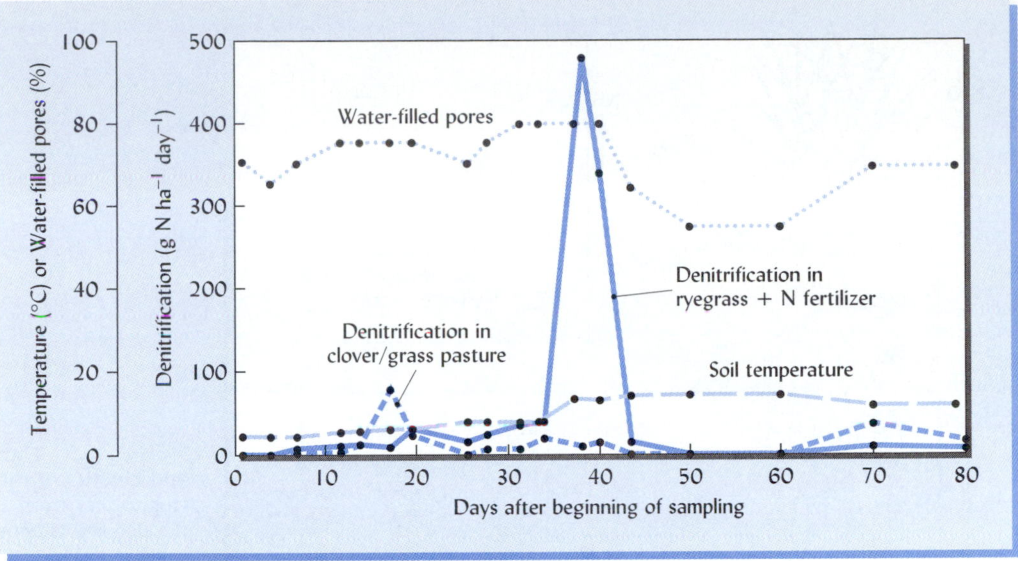

FIGURE 12.5 Changes in denitrification and soil conditions in spring on a well-drained, fine-textured soil (East Keswick silty clay loam, Udalfs) in Wales, U.K. Two pasture systems were studied: ryegrass receiving 150 kg N/ha annually and a clover–grass mixture. The fertilizer nitrogen applied to the ryegrass system was intended to approximately equal the nitrogen fixed from the atmosphere by the clover in the mixed system (see Section 12.11). Note the sporadic nature of the denitrification process, with most of the nitrogen loss occurring during a brief period when the soil was warm, wet, and high in nitrate. This episodic pattern is typical of disturbed agroecosystems and stands in contrast to the slow, steady pattern of denitrification usually found in undisturbed natural forests. [Data from Colbourn (1993)]

FIGURE 12.6 Denitrification can be very efficient in removing nitrogen in flooded systems that combine aerobic and anaerobic zones and have high concentrations of available organic carbon. Some examples of such systems are (*a*) a flooded rice paddy, (*b*) a tidal wetland, (*c*) a site for treating sewage effluent by overland flow (effluent applied by sprinklers, see arrows), and (*d*) a manure storage lagoon. (Photos courtesy of R. Weil)

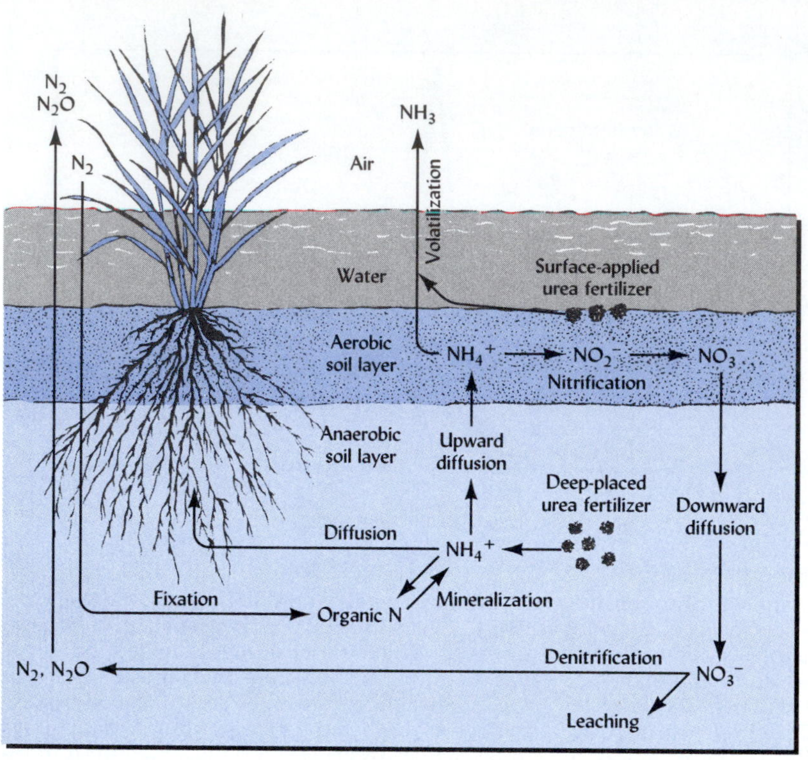

FIGURE 12.7 Nitrification–denitrification reactions and kinetics of the related processes controlling nitrogen loss from the aerobic anaerobic layers of a flooded soil system. Nitrates, which form in the thin aerobic soil layer just below the soil–water interface, diffuse into the anaerobic (reduced) soil layer below and are denitrified to the N_2 and N_2O gaseous forms, which are lost to the atmosphere. Placing the urea or ammonium-containing fertilizers deep in the anaerobic layer prevents the oxidation of ammonium ions to nitrates, thereby greatly reducing the loss. [Modified from Patrick (1982)]

by nitrification during the dry periods are often subject to denitrification when the soils are submerged. Even when submerged, the soil permits both reactions to take place at once—nitrification at the soil–water interface where some oxygen derived from the water is present and denitrification at lower soil depths (see Figure 12.7). However, nitrogen losses can be dramatically reduced by keeping the soil flooded and by deep placement of the fertilizer into the reduced zone of the soil. In this zone there is insufficient oxygen to allow nitrification to proceed, so the nitrogen remains in the ammonium form and is not susceptible to loss by denitrification.

The sequential combination of nitrification and denitrification also operates in natural and artificial wetlands. Tidal wetlands, which become alternately anaerobic and aerated as the water level rises and falls, have particularly high potentials for converting nitrogen to gaseous forms. Often the resulting rapid loss of nitrogen is considered to be a beneficial function of wetlands, in that the process protects estuaries and lakes from the eutrophying effects of too much nitrogen.

Denitrification in Groundwater

Recent studies on the movement of nitrate in groundwater have documented the significance of denitrification taking place in the poorly drained soils under **riparian** vegetation (mainly woodlands adjacent to streams). In most cases studied in humid temperate regions, contaminated groundwater lost most of its nitrate load as it flowed through the riparian zone on its way to the stream. The apparent removal of nitrate may be quite dramatic, whether the nitrate source is septic drainfields or fertilized cropland (Figure 12.8). Most of the nitrate loss is believed to be by denitrification, stimulated by organic compounds leached from the decomposing forest litter and by the anaerobic conditions that prevail in the wet riparian zone soils.

We have just discussed a number of biological processes that lead to losses of nitrogen from the soil system. We now turn our attention to the principal biological process by which soil nitrogen is replenished.

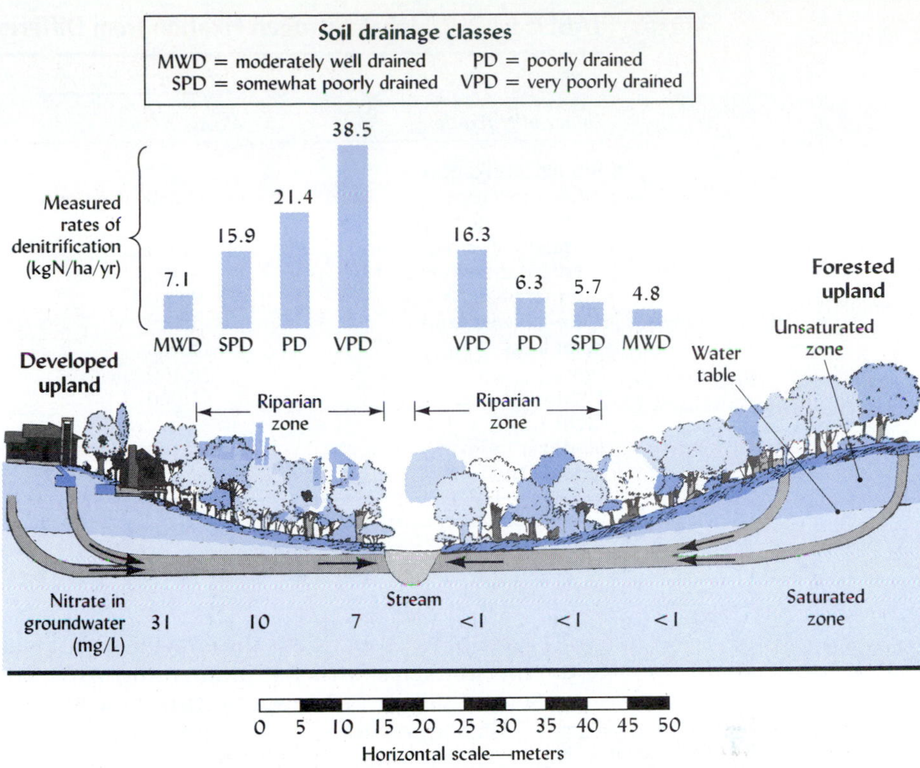

FIGURE 12.8 Denitrification in riparian wetlands receiving groundwater with high- and low-nitrate contents. The high-nitrate groundwater (left) came from sites with heavy development of houses using septic drainfields for sewage disposal (see Section 6.10). The low-nitrate groundwater (right side) came from an area of undeveloped forest. The riparian zones on both sides of the stream were covered with red maple–dominated forest. Within a few meters after entering the riparian wetland, the nitrate content of the contaminated groundwater was reduced by 75%, from 31 ppm nitrate to less than 7 ppm nitrate. The soils in this study site are very sandy Inceptisols and Entisols. In other regions, riparian zones with finer-textured soils have shown even more complete removal of nitrate from groundwater. [Data shown are from Hanson et al. (1994)]

12.11 BIOLOGICAL NITROGEN FIXATION

Soil bacteria in the nitrogen cycle:
http://www.soils.umn.edu/academics/classes/soil2125/doc/s9chap2.htm

Next to plant photosynthesis, **biological nitrogen fixation** is probably the most important biochemical reaction for life on earth. Through this process, certain organisms convert the inert dinitrogen gas of the atmosphere (N_2) to nitrogen-containing organic compounds that become available to all forms of life through the nitrogen cycle. The process is carried out by a limited number of microorganisms, including several species of bacteria, a number of actinomycetes, and certain cyanobacteria (blue-green algae).

Globally, enormous amounts of nitrogen are fixed biologically each year. Terrestrial systems alone fix an estimated 139 million Mg, about twice as much as is industrially fixed in the manufacture of fertilizers (Table 12.2).

The Mechanism. Regardless of the organisms involved, the key to biological nitrogen fixation is the enzyme *nitrogenase,* which catalyzes the reduction of dinitrogen gas to ammonia.

$$N \equiv N + 8H^+ + 8e \xrightarrow[\text{(Fe,Mo)}]{\text{(Nitrogenase)}} 2\,\begin{matrix} H \\ | \\ N\text{-}H \\ | \\ H \end{matrix} + H_2 \qquad (12.7)$$

The ammonia, in turn, is combined with organic acids to form amino acids and, ultimately, proteins.

$$NH_3 + \text{organic acids} \longrightarrow \text{amino acids} \longrightarrow \text{proteins} \qquad (12.8)$$

TABLE 12.2 Global Nitrogen Fixation from Different Sources

Source of N fixation	Nitrogen fixed per year: Area, 10^6 ha	Rate, kg/ha	Total fixed, 10^6 Mg
Biological fixation			
Legume crops	250	140	35
Nonlegume crops	1,150	8	9
Meadows and grassland	3,000	15	45
Forest and woodland	4,100	10	40
Other vegetated land	4,900	2	10
Ice-covered land	1,500	0	0
Total land	14,900		139
Sea	36,100	1	36
Total biological	51,000		175
Lightning			8
Fertilizer industry			77
Grand Total			260

Calculated from Jenkinson (1990) and Burns and Hardy (1975).

The site of N_2 reduction is the enzyme **nitrogenase**, a complex consisting of two proteins, the smaller of which contains iron while the larger contains molybdenum and iron. Several salient facts about nitrogenase and its function are worth noting, for this enzyme is unique and of great ecological importance.

The reduction of N_2 to NH_3 by nitrogenase requires a great deal of energy to break the triple bond between the nitrogen atoms. Therefore, the process is greatly enhanced by association with higher plants, which can supply this energy from photosynthesis.

The reduction reaction is end-product inhibited—an accumulation of ammonia will inhibit nitrogen fixation. Also, too much nitrate in the soil will inhibit the formation of nodules (see Section 12.12). Nitrogen-fixing organisms have a relatively high requirement for molybdenum, iron, phosphorus, and sulfur, because these nutrients are either part of the nitrogenase molecule or are needed for its synthesis and use. The most efficient nitrogen-fixers are able to recapture some of the energy expended in making hydrogen gas, using this energy to fix additional nitrogen.

Fixation Systems. Biological nitrogen fixation occurs through a number of microorganism, with or without direct association with higher plants (Table 12.3). Although the legume–bacteria symbiotic systems have received the most attention, recent findings suggest that the other systems involve many more families of plants worldwide, and may be just as important as the legume-associated systems in suppling nitrogen to the soil. Each major system will be discussed briefly.

TABLE 12.3 Information on Different Systems of Biological Nitrogen Fixation

N-fixing systems	Organisms involved	Plants involved	Site of fixation
Symbiotic			
Obligatory			
Legumes	Bacteria *Rhizobia* and *Bradyrhizobia*	Legumes	Root nodules
Nonlegumes (angiosperms)	Actinomycetes (*Frankia*)	Nonlegumes (angiosperms)	Root nodules
Associative			
Morphological involvement	Cyanobacteria, bacteria	Various higher plants and microorganisms	Leaf and root nodules, lichens
Nonmorphological involvement	Cyanobacteria, bacteria	Various higher plants and microorganisms	Rhizosphere (root environment) Phyllosphere (leaf environment)
Nonsymbiotic	Cyanobacteria, bacteria	Not involved with plants	Soil, water independent of plants

12.12 SYMBIOTIC FIXATION WITH LEGUMES

The **symbiosis** (mutually beneficial relationship) of legumes and bacteria of the genera *Rhizobium* and *Bradyrhizobium* provide the major biological source of fixed nitrogen in agricultural soils. These organisms infect the root hairs and the cortical cells, ultimately inducing the formation of **root nodules** that serve as the site of nitrogen fixation (Figure 12.9). In a mutually beneficial association, the host plant supplies the bacteria with carbohydrates for energy, and the bacteria reciprocate by supplying the plant with fixed-nitrogen compounds.

Organisms Involved. A given *Rhizobium* or *Bradyrhizobium* species will infect some legumes but not others. For example, *Rhizobium trifolii* inoculates *Trifolium* species (most clovers), but not sweet clover, which is in the genus *Melilotus*. Likewise, *Rhizobium phaseoli* inoculates *Phaseolus vulgaris* (dry beans), but not soybeans, which are in the genus *Glycine*. This specificity of interaction is one basis for classifying rhizobia (see Table 12.4). Legumes that can be inoculated by a given *Rhizobium* species are included in the same cross-inoculation group.

In areas where a given legume has been grown for several years, the appropriate species of *Rhizobium* is probably present in the soil and the legume will be well-nodulated (check the roots for nodule that show a red color when cut open, see Plate 59). Often, however, the natural *Rhizobium* population in the soil is too low or the strain of the *Rhizobium* species present is not effective. In such circumstances, special mixtures of the appropriate *Rhizobium* and *Bradyrhizobium* inoculant may be applied, either by coating the legume seeds or by applying the inoculant directly to the soil. Effective and competitive strains of *Rhizobium,* which are available commercially, often give significant yield increases, but only if used on the proper crops.

Quantity of Nitrogen Fixed. The rate of biological fixation is greatly dependent on soil and climatic conditions. High levels of available nitrogen, whether from the soil or added in fertilizers, tend to depress biological nitrogen fixation (Figure 12.10). Apparently, plants make the heavy energy investment required for symbiotic nitrogen fixation only when short supplies of mineral nitrogen make nitrogen fixation necessary.

Although quite variable from site to site, the amount of nitrogen biologically fixed can be quite high, especially for those systems involving nodules, which supply energy from photosynthates and protect the nitrogenase enzyme system (Table 12.5). Non-nodulating or nonsymbiotic systems generally fix relatively small amounts of nitro-

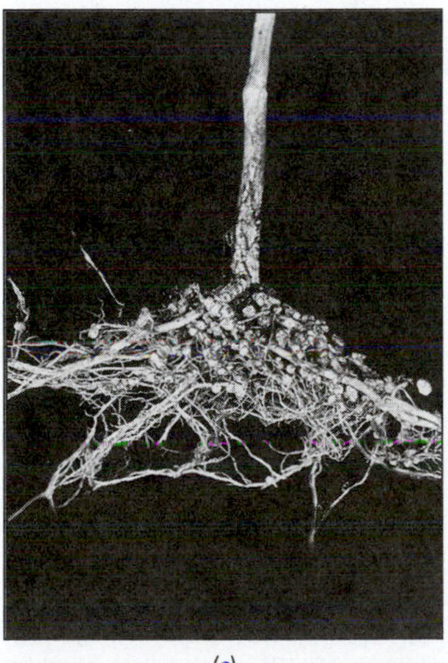

(a)

(b)

(c)

FIGURE 12.9 Photos illustrating soybean nodules. In (*a*) the nodules are seen on the roots of the soybean plant, and a close-up (*b*) shows a few of the nodules associated with the roots. A scanning electron micrograph (*c*) shows a single plant cell within the nodule stuffed with the bacterium *Bradyrhizobium japonicum*. (Courtesy of W. J. Brill, University of Wisconsin)

TABLE 12.4 Classification of Rhizobia Bacteria and Associated Legume Cross-Inoculation Groups

The genus Rhizobium *contains fast-growing, acid-producing bacteria, while those of* Bradyrhizobium *are slow growers that do not produce acid. A third genus,* Azorhizobium, *which is not shown, produces stem nodules on* Sesbania rostrata.

Bacteria		
Genus	*Species/subgroup*	*Host legume*
Rhizobium	*R. leguminosarum*	
	bv. *viceae*	*Vicia* (vetch), *Pisum* (peas), *Lens* (lentils), *Lathyrus* (sweet pea)
	bv. *trifolii*	*Trifolium* spp. (most clovers)
	bv. *phaseoli*	*Phaseolus* spp. (dry bean, runner bean, etc.)
	R. Meliloti	*Melilotus* (sweet clover, etc.), *Medicago* (alfalfa), *Trigonella,* (fenugreek)
	R. loti	*Lotus* (trefoils), *Lupinus* (lupins), *Cicer* (chickpea), *Anthyllis, Leucaena,* and many other tropical trees
	R. Fredii	*Glycine* spp. (e.g., soybean)
Bradyrhizobium	*B. japonicum*	*Glycine* spp. (e.g., soybean)
	B. sp.	*Vigna* (cowpeas), *Arachis* (peanut), *Cajanus* (pigeon pea), *Pueraria* (kudzu), *Crotolaria* (crotolaria), and many other tropical legumes

gen. Nonetheless, many natural plant communities and agricultural systems (generally involving legumes) derive the bulk of their nitrogen needs from biological fixation.

Effect on Soil Nitrogen Level. The data in Table 12.5 show that symbiotic nitrogen fixation is definitely beneficial to both forestry and agriculture. This source of nitrogen is not available to most coniferous and hardwood trees, grasses, cereal crops, and vegetables that lack symbiotic associations, unless they are grown in association with nodulated species. Over time, the presence of nitrogen-fixing species can significantly increase the nitrogen content of the soil and benefit nonfixing species grown in association with fixing species (see Figure 12.11).

Some crops, such as beans and peas, are such weak nitrogen fixers that most of the nitrogen they absorb must come from the soil. Consequently, it should not be assumed that the symbiotic systems always increase soil nitrogen. Only in cases where the soil is

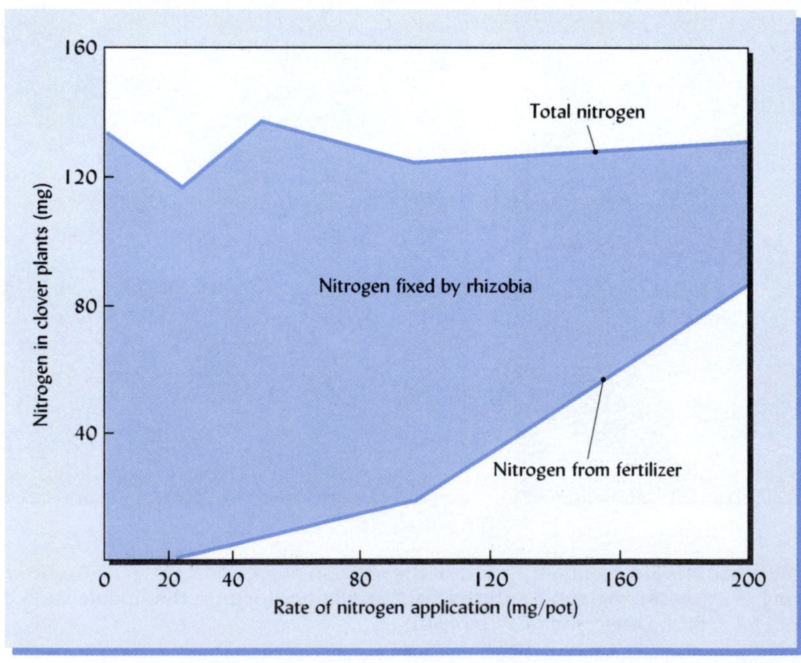

FIGURE 12.10 Influence of added inorganic nitrogen on the total nitrogen in clover plants, the proportion supplied by the fertilizer, and that fixed by the rhizobium organisms associated with the clover roots. Increasing the rate of nitrogen application decreased the amount of nitrogen fixed by the organisms in this greenhouse experiment. [From Walker et al. (1956)]

TABLE 12.5 Typical Levels of Nitrogen Fixation from Different Systems

Crop or plant	Associated organism	Typical levels of nitrogen fixation, kg N/ha/yr
Symbiotic		
Legumes (nodulated)		
Ipil Ipil tree (*Leucena leucocephala*)	Bacteria (*Rhizobium*)	100–500
Locust tree (*Robina* spp.)		75–200
Alfalfa (*Medicago sativa*)		150–250
Clover (*Trifolium pratense L.*)		100–150
Lupine (*Lupinus*)		50–100
Vetch (*Vicia vilbosa*)		50–150
Bean (*Phaseolus vulgaris*)		30–50
Cowpea (*Vigna unguiculata*)	Bacteria (*Bradyrhizobium*)	50–100
Peanut (*Arachis*)		40–80
Soybean (*Glycine max L.*)		50–150
Pigeon pea (*Cajunus*)		150–280
Kudzu (*Pueraria*)		100–140
Nonlegumes (nodulated)		
Alders (Alnus)	Actinomycetes (*Frankia*)	50–150
Species of Gunnera	Cyanobacteria[a] (*Nostoc*)	10–20
Nonlegumes (nonnodulated)		
Pangola grass (*Degetaria decumbens*)	Bacteria (*Azospirillum*)	5–30
Bahia grass (*Paspalum notatum*)	Bacteria (*Azotobacter*)	5–30
Azolla	Cyanobacteria[a] (*Anabena*)	150–300
Nonsymbiotic	Bacteria (*Azotobacter, Clostridium*)	5–20
	Cyanobacteria[a] (various)	10–50

[a]Sometimes referred to as *blue-green algae*.

low in available nitrogen, and vegetation includes strong nitrogen fixers, would this be likely to be true.

In the case of legume crops harvested for seed or hay, most of the nitrogen fixed is removed from the field with the harvest. Nitrogen additions from such crops should be considered as nitrogen *savers* for the soil rather than nitrogen builders. On the other hand, considerable buildup of soil nitrogen can be achieved by perennial legumes (such as alfalfa) and by annual legumes (such as hairy vetch) whose entire growth is returned to the soil as **green manure.** Such nitrogen buildup should be taken into account when estimating nitrogen fertilizer needs for maximum plant production with minimal environmental pollution (see Section 14.3).

Fate of Nitrogen Fixed by Legume Bacteria. Some of the nitrogen fixed in root nodules is used directly by the host plant, which thereby benefits greatly from the symbiosis described in Section 12.11. Second, some fixed nitrogen may become available to nonfixing plants growing in association with nitrogen-fixing plants. The vigorous development of a grass

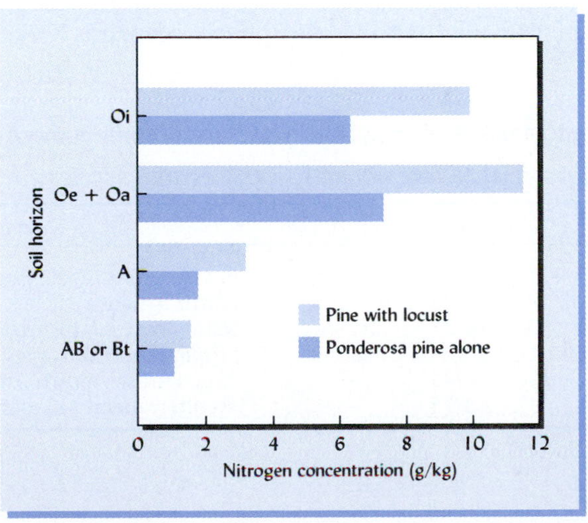

FIGURE 12.11 Nitrogen contents of forest soil horizons showing the effects of New Mexican locust trees (*Robinia neomexicana*) growing in association with ponderosa pine (*Pinus ponderosa*) in a region of Arizona receiving about 670 mm rainfall per year. The data are means from 20 stands of ponderosa pine, half of them with the nitrogen-fixing legume trees (locust) in the understory. The soils are Eutrustalfs and Argiustolls with loam and clay loam textures. [Data from Klemmedson (1994)]

in a legume–grass mixture is evidence of this rapid release, as are the relatively high nitrate concentrations sometimes measured in groundwater under legume crops.

The third pathway for the fixed nitrogen is immobilization by heterotrophic microorganisms and eventual incorporation into the soil organic matter (see Section 12.4).

12.13 SYMBIOTIC FIXATION WITH NONLEGUMES

Nodule-Forming Nonlegumes

Nearly 200 species from more than a dozen genera of nonlegumes are known to develop nodules and to accommodate symbiotic nitrogen fixation. Included are several important groups of angiosperms, listed in Table 12.6. The roots of these plants, which are present in certain forested areas and wetlands, form distinctive nodules when their root hairs are invaded by soil actinomycetes of the genus *Frankia*.

The rates of nitrogen fixation per hectare compare favorably with those of the legume–*Rhizobium* complexes (see Table 12.5). On a worldwide basis, the total nitrogen fixed in this way may even exceed that fixed by agricultural legumes. Because of their nitrogen-fixing ability, certain tree–actinomycete complexes are able to colonize infertile soils and newly forming soils on disturbed lands, which may have extremely low fertility and other conditions that limit plant growth (Figure 12.12). Once nitrogen-fixing plants become established and begin to build up the soil nitrogen supply through leaf litter and root exudation, the land becomes more hospitable for colonization by other species. *Frankia* thus play an important role in the nitrogen economy of areas undergoing succession, as well as in established forest and marshes.

Certain cyanobacteria are known to develop nitrogen-fixing symbiotic relations with green plants. One involves nodule formation on the stems of *Gunnera,* an angiosperm common in marshy areas of the southern hemisphere. In this association, cyanobacteria of the genus *Nostoc* fix 10 to 20 kg N/ha/yr (see Table 12.5).

Symbiotic Nitrogen Fixation Without Nodules

Among the most significant nonnodule nitrogen-fixing systems are those involving cyanobacteria. One system of considerable practical importance is the *Azolla–Anabaena* complex, which flourishes in certain rice paddies of tropical and semitropical areas. The *Anabaena* cyanobacteria inhabit cavities in the leaves of the floating fern *Azolla* and fix quantities of nitrogen comparable to those of the more efficient *Rhizobium*–legume complexes (see Table 12.5).

A more widespread but less intense nitrogen-fixing phenomenon is that which occurs in the *rhizosphere* of certain grasses and other nonlegume plants. The organisms responsible are bacteria, especially those of the *Spirillum* and *Azotobacter* genera. Root exudates supply these microorganisms with energy for their nitrogen-fixing activities.

Scientists have reported a wide range of rates for rhizosphere nitrogen fixation, with the highest values observed in association with certain tropical grasses. Even if typical rates are only 5 to 30 kg N/ha/yr, the vast areas of tropical grasslands suggest that the total quantity of nitrogen fixed by rhizosphere organisms is likely very high (see Table 12.2).

TABLE 12.6 Number and Distribution of Major Actinomycete-Nodulated Nonlegume Angiosperms

In comparison, there are about 13,000 legume species.

Genus	*Family*	*Species*[a] *nodulated*	*Geographic distribution*
Alnus	Betulaceae	33/35	Cool regions of the northern hemisphere
Ceanothus	Rhamnaceae	31/35	North America
Myrica	Myricaceae	26/35	Many tropical, subtropical, and temperate regions
Casuarina	Casuarinaceae	24/25	Tropics and subtropics
Elaeagnus	Elaeagnaceae	16/45	Asia, Europe, North America
Coriaria	Coriariaceae	13/15	Mediterranean to Japan, New Zealand, Chile to Mexico

[a]Number of species nodulated/total number of species in genus.
Selected from Torrey (1978).

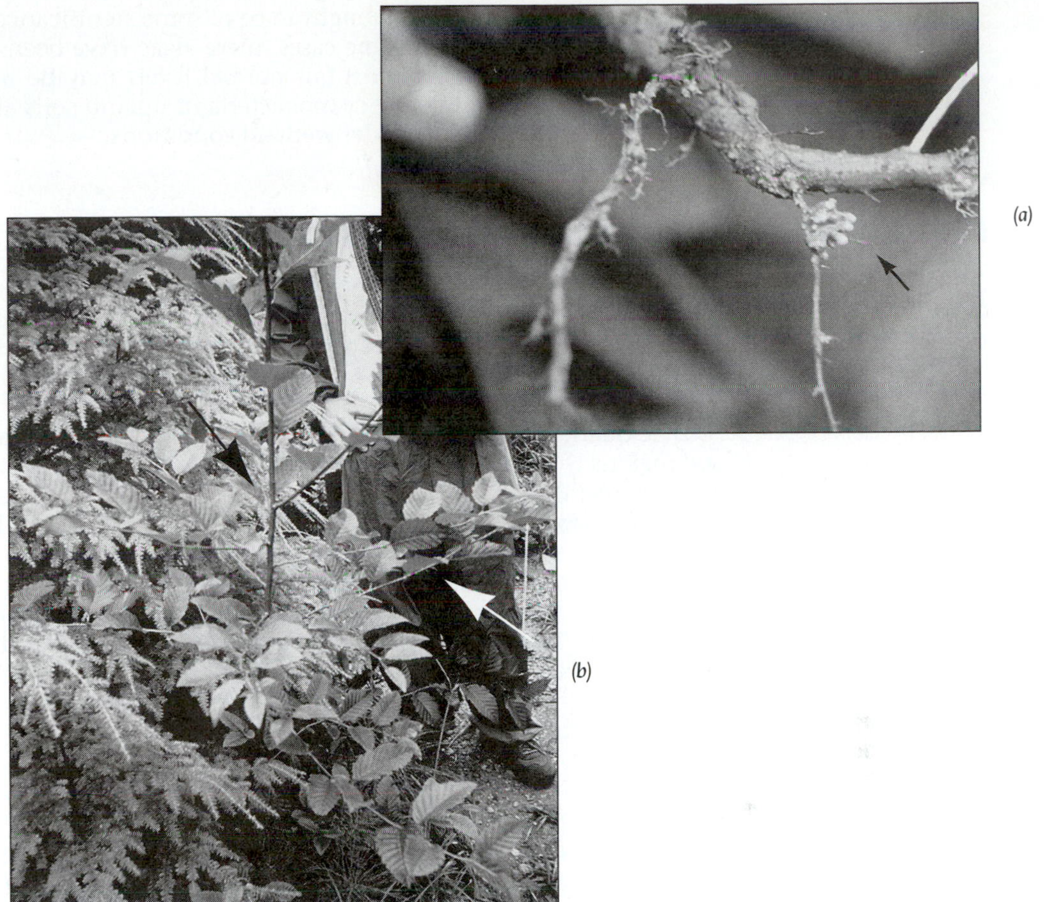

FIGURE 12.12 Soil actinomycetes of the genus *Frankia* can nodulate the roots of certain woody plant species and form a nitrogen-fixing symbiosis that rivals the legume–rhizobia partnership in efficiency. The actinomycete-filled root nodule (*a*) is the site of nitrogen fixation. The red alder tree (*b*) is among the first pioneer tree species to revegetate disturbed or badly eroded sites in high-rainfall areas of the Pacific Northwest in North America. This young alder is thriving despite the nitrogen-poor, eroded condition of the soil, because it is not dependent on soil nitrogen for its needs. (Photos courtesy of R. Weil)

12.14 NONSYMBIOTIC NITROGEN FIXATION

Certain free-living microorganisms present in soils and water are able to fix nitrogen. Because these organisms are not directly associated with higher plants, the transformation is referred to as *nonsymbiotic* or *free-living.*

Fixation by Heterotrophs

Several different groups of bacteria and cyanobacteria are able to fix nitrogen nonsymbiotically. In upland mineral soils, the major fixation is brought about by species of two genera of heterotrophic aerobic bacteria, *Azotobacter* (in temperate zones) and *Beijerinckia* (in tropical soils). Certain aerobic bacteria of the genus *Clostridium* are also able to fix nitrogen.

The amount of nitrogen fixed by these heterotrophs generally ranges from 5 to 20 kg N/ha/yr—only a small fraction of the nitrogen needed by crops. However, in some natural ecosystems this level of nitrogen fixation would make a significant contribution toward meeting nitrogen needs.

Fixation by Autotrophs

Among the autotrophs able to fix nitrogen are certain photosynthetic bacteria and cyanobacteria. In the presence of light, these organisms are able to fix carbon dioxide and nitrogen simultaneously. The contribution of the photosynthetic bacteria is uncertain, but that of cyanobacteria is thought to be of some significance, especially in wetland areas and in rice paddies. In some cases, these algae have been found to fix sufficient nitrogen for moderate rice yields, but normal levels may be no more than 20 to 30 kg N/ha/yr. Nitrogen fixation by cyanobacteria in upland soils also occurs, but the level is much lower than is found under wetland conditions.

12.15 ADDITION OF NITROGEN TO SOIL IN PRECIPITATION

The atmosphere contains ammonia, nitrates, and other nitrogen compounds derived primarily from the combustion of coal and petroleum products. These atmosphere-borne nitrogen compounds are added to the soil through rain, snow, and dust. Although the rates of addition per hectare are typically small, the total quantity of nitrogen added annually is significant.

The quantities of ammonia and nitrates in precipitation are greater in humid tropical regions than in humid temperate regions, and are larger in the latter than in semiarid, temperate climates. Rainfall additions of nitrogen are highest near cities and industrial areas and near large animal feedlots (Table 12.7). The environmental impact of acid rain (see Section 9.7) and its influence on vegetation (forests and crops) and surface waters tends to overshadow the nutrient benefit of precipitation-supplied nitrates.

Typically, the nitrogen in precipitation is about two-thirds ammonium and one-third nitrate. The range of total nitrogen (nitrate + ammonium) added by precipitation annually is 1 to 25 kg N/ha. A figure of 5 to 8 kg N/ha would be typical for nonindustrial temperate regions. This modest annual acquisition of nitrogen is probably of more significance to the nitrogen budget in natural ecosystems than in agriculture.

Effects on Forest Ecosystems

In a mature forest, net plant uptake and microbial immobilization are in such close balance with release of nitrogen by mineralization that many of these systems have very little ability to retain incoming nitrogen. Nitrogen from the atmosphere enters the forest largely as nitric acid (HNO_3) or as ammonium, which is converted to nitric acid upon microbial nitrification (see Section 12.8). As a result, the H^+ ions from HNO_3 and other acids displace nonacid cations on the soil colloids. The displaced cations, especially calcium and magnesium, leach downward with the nitrate (and sulfate; see Section 12.23). This causes imbalances in the nutrition of the trees and a decline in forest productivity (see also Chapter 9 and Section 12.8).

12.16 REACTIONS OF NITROGEN FERTILIZERS

Use of nitrogen as fertilizer: http://pmep.cce.cornell.edu/facts-slides-self/facts/nit-el-grw89.html

Most commercial fertilizers supply nitrogen in soluble forms, such as nitrate or ammonium, or as urea, which rapidly hydrolyzes to form ammonium. Ammonium and nitrate ions from fertilizer are taken up by plants and participate in the nitrogen cycle in exactly the same way as ammonium and nitrate derived from organic matter mineralization or other sources. The main difference is that nitrogen from heavy doses of fertilizer

TABLE 12.7 Amounts of Nitrogen Brought Down in Precipitation Annually in Different Parts of the United States

	Range in annual deposition, kg/ha		
Areas in the United States	*Nitrate nitrogen*	*Ammonium nitrogen*	*Total nitrogen, kg/ha*
Rural area of Ohio[a]	6.6–10.6	6.2–8.2	12.8–18.8
Industrialized Northeast[b]	4.3–7.4	8.6–14.8	12.9–22.2
Borders of NE industrialized areas[b]	2.8–4.1	5.6–8.2	8.4–12.3
Open areas in West[b]	0.4–0.6	0.8–1.2	1.2–1.8

[a]Owens et al. (1992).
[b]Ammonium nitrogen is calculated as twice the nitrate deposition, the approximate ratio from numerous measurements, although it may be incorrect at any specific site [U.S./Canada Work Group No. 2 (1982)].

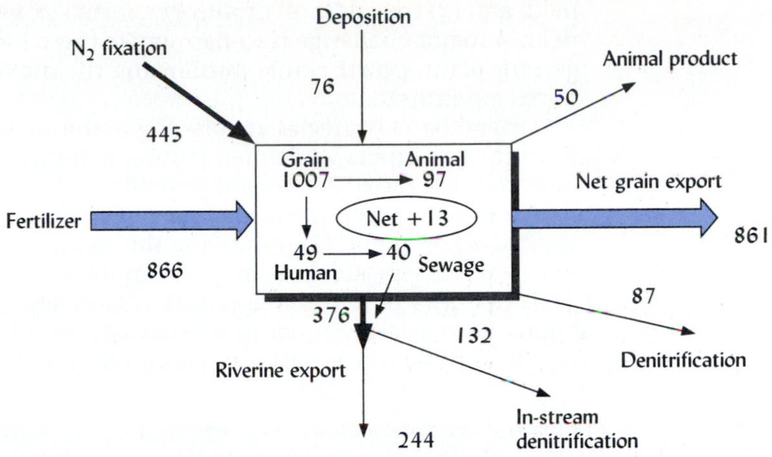

FIGURE 12.13 Diagram showing the annual inputs and outputs of nitrogen for the state of Illinois, averaged for 18 years (1979–96 for terrestrial fluxes and 1980–97 for riverine export). The units are 1000 Mg N/year. The major inputs of nitrogen are commercial fertilizers, nitrogen fixation, and deposition from the atmosphere. Fluxes inside the box are for nitrogen in the total grain harvested, its consumption by animals and humans, and the human sewage that moves into rivers from sewage plants. The major outputs are the nitrogen in the animals sold and the grain exported outside the state, the nitrogen that leaches or runs off into rivers (riverine export), and that lost by denitrification from the soil and from streams as they move toward the sea. Note in the box that a small amount of N (13,000 mg/yr) remains in the system. This may contribute to the level of nitrogen in the soil. Note that animal manure is not considered as either an input or an output since it represents a recycling of nitrogen that comes from the soil and goes back into the soil. This diagram illustrates many of the principles covered in this chapter. [From David and Gentry (2000)]

is far more concentrated in time (available all at once) and space (applied in narrow bands or in a thin layer in the soil) than is nitrogen from other sources. While judicious application of such concentrations may be very beneficial for plant growth, it may also have serious detrimental effects.

First, high concentrations of ammonia gas from ammonium-releasing fertilizers (such as urea and anhydrous ammonia) may be toxic to soil organisms and favor loss of this gas to the atmosphere, especially in alkaline soils. Also, because 2 moles of acidity are formed for every 1 mole of ammonium nitrogen that undergoes nitrification to nitrates, use of excess ammonium fertilizers or manure increases soil acidity (see Sections 9.7 and 12.8).

To save time and money, many land managers in both agriculture and forestry apply most of the year's supply of nitrogen at one time. Unfortunately, the plants and microorganisms cannot assimilate the applied nitrogen fast enough to prevent major losses by leaching, surface runoff, denitrification, and ammonia volatilization (see Figure 12.13). The environmental consequences of these losses have already been discussed. As a general rule, it is more environmentally benign to apply nitrogen fertilizer in multiple small doses that coincide with the plant demand.

12.17 PRACTICAL MANAGEMENT OF SOIL NITROGEN IN AGRICULTURE

Nitrogen management for no-tillage systems: http://muextension.missouri.edu/xplor/agguides/soils/g09175.htm

The goals of nitrogen control are threefold: (1) the maintenance of an adequate nitrogen supply in the soil, (2) the regulation of the soluble forms of nitrogen to ensure that enough is readily available to provide all the nitrogen plants need for optimum growth, and (3) the minimization of environmentally damaging leakage from the soil–plant system.

To understand the problem of nitrogen control, it is useful to consider the nitrogen inputs and outputs for a farm or ecosystem (Figure 12.13). The features that make the cropland balance different from that for a natural ecosystem are mainly: (1) the removal of nitrogen contained in the harvested product, which represents a major loss from the field, and (2) the addition of nitrogen fertilizer, which represents a major input into the field. A major challenge is to harmonize these losses and gains to provide ample nitrogen for plant growth while minimizing the movement of excess nitrogen into the external environment.

Several basic strategies are proving useful in achieving a rational reduction of excessive nitrogen inputs while maintaining or improving production levels and profitability in agriculture (Box 12.4, including Table 12.8). These approaches include (1) taking into account the nitrogen contribution from *all* sources and reducing the amount of fertilizer applied accordingly; (2) improving the efficiency with which fertilizers, as well as such so-called waste products as animal manure, are used; (3) avoiding overly optimistic yield goals that lead to fertilizer application rates designed to meet crop needs that are much higher than actually occur in most years; and (4) improving crop response knowledge, which identifies the lowest nitrogen application that is likely to produce optimum profit.

BOX 12.4 RATIONALIZING NITROGEN INPUTS IN AGRICULTURAL SYSTEMS

The relative importance and absolute magnitudes of the various nitrogen cycle pathways vary greatly with the particular crop grown and with variations in climate and management. A good crop of wheat or cotton may remove only 100 kg N/ha, nearly half of which may be returned to the soil in the stalks or straw. A bumper crop of silage corn, in contrast, may contain over 250 kg N/ha, and a good annual yield of alfalfa or well-fertilized grass hay removes more than 350 kg N/ha. Regardless of the crop, the goal should be to minimize the losses of nitrogen by all pathways except crop removal (this is, after all, what the farmer sells and what we all eat). The need for commercial nitrogen fertilizer will be governed by the degree to which the farmer is able to integrate symbiotic nitrogen fixation, manure and residue recycling, and loss minimization into the farming system.

If crops and livestock are combined in a farming system, then most of the nitrogen removed in crops can be returned in animal manures. Fertilizer should be considered as a supplement to the nitrogen made available from organic matter mineralization, biological nitrogen fixation, rainfall, animal manure, and crop residues. Nitrogen losses by leaching and denitrification generally become a problem only when nitrogen fertilization exceeds the amount needed to fill the gap between crop uptake needs and the supply from these other sources.

When we examine the nitrogen balance for cropland on a national scale, it is clear that more nitrogen is being applied than can be properly used (Table 12.8). In the United States, cropland receives approximately 20.9 million Mg of nitrogen annually, while it exports some 13.5 million Mg of nitrogen as crop harvests and residues. In other words, nitrogen inputs are more than 50% greater than intended outputs, leaving over 7 million Mg of nitrogen to either accumulate in soils (unlikely in most cases) or leak into the environment via leaching, erosion, and runoff, or via gaseous losses to the atmosphere. While some losses from agroecosystems are inevitable, it is becoming widely acknowledged that cropland in the United States and in other industrial countries has been receiving more nitrogen than plants can effectively use, and that this excess will have to be reduced if nitrogen pollution is to be controlled.

TABLE 12.8 Nitrogen Inputs and Outputs for All Cropland in the United States in 1987 and 1977

Only the major categories are considered. These are aggregate data for the entire country and should not be considered representative of any given farm field. The estimates for the 2 years were made independently of each other, so the close agreement lends validity to both.

	Millions of Mg[a]	
Nitrogen output or input	*1987*	*1977*
Outputs		
Harvested crops	10.60	8.9
Crop residues	2.89	3.0
Total outputs	13.50	11.9
Inputs to cropland		
As commercial fertilizer	9.39	9.5
As legume N fixation	6.87	7.2
In crop residues returned	2.89	3.0
Recoverable manure	1.73	1.4
Total inputs	20.9	21.1
Balance = inputs − outputs	7.42	9.2

[a]Data for 1987 from Table 6-3 in National Research Council (1993). Data for 1977 abstracted from Power (1981) as cited in NRC (1993). For a review of nitrogen use in agroecosystems, see Cassman, et al. (2002).

12.18 IMPORTANCE OF SULFUR[9]

Sulfur as a plant nutrient: http://www.soil.ncsu.edu/publications/Soilfacts/AG-439-15/

Sulfur has long been recognized as indispensable for many reactions in living cells. In addition to its vital roles in plant and animal nutrition, sulfur is also responsible for several types of air, water, and soil pollution and is therefore of increasing environmental interest. The environmental problems associated with sulfur include acid precipitation, certain types of forest decline, acid mine drainage, acid sulfate soils, and even some toxic effects in drinking water used by humans and livestock.

Roles of Sulfur in Plants and Animals

Soil test for sulfur: http://www.ianr.unl.edu/pubs/soil/g901.htm

Sulfur is a constituent of the amino acids methionine, cysteine, and cystine, deficiencies of which result in serious human malnutrition. The vitamins biotin, thiamine, and B1 contain sulfur, as do many enzymes that regulate such activities as photosynthesis and nitrogen fixation. It is believed that sulfur-to-sulfur bonds link certain sites on long chains of amino acids, causing proteins to assume the specific three-dimensional shapes that are the key to their catalytic action. Sulfur is closely associated with nitrogen in the processes of protein and enzyme synthesis. Sulfur is also an essential ingredient of the aromatic oils that give the cabbage and onion families of plants their characteristic odors and flavors. It is not surprising that among the plants, the legume, cabbage, and onion families require especially large amounts of sulfur.

Deficiencies of Sulfur

Healthy plant foliage generally contains 0.15 to 0.45% sulfur, or approximately one-tenth as much sulfur as nitrogen. Plants deficient in sulfur tend to become spindly and develop thin stems and petioles. Their growth is slow, and maturity may be delayed. They also have a chlorotic light green or yellow appearance (Plate 49, after page 498). Symptoms of sulfur deficiency are similar to those associated with nitrogen deficiency (see Section 12.1); however, the chlorosis develops first on the youngest leaves as sulfur supplies are depleted. Sulfur-deficient leaves on some plants show interveinal chlorosis or faint striping that distinguishes them from nitrogen-deficient leaves.

As a result of the following three independent trends, sulfur deficiencies in agricultural plants have become increasingly common during the past several decades.

1. Enforcement of clean air standards has led to reduction of sulfur dioxide (SO_2) emissions to the atmosphere from the burning of fossil coal and oil.
2. The sulfur content of today's highly concentrated N-P-K fertilizers are far lower than those of a generation ago. Thus, less sulfur is being added.
3. As crop harvests have increased, larger amounts of sulfur are being removed from the soils. Consequently, the need for sulfur has increased just as the inputs from all sources have declined.

AREAS OF DEFICIENCY. Sulfur deficiencies have been reported in most areas of the world, but are most prevalent in areas where soil parent materials are low in sulfur, where extreme weathering and leaching has removed this element, or where there is little replenishment of sulfur from the atmosphere. In many tropical countries, one or more of these conditions prevail and sulfur-deficient areas are common.

In the United States, deficiencies of sulfur are most common in the Southeast, the Northwest, California, and the Great Plains. In the Northeast and in other areas with heavy industry and large cities, sulfur deficiencies are not yet widespread.

12.19 NATURAL SOURCES OF SULFUR

The three major natural sources of sulfur that can become available for plant uptake are (1) *organic matter,* (2) *soil minerals,* and (3) *sulfur gases in the atmosphere.* In natural ecosystems where most of the sulfur taken up by plants is eventually returned to the same soil, these three sources combined are usually sufficient to supply the needs of growing plants.

[9]For a discussion of sulfur and agriculture, see Tabatabai (1986).

Organic Matter

In surface soils in temperate, humid regions, 90 to 98% of the sulfur is usually present in organic forms (see Figure 12.14). As is the case for nitrogen, the exact forms of the sulfur in the organic matter are not known. Over time, however, soil microorganisms break down the organic sulfur compounds into soluble inorganic forms, mainly sulfate. This mineralization of organic compounds to release sulfate is analogous to the release of ammonium and nitrate from organic matter discussed in Section 12.4.

In arid and semiarid regions, less organic matter is present in the surface soils. Furthermore, the proportion of organic sulfur is not likely to be as high in arid- and semiarid-region soils as it is in humid-region soils, especially in the subsoils, where gypsum ($CaSO_4 \cdot 2H_2O$) may be prominent.

Soil Minerals

The inorganic forms of sulfur are not as plentiful as the organic forms; however, they include sulfates that are reasonably soluble in water and are available for plant uptake. These sulfates are most prominent in the lower horizons of soils of low-rainfall areas. Also included are the sulfide forms found in some poorly drained humid-region areas. If these soils are drained, the sulfides are oxidized to sulfuric acid, and very low pH *acid sulfate* soils result (see Section 9.7).

Another mineral source of sulfur is the clay fraction of some soils high in Fe, Al oxides and kaolinite. These clays are able to strongly adsorb sulfate from soil solution and subsequently release it slowly by anion exchange, especially at low pH. Oxisols and other highly weathered soils of the humid tropics and subtropics may contain large stores of sulfate, especially in their subsoil horizons (see Figure 12.14*c*). Considerable sulfate may also be bound by the metal oxides in the spodic horizons under certain temperate forests.

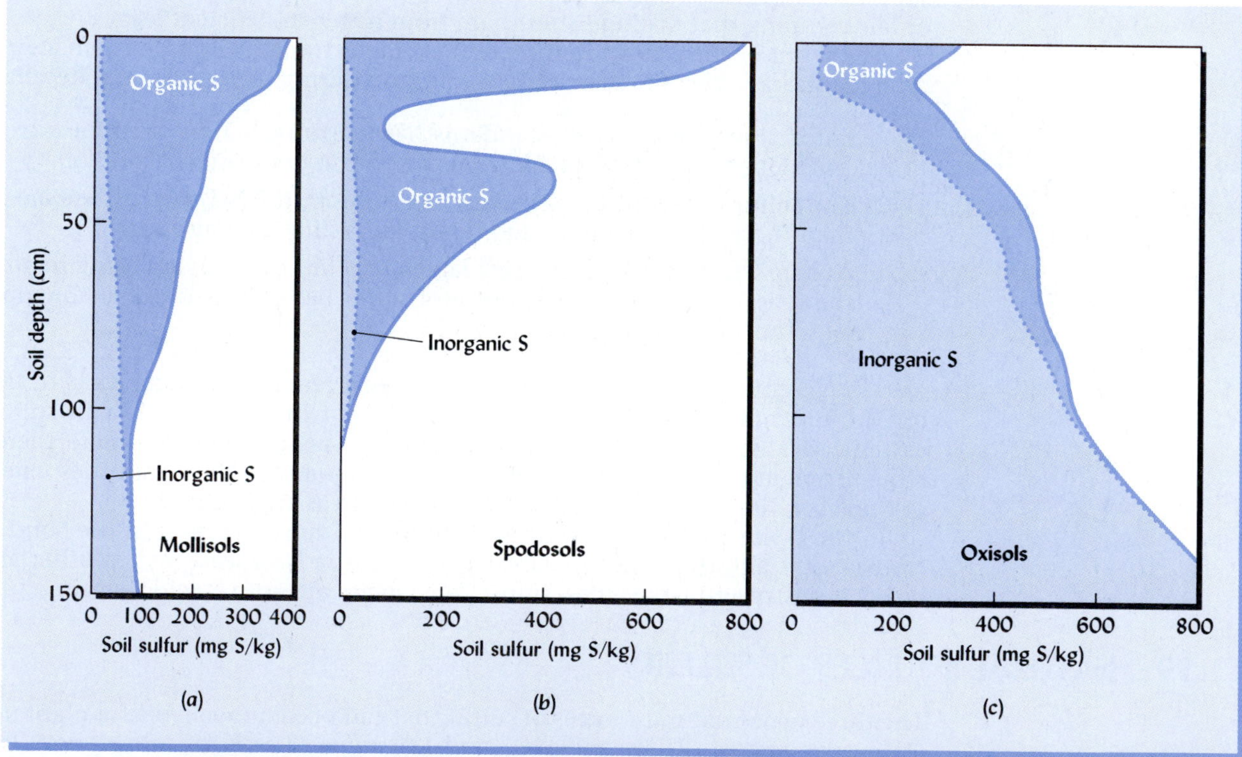

FIGURE 12.14 The distribution of organic and inorganic sulfur in representative soil profiles of the soil orders Mollisols, Spodosols, and Oxisols. In each, soil organic forms dominate the surface horizon. Considerable inorganic sulfur, both as adsorbed sulfate and calcium sulfate minerals, exists in the lower horizons of Mollisols. Relatively little inorganic sulfur exists in Spodosols. However, the bulk of the profile sulfur in the humid tropics (Oxisols) is present as sulfate adsorbed to colloidal surfaces in the subsoil.

Atmospheric Sulfur[10]

Sulfur emissions and air pollution (scroll down page): http://www.iitap.iastate.edu/gcp/chem/nitro/nitro_lecture.html

The atmosphere contains varying quantities of carbonyl sulfide (COS), hydrogen sulfide (H_2S), sulfur dioxide (SO_2), and other sulfur gases, as well as sulfur-containing dust particles. Some of these materials are oxidized in the atmosphere to sulfates, forming H_2SO_4 and sulfate salts, such as $CaSO_4$ and $MgSO_4$. When these solids and gases return to the earth as dry particles and gases, it is called *dry deposition;* when they are brought down with precipitation, it is called *wet deposition.* Although the proportion of these two forms of deposition varies from one place to another, each typically supplies about half of the total.

The industrialized northeastern states have the highest deposition of sulfur in the United States, commonly ranging from 30 to 75 kg S/ha/yr downwind from industrial sites. In rural areas of the western United States, away from industrial cities and smelting plants, only 2 to 5 kg S/ha/yr is deposited. In rural Africa, 1 to 4 kg S/ha/yr is deposited. Regional patterns of air quality and sulfur deposition are evident as the less acute effects of industrial emissions are felt hundreds of kilometers downwind (Figure 12.15).

Atmospheric sulfur, whether in the wet or dry form, is absorbed by the soils or directly by the plant foliage. The quantity that plants can absorb directly is variable, but in some cases 25 to 35% of the plant sulfur can come from this source even if available soil sulfate is adequate. In sulfur-deficient soils, about half of the plant needs can come from the atmosphere.

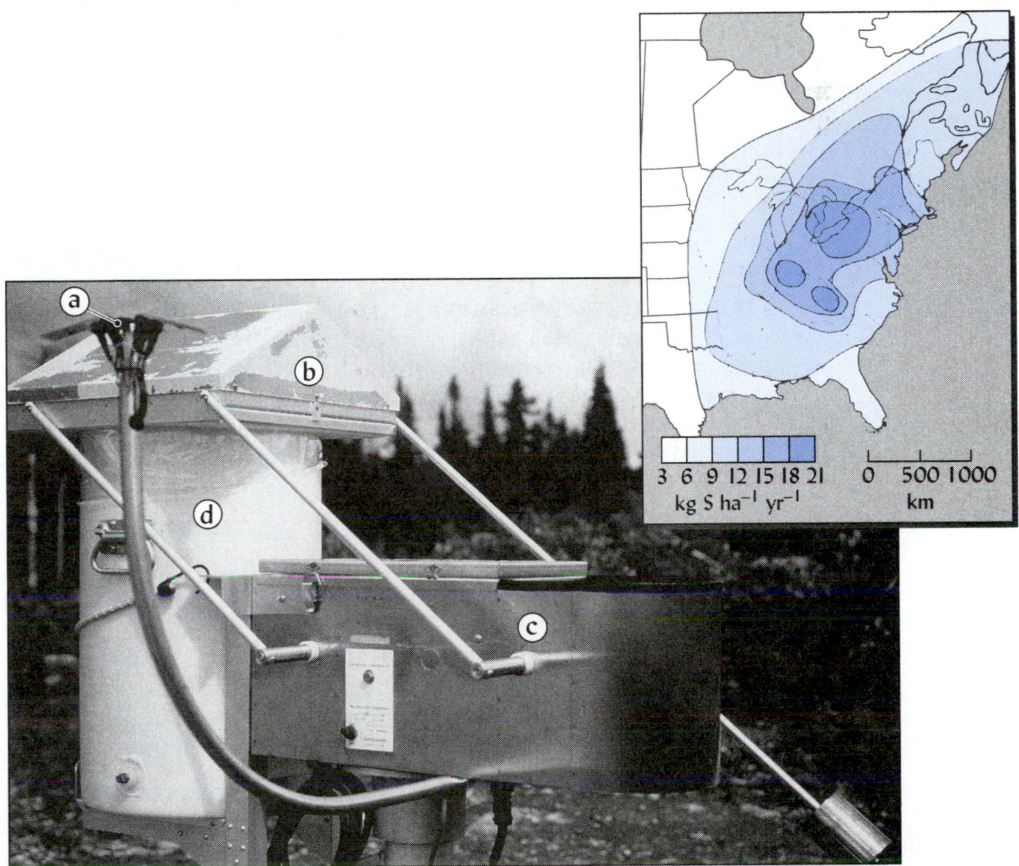

FIGURE 12.15 The apparatus shown in the picture collects both wet and dry sulfur deposition. A sensor (*a*) triggers the small roof (*b*) to move over and cover the dry deposition collection chamber (*c*) at the first sign of precipitation. The wet deposition chamber (*d*) is then exposed to collect precipitation. When the precipitation ceases, the sensor triggers the roof to move back over the wet deposition collection chamber so that dry deposition can again be collected. The map shows the typical geographic distribution of sulfur deposition from the atmosphere in eastern North America. The data are based on measurements of sulfur in wet deposition and the assumption that dry deposition contributes approximately an equal share of the total. [Photo courtesy of R. Weil; map modified from Olsen and Slavich (1986)]

[10]For a discussion of this topic, see National Academy of Sciences (1983).

Environmental concerns about high sulfur levels in the atmosphere and the resulting acid rain are of great practical significance to both forestry and agriculture. The elevated levels of atmospheric sulfur and associated acid effects are causing serious damage to certain types of forest ecosystems. On the other hand, continued reductions in atmospheric sulfur are resulting in plant deficiencies of this element in other areas, especially for high-yield-potential agricultural crops. In these cases, part of the crop's sulfur requirement may have to be supplied by increasing the sulfur content of the fertilizers used, thereby increasing crop production costs.

12.20 THE SULFUR CYCLE

Biogeochemical cycles, including sulfur cycle: http://www2.nau.edu/doetqp/courses/env440/lectures/lec6/lec6.html

The major transformations that sulfur undergoes in soils are shown in Figure 12.16. The inner circle shows the relationships among the four major forms of this element: (1) *sulfides,* (2) *sulfates,* (3) *organic sulfur,* and (4) *elemental sulfur.* The outer portions show the most important sources of sulfur and how this element is lost from the system.

Considerable similarity to the nitrogen cycle is evident (compare Figures 12.1 and 12.16). In each case, the atmosphere is an important source of the element in question. Both elements are held largely in the soil organic matter, both are subject to microbial oxidation and reduction, both can enter and leave the soil in gaseous forms, and both are subject to some degree of leaching in the anionic form. Microbial activities are responsible for many of the transformations that determine the fates of both nitrogen and sulfur.

Figure 12.16 should be referred to frequently in conjunction with the following, more detailed examination of sulfur in plants and soils.

12.21 BEHAVIOR OF SULFUR COMPOUNDS IN SOILS

Mineralization

Sulfur behaves much like nitrogen as it is absorbed by plants and microorganisms and moves through the sulfur cycle. The organic forms of sulfur must be mineralized by soil organisms if the sulfur is to be used by plants. When conditions are favorable for general microbial activity, sulfur mineralization occurs. Some of the more easily decomposed organic compounds in the soil are sulfate esters, from which microorganisms release sulfate ions directly. However, in much of the soil organic matter, sulfur in the reduced state is bonded to carbon atoms in protein and amino acid compounds. In the latter case, the mineralization reaction might be expressed as follows:

$$\underset{\text{Proteins and other organic combinations}}{\text{Organic sulfur}} \longrightarrow \underset{H_2S\text{ and other sulfides are simple examples}}{\text{decay products}} \xrightarrow{O_2} \underset{\text{Sulfates}}{SO_4^{2-}} + 2H^+ \qquad (12.9)$$

Because this release of available sulfate is mainly dependent on microbial processes, the supply of available sulfate in soils fluctuates with seasonal, and sometimes daily, changes in environmental conditions (Figure 12.17). These fluctuations lead to the same difficulties in predicting and measuring the amount of sulfur available to plants as discussed in the case of nitrogen.

Immobilization

Immobilization of inorganic forms of sulfur occurs when low-sulfur, energy-rich organic materials are added to soils. The immobilization mechanism is thought to be the same as for nitrogen—the energy-rich material stimulates microbial growth, and the inorganic sulfate is assimilated into microbial tissue. A C/S ratio greater than 400:1 generally leads to such immobilization of sulfur. When the microbial activity subsides, the inorganic sulfate reappears in the soil solution.

The pattern of S immobilization in soils suggests that, like nitrogen, sulfur in soil organic matter may be associated with organic carbon in a reasonably constant ratio. Data from a number of soils in temperate regions suggest that the C/N/S ratio of 100:8:1 is quite representative.

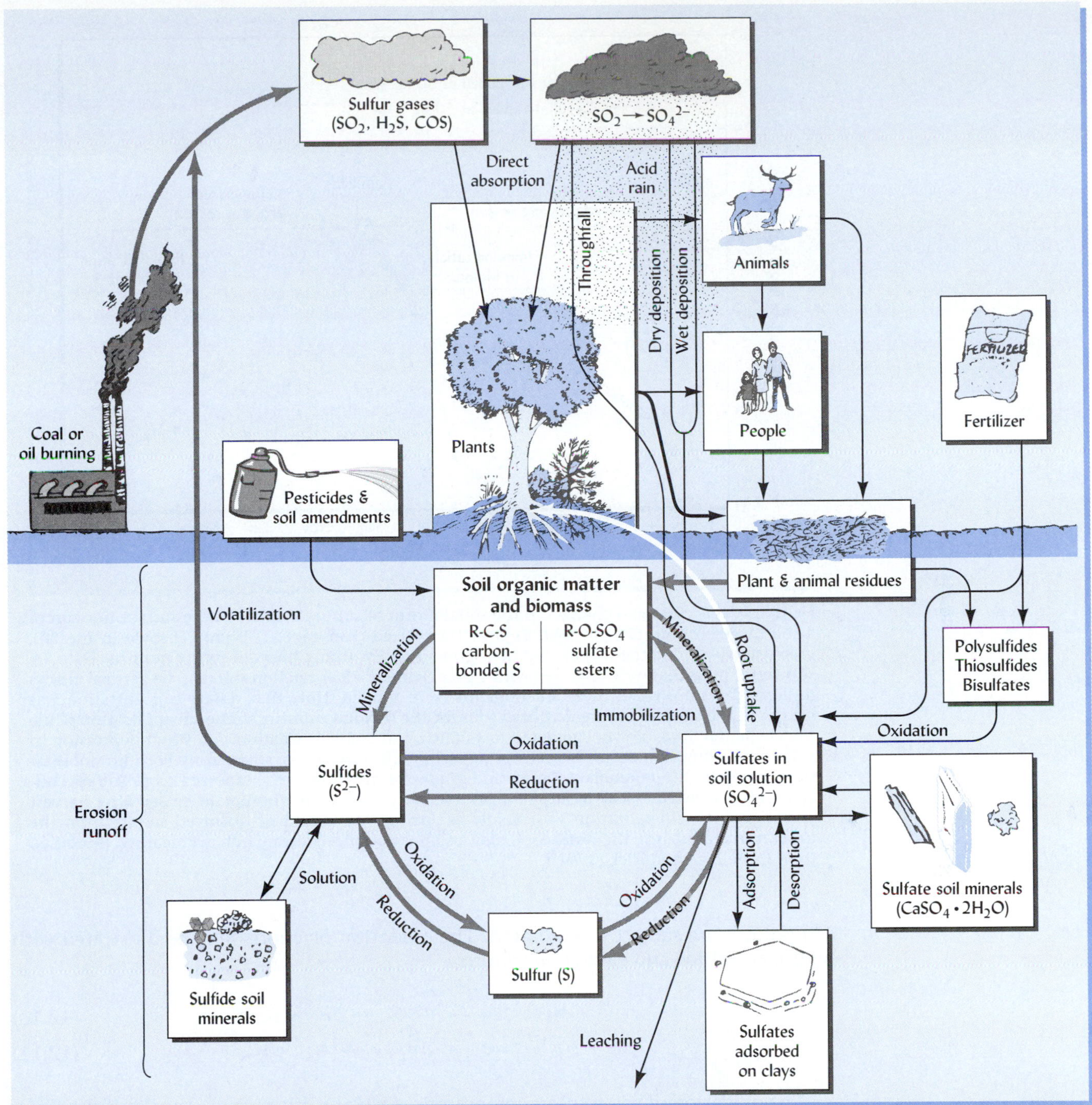

FIGURE 12.16 The sulfur cycle, showing some of the transformations that occur as this element is cycled through the soil–plant–animal–atmosphere system. In the surface horizons of all but a few types of arid-region soils, the great bulk of sulfur is in organic forms. However, in deeper horizons or in excavated soil materials, various inorganic forms may dominate. The oxidation and reduction reactions that transform sulfur from one form to another are mainly mediated by soil microorganisms.

12.22 SULFUR OXIDATION AND REDUCTION

The Oxidation Process

During the microbial decomposition of organic carbon–bonded sulfur compounds, sulfides are formed along with other incompletely oxidized substances, such as elemental sulfur (S^0), thiosulfates ($S_2O_3^{2-}$), and polythionates ($S_{2x}O_{3x}^{2-}$). Like the ammonium compounds formed when nitrogenous materials are decomposed, these reduced sulfur

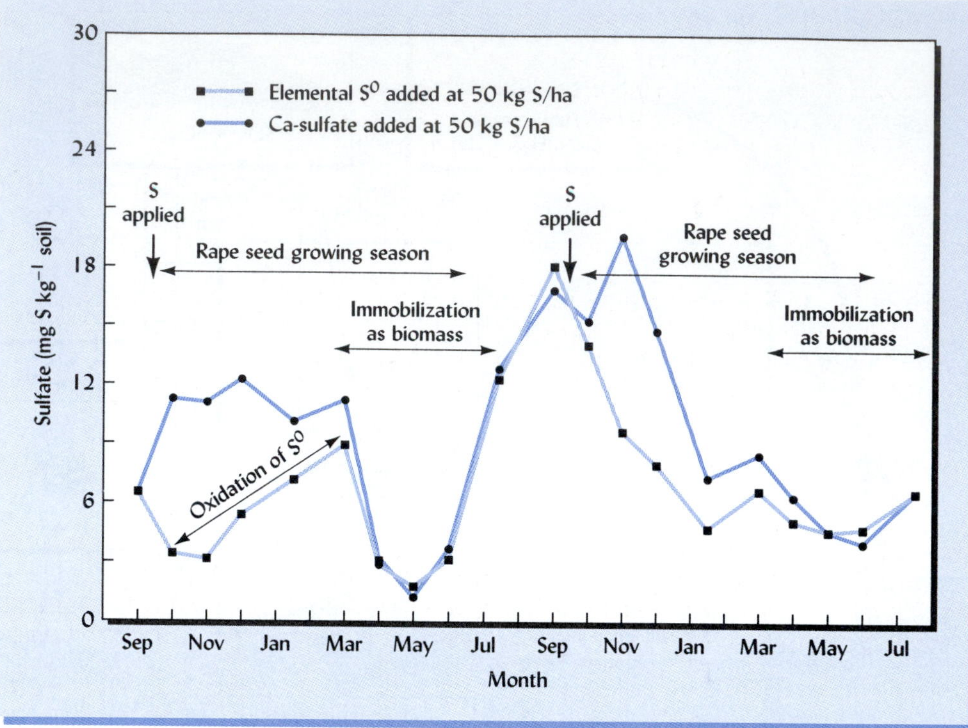

FIGURE 12.17 Seasonal changes in the sulfate form of sulfur available in the surface horizon of a soil (Argixeroll) in Oregon used to grow the oilseed crop, rape. This crop is sown in the fall, grows slowly during the winter, and then grows rapidly during the cool spring months. Data are shown for plots that were fertilized with either elemental S or calcium sulfate. The vertical arrows indicate dates on which these amendments were applied. Note that sulfate concentration was greater in the calcium sulfate–fertilized soils for the first few months after each application while the elemental S was slowly converted to sulfate by microbial oxidation. A distinct depression in sulfate concentration occurred each spring as the soil warmed up, stimulating both immobilization of sulfate into microbial biomass and uptake of sulfate into the rapeseed crop. Sulfate concentrations reached a peak in late summer and early fall, when crop uptake ceased after harvest and microbial mineralization was rapidly occurring. Movement of dissolved sulfate from the lower horizons up into the surface soil may also have occurred during hot, dry weather. [Modified from Castellano and Dick (1991)]

compounds are subject to oxidation. The oxidation reactions may be illustrated with hydrogen sulfide and elemental sulfur:

$$H_2S + 2O_2 \longrightarrow H_2SO_4 \longrightarrow 2H^+ + SO_4^{2-} \tag{12.10}$$

$$2S + 3O_2 + 2H_2O \longrightarrow 2H_2SO_4 \longrightarrow 4H^+ + SO_4^{2-} \tag{12.11}$$

The oxidation of some sulfur compounds, such as sulfites (SO_3^{2-}) and sulfides (S^{2-}), can occur by strictly chemical reactions. However, most sulfur oxidation in soils is *biochemical* in nature, carried out by a number of autotrophic bacteria, which include five species of the genus *Thiobacillus*. Since the environmental requirements and tolerances of these five species vary considerably, the process of sulfur oxidation occurs over a wide range of soil conditions. For example, sulfur oxidation may occur at pH values ranging from <2 to >9. This flexibility is in contrast to the comparable nitrogen oxidation process, nitrification, which requires a rather narrow pH range closer to neutral.

Note that sulfur oxidation is an acidic process which, as we learned in Chapter 9, can be both beneficial and detrimental. The H_2SO_4 produced from sulfur oxidation can be used to help reclaim sodic soils, and to lower the soil pH for acid-loving plants such as rhododendrons. But the process also stimulates the formation of highly toxic *acid sulfate soils* with pH values as low as 1.5 (see Section 9.7). Likewise, it (along with oxides of nitrogen) contributes to the problems of acid rain with its detrimental effects, particularly on forests. Last, it increases the quantities of liming materials that must be added to maintain appropriate pH levels in humid-region soils.

The Reduction Process

Like nitrate ions, sulfate ions tend to be unstable in anaerobic environments. They are reduced to sulfide ions by a number of bacteria of two genera, *Desulfovibro* (five species) and *Desulfotomaculum* (three species). The organisms use the sulfur as an electron acceptor to oxidize organic materials:

$$\underset{\text{Organic alcohol}}{2R{-}CH_2OH} + \underset{\text{Sulfate}}{SO_4^{2-}} \longrightarrow \underset{\text{Organic acid}}{2R{-}COOH} + 2H_2O + \underset{\text{Sulfide}}{S^{2-}} \qquad (12.12)$$

In poorly drained soils, the sulfide ion reacts immediately with iron or manganese, which in anaerobic conditions are typically present in the reduced forms. By tying up the soluble reduced iron, the formation of iron sulfides helps prevent iron toxicity in rice paddies and marshes. This reaction may be expressed as follows:

$$\underset{\substack{\text{Dissolved}\\\text{ferrous iron}}}{Fe^{2+}} + \underset{\text{Sulfide}}{S^{2-}} \longrightarrow \underset{\substack{\text{Iron sulfide}\\\text{(solid)}}}{FeS} \qquad (12.13)$$

The oxidation and reduction reactions of inorganic sulfur compounds play an important role in determining the quantity of sulfate (the plant-available nutrient form of sulfur) present in soils at any one time. Also, the state of sulfur oxidation is an important factor in the acidity of soil and water draining from soils.

12.23 SULFUR RETENTION AND EXCHANGE

The sulfate ion is the form in which plants absorb most of their sulfur from soils. Since many sulfate compounds are quite soluble, the sulfate would be readily leached from the soil, especially in humid regions, were it not for its adsorption by the soil colloids. As pointed out in Chapter 8, most soils have some anion exchange capacity, which is associated with iron and aluminum oxide coatings and clays and, to a limited extent, with 1:1-type silicate clays. Sulfate ions are attracted by the positive charges that characterize acid soils containing these clays. They also react directly with hydroxy groups exposed on the surfaces of these clays. Figure 12.18 illustrates sulfate adsorption mechanisms on the surface of some Fe, Al oxides and 1:1-type clays. Note that adsorption increases at

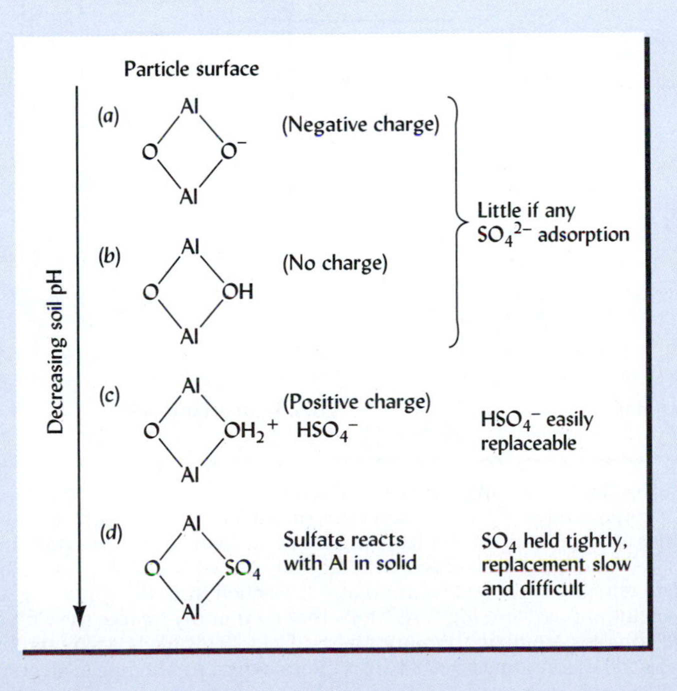

FIGURE 12.18 Effect of decreasing soil pH on the adsorption of sulfates by 1:1-type silicate clays and oxides of Fe and Al (reaction with a surface-layer Al is illustrated). At high pH levels (*a*), the particles are negatively charged, the cation exchange capacity is high, and non-acid cations are adsorbed. Sulfates are repelled by the negative charges. As acidity is increased (*b*), the H^+ ions are attracted to the particle surface, and the negative charge is satisfied but the SO_4^{2-} ions are still not attracted. At still lower pH values (*c*), more H^+ ions are attracted to the particle surface, resulting in a positive charge that attracts the SO_4^{2-} ion. This is easily exchanged with other anions. At still lower pH levels, the SO_4^{2-} reacts directly with Al and becomes a part of the crystal structure. Such sulfate is tightly bound, as an inner-sphere complex, and it is removed very slowly, if at all.

lower pH values as positive charges that become more prominent on the particle surfaces attract the sulfate ions. Some sulfate reacts with the clay particles, becoming tightly bound, and is only slowly available for plant uptake and leaching.

In warm, humid regions, surface soils are typically quite low in sulfur. However, much sulfate may be held by the iron and aluminum oxides and 1:1-type silicate clays that tend to accumulate in the subsoil horizons of the Ultisols and Oxisols of these regions (see Figure 12.14). Symptoms of sulfur deficiency commonly occur early in the growing season on Ultisols in the southeastern United States, especially those with sandy, low-organic-matter surface horizons. However, the symptoms may disappear as the crop matures and its roots reach the deeper horizons where sulfate is retained.

Sulfate Adsorption and Leaching of Nonacid Cations

Sulfur cycling retention, and mobility in soils: A review: http://www.fs.fed.us/ne/newtown_square/publications/technical_reports/pdfs/1998/gtrne250.pdf

When sulfate anions leach from the soil, they are usually accompanied by equivalent quantities of Ca and Mg and other nonacid cations. In soils with high sulfate adsorption capacities, sulfate leaching is low and the loss of companion cations is also low (Figure 12.19). In contrast, much sulfate may leach from low-sulfate-adsorbing soils, taking with it considerable quantities of nonacid cations. The role of sulfate sorption in conserving cations in the soil profile is of considerable importance in forested areas that receive acid rain.

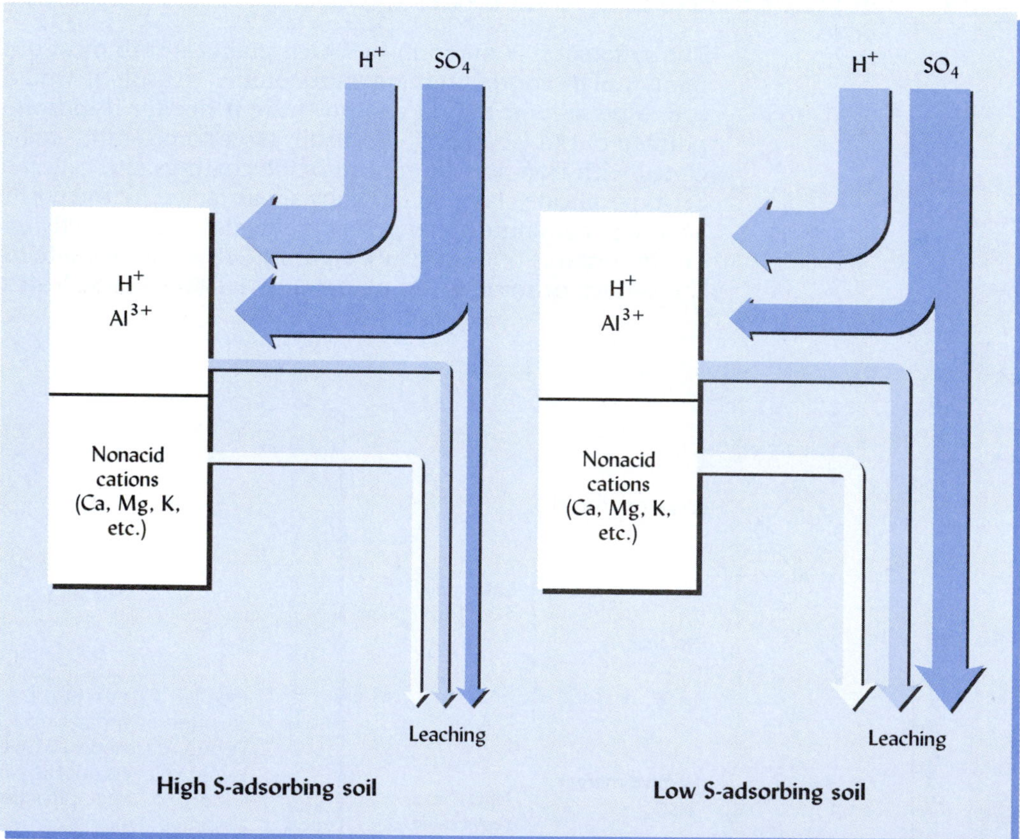

FIGURE 12.19 Diagrams illustrating cation leaching losses as influenced by sulfate adsorption capacities of forest soils. When acid rain containing SO_4^{2-} and H^+ ions falls on soils with high SO_4^{2-}-adsorbing capacities (left), the small quantities of SO_4^{2-} available for leaching are accompanied by correspondingly small amounts of cations such as Ca^{2+} and Mg^{2+}. Where acid rain falls on soils with low SO_4^{2-} adsorbing capacity (right), most of the sulfate remains in the soil solution and is leached from the soil along with equivalent quantities of cations, including Ca^{2+} and Mg^{2+}. Al^{3+} ions that commonly replace the Ca^{2+} and Mg^{2+} lost from the soil exchange complex are toxic to many forest species. This likely accounts for at least part of the negative effects of acid rain on some forested areas. [Redrawn from Mitchell et al. (1992)]

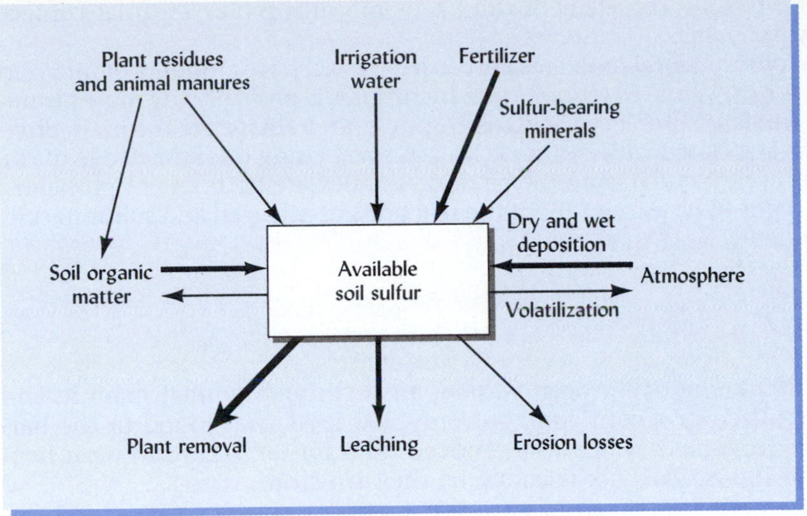

FIGURE 12.20 Major gains and losses of available soil sulfur. The thickness of the arrows indicates the relative amounts of sulfur involved in each process under average conditions. Considerable variation occurs in the field.

12.24 SULFUR AND SOIL FERTILITY MAINTENANCE

Sulfur for vegetables and field crops: http://www.msue.msu.edu/msue/imp/modfl/05209704.html

Figure 12.20 depicts the major gains and losses of sulfur from soils. The problem of maintaining adequate quantities of sulfur for mineral nutrition of plants is becoming increasingly important. Even though chances for widespread sulfur deficiencies are generally less than for nitrogen, phosphorus, and potassium, increasing crop removal of sulfur makes it essential that farmers be attentive to prevent deficiencies of this element. In some parts of the world (especially in certain semiarid grasslands), sulfur is already the next most limiting nutrient after nitrogen.

Crop residues and farmyard manures can help replenish the sulfur removed in crops, but these sources generally can help to recycle only those sulfur supplies that already exist within a farm. In regions with low-sulfur soils, greater dependence must be placed on fertilizer additions. Regular applications of sulfur-containing materials are now necessary for good crop yields in large areas far removed from cities and industrial plants. There will certainly be an increased necessity for the use of sulfur in the future.

12.25 CONCLUSION

Sulfur and nitrogen have much in common concerning the manner in which they cycle in soils. Both are held by soil colloids in slowly available forms. Both are found in proteins and other organic forms as part of the soil organic matter. Their release to inorganic ions (SO_4^{2-}, NH_4^+, and NO_3^-), in which forms they are available to higher plants, is accomplished by soil microorganisms. Anaerobic soil organisms are able to change these elements into gaseous forms, which are then released to the atmosphere to be joined by similar gases released from industrial plants and from vehicle engines. These gases are then deposited on plants, soils, and other objects in forms that are popularly termed *acid rain,* which has serious consequences to forestry, agriculture, and society in general.

There are some significant differences between nitrogen and sulfur. Large amounts of both soil nitrogen and soil sulfur are found in organic compounds; however, a higher proportion of soil sulfur is found in inorganic compounds, especially in some drier areas where gypsum ($CaSO_4 \cdot 2H_2O$) is abundant in the subsoil. Some soil organisms have the ability to fix elemental N_2 gas into compounds usable by plants. No analogous process occurs for sulfur. Also, nitrogen, which is removed in much larger quantities by plants, must be replenished regularly by organic residues, manure, or chemical fertilizers.

In rural areas away from cities and industrial plants, sulfur deficiencies in plants are increasingly common. As a result of increased crop removal and reductions in incidental applications of sulfur to soils, this element will likely join nitrogen as a regular component of soil fertility programs.

Several serious environmental problems are caused by excessive amounts of nitrogen or sulfur in certain forms. While water pollution by nitrates is probably the most prominent of the nitrogen-related problems, acid deposition and drainage are the main environmental concerns associated with sulfur cycling in soils. Using our knowledge of the nitrogen and sulfur cycles, much can now be done to alleviate both sets of problems. However, much remains to be learned about the reactions of nitrogen and sulfur in soils and in the environment.

STUDY QUESTIONS

1. A total of 300 kg/ha of nitrogen was supplied through animal manure and chemical fertilizer for a corn crop, but only 200 kg/ha was found in the harvested silage corn, all of which was removed from the land. Explain what happened to the 100 kg/ha N not taken up by the corn crop.
2. Discuss the benefits and ill effects of sulfur and nitrogen that are added to soils by humans and by natural processes each year.
3. Some 20 kg/ha of nitrogen contained in wheat straw was added to a soil that had a nitrate level of 10 kg/ha. A couple of weeks later a soil test showed no nitrates. How do you explain this situation?
4. Modern domestic animal production systems have some environmental and health implications relating to nitrogen. What are they, and how can they be managed?
5. What differences would you expect in nitrate contents of streams from a forested watershed and one where agricultural crops are grown, and why?
6. What is *acid rain,* what are the sources of this precipitation, and what are its implications for forestry and agriculture?
7. Nitrogen is fixed from the atmosphere and also by vermiculite clays and humus. Differentiate between these two processes and indicate the role of microbes in each process.
8. Chemical fertilizers and manures are commonly added to agricultural soils, and yet these soils are often lower in nitrogen and other nutrients than are nearby forested soils. Explain why this is the case.
9. How do riparian forests help reduce nitrate contamination of streams and rivers?
10. Sulfur deficiencies are more likely in agricultural fields in the United States today than 40 years ago. What are the three major reasons for this?
11. In some tropical regions, agroforestry systems that involve mixed cropping of trees and food crops are used. What advantages in nitrogen management do such systems have over monocropping systems that do not involve trees?

REFERENCES

Burns, R. C., and R. W. F. Hardy. 1975. *Nitrogen Fixation in Bacteria and Higher Plants.* (Berlin: Springer-Verlag).

Cassman, K. G., A. Dobermann, and D. T. Walters. 2002. "Agroecosystems, Nitrogen-use Efficiency, and Nitrogen Management," *AMBIO: A Journal of the Human Environment* **31**:132–140.

CAST. 1999. *Gulf of Mexico Hypoxia: Land and Sea Interactions,* Task Force Report 134. (Ames, Iowa: Council for Agricultural Science and Technology).

Castellano, S. D., and R. P. Dick. 1991. "Cropping and sulfur fertilization influence on sulfur transformations in soil," *Soil Sci. Soc. Amer. J.* **54**:114–121.

Colbourn, P. 1993. "Limits to denitrification in two pasture soils in a maritime temperate climate," *Agriculture, Ecosystems and Environment* **43**:49–68.

David, M. B., and L. E. Gentry. 2000. "Anthropogenic inputs of nitrogens and phosphorus and riverine export for Illinois, USA," *J. Environ. Qual.* **29**:494–508.

De Walle, D. R., et al. 1999. "Dendrochemical detection of base cation depletion and watershed acidification," in W. E. Sharpe and J. R. Brohan (eds.), *Proceedings of the 1998 PA Acidic Deposition Conference,* vol. 1., chap. 10. (University Park, Penn.: Environ. Res. Inst.).

Diaz, R. J. 2001. "Overview of hypoxia around the world," *J. Environ. Qual.* **30**:275–281.

Galloway, J. N., and E. B. Cowling. 2002. "Reactive Nitrogen and the World: 200 Years of Change," *AMBIO: A Journal of the Human Environment* **31**:64–71.

Goolsby, D. A., et al. 1999. *Flux and Sources of Nutrients in the Mississippi—Atchafalaya River Basin: Topic 3 Report for the Integrated Assessment on Hypoxia in the Gulf of Mexico.* NOAA Coastal Ocean Program Decision Analysis Series no. 7. (Silver Spring, Md.: NOAA Coastal Ocean Program).

Hanson, G. C., P. M. Groffman, and A. J. Gold. 1994. "Denitrification in riparian wetlands receiving high and low groundwater nitrate inputs," *J. Environ. Qual.* **23**:917–922.

Jenkinson, D. S. 1990. "An introduction to the global nitrogen cycle," *Soil Use and Management* **6**:56–61.

Klemmedson, J. O. 1994. "New Mexican locust and parent material: Influence on forest floor and soil macronutrients," *Soil Sci. Soc. Amer. J.* **58**:974–980.

Leffelaar, P. A., and W. W. Wessel. 1988. "Denitrification in a homogeneous, closed system: Experimental and simulation," *Soil Sci.* **146**:335–349.

Leuch, D., S. Haley, P. Liapis, and B. McDonald. 1995. *The EU Nitrate Directive and CAP Reform: Effects on Agricultural Production, Trade and Residual Soil Nitrogen.* Report 255. (Washington, D.C.: USDA Economic Research Service).

Mitchell, M. J., et al. 1992. "Sulfur dynamics of forest ecosystems," in Howarth et al. (eds.), *Sulfur Cycling on the Continents, 1992 SCOPE.* (New York: Wiley).

Murphy, D. W., et al. 2000. "Soluble nitrogen in agricultural soils," *Biol. Fertil. Soils* **30**:374–387.

National Academy of Sciences. 1983. *Acid Deposition: Atmospheric Processes in Eastern North America.* (Washington, D.C.: National Academy Press).

National Research Council. 1993. *Soil and Water Quality: An Agenda for Agriculture.* (Washington, D.C.: National Academy of Sciences).

Olsen, A. R., and A. L. Slavich. 1986. *Acid Precipitation in North America: 1984 Annual Data Summary,* from Acid Deposition System Data Base. Environmental Protection Agency Report EPA/600/4-86/033. (Washington, D.C.: U.S. Environmental Protection Agency).

Owens, L. B., W. M. Edwards, and R. W. Van Keuren. 1992. "Nitrate levels in shallow groundwater under pastures receiving ammonium nitrate or slow-release nitrogen fertilizer," *J. Environ. Qual.* **21**(4):607–613.

Patrick, W. H., Jr. 1982. "Nitrogen transformations in submerged soils," in F. J. Stevenson (ed.), *Nitrogen in Agricultural Soils.* Agronomy Series no. 27. (Madison, Wis.: Amer. Soc. Agron., Crop Sci. Soc. Amer., Soil Sci. Soc. Amer.).

Perakis, S. S., and L. O. Hedin. 2002. "Nitrogen loss from unpolluted South American forests mainly via dissolved organic compounds," *Nature* **415**:416–419.

Power, J. F. 1981. "Nitrogen in the cultivated ecosystem," in F. E. Clark and T. Rosswall (eds.), *Terrestrial Nitrogen Cycles—Processes, Ecosystem Strategies and Management Impacts.* Ecological Bulletin no. 33. (Stockholm, Sweden: Swedish National Research Council), pp. 529–546.

Prasad, R., and J. F. Power. 1995. "Nitrification inhibitors for agriculture, health and the environment," *Advances in Agronomy* **54**:233–280.

Smil, V. 1999. "Nitrogen in crop production: an account of global flows," *Global Biogeochem. Cycl.* **13**:647–662.

Tabatabai, S. J. 1986. *Sulfur in Agriculture.* Agronomy Series no. 27 (Madison, Wis.: Amer. Soc. Agron., Crop Sci. Soc. Amer., Soil Sci. Soc. Amer.).

Torrey, J. G. 1978. "Nitrogen fixation by actinomycete-induced angiosperms," *BioScience* **28**:586–592.

U.S./Canada Work Group No. 2. 1982. *Atmospheric Sciences and Analysis.* Final Report; J. L. Ferguson and L. Machata (cochairmen). (Washington, D.C.: U.S. Environmental Protection Agency).

Walker, T. W., et al. 1956. "Fate of labeled nitrate and ammonium nitrogen when applied to grass and clover grown separately and together," *Soil Sci.* **81**:339–352.

13

Soil Phosphorus, Potassium, and Micronutrients

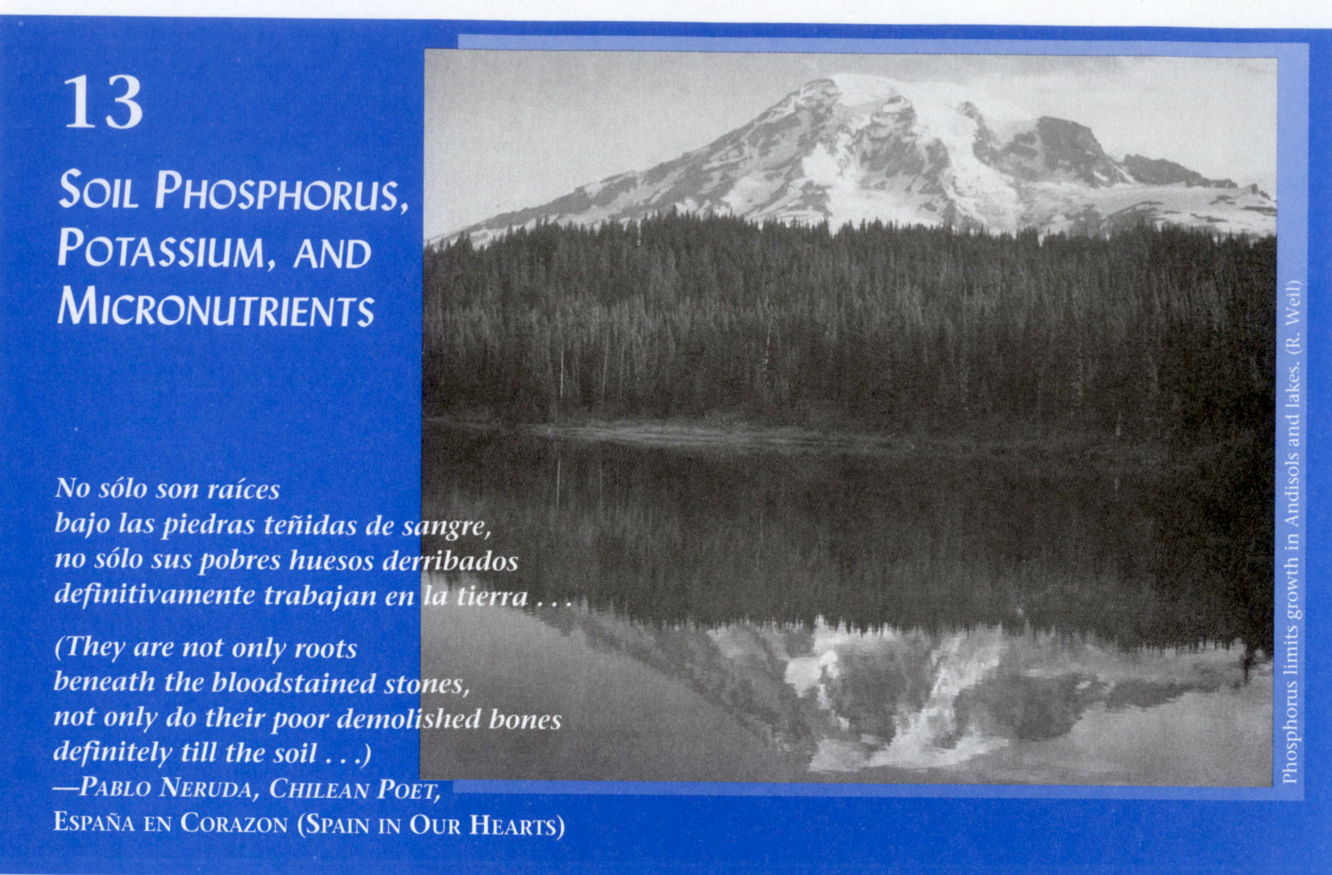

Phosphorus limits growth in Andisols and lakes. (R. Weil)

No sólo son raíces
bajo las piedras teñidas de sangre,
no sólo sus pobres huesos derribados
definitivamente trabajan en la tierra . . .

(They are not only roots
beneath the bloodstained stones,
not only do their poor demolished bones
definitely till the soil . . .)
—*Pablo Neruda, Chilean Poet,*
España en Corazon (Spain in Our Hearts)

Among the nutrient elements, phosphorus is second only to nitrogen in its importance for the productivity and health of both terrestrial and aquatic ecosystems. The total quantity of phosphorus in native soils is low, with most of it present in forms quite unavailable to plants. Phosphorus is so scarce in most ecosystems, that archaeologists often use the presence of high soil phosphorus concentrations as an indication of prehistoric human habitation (discarded bones contain the phosphorus collected by game animals from across the humans' entire hunting territory). We shall see that natural ecosystems have evolved ways of maximizing the efficiency with which plants use and recycle limited phosphorus supplies. In agricultural systems, however, the low availability of phosphorus often leads to major social and environmental problems. Unproductive soils fail to produce adequate crops, forcing poor farmers to clear more land in order to survive. The cleared lands support poor vegetative cover, and so are subject to erosion that further degrades the soils and pollutes rivers and lakes with sediment-laden runoff. This scenario pertains in much of sub-Saharan Africa, where inadequate supplies of soil phosphorus have contributed to the decline of per capita food production in recent decades.

Farmers in most industrialized countries have for many years overcompensated for the initial low phosphorus availability by adding far more P to agricultural soils than crop harvests remove. Excessive applications of P-containing fertilizers (Figure 13.1) and organic wastes have led to a buildup of phosphorus in the surface soil. In many agricultural watersheds, the resulting P-laden runoff contributes to the process of **eutrophication,**[1] which may jeopardize drinking water supplies and restrict the use of aquatic systems for fisheries, recreation, industry and aesthetics. Phosphorus control is therefore a high priority of most national and regional water quality programs.

---

[1]The word *eutrophication* comes from the Greek *eutruphos,* meaning "well nourished or nourishing." Natural accumulation of nutrients over centuries causes lakes to fill in with plants and lose dissolved oxygen needed by fish—the process of eutrophication. Excessive inputs of nutrients under human influence tremendously speeds this process and is called *cultural eutrophication.*

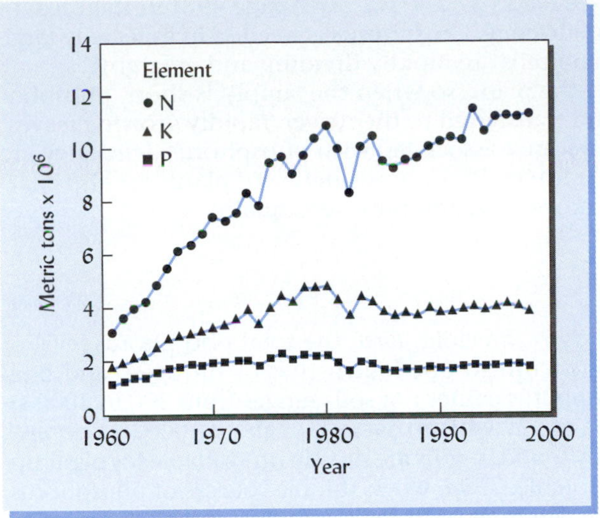

FIGURE 13.1 Estimated annual addition of fertilizer nitrogen, phosphorus, and potassium to cropland in the United States since 1961. Note the very rapid increase in the use of nitrogen fertilizers, probably due to the replacement of legume-based rotations with monocropping and heavy nitrogen applications. The annual applications of phosphorus increased steadily until about 1980, then declined somewhat to level off at about 2 million Mg per year. The decline and leveling off of phosphorus applications reflects the reduced need for phosphorus buildup in agricultural soils since their phosphorus fixing capacity has been largely saturated. The trend toward lower rates of both phosphorus and potassium since 1980 was likely influenced by economic and agronomic factors, not environmental concerns. For example, the oil embargo of 1973–1974 increased fertilizer prices dramatically, resulting in the notable dip in application rates in 1974. The sharp dip in 1983 reflected government policy of "payment in kind" to induce farmers to restrict their corn plantings. In some parts of the world such as sub-Saharan Africa, sustainable crop production will require increasing applications, particularly of phosphorus, for many years to come. (Based on data from the Food and Agriculture Organization)

Potassium does not play as direct a role in environmental quality as does phosphorus, but inadequate supplies of this element commonly limit plant growth, reduce plant resistance to pests and result in low quality of agricultural produce. Even though most soils have large total supplies of this element, most of that present is tied up in the form of insoluble minerals and is unavailable for plant use. Also, plants require potassium in such large amounts that careful management practices are necessary to make this nutrient available rapidly enough to optimize plant growth.

Of the 18 elements known to be essential for plant growth, 9 are required in such small amounts that they are called *micronutrients* or *trace elements*. The ill effects of deficiencies and even toxicities of these elements are becoming more noted throughout the world in terms of stunted growth, low yields, and even plant death. Very small applications of micronutrients can produce dramatic results, while problems of toxicity loom when an oversupply is present.

13.1 ROLE OF PHOSPHORUS IN PLANT NUTRITION AND SOIL FERTILITY[2]

ATP-the molecule and how it works:
http://www.bris.ac.uk/Depts/Chemistry/MOTM/atp/atp1.htm

Neither plants nor animals can live without phosphorus. It is an essential component of **adenosine triphosphate** (ATP) which is the *energy currency* that drives most biochemical processes, including the uptake of nutrients and their transport within the plant. Phosphorus is an essential component of **deoxyribonucleic acid** (DNA), the seat of genetic inheritance, and of **ribonucleic acid** (RNA), which directs protein synthesis in both plants and animals. Phospholipids, which play critical roles in cellular membranes, are another class of universally important phosphorus-containing compounds. For most plant species, the total phosphorus content of healthy leaf tissue is 0.2 and 0.4% of the dry matter.

Phosphorus and Plant Growth

P as an essential nutrient-sources and management:
http://www.soils.wisc.edu/~barak/soilscience326/phosphorus.htm

Aspects of Plant Growth Enhanced. Adequate phosphorus nutrition enhances many aspects of plant physiology, including the fundamental processes of photosynthesis, nitrogen fixation, flowering, fruiting (including seed production), and maturation. Root growth, particularly development of lateral roots and fibrous rootlets, is encouraged by phosphorus as is straw strength in cereal crops. Improvement of crop quality, especially in forages and vegetables, is another benefit attributed to this nutrient.

Symptoms of Phosphorus Deficiency in Plants. A phosphorus-deficient plant is usually stunted, thin stemmed, and spindly, but its foliage is often dark, almost bluish green. In severe cases, phosphorus deficiency can cause yellowing and senescence of leaves.

[2]For a review of the plant availability of this element, see Sharpley (2000).

Some plants develop purple colors (see Plates 47 and 52, after page 498) in their leaves and stems as a result of phosphorus deficiency. Phosphorus is needed in especially large amounts in meristematic tissues, where cells are rapidly dividing and enlarging.

Phosphorus is very mobile within the plant, so when the supply is short, phosphorus in the older leaves is mobilized and transferred to the newer, rapidly growing leaves. Both the purpling and premature senescence associated with phosphorus deficiency are therefore most prominent on the older leaves. Phosphorus-deficient plants are also characterized by delayed maturity, sparse flowering, and poor seed quality.

The Phosphorus Problem in Soil Fertility

Agronomic guide to soil phosphorus:
http://www.extension.umn.edu/distribution/cropsystems/DC0792.html

The phosphorus problem in soil fertility is threefold. *First,* the total phosphorus level of soils is low, usually no more than one-tenth to one-fourth that of nitrogen, and one-twentieth that of potassium. The phosphorus content of soils ranges from 200 to 2000 kg phosphorus in the upper 15 cm of 1 ha of soil, with an average of about 1000 kg P. *Second,* the phosphorus compounds commonly found in soils are mostly unavailable for plant uptake, often because they are highly insoluble. *Third,* when soluble sources of phosphorus, such as those in fertilizers and manures, are added to soils, they are fixed[3] (changed to unavailable forms) and in time form highly insoluble compounds. These three factors that constrain levels of available phosphorus in soils help account for the low crop productivity in regions such as sub-Saharan Africa where replenishment of soil phosphorus is inadequate. They also help us understand the environmental consequences of human mismanagement of the phosphorus, as discussed in the next section.

13.2 EFFECTS OF PHOSPHORUS ON ENVIRONMENTAL QUALITY[4]

Detergents, phosphates and water quality:
http://www.nhm.ac.uk/mineralogy/phos/pfile14.htm

The principal environmental problems related to soil phosphorus are *land degradation* caused by too little available phosphorus and *accelerated eutrophication* caused by too much. Both problems are related to the role of phosphorus as a plant nutrient.

Land Degradation

Many highly weathered soils in the warm, humid, and subhumid regions of the world have little capacity to supply phosphorus for plant growth. The low phosphorus availability is partly a result of extensive losses of phosphorus during long periods of relatively intense weathering and partly due to the low availability of phosphorus in the aluminum and iron combinations that are the dominant forms of phosphorus in these soils.

Undisturbed natural ecosystems in these regions usually contain enough phosphorus in the biomass and soil organic matter to maintain a substantial standing crop of trees or grasses. Most of the phosphorus taken up by the plants is that released from the decomposing residues of other plants. Very little is lost as long as the system remains undisturbed.

Once the land is cleared for agricultural use (by timber harvest or by forest fires), the losses of phosphorus in eroded soil particles, in runoff water, and in biomass removals (harvests) can be substantial. Within just a few years, the system may lose most of the phosphorus that had cycled between the plants and the soils. The remaining inorganic phosphorus in the soil is largely unavailable for plant uptake. In this manner, the phosphorus-supplying capacity of the disturbed soil rapidly becomes so low that regrowth of natural vegetation is sparse and, on land cleared for agricultural use, crops soon fail to produce useful yields (Figure 13.2).

Leguminous plants that might be expected to replenish soil nitrogen supplies are particularly hard hit by phosphorus deficiency, because low phosphorus supply inhibits effective nodulation and retards the biological nitrogen-fixation process. The spindly

[3]Note that the term *fixation* as applied to phosphorus has the same general meaning as the chemical fixation of potassium or ammonium ions; that is, the chemical being fixed is bound, entrapped, or otherwise held tightly by soil solids in a form that is relatively unavailable to plants. In contrast, the fixation of gaseous nitrogen refers to the biological conversion of N_2 gas to combined forms that plants can use.

[4]For a number of excellent research and review papers on practical and innovative measures to control losses of phosphorus from agricultural lands, see *J. Environ. Qual.* **29**:1–181 (2000).

FIGURE 13.2 On many highly weathered soils in the tropics and subtropics, low availability of phosphorus, as well as of other nutrients, may limit the regrowth of natural vegetation or the production of crops after natural vegetation is cleared. (*a*) Regrowth of natural vegetation is very sparse in the demarcated unfertilized plot in this disturbed area of Caribbean rain forest in Puerto Rico. (*b*) and (*c*) Phosphorus has been so depleted in this African woman's cornfield that there will be little or no grain to harvest from her severely phosphorus-deficient crop. In both cases, poor plant growth leaves the soil susceptible to accelerated erosion and further degradation. (Photos courtesy of R. Weil)

plants, deficient in both phosphorus and nitrogen, can provide little vegetative cover to prevent heavy rains from washing away the surface soil. The resulting erosion will further reduce soil fertility and water-holding capacity. The increasingly impoverished soils can support less and less vegetative cover, and so the degradation accelerates. Meanwhile, the soil particles lost by erosion become sediment farther down in the watershed, filling in reservoirs and increasing the turbidity of the rivers.

There are probably 1 to 2 billion ha of land in the world where phosphorus deficiency limits growth of both crops and native vegetation. Much of this land lies in poor countries whose farmers have little money for fertilizers. In much of sub-Saharan Africa,[5] additions of phosphorus fertilizers for food crops are a fraction of the rate of removal of this element in the harvested crops. The soils have been mined for years, and

[5] For a review of soil phosphorus problems in Africa, see Buresh et al. (1997).

BOX 13.1 EUTROPHICATION

Abundant growth of plants in terrestrial systems is usually considered beneficial, but in aquatic systems such growth can cause water-quality problems. The unwanted growth of algae (floating single-celled plants) and of aquatic weeds, termed *eutrophication*, can make ponds and lakes unsatisfactory environments for fish and can make the water unsuitable for drinking.

In natural ecosystems, lakes are commonly clear and free of excess plant growth and have highly diverse communities of organisms, because some factor (such as low levels of nitrogen or phosphorus) is limiting the growth of algae and other plants. Brackish and salty waters are most often nitrogen-limited, and plant growth is stimulated by high-nitrogen pollutants.

Most freshwater lakes and streams, on the other hand, are usually phosphorus-limited, partly because phosphorous levels in the water are typically very low and partly because nitrogen is being supplied through nitrogen fixation by some cyanobacteria (blue-green algae). Critical levels of phosphorus in water, above which eutrophication is likely to be triggered, are approximately 0.03 mg/L of dissolved phosphorus and 0.1 mg/L of total phosphorus.

When phosphorus is added to a phosphorus-limited lake, it stimulates a burst of algal growth (referred to as an *algal bloom*) and, often, a shift in the dominant algal species. The phosphorus-stimulated algae may cover the surface of the water with mats of algal scum. The lake may also be choked with higher plants that are also stimulated by the added phosphorus. When these aquatic weeds and algal mats die, they sink to the bottom, where their decomposition by microorganisms uses up much of the oxygen dissolved in the water. The decrease in oxygen (anoxic conditions) severely limits the growth of many aquatic organisms, especially fish. Such eutrophic lakes often become turbid, limiting growth of submerged aquatic vegetation and benthic (bottom feeding) organisms that serve as food for much of the fish community. In extreme cases eutrophication can lead to massive fish kills (Figure 13.3).

Eutrophication can transform clear, oxygen-rich, good-tasting water into cloudy, oxygen-poor, foul-smelling, bad-tasting, and possibly toxic water. Eutrophic conditions favor the growth of *Cyanobacter*, blue-green algae that are mostly undesirable food for zooplankton, a major food source for fish. These *Cyanobacter* produce toxins and bad-tasting and -smelling compounds that make the water unsuitable for human or animal consumption. Some filamentous algae can clog water treatment intake filters and thereby increase the cost of water remediation. Dense growth of both algae and aquatic weeds may make the water useless for boating and swimming. Furthermore, eutrophic waters generally have a reduced level of biological diversity (fewer species) and fewer fish, of desirable species.

Figure 13.3 In extreme cases of eutrophication, massive fish kills can occur in sensitive lakes and rivers. The kills result from anoxic conditions, which are brought on by the decay of the masses of algae stimulated by elevated inputs of phosphorus (or sometimes of nitrogen). (Photo courtesy of R. Weil)

crop yields remain low. Without properly managed inputs of phosphorus and biological means of enhancing phosphorus availability, there can be little hope of restoring these lands to productivity and their inhabitants to prosperity.

Water-Quality Degradation

Lake Okeechobee's problems: http://www.sfwmd.gov/org/wrp/wrp_okee/projects/factsheet.pdf

Accelerated or cultural eutrophication (Box 13.1 and Figure 13.3) is caused by phosphorus entering streams from both **point sources** and **nonpoint sources.** Point sources, such as sewage treatment plant outflows, industries, and the like, are relatively easy to identify, regulate, and clean up. Nonpoint sources, which are principally runoff water and sediments from soils throughout the affected watershed, are more difficult to identify and control. Consequently, these diffuse sources of phosphorus are major causes of eutrophication in most industrialized countries. In intensively cultivated areas of these countries, phosphorus additions from fertilizers, manures (Figure 13.4), and municipal wastes far exceed the removal of this element in harvested crops. This leads to a buildup of phosphorus in the soils, and to eutrophication (see Box 13.1). Figure 13.5 shows how the phosphorus buildup influences the balance of this element in the surface soils of intensively cultivated areas compared to that in natural forested watersheds.

Phosphorus Losses Runoff

Phosphorus in urban runoff: http://wi.water.usgs.gov/pubs/WRIR-99-4021/index.html

Agricultural management that involves disturbing the soil surface with tillage generally increases the amount of phosphorus carried away on eroded sediment (i.e., **particulate P**). On the other hand, fertilizer or manure that is left unincorporated on the surface of cropland or pastures usually leads to increased losses of phosphorus dissolved in the runoff water (i.e., **dissolved P**). These trends can also be seen in Table 13.1 by comparing phosphorus losses from no-till and conventionally tilled wheat. No-till systems may increase losses of soluble phosphorus, but they generally more than compensate for this by constraining soil erosion and thereby reducing particulate losses of P.

Disturbances to natural vegetation, such as timber harvest or wildfires, also increase the loss of phosphorus, primarily via eroded sediment. Erosion tends to transport predominantly the clay and organic matter fractions of the soil, which are relatively rich in phosphorus, leaving behind the coarser, lower-phosphorus fractions. Thus, compared to the original soil, eroded sediment is often enriched in phosphorus by a ratio of 2 or more (Table 13.2).

Runoff from animal feedlots and holding areas where hundreds or even thousands of cattle, hogs, or poultry are confined are increasingly significant sources of phosphorus that lead to eutrophication. Manure from a herd of 10,000 beef cattle contains as much phosphorus and nitrogen as is found in human wastes of a city of 100,000. Even if animal manure is spread on nearby fields, heavy rains induce runoff and erosion losses that carry significant quantities of soluble and particulate phosphorus. Much of this element can find its way into lakes and ponds, where eutrophication occurs.

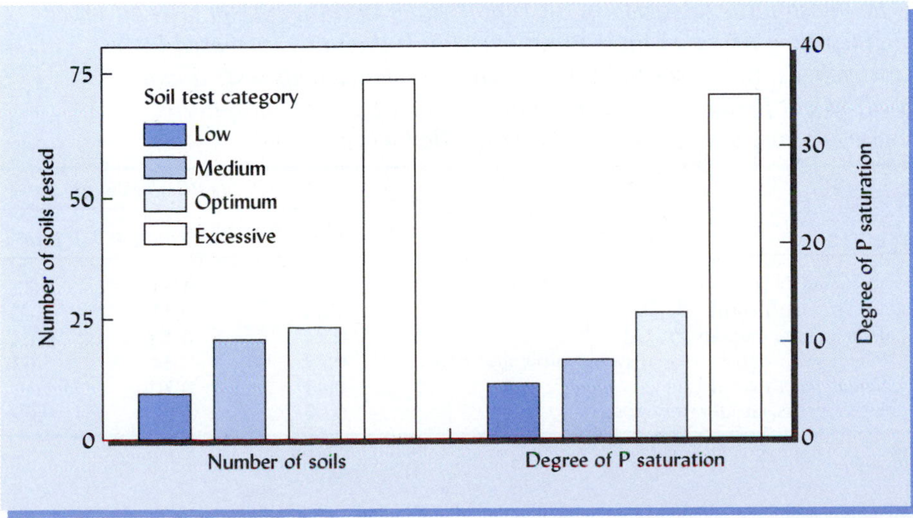

FIGURE 13.4 Chemical analyses of samples of 127 soils from Delaware (122) and the Netherlands (5) showed the phosphorus levels to be excessive in 73 of the soils, and that these soils also had a high degree of P saturation. The risk of phosphorus moving from these soils through erosion, runoff, or leaching to pollute nearby surface waters is high. Furthermore, adding additional fertilizer phosphorus to these soils would certainly make the problem worse. [Drawn from data in Paulter and Sims (2000)]

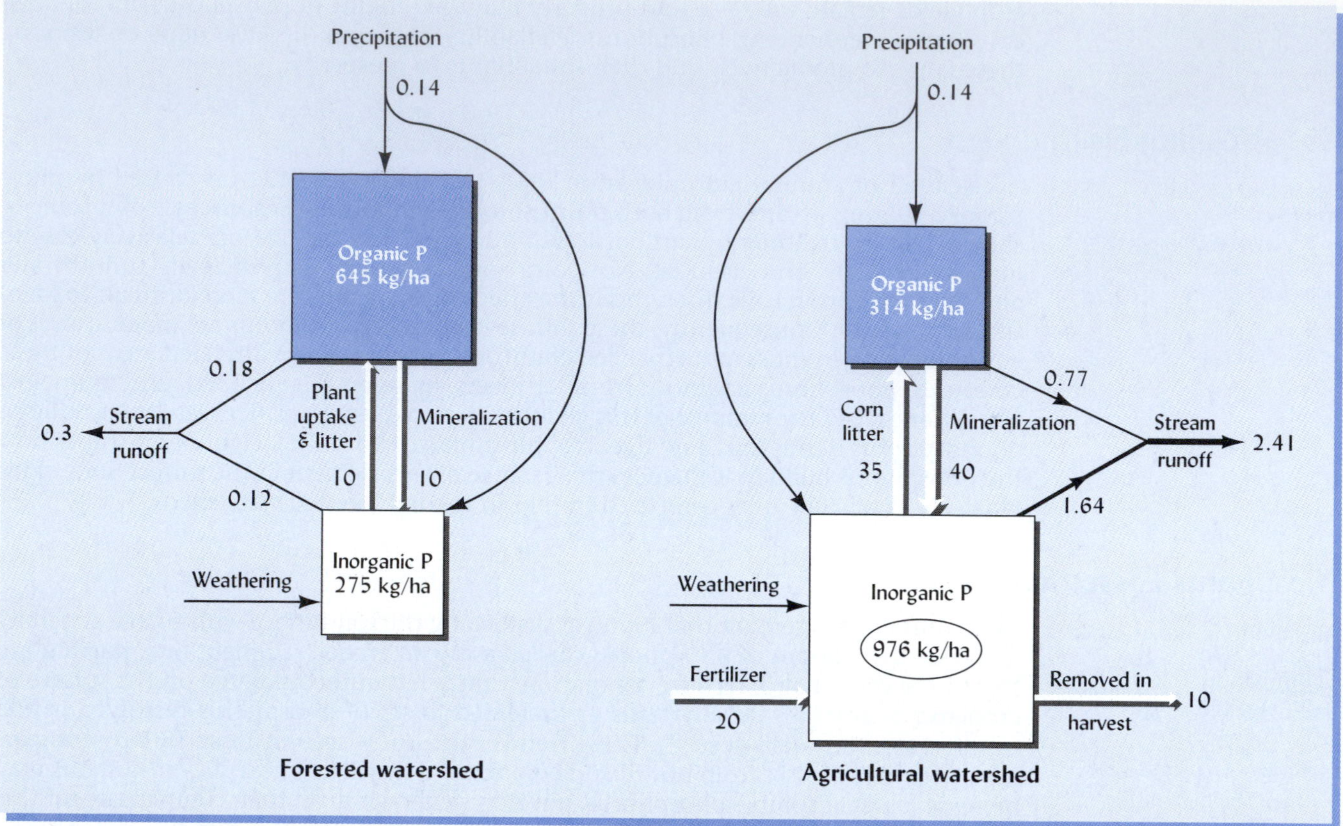

FIGURE 13.5 Phosphorus balance in surface soils (Ultisols) of adjacent forested and agricultural watersheds. The forest consisted primarily of mature hardwoods that had remained relatively undisturbed for 45 or more years. The agricultural land was producing row crops for more than 100 years. It appears that in the agricultural soil about half of the organic phosphorus has been converted into inorganic forms or lost from the system since cultivation began. At the same time, substantial amounts of inorganic phosphorus accumulated from fertilizer inputs. Compared to the forested soil, mineralization of organic phosphorus was about 4 times as great in the agricultural soil, and the amount of phosphorus lost to the stream was 8 times as great. Flows of phosphorus, represented by arrows, are given as kg/ha/yr. Although not shown in the diagram, it is interesting to note that nearly all (95%) of the phosphorus lost from the agricultural soil was in particulate form, while losses from the forest soil were 33% dissolved and 77% particulate. [Data from Vaithiyanathan and Correll (1992)]

TABLE 13.1 Influence of Wheat Production and Tillage on Annual Losses of Phosphorus in Runoff Water and Eroded Sediments Coming from Soils in the Southern Great Plains

The total phosphorus lost includes the phosphorus dissolved in the runoff water and the phosphorus adsorbed to the eroded particles.[a] Although cattle grazing on the natural grasslands probably increased losses of phosphorus from these watersheds, the losses from the agricultural watersheds were about 10 times as great. The no-till wheat fields tended to lose much less particulate phosphorus, but more soluble phosphorus, than the conventionally tilled wheat fields.

		kg P/ha/yr		
Location and soil	*Management*	*Soluble P*	*Particulate P*	*Total P*
El Reno, Okla. Paleustolls, 3% slope	Wheat with conventional plow and disk	0.21	3.51	3.72
	Wheat with no-till	1.04	0.43	1.42
	Native grass, heavily grazed	0.14	0.10	0.24
Woodward, Okla. Ustochrepts, 8% slope	Wheat with conventional sweep plow and disk	0.23	5.44	5.67
	Wheat with no-till	0.49	0.70	1.19
	Native grass, moderately grazed	0.02	0.07	0.09

[a]Wheat was fertilized with up to 23 kg of P each fall.
Data from Smith et al. (1991).

TABLE 13.2 Effect of Forest Fire on Erosion, Phosphorus Loss and Enrichment Ratios During the Following Year from a Scrub Forest in Northwestern Spain with Steeply Sloped Sandy Loam Soils.

Note that the more severe the burn, the greater were the losses of sediment and associated P. In all cases, the enrichment ratios were slightly greater than 2, indicating that the sediments eroded from the plots were more than twice as concentrated in phosphorus as were the surface soils themselves.

	Sediment lost, Mg/ha	*Total P lost, kg/ha*	*Enrichment ratio*[b]
Unburned control	2	1.4	2.6
Moderately burned by prescribed fire	5	4.3	2.1
Severely burned by wildfire	13+[a]	9.1+	2.2

[a]Values for severely burned plots were measured for 10 months, while the other plots were measured for 12 months.
[b]Enrichment ratio = (mg P/kg sediment)/(mg P/kg soil).
Data from Saa et al. (1994).

Intensive animal production and pollution from domestic and commercial development are thought to be responsible for blooms of ***Pfiesteria*** (a dinoflagellate microbe) and algae that are responsible for the death of millions of fish in estuaries and rivers of the coastal plain area of the eastern United States. In particular, researchers in North Carolina believe that elevated phosphorus in streams induces the life stage of *Pfiesteria* that releases a toxin 1,000 times more deadly than cyanide, killing small fish in a matter of minutes and larger ones in a few hours.

The preceding examples make it clear that the control of eutrophication in many rivers and lakes will require improved management of the phosphorus cycle in soils. Specifically, it is critical to reduce the transport of phosphorus from soils to water (see Sections 14.3 and 14.13).

13.3 THE PHOSPHORUS CYCLE[6]

Global phosphorus reservoirs, fluxes and turnover times:
http://www.ess.uci.edu/~reeburgh/fig4.html

To manage phosphorus for economic plant production and for environmental protection, we will have to understand the nature of the different forms of phosphorus found in soils and the manner in which these forms of phosphorus interact within the soil and in the larger environment. The cycling of phosphorus within the soil, from the soil to higher plants and back to the soil, is illustrated in Figure 13.6.

Phosphorus in Soil Solution. Compared to other macronutrients, such as sulfur and calcium, the concentration of phosphorus in the soil solution is very low, generally ranging from 0.001 mg/L in very infertile soils to about 1 mg/L in rich, heavily fertilized soils. Plant roots absorb phosphorus dissolved in the soil solution, mainly as phosphate ions (HPO_4^{-2} and $H_2PO_4^-$), but some soluble organic phosphorus compounds are also taken up. The chemical species of phosphorus present in the soil solution is determined by the solution pH, as shown in Figure 13.7. In strongly acid soils (pH 4 to 5.5), the monovalent anion $H_2PO_4^-$ dominates, while alkaline solutions are characterized by the divalent anion HPO_4^{-2}. Both anions are important in near-neutral soils. Of the two anions, $H_2PO_4^-$ is thought to be slightly more available to plants, but effects of pH on phosphorus reactions with other soil constituents are more important than the particular phosphorus anion present.

Uptake by Roots and Mycorrhizae. Plant uptake of phosphate ions from the soil solution is curtailed by the extremely slow diffusion of these ions to root surfaces. This is overcome in part by the growth of roots into zones where the ions are held. Also, phosphate ions move to plant roots through symbiosis with mycorrhizal fungi (Figure 13.8). The microscopic, threadlike mycorrhizal hyphae extend out into the soil several centimeters from

[6]For a review of the phosphorus cycle, see Stevenson (1986).

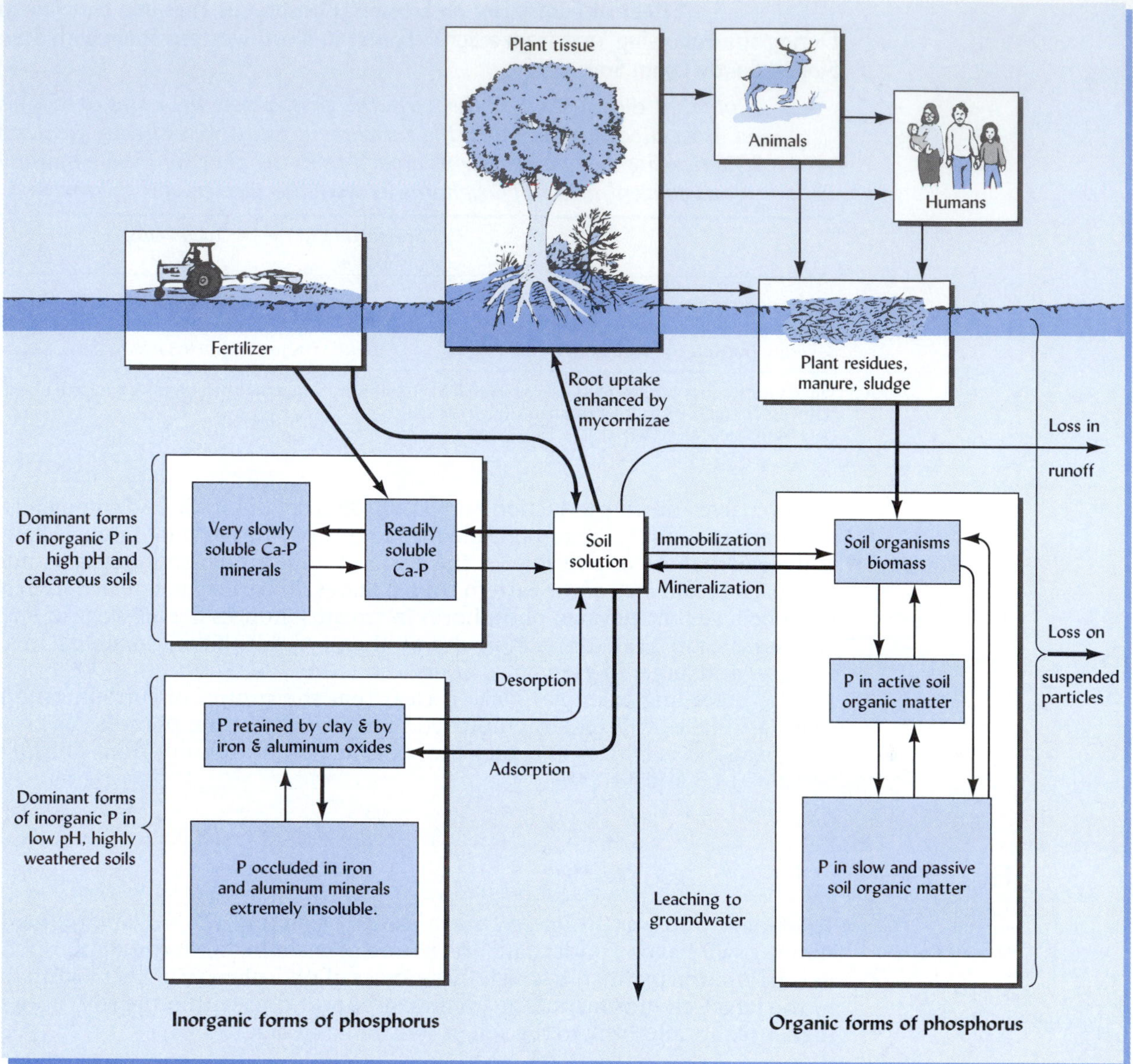

FIGURE 13.6 The phosphorus cycle in soils. The boxes represent pools of the various forms of phosphorus in the cycle, while the arrows represent translocations and transformations among these pools. The three largest white boxes indicate the principal groups of phosphorus-containing compounds found in soils. Within each of these groups, the less soluble, less available forms tend to dominate.

the root surfaces. The hyphae are able to absorb phosphorus ions as the ions enter the soil solution, and may even be able to access some strongly bound forms of phosphorus. The hyphae then bring the phosphate to the root by transporting it inside the hyphal cells, where soil-retention mechanisms cannot interfere with phosphate movement.

Chemical Forms in Soils. In most soils, the amount of phosphorus available to plants from the soil solution at any one time is very low, seldom exceeding about 0.01% of the total phosphorus in the soil. The bulk of the soil phosphorus exists in three general groups of compounds—*organic phosphorus, calcium-bound inorganic phosphorus,* and *iron- or aluminum-bound inorganic phosphorus* (see Figure 13.6). The organic phosphorus is distributed among the active, slow, and passive fractions of soil organic matter (see Section 11.8). Of the inorganic phosphorus, the calcium compounds predominate in most alkaline soils, while the iron and aluminum forms are most important in acidic soils. All three groups of compounds slowly contribute phosphorus to the soil solution, but most of the phosphorus in each group is of very low solubility and not readily available for plant uptake.

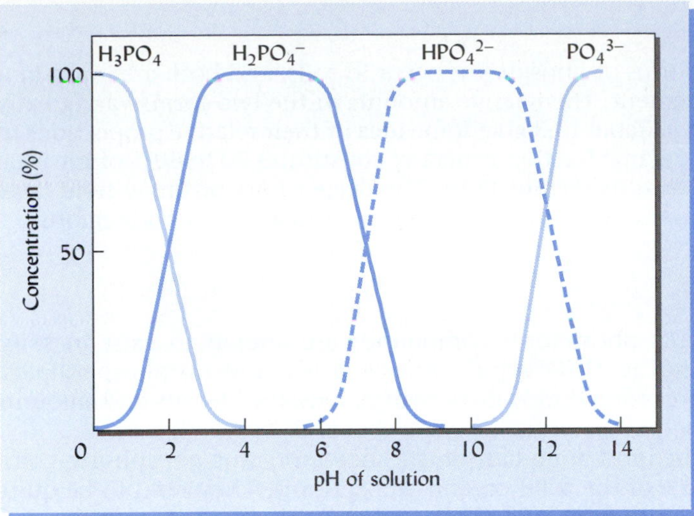

FIGURE 13.7 The effect of pH on the relative concentrations of the three species of phosphate ions. At lower pH values, more H^+ ions are available in the solution, and thus the phosphate ion species containing more hydrogen predominates. In near-neutral soils, HPO_4^{-2} and $H_2PO_4^-$ are found in nearly equal amounts. Both of these species are readily available for plant uptake.

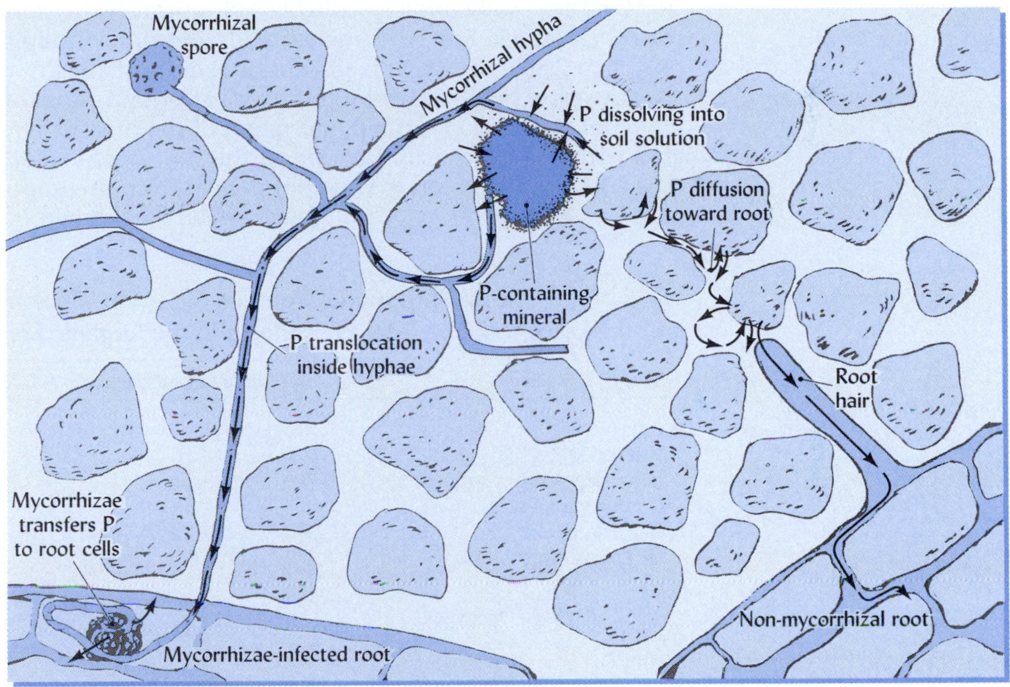

FIGURE 13.8 Roles of diffusion and mycorrhizal hyphae in the movement of phosphate ions to plant roots. In soils with low solution phosphorus concentration and high phosphorus fixation, slow diffusion may seriously limit the ability of roots to obtain sufficient phosphorus. The hyphae of symbiotic mycorrhizal fungi help overcome this problem. They penetrate the soil, absorb the phosphorus, and by cytoplasmic streaming inside the hyphae, transport phosphorus to the plant roots. This makes the plant much less dependent on the diffusion of phosphate ions through the soil.

Gains and Losses. The principal pathways by which phosphorus is lost from the soil system are plant removal (5 to 50 kg/ha annually in harvestable biomass), erosion of phosphorus-carrying soil particles (0.1 to 10 kg/ha annually on organic and mineral particles), phosphorus dissolved in surface runoff water (0.01 to 3.0 kg/ha annually) and leaching to groundwater (0.0001 to 0.4 kg/ha annually). For each pathway, the higher figures cited for annual phosphorus loss would most likely apply to cultivated soils (especially those treated with manure), the lower ones to forested and natural grassland areas (see Figure 13.5).

13.4 ORGANIC PHOSPHORUS IN SOILS

Both inorganic and organic forms of phosphorus occur in soils, and both are important to plants as sources of this element. The relative amounts in the two forms vary greatly from soil to soil, but the data in Table 13.3 give some idea of their relative proportions in a range of mineral soils. The organic fraction generally constitutes 20 to 80% of the total phosphorus in surface soil horizons (Figure 13.9). The deeper horizons may hold large amounts of inorganic phosphorus, especially in soils from arid and semiarid regions.

Organic Phosphorus Compounds

Three broad groups of organic phosphorus compounds are known to exist in soils: (1) phosphate esters of inositol [$C_6H_6(OH)_6$]; (2) nucleic acids; and (3) phospholipids. While other organic phosphorus compounds are present in soils, the identity and amounts present are less well understood.

Inositol phosphates are the most abundant of the known organic phosphorus compounds, making up 10 to 50% of the total organic phosphorus. They tend to be quite stable in acid and alkaline conditions, and they interact with the higher-molecular-weight humic compounds. These properties may account for their relative abundance in soils.

Nucleic acids are adsorbed by humic compounds as well as by silicate clays. Adsorption on these soil colloids probably helps protect the phosphorus in nucleic acids from microbial attack. Still, the nucleic acids and phospholipids together probably make up only 1 to 2% of the organic phosphorus in most soils.

The other chemical compounds that contain most of the soil organic phosphorus have not yet been identified, but much of the organic phosphorus appears to be associated with the fulvic acid fraction of the soil organic matter. Our ignorance of the specific compounds involved does not detract from the importance of these compounds as suppliers of phosphorus through microbial breakdown.[7]

TABLE 13.3 Total Phosphorus Content of Surface Soils from Different Locations and the Percentage of Total Phosphorus in the Organic Form

Soils	*Number of samples*	*Total P, mg/kg*	*Organic fraction, %*
Western Oregon			
Upland soils	4	357	66
Old valley-filling soils	4	1479	30
Recent valley soils	3	848	26
New York			
Histosols (cultivated)	8	1491	52
Iowa			
Mollisols and Alfisols	6	561	44
Arizona	19	703	36
Australia	3	422	75
Texas			
Ustolls	2	369	34
Hawaii			
Andisol	1	4700	37
Oxisol	1	1414	19
Zimbabwe			
Alfisols	22	899	56
Maryland			
Ultisols (silt loams)	6	650	59
Ultisols (forested sandy loams)	3	472	70
Ultisols (sandy loam, cropland)	4	647	25

Data for Oregon, Iowa, and Arizona from sources quoted by Brady (1974); Australia from Fares et al. (1974); New York from Cogger and Duxbury, (1984); Hawaii from Soltanpour et al. (1988); Zimbabwe courtesy of R. Weil and R. Folle; Texas from Raven and Hossner (1993); Maryland (silt loams) from Weil et al. (1988); Maryland (sandy loams) from Vaithiyanathan and Correll (1992).

[7]Although plants are able to absorb some organic phosphorus compounds directly, the level of such absorption is thought to be very low compared to that of inorganic phosphates.

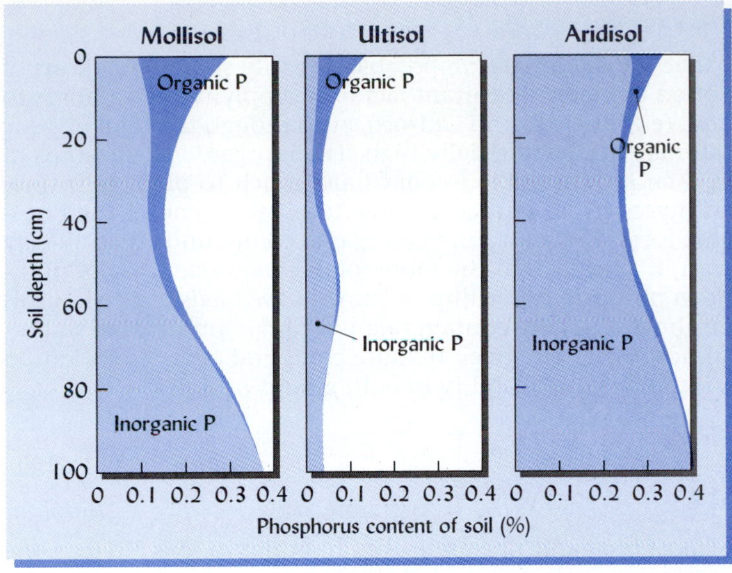

FIGURE 13.9 Phosphorus contents of representative soil profiles from three soil orders. All three soils contain a high proportion of organic phosphorus in their surface horizons. The Aridisol has a high inorganic phosphorus content throughout the profile because rainfall during soil formation was insufficient to leach much of the inorganic phosphorus compounds from the soil. The increased phosphorus in the subsoil of the Ultisol is due to adsorption of inorganic phosphorus by iron and aluminum oxides in the B horizon. In both the Mollisol and Aridisol, most of the subsoil phosphorus is in the form of inorganic calcium phosphate compounds.

Much of the phosphorus in the soil solution and in leachates from land that has received large quantities of animal wastes is present as **dissolved organic phosphorus** (DOP). DOP is generally more mobile than soluble inorganic phosphates, probably because it is not so readily adsorbed by organic matter, clays and $CaCO_3$ in the soil. In the lower horizons of such soils, the DOP commonly makes up more than 50% of the total soil solution phosphorus. As a consequence, in heavily manured areas with sandy soils DOP can leach downward to nearly 2 m. In fields with high water tables, the phosphorus can move with the groundwater to nearby lakes or streams and thereby contribute significantly to eutrophication.

Mineralization of Organic P

Phosphorus held in organic form can be mineralized and immobilized by the same general processes that release nitrogen and sulfur from soil organic matter (see Chapter 12):

$$\xleftarrow{\quad\text{Immobilization}\quad}$$

$$\text{Organic P forms} \underset{\text{Microbes}}{\overset{\text{Microbes}}{\rightleftharpoons}} \underset{\text{Soluble phosphate}}{H_2PO_4^-} \xrightleftharpoons{Fe^{3+},\ Al^{3+},\ Ca^{2+}} \underset{\text{Insoluble fixed P}}{\text{Fe, Al, Ca phosphates}} \tag{13.1}$$

$$\xrightarrow{\quad\text{Mineralization}\quad}$$

Net immobilization of soluble phosphorus is most likely to occur if residues added to the soil have a C/P ratio greater than 300:1, while net mineralization is likely if the ratio is below 200:1. Mineralization of organic phosphorus in soils is subject to many of the same influences that control the general decomposition of soil organic matter—temperature, moisture, tillage, and so forth (see Section 11.3). In temperate regions, mineralization of organic phosphorus in soils typically releases 5 to 20 kg P/ha/yr, most of which is readily absorbed by growing plants. These values can be compared to the annual uptake of phosphorus by most crops, trees, and grasses, which generally ranges from 5 to 30 kg P/ha. When forested soils are first brought under cultivation in tropical climates, the amount of phosphorus released by mineralization may exceed 50 kg/ha/yr, but unless phosphorus is added from outside sources, these high rates of mineralization will soon decline due to the depletion of readily decomposable soil organic matter. In Florida, rapid mineralization of organic matter in Histosols (Saprists) drained for agricultural use is estimated to release about 80 kg P/ha/yr. Unlike most mineral soils, these organic soils possess little capacity to retain dissolved phosphorus, so water draining from them is quite concentrated in phosphorus (0.5 to 1.5 mg P/L) and is thought to be contributing to the degradation of the Everglades wetland system.

Contribution of Organic Phosphorus to Plant Needs

Recent evidence indicates that the readily decomposable or easily soluble fractions of soil *organic phosphorus* are often the most important factor in supplying phosphorus to plants in *highly weathered soils* (e.g., Ultisols and Oxisols), even though the total organic matter content of these soils may not be especially high. The inorganic phosphorus in the highly weathered soils is far too insoluble to contribute much to plant nutrition. Apparently plant roots and mycorrhizal hyphae are able to obtain some of the phosphorus released from organic forms before it forms inorganic compounds that quickly become insoluble. In contrast, it appears that the more soluble *inorganic forms* of phosphorus play the biggest role in phosphorus fertility of *less weathered soils* (e.g., Mollisols and Vertisols), even though these generally contain relatively high amounts of soil organic matter. Cycling phosphorus through green manure crops and cover crops left on the surface as a mulch can improve its availability in both groups of soils.

13.5 INORGANIC PHOSPHORUS IN SOILS

As indicated by Figure 13.6, most inorganic phosphorus compounds in soils fall into one of two groups: (1) those containing calcium, and (2) those containing iron and aluminum, and less frequently, manganese (see Table 13.4). Except for the mono- and dicalcium phosphates supplied by commercial fertilizers, all the inorganic soil phosphorus compounds listed in Table 13.4 are quite insoluble. The calcium phosphates, especially the **apatite** minerals, are stable and extremely insoluble at high pH levels. In contrast, iron- and aluminum-containing phosphates such as strengite ($FePO_4 \cdot 2H_2O$) and variscite ($AlPO_4 \cdot 2H_2O$) are stable only in very acid soils where they are most insoluble.

At intermediate pH levels, phosphates are also found associated with some 1:1-type silicate clays. Apparently, phosphate ions have reacted with or replaced hydroxyl groups at the edges of the clay particles and are bound in relatively unavailable inner sphere complexes. Once again, soil pH is a major factor in determining the availability of the phosphorus, as we shall see in the next Section.

13.6 SOLUBILITY OF INORGANIC PHOSPHORUS IN ACID SOILS

The particular types of reactions that fix phosphorus in relatively unavailable forms are closely related to soil pH (Figure 13.10). In acid soils these reactions involve mostly Al, Fe, or Mn, either as dissolved ions, as oxides, or as hydrous oxides. Many soils contain such hydrous oxides as coatings on soil particles and as interlayer precipitates in silicate clays. At moderate pH values, adsorption on the edges of kaolinite or on the iron oxide coating on kaolinite clays plays an important role. Figure 13.11 illustrates some of the reactions by which phosphate is fixed by iron and aluminum compounds.

TABLE 13.4 Inorganic Phosphorus-Containing Compounds Commonly Found in Soils

In each group, the compounds are listed in order of increasing solubility.

Compound	*Formula*
Iron and aluminum compounds	
Strengite	$FePO_4 \cdot 2H_2O$
Variscite	$AlPO_4 \cdot 2H_2O$
Calcium compounds	
Fluorapatite	$[3Ca_3(PO_4)_2] \cdot CaF_2$
Carbonate apatite	$[3Ca_3(PO_4)_2] \cdot CaCO_3$
Hydroxy apatite	$[3Ca_3(PO_4)_2] \cdot Ca(OH)_2$
Oxy apatite	$[3Ca_3(PO_4)_2] \cdot CaO$
Tricalcium phosphate	$Ca_3(PO_4)_2$
Octacalcium phosphate	$Ca_8H_2(PO_4)_6 \cdot 5H_2O$
Dicalcium phosphate	$CaHPO_4 \cdot 2H_2O$
Monocalcium phosphate	$Ca(H_2PO_4)_2 \cdot H_2O$

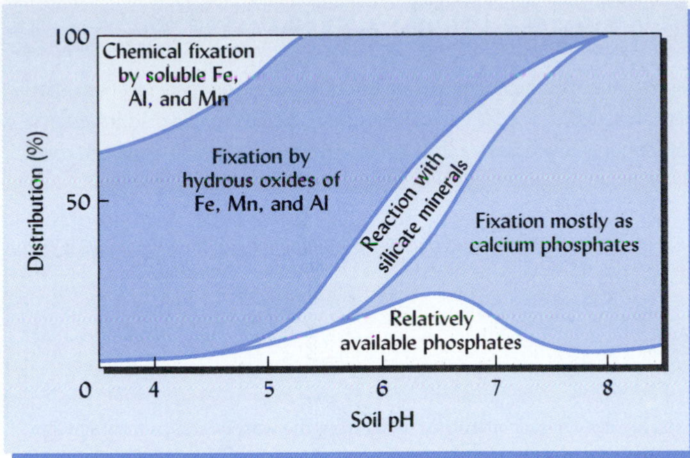

FIGURE 13.10 Inorganic fixation of added phosphates at various soil pH values. Average conditions are postulated, and it is not to be inferred that any particular soil would have exactly this distribution. The actual proportion of the phosphorus remaining in an available form will depend upon contact with the soil, time for reaction, and other factors. It should be kept in mind that some of the added phosphorus may be changed to an organic form in which it would be temporarily unavailable.

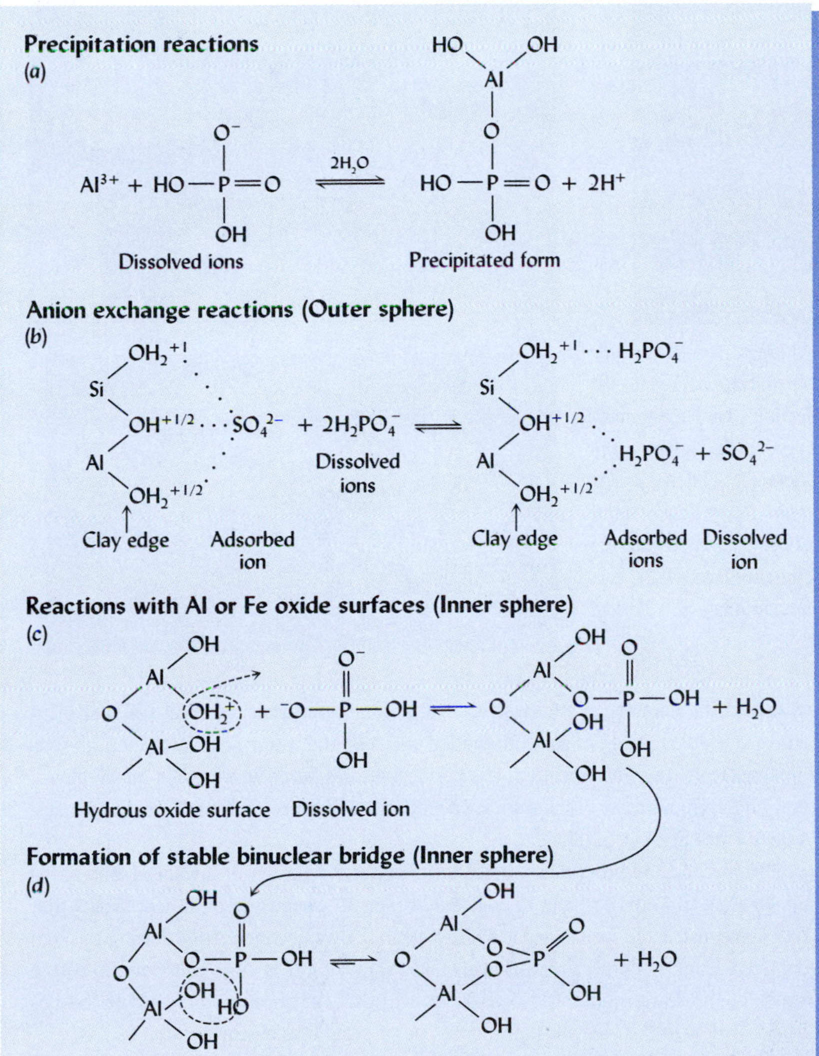

FIGURE 13.11 Several of the reactions by which phosphate ions are removed from soil solution and fixed by reaction with iron and aluminum in various hydrous oxides. Freshly precipitated aluminum, iron, and manganese phosphates (*a*) are relatively available, though over time they become increasingly unavailable. In (*b*) the phosphate is reversibly adsorbed by anion exchange. In reactions of the type shown in (*c*) a phosphate ion replaces an $—OH_2$ or an —OH group in the surface structure of Al or Fe hydrous oxide minerals. In (*d*) the phosphate further penetrates the mineral surface by forming a stable binuclear bridge. The adsorption reactions (*b, c, d*) are shown in order—from those that bind phosphate with the least tenacity (relatively reversible and somewhat more plant-available) to those that bind phosphate most tightly (almost irreversible and least plant-available). It is probable that, over time, phosphate ions added to a soil may undergo an entire sequence of these reactions, becoming increasingly unavailable. Note that (*b*) illustrates an outer-sphere complex, while (*c*) and (*d*) are examples of inner-sphere complexes (see Figure 8.12).

Precipitation by Iron, Aluminum, and Manganese Ions

Probably the easiest type of phosphorus-fixation reaction to visualize is the simple reaction of $H_2PO_4^-$ ions with dissolved Fe^{3+}, Al^{3+}, and Mn^{3+} ions to form insoluble hydroxy phosphate precipitates such as those formed in wastewater treatment plants (see Box 13.2 and Figure 13.12). In acid soils, enough soluble Al, Fe, or Mn is usually present to cause

BOX 13.2 PHOSPHORUS REMOVAL FROM WASTEWATER

Environmental soil scientists and engineers remove phosphorus from municipal wastes by taking advantage of some of the same reactions that bind phosphorus in soils. After primary and secondary sewage treatment that removes solids and oxidizes most of the organic matter, tertiary treatment in huge, specially designed tanks (Figure 13.12 left) causes phosphorus to precipitate through reactions with iron and aluminum compounds, such as the following:

$$\underset{\text{Alum}}{Al_2(SO_4)_3 \cdot 14H_2O} + \underset{\text{Soluble phosphate}}{2PO_4^{3-}} \rightarrow \underset{\text{Insoluble AlP}}{2AlPO_4} + 3SO_4^{2-} + 14H_2O \tag{13.2}$$

$$\underset{\text{Ferric chloride}}{FeCl_3} + \underset{\text{Soluble phosphate}}{PO_4^{3-}} \rightarrow \underset{\text{Insoluble FeP}}{FePO_4} + 3Cl^- \tag{13.3}$$

The insoluble aluminum and iron phosphates settle out of solution and are later mixed with other solids from the wastewater to form sewage sludge. The low-phosphorus water, after minor processing, is returned to the river.

FIGURE 13.12 *Increasingly modern sewage plants (left) are required to include tertiary treatment facilities for phosphorus removal. The chemical reactions in the sewage treatment process, which are similar to those affecting phosphorus availability in soils, occur when the treatment plant effluent is passed through an artificial wetland (above right) to remove nitrogen by denitrification (see Section 12.10). They also occur when the effluent goes through the final stage (lower right) whereby pollutants are removed by natural processes. After being sprayed onto the field by an irrigation system (see arrow), the wastewater percolates through the soil, which in this case is a peat that had to be amended by adding iron before it worked as planned. (Photos courtesy of Ray Weil)*

Other less-expensive tertiary treatment approaches involve the spraying of the wastewater on vegetated soils. Natural soil and plant processes clean the phosphorus and other constituents out of the waste water. In some *infiltration systems,* the water percolates through relatively permeable soils. Other systems, termed *overland flow systems,* use finer-textured, less-permeable soils, over which water slowly flows, permitting the upper few centimeters of soil and the vegetation to remove most of the soluble phosphorus and other contaminants. In both systems, advantage is taken of the soil's phosphate-fixing capacity (see Section 13.8).

As the following example illustrates, knowledge of soil properties and processes is essential for effective design of advanced land-based wastewater treatment systems. A large environmental engineering firm won a contract to build a new type of wastewater treatment facility that would look more like a park than a sewage plant because it used constructed wetlands and soils to clean the wastewater. After flowing through a number of artificial marshes and filtering systems (Figure 13.12, upper right), the wastewater was sprayed into a large field covered with a layer of artificial permeable "soil" several meters thick (Figure 13.12, lower right). The expectation was that the phosphorus would be fixed as it moved through the "soil," leaving the groundwater sufficiently low in this element that it could be released into a nearby estuary.

Unfortunately, the system didn't work. The water coming from the bottom of the artificial profile was higher in phosphorus than before the treatment. The problem? The designers had specified peat (very low in mineral colloids) rather than mineral soil as the artificial "soil" through which the water percolated. Consequently, insufficient iron, aluminum, or calcium compounds were available to fix the phosphorus. Soil scientists brought in to assist the municipality recognized the problem and recommended that a few trainloads of steel wool dust (metallic iron) be incorporated into the peat. As the steel wool rusted (oxidized), ferric iron formed and reacted strongly with the dissolved phosphorus in the wastewater. The phosphorus was precipitated without appreciably decreasing the desirable high permeability of the peat "soil."

the chemical precipitation of nearly all dissolved $H_2PO_4^-$ ions by reactions such as the following (using the aluminum cation as an example):

$$\underset{\text{(soluble)}}{Al^{3+} + H_2PO_4^- + 2H_2O} \rightleftharpoons \underset{\text{(insoluble)}}{2H^+ + Al(OH)_2H_2PO_4} \tag{13.4}$$

Freshly precipitated hydroxy phosphates are slightly soluble and initially at least, somewhat available to plants. Over time, however, as the precipitated hydroxy phosphates age, they become less soluble and the phosphorus in them becomes almost completely unavailable to most plants.

Reaction with Hydrous Oxides and Silicate Clays

Most of the phosphorus fixation in acid soils probably occurs when $H_2PO_4^-$ ions react with, or become adsorbed to, the surfaces of insoluble oxides of iron, aluminum, and manganese, such as gibbsite ($Al_2O_3 \cdot 3H_2O$) and goethite ($Fe_2O_3 \cdot 3H_2O$). The large quantities of Fe, Al oxides, and 1:1 clays present in many soils make possible the fixation of extremely large amounts of phosphorus by these reactions.

Although all the exact mechanisms have not been identified, $H_2PO_4^-$ ions are known to react with iron and aluminum mineral surfaces in several different ways, resulting in different degrees of phosphorus fixation. Some of these reactions are shown diagrammatically in Figure 13.11).

The $H_2PO_4^-$ anion may be attracted to positive charges that develop under acid conditions on the surfaces of iron and aluminum oxides and the broken edges of kaolinite clays (see Figure 13.11*b*). The adsorbed $H_2PO_4^-$ anions are subject to anion exchange with certain other anions, such as OH^-, SO_4^{2-}, and MoO_4^{2-}. Since this type of adsorption of $H_2PO_4^-$ ions is reversible, the phosphorus may slowly become available to plants.

The phosphate ion may also replace a structural hydroxyl to become chemically bound to the oxide (or clay) surface (see Figure 13.11*c*). This reaction, while reversible, binds the phosphate too tightly to allow its ready replacement by other anions. The availability of phosphate bound in this manner is very low. Over time, a second oxygen of the phosphate ion may replace a second hydroxyl, so that the phosphate becomes chemically bound to two adjacent aluminum (or iron) atoms in the hydrous oxide surface (see Figure 13.11*d*). With this step, the phosphate becomes an integral part of the oxide mineral and the likelihood of its release back to the soil solution is extremely small.

Finally, as more time passes, the precipitation of additional iron or aluminum hydrous oxide may bury the phosphate deep inside the oxide particle. Such phosphate is termed *occluded,* and is the least available form of phosphorus in most acid soils.

Effect of Iron Reduction Under Wet Conditions

Phosphorus bound to iron oxides by the mechanisms just discussed is very insoluble under well-aerated conditions. However, prolonged anaerobic conditions can reduce the iron in these complexes from Fe^{3+} to Fe^{2+}, making the iron–phosphate complex much more soluble and causing it to release phosphorus into solution. The release of phosphorus from iron phosphates by means of the reduction and subsequent solubilization of iron improves the phosphorus availability in soils used for paddy rice. Unfortunately, this process also allows phosphorous-containing sediment layers at the bottom of lakes and estuaries to contribute to eutrophication.

13.7 INORGANIC PHOSPHORUS AVAILABILITY AT HIGH pH VALUES[8]

The availability of phosphorus in alkaline soils is determined principally by the solubility of the various calcium phosphate compounds present. In neutral and slightly alkaline soils, soluble $H_2PO_4^-$ quickly reacts with calcium to form a sequence of products of decreasing solubility. For instance, highly soluble monocalcium phosphate [$Ca(H_2PO_4)_2 \cdot H_2O$] added as concentrated superphosphate fertilizer rapidly reacts with

[8]See Sample et al. (1980) for a discussion of this subject.

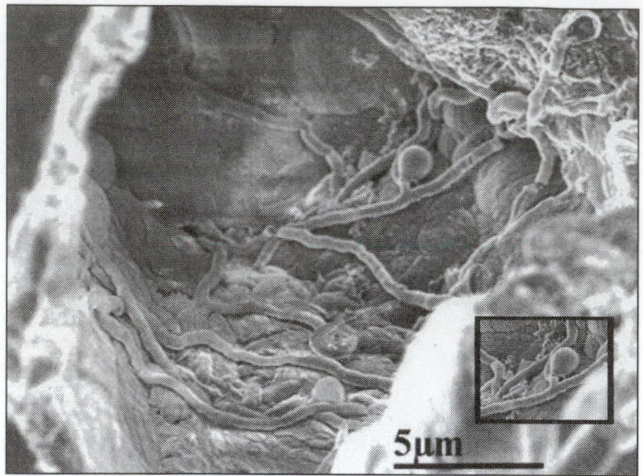

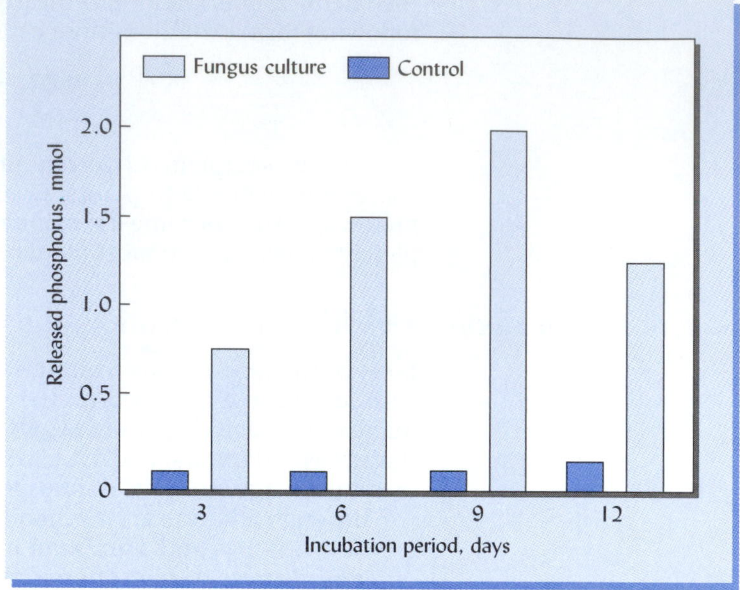

FIGURE 13.13 Certain soil microorganisms can increase the availability of phosphorus in minerals such as rock phosphate and aluminum phosphates that normally hold the phosphorus in very insoluble forms. (*Left*) A micrograph of a fungus growing on the surface of an aluminum phosphate found in soils. The fungus is thought to produce organic acids that help release some soluble phosphorus. (*Right*) In another experiment, phosphorus as released from rock phosphate that was incubated for different periods of time with a liquid culture of a fungus (*Aspergillus niger*) that had been isolated from a tropical soil. [Micrograph (*left*) courtesy Dr. Anne Taunton, University of Wisconsin; graph (*right*) drawn from data in Goenadi et al. (2000)]

calcium carbonate in the soil to form first dicalcium phosphate ($CaHPO_4 \cdot 2H_2O$) and then tricalcium phosphate [$Ca_3(PO_4)_2$], as follows:

$$\underset{\substack{\text{Monocalcium phosphate}\\ \text{(soluble)}}}{Ca(H_2PO_4)_2 \cdot H_2O} + 2H_2O \xrightarrow{CaCO_3} \underset{\substack{\text{Dicalcium phosphate}\\ \text{(slightly soluble)}}}{2(CaHPO_4 \cdot 2H_2O)} + CO_2\uparrow \xrightarrow{CaCO_3} \underset{\substack{\text{Tricalcium phosphate}\\ \text{(very low solubility)}}}{Ca_3(PO_4)_2} + CO_2\uparrow + 5H_2O \qquad (13.5)$$

The solubility of these compounds and, in turn, the plant availability of the phosphorus they contain decrease as the phosphorus changes from the $H_2PO_4^-$ ion to tricalcium phosphate [$Ca_3(PO_4)_2$]. Although this latter compound is quite insoluble, it may undergo further reactions to form even more insoluble compounds, such as the apatites shown in Table 13.4. These compounds are thousands of times less soluble than freshly formed tricalcium phosphates.

Reversion of soluble fertilizer phosphorus to extremely insoluble calcium phosphate forms is most serious in the calcareous soils of low-rainfall regions (e.g., western United States). Iron and aluminum impurities in calcite particles may also adsorb considerable amounts of phosphate in these soils. Because of the various reactions with $CaCO_3$, phosphorus availability tends to be nearly as low in the Aridisols, Inceptisols, and Mollisols of arid regions as in the highly acid Spodosols and Ultisols of humid regions, where iron, aluminum, and manganese limit phosphorus availability.

Recent research has called attention to the importance of certain soil microorganisms (bacteria and fungi) in enhancing the solubility of both calcium and aluminum phosphates. Figure 13.13 illustrates the activity of fungi in releasing organic acids such as citric acid that solubilize the highly insoluble phosphates, making them available for absorption first by the fungi and ultimately by higher plants. These organisms may be responsible for enhancing the meager availability of phosphate minerals to both natural and crop plants.

13.8 PHOSPHORUS-FIXATION CAPACITY OF SOILS

Soils may be characterized by their capacity to fix phosphorus in unavailable, insoluble forms. The phosphorus-fixation capacity of a soil may be conceptualized as the total number of sites on soil particle surfaces capable of reacting with phosphate ions. Phosphorus fixation may also be due to reactive soluble iron, aluminum, or manganese. The different types of fixation mechanisms are illustrated schematically in Figure 13.14.

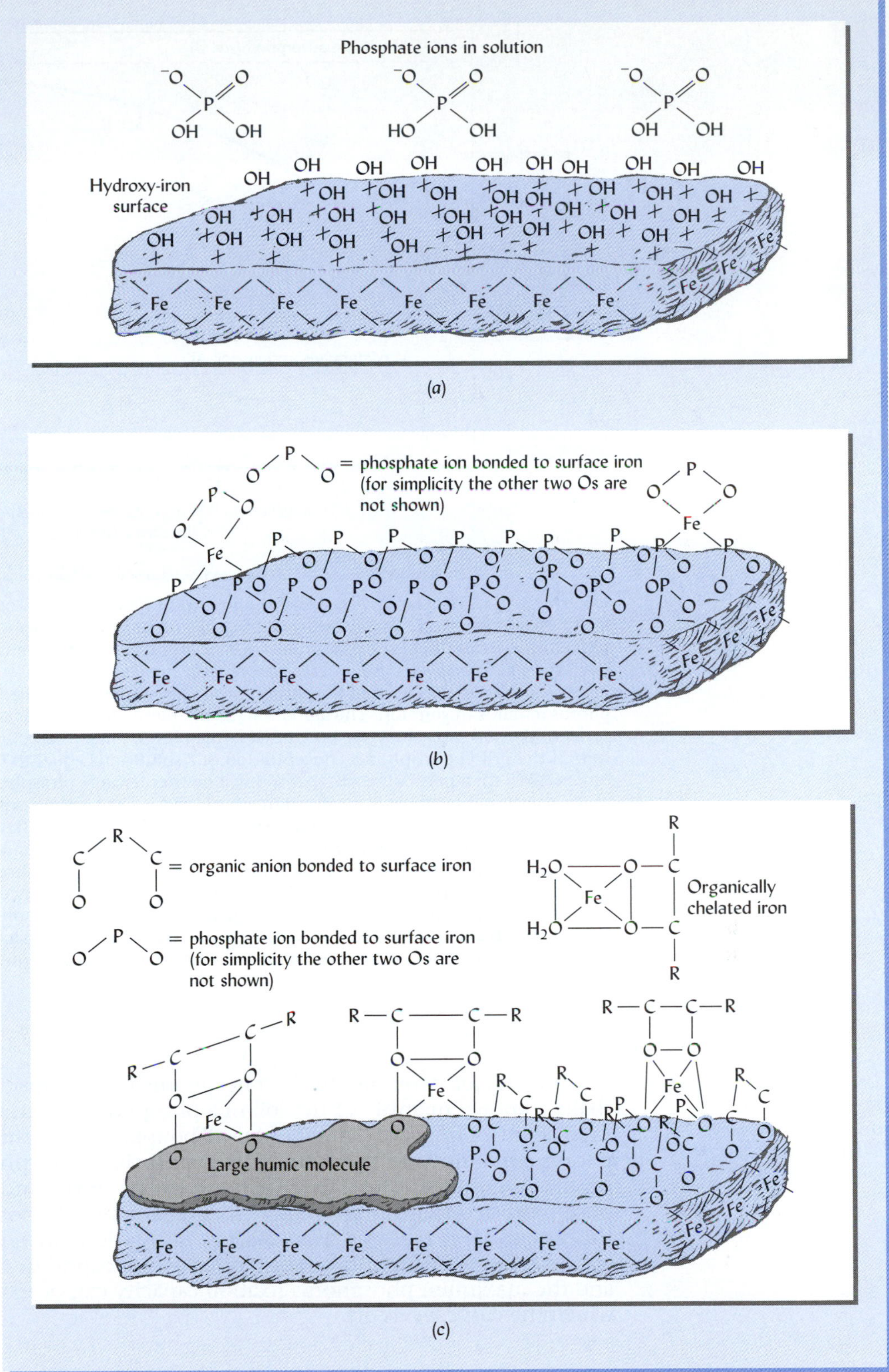

FIGURE 13.14 Schematic illustration of phosphorus-fixation sites on a soil particle surface showing hydrous iron oxide as the primary fixing agent. In part (*a*) the sites are shown as +s, indicating positive charges or hydrous metal oxide sites, each capable of fixing a phosphate ion. In part (*b*) the fixation sites are all occupied by phosphate ions (the soil's fixation capacity is satisfied). Part (*c*) illustrates how organic anions, larger organic molecules, and certain strongly fixed inorganic anions can reduce the sites available for fixing phosphorus. Such mechanisms account for the reduced phosphorus fixation and greater phosphorus availability brought about when organic matter is added to a soil.

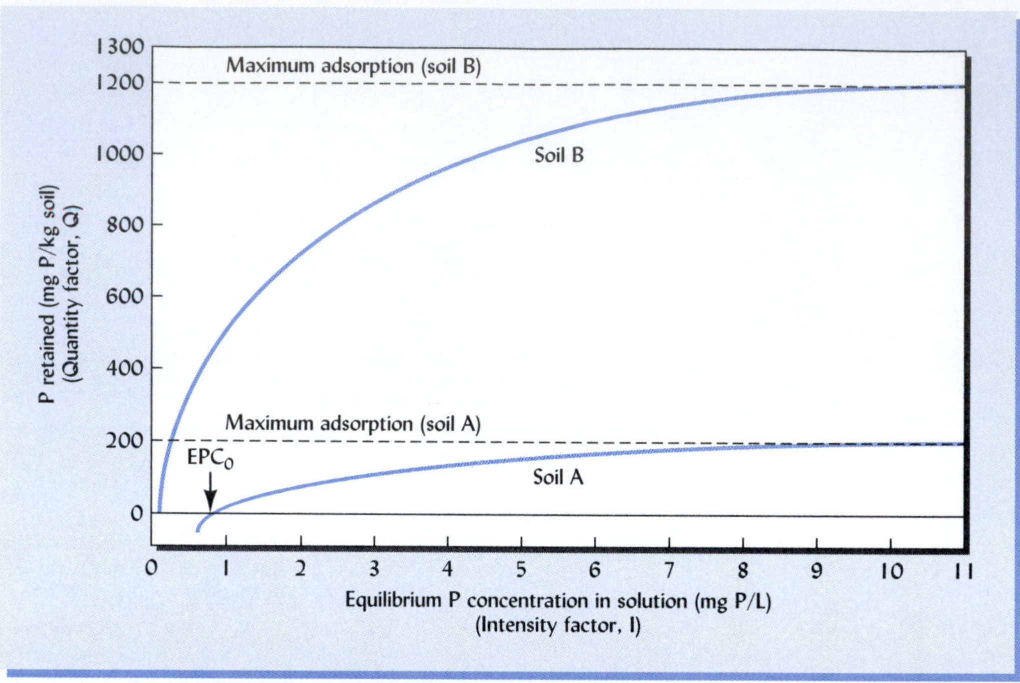

FIGURE 13.15 The relationship between phosphorus fixation and phosphorus in solution when two different soils (A and B) are shaken with solutions of various initial phosphorus concentrations. Initially, each soil removes nearly all of the phosphorus from solution, and as more and more concentrated solutions are used, the soil fixes greater amounts of phosphorus. However, eventually solutions are used that contain so much phosphorus that most of the phosphorus-fixation sites are satisfied, and much of the dissolved phosphorus remains in solution. The amount fixed by the soil levels off as the *maximum phosphorus-fixing capacity* of the soil is reached (see horizontal dashed lines: for soil A, 200 mg P/kg; for soil B, 1200 mg P/kg soil). If the initial phosphorus concentration of a solution is equal to the *equilibrium phosphorus concentration (EPC)* for a particular soil, that soil will neither remove phosphorus from nor release phosphorus to the solution (i.e., phosphorus fixation = 0 and EPC = EPC$_0$). If the solution phosphorus concentration is less than the EPC$_0$, the soil will release some phosphorus (i.e., the fixation will be negative). In this example, soil B has a much *higher* phosphorus-fixing capacity and a much *lower* EPC than does soil A. It can also be said that soil B is highly *buffered,* because much phosphorus must be added to this soil to achieve a small increase in the equilibrium solution phosphorus concentration. On the other hand, if a plant root were to remove a relatively large amount of total phosphorus from the soil, only a small change would occur in the equilibrium solution concentration. For a review of buffering capacity as it relates to the available phosphorus and potassium, as well as other nutrients, see Nair (1996).

Factors affecting phosphorus availability in soils:
http://www.petrik.com/PUBLIC/library/misc_/phosphavailability.htm

One way of determining the phosphorus-fixing capacity of a particular soil is to shake a known quantity of the soil in a phosphorus solution of known concentration. After about 24 hours, an equilibrium will be approached, and the concentration of phosphorus remaining in the solution [the **equilibrium phosphorus concentration (EPC)**] can be determined. The difference between the initial and final (*equilibrium*) solution phosphorus concentrations represents the amount of phosphorus fixed by the soil. If this procedure is repeated using a series of solutions with different initial phosphorus concentrations, the results can be plotted as a phosphorus-fixation curve (Figure 13.15) and the maximum phosphorus-fixation capacity can be extrapolated from the value at which the curve levels off.

Factors Affecting the Extent of Phosphorus Fixation in Soils

Soils that can fix more than 350 mg P/kg of soil (about 700 kg P/ha) are generally considered to be high phosphorus-fixing soils (see Table 13.5). The phosphorus-fixation capacity of a soil is positively related to certain soil properties, such as soil pH, organic matter content, calcium carbonate content, clay content, and contents of iron, aluminum, and manganese, especially the hydrous oxides of these metals.

TABLE 13.5 Maximum Phosphorus-Fixation Capacity of Several Soils of Varied Content and Kinds of Clays

Fe, Al oxides (especially amorphous types) fix the largest quantities and silicate clays (especially 2:1-type) fix the least.

		Clay		
Soil Great Group (and series, if known)	*Location*	*Percent*	*Type*	*Maximum P fixation, mg P/kg soil*
Evesboro (Quartzipsamment)	Maryland	6	Kaolinite, Fe, Al oxides	125
Kandiustalf	Zimbabwe	20	Kaolinite, Fe oxides	394
Kitsap (Xerept)	Washington	12	2:1 clays, allophane	453
Matapeake (Hapludult)	Maryland	15	Chlorite, kaolinite, Fe oxides	465
Rhodustalf	Zimbabwe	53	Kaolinite, Fe oxides	737
Newberg (Haploxeroll)	Washington	38	2:1 clays, Fe oxides	905
Tropohumults	Cameroon	46	Fe, Al oxides, kaolinite	2060

Cameroon data courtesy of V. Ngachie; Washington data from Kuo (1988); Maryland and Zimbabwe data courtesy of R. Weil and F. Folle.

Amount of Clay Present. Most of the compounds with which phosphorus reacts are in the finer soil fractions. Therefore, if soils with similar pH values and mineralogy are compared, phosphorus fixation tends to be more pronounced and ease of phosphorus release tends to be lowest in those soils with higher clay contents (Table 13.5).

Type of Clay Minerals Present. Some clay minerals are much more effective at phosphorus fixation than others. Generally, those clays that possess greater anion exchange capacity (due to positive surface charges) have a greater affinity for phosphate ions. For example, extremely high phosphorus fixation is characteristic of allophane clays typically found in Andisols and other soils associated with volcanic ash. Oxides of iron and aluminum, such as gibbsite and goethite, also strongly attract and hold phosphorus ions. Among the layer silicate clays, kaolinite has a greater phosphorus-fixation capacity than most. The 2:1 clays of less weathered soils have relatively little capacity to bind phosphorus. Differences among clays in their phosphorus-fixing capacities help explain the relatively low phosphorus-fixing capacities of Vertisols and Mollisols (high in 2:1-type clays), and the much higher capacities of Ultisols and Oxisols (high in Fe, Al oxides, and 1:1-type clays) and Andisols (high in allophane; see Figure 13.16).

Effect of Soil pH. The greatest degree of phosphorus fixation occurs at very low and very high soil pH. As pH increases from below 5.0 to about 6.0, the iron and aluminum phosphates become somewhat more soluble. Also, as pH drops from greater than 8.0 to below 6.0, calcium phosphate compounds increase in solubility. Therefore, as a general rule in mineral soils, phosphate fixation is at its lowest (and plant availability is highest) when soil pH is maintained in the 6.0 to 7.0 range (see Figure 13.10). Even in this range, however, phosphate availability may still be low, and added soluble phosphates will be readily fixed by soils.

Effect of Organic Matter. Organic matter generally has little capacity to strongly fix phosphate ions. Soils high in organic matter, especially active fractions of organic matter (see Section 11.7), generally exhibit relatively low levels of phosphorus fixation. Several mechanisms are responsible for the reduced phosphorus fixation associated with high soil organic matter (see Figure 13.14). First, large humic molecules can adhere to the surfaces of clays and metal hydrous oxide particles, masking the phosphorus-fixation sites and preventing them from interacting with phosphorus ions in solution. Second, organic acids produced by plant roots and microbial decay can serve as organic anions, which are attracted to positive charges and hydroxyls on the surfaces of clays and hydrous oxides. These organic anions may compete with phosphorus ions for fixation sites. Third, certain organic acids and similar compounds can entrap reactive Al and Fe in stable organic complexes called *chelates* (see Section 13.21). Once chelated, these metals are unavailable for reaction with phosphorus ions in solution.

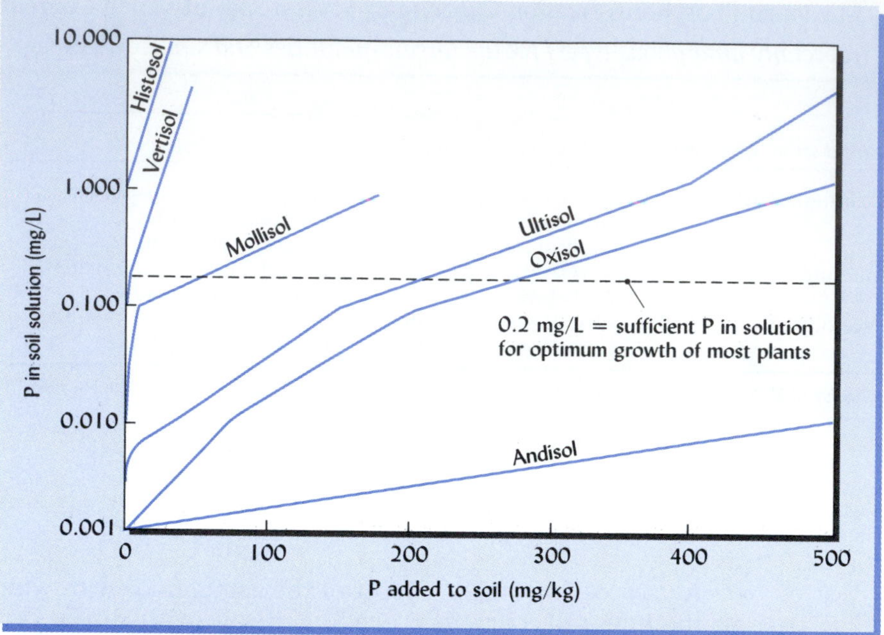

FIGURE 13.16 Typical soil solution levels of phosphorus when increasing amounts of fertilizer phosphorus are added to representative soils from several different orders. Without help from mycorrhizal fungi, most crop plants require a concentration of about 0.2 mg/L in the soil solution for optimum growth (see Table 13.6). The graph suggests very low phosphorus fixing capacities of Histisols and Vertisols, making it possible for soils in these orders when intially cultivated to require little if any fertilizer to support good plant growth. In contrast, because of their high phosphorus-fixing capacities, unfertilized Ultisols and Oxisols would require a minimum of 200–300 Kg P/ha for good plant growth, and Andisols would require many times this amount. Keep in mind that the above relationships would not necessarily apply for a given soil in each order, especially if the soil had been cultivated for a long time without fertilizer additions, or with high rates of phosphorus fertilizers that would have satisfied the soil's P-fixing capacity. (Curves are representative of data from many sources)

13.9 PLANT GENETICS AND PHOSPHORUS AVAILABILITY

Recent developments in plant breeding and genetics suggest some means of controlling the amount of phosphorus available for plant uptake and for movement from soil into groundwater and streams. For example, genes have been inserted into rice and corn that can stimulate the plant root excretion of citric acid which increases the solubility of calcium phosphates. Such developments could be of special interest to low-income farmers of areas such as sub-Saharan Africa for whom ground phosphate rock may be much more affordable than manufactured fertilizers.

Another interesting development is the creation of new low phytin grain varieties whose phosphorus is more readily absorbed by nonruminant animals such as pigs, poultry, and humans, thereby reducing the amount of phosphorus these animals excrete in their wastes. Manure from animals fed the new grains is considerably lower in phosphorus, and when spread on the land should be less prone to cause water pollution. Collaboration between soil and plant scientists offers one pathway to help alleviate problems of deficiency or excess of this important element.

13.10 PRACTICAL CONTROL OF PHOSPHORUS IN SOILS

Soil tests and agronomic and environmental management of phosphorus:
http://www.extension.umn.edu/distribution/cropsystems/DC6797.html

Based on the principles of soil phosphorus behavior discussed in this chapter, a number of approaches can be suggested to ameliorate deficiencies and excesses of phosphorus in soils.

1. *Choice of fertilizer rates to fit soil phosphorus status.* Where phosphorus-fixing capacity is grossly unsaturated, optimum crop yields will likely require fertilizer rates that considerably exceed the plant uptake of this element. As phosphorus

availability increases, however, fertilizer rates should be decreased to supply no more than the amount being taken up by the plants and to prevent phosphorus moving into the drainage waters (see Figure 13.17).

2. *Placement of phosphorus.* By localized (band) placement of the added phosphorus fertilizers, reactions with the bulk of the soil can be minimized, and the maintenance of phosphorus availability enhanced. Banded application of starter fertilizers is standard practice, as is the use of pellets and placement in holes around trees. In untilled systems, a surface band is often created. Such localized placement allows the use of less fertilizer.
3. *Combination of ammonium and phosphorus fertilizers.* The plant uptake of phosphorus can be increased, especially in alkaline soils, if ammonium and phosphorus fertilizers are combined. Apparently, oxidation or uptake of the ammonium ions acidifies the rhizosphere and slows the formation of the more insoluble calcium phosphate compounds.
4. *Cycling of organic matter.* The judicious application of organic materials such as animal manures, green manures, and plant residues can increase phosphorus availability. During the microbial breakdown of these materials, phosphorus is released slowly and can be taken up by plants before it reacts with the soil. Also, organic compounds can protect phosphorus from fixation by masking the fixation sites on the soil colloids, and by forming organic complexes (chelates) with Al, Fe, and Mn ions, thereby limiting the reaction of these ions with phosphorus.
5. *Control of soil pH.* By maintaining pH between 6 and 7, the availability of phosphorus can be optimized in most systems. Proper liming of acid soils and acidifying of alkaline soils (using a material such as elemental sulfur) can help achieve this range.
6. *Enhancement of mycorrhizal symbiosis.* Practices to enhance mycorrhizal symbiosis should be encouraged. In some cases it may be necessary to inoculate tree seedlings with efficient strains of mycorrhizal fungi. For most plants, however, the mycorrhizal symbiosis can best be enhanced by fostering appropriate soil conditions through crop rotations, organic matter additions, and minimum tillage.
7. *Choice of phosphorus-efficient plants.* By choosing plant species known to be efficient in their foraging for phosphorus, the potential for good growth can be enhanced. Table 13.6 suggests that differences in this efficiency exist. While mechanisms for more efficient phosphorus uptake are not always known, advantage should be taken of the differences in efficiency known to exist.
8. *Reduction of runoff and sediment losses, especially from fertilized and manured land.* Never apply manure or fertilizer to frozen land, because much of the soluble compounds will be washed into streams and rivers. Use conservation tillage practices that minimize runoff and erosion. Use cover crops and crop residues to maximize movement of water into the soil rather than off it.

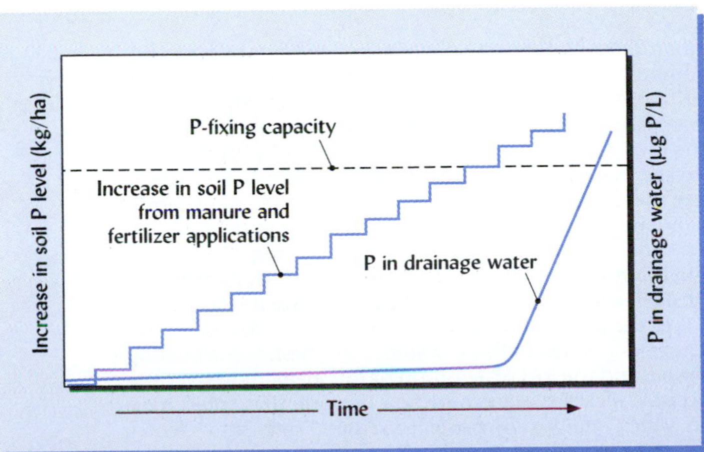

FIGURE 13.17 Generalized relationship between excess phosphorus added in manure and fertilizers, and the concentration of P in the drainage water from agricultural soils in the United States. Note that for many years the total annual additions of phosphorus have exceeded the removal of this element in harvested plants. Initially, little of this excess P leached from the soils since it was bound (fixed) in insoluble compounds such as those containing iron, aluminum, or calcium. Ultimately, however, sufficient excess P has been applied to satisfy the capacity of many soils to fix the phosphorus. This makes it easier for some of the added phosphorus to move into the drainage system, and from there to downstream ponds, reservoirs, lakes, and oceans. This same situation also prevails in most of the industrialized nations of Europe, Asia, and North America.

TABLE 13.6 Concentration of Phosphorus in Soil Solution Required for Near-Optimal Growth (95% of Maximum Yield) of Various Plants

Plant	*Approximate P in soil solution, mg/L*
Cassava	0.005
Peanut	0.01
Corn	0.05
Sorghum	0.06
Cabbage	0.04
Soybean	0.20
Tomato	0.20
Head lettuce	0.30

Data from Fox (1981).

9. *Capturing excess phosphorus before it enters mainstream channels.* The use of natural or constructed wetlands to tie up the phosphorus before it enters streams should be encouraged. Some of this phosphorus will be in soluble forms, but is most likely to be in the sediment. Also, prior to their being applied to the field, some phosphorus-containing waste products and manures can be treated with compounds (e.g., alum) that tie up the phosphorus and prevent its movement off the land in runoff water.

The soil's small pool of available phosphorus is supplemented primarily by the addition of P-containing fertilizers, and depleted mostly by the removal of this element from the land by harvested crops. The balance of these two opposing forces determines not only the ability of a soil to support plant growth, but also the environmental hazards that result from too much phosphorus. Figure 13.18, showing the phosphorus balance for agriculture in the state of Illinois, illustrate these two major inputs and outputs. This figure also suggests the continued buildup of phosphorus in the soil, and warns us of the environmental problems that can result if too much phosphorus moves into natural waters.

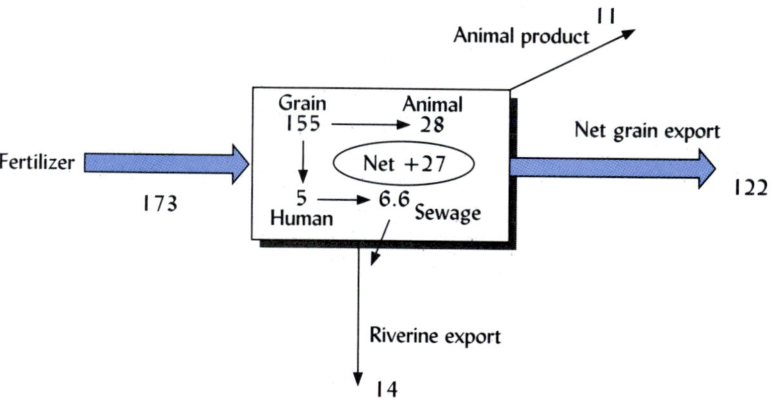

FIGURE 13.18 Diagram showing the annual inputs and outputs of phosphorus for the State of Illinois, averaged for 18 years (1979–1996 for terrestrial fluxes and 1980–1997 for riverine export). The units are 1,000 Mg P/yr. The major input into the system is fertilizer phosphorus. Fluxes inside the box are for phosphorus in the total grain harvested, its consumption by animals and humans, and the sewage from animals and humans that moves into rivers. The major outputs are the net grain export outside the state, the phosphorus in the animals sold, and the phosphorus moved from the state in the river systems (riverine export). Note that within the box some 27,000 Mg of phosphorus accumulates annually, mostly residing in the soils of the state. Phosphorus in manures is not considered as either an input or output, since it is merely recycled from the soil into the plants and back into the soil. This diagram illustrates some of the principles covered in this chapter. It emphasizes the buildup of phosphorus in the soil, and the loss of a portion of this element in river waters, a portion that means little for plant nutrition but one that could have damaging effects on water quality downstream. [From David and Gentry (2000)]

13.11 POTASSIUM: NATURE AND ECOLOGICAL ROLES[9]

Of all the essential elements, potassium is the third most likely, after nitrogen and phosphorus, to limit plant productivity. For this reason it is commonly applied to soils, and is a component of most mixed fertilizers.

The potassium story differs in many ways from that of phosphorus. Unlike phosphorus (or sulphur and, to a large extent, nitrogen), potassium is present in the soil solution only as a positively charged cation, K^+. Like phosphorus, potassium does not form any gases that could be lost to the atmosphere. Its behavior in the soil is influenced primarily by soil cation exchange properties (see Chapter 8) and mineral weathering (Chapter 2), rather than by microbiological processes. Unlike nitrogen and phosphorus, potassium causes no off-site environmental problems when it leaves the soil system. It is not toxic and does not cause eutrophication in aquatic systems.

13.12 POTASSIUM IN PLANT AND ANIMAL NUTRITION

Potassium uptake in plants: battery-like mechanism shows surprising power: http://www.biotech.wisc.edu/Education/biotechnews/potassium.html

Although potassium plays numerous roles in plant and animal nutrition, it is not actually incorporated into the structures of organic compounds. Instead, potassium remains in the ionic form (K^+) in solution in the cell, or acts as an activator for cellular enzymes. Potassium is known to activate over 80 different enzymes responsible for such plant and animal processes as energy metabolism, starch synthesis, nitrate reduction, photosynthesis, and sugar degradation. Certain plants, many of which evolved in sodium-rich semiarid environments, can substitute sodium or other monovalent ions to carry out some, but not all, of the functions of potassium.

Potassium is especially important in helping plants adapt to environmental stresses. Good potassium nutrition is linked to improved drought tolerance, improved winter hardiness, better resistance to certain fungal diseases, and greater tolerance to insect pests. Potassium also enhances the quality of flowers, fruits, and vegetables by improving flavor and color and strengthening stems (thereby reducing lodging). In many of these respects, potassium seems to counteract some of the detrimental effects of excess nitrogen.

In animals, including humans, potassium plays critical roles in regulating the nervous system and in the maintenance of healthy blood vessels. Diets that include such high-potassium foods as bananas, potatoes, orange juice, and leafy green vegetables have been shown to lower human risk of stroke and heart disease. Maintaining a balance between potassium and sodium is especially important in human diets.

Deficiency Symptoms in Plants

Compared to deficiencies of phosphorus and many other nutrients, a deficiency of potassium is relatively easy to recognize in most plants. When potassium is deficient, the tips and edges of the oldest leaves begin to yellow (chlorosis) and then die (necrosis), so that the leaves appear to have been burned on the edges (Figure 13.19 and Plate 44, after page 498). On some plants, the necrotic leaf edges may tear, giving the leaf a ragged appearance. In several important forage and cover-crop legume species, potassium deficiency produces small, white necrotic spots that form a unique pattern along the leaflet margins (Fig 13.19*d*); this easily recognized symptom is one that people often mistake for insect damage.

13.13 THE POTASSIUM CYCLE

Figure 13.20 shows the major forms in which potassium is held in soils and the changes it undergoes as it is cycled through the soil–plant system. The original sources of potassium are the primary minerals, such as micas (biotite and muscovite) and potassium feldspar (orthoclase and microcline). As these minerals weather, their rigid lattice structures become more pliable. For example, potassium held between the 2:1-type crystal

---

[9]For further information on this topic, see Mengel and Kirkby (1987) and Munson (1985).

(a)

(b)

(c)

(d)

FIGURE 13.19 Potassium deficiency often produces easily recognized foliar symptoms, mainly on older leaves: (*a*) chlorotic margins on boxwood leaves; (*b*) chlorotic leaf margins on soybean; (*c*) ragged, necrotic margins of older banana leaves; and (*d*) small, white necrotic spots on hairy vetch leaflets. [Photos (*a*), (*c*), and (*d*) courtesy of R. Weil; photo (*b*) courtesy of Potash & Phosphate Institute]

layers of mica is in time made more available, first as nonexchangeable but slowly available forms near the weathered edges of minerals and, eventually, as readily exchangeable and soil solution forms from which it is absorbed by plant roots.

Potassium is taken up by plants in large quantities. A portion of this potassium is leached from plant foliage by rainwater (through-fall) and returned to the soil, and a portion is returned to the soil with the plant residues. In natural ecosystems, most of the potassium taken up by plants is returned in these ways or as wastes (mainly urine) from animals feeding on the vegetation. Some potassium is lost with eroded soil particles and in runoff water, and some is lost to groundwater by leaching. In agroecosystems, from one-fifth (e.g., in cereal grains) to nearly all (e.g., in hay crops) of the potassium taken up by plants may be exported to distant markets, from which it is unlikely to return.

At any one time, most soil potassium is in primary minerals and nonexchangeable forms. In relatively fertile soils, the release of potassium from these forms to the exchangeable and soil solution forms that plants can use directly may be sufficiently rapid to keep plants supplied with enough potassium for optimum growth. On the other hand, where high yields of agricultural crops or timber are removed from the land, or

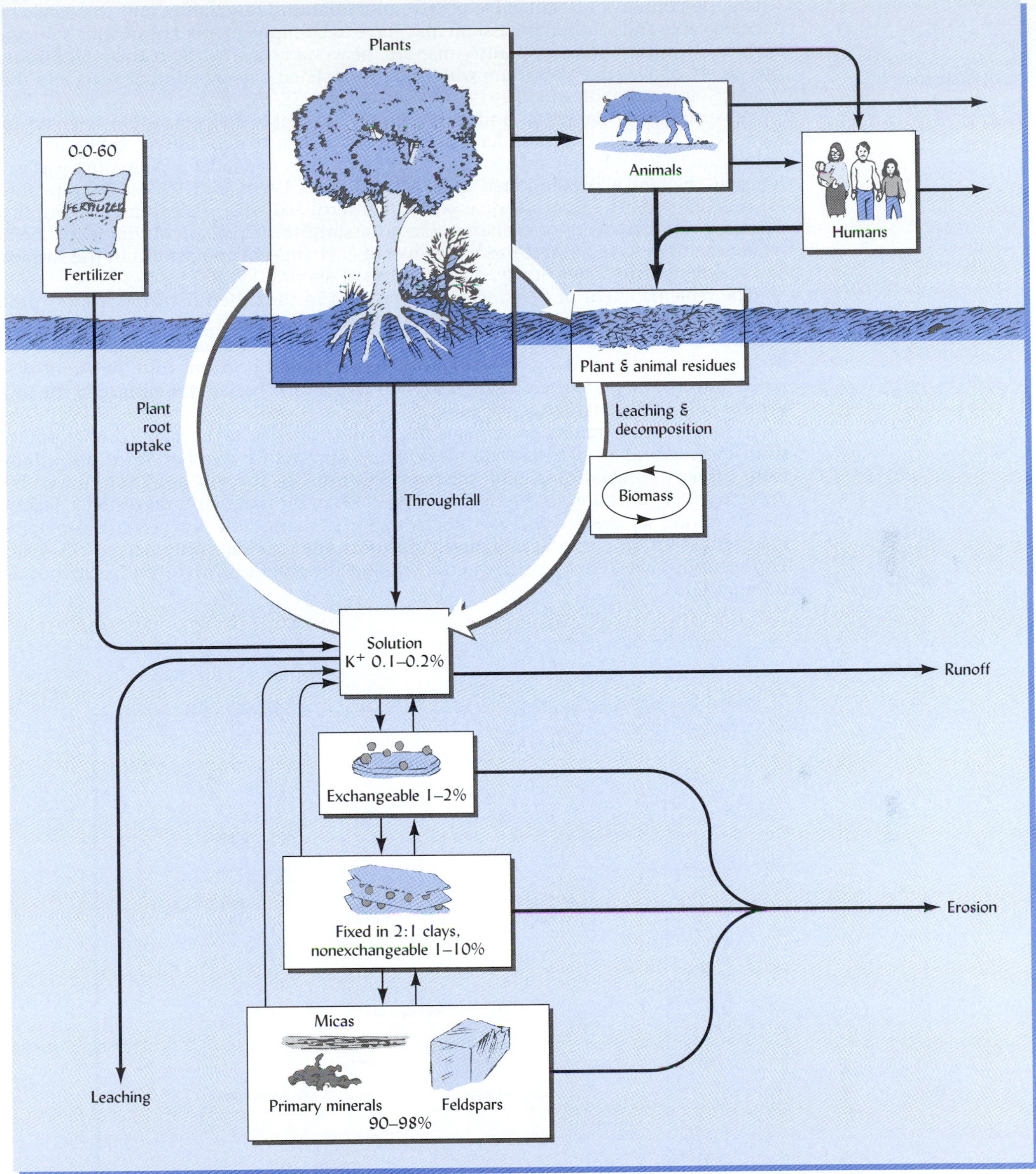

FIGURE 13.20 Major components of the potassium cycle in soils. The inner circle emphasizes the biological cycling of potassium from the soil solution to plants and back to the soil via plant residues or animal wastes. Primary and secondary minerals are the original sources of the element. Exchangeable potassium may include those ions held and released by both clay and humus colloids, but potassium is not a structural component of soil humus. The interactions among solution potassium, exchangeable, nonexchangeable, and structural potassium in primary minerals is shown. The bulk of soil potassium occurs in the primary and secondary minerals, and is released very slowly by weathering processes.

Potassium's ecological benefits in forests of the Pacific Northwest:
http://www.cnr.uidaho.edu/extforest/electron.htm

where the content of weatherable potassium-containing minerals is low, the levels of exchangeable and solution potassium may have to be supplemented by outside sources, such as chemical fertilizers, poultry manure, or wood ashes. Without these additions, the supply of available potassium will likely be depleted over a period of years and the productivity of the soil will likewise decline.

An example of depletion and restoration of available soil potassium is given in Figure 13.21. Farming without fertilizers for over a century depleted the exchangeable potassium in a sandy soil in New York. After the supply of available potassium was exhausted, the land was abandoned in the 1920s. In the 1920s and 1930s, a red pine forest was planted. The trees in some plots were fertilized with potassium, causing the expected rapid recovery of exchangeable potassium to its preagricultural level. Even where the trees were not fertilized, the level of exchangeable potassium in the surface soil was replenished, but slowly, over a period of about 80 years.

The replenishment of exchangeable potassium in the unfertilized forest plots provides an example of the ability of unharvested perennial plants, such as forest trees and pasture legumes, to ameliorate soil fertility over time. Deep-rooted perennials often act as "nutrient pumps," taking potassium from deep subsoil horizons into their root systems, translocating it to their leaves, and then recycling it back to the surface of the soil via leaf fall and throughfall (see Figure 2.17).

In most mature natural ecosystems the small (1 to 5 kg/ha) annual losses of potassium by leaching and erosion are more than balanced by weathering of potassium from primary minerals and nonexchangeable forms in the soil profile, followed by vegetative translocation to the surface of the soil. In most agricultural systems, leaching losses are far greater because much higher exchangeable K levels are generally maintained for crop production, and crop roots are active for only part of the year. The sections that follow give greater details on the reactions involved in the potassium cycle.

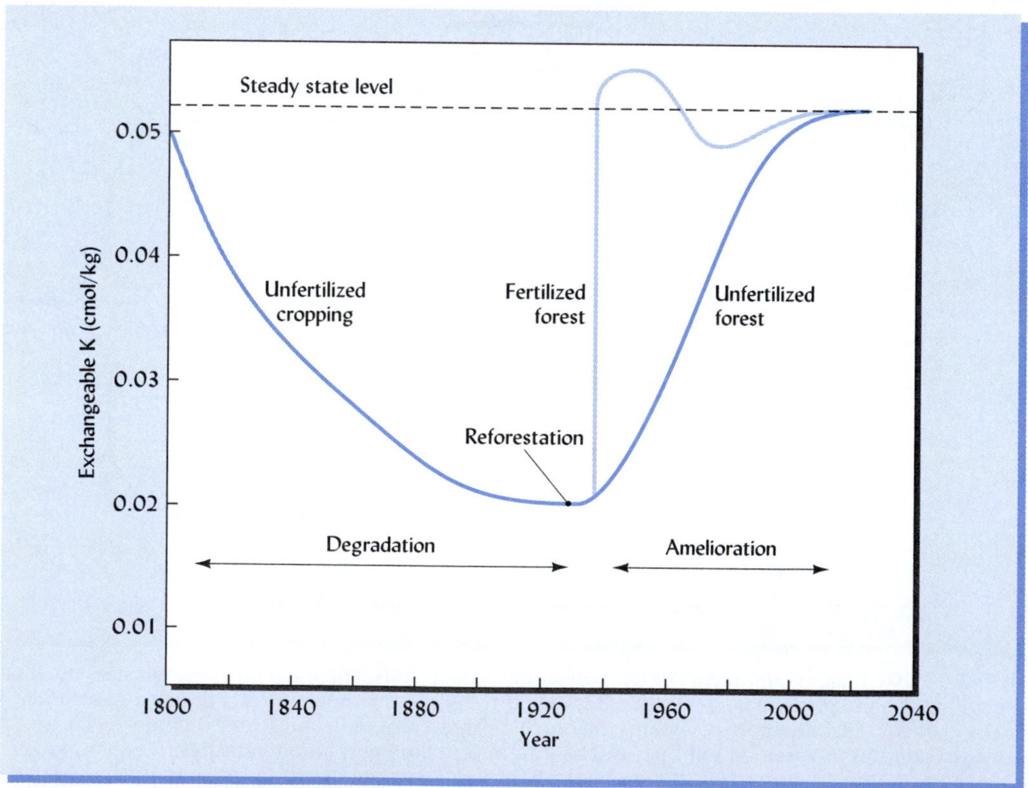

FIGURE 13.21 The general pattern of depletion of A-horizon exchangeable potassium by decades of exploitative farming, followed by its restoration under forest vegetation. The forest consisted of red pine trees planted on a Plainfield loamy sand (Udipsamment) in New York. This soil has a very low cation exchange capacity and low levels of exchangeable K^+. [From Nowak et al. (1991)]

13.14 THE POTASSIUM PROBLEM IN SOIL FERTILITY

Availability of Potassium

The total quantity of potassium is generally greater than that of any other major nutrient element. Amounts as great as 30,000 to 50,000 kg potassium in the upper 15 cm of 1 ha of soil are not at all uncommon (see Table 1.3).

Yet the quantity of potassium held in an easily exchangeable condition at any one time often is very small. Most of this element is held rigidly as part of the primary minerals or is fixed in forms that are, at best, only moderately available to plants. Therefore, the situation with respect to potassium utilization parallels that of phosphorus and nitrogen in at least one way: A very large proportion of all three of these elements in the soil is insoluble and relatively unavailable to growing plants.

Leaching Losses

Potassium is much more readily lost by leaching than is phosphorus. Drainage waters from soils receiving liberal fertilizer applications usually contain considerable quantities of potassium. From representative humid-region soils growing annual crops and receiving only moderate rates of fertilizer, the annual loss of potassium by leaching is usually about 25 to 50 kg/ha, the greater values being typical of acid, sandy soils.

Losses would undoubtedly be much larger were the leaching of potassium not slowed by the attraction of the positively charged potassium ions to the negatively charged cation exchange sites on clay and humus surfaces. Liming an acid soil to raise its pH can reduce the leaching losses of potassium because of the *complementary ion effect* (see Section 8.10 and Figure 13.22). Potassium ions can more easily exchange with the calcium and magnesium ions in the limed soil than with aluminum ions that dominate the exchange complex of acid soils. Thus, in a limed soil, K^+ can be more readily retained on the exchange complex, and leaching of this element is reduced.

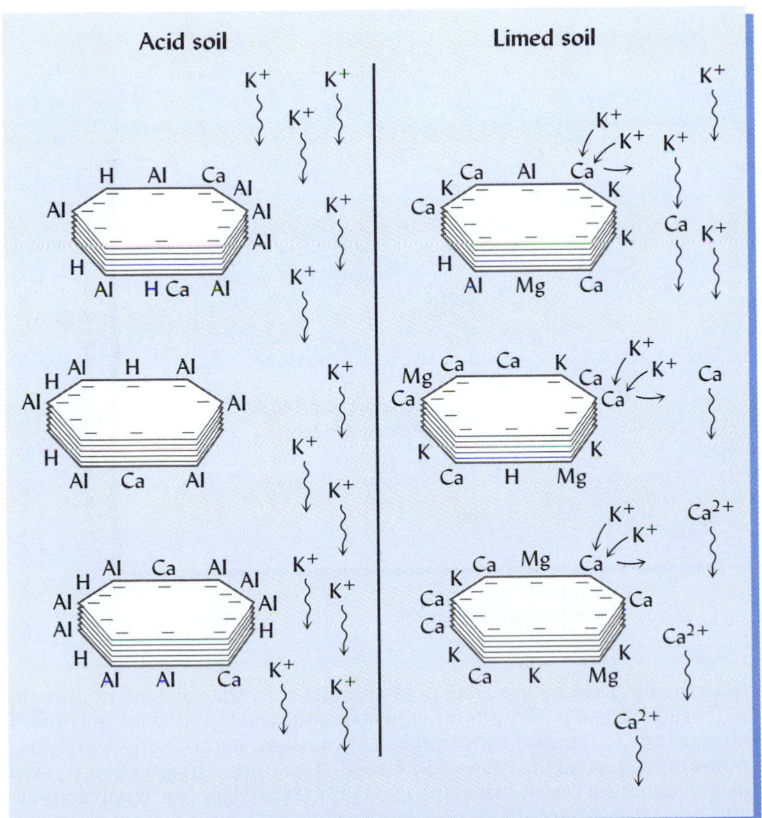

FIGURE 13.22 Diagram illustrating how liming an acid soil can reduce leaching losses of potassium. The fact that the K^+ ions can more easily replace Ca^{2+} ions than they could replace Al^{3+} ions allows more of the K^+ ions to be removed from solution by cation exchange in the limed (high-calcium) soil. The removal of K^+ ions from solution by adsorption on the colloids will reduce their loss by leaching, but they will still be at least moderately available for plant uptake.

Plant Uptake and Removal

Fertilizer industry view on maintaining high levels of soil potassium:
http://www.pda.org.uk/leaflets/28/no28-page1.html

Plants take up very large amounts of potassium, often 5 to 10 times as much as for phosphorus and about the same amount as for nitrogen. If most or all of the aboveground plant parts are removed in harvest, the drain on the soil supply of potassium can be very large. For example, a 60 Mg/ha yield of corn silage may remove 160 kg/ha of potassium. Conventional bolewood timber harvest typically removes about 100 kg/ha of potassium. If the entire tree is chipped and removed, as for paper pulp, the removal of potassium may be twice as great. A high-yielding legume hay crop may remove 400 kg/ha of potassium each year.

Luxury Consumption

Moreover, this situation is made even more critical by the tendency of plants to take up soluble potassium far in excess of their needs if sufficiently large quantities are present. This tendency is termed **luxury consumption**, because the excess potassium absorbed does not increase plant growth to any extent.

The principles involved in luxury consumption are shown in Figure 13.23. For many plants there is a direct relationship between the available potassium (soil plus fertilizer) and the removal of this element by the plants. However, only a certain amount of this element is needed for optimum growth, and this is termed *required potassium*. Potassium taken up by the plant above this critical required level is considered a *luxury*. If plant residues are not returned to the soil, the removal of this excess potassium is decidedly wasteful. In addition, high levels of potassium may depress calcium uptake and cause nutritional imbalances both in the plants and in animals that consume them (a problem especially critical for dairy cows).

In summary, the problem of potassium is at least threefold: (1) A very large proportion is relatively unavailable to higher plants; (2) it is subject to leaching losses; and (3) the removal of potassium by plants is high, especially when luxury quantities of this element are supplied. With these ideas as a background, the various forms and availabilities of potassium in soils will now be considered.

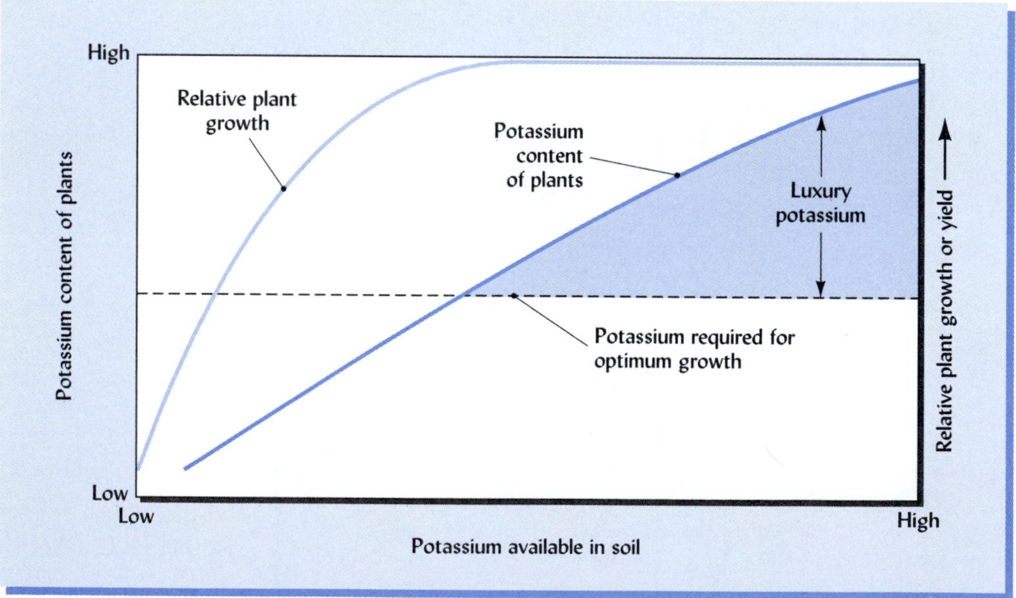

FIGURE 13.23 The general relationship between available potassium level in the soils, plant growth, and plant uptake of potassium. If available soil potassium is raised above the level needed for maximum plant growth, many plants will continue to increase their uptake of potassium without any corresponding increase in growth. The potassium taken up in excess of that needed for optimum growth is termed *luxury consumption*. Such luxury consumption may be wasteful, especially if the plants are completely removed from the soil. It may also cause dietary imbalance in grazing animals.

13.15 FORMS AND AVAILABILITY OF POTASSIUM IN SOILS

Potassium as an essential nutrient:
http://www.soils.wisc.edu/~barak/soilscience326/potassium.htm

Four forms of soil potassium are shown in the potassium-cycle diagram (Figure 13.20, from the bottom upward): (1) K in primary mineral crystal structures, (2) K in nonexchangeable positions in secondary minerals, (3) K in exchangeable form on soil colloid surfaces, and (4) potassium ions soluble in water. The total amount of potassium in a soil and the distribution of potassium among these four major pools is largely a function of the kinds of clay minerals present. Generally, soils dominated by 2:1 clays contain the most potassium; those dominated by kaolinite contain the least (Table 13.7). In terms of availability for plant uptake, the following interpretation applies to the different forms of soil potassium:

K in primary mineral structure	*Unavailable*
Nonexchangeable K in secondary minerals	*Slowly available*
Exchangeable K on soil colloids K soluble in water	*Readily available*

Relatively Unavailable Forms

Some 90 to 98% of all soil potassium in a mineral soil is in relatively unavailable forms (see Figure 13.20), mostly in the crystal structures of feldspars and micas. These minerals are quite resistant to weathering and supply relatively small quantities of potassium during a given growing season. However, their cumulative release of potassium over a period of years undoubtedly is of some importance. As already mentioned, the roots of some plants can obtain a significant portion of their potassium supply from these minerals, apparently by depleting the potassium ions from the solution around the edges of these minerals, thereby favoring the dissolution of the mineral.

Readily Available Forms

Only 1 to 2% of the total soil potassium is readily available. Available potassium exists in soils in two forms: (1) in the soil solution, and (2) in exchangeable potassium adsorbed on the soil colloidal surfaces. Although most of this available potassium is in the exchangeable form (approximately 90%), soil solution potassium is most readily absorbed by higher plants (but, unfortunately, is subject to considerable leaching loss).

As represented in Figure 13.20, these two forms of readily available potassium are in dynamic equilibrium. This equilibrium is of extreme practical importance. When plants absorb potassium from the soil solution, some of the exchangeable potassium immediately moves into the soil solution until the equilibrium is again established. When water-soluble fertilizers are added, the soil solution becomes potassium enriched and the reverse adjustment occurs—potassium from soil solution moves onto the exchange complex. The exchangeable potassium can be seen as an important buffer mechanism for soil solution potassium.

TABLE 13.7 The Influence of Dominant Clay Minerals on the Amounts of Water-Soluble, Exchangeable, Fixed (Nonexchangeable), and Total Potassium in Soils

The values given are means for many soils in 10 soil orders sampled in the United States and Puerto Rico.

	Dominant clay mineralogy of soils, mg K/kg soil		
Potassium pool	*Kaolinitic (26 soils)*	*Mixed (53 soils)*	*Smectitic (23 soils)*
Total potassium	3,340	8,920	15,780
Exchangeable potassium	45	224	183
Water-soluble potassium	2	5	4

Data from Sharpley (1990).

Slowly Available Forms

In the presence of vermiculite, smectite, and other 2:1-type minerals, the K^+ ions (as well as the similarly sized NH_4^+ ions) in the soil solution (or added as fertilizers) not only become adsorbed but also may become definitely fixed by the soil colloids (Figure 13.24). Potassium (and ammonium) ions fit in between layers in the crystals of these normally expanding clays and become an integral part of the crystal. These ions cannot be replaced by ordinary exchange processes and consequently are referred to as *nonexchangeable ions.* As such, they are not readily available to most higher plants. This form is in equilibrium, however, with the more available forms and consequently acts as an extremely important reservoir of slowly available nutrients (Figure 13.25).

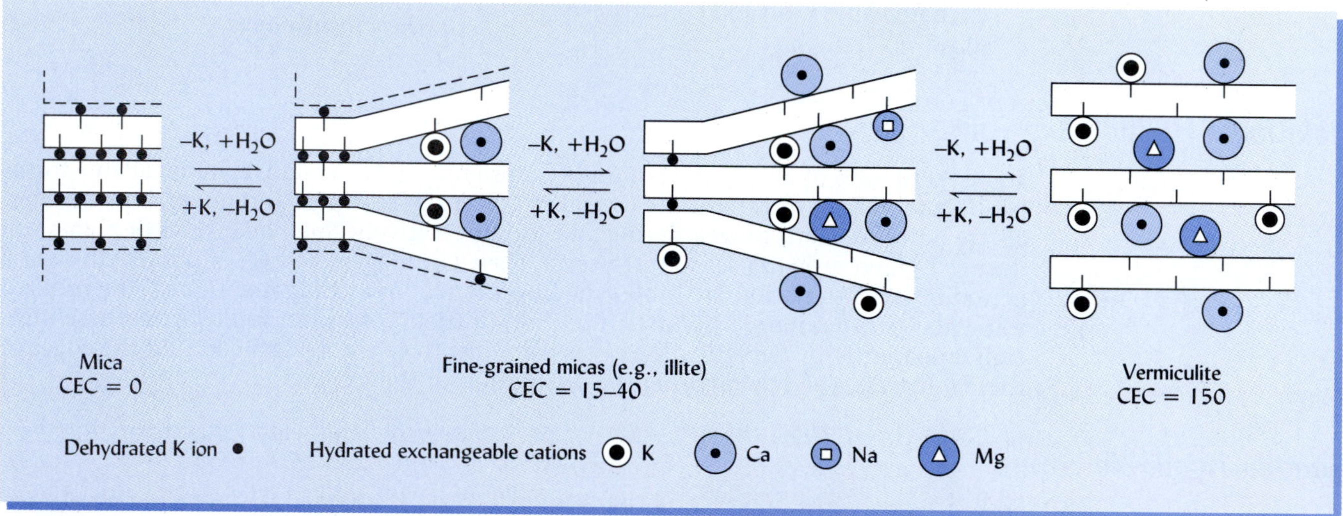

FIGURE 13.24 Diagrammatic illustration of the release and fixation of potassium between primary micas, fine-grained mica (illite clay), and vermiculite. In the diagram, the release of potassium proceeds to the right, while the fixation process proceeds to the left. Note that the dehydrated potassium ion is much smaller than the hydrated ions of Na^+, Ca^{2+}, Mg^{2+}, etc. Thus, when potassium is added to a soil containing 2:1-type minerals such as vermiculite, the reaction may go to the left and potassium ions will be tightly held (fixed) in between layers within the crystal, producing a fine-grained mica structure. Ammonium ions (NH_4^+) are of a similar size and charge to potassium ions and may be fixed by similar reactions. [Modified from McLean (1978)]

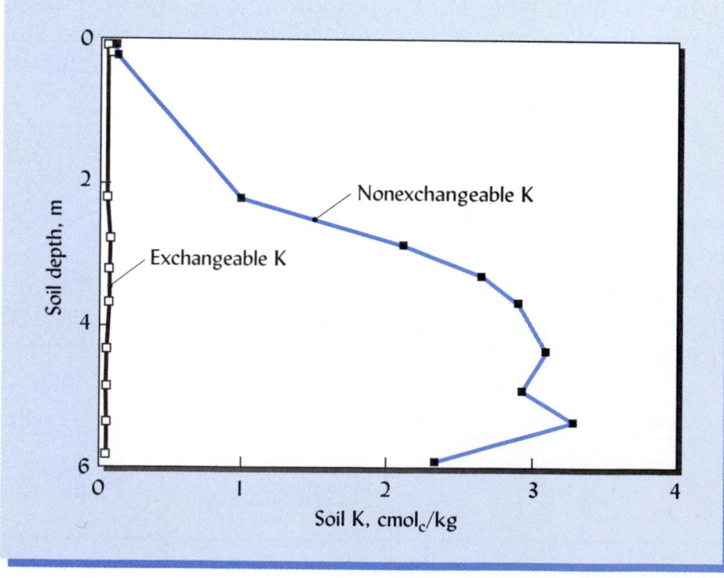

FIGURE 13.25 Exchangeable and nonexchangeable potassium levels in a Ultisol in South Carolina after 30 years growth of a loblolly pine following 150 years of cultivated crops. Although the exchangeable potassium level was quite low, tree growth was not adversely affected, and large quantities of this element were absorbed by the trees over the 30-year period. This was made possible by the conversion of nonexchangeable potassium to the exchangeable form, which was readily taken up by the trees. The upper horizons may have been depleted somewhat of nonexchangeable K, but the deep tree roots were able to use the potassium released from the nonexchangeable form in the lower horizons. [From Richter et al. (1994)]

Release of Fixed Potassium

The quantity of nonexchangeable or fixed potassium in some soils is quite large. The fixed potassium in such soils is continually released to the exchangeable form in amounts large enough to be of great practical importance. The entire equilibrium may be represented for potassium as follows:

$$\text{Nonexchangeable K} \underset{}{\overset{\text{Slow}}{\rightleftharpoons}} \text{Exchangeable K} \underset{}{\overset{\text{Rapid}}{\rightleftharpoons}} \text{Soil solution K} \quad (13.6)$$

As a result of these relationships, very sandy soils with low CEC are poorly buffered with respect to potassium. In them, the potassium ion concentration may be quite high at the beginning of a growing season or just after fertilization, but the soils have little capacity to maintain the potassium concentration, as plants remove the dissolved potassium from the soil solution during the growing season. Late-season potassium deficiency may result. In finer-textured soils with a greater CEC (and therefore greater buffering capacity), the initial solution concentration of potassium may be somewhat lower, but the soil is capable of maintaining a fairly constant supply of solution potassium ions throughout the growing season.

13.16 FACTORS AFFECTING POTASSIUM FIXATION IN SOILS

Four soil conditions markedly influence the amounts of potassium fixed: (1) the nature of the soil colloids, (2) wetting and drying, (3) freezing and thawing, and (4) the presence of excess lime.

Effects of Type of Clay and Moisture

The ability of the various soil colloids to fix potassium varies widely. Kaolinite and other 1:1-type clays fix little potassium. On the other hand, clays of the 2:1 type, such as vermiculite, fine-grained mica (illite), and smectite, fix potassium very readily and in large quantities. Even silt-sized fractions of some micaceous minerals fix and subsequently release potassium (Table 13.8).

Alternate wetting/drying and freezing/thawing have been shown to enhance both the fixation of potassium in nonexchangeable forms and the release of previously fixed potassium to the soil solution. Although the practical importance of this is recognized, its mechanism is not well understood.

Applications of lime sometimes result in an increase in potassium fixation of soils. This is not surprising, since in strongly acid soils the tightly held H^+ and hydroxy aluminum ions prevent the potassium ions from being closely associated with the colloidal surfaces, which reduces their susceptibility to fixation. As the pH increases, the H^+ and hydroxyl aluminum ions are removed or neutralized, and it is easier for potassium ions to move closer to the colloidal surfaces, where they are more susceptible to fixation in 2:1 clays.

TABLE 13.8 Average Uptake of Potassium by the Roots of Ryegrass Grown on 14 High Mica Arable Loess-Derived Soils (Alfisols) with and without the Clay Removed

The mica in the silt and sand fractions provided about 3/4 as much potassium as did the whole soils. It is probable that the silt fraction was a major source of potassium in these soils.

	K uptake by ryegrass root (mg/pot)		
	First harvest	*Second harvest*	*Total*
Whole soils	103.2	41.3	144.5
Silt + sand fraction	73.9	23.7	97.6

Calculated from Mengel et al. (1998).

Furthermore, high calcium and magnesium levels in the soil solution may reduce potassium uptake by the plant because cations tend to compete against one another for uptake by roots. Since the absorption of the potassium ion by plant roots is affected by the activity of other ions in the soil solution, some authorities prefer to use the following ratio:

$$\frac{[K^+]}{\sqrt{[Ca^{2+}] + [Mg^{2+}]}}$$

rather than the potassium concentration to indicate the available potassium level in solution. Finally, potassium deficiency frequently occurs in calcareous soils even when the amount of exchangeable potassium present would be adequate for plant nutrition on other soils. Potassium fixation as well as cation ratios may be responsible for these adverse effects on calcium carbonate–rich soil.

13.17 PRACTICAL ASPECTS OF POTASSIUM MANAGEMENT

Avoiding luxury K in alfalfa that harms cow health: http://www.uwex.edu/ces/crops/TissueK.htm

Although there are differences among soils in their potassium management requirements, a few common suggestions for optimum management of this element are appropriate.

1. Obtain a clear understanding of the sources of gains and losses of available potassium, and of the factors affecting these gains and losses (Figure 13.26). Recognize that the relative importance of each source will vary among soils (e.g., sand versus clays) and among ecosystems (e.g., agricultural versus forests).
2. Take full advantage of the natural potassium-supplying power of soils, remembering that for most soils it may not be necessary to replace all the potassium removed by plants. The natural supplies are commonly adequate for most nonagricultural systems, but generally must by supplemented by outside sources where crops (including trees) are removed from the land.
3. Avoid highly acid soil conditions that may permit extensive losses of potassium.
4. Recycle as much of the plant-absorbed potassium as feasible by returning to the soil plant residues and animal manures, both of which are highly laden with potassium.
5. Where fertilizers are required to meet potassium needs, apply the element in frequent (annual), light applications rather than in heavy dressings every few years. This will reduce the possibilities of luxury consumption, losses by leaching, and fixation in relatively unavailable forms.

The global use of potassium fertilizers will continue to increase as the pools of available soil potassium are depleted and increasing crop yields place demands on the soil that can-

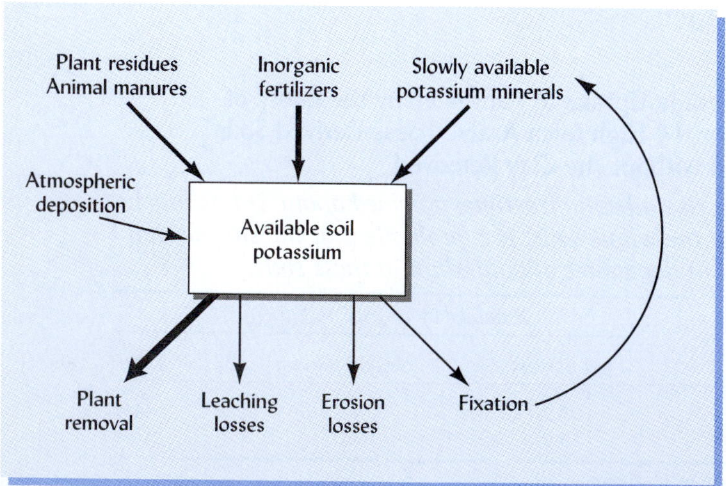

FIGURE 13.26 Gains and losses of *available* soil potassium under average field conditions. The approximate magnitude of the changes is represented by the width of the arrows. For any specific case, the actual amounts of potassium added or lost undoubtedly may vary considerably from this representation. As was the case with nitrogen and phosphorus, fertilizers are important in meeting plant demands in many agroecosystems.

not be met by mineral weathering alone. This is especially true in cash-crop areas and in regions where sandy soils or highly weathered soils are prominent. In some less developed agricultural regions of the world where current rates of potassium fertilizer use are quite low, such rates may have to increase if yields are to be increased or even maintained.

13.18 IMPORTANCE OF MICRONUTRIENT ELEMENTS[10]

The nine so-called *micronutrients* or *trace elements* are no less important to plant growth than are the macronutrients; they are merely needed in much smaller quantities. The trace elements include iron, manganese, zinc, copper, cobalt, nickel, boron, molybdenum, and chlorine. Increased attention is being given to them for the following reasons.

1. Intensive plant production practices have increased crop yields, resulting in greater removal of micronutrients from soils.
2. The trend toward high-analysis fertilizers has reduced the use of impure salts and organic manures, which formerly supplied significant amounts of micronutrients.
3. Increased knowledge of plant nutrition and improved methods of analysis in the laboratory are helping in the diagnosis of micronutrient deficiencies that might formerly have gone unnoticed.
4. Increasing evidence indicates that food grown on soils with low levels of trace elements may provide insufficient human dietary levels of certain elements, even though the crop plants show no signs of deficiency themselves.

Interest in trace elements has also been sparked by problems of toxicity resulting from an oversupply of these elements. Levels of trace elements toxic to plants or to animals may result from natural soil conditions, pollution, or soil management practices. Such toxicities are receiving more critical attention since toxic levels of trace elements in soils commonly lead to water pollution, with subsequent threats to human and animal health.

Micronutrients are required in very small quantities, their concentrations in plant tissue being one or more orders of magnitude lower than for the macronutrients (Figure 13.27). The ranges of plant tissue concentrations considered deficient, adequate, and toxic for several micronutrients are illustrated in Figure 13.28.

13.19 ROLE OF THE MICRONUTRIENTS[11]

Physiological Roles

Diagnosis of micronutrient deficiency in ornamental woody plants:
http://plantclinic.cornell.edu/FactSheets/microchlorosis/microchlorosis.htm

Micronutrients play many complex roles in plant nutrition. While most of the micronutrients participate in the functioning of a number of enzyme systems (Table 13.9), there is considerable variation in the specific functions of the various micronutrients in plant and microbial growth processes. For example, copper, iron, and molybdenum are capable of acting as electron carriers in the enzyme systems that bring about oxidation-reduction reactions in plants. Such reactions are essential steps in photosynthesis and

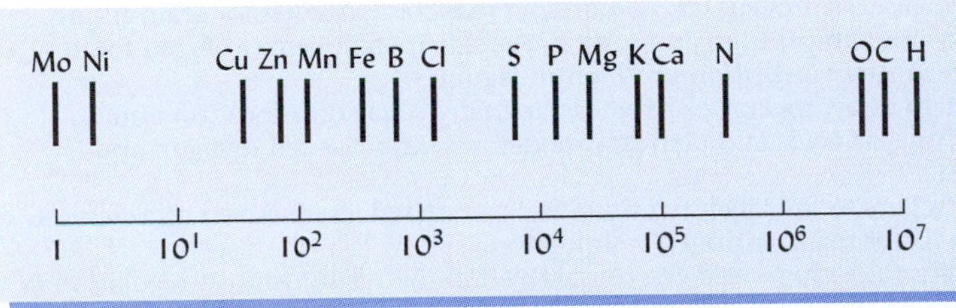

FIGURE 13.27 Relative numbers of atoms of the essential elements in alfalfa at bloom stage, expressed logarithmically. Note that there are more than 10 million hydrogen atoms for each molybdenum atom. Even so, normal plant growth would not occur without molybdenum. [Modified from Viets (1965)]

[10]For review articles on this subject, see Welch (1995), Mortvedt et al. (1991), and Stevenson (1986).
[11]For a review of the role of individual micronutrients, see Mengel and Kirkby (1987) and Marschner (1995). For nickel, see Brown et al. (1987).

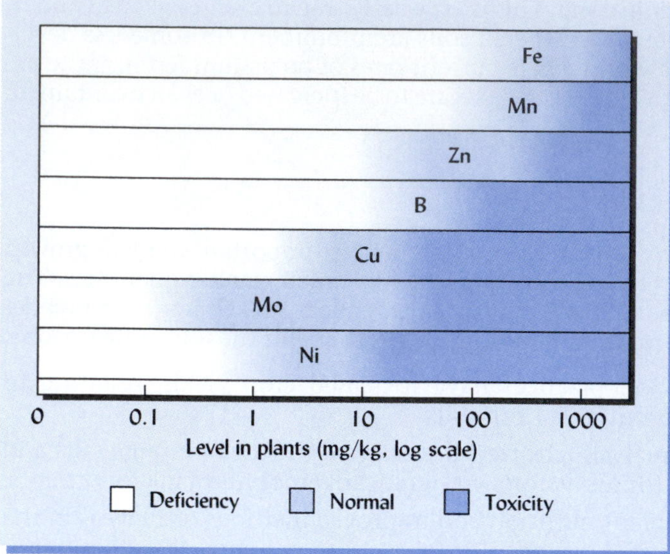

FIGURE 13.28 Deficiency, normal, and toxicity levels in plants for seven micronutrients. Note that the range is shown on a logarithm base and that the upper limit for manganese is about 10,000 times the lower range for molybdenum and nickel. In using this figure, keep in mind the remarkable differences in the ability of different plant species and cultivars to accumulate and tolerate different levels of micronutrients. (Based on data from many sources)

many other metabolic processes. Zinc and manganese function in many plant enzyme systems as bridges to connect the enzyme with the substrate upon which it is meant to act. The functions of other micronutrients are outlined in Table 13.9.

Deficiency Symptoms

Since most of the micronutrients are relatively immobile in the plant, they are not readily transferred from older leaves to younger ones. Therefore, the concentration of the nutrient tends to be lowest, and the symptoms of deficiency most pronounced, in the younger leaves, which develop after the supply of the nutrient has run low.[12] The iron-deficient sorghum in Plate 29 and the zinc-deficient corn and peach leaves in Plates 30 and 31 (following pages 114 and 498) illustrate the pattern of pronounced deficiency symptoms on the younger leaves. Boron deficiency tends to cause aborted or misshapen flowers or leaves as illustrated for roses in Figure 13.29.

TABLE 13.9 Functions of The Nine Micronutrients in Higher Plants

Micronutrient	*Functions in higher plants*
Zinc	Present in several dehydrogenase, proteinase, and peptidase enzymes; promotes growth hormones and starch formation; promotes seed maturation and production.
Iron	Present in several peroxidase, catalase, and cytochrome oxidase enzymes; found in ferredoxin, which participates in oxidation-reduction reactions (e.g., NO_3^- and SO_4^{2-} reduction and N fixation); important in chlorophyll formation.
Copper	Present in lactase and several other oxidase enzymes; important in photosynthesis, protein and carbohydrate metabolism, and probably nitrogen fixation.
Manganese	Activates decarboxylase, dehydrogenase, and oxidase enzymes; important in photosynthesis, nitrogen metabolism, and nitrogen assimilation.
Nickel	Essential for urease, hydrogenases, and methyl reductase; needed for grain filling, seed viability, iron absorption, and urea and ureide metabolism (to avoid toxic levels of these nitrogen-fixation products in legumes).
Boron	Activates certain dehydrogenase enzymes; facilitates sugar translocation and synthesis of nucleic acids and plant hormones; essential for cell division and development.
Molybdenum	Present in nitrogenase (nitrogen fixation) and nitrate reductase enzymes; essential for nitrogen fixation and nitrogen assimilation.
Chloride	Essential for photosynthesis and enzyme activation. Active in osmosis regulation of plants on saline soils.
Cobalt	Essential for nitrogen fixation; found in vitamin B_{12}.

[12]This pattern is in contrast to most of the macronutrients (except sulfur), which are more easily translocated by the plants and so become most deficient in the older leaves.

(a)

(b)

FIGURE 13.29 A misshapen rose, or *bullhead,* on a boron-deficient tea rose (*a*). A well-formed bloom on the same plant after a solution containing boron was applied to the leaves and young buds (*b*). (Photos courtesy of R. Weil)

13.20 SOURCE OF MICRONUTRIENTS

Inorganic Forms

Scroll down page for details on micronutrients in hydroponics: http://www.hydrofarm.com/content/articles/factors_plant.html

Sources of the nine micronutrients vary markedly from area to area. The wide variability in the content of these elements in soils and suggested contents in a representative soil is shown in Table 13.10.

All of the micronutrients have been found in varying quantities in igneous rocks. Two of them, iron and manganese, have prominent structural positions in primary silicate minerals, such as biotite and hornblende. Others, such as cobalt and zinc, also may occupy structural positions as minor replacements for the major constituents of silicate minerals, including clays.

Anions such as borate and molybdate in soils may undergo adsorption or reactions similar to those of the phosphates. Chlorine, by far the most soluble of the group, is added to soils in considerable quantities each year through rainwater. Its incidental addition to soils in fertilizers and in other ways helps prevent the deficiency of chlorine under field conditions. Chlorine is only very weakly adsorbed to soil colloids.

TABLE 13.10 Major Sources of Nine Micronutrients, Ranges and Representative Contents of These Nutrients in Soils, Along with Their Representative Contents in Harvested Crops

The ratio of soil to crop contents emphasizes the primary need to increase the efficiency of plants in absorbing these nutrients from the soil.

			Representative values		
Element	*Major sources*	*Range, kg/ha/15 cm*	*Soil, kg/ha-15 cm*	*Crop, kg/ha*	*Soil/crop ratio*
Fe	Oxides, sulfides, silicates	20,000–220,000	56,000	2	28,000
Mn	Oxides, silicates, carbonates	45–9,000	2200	0.5	4,400
Zn	Sulfides, carbonates, silicates	25–700	110	0.3	366
Cu	Sulfides, hydroxy carbonates, oxides	4–2,000	45	0.1	450
Ni	Silicates (especially serpentine)	10–2,200	45	0.02	2,250
B	Borosilicates, borates	8–200	22	0.2	110
Mo	Sulfides, oxides, molybdates	0.4–10	5	0.02	250
Cl	Chlorides	15–100	22	2.5	0.9
Co	Silicates	2–90	18	0.02	900

TABLE 13.11 Forms of Micronutrients Dominant in the Soil Solution

Micronutrient	*Dominant soil solution forms*
Iron	Fe^{2+}, $Fe(OH)_2^{+}$, $Fe(OH)^{2+}$, Fe^{3+}
Manganese	Mn^{2+}
Zinc	Zn^{2+}, $Zn(OH)^{+}$
Copper	Cu^{2+}, $Cu(OH)^{+}$
Molybdenum	MoO_4^{2-}, $HMoO_4^{-}$
Boron	H_3BO_3, $H_2BO_3^{-}$
Cobalt	Co^{2+}
Chlorine	Cl^{-}
Nickel	Ni^{2+}, Ni^{3+}

From data in Lindsay (1972).

Organic Forms

Organic matter is an important secondary source of some of the trace elements. Several of them tend to be held as complex combinations by organic (humus) colloids. Copper is especially tightly held by organic matter, so much so that its availability can be very low in organic soils (Histosols). In uncultivated profiles, there is a somewhat greater concentration of micronutrients in the surface soil, much of it presumably in the organic fraction. Correlations between soil organic matter and contents of copper, molybdenum, and zinc have been noted. Although the elements thus held are not always readily available to plants, their release through decomposition is undoubtedly an important fertility factor. Animal manures are a good source of micronutrients, much of it present in organic forms.

Forms in Soil Solution

The dominant forms of micronutrients that occur in the soil solution are listed in Table 13.11. The specific forms present are determined largely by the pH and by soil aeration (i.e., redox potential). Note that the cations are present in the form of either simple cations or hydroxy metal cations. The simple cations tend to be dominant under highly acid conditions. The more complex hydroxy metal cations become more prominent as the soil pH is increased.

Molybdenum is present mainly as MoO_4^{2-}, an anionic form that reacts at low pH in ways similar to those of phosphorus. Although boron also may be present in anionic form at high pH levels, research suggests that undissociated boric acid (H_3BO_3) is the form that is dominant in the soil solution and is absorbed by higher plants.

The cycling of micronutrients through the soil–plant–animal system is illustrated in a generalized way by Figure 13.30. Although not every micronutrient will participate in every pathway shown therein, it can be seen that organic chelates, soil colloids, soil organic matter, and soil minerals all contribute micronutrients to the soil solution and, in turn, to growing plants. As we turn our attention to micronutrient availability, it will be helpful to refer back to Figure 13.30 to see the relationships among the processes involved.

Land disposal of industrial and domestic wastes can affect the forms and quantities of macronutrients and other trace elements found in the soil solution. The macronutrient contents of these wastes (particularly those of nitrogen and phosphorus) have long been appreciated, but only recently have we recognized the significance of their trace element contents. Small quantities of trace elements applied in these wastes can help alleviate nutrient deficiencies. However, repeated land applications of large quantities of wastes have increased the soil levels of some trace elements to their toxicity ranges. These levels are adversely affecting not only the plants and the animals that consume them, but also the quality of the water running off and through the soil.

13.21 FACTORS AFFECTING THE AVAILABILITY OF MICRONUTRIENTS

Micronutrient deficiencies are most apt to occur where the total contents of the nutrients in the soil are low, or where soil conditions constrain the availability of whatever nutrients are present. Highly leached, acid, sandy soils are commonly low in micronutrients,

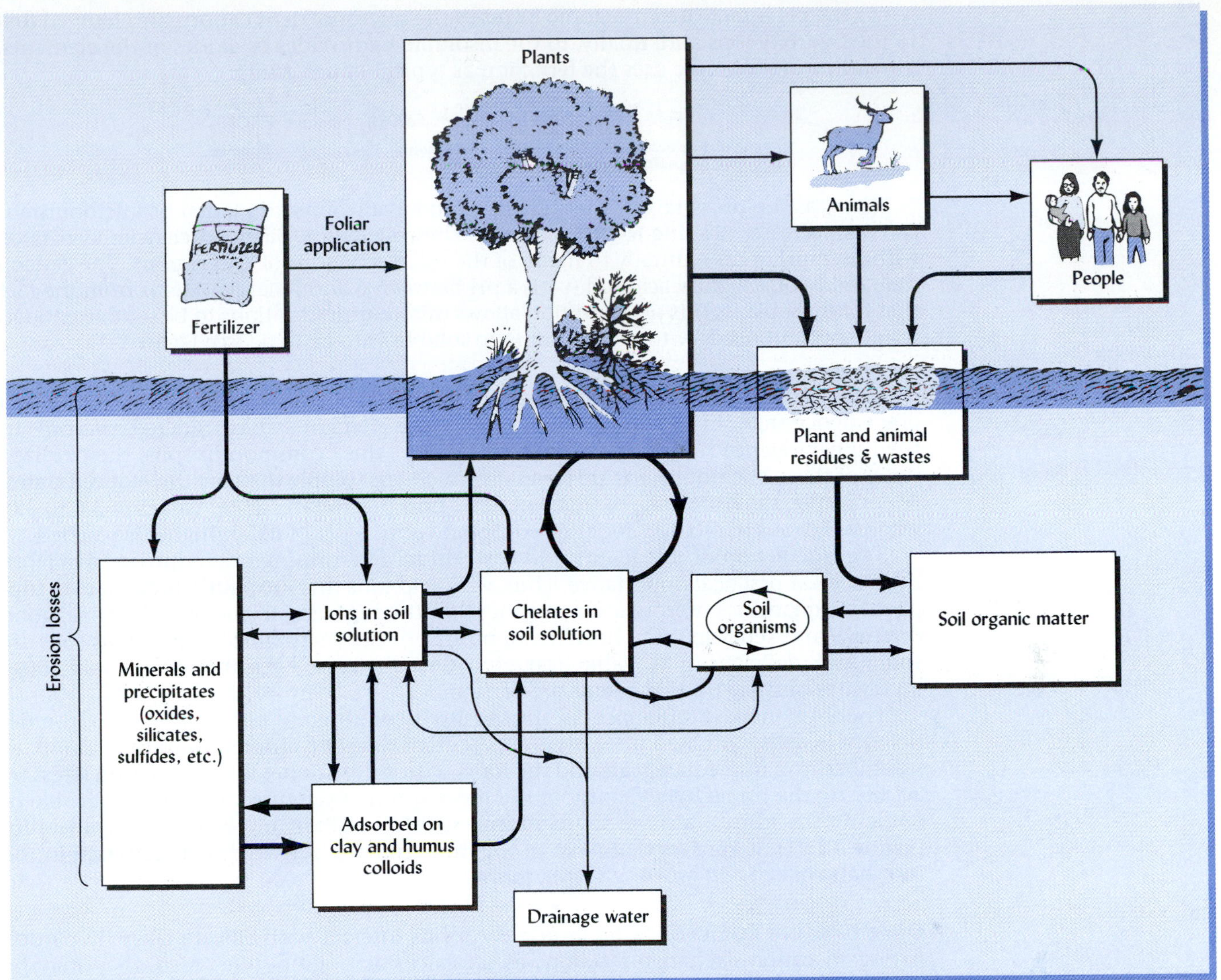

FIGURE 13.30 Cycling and transformations of micronutrients in the soil–plant–animal system. Although all micronutrients may not follow each of the pathways shown, most are involved in the major components of the cycle. The formation of chelates, which keep most of these elements in soluble forms, is a unique feature of this cycle.

as are most organic soils. Likewise, micronutrient deficiencies are most apt to occur in intensively cropped areas where the harvested crops are removed, and where macronutrients are being replenished through added fertilizers. Also, eroded soils that have lost much of their upper horizons and associated organic matter are more apt to be low in most micronutrients. In all cases, however, the availability of micronutrients is influenced by several general soil conditions which we will now consider.

Micronutrient Cation Availability

Zinc and Iron in Nebraska: http://www.ianr.unl.edu/pubs/soil/g596.htm

The availability of the six micronutrient cations is affected in a similar way by several soil characteristics.

Soil pH. The micronutrient cations are most soluble and available under acid conditions. In very acid soils, there is a relative abundance of the ions of iron, manganese, zinc, copper, and nickel. In some cases, concentrations are sufficiently high to be toxic to common plants.

As the pH is increased, the ionic forms of the micronutrient cations are changed first to the hydroxy ions and, finally, to the insoluble hydroxides or oxides of the elements. The following example uses the ferric ion as typical of the group:

$$\underset{\substack{\text{Simple cation}\\ \text{(soluble)}}}{Fe^{3+}} \xrightarrow{OH^-} Fe(OH)^{2+} \xrightarrow{OH^-} \underset{\substack{\text{Hydroxy metal cations}\\ \text{(soluble)}}}{Fe(OH)_2{}^+} \xrightarrow{OH^-} \underset{\substack{\text{Hydroxide}\\ \text{(insoluble)}}}{Fe(OH)_3} \qquad (13.7)$$

Thus, as the pH is raised, the solubility (and availability to plants) of micronutrient cations decreases, leading to deficiencies of these elements. Such deficiencies associated with high pH occur naturally in many of the calcareous soils of arid regions. The general desirability of a slightly acid soil (with a pH between 6 and 7) largely stems from the fact that for most plants this pH condition allows micronutrient cations to be soluble enough to satisfy plant needs without becoming so soluble as to be toxic (see Figure 9.11).

OXIDATION STATE AND pH. Iron, manganese, nickel, and copper occur in soils in more than one valence state. In the lower valence states, the elements are considered *reduced;*[13] in the higher valence state they are *oxidized.* At pH values common in soils, the oxidized states of these micronutrients are generally much less soluble than are the reduced states. For example, the hydroxide of trivalent ferric iron precipitates at pH values of 3.0 to 4.0, whereas ferrous hydroxide does not precipitate until a pH of 6.0 or higher is reached.

The interaction of soil acidity and aeration in determining micronutrient availability is of great practical importance. Thus, very acid soils that are poorly drained may supply toxic quantities of iron and manganese. At the high end of the soil pH range, good drainage and aeration often have the opposite effect. Well-oxidized calcareous soils are sometimes deficient in available iron, zinc, or manganese even though adequate total quantities of these trace elements are present.

There are marked differences in the sensitivity of different plant varieties to iron deficiency in soils with high pH. This is apparently caused by differences in their ability to solubilize iron immediately around the roots. Efficient varieties respond to iron stress by acidifying the immediate vicinity of the roots and by excreting compounds capable of reducing the iron to a more soluble form, with a resultant increase in its availability (Figure 13.31). It appears that most of the iron-reducing activity is concentrated in the root hairs of actively growing young roots.

OTHER INORGANIC REACTIONS. Micronutrient cations interact with silicate clays by participating in cation exchange reactions as do calcium or aluminum. Also, they may be tightly bound or fixed to certain silicate clays, especially the 2:1 type. Zinc, manganese, cobalt, and iron ions sometimes occur in the crystal structure of these clays. Depending on conditions, they may be released from the clays or fixed by them in a manner similar

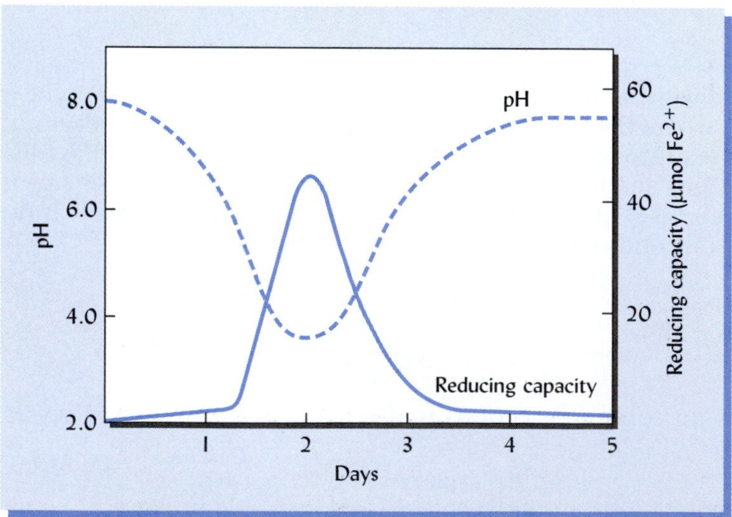

FIGURE 13.31 Response of one variety of sunflowers to iron deficiency. When the plant became stressed owing to iron deficiency, plant exudates lowered the pH and increased the reducing capacity immediately around the roots. Iron is solubilized and taken up by the plant, the stress is alleviated, and conditions return to normal. [From Marschner et al. (1974) as reported by Olsen et al. (1981); used with permission of *American Scientist*]

[13]For a review of microbial reduction of certain trace elements, see Lovely (1996).

to that by which potassium is fixed (see Section 13.15). The fixation may cause serious deficiency in the case of cobalt, and sometimes zinc, because these elements are present in soil in such small quantities (see Table 13.10).

The application of large quantities of phosphate fertilizers can adversely affect the supply of some of the micronutrients. The uptake of both iron and zinc may be reduced in the presence of excess phosphates. For environmental quality reasons, as well as from a practical standpoint, phosphate fertilizers should be used in only those quantities required for good plant growth.

Organic Matter. Organic matter, organic residues, and manure applications affect the immediate and potential availability of micronutrient cations. Some organic compounds react with these cations to form water-insoluble complexes that can protect the nutrients from interactions with mineral particles that can bind them in even more insoluble forms. Other complexes provide slowly available nutrients as they undergo microbial breakdown. Organic complexes that enhance micronutrient availability are considered later in this section.

Deficiencies of copper and, to a lesser extent, manganese are often found on poorly drained soils that are high in organic matter (e.g., peats and marshes). Zinc is also retained by organic matter, but deficiencies stemming from this retention are not common.

Micronutrient-enriched organic products have been used as a nutrient source on soils deficient in available trace elements. For example, composts of iron-enriched organic materials, such as forest by-products, peat, animal manures, and plant residues, have been found to be effective on iron-deficient soils.

Role of Mycorrhizae. A symbiosis between most higher plants and certain soil fungi produces mycorrhizae (fungus roots), which are far more efficient than normal plant roots in several respects. The nature of the mycorrhizal symbiosis and its importance in phosphorus nutrition were described in Sections 10.9 and 13.3, but it is worth mentioning here that mycorrhizae have also been shown to increase plant uptake of micronutrients (Table 10.6). Crop rotations and other practices that encourage a diversity of mycorrhizal fungi may thereby improve micronutrient nutrition of plants.

Surprisingly, mycorrhizae also appear to protect plants from excessive uptake of micronutrients and other trace elements where these elements are present in potentially toxic concentrations. Seedlings of such trees as birch, pine, and spruce are able to grow well on sites contaminated with high levels of zinc, copper, nickel, and aluminum only if their roots are sheathed by ectomycorrhizae. The mycorrhizae apparently help exclude these metallic cations from the root stele and prevent long-distance transport of metal cations within the plant.

Organic Compounds as Chelates

Brief chemistry of chelates and chelating agents:
http://scifun.chem.wisc.edu/CHEMWEEK/chelates/chelates.html

The cationic micronutrients react with certain organic molecules to form the organometallic complexes called **chelates** (from the Greek *chele,* claw). In these complexes, a given micronutrient is surrounded by an envelope of C, H, O, and other atoms that protect it from reactions with the soil particles (see Figure 13.32). These organic molecules may be synthesized by plant roots and released to the surrounding soil, may be present in the soil humus, or may be commercial synthetic compounds added to the soil to increase micronutrient availability. Examples of iron chelate ring structures are shown in Figure 13.32.

Advantages of chelated micronutrients:
http://www.agrisol.co.nz/chelate.htm

The effect of chelation can be illustrated with iron. In the absence of chelation, when an inorganic iron salt such as ferric sulfate is added to a calcareous soil, most of the iron is quickly rendered unavailable by reaction with hydroxide, as follows:

$$\underset{\text{(available)}}{Fe^{3+} + 3OH^-} \rightleftharpoons \underset{\text{(unavailable)}}{FeOOH + H_2O} \qquad (13.8)$$

In contrast, if the iron is chelated, it largely remains in the chelate form, which is available for uptake by plants. In this reaction, the available iron chelate reactant is favored:

$$\underset{\text{(available)}}{\text{Fe chelate} + 3OH^-} \rightleftharpoons \underset{\text{(unavailable)}}{FeOOH + \text{chelate}^{3-}} + H_2O \qquad (13.9)$$

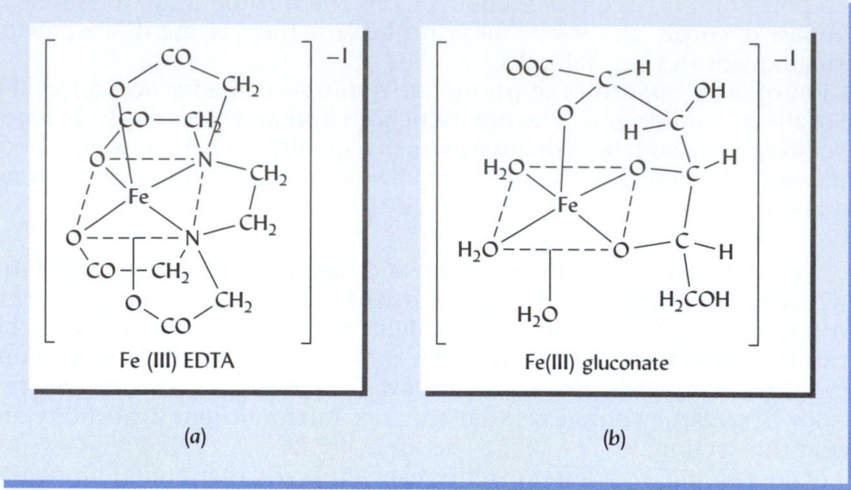

FIGURE 13.32 Structural formula for two common iron chelates, ferric ethylenediaminetetraacetate (Fe-EDTA) (*a*) and ferric gluconate (*b*). In both chelates, the iron is protected, and yet can be used by plants. [Diagrams from Clemens et al. (1990); reprinted by permission of Kluwer Academic Publishers]

The mechanism by which micronutrients from chelates are absorbed by plants is different for different plants. Many dicots appear to remove the metallic cation from the chelate at the root surface, chemically reducing (in the case of iron) and taking up the cation while releasing the organic chelating agent in the soil solution. Roots of certain grasses have been shown to take in the entire chelate–metal complex, reducing and removing the metallic cation inside the root cell, then releasing the organic chelate back to the soil solution (Figure 13.33). In both cases, it appears that the primary role of the chelate is to allow metallic cations to remain in solution so they can diffuse through the soil to the root. Once the micronutrient cations are inside the plant, other organic chelates (such as citrates) may be carriers of these cations to different parts of the plant.

The chelates of different micronutrients vary in their stability and therefore as sources of micronutrients. If a mixture of Fe^{2+}, Mn^{2+}, and Zn^{2+} occurs in the soil solution and the Fe^{2+} forms the most stable chelate, it will be most protected from the reaction with the soil and will remain most available for plant uptake. Knowledge of the stabilities of different commercial chelates helps in determining which to use to alleviate the deficiency of a given nutrient. In some cases, mixtures of such chelates are provided since more than one nutrient may be in deficiency status. Commercial chelates are used substantially in industrialized countries, particularly to ameliorate micronutrient deficiencies of fruit trees and ornamentals.

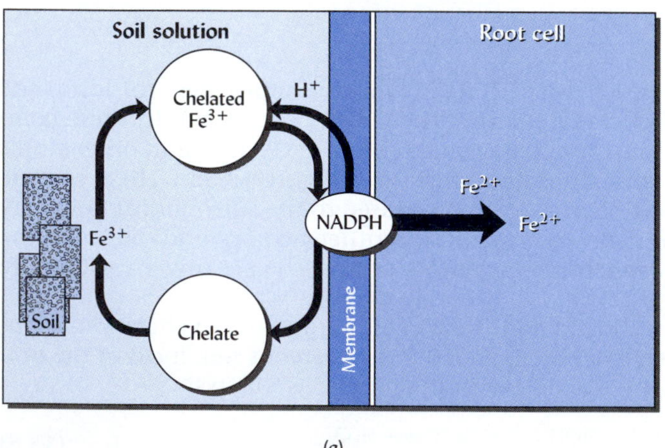

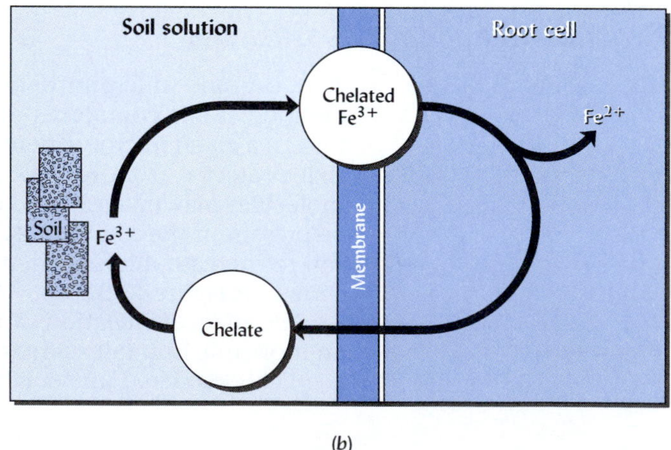

FIGURE 13.33 Two ways in which plants utilize micronutrients held in chelated form. (*a*) Dicotyledonous plants such as cucumber and peanuts produce strong reducing agents (NADPH) that reduce iron at the outer surface of the root membrane. They then take in only the reduced iron, leaving the organic chelate in the soil solution where it can complex another iron atom. (*b*) Grass plants such as wheat or corn apparently take the entire chelate–metal complex into their root cells. They then remove the iron, reduce it, and return the chelate to the soil solution.

Micronutrient Anion Availability

Managing boron in a temperate humid region: http://www.extension.umn.edu/distribution/cropsystems/DC0723.html

Unlike the cations needed in trace quantities by plants, the anion micronutrients seem to have relatively little in common with each other. Chlorine, molybdenum, and boron are quite different chemically, so little similarity would be expected in their reaction in soils.

CHLORINE. Chlorine is absorbed in larger quantities by most crop plants than any of the micronutrients except iron. Most of the chlorine in soils is in the form of the chloride ion, which leaches rather freely from humid-region soils. Except where toxic quantities of chlorine are found in saline soils, there are no known natural soil conditions that reduce the availability and use of this element. Accretions of chlorine from the atmosphere, along with those from fertilizer salts such as potassium chloride, are sufficient to meet most crop needs.

BORON. Boron is one of the most commonly deficient of all the micronutrients. The availability of boron is related to the soil pH, this element being most available in acid soils. Soluble boron is present in soils mostly as boric acid $[B(OH)_3]$ or as $B(OH)_4^-$. These compounds can exchange with the OH groups on the edges and surfaces of variable charge clays such as kalonite, and especially with the oxides of iron and aluminum. The boron so adsorbed is quite tightly bound, especially between pH 7 and 9, the range of lowest availability of this element (see Figure 9.12).

Since boron is adsorbed by humus, organic matter serves as a major reservoir of this element in many soils, and exerts considerable control over its availability. Boron availability is impaired by long dry spells that may reduce the rate of mineralization of the organically held boron. Boron deficiencies are also common in calcareous Aridisols and in other soils with high pH.

MOLYBDENUM. Soil pH is the most important factor influencing the availability and plant uptake of molybdenum. The following equations show the forms of this element present at low and high soil pH:

$$H_2MoO_4 \underset{+H^+}{\overset{+OH^-}{\rightleftharpoons}} HMoO_4^- + H_2O \qquad (13.10)$$

$$HMoO_4^- \underset{+H^+}{\overset{+OH^-}{\rightleftharpoons}} MoO_4^{2-} + H_2O \qquad (13.11)$$

At low pH values, the molybdenum is adsorbed by silicate clays and, more especially, by oxides of iron and aluminum through ligand exchange with hydroxide ions on the surface of the collodial particles.

The liming of acid soils will usually increase the availability of molybdenum. The effect is so striking that some researchers, especially those in Australia and New Zealand, argue that the primary reason for liming very acid soils is to supply molybdenum. Furthermore, in some instances 30 g or so of molybdenum added to acid soils has given about the same increase in the yield of legumes as has the application of several megagrams of lime.

SELENIUM. Although not a plant nutrient, selenium is important to animal and human health. Selenium is found in nature in four major solid forms—selenate (SeO_4^{2-}), selenite (SeO_3^{2-}), elemental Se (Se^0), and selenide (Se^{2-}). It also exists in volatile compounds, such as dimethyl selenide and dimethyl diselenide, which can be released as gases to the atmosphere.

Under acid conditions (pH 4.5 to 6.5), *selenite* is usually dominant. This form is only slowly available, but may be added to alleviate selenium deficiencies, as it will not likely induce selenium toxicities. At higher pH values (7.5 to 8.5) and in well-aerated soils, the more soluble *selenate* dominates and often gives rise to such toxicities.

Under anaerobic conditions, soil microorganisms can reduce the selenate and selenite to solid selenide and elemental selenium forms, both of which are quite insoluble. Advantage can be taken of such reduction to remove toxic levels of soluble selenium from soil and water, a promising process of bioremediation (see Box 13.3 and Figure 13.34). This microbial reduction also produces relatively nonphytoxic volatile selenium compounds

BOX 13.3 SELENIUM—BOUND AND VAPORIZED

Selenium is an essential element for humans and animals, but is not generally considered to be essential for plant processes. However, this element may exist in some soils at levels that are toxic, not only to plants, but to animals that consume these plants. For example, it is thought to be responsible for "blind staggers" in ruminants and for embryonic deformities in waterfowl. Natural selenium toxicities are found notably in soils developed on marine sedimentary parent material that is high in selenium. But human activities such as irrigation and mining can significantly increase the levels of selenium in soils, sometimes to toxic levels.

Research has shown that selenium exists in several forms in soils, and that the particular forms present determine the degree of toxicity much more than does the total amount of selenium in the soil. Similar to sulfur, selenium is found in four oxidation states in soils: 2−, 0, 4+, and 6+. The particular compounds present at any one time are determined largely by factors such as redox potential, pH, organic ligands present, and the activities of soil microorganisms.

Figure 13.34 illustrates the chemical forms present and shows the reactions and solubility of selenium in soils.

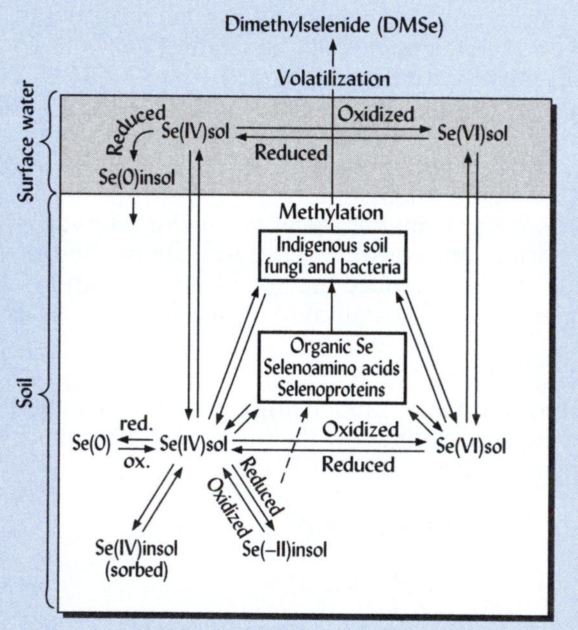

FIGURE 13.34 Selenium transformations in wetland soils that produce both highly insoluble compounds and soluble compounds that are readily taken up by plants, often at such high levels that they are toxic to the animals that consume them.

Irrigation waters carry two relatively soluble forms of selenium, six-valent selenates [Se(VI) or SeO_4^{2-}] and four-valent selenites [Se(IV) or SeO_3^{2-}]. When the selenium first moves into the soil, some of it is reduced quickly to the very insoluble elemental selenium (Se), which is largely unavailable to plants and is nontoxic. Further transformations take place as both soluble forms move downward into the soil. Reducing conditions shift the equilibrium between the two forms toward the selenites (IV), which tend to be tightly sorbed by iron oxides. Further microbially induced reduction leads to the formation of not only elemental selenium [Se(O)] but to selenides [Se(II) or Se^{2-}], both of which are quite insoluble. Such reduced conditions encourage the formation of insoluble forms, thereby reducing the toxicity of the selenium present.

As the microbes and higher plants metabolize selenium, it is assimilated in organic forms such as selenoamino acids and selenoproteins, most of which are also quite insoluble. But some soluble organic forms of selenium are absorbed by plants and microorganisms, along with inorganic forms, particularly the more soluble selenates (VI) present under oxidized conditions. Some plant species such as rice and members of the *Brassica* family (generally in association with soil micro organisms) are able to produce volatile selenium compounds that are released to the atmosphere. The exact mechanisms involved in the formation of the volatile compounds are not as yet understood, but the diagram shows how fungi and bacteria can attach methyl groups (methylation) to organoselenium compounds to form volatile gases such as dimethylselenide (DMSe) that can be released into the atmosphere. DMSe is 700 times less toxic than the selenates, and can be dispersed into the atmosphere without any environmental damage. The process seems to work best in moist but not flooded soils, which also contain organic residues and animal manures as sources of metabolic energy.

These processes that either change selenium to insoluble forms or release the element into the atmosphere as gases seem to have the greatest potential of reducing selenium toxicity. They are being pursued vigorously to allow continued high crop productivity without damaging the environment.

Concepts for the diagram are from Hayes and Traina (1998) and Frankenberger and Losi (1995). An excellent review of research on phytovolatilization of selenium is found in Zayed et al. (2000).

(e.g., dimethyl selenide) that can be released to the atmosphere.[14] The production of such volatile compounds is also being considered as a means of reducing toxic levels of selenium in water and soils.

A number of other trace elements found in the anionic form have been reported to be essential for either plants or animals, but we will consider only arsenic and chromium because of their toxic effects on humans and other animals.

Arsenic. Some studies have shown arsenic to be essential for both plant and animal life, but its negative effect, particularly on humans, is most widely recognized. Its presence in soils, groundwater, and well water is of concern to people around the world, but especially in Bangladesh, India, China, Chile, and Slovakia. In Bangladesh, for example, more than 20 million of its 126 million people are believed to be drinking arsenic-contaminated water. The well water in the United States is generally safe, but the arsenic content of some wells exceeds the current maximum contaminant level (MCL), which may become even more stringent in the future (Figure 13.35).

Arsenic is associated with the soil in two major forms, arsenite (AsO_3^{3-}, or three-valent AsIII) and arsenate (AsO_4^{3-}, or five-valent AsV). Arsenites (AsIII) are generally more mobile, and move more easily into groundwater, the source of drinking water in many parts of the world. For this reason, wet and reduction-prone conditions are to be avoided to minimize dissolution and movement of the most toxic forms of arsenic.

Scientists are evolving methods for the remediation of arsenic-contaminated waters. For example, they are trying to use hydrous oxides of iron as sorbing agents for

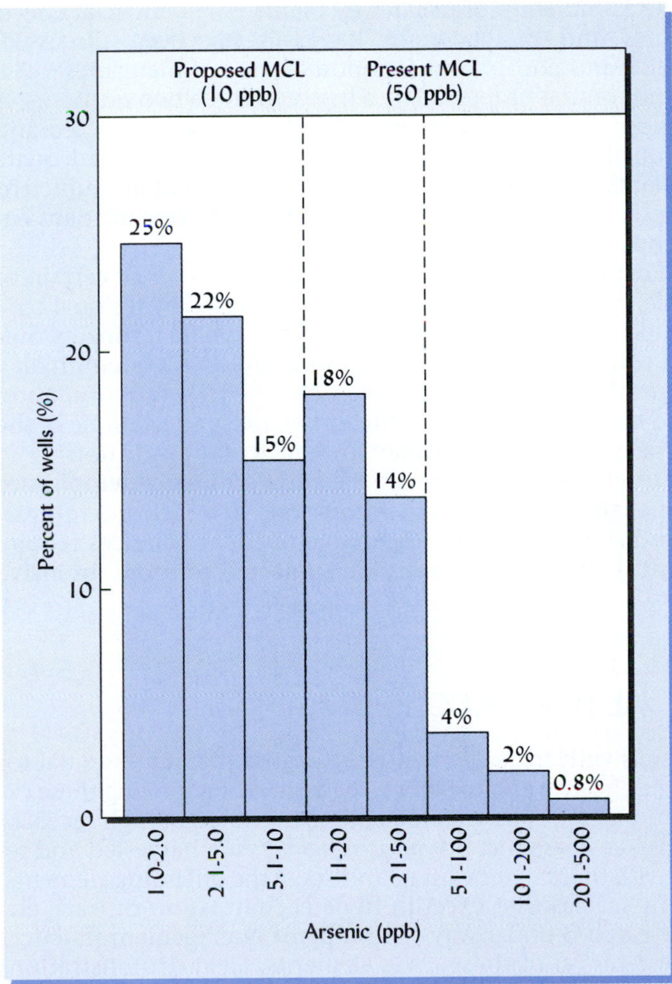

FIGURE 13.35 The percentage of 247 Arizona wells with increasing levels of arsenic in the drinking water. Note that some 7% of the wells had arsenic levels that exceed the current maximum contaminant level (MCL) of 50 ppb established by the U.S. Environmental Protection Agency (EPA). More than 38% of the wells would not meet a new MCL standard that was proposed by the EPA in 2001 (10 ppb). The health risks of arsenic in soils and water are well illustrated by these data. [Data from the U.S. Geological Survey National Water Quality Assessment; see Spencer (2000)]

[14]Haygarth et al. (1995) attributed the source of 30 to 50% of the selenium taken up by pasture plants in one experiment to the atmosphere.

Chemical transformations of chromium in soils. Relevance to mobility, bioavailability and remediation: http://www.chromium-asoc.com/publications/crfile8feb02.htm

the arsenic in drinking water. They have discovered that certain plants are *hyperaccumulators* of arsenic, an example being a fern that is known to have accumulated as much as 2% arsenic in research plots in Florida. By growing and harvesting such plants, it may be possible to reduce the level of soluble arsenic in the soils and the groundwater.

Chromium. In trace amounts, chromium appears to be essential for human life, but, like arsenic, it is a carcinogen when absorbed in larger doses. When chromium-containing wastes from industrial plants making products such as steel are applied to the land, the soil as well as groundwater and drinking water can be polluted.

Like arsenic, chromium is found in soils in two major oxidation states: a trivalent form [Cr(III)] and a hexavalent form [Cr(VI)]. In contrast to arsenic, however, the more highly oxidized state [Cr(VI)] is the more soluble and mobile, especially if the pH values are above about 5.5.

To remediate Cr(VI)-contaminated soil and water, the chromium must be reduced to the Cr(III) state. This process can be enhanced if the soil is kept wet to reduce the level of oxygen, and if an abundance of organic material is applied.

13.22 NEED FOR NUTRIENT BALANCE

Guidelines for micronutrient fertilizers and their use: http://ohioline.osu.edu/hyg-fact/1000/1252.html

Nutrient balance among the trace elements is as essential as, but even more difficult to maintain than, macronutrient balance. Some of the plant enzyme systems that depend on micronutrients require more than one element. For example, both manganese and molybdenum are needed for the assimilation of nitrates by plants. The beneficial effects of combinations of phosphates and molybdenum have already been discussed. Apparently, some plants need zinc and phosphorus for optimum use of manganese. The use of boron and calcium depends on the proper balance between these two nutrients. A similar relationship exists between potassium and copper and between potassium and iron in the production of good-quality potatoes. Copper utilization is favored by adequate manganese, which in some plants is assimilated only if zinc is present in sufficient amounts. These nutrient interactions are examples of **synergism**, where one nutrient enhances the utilization of a second nutrient.

Other *antagonistic* effects may be used effectively in reducing toxicities of certain of the micronutrients. For example, copper toxicity of citrus groves caused by residual copper from fungicidal sprays may be reduced by adding iron and phosphate fertilizers. Sulfur additions to calcareous soils containing toxic quantities of soluble molybdenum may reduce the availability, and hence the toxicity, of molybdenum. The hyperaccumulation of phosphorus, manganese, and magnesium in zinc-deficient plants is an example of the complex interactions among essential elements as they influence plant nutrition.

These examples of nutrient interactions, both beneficial and detrimental, emphasize the highly complicated nature of the biological transformations in which micronutrients are involved. The total land area on which unfavorable nutrient balances require special micronutrient treatment is increasing as soils are subjected to more intensive cropping methods.

13.23 TRACE ELEMENT CLEANUP AND METAL HYPERACCUMULATORS[15]

Metal-scavenging plants to cleanse the soil: http://www.soils.wisc.edu/~barak/soilscience326/agres.htm

Human activities may pollute soils with trace elements, both those that are essential for plants and those that are not. One effective and inexpensive method of removing these excess chemicals is to grow certain plant species that are able to accumulate high levels of trace elements in their tissues. When these metal *hyperaccumulators* are harvested and removed from the land, they take with them significant quantities of the offending elements.

Significant numbers of plant species that excel in hyperaccumulation of trace elements have been identified. Research is under way to find plant management practices that will maximize the rates of removal of these toxic elements. Field demonstrations have already shown their potential.

[15]For a recent review of progress in using biological means of removing pollutants from soil and water, see Terry and Banuelos (2000).

13.24 SOIL MANAGEMENT AND MICRONUTRIENT NEEDS

In seeking the cause of plant abnormalities, one should keep in mind the conditions under which micronutrient deficiencies or toxicities are likely to occur. Sandy soils, mucks, and soils having very high or very low pH values are prone to micronutrient deficiencies. Areas of intensive cropping and heavy macronutrient fertilization may be deficient in the micronutrients.

The optimum management of these soils may differ from one situation to another, but the following suggestions are widely applicable.

1. Recognize and evaluate the major sources of the micronutrients and the general reactions that make them available (Figure 13.36). Give highest priority to those sources, such as organic residues, over which some control can be exerted.
2. Avoid extremes in soil pH to minimize deficiencies and toxicities of the different micronutrients. Generally, pH values between 5.5 and 7 are most appropriate.
3. Avoid extremes in soil drainage status to minimize these micronutrient constraints. If controlling the drainage status is not practicable, turn to pH modification or choice of plants to alleviate the constraint.
4. Choose plant species or varieties known to tolerate excesses or deficiencies of the micronutrients (see Table 13.12).
5. Utilize organic residues or farm manures to the extent feasible to overcome micronutrient deficiencies. Where these sources are inadequate, turn to commercial fertilizers.
6. Use great care in selecting and applying fertilizer sources of micronutrients to avert risks of overdose and/or upsetting nutrient balances. (See Table 14.9 for examples of micronutrient fertilizer products.)

Economic responses to micronutrients are becoming more widespread as intensity of plant production increases. For example, responses of fruits, vegetables, and field crops to zinc and iron applications are common in areas with neutral to alkaline soils. Even on acid soils, deficiencies of these elements are increasingly encountered. Molybdenum, which has been used for some time for forage crops and for cauliflower and other vegetables, has received attention in recent years for forest nurseries and soybeans, especially on acid soils. These examples, along with those from muck areas and sandy soils where micronutrients have been used for decades, illustrate the need for these elements if optimum yields are to be maintained. Limited research suggests that there may be large areas in developing countries with micronutrient deficiencies or toxicities. As macronutrient deficiencies are addressed and yields are increased, more micronutrient deficiencies will undoubtedly come to the fore.

A soil will often produce a micronutrient deficiency in some plants but not in others. Plant species, and varieties within species, differ widely in their susceptibility to

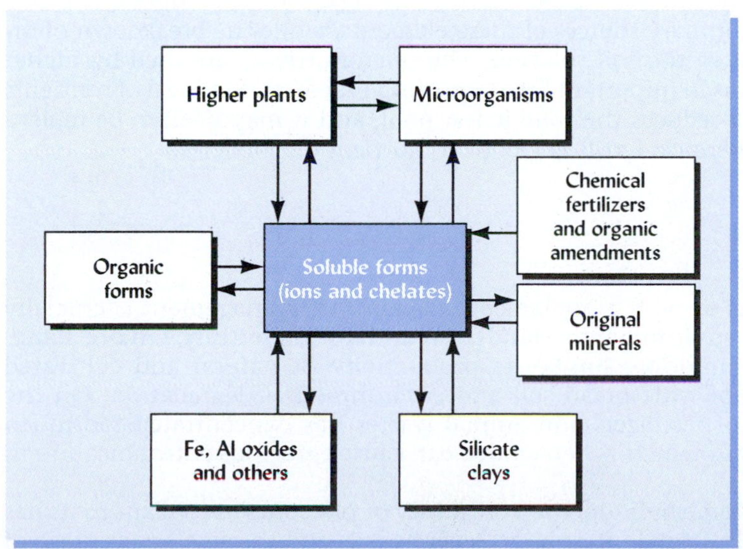

FIGURE 13.36 Diagram of soil sources of soluble forms of micronutrients and their utilization by plants and microorganisms.

TABLE 13.12 Plants Known to Be Especially Susceptible or Tolerant to, and Soil Conditions Conducive to, Micronutrient Deficiencies

Plants which are most susceptible to deficiency of a micronutrient often have a relatively high requirement for that nutrient and may be relatively tolerant to levels of that nutrient that would be high enough to cause toxicity to other plants.

Micronutrient	*Common range in rates recommended for soil application[a], kg/ha*	*Plants most commonly deficient (high requirement or low efficiency of uptake)*	*Plants rarely deficient (low requirement or high efficiency of uptake)*	*Soil conditions commonly associated with deficiency*
Iron	0.5–10.0	Blueberries, azaleas, roses, holly, grapes, nut trees, maple, bean, sorghum, oaks	Wheat, alfalfa, sunflower, cotton	Calcareous, high pH, waterlogged alkaline soils
Manganese	2–20	Peas, oats, apple, sugar beet, raspberry, citrus	Cotton, soybean, rice, wheat	Calcareous, high pH, drained wetlands, low organic matter, sandy soils
Zinc	0.5–20	Corn, onion, pines, soybeans, beans, pecans, rice, peach, grapes	Carrots, asparagus, safflower, peas, oats, crucifers, grasses	Calcareous soils, acid, sandy soils, high phosphorus
Copper	0.5–15	Wheat, corn, onions, citrus, lettuce, carrots	Beans, potato, peas, pasture grasses, pines	Histosols, very acid, sandy soils
Boron	0.5–5	Alfalfa, cauliflower, celery, grapes, conifers, apples, peanut, beets, rapeseed, pines	Barley, corn, onion, turf grass, blueberry, potato, soybean	Low organic matter, acid, andy soils, recently limed soils, droughty soils, soils high in 2:1 clays
Molybdenum	0.05–0.5	Alfalfa, sweet clover, crucifers (broccoli, cabbage, etc.), citrus, most legumes	Most grasses	Acid sandy soils, highly weathered soils with amorphous Fe and Al

[a]The lower end of each range is typical for banded application; the higher end is typical for broadcast applications.

micronutrient deficiency or toxicity. Table 13.12 provides examples of plant species known to be particularly susceptible or tolerant to several micronutrient deficiencies.

Both soil management and plant breeding can improve the nutritional value of food crops to help alleviate suffering from human micronutrient deficiencies. Soil and plant scientists are collaborating to enhance the micronutrient nutrition of hundreds of millions of people who can afford to eat only a meager diet consisting mainly of staple grains with few vegetables or fruits that could provide ample micronutrients. For example, scientists are using traditional and transgenic techniques to reduce the occurrence of iron deficiency anemia among the poor of Asia by increasing the content of readily assimilated iron in certain high-yielding cultivars of rice.

Micronutrient Availability

Major sources of micronutrients and the general reactions that make them available to higher plants and microorganisms are summarized in Figure 13.36. Original and secondary minerals are the primary sources of these elements, while the breakdown of organic forms releases ions to the soil solution. The micronutrients are used by higher plants and microorganisms in important life-supporting processes. Removal of nutrients in crop or timber harvest reduces the soluble ion pool, and it may need to be replenished with manures or chemical fertilizers to avoid nutrient deficiencies.

13.25 CONCLUSION

Soil phosphorus presents us with a double-edged sword: Its management is crucially important from both an environmental standpoint and for soil fertility. On one hand, too little phosphorus commonly limits the productivity of natural and cultivated plants and is the cause of widespread soil and environmental degradation. On the other hand, excess use of fertilizers and animal wastes has concentrated too much phosphorous in some soils, which, when eroded can cause egregious eutrophication in surface waters.

Except in cases of extreme buildup, the availability of phosphorus to plant roots has a double constraint: the low total phosphorus level in soils and the small percentage of this level that is present in available forms. Furthermore, even when soluble phosphates are added to soils, they are quickly fixed into insoluble forms that in time become quite unavailable to growing plants. In acid soils, the phosphorus is fixed primarily by iron, aluminum, and manganese; in alkaline soils, by calcium and magnesium. This fixation greatly reduces the efficiency of phosphate fertilizers, with little of the added phosphorus being taken up by plants.

Potassium is generally abundant in soils, but it, too, is present mostly in forms that are quite unavailable for plant absorption. Fortunately, however, some soils contain considerable nonexchangeable but slowly available forms of this element. Over time this potassium can be released to exchangeable and soil solution forms that can be quickly absorbed by plant roots. This is fortunate, because the plant requirements for potassium are high—5 to 10 times that of phosphorus and similar to that of nitrogen.

Micronutrients are becoming increasingly important to world agriculture as crop removal of these essential elements increases. Deficiencies of these elements are due not only to low contents of these elements in soils but more often to their unavailability to growing plants. They are adsorbed by inorganic constituents, such as Fe, Al oxides, and form complexes with organic matter, some of which are only sparingly available to plants. Other such organic complexes, known as chelates, protect some of the micronutrient cations from inorganic adsorption and make them available for plant uptake.

Toxicities of micronutrients and other trace elements are also quite common. They retard both plant and animal growth, and, through groundwater pollution, can adversely affect human health. Removing excesses of these elements from soil and water, or rendering them unavailable for plant uptake, is one of the challenges facing soil and plant scientists. Soil management practices to overcome deficiencies and toxicities of trace elements are being sought by soil and plant scientists.

STUDY QUESTIONS

1. You have learned that nitrogen, potassium, and phosphorus are all "fixed" in the soil. Compare the processes of these fixations and the benefits and constraints they each provide.
2. What is meant by *eutrophication,* and how is it influenced by farm practices involving phosphorus?
3. In the spring a certain surface soil showed the following soil test: soil solution K = 20 kg/ha; exchangeable K = 200 kg/ha. After two crops of alfalfa hay that contained 250 kg/ha of potassium were harvested and removed, a second soil test showed soil solution K = 15 kg/ha and exchangeable K = 150 kg/ha. Explain why there was not a greater reduction in soil solution and exchangeable K levels.
4. What is the effect of soil pH on the availability of phosphorus, and what are the unavailable forms at the different pH levels?
5. What is *luxury consumption* of plant nutrients, and what are its advantages and disadvantages?
6. How does phosphorus that forms relatively insoluble inorganic compounds in soils find its way into streams and other waterways?
7. During a year's time some 250 kg nitrogen and only 30 g molybdenum have been taken up by the trees growing on a hectare of land. Would you therefore conclude that the nitrogen was more essential for the tree growth? Explain.
8. Soybeans growing on a recently limed soil show evidence of a deficiency of a nutrient, thought by some to be molybdenum. Do you agree with this diagnosis? If not, what is your explanation?
9. What are *chelates,* how do they function, and what are their sources?
10. The addition of only 1 kg/ha of a micronutrient to an acid soil on which lime-loving cauliflower was being grown gave considerable growth response. Which of the micronutrients would it likely have been? Explain.
11. Two Aridisols, both at pH 8, were developed from the same parent material, one having restricted drainage, the other being well drained. Plants growing on the well-drained soils showed iron-deficiency symptoms; those on the less-well-drained soil did not. What is the likely explanation for this?

REFERENCES

Baker, A. J. M., et al. 2000. "Metal hyperaccumulator plants: A review of the ecology and physiology of a biological resource for phytoremediation of metal-polluted soils," in N. Terry and G. Banuelos (eds.), *Photoremediation of Contaminated Soil and Water.* (New York: Lewis Publishers), ch. 5.

Brady, N. C. 1974. *The Nature and Properties of Soils,* 8th ed. (New York: Macmillan).

Brown, P. H., R. M. Welch, and E. E. Cary. 1987. "Nickel: A micronutrient essential for higher plants," *Plant Physiol.* **85**:801–803.

Buresh, R. J., P. C. Smithson, and D. T. Hellums. 1997. "Building soil phosphorus capital in Africa," in R. J. Buresh, P. A. Sanchez, and F. Calhoun (eds.), *Replenishing Soil Fertility in Africa.* SSSA Special Publication no. 51. (Madison, Wis.: Soil Sci. Soc. Amer.).

Clemens, D. F., B. M. Whitehurst, and G. B. Whitehurst. 1990. "Chelates in agriculture," *Fertilizer Research* **25**:127–131.

Cogger, C., and J. M. Duxbury. 1984. "Factors affecting phosphorus loss from cultivated organic soils," *J. Environ. Qual.* **13**:111–114.

David, M. B., and L. E. Gentry. 2000. "Anthropogenic inputs of nitrogen and phosphorus and riverine export for Illinois, USA," *J. Environ. Qual.* **29**:494–508.

Fares, F., et al. 1974. "Quantitative survey of organic phosphorus in different soil types," *Phosphorus in Agriculture.* (Madison, Wis.: Amer. Soc. Agron.), pp. 25–40.

Fox, R. L. 1981. "External phosphorus requirements of crops," in *Chemistry in the Soil Environment.* ASA Special Publication no. 40. (Madison, Wis.: Amer. Soc. Agron. and Soil Sci. Soc. Amer.), pp. 223–239.

Frankenberger, W. T., and M. E. Losi. 1995. "Applications of bioremediation in the cleanup of heavy metals and metalloids," SSSA Special Publication no. 43. (Madison, Wis: Soil Sci. Soc. of Amer.).

Goenadi, D., H. Siswanto, and Y. Sugiarto. 2000. "Bioactiviation of poorly soluble phosphate rocks with a phosphorus-solubilizng fungus," *Soil Sci. Amer. J.* **64**:927–932.

Hayes, K. F., and S. J. Traina. 1998. "Metal ion speciation and its significance in ecosystem health," in P. M. Huang et al. (eds.), *Soil Chemistry and Ecosystem Health.* SSSA Special Publication no. 52. (Madison, Wis: Soil Sci. Soc. of Amer.).

Haygarth, P. M., A. F. Harrison, and K. C. Jones. 1995. "Plant selenium from soil and the atmosphere," *J. Environ. Qual.* **24**:768–771.

Kuo, S. 1988. "Application of modified Langmuir isotherm to phosphate sorption by some acid soils," *Soil Sci. Soc. Amer. J.* **52**:97–102.

Lindsay, W. L. 1972. "Inorganic phase equilibria of micronutrients in soils," in J. J. Mortvedt, P. M. Giordano, and W. L. Lindsay (eds.), *Micronutrients in Agriculture* (Madison, Wis.: Soil Sci. Soc. Amer.).

Lovley, D. R. 1996. "Microbial reduction of iron, manganese and other metals," *Advances in Agronomy* **54**:175–231.

Marschner, H. 1995. *Mineral Nutrition of Higher Plants,* 2d ed. (New York: Academic Press).

Marschner, H., A. Kalisch, and V. Romheld. 1974. "Mechanism of iron uptake in different plant species," *Proc. 7th Int. Colloquium on Plant Analysis and Fertilizer Problems,* Hanover, West Germany.

McLean, E. O. 1978. "Influence of clay content and clay composition on potassium availability," in G. S. Sekhon (ed.), *Potassium in Soils and Crops.* (New Delhi, India: Potash Research Institute of India), pp. 1–19.

Mengel, K., and E. A. Kirkby. 1987. *Principles of Plant Nutrition,* 4th ed. (Bern, Switzerland: International Potash Institute).

Mengel, K., Rahmutullah, and H. Dou. 1998. "Release of potassium from the silt and sand fractions of loess-derived soils," *Soil Science* **163**:805–813.

Mortvedt, J. J., F. R. Cox, L. M. Shuman, and R. M. Welch (eds.). 1991. *Micronutrients in Agriculture.* SSSA Book Series no. 4. (Madison, Wis.: Soil Sci. Soc. Amer.).

Munson, R. D. (ed.). 1985. *Potassium in Agriculture.* (Madison, Wis.: Amer. Soc. Agron.).

Nair, K. P. P. 1996. "The buffering power of plant nutrients and effects on availability," *Advances in Agronomy* **57**:237–287.

Nowak, C. A., R. B. Downard, Jr., and E. H. White. 1991. "Potassium trends in red pine plantations at Pack Forest, New York," *Soil Sci. Soc. J.* **55**:847–850.

Olsen, R. A., R. B. Clark, and J. H. Bennett. 1981. "The enhancement of soil fertility by plant roots," *Amer. Scientist* **69**:378–384.

Olsen, S. R., and F. E. Khasawneh. 1980. "Use and limitations of physical-chemical criteria for assessing the status of phosphorus in soils," in F. E. Khasawneh et al. (eds.), *The Role of Phosphorus in Agriculture.* (Madison, Wis.: Amer. Soc. of Agron.).

Paulter, M. C., and J. T. Sims. 2000. "Relationship between soil test phosphorus, soluble phosphorus, and phosphorus saturation in Delaware soils," *Soil Sci. Soc. Am. J.* **64**:765–773.

Raven, K. P., and L. R. Hossner. 1993. "Phosphorus desorption quantity-intensity relationships in soils," *Soil Sci. Soc. Amer. J.* **57**:1501–1508.

Richter, D. D., et al. 1994. "Soil chemical change during three decades in an old-field Loblolly pine (*Pinus taeda L.*) ecosystem," *Ecology* **75**:1463–1473.

Saa, A., et al. 1994. "Forms of phosphorus in sediments eroded from burnt soils," *J. Environ. Qual.* **23**:739–746.

Sample, E. C., et al. 1980. "Reactions of phosphate fertilizers in soils," in F. E. Khasawneh et al. (eds.), *The Role of Phosphorus in Agriculture.* (Madison, Wis.: Amer. Soc. Agron.).

Sharpley, A. N. 1990. "Reaction of fertilizer potassium in soils of differing mineralogy," *Soil Sci.* **49**:44–51.

Sharpley, A. N. 2000. "Phosphorus availability," in M. E. Summer (ed), *Handbook of Soil Science.* (New York: CRC Press), pp. D-18–D-38.

Smith, S. J., et al. 1991. "Water quality impacts associated with wheat culture in the Southern Plains," *J. Environ. Qual.* **20**:244–249.

Soltanpour, P. N., R. L. Fox, and R. C. Jones. 1988. "A quick method to extract organic phosphorus from soils," *Soil Sci. Soc. Amer. J.* **51**:255–256.

Spencer, J. E. 2000. "Arsenic in groundwater," *Arizona Geology* **30**(3).

Stevenson, F. J. 1986. *Cycles of Soil Carbon, Nitrogen, Phosphorus, Sulfur, and Micronutrients.* (New York: Wiley).

Terry, N., and G. Banuelos (eds.). 2000. *Phytoremediation of Contaminated Soil and Water.* (New York: Lewis Publishers).

Vaithiyanathan, P., and D. L. Correll. 1992. "The Rhode River watershed: Phosphorus distribution and export in forest and agricultural soils," *J. Environ. Qual.* **21**:280–288.

Viets, F. J., Jr. 1965. "The plants' need for and use of nitrogen," in *Soil Nitrogen* (*Agronomy,* no. 10). (Madison, Wis.: Amer. Soc. Agron.).

Weil, R. R., P. W. Benedetto, L. J. Sikora, and V. A. Bandell. 1988. "Influence of tillage practices on phosphorus distribution and forms in three Ultisols," *Agron. J.* **80**:503–509.

Welch, R. M. 1995. "Micronutrient nutrition of plants," *Critical Reviews in Plant Science* **14**(1):49–82.

Zayed, A. E., et al. 2000. "Remediation of selenium-polluted soils and waters by phytovolatilization," in N. Terry and G. Banuelos (eds.), *Phytoremediation of Contaminated Soil and Water.* (New York: Lewis Publishers), ch. 4.

14

Practical Nutrient Management

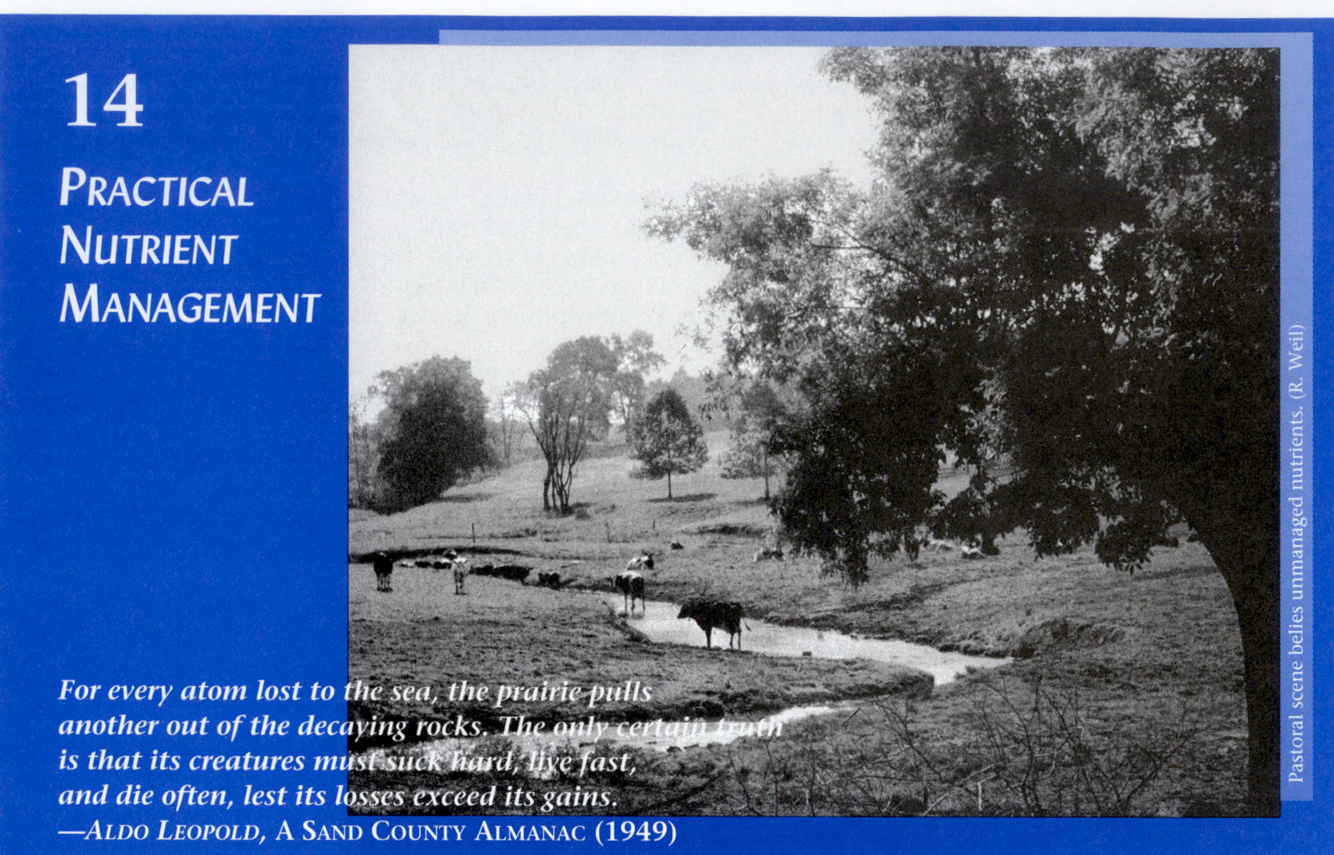

Pastoral scene belies unmanaged nutrients. (R. Weil)

For every atom lost to the sea, the prairie pulls another out of the decaying rocks. The only certain truth is that its creatures must suck hard, live fast, and die often, lest its losses exceed its gains.
—*Aldo Leopold,* A Sand County Almanac (1949)

As stewards of the land, soil managers must keep nutrient cycles in balance. By doing so they maintain the soil's capacity to supply the nutritional needs of plants and, indirectly, of us all. While undisturbed ecosystems may need no intervention, few ecosystems are so undisturbed. More often, human hands have directed the output of the ecosystem for human ends. Forests, farms, fairways, and flower gardens are ecosystems modified to provide us with lumber, food, recreational opportunities, and aesthetic satisfaction. By their very nature, managed ecosystems need management.

In managed ecosystems, nutrient cycles can become unbalanced through increased removals (e.g., harvest of timber and crops), through increased system leakage (e.g., leaching and runoff), through simplification (e.g., monoculture, be it of pine tree or sugarcane), through increased demands for rapid plant growth (whether the soil is naturally fertile or not), and through increased animal density (especially if imported feed brings in nutrients from outside the ecosystem). Some of the greatest impacts of land management are felt not on the land but in the water, where excess nutrients play havoc with aquatic ecosystems. The land manager is, of necessity, also a nutrient and environmental manager.

In this chapter we will discuss methods of enhancing nutrient recycling, as well as sources of additional nutrients that can be applied to soils or plants. We will learn how to diagnose nutritional disorders of plants and correct soil fertility problems. Building on the principles set out earlier, this chapter contains much practical information on profitable production of abundant, high-quality plant products and on maintaining the quality of both the soil and the rest of the environment.

14.1 GOALS OF NUTRIENT MANAGEMENT[1]

Nutrient management aims to achieve four broad, interrelated goals: (1) cost-effective production of high-quality plants, (2) efficient use and conservation of nutrient resources, (3) maintenance or enhancement of soil quality, and (4) protection of the environment beyond the soil.

Plant Production

Three of the primary types of plant production in which people engage are: (1) *agriculture,* (2) *forestry,* and (3) *ornamental landscaping.* The main nutrient management goal for agriculture is to increase plant yields, thereby helping the subsistent farmers to feed their families and the commercial farmers to enhance their incomes. Unfortunately, farmers tend to judge the success or failure of their management schemes in only 3 to 6 months, the period of growth of the crops they are producing. This is nothing like the span of several human generations that may be needed to fully evaluate the effectiveness of the practices they choose to use.

In forestry, the principal plant product may be measured in terms of volume of lumber or paper produced. In these instances, nutrient management aims to enhance rate of growth so that the time between investment and payoff can be minimized. The survival rate of tree seedlings is also important. Wildlife habitat and recreational values may also be primary or secondary products. The time frame in forestry, measured in decades or even centuries, tends to limit the intensity of nutrient management interventions that can be profitably undertaken.

When soils are used for ornamental landscaping purposes, the principal objective is to produce quality, aesthetically pleasing plants. Whether the plants are produced for sale or not, relatively little attention is paid to yield of biomass produced. Labor costs and convenience are generally of more concern than fertilizer costs; hence, expensive, slow-release fertilizers are widely used.

Conservation of Nutrient Resources

Two concepts that are key to the goal of conserving nutrient resources are (1) renewal or reuse of the resources, and (2) nutrient budgeting that reflects a balance between system inputs and outputs.

The first law of thermodynamics suggests that all material resources are ultimately renewable, since the elements are not destroyed by use but are merely recombined and moved about in space. In practical terms, however, once a nutrient has been removed from a plot of land and dispersed into the larger environment, it may be difficult if not impossible to use it again as a nutrient for that particular soil. For instance, phosphorus deposited in a lake bottom with eroded sediment and nitrogen buried in a landfill as a component of garbage are not available for reuse. In contrast, land application of composted municipal garbage and irrigation with sewage effluent are examples of practices that treat nutrients as reusable resources.

Recycling is a form of reuse in which nutrients are returned to the same land from which they were previously removed. Litter fall from perennial vegetation recycles nutrients to the soil naturally. Leaving crop residues in the field and spreading barnyard manure onto the land from which the cattle feed was harvested are both examples of managed nutrient recycling. The term *renewable resource* best applies to soil nitrogen, which can be replenished from the atmosphere by biological nitrogen fixation (see Sections 12.11 to 12.13). Manufacture of nitrogen fertilizers also fixes atmospheric nitrogen, but at a large cost in nonrenewable fossil fuel energy. Other fertilizer nutrients, such as potassium and phosphorus, are mined or extracted from nonrenewable mineral deposits or from mineral-laden seas and oceans. Therefore it would be wise to make careful husbandry of the Earth's limited nutrient resources a part of any long-term nutrient management program.

---

[1]For an overview of issues and advances in nutrient management for agriculture and environmental quality, see Magdoff et al. (1997). A standard textbook on management of agricultural soil fertility and fertilizers is Havlin et al. (1999).

A useful step in planning nutrient management is to conceptualize the nutrient flows for the particular system under consideration. Such a flowchart should attempt to account for all the major inputs and outputs of nutrients. Simplified examples of such nutrient budgets are shown in Figure 14.1.

Nutrient Budgets

N, P, K content of some natural materials:
http://www.ces.ncsu.edu/depts/hort/consumer/quickref/fertilizer/nutrient_content_natural.html

Addressing nutrient imbalances, shortages, and surpluses may call for analysis of nutrient flows, not just on a single farm or enterprise, but on a watershed or regional scale. For example, in many African countries exports of agricultural and forest products carry with them more nutrients than are imported into the country as fertilizers, food, or animal feed. During the past 30 years, an average of 22 kg/ha N, 2.5 kg/ha P, and 15 kg/ha K have been lost *annually* from about 200 million ha of cultivated land in sub-Saharan Africa (excluding South Africa). This net *negative* nutrient balance for African soils[2] appears to be a contributing factor in their impoverishment, the reduction of agricultural productivity, and the stagnation or decline of national economies.

By contrast, in temperate regions, the cultivated land on average receives nutrients in excess of those removed in crops, runoff, and erosion. During the same 30-year period, the nearly 300 million ha of cultivated temperate-region soils had a net *positive* annual nutrient balance of at least 60 kg/ha N, 20 kg/ha P, and 30 kg/ha K. While nutrient deficiencies still occur in some fields, the great majority of soils have experienced a nutrient buildup. Some excess nutrients move into streams, lakes, or the atmosphere, where they can contribute to environmental damage (Section 14.2).

In some countries the concentration of livestock in huge confinement facilities using feed imported from other areas has led to *regional* nutrient imbalances and concomitant serious water pollution problems. The animal manure produced in these production sites contains nutrients far in excess of the amounts that can be used in an efficient and environmentally safe manner by crops in nearby fields (for example, see Table 14.1).

Soil Quality and Productivity

Feed composition tables for nutrient balance in animal farms:
http://books.nap.edu/books/0309032458/html/index.html

The concept of using nutrient management to enhance soil quality goes far beyond simply supplying nutrients for the current year's plant growth. Rather, it includes the long-term nutrient-supplying and -cycling capacity of the soil, improvement of soil physical properties or tilth, maintenance of above- and belowground biological functions and diversity, and the avoidance of chemical toxicities. Likewise, the management tools employed go far beyond the application of various fertilizers (although this may be an important component of nutrient management). Nutrient management requires the integrated management of physical, chemical, and biological processes. The effects of tillage on organic matter accumulation (Chapter 11), the increase in nutrient

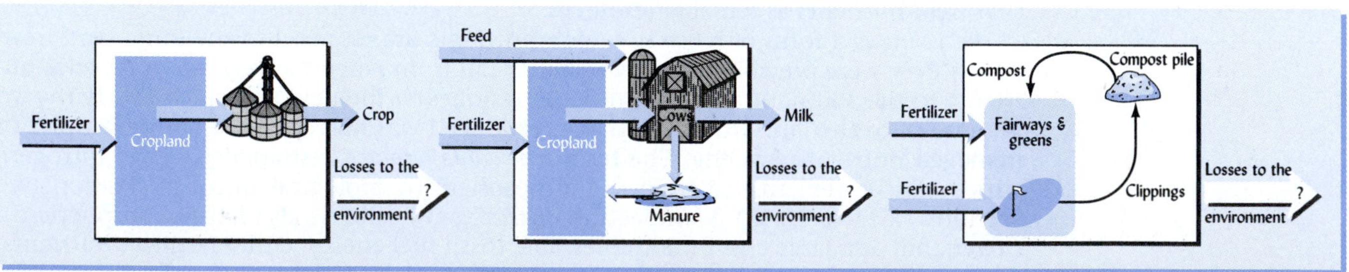

FIGURE 14.1 Representative conceptual nutrient flowcharts for a cash-grain farm, a dairy farm, and a golf course. Only the managed inputs, key recycling flows, and outputs are shown. Unmanaged inputs, such as nutrient deposition in rainfall, are not shown. Outputs that are difficult to manage, such as leaching and runoff losses to the environment, are shown as being variable. Although information on unmanaged inputs and outputs is not always readily available, it must be taken into consideration in developing a complete nutrient management plan. Such flowcharts are a starting point in identifying imbalances between inputs and outputs that could lead to wasted resources, reduced profitability, and environmental damage.

[2] For discussions of the negative nutrient balance in African soils, see Sanchez et al. (1997) and Smaling et al. (1997).

TABLE 14.1 Statewide Nitrogen and Phosphorus Balance for Agriculture in Delaware, a State with Concentrated Poultry Production

Each year about 10,000 more Mg of N were applied to Delaware's soils than were needed by the crops grown. This is nearly 50 kg N/ha. Much of this excess N leaves these farms as nitrates in surface runoff and groundwater. Phosphorus was proportionally even further out of balance, resulting in the buildup of this element in surface soils and increased losses in runoff.

Nitrogen requirement or source	*Farmland area, ha*	*Amount of phosphorus, Mg/yr*	*Amount of nitrogen, Mg/yr*
Requirements			
Corn	69,600	940	9,800
Soybeans	80,600	1,085	0
Wheat	24,300	330	2,200
Barley	11,000	150	1,000
Other crops	32,400	435	3,600
Total	217,900	2,940	16,600
Nutrient Sources			
Poultry manure	—	3,495	7,865
Fertilizer sales	—	2,955	19,275
Other wastes	—	Unknown	Unknown
Total sources of nutrient	—	6,450+	27,140+
Balance (excess N or P)	—	3,510+	10,580+

From Sims and Wolf (1994).

availability brought about by earthworm activity (Chapter 10), the role of mycorrhizal fungi in phosphorus uptake by plants (Chapter 13), and the impact of fire on soil nutrient and water supplies (Chapter 7) are all examples of components of integrated nutrient management.

14.2 ENVIRONMENTAL QUALITY

WATERSHEDS program to assess nutrients and other water quality problems: http://h2osparc.wq.ncsu.edu/

Nutrient management impacts the environment most directly with water-quality problems caused by nitrogen (N) and phosphorus (P). Together, these two nutrients are the most widespread cause of water-quality impairment in lakes and estuaries; and in rivers and streams, they are second only to sediment (see Chapter 15). As explained in Box 13.1, the supply of either N or P usually limits the growth of seaweeds, algae, and phytoplankton. If more of the limiting nutrient is added, growth of these aquatic organisms explodes leading to eutrophication (Table 14.2). In most freshwater lakes and streams, as little as 0.025 mg P/L can accelerate eutrophication. In salty estuaries and coastal waters, N is more likely to cause eutrophication. In addition to stimulating eutrophication, ammonium-N (NH_4^+-N), which is in equilibrium with ammonia gas (NH_3), can be directly toxic to fish at levels above 2 mg N/L. The nitrate form of N (NO_3^-) is also of concern, as levels above 10 mg N/L are considered potentially toxic in drinking water (see Box 12.3).

Most industrialized countries have made great strides in reducing nutrient pollution from factory and municipal sewage outfalls (called **point sources**, because they are clearly localized). However, there has been much less progress in controlling nutrients in runoff water coming from the landscape (called **nonpoint sources**, because they are

TABLE 14.2 Some Adverse Effects of Eutrophication on Various Aquatic Ecosystems

Excessive nitrogen or phosphorus stimulates algal growth and species changes leading to the other adverse effects.

- Increased phytoplankton growth
- Increases in bloom-forming or toxic phytoplankton species
- Increased coastal blooms of toxic dinoflagellate or gelatinous zooplankton
- Increased masses of algae, covering bottom sediments and submerged aquatic vegetation
- Shifts in species of macroscopic plants
- Death of coral reef communities
- Increased water turbidity and decreased light transmission
- Decreased growth of submerged aquatic vegetation (SAV) that serves as fish habitat and underpins some aquatic food webs
- Oxygen depletion
- Fish kills
- Loss of desirable fish species
- Reduced harvests of marketable fish and shellfish
- Taste, odor, and water treatment problems

TABLE 14.3 Typical Components of a Nutrient Management Plan

The plan provides practical guidelines for nutrient application to land in a manner that maximizes nutrient use efficiency and minimizes water pollution risks. This chapter includes concepts underlying many of the components.

- Aerial site photographs or maps and a soil map
- Current and/or planned plant production sequence or crop rotation
- Soil test results and recommended nutrient application rates
- Plant tissue analysis results
- Nutrient analysis of manure or other soil amendments
- Realistic yield goals and a description of how they were determined
- A complete nutrient budget for N, P, and K in the production system
- An accounting of all nutrient inputs such as fertilizers, animal manure, sewage sludge, irrigation water, compost, and atmospheric deposition
- Planned rates, methods, and timing of nutrient applications.
- Location of environmentally sensitive areas or resources, if present.
- The potential risk of N and/or P water pollution as assessed by a N leaching index, P site index, or other acceptable assessment tools

diffuse and not easily identified). Streams draining forestland and rangeland are generally far lower in nutrients than are those draining agricultural land. See Chapters 12 and 13 to review the pathways by which N and P can be lost from soils.

Nutrient Management Plans

Managing nonpoint source pollution from agriculture: http://www.epa.gov/owow/nps/facts/point6.htm

One tool for reducing nonpoint source N and P pollution is a nutrient management plan—a document that records an integrated strategy and specific practices for how nutrients will be used in plant production (Table 14.3). Increasingly, these plans are seen as legal documents used to implement government regulations on nonpoint source nutrient pollution. Generally, a specially trained soil scientist, who consults closely with the landowner to meet both environmental goals and practical needs, prepares the document. The plan attempts to balance the inputs of N and P (and other nutrients) with their desirable outputs (i.e., removal in harvested products) to prevent undesirable outputs (runoff, leaching) that exceed **maximum allowable daily loadings**, the largest amount of nutrient runoff and leaching (g/ha/day) permitted from an area of land. An important component of many nutrient management plans, the *P site index,* is explained in Section 14.13.

In the United States, practices officially sanctioned to minimize water pollution with excess N and P are known as **best management practices (BMPs).** Four general types of practices will now be briefly considered: (1) buffer strips, (2) cover crops, (3) conservation tillage, and (4) forest stand management.

Riparian Buffer Strips[3]

Stream Buffer Newsletter: http://nacdnet.org/buffers/

Buffer strips of dense vegetation situated along the bank of a stream or other body of water (the **riparian zone**) are a simple and generally cost-effective method to protect water from the polluting effects of a nutrient-generating land use. Buffer strips may consist of natural or planted species, including grasses, shrubs, trees (Plate 25, after page 114), or a combination of these vegetation types (Figure 14.2).

How They Work. Water running off the surface of the nutrient-rich land passes through the riparian buffer strip before it reaches the stream. Plants and litter reduce water velocity and cause most of the sediment, and attached nutrients to settle out. Some dissolved nutrients are adsorbed by the organic mulch and mineral soil or are taken up by the buffer strip plants. The decreased flow velocity also increases the retention time—the length of time during which microbial action can work to break down pesticides before they reach the stream. Under some circumstances, buffer strips along streams can also reduce the nitrate levels in the groundwater flowing under them (see Figure 12.8).

Design and Management. To preserve the dense vegetation and litter, riparian zones should be kept off limits to machinery and cattle. The width needed for optimum cleanup may vary from 6 to 60 m, although a width of 10 m is usually sufficient to

[3] For details about riparian buffer strips in forested areas, see Belt and O'Laughlin (1994); for information related to buffers in agricultural settings, see Chesapeake Bay Program (1995).

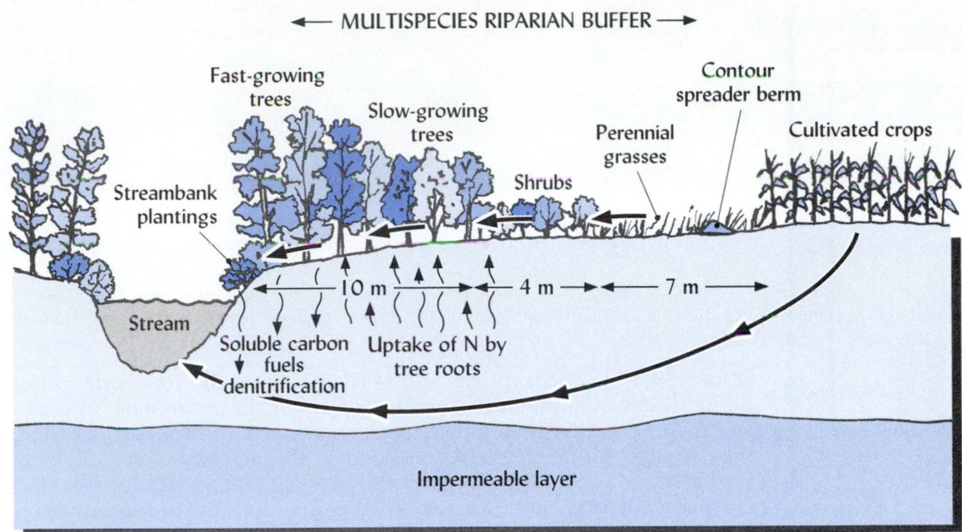

FIGURE 14.2 A multispecies riparian buffer strip designed to protect the stream from nutrients and sediment in runoff while also providing wildlife habitat benefits. A grass-covered level berm spreads runoff water evenly to avoid gullies. The perennial grasses filter out sediments and take up dissolved nutrients. Deep tree roots remove some nutrients from shallow groundwater. Soluble carbon from tree litter percolates downward to provide an energy source for anaerobic denitrifying bacteria that can remove additional nitrogen from shallow groundwater. The woody vegetation also provides wildlife habitat, shades the stream, and provides in-stream woody debris for fish habitat. A total buffer width of 20 to 30 m is usually sufficient.

obtain most of the nutrient- and sediment-removal benefits on slopes of less than 8%. A well-designed buffer strip can compensate the landowner for loss of productive land by providing some real benefits: turnaround space for field equipment, valuable hay from grass buffers, improved fishing by shading the stream and providing large woody debris for fish habitat, and enhanced recreational values associated with increased wildlife populations. Thus, installation of buffer strips can often be a win-win situation.

Cover Crops

Instead of being harvested, a cover crop is grown to provide vegetative cover for the soil and then is either killed and left on the surface as a mulch, or tilled into the soil as a *green manure*. In humid regions where some water use by the cover crop is tolerable, growing a cover crop provides numerous benefits compared to leaving the soil bare for the off-season. Most relevant to the topic of this chapter, cover crops may reduce the loss of nutrients and sediment in surface runoff, and they may conserve nutrients that would otherwise leach below the root zone.

Cover Crops Reduce Runoff Losses. A cover crop reduces nutrient losses in runoff for two reasons. First, the protective foliage and litter prevent the formation of a crust at the soil surface, thus maintaining a high rate of infiltration (see Figure 6.5*b*) and reducing runoff. Second, for the runoff that does occur, the cover crop helps remove both sediment and nutrients by the same mechanisms that operate in a buffer strip, as previously described.

Cover Crops Reduce Leaching Losses. Cover crops can reduce the leaching losses of nutrients, principally nitrogen (Figure 14.3). In many temperate humid regions, the greatest potential for leaching occurs during the fall and winter, after harvest and before planting of the main crop in the spring. During this time of vulnerability, an actively growing cover crop will slow percolation and remove much of the nitrogen from the soil solution, incorporating this nutrient into plant tissue. For this purpose, an ideal cover crop should rapidly produce as extensive a root system as possible. Winter annual cereals (rye, wheat, oats, etc.) and brassicas (radish, mustard, rapseed, etc.) have proven to be more efficient than legumes (vetch, clover, etc.) at mopping up leftover soluble nitrogen.

Conservation Tillage

The term **conservation tillage** applies to agricultural practices that keep at least 30% of the soil surface covered by plant residues. The effects of conservation tillage on soil properties and on the prevention of soil erosion are discussed in Section 15.5. Here, we emphasize the effects on nutrient losses.

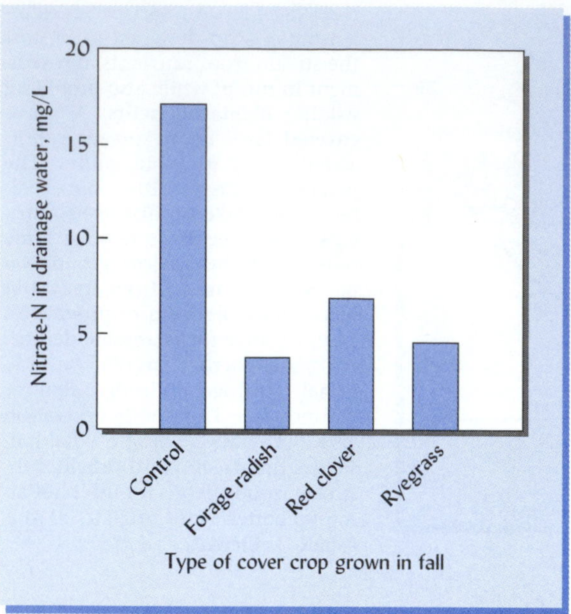

FIGURE 14.3 Cover crops can substantially reduce the concentration of nitrate-N in winter drainage water by soaking up excess mineral nitrogen left in the soil after the summer crop is harvested. In the case illustrated, the cover crops were grown from September through November following the harvest of sweet corn. The data show averages for 2 years and two sites in Quebec, Canada. Both sites had fine-textured soils in the Inceptisols order. Among the cover crops, red clover was the least effective in reducing nitrogen leaching, partly because it produced the least growth and partly because it is a nitrogen-fixing legume that is relatively inefficient at scavenging leftover nitrogen. [Data from Isse et al. (1999)]

Compared to plowed fields with little residue cover, conservation tillage usually reduces the total amount of water running off the land surface, and reduces even more the load of nutrients and sediment carried by that runoff (Table 13.1). When combined with a cover crop, the reductions are greater still. In most situations, the less the soil surface is disturbed by tillage, even when manure or sewage sludge is spread on it, the smaller are the losses of nutrients in surface runoff. The relatively small amounts of nutrients lost from untilled land (no-till, pastures, and forests) tend to be mostly dissolved in the water rather than attached to sediment particles, while the reverse is true for tilled land.

On the other hand, the loss of nutrients by leaching may be somewhat greater with conservation tillage than with conventional tillage. In conservation tillage systems, a higher percentage of the precipitation or irrigation water infiltrates into the soil, where it may carry nutrients downward. However, over time, many no-till soils develop large macropores (such as worm burrows) that are open to the soil surface. Water may move down rapidly through these large macropores, with the result that nutrients held in the finer pores of the soil matrix are bypassed and do not leach into the lower horizons (see Figure 6.21). In such situations, nutrient leaching may actually be less where conservation tillage is used.

Nutrient Losses Associated with Forest Management

Major sources of nutrients to streams and groundwater: http://water.usgs.gov/nawqa/circ-1136/h7.html

Undisturbed forests lose nutrients primarily by (1) leaching and runoff of dissolved ions and organic compounds, (2) erosion of nutrient-containing organic litter and mineral particles, and (3) volatilization of certain nutrients during fires. Most often, the output of such elements as calcium, magnesium, and potassium (but usually not nitrogen) in streams is greater than the input from atmospheric deposition, but weathering from soil and rock minerals, combined with atmospheric deposition, usually can maintain the plant-available supply of these elements. Nitrogen losses most commonly range from about 1 to 5 kg/ha each year, but atmospheric inputs of nitrogen (not including biological nitrogen fixation) are usually 2 to 3 times as great. Annual losses of phosphorus are typically very low (< 0.2 kg/ha), and are closely balanced by inputs of this element from the atmosphere and by slow release from mineral weathering.

Management of forests to produce marketable wood products tends to increase losses by all three pathways just mentioned, plus it adds a fourth very significant pathway, namely removal of nutrients in forest products (whole trees, logs, or pine straw). Forest management practices that physically disturb the soil (and therefore tend to increase nutrient losses mainly by erosion) include building roads for timber harvesting, dragging logs on the ground (skidding), and after-harvest site preparation for planting new trees (see Section 15.6).

FIGURE 14.4 Clear-cut harvest of loblolly pine on Coastal Plain soils. A few mature trees were left standing as seed trees. The forest floor is covered with slash of all sizes, but little living vegetation has appeared (1 month after harvest in spring). If the site is replanted and not treated with herbicides, nutrient uptake by roots should resume within a year or so. Removal of the tree canopy allows the sun to warm the soil, accelerating mineralization of nutrients in the forest floor, but decomposition of the high-C/N-ratio slash may immobilize much of the nitrogen released. (Photo courtesy of R. Weil)

Soil Disturbance. Disturbance of forest soil not only leaves it more vulnerable to erosion, but also can alter the nutrient balance in several other ways. Carefully planned tree harvesting and regeneration using methods that minimize soil disturbance and hasten revegetation can keep nutrient losses to low levels (Figure 14.4). Two practices should generally be avoided, as they can be particularly damaging to the forest nutrient cycle: (1) extended suppression of unwanted vegetation with herbicides, and (2) windrowing of stumps and slash (tree branches and tops left after harvesting logs) to clear land for easy planting after tree harvesting.

Nutrient Losses. Because of increased litter decomposition and interrupted nutrient uptake, streams draining many clear-cut watersheds periodically carry somewhat elevated levels of nitrogen and other nutrients for several years after the timber harvest has occurred (Figure 14.5, left). The concentrations of nitrogen are usually too small (<4 mg N/L) to immediately threaten streamwater quality. However, this loss of nutrients, combined with nutrient removal in the harvested trees, raises concerns about soil depletion and site productivity, especially on sites that are originally nutrient-poor.

Fertilization. Fertilizer application to forests is becoming an increasingly common practice. The principal nutrients applied are nitrogen and/or phosphorus, along with some sulfur and boron. Fertilizers prevent soil impoverishment in the face of the nutrient losses just discussed, and can substantially accelerate plant growth and raise productivity. As might be expected, increases in peak nutrient exports can often be detected when forested watersheds undergo fertilization (Figure 14.5, right). Although the effects of forest fertilization on water quality do not yet approach those associated with fertilizer use in agriculture, foresters would be well served to study the lessons learned from agricultural use of fertilizers and so avoid making the nutrient management mistakes that have plagued agriculture.

14.3 NUTRIENT RESOURCES AND CYCLES

Internal nutrient resources come from within the ecosystem, be it a forest, a farm, or a home in the suburbs, and are generally preferred, since their financial and environmental costs are minimized. These resources include mineral weathering within the soil profile, biological nitrogen fixation, atmospheric deposition, and various forms of internal recycling, such as animal manure application, utilization of cover crops, and plant-residue management.

If internal resources prove inadequate, nutrients must be imported from resources external to the system, such as purchased inorganic or organic fertilizers and (so-called) wastes.

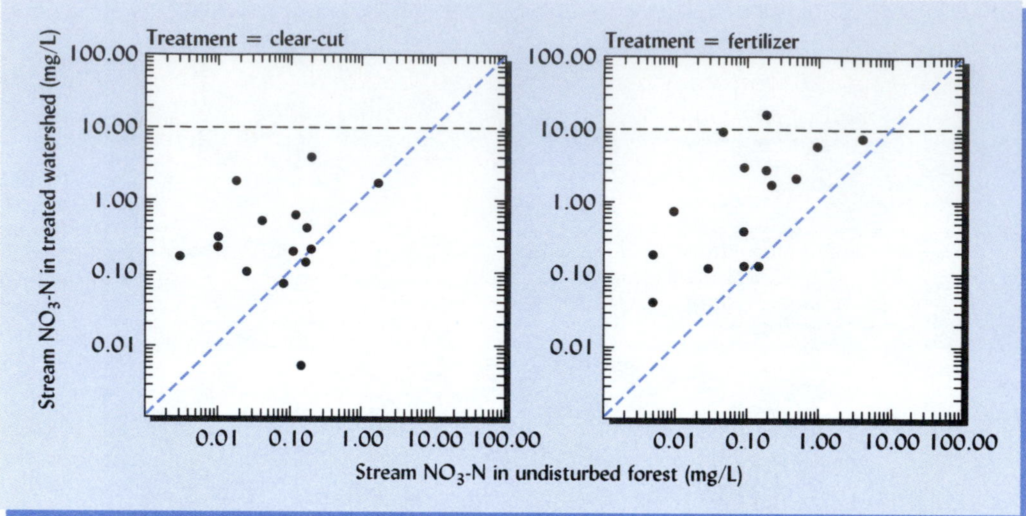

FIGURE 14.5 Effects of clear-cutting or fertilizer treatments on nitrate-N in streams draining forested watersheds. Each data point represents one pair of watersheds (treated and undisturbed control). Points above the diagonal dashed line indicate increased nitrate-N concentration compared to the control. Note that log scales are used. (*Left*) Nitrate-N concentrations in streams from watersheds that were either completely clear-cut or left undisturbed. In 10 of 15 watersheds studied, clear-cutting increased average stream nitrate-N. It is likely that short-term peak nitrate-N concentrations increased more than did the annual averages shown. (*Right*) Short-term peak nitrate-N concentrations with or without application of 200 to 250 kg N/ha as urea fertilizer. In all but two of the watersheds, fertilizing the forest substantially increased peak nitrate concentrations in the streamwater, in some cases to more than 10 mg/L. [Data from Brown and Binkley (1994); see also Binkley et al. (1999)]

Water quality standards and concerns about nutrients in water:
http://water.usgs.gov/nawqa/circ-1136/h6.html#NIT

Depending on the parent materials and climate, weathering of minerals (see Section 2.3) may release significant quantities of nutrients (Table 14.4). For timber production, most nutrients are released fast enough from either parent materials or decaying organic matter to supply adequate nutrients. In agricultural systems, however, some nutrients generally must be added, since nutrients are removed from the land annually in harvested crops. Negligible amounts of nitrogen are released by weathering of mineral parent materials, so other mechanisms (including biological nitrogen fixation) are needed to resupply this important nutrient. Release of nutrients by mineralization of soil organic matter is important in short-term nutrient cycling, but in the long run, the organic matter and the nutrients it contains must be replenished or soil fertility will be depleted.

Nutrient Economy of Forests[4]

Forest fertilization in future forestry planning:
http://cru.cahe.wsu.edu/CEPublications/eb1800/eb1800.html

Recycling. In most forests, organic matter mineralization is the main source of nutrients for tree growth, and the rates of nutrient uptake from the soil by the trees closely match the rates of release by mineralization. Surface plant litter makes up but a small proportion of the nutrients cycled from trees to the soil decomposers each year. The biggest portion comes from a combination of fine roots and mycorrhizal hypha (see Sections 10.7 and 10.9). A second major source of nutrients for each season's tree growth is the recycling process within the trees themselves. Nutrients are translocated from the leaves to the twigs and branches just prior to litter fall. The translocated nutrients are then remobilized early in the next growing season to produce new leaf growth and wood increment. Additional sources are atmospheric deposition and weathering of soil minerals.

Surface Soil Enrichment. Trees may obtain most of their nutrients from the surface horizons, but they are also well adapted to gathering nutrients from deep in the soil profile, where much of the nutrient release from parent material takes place. Overall nutrient-

[4]For in-depth treatments of this topic, see Attiwill and Leeper (1987), Perry (1994), and Likens and Bormann (1995). Aber et al. (2000) provide insights on ecological challenges in forestry.

TABLE 14.4 Amounts of Selected Nutrients Released by Mineral Weathering in a Representative Humid, Temperate Climate, Compared with Amounts Removed by Silvicultural and Agricultural Harvests

For the forest, weathering release and harvest removal are roughly balanced, but cropping removes much more of some nutrients than can be released by weathering. Leaching losses are not shown.

	Amount released or removed, kg/ha			
	P	*K*	*Ca*	*Mg*
Weathered from igneous parent material over 50 years	5–25	250–1000	150–1500	50–500
Removed in harvest of 50-year-old deciduous bole wood	10–20	60–150	175–250	25–100
Removed by 50 annual harvests of a corn–wheat–soybean rotation	1200	2000	550	500

Estimated from many sources.

(a)

(b)

FIGURE 14.6 Two examples of agroforestry systems. (*a*) The deep-rooted *Acacia albida* enriches the soil under its spreading branches (arrow points to a man). Conveniently, these trees leaf out in the dry season and lose their leaves during the rainy season when crops are grown. It is an African tradition to leave these trees standing when land is cleared for crop production. Crops growing under the trees yield more and have higher contents of sulfur, nitrogen, and other nutrients. (*b*) Branches pruned from widely spaced rows of leguminous trees are spread as a mulch on the soil surface in the alleys between the tree rows, thus enriching the alleys with nutrients from the leaves as well as conserving soil moisture. The crops grown in this alley-cropping system may yield better than crops grown alone, but only if competition between trees and crop plants for light and water can be kept to a minimum. (Photos courtesy of R. Weil)

use efficiency can sometimes be increased by combining trees and agricultural crops into what are known as *agroforestry systems* (Figure 14.6).

In Chapter 2 (Figure 2.17) we saw that trees can act as nutrient pumps, taking up nutrients that occur deep in the profile because of weathering or leaching, and depositing them at the soil surface as litter that will decompose and release the nutrients where they can be of use to relatively shallow-rooted agricultural crops. Nitrogen-fixing trees (mostly legumes) can also add nitrogen to the surface soil with their nitrogen-rich leaf litter. Trees may also enhance the fertility of the soil in their vicinity by trapping windblown dust, thus increasing the deposition of such nutrients as calcium, phosphorus, and sulfur.

Some Effects of Fire

Hot-burning forest wildfires convert a great deal of nitrogen and sulfur, and some phosphorus, to gaseous forms in which they are lost from the site. Burned-over land continues to lose more phosphorus in runoff for several years after a high-intensity burn (see Table 13.2). Ashes, both from wildfires and from prescribed burns (low-intensity intentional fires), contain high levels of soluble K, Mg, Ca, and P, increasing the short-term availability of these nutrients, but also increasing the rate of loss of these nutrients from the forest ecosystem. Wildfires usually result in greater nutrient losses than do prescribed burns, because the high heat associated with wildfires destroys some of the soil organic matter as well as the aboveground biomass. Another nutrient aspect of forest fires is that fire retardants used to fight wildfires are mostly fertilizer-type materials containing N, P and B.

Rangeland Nutrient Cycling

The burning of rangeland grasses (or crop residues) usually produces much less volatilization of nitrogen and sulfur than do forest wildfires, as grass fires move quickly and burn at relatively low temperatures. While the loss of organic matter consumed by the fire is undeniable, the nutrients released may stimulate enough extra plant biomass production that soil organic matter may actually accumulate to higher levels under grasslands subject to occasional burns than under those where fire is completely controlled. Grazing by large mammals, if not too frequent and intense, can also stimulate increased plant production and quality (Box 14.1 and Figure 14.7). Fire and grazing, both natural components in rangeland ecosystems, can be important tools in nutrient management.

Leguminous Cover Crops to Supply Nitrogen

In temperate regions, winter annual legumes such as vetch, clovers, and peas can be established in fall before the weather becomes too cold for cover crop growth. Then, in spring, the cover crop will resume growth and associated microorganisms will fix as

BOX 14.1 GRAZING MAMMALS AND NUTRIENT CYCLING IN SOILS

In the Serengeti plains of East Africa, the spatial distribution of nonmigratory antelopes and gazelles (and by inference, the lions and cheetahs that prey on them) is influenced by variations in soil fertility. The graphs in Figure 14.7 describe a study in which animals were found grazing most frequently in areas where the soils were high in nitrogen and sodium, two mineral nutrients critical to the health and survival of pregnant or lactating females and their young. Not only did the animals seek out the more fertile soils, but their activities also actually enhanced the cycling of nutrients, making the soils they frequented more fertile. First, the animals left manure and urine that contained most of the nutrients they consumed, but in a more easily decomposable form than in the original plant material. Second, the animals' grazing action seems to have stimulated vigorous growth of the palatable, easily decomposed plant species, thus speeding the cycling of nutrients. Leaf nitrogen concentrations in plant regrowth after grazing generally were higher than in ungrazed plants, making the resulting plant residues more readily recyclable in the soil.

Overgrazing occurs when animal density exceeds the carrying capacity of the land. Improvements in the condition of vegetation from grazing by free-ranging wild animals contrast with the negative impacts caused by overgrazing. Continual grazing by livestock may kill off the most palatable species, so that the vegetation becomes relatively sparse and dominated by less palatable plants. In some cases, the residues of these plants are also less decomposable. The resulting impairment of the nutrient cycling processes accelerates the deterioration of the vegetative cover and exposes the soil to the erosive action of wind and water (see Section 15.8).

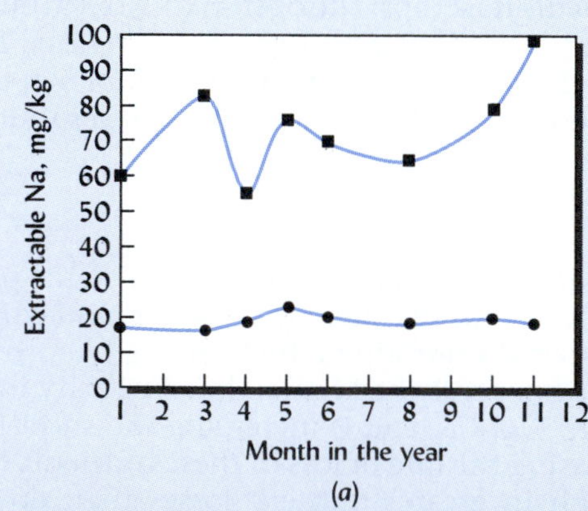

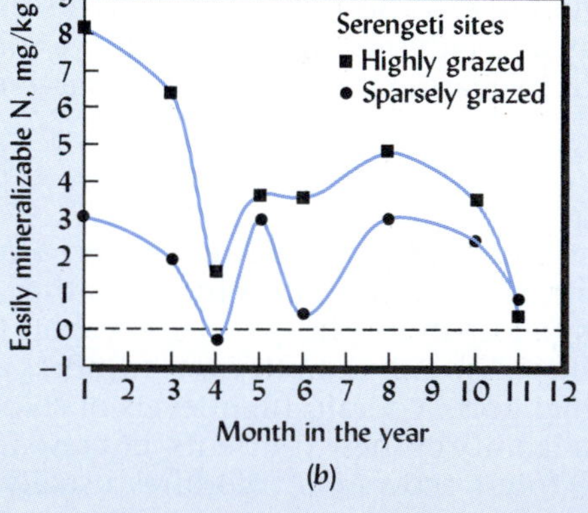

FIGURE 14.7 Plant-available sodium (a) and nitrogen (b) in soils where gazelles habitually graze heavily or almost not at all. Averages for two pairs of sites and two years of study in Tanzania's Serengeti National Park. [Data recalculated from McNaughton et al. (1997)]

much as 3 kg/ha of nitrogen daily. The cover crop is usually killed chemically or mechanically a week or two before the main crop is planted. The cover crop biomass can be left on the surface as a mulch in a no-till system, or plowed under in a conventional tillage system. In a no-till system, it is also possible to plant the main crop into the living cover crop and kill the cover crop a few days later, thus extending the mulch effect and maximizing (but somewhat delaying) the nitrogen contribution.

A cover crop system may be able to replace part or all of the nitrogen fertilizer normally used to grow the main crop (Figure 14.8). An economic evaluation of cover crop use should include the many benefits in addition to provision of nitrogen (see Section 14.2). The use of winter legume cover crops to provide nitrogen for nonlegume crops is becoming increasingly popular in orchards, rice paddies (drained during the dry season), corn fields, vegetable fields, and gardens. Environmental restrictions on phosphorus applications (Sections 14.2 and 14.13) are making cover crops increasingly important in **organic farming** systems, which have to reduce their dependence on external organic nitrogen sources such as manure and compost because of the phosphorus contents of these materials.

Crop Rotations

Growing one particular crop year after year on the same land generally produces lower yields of that crop and engenders more negative impacts on the soil and environment than if that crop is grown in sequence (rotation) with other crops. Crop rotation interrupts weed, disease, and insect pest cycles; uses nutrients more effectively because of different rooting patterns, types of residues, and nutrient requirements; and, may have positive effects on mycorrhizal diversity (see Section 10.9). These and other phenomena may explain why, even where pests and nutrient supply are optimally controlled, crops consistently yield 10 to 20% more in rotations than in continuous culture.

A detailed consideration of different cropping systems and rotations is beyond the scope of this book, but the nutrient management aspects of rotating legumes with nonlegumes deserve mention here. Similar to the cover crop effects just discussed, a legume main crop may substantially reduce the amount of nitrogen that needs to be added to grow a subsequent nonleguminous crop, even though much of the biomass and accumulated nitrogen in the legume are removed with the crop harvest. Perennial forage legumes, such as alfalfa, tend to produce the greatest effects in this regard, but the nitrogen contributions of such grain legumes as soybeans and peanuts should also be taken into account when planning nitrogen application to nonlegumes such as cereal grains (Figure 14.9).

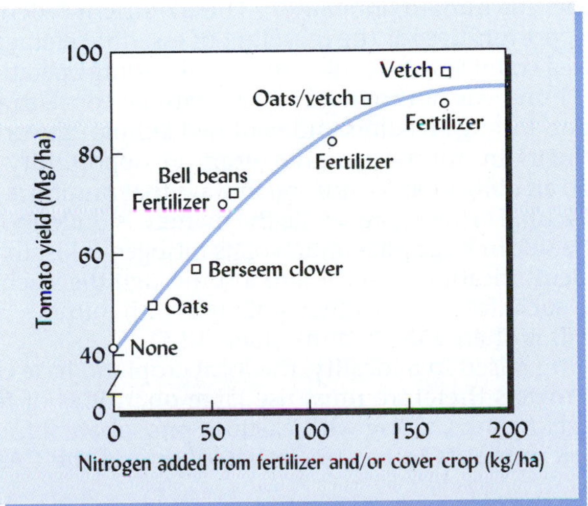

FIGURE 14.8 Comparative effects of winter cover crops and inorganic fertilizer on the yield of the following main crop of processing tomatoes grown on a Xeralf soil in California. The amount of nitrogen shown on the *x*-axis was added either as inorganic fertilizer or as aboveground residues of the cover crops, or a combination of both. Note that nitrogen from either source seemed to be equally effective. For tomato, which requires nitrogen over a long period of time, vetch alone or vetch mixed with oats produced enough nitrogen for near-optimal yields. [Data abstracted from Stivers et al. (1993) ©Lewis Publishers, an imprint of CRC Press, Boca Raton, Florida]

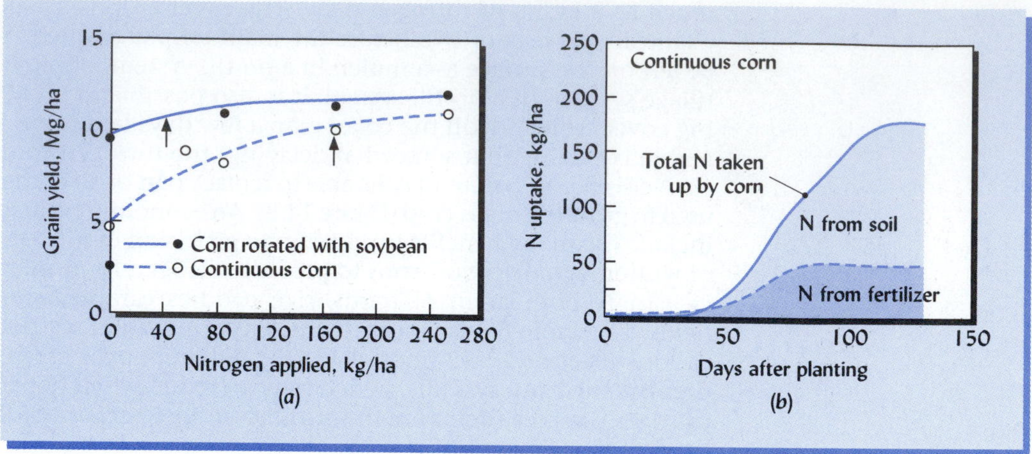

FIGURE 14.9 Crop rotation can increase yields and reduce fertilizer requirements. (*a*) The amount of fertilizer nitrogen required (↑) for high corn yields was less than 50 kg N/ha in the corn–soybean rotation (solid line), but over 150 kg N/ha in the continuous corn system (dashed line). These soils supported either corn every year or corn every second year alternating with soybeans for the 12 years. The rotation corn maintained some yield advantage even when high amounts of nitrogen were applied, indicating that in addition to increased nitrogen availability after soybeans, the crop rotation conferred other benefits, such as pest reduction or improved soil microbial activity. Crop rotations involving more than two crops usually produce even greater benefits, especially if several years in the rotation are devoted to perennial grass-legume hay or pasture. (*b*) The use of isotopically labeled fertilizer ($^{15}NH_4NO_3$) showed that even when a relatively high rate was applied (168 kg N/ha), most of the corn's N needs came not directly from the fertilizer, but from the mineralization of soil organic matter. This was especially true in the latter part of the growing season. By harvest time, the fertilizer-derived nitrogen taken up by the corn represented less than one-third of the nitrogen taken up by the plants, and accounted for less than one-third of the fertilizer nitrogen applied to the soil. Corn grown in rotation with soybeans (not shown) took up more nitrogen from both soil and fertilizer sources than did corn grown after corn. These data are averages for 2 years and two soils (a silt loam Ustoll and a loam Udoll) in Kansas, but the results are typical of those found in many parts of the world (compare Table 14.10). [Data recalculated from Omay et al. (1998)]

14.4 RECYCLING NUTRIENTS THROUGH ANIMAL MANURES[5]

Maryland's manure matching service helps manure do the most good: http://www.mda.state.md.us/nutrient/manure3.pdf

For centuries, the use of farm manure has been synonymous with a successful and stable agriculture. Not only does manure supply organic matter and plant nutrients to the soil, but it is often associated with the production of soil-conserving forage crops used to feed animals. Huge quantities of farm manure are available each year for possible return to the land. For each kilogram of liveweight of farm animals, about 4 kg dry weight of manure is produced in a year. In the United States the farm animal population voids about 10 times as much manure solids as does the human population. These nutrient-laden organic manures provide remarkable opportunities for the recycling of essential elements.

Unfortunately, in most industrialized countries, concentrated animal feeding operations (CAFOs) have emerged—tens or even hundreds of thousands of animals are concentrated in a single facility. Livestock production in huge feedlots and confined animal "factories" has changed the perception of nutrients in animal manures from an opportunity for *recycling* to enhance crop production to an obligation for *disposal* in ways that minimize environmental contamination (Figure 14.10). Furthermore, while the manure is underfoot of the animals and after it is moved to the side in huge piles, much of its nitrogen is lost to the atmosphere as ammonia or through denitrification. Some is also lost through the leaching of nitrates. The groundwater under such feedlots is often polluted with nitrates and pathogens, and water from nearby wells is often unfit to drink (Table 14.5).

As a result of so many animals being raised in a locality, the local cropland base cannot provide all the feed required. Growers therefore must use large quantities of feed grains grown elsewhere. These imported grains, along with calcium phosphate mineral supplements added to the feed, cause nutrients to be concentrated around the CAFO.

[5] Some of the challenges of such recycling are discussed in Gardner (1997).

FIGURE 14.10 A large feedlot in Colorado where 100,000 cattle are fed on grain imported from distant farms. It is difficult for such concentrated animal feeding operations to recycle the nutrients in the manure back to the land on which the cattle feed was grown. Instead of being seen as a valued resource, the manure in this situation may be considered a waste to be disposed of. Agriculture is challenged to structure itself in a more ecologically balanced manner that better integrates its animal and crop components. (Photo courtesy of R. Weil)

Feeding strategies to optimize dairy cow performance while loss of N, P and K to the environmental: http://www.ianr.unl.edu/pubs/dairy/g1306.htm

The manure produced therefore contains far more nitrogen and phosphorus than the local cropland base can properly utilize (see Table 14.1).

In the past, all the manure was applied to nearby fields, resulting in very high accumulations of nitrogen and phosphorus and pollution of surface and groundwater. As the pollution problem and its causes have been recognized, regulations (and common sense) require that manure not be applied to fields already saturated with nutrients. Instead, much of this manure must be transported to distant farms whose soils can make use of added nitrogen and phosphorus.

In recognition of the public benefits of reduced water pollution, governments may invest in programs designed to (1) promote composting (to reduce the volume of manure and the solubility of the nutrients in it); (2) eliminate the overfeeding of phosphorus supplements to reduce the concentration of phosphorus in the manure; (3) mix iron or aluminum compounds with the manure to reduce the solubility of the phosphorus; and (4) facilitate transportation of manure to areas with soils that can effectively utilize the nutrients.

Nutrient Composition of Animal Manures

Animals excrete large portions of the nutrient elements consumed by them in their feeds. For this reason, animal manures are valuable sources of both macro- and micronutrients (Table 14.6). However, one has to be cautious in interpreting general statements about the value and use of manure. For a particular type of animal, the actual water and nutrient content of a load of manure will depend on the nutritional quality of the animals' feed, how the manure was handled, and the conditions under which it has been stored. The variability from one type of animal to another (e.g., broiler chicken manure compared to horse manure) is even greater.

TABLE 14.5 The Effect of Soil and Site Characteristics on the Percentage of Wells in Five Midwestern States Having Nitrate-N Concentrations Greater than 10 mg/L, the Upper Limit Considered Suitable for Human Consumption

Sites with sandy soils, near cropland, near barnyards, and having shallow wells had the highest nitrate-N levels.

	Texture of soils		*Proximity to cropland*		*Proximity to feedlot or barnyard*		*Well depth*		
Characteristics	*Sandy*	*Clayey*	*<6 m*	*Out of sight*	*<6 m*	*Out of sight*	*Deep >30 m*	*Shallow <15 m*	*Shallow wells near barnyard or feedlot*
No. of wells	2412	6415	1684	3098	704	7520	5106	3467	158
Percentage with nitrate-N levels >10 mg/L	7.2	3.1	6.4	1.8	12.2	2.8	1.1	9.7	25.3

Data from Richards et al. (1996).

TABLE 14.6 Commonly Used Organic Nutrient Sources: Their Approximate Nutrient Contents and Other Characteristics

Along with nitrogen-fixing legumes grown in rotation and as cover crops, materials such as these (excepting sewage sludge and municipal solid wastes) provide the mainstay of nutrient supply in organic farming. The nutrient contents shown for animal manures are typical of well-fed livestock in confinement production systems. Manure from free-range animals not given feed supplements may be considerably lower in both nitrogen and phosphorus.

		Percent of dry weight						*g/Mg of dry weight*						
Material	*Water,[a] %*	*Total N*	*P*	*K*	*Ca*	*Mg*	*S*	*Fe*	*Mn*	*Zn*	*Cu*	*B*	*Mo*	
Activated sewage sludge	<10	6	1.5	0.5	—	—	—	—	—	450	—	—	—	Most common form is Milorganite®, N available over 2 to 6 months.
Cottonseed meal	<15	7	1.5	1.5	—	—	—	—	—	—	—	—	—	Acidifies the soil. Commonly used as livestock feed.
Dairy cow manure	75	2.4	0.7	2.1	1.4	0.8	0.3	1,800	165	165	30	20		May contain high-C bedding.
Dried blood	<10	13	1	1	—	—	—	—	—	—	—	—	—	Slaughterhouse by-product, N is available quickly.
Dried fish meal	<15	10	3	3	—	—	—	—	—	—	—	—	—	Incorporate or compost because of bad odors. Can feed to livestock.
Feedlot cattle manure[d]	80	1.9	0.7	2.0	1.3	0.7	0.5	5,000	40	8	2	14	1	May contain soil and high soluble salts.
Horse manure	63	1.4	0.4	1.0	1.6	0.6	0.3	—	200	125	25	—	—	May contain high-C bedding.
Municipal solid waste compost[e]	40	1.2	0.3	0.4	3.1	0.3	0.2[f]	14,000	500	650	280	60	7	May have high C/N and contain heavy metals, plastic, and glass.
Poultry (broiler) manure[b]	35	4.4	2.1	2.6	2.3	1.0	0.6	1,000	413	480	172	40	0.7	May contain high-C bedding, high soluble salts, arsenic, and ammonia.
Sewage sludge	80	4.5	2.0	0.3	1.5[g]	0.2	0.2	16,000[g]	200	700	500	100	15	May contain high soluble salts and toxic heavy metals.
Sheep manure	68	3.5	0.6	1.0	0.5	0.2	0.2	—	150	175	30	30	—	
Spoiled legume hay	40	2.5	0.2	1.8	0.2	0.2	0.2	100	100	50	10	1,500	3	May contain weed seeds.
Swine manure[c]	72	2.1	0.8	1.2	1.6	0.3	0.3	1,100	182	390	150	75	0.6	May contain elevated Cu levels.
Wood wastes	—	—	0.2	0.2	0.2	1.1	0.2	2,000	8,000	500	50	30	—	Very high C/N ratio; must be supplemented by other N.
Young rye green manure	85	2.5	0.2	2.1	0.1	0.05	0.04	100	50	40	5	5	.05	Nutrient content decreases with advanced growth stage.

[a]Water content given for fresh materials. Processing and storage methods may alter water content to less than 5% (heat-dried) or to more than 93% (slurry).
[b]Broiler and dairy manure composition estimated from means of approximately 800 and 400 samples analyzed by the University of Maryland manure analysis program 1985–1990.
[c]Composition of swine, sheep, and horse manure calculated from North Carolina Cooperative Extension Service Soil Fact Sheets prepared by Zublena et al. (1993).
[d]Feedlot manure composition is based on average analysis reported in Eghball and Power (1994).
[e]Composition of municipal solid waste compost based on mean values for the products of 10 composting facilities in the United States as reported by He et al. (1995).
[f]Sulfate-S.
[g]Sludge contents of Ca and Fe may vary 10-fold depending on the wastewater treatment processes used.
Data derived from many sources.

FIGURE 14.11 Spreading solid dairy manure on cropland recycles nutrients efficiently but is labor-intensive and time-consuming for the farmer. Many loads of manure will have to be hauled to fertilize this field. The manure should be incorporated as soon as possible after spreading, and spreading should not be done on frozen soils. Calibration of spreaders is important to prevent unintentional overapplication of nutrients. (Photo courtesy of R. Weil)

On a dry-weight basis, animal manures contain from 2 to 5% N, 0.5 to 2% P, and 1 to 3% K. These values are one-half to one-tenth as great as are typical for commercial fertilizers. However, unless it has been specially processed or stored outdoors in a dry climate, manure is rarely spread in the dry form, but contains a great deal of water. As it comes from the animal, manure has a high water content (see Table 14.6). If the fresh manure is handled as a solid and spread directly on the land (Figure 14.11), the high water content is a nuisance that adds to the expense of hauling. If the manure is handled and digested in a liquid form or slurry, even more water dilutes the nutrient content to values much lower than those cited for dry manure in Table 14.6. The high water and low nutrient content make it difficult to economically justify transporting manure to distant fields where it might do the most good. However, the value of the micronutrients in manure (Table 14.6) and the nonnutrient benefits of its organic matter (see Section 11.7) may be even greater than that of its N-P-K content, and should be included in any economic evaluation of manure transport.

Anaerobic Digestion with Biogas Production

If manure is made into a liquid slurry and allowed to become anaerobic (air is not stirred in), methane is a major gaseous product of the resulting anaerobic decomposition. The *biogas* produced by anaerobic manure digestion is about 80% methane and 20% carbon dioxide and can be burned much like commercial natural gas. Developing countries have attempted to use small-scale manure digesters to supply cooking and heating fuel for remote villages.

Recently, commercial ventures and some farms have established large-scale production of biogas from manure. The gas is most commonly used to generate electricity, which is sold to local utility companies. The slurry remaining after digestion still contains most of the nutrients from the manure (though much organic carbon and some nitrogen has been transformed to gases), and it can be pumped to nearby fields or further processed to allow for wider sale and distribution.

In the future such technologies, along with more integrated animal farming systems, may help to redress the serious nutrient imbalances that have developed with regard to manure production in industrialized agriculture.

14.5 INDUSTRIAL AND MUNICIPAL BY-PRODUCTS[6]

In addition to farm manure, four major types of organic wastes are of significance in land application: (1) municipal garbage, (2) sewage effluents and sludges (see Section 14.6), (3) food-processing wastes, and (4) wastes of the lumber industry. Although once seen as

[6]For a collection of technical papers discussing the potential benefits and problems associated with the land application of these by-products, see Powers and Dick (2000).

mere waste products to be flushed into rivers and out to sea, these materials are increasingly seen as sources of nutrients and organic matter that can be used beneficially to promote soil productivity in agriculture, forestry, and disturbed-land reclamation. However, because of their uncertain content of toxic chemicals, these and other industrial wastes may or may not be acceptable for land application.

Garbage

Municipal garbage, which has traditionally been used extensively to enhance crop production in China and other Asian countries, is becoming more widely used in the United States and Europe. About 50 to 60% of municipal solid waste (MSW) consists of decomposable materials (paper, yard waste, etc.). Once the inorganic glass, metals, and so forth are removed, MSW can be composted, sometimes in conjunction with sewage sludge, poultry manure, or other nutrient-rich materials, to produce MSW compost that is then applied to the land. Most MSW in the United States is incinerated or landfilled, but growing concerns about air quality and scarcity of landfill space are raising the level of interest in using soil application as a means of disposal. The very low nutrient content of MSW compost (see Table 14.6) makes the transportation of a given quantity of nutrients in MSW compost even more expensive than for animal manures. Potentially, the entire annual production of organic-material MSW, some 80 million cubic meters in the United States, could be recycled as a soil amendment using less than 10% of the agricultural land in the country.

Food-Processing Wastes

Land application of food-processing wastes is being practiced in selected locations, but the practice is focused almost entirely on pollution abatement and not on crop production. Liquid wastes are commonly applied through sprinkle irrigation to permanent grass. Plant-processing schedules dictate the timing and rate of application, which may not be suitable for optimum crop production.

Wood Wastes

Sawdust, wood chips, and shredded bark from the lumber industry have long been sources of soil amendments and mulches, especially for home gardeners and landscapers. These wastes are high in lignin and related materials, have very high C/N ratios, and therefore decompose very slowly—desirable properties in a mulch material. However, they do not readily supply plant nutrients. In fact, sawdust incorporated into soils to improve soil physical properties may cause plants to become nitrogen deficient unless an additional source of nitrogen is applied with the sawdust (see Section 11.3).

Recycling by Composting

In the past, most of the organic wastes in the United States found their way into the thousands of landfills that dotted the countryside. But the number of landfill sites has decreased by 75% in the past 20 years, and the environmental consequences of dumping organic materials in these sites have become more apparent.

Among the alternatives that are being more widely used are composting operations that combine yard trimmings, food scraps, supermarket wastes, and paper in compost piles with such materials as sawdust, manure, and some industrial wastes (e.g., coal ashes). When appropriate moisture is supplied, decomposition takes place, and in time composted materials that will support plant growth are available for recycling to the land (see Section 11.5). The compost sites may be large or small operations run by communities, small businesses, or even individuals. The number of such facilities in the United States has increased more than fourfold since 1990.

Wastewater Treatment By-Products

Sewage treatment has evolved over the past century in response to society's desire (and resultant regulations) to avoid polluting our rivers and oceans with pathogens, oxygen-demanding organic debris, and eutrophying nutrients. The ever-more-stringent efforts to clean up the wastewater before returning it to natural waters has two basic consequences. First, the amount of material *removed* from the wastewater during the treatment process has increased tremendously. This solid material, known as *sewage sludge,* must also be disposed of safely. Second, a goal of advanced wastewater treatment is to remove nutrients (mainly phosphorus, but increasingly also nitrogen) from the *effluent* (the treated water that is returned to the stream), a job that can be accomplished safely and economically by allowing the partially treated effluent to interact with a soil–plant system. Therefore, there is a growing interest in using soils to assist with the sewage problem in two ways: (1) as a system of assimilating, recycling, or disposing of the solid sludge; and (2) as a means of carrying out the final removal of nutrients and organics from the liquid effluent.

Sewage Effluent. Some cities operate sewage farms on which they produce crops, usually animal feeds and forages, that offset part of the expense of effluent disposal. Another beneficial use of nutrient-rich sewage effluent is the irrigation of forest land (Figure 14.12) as a method of advanced wastewater treatment. The rate of wood production is also greatly increased as a result of both the water and nutrients in the effluent.

In a carefully planned and managed effluent irrigation system, the combination of (1) nutrient uptake by the plants, (2) adsorption of inorganic and organic constituents by soil colloids, and (3) degradation of organic compounds by soil microorganisms results in the purification of the wastewater. Percolation of the purified water eventually replenishes the groundwater supply.

Using biosolids as a plant nutrient source, including a rate worksheet:
www.agweb.okstate.edu/pearl/plantsoil/soils/cr-2201.pdf

Sewage Sludge or Biosolids. Sewage sludge has been spread on the land for decades, and its use will likely increase in the future. If sewage sludge has been treated to meet certain land-application standards (low pathogen and contaminant levels), the term **biosolids** may be applied. The product Milorganite®, a dried, activated (oxygenated) sludge sold by the Milwaukee Sewerage Commission, has been used widely in North America since 1927, especially on turf grass. Numerous other cities market composted sludge products to landscaping and other specialty users. The great bulk of sewage sludge (biosolids) used on land is applied as liquid slurry or as partially dried cake.

Composition of Sewage Sludge. As might be expected, the composition of sludge varies from one sewage treatment plant to another. Representative values for plant nutrients are given in Table 14.6. Like manure and other organic nutrient sources, sewage sludge contributes micronutrients as well as macronutrients. Levels of plant micronutrient metals (zinc, copper, iron, manganese, and nickel) as well as other heavy metals (cadmium, chromium, lead) are determined largely by the degree to which industrial wastes have been mixed in with domestic wastes. In the United States, the levels of metals in sewage are far lower than they were in the past, because of source-reduction programs that require industrial facilities to remove pollutants *before* sending their sewage to municipal treatment plants. Nonetheless, vigilance must be maintained to avoid sludges too contaminated for safe land application.

FIGURE 14.12 Final treatment of sewage effluent and recharge of groundwater are being accomplished by natural soil and plant processes in this effluent-irrigated forest on Ultisols near Atlanta, Georgia. Nutrient flows, groundwater quality, and tree growth are carefully monitored. Some of the greatly increased production of wood is used as an energy source to run the sewage treatment plant. (Photo courtesy of R. Weil)

In comparison with inorganic fertilizers, sludges are generally low in nutrients, especially potassium. Most of the potassium in sewage is soluble and remains with the effluent during treatment. The phosphorus content is higher, because advanced sewage treatment is designed to remove phosphorus from the effluent and deposit it in the sludge (see Box 13.2). If the sewage treatment precipitates phosphorus by reactions with iron or aluminum compounds, the phosphorus in the sludge will likely have a very low availability to plants.

14.6 PRACTICAL UTILIZATION OF ORGANIC NUTRIENT SOURCES

Using biosolids as a plant nutrient source:
http://www.agweb.okstate.edu/pearl/plantsoil/soils/cr-2201.pdf

In Section 11.7 we discussed the many beneficial effects on soil physical and chemical properties that can result from amendment of soils with decomposable organic materials. Here we will focus on the nutrient management aspects of organic waste utilization. Whether the material is sewage sludge, farm manure, or MSW compost, several general principles apply to the ecologically sound application of the material to soils:

The rate of application is generally governed by the amount of nitrogen or phosphorus that the organic material will make available to plants. Nitrogen usually is the first criterion because it is needed in the largest quantity by most plants, and because excess nitrogen can present a pollution problem (see Section 12.9). However, the P/N ratio in most organic sources is higher than in plant tissue. Consequently, if the organic materials supply sufficient nitrogen to meet plant needs, excessive levels of soil phosphorus will result (see Section 13.2), and must be taken into account in the long run.

For soils that already have high levels of phosphorus, the amount of an organic nutrient source applied may be limited by its phosphorus content. One of the reasons that animals manures have caused problems with excessive soil phosphorus levels is that commercial animal feeds are commonly fortified with 25 to 50% more phosphorus than the animals actually need. Potentially, toxic heavy metals in some materials may also limit the rate of application.

A small fraction of the nitrogen in manure or sludge may be soluble and immediately available, but the bulk of the nitrogen must be released by microbial mineralization of organic compounds. Those materials that have been partially decomposed during treatment and handling (e.g., composts and digested sludge) release a lower percentage of the nitrogen (Table 14.7).

If a field is treated annually with an organic material, the application rate needed will become progressively smaller because, after the first year, the amount of nitrogen released from material applied in previous years must be subtracted from the total to be applied afresh (see Box 14.2). This is especially true for composts for which the initial availability of the nitrogen is quite low. Instead of making progressively smaller applications, another practical strategy is to use a moderate application every year, but supplement the nitrogen from other sources in the first few years until nitrogen release from residual previous applications can supply the entire requirement.

TABLE 14.7 Release of Mineral Nitrogen from Various Organic Materials Applied to Soils, as Percent of the Organic Nitrogen Originally Present[a]

For example, if 10 Mg of poultry litter initially contains 300 kg (3.0%) nitrogen in organic forms, 50% or 150 kg of nitrogen would be mineralized in the first year. Another 15% (0.15 × 300) or 45 kg of nitrogen would be released in the second year.

Organic nitrogen source	*Year 1*	*Year 2*	*Year 3*	*Year 4*
Poultry floor litter	50	15	8	3
Dairy manure (fresh solid)	35	18	9	4
Swine manure lagoon liquid	50	15	8	3
Feedlot cattle manure	40	18	8	2
Composted fedlot manure	20	8	4	1
Lime-stabilized, aerobically digested sewage sludge	40	12	5	2
Anaerobically digested sewage sludge	20	8	4	1
Composted sewage sludge	10	5	3	2
Activated, unstabilized sewage sludge	45	15	4	2

[a]These values are approximate and may need to be increased for warm climates or sandy soils and decreased for cold or dry climates or heavy clay soils.

BOX 14.2 CALCULATION OF AMOUNT OF ORGANIC NUTRIENT SOURCE NEEDED TO SUPPLY NITROGEN FOR A CROP

Our example is a field producing corn for 2 consecutive years. The goal is to produce 7000 kg/ha of grain each year. This yield normally requires the application of 120 kg/ha of available nitrogen, which we expect to obtain from a lime-stabilized sewage sludge containing 4.5% total N and 0.2% mineral N (ammonium and nitrate).

Year 1

% organic N in sludge = *total N − mineral N* = 4.5% − 0.2% = 4.3%.
Organic N in 1 Mg of sludge = 0.043 × 1000 *kg* = 43 *kg* N.
Mineral N in 1 Mg of sludge = 0.002 × 1000 *kg* = 2 *kg* N.
Mineralization rate for lime-stabilized sludge in first year (Table 14.7) = 40% *of organic* N.
Available N mineralized from 1 Mg sludge in first year = 0.40 × 43 *kg* N = 17.2 *kg* N.
Total available N from 1 Mg sludge = *mineral N* + *mineralized N* = 2.0 + 17.2 = 19.2 *kg* N.
Amount of (dry) sludge needed = 120 *kg N/(19.2 kg available N/Mg dry sludge)* = 6.25 Mg *dry sludge*
Adjust for moisture content of sludge (e.g., assume sludge has 25% solids and 75% water).
Mg wet sludge to apply: 6.25 Mg dry sludge/(0.25 Mg dry sludge/Mg wet sludge) = 25 Mg wet sludge.

Year 2

Second year mineralization rate (Table 14.7) = 12% *of original organic* N.
N mineralized in year 2 from sludge applied in year = 0.12 × 43 *Kg N/Mg* × 6.25 *Mg dry sludge* = 32.25 *kg N from sludge in year* 2.
N needed from sludge applied in year 2 = *N needed by corn* − *N from sludge applied in year 1*
= 120 *kg* − 32.25 *kg* = 87.75 *kg N needed/ha.*
Amount of (dry) sludge to apply in year 2 = 87.75 *kg N/(19.2 kg available N/Mg dry sludge)* = 4.57 *Mg dry sludge/ha.*
Adjust for moisture content of sludge (e.g., assume sludge has 25% solids and 75% water).
Mg of wet sludge to apply: 4.57 Mg dry sludge/(0.25 Mg dry sludge/Mg wet sludge) = 18.3 Mg wet sludge

Note that the 10.82 Mg sludge (6.25 + 4.57) also provided 216 kg P/ha (assuming 2% P; see Table 14.6), an amount that greatly exceeds the crop requirement and will soon lead to excessive buildup of P.

The nutrient and moisture contents of organic amendments vary widely among sources, and even from batch to batch, depending on how the material has been stored and treated. Therefore, general values such as those in Table 14.6 should not be relied upon when calculating rates of material to apply. Instead, representative samples of the material should be analyzed in a laboratory. Analyses are generally required by laws regulating land application of sewage sludge, but other materials are not as widely regulated in this regard.

Nutrients from organic sources will produce the highest return and cause the least environmental damage if applied to fields that are relatively poorly supplied with nitrogen and phosphorus. Unfortunately, cost of transport and considerations of convenience provide incentives to apply organic materials on land close to their source, often leading to the nutrient imbalances discussed in Section 14.4.

Special Uses

There are a number of special uses of organic nutrient sources for which the organic matter component plays a special role. Examples include applications to denuded soil areas resulting from erosion, from land-leveling for irrigation, or from mining operations (Table 14.8). Improvements in water-holding capacity and soil structure brought about by organic material application may be just as important as the long-term, slow-release, nutrient-supplying features of the organic wastes. For example, in the mine reclamation project featured in Table 14.8, not only did planted species grow faster, but woody and herbaceous native perennial plants also preferentially invaded the sludge treated plots. Initial applications of 50 to more than 100 Mg/ha may be justified to supply organic matter as well as nutrients, provided the material is not so high in nitrogen as to create the potential for nitrate leaching.

TABLE 14.8 Heavy Application of Sewage Sludge Helps Restoration of Drastically Disturbed Site

Acid, infertile, shaley soils forming from coal mine spoil were seeded in year 1 with a mix of grass and broadleaf species. The revegetation effort benefited from both the nutrients and the organic matter in the sludge. Compared to fertilizer, sludge application increased the concentration of some potentially toxic metals in grass tissue, but not above normal levels.

	Standing plant biomass		*Metals in plant tissue in year 2, mg/kg*		
	Year 2	*Year 5*	*Cr*	*Ni*	*Cu*
Control (no amendment)	1.55	1.49	0.9	5.7	12.1
Commercial fertilizer	4.79	3.10	0.5	1.6	14.8
Sewage sludge (368 Mg/ha)	5.68	3.07	1.8	0.6	24.8

Data from Haering et al. (2000)

Special cases of micronutrient deficiency can be ameliorated with manure application. Such treatments are sometimes used when there is some uncertainty about which specific nutrient is lacking. Manure applications can usually be made with little concern for adding toxic quantities of the micronutrients. A well-documented exception is the use of manure from swine that have been fed copper supplements to stimulate faster growth. Application of such manure can result in the buildup of phytotoxic levels of copper in amended soils.

14.7 INORGANIC COMMERCIAL FERTILIZERS

The worldwide use of fertilizers increased dramatically during the latter half of the 20th century (Figure 14.13), accounting for a significant part of the equally dramatic increases in crop yields during the same period. The need to supplement forest soil fertility is also increasing as the demands for forest products increase the removal of nutrients and competing uses of land leave forestry with more infertile, marginal sites.

Regional Use of Fertilizers

Fertilizer use statistics within a region tell whether fertilizers are contributing to soil quality enhancement or to environmental degradation. For example, in the Netherlands, nitrogen additions from fertilizers and manures are more than 4 times the removal of this

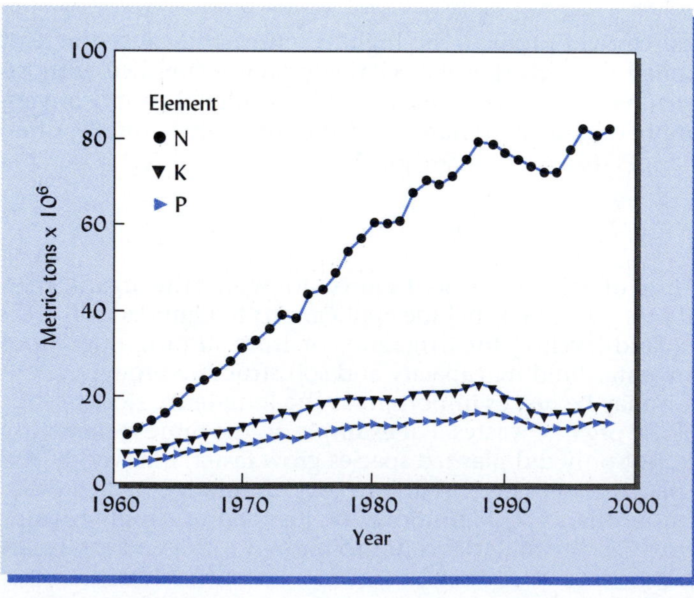

FIGURE 14.13 World fertilizer use since 1960 by major nutrient element. The use of nitrogen has increased much faster than that of phosphorus and potassium. The dip in world fertilizer use in the early 1990s was due mainly to drastic reductions in Russia and Ukraine after the collapse of the Soviet Union. The apparent leveling off since 1990 of the world total is the result of decreases in most of the industrialized countries combined with continued increases in China, India, and much of the developing world. In the latter countries, the removal of nutrients in crop harvests still exceeds the amounts returned to the soil. [Data from the United Nations Food and Agriculture Organization]

element in harvested crops. In contrast, soils in sub-Saharan Africa are literally being mined—far less nutrients are being added from all sources than are being removed in the crops. The fertilizer nutrient rate in this region is only about 10% of the world average. We must be as site specific as possible in evaluating the role of fertilizers in meeting humanitarian and environmental goals.

Origin and Processing of Inorganic Fertilizers

Most fertilizers are inorganic salts containing readily available plant nutrient elements. Some are manufactured, but others, such as phosphorus and potassium, are found in natural geological deposits. Relicts from ancient seas, **beds of solid salts** located far beneath the earth's surface are the primary sources of potassium. The salts are mined and then purified, yielding such compounds as potassium chloride and potassium sulfate. Apatite found in **phosphate rock deposits** is the primary source of phosphorus fertilizers. Since apatite is extremely insoluble, however, it is treated with sulfuric, phosphoric, or nitric acid to produce materials, such as triple superphosphate, that yield phosphorus in a readily available form.

Under very high temperatures and pressures, the **nitrogen gas in the atmosphere** is fixed with hydrogen from natural gas to produce ammonia gas, NH_3. This fixation consumes tremendous quantities of energy—usually from fossil fuels. When the ammonia gas is then put under moderate pressure, it liquifies, forming anhydrous ammonia. Because of its low cost and ease of application, more nitrogen is applied directly to soils in anhydrous ammonia than in any other fertilizer. More important, it is the starting point for the manufacture of most of the other nitrogen carriers, including urea, ammonium nitrate, ammonium sulfate, sodium nitrate, and liquid mixtures known as "nitrogen solutions."

Many commercial fertilizers contain two or more of the primary nutrient elements. In some cases, mixtures of the primary nutrient carriers are used. But in others, a given compound may carry two nutrients, examples being monoammonium phosphate, diammonium phosphate, and potassium nitrate. Care must be used in selecting the components of a mixed fertilizer since some compounds are not compatible with others, resulting in poor physical condition and reduced nutrient availability in the mixture.

Properties and Use of Inorganic Fertilizers

The composition of inorganic commercial fertilizers is much more precisely defined (Table 14.9) than is the case for the organic material discussed above. Most fertilizers supply plants with nitrogen, phosphorus, and/or potassium—sometimes called the *primary fertilizer elements*. Fertilizers that supply sulfur, magnesium, and the micronutrients are also manufactured.

A particular nutrient (say, nitrogen) can be supplied by many different *carriers,* or fertilizer compounds (Table 14.9). Decisions as to which fertilizers to use must take into account not only the nutrients they contain, but also other characteristics of the individual carriers. Table 14.9 provides information about some of these characteristics, such as the salt hazard (see also Section 9.7), acid-forming tendency (see also Section 12.16), tendency to volatilize, ease of solubility, and content of nutrients other than the principal one.

Physical Forms of Marketed Fertilizer

Increases in fertilizer use rates and labor costs, along with improved means of transporting and handling the fertilizer and increased availability of custom applicators, have favored two types of fertilizer application in the United States: (1) unbagged, dry solids handled in **bulk** form, and (2) **liquid** or fluid forms stored, transported, and applied from tanks. Bulk spreading, the favorite for applying multinutrient fertilizers, is often done for the grower by a fertilizer dealer or private custom applicator who also blends the desired nutrients. Liquid fertilizers comprise more than half of the single-nutrient carriers sold in the United States and about 40% of all fertilizers.

Fertilizer Grade

Every fertilizer label states the *grade* as a three-number code, such as 10-5-10 or 6-24-24. These numbers stand for percentages indicating the *total* nitrogen (N) content, the *available* phosphate (P_2O_5) content, and the *soluble* potash (K_2O) content. Plants do not

TABLE 14.9 Commonly Used Inorganic Fertilizer Materials: Their Nutrient Contents and Other Characteristics

Fertilizer	N	P	K	S	Salt hazard	Acid formation[b]	Other nutrients & comments
	Percent by weight						
Primarily sources of nitrogen							
Anhydrous ammonia (NH_3)	82				Low	−148	Pressurized equipment needed; toxic gas; must be injected into soil.
Urea [$CO(NH_2)_2$]	45				Moderate	−84	Soluble; hydrolyses to ammonium forms. Volatilizes if left on soil surface.
Ammonium nitrate (NH_4NO_3)	33				High	−59	Absorbs moisture from air; can be left on soil surface. Can explode if mixed with organic dust or S.
Sulfur-coated urea	30–40			13–16	Low	−110	Variable slow rate of release.
UF (ureaformaldehyde)	30–40				Very low	−68	Slowly soluble; faster with warm temperatures.
UAN solution	30				Moderate	−52	Most commonly used liquid N.
IBDU (isobutylidene diurea)	30				Very low	—	Slowly soluble.
Ammonium sulfate [$(NH_4)_2SO_4$]	21			24	High	−110	Rapidly lowers soil pH; very easy to handle.
Sodium nitrate ($NaNO_3$)	16				Very high	+29	Hardens, disperses soil structure.
Potassium nitrate (KNO_3)	13		36	0.2	Very high	+26	Very rapid plant response.
Primarily sources of phosphorus							
Monoammonium phosphate ($NH_4H_2PO_4$)	11	21–23		1–2	Low	−65	Best as starter.
Diammonium phosphate [$(NH_4)_2HPO_4$]	18–21	20–23		0–1	Moderate	−70	Best as starter.
Triple superphosphate		19–22		1–3	Low	0	15% Ca
Phosphate rock [$Ca_3(PO_4)_2 \cdot CaX$]		8–18[a]			Very low	Variable	Low to extremely low availability. Best as fine powder on acid soils. 30% Ca Contains some Cd, F, etc.
Single superphosphate		7–9		11	Low	0	Nonburning, can place with seed. 20% Ca.
Bonemeal	1–3[a]	10[a]	0.4		Very low	—	Slow availability of N, P as for phosphate rock. 20% Ca.
Colloidal phosphate		8[a]			Very low	—	P availability as for phosphate rock. 20% Ca.
Primarily sources of potassium							
Potassium chloride (KCl)			50		High	0	47% Cl–may reduce some diseases.
Potassium sulfate (K_2SO_4)			42	17	Moderate	0	Use where Cl not desirable.
Wood ashes		0.5–1	1–4		Moderate to high	+40	About ½ the liming value of limestone; caustic. 10–20% Ca 2–5% Mg; 0.2% Fe, 0.8% Mn.
Greensand		0.6	6		Very low	0	Very low availability.
Granite dust			4		Very low	0	Very slow availability.
Primarily sources of other nutrients							
Basic slag		1–7			Low	+70	Slow availability; best on acid soils. 3–30% Ca, 3% Mg, 10% Fe, 2% Mn
Gypsum ($CaSO_4 \cdot 2H_2O$)				19	Low	0	Stabilizes soil structure; no effect on pH; Ca and S readily available. 23% Ca.
Calcitic limestone ($CaCO_3$)					Very low	+95	Slow availability; raises pH. 36% Ca
Dolomitic limestone [$CaMg(CO_3)_2$]					Very low	+95	Very slow availability; raises pH. ~24% Ca, ~12% Mg.
Epsom salts ($MgSO_4 \cdot 7H_2O$)				13	Moderate	0	No effect on pH; water soluble. 2% Ca, 10% Mg
Sulfur, flowers (S)				95	—	−300	Irritates eyes; very acidifying; slow acting; requires microbial oxidation.
Solubor					Moderate	—	Very soluble; compatible with foliar sprays. 20.5% B
Borax ($Na_2B_4O_7 \cdot 10H_2O$)					Moderate	—	Very soluble. 11% B; 9% Na
EDTA chelates					—	—	See label. Usually 13% Cu or 10% Fe or 12% Mn or 12% Zn
Cu, Fe, Mn, or Zn sulfates				13–20			25% Cu, 19% Fe, 27% Mn, or 35% Zn, very soluble.

[a]Highly variable contents.

[b]A negative number indicates that acidity is produced; a positive number indicates that alkalinity is produced; kg $CaCO_3$ needed to neutralize acidity per 100kg of material applied.

take up phosphorus and potassium in these chemical forms, nor do any fertilizers actually contain P_2O_5 or K_2O. Unfortunately, these "oxide" expressions found their ways into state laws governing the sale of fertilizers, and there is considerable resistance to changing them, although some progress is being made. In scientific work and in this textbook, the simple elemental contents are used (P and K) wherever possible. Box 14.3 and Figure 14.14 explain how to convert between the elemental and oxide forms of expression.

The grade is important from an economic standpoint because it conveys the analysis or concentration of the nutrient in a carrier. When properly applied, most fertilizer carriers give equally good results for a given amount of nutrient element. The more concentrated carriers are usually the most economical to use, because less weight of fertilizer must be transported to supply the needed quantity of a given nutrient. Hence, economic comparisons among different equally suitable fertilizers should be based on the price per kilogram of nutrient, not the price per kilogram of fertilizer.

Fate of Fertilizer Nutrients

A common *myth* about fertilizers suggests that inorganic fertilizers applied to soil directly feed the plant, and that therefore the biological cycling of nutrients are of little consequence where inorganic fertilizers are used. The reality is that nutrients added by normal application of fertilizers, whether organic or inorganic, are incorporated into complex soil nutrient cycles, and that relatively little of the fertilizer nutrient (from 10 to 60%) actually winds up in the plant being fertilized during the year of application (Table 14.10). Even when the application of fertilizer greatly increases both plant growth and nutrient uptake, the fertilizer stimulates increased cycling of the nutrients, and the nutrient ions taken up by the plant come largely from various pools in the soil and not directly from the fertilizer. Generally, as fertilizer rates are increased, the efficiency of fertilizer nutrient use decreases, leaving behind in the soil an increasing proportion of the added nutrient.

BOX 14.3 HOW MUCH NITROGEN, PHOSPHORUS, AND POTASSIUM IS IN A BAG OF 6-24-24?

Conventional labeling of fertilizer products reports percentage N, P_2O_5, and K_2O. Thus, a fertilizer package (Figure 14.14) labeled as 6-24-24 (6% nitrogen, 24% P_2O_5, 24% K_2O) actually contains 6% N, 10.5% P, and 19.9% K (see calculations below). To determine the amount of fertilizer needed to supply the recommended amount of a given nutrient, first convert percent P_2O_5 and percent K_2O to percent P and K, by calculating the proportion of P_2O_5 that is P and the proportion of K_2O that is K. The following calculations may be used:

Given that the molecular weights of P, K, and O are as follows: P = 31, K = 39, 0 = 16 *g/mol*

Molecular weight of $P_2O_5 = 2(31) + 5(16) = 142$ *g/mol*

$$\text{Proportion P in } P_2O_5 = \frac{2P}{P_2O_5} = \frac{2(31)}{2(31) + 5(16)} = 0.44$$

To convert P_2O_5 to P multiply percent P_2O_5 by 0.44

Molecular weight of $K_2O = 2(39) + 16 = 94$

$$\text{Proportion K in } K_2O = \frac{2K}{K_2O} = \frac{2(39)}{2(39) + 16} = 0.83$$

To convert K_2O to K multiply percent K_2O by 0.83

Thus, if the bag in Figure 14.14 contains 25 kg of the 6-24-24 fertilizer, it will supply 1.5 kg N (0.06 × 25); 2.6 kg P (0.24 × 0.44 × 25); and 5 kg K (0.24 × 0.83 × 25).

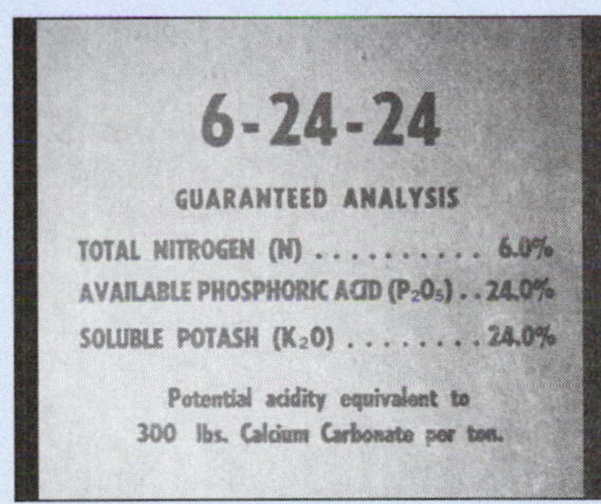

FIGURE 14.14 *A typical commercial fertilizer label. Note that a calculation must be performed to determine the percentage of the nutrient elements P and K in the fertilizer since the contents are expressed as if the nutrients were in the forms of P_2O_5 and K_2O. Also note that after interacting with the plant and soil, this material would cause an increase in soil acidity that could be neutralized by 300 units of $CaCO_3$ per 2000 units (1 ton = 2000 lbs) of fertilizer material.*

TABLE 14.10 Source of Nitrogen in Corn Plants Grown in North Carolina on an Enon Sandy Loam Soil (Ultic Hapludalf) Fertilized with Three Rates of Nitrogen as Ammonium Nitrate

The source of the nitrogen in the corn plant was determined by using fertilizer tagged with the isotope ^{15}N. Moderate use of nitrogen fertilizer increased the uptake of N already in the soil system as well as that derived from the fertilizer.

Fertilizer nitrogen applied, kg/ha	*Corn grain yield, Mg/ha*	*Total N in corn plant, kg/ha*	*Fertilizer-derived N in corn, kg/ha*	*Soil-derived N in corn, kg/ha*	*Fertilizer-derived N in corn as percent of total N in corn*	*Fertilizer-derived N in corn as percent of N applied*
50	3.9	85	28	60	33	56
100	4.6	146	55	91	38	55
200	5.5	157	86	71	55	43

Calculated from Reddy and Reddy (1993).

The Concept of the Limiting Factor

A famous German chemist named Justus von Liebig is credited with first publishing the concept that *plant production can be no greater than that level allowed by the growth factor present in the lowest amount relative to the optimum amount for that factor.* This growth factor, be it temperature, nitrogen, or water supply, will limit the amount of growth that can occur and is therefore called the **limiting factor** (Figure 14.15).

If a factor is not the limiting one, increasing it will do little or nothing to enhance plant growth. In fact, increasing the amount of a nonlimiting factor may actually reduce plant growth by throwing the system further out of balance. For example, if a plant is limited by lack of phosphorus, adding more nitrogen may only aggravate the phosphorus deficiency.

Looked at another way, applying available phosphorus (the first limiting nutrient in this example) may allow the plant to respond positively to a subsequent addition of nitrogen. Thus, the increased growth obtained by applying two nutrients together often is much greater than the sum of the growth increases obtained by applying each of the two nutrients individually. Such an *interaction* or *synergy* between two nutrients can be seen in the data of Figure 14.16.

14.8 FERTILIZER APPLICATION METHODS

Wise, effective fertilizer use involves making correct decisions regarding *which* nutrient element(s) to apply, *how much* of each needed nutrient to apply, *what type* of material or carrier to use (Tables 14.6 and 14.9 list some of the choices), *in what manner* to apply the

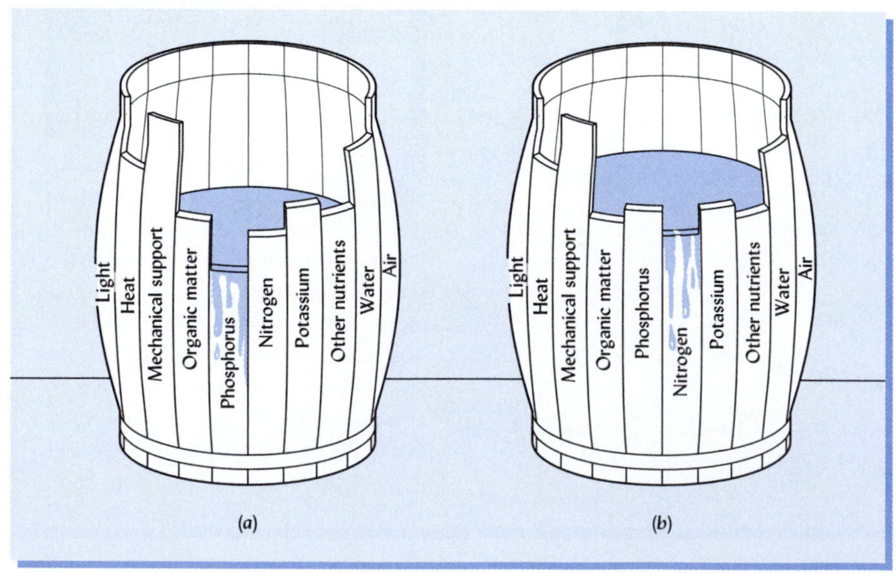

FIGURE 14.15 An illustration of the law of the minimum and the concept of the limiting factor. Plant growth is constrained by the essential element (or other factor) that is most limiting. The level of water in the barrel represents the level of plant production. (*a*) Phosphorus is represented as being the factor that is most limiting. Even though the other elements are present in more than adequate amounts, plant growth can be no greater than that allowed by the level of phosphorus available. (*b*) When phosphorus is added, the level of plant production is raised until another factor becomes most limiting—in this case, nitrogen.

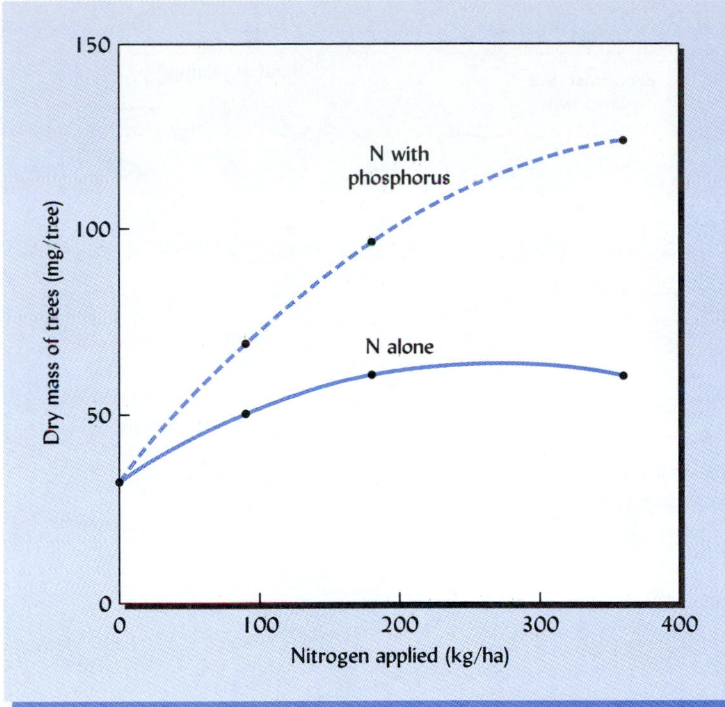

FIGURE 14.16 Increase in biomass of year-old white pine seedlings in response to nitrogen fertilizer given with or without phosphorus on a sandy soil in southern Ontario, Canada. In the unfertilized soil, apparently nitrogen was most limiting, and therefore no response to phosphorus was obtained unless the nitrogen level was raised (compare curves at zero nitrogen rate). Once nitrogen was applied, the trees did respond to added phosphorus. After about 150kg/ha of nitrogen was supplied, phosphorus became the most limiting nutrient, with further response to more nitrogen occurring only where phophorus was also applied. This is an example of a nitrogen-by-phosphorus interaction. [Redrawn from Teng and Timmer (1994)]

material, and, finally, *when* to apply it. We will leave information on the first two decisions until Section 14.10. Here we will discuss the alternatives available with regard to the last two decisions.

There are three general approaches to applying fertilizers: (1) *broadcast application,* (2) *localized placement,* (3) and *foliar application.* Each method has some advantages and disadvantages and may be particularly suitable for different situations. Often some combination of the three methods is used.

Broadcasting

In many instances fertilizer is spread evenly over the entire field or area to be fertilized. This method is called **broadcasting.** Often the broadcast fertilizer is mixed into the plow layer by means of tillage, but in some situations it is left on the soil surface and allowed to be carried into the root zone by percolating rain or irrigation water (Fig 14.17*a-c*). The broadcast method is most appropriate when a large amount of fertilizer is being applied with the aim of raising the fertility level of the soil over a long period of time.

For phosphorus, zinc, manganese, and other nutrients that tend to be strongly retained by the soil, broadcast applications are usually much less efficient than localized placement. Often 2 to 3 kg of fertilizer must be broadcast to achieve the same response as from 1 kg that is placed in a localized area.

A heavy one-time application of phosphorus and potassium fertilizer, broadcast and worked into the soil, is a good preparation for establishing perennial plants such as lawns, pastures, and orchards. It may be necessary to broadcast a top dressing in subsequent years, being careful not to allow fertilizers with high salt hazards (see Table 14.9) to remain in contact with the foliage long enough to cause salt burn. Because of its mobility in the soil, nitrogen does not suffer from reduced availability when broadcast, but if left on the soil surface much may be lost by volatilization. Volatilization losses are especially troublesome for urea and ammonium fertilizers applied to soils with a high pH. Nitrogen is commonly broadcast (sprayed) in a liquid form, often in a solution that also contains other nutrients or chemicals. Runoff studies have shown that most of the annual loss of nutrients (or herbicides, if surface broadcast) usually occurs during the first one or two heavy-rainfall events after the broadcast application. Where irrigation is practiced, liquid fertilizers can be applied in the irrigation water, a practice sometimes called **fertigation.**

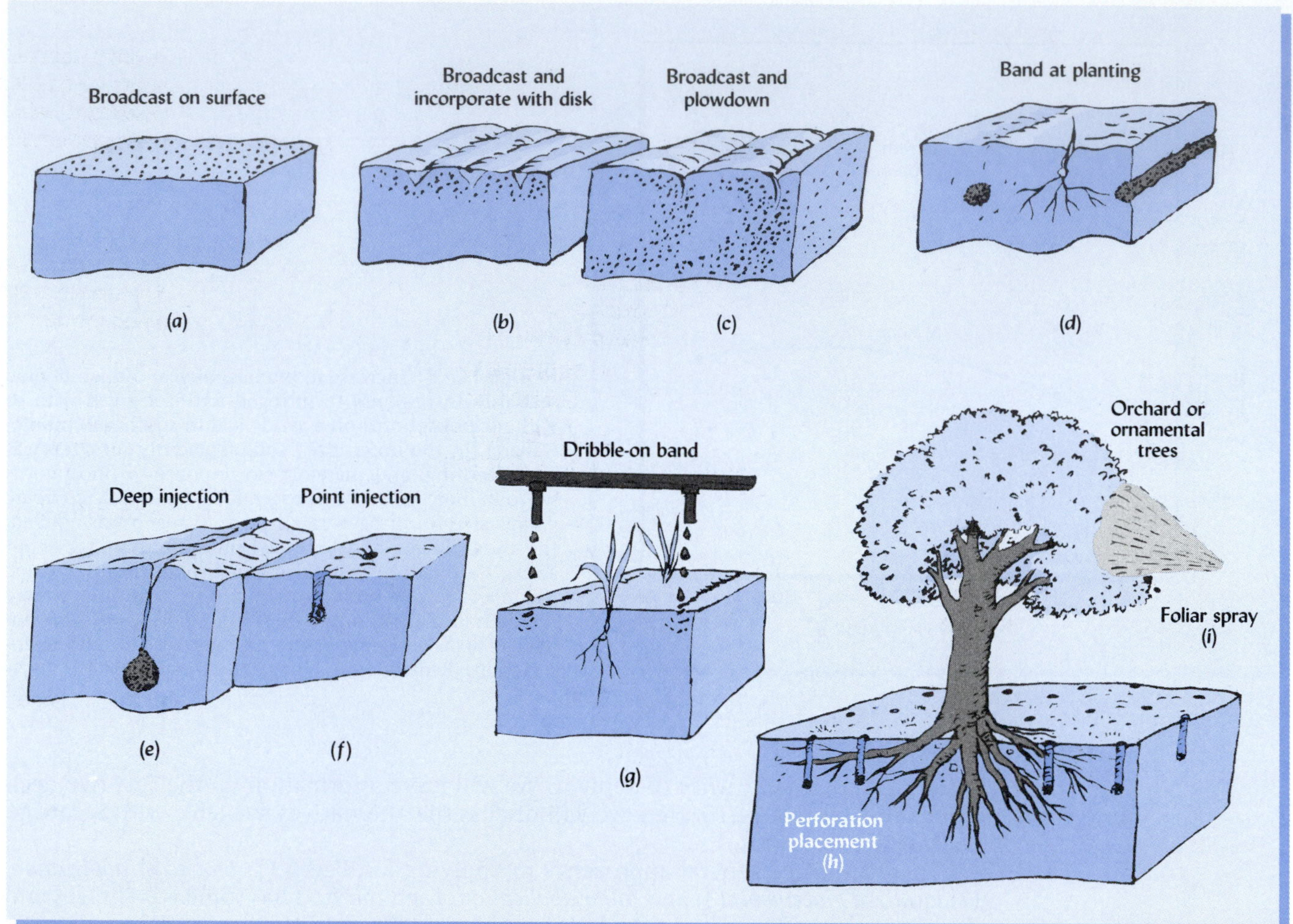

FIGURE 14.17 Fertilizers may be applied by many different methods, depending on the situation. Methods (*a*) to (*c*) represent broadcast fertilizer, with or without incorporation. Methods (*d*) to (*h*) are variations of localized placement. Method (*i*) is foliar application and has special advantages, but also limitations. Commonly, two or three of these methods may be used in sequence. For example, a field may be prepared with (*c*) before planting; (*d*) may be used during the planting operation; (*g*) may be used as a **side-dressing** early in the growing season; and, finally, (*i*) may used to correct a micronutrient deficiency that shows up in the middle of the season.

Localized Placement

Although it is commonly thought that nutrients must be thoroughly mixed throughout the root zone, research has clearly shown that a plant can easily obtain its entire supply of a nutrient from a concentrated localized source in contact with only a small fraction of its root system. There are at least two reasons why fertilizer is often more effectively used by plants if it is placed in a localized concentration rather than mixed with soil throughout the root zone. First, localized placement reduces the amount of contact between soil particles and the fertilizer nutrient, thus minimizing the opportunity for adverse fixation reactions. Second, in the fertilized zone the concentration of the nutrient in the soil solution at the root surface will be very high, resulting in greatly enhanced uptake by the roots.

STARTER APPLICATION. Localized placement is especially effective for young seedlings, in cool soils in early spring, and for plants that grow rapidly with a big demand for nutrients early in the season. For these reasons, starter fertilizer is often applied in bands on either side of the seed as the crop is planted. Since germinating seeds can be injured by fertilizer salts, and since these salts tend to move upward as water evaporates from the soil surface, the best placement for starter fertilizer is approximately 5 cm below and 5 cm off to the side from the seed row (see Figure 14.17*d*).

LIQUID FERTILIZERS. Liquid fertilizers and slurries of manure and sewage sludge can also be applied in bands rather than broadcast. Bands of these liquids are placed 10 to 30 cm

PLATE 31 Urban soils (referred to by some as *Urbents*) often hold surprises in their profiles. Here a tree planting hole reveals a buried A horizon (*top*) and buried asphalt (*lower*).

PLATE 34 Mass wasting of clayey soils on steep slope may occur when saturated with water, as in this rotational block slide in East Africa. Note man in center for scale.

PLATE 35 Two large soil stockpiles on a construction site. The brown A horizon was set aside for landscaping topsoil; the redder B horizon (*back*) was stockpiled for use as fill and road base.

PLATE 32 Soil saturated beyond its liquid limit by torrential rains caused a landslide and mudflow that pushed huge tropical cloud forest trees downslope like toothpicks and demolished a village at the foot of this mountain in Honduras.

PLATE 33 The red, kaolinitic soil in the foreground was hauled in to build up a stable roadbed across this low-lying landscape in South Central Tanzania. The black soils are rich in expansive clays, which would break up the pavement if used for the subbase.

PLATE 36 An uneven layer of silt blankets a glacial deposit of coarse sand and gravel in this Rhode Island Inceptisol. The profile water-holding capacity varies with the thickness of the silt, causing an irregular pattern of drought-stricken turf grass (*inset*).

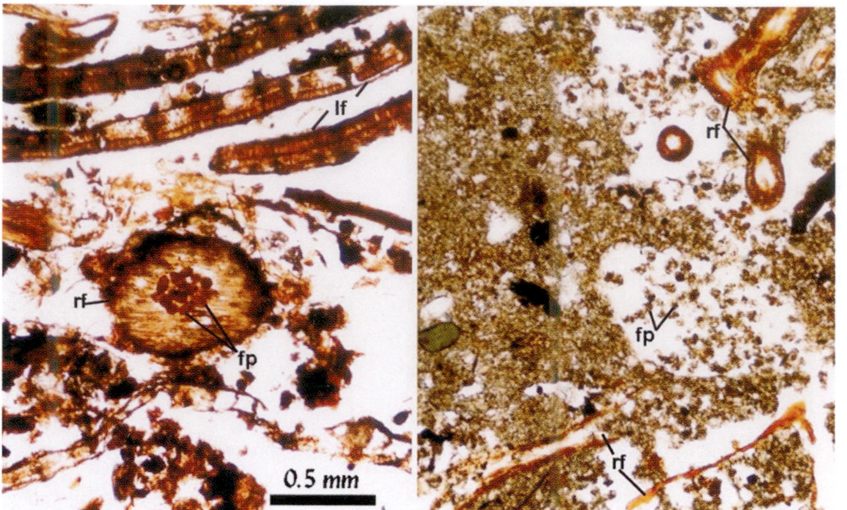

PLATE 37 Influence of soil fauna on microstructure in O (*left*) and A (*right*) horizons of a forested Ultisol in Tennessee. Particulate organic matter (POM) includes leaf fragments (lf), fecal pellets (fp) and root fragments (rf).

PLATE 38 Roots from sweet pepper plants follow organic-matter-lined earthworm burrows through the compacted B horizon of a Pennsylvania Inceptisol. Earthworm activity was encouraged by 15 years of no-till practices and cover crops.

PLATE 39 Plant roots grow along the gleyed coating of a fragipan prism whose reddish interior is too dense for roots to grow. The roots are squeezed flat between the prisms.

PLATE 40 This cicada nymph, nestled 60 cm deep in the B horizon of a forested Ultisol, will feed on oak tree roots for several years before emerging as an adult. Free water in the macropore bathes the cicada, whose burrowing promotes drainage and enhances root growth.

PLATE 41 Traditional humid-region farmers use slash-and-burn systems. They chop down patches of forest and burn the dead vegetation, returning many nutrients in the ash. Note fire in the background and burned logs in the foreground of this Sri Lankan woman's new clearing.

PLATE 42 Drought-stressed grass shows importance of soil depth. The rectangular area of brown grass is underlain by a shallow (25-cm) layer of soil atop the roof of an underground library. The trees and green grass at right grow in deeper soil.

PLATE 43 The soil on the left of this hydrangea was limed, that on the right was acidified (with $FeSO_4$). After a year, blue flowers formed on the low-pH side, pink on the high-pH side.

PLATE 44 Leaves from near the bottom of nitrogen-deficient (yellow tip and midrib), potassium-deficient (necrotic leaf edges), and normal corn plants. All the leaves came from the same field.

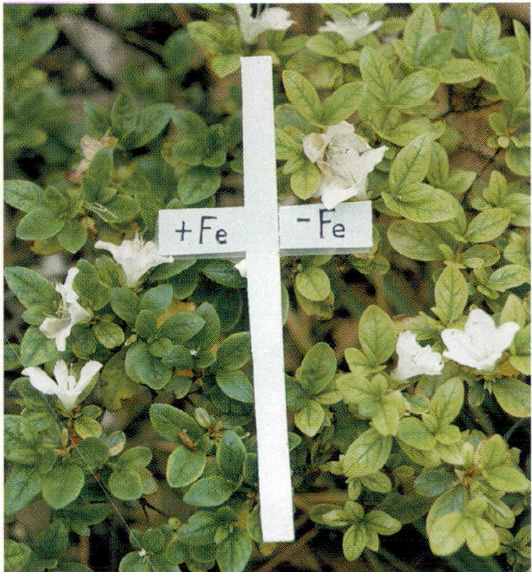

PLATE 45 This iron-deficient azalea was sprayed with $FeSO_4$ on one side 3 days before being photographed. Soil pH higher than 5.5 can induce such iron deficiency.

PLATE 46 Magnesium deficiency causes interveinal chlorosis on the older leaves of this poinsettia.

PLATE 47 Phosphorus deficiency causes severe stunting and purpling of older leaves of this tomato.

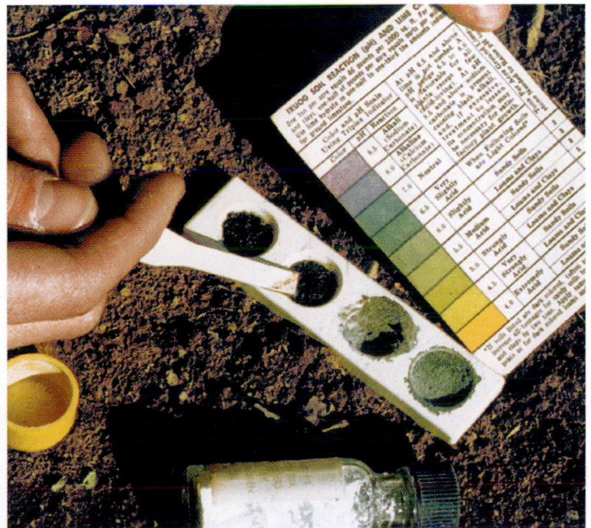

PLATE 48 The color developed after wetting soil with a pH-sensitive dye can be compared to a color chart in order to estimate soil pH in the field. This soil has a pH of about 7. See page 280.

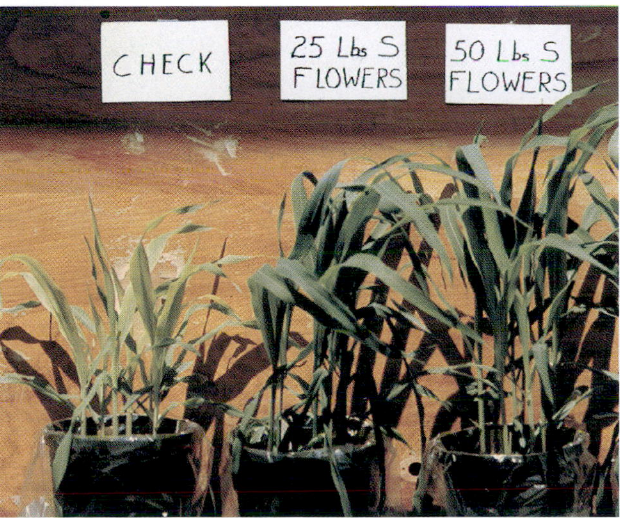

PLATE 49 Sulfur deficiency typically causes chlorosis (yellowing) on the *youngest* leaves first, or on all the foliage, as in the sorghum plant on the left. This contrasts with nitrogen deficiency, which causes chlorosis first on the *oldest* leaves. Plants growing on low-sulfur soil responded to sulfur addition.

PLATE 50 Nitrogen-deficient corn on Udolls in central Illinois. Ponded water after heavy rains resulted in nitrogen loss by denitrification and leaching.

PLATE 51 Slow-moving coastal plain stream choked with algal bloom caused by nitrogen and phosphorus from upstream farmland.

PLATE 52 Normal (*left*) and phosphorus-deficient (*right*) corn plants. Note stunting and purple color.

PLATE 53 Zinc deficiency on peach tree. Note whorl of small, misshaped leaves.

PLATE 54 Zinc deficiency on sweet corn. Note broad whitish bands.

PLATE 55 Eroded calcareous soil (Ustolls) with iron-deficient sorghum.

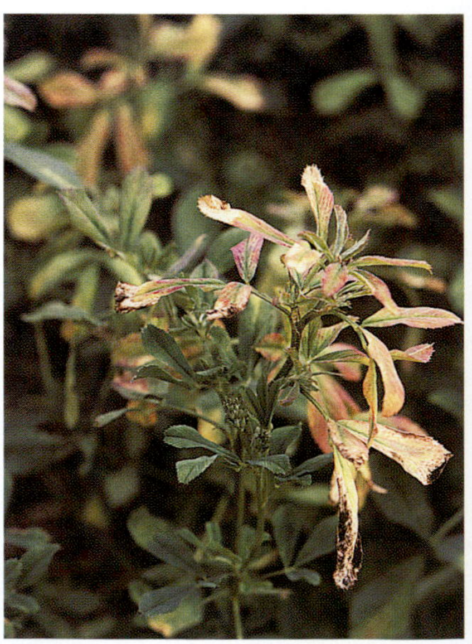

PLATE 56 Boron deficiency on alfalfa. Note reddish foliage.

PLATE 57 Pink blooms belong to pioneering redbud (*Cercis canadensis* L) trees, a nitrogen-fixing legume that enriches the soil for the other species (which eventually will take over as the forest matures).

PLATE 58 Nitrogen deficiency on a pine tree. The yellowing (chlorosis) occurs mainly on the older needles.

PLATE 59 The large soybean root nodule was cut open to show its red interior indicative of active nitrogen fixation. The red comes from an iron-coordinated compound very similar to the hemoglobin that makes human blood turn red when oxygenated.

PLATE 60 Looking like snow, the salt crust covering this soil formed when salt-laden groundwater in this salt marsh near the Great Salt Lake in Utah rose by capillarity and evaporated from the soil surface, leaving the dissolved salt behind.

PLATE 61 Sea spray has caused salt injury (brown leaves) despite the high salt tolerance of this Bermuda grass at Pebble Beach Golf Course in California.

PLATE 62 It's a good thing this homeowner readjusted his spreader before he finished fertilizing the lawn. Salt "burn" from too much nitrogen fertilizer.

PLATE 63 Iron deficiency causes yellowing with sharply contrasting green veins on the younger leaves. Rose growing in soil with pH 6.8.

PLATE 64 Stream polluted by acid drainage caused by sulfurization of soils forming in coal mine spoil. $FeSO_4$ in the acid drainage oxidizes in the stream to cause the orange color.

PLATE 65 Early stages of soil formation in material dredged from Baltimore Harbor. Sulfidic materials (black), acid drainage (orange liquid), salt accumulations (whitish crust), and initiation of prismatic structure (cracks) are all evident.

PLATE 66 Landsat Thematic Mapper image of Washington, D.C. (*upper left*), and sediment-laden Potomac River (*center*). Composite image with natural colors.

PLATE 67 Landsat Thematic Mapper image of Palo Verde Valley, California, irrigation scheme. Composite image using bands 2, 3, and 4. Lush vegetation appears bright red.

Plate 1 courtesy of Bill Waltman and D. J. Lathwell, Cornell University; Plates 2, 10, 11, 13, 15, 16, 18–23, 25–36, and 38–65 courtesy of Ray Weil; Plates 3, 4, 8 and 12 courtesy of R. W. Simonson; Plates 5 and 14 courtesy of Chien-Lu Ping, Agriculture and Forestry Experimental Station, University of Alaska—Fairbanks; Plates 6, 7 and 9 courtesy of Soil Science Society of America; Plate 17 courtesy of Stefan Doerr, University of Swansea, Wales; Plate 24 courtesy of Carlos F. Dorronsoro Fernandez, University of Granada, Spain; Plate 37 courtesy of Debra Phillips, Oak Ridge National Laboratory, Tenn.; Plate 66 copyright Space Imaging, Inc.; Plate 67 courtesy of Earth Satellite Corp, Rockville, Md.

deep in the soil by a process known as *knife injection* (Figures 14.17*e*). In addition to the advantages mentioned for banding fertilizer, injection of these organic slurries reduces runoff losses and odor problems. Anhydrous ammonia and pressurized nitrogen solutions must be injected at depths of 15 and 5 cm, respectively, into the soil to prevent losses by volatilization.

Dribble Application. Another approach to banding liquids (though not slurries) is to dribble a narrow stream of liquid fertilizer alongside the crop row as a side dressing (Figure 14.17*g*). The use of a stream instead of a fine spray changes the application from broadcast to banding and results in enough liquid in a narrow zone to cause the fertilizer to soak into the soil. This action greatly reduces volatilization loss of nitrogen.

Drip Irrigation. The use of drip irrigation systems (see Section 6.11) has greatly facilitated the localized application of nutrients in irrigation water. Because drip fertigation is applied at frequent intervals, the plants are essentially spoonfed, and the efficiency of nutrient use is quite high.

Perforation Method for Trees. Trees in orchards and ornamental plantings are best treated individually, the fertilizer being applied around each tree within the spread of the branches but beginning approximately 1 m from the trunk (see Figure 14.17*h*). The fertilizer is best applied by what is called the *perforation* method. Numerous small holes are dug around each tree within the outer half of the branch-spread zone and extending down into the upper subsoil where the fertilizer is placed. This method of application places the nutrients within the tree root zone and avoids an undesirable stimulation of the grass or cover that may be growing around the trees.

Foliar Application

Plants are capable of absorbing nutrients through their leaves in limited quantities. Under certain circumstances, the best way to apply a nutrient is *foliar application*—spraying a dilute nutrient solution directly onto the plant leaves (Figure 14.17*i*). Diluted NPK fertilizers, micronutrients, or small quantities of urea can be used as foliar sprays, although care must be taken to avoid significant concentrations of Cl^- or NO_3^-, which can be toxic to some plants. Foliar fertilization may conveniently fit in with other field operations for horticultural crops, because the fertilizer is often applied simultaneously with pesticide sprays.

The amount of nutrients that can be sprayed on leaves in a single application is quite limited. Therefore, while a few spray applications may deliver the entire season's requirement for a micronutrient, only a small portion of the macronutrient needs can be supplied in this manner. The danger of leaf injury is especially high during dry, hot weather, when the solution quickly evaporates from the leaf surface, leaving behind the fertilizer salts. Spraying on cool, overcast days or during early morning or late evening hours reduces the risk of injury.

14.9 TIMING OF FERTILIZER APPLICATION

Availability When the Plants Need It

For mobile nutrients such as nitrogen (and to some degree potassium), the general rule is to make applications as close as possible to the period of rapid plant nutrient uptake. For rapid-growing summer annuals, such as corn, this means making only a small starter application at planting time and applying most of the needed nitrogen as a side-dressing just before the plants enter the rapid nutrient accumulation phase, usually about 4 to 6 weeks after planting. For cool-season plants, such as winter wheat or certain turf grasses, most of the nitrogen should be applied about the time of spring "green-up," when the plants resume a rapid growth rate. For trees, the best time is when new leaves are forming. With slow-release organic sources, time should be allowed for mineralization to take place prior to the plants' period of maximum uptake.

Environmentally Sensitive Periods

In temperate (Udic and Xeric) climates, most leaching takes place in the winter and early spring when precipitation is high and evapotranspiration is low. Nitrates left over after plant uptake has ceased have the potential for leaching during this period. In this regard it should be noted that, for grain crops, the rate of nutrient uptake begins to decline during grain-filling stages and has virtually ceased long before the crop is ready for harvest. With inorganic nitrogen fertilizers, avoiding leftover nitrates is largely a matter of limiting the amount applied to what the plants are expected to take up. However, for slow-release organic sources applied in late spring or early summer, mineralization is likely to continue to release nitrates after the crop has matured and ceased taking them up. To the extent that this timing of nitrate release is unavoidable, nonleguminous cover crops should be planted in the fall to absorb the excess nitrate being released.

SPLIT APPLICATIONS. In high-rainfall conditions and on permeable soils, dividing a large dose of fertilizer into two or more split applications may avoid leaching losses prior to the crop's establishment of a deep root system. In cold climates, another environmentally sensitive period occurs during early spring, when snowmelt over frozen or saturated soils results in torrents of runoff water that may carry to rivers and streams soluble nutrients from manure or fertilizer that may be at or near the soil surface.

Application of urea fertilizers to mature forests is usually carried out when rains can be expected to wash the nutrients into the soil and minimize volatilization losses. Furthermore, nitrogen fertilization of forests should occur just prior to the onset of the growing season (early spring, or in warm climates, winter) so that the tree roots will have the entire growing season to utilize the nitrogen. Fertilizer applications commonly result in a pulse of nitrate leaving the watershed for several weeks following the fertilizer application. The nitrogen loss can be reduced if a 10 to 15 m unfertilized buffer is maintained along all streams

Physiologically Appropriate Timing

It is important to make nutrients available when they will strengthen plants and improve their quality. For example, too much nitrogen in the summer may stress a cool-season turf grass, while high nitrogen late in the season will reduce the sugar content of sugar crops. A good supply of potassium is particularly important in the fall to enable plants to improve their winter hardiness. Planting-hole or broadcast application of phosphorus at the time of the tree planting brings good results on phosphorus-poor sites. However, broadcast application of nitrogen to trees soon after the seedlings are planted may benefit fast-growing weeds more than the desired trees.

Practical Field Limitations

Sometimes it is simply not possible to apply fertilizers at the ideal time of the year. For example, although a crop may respond to a late-season side-dressing, such an application will be difficult if the plants are too tall to drive over without damaging them. Early spring applications may be limited by the need to avoid compacting wet soils. Application by airplane or helicopter may increase the flexibility possible. Economic costs or the time demands of other activities may also require that compromises be made in the timing of nutrient application.

14.10 DIAGNOSTIC TOOLS AND METHODS[7]

Three basic tools are available for diagnosing soil fertility problems: (1) *field observations,* (2) *plant tissue analysis,* and (3) *soil analysis* (soil testing). To effectively guide the application of nutrients, as well as to diagnose problems as they arise in the field, all three approaches should be integrated. There is no substitute for careful observation and *recording* of circumstantial evidence and symptoms in the field. Effective observation

[7]For a detailed discussion of both plant analysis and soil testing, see Westerman (1990).

Nutrient removal by crops: http://www.soil.ncsu.edu/publications/Soilfacts/AG-439-16/#Micronutrients

and interpretation requires skill and experience, as well as an open mind. It is not uncommon for a supposed soil fertility problem to actually be caused by soil compaction, weather conditions, pest damage, or human error. The task of the diagnostician is to use all the tools available in order to identify the factor that is limiting plant growth, and then devise a course of action to alleviate the limitation.

Plant Symptoms and Field Observations

Nutrient deficiencies in trees, with color photos: http://www.utextension.utk.edu/spfiles/SP543.pdf

This detectivelike work can be one of the more exciting and challenging aspects of nutrient management. To be an effective soil fertility diagnostician, several general guidelines are helpful.

First, develop an organized way to *record your observations* for use in making interpretations at a later date. Sketch a map of the site showing features you have observed and the distribution of symptoms.

Second, *talk to the person* who owns or manages the land. Ask when the problem was first observed and if any recent changes have taken place. *Obtain records* on plant growth or crop yield from previous years, and ascertain the history of management of the site for as many years as possible.

Look for spatial patterns—how the problem seems to be distributed in the landscape and in individual plants. Linear patterns across a field may indicate a problem related to tillage, drain tiles, or the incorrect spreading of lime or fertilizer. Poor growth concentrated in low-lying areas may relate to the effects of soil aeration. Poor growth on the high spots in a field may reflect the effects of erosion and possibly exposure of subsoil material with an unfavorable pH.

Closely examine individual plants for characteristic nutrient deficiency symptoms on leaves and other plant parts. Examples of such symptoms are shown in Plates 44 to 56. Determine if the symptoms are most pronounced on the younger leaves (as is the case for most of the micronutrient cations) or on the older leaves (as is the case for nitrogen, potassium, and magnesium). Some nutrient deficiencies are quite reliably identified from foliar symptoms, while others produce symptoms that may be confused with disease symptoms or damage from herbicides, insects, or poor aeration.

Observe and *measure differences in plant growth* and crop yield that may reflect different levels of soil fertility, even though no leaf symptoms are apparent. Check *both* aboveground and belowground growth. Are mycorrhizae associated with tree roots? Are legumes well nodulated? Is root growth restricted?

Plant Tissue Analysis

Color images of foliar symptoms of plant mineral nutrient deficiencies: http://www.nrsl.umd.edu/courses/nrsc411/

NUTRIENT CONCENTRATIONS. The concentration of essential elements in plant tissue is related to plant growth or crop yield, as shown in Figure 14.18. The range of tissue concentrations at which the supply of a nutrient is sufficient for optimal plant growth is termed the **sufficiency range.** At the upper end of the sufficiency range, plants may be participating in luxury consumption, additional nutrient uptake without additional growth (see Figure 13.23). At concentrations above the sufficiency range, plant growth may decline as nutrient elements reach concentrations that are toxic to plant cells or interfere with the use of other nutrients. If tissue concentrations are in the **critical range**, the supply is just marginal and growth is expected to decline if the nutrient becomes any less available, even though visible foliar symptoms may not be exhibited ("hidden hunger"). Plants with tissue concentrations below the sufficiency range for a nutrient are likely to respond to additions of that nutrient if no other factor is more limiting. The sufficiency range and critical range have been well characterized for many plants, especially for agronomic and major horticultural crops. Less is known about forest trees and ornamentals.[8] Sufficiency ranges for 11 essential elements in a variety of plants are listed in Table 14.11.

TISSUE ANALYSIS. Tissue analysis can be a powerful tool for identifying plant nutrient problems if several simple precautions are taken. First, it is critical that the correct plant part be sampled. Second, the plant part must be sampled at the specified stage of

[8]Detailed information on tissue analysis for a large number of plant species can be found in Reuter and Robinson (1986).

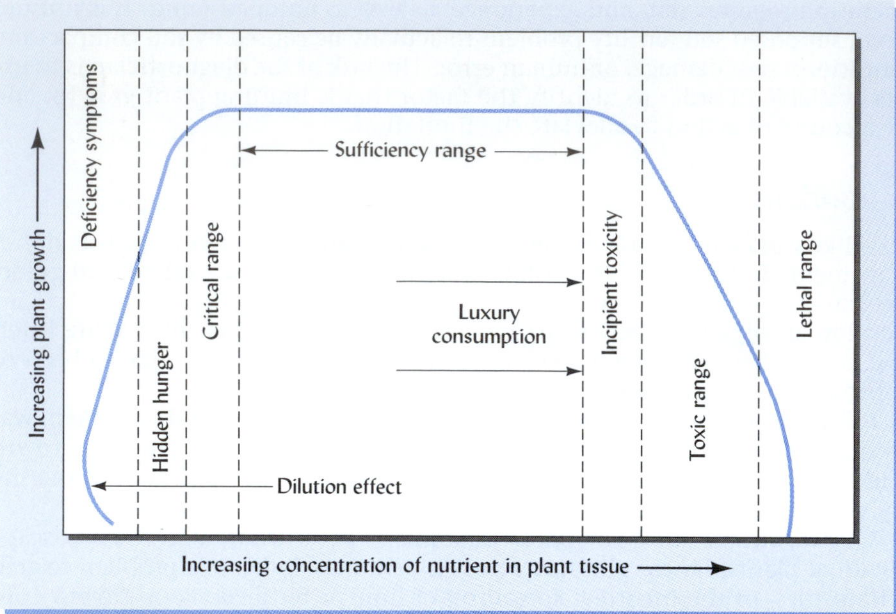

FIGURE 14.18 The relationship between plant growth or yield and the concentration of an essential element in the plant tissue. For most nutrients there is a relatively wide range of values associated with normal, healthy plants (the sufficiency range). Beyond this range, plant growth suffers from either too little or too much of the nutrient. The critical range (CR) is commonly used for the diagnosis of nutrient deficiency. Nutrient concentrations below the CR are likely to reduce plant growth even if no deficiency symptoms are visible. This moderate level of deficiency is sometimes called *hidden hunger.* The odd hook at the lower left of the curve is the result of the so-called dilution effect that is often observed when extremely stunted, deficient plants are given a small dose of the limiting nutrient. The growth response may be so great that even though somewhat more of the element is taken up, it is diluted in a much greater plant mass.

growth, because the concentrations of most nutrients decrease considerably as the plant matures. Third, it must be recognized that the concentration of one nutrient may be affected by that of another nutrient, and that sometimes the ratio of one nutrient to another (e.g., Mg/K, N/S, or Fe/Mn) may be the most reliable guide to plant nutritional status. In fact, several elaborate mathematical systems for assessing the ratios or balance among nutrients have proven useful for certain plant species.[9] Because of the uncertainties and complexities in interpreting tissue concentration data, it is wise to sample plants from the best and worst areas in a field or stand. The difference between samples may provide valuable clues concerning the nature of the nutrient problem.

[9]The best developed of the multinutrient ratio systems is known as the Diagnostic Recommendation Integrated System (DRIS). For details, see Walworth and Sumner (1987).

TABLE 14.11 A Guide to Sufficiency Ranges for Tissue Analysis of Selected Plant Species

Plant species and part to sample	*Content, %*						*Content, μg/g*				
	N	*P*	*K*	*Ca*	*Mg*	*S*	*Fe*	*Mn*	*Zn*	*B*	*Cu*
Pine trees (*Pinus* spp.) Current-year needles near terminal	1.2–1.4	0.10–0.18	0.3–0.5	0.13–0.16	0.05–0.09	0.08–0.12	20–100	50–600	20–50	3–9	2–6
Turf grasses, warm season Clippings	2.8–3.5	0.30–0.55	1.0–2.5	0.50–1.2	0.15–0.60	0.15–0.45	35–500	25–150	20–55	10–30	5–30
Cool season turf grasses, Clippings	3.0–6.0	0.3–0.4	2–4	0.4–0.8	0.2–0.4	0.25–0.4	40–500	20–100	20–50	8–20	6–30
Corn (*Zea mays*) Ear-leaf at tasseling	2.5–3.5	0.20–0.50	1.5–3.0	0.2–1.0	0.16–0.40	0.16–0.50	25–300	20–200	20–70	6–40	6–40
Soybean Recently matured trifoliate, flowering stage	4.0–5.0	0.31–0.50	2.0–3.0	0.45–2.0	0.25–0.55	0.25–0.55	50–250	30–200	25–50	25–60	8–20
Apple Leaf from base of non-fruiting shoots	1.8–2.4	0.15–0.30	1.2–2.0	1.0–1.5	0.25–0.50	0.13–0.30	50–250	35–100	20–50	20–50	5–20
Wheat First leaf blade from top of plant	2.2–3.3	0.24–0.36	2.0–3.0	0.28–0.42	0.19–0.30	0.20–0.30	35–55	30–50	20–35	5–10	6–10
Tomato Most recently matured leaf at early bloom	3.2–4.8	0.32–0.48	2.5–4.2	1.7–4.0	0.45–0.70	0.60–1.0	120–200	80–180	30–50	35–55	8–12
Alfalfa Upper third of plant at first flower	3.0–4.5	0.25–0.50	2.5–3.8	1.0–2.5	0.3–0.8	0.3–0.5	50–250	25–100	25–70	6–20	30–80

Data derived from many sources.

Cornstalk Nitrate. The *end of season cornstalk nitrate test* (CSNT) measures the nitrate content of the lower portion of mature cornstalks to improve future nitrogen management for corn. This test is particularly helpful in identifying excessive levels of nitrogen in the plant at harvest time, which, in turn, is an indication of excessive levels in the soil at the end of the season.

14.11 SOIL ANALYSIS

Since the total amount of an element in a soil tells us very little about the ability of that soil to supply that element to plants, more meaningful *partial soil analyses* have been developed. *Soil testing* is the routine partial analysis of soils for the purpose of guiding nutrient management. The soil testing process consists of three critical phases: (1) *sampling* the soil, (2) chemically *analyzing* the sample, and (3) *interpreting* the analytical result to make a recommendation on the kind and amounts of nutrients to apply.

Sampling the Soil

Because of the variability in nutrient levels from spot to spot, it is always advisable to divide a given field or property into as many distinct areas as practical when taking soil samples to determine nutrient needs. For example, suppose a 20 ha field has a 2 ha low spot in the middle, and 5 ha at one end that used to be a permanent pasture. These two areas should be sampled, and later managed, separately from the remainder of the field. Similarly, a homeowner should sample flower beds separately from lawn areas, low spots separately from sloping areas, and so on. On the other hand, known areas of unusual soil that are too small or irregular to be managed separately should be avoided and not included in the composite sample from the whole field.

Composite Sample. Usually, a soil probe is used to remove a thin cylindrical core of soil from at least 12 to 15 randomly scattered places within the land area to be represented (Figure 14.19). The 12 to 15 subsamples are thoroughly mixed in a plastic bucket, and about 0.5 L of the soil is placed in a labeled container and sent to the lab. If the soil is moist, it should be air-dried without sun or heat prior to packaging for routine soil tests. Heating the sample might cause a falsely high result for certain nutrients. However, samples for analysis of soil nitrates should be rapidly dried under a fan or in an oven at about 65°C.

How to get a good soil sample:
http://www.agweb.okstate.edu/pearl/plantsoil/soils/f-2207.pdf

Depth and Time to Sample. The standard depth of sampling for a plowed soil is the depth of the plowed layer, about 15 to 20 cm, but various other depths are also used (see Figure 14.19). Because in many unplowed soils nutrients are stratified in contrasting layers, the depth of sampling can greatly alter the results obtained. Seasonal changes are often observed in soil test results for a given area. A good practice is to sample each area every year or two (always at the same time of year), so that the soil test levels can be tracked over the years to determine whether nutrient levels are being maintained, increased, or depleted.

Timing for Special Nitrogen Tests. Timing is especially critical to determine the amount of mineralized nitrogen in the root zone. In relatively dry, cold regions (e.g., the Great Plains), the *residual nitrate* test is done on 60 cm deep samples obtained sometime between fall and before planting in spring. In humid regions, where nitrate leaching is more pronounced, samples for the *Preside-dress nitrate test* (PSNT) are taken from the upper 30 cm when the corn is about 30 cm tall, just in time to determine how much nitrogen to apply as the crop enters its period of most rapid nitrogen uptake. In this case, the soil must be sampled during the narrow window of time when spring mineralization has peaked, but plant uptake has not yet begun to deplete the nitrate produced (see Figure 12.3).

Chemical Analysis of the Sample

In general, soil tests attempt to extract from the soil amounts of essential elements that are correlated with the nutrients taken up by plants. Different extraction solutions are employed by various laboratories.

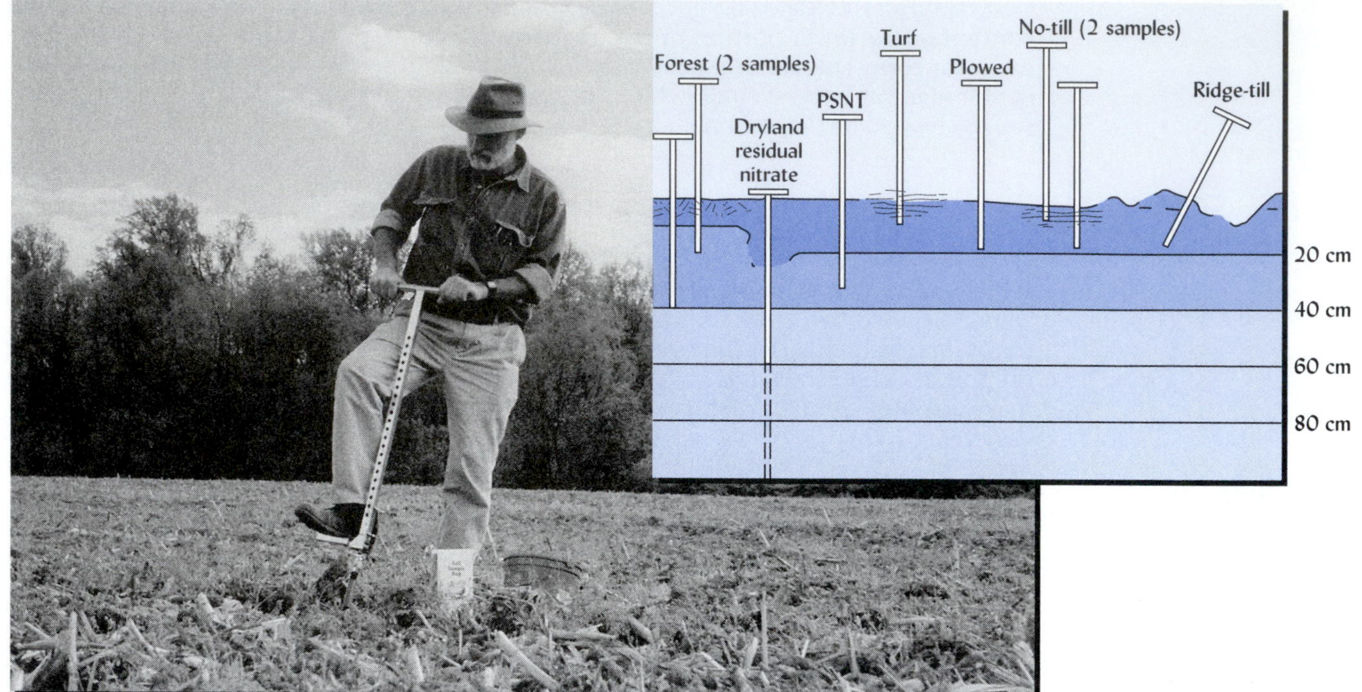

FIGURE 14.19 (Left) Taking the soil sample in the field is often the most error-prone step in the soil testing process because soil properties vary greatly both with depth and from place to place in even a uniform-appearing field. Areas that are obviously unusual (wet spots, places where manure was piled, eroded spots, etc.) should be avoided. (Right) The proper depth to sample depends on the purpose of the soil test and the nature of the soil. Some suggested depths for different situations are shown. (Photo courtesy of R. Weil)

Soil cation nutrient balancing—missing link, or red herring?:
http://www.vabf.org/soilrel.php

The most common and reliable tests are those for soil pH, potassium, phosphorus, and magnesium. Micronutrients are sometimes extracted using chelating agents, especially for calcareous soils in the more arid regions. While the nitrate and sulfate present in the soil at the time of sampling can be measured, predicting the availability of nitrogen and sulfur is considerably more difficult because of the many biological factors involved.

Because the methods used by different labs may be appropriate for different types of soils, it is advisable to send a soil sample to a lab in the same region from which the soil originated. Such a lab is likely to use procedures appropriate for the soils of the region and should have access to data correlating the analytical results to plant responses on soils of a similar nature.

Soil tests designed for use on soils or soil-based potting media generally do not give meaningful results when used on artificial peat-based soilless potting media. Special extraction procedures must be used for the latter, and the results must then be correlated to nutrient uptake and growth of plants grown in similar media.

These examples should further emphasize the importance of providing the soil testing lab with complete information concerning the nature of your soil, its management history, and your plans for its future use.

Interpreting the Results to Make a Recommendation

Interpretation is, perhaps, the most controversial aspect of soil testing. The soil test values themselves are merely *indices* of nutrient-supplying power. They do not indicate the actual amount of nutrient that will be supplied. For this reason, it is best to think of soil test reports as more indicative than quantitative.

Many years of field experimentation at many sites are needed to determine which soil test level indicates a low, medium, or high capacity to supply the nutrient tested. Such categories are used to predict the likelihood of obtaining a profitable response from the application of the nutrient tested (Figure 14.20). Since the actual units of measurement (ppm or lb/acre) have little actual meaning to the end user, soil test results are increasingly reported as index values on a relative scale (often with 75 to 100 considered optimal) as in Figure 14.21.

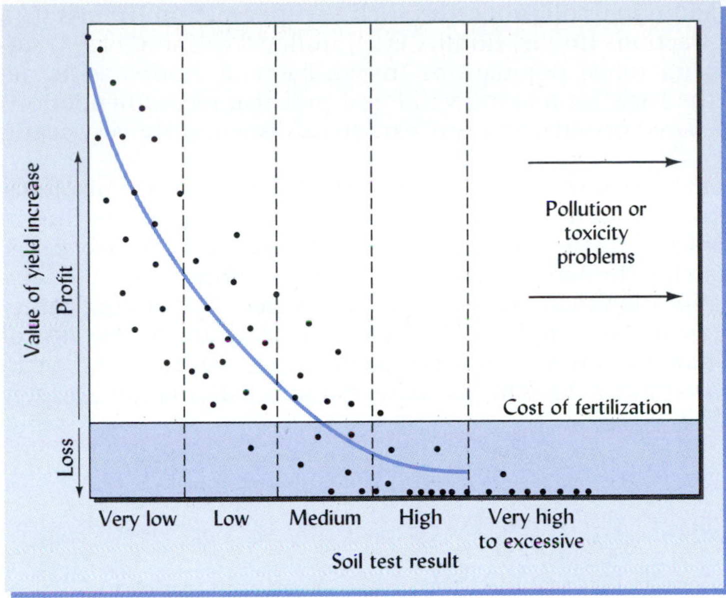

FIGURE 14.20 The relationship between soil test results for a nutrient and the extra yield obtained by fertilizing with that nutrient. Each data point represents the *difference* in plant yield between the fertilized and the unfertilized soil. Because many factors affect yield and because soil tests can only approximately predict nutrient availability, the relationship is not precise, but the data points are scattered about the trend line. If the point falls above the fertilizer cost line, the extra yield was worth more than the cost of the fertilizer and a profit would be made. For a soil testing in the very low and low categories, a profitable response to fertilizer is very likely. For a soil testing medium, a profitable response is a 50:50 proposition. For soils testing in the high category, a profitable response is unlikely.

Recommendations for nutrient applications take into consideration practical knowledge of the plants to be grown, the characteristics of the soil under study, and other environmental conditions. Management history and field observations can help relate soil test data to fertilizer needs.

Merits of Soil Testing

Dependability in predicting how plants will respond to soil amendments varies with the particular soil test, the tests for some parameters being much more reliable than others because of the consistency and breadth of field correlation data. In general, the tests for pH (need for liming or acidification), P, K, Mg, B, and Zn are quite reliable.

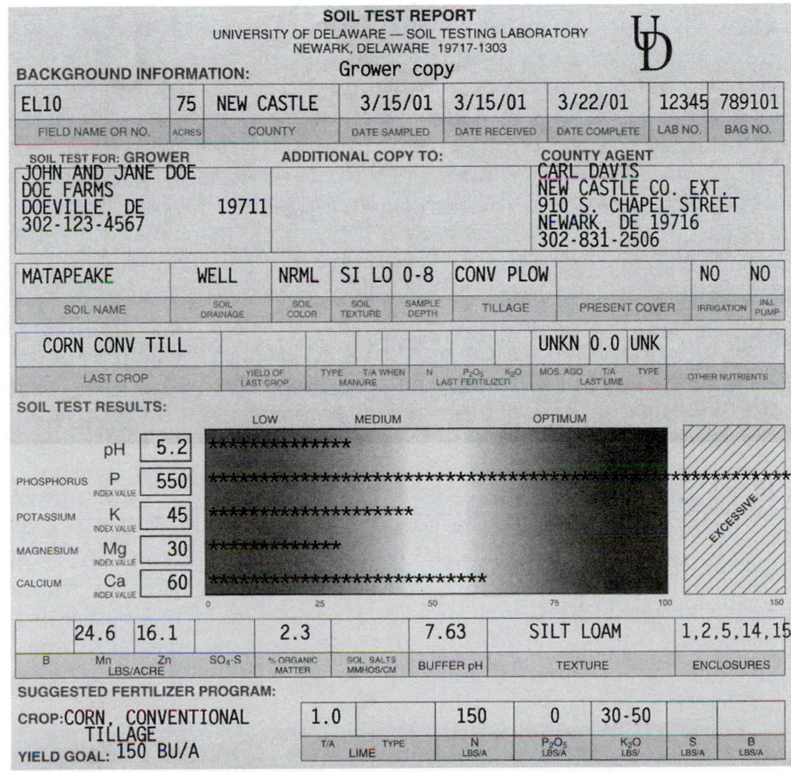

SOIL TEST REPORT
UNIVERSITY OF DELAWARE — SOIL TESTING LABORATORY
NEWARK, DELAWARE 19717-1303

BACKGROUND INFORMATION: Grower copy

FIELD NAME OR NO.	ACRES	COUNTY	DATE SAMPLED	DATE RECEIVED	DATE COMPLETE	LAB NO.	BAG NO.
EL10	75	NEW CASTLE	3/15/01	3/15/01	3/22/01	12345	789101

SOIL TEST FOR: GROWER
JOHN AND JANE DOE
DOE FARMS
DOEVILLE, DE 19711
302-123-4567

ADDITIONAL COPY TO:

COUNTY AGENT
CARL DAVIS
NEW CASTLE CO. EXT.
910 S. CHAPEL STREET
NEWARK, DE 19716
302-831-2506

SOIL NAME	SOIL DRAINAGE	SOIL COLOR	SOIL TEXTURE	SAMPLE DEPTH	TILLAGE	PRESENT COVER	IRRIGATION	INJ. PUMP
MATAPEAKE	WELL	NRML	SI LO	0-8	CONV PLOW		NO	NO

LAST CROP	YIELD OF LAST CROP	MANURE TYPE	MANURE T/A	MANURE WHEN	LAST FERTILIZER N	LAST FERTILIZER P_2O_5	LAST FERTILIZER K_2O	LAST LIME MOS. AGO	LAST LIME T/A	LAST LIME TYPE	OTHER NUTRIENTS
CORN CONV TILL								UNKN	0.0	UNK	

SOIL TEST RESULTS: LOW — MEDIUM — OPTIMUM — EXCESSIVE

		Value	Bar
	pH	5.2	**************
PHOSPHORUS	P (INDEX VALUE)	550	**
POTASSIUM	K (INDEX VALUE)	45	********************
MAGNESIUM	Mg (INDEX VALUE)	30	*************
CALCIUM	Ca (INDEX VALUE)	60	*****************************

Scale: 0, 25, 50, 75, 100, 150

B	Mn (LBS/ACRE)	Zn (LBS/ACRE)	SO_4-S	% ORGANIC MATTER	SOL. SALTS MMHOS/CM	BUFFER pH	TEXTURE	ENCLOSURES
	24.6	16.1		2.3		7.63	SILT LOAM	1,2,5,14,15

SUGGESTED FERTILIZER PROGRAM:
CROP: CORN, CONVENTIONAL TILLAGE
YIELD GOAL: 150 BU/A

LIME T/A	LIME TYPE	N LBS/A	P_2O_5 LBS/A	K_2O LBS/	S LBS/A	B LBS/A
1.0		150	0	30-50		

FIGURE 14.21 A typical report sheet from a soil test lab. The report records the nature and history of the field, reports the levels of certain nutrients and soil properties that were measured in the lab, and gives recommendations for the amounts and types of soil amendments to apply for a specific crop. For the particular field sampled here, the following interpretations can be made. The pH is somewhat low and should be raised to near 5.8 to 6.0. At least some of the liming material should be dolomitic limestone to add magnesium, which is in low to medium supply. Moderate amounts of potassium are likely to give an economic response. Sulfur and boron are not part of the standard analysis package in Delaware, but might be included if requested. The phosphorus level is highly excessive, so phosphorus additions of any kind should be avoided, and steps should be taken to assess and manage potential transport of phosphorus to waterways. As Delaware is in a humid region where most soluble nitrogen may leach away over the winter, nitrogen application is recommended based on the yield goal and soil properties that affect mineralization, but not on an analytical test for available nitrogen. In an arid or semiarid region, a soil test lab would likely also measure and report nitrate and electrical conductivity in the soil and use these values in making recommendations concerning nitrogen fertilization and salt management, respectively. (Soil test report courtesy of K. L. Gartley, University of Delaware)

Some soil test labs report additional soil properties such as other micronutrients (Cu, Mn, Fe, Mo, etc.), humus fractions (fulvic, humic, etc.), nonacid cation ("base") saturation, and even various microbial populations (fungi, bacteria, nonparasitic nematodes, etc.). However, the basis for making valid and practical recommendations for managing soil fertility based on such soil test parameters is currently very scant, at best.

Generally, soil testing has been most relied upon in agricultural systems, while foliar analysis has proved more widely useful in forestry. The limited use of soil testing in forestry is partly due to the fact that trees, with their extensive perennial root systems, integrate nutrient bioavailability throughout the profile. This, combined with the typically complex nutrient stratification in forested soils, creates a great deal of uncertainty about how to obtain a representative sample of soil for analysis. In addition, because of the comparably long time frame in forestry, limited information is available on the correlation of soil test levels with timber yields in the sense that such information is widely available for agronomic crops.

14.12 SITE-SPECIFIC NUTRIENT MANAGEMENT

Further information on precision farming: http://www.ag.arizona.edu/precisionag/

Recent advances in technology have combined the global positioning system (GPS) of earth-orbiting satellites, computer programs capable of making detailed maps that integrate information about many soil properties, and technologies that allow farm equipment to alter the rate of fertilizer, seed, or chemical delivery on the go. The combined technology allows farmers to make nutrient management more site-specific than was previously practical for large farming operations (Figure 14.22). Portable GPS receivers can plot one's exact location (to within a few meters) as one moves across a large field. Therefore, if one is taking soil samples, the location of each sample can be georeferenced with north–south and east–west coordinates.

In practice, a large field is divided into cells in a grid pattern, each cell usually being about 1 ha in area. For example, 18 separate georeferenced soil samples (each being a composite of about 15 to 20 soil cores) may be collected from a single 18 ha field (Figure 14.22). These soil samples can be analyzed for such properties as texture, organic matter content, pH, and soil test levels of phosphorus and potassium. A computer program can then produce a map of the spatial distribution of each soil property measured. Special statistical methods are used to estimate the location of continuous boundaries between areas differing in regard to each soil property. Thus, one map might show areas of low, medium, and high soil test phosphorus levels. Another map might show areas of high, medium, and low clay content, and so on. Information on other spatial variables, such as soil classification, drainage class, past management, crop cultivars to be planted, and so forth, can be fed into the program to generate additional maps. A sophisticated computer program (known as a geographic information system, GIS) then integrates the information from these individual maps to create a new combined map showing the different application rates of fertilizer (or other material) that are recommended for different parts of the field.

For example, areas mapped as testing low in phosphorus might be slated to receive higher-than-average application rates of phosphorus fertilizer, while areas mapped as being high in phosphorus might receive no phosphorus fertilizer at all. Similarly, reduced rates of nitrogen fertilizer application might be mapped for areas of sandy soils with high leaching potential but low yield potential. The application-rate recommendation maps are then programmed into a computer on board the machine that spreads the fertilizer. The fertilizer spreader also has a GPS receiver installed, so its computer tracks its location on the map as it moves across the field. The computer then signals small motors on the fertilizer spreader to increase or decrease the rate of application, as called for on the map. Thus, the system is designed to reduce both overapplication on the high-fertility areas (which should save money and reduce nutrient pollution potential) and underapplication on the low-fertility sites (which should make money by producing a higher yield). The total amount of fertilizer added may not be greatly different from that used in the past, but application rates should be more in tune with plant needs and environmental cautions.

At harvest time, similar computerized satellite linkages are used to monitor yields in different parts of the field and to create maps showing the yield differences. The yield

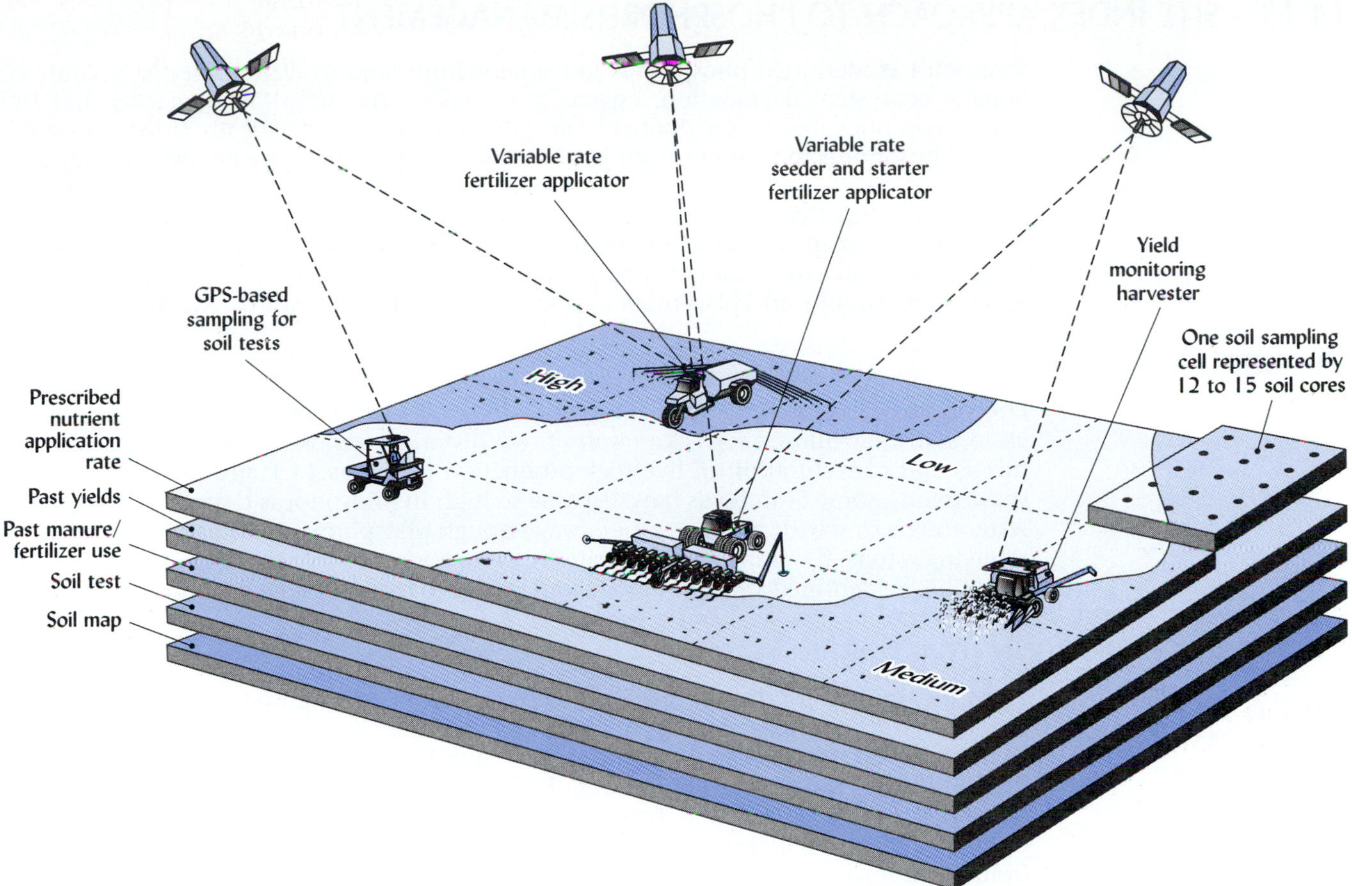

FIGURE 14.22 Space age technology is being used to facilitate site-specific nutrient management systems. Earth-orbiting satellites and appropriate computer software are the bases for the global positioning system (GPS), which can plot the location of many soil sampling and plant production sites on a grid basis within a field. A soil sample (composite of 20 cores) was taken from each cell (about 1 ha) in an 18 ha field and was analyzed. Using the soil analysis data from these samples, information on past yields and amendments, and soil survey maps, geographic information system (GIS) computer programs can create a map showing areas that should receive different rates of nutrient application (top layer in diagram). The satellite/computer systems can then be used to control variable-rate fertilizer applicators that apply only the amounts of nutrients that the soil tests and other information suggest are needed. At harvest time, similar satellite/computer connections make possible the monitoring of crop yields as the harvest machine traverses the field. The yield data are used to create yield maps, which can then be used to further refine nutrient-management systems. [Modified from PPI (1996)]

Sensors to control fertilizer application rate for improved nitrogen use efficiency:
http://www.dasnr.okstate.edu/nitrogen_use/sensor_based_improvement_of_nue.htm

Sensing soil properties for site-specific management:
http://www.precisionag.org/PDF/ch10.pdf

maps produced by these GPS-equipped combines are proving to be the most popular component of the entire system, as they help farmers pinpoint problem areas for investigation. Yield levels in some parts of a field, even in one that is uniform in appearance, commonly are 2 to 3 times as high as in other parts of the field. By overlaying the yield map on the soil nutrient maps, it is possible to determine the extent to which nutrient deficiencies are constraining yields, but other problems such as poor drainage, low organic matter, or even feeding by wildlife may also come to light.

This site-specific nutrient management system is an integral part of what is commonly referred to as *precision farming*. Opportunities for the control of insects and weeds and for modifying plant seeding rates and depths can be utilized on a site-specific basis rather than on a field basis. Such computerized GPS-based systems are expensive in terms of equipment costs and costs to collect all the requisite information. Costs may be reduced if sensing instruments that estimate soil fertility properties "on the go" can replace laborious collection of grid soil samples and time-consuming laboratory analyses (see Figure 9.21 for an example).

The technology may not be economically viable for all farms, but its use is spreading, particularly among growers who have access to custom operators with computer expertise and the appropriate equipment. Precision farming may help assure that nutrients will be used only where they are needed to enhance plant production. The technology may also be helpful in implementing environmental nutrient management based on a site-specific index of pollution potential, the topic we turn to next.

14.13 SITE-INDEX APPROACH TO PHOSPHORUS MANAGEMENT

Research has identified phosphorus movement from land to water as a major cause of aquatic ecosystem degradation, especially for lakes. The situation demands that the amounts of phosphorus contributed from agricultural land be drastically reduced in some areas. Phosphorus movement from land to water is determined by *phosphorus transport, phosphorus source,* and *phosphorus management* parameters. Transport parameters include the mechanisms that govern how rain and irrigation water cause phosphorus to move across the landscape in runoff and sediment. Source parameters include the amount and forms of phosphorus on and in the soil. Management parameters include the method of application, timing, and placement of phosphorus and such disturbances as tillage.

Overenrichment of Soils

Phosphorus so scarce in nature, has accumulated over a period of decades to excessive levels in many agricultural soils as a result of two historical trends: overapplication of P fertilizers and concentration of livestock production (Sections 14.1 and 14.4). Because of these trends, some farm fields have become so high in phosphorus that rain or irrigation water interacting with the soil carries away enough phosphorus to impair the ecology of receiving waters. It is imperative that sites with such a high potential to cause phosphorus pollution be identified and appropriately managed to reduce their environmental impact.

Transport of Phosphorus from Land to Water

Two fundamental conditions must be met for a soil to cause significant phosphorus pollution. First, the supply of phosphorus in the soil must be relatively large; second, the characteristics of the site and soil must allow significant transport of phosphorus from the soil in the field to a receiving water body such as a lake or stream. A review of the phosphorus cycle (Section 13.3) reminds us that phosphorus can move from land to water by three principal pathways: attached to eroded soil particles (the main pathway from tilled soils); dissolved in water running off the surface of the land (a major pathway for pastures, woodland, and no-till cropland); and dissolved or attached to suspended colloidal particles in water percolating down the profile (a significant pathway in very sandy soils with shallow water tables and possibly in heavily manured soils).

Phosphorus Soil Test Level as Indicator of Potential Losses

Phosphorus in Eroded Sediment. Phosphorous attached to eroded soil particles accounts for the greatest amount of phosphorus contributed to receiving waters from most agricultural land. Soils that are both highly erodible and highly enriched with phosphorus (as indicated by excessive phosphorus soil test levels) represent the most serious potential for phosphorus pollution.

Phosphorus in Runoff Water. As described in Section 14.11, soil testing was designed to determine how well a soil can supply nutrients to plant roots. However, routinely used phosphorus soil tests have been shown to also provide a useful (though not perfect) indication of how readily phosphorus will desorb from a soil and dissolve in runoff water. If phosphorus soil test levels of the upper few centimeters of soil are found to be near or below the critical range for optimum plant growth, little phosphorus is lost in runoff water (see Figure 14.23). But if the soil test levels are much above this critical range, the phosphorus level in the runoff water will likely increase exponentially. Research suggests that soils can be maintained at phosphorus levels sufficient for optimal plant growth without undue losses of phosphorus in runoff water. As is also the case with nitrogen, significant environmental damages are associated with excess—the application of more nutrient than is really needed. Therefore, the most profitable level of phosphorus fertility should also be an environmentally sound level, and one that also conserves the world's finite stocks of this essential element.

Phosphorus in Drainage Water. In most soils under most circumstances, significant quantities of phosphorus are unlikely to be lost in drainage water, because as phosphorus-laden water percolates from the surface down through the subsurface soil horizons, dissolved $H_2PO_4^-$ is strongly sorbed by inner-sphere complexation onto Fe-, Al-, Mn-, and Ca-containing mineral surfaces (Sections 8.7 and 13.8). However, under certain cir-

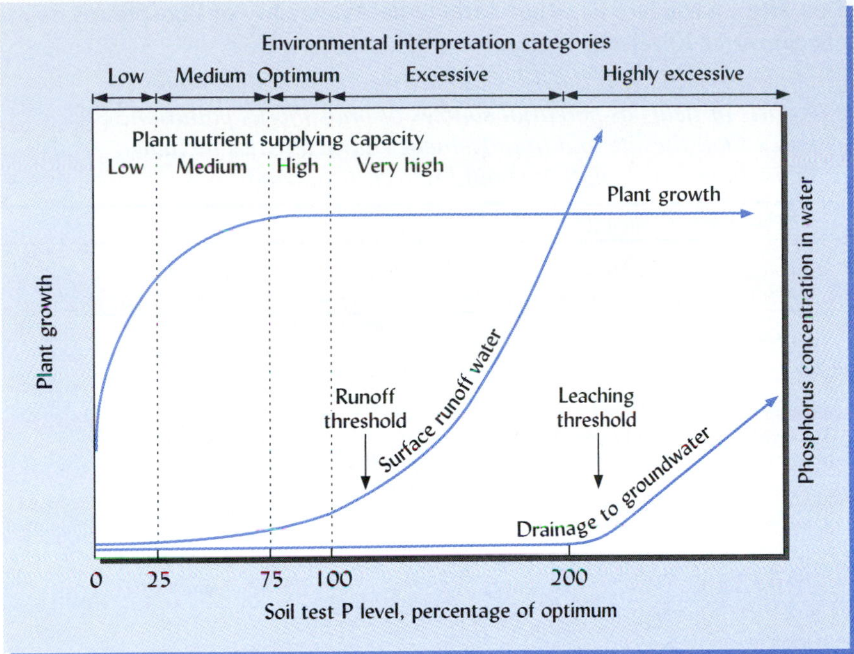

FIGURE 14.23 The generalized relationship between levels of plant-available phosphorus in soils (soil test P level) and environmental losses of P dissolved in surface runoff and subsurface drainage waters. The generalized relationship between traditional plant nutrient supply interpretation categories and environmental interpretation categories is also indicated. The diagram suggests that, fortunately, soil P levels can be achieved that are both conducive to optimum plant growth and protective of the environment. If P losses by soil erosion are controlled (although not shown, this is a big if), significant quantities of dissolved P would be lost only when soils contain P levels in excess of those needed for optimum plant growth. Losses of P by leaching in drainage water would be significant only at very high soil P levels, as this pathway rapidly increases only after the P-sorption capacity of the soil is substantially saturated. Note the threshold levels (vertical arrows) for P losses. The vertical axes are not to scale. [Figure based on data and concepts discussed in Sharpley (1997), Beegle et al. (2000), and Higgs et al. (2000)]

cumstances this mechanism is not so effective in removing phosphorus from drainage water. These circumstances include the following:

The percolating water is flowing through macropores and has little contact with soil surfaces (Section 6.8). This may occur under long-term no-till management or in pastures.

The dissolved phosphorus is not in mineral ($H_2PO_4^-$ or HPO_4^{2-}) forms, but is part of soluble organic molecules, which are not strongly sorbed by mineral surfaces. This situation is common under forests or heavily manured soils.

The phosphorus is sorbed onto the surface of dispersed colloidal particles that are so small that they remain suspended in the percolating water.

The phosphorus-fixation capacity of the soil is so small (as in sands, Histosols, and some waterlogged soils with reduced iron) or already so saturated with phosphorus (as in soils overloaded with phosphorus after many years of excessive manure and fertilizer applications) that little more phosphorus can be sorbed (as illustrated in Figure 14.23).

Site Characteristics Influencing Transport of Phosphorus

To cause environmental damage, the phosphorus-rich sites must also have characteristics conducive to the transport of phosphorus to sensitive bodies of water. Such characteristics might include close proximity to the water body, the absence of a protective vegetated buffer, erodible soils, or high runoff rates caused by low permeability and steep slopes. Control of soil erosion by conservation tillage, mulches, and other practices discussed in Chapter 15 can greatly reduce the transport of sediment-associated phosphorus, but will have little effect on, or may even increase, the loss of phosphorus dissolved in runoff or drainage water. In fact, when no-till fields are fertilized, the phosphorus tends to accumulate in the upper 1 to 2 cm of soil, where it is most susceptible to desorbing into runoff water. The beneficial influence of buffers in removing both dissolved and sediment-associated phosphorus from runoff was discussed in Section 14.2.

Phosphorus Site Index

Usually, most of the phosphorus entering a river or lake originates from only a small fraction of the watershed land area. Effective environmental management requires identification of the high-risk sites. The *phosphorus site index* (Table 14.12) integrates phosphorus source, transport, and management characteristics of a site into an index of phosphorus pollution risk. Box 14.4 illustrates how such an index can be calculated for a specific farm field (or a separately identified part of a field).

TABLE 14.12 Phosphorus Site Index Based on Site Characteristics That Affect the Availability of Phosphorus at the Source and the Transport of Phosphorus to Sensitive Receiving Waters[a]

This example illustrates the types of parameters and relationships among parameters that might be used to rank individual farm fields or parts of fields as potential sources of phosphorus pollution. In adapting this type of phosphorus index tool, the site and management characteristics included, the weights used for each, and the method of calculation should reflect local conditions.

		Phosphorus loss rating (B)					
Parameter	*Weight factor (A)*	*None 0*	*Low 1*	*Medium 2*	*High 4*	*Very high 8*	*Score (A × B)*
				P transport characteristics			
1. Annual soil erosion by rain[b]	1.5	<2	2–8	8–12	12–32	>32	
2. Erosion by irrigation water	1.0	Not irrigated	Tailwater return or no visible runoff	*QS*[c] >38 for erosion-resistant soils	*QS* >38 for erodible soils or visible runoff at field border	*QS* >38 for very erodible soils or rills visible	
3. Runoff class[d]	1	Very low	Low	Medium	High	Very high	
4. Distance from field edge to lake or stream	1	>30 m	15–30 m	8–15 m	4–8 m	<4 m	
5. Vegetated buffer between field and lake or stream[e]	0.5	>30 m	15–30 m	8–15 m	4–8 m	<4 m	
6. No P application zone (above any vegetated buffer)	0.5	>15 m	10–15 m	5–10 m	2–5 m	<2 m	
7. Receiving water priority rating[f]	1	Very low	Low	Medium	High	Very high	
				Score 1 + score 2 + . . . + score 7 = **total score for P transport→**			
				P source characteristics			
1. Soil test P category where >100 is considered excessive[g]	1	Low 0–25	Medium 26–75	Optimum 76–100	Excessive 101–200	Highly excessive >201	
2. P fertilizer application rate, kg P/ha	0.75	None	1–15	16–45	46–75	>75	
3. P fertilizer application method	0.5	None applied	Inject or band deeper than 5 cm	Incorporate within 5 days of application	Surface apply in summer or incorporate after >5 days	Surface apply during winter and not incorporate	
4. Organic P source application rate, kg P/ha[h]	1.0	None	1–15	16–30	31–45	>46	
5. Organic P source application method	1.0	None applied	Inject or band deeper than 5 cm	Incorporate within 5 days of application	Surface applied in summer or incorporate after >5 days	Surface apply during winter and not incorporate	
				Score 1 + score 2 +. . . + score 5 = **total score for P source→**			
				(P transport score) × (P source score) = **combined P site index for risk of water pollution→**			

Interpretation of index

Combined P site index:	0–10	11–50	51–250	251–1000	>1000
Risk of P pollution:	Negligible	Low	Medium	High	Very High

[a]Adapted from models and concepts discussed in USDA/NRSC (2000) and Gburek et al. (2000). In any given region, the appropriate phosphorus index matrix table would differ somewhat from the general model shown here. English units are commonly used rather than metric for communication with farmers in the United States.

[b]Mg/ha As determined by RUSLE—see Section 15.4.

[c]Q = flow rate (L/min) of water applied to a furrow; S = slope (%). $Q \times S = QS$.

[d]Ranging from permeable soils on nearly level sites (none) to impermeable soils on slopes >20% (very high).

[e]See Section 14.2. Perennial grass nearest the field combined with forest vegetation nearest the stream is ideal.

[f]Not all areas have official priority rankings. Low priority would typically be an already-polluted artificial holding pond, while highest priority would be a pristine lake very sensitive to eutrophication.

[g]Not all soil test reports give results on this scale. "Excessive" may be termed "very high" by some labs (see Figure 14.23).

[h]The P from an organic source may be multiplied by an availability factor to estimate the fraction of total P in an organic material that will be released in a year.

14.14 SOME BROADER ASPECTS OF FERTILIZER PRACTICE

Nitrogen Credits. Because of the prominent response by most nonlegume plants, and because of environmental quality considerations, the initial focus in most soil fertility plans is on nitrogen. Agriculturalists usually determine the optimal level of nitrogen fertilizer from the shape of the response curve and economic considerations (see following). This optimal fertilizer rate should be adjusted by the amount of any additional gains or losses not taken into account in the standard response curve. For example, nitrogen contributions from previous or current manure applications (see Box 14.2), legume cover crop (see Figure 14.8), previous legume in the rotation (see Figure 14.9), or nitrate in irrigation water should be subtracted from the amount of fertilizer recommended.

Profitability. Farmers use fertilizers to make a living. The most profitable rate is determined by the ratio of the value of the extra yield expected to the cost of the fertilizer applied. The law of diminishing returns applies. Therefore, the most profitable fertilizer rate will be somewhat less than the rate that would produce the very highest yield (Figure 14.24).

BOX 14.4 HOW A PHOSPHORUS SITE INDEX WORKS

The phosphorus site index table assigns a rating of 0 to 8 for different degrees of risk from each parameter. In Table 14.12, the parameters are grouped into phosphorus source parameters and phosphorus transport parameters. The weights listed in the second column represent professional judgments as to how much each parameter is likely to influence phosphorus pollution. The weight multiplied by the rating gives a partial score for each parameter. These partial scores are then added together to give two total scores, one for phosphorus source and one for phosphorus transport. The product obtained by multiplying these two scores yields an overall combined score that indicates to which of five categories of risk the site belongs. Consider the following example:

A particular field is located only 12 m from the edge of a sensitive lake (priority high) and there is no zone left at the field edge without phosphorus application. The field is bordered by a 12-m-wide vegetated buffer strip. The field is not irrigated and is moderately erodible (10 Mg/ha estimated annual erosion). The runoff class is medium because although the soil is moderately permeable, the site is steeply sloping. The upper layer of soil tests in the highly excessive range (300) because of a long history of phosphorus applications. No inorganic phosphorus fertilizer is now applied, but plans are to surface-apply, in winter, enough liquid manure to supply about 40 kg P/ha. The parameter scores in Table 14.12 are calculated here. For example, for the first parameter (erosion), the weight factor from Table 14.12 is 1.5 and the rating is 2 (8 to 10 Mg/ha class), so $1.5 \times 2 = 3.0$, the partial score for this parameter.

Parameter	*Calculations (source parameters)*	*Partial score*	*Parameter*	*Calculations (transport parameters)*	*Partial score*
1	$1.5 \times 2 =$	3.0	1	$1.0 \times 8 =$	8.0
2	$1.0 \times 0 =$	0.0	2	$0.75 \times 0 =$	0
3	$1.0 \times 2 =$	2.0	3	$0.5 \times 0 =$	0
4	$1.0 \times 2 =$	2.0	4	$1.0 \times 4 =$	4.0
5	$0.5 \times 2 =$	1.0	5	$1.0 \times 8 =$	8.0
6	$0.5 \times 8 =$	4.0	Total P source score		20.0
7	$1.0 \times 4 =$	4.0			
Total P transport score		16.0			

Combined P pollution risk index: $16 \times 20 = 320$, indicates high risk

Based on the combined score of 320, this site has a high risk for phosphorus movement from the site (see interpretation chart at bottom of Table 14.12). There is a high probability for an adverse impact to surface-water resources unless remedial action is taken. Soil and water conservation as well as phosphorus management practices are necessary to reduce the risk of phosphorus movement and probable water-quality degradation. The partial scores suggest that the situation could be corrected by addressing transport parameter 6 (create a no-phosphorus-application zone along the lower edge of the field), source parameter 4 (reduce the amount of organic nutrient source applied), and source parameter 5 (incorporate the organic nutrient source rather than applying it to the surface in winter). Transport parameter 7 (the priority ranking of the receiving water) and source parameter 1 (the current phosphorus soil test level) also contribute substantially to the combined index, but these factors are not subject to management change.

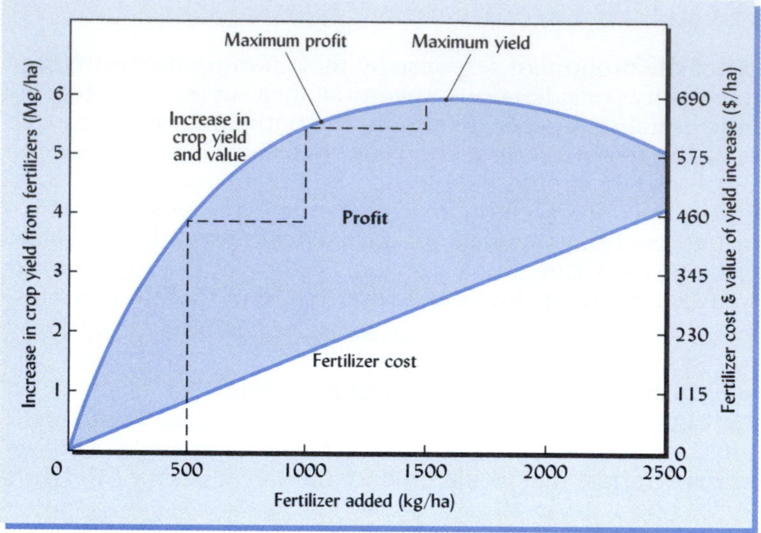

FIGURE 14.24 Relationships among the rate of fertilizer addition, crop yield increase, fertilizer costs, and profit from adding fertilizers. Note that the yield increase (and profit) from the first 500 kg of fertilizer is much greater than from the second and third 500 kg. Also note that the maximum profit is obtained at a lower fertilizer rate than that needed to give a maximum yield. The calculations assume that the yield response to added fertilizer takes the form of a smooth quadratic curve, an assumption which often leads to an overestimate of the amount of fertilizer needed to optimize profit.

Response Curves. Traditionally, economic analysis of optimum fertilizer rates has assumed that the plant response to fertilizer inputs was represented by a smooth curve following a quadratic function ($y = a + bx + cx^2$). In fact, actual data obtained can be just as well described by a number of other mathematical functions (Figure 14.25). This seemingly esoteric observation can have a great effect on the amount of fertilizer recommended and, in turn, on the likelihood of environmental harm from excessive fertilizer use. Among the various models studied, the linear-plateau approach (Figure 14.25*a*) usually leads to the lowest fertilizer recommendation, least wasted fertilizer, and least environmental damage. Nonetheless, the uncertainties related to nitrogen supply are still sufficiently high, and the cost of nitrogen fertilizer sufficiently low, that farmers tend to err on the side of oversupply, just for "insurance." In most years, this philosophy can be disadvantageous to both the farmer and the environment.

Anyone involved with the actual production of plants can testify to the enormous improvements that accrue from judicious use of organic and inorganic nutrient supplements. The preceding discussion should not imply that such supplements are not needed, but only that their optimum levels are difficult to determine with precision.

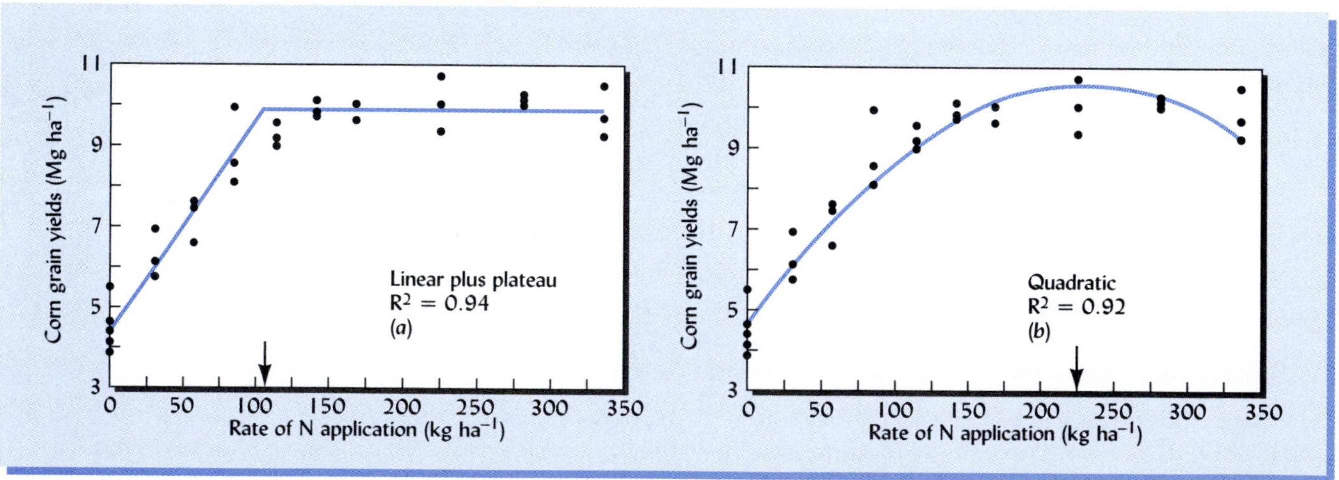

FIGURE 14.25 An example of how the mathematical function chosen to represent fertilizer-response data can effect the amount of fertilizer recommended. The data in both graphs are exactly the same and represent the response of corn yields in Iowa to increasing levels of nitrogen fertilization. The vertical arrows indicate the recommended, most profitable rate of nitrogen to apply according to each mathematical model. Note that both models fit the data equally well (as indicated by the very similar R^2 values), but that the linear-plus-plateau model predicts an optimum nitrogen rate of 104 kg/ha, while the standard quadratic model suggests that 222 kg/ha is the optimum. The extra 118 kg/ha of nitrogen may have no effect on crop yield, but may greatly increase the risk of environmental damage. [Redrawn from Cerrato and Blackmer (1990)]

14.15 CONCLUSION

Organic-biodynamic v. conventional farming comparison. Check soil fertility and nutrient balance:
http://www.fibl.ch/english/research/index.htm

The continuous availability of plant nutrients is critical for the sustainability of most ecosystems. The challenge of nutrient management is threefold: (1) to provide adequate nutrients for plants in the system; (2) to simultaneously ensure that inputs are in balance with plant utilization of nutrients, thereby conserving nutrient resources; and (3) to prevent contamination of the environment with unutilized nutrients.

The recycling of plant nutrients must receive primary attention in any ecologically sound management system. This can be accomplished in part by returning plant residues to the soil. These residues can be supplemented by judicious application of the organic wastes that are produced in abundance by municipal, industrial, and agricultural operations worldwide. The use of cover crops grown specifically to be returned to the soil is an additional organic means of recycling nutrients.

For sites from which crops or forest products are removed, nutrient losses commonly exceed the inputs from recycling. Inorganic fertilizers will continue to supplement natural and managed recycling to replace these losses and to increase the level of soil fertility so as to enable humankind to not only survive, but to flourish on this planet. In extensive areas of the world, fertilizer use will have to be increased above current levels to avoid soil and ecosystem degradation, to remediate degraded soils, and to enable profitable production of food and fiber.

The use of fertilizers, both inorganic and organic, should not be done in a simply habitual manner or for so-called insurance purposes. Rather, soil testing and other diagnostic tools should be used to determine the true need for added nutrients. In managing nitrogen and phosphorus, increasing attention will have to be paid to the potential for transport of these nutrients from soils where they are applied to waterways where they can become pollutants. If soils are low in available nutrients, fertilizers often return several dollars' worth of improved yield for every dollar invested. However, where the nutrient-supplying power of the soil is already sufficient, adding fertilizers is likely to be damaging both to the bottom line and to the environment.

STUDY QUESTIONS

1. The groundwater under a heavily manured field is high in nitrates, but by the time it reaches a stream the nitrate concentration has declined considerably. What are likely explanations for the reduction in nitrate?
2. A cornstalk nitrate test at harvest time shows a farmer that the soil likely contains considerable unused nitrates. To minimize nitrate leaching, the farmer wants to grow a cover crop. What characteristics would you look for in choosing a cover crop to ameliorate this situation?
3. What effect do forest fires have on nutrient availabilities and toxicities?
4. Compare the resource-conservation and environmental-quality issues related to each of the three so-called fertilizer elements, N, P, and K.
5. Compare the relative advantages and disadvantages of organic and inorganic nutrient sources.
6. Traditionally, animal manures have been considered of major benefit to farms in maintaining soil quality. Today, in many situations they are a liability. Explain and indicate some means of alleviating the problems.
7. Discuss the concept of the *limiting factor* and indicate its importance in enhancing or constraining plant growth.
8. Why are nutrient cycling problems in agricultural systems more prominent than in those in forested areas?
9. Discuss how GIS-based site-specific nutrient-application technology might improve profitability and reduce environmental degradation. Suggest factors that might prevent degradation and improve profitability.
10. Discuss the value and limitations of soil tests as indicators of plant nutrient needs and water pollution risks.

REFERENCES

Aber, J., et al. 2000. "Applying ecological principles to management of the U.S. National Forests," *Issues in Ecology 6: http://esa.sdsc.edu/issues6.htm* (confirmed December 12, 2000). (Washington D.C.: Ecological Society of America).

Attiwill, P. M., and G. W. Leeper. 1987. *Forest Soils and Nutrient Cycles.* (Melbourne, Australia: Melbourne University Press).

Beegle, D. B., O. T. Carton, and J. S. Bailey. 2000. "Nutrient management planning: Justification, theory, practice," *J. Environ. Quality* **29**:72–79.

Belt, G. H., and J. O'Laughlin. 1994. "Buffer strip design for protecting water quality and fish habitat," *Western J. Applied Forestry* **9**(2):41–45.

Binkley, D., H. Burnham, and H. L. Allen. 1999. "Water quality impacts of forest fertilization with nitrogen and phosphorus, " *Forest Ecology and Management* **121**:191–213.

Brown, T. C., and D. Binkley. 1994. "Effect of management on water quality in North American forests," USDA Forest Service General Technical Report RM-248. (Ft. Collins, Colo.: USDA Forest Service).

Cerrato, M. E., and A. M. Blackmer. 1990. "Comparison of models from describing corn yield response to nitrogen fertilizer," *Agron. J.* **98**:138–143.

Chesapeake Bay Program. 1995. "Water quality functions of riparian forest buffer systems in the Chesapeake Bay watershed," EPA 903-R-95-004. (Washington, D.C.: USEPA).

Eghball, B., and J. F. Power. 1994. "Beef cattle feedlot manure management," *J. Soil and Water Conservation* **49**:113–122.

Gardner, G. 1997. *Recycling Organic Waste: From Urban Pollutant to Farm Resource.* Worldwatch Paper 135. (Washington, D.C.: Worldwatch Institute).

Gburek, W. J., A. N. Sharpley, L. Heatherwaite, and G. J. Folmar. 2000. "Phosphorus management at the watershed scale: A modification of the phosphorus index," *J. Environ. Quality* **29**:130–144.

Haering, K. C., W. L. Daniels, and S. E. Feagley. 2000. "Reclaiming mined lands with biosolids, manures and papermill sludges," in R. I. Barnhisel, W. L. Daniels, and R. G. Darmody (eds.), *Reclamation of Drastically Disturbed Lands,* Agronomy Monograph 41. (Madison, Wis.: American Society of Agronomy), pp. 615–644.

Havlin, J., D. Beaton, and S. L. Tisdale. 1999. *Soil Fertility and Fertilizers: An Introduction to Nutrient Management.* (Upper Saddle River, N.J.: Prentice Hall).

He, Xin-Tao, T. Logan, and S. Traina. 1995. "Physical and chemical characteristics of selected U.S. municipal solid waste composts," *J. Environ. Qual.* **24**:543–552.

Higgs, B., A. E. Johnston, J. L. Salter, and C. J. Dawson. 2000. "Some aspects of achieving sustainable phosphorus use in agriculture," *J. Environ. Quality* **29**:80–87.

Isse, A., A. MacKenzie, K. Stewart, D. Cloutier, and D. Smith. 1999. "Cover crops and nutrient retention for subsequent sweet corn production," *Agron. J.* **91**:934–939.

Likens, G. E., and F. H. Bormann. 1995. *Biogeochemistry of a Forested Ecosystem,* 2d ed. (New York: Springer-Verlag).

Magdoff, F., L. Lanyon, and B. Liebhardt. 1997. "Nutrient cycling, transformations, and flows: Implications for a more sustainable agriculture," *Advances in Agronomy* **60**:2–73.

McNaughton, S. J., F. F. Banyikwa, and M. M. McNaughton. 1997. "Promotion of the cycling of diet-enhancing nutrients by African grazers," *Science* **278**:1798–1800.

National Research Council. 1993. *Soil and Water Quality: An Agenda for Agriculture.* (Washington, D.C.: National Academy of Sciences).

Olsen, R.A., K.D. Frank, P. Graboushi, and G. Rehm. 1982. "Economic and agronomic impacts of varied philosophies of soil testing," *Agron.* J. **74**:492–499.

Omay, A. B., C. W. Rice, L. D. Maddux, and W. B. Gordon. 1998. "Corn yield and nitrogen uptake in monoculture and in rotation with soybean," *Soil Sci. Soc. Amer. J.* **62**:1596–1603.

Perry, D. A. 1994. *Forest Ecosystems.* (Baltimore: Johns Hopkins University Press).

Powers, J. F., and W. P. Dick (eds.). 2000. *Land Application of Agricultural, Industrial, and Municipal By-Products.* Soil Science Society of America Book Series no. 6. (Madison, Wis.: Soil Sci. Soc. Amer.).

PPI. 1996. "Site-specific nutrient management systems for the 1990's." Pamphlet. (Norcross, Ga.: Potash and Phosphate Institute and Foundation for Agronomic Research; Saskatoon, Sask.: Potash Phosphate Institute of Canada).

Reddy, G. B., and K. R. Reddy. 1993. "Fate of nitrogen-15 enriched ammonium nitrate applied to corn," *Soil Sci. Soc. Amer. J.* **57**:111–115.

Reuter, D. J., and J. B. Robinson. 1986. *Plant Analysis: An Interpretation Manual.* (Melbourne, Australia: Inkata Press).

Richards, R. P., et al. 1996. "Well water, well vulnerability, and agricultural contamination in the midwestern United States," *J. Environ. Quality* **25**:389–402.

Robinson, C. A., M. Ghaffarzadeh, and R. M. Cruze. 1996. "Vegetative filter strip effects on sediment concentration in cropland runoff," *J. Soil and Water Conserv.* **50**:227–230.

Sanchez, P. A., et al. 1997. "Soil fertility replenishment in Africa: An investment in natural resource capital," in R. J. Buresh, P. A. Sanchez, and F. Calhoun (eds.), *Replenishing Soil Fertility in Africa.* ASA/SSSA Publication. (Madison, Wis.: Soil Sci. Soc. Amer.).

Sharpley, A. N. 1997. "Rainfall frequency and nitrogen and phosphorus runoff from soil amended with poultry litter," *J. Environ. Quality* **26**:1127–1132.

Sims, J. T., and D. C. Wolf. 1994. "Poultry waste management: Agricultural and environmental issues," *Advances in Agronomy* **52**:1–83.

Smaling, E. M. A., S. M. Nwanda, and B. H. Jensen. 1997. "Soil fertility in Africa is at stake," in R. J. Buresh, P. A. Sanchez, and F. Calhoun (eds.), *Replenishing Soil Fertility in Africa.* ASA/SSSA Publication. (Madison, Wis.: Soil Sci. Soc. Amer.).

Stivers, L. J., C. Shennen, L. E. Jackson, K. Groody, C. J. Griffin, and P. R. Miller. 1993. "Winter cover cropping in vegetable production systems in California," in M. G. Paoletti, W. Foissner, and D. Coleman (eds.), *Soil Biota, Nutrient Cycling and Farming Systems.* (Boca Raton, Fla.: Lewis Publishers).

Teng, Y., and V. R. Timmer. 1994. "Nitrogen and phosphorus in an intensely managed nursery soil–plant system," *Soil Sci. Soc. Amer. J.* **58**:232–238.

USDA Natural Resources Conservation Service. 2000. *The P Index: A Phosphorus Assessment Tool: www.nhq.nrcs.usda.gov/BCS/nutri/piframe.html* (confirmed December 2000).

Walworth, J. L., and M. E. Sumner. 1987. "The diagnosis and recommendation integrated system (DRIS)," *Advances in Soil Science* **6**:149–187.

Westerman, R. L. (ed.). 1990. *Soil Testing and Plant Analaysis,* 3d ed. (Madison, Wis.: Soil Sci. Soc. Amer.).

Zublena, J. P., J. C. Barker, and T. A. Carter. 1993. "Poultry manure as a fertilizer source," *Soil Facts.* (Raleigh, N.C.: North Carolina Cooperative Extension Service, North Carolina State University).

15

SOIL EROSION AND ITS CONTROL

The wind crosses the brown land, unheard . . .
—*T.S. ELIOT,* THE WASTE LAND

Water lays waste unprotected soil. (R. Weil)

No soil phenomenon is more destructive worldwide than the erosion caused by wind and water. Since prehistoric times people have brought the scourge of soil erosion upon themselves, suffering impoverishment and hunger in its wake. Past civilizations have disintegrated as their soils, once deep and productive, washed away, leaving only thin, infertile, rocky relics of the past. When viewing the nearly barren hills in central India, or in parts of Greece, Lebanon, and Syria, it is hard to imagine that agricultural communities once flourished in these places.

The current world threat of soil erosion is more ominous than at any time in history. Since 1960, farmers have had to more than double world food output to feed the unprecedented numbers of people on Earth. In the low-income countries, the ratio of people to available cropland is already high and rising rapidly. While intensified cultivation of fertile, relatively level lands has helped produce much of the needed food, many nations have had to expand the area of land under cultivation, clearing and burning steep, forested slopes and plowing grasslands. Population pressures have also led to overgrazing of rangelands and overexploiting of timber resources. All these activities degrade or remove natural vegetation, causing the underlying soils to become much more susceptible to the destructive action of erosion. This results in a downward cycle of deterioration—land degradation leads to poor crops, human poverty, and reduced vegetative protection for the soil, which in turn accelerates erosion and drives ever more desperate people to clear, cultivate, and degrade still more land.

The degraded productivity of farm-, forest-, and rangelands tells only part of the sad erosion story. Soil particles washed or blown from the eroding areas are subsequently deposited elsewhere—in nearby low-lying landscape sites; in streams and rivers; or in downstream reservoirs, lakes, and harbors. The environmental and economic damages suffered by sites on which the eroded soil materials are deposited may be as great or greater than that incurred on the sites from which the soil material was removed. The displaced soil materials (sediment and dust) lead to major water and air pollution problems, bringing enormous economic and social costs to society.

Combating soil erosion is everybody's business. Fortunately, recent decades have seen much progress in understanding the mechanisms of erosion and in developing techniques that can effectively and economically control soil loss in most situations. This chapter will equip you with some of the concepts and tools you will need to do your part in solving this pressing world problem.

15.1 SIGNIFICANCE OF SOIL EROSION AND LAND DEGRADATION[1]

Land Degradation

Extent of soil degradation-world map:
http://www.terrestris.de/soilerosion/extent.html

During the past half century, human land use and associated activities have degraded some 5 billion ha (about 43%) of earth's vegetated land. Such **land degradation** results in a reduced productive potential and a diminished capacity to provide benefits to humanity. Much of this degradation (on about 3.6 billion ha) is linked to **desertification,** the spreading of desert conditions that disrupt semiarid and arid ecosystems (including agroecosystems). A major cause of desertification is overgrazing by cattle, sheep, and goats, a factor that likely accounts for about a third of all land degradation, mainly in such dry regions as the Sahel in northern Africa, East Asia, and the rangeland of the American Southwest. Likewise, the indiscriminate felling of rain forest trees has already degraded nearly 0.5 billion ha in the humid tropics. Additionally, inappropriate agricultural practices continue to degrade land in all climatic regions.

Soil–Vegetation Interdependency

The "herd effect" in intensively managed grazing for sustainable use of arid rangeland:
http://www.holisticmanagement.org/ahm_graze.cfm

Degraded lands may suffer from destruction of native vegetation communities, reduced agricultural yields, lowered animal production, and simplification of once-diverse natural ecosystems with or without accompanying degradation of the soil resource. On about 2 billion of the 5 billion ha of degraded lands in the world, soil degradation is a major part of the problem (see Figure 15.1). In some cases the soil degradation occurs mainly as deterioration of physical properties by compaction or surface crusting (see Sections 4.5 and 4.8), or as deterioration of chemical properties by acidification (see Section 9.7) or salt accumulation (see Section 9.13). However, most (~85%) soil degradation stems from erosion—the destructive action of wind and water.

The two main components of land degradation—damage to soils and to plant communities—are not independent, but interact to cause a downward spiral of accelerating deterioration. Overgrazing, fires, deforestation, and inappropriate farming diminish the ability of the vegetation to protect the soil. Likewise, soil erosion, crusting, salinization, and nutrient depletion diminish the ability of the soil to support vegetation. Poor farmers, trying to grow food for their families, are often caught up in this cycle of impoverishment as they clear and till ever more land in a desperate attempt to make up for declining soil productivity. Improvements in both soil and vegetation management must go hand in hand if the productive potential of the land is to be protected—or even restored by moving *up* rather than *down* the spiral.

Geological Versus Accelerated Erosion

GEOLOGICAL EROSION. **Erosion** is a process that transforms soil into **sediment.** Soil erosion that takes place naturally, without the influence of human activities, is termed **geological erosion.** It is a natural leveling process. It inexorably wears down hills and mountains, and through subsequent deposition of the eroded sediments, it fills in valleys, lakes, and bays. Many of the landforms we see around us—canyons, buttes, rounded hills, river valleys, deltas, plains, and pediments—are the result of geological erosion and deposition. The vast deposits that now appear as sedimentary rocks originated in this way.

[1]For a readable account of historical degradation of land and water resources, see Hillel (1991). Overviews of the current extent of soil erosion and other forms of land degradation are given by Oldeman (1994), Daily (1997), and Rosensweig and Hillel (1998).

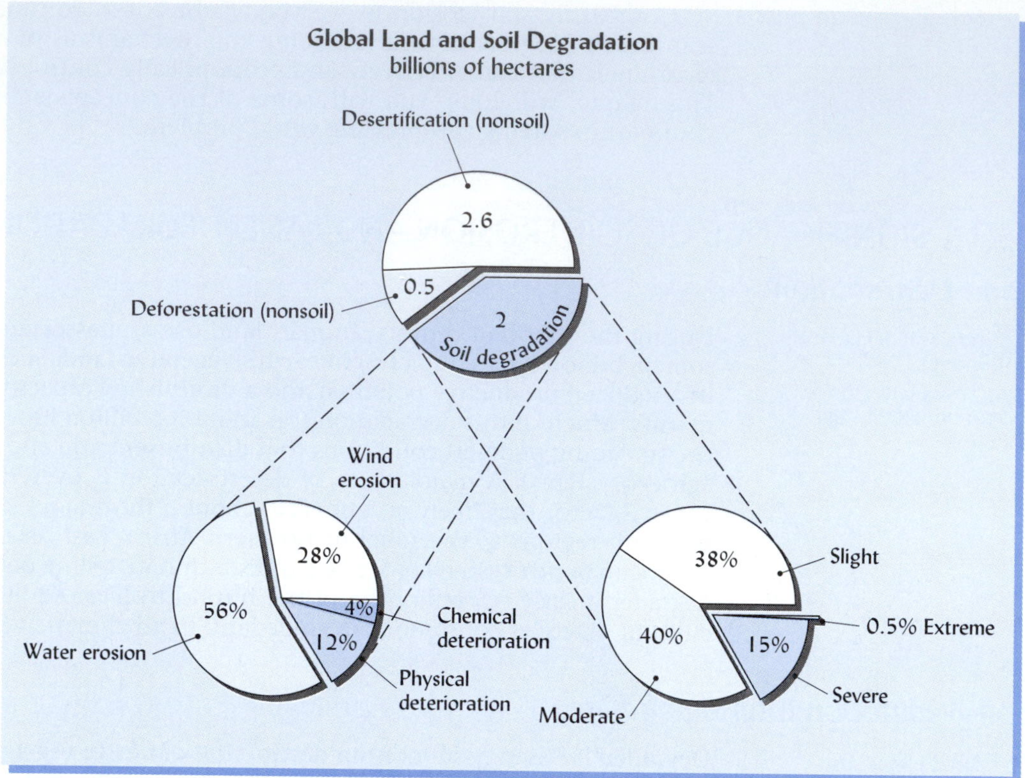

FIGURE 15.1 Soil degradation as a part of global land degradation caused by overgrazing, deforestation, inappropriate agricultural practices, fuel wood overexploitation, and other human activities. About 60% of degraded land has not suffered soil degradation. Of the 2 billion ha of land with degraded soils, most could be restored easily (*slight* degradation) or with considerable financial and technical investments (*moderate* degradation). *Severely* degraded soils are currently useless for agriculture and would require major international assistance for restoration. About 9 million ha (0.5% of degraded soils) are *extremely* degraded and incapable of restoration. About 85% of the soil degradation is caused by erosion by wind and water. [FAO data selected from Oldeman (1994) and Daily (1997)]

In most settings, geological erosion wears down the land slowly enough that new soil forms from the underlying rock or regolith faster than the old soil is lost from the surface. The very existence of soil profiles bears witness to the net accumulation of soil and the effectiveness of undisturbed natural vegetation in protecting the land surface from erosion.

The rate of geological soil erosion varies greatly with both rainfall and type of material comprising the regolith. Geological erosion by water tends to be greatest in semiarid regions where rainfall is enough to be damaging, but not enough to support dense, protective vegetation (Figure 15.2).

Accelerated Erosion. **Accelerated erosion** occurs when people disturb the soil or the natural vegetation by grazing livestock, cutting forests for agricultural use (Figure 15.3), plowing hillsides, or tearing up land for construction of roads and buildings. Accelerated erosion is often 10 to 1000 times as destructive as geological erosion, especially on sloping lands in regions of high rainfall. Rates of erosion by wind and water on agricultural land in Africa, Asia, and South America are thought to average about 30 to 40 Mg/ha annually. In the United States, the average erosion rate on cropland is about 12 Mg/ha—7 Mg by water and 5 Mg by wind. Some cultivated soils are eroding at 10 times these average rates. In comparison, erosion on undisturbed humid-region grasslands and forests generally occurs at rates considerably below 0.1 Mg/ha.

About 4 billion Mg of soil is moved annually by soil erosion in the United States, some two-thirds by water and one-third by wind. More than half of the movement by water and about 60% of the movement by wind takes place on croplands that produce most of the country's food. Much of the remainder comes from semiarid rangelands, from road building and timber harvest on forest lands, and from soils disturbed for highway and building construction. Although progress has been made in reducing erosion

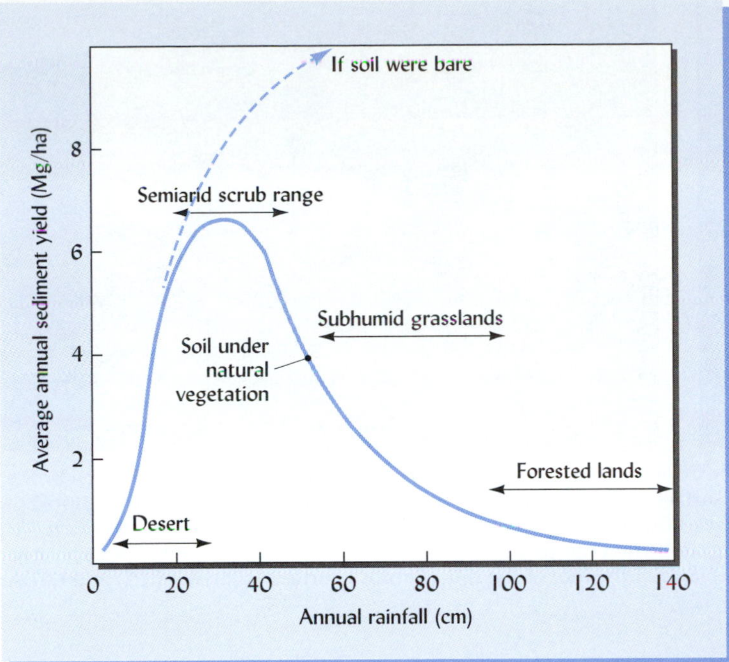

FIGURE 15.2 Generalized relationship between annual rainfall and soil loss from geologic erosion by water. The actual amount of sediment lost annually per hectare will depend on other climatic variables, topography, and the type of soils in the watershed. Note that sediment yields are greatest in semiarid regions. Here a number of severe runoff–generating storms occur in most years, but the total rainfall is too little to support much protective plant cover. By comparison, the very dry deserts have too little rain to cause much erosion, and the well-watered regions support dense forests that effectively protect the soil. Where the natural vegetation is destroyed by plowing, erosion from the bare soils is much higher with increasing rainfall, as indicated by the dashed curve.

(see Section 15.11), current high losses are simply not acceptable for long-term sustainability and must be further reduced.

Under the influence of accelerated erosion, soil is commonly washed or blown away faster than new soil can form by weathering or deposition. As a result, the soil depth suitable for plant roots is often reduced. In severe cases, gently rolling terrain may become scarred by deep gullies, and once-forested hillsides may be stripped down to bare rock.

Accelerated erosion often makes the soils in a landscape more heterogenous. One example of increased heterogeneity occurs as overgrazed semiarid grasslands degrade into

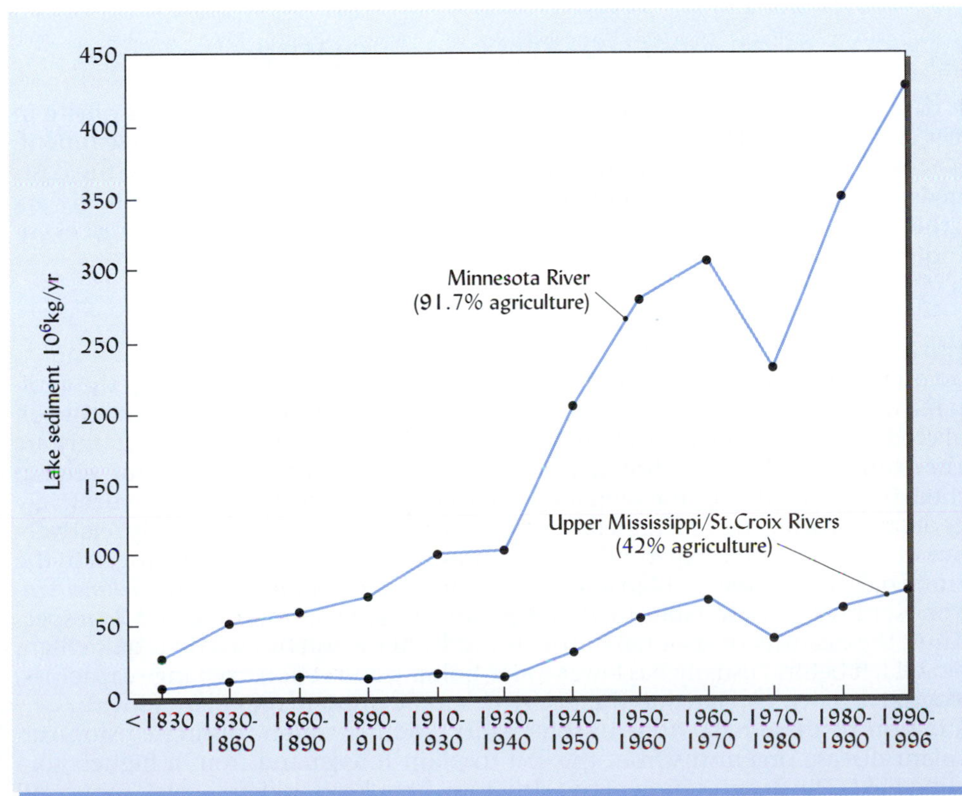

FIGURE 15.3 Sediment deposition rates into Lake Pepin (Minnesota-Wisconsin border) from watersheds of the Minnesota River (92% cultivated) and the upper Mississippi and Croix Rivers (42% cultivated). Neither watershed deposited much sediment before about 1830 because little of the land had been cleared for farming. But as agriculture expanded, deposition rates increased, especially from the Minnesota watershed where intensive row-crop agriculture became dominant. [Redrawn from Kelley and Nater (2000)]

FIGURE 15.4 Erosion and deposition occur simultaneously across a landscape. (Left) The soil on this ridgetop was worn down by erosion during nearly 300 years of cultivation. The surface soil exposed on the ridgetop consists mainly of light-colored C horizon material. At sites lower down the slope the surface horizon shows mainly A and B horizon material, some of which has been deposited after eroding from locations upslope. (Right) Erosion on the sloping wheat field in the background has deposited a thick layer of sediment in the foreground, burying the plants at the foot of the hill. [Photos courtesy of R. Weil (left) and USDA Natural Resources Conservation Service (right)]

scrub brush ecosystems. During this degradation "islands of fertility" form under shrubs, while the soils between shrubs become increasingly thin and infertile. Another example can be seen in the striking differences in surface soil color that develop as ridgetop soils are truncated, exposing material from the B or C horizon at the land surface, while soils lower in the landscape are buried under organic-matter-enriched sediment (Figure 15.4 and Plate 28 following page 114).

It should not be assumed that all accelerated erosion comes from cropland. Nearly half of the soil lost by erosion comes from building and highway construction sites, and from other land uses such as the grazing of rangelands and the harvesting of forests. In any case, however, the losses come from human uses or misuses of land resources.

15.2 ONSITE AND OFFSITE EFFECTS OF ACCELERATED SOIL EROSION

Sediment, nutrient and chemical loading. Click on "Ecosystem Threats": http://www.bigdarby.org/

Erosion damages the site on which it occurs and also has undesirable effects offsite in the larger environment. The offsite costs relate to the effects of excess water, sediment, and associated chemicals on downhill and downstream environments. While the costs associated with either or both of these types of damages may not be immediately apparent, they are real and grow with time. Landowners and society as a whole must eventually foot the bill.

Types of Onsite Damages

The most obviously damaging aspect of erosion is the loss of soil itself. In reality, the damage done to the soil is greater than the amount of soil lost would suggest, because the soil material eroded away is almost always more valuable than that left behind. Not only are surface horizons eroded while subsurface horizons (which are usually less useful) remain untouched, but the quality of the remaining topsoil is also impaired. Erosion selectively removes organic matter and fine mineral particles, while leaving behind mainly relatively less active, coarser fractions. Experiments have shown organic matter and nitrogen in the eroded material to be 5 times as high as in the original topsoil. Comparable *enrichment ratios* for phosphorus (review Table 13.2) and potassium are commonly 2 and 3, respectively. Thus, the quantity of essential nutrients lost from the soil by erosion is quite high. Also, the soil left behind usually has lower water-holding and cation exchange capacities, less biological activity, and a reduced capacity to supply nutrients for plant growth.

In addition to these reductions in soil-quality, soil movement during erosion can spread plant disease organisms from the soil to plant foliage and from a higher- to a lower-lying field. The deterioration of soil structure often leaves a dense crust on the soil

Sediment Happens. Problems in aquatic systems: http://www.bayjournal.com/02-07/soil.htm

surface, which in turn greatly reduces water infiltration and increases water runoff. Newly planted seeds and seedlings may be washed downhill, trees may be uprooted, and small plants may be buried in sediment. In the case of wind erosion, fruits and foliage may be damaged by the sandblasting effect of blowing soil particles.

Finally, gullies that carve up badly eroded land may make the use of tractors impossible and may undercut pavements and building foundations, causing unsafe conditions and expensive repairs.

Types of Offsite Damages

Erosion moves sediment and nutrients off the land, creating the two most widespread water pollution problems in our rivers and lakes. The nutrients impact water quality largely through the process of eutrophication caused by excessive nitrogen and phosphorus, as discussed in Section 13.2. In addition to nutrients, sediment and runoff water may also carry toxic metals and organic compounds, such as pesticides. The sediment itself is a major water pollutant, causing a wide range of environmental damages.

Damages from Sediment. Sediment deposited on the land may smother crops and other low-growing vegetation (Figure 15.4). It fills in roadside drainage ditches and creates hazardous driving conditions where mud covers the roadway.

Sediment that washes into streams makes the water cloudy or turbid (Figure 15.5*a* and Plate 66). High **turbidity** prevents sunlight from penetrating the water and thus reduces

(*a*)

(*b*)

FIGURE 15.5 Offsite damages caused by soil erosion include the effect of sediment on aquatic systems. (*a*) A sediment-laden tributary stream empties into the relatively clear waters of a larger river. The turbid water will foul fish gills, inhibit submerged aquatic vegetation, and clog water-purification systems. Part of the sediment will settle out on the river bottom, covering fish-spawning sites and raising the river bed enough to aggravate the severity of future flooding episodes. (*b*) Expensive dredging and excavation (by the dragline in the foreground) is being undertaken to restore the beauty, recreational value, and flood-control function of this pond in a neighborhood park after accumulated sediment had transformed it into a mere mudflat. The watershed upstream from the pond had undergone a period of rapid suburban development, during which adequate sediment-control practices were not used on the construction sites. [Photos courtesy of USDA Natural Resources Conservation Service (*a*) and R. Weil (*b*)].

photosynthesis and survival of the *submerged aquatic vegetation (SAV)*. The demise of the SAV in turn degrades the fish habitat and upsets the aquatic food chain. The muddy water also fouls the gills of some fish. Sediment deposited on the stream bottom can have a disastrous effect on many freshwater fish by burying the pebbles and rocks among which they normally spawn. The buildup of bottom sediments can actually raise the level of the river, so that flooding becomes more frequent and more severe. For example, to counter the rising river bottom, flood-control levees along the Mississippi must be constantly enlarged.

A number of major problems occur when the sediment-laden rivers reach a lake, reservoir, or estuary. Here, the water slows and drops its load of sediment. Eventually reservoirs—even those formed by giant dams—become mere mudflats, completely filled in with sediment (see Figure 15.5*b*). Prior to that, the capacity of the reservoir to store water for irrigation or municipal water systems is progressively reduced, as is the capacity for floodwater retention or hydroelectric generation. It is estimated that 1.5 billion Mg of sediment is deposited each year in the nation's reservoirs, harbors, and shipping channels.

WINDBLOWN DUST. Wind erosion also has its offsite effects. The blowing sands may bury roads and fill in drainage ditches, necessitating expensive maintenance. The sandblasting effect of windborne soil particles may damage the fruits and foliage of crops in neighboring fields, as well as the paint on vehicles and buildings many kilometers downwind from the eroding site.

The finest particles blow the farthest, and may present a major human health hazard. While silt-sized particles are generally filtered by nose hairs or trapped in the mucus of the windpipe and bronchial tubes, smaller clay-sized particles often pass through these defenses and lodge in the alveoli (air sacs) of the lungs. Long-term epidemiological studies suggest that the number of deaths resulting from people inhaling this fine *fugitive dust* could even exceed the number of deaths from car highway accidents.

ESTIMATED COSTS OF EROSION. The total onsite costs of erosion in the United States, including costs of replacing nutrients and water lost through accelerated erosion, as well as crop yield reductions due to reduced soil depth, have been estimated at between $4 and $27 billion per year.

The offsite costs of erosion are likely even greater, especially because of the health effects of windblown particles and the reduced recreational (fishing, swimming, and aesthetic) value of muddy waters. The total of these annual offsite costs has been estimated at $5 to $17 billion. The grand total annual cost of erosion in the United States is therefore likely between $9 and $44 billion. Such high costs are a sobering reminder of the burden that society bears as a result of poor land management and would seem to justify increasing the sums allocated to the battle against erosion.

MAINTENANCE OF SOIL PRODUCTIVITY. Over the long term, accelerated soil erosion that exceeds the rate of soil formation leads to declining productivity on most soils. In the United States, crop yields on severely eroded soils are often 20 to 40% lower than on similar soils with only slight erosion. Ultimately, the rate of decline of soil productivity, or the cost of maintaining constant crop-yield levels, is determined by such soil properties as *depth to a root-restricting layer* and *permeability of the subsoil* (Figure 15.6).

Soil-Loss Tolerance

A tolerable soil loss (***T* value**) is the maximum amount of soil that can be lost annually by the combination of water and wind erosion on a particular soil without degrading that soil's long-term productivity. Currently, *T* values are based on the best judgment of informed soil scientists, rather than on rigorous research data. The *T* values for soils in the United States commonly range from 5 to 11 Mg/ha. They depend on a number of soil-quality and management factors, including soil depth, organic matter content, and the use of water-control practices. Soils with shallow layers of infertile, impermeable, or rocky material are generally assigned *T* values near the low end of the range. In some tropical areas, the low nutrient-supplying power of the subsoil suggests that appropriate *T* values may be considerably lower than those used in the United States.

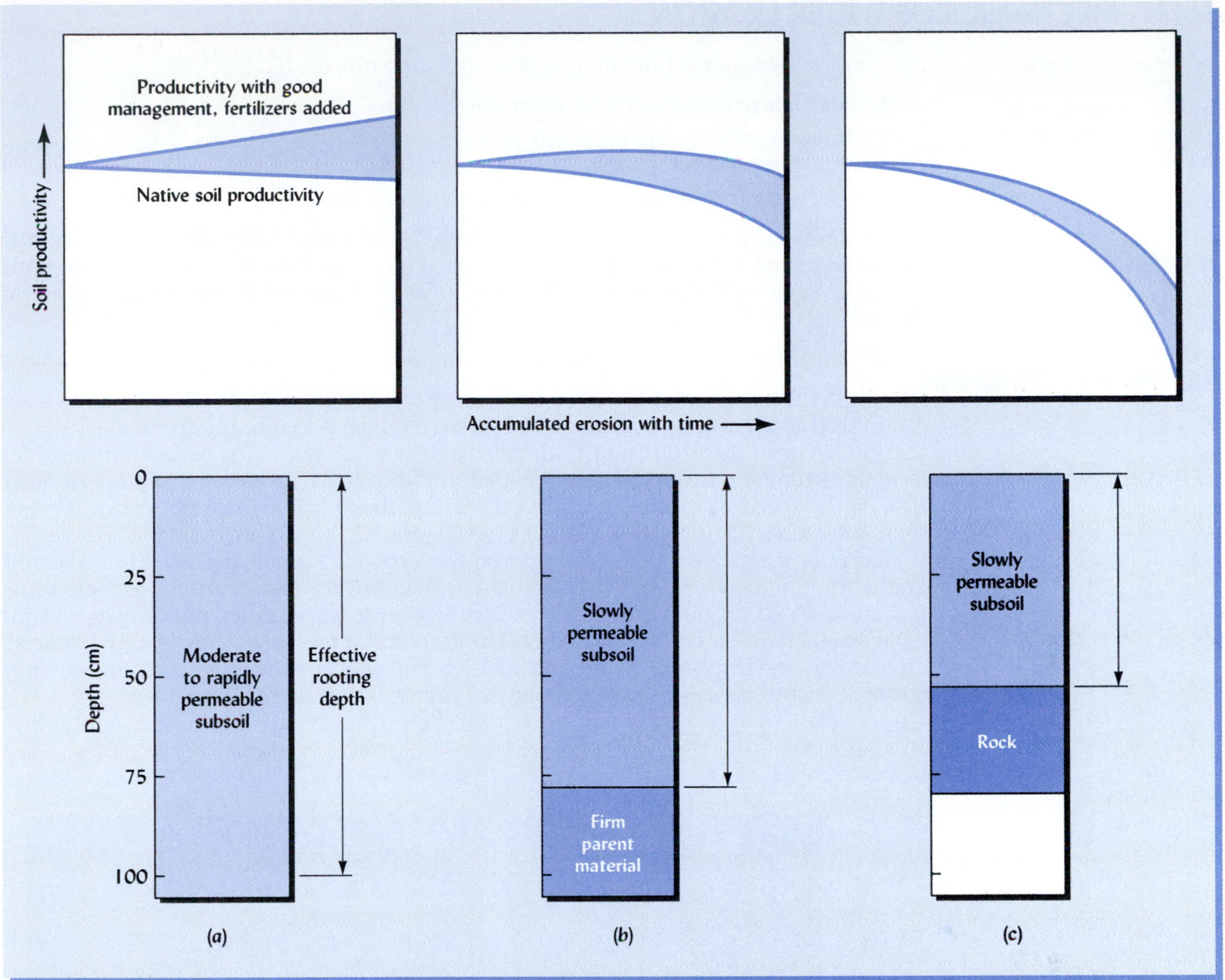

FIGURE 15.6 Effect of erosion over time on the productivity of three soils differing in depth and permeability. Productivity on soil (*a*) actually increases with time because of good management practices and fertilizer additions, even though the native soil productivity declines as a result of erosion. Because soil (*c*) is shallow and has restricted permeability, its productivity declines rapidly as a result of erosion, a decline that good management and fertilizers cannot prevent. Soil (*b*), which is intermediate in both depth and permeability, suffers only a slight decline in productivity due to erosion. Soil characteristics clearly influence the effect of erosion on soil productivity.

The majority of agricultural soils in the United States are currently assigned the highest *T* value, 11 Mg/ha. This represents a maximum allowable loss of about 0.9 mm of soil depth annually, a rate at which it would take about 225 years to lose the equivalent of an entire Ap horizon.[2] Under good agricultural management in a deep, permeable soil profile, this may be sufficient time to allow replacement of the lost material as new Ap horizon material forms from underlying subsoil material. Development of horizons in undisturbed soils under natural vegetation is most likely much slower than this, so the *T* value assigned to a particular type of soil probably would not provide sufficient protection for many rangeland or forest sites.

The concept of *T* values is useful in focusing attention on the soils where improved practices are needed to maintain long-term productivity. The 1992 U.S. National Resource Inventory suggests that 33% of all cropland is eroding at rates in excess of the *T* value, and about 15% at more than twice the *T* value. Clearly, much work remains to be done in controlling excessive soil loss.

[2]This calculation assumes that the Ap horizon is 200 mm thick and the bulk density of the soil is 1.25 Mg/m^3.

15.3 MECHANICS OF WATER EROSION

Details on raindrop impact and the soil erosion process: http://www.terrestris.de/soilerosion/processesmain.html

Soil erosion by water is fundamentally a three-step process (Figure 15.7):

1. *Detachment* of soil particles from the soil mass
2. *Transportation* of the detached particles downhill by floating, rolling, dragging, and splashing
3. *Deposition* of the transported particles at some place lower in elevation

On comparatively smooth soil surfaces, the beating action of raindrops causes most of the detachment. Where water is concentrated into channels, the cutting action of turbulent flowing water detaches soil particles. In some situations, freezing–thawing action also contributes to soil detachment.

Influence of Raindrops

A raindrop accelerates as it falls until it reaches *terminal velocity*—the speed at which the friction between the drop and the air balances the force of gravity. Larger raindrops fall faster, reaching a terminal velocity of about 30 km/h, a little faster than a person can run. As the speeding raindrops impact the soil with explosive force, they transfer their high kinetic energy to the soil particles (see Figure 15.7).

Raindrop impact exerts three important effects: (1) it detaches soil; (2) it destroys granulation; and (3) its splash, under certain conditions, causes an appreciable transportation of soil. So great is the force exerted by raindrops that they not only loosen and detach soil granules, but may even beat the granules to pieces. As the dispersed material dries it may develop into a hard crust, which will prevent the emergence of seedlings and will encourage runoff from subsequent precipitation (see Section 4.8).

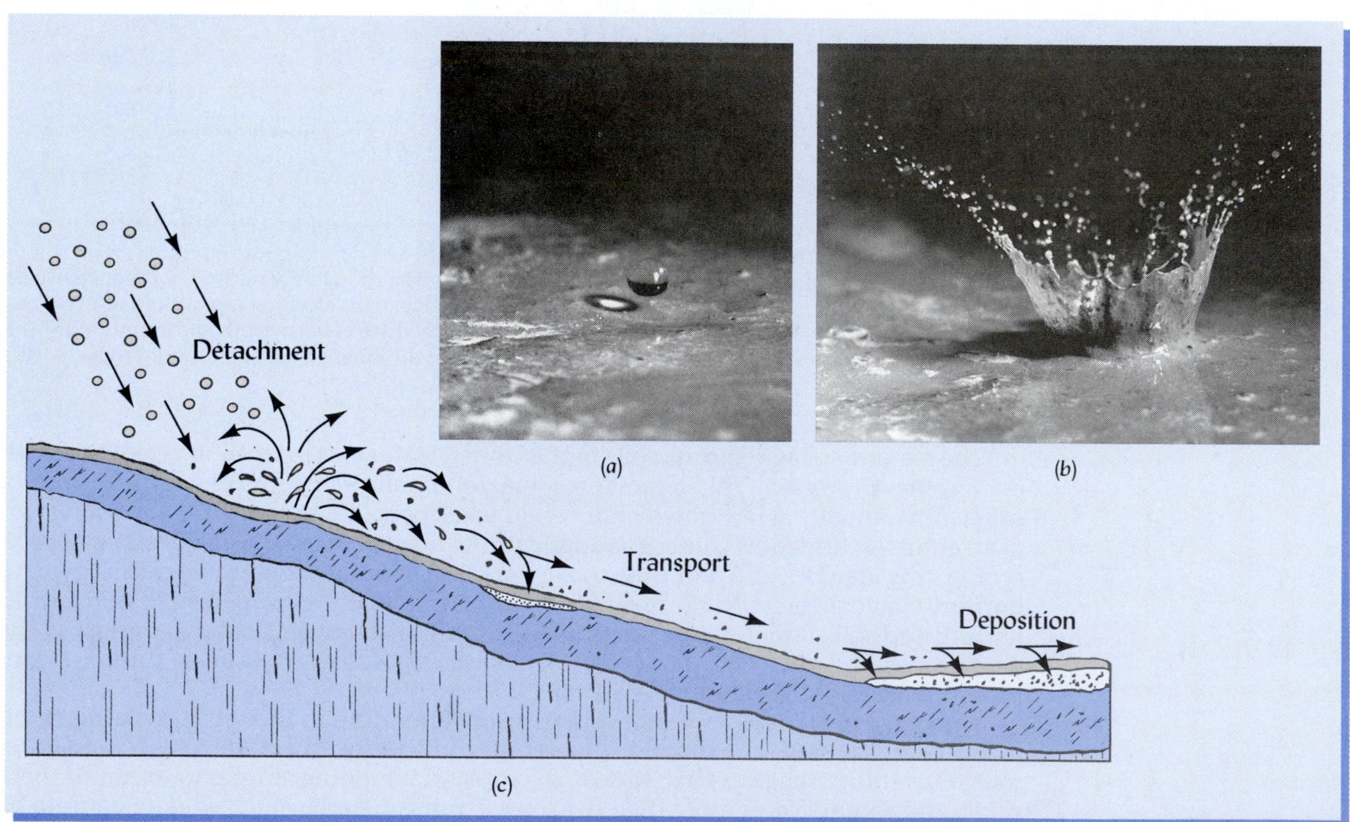

FIGURE 15.7 The three-step process of soil erosion by water begins with the impact of raindrops on wet soil. (*a*) A raindrop speeding toward the ground. (*b*) The splash that results when the drop strikes a wet, bare soil. Such raindrop impact destroys soil aggregates, encouraging sheet and interill erosion. Also, considerable soil may be moved by the splashing process itself. The raindrop affects the detachment of soil particles, which are then transported and eventually deposited in locations downhill (*c*).

History may someday record that one of the truly significant scientific advances of the 20th century was the realization that most erosion is initiated by the impact of raindrops, rather than the flow of running water. For centuries prior to this realization, soil conservation efforts aimed at controlling the more visible flow of water across the land, rather than protecting the soil surface from the impact of raindrops.

Transportation of Soil

Raindrop Splash Effects. When raindrops strike a wet soil surface, they detach soil particles and send them flying in all directions (see Figure 15.7). On a soil that is subject to easy detachment, a very heavy rain may splash as much as 225 Mg/ha of soil, some of the particles splashing as much as 0.7 m vertically and 2 m horizontally. If the land is sloping or if the wind is blowing, this splashing may be greater in one direction, leading to considerable net horizontal movement of soil.

Role of Running Water. Runoff water plays the major role in the transportation step of soil erosion. If the rate of rainfall exceeds the soil's infiltration capacity, water will pond on the surface and begin running downslope. The soil particles sent flying by raindrop impact will then land in flowing water, which will carry them down the slope. So long as the water is flowing smoothly in a thin layer (sheet flow), it has little power to detach soil; however, in most cases the water is soon channeled by irregularities in the soil surface which causes it to increase in both velocity and turbulence. The channelized flow then not only carries along soil splashed by raindrops, but also begins to detach particles as it cuts into the soil mass.

Types of Water Erosion

Three types of water erosion are generally recognized: (1) *sheet,* (2) *rill,* and (3) *gully* (Figure 15.8). In **sheet erosion,** splashed soil is removed more or less uniformly, except that tiny columns of soil often remain where pebbles intercept the raindrops (see Figure 15.8*a*). However, as the sheet flow is concentrated into tiny channels (termed **rills**), **rill erosion** becomes dominant. Rills are especially common on bare land whether newly planted or in fallow (see Figure 15.8*b*). Rills are channels small enough to be smoothed by normal tillage, but the damage is already done—the soil is lost. When sheet erosion takes place primarily between irregularly spaced rills, it is called **interrill erosion.**

Where the volume of runoff is further concentrated, the rushing water cuts deeper into the soil, deepening and coalescing the rills into larger channels, termed **gullies** (see Figure 15.8*c*). This is **gully erosion.** Gullies on cropland are obstacles to tractors and cannot be removed by ordinary tillage practices. All three types may be serious, but sheet and rill erosion, although less noticeable than gully erosion, are responsible for most of the soil moved.

Deposition of Eroded Soil

Erosion may send soil particles on a journey of a thousand kilometers or more—off the hills, into creeks, and down great muddy rivers to the ocean. On the other hand, eroded soil may travel only a meter or two before coming to rest in a slight depression on a hillside or at the foot of a slope (as shown in Figure 15.4). The amount of soil delivered to a stream, divided by the amount eroded, is termed the **delivery ratio.** As much as 60% of eroded soil may reach a stream (delivery ratio = 0.60) in certain watersheds where valley slopes are very steep. As little as 1% may reach the streams draining a gently sloping coastal plain. Typically, the delivery ratio is larger for small watersheds than for large ones, because the latter provide many more opportunities for deposition before a major stream is reached. It is estimated that about 5 to 10% of all eroded soil in North America is washed out to sea. The remainder is deposited in reservoirs, river beds, on flood plains, or on relatively level land farther up the watershed.

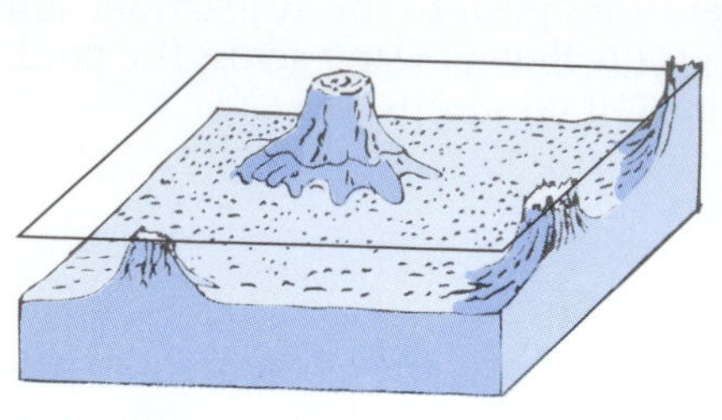

(*a*) Sheet erosion

(*b*) Rill erosion

(*c*) Gully erosion

FIGURE 15.8 Three major types of soil erosion. *Sheet erosion* is relatively uniform erosion from the entire soil surface. Note that the perched stones and pebbles have protected the soil underneath from sheet erosion. The pencil gives a sense of scale. *Rill erosion* is initiated when the water concentrates in small channels (rills) as it runs off the soil. Subsequent cultivation may erase rills, but it does not replace the lost soil. *Gully erosion* creates deep channels that cannot be erased by cultivation. Although gully erosion looks the most catastrophic of the three, far more total soil is lost by the less obvious sheet and rill erosion. [Drawings from FAO (1987); photos courtesy USDA Natural Resources Conservation Service]

15.4 FACTORS AFFECTING EROSION BY WATER

The Universal Soil Loss Equation[3]

The major factors that affect erosion by water are quantified in the **universal soil-loss equation (USLE):**

$$A = RKLSCP$$

where A, the predicted soil loss, is the product of:

R = rainfall erosivity	Rain-related factor
K = soil erodibility L = slope length S = slope gradient or steepness	Soil-related factors

[3]For discussion of the original USLE, see Wischmeier and Smith (1978), and for the RUSLE, see Renard et al. (1997). In this textbook, we use the scientifically acceptable SI units for the R and K factors in our discussion of these erosion equations. However, since these soil-loss equations were published in the United States for use by landowners and the general public, most maps, tables, and computer programs available supply values for the R and K factors in customary English units, rather than in SI units. When using English units for the R and K factors, the soil loss A is expressed in tons (2000 lb) per acre, which can be easily converted to Mg/ha by multiplying by 2.24. For details on converting the customary English units to SI units, see Foster et al. (1981).

Erosion principles and control strategies with focus on Africa:
http://www.fao.org/docrep/T1765E/t1765e00.htm#Contents

C = cover and management
P = erosion-control practices
} Land management factors

The USLE was designed to predict only the amount of soil loss by sheet and rill erosion in an average year for a given location. It cannot predict erosion from a specific year or storm, nor can it predict the extent of gully erosion and sediment delivery to streams. It can, however, show how varying any combination of the soil- and land-management-related factors might be expected to influence soil erosion, and therefore can be used as a decision-making aid in choosing the most effective strategies to conserve soil.

The Revised Universal Soil Loss Equation (RUSLE)

Background, use and download for RUSLE and associated data:
www.sedlab.olemiss.edu/rusle/

The USLE has been used widely since the 1970s. In the early 1990s, the basic USLE was updated and computerized to create an erosion-prediction tool called the **revised universal soil-loss equation (RUSLE).** The RUSLE uses the same basic factors of the USLE just shown, although some are better defined and interrelationships among them improve the accuracy of soil-loss prediction. The RUSLE is a computer software package that is constantly being improved and modified as experience is gained from its use around the world.

As we are about to see, the five factors that comprise the USLE provide a useful framework for understanding soil erosion and its control.

Rainfall Erosivity Factor R

The rainfall **erosivity** factor R represents the driving force for sheet and rill erosion. It takes into consideration the total rainfall and, more important, the intensity and seasonal distribution of the rain. Rain intensity is of great importance for two reasons: (1) intense rains have a large drop size, which results in much greater kinetic energy being available to detach soil particles; and (2) the higher the rate of rainfall, the more runoff that occurs, providing the means to transport detached particles. Gentle rains of low intensity may cause little erosion, even if the total annual precipitation is high. In contrast, a few torrential downpours may result in severe damage, even in areas of low annual rainfall. Likewise, soil losses are heavy if the rain falls when the soil is just thawing or is relatively bare because of recent disturbance.

Rainfall index values for locations in the United States are shown in Figure 15.9. Note that they vary from less than 10 in areas of the west to more than 700 along the coasts of Louisiana [the map gives R values in English units that can be converted to Sl units of (MJ · mm)/(ha · h · yr) if multiplied by 17.02]. Similar data have been generated in other parts of the world. Generally, rainfall tends to be more intense and more erosive in subtropical and tropical regions than in temperate regions.

Rainfall intensity in most locations is so highly variable that actual erosivity in any single year is commonly 2 to 5 times greater or smaller than the long-term average. In fact, a few unusually intense, heavy storms often account for most of the erosion that takes place during a typical decade. Conservation practices based on the predictions of the USLE or RUSLE using long-term average R factors may not be sufficient to limit erosion damages from these relatively rare, but extremely damaging, storms.

Soil Erodibility Factor K

The soil **erodibility** factor K indicates a soil's inherent susceptibility to erosion. The K value assigned to a particular type of soil indicates the amount of soil lost per unit of erosive energy in the rainfall, assuming a standard research plot (22 m long, 9% slope) on which the soil is kept continuously bare by tillage.

The two most significant and closely related soil characteristics influencing erodibility are (1) *infiltration capacity* and (2) *structural stability.* High infiltration means that less water will be available for runoff, and the surface is less likely to be ponded (which would make it more susceptible to splashing). Stable soil aggregates resist the beating action of rain, and thereby save soil even though runoff may occur. Certain tropical clay soils high in hydrous oxides of iron and aluminum are known for their highly stable aggregates that resist the action of torrential rains. Downpours of a similar magnitude on swelling-type clays would be disastrous.

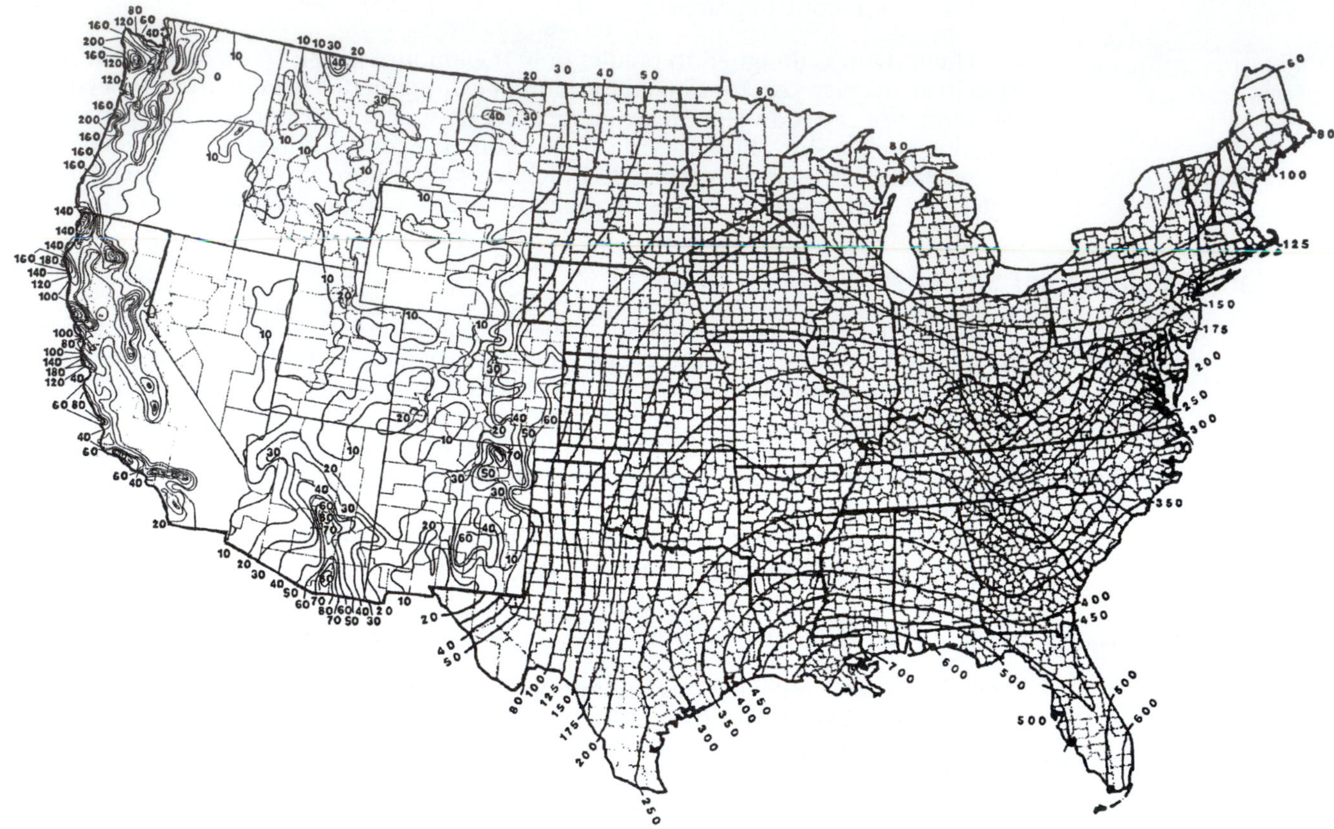

FIGURE 15.9 The geographic distribution of R values for rainfall erosivity in the continental United States. Note the very high values in the humid, subtropical Southeast, where annual rainfall is high and intense storms are common. Similar amounts of annual rainfall along the coast of Oregon and Washington in the Northwest result in much lower R values because there the rain mostly falls gently over long periods. The complex patterns in the West are mainly due to the effects of mountain ranges. Values on map are in units of 100 (ft · ton· in.)/(acre · yr). To convert to SI units of (MJ · mm)/(ha · h · yr), multiply by 17.02. [Redrawn from USDA/ARS (1997)]

Basic soil properties that tend to result in high K values include high contents of silt and very fine sand; expansive types of clay minerals; a tendency to form surface crusts; the presence of impervious soil layers; and blocky, platy, or massive soil structure. Soil properties that tend to make the soil more *resistant* to erosion (low K values) include high soil-organic-matter content, nonexpansive types of clays, and strong granular structure. Approximate K values for soils at selected locations are shown in Table 15.1. Site-specific K values obtained with the RUSLE program will vary somewhat from these values. Note that the K factor in SI units normally varies from near zero to about 0.1. Soils with high rates of water infiltration commonly have K values of 0.025 or below, while more easily eroded soils with low infiltration rates have K factors of 0.04 or higher.

Topographic Factor LS

The topographic factor LS reflects the influence of length and steepness of slope on soil erosion. It is expressed as a unitless ratio with soil loss from the area in question in the numerator, and that from a standard plot (9% slope, 22 m long) in the denominator. The longer the slope, the greater the opportunity for concentration of the runoff water.

The LS factors in Table 15.2 illustrate the increases that occur as slope length and steepness increase. Three subtables are given for sites with low, moderate, and high ratios of rill to interrill (sheet) erosion.

TABLE 15.1 Computed K Values for Soils at Different Locations

The values listed are in SI units. The computerized RUSLE model may give more accurate values for some locations.

Soil	*Location*	*Compounds[a] K*
Udalf (Dunkirk silt loam)	Geneva, N.Y.	0.091
Udalf (Keene silt loam)	Zanesville, Ohio	0.063
Udult (Lodi loam)	Blacksburg, Va.	0.051
Udult (Cecil sandy clay loam)	Watkinsville, Ga.	0.048
Udoll (Marshall silt loam)	Clarinda, Iowa	0.044
Udalf (Hagerstown silty clay loam)	State College, Pa.	0.041
Ustoll (Austin silt)	Temple, Tex.	0.038
Aqualf (Mexico silt loam)	McCredie, Mo.	0.034
Udult (Cecil sandy loam)	Clemson, S.C.	0.034
Udult (Cecil sand loam)	Watkinsville, Ga.	0.030
Alfisols	Indonesia	0.018
Alfisols	Benin	0.013
Oxisols	Ivory Coast	0.013
Udult (Tifton loamy sand)	Tifton, Ga.	0.013
Ultisols	Hawaii	0.012
Alfisols	Nigeria	0.008
Udept (Bath flaggy silt loam)	Arnot, N.Y.	0.007
Ultisols	Nigeria	0.005
Oxisols	Puerto Rico	0.001

[a] To convert these *K* values from (Mg · ha · h)/(ha · MJ · mm) to English units of (ton · acre · h)/(100 acres · ft-ton · in.), simply multiply the values in this table by 7.6.

From Wischmeier and Smith (1978); data for tropical soils cited by Cassel and Lal (1992).

Cover and Management Factor C

Erosion and runoff are markedly affected by different types of vegetative cover and cropping systems (Table 15.3). Undisturbed forests and dense grass provide the best soil protection and are about equal in their effectiveness. Forage crops (both legumes and grasses) are next in effectiveness because of their relatively dense cover. Small grains, such as wheat and oats, are intermediate and offer considerable obstruction to surface

TABLE 15.2 Some Representative Values of the Topographic Factor LS on Three Types of Sites with Various Slope Gradients and Lengths

	Approximate slope length, m				
Slope gradient, %	*25*	*50*	*100*	*150*	*200*
Sites with low ratios of rill to interrill erosion (e.g., many rangelands)					
2	0.25	0.27	0.30	0.32	0.33
4	0.47	0.55	0.64	0.70	0.75
8	0.93	1.16	1.44	1.65	1.80
12	1.58	2.05	2.65	3.08	3.42
16	2.30	3.05	4.06	4.80	5.40
Sites with moderate ratios of rill to interrill erosion (e.g., most tilled row-crop land)					
2	0.25	0.29	0.35	0.38	0.41
4	0.48	0.62	0.79	0.92	1.02
8	0.94	1.32	1.84	2.24	2.57
12	1.62	2.37	3.48	4.34	5.09
16	2.35	3.54	5.33	6.77	8.03
Sites with high ratios of rill to interrill erosion (e.g., freshly disturbed construction sites and new seedbeds)					
2	0.25	0.33	0.44	0.51	0.58
4	0.49	0.71	1.02	1.27	1.48
8	0.96	1.52	2.38	3.10	3.74
12	1.65	2.70	4.43	5.91	7.24
16	2.39	4.00	6.69	9.03	11.17

Modified from Renard et al. (1997).

TABLE 15.3 Effect of Plant Cover on Soil Erosion by Water in the Humid Zone of West Africa

The data are averaged over many nearby sites, all with similar slopes, soils, and amounts of rainfall. The erosive influence of cultivation (with various types of tillage), especially bare fallows, is illustrated, as is the protective effect of undisturbed forest vegetation.

Type of cover	*Number of sites*	*Mean rainfall, mm/y*	*Runoff, % of rainfall*	*Erosion, Mg/ha*
Forest protected from fire	11	1293	0.9	0.10
Forest with light fires	13	1289	1.1	0.27
Natural bunch-grass fallow	7	1203	16.6	4.88
Groundnut (peanut)	32	1329	20.7	7.70
Upland rice	17	946	23.3	5.52
Maize (corn)	17	1405	17.7	7.63
Failed crops and bare soil	11	1154	39.5	21.28

Data selected from that cited by Pierre (1992).

Soil management to control erosion and USLE factors in the U.S. Corn Belt:
http://www.agron.iastate.edu/courses/agron212/Readings/Soil_erosion.htm

wash. Row crops, such as corn, soybeans, and potatoes, offer relatively little living cover during the early growth stages and thereby leave the soil susceptible to erosion unless residues from previous crops cover the soil surface.

Cover crops consist of plants that are similar to the forage crops just mentioned. They can provide soil protection during the time of year between the growing seasons for annual crops. For widely spaced perennial plantings, such as orchards and vineyards, cover crops can permanently protect the soil between rows of trees or vines. A mulch of plant residue or applied materials is also effective in protecting soils. Research on all continents has shown that a surface mulch does not have to be thick or cover the soil completely to make a major contribution to soil conservation. Even small increases in surface cover result in large reductions in soil erosion, particularly interrill erosion (Figure 15.10).

Regulation of grazing to maintain a dense vegetative cover on range and pasture land and the inclusion of close-growing hay crops in rotation with row crops on arable land

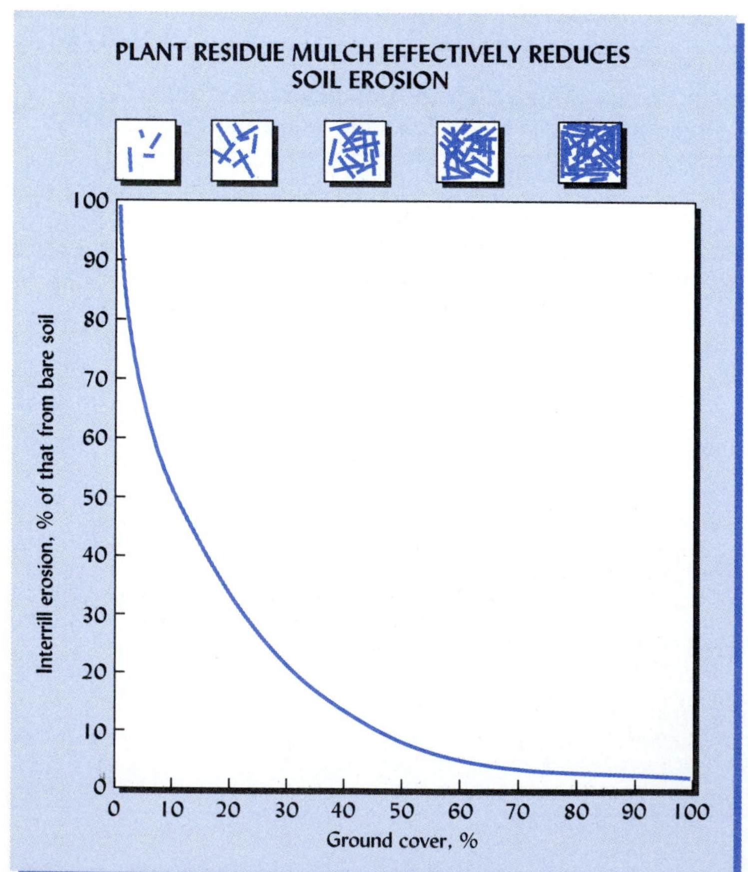

FIGURE 15.10 Reduction in interrill erosion achieved by increasing ground cover percentage. The diagrams above the graph illustrate 5, 20, 40, 60, and 80% ground cover. Note that even a light covering of mulch has a major effect on soil erosion. The graph applies to interrill erosion. On steep slopes, some rill erosion may occur even if the soil is well covered. [Generalized relationship based on results from many studies]

TABLE 15.4 **Examples of C Values for the Cover and Vegetation Management Factor**

The C values indicate the ratio of soil eroded from a particular vegetation system to that expected if the soil were kept completely bare. Note the effects of canopy cover, surface litter (residue) cover, tillage, and crop rotation. The C values are site- and situation-specific and must be calculated from local information on plant growth habit, climate, and so on. In the United States, specific values may be obtained from the RUSLE computer program or from local offices of the USDA Natural Resources Conservation Service.

Vegetation	*Management/condition*	*C value*
Range grasses and low (<1 m) shrubs	75% canopy cover, no surface litter	0.17
	75% canopy cover, 60% cover with decaying litter	0.032
Scrub brush about 2 m tall	25% canopy cover, no litter	0.40
	75% canopy cover, no litter	0.28
Trees with no understory, about 4 m drop fall	75% canopy cover, no litter	0.36
	75% canopy cover, 40% leaf litter cover	0.09
	75% canopy cover, 100% leaf litter cover	0.003
Woodland with understory	90% canopy cover, 100% litter cover	0.001
Permanent pasture	Dense stand of grass sod	0.003
Corn–soybean rotation	Fall plowing, conventional tillage, residues removed	0.53
	Spring chisel plow–plant conservation tillage, 2500 kg/ha surface residues after planting	0.22
	No-till planting, 5000 kg/ha surface residues after planting	0.06
Corn–soybean–wheat–hay rotation	Fall plowing, conventional tillage, residues removed	0.20
	Spring chisel plow–plant conservation tillage, 2500 kg/ha surface residues after planting	0.13
	No-till planting, 5000 kg/ha surface residues after planting	0.05
Corn–oats–hay–hay rotation	Spring conventional plowing before planting	0.05
	No-till planting	0.03

Values typical of midwestern United States. Based on Wischmeier and Smith (1978) and Schwab et al. (1996).

will help control both erosion and runoff. Likewise, the use of conservation tillage systems, which leave most of the plant residues on the surface, greatly decreases erosion hazards.

The *C* factor in the USLE or RUSLE will approach 1.0 where there is little soil cover (e.g., a bare seedbed in the spring or freshly graded bare soil on a construction site). It will be low (e.g., < 0.10) where large amounts of plant residues are left on the land or in areas of dense perennial vegetation (Table 15.4).

Values of *C* are specific to each region and type of vegetation or soil management. Estimates based on experiment data and field experience are available from conservation offices and the RUSLE program.

Support Practice Factor *P*[4]

On some sites with long and/or steep slopes, erosion control achieved by management of vegetative cover, residues, and tillage must be augmented by the construction of physical structures or other steps aimed at guiding and slowing the flow of runoff water. These **support practices** determine the value of the *P* factor in the USLE. The *P* factor is the ratio of soil loss with a given support practice to the corresponding loss if row crops were planted up and down the slope. If there are no support practices, the *P* factor is 1.0. The support practices include tillage on the contour (Figure 15.11), contour strip-cropping, terrace systems, and grassed waterways (Figure 15.12), all of which will tend to reduce the *P* factor.

Narrow strips of permanent vegetation (usually grasses or shrubs) planted on the contour can act as barriers to slow runoff, trap sediment, and eventually build up "natural" or "living" terraces. In some situations, tropical grasses (e.g., a deep-rooted, drought-tolerant species called *vetiver grass*) have shown considerable promise as an affordable alternative to the construction of terraces (Figure 15.13). The deep-rooted vetiver plants have dense, stiff stems that tend to filter out soil particles from the muddy

[4]Many of the erosion-control practices or management techniques discussed with regard to the *C* and *P* factors and in later sections of this chapter are considered to be **best management practices (BMPs)** under provisions of the Clean Water Act in the United States. The act defines BMPs as "optimal operating methods and practices for reducing or eliminating water pollution" from land-use activities.

FIGURE 15.11 Contour ridges must be carefully laid out with sufficient height to hold back water from even heavy rainfall. Here, surface retention is fast becoming surface runoff. (Photo courtesy of R. Weil)

Fact sheet summarizing erosion processes and practices in Canada:
http://www.gov.on.ca/OMAFRA/english/engineer/facts/87-040.htm

runoff. This sediment builds up on the upslope side of the grass barrier and, in time, actually creates a terrace that may be more than 1 m above the soil surface on the downslope side of the plants.

Examples of *P* values for contour tillage and strip-cropping at different slope gradients are shown in Table 15.5. Note that *P* values increase with slope and that they are lower for strip-cropping, illustrating the importance of this practice for erosion control. Terracing also reduces the *P* values. Unlike the USLE, the RUSLE takes into account interactions between support practices and subfactors such as slope and soil water infiltration.

The five factors of the USLE (*R, K, LS, C,* and *P*) have suggested many approaches to the practical control of soil erosion. A sample calculation is shown in Box 15.1 to show how the USLE can help evaluate erosion-control options.

More than half of the soil eroded in the United States comes from agricultural land, with the remainder coming from forests, rangeland, and construction sites. We will now focus on several specific erosion-control technologies appropriate for these various types of land uses.

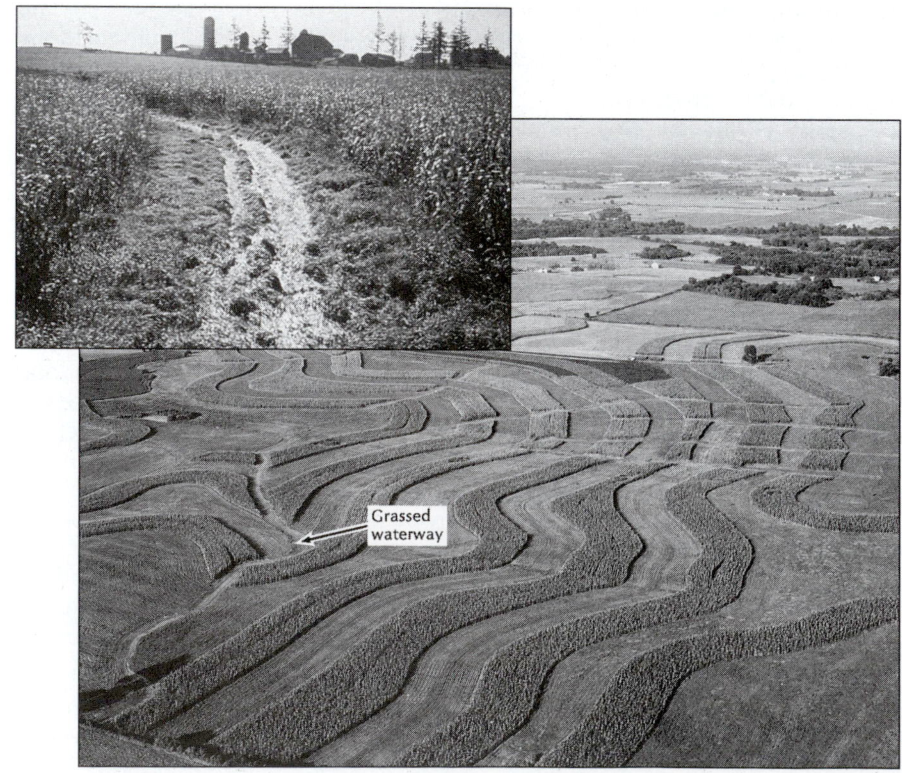

FIGURE 15.12 An aerial photo of farmland in Kentucky where contour strip-cropping is practiced and grassed waterways (see arrow) control the flow of water off cropland and prevent gully erosion. (Inset) A grassed waterway in action, showing how the permanent grass sod can resist the scouring force of water and safely conduct water off the field. [Photos courtesy of USDA National Resources Conservation Service]

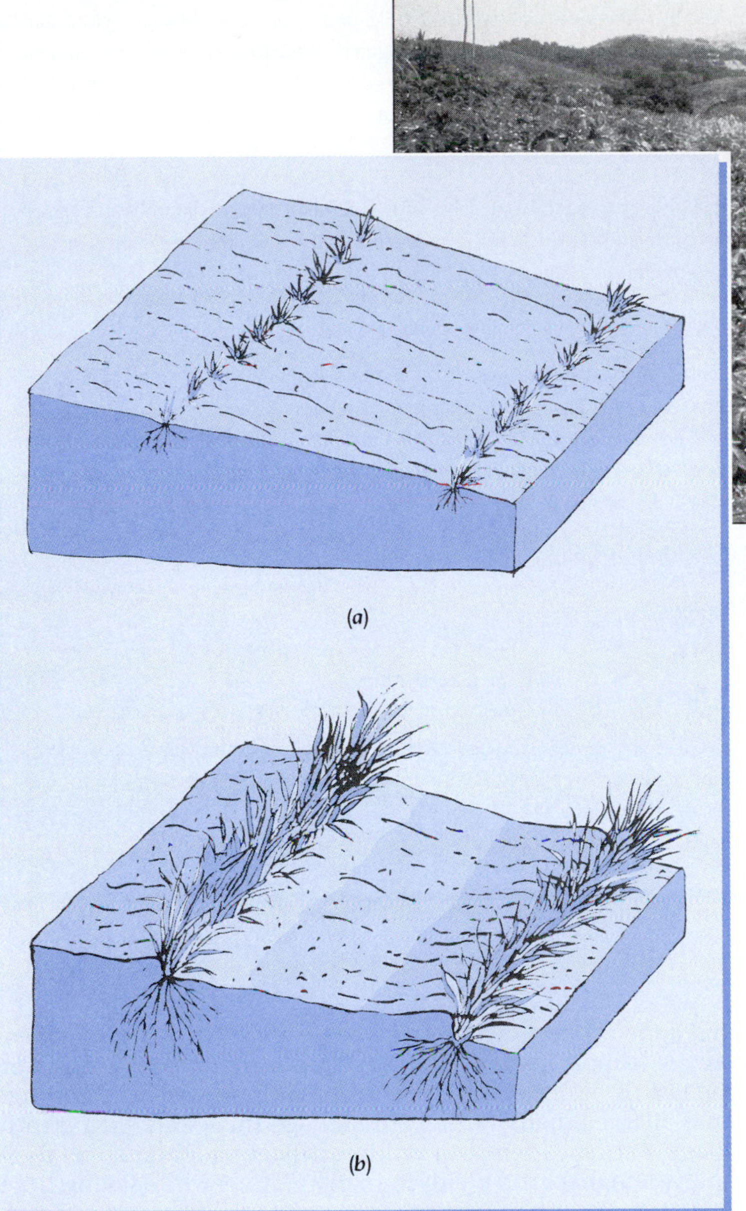

FIGURE 15.13 The use of vegetative barriers to create natural terraces. (Photo) A tropical grass (vetiver) has been planted vegetatively on the contour in a cassava field by pushing root sprigs into the soil. (*a*) The root cuttings are planted perpendicular to the slope direction. In a year or so the grass will be well established and its dense root and shoot growth will serve as a barrier to hold soil particles while permitting some water to pass on through. (*b*) Note the buildup of soil above the grass, basically forming a terrace wall. Perennial tall wheatgrass is being used experimentally in the Northern Plains area of the United States to serve as a barrier against snow movement and later against wind erosion. (Photo courtesy of Centro Internacional Agricultura Tropical in Cali, Colombia)

TABLE 15.5 P Factors for Contour and Strip-Cropping at Different Slopes and the Terrace Subfactor at Different Terrace Intervals

The product of the contour or strip-cropping factors and the terrace subfactor gives the P value for terraced fields.

				Terrace subfactor	
Slope, %	*Contour* P *factor*	*Strip-cropping* P *factor*	*Terrace interval, m*	*Closed outlets*	*Open outlets*
1–2	0.60	0.30	33	0.5	0.7
3–8	0.50	0.25	33–44	0.6	0.8
9–12	0.60	0.30	43–54	0.7	0.8
13–16	0.70	0.35	55–68	0.8	0.9
17–20	0.80	0.40	69–60	0.9	0.9
21–25	0.90	0.45	90	1.0	1.0

Contour and strip-cropping factors from Wischmeier and Smith (1978); terrace subfactor from Foster and Highfill (1983).

BOX 15.1 CALCULATIONS OF EXPECTED SOIL LOSS USING USLE

The RUSLE computer software is designed to calculate the expected soil loss from specific cropping systems at a given location.

The principles involved in both USLE and RUSLE can be verified by making calculations using USLE and its associated factors. Note that the factors in the USLE are related to each other in a multiplicative fashion. Therefore, if any one factor can be made to be near zero, the amount of soil loss A will be near zero.

Assume, for example, a location in Iowa on a Marshall silt loam with an average slope of 4% and an average slope length of 50 m. Assume further that the land is clean-tilled and fallowed.

Figure 15.9 shows that the R factor for this location is about 150 in English units or (150 × 17) 2550 in SI units. The K factor for a Marshall silt loam in central Iowa is 0.044 (Table 15.1) and the topographic factor *LS* from Table 15.2 is 0.71 (high rill to interrill ratio on soil kept bare). The C factor is 1.0, since there is no cover or other management practice to discourage erosion. If we assume the tillage is up and down the hill, the P value is also 1.0. Thus, the anticipated soil loss can be calculated by the USLE ($A = RKLSCP$):

$$A = (2550)(0.044)(0.71)(1.0)(1.0) = 79.7 \text{ Mg/ha or } 35.6 \text{ tons/acre}$$

If the crop rotation involved corn–soybean–wheat–hay and conservation tillage practices (e.g., spring chisel plow tillage) were used, a reasonable amount of residue would be left on the soil surface. Under these conditions, the C factor may be reduced to about 0.13 (Table 15.4). Likewise, if the tillage and planting were done on the contour, the *P* value would drop to about 0.5 (Table 15.5). P could be reduced further to 0.4(0.5 × 0.8) if terraces with open outlets were installed about 40 m apart. Furthermore, with crops on the land the site would have a moderate rill to interrill erosion ratio, so the *LS* factor would be only 0.62 (middle of Table 15.2). With these figures the soil loss becomes:

$$A = (2550)(0.044)(0.62)(0.13)(0.4) = 3.6 \text{ Mg/ha or } 1.5 \text{ tons/acre}$$

The units for the calculation are not normally shown, but they are:

$$\frac{\text{MJ} \cdot \text{mm}}{\text{ha} \cdot \text{h} \cdot \text{yr}} \times \frac{\text{Mg} \cdot \text{ha} \cdot \text{h}}{\text{ha} \cdot \text{MJ} \cdot \text{mm}} = \frac{\text{Mg}}{\text{ha} \cdot \text{yr}} (\text{LS, C, and P are unitless ratios}).$$

The benefits of good cover and management and support practices are obvious. The figures cited were chosen to provide an example of the utility of the universal soil-loss equation, but calculations can be made for any specific location. In the United States, pertinent factor values that can be used for erosion prediction in specific locations are generally available from state offices of the USDA Natural Resource Conservation Service. The necessary factors are also built into the RUSLE software.

15.5 CONSERVATION TILLAGE[5]

Core 4 conservation tillage practices:
http://www.ctic.purdue.edu/

For centuries, conventional agricultural practice around the world encouraged extensive soil tillage that leaves the soil bare and unprotected from the ravages of erosion (Figure 15.14*a*). During the last three decades, the development of herbicides and special planting equipment has allowed many farmers to manage their soils with greatly reduced tillage—or no tillage. Farmer interest in reduced tillage heightened as these systems produced equal or even higher crop yields in many regions while saving time, fuel, money—and soil. The latter attribute earned these systems the name of **conservation tillage** (Table 15.6).

Conservation tillage for organic farming where herbicides ar not used:
http://www.attra.org/attra-pub/organicmatters/conservationtillage.html

Conservation tillage systems range from those that merely reduce excess tillage to the no-tillage system, which uses no tillage beyond the slight soil disturbance that occurs as the planter cuts a planting slit through the residues and several cm into the soil (Figure 15.14*c*). The conventional **moldboard plow** was designed to leave the field "clean," that is, free of surface residues. In contrast, conservation tillage systems, such as **chisel plowing** (Figure 15.14*b*), stir the soil but only partially incorporate surface residues, leaving more than 30% of the soil covered. **Stubble mulching**, whose water-conserving attributes were highlighted in Section 6.6, is another example. **Ridge tillage** is a conservation system in which crops are planted atop permanent 15 to 20 cm high ridges. About 30% soil coverage is maintained, even though the ridges are scraped off a bit for planting and then built up again by shallow tillage to control weeds.

[5]For reviews of conservation tillage technology, see Blevins and Frye (1992) and Carter (1994).

(a)

(b)

(c)

FIGURE 15.14 Conventional and conservation tillage practices. (*a*) In conventional tillage, a moldboard plow inverts the upper soil horizon, burying all plant residues and producing a bare soil surface. (*b*) A chisel plow, one type of conservation tillage implement, stirs the soil but leaves a good deal of the crop residues on the soil surface. (*c*) In no-till systems, one crop is planted directly into a cover crop or the residue of a previous cash crop, with only a narrow band of soil disturbed. No-till systems leave virtually all of the residue on the soil surface, providing up to 100% cover and nearly eliminating erosion losses. Here soybeans are planted into a cover crop that will be killed with a herbicide (weed-killing chemical) to form a surface mulch. (Photos courtesy of R. Weil)

With **no-tillage** systems, we can expect 50 to 100% of the surface to remain covered (Table 15.7). Well-managed continuous no-till systems in humid regions include cover crops during the winter and high-residue-producing crops in the rotation. Such systems keep the soil completely covered at all times and build up organic surface layers somewhat like those found in forested soils.

Conservation tillage systems generally provide yields equal to or greater than those from conventional tillage, provided the soil is not poorly drained and in a cool region. However, during the transition from conventional tillage to no tillage, crop yields may decline slightly for several years for reasons associated with some of the effects outlined in the following subsections.

Adaptation by Farmers[6]

In recent years conservation tillage has become increasingly popular, being used on nearly two-fifths of the nation's cropland. Conservationists project that as much as 60% of the cropland in the United States will be managed with some kind of conservation

[6]For the inspiring story of one Chilean farmer's struggle to conquer the forces of erosion and degradation and restore the health of his soil, see Crovetto (1996).

TABLE 15.6 General Classification of Different Conservation Tillage Systems

All systems maintain at least 30% of the crop residues on the surface.

Tillage system	*Operation involved*
No-till	Soil undisturbed prior to planting, which occurs in narrow seedbed, 2.5 to 7.5 cm wide. Weed control primarily by herbicides.
Ridge till (till, plant)	Soil undisturbed prior to planting, which is done on ridges 10 to 15 cm higher than row middles. Residues moved aside or incorporated on about one-third of soil surface. Herbicides and cultivation to control weeds.
Strip till	Soil undisturbed prior to planting. Narrow and shallow tillage in row using rotary tiller, in-row chisel, and so on. Up to one-third of soil surface is tilled at planting time. Herbicides and cultivation to control weeds.
Mulch till	Soil surface disturbed by tillage prior to planting, but at least 30% of residues left on or near soil surface. Tools such as chisels, field cultivators, disks, and sweeps are used (e.g., stubble mulch). Herbicides and cultivation to control weeds.
Reduced till	Any other tillage and planting system that keeps at least 30% of residues on the surface.

Definitions used by Conservation Technology Information Center, West Lafayette, Ind.

TABLE 15.7 The Effect of Tillage Systems in Nebraska on the Percentage of Land Surface Covered by Crop Residues

Note that the conventional moldboard plow system provided essentially no cover, while no-till and planting on a ridge (ridge till) provided best cover.

		Percentage of fields with residue cover greater than			
Tillage system	*Number of fields*	*15%*	*20%*	*25%*	*30%*
Moldboard	33	3	0	0	0
Chisel	20	40	15	5	0
Disk	165	40	20	9	4
Field cultivate	13	46	23	0	0
Ridge till (till, plant)	2	100	50	50	0
No-till	3	100	100	100	100
All systems	236	36	18	9	4

From Dickey et al. (1987).

tillage by about 2010. Already, 80% or more of the cropland in a few areas within the United States is managed with conservation tillage. No-tillage and other conservation systems are also being used in other parts of the world (Figure 15.15). One of the most significant examples of no-tillage expansion has been in southern Brazil. Thousands of small-scale soybean and corn farmers have successfully adapted cover-crop-based no-tillage systems using animal traction or small tractors.

Erosion Control by Conservation Tillage

Since conservation tillage systems were initiated, hundreds of field trials have demonstrated that these tillage systems allow much less soil erosion than do conventional tillage methods. Surface runoff is also decreased, although the differences are not as pronounced as with soil erosion. These differences are reflected in the much lower *C* factor values assigned to conservation tillage systems (see Table 15.4).

The erosion-control value of an undisturbed surface residue mulch was discussed in the previous section. Conservation tillage also significantly reduces the loss of nutrients dissolved in runoff water or attached to sediment (review Table 13.1).

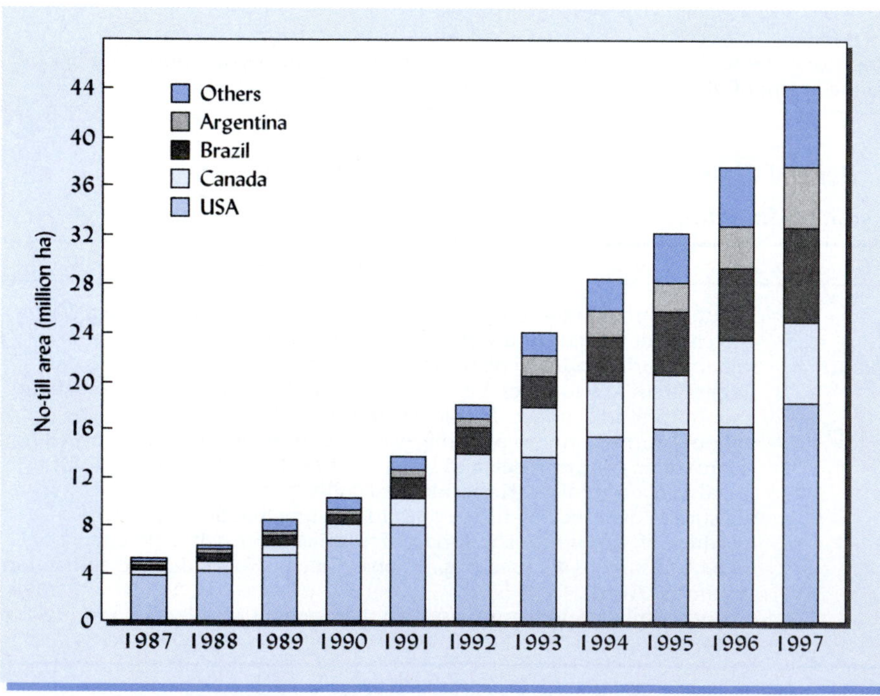

FIGURE 15.15 Spread of conservation tillage has been dramatic in the past 10 to 15 years, not only in the United States, but globally. As the chart illustrates, no-till practices in other countries such as Canada, Brazil, and Argentina have expanded very rapidly. By 1997 nearly 60% of the land on which no-till practices were used worldwide was found outside the United States. (Modified from charts provided by the Conservation Technology Center, Lafayette, Indiana)

Effect on Soil Properties

Erosion of soils irrigated with effluent water—a technical paper:
http://www.toprak.org.tr/isd/isd_54.htm

When soil management is converted from plow tillage to conservation tillage (especially no-tillage), numerous soil properties are affected, mostly in favorable ways. The changes are most pronounced in the upper few centimeters of soil. Generally, the changes are greatest for systems that produce the most plant residue (especially corn and small grains in humid regions), retain the most residue coverage, and cause the least soil disturbance. Many of these changes are illustrated in other chapters of this textbook, so the discussion here will be brief.

Physical Properties. Macroporosity and aggregation (see Sections 4.7 and 4.8) are increased as active organic matter builds up and earthworms and other organisms establish themselves. Infiltration and internal drainage are generally improved, as is soil water-holding capacity. Some of these effects are illustrated in Figure 15.16. The enhanced infiltration capacity of no-till-managed soils is generally quite desirable, but in some cases it may lead to more rapid leaching of nitrates and other water-soluble chemicals.

In cool regions, soils with restricted drainage may yield somewhat less using conservation tillage, because soils are wetter and cooler than with conventional tillage. Reduced yields have discouraged the adoption of conservation tillage in these regions. However, limited preplanting tillage over the crop row (see Section 7.11), and ridge tillage are conservation tillage systems that allow at least part of the soil to warm faster and largely overcome these problems.

Chemical Properties. No-tillage systems significantly increase the organic matter content of the upper few centimeters of soil (see Figures 15.16 and 11.17). During the initial 4 to 6 years of no-till management, the buildup of organic matter results in the immobilization of nutrients (see Section 11.3), especially nitrogen. This is in contrast to the mineralization of nutrients that is encouraged by the decline of soil organic matter under conventional tillage. Eventually, when soil organic matter stabilizes at a new higher level, nutrient mineralization rates under no-till increase.

In no-tillage systems, nutrient elements tend to accumulate in the upper few centimeters of soil as they are added to the soil surface in crop residues, animal manures,

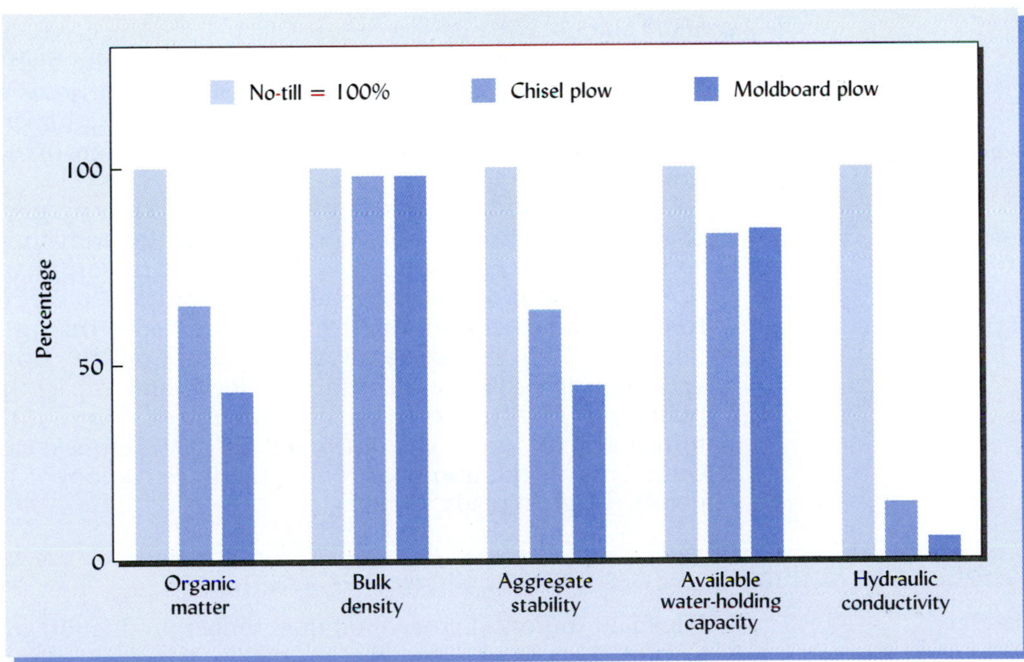

FIGURE 15.16 The comparative effects of 28 years of three tillage systems on soil organic matter content and a number of soil physical properties of an Alfisol in Ohio. Values for the no-till system were taken as 100, and the others are shown in comparison. Bulk density was about the same for each tillage system, but for all other properties the no-till system was decidedly more beneficial then either of the other two systems. The saturated hydraulic conductivity was especially high with no-till. [From Mahboubi et al. (1993); used with permission of the Soil Science Society of America]

chemical fertilizers, and lime. However, research indicates that because of the surface mulch, crop roots (like those of trees in the untilled forest soil environment) have no trouble obtaining nutrients from the near-surface soil layers.

Without tillage to mix the soil, the acidifying effects of nitrogen oxidation, residue decomposition, and rainfall are concentrated in the upper few centimeters of soil, the pH of which may drop more rapidly than that of the whole plow layer in conventional systems (see Sections 9.6 and 9.7). In humid regions this acidity must be countered by application of liming materials to the soil.

Biological Effects. The abundance, activity, and diversity of soil organisms tend to be greatest in conservation tillage systems characterized by high levels of surface residue and little physical soil disturbance (see Section 10.15). Earthworms and fungi, both important for soil structure, are especially favored.

15.6 CONTROL OF ACCELERATED EROSION ON RANGELAND AND FORESTLAND

Rangeland Problems

Interactive erosion calculator based on USLE-you can test effects of soil, land-use, management and climate:
http://www.interwet.psu.edu/S1.htm

Many semiarid rangelands lose large amounts of soil under natural conditions, but accelerated erosion can lead to even greater losses if human influences are not carefully managed. Overgrazing, which leads to the deterioration of the vegetative cover on rangelands, is a prime example. Grass cover generally protects the soil better than the scattered shrubs that usually replace it under the influence of poorly managed livestock grazing. In addition, cattle congregating around poorly distributed water sources and salt licks may completely denude the soil around these facilities. Cattle trails, as well as ruts from off-road vehicles, can channelize runoff water and spawn gullies that eat into the landscape. Because of the prevalence of dry conditions, wind erosion (to be discussed in Sections 15.8 and 15.9) also plays a major role in the deterioration of rangeland soils.

Erosion on Forestlands

In contrast to deserts and rangelands, land under healthy, undisturbed forests loses very small amounts of soil. However, accelerated erosion can be a serious problem on forested land, both because the rates of soil loss may be quite high and because the amount of land involved is often enormous. The main cause of accelerated erosion in forested watersheds is usually the construction of logging roads, timber-harvest operations, and the trampling of trails and off-trail areas by large numbers of recreational users (or cattle, in some areas).

To understand and correct these problems, it is necessary to realize that the secret of low natural erosion from forested land is the undisturbed forest floor, the O horizons that protect the soil from the impact of raindrops and allow such high infiltration rates that surface runoff is minimal. Contrary to the common perception, it is the forest floor, rather than the tree canopy or roots, that protects the soil from erosion (Figure 15.17*a*). In fact, rainwater dripping from the leaves of tall trees often forms very large drops that reach terminal velocity and impact the ground with more energy than direct rain from even the most intense of storms. If the forest floor has been disturbed and mineral soil exposed, serious splash erosion can result (see Figure 15.17*b*). Gully erosion can also occur under the forest canopy if water is concentrated, as by poorly designed roads.

Practices to Reduce Soil Loss Caused by Timber Production[7]

The main sources of eroded soil from timber production are *logging roads* (that are built to provide access to the area by trucks), *skid trails* (the paths along which logs are dragged), and *yarding areas* (the areas where collected logs are sized and loaded onto trucks). Relatively little erosion results directly from the mere felling of the trees (except

[7]For a general introduction to this topic, see Nyland (1996).

(a)

(b)

FIGURE 15.17 The leaf mulch on the forest floor, rather than the tree roots or canopy, provides most of the protection against erosion in a wooded ecosystem. (*a*) An undisturbed temperate deciduous forest floor (as seen through a rotten stump). The leafless canopy will do little to intercept rain during winter months. During the summer, rainwater dripping from the foliage of tall trees may impact the forest floor with as much energy as unimpeded rain. (*b*) Severe erosion has taken place under the tree canopy in a wooded area where the protective forest floor has been destroyed by foot traffic. The exposed tree roots indicate that nearly 25 cm of the soil profile has washed away. (Photos courtesy of R. Weil)

FIGURE 15.18 An open-top culvert in a well-designed logging road in Montana. This simple structure, along with proper road bed alignment, can greatly reduce gully erosion caused by water flowing unimpeded along roads and trails in forested areas. Water bars or open-top culverts placed at frequent intervals lead runoff water, a little at a time, off the road and into densely vegetated areas. (Photos courtesy of R. Weil)

where large tree roots are needed to anchor the soil against landslides. Strategies to control erosion should include consideration of the following: (1) intensity of timber harvest, (2) methods used to remove logs, (3) scheduling of timber harvests, and (4) design and management of roads and trails (see Figure 15.18). Soil disturbance in preparation for tree regeneration (such as tillage to eliminate weed competition or provide better seed-to-soil contact) must also be limited to sites with low susceptibility to erosion.

When forests are harvested, buffer strips as wide as 1.5 times the height of the tallest trees should generally be left untouched along all streams. As discussed in Section 14.2, buffer strips of dense vegetation have a high capacity to remove sediment and nutrients from runoff water. Forested buffers also protect the stream from excessive logging debris. In addition, streamside trees shade the water, protecting it from the undesirable heating that would result from exposure to direct sunlight.

15.7 EROSION AND SEDIMENT CONTROL ON CONSTRUCTION SITES

Video on soil erosion and sediment pollution, how they affect waterways, and what can be done. Focus on urban soils:
http://www.greenworks.tv/waterquality/erosion.htm

Although active construction sites cover relatively little land in most watersheds, they may still be a major source of eroded sediment, because the potential erosion per hectare on drastically disturbed land is commonly 100 times that on agricultural land. Heavy sediment loads are characteristic of rivers draining watersheds in which land use is changing from farm and forest to built-up land. Historically, once urbanization of a watershed is complete (all land being either paved over or covered by well-tended lawns), sedimentation rates return to levels as low (or lower) than before the development took place.

To prevent serious sediment pollution from construction sites, government agencies in the United States (e.g., through state laws and the federal Clean Water Act of 1992) and in many other industrialized countries require that contractors develop detailed erosion or sediment control plans before initiating construction projects that will disturb a significant amount of land. The goals of erosion control on construction sites are (1) to avoid onsite damage such as undercutting of foundations or finished grades and loss of topsoil needed for eventual landscaping; and (2) to retain eroded sediment onsite so as to avoid the environmental damages (and liabilities) that would result from deposition of sediment on neighboring land and roads, and in ditches, reservoirs, and streams.

Principles of Erosion Control on Construction Sites

Liquefaction of saturated soils—see collapsing buildings:
http://www.ce.washington.edu/~liquefaction/html/what/what1.html

Five basic steps are useful in developing plans to meet the aforementioned goals:

1. When possible, schedule the main excavation activities for low-rainfall periods of the year.
2. Divide the project into as many phases as possible, so that only a few small areas must be cleared of vegetation and graded at any one time.
3. Cover disturbed soils as completely as possible, using vegetation or other materials.
4. Control the flow of runoff to move the water safely off the site without destructive gully formation.
5. Trap the sediment before releasing the runoff water offsite.

The last three steps bear further elaboration. They are best implemented as specific practices integrated into an overall erosion control plan for the site.

Keeping the Disturbed Soil Covered

Soils freshly disturbed by excavation or grading operations are characterized by very high erodibility (*K* values). This is especially true for low-organic-matter subsoil materials. Potential erosion can be extremely high (200 to 400 Mg/ha is not uncommon) unless the *C* value is made very low by providing good soil cover. This is best accomplished by allowing the natural vegetation to remain undisturbed for as long as possible, rather than clearing and grading the entire project area at the beginning of construction (see step 1, preceding). Once a section of the site is graded, any sloping areas not directly involved in the construction should be sodded or sown to fast-growing grass species adapted to the soil and climatic conditions. (Plate 33 shows erosion on an unprotected road bank.)

Seeded areas should be covered with **mulch** or specially manufactured **erosion blankets** (Figure 15.19, bottom). Erosion blankets, made of various biodegradable or nonbiodegradable materials, provide instant soil cover and protect the seed from being washed away (Table 15.8).

A commonly used technology to protect steep slopes and areas difficult to access, such as road cuts, is the *hydroseeder* (Figure 15.19, top) that sprays out a mixture of seed,

FIGURE 15.19 Two methods of establishing vegetative cover on steep, unstable slopes. (Top) A hydroseeder allows vegetative cover to be efficiently established on difficult-to-reach areas. The machine is spraying a mixture of water, chopped straw, grass seed, fertilizer, and sticky polymers that hold the mulch in place until the grass seed can take root. (Bottom) Erosion-control mats made of plastic netting or natural materials like jute are laid down over newly seeded grass to hold the seed and soil in place until the vegetative cover is established. (Photos courtesy of R. Weil)

Water pollution from urban erosion and storm water: http://www.epa.gov/reg3wapd/stormwater/construction.htm

fertilizer, lime (if needed), mulching material, and sticky polymers. Good construction-site management includes removal and stockpiling of the A-horizon material before an area is graded (see Plate 35 following page 498). This soil material is often quite high in fertility and is a potential source of sediment and nutrient pollution. The stockpile should therefore be given a grass cover to protect it from erosion until it is used to provide topsoil for landscaping around the finished structures.

Controlling the Runoff

Freshly exposed and disturbed subsoil material is highly susceptible to the cutting action of flowing water. The gullies so formed may ruin a grading job, undercut pavements and foundations, and produce enormous sediment loads. The flow of runoff water must be controlled by carefully planned grading, terracing, and channel construction. Most construction sites require a perimeter waterway to catch runoff before it leaves the site and to channel it to a retention basin.

The sides and bottom of such channels must be covered with "armor" to withstand the cutting force of flowing water. Where high water velocities are expected, the soil must be protected with **hard armor** such as **riprap** (large angular rocks, as shown in Figure 15.20), **gabions** (rectangular wire-mesh containers filled with hand-sized stone), or interlocking concrete blocks. The soil is first covered with a **geotextile** filter cloth (a tough nonwoven material) to prevent mixing of the soil into the rock or stone.

TABLE 15.8 Effectiveness of Various Erosion-Control Materials in Reducing Runoff and Soil Loss and in Establishing Vegetative Cover on Disturbed Sites

All plots (except sod) were seeded with grass and covered with the materials listed. Note that all treatments greatly reduced erosion and, to a lesser degree, runoff losses. The sod was by far the most effective material, followed by the straw mulch. None of the materials had a negative effect on grass establishment, while the straw mulch may have had a positive effect. In other applications (such as lining of channels) where high velocity of runoff is encountered, the manufactured erosion blankets would be more effective than straw.

Material	*Soil loss,*[a] *kg/ha*	*Runoff loss, % of rain*	*Time to initiation of runoff, s*	*Ground cover established,*[b] *%*
Bare soil	6650	83	34	50
Jute (woven mesh)	410	68	62	61
Wood shavings in nonwoven polyester netting	810	74	74	69
Coconut fiber mat	1070	76	86	58
Straw (4.5 Mg/ha)	590	60	102	76
Grass sod	100	28	341	NA[c]

[a] Soil loss from a single simulated rain event at a rate of 96 mm/h. Means of two soils: a Sassafras loamy sand with 8% slope and a Matapeake sandy clay loam with 15% slope.

[b] Percent vegetative cover established one year after Kentucky 31 tall fescue grass was seeded and covered by the various materials, in a separate study.

[c] Not applicable since sod provided 100% cover and was not seeded.

Data from Krenisky et al. (1998)

FIGURE 15.20 The flow of runoff from large areas of bare soil must be carefully controlled if offsite pollution is to be avoided. Here, a carefully designed channel with a grass sod bottom and sides lined with large rocks (riprap) prevents gully erosion, reduces soil loss, and guides runoff around the perimeter of a construction site. (Photo courtesy of R. Weil)

In smaller channels, and on more gentle slopes where relatively low water velocities will be encountered, **soft armor,** such as grass sod or erosion blankets, can be used. Generally, soft armor is cheaper and more aesthetically appealing than hard armor. Newer approaches to erosion control often involve reinforced vegetation (e.g., trees or grasses planted in openings between concrete block or in tough erosion mats).

The term **bioengineering** describes techniques that use vegetation (locally native, noninvasive species are preferred) and natural biodegradable materials to protect channels subject to rather high water velocities. The so-called **live stake** technique (Figure 15.21) is an example of a bioengineering approach commonly used to stabilize soil along channels subject to high velocity water.

FIGURE 15.21 An example of *bioengineering* along a stream bed that had been disturbed during the construction of a commercial airport in Illinois. Living willow branches are pounded into the soft erodible stream bank to anchor the soil and reduce the scouring power of the water during high flows. Eventually the willow branches will take root and sprout (inset), providing trees that will permanently stabilize the stream bank and improve wildlife habitat. (Photos courtesy of R. Weil)

Trapping the Sediment

Indigenous Ethiopians outwit a river using "devil's tie" armor:
http://www.nuffic.nl/ciran/ikdm/7-2/abay.html

For small areas of disturbed soils, several forms of sediment barriers can be used to filter the runoff before it is released. The most commonly used types of silt barriers are straw bales and woven fabric silt fences. If installed properly, both can effectively slow the water flow so that most of the sediment is deposited on the uphill side of the barrier (Figure 15.22), while relatively clear water passes through.

On large construction sites, a system of protected slopes and channels leads storm runoff water to one or more retention and sedimentation ponds located at the lowest elevation of the site. As the flowing water meets the still water in the pond it drops most of its sediment load, allowing the relatively clear water to be skimmed off the top and released to the next pond or off the site. Wetlands (Section 7.7) are often constructed to help purify the overflow from sedimentation ponds before the water is released into the natural stream or river.

Construction site erosion-control measures are commonly designed to retain the runoff from small storms onsite. The retention ponds must also be able to deal with runoff generated from intense rainstorms—the kind that may be expected to occur on a site only once in every 10 to even 100 years. While expensive to construct, well-designed sediment-retention ponds can be incorporated as permanent aesthetic water features that enhance the value of the final project.

15.8 WIND EROSION: IMPORTANCE AND FACTORS AFFECTING IT

Up to this point we have focused on soil erosion by water, but wind, too, causes much soil loss. Wind erosion is most common in arid and semiarid regions, and it is a problem on some soils in humid climates as well. It occurs when strong winds blow across soils with relatively dry surface layers. All kinds of soils and soil materials are affected. The finer soil particles may be carried to great heights and for thousands of kilometers—even from one continent, across the ocean, to another.

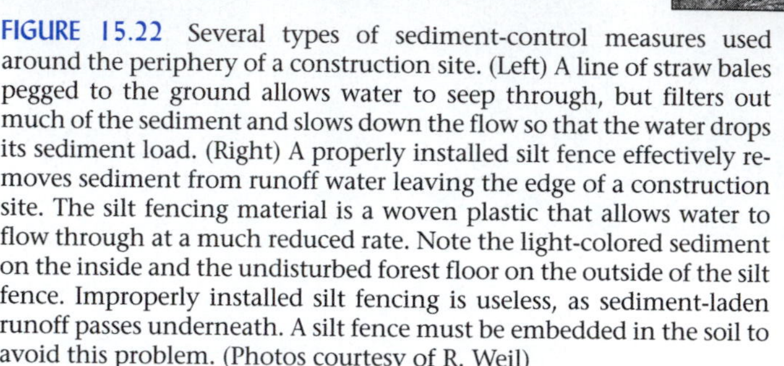

FIGURE 15.22 Several types of sediment-control measures used around the periphery of a construction site. (Left) A line of straw bales pegged to the ground allows water to seep through, but filters out much of the sediment and slows down the flow so that the water drops its sediment load. (Right) A properly installed silt fence effectively removes sediment from runoff water leaving the edge of a construction site. The silt fencing material is a woven plastic that allows water to flow through at a much reduced rate. Note the light-colored sediment on the inside and the undisturbed forest floor on the outside of the silt fence. Improperly installed silt fencing is useless, as sediment-laden runoff passes underneath. A silt fence must be embedded in the soil to avoid this problem. (Photos courtesy of R. Weil)

The American "Dust Bowl" experience:
http://www.pbs.org/wgbh/amex/dustbowl/

The erosive power of wind is demonstrated in Figure 15.23. The erosive force of the wind tends to be much greater in some regions than in others. For example, the semiarid Great Plains region of the United States is subject to winds with 5 to 10 times the erosive force of the winds common in the humid East. In the Great Plains, the winds are most powerful in the winter season. In other regions, high winds occur most commonly during the hot summers.

Wind erosion is a worldwide problem. Large areas of the former Soviet Union and sub-Saharan Africa have been badly damaged by wind erosion. Overgrazing and other misuses of the fragile lands of arid and semiarid areas have allowed wind erosion to depress soil productivity and bring starvation and misery to millions of people in these areas.

Mechanics of Wind Erosion

Like water erosion, wind erosion involves three processes: (1) *detachment,* (2) *transportation,* and (3) *deposition.* The moving air, itself, results in some detachment of tiny soil grains from the granules or clods of which they are a part. However, when the moving air is laden with soil particles, its abrasive power is greatly increased. The impact of these rapidly moving grains dislodges other particles from soil clods and aggregates. These dislodged particles are then ready for one of the three modes of wind-induced transportation, depending mostly on their size.

SALTATION. The first and most important mode of particle transportation is that of **saltation,** or the movement of soil by a series of short bounces along the ground surface (Figure 15.24). The particles remain fairly close to the ground as they bounce, seldom rising more than 30 cm. Depending on conditions, this process may account for 50 to 90% of the total movement of soil.

SOIL CREEP. Saltation also encourages **soil creep,** or the rolling and sliding along the surface of the larger particles. The bouncing particles carried by saltation strike the large aggregates and speed up their movement along the surface. Soil creep accounts for the movement of particles up to about 1.0 mm in diameter, which may amount to 5 to 25% of the total movement.

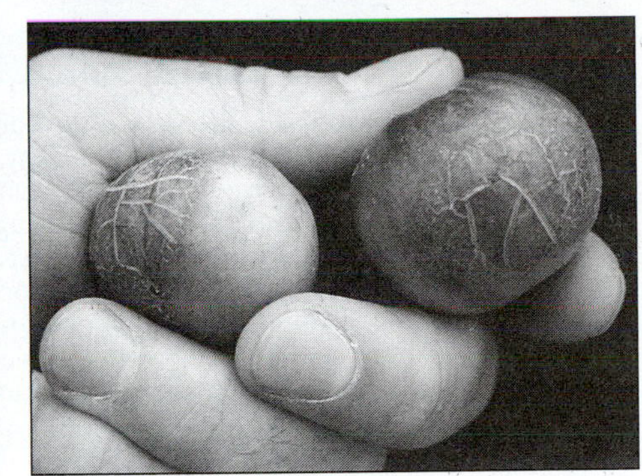

FIGURE 15.23 Wind erosion in action. (Upper left) Much wind erosion comes from dust storms such as this one moving across the High Plains of Texas. The swirling black cloud consists of fine particles eroded from the soil by high winds sweeping across the flat farmland and range. Apparently, most of the land was not as well covered as the wheat field in the foreground. (Upper right) Soil eroded by wind during a single dust storm has piled up to a depth of nearly one m along a fencerow in Idaho. The existing soil and plants are covered by deposits that are quite unproductive since the soil structure has been destroyed. Also, these deposits are subject to further movement when the wind direction changes. (Lower left) Direct wind damage to a tomato crop in a sandy field in Delaware. The tomato plants are buried beneath windblown sandy soil. (Lower right) Young tomato plants show damage incurred by sandblasting during a windstorm. (Upper left photo courtesy of Dr. Chen Weinan, USDA Agricultural Research Service, Warm Springs, TX; other photos courtesy R. Weil)

Suspension. The most spectacular method of transporting soil particles is by movement in *suspension*. Here, dust particles of a fine-sand size and smaller are moved parallel to the ground surface and upward. Although some of them are carried at a height no greater than a few meters, the turbulent action of the wind results in others being carried kilometers upward into the atmosphere and many hundreds of kilometers horizontally. These particles return to the earth only when the wind subsides and/or when precipitation washes them down. Although it is the most striking manner of transportation, suspension seldom accounts for more than 40% of the total and is generally no more than about 15%.

Factors Affecting Wind Erosion

Soil Moisture. Susceptibility to wind erosion is related to the moisture content of soils. Wet soils do not blow because of the adhesion between water and soil particles. Dry winds generally lower the moisture content of the surface soil to below the wilting point before wind erosion takes place.

Wind Velocity and Turbulence. The rate of wind movement, especially gusts having greater than average velocity, will influence erosion. Tests have shown that wind speeds of about 25 km/h (7 m/s) are required to initiate soil movement. At higher wind speeds,

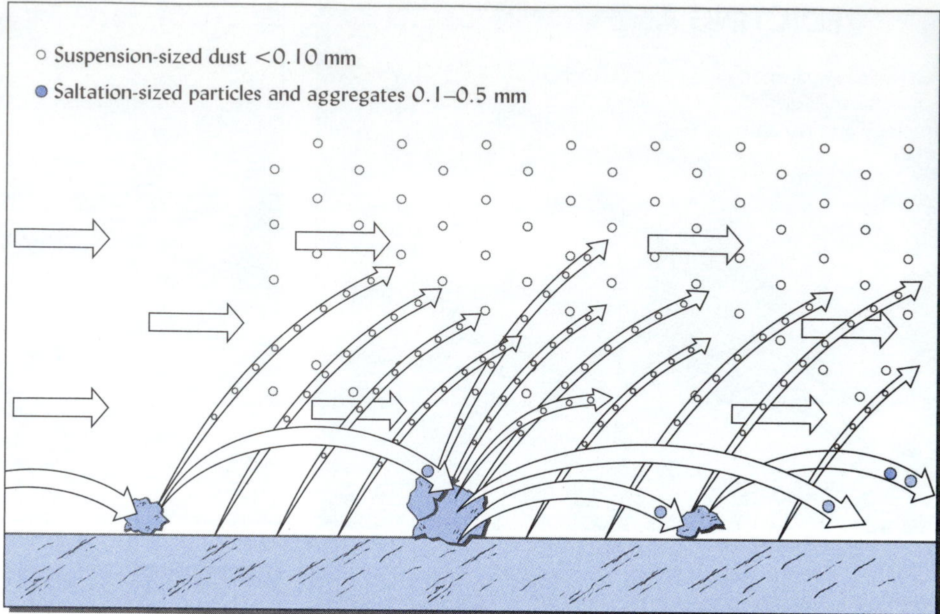

FIGURE 15.24 An illustration of the generation of dust particles during wind erosion. As the wind blows from left to right, some fine particles are picked up from the soil surface and carried into the atmosphere, where they remain suspended until the wind velocity is reduced. Other midsize particles or aggregates, being too large to be carried up in suspension, are bounced along the soil surface. When they strike the surfaces of the crusted areas of soil or larger soil aggregates, they break loose particles of various sizes. The finer particles move in suspension up into the air above, and the medium-sized particles continue to bounce along the soil surface. This process of particle movement stimulated by the medium-sized particles as they skim along the surface is termed **saltation.** [Modified from Hughes (1980); used with permission of Deere & Company, Moline, Ill.]

soil movement is proportional to the cube of the wind velocity. Thus, the quantity of soil carried by wind goes up very rapidly as speeds above 30 km/h are reached. Wind turbulence also influences the capacity of the atmosphere to transport matter. Although the wind itself has some direct influence in picking up fine soil, the impact of wind-carried particles as they strike the soil is probably more important.

Soil Properties. Wind erosion is less severe where soil surface roughness is enhanced by tillage methods that leave large clods or ridges on the soil surface. Leaving a stubble mulch (see Section 6.6) is an even more effective way of reducing windborne soil losses.

Several other soil characteristics influencing wind erosion are (1) mechanical stability of soil clods and aggregates, (2) stability of soil crusts, and (3) bulk density and size of erodible soil fractions. Some clods resist the abrasive action of wind-carried particles. If a soil crust resulting from a previous rain is present, it, too, may be able to withstand the wind's erosive power. The presence of clay, organic matter, and other cementing agents is also important in helping clods and aggregates resist abrasion. This is one reason why sandy soils, which are low in such agents, are so easily eroded by wind.

Soil particles or aggregates about 0.1 mm in diameter are more erodible than those larger or smaller in size. Thus, fine sandy soils are quite susceptible to wind erosion. Particles or aggregates about 0.1 mm in size are also partly responsible for the movement of other particles. Saltating particles bounce against larger particles, causing surface creep. The saltating particles also collide with smaller particles, sending them up into suspension.

Vegetation. Vegetation or a stubble mulch will reduce wind erosion hazards, especially if the rows run perpendicular to the prevailing wind direction. This effectively slows wind movement near the soil surface. In addition, plant roots help bind the soil and make it less susceptible to wind damage.

15.9 PREDICTING AND CONTROLLING WIND EROSION

More on wind erosion processes and models: http://www.weru.ksu.edu/weps.html

A Wind Erosion Prediction Equation (WEQ) has been in use since the late 1960s:

$$E = f(ICKLV)$$

Where the predicted wind erosion E is a function f of:

I = soil erodibility factor
C = climate factor
K = soil-ridge-roughness factor
L = width of field factor
V = vegetative cover factor

The WEQ involves the major factors that determine the severity of the erosion, but it also considers how these factors interact with each other. Consequently, it is not as simple as is the USLE for water erosion. The **soil erodibility factor** I relates to the properties of the soil and to the degree of slope of the site in question. The **soil-ridge-roughness** factor K takes into consideration the clodiness of the soil surface, vegetative cover V, and ridges on the soil surface. The **climatic factor** C involves wind velocity, soil temperature, and precipitation (which helps control soil moisture). The **width of field factor** L is the width of a field in the downwind direction. Naturally, the width changes as the direction of the wind changes, so the prevailing wind direction is generally used. The **vegetative cover** V relates not only to the degree of soil surface covered with residues, but to the nature of the cover—whether it is living or dead, still standing, or flat on the ground.

A revised, more complex, and more accurate computer-based prediction model has been developed, known as the Revised Wind Erosion Equation (RWEQ). It is still an empirical model based on many years of research to characterize the relationship between observable conditions and resulting wind erosion severity. Table 15.9 outlines the factors taken into consideration by RWEQ.

TABLE 15.9 Some Factors Integrated in the Revised Wind Erosion Equation (RWEQ) Model

The RWEQ program calculates values for each factor for each 15-day period during the year. It also calculates interactions; that is, it uses the value of one factor to modify other factors.

Model factor	*Subfactor terms in the model*	*Comments*
Weather factor	Weather factor *WF;* includes terms for wind velocity, direction, air temperature, solar radiation, rainfall, and snow cover.	Modified by other factors such as soil wetness.
Soil factors	Erodible fraction *EF;* fraction smaller than 0.84 mm diameter.	Based on sand, silt, organic matter, and rock cover.
	Soil crust factor *SCF.*	Induced by rainfall, eliminated by tillage.
	Surface roughness *SR;* a random component due to clods and/or an oriented component due to ridges.	Interacts with rainfall and tillage.
	Soil wetness *SW.*	Computed from rainfall minus evapotranspiration.
Tillage factor	Tillage factor *TF;* depends on type of implement, soil conditions, timing, and so forth.	Modifies surface roughness, crust factor, and so forth.
Hill factor	Hill factor *HF;* slope gradients and length input.	Affects wind speed (high going upslope, lower going down).
Irrigation factor	Irrigation factor *IR.*	Equivalent to added rainfall.
Crops factor	Flat residue *SLR;* factor.	Soil cover by residues lying on the surface is estimated for each crop, including changes over time due to decay, and so forth.
	Standing residue SLR_s; depends on crop, harvest height, plant density, and so forth.	Standing residues reduce the wind speed at the soil surface. Decay is slower than for flat residues.
	Crop canopy factor SLR_c.	Changes daily with crop growth.
Barriers	Barriers (e.g., tree windbreaks); includes orientation, density, height, spacing.	Reduce leeward wind velocity.

Based on information in Fryrear et al. (2000).

Control of Wind Erosion

Soil Moisture and Vegetation. The factors of the WEQ give clues to methods of reducing wind erosion. For example, since soil moisture increases cohesiveness, the wind speed required to detach soil particles increases dramatically as soil moisture increases. Therefore, where irrigation water is available, it is common to moisten the soil surface when high winds are predicted (Figure 15.25). Unfortunately, most wind erosion occurs in dry regions without available irrigation. A vegetative cover also discourages soil blowing, especially if the plant roots are well established. In dry farming areas, however, sound moisture-conserving practices require summer fallow on some of the land, and hot, dry winds reduce the moisture in the soil surface. Consequently, other means must be employed on cultivated lands of these areas.

Tillage. Keeping the soil surface rough and maintaining some vegetative cover is accomplished by using appropriate tillage practices. However, the vegetation should be well anchored into the soil to prevent it from blowing away. Stubble mulch has proven to be effective for this purpose (see Section 6.6).

The effect of tillage depends not only on the type of implement used, but also on the timing of the tillage operation. Tillage can greatly reduce wind erosion if it is done while there is sufficient soil water to cause large clods to form. Tillage on a dry soil may produce a fine, dusty surface that aggravates the erosion problem. Tillage to provide for a cloddy surface condition should be at right angles to the prevailing winds. Likewise, strip-cropping and alternate strips of cropped and fallowed land should be perpendicular to the wind.

Barriers. Barriers such as shelterbelts (Figure 15.26*a,b*) are effective in reducing wind velocities for short distances and for trapping drifting soil. Various devices are used to control blowing of sands, sandy loams, and cultivated peat soils (even in humid regions). Windbreaks and tenacious grasses and shrubs are especially effective. Picket

FIGURE 15.25 Wind erosion control in an area of productive Histosols (Saprists) in central Michigan. Prior to being cleared and drained, this area was a partially forested bog. When dry, cultivated Histosols are very light and fluffy and susceptible to wind erosion. The rows of trees (mainly willows) were planted perpendicular to the prevailing winds to slow the wind velocity and protect these valuable organic soils from erosion. Wetting the soil surface is another effective means of reducing wind erosion, as seen by the darker-colored field in the background (where the water table was raised) and the darker circles in the foreground where sprinkler irrigation was used. Note that the photo was taken in early spring before most crops were planted and before the trees had fully leafed out. (Photo courtesy of USDA Natural Resources Conservation Service)

(a)

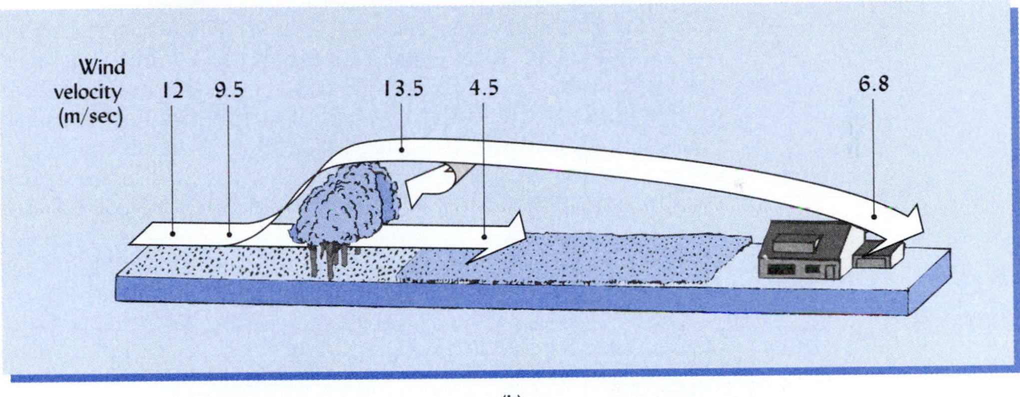

(b)

(c)

FIGURE 15.26 Windbreaks to reduce wind erosion. (*a*) Shrubs and trees make good windbreaks and add beauty to a North Dakota farm homestead. (*b*) The effect of a windbreak on wind velocity. The wind is deflected upward by the trees and is slowed down even before reaching them. On the leeward side further reduction occurs; the effect being felt as far as 20 times the height of the trees. (*c*) Narrow strips of cereal rye act as miniature windbreaks to protect watermelons from wind erosion on a loamy sand on the mid-Atlantic coastal plain. [Photo (*a*) courtesy of USDA Natural Resources Conservation Service; (*b*) from FAO (1987); (*c*) courtesy of R. Weil]

fences and burlap screens, though less efficient as windbreaks than such trees as willows, are often preferred because they can be moved from place to place as crops and cropping practices are varied. Rye, planted in narrow strips across the field, is sometimes used on peat lands and on sandy soils (see Figure 15.26*c*). Narrow rows of such perennial grasses as tall wheatgrass are being evaluated for a combination of wind erosion control and capturing of winter snows in the Northern Plains states.

15.10 LAND CAPABILITY CLASSIFICATION AS A GUIDE TO CONSERVATION

The land capability classification system devised by the U.S. Department of Agriculture uses eight **land capability classes** to indicate the *degree* of limitation imposed on land uses (Figure 15.27), with Class I the least limited and Class VIII the most limited. Each land use class may have four subclasses that indicate the *type* of limitation encountered: risks of erosion (e); wetness, drainage, or flooding (w); root-zone limitations, such as acidity, density, and shallowness (s); and climatic limitations, such as short growing season (c). The erosion (e) subclasses are the most common, and they will be the focus of our attention here. For example, Class IIe land is slightly susceptible to erosion, while Class VIIIe is extremely susceptible. Figure 15.28 shows the appropriate intensity of use allowable for each of these land capability classes if erosion losses (or problems associated with the other subclasses) are to be avoided.

In the United States, Classes I, V, and VIII each account for less than 3% of the nonfederal land. The other classes are much more common, with about 20% of the land in each of Classes II, III, VI, and VII and about 14% in Class IV. This means that about 43% of the land in the United States (242 million ha) is suitable for regular cultivation (Classes I, II, and III). Another 14% (78 million ha) is marginal for growing cultivated crops. The remainder is suited primarily for grasslands and forests, and is used mostly for those purposes. A brief description of the land-use capability classes follows.

Class I. These soils are deep and well drained, have nearly level topography (see Figure 15.27), and have few limitations on their use. They can be used continuously

FIGURE 15.27 Several land capability classes in San Mateo County, California. A range is shown from the nearly level land in the foreground (Class I), which can be cropped intensively, to the badly eroded hillsides (Classes VII and VIII). Although topography and erosion hazards are emphasized here, it should be remembered that other factors—drainage, stoniness, droughtiness—also limit soil usage and help determine the land capability class. (Courtesy USDA Natural Resources Conservation Service)

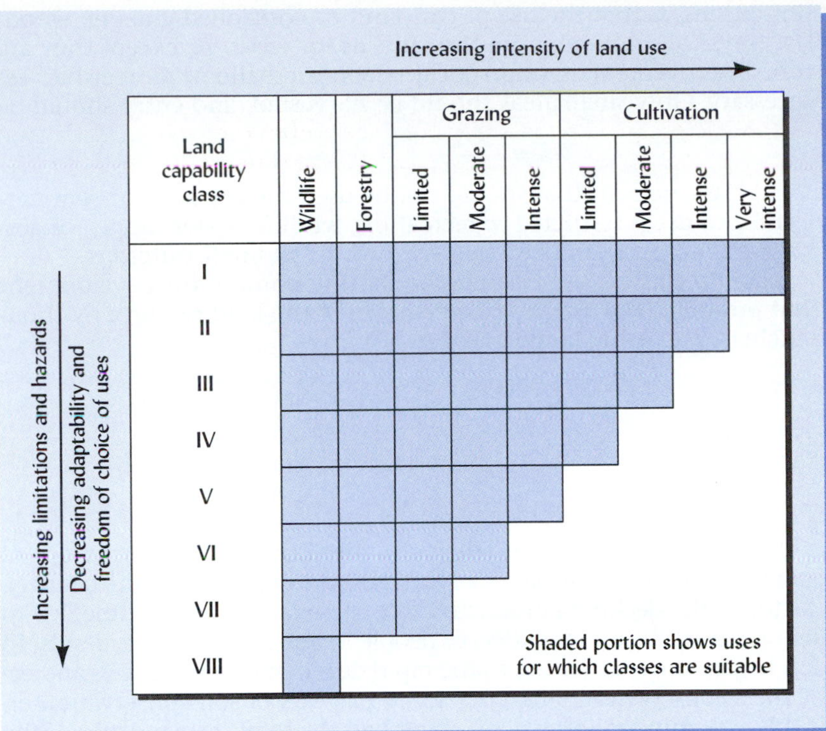

FIGURE 15.28 Intensity with which each land capability class can be used with safety. Note the increasing limitations on the safe uses of the land as one moves from Class I to Class VIII. [Modified from Hockensmith and Steel (1949)]

for row crops, tree nurseries, or for any of the less-intensive uses shown in Figure 15.28. For most purposes, these soils provide the greatest productivity at the least cost.

CLASS II. Soils in Class II have some limitations that reduce the choice of uses or require moderate conservation practices. Tillage and row-crop production should be restricted, or some conservation practices (conservation tillage, grassed waterways, contour strips, etc.) should be installed.

CLASS III. Soils in this class have severe limitations that reduce the choice of plants or require special conservation practices, or both. The same crops may be grown on Class III land as on Classes I and II land, but crops that provide soil cover, such as grasses and legumes, must be more prominent in the rotations used. The amount of land used for row crops is severely restricted. For the w subclass, drainage systems may be needed. Special care and scheduling for timber-harvest operation may be required.

CLASS IV. Soils in this class can be used for cultivation, but there are severe limitations on the choice of crops and management practices. If the land is to be cultivated, crops should be mostly close-growing types (e.g., wheat or barley) and sod or hay crops must be used extensively. Row crops cannot be grown safely, except in no-till systems. The choice of crops may be limited by excess moisture (subclass w), as well as by erosion hazards. For tilled cropland, terracing, contour strips, or other special conservation practices are required. For timber harvest on forestland, special techniques may be required.

CLASS V. These lands are generally not suited to crop production because of factors other than erosion hazards. Such limitations include (1) frequent stream overflow, (2) growing season too short for crop plants, (3) stony or rocky soils, and (4) ponded areas where drainage is not feasible. Often, pastures can be improved on this class of land.

CLASS VI. Soils in this class have extreme limitations (typically steep slopes and high erosion hazards) that restrict use largely to pasture, range, woodland, or wildlife. Forest entry should be limited, special care should be used in path construction, and alternative timber-harvest techniques should be used.

CLASS VII. Severe limitations restrict the use of this land to controlled grazing, woodland, or wildlife. The physical limitations are the same as for Class VI, except they are so strict that pasture improvement is impractical. Cable or balloon timber-harvest methods may be necessary. Only small areas should be harvested, and entry should be strictly limited.

CLASS VIII. In this land class are soils that should not be used for any kind of commercial plant production. Land use is restricted to recreation, wildlife, water supply, or aesthetic purposes. This land includes sand beaches, river wash, and rock outcrops.

The eight land capability classes have become the starting point in the development of land-use plans that promote wise use and conservation of the land resource by thousands of farmers, ranchers, and other landowners.

15.11 PROGRESS IN SOIL CONSERVATION

Soil Erosion

Soil erosion in the United States accelerated when the first European settlers chopped down trees and began to farm the sloping lands of the humid eastern part of the country. Soil erosion was a factor in the declining productivity of these lands that, in time, led to their abandonment and the westward migration of people in search of new farmlands. In the 1930s, the nation began to pay attention to the rapid degeneration of its soils and organized support for their conservation. But, after some 70 years of soil conservation efforts, soil erosion is still a prominent problem on about half the total cropland area of the country.

Since 1982, rather remarkable progress has been made in reducing soil erosion in the U.S. (Figure 15.29), largely as a result of two factors: (1) the spread of conservation tillage (see Section 15.5), and (2) the implementation of land-use changes as part of the conservation reserve program. Progress continues to be made on both fronts; however, about one-third of the nation's cultivated cropland is still losing more than 11 Mg/ha/yr, the maximum loss that can be sustained without serious loss of productivity on most soils.

The Conservation Reserve Program

A major part (about 60%) of the reduction in soil erosion experienced in the United States since 1982 is due to government programs that have paid farmers to shift some land from crops to grasses and forests. Establishing grass or trees on these former croplands reduced the sheet and rill erosion from an average of 19.3 to 1.3 Mg/ha and the wind erosion from 24 to 2.9 Mg/ha. Between 1982 and 1992, 13 million ha of cropland were diverted to such noncultivated uses through the *Conservation Reserve Program*

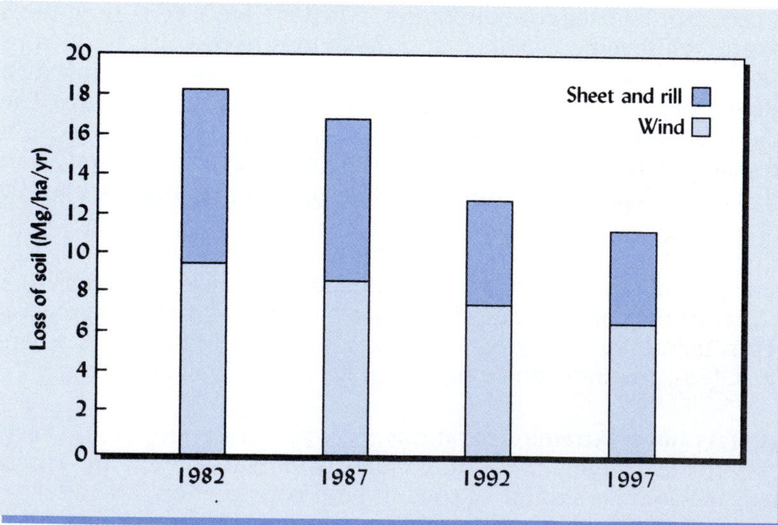

FIGURE 15.29 Rates of soil loss by water and wind erosion in the United States from 1982 to 1997. Farmer adoption of conservation tillage practices along with the Conservation Reserve Program likely account for these remarkable reduction (Data from USDA 1997 National Resources Inventory)

(CRP). Only about half (7 million ha) of the CRP land was on what is considered to be **highly erodible land** (HEL).[8] In 1995, the CRP was reoriented to better target these and other environmentally sensitive areas.

The CRP is basically an arrangement by which the nation's taxpayers pay rent to farmers to forego cropping part of their farmland and instead plant grass or trees on it. The benefits to the nation have been evident, not only in greatly reduced soil losses and sediment pollution, but also in a dramatic upswing in wildlife as bird and animal populations take advantage of the newly restored habitat. Where strips of land along streams (riparian zone buffers) have been incorporated into the CRP, water-quality benefits have also been amplified.

Conservation Management to Enhance Soil Quality

In a broader sense, conservation management practices are those that improve soil quality in more ways than just by protecting the soil from erosion. Properties that indicate the level of soil quality, especially those associated with soil organic matter, can be enhanced by such conservation measures as minimizing tillage, maximizing residue cover of the soil surface, providing for diversity of plant types, keeping soil under grass sod vegetation for at least part of the time, adding organic amendments where practical, and maintaining balanced soil fertility. Improved soil quality, in turn, enhances the soil's capacity to support plants, resist erosion, prevent environmental contamination, and conserve water. Conversation management therefore can lead to the upward spiral of soil and environmental improvement referred to in Section 15.1.

Adapting Soil Conservation to the Needs of Resource-Poor Farmers[9]

In much of the world, so little land is available to each farmer for food production that nations cannot afford the luxury of following the land-use capability classification recommendations as outlined in Section 15.10. Many farmers must use *all* land capable of food production simply to stave off starvation and impoverishment. These farmers often realize that farming erodible land jeopardizes their future livelihood and that of their children, but they see no choice. It is imperative to either find nonagricultural employment for these people or to find farming systems that are sustainable on these erodible lands.

Fortunately, some farmers and scientists have developed, through long traditions of adaptation or through innovation and research, farming systems that *can* produce food and profits while conserving such erodible soil resources. Examples include the traditional Kandy Home Gardens of Sri Lanka's humid mountains, in which a rain forest–like mixed stand of tall fruit and nut trees is combined with an understory of pepper vines, coffee bushes, and spice plants to provide valuable harvests while keeping the soil under perennial vegetative protection. Another example comes from Central America, where farmers have learned to plant thick stands of velvet bean (*Mucuna*) or other viney legumes that can be chopped down by machete to leave a soil-protecting, water-conserving, weed-inhibiting mulch on steep farmlands. In Asia, steep lands have been carefully terraced in ways that allow production of food, even paddy rice, on very steep land without causing significant erosion.

Many examples from the United States and around the world make it clear that when governments cajole, pay, or force farmers into installing soil conservation measures on their land, the results are unlikely to be long-lasting. Usually, farmers will abandon the unwanted practices as soon as the pressure is off. On the other hand, if scientists and conservationists work *with* farmers to help them develop and adapt conservation systems that the farmers feel are of benefit to them and their land, then effective and lasting progress can be made. Experience with conservation tillage systems in the United States, mulch farming systems in Central America, and vegetative contour barriers in Asia have shown that farmers can help develop practices that are good for their land and for their profits: a win-win situation.

---

[8]Highly erodible land is defined by the USDA Natural Resources Conservation Service as land for which the USLE predicts that soil loss on bare, unprotected soil would be 8 times the T value for that soil: $R \times K \times LS \geq 8T$.

[9]For additional perspectives on this topic, see Pieri et al. (2002).

TABLE 15.10 Proportion of the Potential Income from Ecosystem Goods and Services Provided by Five Hypothetical Land-Based Businesses that Vary Greatly in their Soil-Conserving Capabilities

*Note that only 30 to 65% of the value derives from traditional sources of income (**in bold**). Society will have to develop innovative political, social, and marketing means of appropriately rewarding each business for the total revenue or value it generates.*

	Proportion of net income (%)				
Commodity or service	*Forestry business*	*Dairy farm business*	*Ranch business*	*Golf course*	*Grain business*
Milk	0	**20**	0	0	0
Grains	0	**5**	0	0	**60**
Hay	5	**5**	**10**	3	0
Cattle	0	**5**	**30**	0	0
Timber	**30**	**5**	**5**	0	0
Carbon sequestration	15	10	**5**	2	12
Water filtration/collection	15	10	10	10	8
Air purification	5	5	5	5	5
Soil fertility	5	5	5	5	3
Scenic beauty	5	10	5	15	5
Recreation	**5**	5	5	**50**	**5**
Biodiversity	10	10	10	5	2
Salinity control	0	0	5	5	0
Medicinal plants	5	5	5	0	0

Based on concepts summarized in Daily et al. (2000).

15.12 REAL VALUE OF SOIL CONSERVATION

Some soil conservation practices are not adopted because they are considered to be uneconomical; that is, they do not pay the landowner enough to justify their use. There is growing recognition, however, that the true value to society of many conservation practices and systems far exceeds their costs. This value includes factors such as the abatement of downstream siltation, water quality, carbon sequestration, biodiversity, watershed protection, and scenic beauty. Each of these has intrinsic value, but that value cannot always be easily measured, nor are their simple ways of rewarding those who enhance it.

To better understand the true values of any commodity or service, the ecosystems providing these outcomes must be considered as capital assets with various intrinsic values. For example, in determining the total income from a forest ecosystem we should add to the timber sales the income equivalent attributable to such values as scenic beauty, watershed protection, and downstream water quality. Such methodology clearly demonstrates that the monetary income from ecosystems is often far exceeded by the "income" from other values of the ecosystem (see Table 15.10).

Recognition of the benefits to society of various conservation practices has taken two forms: economic incentives for adoption and use of the practices, and penalties and fines for not doing so. Both options seem to be effective depending upon the circumstances. In either case, it is to the economic advantage of the landowner that appropriate soil management practices are followed, and that the loss of soil and water is minimized. Thus, the success of soil conservation systems require political and social actions to ensure the utilization of scientific findings discussed in this textbook.

15.13 CONCLUSION

Accelerated soil erosion is one of the most critical environmental and social problems facing humanity today. Erosion degrades soils, making them less productive of the plants on which animals and people depend. Equally important, erosion causes great damages downstream in reservoirs, lakes, waterways, harbors, and municipal water supplies. Wind erosion also causes fugitive dust that may be harmful to human health.

Nearly 4 billion Mg of soil is eroded each year on land in the United States alone. Half of this erosion occurs on the nation's croplands, the remainder on harvested timber areas, rangelands, and construction sites. Some one-third of the cropland still suffers from erosion that exceeds levels thought to be tolerable.

By sheet and rill erosion, water carries most of the sediment in humid areas. Gullies created by infrequent, but violent, storms account for much of the erosion in drier areas. Wind is the primary erosion agent in many drier areas, especially where the soil is bare and low in moisture during the season when strong winds blow.

Protecting soil from the ravages of wind or water is by far the most effective way to constrain erosion. In croplands and forests, such protection is due mainly to the cover of plants and their residues. Conservation tillage practices maintain vegetative cover on at least 30% of the soil surface, and the widening adoption of these practices has contributed to the significant reductions in soil erosion achieved over the past two decades. Crop rotations that include sod and close-growing crops, coupled with such practices as contour tillage, strip cropping, and terracing, also help combat erosion on farmland.

In forested areas, most erosion is associated with timber-harvesting practices and forest road construction. For the sake of future forest productivity and current water quality, foresters must become more selective in their harvest practices and invest more in proper road construction.

Construction sites for roads, buildings, and other engineering projects lay bare many scattered areas of soils that add up to a serious erosion problem. Control of sediment from construction sites requires carefully phased land clearing, along with vegetative and artificial soil covers, and installation of various barriers and sediment-holding ponds. These measures may be expensive to implement, but the costs to society that result when sediment is not controlled are too high to ignore. Once construction is completed, erosion rates on urban areas are commonly as low as those on areas under undisturbed native vegetation.

Erosion-control systems must be developed in collaboration with those who use the land, and especially the poor, for whom immediate needs must overshadow concerns for the future. As put succinctly in *The River,* a classic 1930s documentary film produced during the rebirth of American soil erosion consciousness, "Poor land makes poor people, and poor people make poor land."

STUDY QUESTIONS

1. Explain the distinction between *geologic erosion* and *accelerated erosion.* Is the difference between the two greater in humid or arid regions?
2. When erosion takes place by wind or water, what are three important types of damages that result on the land whose soils is eroding? What are five important types of damages that erosion causes in locations away from the eroding site?
3. What is a common *T* value, and what is meant by this term? Explain why certain soils have been assigned a higher *T* value than other soils.
4. Describe the three main steps in the water erosion process.
5. Many people assume that the amount of soil eroded on the land in a watershed (*A* in the universal soil-loss equation) is the same as the amount of sediment carried away by the stream draining that watershed. What factor is missing that makes this assumption incorrect? Do you think that this means the USLE should be renamed?
6. Why is the total annual rainfall in an area *not* a very good guide to the amount of erosion that will take place on a particular type of bare soil?
7. Contrast the properties you would expect in a soil with either a very high *K* value or a very low *K* value.
8. How much soil is likely to be eroded from a Keene silt loam in central Ohio, on a 12% slope, 100 m long, if it is in dense permanent pasture and has no support practices applied to the land? Use the information available in this chapter to calculate an answer.
9. What type of conservation tillage leaves the greatest amount of soil cover by crop residues? What are the advantages and disadvantages of this system?

10. Why are narrow strips of grass planted on the contour sometimes called a *living terrace?*
11. In most forests, which component of the ecosystem provides the primary protection against soil erosion by water, the *tree canopy, tree roots,* or *leaf litter?*
12. Certain soil properties generally make land susceptible to erosion by wind or erosion by water. List four properties that characterize soils highly susceptible to wind erosion. Indicate which two of these properties should also characterize soils highly susceptible to water erosion, and which two should not.
13. Which three factors in the wind erosion prediction equation (WEQ) can be affected by tillage? Explain.
14. Describe a soil in land capability Class IIw soil in comparison with one in Class IVe.
15. Why is it important that there be a close relationship between land in the CRP and that considered to be HEL?

REFERENCES

Blevins, R. L., and W. W. Frye. 1992. "Conservation tillage: An ecological approach to soil management," *Advances in Agronomy* **51**:33–78.

Carter, M. R. 1994. *Conservation Tillage in Temperate Agroecosystems.* (Boca Raton, Fla.: Lewis Publishers).

Cassel, D. K., and R. Lal. 1992. "Soil physical properties of the tropics: Common beliefs and management constraints," in R. Lal and P. A. Sanchez (eds.), *Myths and Science of Soils of the Tropics.* SSA Special Publication no. 29. (Madison, Wis.: Soil Sci. Soc. of Amer.), pp. 61–89.

Crovetto, C. 1996. *Stubble Over the Soil.* (Madison, Wis.: Amer. Soc. Agron.).

Daily, G. 1997. "Restoring value to the world's degraded lands," *Science* **269**:350–354.

Daily, G. C., et al. 2000. "The value of nature and the nature value," *Science* **289**:395–396.

Dickey, E. C., et al. 1987. "Conservation tillage: Perceived and actual use," *J. Soil Water Cons.* **42**:431–434.

FAO. 1987. *Protect and Produce.* (Rome: U.N. Food and Agriculture Organization).

Foster, G. R., and R. E. Highfill. 1983. "Effect of terraces on soil loss: USLEP factor values for terraces," *J. Soil Water Cons.* **38**:48–51.

Foster, G. R., D. K. McCool, K. G. Renard, and W. C. Moldenhauer. 1981. "Conversion of the universal soil loss equation to SI metric units," *J. Soil Water Cons.* **36**:355–359.

Fryrear, D. W. 2000. "RWEQ: Improved wind erosion technology," *J. Soil Water Cons.* **55**:183–189.

Hillel, D. 1991. *Out of the Earth: Civilization and the Life of the Soil.* (New York: The Free Press).

Hockensmith, R. D., and J. G. Steel. 1949. "Recent trends in the use of the land-capability classification," *Soil Sci. Soc. Amer. Proc.* **14**:383–388.

Hudson, N. 1995. *Soil Conservation,* 3rd ed. (Ames, Iowa: Iowa State University Press).

Hughes, H. A. 1980. *Conservation Farming.* (Moline, Ill.: John Deere and Company).

Kelley, D. W., and E. A. Nater. 2000. "Historical sediment flux from three watersheds into Lake Pepin, Minnesota, USA," *J. Environ. Qual.* **29**:561–568.

Krenisky, E. C., M. J. Carroll, R. L. Hill, and J. M. Krouse. 1998. "Runoff and sediment losses from natural and man-made erosion control materials," *Crop Sci.* **38**:1042–1046.

Mahboubi, A. A., R. Lal, and N. R. Faussey. 1993. "Twenty-eight years of tillage effects on two soils in Ohio," *Soil Sci. Soc. Amer. J.* **57**:506–512.

Nyland, R. D. 1996. *Silviculture: Concepts and Applications.* (New York: McGraw-Hill).

Oldeman, L. R. 1994. "The global extent of soil degradation," in D. J. Greenland and I. Szabolcs (eds.), *Soil Resilience and Sustainable Land Use.* (Wallingford, U.K.: CAB International).

Pierre, C. 1992. *Fertility of Soils: A Future for Farming in the West African Savannah.* (Berlin: Springer-Velag).

Pieri, C., G. Evers, J. Landers, P. O'Connell, and E. Terry. 2002. No-Till Farming for Sustainable Rural Development. Agriculture & Rural Development Working Paper 24536. Washington, DC: The International Bank for Reconstruction and Development.

Renard, K. G., G. Foster, G. Weesies, D. McCool, and D. Yoder. 1997. *Predicting Soil Erosion by Water: A Guide to Conservation Planning with the Revised Universal Soil Loss Equation (RUSLE).* Agricultural Handbook no. 703. (Washington, D.C.: USDA).

Rosensweig, C., and D. Hillel. 1998. *Climate Change and the Global Harvest: Potential Impacts of the Greenhouse Effect on Agriculture.* (Cary, N.C.: Oxford University Press).

Schwab, G. O., D. D. Fangmeirer, and W. J. Elliot. 1996. *Soil and Water Management Systems,* 4th ed. (New York: Wiley).

USDA/ARS. 1997. *Predicting Soil Erosion By Water: A Guide to Conservation Planning with the Revised Universal Soil Loss Equation (RUSLE).* Agricultural Handbook 703 (Washington, DC: USDA Agriculture Research Service).

USDA. 1997. *1997 National Resources Inventory.* (Washington, D.C.: USDA Natural Resources Conservation Service).

Wischmeier, W. J., and D. D. Smith. 1978. *Predicting Rainfall Erosion Loss—A Guide to Conservation Planning.* Agricultural Handbook no. 537. (Washington, D.C.: USDA).

Appendix A
Canadian and FAO Soil Classification Systems

The Canadian Soil Classification System is one of many national soil classification systems used in various countries around the world. Of these, it is perhaps the most closely aligned with the U.S. *Soil Taxonomy.* It includes five hierarchical categories: order, great group, subgroup, family, and series. The system is designed to apply principally to the soils of Canada. The soil orders of the Canadian System of Soil Classification are described in Table A.1 and soil orders and some great groups are compared to the U.S. *Soil Taxonomy* in Table A.2.

TABLE A.1 Summary with Brief Descriptions of the Soil Orders in the Soil Classification System of Canada

Brunisolic	Soils with sufficient development to exclude them from the Regosolic order, but lacking the degree or kind of horizon development specified for other soil orders.
Chernozemic	Soils with high base saturation and surface horizons darkened by the accumulation of organic matter from the decomposition of plants from grassland or grassland-forest ecosystems.
Cryosolic	Soils formed in either mineral or organic materials that have permafrost either within 1 m of the surface or within 2 m if more than one-third of the pedon has been strongly cryoturbated, as indicated by disrupted, mixed, or broken horizons.
Gleysolic	Gleysolic soils have features indicative of periodic or prolonged saturation (i.e., gleying, mottling) with water and reducing conditions.
Luvisolic	Soils with light-colored, eluvial horizons that have illuvial B horizons in which silicate clay has accumulated.
Organic	Organic soils developed on well- to undecomposed peat or leaf litter.
Podzolic	Soils with a B horizon in which the dominant accumulation product is amorphous material composed mainly of humified organic matter combined in varying degrees with Al and Fe.
Regosolic	Weakly developed soils that lack development of genetic horizons.
Solonetzic	Soils that occur on saline (often high in sodium) parent materials, which have B horizons that are very hard when dry and swell to a sticky mass of very low permeability when wet. Typically the solonetzic B horizon has prismatic or columnar macrostructure that breaks to hard to extremely hard, blocky peds with dark coatings.
Vertisolic	Soils with high contents of expanding clays that have large cracks during the dry parts of the year and show evidence of swelling, such as gilgae and slickensides.

TABLE A.2 Comparison of U.S. Soil Taxonomy and the Canadian Soil Classification System

Note that because the boundary criteria differ between the two systems, certain U.S. Soil Taxonomy *soil orders have equivalent members in more than one Canadian Soil Classification System soil order.*[a]

U.S. Soil Taxonomy *soil order*	*Canadian system soil order*	*Canadian system great group*	*Equivalent lower-level taxa in U.S.* Soil Taxonomy
Alfisols	Luvisolic	Gray Brown Luvisols	Hapludalfs
		Gray Luvisols	Haplocryalfs, Eutrocryalfs, Fragudalfs, Glossocryalfs, Palecryalfs, and some subgroups of Ustalfs and Udalfs
	Solonetzic	Solonetz	Natrudalfs and Natrustalfs
		Solod	Glossic subgroups of Natraqualfs, Natrudalfs, and Natrustalfs
Andisols	Components of Brunisolic and Cryosolic		
Aridisols	Solonetzic		Frigid families of Natrargids
Entisols	Regosolic		Cryic great groups and frigid families of Entisols, except Aquents
		Regosol	Cryic great groups and frigid families of Folists, Fluvents, Orthents, and Psamments
Gelisols	Cryosolic	Turbic Cryosol	Turbels
		Organic Cryosol	Histels
		Stagnic Cryosol	Orthels
Histosols	Organic	Fibrisol	Cryofibrists, Sphagnofibrists
		Mesisol	Cryohemists
		Humisol	Cryosaprists
Inceptisols	Brunisolic	Melanic Brunisol	Some Eutrustepts
		Eutric Brunisol	Subgroups of Cryepts; frigid and mesic families of Haplustepts
		Sombric Brunisol	Frigid and mesic families of Udepts, and Ustept and Humic Dystrudepts
		Dystric Brunisol	Frigid families of Dystrudepts and Dystrocryepts
	Gleysolic		Cryic subgroups and frigid families of Aqualfs, Aquolls, Aquepts, Aquents, and Aquods
		Humic Gleysol	Humaquepts
		Gleysol	Cryaquepts and frigid families of Fragaquepts, Epiaquepts, and Endoaquepts
Mollisols	Chernozemic	Brown	Xeric and Ustic subgroups of Argicryolls and Haplocryolls
		Dark Brown	Subgroups of Argicryolls and Haplocryolls
		Black	Typic subgroups of Argicryolls and Haplocryolls
		Dark Gray	Alfic subgroups of Argicryolls
	Solonetzic	Solonetz	Natricryolls and frigid families of Natraquolls and Natralbolls
		Solod	Glossic subgroups of Natricryolls
Oxisols	Not relevant in Canada		
Spodosols	Podzolic	Humic Podzol	Humicryods, Humic Placocryods, Placohumods, and frigid families of other Humods
		Ferro-Humic Podzol	Humic Haplocryods, some Placorthods, and frigid families of humic subgroups of other Orthods
		Humo-Ferric Podzol	Haplorthods, Placorthods, and frigid families of other Orthods and Cryods except humic subgroups
Ultisols	Not relevant in Canada		
Vertisols	Vertisolic	Vertisol	Haplocryerts
		Humic Vertisol	Humicryerts

[a] Based on information in Soil Classification Working Group, 1998, *The Canadian System of Soil Classification,* 3d ed. (Ottawa: Agriculture and Agri-Food Canada). Publication No. A53-1646/1997E.

The FAO Soil Classification System is really a map legend for the world soil map made by FAO/UNESCO. A brief description of the map units is given in Table A.3. The FAO map legend terms are used by many scientists around the world to describe different kinds of soils. Global and regional soil databases are often organized according to this system. The soil characteristics recognized and the terminology used borrow heavily from the Russian soil classification system, as well as from U.S. *Soil Taxonomy*.

TABLE A.3 Soil Map Units for the FAO/UNESCO Soil Map of the World

Acrisols	Low base status soils with argillic horizons
Andosols	Soils formed in volcanic ash that have dark surfaces
Arenosols	Soils formed from sand
Cambisols	Soils with slight color, structure, or consistency change due to weathering
Chernozems	Soils with black surface, high humus under prairie vegetation
Cryosols	Soils of cold climates with permafrost
Ferralsols	Highly weathered soils with sesquioxide-rich clays
Fluvisols	Water-deposited soils with little alteration
Gleysols	Soils with mottled or reduced horizons due to wetness
Greyzems	Soils with dark surface, bleached E horizon, and textural B horizon
Histosols	Organic soils
Kastanozems	Soils with chestnut surface color, steppe vegetation
Lithosols	Shallow soils over hard rock
Luvisols	Medium to high base status soils with argillic horizons
Nitosols	Soils with low CEC clay in argillic horizons
Planosols	Soils with abrupt A–B horizon contact
Phaeozems	Soils with dark surface, more leached than Kastanozem or Chernozem
Podzols	Soils with light-colored eluvial horizon and subsoil accumulation of iron, aluminum, and humus
Podzoluvisols	Soils with leached horizons tonguing into argillic B horizons
Rankers	Thin soils over siliceous material
Regosols	Thin soil over unconsolidated material
Rendzinas	Shallow soils over limestone
Solonchaks	Soils with soluble salt accumulation
Solonetz	Soils with high sodium content
Vertisols	Self-mulching, inverting soils, rich in smectite clay
Xerosols	Dry soils of semiarid regions
Yermosols	Desert soils

Appendix B

SI Units, Conversion Factors, and Periodic Table of the Elements

Basic SI Units of Measurement

Parameter	*Basic unit*	*Symbol*
Amount of substance	mole	mol
Electrical current	ampere	A
Length	meter	m
Luminous intensity	candela	cd
Mass	gram (kilogram)	g (kg)
Temperature	kelvin	K
Time	second	s

Prefixes Used to Indicate Order of Magnitude

Prefix	*Multiple*	*Abbreviation*	*Multiplication factor*
exa	10^{18}	E	1,000,000,000,000,000,000
peta	10^{15}	P	1,000,000,000,000,000
tera	10^{12}	T	1,000,000,000,000
giga	10^{9}	G	1,000,000,000
mega	10^{6}	M	1,000,000
kilo	10^{3}	k	1,000
hecto	10^{2}	h	100
deca	10	da	10
deci	10^{-1}	d	0.1
centi	10^{-2}	c	0.01
milli	10^{-3}	m	0.001
micro	10^{-6}	μ	0.000 001
nano	10^{-9}	n	0.000 000 001
pico	10^{-12}	P	0.000 000 000 001
femto	10^{-15}	f	0.000 000 000 000 001
atto	10^{-18}	a	0.000 000 000 000 000 001

Factors for Converting Non-SI Units to SI Units

Non-SI Unit	*Multiply by*[a]	*To obtain SI Unit*
	Length	
inch, in.	2.54	centimeters, cm (10^{-2} m)
foot, ft	0.304	meter, m
mile,	1.609	kilometer, km (10^{3} m)
micron,µ	1.0	micrometer, µm (10^{-6} m)
Ångstrom unit, Å	0.1	nanometer, nm (10^{-9} m)
	Area	
acre, ac	0.405	hectare, ha (10^{4} m^2)
square foot, ft^2	9.29×10^{-2}	square meter, m^2
square inch, in^2	645	square millimeter, mm^2
square mile, mi^2	2.59	square kilometer, km^2
	Volume	
bushel, bu	35.24	liter, L
cubic foot, ft^3	2.83×10^{-2}	cubic meter, m^3
cubic inch, $in.^3$	1.64×10^{-5}	cubic meter, m^3
gallon (U.S.), gal	3.78	liter, L
quart, qt	0.946	liter, L
acre-foot, ac-ft	12.33	hectare-centimeter, ha-cm
acre-inch, ac-in.	1.03×10^{-2}	hectare-meters, ha-m
ounce (fluid), oz	2.96×10^{-2}	liter, L
pint, pt	0.473	liter, L
	Mass	
ounce (avdp), oz	28.4	gram, g
pound, lb	0.454	kilogram, kg (10^{3} g)
ton (2000 lb)	0.907	megagram, Mg (10^{6} g)
tonne (metric), t	1000	kilogram, kg
	Radioactivity	
curie, Ci	3.7×10^{10}	becquerel, Bq
picocurie per gram, pCi/g	37	becquerel per kilogram, Bq/kg
	Yield and Rate	
pound per acre, lb/ac	1.121	kilogram per hectare, kg/ha
bushel per acre (60 lb), bu/ac	67.19	kilogram per hectare, kg/ha
bushel per acre (56 lb), bu/ac	62.71	kilogram per hectare, kg/ha
bushel per acre (48 lb), bu/ac	53.75	kilogram per hectare, kg/ha
gallon per acre (U.S.), gal/ac	9.35	liter per hectare, L/ha
ton (2000 lb) per acre	2.24	megagram per hectare, Mg/ha
miles per hour, mph	0.447	meter per second, m/s
gallon per minute (U.S.), gpm	0.227	cubic meter per hour, m^3/h
cubic feet per second, cfs	101.9	cubic meter per hour, m^3/h
	Pressure	
atmosphere, atm	0.101	megapascal, MPa (10^{6} Pa)
bar	0.1	megapascal, MPa
pound per square foot, lb/ft^2	47.9	pascal, Pa
pound per square inch, lb/in^2	6.9×10^{3}	pascal, Pa
	Temperature	
degrees Fahrenheit (°F − 32)	0.556	degrees, °C
degrees Celsius (°C + 273)	1	Kelvin, K
Energy		
British thermal unit, Btu	1.05×10^{3}	joule, J
calorie, cal	4.19	joule, J
dyne, dyn	10^{-5}	newton, N
erg	10^{-7}	joule, J
foot-pound, ft-lb	1.36	joule, J
	Concentrations	
percent, %	10	gram per kilogram, g/kg
part per million, ppm	1	milligram per kilogram, mg/kg
milliequivalents per 100 grams	1	centimole per kilogram, cmol/kg

[a] To convert from SI to non-SI units, *divide* by the factor given.

Periodic Table of the Elements with Notes Concerning Relevance to Soil Science

Based on atomic mass of $^{12}C = 12.0$. Numbers in parentheses are the mass numbers of the most stable isotopes of radioactive elements.

Group IA	Group IIA	Group IIIB	Group IVB	Group VB	Group VIB	Group VIIB	← Group	Group VIIIB	Group →	Group IB	Group IIB	Group IIIA	Group IVA	Group VA	Group VIA	Group VIIA	Group VIIIA
1 H 1.01 Hydrogen																	2 He 4.00 Helium
3 Li 6.94 Lithium	4 Be 9.01 Beryllium											5 B 10.81 Boron	6 C 12.01 Carbon	7 N 14.01 Nitrogen	8 O 16.00 Oxygen	9 F 19.00 Fluorine	10 Ne 20.18 Neon
11 Na 22.99 Sodium	12 Mg 24.30 Magnesium											13 Al 26.98 Aluminum	14 Si 28.09 Silicon	15 P 30.97 Phosphorus	16 S 32.07 Sulfur	17 Cl 35.45 Chlorine	18 Ar 39.95 Argon
19 K 39.10 Potassium	20 Ca 40.08 Calcium	21 Sc 44.96 Scandium	22 Ti 47.88 Titanium	23 V 50.94 Vanadium	24 Cr 52.00 Chromium	25 Mn 54.94 Manganese	26 Fe 55.85 Iron	27 Co 58.93 Cobalt	28 Ni 58.69 Nickel	29 Cu 63.55 Copper	30 Zn 65.38 Zinc	31 Ga 69.72 Gallium	32 Ge 72.59 Germanium	33 As 74.92 Arsenic	34 Se 78.96 Selenium	35 Br 79.90 Bromine	36 Kr 83.80 Krypton
37 Rb 85.47 Rubidium	38 Sr 87.62 Strontium	39 Y 88.91 Yttrium	40 Zr 91.22 Zirconium	41 Nb 92.91 Niobium	42 Mo 95.94 Molybdenum	43 Tc (98) Technetium	44 Ru 101.07 Ruthenium	45 Rh 102.91 Rhodium	46 Pd 106.42 Palladium	47 Ag 107.87 Silver	48 Cd 112.41 Cadmium	49 In 114.82 Indium	50 Sn 118.71 Tin	51 Sb 121.75 Antimony	52 Te 127.60 Tellurium	53 I 126.90 Iodine	54 Xe 131.29 Xenon
55 Cs 132.91 Cesium	56 Ba 137.33 Barium	57 La 138.91 Lanthanum	72 Hf 178.49 Hafnium	73 Ta 180.95 Tantalum	74 W 183.85 Tungsten	75 Re 186.21 Rhenium	76 Os 190.2 Osmium	77 Ir 192.22 Iridium	78 Pt 195.08 Platinum	79 Au 196.97 Gold	80 Hg 200.59 Mercury	81 Tl 204.38 Thallium	82 Pb 207.2 Lead	83 Bi 208.98 Bismuth	84 Po (209) Polonium	85 At (210) Astatine	86 Rn (222) Radon
87 Fr (223) Francium	88 Ra (226) Radium	89 Ac (227) Actinium	104 Unq (261) Unnilquadium	105 Unp (262) Unnilpentium	106 Unh (263) Unnilhexium	107 Uns (262) Unnilseptium	108 Uno (265) Unniloctium										

Metals ← → Nonmetals

58 Ce 140.12 Cerium	59 Pr 140.91 Praseodymium	60 Nd 144.24 Neodymium	61 Pm (145) Promethium	62 Sm 150.36 Samarium	63 Eu 151.96 Europium	64 Gd 157.25 Gadolinium	65 Tb 158.93 Terbium	66 Dy 162.50 Dysprosium	67 Ho 164.93 Holmium	68 Er 167.26 Erbium	69 Tm 168.93 Thulium	70 Yb 173.04 Ytterbium	71 Lu 174.97 Lutetium
90 Th (232) Thorium	91 Pa (231) Protactinium	92 U (238) Uranium	93 Np (237) Neptunium	94 Pu (244) Plutonium	95 Am (243) Americium	96 Cm (247) Curium	97 Bk (247) Berkelium	98 Cf (251) Californium	99 Es (252) Einsteinium	100 Fm (257) Fermium	101 Md (258) Mendelevium	102 No (259) Nobelium	103 Lr (260) Lawrencium

Key: 87 — Atomic number; Fr — Symbol; (223) — Atomic mass; Francium

- Elements known to be nutrients for animals or plants. Some are also toxic in excessive amounts.
- Elements toxic to organisms in small amounts, and not known to serve as nutrients.
- Other elements commonly studied in Soil Science because of soil-environmental impacts or because of their use as tracers or electrodes. (Br is used to trace anionic solutes such as nitrate. Isotopes of Rb and Sr are used to trace K and Ca in plants and soils. Cs and Ti are used to trace geological processes such as soil erosion. Pt and Ag are used in electrodes for measuring soil redox potential and pH, respectively.)

Glossary of Soil Science Terms[1]

A horizon The surface horizon of a mineral soil having maximum organic matter accumulation, maximum biological activity, and/or eluviation of materials such as iron and aluminum oxides and silicate clays.

abiotic Nonliving; often in reference to environmental influences such as rainfall, heat and minerals.

accelerated erosion Erosion much more rapid than natural, geological erosion; primarily as a result of human activities.

acid cations Cations, principally Al^{3+}, Fe^{3+}, and H^+, that contribute to H^+ ion activity either directly or through hydrolysis reactions with water. *See also* nonacid cations.

acid rain Atmospheric precipitation with pH values less than about 5.6, the acidity being due to inorganic acids such as nitric and sulfuric that are formed when oxides of nitrogen and sulfur are emitted into the atmosphere.

acid saturation The proportion or percentage of a cation-exchange site occupied by acid cations.

acid soil A soil with a pH value <7.0. Usually applied to surface layer or root zone, but may be used to characterize any horizon. *See also* reaction, soil.

acid sulfate soils Soils that are potentially extremely acid (pH <3.5) because of the presence of large amounts of reduced forms of sulfur that are oxidized to sulfuric acid if the soils are exposed to oxygen when they are drained or excavated. A sulfuric horizon containing the yellow mineral jarosite is often present. *See also* cat clays.

acidity, active The activity of hydrogen ions in the aqueous phase of a soil. It is measured and expressed as a pH value.

acidity, residual Soil acidity that can be neutralized by lime or other alkaline materials but cannot be replaced by an unbuffered salt solution.

acidity, salt replaceable Exchangeable hydrogen and aluminum that can be replaced from an acid soil by an unbuffered salt solution such as KCl or NaCl.

acidity, total The total acidity in a soil. It is approximated by the sum of the salt-replaceable acidity plus the residual acidity.

actinomycetes A group of bacteria that usually produce a characteristic branched mycelium.

activated sludge Sludge that has been aerated and subjected to bacterial action.

active layer The upper portion of a Gellisol that is subject to freezing and thawing and is underlain by permafrost.

active organic matter A portion of the soil organic matter that is relatively easily metabolized by microorganisms and cycles with a half-life in the soil of a few days to a few years.

[1] This glossary was compiled and modified from several sources, including *Glossary of Soil Science Terms* [Madison, Wis.: Soil Sci. Soc. Amer. (1997)]; *Resource Conservation Glossary* [Ankeny, Iowa: Soil Cons. Soc. Amer. (1982)]; and *Soil Taxonomy* [Washington, D.C.: U.S. Department of Agriculture (1999)].

adhesion Molecular attraction that holds the surfaces of two substances (e.g., water and sand particles) in contact.

adsorption The attraction of ions or compounds to the surface of a solid. Soil colloids adsorb large amounts of ions and water.

adsorption complex The group of organic and inorganic substances in soil capable of adsorbing ions and molecules.

aerate To impregnate with gas, usually air.

aeration, soil The process by which air in the soil is replaced by air from the atmosphere. In a well-aerated soil, the soil air is similar in composition to the atmosphere above the soil. Poorly aerated soils usually contain more carbon dioxide and correspondingly less oxygen than the atmosphere above the soil.

aerobic (1) Having molecular oxygen as a part of the environment. (2) Growing only in the presence of molecular oxygen, as aerobic organisms. (3) Occurring only in the presence of molecular oxygen (said of certain chemical or biochemical processes, such as aerobic decomposition).

aerosolic dust A type of eolian material that is very fine (about 1 to 10 μm) and may remain suspended in the air over distances of thousands of kilometers. Finer than most *loess*.

aggregate (soil) Many soil particles held in a single mass or cluster, such as a clod, crumb, block, or prism.

agric horizon A diagnostic subsurface horizon in which clay, silt, and humus derived from an overlying cultivated and fertilized layer have accumulated. Wormholes and illuvial clay, silt, and humus occupy at least 5% of the horizon by volume.

agronomy The theory and practice of field-crop production and scientific management of soil and land.

air-dry (1) The state of dryness (of a soil) at equilibrium with the moisture content in the surrounding atmosphere. The actual moisture content will depend upon the relative humidity and the temperature of the surrounding atmosphere. (2) To allow to reach equilibrium in moisture content with the surrounding atmosphere.

air porosity The proportion of the bulk volume of soil that is filled with air at any given time or under a given condition, such as a specified moisture potential; usually the large pores.

albic horizon A diagnostic subsurface horizon from which clay and free iron oxides have been removed or in which the oxides have been segregated to the extent that the color of the horizon is determined primarily by the color of the primary sand and silt particles rather than by coatings on these particles.

Alfisols An order in *Soil Taxonomy.* Soils with gray to brown surface horizons, medium to high supply of bases, and B horizons of illuvial clay accumulation. These soils form mostly under forest or savanna vegetation in climates with slight to pronounced seasonal moisture deficit.

algal bloom A population explosion of algae in surface waters, such as lakes and streams, often resulting in high turbidity and green or red water, and commonly stimulated by nutrient enrichment with phosphorus and nitrogen.

alkaline soil Any soil that has pH >7. Usually applied to the surface layer or root zone but may be used to characterize any horizon or a sample thereof. *See also* reaction, soil.

allelochemical An organic chemical by which one plant can influence another. *See* allelopathy.

allelopathy The process by which one plant may affect other plants by biologically active chemicals introduced into the soil, either directly by leaching or exudation from the source plant, or as a result of the decay of the plant residues. The effects, though usually negative, may also be positive.

allophane A poorly defined aluminosilicate mineral whose structural framework consists of short runs of three-dimensional crystals interspersed with amorphous noncrystalline materials. It is prevalent in volcanic ash materials.

alluvial fan Fan-shaped alluvium deposited at the mouth of a canyon or ravine where debris-laden waters fan out, slow down, and deposit their burden.

alluvium A general term for all detrital material deposited or in transit by streams, including gravel, sand, silt, clay, and mixtures of these. Unless otherwise noted, alluvium is unconsolidated.

alpha particle A positively charged particle (consisting of two protons and two neutrons) that is emitted by certain radioactive compounds.

aluminosilicates Compounds containing aluminum, silicon, and oxygen as main constituents. An example is microcline, $KAlSi_3O_8$.

amendment, soil Any substance other than fertilizers, such as lime, sulfur, gypsum, and sawdust, used to alter the chemical or physical properties of a soil, generally to make it more productive.

amino acids Nitrogen-containing organic acids that couple together to form proteins. Each acid molecule contains one or more amino groups ($—NH_2$) and at least one carboxyl group (—COOH). In addition, some amino acids contain sulfur.

ammonification The biochemical process whereby ammoniacal nitrogen is released from nitrogen-containing organic compounds.

ammonium fixation The entrapment of ammonium ions by the mineral or organic fractions of the soil in forms that are insoluble in water and are at least temporarily nonexchangeable.

amorphous material Noncrystalline constituents of soils.

anaerobic (1) Without molecular oxygen. (2) Living or functioning in the absence of air or free oxygen.

andic properties Soil properties related to volcanic origin of materials, including high organic carbon content, low bulk density, high phosphate retention, and extractable iron and aluminum.

Andisols An order in *Soil Taxonomy.* Soils developed from volcanic ejecta. The colloidal fraction is dominated by allophane and/or Al-humus compounds.

angle of repose The maximum slope steepness at which loose, cohesionless material will come to rest.

anion Negatively charged ion; during electrolysis it is attracted to the positively charged anode.

anion exchange Exchange of anions in the soil solution for anions adsorbed on the surface of clay and humus particles.

anion exchange capacity The sum total of exchangeable anions that a soil can adsorb. Expressed as centimoles of charge per kilogram ($cmol_c/kg$) of soil (or of other adsorbing material, such as clay).

anoxic *See* anaerobic.

anthropic epipedon A diagnostic surface horizon of mineral soil that has the same requirements as the mollic epipedon but that has more than 250 mg/kg of P_2O_5 soluble in 1% citric acid, or is dry more than 10 months (cumulative) during the period when not irrigated. The anthropic epipedon forms under long-continued cultivation and fertilization.

antibiotic A substance produced by one species of organism that, in low concentrations, will kill or inhibit growth of certain other organisms.

Ap The surface layer of a soil disturbed by cultivation or pasturing.

apatite A naturally occurring complex calcium phosphate that is the original source of most of the phosphate fertilizers. Formulas such as $[3Ca_3(PO_4)_2] \cdot CaF_2$ illustrate the complex compounds that make up apatite.

aquic conditions Continuous or periodic saturation (with water) and reduction, commonly indicated by redoximorphic features.

aquiclude A saturated body of rock or sediment that is incapable of transmitting significant quantities of water under ordinary water pressures.

aquifer A saturated, permeable layer of sediment or rock that can transmit significant quantities of water under normal pressure conditions.

arbuscular mycorrhiza (AM) A common endomycorrhizal association produced by phycomycetous fungi and characterized by the development, within root cells, of small structures known as *arbuscules*. Some also form, between root cells, storage organs known as *vesicles*. Host range includes many agricultural and horticultural crops. Formerly called vesicular arbuscular mycorrhiza (VAM). *See also* endotrophic mycorrhiza.

arbuscule Specialized branched structure formed within a root cortical cell by endotrophic mycorrhizal fungi.

argillan A thin coating of well-oriented clay particles on the surface of a soil aggregate, particle, or pore. A clay film.

argillic horizon A diagnostic subsurface horizon characterized by the illuvial accumulation of layer-lattice silicate clays.

arid climate Climate in regions that lack sufficient moisture for crop production without irrigation. In cool regions annual precipitation is usually less than 25 cm. It may be as high as 50 cm in tropical regions. Natural vegetation is desert shrubs.

Aridisols An order in *Soil Taxonomy.* Soils of dry climates. They have pedogenic horizons, low in organic matter, that are never moist for as long as 3 consecutive months. They have an ochric epipedon and one or more of the following diagnostic horizons: argillic, natric, cambic, calcic, petrocalcic, gypsic, petrogypsic, salic, or a duripan.

aspect (of slopes) The direction (e.g., south or north) that a slope faces with respect to the sun.

association, soil *See* soil association.

Atterberg limits Water contents of fine-grained soils at different states of consistency.

- **liquid limit (LL)** The water content corresponding to the arbitrary limit between the liquid and plastic states of consistency of a soil.
- **plastic limit (PL)** The water content corresponding to an arbitrary limit between the plastic and semisolid states of consistency of a soil.
- **shrinkage limit (SL)** The water content above which a mass of soil material will swell in volume, but below which it will shrink no further.

autochthonous organisms Those microorganisms thought to subsist on the more resistant soil organic matter and little affected by the addition of fresh organic materials. *Contrast with* zymogenous organisms.

autotroph An organism capable of utilizing carbon dioxide or carbonates as the sole source of carbon and obtaining energy for life processes from the oxidation of inorganic elements or compounds such as iron, sulfur, hydrogen, ammonium, and nitrites, or from radiant energy. *Contrast with* heterotroph.

available nutrient That portion of any element or compound in the soil that can be readily absorbed and assimilated by growing plants. ("Available" should not be confused with "exchangeable.")

available water The portion of water in a soil that can be readily absorbed by plant roots. The amount of water released between the field capacity and the permanent wilting point.

B horizon A soil horizon, usually beneath the A horizon, that is characterized by one or more of the following: (1) a concentration of silicate clays, iron and aluminum oxides, and humus, alone or in combination; (2) a blocky or prismatic structure; and (3) coatings of iron and aluminum oxides that give darker, stronger, or redder color.

bar A unit of pressure equal to 1 million dynes per square centimeter (10^6 dynes/cm^2). It approximates the pressure of a standard atmosphere.

base-forming cations (Obsolete) Those cations that form strong (strongly dissociated) bases by reaction with hydroxyl; e.g., K^+ forms potassium hydroxide ($K^+ + OH^-$). *See* nonacid cations.

base saturation percentage The percentage of the total cation exchange capacity saturated with exchangeable cations other than hydrogen and aluminum. *See* nonacid saturation.

bedding (Engineering) Arranging the surface of fields by plowing and grading into a series of elevated beds separated by shallow depressions or ditches for drainage.

bedrock The solid rock underlying soils and the regolith in depths ranging from zero (where exposed by erosion) to several hundred feet.

bench terrace An embankment constructed across sloping fields with a steep drop on the downslope side.

bioaccumulation A buildup within an organism of specific compounds due to biological processes. Commonly applied to heavy metals, pesticides, or metabolites.

bioaugmentation The cleanup of contaminated soils by adding exotic microorganisms that are especially efficient at breaking down an organic contaminant. A form of *bioremediation.*

biodegradable Subject to degradation by biochemical processes.

biological nitrogen fixation Occurs at ordinary temperatures and pressures. It is commonly carried out by certain bacteria, algae, and actinomycetes, which may or may not be associated with higher plants.

biomass The total mass of living material of a specified type (e.g., microbial biomass) in a given environment (e.g., in a cubic meter of soil).

biopores Soil pores, usually of relatively large diameter, created by plant roots, earthworms, or other soil organisms.

bioremediation The decontamination or restoration of polluted or degraded soils by means of enhancing the chemical degradation or other activities of soil organisms.

biosequence A group of related soils that differ, one from the other, primarily because of differences in kinds and numbers of plants and soil organisms as a soil-forming factor.

biosolids Sewage sludge that meets certain regulatory standards, making it suitable for land application. *See* sewage sludge.

biostimulation The cleanup of contaminated soils through the manipulation of nutrients or other soil environmental factors to enhance the activity of naturally occurring soil microorganisms. A form of *bioremediation.*

blocky soil structure Soil aggregates with blocklike shapes; common in B horizons of soils in humid regions.

broad-base terrace A low embankment with such gentle slopes that it can be farmed, constructed across sloping fields to reduce erosion and runoff.

broadcast Scatter seed or fertilizer on the surface of the soil.

brownfields Abandoned, idled, or underused industrial and commercial facilities where expansion or redevelopment is complicated by real or perceived environmental contamination.

buffering capacity The ability of a soil to resist changes in pH. Commonly determined by presence of clay, humus, and other colloidal materials.

bulk blended fertilizers Solid fertilizer materials blended together in small blending plants, delivered to the farm in bulk, and usually spread directly on the fields by truck or other special applicator.

bulk blending Mixing dry individual granulated fertilizer materials to form a mixed fertilizer that is applied promptly to the soil.

bulk density, soil The mass of dry soil per unit of bulk volume, including the air space. The bulk volume is determined before drying to constant weight at 105°C.

buried soil Soil covered by an alluvial, loessal, or other deposit, usually to a depth greater than the thickness of the solum.

bypass flow *See* preferential flow.

C horizon A mineral horizon, generally beneath the solum, that is relatively unaffected by biological activity and pedogenesis and is lacking properties diagnostic of an A or B horizon. It may or may not be like the material from which the A and B have formed.

calcareous soil Soil containing sufficient calcium carbonate (often with magnesium carbonate) to effervesce visibly when treated with cold 0.1 N hydrochloric acid.

calcic horizon A diagnostic subsurface horizon of secondary carbonate enrichment that is more than 15 cm thick, has a calcium carbonate equivalent of more than 15%, and has at least 5% more calcium carbonate equivalent than the underlying C horizon.

caliche A layer near the surface, more or less cemented by secondary carbonates of calcium or magnesium precipitated from the soil solution. It may occur as a soft, thin soil horizon; as a hard, thick bed just beneath the solum; or as a surface layer exposed by erosion.

cambic horizon A diagnostic subsurface horizon that has a texture of loamy very fine sand or finer, contains some weatherable minerals, and is characterized by the alteration or removal of mineral material. The cambic horizon lacks cementation or induration and has too few evidences of illuviation to meet the requirements of the argillic or spodic horizon.

capillary conductivity (Obsolete) *See* hydraulic conductivity.

capillary fringe A zone in the soil just above the plane of zero water pressure (water table) that remains saturated or almost saturated with water.

capillary water The water held in the capillary or *small* pores of a soil, usually with a tension >60 cm of water. *See also* soil water potential.

carbon cycle The sequence of transformations whereby carbon dioxide is fixed in living organisms by photosynthesis or by chemosynthesis, liberated by respiration and by the death and decomposition of the fixing organism, used by heterotrophic species, and ultimately returned to its original state.

carbon/nitrogen ratio The ratio of the weight of organic carbon (C) to the weight of total nitrogen (N) in a soil or in organic material.

carnivore An organism that feeds on animals.

casts, earthworm Rounded, water-stable aggregates of soil that have passed through the gut of an earthworm.

cat clays Wet clay soils high in reduced forms of sulfur that, upon being drained, become extremely acid because of the oxidation of the sulfur compounds and the formation of sulfuric acid. Usually found in tidal marshes. *See* acid sulfate soils.

catena A group of soils that commonly occur together in a landscape, each characterized by a different slope position and resulting set of drainage-related proprieties. *See also* toposequence.

cation A positively charged ion; during electrolysis it is attracted to the negatively charged cathode.

cation exchange The interchange between a cation in solution and another cation on the surface of any surface-active material, such as clay or organic matter.

cation exchange capacity (CEC) The sum total of exchangeable cations that a soil can adsorb. Sometimes called *total-exchange capacity, base-exchange capacity,* or *cation-adsorption capacity.* Expressed in centimoles of charge per kilogram ($cmol_c/kg$) of soil (or of other adsorbing material, such as clay).

cemented Indurated; having a hard, brittle consistency because the particles are held together by cementing substances, such as humus, calcium carbonate, or the oxides of silicon, iron, and aluminum.

channery Thin, flat fragments of limestone, sandstone, or schist up to 15 cm (6 in.) in major diameter.

chelate (Greek, claw) A type of chemical compound in which a metallic ion is firmly combined with an organic molecule by means of multiple chemical bonds.

chert A structureless form of silica, closely related to flint, that breaks into angular fragments.

chisel, subsoil A tillage implement with one or more cultivator-type feet to which are attached strong knifelike units used to shatter or loosen hard, compact layers, usually in the subsoil, to depths below normal plow depth. *See also* subsoiling.

chlorite A 2:1:1-type layer-structured silicate mineral having 2:1 layers alternating with a magnesium-dominated octahedral sheet.

chlorosis A condition in plants relating to the failure of chlorophyll (the green coloring matter) to develop. Chlorotic leaves range from light green through yellow to almost white.

chroma (color) *See* Munsell color system.

chronosequence A sequence of related soils that differ, one from the other, in certain properties primarily as a result of time as a soil-forming factor.

classification, soil *See* soil classification.

clay (1) A soil separate consisting of particles <0.002 mm in equivalent diameter. (2) A soil textural class containing >40% clay, <45% sand, and <40% silt.

clay mineral Naturally occurring inorganic material (usually crystalline) found in soils and other earthy deposits, the particles being of clay size, that is, <0.002 mm in diameter.

claypan A dense, compact, slowly permeable layer in the subsoil having a much higher clay content than the overlying material, from which it is separated by a sharply defined boundary. Claypans are usually hard when dry and plastic and sticky when wet. *See also* hardpan.

climosequence A group of related soils that differ, one from another, primarily because of differences in climate as a soil-forming factor.

clod A compact, coherent mass of soil produced artificially, usually by such human activities as plowing and digging, especially when these operations are performed on soils that are either too wet or too dry for normal tillage operations.

coarse fragments Mineral (rock) soil particles larger than 2 mm in diameter. Compare to fine earth fraction.

coarse texture The texture exhibited by sands, loamy sands, and sandy loams (except very fine sandy loam).

cobblestone Rounded or partially rounded rock or mineral fragments 7.5 to 25 cm (3 to 10 in.) in diameter.

cocomposting A method of composting in which two materials of differing but complementary nature are mingled together and enhance each other's decomposition in a compost system.

cohesion Holding together: force holding a solid or liquid together, owing to attraction between like molecules. Decreases with rise in temperature.

collapsible soil Certain soil that may undergo a sudden loss in strength when wetted.

colloid, soil (Greek, *gluelike*) Organic and inorganic matter with very small particle size and a correspondingly large surface area per unit of mass.

colluvium A deposit of rock fragments and soil material accumulated at the base of steep slopes as a result of gravitational action.

color The property of an object that depends on the wavelength of light it reflects or emits.

columnar soil structure *See* soil structure types.

companion planting The practice of growing certain species of plants in close proximity because one species has the effect of improving the growth of the other, sometimes by positive *allelopathic* effects.

compost Organic residues, or a mixture of organic residues and soil, that have been piled, moistened, and allowed to undergo biological decomposition. Mineral fertilizers are sometimes added. Often called *artificial manure* or *synthetic manure* if produced primarily from plant residues.

concretion A local concentration of a chemical compound, such as calcium carbonate or iron oxide, in the form of grains or nodules of varying size, shape, hardness, and color.

conduction The transfer of heat by physical contact between two or more objects.

conductivity, hydraulic *See* hydraulic conductivity.

conservation tillage *See* tillage, conservation.

consistence The combination of properties of soil material that determines its resistance to crushing and its ability to be molded or changed in shape. Such terms as *loose, friable, firm, soft, plastic,* and *sticky* describe soil consistence.

consistency The interaction of adhesive and cohesive forces within a soil at various moisture contents as expressed by the relative ease with which the soil can be deformed or ruptured.

consociation *See* soil consociation.

consolidation test A laboratory test in which a soil mass is laterally confined within a ring and is compressed with a known force between two porous plates.

constant charge The net surface charge of mineral particles, the magnitude of which depends only on the chemical and structural composition of the mineral. The charge arises from isomorphous substitution and is not affected by soil pH.

consumptive use The water used by plants in transpiration and growth, plus water vapor loss from adjacent soil or snow, or from intercepted precipitation in any specified time. Usually expressed as equivalent depth of free water per unit of time.

contour An imaginary line connecting points of equal elevation on the surface of the soil. A contour terrace is laid out on a sloping soil at right angles to the direction of the slope and nearly level throughout its course.

contour strip-cropping Layout of crops in comparatively narrow strips in which the farming operations are performed approximately on the contour. Usually strips of grass, close–growing crops, or fallow are alternated with those of cultivated crops.

controlled traffic A farming system in which all wheeled traffic is confined to fixed paths so that repeated compaction of the soil does not occur outside the selected paths.

convection The transfer of heat through a gas or solution because of molecular movement.

cover crop A close-growing crop grown primarily for the purpose of protecting and improving soil between periods of regular crop production or between trees and vines in orchards and vineyards.

creep Slow mass movement of soil and soil material down relatively steep slopes, primarily under the influence of gravity, but facilitated by saturation with water and by alternate freezing and thawing.

crop rotation A planned sequence of crops growing in a regularly recurring succession on the same area of land, as contrasted to continuous culture of one crop or growing different crops in haphazard order.

crotovina A former animal burrow in one soil horizon that has been filled with organic matter or material from another horizon (also spelled *krotovina*).

crumb A soft, porous, more or less rounded natural unit of structure from 1 to 5 mm in diameter. *See also* soil structure types.

crushing strength The force required to crush a mass of dry soil or, conversely, the resistance of the dry soil mass to crushing. Expressed in units of force per unit area (pressure).

crust (soil) (i) physical A surface layer on soils, ranging in thickness from a few millimeters to as much as 3 cm, that physical-chemical processes have caused to be much more compact, hard and brittle when dry than the material immediately beneath it. **(ii) microbiotic** An assemblage of cyanobacteria, algae, lichens, liverworts and mosses that commonly forms an irregular crust on the soil surface, especially on otherwise barren, arid-region soils. Also referred to as cryptogamic, cryptobiotic or biological crusts.

cryophilic Pertaining to low temperatures in the range of 5° to 15°C, the range in which cryophilic organisms grow best.

cryoturbation Physical disruption and displacement of soil material within the profile by the forces of freezing and thawing. Sometimes called *frost churning,* it results in irregular, broken horizons, involutions, oriented rock fragments, and accumulation of organic matter on the permafrost table.

crystal A homogeneous inorganic substance of definite chemical composition bounded by plane surfaces that form definite angles with each other, thus giving the substance a regular geometrical form.

crystal structure The orderly arrangement of atoms in a crystalline material.

cultivation A tillage operation used in preparing land for seeding or transplanting or later for weed control and for loosening the soil.

cutans A modification of the texture, structure, or fabric at natural surfaces in soil materials due to concentration of particular soil constituents.

cyanobacteria Chlorophyll-containing bacteria that accommodate both photosynthesis and nitrogen fixation. Formerly called blue-green algae.

deciduous plant A plant that sheds all its leaves every year at a certain season.

decomposition Chemical breakdown of a compound (e.g., a mineral or organic compound) into simpler compounds, often accomplished with the aid of microorganisms.

deflocculate (1) To separate the individual components of compound particles by chemical and/or physical means. (2) To cause the particles of the *disperse phase* of a colloidal system to become suspended in the *dispersion medium.*

delineation (soil) An individual polygon shown by a closed boundary on a soil map that defines the area, shape, and location of a map unit within a landscape.

delta An alluvial deposit formed where a stream or river drops its sediment load upon entering a quieter body of water.

denitrification The biochemical reduction of nitrate or nitrite to gaseous nitrogen, either as molecular nitrogen or as an oxide of nitrogen.

density *See* particle density; bulk density.

desalinization Removal of salts from saline soil, usually by leaching.

desert crust A hard layer, containing calcium carbonate, gypsum, or other binding material, exposed at the surface in desert regions.

desert varnish A thin, dark, shiny film or coating of iron oxide and lesser amounts of manganese oxide and silica formed on the surfaces of pebbles, boulders, rock fragments, and rock outcrops in arid regions.

desert pavement A natural residual concentration of closely packed pebbles, boulders, and other rock fragments on a desert surface where wind and water action has removed all smaller particles.

desorption The removal of sorbed material from surfaces.

detritivore An organism that subsists on detritus.

detritus Debris from dead plants and animals.

diagnostic horizons (As used in *Soil Taxonomy*): Horizons having specific soil characteristics that are indicative of certain classes of soils. Horizons that occur at the soil surface are called *epipedons;* those below the surface, *diagnostic subsurface horizons.*

diatomaceous earth A geologic deposit of fine, grayish, siliceous material composed chiefly or wholly of the remains of diatoms. It may occur as a powder or as a porous, rigid material.

diatoms Algae having siliceous cell walls that persist as a skeleton after death; any of the microscopic unicellular or colonial algae constituting the class Bacillariaceae. They occur abundantly in fresh and salt waters and their remains are widely distributed in soils.

diffusion The movement of atoms in a gaseous mixture or of ions in a solution, primarily as a result of their own random motion.

dioctahedral sheet An octahedral sheet of silicate clays in which the sites for the six-coordinated metallic atoms are mostly filled with trivalent atoms, such as Al^{3+}.

disintegration Physical or mechanical breakup or separation of a substance into its component parts (e.g., a rock breaking into its mineral components).

disperse (1) To break up compound particles, such as aggregates, into the individual component particles. (2) To distribute or suspend fine particles, such as clay, in or throughout a dispersion medium, such as water.

distribution coefficient (Kd) The distribution of a chemical between soil and water.

dissolution Process by which molecules of a gas, solid, or another liquid dissolve in a liquid, thereby becoming completely and uniformly dispersed throughout the liquid's volume.

diversion terrace *See* terrace.

drain (1) To provide channels, such as open ditches or drain tile, so that excess water can be removed by surface or by internal flow. (2) To lose water (from the soil) by percolation.

drain field, septic tank An area of soil into which the effluent from a septic tank is piped so that it will drain through the lower part of the soil profile for disposal and purification.

drainage, soil The frequency and duration of periods when the soil is free from saturation with water.

drift Material of any sort deposited by geological processes in one place after having been removed from another. Glacial drift includes material moved by the glaciers and by the streams and lakes associated with them.

drumlin Long, smooth, cigar-shaped low hills of glacial till, with their long axes parallel to the direction of ice movement.

dryland farming The practice of crop production in low-rainfall areas without irrigation.

duff The matted, partly decomposed organic surface layer of forest soils.

duripan A diagnostic subsurface horizon that is cemented by silica, to the point that air-dry fragments will not slake in water or HCl. Hardpan.

dust mulch A loose, finely granular or powdery condition on the surface of the soil, usually produced by shallow cultivation.

E horizon Horizon characterized by maximum illuviation (washing out) of silicate clays and iron and aluminum oxides; commonly occurs above the B horizon and below the A horizon.

earthworms Animals of the Lumbricidae family that burrow into and live in the soil. They mix plant residues into the soil and improve soil aeration.

ectotrophic mycorrhiza (ectomycorrhiza) A symbiotic association of the mycelium of fungi and the roots of certain plants in which the fungal hyphae form a compact mantle on the surface of the roots and extend into the surrounding soil and inward between cortical cells, but not into these cells. Associated primarily with certain trees. *See also* endotrophic mycorrhiza.

edaphology The science that deals with the influence of soils on living things, particularly plants, including human use of land for plant growth.

effective cation exchange capacity The amount of cation charges that a material (usually soil or soil colloids) can hold at the pH of the material, measured as the sum of the exchangeable Al^{3+}, Ca^{2+}, Mg^{2+}, K^{+}, and Na^{+}, and expressed as moles or cmol of charge per kg of material.

effective precipitation That portion of the total precipitation that becomes available for plant growth or for the promotion of soil formation.

Eh In soils, it is the potential created by oxidation-reduction reactions that take place on the surface of a platinum electrode measured against a reference electrode, minus the Eh of the reference electrode. This is a measure of the oxidation-reduction potential of electrode-reactive components in the soil. *See also* pe.

electrical conductivity (EC) The capacity of a substance to conduct or transmit electrical current. In soils or water, measured in siemens/meter, and related to dissolved solutes.

eluviation The removal of soil material in suspension (or in solution) from a layer or layers of a soil. (Usually, the loss of material in *solution* is described by the term *leaching*.) *See also* leaching.

endoaquic (endosaturation) A condition or moisture regime in which the soil is saturated with water in all layers from the upper boundary of saturation (water table) to a depth of 200 cm or more from the mineral soil surface. *See also* epiaquic.

endotrophic mycorrhiza (endomycorrhiza) A symbiotic association of the mycelium of fungi and roots of a variety of plants in which the fungal hyphae penetrate directly into root hairs, other epidermal cells, and occasionally into cortical cells. Individual hyphae also extend from the root surface outward into the surrounding soil. *See also* arbuscular mycorrhiza.

Entisols An order in *Soil Taxonomy.* Soils have no diagnostic pedogenic horizons. They may be found in virtually any climate on very recent geomorphic surfaces.

eolian soil material Soil material accumulated through wind action. The most extensive areas in the United States are silty deposits (loess), but large areas of sandy deposits also occur.

epipedon A diagnostic surface horizon that includes the upper part of the soil that is darkened by organic matter, or the upper eluvial horizons, or both. *(Soil Taxonomy.)*

equilibrium phosphorus concentration (EPC) The concentration of phosphorus in a solution in equilibrium with a soil, the EPC_O being the concentration of phosphorus achieved by desorption of phosphorus from a soil to phosphorus-free distilled water.

epiaquic (episaturation) A condition in which the soil is saturated with water due to a perched water table in one or more layers within 200 cm of the mineral soil surface, implying that there are also one or more unsaturated layers within 200 cm below the saturated layer. *See also* endoaquic.

erosion (1) The wearing away of the land surface by running water, wind, ice, or other geological agents, including such processes as gravitational creep. (2) Detachment and movement of soil or rock by water, wind, ice, or gravity.

esker A narrow ridge of gravelly or sandy glacial material deposited by a stream in an ice-walled valley or tunnel in a receding glacier.

essential element A chemical element required for the normal growth of plants.

eukaryote An organism whose cells each have a visibly evident nucleus.

eutrophic Having concentrations of nutrients optimal (or nearly so) for plant or animal growth. (Said of nutrient solutions or of soil solutions.)

eutrophication Nutrient enrichment of lakes, ponds, and other such waters that stimulates the growth of aquatic organisms, which leads to a deficiency of oxygen in the water body.

evapotranspiration The combined loss of water from a given area, and during a specified period of time, by evaporation from the soil surface and by transpiration from plants.

exchange capacity The total ionic charge of the adsorption complex active in the adsorption of ions. *See also* anion exchange capacity; cation exchange capacity.

exchangeable ions Positively or negatively charged atoms or groups of atoms that are held on or near the surface of a solid particle by attraction to charges of the opposite sign, and which may be replaced by other like-charged ions in the soil solution.

exchangeable sodium percentage The extent to which the adsorption complex of a soil is occupied by sodium. It is expressed as follows:

$$ESP = \frac{\text{exchangeable sodium (cmol}_c\text{/kg soil)}}{\text{cation exchange capacity (cmol}_c\text{/kg soil)}} \times 100$$

exfoliation Peeling away of layers of a rock from the surface inward, usually as the result of expansion and contraction that accompany changes in temperature.

expansive soil Soil that undergoes significant volume change upon wetting and drying, usually because of a high content of swelling-type clay minerals.

external surface The area of surface exposed on the top, bottom, and sides of a clay crystal or micelle.

facultative organism An organism capable of both aerobic and anaerobic metabolism.

fallow Cropland left idle to restore productivity, mainly through accumulation of nutrients, water, and/or organic matter. Preceding a cereal grain crop in semiarid regions, land is commonly left in *summer fallow* for a period during which weeds are controlled by chemicals or tillage and water is allowed to accumulate in the soil profile. In humid regions, fallow land may be allowed to grow up in natural vegetation for a period ranging from a few months to many years. *Improved fallow* involves the purposeful establishment of plant species capable of restoring soil productivity more rapidly than a natural plant succession.

family, soil In *Soil Taxonomy,* one of the categories intermediate between the great group and the soil series. Families are defined largely on the basis of physical and mineralogical properties of importance to plant growth.

fauna The animal life of a region or ecosystem.

fen A calcium-rich, peat-accumulating wetland with relatively stagnant water.

ferrihydrite, $Fe_5HO_8 \cdot 4H_2O$ A dark red-brown poorly crystalline iron oxide that forms in wet soils.

fertigation The application of fertilizers in irrigation waters, commonly through sprinkler systems.

fertility, soil The quality of a soil that enables it to provide essential chemical elements in quantities and proportions for the growth of specified plants.

fertilizer Any organic or inorganic material of natural or synthetic origin added to a soil to supply certain elements essential to the growth of plants.

fertilizer requirement The quantity of certain plant nutrient elements needed, in addition to the amount supplied by the soil, to increase plant growth to a designated optimum.

fibric materials *See* organic soil materials.

field capacity (field moisture capacity) The percentage of water remaining in a soil 2 or 3 days after its having been saturated and after free drainage has practically ceased.

fine earth fraction That portion of the soil that passes through a 2mm diameter sieve opening. *Compare to* coarse fragments.

fine texture The texture exhibited by soils in the clay loam, sandy clay loam, silty clay loam, sandy clay, silty clay, and clay textural classes.

fine-grained mica A silicate clay having a 2:1-type lattice structure with much of the silicon in the tetrahedral sheet having been replaced by aluminum and with considerable interlayer potassium, which binds the layers together, prevents interlayer expansion and swelling, and limits interlayer cation exchange capacity.

fixation (1) For other than elemental nitrogen: the process or processes in a soil by which certain chemical elements essential for plant growth are converted from a soluble or exchangeable form to a much less soluble or to a nonexchangeable form; for example, potassium, ammonium, and phosphate fixation. (2) For elemental nitrogen: process by which gaseous elemental nitrogen is chemically combined with hydrogen to form ammonia. *See* biological nitrogen fixation.

flagstone A relatively thin rock or mineral fragment 15 to 38 cm in length commonly composed of shale, slate, limestone, or sandstone.

flocculate To aggregate or clump together individual, tiny soil particles, especially fine clay, into small clumps or floccules. Opposite of *deflocculate* or *disperse.*

floodplain The land bordering a stream, built up of sediments from overflow of the stream and subject to inundation when the stream is at flood stage. Sometimes called *bottomland.*

flora The sum total of the kinds of plants in an area at one time. The organisms loosely considered to be of the plant kingdom.

fluorapatite A member of the apatite group of minerals containing fluorine. Most common mineral in phosphate rock.

fluvial deposits Deposits of parent materials laid down by rivers or streams.

fluvioglacial *See* glaciofluvial deposits.

foliar diagnosis An estimation of mineral nutrient deficiencies (excesses) of plants based on examination of the chemical composition of selected plant parts, and the color and growth characteristics of the foliage of the plants.

forest floor The forest soil O horizons, including litter and unincorporated humus, on the mineral soil surface.

fragipan Dense and brittle pan or subsurface layer in soils that owes its hardness mainly to extreme density or compactness rather than high clay content or cementation. Removed fragments are friable, but the material in place is so dense that roots penetrate and water moves through it very slowly.

friable A soil consistency term pertaining to soils that crumble with ease.

frigid A soil temperature class with mean annual temperature below 8°C.

fritted micronutrients Sintered silicates having total guaranteed analyses of micronutrients with controlled (relatively slow) release characteristics.

fulvic acid A term of varied usage but usually referring to the mixture of organic substances remaining in solution upon acidification of a dilute alkali extract from the soil.

functional diversity The characteristic of an ecosystem exemplified by the capacity to carry out a large number of biochemical transformations and other functions.

functional group An atom, or group of atoms, attached to a large molecule. Each functional group (e.g., —OH, —CH_3, —COOH) has a characteristic chemical reactivity.

fungi Eukaryote microorganisms with a rigid cell wall. Some form long filaments of cells called *hyphae* that may grow together to form a visible body.

furrow slice The uppermost layer of an arable soil to the depth of primary tillage; the layer of soil sliced away from the rest of the profile and inverted by a moldboard plow.

gamma ray A high-energy ray (photon) emitted during radioactive decay of certain elements.

Gelisols An order in *Soil Taxonomy.* Soils that have permafrost within the upper 1 m, or upper 2 m if cryoturbation is also present. They may have an ochric, histic, mollic, or other epipedon.

gellic materials Mineral or organic soil materials that have *cryoturbation* and/or ice in the form of lenses, veins, or wedges and the like.

genesis, soil The mode of origin of the soil, with special reference to the processes responsible for the development of the solum, or true soil, from the unconsolidated parent material.

genetic horizon Soil layers that resulted from soil-forming (pedogenic) processes, as opposed to sedimentation or other geologic processes.

geographic information system (GIS) A method of overlaying, statistically analyzing and integrating large volumes of spatial data of different kinds. The data are referenced to geographical coordinates and encoded in a form suitable for handling by computer.

geological erosion Wearing away of the earth's surface by water, ice, or other natural agents under natural environmental conditions of climate, vegetation, and so on, undisturbed by man. Synonymous with *natural erosion.*

gibbsite, $Al(OH)_3$ An aluminum trihydroxide mineral most common in highly weathered soils, such as oxisols.

gilgai The microrelief of soils produced by expansion and contraction with changes in moisture. Found in soils that contain large amounts of clay that swells and shrinks considerably with wetting and drying. Usually a succession of microbasins and microknolls in nearly level areas or of microvalleys and microridges parallel to the direction of the slope.

glacial drift Rock debris that has been transported by glaciers and deposited, either directly from the ice or from the meltwater. The debris may or may not be heterogeneous.

glacial till *See* till.

glaciofluvial deposits Material moved by glaciers and subsequently sorted and deposited by streams flowing from the melting ice. The deposits are stratified and may occur in the form of outwash plains, deltas, kames, eskers, and kame terraces.

gleyed A soil condition resulting from prolonged saturation with water and reducing conditions that manifest themselves in greenish or bluish colors throughout the soil mass or in mottles.

glomalin A protein-sugar complex substance secreted by certain fungi resulting in a sticky hyphal surface thought to contribute to aggregate stability.

goethite, FeOOH A yellow-brown iron oxide mineral that accounts for the brown color in many soils.

granular structure Soil structure in which the individual grains are grouped into spherical aggregates with indistinct sides. Highly porous granules are commonly called *crumbs.* A well-granulated soil has the best structure for most ordinary crop plants. *See also* soil structure types.

granulation The process of producing granular materials. Commonly used to refer to the formation of soil structural granules, but also used to refer to the processing of powdery fertilizer materials into granules.

grassed waterway A natural or constructed waterway covered with erosion-resistant grasses that permits removal of runoff water without excessive erosion.

gravitational potential That portion of the total *soil water potential* due to differences in elevation of the reference pool of pure water and that of the soil water. Since the soil water elevation is usually chosen to be higher than that of the reference pool, the gravitational potential is usually positive.

gravitational water Water that moves into, through, or out of the soil under the influence of gravity.

great group A category in *Soil Taxonomy.* The classes in this category contain soils that have the same kind of horizons in the same sequence and have similar moisture and temperature regimes.

green manure Plant material incorporated with the soil while green, or soon after maturity, for improving the soil.

greenhouse effect The entrapment of heat by upper atmosphere gases, such as carbon dioxide, water vapor, and methane, just as glass traps heat for a greenhouse. Increases in the quantities of these gases in the atmosphere will likely result in global warming that may have serious consequences for humankind.

groundwater Subsurface water in the zone of saturation that is free to move under the influence of gravity, often horizontally to stream channels.

gully erosion The erosion process whereby water accumulates in narrow channels and, over short periods, removes the soil from this narrow area to considerable depths, ranging from 1 to 2 ft to as much as 23 to 30 m (75 to 100 ft).

gypsic horizon A diagnostic subsurface horizon of secondary calcium sulfate enrichment that is more than 15 cm thick.

gypsum requirement The quantity of gypsum required to reduce the exchangeable sodium percentage in a soil to an acceptable level.

halophyte A plant that requires or tolerates a saline (high salt) environment.

hardpan A hardened soil layer, in the lower A or in the B horizon, caused by cementation of soil particles with organic matter or with such materials as silica, sesquioxides, or calcium carbonate. The hardness does not change appreciably with changes in moisture content and pieces of the hard layer do not slake in water. *See also* caliche; claypan.

harrowing A secondary broadcast tillage operation that pulverizes, smooths, and firms the soil in seedbed preparation, controls weeds, or incorporates material spread on the surface.

heaving The partial lifting of plants, buildings, roadways, fenceposts, etc., out of the ground, as a result of freezing and thawing of the surface soil during the winter.

heavy metals Those metals that have densities of 5.0 Mg/m or greater. Elements in soils include Cd, Co, Cr, Cu, Fe, Hg, Mn, Mo, Pb, and Zn.

heavy soil (Obsolete in scientific use) A soil with a high content of the fine separates, particularly clay, or one with a high drawbar pull, hence difficult to cultivate.

hematite, Fe_2O_3 A red iron oxide mineral that contributes red color to many soils.

hemic material Organic soil material at an intermediate degree of decomposition that contains 1/6 to 3/4 recognizable fibers (after rubbing) of undecomposed plant remains. Bulk density is usually very low, and water holding capacity very high.

herbicide A chemical that kills plants or inhibits their growth; intended for weed control.

herbivore A plant-eating animal.

heterotroph An organism capable of deriving energy for life processes only from the decomposition of organic compounds and incapable of using inorganic compounds as sole sources of energy or for organic synthesis. *Contrast with* autotroph.

histic epipedon A diagnostic surface horizon consisting of a thin layer of organic soil material that is saturated with water at some period of the year unless artificially drained and that is at or near the surface of a mineral soil.

Histosols An order in *Soil Taxonomy.* Soils formed from materials high in organic matter. Histosols with essentially no clay must have at least 20% organic matter by weight (about 78% by volume). This minimum organic matter content rises with increasing clay content to 30% (85% by volume) in soils with at least 60% clay.

horizon, soil A layer of soil, approximately parallel to the soil surface, differing in properties and characteristics from adjacent layers below or above it. *See also* diagnostic horizons.

horticulture The art and science of growing fruits, vegetables, and ornamental plants.

hue (color) *See* Munsell color system.

humic acid A mixture of variable or indefinite composition of dark organic substances, precipitated upon acidification of a dilute alkali extract from soil.

humic substances A series of complex, relatively high molecular weight, brown to black organic substances that make up 60 to 80% of the soil organic matter and are generally quite resistant to ready microbial attack.

humid climate Climate in regions where moisture, when distributed normally throughout the year, should not limit crop production. In cool climates annual precipitation may be as little as 25 cm; in hot climates, 150 cm or even more. Natural vegetation in uncultivated areas is forests.

humification The processes involved in the decomposition of organic matter and leading to the formation of humus.

humin The fraction of the soil organic matter that is not dissolved upon extraction of the soil with dilute alkali.

humus That more or less stable fraction of the soil organic matter remaining after the major portions of added plant and animal residues have decomposed. Usually it is dark in color.

hydration Chemical union between an ion or compound and one or more water molecules, the reaction being stimulated by the attraction of the ion or compound for either the hydrogen or the unshared electrons of the oxygen in the water.

hydraulic conductivity An expression of the readiness with which a liquid, such as water, flows through a solid, such as soil, in response to a given potential gradient.

hydric soils Soils that are water-saturated for long enough periods to produce reduced conditions and affect the growth of plants.

hydrogen bonding Relatively low energy bonding exhibited by a hydrogen atom located between two highly electronegative atoms, such as nitrogen or oxygen.

hydrologic cycle The circuit of water movement from the atmosphere to the earth and back to the atmosphere through various stages or processes, as precipitation, interception, runoff, infiltration, percolation, storage, evaporation, and transpiration.

hydrolysis A reaction with water that splits the water molecule into H^+ and OH^- ions. Molecules or atoms participating in such reactions are said to *hydrolyze*.

hydronium A hydrated hydrogen ion (H_3O^+), the form of hydrogen ion usually found in aqueous systems.

hydroponics Plant-production systems that use nutrient solutions and no solid medium to grow plants.

hydrous mica *See* fine-grained mica.

hydroxyapatite A member of the apatite group of minerals rich in hydroxyl groups. A nearly insoluble calcium phosphate.

hygroscopic coefficient The amount of moisture in a dry soil when it is in equilibrium with some standard relative humidity near a saturated atmosphere (about 98%), expressed in terms of percentage on the basis of oven-dry soil.

hyperthermic A soil temperature class with mean annual temperatures >22°C.

hypha (pl. hyphae) Filament of fungal cells. Actinomycetes also produce similar, but thinner, filaments of cells.

hypoxia State of oxygen deficiency in an environment so low as to restrict biological respiration (in water, typically less than 2 to 3 mg O_2/L).

hysteresis A relationship between two variables that changes depending on the sequences or starting point. An example is the relationship between soil water content and water potential, for which different curves describe the relationship when a soil is gaining water or losing it.

igneous rock Rock formed from the cooling and solidification of magma that has not been changed appreciably since its formation.

illite *See* fine-grained mica.

illuvial horizon A soil layer or horizon in which material carried from an overlying layer has been precipitated from solution or deposited from suspension. The layer of accumulation.

illuviation The process of deposition of soil material removed from one horizon to another in the soil; usually from an upper to a lower horizon in the soil profile. *See also* eluviation.

immature soil A soil with indistinct or only slightly developed horizons because of the relatively short time it has been subjected to the various soil-forming processes. A soil that has not reached equilibrium with its environment.

immobilization The conversion of an element from the inorganic to the organic form in microbial tissues or in plant tissues, thus rendering the element not readily available to other organisms or to plants.

imogolite A poorly crystalline aluminosilicate mineral with an approximate formula $SiO_2Al_2O_3 \cdot 2.5H_2O$; occurs mostly in soils formed from volcanic ash.

impervious Resistant to penetration by fluids or by roots.

improved fallow *See* fallow.

Inceptisols An order in *Soil Taxonomy.* Soils that are usually moist with pedogenic horizons of alteration of parent materials but not of illuviation. Generally, the direction of soil development is not yet evident from the marks left by various soil-forming processes or the marks are too weak to classify in another order.

induced systemic resistance Plant defense mechanisms activated by a chemical signal produced by a rhizosphere bacteria. Although the process begins in the soil, it may confer disease resistance to leaves or other aboveground tissues.

indurated (soil) Soil material cemented into a hard mass that will not soften on wetting. *See also* consistence; hardpan.

infiltration The downward entry of water into the soil.

infiltration capacity A soil characteristic determining or describing the *maximum* rate at which water *can* enter the soil under specified conditions, including the presence of an excess of water.

inner-sphere complex A relatively strong (not easily reversed) chemical association or bonding directly between a specific ion and specific atoms or groups of atoms in the surface structure of a soil colloid.

inoculation The process of introducing pure or mixed cultures of microorganisms into natural or artificial culture media.

inorganic compounds All chemical compounds in nature except compounds of carbon other than carbon monoxide, carbon dioxide, and carbonates.

insecticide A chemical that kills insects.

intergrade A soil that possesses moderately well-developed distinguishing characteristics of two or more genetically related great soil groups.

interlayer (mineralogy) Materials between layers within a given crystal, including cations, hydrated cations, organic molecules, and hydroxide groups or sheets.

internal surface The area of surface exposed within a clay crystal or micelle between the individual crystal layers. *Compare with* external surface.

interped pores Voids and cracks between aggregates or other structural units of soil.

interstratification Mixing of silicate layers within the structural framework of a given silicate clay.

ions Atoms, groups of atoms, or compounds that are electrically charged as a result of the loss of electrons (cations) or the gain of electrons (anions).

iron-pan An indurated soil horizon in which iron oxide is the principal cementing agent.

irrigation efficiency The ratio of the water actually consumed by crops on an irrigated area to the amount of water diverted from the source onto the area.

isomorphous substitution The replacement of one atom by another of similar size in a crystal lattice without disrupting or changing the crystal structure of the mineral.

isotopes Two or more atoms of the same element that have different atomic masses because of different numbers of neutrons in the nucleus.

joule The SI energy unit defined as a force of 1 newton applied over a distance of 1 meter.

kame A conical hill or ridge of sand or gravel deposited in contact with glacial ice.

kandic horizon A subsurface diagnostic horizon having a sharp clay increase relative to overlying horizons and having low-activity clays.

kaolinite An aluminosilicate mineral of the 1:1 crystal lattice group; that is, consisting of single silicon tetrahedral sheets alternating with single aluminum octahedral sheets.

Kd *See* distribution coefficient, Kd.

keystone species A few species of organisms which play unique roles that greatly influence the habitat and function of many other species in a community or ecosystem. For example, nitrifying bacteria, or burrowing earthworms.

Koc The distribution coefficient, Kd, calculated based on organic carbon content. Koc = Kd/foc where foc is the fraction of organic carbon.

labile A substance that is readily transformed by microorganisms or is readily available for uptake by plants.

lacustrine deposit Material deposited in lake water and later exposed either by lowering of the water level or by the elevation of the land.

land A broad term embodying the total natural environment of the areas of the earth not covered by water. In addition to soil, its attributes include other physical conditions, such as mineral deposits and water supply; location in relation to centers of commerce, populations, and other land; the size of the individual tracts or holdings; and existing plant cover, works of improvement, and the like.

land capability classification A grouping of kinds of soil into special units, subclasses, and classes according to their capability for intensive use and the treatments required for sustained use. One such system has been prepared by the USDA Natural Resources Conservation Service.

land classification The arrangement of land units into various categories based upon the properties of the land or its suitability for some particular purpose.

land forming Shaping the surface of the land by scraping off the high spots and filling in the low spots with precision grading machinery to create a uniform, smooth slope, often for irrigation purposes. Also called *land smoothing.*

land-use planning The development of plans for the uses of land that, over long periods, will best serve the general welfare, together with the formulation of ways and means for achieving such uses.

laterite An iron-rich subsoil layer found in some highly weathered humid tropical soils that, when exposed and allowed to dry, becomes very hard and will not soften when rewetted. When erosion removes the overlying layers, the laterite is exposed and a virtual pavement results. *See also* plinthite.

layer (Clay mineralogy) A combination in silicate clays of (tetrahedral and octahedral) sheets in a 1:1, 2:1, or 2:1:1 combination.

leaching The removal of materials in solution from the soil by percolating waters. *See also* eluviation.

leaching requirement The leaching fraction of irrigation water necessary to keep soil salinity from exceeding a tolerance level of the crop to be grown.

legume A pod-bearing member of the Leguminosae family, one of the most important and widely distributed plant families. Includes many valuable food and forage species, such as peas, beans, peanuts, clovers, alfalfas, sweet clovers, lespedezas, vetches, and kudzu. Nearly all legumes are associated with nitrogen-fixing organisms.

lichen A symbiotic relationship between fungi and cyanobacteria (blue-green algae) that enhances colonization of bare minerals and rocks. The fungi supply water and nutrients, the cyanobacteria the fixed nitrogen and carbohydrates from photosynthesis.

Liebig's law The growth and reproduction of an organism are determined by the nutrient substance (oxygen, carbon dioxide, calcium, etc.) that is available in minimum quantity with respect to organic needs; the *limiting factor.*

light soil (Obsolete in scientific use) A coarse-textured soil; a soil with a low drawbar pull and hence easy to cultivate. *See also* coarse texture; soil texture.

lignin The complex organic constituent of woody fibers in plant tissue that, along with cellulose, cements the cells together and provides strength. Lignins resist microbial attack and after some modification may become part of the soil organic matter.

lime (agricultural) In strict chemical terms, calcium oxide. In practical terms, a material containing the carbonates, oxides, and/or hydroxides of calcium and/or magnesium used to neutralize soil acidity.

lime requirement The mass of specified liming material, required to raise the pH of the soil to a desired value under field conditions.

limestone A sedimentary rock composed primarily of calcite ($CaCO_3$). If dolomite ($CaCO_3 \cdot MgCO_3$) is present in appreciable quantities, it is called a *dolomitic limestone.*

limiting factor *See* Liebig's law.

liquid limit (LL) *See* Atterberg limits.

lithosequence A group of related soils that differ, one from the other, in certain properties primarily as a result of parent material as a soil-forming factor.

loam The textural-class name for soil having a moderate amount of sand, silt, and clay. Loam soils contain 7 to 27% clay, 28 to 50% silt, and 23 to 52% sand.

loamy Intermediate in texture and properties between fine-textured and coarse-textured soils. Includes all textural classes with the words *loam* or *loamy* as a part of the class name, such as clay loam or loamy sand. *See also* loam; soil texture.

lodging Falling over of plants, either by uprooting or stem breakage.

loess Material transported and deposited by wind and consisting of predominantly silt-sized particles.

luxury consumption The intake by a plant of an essential nutrient in amounts exceeding what it needs. For example, if potassium is abundant in the soil, alfalfa may take in more than it requires.

lysimeter A device for measuring percolation (leaching) and evapotranspiration losses from a column of soil under controlled conditions.

macrofauna Animals of greater than approximately 2 mm diameter.

macronutrient A chemical element necessary in large amounts (usually 50 mg/kg in the plant) for the growth of plants. Includes C, H, O, N, P, K, Ca, Mg, and S. (*Macro* refers to quantity and not to the essentiality of the element.) *See also* micronutrient.

macropores Larger soil pores, generally having a diameter greater than 0.06 mm, from which water drains readily by gravity.

map unit (mapping unit), soil A conceptual group of one to many component soils, delineated or identified by the same name in a soil survey, that represent similar landscape areas. *See also* delineation, soil consociation, soil complex, soil association, and undifferentiated group.

marl Soft and unconsolidated calcium carbonate, usually mixed with varying amounts of clay or other impurities.

marsh Periodically wet or continually flooded area with the surface not deeply submerged. Covered dominantly with sedges, cattails, rushes, or other hydrophytic plants. Subclasses include freshwater and saltwater marshes.

mass flow Movement of nutrients with the flow of water to plant roots.

matric potential That portion of the total *soil water potential* due to the attractive forces between water and soil solids as represented through adsorption and capillarity. It will always be negative.

mature soil A soil with well-developed soil horizons produced by the natural processes of soil formation and essentially in equilibrium with its present environment.

maximum retentive capacity The average moisture content of a disturbed sample of soil, 1 cm high, which is at equilibrium with a water table at its lower surface.

medium texture Intermediate between fine-textured and coarse-textured, including soils classed as: very fine sandy loam, loam, silt loam, and silt.

melanic epipedon A diagnostic surface horizon formed in volcanic parent material, that contains more than 6% organic carbon, is dark in color, and has a very low bulk density and high anion adsorption capacity.

mellow soil A very soft, very friable, porous soil without any tendency toward hardness or harshness. *See also* consistence.

mesic A soil temperature class with mean annual temperature 8° to 15°C.

mesofauna Animals of medium size, between approximately 2 and 0.2 mm in diameter.

mesophilic Pertaining to moderate temperatures in the range of 15° to 35°C, the range in which mesophilic organisms grow best and in which mesophilic composting takes place.

metamorphic rock A rock that has been greatly altered from its previous condition through the combined action of heat and pressure. For example, marble is a metamorphic rock produced from limestone, gneiss is produced from granite, and slate is produced from shale.

methane, CH_4 An odorless, colorless gas commonly produced under anaerobic conditions. When released to the upper atmosphere, methane contributes to global warming. *See also* greenhouse effect.

micas Primary aluminosilicate minerals in which two silica tetrahedral sheets alternate with one alumina/magnesia octahedral sheet with entrapped potassium atoms fitting between sheets. They separate readily into visible sheets or flakes.

microfauna That part of the animal population which consists of individuals too small to be clearly distinguished without the use of a microscope. Includes protozoans and nematodes.

microflora That part of the plant population which consists of individuals too small to be clearly distinguished without the use of a microscope. Includes actinomycetes, algae, bacteria, and fungi.

micronutrient A chemical element necessary in only extremely small amounts (<50 mg/kg in the plant) for the growth of plants. Examples are B, Cl, Cu, Fe, Mn, and Zn. (*Micro* refers to the amount used rather than to its essentiality.) *See also* macronutrient.

micropores Relatively small soil pores, generally found within structural aggregates and having a diameter less than 0.06 mm. *Contrast to* macropores.

microrelief Small-scale local differences in topography, including mounds, swales, or pits that are only 1 m or so in diameter and with elevation differences of up to 2 m. *See also* gilgai.

mineral soil A soil consisting predominantly of, and having its properties determined predominantly by, mineral matter. Usually contains <20% organic matter, but may contain an organic surface layer up to 30 cm thick.

mineralization The conversion of an element from an organic form to an inorganic state as a result of microbial decomposition.

minimum tillage *See* tillage, conservation.

minor element (Obsolete) *See* micronutrient.

moisture potential *See* soil water potential.

mole drain Unlined drain formed by pulling a bullet-shaped cylinder through the soil.

mollic epipedon A diagnostic surface horizon of mineral soil that is dark colored and relatively thick, contains at least 0.6% organic carbon, is not massive and hard when dry, has a base saturation of more than 50%, has less than 250 mg/kg P_2O_5 soluble in 1% citric acid, and is dominantly saturated with bivalent cations.

Mollisols An order in *Soil Taxonomy.* Soils with nearly black, organic-rich surface horizons and high supply of bases. They have mollic epipedons and base saturation greater than 50% in any cambic or argillic horizon. They lack the characteristics of Vertisols and must not have oxic or spodic horizons.

montmorillonite An aluminosilicate clay mineral in the smectite group with a 2:1 expanding crystal lattice, with two silicon tetrahedral sheets enclosing an aluminum octahedral sheet. Isomorphous substitution of magnesium for some of the aluminum has occurred in the octahedral sheet. Considerable expansion may be caused by water moving between silica sheets of contiguous layers.

mor Raw humus; type of forest humus layer of unincorporated organic material, usually matted or compacted or both; distinct from the mineral soil, unless the latter has been blackened by washing in organic matter.

moraine An accumulation of drift, with an initial topographic expression of its own, built within a glaciated region chiefly by the direct action of glacial ice. Examples are ground, lateral, recessional, and terminal moraines.

morphology, soil The constitution of the soil, including the texture, structure, consistence, color, and other physical, chemical, and biological properties of the various soil horizons that make up the soil profile.

mottling Spots or blotches of different color or shades of color interspersed with the dominant color.

mucigel The gelatinous material at the surface of roots grown in unsterilized soil.

muck Highly decomposed organic material in which the original plant parts are not recognizable. Contains more mineral matter and is usually darker in color than peat. *See also* muck soil; peat.

muck soil (1) A soil containing 20 to 50% organic matter. (2) An organic soil in which the organic matter is well decomposed.

mulch Any material such as straw, sawdust, leaves, plastic film, and loose soil that is spread upon the surface of the soil to protect the soil and plant roots from the effects of raindrops, soil crusting, freezing, evaporation, etc.

mulch tillage *See* tillage, conservation.

mull A humus-rich layer of forested soils consisting of mixed organic and mineral matter. A mull blends into the upper mineral layers without an abrupt change in soil characteristics.

Munsell color system A color designation system that specifies the relative degrees of the three simple variables of color:

- **chroma** The relative purity, strength, or saturation of a color.
- **hue** The chromatic gradation (rainbow) of light that reaches the eye.
- **value** The degree of lightness or darkness of the color.

mycelium A stringlike mass of individual fungal or actinomycetes hyphae.

myco Prefix designating an association or relationship with a fungus (e.g., mycotoxins are toxins produced by a fungus).

mycorrhiza The association, usually symbiotic, of fungi with the roots of seed plants. *See also* ectotrophic mycorrhiza; endotrophic mycorrhiza; arbuscular mycorrhiza.

natric horizon A diagnostic subsurface horizon that satisfies the requirements of an argillic horizon, but that also has prismatic, columnar, or blocky structure and a subhorizon having more than 15% saturation with exchangeable sodium.

necrosis Death associated with discoloration and dehydration of plant tissues such as leaf tissue.

nematodes Very small round worms abundant in many soils and important because some of them attack plant roots and others play critical food web roles.

neutral soil A soil in which the surface layer, at least to normal plow depth, is neither acid nor alkaline in reaction. In practice this means the soil is within the pH range of 6.6 to 7.3. *See also* acid soil; alkaline soil; pH; reaction, soil.

nitrate depression period A period of time, beginning shortly after the addition of fresh, highly carbonaceous organic materials to a soil, during which decomposer microorganisms have removed most of the soluble nitrate from the soil solution.

nitrification The biochemical oxidation of ammonium to nitrate, predominantly by autotrophic bacteria.

nitrogen assimilation The incorporation of nitrogen into organic cell substances by living organisms.

nitrogen cycle The sequence of chemical and biological changes undergone by nitrogen as it moves from the atmosphere into water, soil, and living organisms, and upon death of these organisms (plants and animals) is recycled through a part or all of the entire process.

nitrogen fixation The biological conversion of elemental nitrogen (N_2) to organic combinations or to forms readily utilized in biological processes.

nodule bacteria *See* rhizobia.

nonacid cations Those cations that do not react with water by hydrolysis to release H^+ ions to the soil solution. These cations do not remove hydroxyl ions from solution, but form strongly dissociated bases such as potassium hydroxide ($K^+ + OH^-$). Formerly called *base cations* or *base-forming cations* in soil science literature.

nonacid saturation The proportion or percentage of cation-exchange sites occupied by nonacid cations. Formerly termed *base saturation.*

nonhumic substances The portion of soil organic matter comprised of relatively low molecular weight organic substances; mostly identifiable biomolecules.

nonlimiting water range The region bounded by the upper and lower soil water content over which water, oxygen, and mechanical resistance are not limiting to plant growth. *Compare with* available water.

nonpoint source Pollution arising from an ill-defined and diffuse source such as runoff from cultivated soils or from urban areas.

no-tillage *See* tillage, conservation.

nucleic acids Complex organic acids found in the nuclei of plant and animal cells; may be combined with proteins as nucleoproteins.

O horizon Organic horizon of mineral soils.

ochric epipedon A diagnostic surface horizon of mineral soil that is too light in color, too high in chroma, too low in organic carbon, or too thin to be a plaggen, mollic, umbric, anthropic, or histic epipedon, or that is both hard and massive when dry.

octahedral sheet Sheet of horizontally linked, octahedral-shaped units that serve as the basic structural components of silicate (clay) minerals. Each unit consists of a central, six-coordinated metallic atom (e.g., Al, Mg, or Fe) surrounded by six hydroxyl groups that, in turn, are linked with other nearby metal atoms, thereby serving as interunit linkages that hold the sheet together.

oligotrophic Environments, such as soils or lakes, which are poor in nutrients.

order, soil The category at the highest level of generalization in *Soil Taxonomy.* The properties selected to distinguish the orders are reflections of the degree of horizon development and the kinds of horizons present.

organic farming A system/philosophy of agriculture that does not allow the use of synthetic chemicals to produce plant and animal products, but instead emphasizes the management of soil organic matter and biological processes. In many countries, products are officially certified as being organic if inspections confirm that they were grown by these methods.

organic fertilizer By-product from the processing of animal or vegetable substances that contain sufficient plant nutrients to be of value as fertilizers.

organic soil A soil in which more than half of the profile thickness is comprised of organic soil materials.

organic soil materials (As used in *Soil Taxonomy*): (1) Saturated with water for prolonged periods unless artificially drained and having 18% or more organic carbon (by weight) if the mineral fraction is more than 60% clay, more than 12% organic carbon if the mineral fraction has no clay, or between 12 and 18% carbon if the clay content of the mineral fraction is between 0 and 60%. (2) Never saturated with water for more than a few days and having more than 20% organic carbon. Histosols develop on these organic soil materials. There are three kinds of organic materials:

- **fibric materials** The least decomposed of all the organic soil materials, containing very high amounts of fiber that are well preserved and readily identifiable as to botanical origin.
- **hemic materials** Intermediate in degree of decomposition of organic materials between the less decomposed fibric and the more decomposed sapric materials.
- **sapric materials** The most highly decomposed of the organic materials, having the highest bulk density, least amount of plant fiber, and lowest water content at saturation.

osmotic potential That portion of the total *soil water potential* due to the presence of solutes in soil water. It will generally be negative.

osmotic pressure Pressure exerted in living bodies as a result of unequal concentrations of salts on both sides of a cell wall or membrane. Water moves from the area having the lower salt concentration through the membrane into the area having the higher salt concentration and, therefore, exerts additional pressure on the side with higher salt concentration.

outer-sphere complex A relatively weak (easily reversed) chemical association or general attraction between an ion and an oppositely charged soil colloid via mutual attraction for intervening water molecules.

outwash plain A deposit of coarse-textured materials (e.g., sands and gravels) left by streams of meltwater flowing from receding glaciers.

oven-dry soil Soil that has been dried at 105°C until it reaches constant weight.

oxic horizon A diagnostic subsurface horizon that is at least 30 cm thick and is characterized by the virtual *absence* of weatherable primary minerals or 2:1 lattice clays and the *presence* of 1:1 lattice clays and highly insoluble minerals, such as quartz sand, hydrated oxides of iron and aluminum, low cation exchange capacity, and small amounts of exchangeable bases.

oxidation The loss of electrons by a substance; therefore, a gain in positive valence charge and, in some cases, the chemical combination with oxygen gas.

oxidation-reduction potential *See* Eh and pe.

Oxisols An order in *Soil Taxonomy.* Soils with residual accumulations of low-activity clays, free oxides, kaolin, and quartz. They are mostly in tropical climates.

pans Horizons or layers in soils that are strongly compacted, indurated, or very high in clay content. *See also* caliche; claypan; fragipan; hardpan.

parent material The unconsolidated and more or less chemically weathered mineral or organic matter from which the solum of soils is developed by pedogenic processes.

particle density The mass per unit volume of the soil particles. In technical work, usually expressed as metric tons per cubic meter (Mg/m^3) or grams per cubic centimeter (g/cm^3).

particle size The effective diameter of a particle measured by sedimentation, sieving, or micrometric methods.

particle size analysis Determination of the various amounts of the different separates in a soil sample, usually by sedimentation, sieving, micrometry, or combinations of these methods.

particle size distribution The amounts of the various soil separates in a soil sample, usually expressed as weight percentages.

particulate organic matter A microbially active fraction of soil organic matter consisting largely of fine particles of partially decomposed plant tissue.

partitioning The distribution of organic chemicals (such as pollutants) into a portion that dissolves in the soil organic matter and a portion that remains undissolved in the soil solution.

pascal An SI unit of pressure equal to 1 newton per square meter.

peat Unconsolidated soil material consisting largely of undecomposed to slightly decomposed, organic matter accumulated under conditions of excessive moisture. *See also* organic soil materials; peat soil.

peat soil An organic soil containing more than 50% organic matter. Used in the United States to refer to the stage of decomposition of the organic matter, *peat* referring to the slightly decomposed or undecomposed deposits and *muck* to the highly decomposed materials. *See also* muck; muck soil; peat.

ped A unit of soil structure such as an aggregate, crumb, prism, block, or granule, formed by natural processes (in contrast to a *clod,* which is formed artificially).

pedology The science that deals with the formation, morphology, and classification of soil bodies as landscape components.

pedon The smallest volume that can be called *a soil.* It has three dimensions. It extends downward to the depth of plant roots or to the lower limit of the genetic soil horizons. Its lateral cross section is roughly hexagonal and ranges from 1 to 10 m^2 in size, depending on the variability in the horizons.

pedosphere The conceptual zone within the ecosystem consisting of soil bodies or directly influenced by them. A zone or sphere of activity in which mineral, water, air and biological components come together to form soils. Usage is parallel to that for "atmosphere" or "biosphere."

pedoturbation Physical disturbance and mixing of soil horizons by such forces as burrowing animals (faunal pedoturbation) or frost churning (cryoturbation).

peneplain A once high, rugged area that has been reduced by erosion to a lower, gently rolling surface resembling a plain.

penetrability The ease with which a probe can be pushed into the soil. May be expressed in units of distance, speed, force, or work depending on the type of penetrometer used.

penetrometer An instrument consisting of a rod with a cone-shaped tip and a means of measuring the force required to push the rod into a specified increment of soil.

perc test *See* percolation test.

percolation, soil water The downward movement of water through soil. Especially, the downward flow of water in saturated or nearly saturated soil at hydraulic gradients of the order of 1.0 or less.

percolation test A measurement of the rate of percolation of water in a soil profile, usually to determine the suitability of a soil for use as a septic drain field.

perforated plastic pipe Pipe, sometimes flexible, with holes or slits in it that allow the entrance and exit of air and water. Used for soil drainage and for septic effluent spreading into soil.

permafrost (1) Permanently frozen material underlying the solum. (2) A perennially frozen soil horizon.

permanent charge *See* constant charge.

permanent wilting point *See* wilting point.

permeability, soil The ease with which gases, liquids, or plant roots penetrate or pass through a bulk mass of soil or a layer of soil.

petrocalcic horizon A diagnostic subsurface horizon that is a continuous, indurated calcic horizon cemented by calcium carbonate and, in some places, with magnesium carbonate. It cannot be penetrated with a spade or auger when dry; dry fragments do not slake in water; and it is impenetrable to roots.

petrogypsic horizon A diagnostic subsurface horizon that is a continuous, strongly cemented, massive gypsic horizon that is cemented by calcium sulfate. It can be chipped with a spade when dry. Dry fragments do not slake in water and it is impenetrable to roots.

pH, soil The negative logarithm of the hydrogen ion activity (concentration) of a soil. The degree of acidity (or alkalinity) of a soil as determined by means of a glass or other suitable electrode or indicator at a specified moisture content or soil-to-water ratio, and expressed in terms of the pH scale.

phase, soil A subdivision of a soil series or other unit of classification having characteristics that affect the use and management of the soil but do not vary sufficiently to differentiate it as a separate series. Included are such characteristics as degree of slope, degree of erosion, and content of stones.

pH-dependent charge That portion of the total charge of the soil particles that is affected by, and varies with, changes in pH.

photomap A mosaic map made from aerial photographs to which place names, marginal data, and other map information have been added.

phyllosphere The leaf surface.

physical weathering The breakdown of rock and mineral particles into smaller particles by physical forces such as frost action. *See also* weathering.

phytotoxic substances Chemicals that are toxic to plants.

placic horizon A diagnostic subsurface horizon of a black to dark reddish mineral soil that is usually thin but that may range from 1 to 25 mm in thickness. The placic horizon is commonly cemented with iron and is slowly permeable or impenetrable to water and roots.

plaggen epipedon A diagnostic surface horizon that is human-made and more than 50 cm thick. Formed by long-continued manuring and mixing.

plant-available water Water held with tensions between field capacity (10 to 30 kPa) and wilting point (15 kPa).

plant nutrients *See* essential element.

plastic limit (PL) *See* Atterberg limits.

plastic soil A soil capable of being molded or deformed continuously and permanently, by relatively moderate pressure, into various shapes. *See also* consistence.

platy Consisting of soil aggregates that are developed predominantly along the horizontal axes; laminated; flaky.

plinthite (brick) A highly weathered mixture of sesquioxides of iron and aluminum with quartz and other diluents that occurs as red mottles and that changes irreversibly to hardpan upon alternate wetting and drying.

plow layer The soil ordinarily moved when land is plowed; equivalent to *surface soil*.

plowpan A subsurface soil layer having a higher bulk density and lower total porosity than layers above or below it, as a result of pressure applied by normal plowing and other tillage operations.

plowing A primary broad-base tillage operation that is performed to shatter soil uniformly with partial to complete inversion.

point of zero charge The pH value of a solution in equilibrium with a particle whose net charge, from all sources, is zero.

point source Pollution arising from a well defined source such as an industrial plant or a cattle feedlot.

polypedon (As used in *Soil Taxonomy*) Two or more contiguous pedons, all of which are within the defined limits of a single soil series; commonly referred to as a *soil individual*.

pore size distribution The volume of the various sizes of pores in a soil. Expressed as percentages of the bulk volume (soil plus pore space).

porosity, soil The volume percentage of the total soil bulk not occupied by solid particles.

potential acidity The acidity that could potentially be formed if reduced sulfur compounds in a potential acid sulfate soil were to become oxidized.

predator An animal which kills and eats other animals.

preferential flow Nonuniform movement of water and its solutes through a soil along certain pathways, which are often macropores.

primary consumer An organism that subsists on plant material.

primary mineral A mineral that has not been altered chemically since deposition and crystallization from molten lava.

primary producer An organism (usually a photosynthetic plant) that creates organic, energy-rich material from inorganic chemicals, solar energy, and water.

primary tillage *See* tillage, primary.

priming effect The increased decomposition of relatively stable soil humus under the influence of much enhanced, generally biological, activity resulting from the addition of fresh organic materials to a soil.

prismatic soil structure A soil structure type with prismlike aggregates that have a vertical axis much longer than the horizontal axes.

prokaryote An organism whose cells do not have a distinct nucleus.

Proctor test A laboratory procedure that indicates the maximum achievable bulk density for a soil and the optimum water content for compacting a soil.

productivity, soil The capacity of a soil for producing a specified plant or sequence of plants under a specified system of management. Productivity emphasizes the capacity of soil to produce crops and should be expressed in terms of yields.

profile, soil A vertical section of the soil through all its horizons and extending into the parent material.

protein Any of a group of nitrogen-containing organic compounds formed by the polymerization of a large number of amino acid molecules and that, upon hydrolysis, yield these amino acids. They are essential parts of living matter and are one of the essential food substances of animals.

protonation Attachment of protons (H^+ ions) to exposed OH groups on the surface of soil particles, resulting in an overall positive charge on the particle surface.

protozoa One-celled eukaryotic organisms, such as amoeba.

puddled soil Dense, massive soil artificially compacted when wet and having no aggregated structure. The condition commonly results from the tillage of a clayey soil when it is wet.

rain, acid *See* acid rain.

reaction, soil (No longer used in soil science) The degree of acidity or alkalinity of a soil, usually expressed as a pH value or by terms ranging from extremely acid for pH values <4.5 to very strongly alkaline for pH values >9.0.

recharge area A geographic area in which an otherwise confined aquifer is exposed to surficial percolation of water to recharge its groundwater.

redox concentrations Zones of apparent accumulations of Fe-Mn oxides in soils.

redox depletions Zones of low chroma (<2) where Fe-Mn oxides, and in some cases clay, have been stripped from the soil.

redox potential The electrical potential (measured in volts or millivolts) of a system due to the tendency of the substances in it to give up or acquire electrons.

redoximorphic features Soil properties associated with wetness that result from reduction and oxidation of iron and manganese compounds after saturation and desaturation with water. *See also* redox concentrations; redox depletions.

reduction The gain of electrons, and therefore the loss of positive valence charge, by a substance. In some cases, a loss of oxygen or a gain of hydrogen is also involved.

regolith The unconsolidated mantle of weathered rock and soil material on the earth's surface; loose earth materials above solid rock. (Approximately equivalent to the term *soil* as used by many engineers.)

relief The relative differences in elevation between the upland summits and the lowlands or valleys of a given region.

residual material Unconsolidated and partly weathered mineral materials accumulated by disintegration of consolidated rock in place.

resilience The capacity of a soil (or other ecosystem) to return to its original state after a disturbance.

revised universal soil-loss equation (RUSLE) A computerized, expanded version of the universal soil-loss equation.

rhizobacteria Bacteria specially adapted to colonizing the surface of plant roots and the soil immediately around plant roots. Some have effects that promote plant growth, while others have effects that are deleterious to plants.

rhizobia Bacteria capable of living symbiotically with higher plants, usually in nodules on the roots of legumes, from which they receive their energy, and capable of converting atmospheric nitrogen to combined organic forms; hence the term *symbiotic nitrogen-fixing bacteria.* (Derived from the generic name *Rhizobium.*)

rhizoplane The root surface–soil interface. Used to describe the habitat of root-surface-dwelling microorganisms.

rhizosphere That portion of the soil in the immediate vicinity of plant roots in which the abundance and composition of the microbial population are influenced by the presence of roots.

rill A small, intermittent water course with steep sides; usually only a few centimeters deep and hence no obstacle to tillage operations.

rill erosion An erosion process in which numerous small channels of only several centimeters in depth are formed; occurs mainly on recently cultivated soils. *See also* rill.

riparian zone The area, both above and below the ground surface, that borders a river.

riprap Broken rock, cobbles, or boulders placed on earth surfaces, such as the face of a dam or the bank of a stream, for protection against the action of water (waves); also applied to brush or pole mattresses, or brush and stone, or other similar materials used for soil erosion control.

rock The material that forms the essential part of the earth's solid crust, including loose incoherent masses such as sand and gravel, as well as solid masses of granite and limestone.

root interception Acquisition of nutrients by a root as a result of the root growing into the vicinity of the nutrient source.

root nodules Swollen growths on plant roots, including those in which symbiotic microorganisms live.

runoff The portion of the precipitation on an area that is discharged from the area through stream channels. That which is lost without entering the soil is called *surface runoff* and that which enters the soil before reaching the stream is called *groundwater runoff* or *seepage flow* from groundwater. (In soil science *runoff* usually refers to the water lost by surface flow; in geology and hydraulics *runoff* usually includes both surface and subsurface flow.)

salic horizon A diagnostic subsurface horizon of enrichment with secondary salts more soluble in cold water than gypsum. A salic horizon is 15 cm or more in thickness.

saline seep An area of land in which saline water seeps to the surface, leaving a high salt concentration behind as the water evaporates.

saline soil A nonsodic soil containing sufficient soluble salts to impair its productivity. The conductivity of a saturated extract is >4 dS/m, the exchangeable sodium adsorption ratio is less than about 13, and the pH is <8.5.

saline-sodic soil A soil containing sufficient exchangeable sodium to interfere with the growth of most crop plants and containing appreciable quantities of soluble salts. The exchangeable sodium adsorption ratio is >13, the conductivity of the saturation extract is >4 dS/m (at 25°C), and the pH is usually 8.5 or less in the saturated soil.

salinization The process of accumulation of salts in soil.

saltation Particle movement in water or wind where particles skip or bounce along the stream bed or soil surface.

sand A soil particle between 0.05 and 2.0 mm in diameter; a soil textural class.

sapric materials *See* organic soil materials.

saprolite Soft; friable, weathered bedrock that retains the fabric and structure of the parent rock while exhibiting extensive inter- and intra-crystal weathering.

saprophyte An organism that lives on dead organic material.

saturated (soil) A state in which all pore space is filled with water.

saturated paste extract The extract from a saturated soil paste, the electrical conductivity E_c of which gives an indirect measure of salt content in a soil.

saturation percentage The water content of a saturated soil paste, expressed as a dry weight percentage.

savanna (savannah) A grassland with scattered trees, either as individuals or clumps. Often a transitional type between true grassland and forest.

secondary mineral A mineral resulting from the decomposition of a primary mineral or from the reprecipitation of the products of decomposition of a primary mineral. *See also* primary mineral.

sediment Transported and deposited particles or aggregates derived from soils, rocks, or biological materials.

sedimentary rock A rock formed from materials deposited from suspension or precipitated from solution and usually being more or less consolidated. The principal sedimentary rocks are sandstones, shales, limestones, and conglomerates.

seedbed The soil prepared to promote the germination of seed and the growth of seedlings.

self-mulching soil A soil in which the surface layer becomes so well aggregated that it does not crust and seal under the impact of rain but instead serves as a surface mulch upon drying.

semiarid Term applied to regions or climates where moisture is more plentiful than in arid regions but still definitely limits the growth of most crop plants. Natural vegetation in uncultivated areas is short grasses.

separate, soil One of the individual-sized groups of mineral soil particles—sand, silt, or clay.

septic tank An underground tank used in the deposition of domestic wastes. Organic matter decomposes in the tank, and the effluent is drained into the surrounding soil.

series, soil The soil series is a subdivision of a family in *Soil Taxonomy* and consists of soils that are similar in all major profile characteristics.

sewage effluent The liquid part of sewage or wastewater; it is usually treated to remove some portion of the dissolved organic compounds and nutrients present from the original sewage.

sewage sludge Settled sewage solids combined with varying amounts of water and dissolved materials, removed from sewage by screening, sedimentation, chemical precipitation, or bacterial digestion. Also called *biosolids* if certain quality standards are met.

shear Force, as of a tillage implement, acting at right angles to the direction of movement.

sheet (Mineralogy) A flat array of more than one atomic thickness and composed of one or more levels of linked coordination polyhedra. A sheet is thicker than a plane and thinner than a layer. Examples: tetrahedral sheet, octahedral sheet.

sheet erosion The removal of a fairly uniform layer of soil from the land surface by runoff water.

shelterbelt A wind barrier of living trees and shrubs established and maintained for protection of farm fields. Syn. *windbreak*.

shifting cultivation A farming system in which land is cleared, the debris burned, and crops grown for 2 to 3 years. When the farmer moves on to another plot, the land is then left idle for 5 to 15 years; then the burning and planting process is repeated.

short-range order minerals Minerals, such as allophane, whose structural framework consists of short distances of well-ordered crystalline structure interspersed with distances of noncrystalline amorphous materials.

shrinkage limit (SL) see Atterberg limits.

side-dressing The application of fertilizer alongside row-crop plants, usually on the soil surface. Nitrogen materials are most commonly side-dressed.

siderophore A nonporphyrin metabolite secreted by certain microorganisms that forms a highly stable coordination compound with iron.

silica/alumina ratio The molecules of silicon dioxide (SiO_2) per molecule of aluminum oxide (Al_2O_3) in clay minerals or in soils.

silica/sesquioxide ratio The molecules of silicon dioxide (SiO_2) per molecule of aluminum oxide (Al_2O_3) plus ferric oxide (Fe_2O_3) in clay minerals or in soils.

silt (1) A soil separate consisting of particles between 0.05 and 0.002 mm in equivalent diameter. (2) A soil textural class.

silting The deposition of waterborne sediments in stream channels, lakes, reservoirs, or on floodplains, usually resulting from a decrease in water velocity.

site index A quantitative evaluation of the productivity of a soil for forest growth under the existing or specified environment.

slash and burn *See* shifting cultivation.

slick spots Small areas in a field that are slick when wet because of a high content of alkali or exchangeable sodium.

slickensides Stress surfaces that are polished and striated and are produced by one mass sliding past another.

slope The degree of deviation of a surface from horizontal, measured in a numerical ratio, percent, or degrees.

slow fraction (of soil organic matter) That portion of soil organic matter that can be metabolized with great difficulty by the microorganisms in the soil and therefore has a slow turnover rate with a half-life in the soil ranging from a few years to a few decades. Often this fraction is the product of some previous decomposition.

smectite A group of silicate clays having a 2:1-type lattice structure with sufficient isomorphous substitution in either or both the tetrahedral and octahedral sheets to give a high interlayer negative charge and high cation exchange capacity and to permit significant interlayer expansion and consequent shrinking and swelling of the clay. Montmorillonite, beidellite, and saponite are in the smectite group.

sodic soil A soil that contains sufficient sodium to interfere with the growth of most crop plants, and in which the sodium adsorption ratio is 13 or greater.

sodium adsorption ratio (SAR)

$$SAR = \frac{[Na^+]}{\sqrt{\frac{1}{2}([Ca^{2+}] + [Mg^{2+}])}}$$

where the cation concentrations are in millimoles of charge per liter ($mmol_c/L$).

soil (1) A dynamic natural body composed of mineral and organic solids, gases, liquids and living organisms which can serve as a medium for plant growth. (2) The collection of natural bodies occupying parts of the earth's surface that is capable of supporting plant growth and that has properties resulting from the integrated effects of climate and living organisms acting upon parent material, as conditioned by topography, over periods of time.

soil air The soil atmosphere; the gaseous phase of the soil, being that volume not occupied by soil or liquid.

soil alkalinity The degree or intensity of alkalinity of a soil, expressed by a value >7.0 on the pH scale.

soil amendment Any material, such as lime, gypsum, sawdust, or synthetic conditioner, that is worked into the soil to make it more amenable to plant growth.

soil association A group of defined and named taxonomic soil units occurring together in an individual and characteristic pattern over a geographic region, comparable to plant associations in many ways.

soil auger A tool used to bore small holes up to several meters deep in soils to bring up samples of material from various soil layers. It consists of a long T-handle attached to either a cylinder with twisted teeth or a screwlike bit.

soil classification (*Soil Taxonomy*) The systematic arrangement of soils into groups or categories on the basis of their characteristics. *See* order; suborder; great group; subgroup; family; and series.

soil complex A mapping unit used in detailed soil surveys where two or more defined taxonomic units are so intimately intermixed geographically that it is undesirable or impractical, because of the scale being used, to separate them. A more intimate mixing of smaller areas of individual taxonomic units than that described under *soil association.*

soil compressibility The property of a soil pertaining to its capacity to decrease in bulk volume when subjected to a load.

soil conditioner Any material added to a soil for the purpose of improving its physical condition.

soil conservation A combination of all management and land-use methods that safeguard the soil against depletion or deterioration caused by nature and/or humans.

soil consociation A kind of soil map unit named for the dominant soil taxon in the delineation, and in which at least half of the pedons are of the named soil taxon, and most of the remaining pedons are so similar as to not affect most interpretations.

soil correlation The process of defining, mapping, naming, and classifying the kinds of soils in a specific soil survey area, the purpose being to ensure that soils are adequately defined, accurately mapped, and uniformly named.

soil erosion *See* erosion.

soil fertility *See* fertility, soil.

soil genesis *See* genesis, soil.

soil horizon *See* horizon, soil.

soil loss tolerance (*T*) The maximum rate of annual soil loss that will permit plant productivity to be maintained economically and indefinitely.

soil management The sum total of all tillage operations, cropping practices, fertilizer, lime, and other treatments conducted on or applied to a soil for the production of plants.

soil map A map showing the distribution of soil types or other soil mapping units in relation to prominent physical and cultural features of the earth's surface.

soil mechanics and engineering A subspecialization of soil science concerned with the effect of forces on the soil and the application of engineering principles to problems involving the soil.

soil moisture potential *See* soil water potential.

soil monolith A vertical section of a soil profile removed from the soil and mounted for display or study.

soil morphology The physical constitution, particularly the structural properties, of a soil profile as exhibited by the kinds, thicknesses, and arrangement of the horizons in the profile, and by the texture, structure, consistence, and porosity of each horizon.

soil order *See* order, soil.

soil profile *See* profile, soil.

soil organic matter (SOM) The organic fraction of the soil that includes plant and animal residues at various stages of decomposition, cells and tissues of soil organisms, and substances synthesized by the soil population. Commonly determined as the amount of organic material contained in a soil sample passed through a 2 mm sieve.

soil porosity *See* porosity, soil.

soil productivity *See* productivity, soil.

soil quality The capacity of a specific kind of soil to function, within natural or managed ecosystem boundaries, to sustain plant and animal productivity, maintain or enhance water and air quality, and support human health and habitation. Sometimes considered in relation to this capacity in the undisturbed, natural state.

soil reaction *See* reaction, soil; pH, soil.

soil salinity The amount of soluble salts in a soil, expressed in terms of percentage, milligrams per kilogram, parts per million (ppm), or other convenient ratios.

soil separates *See* separate, soil.

soil series *See* series, soil.

soil solution The aqueous liquid phase of the soil and its solutes, consisting of ions dissociated from the surfaces of the soil particles and of other soluble materials.

soil strength A transient soil property related to the soil's solid phase cohesion and adhesion.

soil structure The combination or arrangement of primary soil particles into secondary particles, units, or peds. These secondary units may be, but usually are not, arranged in the profile in such a manner as to give a distinctive characteristic pattern. The secondary units are characterized and classified on the basis of size, shape, and degree of distinctness into classes, types, and grades, respectively.

soil structure classes A grouping of soil structural units or peds on the basis of size from the very fine to very coarse.

soil structure grades A grouping or classification of soil structure on the basis of inter- and intraaggregate adhesion, cohesion, or stability within the profile. Four grades of structure, designated from 0 to 3, are recognized: *structureless, weak, moderate,* and *strong.*

soil structure types A classification of soil structure based on the shape of the aggregates or peds and their arrangement in the profile, including platy, prismatic, columnar, blocky, subangular blocky and granular.

soil survey The systematic examination, description, classification, and mapping of soils in an area. Soil surveys are classified according to the kind and intensity of field examination.

soil temperature classes A criterion used to differentiate soil in *Soil Taxonomy,* mainly at the family level. Classes are based on mean annual soil temperature and on differences between summer and winter temperatures at a depth of 50 cm.

soil textural class A grouping of soil textural units based on the relative proportions of the various soil separates (sand, silt, and clay). These textural classes, listed from the coarsest to the finest in texture, are sand, loamy sand, sandy loam, loam, silt loam, silt, sandy clay loam, clay loam, silty clay loam, sandy clay, silty clay, and clay. There are several subclasses of the sand, loamy sand, and sandy loam classes based on the dominant particle size of the sand fraction (e.g., loamy fine sand, coarse sandy loam).

soil texture The relative proportions of the various soil separates in a soil.

soil water potential (total) A measure of the difference between the free energy state of soil water and that of pure water. Technically it is defined as "that amount of work that must be done per unit quantity of pure water in order to transport reversibly and isothermically an infinitesimal quantity of water from a pool of pure water, at a specified elevation and at atmospheric pressure to the soil water (at the point under consideration)." This *total* potential consists of *gravitational, matric,* and *osmotic* potentials.

solarization The process of heating a soil in the field by covering it with clear plastic sheeting during sunny conditions. The heat is meant to partially sterilize the upper 5 to 15 cm of soil to reduce pest and pathogen populations.

solum (pl: sola) The upper and most weathered part of the soil profile; the A, E, and B horizons.

sombric horizon A diagnostic subsurface horizon that contains illuvial humus but has a low cation exchange capacity and low percentage base saturation. Mostly restricted to cool, moist soils of high plateaus and mountainous areas of tropical and subtropical regions.

sorption The removal from the soil solution of an ion or molecule by adsorption and absorption. This term is often used when the exact mechanism of removal is not known.

species diversity The variety of different biological species present in an ecosystem. Generally, high diversity is marked by many species with few individuals in each.

species richness The number of different species present in an ecosystem, without regard to the distribution of individuals among those species.

specific gravity The ratio of the density of a mineral to the density of water at standard temperature and pressure.

specific heat capacity The amount of kinetic (heat) energy required to raise the temperature of 1 g of a substance (usually in reference to soil or soil components).

specific surface The solid particle surface area per unit mass or volume of the solid particles.

splash erosion The spattering of small soil particles caused by the impact of raindrops on very wet soils. The loosened and separated particles may or may not be subsequently removed by surface runoff.

spodic horizon A diagnostic subsurface horizon characterized by the illuvial accumulation of amorphous materials composed of aluminum and organic carbon with or without iron.

Spodosols An order in *Soil Taxonomy.* Soils with subsurface illuvial accumulations of organic matter and compounds of aluminum and usually iron. These soils are formed in acid, mainly coarse-textured materials in humid and mostly cool or temperate climates.

stem flow The process by which rain or irrigation water is directed by a plant canopy toward the plant stem so as to wet the soil unevenly under the plant canopy.

stratified Arranged in or composed of strata or layers.

strip-cropping The practice of growing crops that require different types of tillage, such as row and sod, in alternate strips along contours or across the prevailing direction of wind.

structure, soil *See* soil structure.

stubble mulch The stubble of crops or crop residues left essentially in place on the land as a surface cover before and during the preparation of the seedbed and at least partly during the growing of a succeeding crop.

subgroup, soil The *great groups* in *Soil Taxonomy* are subdivided into central concept subgroups that show the central properties of the great group, intergrade subgroups that show properties of more than one great group, and other subgroups for soils with atypical properties that are not characteristic of any great group.

submergence potential The positive hydrostatic pressure that occurs below the water table.

suborder, soil A category in *Soil Taxonomy* that narrows the ranges in soil moisture and temperature regimes, kinds of horizons, and composition, according to which of these is most important.

subsoil That part of the soil below the plow layer.

subsoiling Breaking of compact subsoils, without inverting them, with a special knifelike instrument (chisel), which is pulled through the soil at depths usually of 30 to 60 cm and at spacings usually of 1 to 2 m.

sulfuric horizon A diagnostic subsurface horizon in either mineral or organic soils that has a pH <3.5 and fresh straw-colored mottles (called *jarosite mottles*). Forms by oxidation of sulfide-rich materials and is highly toxic to plants.

summer fallow *See* fallow.

surface runoff *See* runoff.

surface seal A thin layer of fine particles deposited on the surface of a soil that greatly reduces the permeability of the soil surface to water.

surface soil The uppermost part of the soil, ordinarily moved in tillage, or its equivalent in uncultivated soils. Ranges in depth from 7 to 25 cm. Frequently designated as the *plow layer,* the *Ap layer,* or the *Ap horizon.*

surface tension The elasticlike phenomenon resulting from the unbalanced attractions among liquid molecules (usually water) and between liquid and gaseous molecules (usually air) at the liquid–gas interface.

swamp An area of land that is usually wet or submerged under shallow freshwater and typically supports hydrophilic trees and shrubs.

symbiosis The living together in intimate association of two dissimilar organisms, the cohabitation being mutually beneficial and obligatory.

synergism (1) The nonobligatory association between organisms that is mutually beneficial. Both populations can survive in their natural environment on their own although, when formed, the association offers mutual advantages. (2) The simultaneous actions of two or more factors that have a greater total effect together than the sum of their individual effects.

talus Fragments of rock and other soil material accumulated by gravity at the foot of cliffs or steep slopes.

taxonomy, soil The science of classification of soils; laws and principles governing the classifying of soil. Also a specific *soil classification* system developed by the U.S. Department of Agriculture.

tensiometer A device for measuring the negative pressure (or tension) of water in soil in situ; a porous, permeable ceramic cup connected through a tube to a manometer or vacuum gauge.

tension, soil-moisture *See* soil water potential.

terrace (1) A level, usually narrow, plain bordering a river, lake, or the sea. Rivers sometimes are bordered by terraces at different levels. (2) A raised, more or less level or horizontal strip of earth usually constructed on or nearly on a contour and designed to make the land suitable for tillage and to prevent accelerated erosion by diverting water from undesirable channels of concentration; sometimes called *diversion terrace.*

tetrahedral sheet Sheet of horizontally linked, tetrahedron-shaped units that serve as one of the basic structural components of silicate (clay) minerals. Each unit consists of a central four-coordinated atom (e.g., Si, Al, Fe) surrounded by four oxygen atoms that, in turn, are linked with other nearby atoms (e.g., Si, Al, Fe), thereby serving as interunit linkages to hold the sheet together.

texture *See* soil texture.

thermal analysis (differential thermal analysis) A method of analyzing a soil sample for constituents, based on a differential rate of heating of the unknown and standard samples when a uniform source of heat is applied.

thermic A soil temperature class with mean annual temperature 15° to 22°C.

thermophilic Pertaining to temperatures in the range of 45° to 90°C, the range in which thermophilic organisms grow best and in which thermophilic composting takes place.

thixotrophy The property of certain clay soils of becoming fluid when jarred or agitated and then setting again when at rest. Similar to *quick,* as in quick clays or quicksand.

till (1) Unstratified glacial drift deposited directly by the ice and consisting of clay, sand, gravel, and boulders intermingled in any proportion. (2) To plow and prepare for seeding; to seed or cultivate the soil.

tillage The mechanical manipulation of soil for any purpose; but in agriculture it is usually restricted to the modifying of soil conditions for crop production.

tillage, conservation Any tillage sequence that reduces loss of soil or water relative to conventional tillage, which generally leaves at least 30% of the soil surface covered by residues, including the following systems:

- **minimum tillage** The minimum soil manipulation necessary for crop production or meeting tillage requirements under the existing soil and climatic conditions.
- **mulch tillage** Tillage or preparation of the soil in such a way that plant residues or other materials are left to cover the surface; also called *mulch farming, trash farming* and *stubble mulch tillage.*
- **no-tillage system** A procedure whereby a crop is planted directly into a seedbed not tilled since harvest of the previous crop; also called *zero tillage.*
- **ridge till** Planting on ridges formed by cultivation during the previous growing period.
- **strip till** Planting is done in a narrow strip that has been tilled and mixed, leaving the remainder of the soil surface undisturbed.

tillage, conventional The combined primary and secondary tillage operations normally performed in preparing a seedbed for a given crop grown in a given geographic area.

tillage, primary Tillage that contributes to the major soil manipulation, commonly with a plow.

tillage, secondary Any tillage operations following primary tillage designed to prepare a satisfactory seedbed for planting.

tilth The physical condition of soil as related to its ease of tillage, fitness as a seedbed, and its impedance to seedling emergence and root penetration.

top-dressing An application of fertilizer to a soil after the crop stand has been established.

toposequence A sequence of related soils that differ, one from the other, primarily because of *topography* as a soil-formation factor.

topsoil (1) The layer of soil moved in cultivation. *See also* surface soil. (2) Presumably fertile soil material used to top-dress roadbanks, gardens, and lawns.

trace elements Elements present in the Earth's crust in concentrations less than 1000 mg/kg. When referring to plant nutrients, the term *micronutrients* is preferred.

trioctahedral An octahedral sheet of silicate clays in which the sites for the six-coordinated metallic atoms are mostly filled with divalent cations, such as Mg^{2+}.

trophic level Levels in a food chain that pass nutrients and energy from one group of organisms to another.

truncated Having lost all or part of the upper soil horizon or horizons.

tuff Volcanic ash usually more or less stratified and in various states of consolidation.

tundra A level or undulating treeless plain characteristic of arctic regions.

Ultisols An order in *Soil Taxonomy.* Soils that are usually moist, low in nonacid cations and have subsurface horizons of illuvial clay accumulations.

umbric epipedon A diagnostic surface horizon of mineral soil that has the same requirements as the mollic epipedon with respect to color, thickness, organic carbon content, consistence, structure, and P_2O_5 content, but that has a base saturation of less than 50%.

universal soil loss equation (USLE) An equation for predicting the average annual soil loss per unit area per year; $A = RKLSPC$, where R is the climatic erosivity factor (rainfall plus runoff), K is the soil erodibility factor, L is the length of slope, S is the percent slope, P is the soil erosion practice factor, and C is the cropping and management factor.

unsaturated flow The movement of water in a soil that is not filled to capacity with water.

vadose zone The aerated region of soil above the permanent water table.

value (color) *See* Munsell color system.

variable charge *See* pH-dependent charge.

varnish, desert A glossy sheen or coating on stones and gravel in arid regions.

vermiculite A 2:1-type silicate clay, usually formed from mica, that has a high net negative charge stemming mostly from extensive isomorphous substitution of aluminum for silicon in the tetrahedral sheet.

vermicompost Compost made by earthworms eating raw organic materials in moist aerated piles, which are kept shallow to avoid heat buildup that could kill the worms.

Vertisols An order in *Soil Taxonomy.* Clayey soils with high shrink–swell potential that have wide, deep cracks when dry. Most of these soils have distinct wet and dry periods throughout the year.

vesicles (1) Unconnected voids with smooth walls. (2) Spherical structures formed inside root cortical cells by vesicular arbuscular mycorrhizal fungi.

virgin soil A soil that has not been significantly disturbed from its natural environment.

water potential, soil *See* soil water potential.

water table The upper surface of groundwater or that level below which the soil is saturated with water.

water table, perched The surface of a local zone of saturation held above the main body of groundwater by an impermeable layer of stratum, usually clay, and separated from the main body of groundwater by an unsaturated zone.

water use efficiency Dry matter or harvested portion of crop produced per unit of water consumed.

waterlogged Saturated with water.

watershed All the land and water within the geographical confines of a drainage divide or surrounding ridges that separate the area from neighboring watersheds.

water-stable aggregate A soil aggregate stable to the action of water, such as falling drops or agitation, as in wet-sieving analysis.

weathering All physical and chemical changes produced in rocks, at or near the earth's surface, by atmospheric agents.

wetland An area of land that has hydric soil and hydrophytic vegetation, typically flooded for part of the year, and forming a transition zone between aquatic and terrestrial systems.

wetting front The boundary between the wetted soil and dry soil during infiltration of water.

wilting point (permanent wilting point) The moisture content of soil, on an oven-dry basis, at which plants wilt and fail to recover their turgidity when placed in a dark, humid atmosphere.

windbreak Planting of trees, shrubs, or other vegetation perpendicular, or nearly so, to the principal wind direction to protect soils, crops, homesteads, etc., from wind and snow.

xenobiotic Compounds foreign to biological systems. Often refers to compounds resistant to decomposition.

xerophytes Plants that grow in or on extremely dry soils or soil materials.

zero tillage *See* tillage, conservation.

zymogenous organisms So-called opportunist organisms found in soils in large numbers immediately following addition of readily decomposable organic materials.

INDEX

F

G

P

T

U

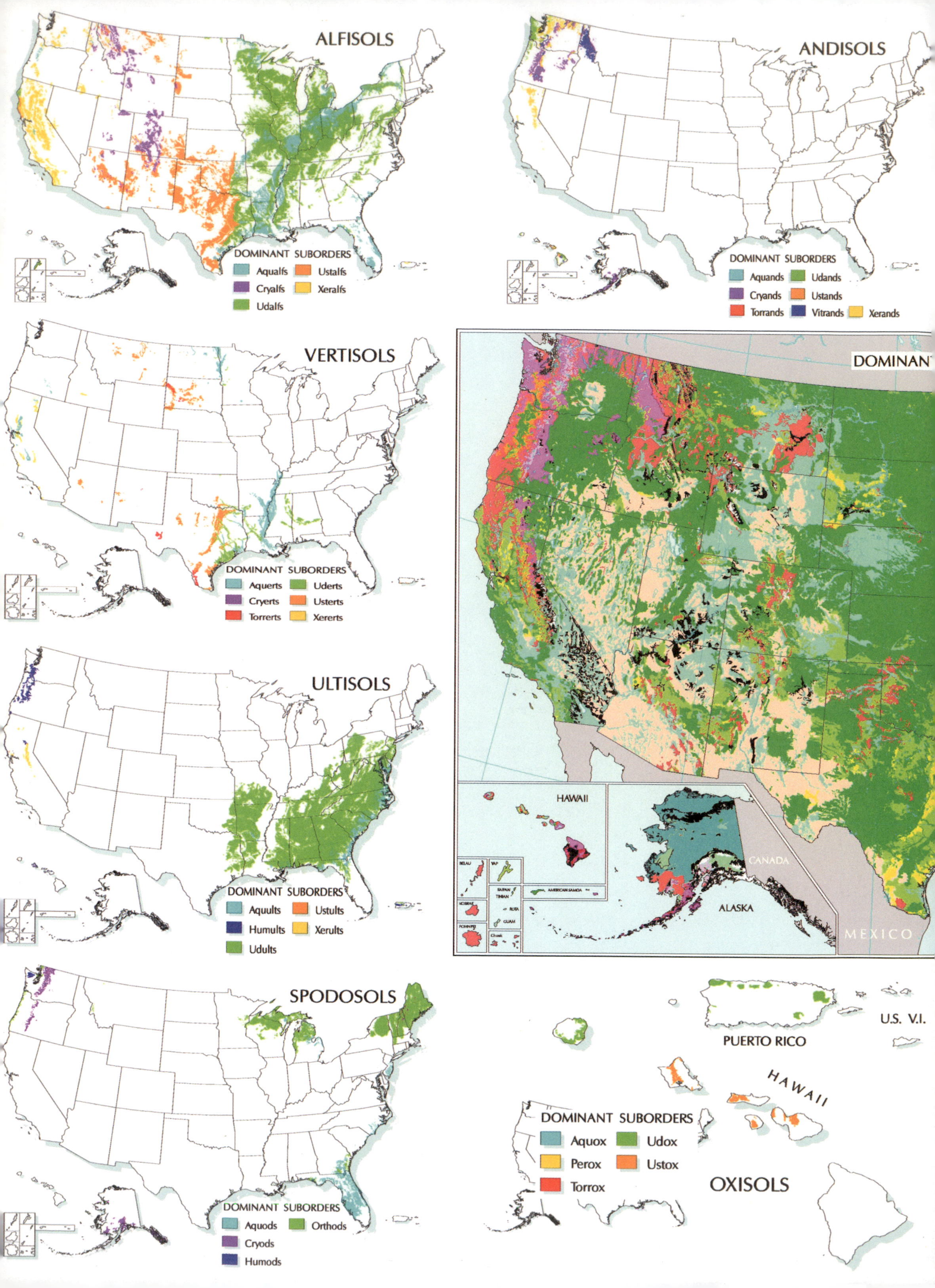

ALFISOLS
DOMINANT SUBORDERS
Aqualfs
Ustalfs
Cryalfs
Xeralfs
Udalfs
ANDISOLS
DOMINANT SUBORDERS
Aquands
Udands
Cryands
Ustands
Torrands
Vitrands
Xerands
VERTISOLS
DOMINANT SUBORDERS
Aquerts
Uderts
Cryerts
Usterts
Torrerts
Xererts
DOMINAN
HAWAII
CANADA
ALASKA
MEXICO
ULTISOLS
DOMINANT SUBORDERS
Aquults
Ustults
Humults
Xerults
Udults
SPODOSOLS
DOMINANT SUBORDERS
Aquods
Orthods
Cryods
Humods
U.S. V.I.
PUERTO RICO
HAWAII
DOMINANT SUBORDERS
Aquox
Udox
Perox
Ustox
Torrox
OXISOLS